The Induction Machine Handbook

The ELECTRIC POWER ENGINEERING Series
Series Editor Leo Grigsby

Published Titles

Electromechanical Systems, Electric Machines, and Applied Mechatronics
Sergey E. Lyshevski

Electrical Energy Systems
Mohamed E. El-Hawary

Electric Drives
Ion Boldea and Syed Nasar

Distribution System Modeling and Analysis
William H. Kersting

Linear Synchronous Motors: Transportation and Automation Systems
Jacek Gieras and Jerry Piech

The Induction Machine Handbook
Ion Boldea and Syed Nasar

Forthcoming Titles

Power System Operations in a Restructured Business Environment
Fred I. Denny and David E. Dismukes

Power Quality
C. Sankaran

The Induction Machine Handbook

Ion Boldea
IEEE Fellow

Syed A. Nasar
IEEE Life Fellow

CRC Press
Boca Raton London New York Washington, D.C.

Library of Congress Cataloging-in-Publication Data

Boldea, I.
Induction Machines Handbook / Ion Boldea, Syed A. Nasar
p. cm. -- (Electric power engineering series)
Includes bibliographical references and index.
ISBN 0-8493-0004-5 (alk. paper)
1. Electric machinery, Induction--Handbooks, manuals, etc. I. Nasar, S.A. II Title.
III. Series.
TK2711.B65 2001
621.31'042—dc21 2001043027
CIP

International Standard Book Number 0-8493-0004-5
Library of Congress Card Number 2001043027
Printed in the United States of America 2 3 4 5 6 7 8 9 0
Printed on acid-free paper

A humble tribute to Nikola Tesla and Galileo Ferraris

PREFACE

The well being of the environmentally-conscious contemporary world is strongly dependent on its efficient production and use of electric energy.

Electric energy is produced with synchronous generators, but for the flexible, distributed, power systems of the near future, the induction machine as an electric generator/motor at variable speeds, is gaining more and more ground.

Pump storage hydroelectric plants already use doubly fed induction machines up to 300MW/unit with the power electronic converter on the rotor side.

On the other hand, the induction motor is the **work horse** of industry due to its ruggedness, low cost, and good performance, when fed from the standard a.c. power grid.

The developments in power electronics and digital control have triggered the widespread use of induction motors for variable speed drives in most industries.

We can safely say that in the last decade the variable speed induction motor has already become the **race horse** of industry.

While in constant speed applications better efficiency and lower costs are the main challenges, in variable speed drives, motion control response quickness, robustness, and precision for ever wider speed and power ranges are the new-added performance indexes.

The recent development of super-high-speed variable drives for speeds up to 60,000 rpm and power in the tens of kW also pose extraordinary technological challenges.

In terms of research and development, new analysis tools such as finite element field methods, design optimization methods, advanced nonlinear circuit models for transients, vector and direct torque control, and wide frequency spectrum operation with power electronics have been extensively used in connection with induction machines in the last 20 years.

There is a dynamic worldwide market for induction machines for constant and variable speed applications; however, an up-to-date comprehensive and coherent treatise in English, dedicated to the induction machine (three phase and single phase)-embracing the wide variety of complex issues of analysis and synthesis (design)-is virtually nonexistent as of this writing. The chapters on IMs in Electric Motor handbooks contain interesting information but lack coherency and completeness.

This book is a daring attempt to build on the now classical books of R. Richter, P. Alger, Heller & Hamata, C. Veinott, and P. Cochran, on induction machines.

It treats in 28 chapters a wide spectrum of issues such as induction machine applications, principles and topologies, materials, windings, electric circuit parameter computation, equivalent circuits (standard and new) and steady state performance, starting and speed control methods, flux harmonics and parasitic torques, skin and saturation effects, fundamental and additional (space and time harmonics) losses, thermal modeling and cooling, transients, specifications and design principles, design below 100 kW, design above 100 kW, design for variable speed, design optimization, three-phase induction generators, linear induction motors, super-high-frequency modelling and behaviour of IMs, testing of three-phase induction machines and single-phase IMs (basics, steady state, transients, generators, design, testing).

Numerical examples, design case studies, and transient behavior waveforms are presented throughout the chapters to make the book self sufficient and easy to use by either young or experienced readers from academia and industry.

It is aimed to be usable and inspiring at the same time it treats both standard and new subjects in induction machines. The rather detailed table of contents is self explanatory in this regard.

Input from readers is most welcome.

The authors

June 2001
Timisoara, RO-1900
Lexington, KY-40506

CONTENTS

Chapter 1

INDUCTION MACHINES: AN INTRODUCTION

1.1. ELECTRIC ENERGY AND INDUCTION MOTORS

The level of prosperity of a community is related to its capability to produce goods and services. But producing goods and services is strongly related to the use of energy in an intelligent way.

Motion and temperature (heat) control are paramount in energy usage. Energy comes into use in a few forms such as thermal, mechanical and electrical.

Electrical energy, measured in kWh, represents more than 30% of all used energy and it is on the rise. Part of electrical energy is used directly to produce heat or light (in electrolysis, metallurgical arch furnaces, industrial space heating, lighting, etc.).

The larger part of electrical energy is converted into mechanical energy in electric motors. Among electric motors, induction motors are most used both for home appliances and in various industries [1-11].

This is so because they have been traditionally fed directly from the three-phase a.c. electric power grid through electromagnetic power switches with adequate protection. It is so convenient.

Small power induction motors, in most home appliances, are fed from the local single phase a.c. power grids. Induction motors are rugged and have moderate costs, explaining their popularity.

In developed countries today there are more than 3 kW of electric motors per person, today and most of it is from induction motors.

While most induction motors are still fed from three-phase or single-phase power grids, some are supplied through frequency changers (or power electronics converters) to provide variable speed.

In developed countries, 10% of all induction motor power is converted in variable speed drives applications. The annual growth rate of variable speed drives has been 9% in the last decade while the electric motor markets showed an average annual growth rate of 4% in the same time.

Variable speed drives with induction motors are used in transportation, pumps, compressors, ventilators, machine tools, robotics, hybrid or electric vehicles, washing machines, etc.

The forecast is that, in the next decade, up to 50% of all electric motors will be fed through power electronics with induction motors covering 60 to 70% of these new markets.

The ratings of induction motors vary from a few tens of watts to 33120 kW (45000 HP). The distribution of ratings in variable speed drives is shown in Table 1.1. [1]

Table 1.1. Variable speed a.c. drives ratings

Power (kW)	1 - 4	5 - 40	40 - 200	200 - 600	>600
Percentage	21%	26%	26%	16%	11%

Intelligent use of energy means higher productivity with lower active energy and lower losses at moderate costs. Reducing losses leads to lower environmental impact where the motor works and lower thermal and chemical impact at the electric power plant that produces the required electrical energy. Variable speed through variable frequency is paramount in achieving such goals. As a side effect, the use of variable speed drives leads to current harmonics pollution in the power grid and to electromagnetic interference (EMI) with the environment. So power quality and EMI have become new constraints on electric induction motor drives.

Digital control is now standard in variable speed drives while autonomous intelligent drives to be controlled and repaired via Internet are on the horizon. And new application opportunities abound: from digital appliances to hybrid and electric vehicles and more electric aircraft.

So much in the future, let us now go back to the first two invented induction motors.

1.2. A HISTORICAL TOUCH

Faraday discovered the electromagnetic induction law around 1831 and Maxwell formulated the laws of electricity (or Maxwell's equations) around 1860. The knowledge was ripe for the invention of the induction machine which has two fathers: Galileo Ferraris (1885) and Nicola Tesla (1886). Their induction machines are shown in Figure 1.1 and Figure 1.2.

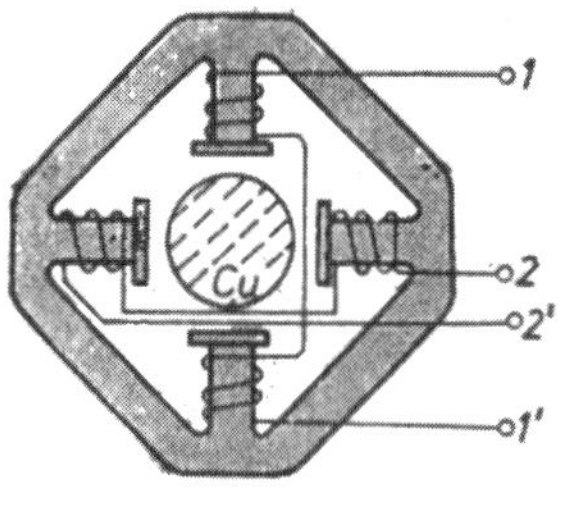

Figure 1.1 Ferrari's induction motor (1885)

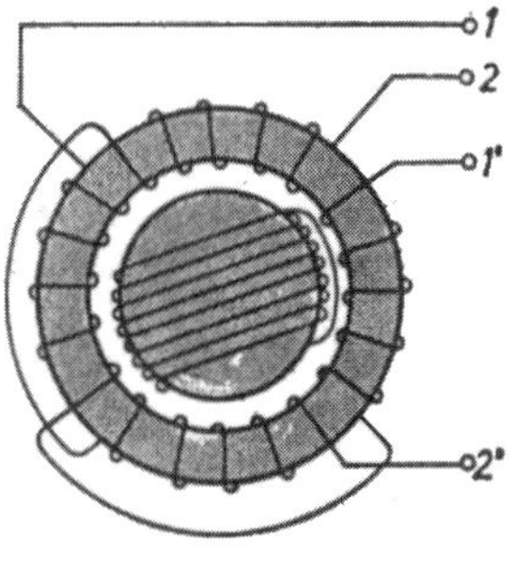

Figure 1.2 Tesla's induction motor (1886)

Both motors have been supplied from a two-phase a.c. power source and thus contained two phase concentrated coil windings 1-1' and 2-2' on the ferromagnetic stator core.

In Ferrari's patent the rotor was made of a copper cylinder, while in the Tesla's patent the rotor was made of a ferromagnetic cylinder provided with a short-circuited winding.

Though the contemporary induction motors have more elaborated topologies (Figure 1.3) and their performance is much better, the principle has remained basically the same.

That is, a multiphase a.c. stator winding produces a traveling field which induces voltages that produce currents in the short-circuited (or closed) windings of the rotor. The interaction between the stator produced field and the rotor induced currents produces torque and thus operates the induction motor. As the torque at zero rotor speed is nonzero, the induction motor is self-starting. The three-phase a.c. power grid capable of delivering energy at a distance to induction motors and other consumers has been put forward by Dolivo-Dobrovolsky around 1880.

In 1889, Dolivo-Dobrovolsky invented the induction motor with the wound rotor and subsequently the cage rotor in a topology very similar to that used today. He also invented the double-cage rotor.

Thus, around 1900 the induction motor was ready for wide industrial use. No wonder that before 1910, in Europe, locomotives provided with induction motor propulsion, were capable of delivering 200 km/h.

However, at least for transportation, the d.c. motor took over all markets until around 1985 when the IGBT PWM inverter was provided for efficient frequency changers. This promoted the induction motor spectacular comeback in variable speed drives with applications in all industries.

Energy efficient, totally enclosed squirrel cage three phase motor
Type M2BA 280 SMB, 90 kW, IP 55, IC 411, 1484 r/min, weight 630 kg

Figure 1.3 A state-of-the-art three-phase induction motor (source ABB motors)

Mainly due to power electronics and digital control, the induction motor may add to its old nickname of "the workhorse of industry" the label of "the racehorse of high-tech".

A more complete list of events that marked the induction motor history follows.

- Better and better analytical models for steady state and design purposes
- The orthogonal (circuit) and space phasor models for transients
- Better and better magnetic and insulation materials and cooling systems
- Design optimization deterministic and stochastic methods
- IGBT PWM frequency changers with low losses and high power density (kW/m^3) for moderate costs
- Finite element methods (FEMs) for field distribution analysis and coupled circuit-FEM models for comprehensive exploration of IMs with critical (high) magnetic and electric loading
- Developments of induction motors for super-high speeds and high powers
- A parallel history of linear induction motors with applications in linear motion control has unfolded
- New and better methods of manufacturing and testing for induction machines
- Integral induction motors: induction motors with the PWM converter integrated into one piece

1.3. INDUCTION MACHINES IN APPLICATIONS

Induction motors are, in general, supplied from single-phase or three-phase a.c. power grids.

Figure 1.4 Start-run capacitor single phase induction motor (Source ABB)

Single-phase supply motors, which have two phase stator windings to provide selfstarting, are used mainly for home applications (fans, washing machines, etc.): 2.2 to 3 kW. A typical contemporary single-phase induction motor with dual (start and run) capacitor in the auxiliary phase is shown in Figure 1.4.

Three-phase induction motors are sometimes built with aluminum frames for general purpose applications below 55 kW (Figure 1.5).

Figure 1.5 Aluminum frame induction motor (Source: ABB)

Table 1.2. EU efficiency classes

EU efficiency classes

Output kW	2-pole Boarderline EFF2/EFF3	EFF1/EFF2	4-pole Boarderline EFF2/EFF3	EFF1/EFF2
1.1	76.2	82.8	76.2	83.8
1.5	78.5	84.1	78.5	85.0
2.2	81.0	85.6	81.0	86.4
3	82.6	86.7	82.6	87.4
4	84.2	87.6	84.2	88.3
5.5	85.7	88.6	85.7	89.2
7.5	87.0	89.5	87.0	90.1
11	88.4	90.5	88.4	91.0
15	89.4	91.3	89.4	91.8
18.5	90.0	91.8	90.0	92.2
22	90.5	92.2	90.5	92.6
30	91.4	92.9	91.4	93.2
37	92.0	93.3	92.0	93.6
45	92.5	93.7	92.5	93.9
55	93.0	94.0	93.0	94.2
75	93.6	94.6	93.6	94.7
90	93.9	95.0	93.9	95.0

Besides standard motors (class B in the U.S.A. and EFF1 in EU), high efficiency classes (class E in U.S.A. and EFF2 and EFF3 in EU) have been developed. Table 1.2. shows data on EU efficiency classes EFF1, EFF2 and EFF3.

Even, 1 to 2% increase in efficiency produces notable energy savings, especially as the motor ratings go up.

Cast iron finned frame efficient motors up to 2000 kW are built today with axial exterior air cooling. The stator and rotor have laminated single stacks.

Typical values of efficiency and sound pressure for such motors built for voltages of 3800 to 11,500 V and 50 to 60 Hz are shown on Table 1.3 (source: ABB). For large starting torque, dual cage rotor induction motors are built (Figure 1.6).

Table 1.3.

Typical values of high voltage 4-pole machines.

Output kW	Efficiency % 4/4 load	3/4 load	1/2 load
500	96.7	96.7	96.1
630	97.0	97.0	96.4
710	97.1	97.1	96.5
800	97.3	97.2	96.8
900	97.4	97.4	96.9
1000	97.4	97.4	97.1
1250	97.6	97.7	97.5
1400	97.8	97.8	97.5
2000	97,9	97,8	97.5

Typical sound pressure levels in dB(A) at 1 meter distance

frame \ rpm	3000	1500	1000	≤750
315	79	78	76	-
355	79	78	76	-
400	79	78	76	75
450	80	78	76	75
500	80	78	76	75
560	80	78	76	75

The variation and measuring tolerance of the figures is ±3 dB(A).

Figure 1.6 Dual cage rotor induction motors for large starting torque (source: ABB)

There are applications (such as overhead cranes) where for safety reasons, the induction motor should be braked quickly, when the motor is turned off. Such an induction motor with integrated brake is shown on Figure 1.7.

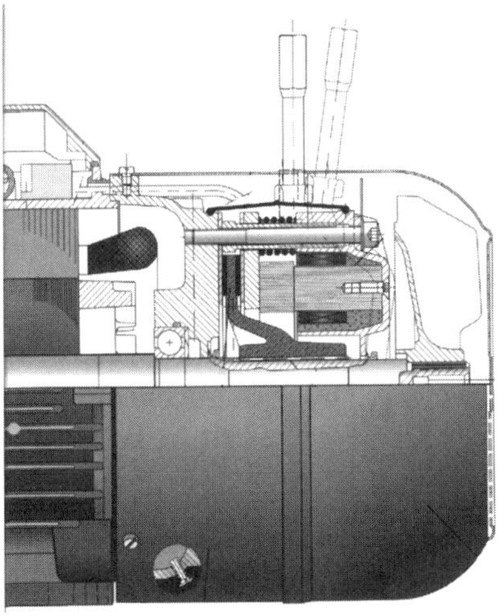

Figure 1.7 Induction motor with integrated electromagnetic brake (source: ABB)

Figure 1.8 Induction motor in pulp and paper industries (source: ABB)

Induction motors used in the pulp and paper industry need to be kept clean from excess pulp fibres. Rated to IP55 protection class, such induction motors prevent the influence of ingress, dust, dirt, and damp (Figure 1.8).

Aluminum frames offer special corrosion protection. Bearing grease relief allows for greasing the motor while it is running.

Induction machines are extensively used for wind turbines up to 750 kW per unit and more. A typical dual winding (speed) induction generator with cage rotor is shown in Figure 1.9.

Generator type M2BA 355 MLA 6/8 B3 E

P_n = 225/50 kW	U_n = 400/400 V D/D	f_n = 50 Hz
n_n = 1007/756 r/min	I_n = 410/95 A	I_s/I_n = 5.2/3.7
T_n = 2230/678 Nm	T_s/T_n = 1.6/1.4	T_{max}/T_n = 4.0/2.1
$\cos\varphi$ = 0.80/0.73	η = 95.7/93.1%	
Q_0 = 120/29.8 kVar	Q_n = 169/46.8 kVAr	

Figure 1.9 Dual stator winding induction generator for wind turbines (source: ABB)

Wind power to electricity conversion has shown a steady growth since 1985. [2] The EU is planning to have 8000 MW of wind power plants by the year 2008. Today, about 3000 MW of wind power generators are at work worldwide, a good part in California.

The environmentally clean solutions to energy conversion are likely to grow in the near future. A 10% coverage of electrical energy needs in many countries of the world seems within reach in the next 20 years. Also small power hydropower plants with induction generators may produce twice as that amount.

Induction motors are used more and more for variable speed applications in association with PWM converters.

Up to 2500 kW at 690 V (line voltage, RMS) PWM voltage source IGBT converters are used to produce variable speed drives with induction motors. A typical frequency converter with a special induction motor series are shown on Figure 1.10.

Constant cooling by integrated forced ventilation independent of motor speed provides high continuous torque capability at low speed in servodrive applications (machine tools, etc.).

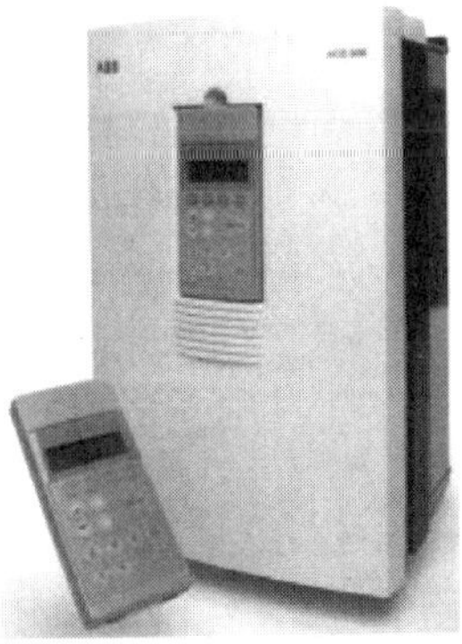

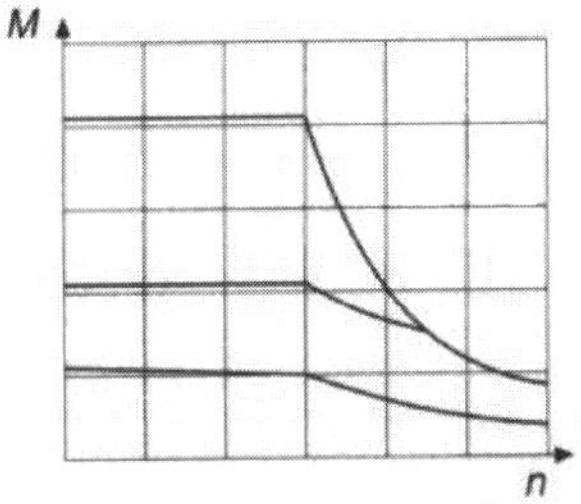

SDM 602 motors

- 1.1 to 75 kW
- Maximum speed 6000 rpm
- Thermal reserves for high pull-out torque and good inverter efficiency
- Enhanced protection against voltage peaks
- Type of enclosure IP 54
- Constant cooling by integrated forced ventilation, independent of motor speed

Figure 1.10 Frequency converter with induction motor for variable speed applications (source: ABB)

Roller table motor,
frame size 355 SB, 35 kW

Roller table motor with a gear,
frame size 200 LB, 9.5 kW

Figure 1.11 Roller table induction motors without a.) and with b.) a gear (source: ABB)

Roller tables use several low speed, ($2p_1$ = 6-12 poles) induction motors with or without mechanical gears, supplied from one or more frequency converters for variable speeds.

The high torque load and high ambient temperature, humidity and dust may cause damage to induction motors unless they are properly designed and built.

Totally enclosed induction motors are fit for such demanding applications (Figure 1.11). Mining applications (hoists, trains, conveyors, etc.) are somewhat similar.

Induction motors are extensively used in marine environments for pumps, fans, compressors, etc. for power up to 700 kW or more. Due to the aggressive environment, they are totally enclosed and may have aluminum (at low power), steel, or cast iron frames (Figure 1.12).

Figure 1.12 Induction motor driving a pump aboard a ship (source: ABB)

Aboard ship, energy consumption reduction is essential, especially money-wise, as electric energy is produced through a diesel engine electrical generator system.

Suppose that electric motors aboard a ship amount to 2000 kW running 8000 hours/year. With energy cost of U.S.$ 0.15/kWh, the energy bill difference per year between two induction motor supplies with 2% difference in motor efficiency is $0.02 \times 2000 \times 8000h \times 0.15$ = U.S.$ 55,200 per year.

Electric trains, light rail people movers in or around town, or trolleybuses of the last generation are propulsed by variable speed induction motor drives.

Most pumps, fans, conveyors, or compressors in various industries are driven by constant or variable speed induction motor drives.

Figure 1.13 2500 kW, 3 kV, 24,000 rpm induction motor (source: ABB)

Figure 1.14 The BC transit system in Vancouver: with linear motion induction motor propulsion (source: UTDC)

The rotor of a 2500 kW, 3 kV, 400 Hz, 2 pole (24,000 rpm) induction motor in different stages of production as shown on Figure 1.13, proves the suitability of induction motors to high speed and high power applications.

Figure 1.15 shows a 3.68 kW (5 HP), 3200 Hz (62,000 rpm) induction motor, with direct water stator cooling, which weighs only 2.268 Kg (5 Pds). This is to show that it is the rather torque than the power that determines the electric motor size.

In parallel with the development of rotary induction motor, power electronics drives linear motion induction motors have witnessed intense studies with quite a few applications. [9, 10] Among them Figure 1.14 shows the UTDC-built linear induction motor people mover (BC transit) in Vancouver now in use for more than a decade.

The panoramic view of induction motor applications sketched above is only to demonstrate the extraordinary breadth of induction machine speed and power ratings and of its applications both for constant and variable speeds.

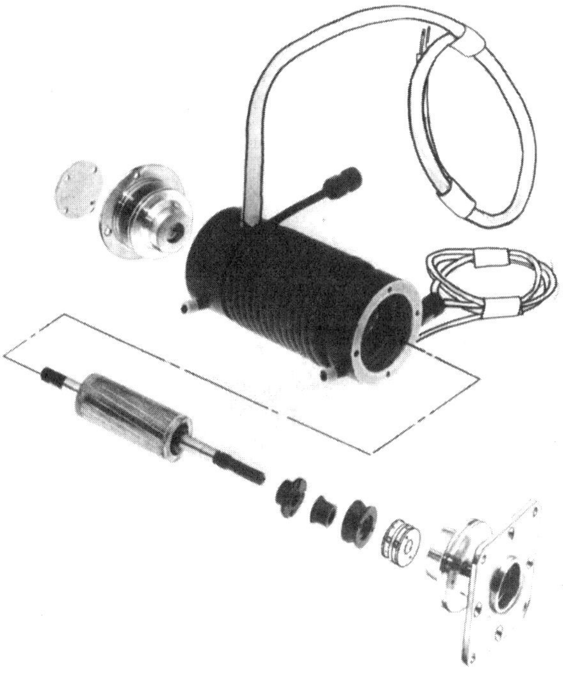

Figure 1.15 3.68 kW (5 HP), 3200 Hz (62,000 rpm) induction motor with forced liquid cooling

1.4. CONCLUSION

After 1885, more than one century from its invention, the induction motor steps into the 21st century with a vigour unparalleled by any other motor.

Power electronics, digital control, computer-added design, and new and better materials have earned the induction motor the new sobriquet of "the racehorse of industry" in addition to the earlier one of "the workhorse of industry".

Present in all industries and in home appliances in constant and variable speed applications, the induction motor seems now ready to make the electric starter/generator system aboard the hybrid vehicles of the near future.

The new challenges in modeling, and optimization design in the era of finite element methods, its control as a motor and generator for even better performance when supplied from PWM converters, and its enormous application potential hopefully justifies this rather comprehensive book on induction machines at the dawn of 21st century.

1.5. REFERENCES

1. R.J. Kerkman, G.L. Skibinski, D.W. Schlegel, AC drives; Year 2000 and Beyond, Record of IEEE-APEC '99, March, 1999.
2. P. Gipe, Wind Energy Comes of Age, Wiley & Sons Inc., New York, 1995.
3. H. Sequenz, The Windings of Electric Machines, vol.3.: A.C. Machines, Springer Verlag, Vienna, 1950 (in German).
4. R. Richter, Electric Machines-Vol. 4-induction machines, Verlag Birkhauser, Bassel/Stuttgart, 1954 (in German).
5. P. Alger, The Nature of Induction Machines, 2nd edition, Gordon & Breach, New York, 1970.
6. C. Veinott, Theory and Design of Small Induction Motors", McGraw-Hill, New York, 1959.
7. J. Stepina, Single Phase Induction Motors, Springer Verlag, 1981 (in German).
8. B. Heller and V. Hamata, Harmonic Field Effects in Induction Machines, Elsevier Scientific, Amsterdam, 1977.
9. E. Laithwaite, Induction machines for special purposes, Newness, 1966.
10. Boldea & S.A. Nasar, Linear Motion Electromagnetic Systems, Wiley Interscience, 1985.
11. EURODEEM by European Commission on Internet: http://iamest.jrc.it/projects/eem/eurodeem.htm

Chapter 2

CONSTRUCTION ASPECTS AND OPERATION PRINCIPLES

The induction machine is basically an a.c. polyphase machine connected to an a.c. power grid, either in the stator or in the rotor. The a.c. power source is, in general, three phase but it may also be single phase. In both cases the winding arrangement on the part of the machine–the primary–connected to the grid (the stator in general) should produce a traveling field in the machine airgap. This traveling field will induce voltages in conductors on the part of the machine not connected to the grid (the rotor, or the mover in general), - the secondary. If the windings on the secondary (rotor) are closed, a.c. currents occur in the rotor.

The interaction between the primary field and secondary currents produces torque from zero rotor speed onward. The rotor speed at which the rotor currents are zero is called the ideal no-load (or synchronous) speed. The rotor winding may be multiphase (wound rotors) or made of bars shortcircuited by end rings (cage rotors).

All primary and secondary windings are placed in uniform slots stamped into thin silicon steel sheets called laminations.

The induction machine has a rather uniform airgap of 0.2 to 3 mm. The largest values correspond to large power, 1 MW or more. The secondary windings may be short-circuited or connected to an external impedance or to a power source of variable voltage and frequency. In the latter case however the IM works as a synchronous machine as it is doubly fed and both stator and rotor-slip frequencies are imposed.

Though historically double stator and double rotor machines have also been proposed to produce variable speed more conveniently, they did not make it to the markets. Today's power electronics seem to move such solutions even further into oblivion.

In this chapter we discuss construction aspects and operation principles of induction machines. A classification is implicit.

The main parts of any IM are

- The stator slotted magnetic core
- The stator electric winding
- The rotor slotted magnetic core
- The rotor electric winding
- The rotor shaft
- The stator frame with bearings
- The cooling system
- The terminal box

The induction machines may be classified many ways. Here are some of them:

- With rotary or linear motion
- Three phase supply or single-phase supply
- With wound or cage rotor

In very rare cases the internal primary is the mover and the external secondary is at a standstill. In most rotary IMs, the primary is the stator and the secondary is the rotor. Not so for linear induction machines. Practically all IMs have a cylindrical rotor and thus a radial airgap between stator and rotor, though, in principle, axial airgap IMs with disk-shaped rotor may be built to reduce volume and weight in special applications.

First we discuss construction aspects of the above mentioned types of IMs and than essentials of operation principles and modes.

2.1. CONSTRUCTION ASPECTS OF ROTARY IMs

Let us start with the laminated cores.

2.1.1. The magnetic cores

The stator and rotor magnetic cores are made of thin silicon steel laminations with unoriented grain-to reduce hysteresis and eddy current losses. The stator and rotor laminations are packed into a single stack (Figure 2.1) or in a multiple stack (Figure 2.2). The latter has radial channels (5-15 mm wide) between elementary stacks (50 to 150 mm long) for radial ventilation.

Single stacks are adequate for axial ventilation.

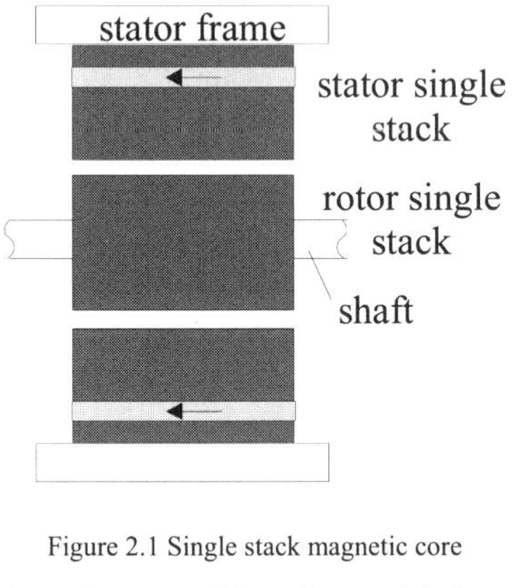

Figure 2.1 Single stack magnetic core

Single-stack IMs have been traditionally used below 100 kW but recently have been introduced up to 2 MW as axial ventilation has been improved drastically. The multistack concept is necessary for large power (torque) with long stacks.

The multiple stacks lead to additional winding losses, up to 10%, in the stator and in the rotor as the coils (bars) lead through the radial channels without producing torque. Also, the electromagnetic field energy produced by the coils (bar) currents in the channels translate into additional leakage inductances which tend to reduce the breakdown torque and the power factor. They also reduce the starting current and torque. Typical multistack IMs are shown in Figure 2.2.

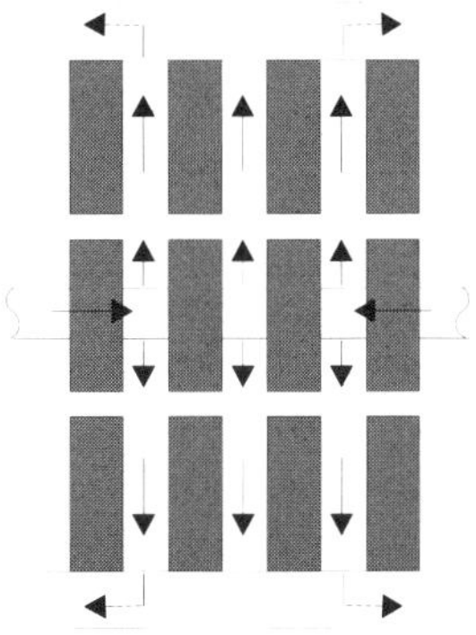

Figure 2.2 Multiple stack IM

For IMs of fundamental frequency up to 300 Hz, 0.5 mm thick silicon steel laminations lead to reasonable core losses 2 to 4 W/Kg at 1T and 50 Hz.

For higher fundamental frequency, thinner laminations are required. Alternatively, anisotropic magnetic powder materials may be used to cut down the core losses at high fundamental frequencies, above 500 Hz, however at lower power factor (see Chapter 3 on magnetic materials).

2.1.2. Slot geometry

The airgap, or the air space between stator and rotor, has to be traveled by the magnetic field produced by the stator. This in turn will induce voltages and produce currents in the rotor windings. Magnetizing air requires large magnetomotive forces (mmfs) or amperturns. The smaller the air (nonmagnetic) gap, the smaller the magnetization mmf. The lower limit of airgap g is determined by mechanical constraints and by the ratio of the stator and slot openings b_{os}, b_{or} to airgap g in order to keep additional losses of surface core and tooth flux pulsation within limits. The tooth is the lamination radial sector between two neighbouring slots.

Putting the windings (coils) in slots has the main merit of reducing the magnetization current. Second, the winding manufacture and placement in slots becomes easier. Third, the winding in slots are better off in terms of mechanical rigidity and heat transmission (to the cores). Finally the total mmf per unit length of periphery (the coil height) could be increased and thus large power IMs could be built efficiently. What is lost is the possibility to build windings (coils) that can produce purely sinusoidal distributed amperturns (mmfs) along the periphery of the machine airgap. But this is a small price to pay for the incumbent benefits.

The slot geometry depends mainly on IM power (torque) level and thus on the type of magnetic wire–with round or rectangular cross section–from which the coils of windings are made. With round wire (random wound) coils for small power IMs (below 100 kW in general), the coils may be introduced in slots wire by wire and thus the slot openings may be small (Figure 2.3a). For preformed coils (in large IMs), made, in general, of rectangular cross-section wire, open or semiopen slots are used (Figure 2.3b, c).

In general, the slots may be rectangular, straight trapezoidal, or rounded trapezoidal. Open and semiopen slots tend to be rectangular (Figure 2.3b, c) in shape and the semiclosed are trapezoidal or rounded trapezoidal (Figure 2.3a).

In an IM, only slots on one side are open, while on the other side, they are semiclosed or semiopen.

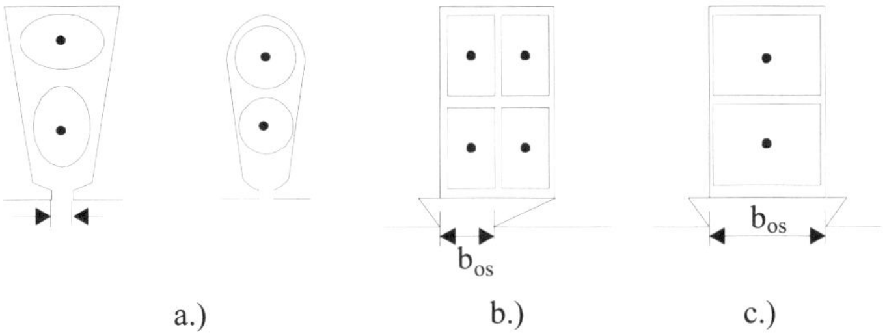

Figure 2.3 Slot geometrics to locate coil windings
a.) semiclosed b.) semiopen c.) open

The reason is that a large slot opening, b_{os}, per gap, g, ratio ($b_{os}/g > 6$) leads to lower average flux density, for given stator mmf and to large flux pulsation in the rotor tooth, which will producc large additional core losses. In the airgap flux density harmonics lead to parasitic torques, noise, and vibration as presented in subsequent, dedicated, chapters. For semiopen and semiclosed slots, $b_{os}/g \cong (4\text{-}6)$ in general. For the same reasons, the rotor slot opening per airgap $b_{or}/g \cong 3\text{-}4$ wherever possible. Too small a slot opening per gap ratio leads to a higher magnetic field in the slot neck (Figure 2.3) and thus to a higher slot leakage inductance, which causes lower starting torque and current and lower breakdown torque.

Slots as in Figure 2.3 are used both for stator and wound rotors. Rotor slot geometry for cage-rotors is much more diversified depending upon

- Starting and rated load constraints (specifications)
- Constant voltage/frequency (V/f) or variable voltage/frequency supply operation
- Torque range.

Less than rated starting torque, high efficiency IMs for low power at constant V/f or for variable V/f may use round semiclosed slots (Figure 2.4a).

Rounded trapezoidal slots with rectangular teeth are typical for medium starting torque (around rated value) in small power IMs (Figure 2.4b).

Closed rotor slots may be used to reduce noise and torque pulsations for low power circulating fluid pumps for homes at the expense of large rotor leakage inductance; that is, lower breakdown torque. In essence the iron bridge (0.5 to 1 mm thick), above the closed rotor slot, already saturates at 10 to 15% of rated current at a relative permeability of 50 or less that drops further to 15 to 20 for starting conditions (zero speed, full voltage).

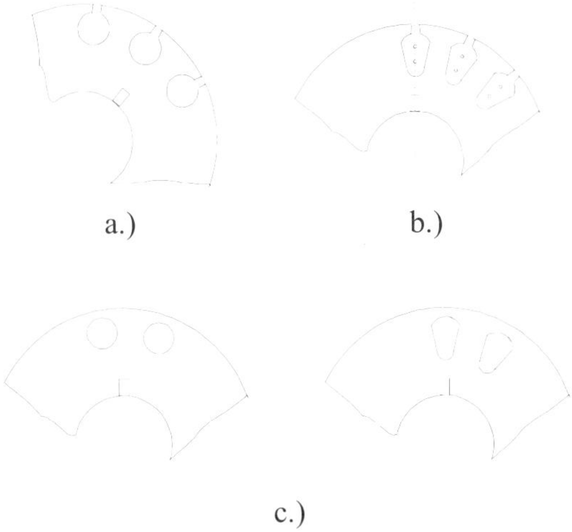

Figure 2.4 Rotor slots for cage rotors
a.) semiclosed and round b.) semiclosed and round trapezoidal c.) closed slots

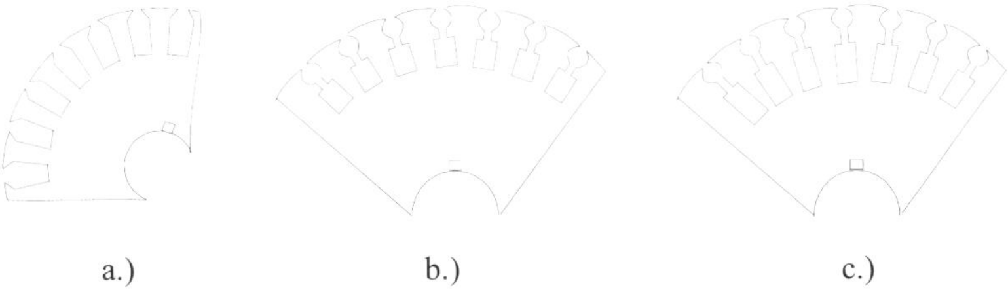

Figure 2.5 Rotor slots for low starting current IMs
a.) high slip, high starting torque b.) moderate starting torque c.) very high starting torque

For high starting torque, high rated slip (lower rated speed with respect to ideal no-load speed), rectangular deep bar rotor slots are used (Figure 2.5a). Inverse trapezoidal or double cage slots are used for low starting current and moderate and large starting torque (Figure 2.5b, c). In all these cases, the rotor slot leakage inductance increases and thus the breakdown torque is reduced to as low as 150 to 200% rated torque.

More general, optimal shape cage rotor slots may be generated through direct FEM optimization techniques to meet desired performance constraints for special applications. In general, low stator current and moderate and high starting torque rely on the rotor slip frequency effect on rotor resistance and leakage inductance.

At the start, the frequency of rotor currents is equal to stator (power grid) frequency f_1, while at full load $f_{sr} = S_n f_1$; S_n, the rated slip, is about 0.08 and less than 0.01 in large IMs:

$$S = \frac{f_1 - np_1}{f_1}; \quad \text{n - speed in rps} \tag{2.1}$$

p_1 is the number of spatial periods of airgap traveling field wave per revolution produced by the stator windings:

$$B_{g_0}(x,t) = B_{g_{m0}} \cos(p_1\theta_1 - \omega_1 t) \tag{2.2}$$

θ_1-mechanical position angle; $\omega_1 = 2\pi f_1$.

Remember that, for variable voltage and frequency supply (variable speed), the starting torque and current constraints are eliminated as the rotor slip frequency Sf_1 is always kept below that corresponding to breakdown torque.

Very important in variable speed drives is efficiency, power factor, breakdown torque, and motor initial or total costs (with capitalized loss costs included).

2.1.3. IM windings

The IM is provided with windings both on the stator and on the rotor. Stator and rotor windings are treated in detail in Chapter 4.

Here we refer only to a primitive stator winding with 6 slots for two poles (Figure 2.6).

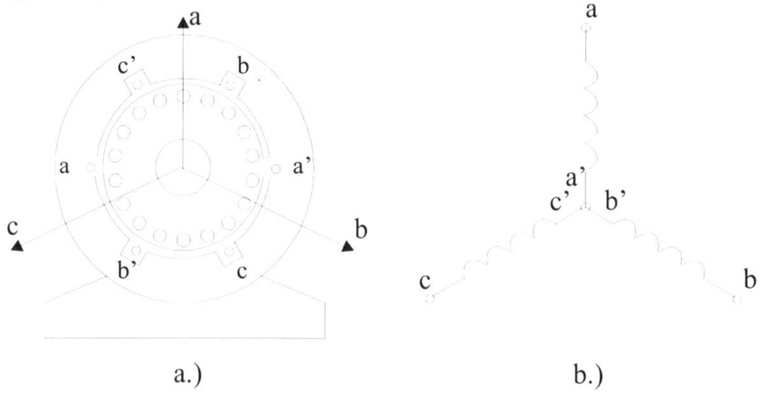

Figure 2.6 Primitive IM with 6 stator slots and cage rotor

Each phase is made of a single coil whose pitch spans half of rotor periphery. The three phases (coils) are space shifted by 120°. For our case there are 120° mechanical degrees between phase axes as $p_1 = 1$ pole pair. For $p_1 = 2$, 3, 4, 5, 6, there will be $120°/p_1$ mechanical degrees between phase axes.

The airgap field produced by each phase has its maximum in the middle of the phase coil (Figure 2.6) and, with the slot opening eliminated, it has a rectangular spatial distribution whose amplitude varies sinusoidally in time with frequency f_1 (Figure 2.7).

It is evident from Figure 2.7 that when the time angle θ_t electrically varies by $\pi/6$, so does the fundamental maximum of airgap flux density with space harmonics neglected, a travelling wavefield in the airgap is produced. Its direction of motion is from phase a to phase b axis, if the current in phase a leads (in time) the current in phase b, and phase b leads phase c. The angular speed of this field is simply ω_1, in electrical terms, or ω_1/p_1 in mechanical terms (see also Equation (2.2)).

$$\Omega_1 = 2\pi n_1 = \omega_1 / p_1 \tag{2.3}$$

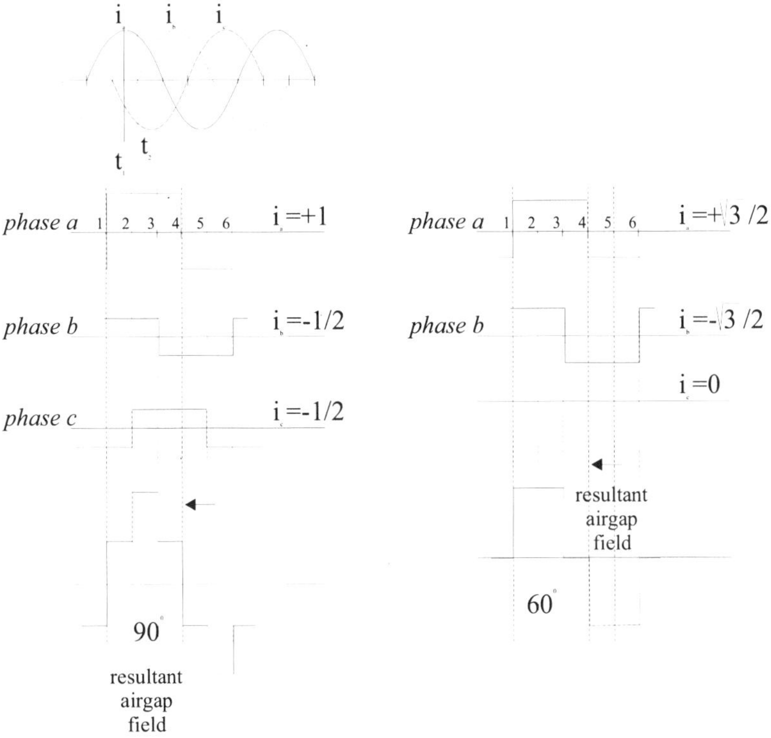

Figure 2.7 Stator currents and airgap field at times t_1 and t_2.

So n_1, the traveling field speed in rps, is

$$n_1 = f_1 / p_1 \tag{2.4}$$

This is how the ideal no load speed for 50(60) Hz is 3000/3600 rpm for $p_1 = 1$, 1500/1800 rpm for $p_1 = 2$ and so on.

As the rated slip S_n is small (less than 10% for most IMs), the rated speed is only slightly lower than a submultiple of f_1 in rps. The crude configuration in Figure 2.7 may be improved by increasing the number of slots, and by using two layers of coils in each slot. This way the harmonics content of airgap flux density diminishes, approaching a better pure traveling field, despite the inherently discontinuous placement of conductors in slots.

A wound stator is shown in Figure 2.8. The three phases may be star or delta connected. Sometimes, during starting, the connection is changed from star to delta (for delta-designed IMs) to reduce starting currents in weak local power grids.

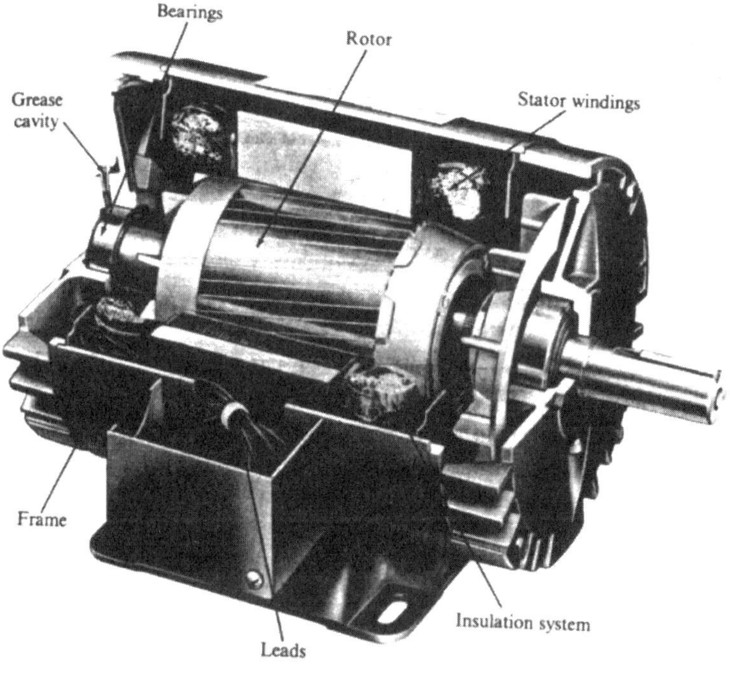

Figure 2.8 IM wound three-phase stator winding with cage rotor

Wound rotors are built in a similar way (Figure 2.9). The slip rings are visible to the right. The stator-placed brush system is not. Single-phase-supply IMs have, on the other hand, in general, two windings on the stator.

The main winding (m) and the auxiliary (or starting) one (a) are used to produce a traveling field in the airgap. Similar to the case of three phases, the two windings are spatially phase shifted by 90° (electrical) in general. To phase shift the current in the auxiliary winding, a capacitor is used.

Figure 2.9 Three-phase wound rotor

In reversible motion applications, the two windings are identical. The capacitor is switched from one phase to the other to change the direction of traveling field.

When auxiliary winding works continuously, each of the two windings uses half the number of slots. Only the number of turns and the wire cross-section differ.

The presence of auxiliary winding with capacitance increases the torque, efficiency, and power factor. In capacitor-start low-power (below 250 W) IMs, the main winding occupies 2/3 of the stator slots and the auxiliary (starting) winding only 1/3. The capacitor winding is turned off in such motors by a centrifugal or time relay at a certain speed (time) during starting. In all cases, a cage winding is used in the rotor (Figure 2.10). For very low power levels (below 100 W in general), the capacitor may be replaced by a resistance to cut cost at the expense of a lower efficiency and power factor.

Finally, it is possible to produce a traveling field with a single phase concentrated coil winding with shaded poles (Figure 2.11). The short-circuiting ring is retarding the magnetic flux of the stator in the shaded pole area with respect to the unshaded pole area.

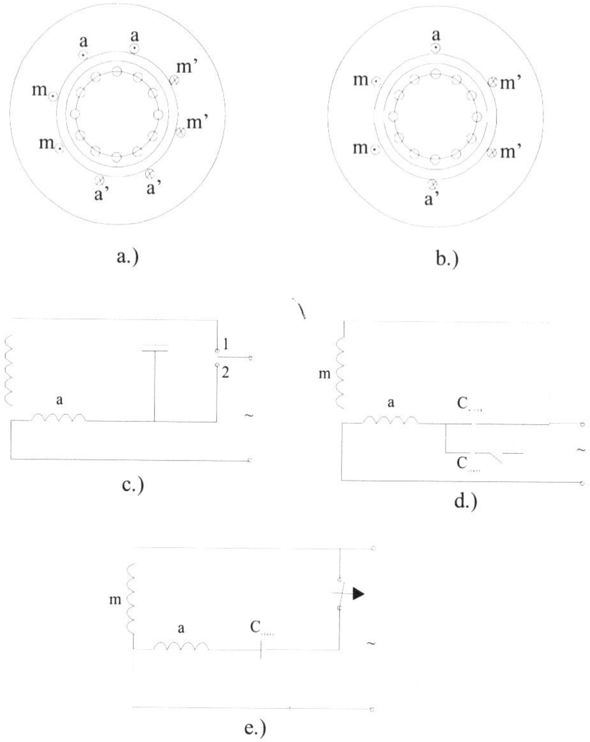

Figure 2.10 Single phase supply capacitor IMs
a.) primitive configuration with equally strong windings
b.) primitive configuration with 2/3, 1/3 occupancy windings
c.) reversible motor d.) dual capacitor connection
e.) capacitor start-only connection

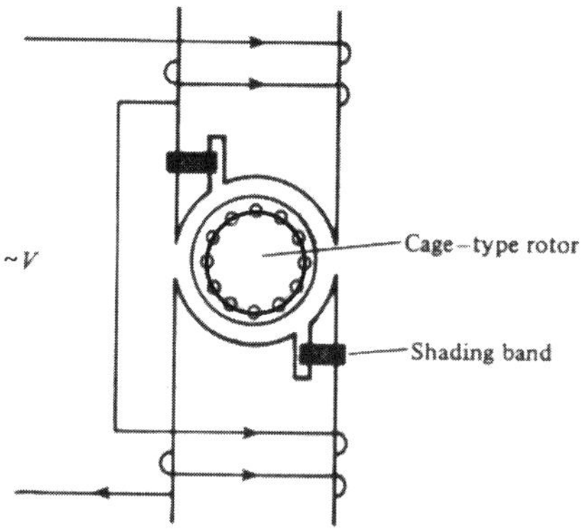

Figure 2.11 Single phase (shaded pole) IM

The airgap field has a traveling component and the motor starts rotating from the unshaded to the shaded pole zone. The rotor has a cage winding. The low cost and superior ruggedness of shaded pole single phase IM is paid for by a lower efficiency and power factor. This motor is more of historical importance and is seldom used, usually below 100 W, where cost is the prime concern.

2.1.4. Cage rotor windings

As mentioned above, the rotor of IMs is provided with single or double cage windings (Figure 2.12), in addition to typical three phase windings on wound rotors.

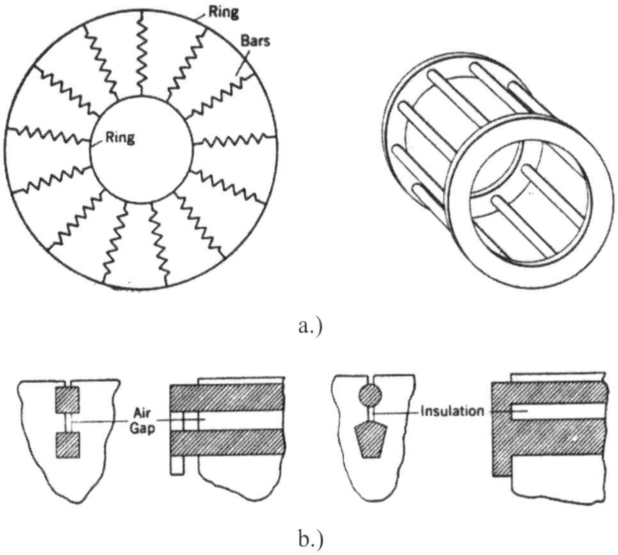

Figure 2.12 Cage rotor windings
a.) single cage b.) double cage

The cage bars and end rings are made of diecast aluminum for low and medium power and from brass or copper for high powers. For medium and high powers, the bars are silver-rings–welded to end to provide low resistance contact.

For double cages brass may be used (higher resistivity) for the upper cage and copper for the lower cage. In this case, each cage has its own end ring, mainly due to thermal expansion constraints.

For high efficiency IMs copper tends to be preferred due to higher conductivity, larger allowable current density, and working temperatures.

The diecasting of aluminum at rather low temperatures results in low rotor mass production costs for low power IMs.

Energy efficient, totally enclosed squirrel cage three phase motor
Type M2BA 280 SMB, 90 kW, IP 55, IC 411, 1484 r/min, weight 630 kg

Figure 2.13 Cutaway view of a modern induction motor

The debate over aluminum or copper is not yet decided and both materials are likely to be used depending on the application and power (torque) level.

Although some construction parts such as frames, cooling system, shafts, bearings, and terminal boxes have not been described here, we will not dwell on them at this time as they will be discussed again in subsequent chapters. Instead, Figure 2.13 presents a rather complete cutaway view of a fairly modern induction motor. It has a single stack magnetic core, thus axial ventilation is used by a fan on the shaft located beyond the bearings. The heat evacuation area is increased by the finned stator frame. This technology has proven practical up to 2 MW in low voltage IMs.

The IM in Figure 2.13 has a single-cage rotor winding. The stator winding is built in two layers out of round magnetic wire. The coils are random wound. The stator and rotor slots are of the semiclosed type. Configuration in Figure 2.13 is dubbed as totally enclosed fan cooled (TEFC), as the ventilator is placed outside bearings on the shaft.

It is a low voltage IM (below 690 V RMS–line voltage).

2.2. CONSTRUCTION ASPECTS OF LINEAR INDUCTION MOTORS

In principle, for each rotary IM there is a linear motion counter-part. The imaginary process of cutting and unrolling the rotary machine to obtain the linear induction motor (LIM) is by now classic (Figure 2.14). [1]

The primary may now be shorter or larger than the secondary. The shorter component will be the mover. In Figure 2.14 the primary is the mover. The primary may be double sided (Figure 2.14d) or single sided (Figure 2.14 c, e).

The secondary material is copper or aluminum for the double-sided LIM and it may be aluminum (copper) on solid iron for the single-sided LIM. Alternatively, a ladder conductor secondary placed in the slots of a laminated core may be used, as for cage rotor rotary IMs (Figure 2.14c). This latter case is typical for short travel (up to a few meters) low speed (below 3 m/s) applications.

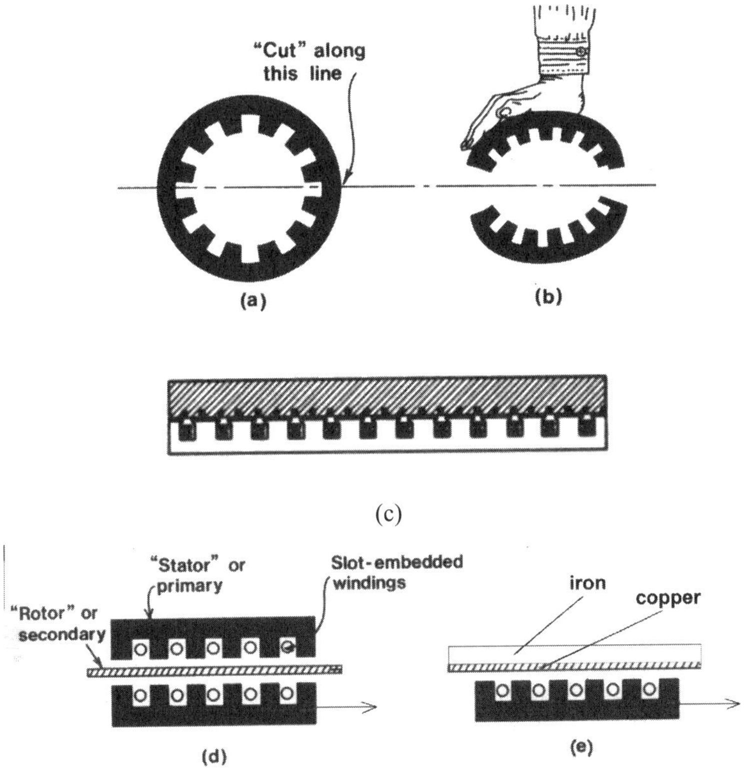

Figure 2.14 Cutting and unrolling process to obtain a LIM

Finally, the secondary solid material may be replaced by a conducting fluid (liquid metal), when a double sided linear induction pump is obtained. [2]

All configurations on Figure 2.14 may be dubbed as flat ones. The primary winding produces an airgap field with a strong traveling component at the linear speed u_s.

$$u_s = \tau \cdot \frac{\omega_1}{\pi} = 2\tau f_1 \tag{2.5}$$

The number of pole pairs does not influence the ideal no-load linear speed u_s. Incidentally, the peripheral ideal no-load speed in rotary IMs has the same formula (2.5) where τ is the pole pitch (the spatial semiperiod of the traveling field).

In contrast to rotary IMs, the LIM has an open magnetic structure along the direction of motion. Additional phenomena called longitudinal effects occur due to this. They tend to increase with speed, deteriorating the thrust, efficiency, and power factor performance. Also, above 3 to 5 m/s speed, the airgap has to be large (due to mechanical clearance constraints): in general, 3 to 12 mm. This leads to high magnetization currents and thus a lower power factor than in rotary IMs.

However, the LIM produces electromagnetic thrust directly and thus eliminates any mechanical transmission in linear motion applications (transportation).

Apart from flat LIM, tubular LIMs may be obtained by rerolling the flat LIM around the direction of motion (Figure 2.15).

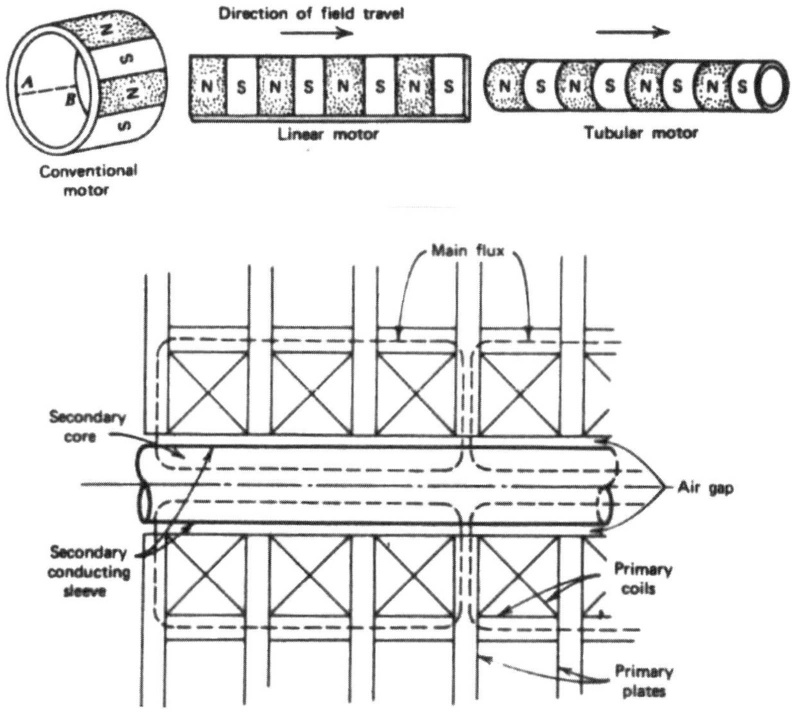

Figure 2.15 The tubular linear induction motor

The coils of primary winding are now of circular shape. The rotor may be an aluminum cylinder on solid iron. Alternatively, a secondary cage may be built. In this case the cage is made of ring shape-bars. Transverse laminations of disc shape may be used to make the magnetic circuit easy to manufacture, but care must be excersized to reduce the core losses in the primary. The blessing of circularity renders this LIM more compact and proper for short travels (1 m or less).

In general, LIMs are characterized by a continuous thrust density (N/cm^2 of primary) of up to 2 (2.5) N/cm^2 without forced cooling. The large values correspond to larger LIMs. The current LIM use for a few urban transportation systems in North America, Middle East, and East Asia has proved that they are rugged and almost maintenance free. More on LIMs in Chapter 20.

2.3. OPERATION PRINCIPLES OF IMs

The operation principles are basically related to torque (for rotary IMs) and, respectively, thrust (for LIMs) production. In other words, it is about forces in traveling electromagnetic fields. Or even simpler, why the IM spins and the LIM moves linearly. Basically the torque (force) production in IMs and LIMs may be approached via

- Forces on conductors in a travelling field
- The Maxwell stress tensor [3]
- The energy (coenergy) derivative
- Variational principles (Lagrange equations) [4]

The electromagnetic traveling field produced by the stator currents exists in the airgap and crosses the rotor teeth to embrace the rotor winding (rotor cage)–Figure 2.16. Only a small fraction of it radially traverses the top of the rotor slot which contains conductor material.

It is thus evident that, with rotor and stator conductors in slots, there are no main forces experienced by the conductors themselves. Therefore, the method of forces experienced by conductors in fields does not apply directly to rotary IMs with conductors in slots.

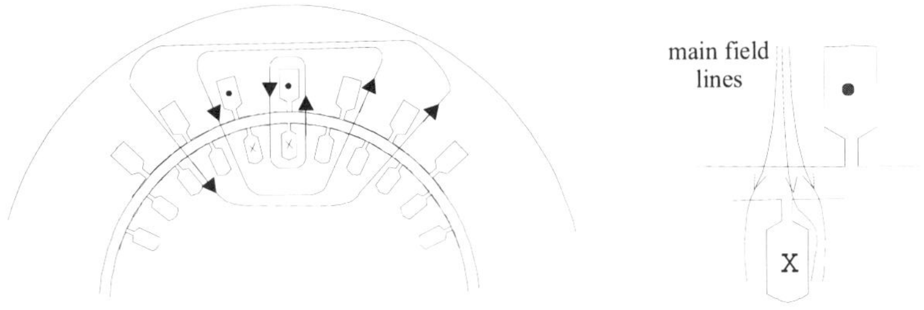

Figure 2.16 Flux paths in IMs

The current occurs in the rotor cage (in slots) because the magnetic traveling flux produced by the stator in any rotor cage loop varies in time even at zero speed (Figure 2.17). If the cage rotor rotates at speed n (in rps), the stator-produced traveling flux density in the airgap (2.2) moves with respect to the rotor with the relative speed n_{sr}.

$$n_{sr} = \frac{f_1}{p_1} - n = S \cdot \frac{f_1}{p_1} \tag{2.6}$$

In rotor coordinates, (2.2) may be written as.

$$B(\theta_r, t) = B_m \cos(p_1\theta_2 - S\omega_1 t) \tag{2.7}$$

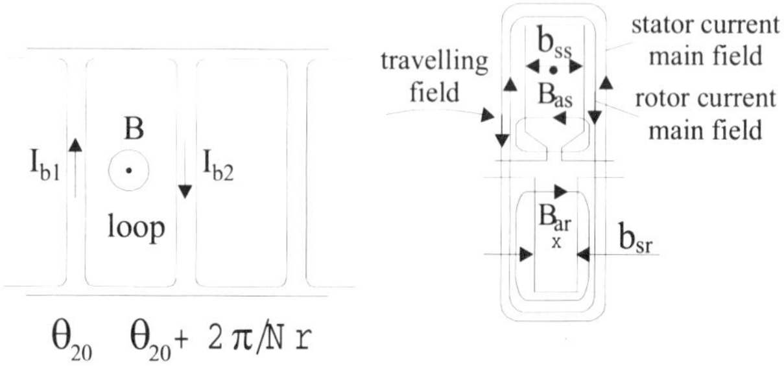

Figure 2.17 Traveling flux crossing the rotor cage loops (a), leakage and main fields (b)

Consequently, with the cage bars in slots, according to electromagnetic induction law, a voltage is induced in loop 1 of the rotor cage and thus a current occurs in it such that its reaction flux opposes the initial flux.

The current which occurs in the rotor cage, at rotor slip frequency Sf_1 (see (2.7)), produces a reaction field which crosses the airgap. This is the main reaction field. Thus the resultant airgap field is the product of both stator and rotor currents. As the two currents tend to be more than $2\pi/3$ when shifted, the resultant (magnetization) current is reasonably low; in fact, it is 25 to 60% of the rated current depending on the machine airgap g to pole pitch τ ratio. The higher the ratio τ/g, the smaller the magnetization current in p.u.

The stator and rotor currents also produce leakage flux paths crossing the slots: B_{as} and B_{ar}.

According to the Maxwell stress tensor theory, at the surface border between mediums with different magnetic fields and permeabilities (μ_0 in air, $\mu \neq \mu_0$ in the core), the magnetic field produces forces. The interaction force component perpendicular to the rotor slot wall is. [3]

$$F_{tn} = \frac{B_{ar}(\theta_r, t) B_{tr}(\theta_r, t)}{\mu_0} \bar{n}_{tooth} \tag{2.8}$$

The magnetic field has a radial (B_{tr}) and a tangential (B_{ar}) component only.

Now, for the rotor slot, $B_{ar}(\theta_r)$–tangential flux density–is the slot leakage flux density.

$$B_{ar}(\theta_r) = \frac{\mu_0 I_b(\theta_r, t)}{b_{sr}} \tag{2.9}$$

The radial flux density that counts in the torque production is that produced by the stator currents, $B_{tr}(\theta_r)$, in the rotor tooth.

$$B_{tr}(\theta_r, t) = B(\theta_r, t) \cdot \frac{b_{tr} + b_{sr}}{b_{sr}} \tag{2.10}$$

In (2.10), b_{tr} is the mean rotor tooth width while $B(\theta_r,t)$ is the airgap flux density produced by the stator currents in the airgap.

When we add the specific Maxwell stress tensors [1] on the left and on the right side walls of the rotor slot we should note that the normal direction changes sign on the two surfaces. Thus the addition becomes a subtraction.

$$f_{tooth}(N/m^2) = -\left(\frac{B_{ar}(\theta_r + \Delta\theta, t)B_{tr}(\theta_r + \Delta\theta, t)}{\mu_0} - \frac{B_{ar}(\theta_r, t)B_{tr}(\theta_r, t)}{\mu_0}\right) \tag{2.11}$$

Essentially the slot leakage field B_{ar} does not change with $\Delta\theta$–the radial angle that corresponds to a slot width.

$$f_{tooth}(N/m^2) = -\frac{B_{ar}(\theta_r, t)}{\mu_0}(B_{tr}(\theta_r + \Delta\theta, t) - B_{tr}(\theta_r, t)) \tag{2.12}$$

The approximate difference may be replaced by a differential when the number of slots is large. Also from (2.9):

$$f_{tooth}(N/m^2) = -\frac{-I_b(\theta_r, t)}{b_{sr}} \cdot \frac{\Delta B(\theta_r, t)}{\Delta\theta_{slot}} \cdot \frac{(b_{tr} + b_{sr})}{b_{sr}} \cdot \Delta\theta_{slot} \tag{2.13}$$

Therefore it is the change of stator produced field with θ_r, the traveling field existence, that produces the tangential force experienced by the walls of each slot. The total force for one slot may be obtained by multiplying the specific force in (2.13) by the rotor slot height and by the stack length.

It may be demonstrated that with a pure traveling field and rotor current traveling wave $I_b(\theta_r,t)$, the tangential forces on each slot pair of walls add up to produce a smooth torque. Not so if the field is not purely traveling.

Based on same rationale, an opposite direction tangential force on stator slot walls may be calculated. It is produced by the interaction of stator leakage field with rotor main reaction field. This is to be expected as action equals reaction according to Newton's third law. So the tangential forces that produce the torque occur on the tooth radial walls. Despite this reality, the principle of IM is traditionally explained by forces on currents in a magnetic field.

It may be demonstrated that, mathematically, it is correct to "move" the rotor currents from rotor slots, eliminate the slots and place them in an infinitely thin conductor sheet on the rotor surface to replace the actual slot-cage rotor configuration (Figure 2.18). This way the tangential force will be exerted directly on the "rotor" conductors. Let us use this concept to explain further the operation modes of IM.

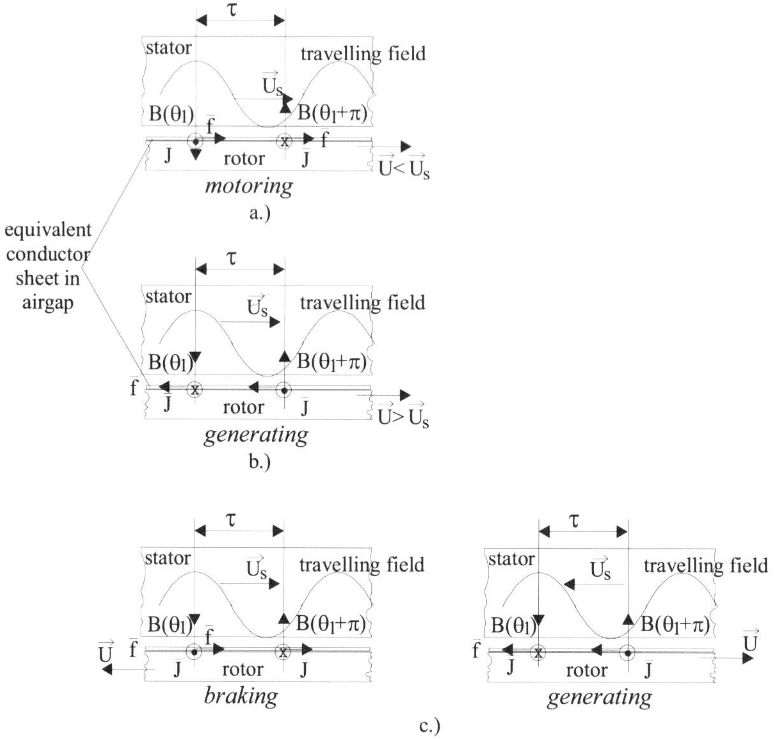

Figure 2.18 Operation modes of IMs

a.) motoring: $\vec{U} < \vec{U}_s$ both in the same direction

b.) generating: $\vec{U} > \vec{U}_s$; both in the same direction

c.) braking: $\vec{U}$ opposite to $\vec{U}_s$; either $\vec{U}$ or $\vec{U}_s$ changes direction.

The relative speed between rotor conductors and stator traveling field is $\overline{U} - \overline{U}_s$, so the induced electrical field in the rotor conductors is.

$$\overline{E} = \left(\overline{U} - \overline{U}_s\right) \times \overline{B} \tag{2.14}$$

As the rotor cage is short-circuited, no external electric field is applied to it, so the current density in the rotor conductor $\overline{J}$ is

$$\overline{J} = \sigma_{Al}\overline{E} \tag{2.15}$$

Finally, the force (per unit volume) exerted on the rotor conductor by the traveling field, f_t, is

$$f_t = \overline{J} \times \overline{B} \tag{2.16}$$

Applying these fundamental equations for rotor speed $\overline{U}$ and field speed $\overline{U}_s$ in the same direction we obtain the motoring mode for $\vec{U} < \vec{U}_s$, and, respectively, the generating mode for $\vec{U} > \vec{U}_s$, as shown on Figure 2.17a, b. In the motoring mode, the force on rotor conductors acts along the direction of motion while, in the generating mode, it acts against it. In the same time, the electromagnetic (airgap) power P_e is negative.

$$P_e = \overline{f}_t \cdot \overline{U}_s > 0\text{ ; for motoring } (f_t > 0, U > 0)$$

$$P_e = \overline{f}_t \cdot \overline{U}_s < 0\text{; for generating } (f_t < 0, U > 0) \quad (2.17)$$

This simply means that in the generating mode the active power travels from rotor to stator to be sent back to the grid after losses are covered. In the braking mode ($U < 0$ and $U_s > 0$ or $U > 0$ and $U_s < 0$), as seen in Figure 2.18c, the torque acts against motion again but the electromagnetic power P_e remains positive ($\overline{U}_s > 0$ and $\overline{f}_t > 0$ or $\overline{U}_s < 0$ and $\overline{f}_t < 0$). Consequently, active power is drawn from the power source. It is also drawn from the shaft. The summation of the two is converted into induction machine losses.

The braking mode is thus energy intensive and should be used only at low frequencies and speeds (low U_s and U), in variable speed drives, to "lock" the variable speed drive at standstill under load.

The linear induction motor operation principles and operation modes are quite similar to those presented for rotary induction machines.

2.4. SUMMARY

- The IM is an a.c. machine. It may be energized directly from a three phase a.c. or single phase a.c. power grid. Alternatively it may be energized through a PWM converter at variable voltage (V) and frequency (f).
- The IM is essentially a traveling field machine. It has an ideal no-load speed $n_1 = f_1/p_1$; p_1 is the number of traveling field periods per one revolution.
- The IM main parts are the stator and rotor slotted magnetic cores and windings. The magnetic cores are, in general, made of thin silicon steel sheets (laminations) to reduce the core losses to values such as 2 to 4 W/Kg at 60 Hz and 1 T.
- Three or two phase windings are placed in the primary (stator) slots. Windings are coil systems connected to produce a travelling mmf (amperturns) in the airgap between the stator and the rotor cores.
- The slot geometry depends on power (torque) level and performance constraints.

  Starting torque and current, breakdown torque, rated efficiency, and power factor are typical constraints (specifications) for power grid directly energized IMs.

- Two phase windings are used for capacitor IMs energised from a single phase a.c. supply to produce traveling field. Single phase a.c. supply is typical for home appliances.
- Cage windings made of solid bars in slots with end rings are used on most IM rotors. The rotor bar cross-section is tightly related to all starting and running performances. Deep-bar or double-cage windings are used for high starting torque, low starting current IMs fed from the power grid (constant V and f).
- Linear induction motors are obtained from rotary IMs by the cut-and-unroll process. Flat and tubular configurations are feasible with single sided or double sided primary. Either primary or secondary may be the mover in LIMs. Ladder or aluminum sheet or iron are typical for single sided LIM secondaries. Continuous thrust densities up to 2 to 2.5 N/cm^2 are feasible with air cooling LIMs.
- In general, the airgap g per pole pitch τ ratio is larger than for rotary IM and thus the power factor and efficiency are lower. However, the absence of mechanical transmission in linear motion applications leads to virtually maintenance-free propulsion systems. Urban transportation systems with LIM propulsion are now in use in a few cities from three continents.
- The principle of operation of IMs is related to torque production. By using the Maxwell stress tensor concept it has been shown that, with windings in slots, the torque (due to tangential forces) is exerted mainly on slot walls and not on the conductors themselves.

  Stress analysis during severe transients should illustrate this reality. It may be demonstrated that the rotor winding in slots can be "mathematically" moved in the airgap and transformed into an equivalent infinitely thin current sheet. The same torque is now exerted directly on the rotor conductors in the airgap. The LIM with conductor sheet on iron naturally resembles this situation.
- Based on the $\bar{J} \times \bar{B}$ force principle, three operation modes of IM are easily identified.
  - Motoring: $|U| < |U_s|$; U and U_s either positive or negative
  - Generating: $|U| > |U_s|$; U and U_s either positive or negative
  - Braking: $(U > 0 \;\&\; U_s < 0)$ or $(U < 0 \;\&\; U_s > 0)$
- For the motoring mode, the torque acts along the direction motion while, for the generator mode, it acts against it as it does during braking mode. However, during generating the IM returns some power to the grid, after covering the losses, while for braking it draws active power also from the power grid.
- Generating is energy–conversion advantageous while braking is energy intensive. Braking is recommended only at low frequency and speed, with variable V/f PWM converter supply, to stall the IM drive on load (like in overhead cranes). The energy consumption is moderate in such cases (as the frequency is small).

2.5. REFERENCES

1. I. Boldea and S.A. Nasar, Linear Motion Electric Machines, J.Wiley Interscience, 1976.
2. I. Boldea & S.A. Nasar, Linear Motion Electromagnetic Systems, Wiley Interscience, New York, 1985, Chapter 5.
3. M. Schwartz, "Principles of Electrodynamics", Dower Publ. Inc., New York, 1972, pp.180.
4. D.C. White, H.H. Woodson, "Electromechanical Energy Conversion", John Wiley and Sons. Inc., London, 1959.

Chapter 3

MAGNETIC, ELECTRIC, AND INSULATION MATERIALS FOR IM

3.1. INTRODUCTION

Induction machines contain magnetic circuits traveled by a.c. and traveling magnetic fields and electric circuits flowed by alternative currents. The electric circuits are insulated from the magnetic circuits (cores). The insulation system comprises the conductór, slot and interphase insulation.

Magnetic, electrical, and insulation materials are characterized by their characteristics (B(H) curve, electrical resistivity, dielectric constant, and breakdown electric field (V/m)) and their losses.

At frequencies encountered in IMs (up to tens of kHz, when PWM inverter fed), the insulation losses are neglected. Soft magnetic materials are used in IM as the magnetic field is current produced. The flux density (B)/magnetic field (H) curve and cycle depend on the soft material composition and fabrication process. Their losses in W/kg depend on the B-H hysteresis cycle, frequency, electrical resistivity, and the a.c. (or) traveling field penetration into the soft magnetic material.

Silicon steel sheets are standard soft magnetic materials for IMs. Amorphous soft powder materials have been introduced recently with some potential for high frequency (high speed) IMs. The pure copper is the favorite material for the stator electric circuit (windings), while aluminum or brass is used for rotor squirrel cage windings.

Insulation materials are getting thinner and better and are ranked into a few classes: A (105^0C), B (130^0C), F (155^0C), H (180^0C).

3.2. SOFT MAGNETIC MATERIALS

In free space the flux density B and the magnetic field H are related by the permeability of free space $\mu_0 = 4\pi 10^{-7}$H/m (S.I.)

$$B\left[\frac{Wb}{m^2}\right] = \mu_0\left[\frac{H}{m}\right] \cdot H\left[\frac{A}{m}\right] \tag{3.1}$$

Within a certain material a different magnetization process occurs.

$$B = \mu \cdot H; \quad \mu = \mu_0 \mu_R \tag{3.2}$$

In (3.2) μ is termed as permeability and μ_R relative permeability (nondimensional).

Permeability is defined for homogenous (uniform quality) and isotropic (same properties in all directions) materials. In nonhomogeneous or (and) nonisotropic materials, μ becomes a tensor. Most common materials are nonlinear: μ varies with B.

A material is classified according to the value of its relative permeability, μ_R, which is related to its atomic structure.

Most nonmagnetic materials are either paramagnetic-with μ_R slightly greater than 1.0, or diamagnetic with μ_R slightly less than 1.0. Superconductors are perfect diamagnetic materials. In such materials when $B \rightarrow 0$, $\mu_R \rightarrow 0$.

Magnetic properties are related to the existence of permanent magnetic dipoles within the matter.

There are quite a few classes of more magnetic materials ($\mu_R >> 1$). Among, them we will deal here with soft ferromagnetic materials. Soft magnetic materials include alloys made of iron, nickel, cobalt and one rare earth element and/or soft steels with silicon.

There is also a class of magnetic materials made of powdered iron particles (or other magnetic material) suspended in an epoxy or plastic (nonferrous) matrix. These softpowder magnetic materials are formed by compression or injection, molding or other techniques.

There are a number of properties of interest in a soft magnetic material such as permeability versus B, saturation flux density, H(B), temperature variation of permeability, hysteresis characteristics, electric conductivity, Curie temperature, and loss coefficients.

The graphical representation of nonlinear B(H) curve (besides the pertinent table) is of high interest (Figure 3.1). Also of high interest is the hysteresis loop (Figure 3.2).

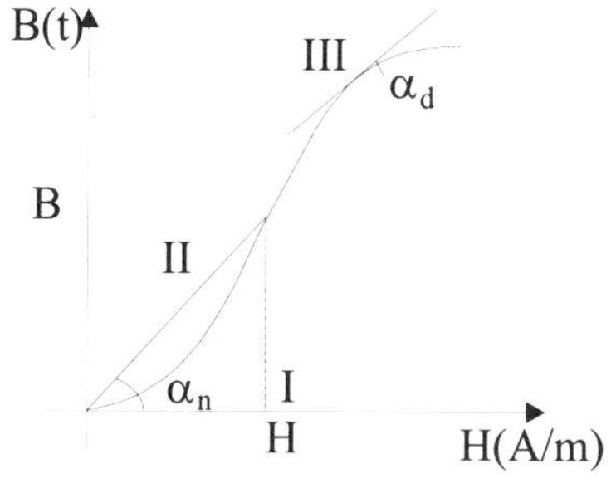

Figure 3.1 Typical B-H curve

There are quite a few standard laboratory methods to obtain these two characteristics. The B-H curve can be obtained two ways: the virgin (initial) B-H curve, obtained from a totally demagnetized sample; the normal (average) B-H curve, obtained as the tips of hysteresis loops of increasing magnitude. There is only a small difference between the two methods.

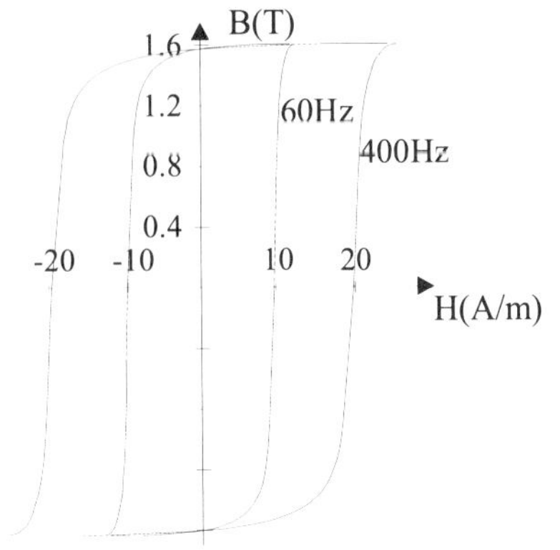

Figure 3.2 Deltamax tape-wound core 0.5 mm strip hysteresis loop

The B-H curve is the result of domain changes within the magnetic material. The domains of soft magnetic materials are 10^{-4}-10^{-7}m in size. When completely demagnetized, these domains have random magnetization with zero flux in all finite samples.

When an external magnetic field H is applied, the domains aligned to H tend to grow when increasing B (region I on Figure 3.1). In region II, H is further increased and the domain walls move rapidly until each crystal of the material becomes a single domain. In region III, the domains rotate towards alignment with H. This results in magnetic saturation B_s. Beyond this condition, the small increase in B is basically due to the increase in the space occupied by the material for $B = \mu_0 H_{r0}$.

This "free space" flux density may be subtracted to obtain the intrinsic magnetization curve. The nonlinear character of B-H curve (Figure 3.1) leads to two different definitions of relative permeability.

- The normal permeability μ_{Rn}:

$$\mu_{Rn} = \frac{B}{\mu_0 H} = \frac{\tan \alpha_n}{\mu_0} \tag{3.3}$$

- The differential relative permeability μ_{Rd}:

$$\mu_{Rd} = \frac{dB}{\mu_0 dH} = \frac{\tan \alpha_d}{\mu_0} \tag{3.4}$$

Only in region II, $\mu_{Rn} = \mu_{Rd}$. In region I and III, in general, $\mu_{Rn} > \mu_{Rd}$ (Figure 3.3). The permeability is maximum in region II. For M19 silicon steel sheets (B_s = 2T, H_s = 40,000 A/m, μ_{Rmax}= 10,000).

So the minimum relative permeability is

$$(\mu_{Rn})_{B_s = 2.0T} = \frac{2.0}{4\pi 10^{-7} \cdot 40000} = 39.8! \tag{3.5}$$

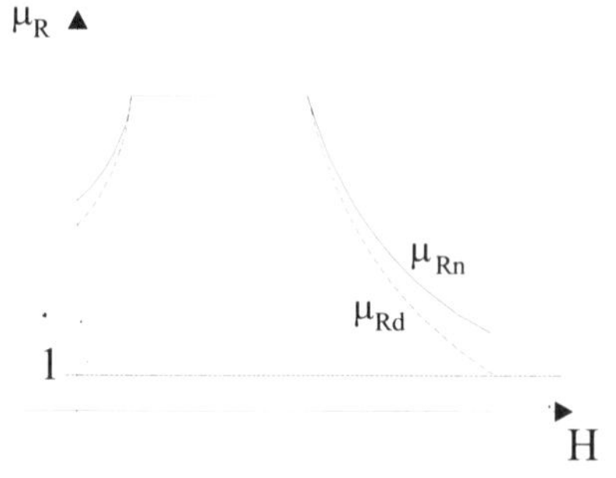

Figure 3.3 Relative permeability versus H

The second graphical characteristic of interest is the hysteresis loop (Figure 3.2). This is a symmetrical hysteresis loop obtained after a number of reversals of magnetic field (force) between $\pm H_c$. The area within the loop is related to the energy required to reverse the magnetic domain walls as H is reversed. This nonreversible energy is called hysteresis loss, and varies with temperature and frequency of H reversals in a given material (Figure 3.2). A typical magnetization curve B-H for silicon steel nonoriented grain is given in Table 3.1.

Table 3.1. B-H curve for silicon (3.5%) steel (0.5mm thick) at 50Hz

B(T)	0.05	0.1	0.15	0.2	0.25	0.3	0.35	0.4	0.45	0.5
H(A/m)	22.8	35	45	49	57	65	70	76	83	90
B(T)	0.55	0.6	0.65	0.7	0.75	0.8	0.85	0.9	0.95	1
H(A/m)	98	106	115	124	135	148	162	177	198	220

B(T)	1.05	1.1	1.15	1.2	1.25	1.3	1.35	1.4	1.45	1.5
H(A/m)	237	273	310	356	417	482	585	760	1050	1340
B(T)	1.55	1.6	1.65	1.7	1.75	1.8	1.85	1.9	1.95	2.0
H(A/m)	1760	2460	3460	4800	6160	8270	11170	15220	22000	34000

Table 3.2.

Induction	Typical DC and Derived AC Magnetizing Force (Oe) of As-Sheared 29 Gage M19 Fully Processed CRNO at Various Frequencies										
(kG)	DC	50 Hz	60 Hz	100 Hz	150 Hz	200 Hz	300 Hz	400 Hz	600 Hz	1000 Hz	1500 Hz	2000 Hz
1.0		0.333	0.334	0.341	0.349	0.356	0.372	0.385	0.412	0.485	0.564	0.642
2.0	0.401	0.475	0.480	0.495	0.513	0.533	0.567	0.599	0.661	0.808	0.955	1.092
4.0	0.564	0.659	0.669	0.700	0.739	0.777	0.846	0.911	1.040	1.298	1.557	1.800
7.0	0.845	0.904	0.916	0.968	1.030	1.094	1.211	1.325	1.553	2.000	2.483	2.954
10.0	1.335	1.248	1.263	1.324	1.403	1.481	1.648	1.822	2.169	2.867	3.697	4.534
12.0	2.058	1.705	1.718	1.777	1.859	1.942	2.129	2.326	2.736	3.657	4.769	5.889
13.0	2.951	2.211	2.223	2.273	2.342	2.424	2.609	2.815	3.244	4.268	5.499	
14.0	5.470	3.508	3.510	3.571	3.633	3.691	3.855	4.132				
15.0	13.928	8.276	8.313	8.366	8.366	8.478	8.651	9.737				
15.5	22.784	13.615	13.587	13.754	13.725	13.776	14.102	16.496				
16.0	35.201	21.589	21.715	21.800	21.842	21.884						
16.5	50.940	32.383	32.506	32.629	32.547	32.588						
17.0	70.260	46.115	46.234	46.392	46.644	46.630						
18.0	122.01											
19.0	201.58											
20.0	393.50											
21.0	1111.84											

It has been shown experimentally that the magnetization curve varies with frequency as in Table 3.2. This time the magnetic field is kept in original data (Oe = 79.55A/m). [1]

In essence the magnetic field increases with frequency for same flux density B. Reduction of the design flux density is recommended when the frequency increases above 200 Hz as the core losses grow markedly with frequency.

3.3. CORE (MAGNETIC) LOSSES

Energy loss in the magnetic material itself is a very significant characteristic in the energy efficiency of IMs. This loss is termed core loss or magnetic loss.

Traditionally, core loss has been divided into two components: hysteresis loss and eddy current loss. The hysteresis loss is equal to the product between the hysteresis loop area and the frequency of the magnetic field in sinusoidal systems.

$$P_h \approx k_h f B_m{}^2 [W/kg]; \quad B_m \text{ - maximum flux density} \tag{3.6}$$

Hysteresis losses are 10 to 30% higher in traveling fields than in a.c. fields for $B_m < 1.5(1.6)$T. However, in a traveling field they have a maximum, in general, between 1.5 to 1.6T and then decrease to low values for B > 2.0T. The computation of hysteresis losses is still an open issue due to the hysteresis cycle complex shape, its dependence on frequency and on the character of the magnetic field (traveling or a.c.) [2].

Preisach modelling of hysteresis cycle is very popular [3] but neural network models have proved much less computation time consuming. [4]

Eddy current losses are caused by induced electric currents in the magnetic material by an external a.c. or traveling magnetic field.

$$P_e \approx k_e f^2 B_m{}^2 [W/kg] \tag{3.7}$$

Finite elements are used to determine the magnetic distribution-with zero electrical conductivity, and then the core losses may be calculated by some analytical approximations as (3.6)-(3.7) or [5]

$$P_{core} \approx k_h f B_m{}^{\alpha} K(B_m) + \frac{\sigma_{Fe}}{12}\frac{d^2 f}{\gamma_{Fe}} \int_{1/f} \left(\frac{dB}{dt}\right)^2 dt + K_{ex} f \int_{1/f} \left(\frac{dB}{dt}\right)^{1.5} dt$$
$$\text{where} \quad K = 1 + \frac{0.65}{B_m}\sum^{n} \Delta B_i \tag{3.8}$$

B_m -maximum flux density
F -frequency
γ_{Fe} -material density
d -lamination thickness
K_h -hysteresis loss constant
K_{ex} -excess loss constant
ΔB_I -change of flux density during a time step

n -total number of time steps

Equation (3.8) is a generalization of Equations (3.6) and (3.7) for nonsinusoidal time varying magnetic fields as produced in PWM inverter IM drives.

For sinusoidal systems, the eddy currents in a thin lamination may be calculated rather easily by assuming the external magnetic field $H_0e^{j\omega_1 t}$ acting parallel to the lamination plane (Figure 3.4).

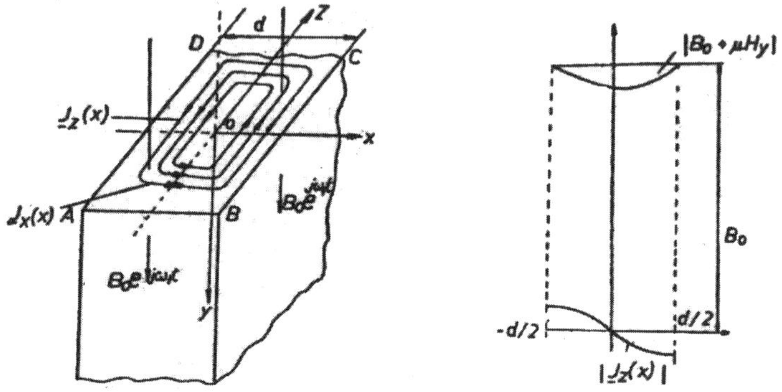

Figure 3.4 Eddy currents paths in a soft material lamination

Maxwell's equations yield

$$\frac{\partial H_y}{\partial x} = J_z; \quad H_{0y} = H_0 e^{j\omega_1 t}$$
$$-\frac{\partial E_z}{\partial x} = -j\omega_1\mu\left(H_{0y} + H_y\right); \ \sigma_{Fe}E_z = J_z \tag{3.9}$$

where J is current density and E is electric field.

As the lamination thickness is small in comparison with its length and width, J_x contribution is neglected. Consequently (3.9) is reduced to

$$\frac{\partial^2 H_y}{\partial x^2} - j\omega_1\mu\sigma_{Fe}H_y = j\omega_1\sigma_{Fe}B_0 \tag{3.10}$$

$B_0 = \mu_0 H_0$ is the initial flux density on the lamination surface.

The solution of (3.10) is

$$H_y(x) = A_1 e^{\gamma x} + A_2 e^{-\gamma x} + \frac{B_0}{\mu_0} \tag{3.11}$$

$$\gamma = \beta(1 + j); \ B = \sqrt{\frac{\omega_1\mu\sigma_{Fe}}{2}} \tag{3.12}$$

The current density $J_z(x)$ is

$$J_z(x) = \frac{\partial H_y}{\partial x} = \gamma\left(A_1 e^{\gamma x} + A_2 e^{-\gamma x}\right) \tag{3.13}$$

The boundary conditions are

$$H_y\left(\frac{d}{2}\right) = H_y\left(-\frac{d}{2}\right) = 0 \tag{3.14}$$

Finally

$$A_1 = A_2 = \frac{B_0}{2\mu \cosh\beta\frac{d}{2}(1+j)} \tag{3.15}$$

$$J_z(x) = -\frac{\beta(1+j)}{\mu}\frac{B_0 \sinh(1+j)\beta x}{\cosh\beta\frac{d}{2}(1+j)} \tag{3.16}$$

The eddy current loss per unit weight P_e is

$$P_e = \frac{2\gamma_{Fe}}{d\sigma_{Fe}}\frac{1}{2}\int_0^{d/2}(J_z(x))^2 dx = \frac{\beta\gamma_{Fe} d\omega_1}{\mu}B_0^2\left[\frac{\sinh(\beta d) - \sin(\beta d)}{\cosh(\beta d) + \cos(\beta d)}\right]\left[\frac{W}{kg}\right] \tag{3.17}$$

The iron permeability has been considered constant within the lamination thickness although the flux density slightly decreases.

For good utilization of the material, the flux density reduction along lamination thickness has to be small. In other words $\beta d << 1$. In such conditions, the eddy-current losses increase with the lamination thickness.

The electrical conductivity σ_{Fe} is also influential and silicon added to soft steel reduces σ_{Fe} to $(2\text{-}2.5)10^6\ (\Omega m)^{-1}$. This is why 0.5-0.6 mm thick laminations are used at 50(60) Hz and, in general, up to 200-300Hz IMs.

For such laminations, eddy current losses may be approximated to

$$P_e \approx K_w B_m^2\left[\frac{W}{kg}\right];\ K_w = \frac{\omega_1^2 \sigma_{Fe} d^2}{24}\gamma_{Fe} \tag{3.18}$$

The above loss formula derivation process is valid for a.c. magnetic field excitation. For pure traveling field the eddy current losses are twice as much for same laminations, frequency, and peak flux density.

In view of the complexity of eddy current and hysteresis losses, it is recommended tests be run to measure them in conditions very similar to those encountered in the particular IM.

Soft magnetic material producers manufacture laminations for many purposes. They run their own tests and provide data on core losses for practical values of frequency and flux density.

Besides Epstein's traditional method, made with rectangular lamination samples, the wound toroidal cores method has also been introduced [6] for a.c. field losses. For traveling field loss measurement, a rotational loss tester may be used. [7]

Typical core loss data for M15–3% silicon 0.5 mm thick lamination material–used in small IMs, is given in Figure 3.5. [8]

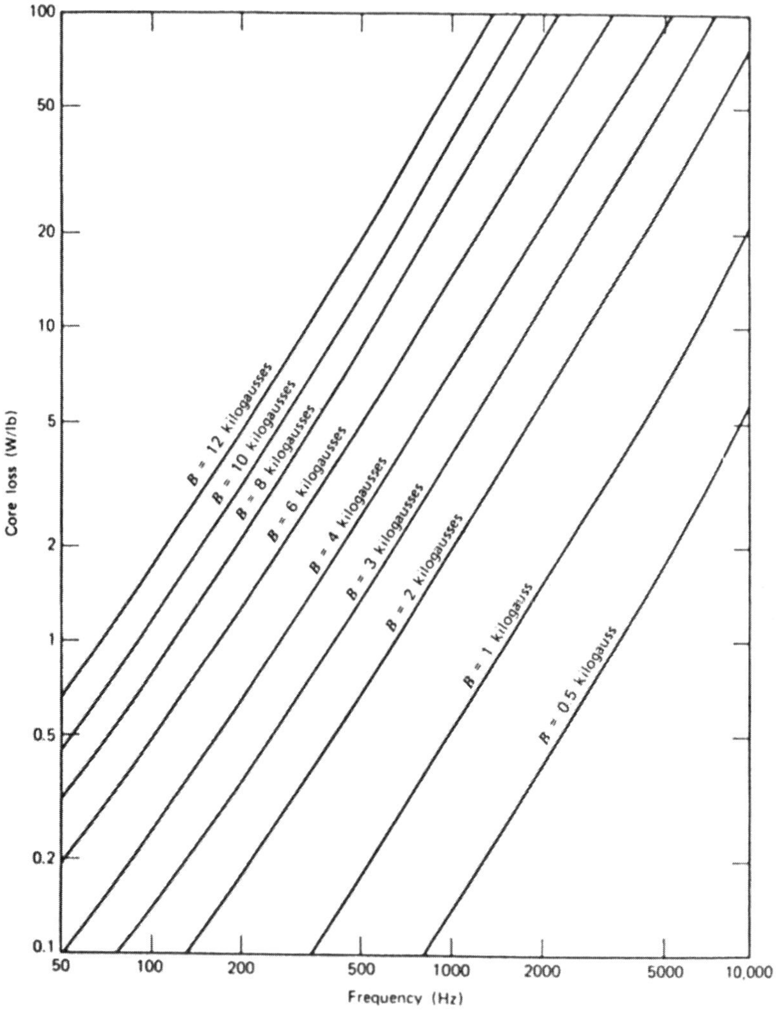

Figure 3.5 Core losses for M15–3% silicon 0.5 mm thick laminations [8]

Table 3.3. [1]

Typical Core Loss (W/lb) of As-Sheared 29 Gage M19 Fully Processed CRNO at Various Frequencies

Induction (kG)	50 Hz	60 Hz	100 Hz	150 Hz	200 Hz	300 Hz	400 Hz	600 Hz	1000 Hz	1500 Hz	2000 Hz
1.0	0.008	0.009	0.017	0.029	0.042	0.074	0.112	0.205	0.465	0.900	1.451
2.0	0.031	0.037	0.072	0.119	0.170	0.300	0.461	0.812	1.786	3.370	5.318
4.0	0.109	0.134	0.252	0.424	0.621	1.085	1.635	2.960	6.340	11.834	18.523
7.0	0.273	0.340	0.647	1.106	1.640	2.920	4.450	8.180	17.753	33.720	53.971
10.0	0.494	0.617	1.182	2.040	3.060	5.530	8.590	16.180	36.303	71.529	116.702
12.0	0.687	0.858	1.648	2.860	4.290	7.830	12.203	23.500	54.258	108.995	179.321
13.0	0.812	1.014	1.942	3.360	5.060	9.230	14.409	27.810	65.100	131.918	
14.0	0.969	1.209	2.310	4.000	6.000	10.920	17.000				
15.0	1.161	1.447	2.770	4.760	7.150	13.000	20.144				
15.5	1.256	1.559	2.990	5.150	7.710	13.942	21.619				
16.0	1.342	1.667	3.179	5.466	8.189						
16.5	1.420	1.763	3.375	5.788	8.674						
17.0	1.492	1.852	3.540	6.089	9.129						

As expected, core losses increase with frequency and flux density. A similar situation occurs with a superior but still common material: steel M19 FP (0.4 mm) 29 gauge (Table 3.3). [1]

A rather complete up-to-date data source on soft magnetic materials characteristics and losses may be found in Reference 1.

Core loss represents 25 to 35% of all losses in low power 50(60) Hz IMs and slightly more in medium and large power IMs at 50(60) Hz. The development of high speed IMs, up to more than 45,000 rpm at 20 kW [9], has caused a new momentum in the research for better magnetic materials as core losses are even larger than winding losses in such applications.

Thinner (0.35 mm or less) laminations of special materials (3.25% silicon) with special thermal treatment are used to strike a better compromise between low 60 Hz and moderate 800/1000 Hz core losses (1.2 W/kg at 60 Hz, 1T; 28 W/kg at 800 Hz, 1T).

6.5% silicon nonoriented steel laminations for low power IMs at 60 Hz have been shown capable of a 40% reduction in core losses. [10] The noise level has also been reduced this way. [10] Similar improvements have been reported with 0.35mm thick oriented grain laminations by alternating laminations with perpendicular magnetization orientation or crossed magnetic structure (CMS). [11]

Soft magnetic composites (SFC) have been produced by powder metallurgy technologies. The magnetic powder particles are coated by insulation layers and a binder which are compressed to provide

- Large enough magnetic permeability
- Low enough core losses
- Densities above 7.1 g/cm^3 (for high enough permeability)

The hysteresis loss tends to be constant with frequency while the eddy current loss increases almost linearly with frequency (up to 1 kHz or so).

At 400 to 500 Hz and above, the losses in SFC become smaller than for 0.5 mm thick silicon steels. However the relative permeability is still low: 100 to 200. Only for recent materials, fabricated by cold compression, the relative permeability has been increased above 500 for flux densities in the 1T range. [12, 13]

Added advantages such as more freedom in choosing the stator core geometry and the increase of slot-filling factor by coil in slot magnetic compression embedded windings [14] may lead to a wide use of soft magnetic composites in induction motors. The electric loading may be thus increased. The heat transmissivity also increases. [12]

In the near future, better silicon 0.5 mm (0.35 mm) thick steel laminations with nonoriented grain seem to remain the basic soft magnetic materials for IM fabrication. For high speed (frequency above 300 Hz) thinner laminations are to be used. The insulation coating layer of each lamination is getting thinner and thinner to retain a good stacking factor (above 85%).

3.4. ELECTRICAL CONDUCTORS

Electric copper conductors are used to produce the stator three (two) phase windings. The same is true for wound rotor windings.

Electrical copper has a high purity and is fabricated by an involved electrolysis process. The purity is well above 99%. The cross-section of copper conductors (wires) to be introduced in stator slots is either circular or rectangular (Figure 3.6). The electrical resistivity of magnetic wire (electric conductor) $\rho_{Co} = (1.65\text{-}1.8) \times 10^{-8}\Omega m$ at 20^0C and varies with temperature as

$$\rho_{Co}(T) = (\rho_{Co})_{20^0}\left[1 + (T - 20)/273\right] \tag{3.19}$$

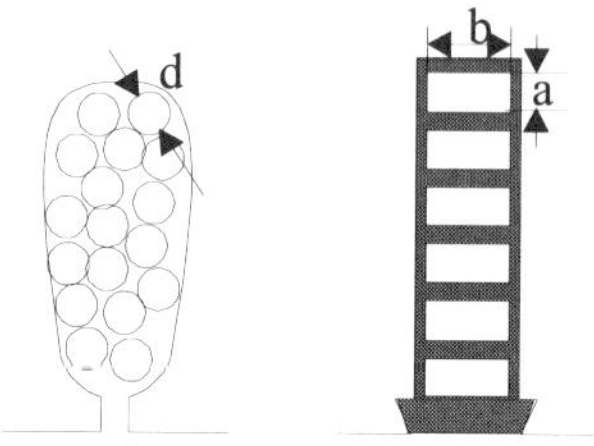

Figure 3.6 Stator slot with round (a) and rectangular (b) conductors

Round magnetic wires come in standardized gauges up to a bare copper diameter of about 2.5mm (3mm) (or 0.12inch), in general (Tables 3.4 and 3.5).

The total cross-section A_{con} of the coil conductor depends on the rated phase current I_{1n} and the design current density J_{con}.

$$A_{con} = I_{1n} / J_{con} \tag{3.20}$$

The design current density varies between 3.5 and 15 A/mm^2 depending on the cooling system, service duty cycle, and the targeted efficiency of the IM. High efficiency IMs are characterized by lower current density (3.5 to 6 A/mm^2). If the A_{con} in (3.19) is larger than the cross section of the largest round wire gauge available, a few conductors of lower diameter are connected in parallel and wound together. Up to 6 to 8 elementary conductors may be connected together.

Table 3.4. Round magnetic wire gauges in inches

Awg Size	Bare Wire Diameter Nominal (Inches)	FILM ADDITIONS (Inches)		OVERALL DIAMETER (Inches)			WEIGHT AT 20°C-68°F		RESISTANCE AT 20°C-68°F		Wires/In. Nom.	Awg Size
		Min.	Max.	Min.	Nom.	Max.	Lbs./M Ft. Nom.	Ft./Lb. Nom.	Ohms/M Ft. Nom.	Ohms./Lb. Nom.		
8	.1285	.0016	.0026	.1288	.1306	.1324	50.20	19.92	.6281	.01251	7.66	8
9	.1144	.0016	.0026	1149	.1165	.1181	39.81	25.12	.7925	.01991	8.58	9
10	.1019	.0015	.0025	1024	.1039	.1054	31.59	31.66	.9988	.03162	9.62	10
11	.0907	.0015	.0025	.0913	.0927	.0941	25.04	39.94	1.26	.05032	10.8	11
12	.0808	.0014	.0024	.0814	.0827	.0840	19.92	50.20	1.59	.07982	12.1	12
13	.0720	.0014	.0023	.0727	.0738	.0750	15.81	63.25	2.00	.1265	13.5	13
14	.0641	.0014	.0023	.0649	.0659	.0670	12.49	80.06	2.52	.2018	15.2	14
15	.0571	.0013	.0022	.0578	.0588	.0599	9.948	100.5	3.18	.3196	17.0	15
16	.0508	.0012	.0021	.0515	.0525	.0534	7.880	126.9	4.02	.5101	19.0	16
17	.0453	.0012	.0020	.0460	.0469	.0478	6.269	159.5	5.05	.8055	21.3	17
18	.0403	.0011	.0019	.0410	.0418	.0426	4.970	201.2	6.39	1.286	23.9	18
19	.0359	.0011	.0019	.0366	.0374	.0382	3.943	253.6	8.05	2.041	26.7	19
20	.0320	.0010	.0018	.0327	.0334	.0341	3.138	318.7	10.1	3.219	29.9	20
21	.0285	.0010	.0018	0292	.0299	.0306	2.492	401.2	12.8	5.135	33.4	21
22	.0253	.0010	.0017	.0260	.0267	.0273	1.969	507.9	16.2	8.228	37.5	22
23	.0226	.0009	.0016	.0233	.0238	.0244	1.572	636.1	20.3	12.91	42.0	23
24	.0201	.0009	.0015	.0208	.0213	.0218	1.240	806.5	25.7	20.73	46.9	24
25	.0179	.0009	.0014	.0186	.0191	.0195	.988	1012	32.4	32.79	52.4	25
26	.0159	.0008	.0013	.0165	.0169	.0174	.779	1284	41.0	52.64	59.2	26
27	.0142	.0008	.0013	.0149	.0153	.0156	.623	1605	51.4	82.50	65.4	27
28	.0126	.0007	.0012	.0132	.0136	.0139	.491	2037	65.3	133.0	73.5	28
29	.0113	.0007	.0012	.0119	.0122	.0126	.395	2532	81.2	205.6	82.0	29
30	.0100	.0006	.0011	.0105	.0109	.0112	.310	3226	104	335.5	91.7	30
31	.0089	.0006	.0011	.0094	.0097	.0100	.246	4065	131	532.5	103	31
32	.0080	.0006	.0010	.0085	.0088	.0091	.199	5025	162	814.1	114	32
33	.0071	.0005	.0009	.0075	.0078	.0081	.157	6394	206	1317	128	33
34	.0063	.0005	.0008	.0067	.0070	.0072	123	8130	261	2122	143	34
35	.0056	.0004	.0007	.0059	.0062	.0064	.0977	10235	331	3388	161	35
36	.0050	.0004	.0007	.0053	.0056	.0058	.0783	12771	415	5300	179	36
37	.0045	.0003	.0006	.0047	.0050	.0052	.0632	15823	512	8101	200	37
38	.0040	.0003	.0006	.0042	.0045	.0047	.0501	19960	648	12934	222	38
39	.0035	.0002	.0005	.0036	.0039	.0041	.0383	26110	847	22115	256	39
40	.0031	.0002	.0005	.0032	.0035	.0037	.0301	33222	1080	35880	286	40
41	.0028	.0002	.0004	.0029	.0031	.0033	.0244	40984	1320	54099	323	41
42	.0025	.0002	.0004	.0026	.0028	.0030	.0195	51282	1660	85128	357	42
43	.0022	.0002	.0003	.0023	.0025	.0026	.0153	65360	2140	139870	400	43
44	.0020	.0001	.0003	.0020	.0022	.0024	.0124	80645	2590	208870	455	44

Table 3.5. Typical round magnetic wire gauges in mm

Rated diameter [mm]	Insulated wire diameter [mm]
0.3	0.327
0.32	0.348
0.33	0.359
0.35	0.3795
0.38	0.4105
0.40	0.4315
0.42	0.4625
0.45	0.4835
0.48	0.515
0.50	0.536
0.53	0.567
0.55	0.5875
0.58	0.6185
0.60	0.639
0.63	0.6705
0.65	0.691
0.67	0.7145
0.70	0.742
0.71	0.7525
0.75	0.7949
0.80	0.8455
0.85	0.897
0.90	0.948
0.95	1.0

1.00	1.051
1.05	1.102
1.10	1.153
1.12	1.173
1.15	1.2035
1.18	1.2345
1.20	1.305
1.25	1.325
1.30	1.356
1.32	1.3765
1.35	1.407
1.40	1.4575
1.45	1.508
1.50	1.559

If A_{con} is larger than 30 to 40 mm^2 (that is 6 to 8, 2.5 mm diameter wires in parallel), rectangular conductors are recommended.

In many countries rectangular conductor cross sections are also standardized. In some cases small cross sections such as (0.8 to 2)·2 mm × mm or (0.8 to 6) × 6 mm × mm.

In general the rectangular conductor height a is kept low (a < 3.55 mm) to reduce the skin effect; that is, to keep the a.c. resistance low. A large cross section area of 3.55 × 50 mm × mm would be typical for large power IMs.

The rotor cage is generally made of aluminum: die-casted aluminum in low power IMs (up to 300 kW or so) or of aluminum bars attached through brazing or welding processes to end rings.

Fabricated rotor cages are made of aluminum or copper alloys and of brass (the upper cage of a double cage) for powers above 300 kW in general. The casting process of aluminum uses the rotor lamination stack as a partial mold because the melting point of silicon steel is much higher than that of aluminum. The electrical resistivity of aluminium $\rho_{Al} \cong (2.7\text{-}3.0)10^{-8}\ \Omega m$ and varies with temperature as shown in (3.19).

Although the rotor cage bars are insulated from the magnetic core, most of the current flows through the cage bars as their resistivity is more than 20 to 30 times smaller than that of the laminated core.

Insulated cage bars would be ideal, but this would severely limit the rotor temperature unless a special high temperature (high cost) insulation coating is used.

3.5. INSULATION MATERIALS

The primary purpose of stator insulation is to withstand turn-to-turn, phase-to-phase and phase-to-ground voltage such that to direct the stator phase currents through the desired paths of stator windings.

Insulation serves a similar purpose in phase-wound rotors whose phase leads are connected to insulated copper rings and then through brushes to stationary devices (resistances or/and special power electronic converters). Insulation is required to withstand voltages associated to: brush rigging (if any) winding connections, winding leads and auxiliaries such as temperature probes and bearings (especially for PWM inverter drives).

The stator laminations are insulated from each other by special coatings (0.013 mm thick) to reduce eddy current core losses.

In standard IMs the rotor (slip) frequency is rather small and thus interlamination insulation may not be necessary, unless the IM is to work for prolonged intervals at large slip values.

For all wound-rotor motors, the rotor laminations are insulated from each other. The bearing sitting is insulated from the stator to reduce the bearing (shaft) voltage (current), especially for large power IMs whose stator laminations are made of a few segments thus allowing a notable a.c. axial flux linkage. This way premature bearing damage may be prevented and even so in PWM inverter fed IMs, where additional common voltage mode superhigh frequency capacitor currents through the bearings occur (Chapter 21).

Stator winding insulation systems may be divided in two types related to power and voltage level.

- Random-wound conductor IMs-with small and round conductors
- Form-wound conductor IMs-with relatively large rectangular conductors

Insulation systems for IMs are characterized by voltage and temperature requirements. The IM insulation has to withstand the expected operating voltages between conductors, (phase) conductors and ground, and phase to phase.

The American National Standards Institute (ANSI) specifies that the insulation test voltage shall be twice the rated voltage plus 100 V applied to the stator winding for 1 minute.

The heat produced by the winding currents and the core losses causes hot-spot temperatures that have to be limited in accordance to the thermal capability of the organic (resin) insulation used in the machine, and to its chemical stability and capability to prevent conductor to conductor, conductor to ground short-circuits during IM operation.

There is continuous, but slow deterioration of the organic (resin) insulation by internal chemical reaction, contamination, and chemical interactions. Thermal degradation develops cracks in the enamel, varnish, or resin, reducing the dielectric strength of insulation.

Insulation materials for electric machines have been organized in stable temperature classes at which they are able to perform satisfactorily for the expected service lifetime.

The temperature classes are (again)

Class A: 105°C	Class F: 155°C
Class B: 130°C	Class G: 180°C

The main insulation components for the random-wound coil windings are the enamel insulation on the wire, the insulation between coils and ground/slot walls-slot liner insulation, and between phases (Figure 3.7).

The connections between the coils of a phase and the leads to the terminal box have to be insulated. Also the binding cord used to tie down endwindings to reduce their vibration is made of insulation materials.

Random-wound IMs are built for voltages below 1 kV. The moderate currents involved can be handled by wound conductors (eventually a few in parallel) where enamel insulation is the critical component. To apply the enamel, the wire is passed through a solution of polymerizable resin and into the high-curing temperature tower where it turns into a thin, solid, and flexible coating.

3.5.1. Random-wound IM insulation

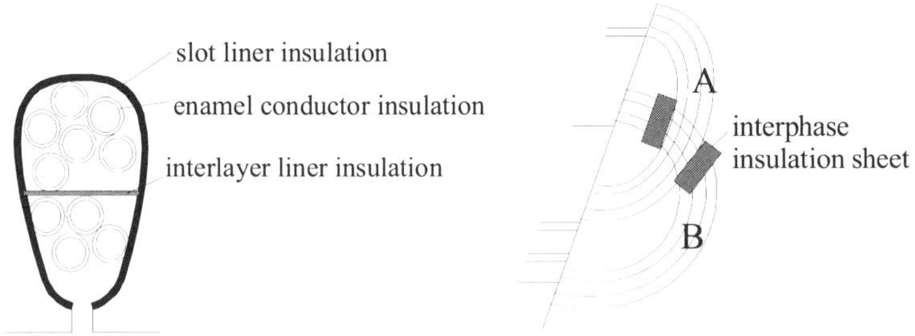

Figure 3.7 Random-wound coils insulation

Several passes are required for the desired thickness (0.025 mm thick or so). There are dedicated standards that mention the tests on enamel conductors (ASTMD-1676; ASTM standards part 39 electric insulation-test methods: solids and solidifying liquids should be considered for the scope.

Enamel wire, stretched and scraped when the coils are introduced into the slots, should survive this operation without notable damage to the enamel. Some insulation varnish is applied over the enamel wire after the stator winding is completed. The varnish provides additional enamel protection against moisture, dirt, and chemical contamination and also provides mechanical support for the windings.

Slot and phase-to-phase insulation for class A temperatures is a somewhat flexible sheet material (such as cellulose paper), 0.125 to 0.25 mm thick, or a polyester film. In some cases fused resin coatings are applied to stator slot walls by electrostatic attraction of a polymerizable resin powder. The stator is heated to fuse and cure the resin to a smooth coating.

For high temperature IMs (class F, H), glass cloth mica paper or asbestos treated with special varnishes are used for slot and phase-to-phase insulation. Varnishes may interact with the emanel to reduce the thermal stability. Enamels and varnishes are tested separately according to ASTM (D2307, D1973, D3145) and IEC standards and together.

Model motor insulation systems (motorettes) are tested according to IEEE standards for small motors.

All these insulation accelerated life tests involve the ageing of insulation test specimens until they fail at temperatures higher than the operating temperature of the respective motor. The logarithms of the accelerated ageing times are then graphed against their reciprocal Kelvin test temperatures (Arhenius graph). The graph is then extrapolated to the planed (reduced) temperature to predict the actual lifetime of insulation.

3.5.2. Form-wound windings

Form-wound windings are employed in high power IMs. The slots are rectangular and so are the conductors. The slot filling factor increases due to this combination.

The insulation of the coil conductors (turns) is applied before inserting the coils in slots. The coils are vacuum also impregnated outside the machine. The slot insulation is made of resin-bonded mica applied as a wrapper or tape with a fibrous sheet for support (in high voltage IMs above 1 to 2 kV).

Vacuum impregnation is done with polymerizable resins which are then cured to solids by heating. During the cure, the conductors may be constrained to size to enter the slot as the epoxy-type resins are sufficiently elastic for the scope.

Voltage, through partial discharges, may cause insulation failure in higher voltage IMs. Incorporating mica in the major insulation schemes solves this problem to a large degree.

A conducting paint may be applied over the slot portion of the coils to fill the space between the insulated preformed coil and slot wall, to avoid partial discharges. Lower and medium voltage coil insulation is measured in accelerated higher temperature tests (IEEE standard 275) by using the model system called formette. Formette testing is similar to motorette testing for random-wound IMs [15].

Diagnostic nondestructive tests to check the integrity and capability of large IM insulation are also standardized [15 - 17].

3.6. SUMMARY

- The three main materials used to build IMs are of magnetic, electric, and insulation type.
- As the IM is an a.c. machine, reducing eddy current losses in its magnetic core is paramount.
- It is shown that these losses increase with the soft magnetic sheet thickness parallel to the external a.c. field.
- Soft magnetic materials (silicon steel) used in thin laminations (0.5 mm thick up to 200Hz) have low hysteresis and eddy current losses (about or less 2 W/kg at 1 T and 60 Hz).
- Besides losses, the B-H (magnetization) curve characterizes a soft magnetic material.

- The magnetic permeability $\mu = B/H$ varies from (5000 to 8000) μ_0 at 1T to (40 to 60)μ_0 at 2.0 T in modern silicon steel laminations. High permeability is essential to low magnetization (no load) current and losses.
- High speed IMs require frequencies above 300 Hz (and up to 800 Hz and more). Thinner silicon lamination steels with special thermal treatments are required to secure core losses in the order of 30-50 W/kg at 800 Hz and 1 T.
- 6.5% silicon steel lamination for small IMs have proven adequate to reduce core losses by as much as 40% at 50 (60) Hz.
- Also, interspersing oriented grain (transformer) laminations (0.35 mm thick) with orthogonal orientation laminations has been shown to produce a 30 to 40% reduction in core losses at 50 (60) Hz and 1T in comparison with 0.5 mm thick nonoriented grain silicon steel used in most IMs.
- Soft magnetic composites have been introduced and shown to produce lower losses than silicon steel laminations only above 300Hz, but at the expense of lower permeability ((100 to 200) μ_0 in general). Cold compression methods are expected to increase slot filling factor notably and thus increase the current loading. Size reduction is obtained also due to the increase of heat transmissivity through soft magnetic composites.
- Electric conductors for stator windings and for wound rotors are made of pure (electrical) copper.
- Cast aluminum is used for rotor cage windings up to 300kW.
- Fabricated aluminum or copper bars and rings are used for higher power IM cage rotors.
- The rotor cage bars are not, in general, insulated from the rotor lamination core. Interbar currents may thus occur.
- The windings are made out of random-wound coils with round wire, and of form-wound coils for large IMs with rectangular wire.
- The windings are insulated from the magnetic core through insulation materials. Also, the conductors are enameled to insulate one conductor from another.
- Insulation systems are classified according temperature limits in four classes: Class A-105^{0}C, Class B-130^{0}C, Class F-155^{0}C, Class G-180^{0}C.
- Insulation testing is thoroughly standardised as the insulation breakdown finishes the operation life of an IM through short-circuit.
- Thinner and better insulation materials keep surfacing as they are crucial to better performance IMs fed from the power grid and from PWM inverters.

3.7. REFERENCES

1. S. Sprague, D. Jones, Using the New Lamination Steels Database in Motor Design, Proceedings of SMMA-2000 Fall Conference in Chicago, pp.1-12.
2. M. Birkfeld, K.A. Hempel, Calculation of the Magnetic Behaviour of Electrical Steel Sheet Under Two Dimensional Excitation by Means of the Reluctance Tensor, IEEE Trans. Vol. MAG-33, No. 5, 1997, pp.3757-3759.

3. I.D. Mayergoyz, Mathematical Models for Hysteresis, Springer Verlag, 1991.
4. H.H. Saliah, D.A. Lowther, B. Forghani, A Neural Network Model of Magnetic Hysteresis for Computational Magnetics", IEEE Trans Vol. MAG-33, No. 5, 1997, pp.4146-4148.
5. M.A. Mueller at al., Calculation of Iron Losses from Time-Stepped Finite Element Models of Cage Induction Machines, Seventh International Conference on EMD, IEE Conf. publication No. 412.
6. A.J. Moses and N. Tutkun, Investigation of Power Losses in Wound Toroidal Cores Under PWM Excitation, IEEE Trans Vol. MAG-33, No. 5, 1997, pp.3763-3765.
7. M. Ehokizono, T. Tanabe, Studies on a new simplified rotational loss tester, IBID, pp.4020-4022.
8. S.A. Nasar, "Handbook of Electrical Machines", Chapter 2, pp.211, 1987, McGraw Hill Inc.
9. W.L.Soong, G.B.Kliman, R.N.Johnson, R.White, J.MIller, Novel High Speed Induction Motor for a Commercial Centrifugal Compressor, IEEE Trans. Vol. IA-36, No. 3, 2000, pp.706-713.
10. M. Machizuki, S. Hibino, F. Ishibashi, Application of 6.5% Silicon Steel Sheet to Induction Motor and to Magnetic Properties, EMPS–Vol. 22, No. 1, 1994, pp.17-29.
11. A. Boglietti, P. Ferraris, M. Lazzari, F. Profumo, Preliminary Consideration About the Adoption of Unconventional Magnetic Materials and Structures for Motors, IBID Vol. 21, No. 4, 1993, pp.427-436.
12. D. Gay, Composite Iron Powder For A.C. Electromagnetic Applications: History and Use, Record of SMMA-2000 Fall Conference, Chicago Oct. 4-6, 2000.
13. M. Persson, P. Jansson, A.G. Jack, B.C. Mecrow, Soft Magnetic Materials–Use for Electric Machines, IEE 7th International Conf. on EMD, 1995, pp.242-246.
14. E.A. Knoth, Motors for the 21st Century, Record of SMMA-2000 Fall Conference Chicago, Oct.4-6, 2000.
15. T.W. Dakin, "Electric machine insulation", Chapter 13 in Electric Machine Handbook, S.A. Nasar, McGraw Hill, 1987.
16. P.L. Cochran, Polyphase Induction Motors, Marcell Dekker, Inc. 1989, Chapter 11.
17. R.M. Engelmann, W.H. Middendorf, "Handbook of Electric Machines, Marcel Dekker, 1995.
18. R. Morin, R. Bartnikas, P. Menard, "A Three Phase Multistress Accelerated Electrical Aging Test Facility for Stator Bars, IEEE Trans Vol. EC-15, No. 2, 2000, pp.149-156.

Chapter 4

INDUCTION MACHINE WINDINGS AND THEIR M.M.Fs

4.1. INTRODUCTION

As shown in Chapter 2, the slots of the stator and rotor cores of induction machines are filled with electric conductors, insulated (in the stator) from cores, and connected in a certain way. This ensemble constitutes the windings. The primary (or the stator) slots contain a polyphase (triple phase or double phase) a.c. winding. The rotor may have either a 3(2) phase winding or a squirrel cage. Here we will discuss the polyphase windings.

Designing a.c. windings means, in fact, assigning coils in the slots to various phases, establishing the direction of currents in coil sides and coil connections per phase and between phases, and finally calculating the number of turns for various coils and the conductor sizing.

We start with single pole number three-phase windings as they are most commonly used in induction motors. Then pole changing windings are treated in some detail. Such windings are used in wind generators or in doubly fed variable speed configurations. Two phase windings are given special attention. Finally, squirrel cage winding m.m.fs are analyzed.

Keeping in mind that a.c. windings are a complex subject having books dedicated to it [1,2] we will treat here first its basics. Then we introduce new topics such as "pole amplitude modulation," "polyphase symmetrization" [4], "intersperse windings" [5], "simulated annealing" [7], and "the three-equation principle" [6] for pole changing. These are new ways to produce a.c. windings for special applications (for pole changing or m.m.f. chosen harmonics elimination). Finally, fractional multilayer three-phase windings with reduced harmonics content are treated in some detail [8,9]. The present chapter is structured to cover both the theory and case studies of a.c. winding design, classifications, and magnetomotive force (mmf) harmonic analysis.

4.2. THE IDEAL TRAVELING M.M.F. OF A.C. WINDINGS

The primary (a.c. fed) winding is formed by interconnecting various conductors in slots around the circumferential periphery of the machine. As shown in Chapter 2, we may have a polyphase winding on the so-called wound rotor. Otherwise, the rotor may have a squirrel cage in its slots. The objective with polyphase a.c. windings is to produce a pure traveling m.m.f., through proper feeding of various phases with sinusoidal symmetrical currents. And all this in order to produce constant (rippleless) torque under steady state:

$$F_{s1}(x,t) = F_{s1m}\cos\left(\frac{\pi}{\tau}x - \omega_1 t - \theta_0\right) \tag{4.1}$$

where

x - coordinate along stator bore periphery

τ - spatial half-period of m.m.f. ideal wave

ω_1 - angular frequency of phase currents

θ_0 - angular position at t = 0

We may decompose (4.1) into two terms

$$F_{s1}(x,t) = F_{s1m}\left[\cos\left(\frac{\pi}{\tau}x - \theta_0\right)\cos\omega_1 t + \sin\left(\frac{\pi}{\tau}x - \theta_0\right)\sin\omega_1 t\right] \tag{4.2}$$

Equation (4.2) has a special physical meaning. In essence, there are now two mmfs at standstill (fixed) with sinusoidal spatial distribution and sinusoidal currents. The space angle lag and the time angle between the two mmfs is $\pi/2$. This suggests that a pure traveling mmf may be produced with two symmetrical windings $\pi/2$ shifted in time (Figure 41.a). This is how the two phase induction machine evolved.

Similarly, we may decompose (4.1) into 3 terms

$$F_{s1}(x,t) = \frac{2}{3}F_{s1m}\left[\cos\left(\frac{\pi}{\tau}x - \theta_0\right)\cos\omega_1 t + \cos\left(\frac{\pi}{\tau}x - \theta_0 - \frac{2\pi}{3}\right)\cos\left(\omega_1 t - \frac{2\pi}{3}\right) + \right.$$
$$\left. + \cos\left(\frac{\pi}{\tau}x - \theta_0 + \frac{2\pi}{3}\right)\cos\left(\omega_1 t + \frac{2\pi}{3}\right)\right] \tag{4.3}$$

Consequently, three mmfs (single-phase windings) at standstill (fixed) with sinusoidal spatial (x) distribution and departured in space by $2\pi/m$ radians, with sinusoidal symmetrical currents–equal amplitude, $2\pi/3$ radians time lag angle–are also able to produce also a traveling mmf (Figure 4.1.b).

In general, m phases with a phase lag (in time and space) of $2\pi/3$ can produce a traveling wave. Six phases (m = 6) would be a rather practical case besides m = 3 phases. The number of mmf electrical periods per one revolution is called the number of pole pairs p_1

$$p_1 = \frac{\pi D}{2\tau}; \quad 2p_1 = 2,4,6,8,... \tag{4.4}$$

where D is the stator bore diameter.

It should be noted that, for $p_1 > 1$, according to (4.4), the electrical angle α_e is p_1 times larger than the mechanical angle α_g

$$\alpha_e = p_1\alpha_g \tag{4.5}$$

A sinusoidal distribution of mmfs (ampereturns) would be feasible only with the slotless machine and windings placed in the airgap. Such a solution is hardly practical for induction machines because the magnetization of a large total airgap would mean very large magnetization mmf and, consequently, low power factor and efficiency. It would also mean problems with severe mechanical stress acting directly on the electrical conductors of the windings.

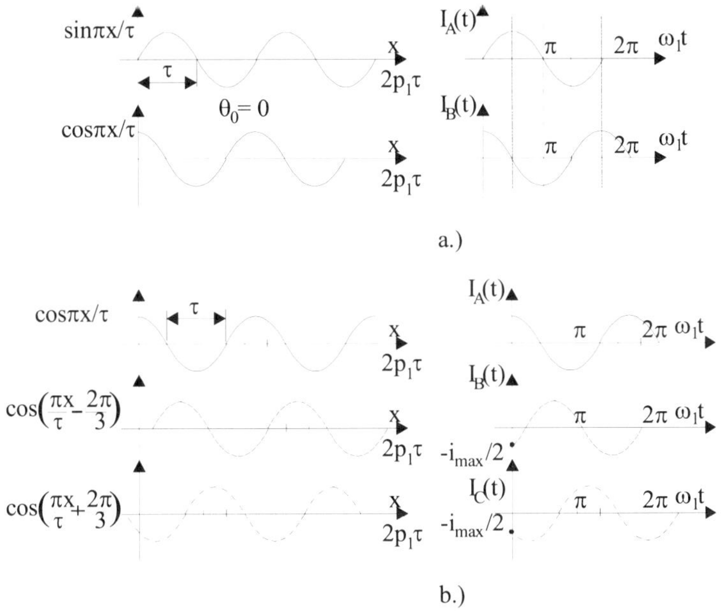

Figure 4.1 Ideal multiphase mmfs
a.) two-phase machine b.) three-phase machine

In practical induction machines, the coils of the windings are always placed in slots of various shapes (Chapter 2).

The total number of slots per stator N_s should be divisible by the number of phases m so that

$$N_s / m = \text{integer} \tag{4.6}$$

A parameter of great importance is the number of slots per pole per phase q:

$$q = \frac{N_s}{2p_1 m} \tag{4.7}$$

The number q may be an integer (q = 1,2, … 12) or a fraction.

In most induction machines, q is an integer to provide complete (pole to pole) symmetry for the winding.

The windings are made of coils. Lap and wave coils are used for induction machines (Figure 4.2).

The coils may be placed in slots in one layer (Figure 4.2a) or in two layers (Figure 4.3.b).

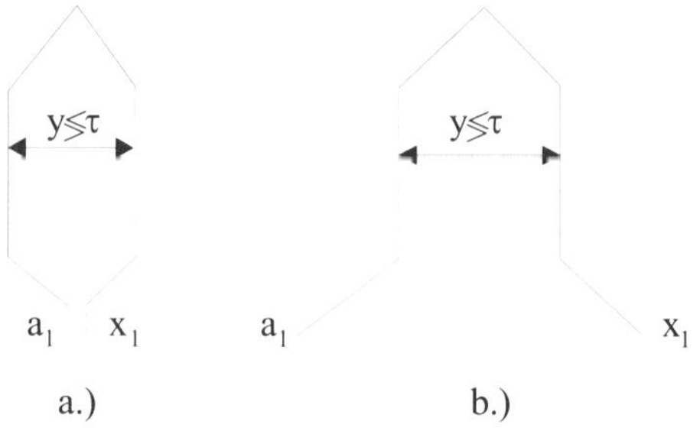

Figure 4.2 Lap a.). and wave b.) single-turn (bar) coils

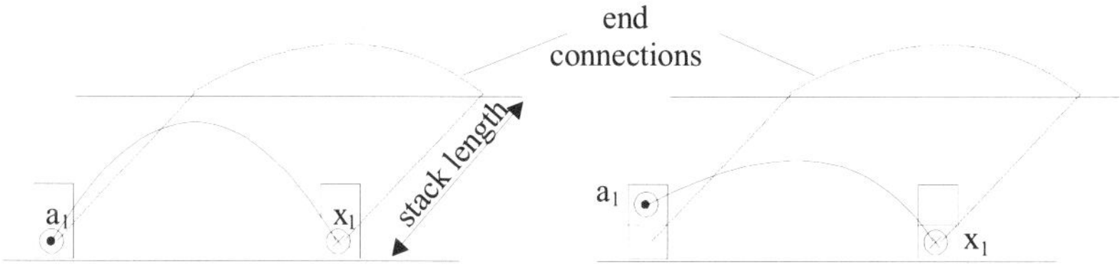

Figure 4.3 Single-layer a.) and double-layer b.) coils (windings)

Single layer windings imply full pitch ($y = \tau$) coils to produce an mmf fundamental with pole pitch τ.

Double layer windings also allow chorded (or fractional pitch) coils ($y < \tau$) such that the end connections of coils are shortened and thus copper loss is reduced. Moreover, as shown later in this chapter, the harmonics content of mmf may be reduced by chorded coils. Unfortunately, so is the fundamental.

4.3. A PRIMITIVE SINGLE-LAYER WINDING

Let us design a four pole ($2p_1 = 4$) three-phase single-layer winding with q = 1 slots/pole/phase. $N_s = 2p_1qm = 2{\cdot}2{\cdot}1{\cdot}3 = 12$ slots in all.

From the previous paragraph, we infer that for each phase we have to produce an mmf with $2p_1 = 4$ poles (semiperiods). To do so, for a single layer winding, the coil pitch $y = \tau = N_s/2p_1 = 12/4 = 3$ slot pitches.

For 12 slots there are 6 coils in all. That is, two coils per phase to produce 4 poles. It is now obvious that the 4 phase A slots are $y = \tau = 3$ slot pitches apart. We may start in slot 1 and continue with slots 4, 7, and 10 for phase A (Figure 4.4a).

Phases B and C are placed in slots by moving 2/3 of a pole (2 slots pitches in our case) to the right. All coils/phases may be connected in series to form one current path (a = 1) or they may be connected in parallel to form two current paths in parallel (a = 2). The number of current paths a is obtained in general by connecting part of coils in series and then the current paths in parallel such that all the current paths are symmetric. Current paths in parallel serve to reduce wire gauge (for given output phase current) and, as shown later, to reduce uncompensated magnetic pull between rotor and stator in presence of rotor eccentricity.

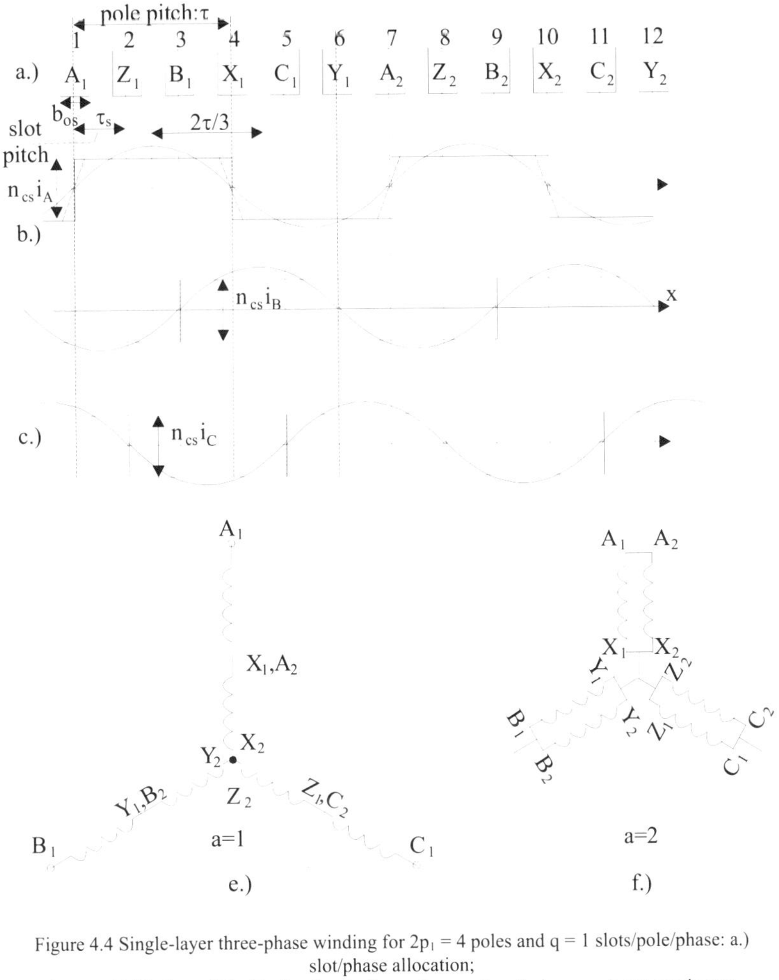

Figure 4.4 Single-layer three-phase winding for $2p_1 = 4$ poles and q = 1 slots/pole/phase: a.) slot/phase allocation;
b.), c.), d.) ideal mmf distribution for the three phases when their currents are maximum;
e.) star series connection of coils/phase; f.) parallel connection of coils/phase

If the slot is considered infinitely thin (or the slot opening $b_{os} \approx 0$), the mmf (ampereturns) jumps, as expected, by $n_{cs} \cdot i_{A,B,C}$, along the middle of each slot.

For the time being, let us consider $b_{os} = 0$ (a virtual closed slot).

The rectangular mmf distribution may be decomposed into harmonics for each phase. For phase A we simply obtain

$$F_{A1}(x,t) = \frac{2}{\pi} \cdot \frac{n_{cs} I\sqrt{2}\cos\omega_1 t}{\nu} \cos\frac{\nu\pi x}{\tau} \tag{4.8}$$

For the fundamental, $\nu = 1$, we obtain the maximum amplitude. The higher the order of the harmonic, the lower its amplitude in (4.8).

While in principle such a primitive machine works, the harmonics content is too rich.

It is only intuitive that if the number of steps in the rectangular ideal distribution would be increased, the harmonics content would be reduced. This goal could be met by increasing q or (and) via chording the coils in a two-layer winding. Let us then present such a case.

4.4. A PRIMITIVE TWO-LAYER CHORDED WINDING

Let us still consider $2p_1 = 4$ poles, m = 3 phases, but increase q from 1 to 2. Thus the total number of slots $N_s = 2p_1qm = 2\cdot2\cdot2\cdot3 = 24$.

The pole pitch τ measured in slot pitches is $\tau = N_s/2p_1 = 24/4 = 6$. Let us reduce the coil throw (span) y such that $y = 5\tau/6$.

We still have to produce 4 poles. Let us proceed as in the previous paragraph but only for one layer, disregarding the coil throw.

In a two, layer winding, the total number of coils is equal to the number of slots. So in our case there are $N_s/m = 24/3$ coils per phase. Also, there are 8 slots occupied by one phase in each layer, four with inward and four with outward current direction. With each layer each phase has to produce four poles in our case. So slots 1, 2; 7', 8'; 13, 14; 19', 20' in layer one belong to phase A. The superscript prime refers to outward current direction in the coils. The distance between neighbouring slot groups of each phase in one layer is always equal to the pole pitch to preserve the mmf distribution half-period (Figure 4.5).

Notice that in Figure 4.5, for each phase, the second layer is displaced to the left by τ-y = 6-5 = 1 slot pitch with respect to the first layer. Also, after two poles, the situation repeats itself. This is typical for a fully symmetrical winding.

Each coil has one side in one layer, say, in slot 1, and the second one in slot y + 1 = 5 + 1 = 6. In this case all coils are identical and thus the end connections occupy less axial room and are shorter due to chording. Such a winding is typical with random wound coils made of round magnetic wire.

For this case we explore the mmf ideal resultant distribution for the situation when the current in phase A is maximum ($i_A = i_{max}$). For symmetrical currents, $i_B = i_C = -i_{max}/2$ (Figure 4.1b).

Each coil has n_c conductors and, again with zero slot opening, the mmf jumps at every slot location by the total number of ampereturns. Notice that half the slots have coils of same phase while the other half accommodate coils of different phases.

The mmf of phase A, for maximum current value (Figure 4.5b) has two steps per polarity as q = 2. It had only one step for q = 1 (Figure 4.4). Also, the resultant mmf has three unequal steps per polarity (q + τ-y = 2 + 6-5 = 3). It is indeed closer to a sinusoidal distribution. Increasing q and using chorded coils reduces the harmonics content of the mmf.

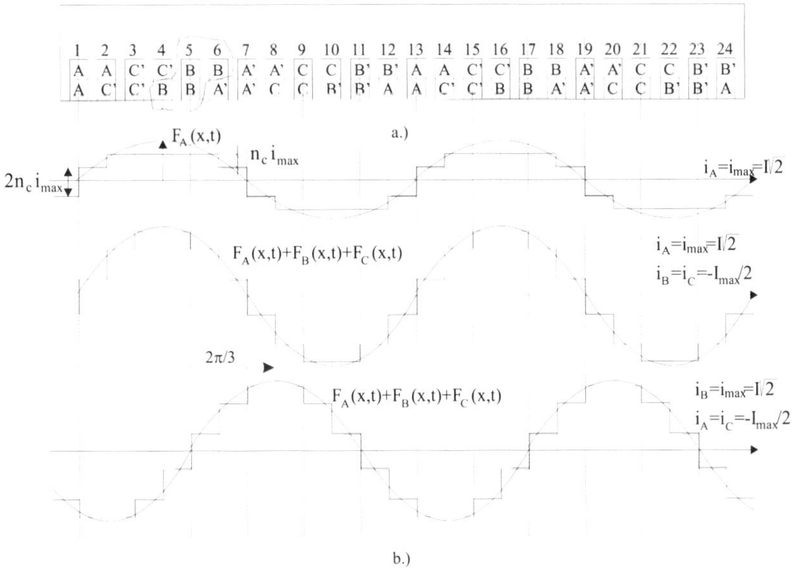

Figure 4.5 Two-layer winding for Ns = 24 slots, 2 p_1 = 4 poles, $y/\tau = 5/6$
a.) slot/phase allocation, b.) mmfs distribution

Also shown in Figure 4.5 is the movement by $2\tau/3$ (or $2\pi/3$ electrical radians) of the mmf maximum when the time advances with $2\pi/3$ electrical (time) radians or T/3 (T is the time period of sinusoidal currents).

4.5. THE MMF HARMONICS FOR INTEGER q

Using the geometrical representation in Figure 4.5, it becomes fairly easy to decompose the resultant mmf in harmonics noticing the step-form of the distributions.

Proceeding with phase A we obtain (by some extrapolation for integer q),

$$F_{A1}(x,t) = \frac{2}{\pi} n_c q I \sqrt{2} K_{q1} K_{y1} \cos\frac{\pi}{\tau} x \cos\omega_1 t \tag{4.9}$$

with
$$K_{q1} = \sin\pi/6/(q\sin\pi/6q) \le 1; K_{y1} = \sin\frac{\pi}{2} y/\tau \le 1 \tag{4.10}$$

K_{q1} is known as the zone (or spread) factor and K_{y1} the chording factor. For q = 1, Kq1 = 1 and for full pitch coils, $y/\tau = 1$, $K_{y1} = 1$, as expected.

To keep the winding fully symmetric $y/\tau \ge 2/3$. This way all poles have a similar slot/phase allocation.

Assuming now that all coils per phase are in series, the number of turns per phase W_1 is

$$W_1 = 2p_1qn_c \tag{4.11}$$

With (4.11), Equation (4.9) becomes

$$F_{A1}(x,t) = \frac{2}{\pi p_1} W_1 I\sqrt{2} K_{q1} K_{y1} \cos\frac{\pi}{\tau} x \cos\omega_1 t \tag{4.12}$$

For three phases we obtain

$$F_1(x,t) = F_{1m}\cos\left(\frac{\pi}{\tau}x - \omega_1 t\right) \tag{4.13}$$

with $$F_{1m} = \frac{3W_1 I\sqrt{2}K_{q1}K_{y1}}{\pi p_1} \text{ (ampereturns per pole)} \tag{4.14}$$

The derivative of pole mmf with respect to position x is called linear current density (or current sheet) A (in Amps/meter)

$$A_1(x,t) = \frac{\partial F_1(x,t)}{\partial x} = A_{1m}\sin\left(-\frac{\pi}{\tau}x + \omega_1 t\right) \tag{4.15}$$

$$A_{1m} = \frac{3\sqrt{2}W_1 I\sqrt{2}K_{q1}K_{y1}}{p_1\tau} = \frac{\pi}{\tau}F_{1m} \tag{4.16}$$

A_{1m} is the maximum value of the current sheet and is also identified as current loading. The current loading is a design parameter (constant) $A_{1m} \approx$ 5,000A/m to 50,000 A/m, in general, for induction machines in the power range of kilowatts to megawatts. It is limited by the temperature rise and increases with machine torquc (sizc).

The harmonics content of the mmf is treated in a similar manner to obtain

$$F(x,t) = \frac{3W_1 I\sqrt{2}K_{q\nu}K_{y\nu}}{\pi p_1 \nu}\cdot$$
$$\cdot\left[K_{BI}\cos\left(\frac{\nu\pi}{\tau}x - \omega_1 t - (\nu-1)\frac{2\pi}{3}\right) - K_{BII}\cos\left(\frac{\nu\pi}{\tau}x + \omega_1 t - (\nu+1)\frac{2\pi}{3}\right)\right] \tag{4.17}$$

with

$$K_{q\nu} = \frac{\sin\nu\pi/6}{q\sin\nu\pi/6q}; K_{y\nu} = \sin\left(\frac{\nu\pi y}{2\tau}\right) \tag{4.18}$$

$$K_{BI} = \frac{\sin(\nu-1)\pi}{3\sin(\nu-1)\pi/3}; K_{BII} = \frac{\sin(\nu+1)\pi}{3\sin(\nu+1)\pi/3} \tag{4.19}$$

Due to mmf full symmetry (with q = integer), only odd harmonics occur. For three-phase star connection, 3K harmonics may not occur as the current sum is zero and their phase shift angle is $3K\cdot 2\pi/3 = 2\pi K$.

We are left with harmonics $\nu = 3K \pm 1$; that is $\nu = 5, 7, 11, 13, 17, \ldots$.

We should notice in (4.19) that for $\nu_d = 3K + 1$, $K_{BI} = 1$ and $K_{BII} = 0$. The first term in (4.17) represents however a direct (forward) traveling wave as for a constant argument under cosinus, we do obtain

$$\left(\frac{dx}{dt}\right) = \frac{\omega_1\tau}{\pi\nu} = \frac{2\tau f_1}{\nu}; \omega_1 = 2\pi f_1 \tag{4.20}$$

On the contrary, for $\nu = 3K-1$, $K_{BI} = 0$, and $K_{BII} = 1$. The second term in (4.17) represents a backward traveling wave. For a constant argument under cosinus, after a time derivative, we have

$$\left(\frac{dx}{dt}\right)_{\nu=3K-1} = \frac{-\omega_1\tau}{\pi\nu} = \frac{-2\tau f_1}{\nu} \tag{4.21}$$

We should also notice that the traveling speed of mmf space harmonics, due to the placement of conductors in slots, is ν times smaller than that of the fundamental ($\nu = 1$).

The space harmonics of the mmf just investigated are due both to the placement of conductors in slots and to the placement of various phases as phase belts under each pole. In our case the phase belts spread is $\pi/3$ (or one third of a pole). There are also two layer windings with $2\pi/3$ phase belts but the $\pi/3$ (60^0) phase belt windings are more practical.

So far the slot opening influences on the mmf stepwise distribution have not been considered. It will be discussed later in this chapter.

Notice that the product of zone (spread or distribution) factor $K_{q\nu}$ and the chording factor $K_{y\nu}$ is called the stator winding factor $K_{w\nu}$.

$$K_{w\nu} = K_{q\nu}K_{y\nu} \tag{4.22}$$

As in most cases, only the mmf fundamental ($\nu = 1$) is useful, reducing most harmonics and cancelling some is a good design attribute. Chording the coils to cancel $K_{y\nu}$ leads to

$$\sin\left(\frac{\nu\pi y}{2\tau}\right) = 0; \frac{\nu\pi y}{2\tau} = n\pi; \frac{y}{\tau} > \frac{2}{3} \tag{4.23}$$

As the mmf harmonic amplitude (4.17) is inversely proportional to the harmonic order, it is almost standard to reduce (cancel) the fifth harmonic (ν = 5) by making n = 2 in (4.23).

$$\frac{y}{\tau} = \frac{4}{5} \tag{4.23'}$$

In reality, this ratio may not be realized with an integer q (q = 2) and thus $y/\tau = 5/6$ or 7/9 is the practical solution which keeps the 5th mmf harmonic low. Chording the coils also reduces K_{y1}. For $y/\tau = 5/6$, $\sin\frac{\pi}{2}\frac{5}{6} = 0.966 < 1.0$ but a 4% reduction in the mmf fundamental is worth the advantages of reducing the coil end connection length (lower copper losses) and a drastical reduction of 5th mmf harmonic.

Mmf harmonics, as will be shown later in the book, produce parasitic torques, radial forces, additional core and winding losses, noise, and vibration.

Example 4.1.

Let us consider an induction machine with the following data: stator core diameter D = 0.1 m, number of stator slots $N_s = 24$, number of poles $2p_1 = 4$, $y/\tau = 5/6$, two-layer winding; slot area $A_{slot} = 100\ mm^2$, total slot fill factor $K_{fill} = 0.5$, current density $j_{Co} = 5\ A/mm^2$, number of turns per coil $n_c = 25$. Let us calculate

a.) The rated current (RMS value), wire gauge
b.) The pole pitch τ
c.) K_{q1} and K_{y1}, K_{w1}
d.) The amplitude of the mmf F_{1m} and of the current sheet A_{1m}
e.) K_{q7}, K_{y7} and F_{7m} ($\nu = 7$)

Solution

Part of the slot is filled with insulation (conductor insulation, slot wall insulation, layer insulation) because there is some room between round wires. The total filling factor of a slot takes care of all these aspects. The mmf per slot is

$$2n_c I = A_{slot} \cdot K_{fill} \cdot J_{Co} = 100 \cdot 0.5 \cdot 5 = 250 \text{Aturns}$$

As $n_c = 25$; I = 250/(2·25) = 5A (RMS). The wire gauge d_{Co} is:

$$d_{Co} = \sqrt{\frac{4}{\pi}\frac{I}{J_{Co}}} = \sqrt{\frac{4}{\pi}\frac{5}{5}} = 1.128\ \text{mm}$$

The pole pitch τ is

$$\tau = \frac{\pi D}{2p_1} = \frac{\pi \cdot 0.15}{2 \cdot 2} = 0.11775\ \text{m}$$

From (4.10)

$$K_{q1} = \frac{\sin\frac{\pi}{6}}{2\sin\frac{\pi}{6 \cdot 2}} = 0.9659$$

$$K_{y1} = \sin\frac{\pi}{2}\cdot\frac{5}{6} = 0.966;\ K_{w1} = K_{q1}K_{y1} = 0.9659\cdot 0.966 = 0.933$$

The mmf fundamental amplitude, (from 4.14), is

$$W_1 = 2p_1qn_c = 2\cdot 2\cdot 2\cdot 25 = 200\text{turns / phase}$$

$$F_{1m} = \frac{3W_1I\sqrt{2}K_{w1}}{\pi p_1} = \frac{3\cdot 200\cdot 5\sqrt{2}\cdot 0.933}{\pi\cdot 2} = 628\text{Aturns / pole}$$

From (4.16) the current sheet (loading) A_{1m} is,

$$A_{1m} = F_{1m}\frac{\pi}{\tau} = 628\cdot\frac{\pi}{0.15} = 13155.3\text{Aturns / m}$$

From (4.18),

$$K_{q7} = \frac{\sin(7\pi/6)}{2\sin(7\pi/6\cdot 2)} = -0.2588;\qquad K_{y7} = \sin\frac{7\pi}{6}\cdot\frac{5}{6} = 0.2588$$

$$K_{w7} = -0.2588\cdot 0.2588 = -0.066987$$

From (4.18),

$$F_{7m} = \frac{3W_1I\sqrt{2}K_{q7}K_{y7}}{\pi p_1 7} = \frac{3\cdot 200\cdot 5\sqrt{2}\cdot 0.066987}{\pi\cdot 2\cdot 7} = 6.445\text{Aturns / pole}$$

This is less than 1% of the fundamental F_{1m} = 628Aturns/pole.

It may be shown that for 120^0 phase belts [10], the distribution (spread) factor K_{qv} is

$$K_{qv} = \frac{\sin v(\pi/3)}{q\sin(v\pi/3\cdot q)} \tag{4.24}$$

For the same case q = 2 and v = 1, we find K_{q1} = sinπ/3 = 0.867. This is much smaller than 0.9659, the value obtained for the 60^0 phase belt, which explains in part why the latter case is preferred in practice.

Now that we introduced ourselves to a.c. windings through two case studies, let us proceed and develop general rules to design practical a.c. windings.

4.6. RULES FOR DESIGNING PRACTICAL A.C. WINDINGS

The a.c. windings for induction motors are usually built in one or two layers.

The basic structural element is represented by coils. We already pointed out (Figure 4.2) that there may be lap and wave coils. This is the case for single turn (bar) coils. Such coils are made of continuous bars (Figure 4.6a) for open slots or from semibars bent and welded together after insertion in semiclosed slots (figure 4.6b).

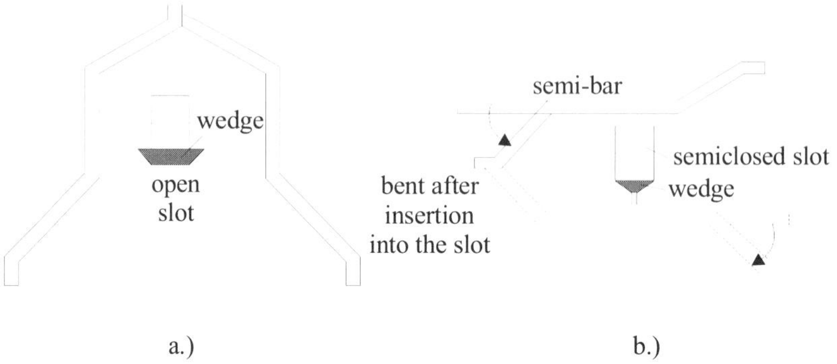

Figure 4.6 Bar coils: a.) continuous bar, b.) semi bar

These are preformed coils generally suitable for large machines.

Continuous bar coils may also be made from a few elementary conductors in parallel to reduce the skin effect to acceptable levels.

On the other hand, round-wire, mechanically flexible coils forced into semiclosed slots are typical for low power induction machines.

Such coils may have various shapes such as shown in Figure 4.7.

A few remarks are in order.

- Wire-coils for single layer windings, typical for low power induction motors (kW range and $2p_1$ = 2pole) have in general wave-shape;
- Coils for single layer windings are always full pitch as an average
- The coils may be concentrated or identical
- The main concern should be to produce equal resistance and leakage inductance per phase
- From this point of view, rounded concentrated or chain-shape identical coils are to be preferred for single layer windings

Double-layer winding coils for low power induction machines are of trapezoidal shape and round shape wire type (Figure 4.8a, b).

For large power motors, preformed multibar (rectangular wire) (Figure 4.8c) or unibar coils (Figure 4.6) are used.

Now to return to the basic rules for a.c. windings design let us first remember that they may be integer q or fractional q ($q = a+b/c$) windings with the total number of slots $N_s = 2p_1qm$. The number of slots per pole could be only an integer. Consequently, for a fractional q, the latter is different and integer for a phase under different poles. Only the average q is fractional. Single-layer windings are built only with an integer q.

As one coil sides occupy 2 slots, it means that $N_s/2m$ = an integer (m– number of phases; m = 3 in our case) for single-layer windings. The number of inward current coil sides is evidently equal to the number of outward current coil sides.

For two-layer windings the allocation of slots per phase is performed in one (say, upper) layer. The second layer is occupied "automatically" by observing the coil pitch whose first side is in one layer and the second one in the second layer. In this case it is sufficient to have N_s/m – an integer.

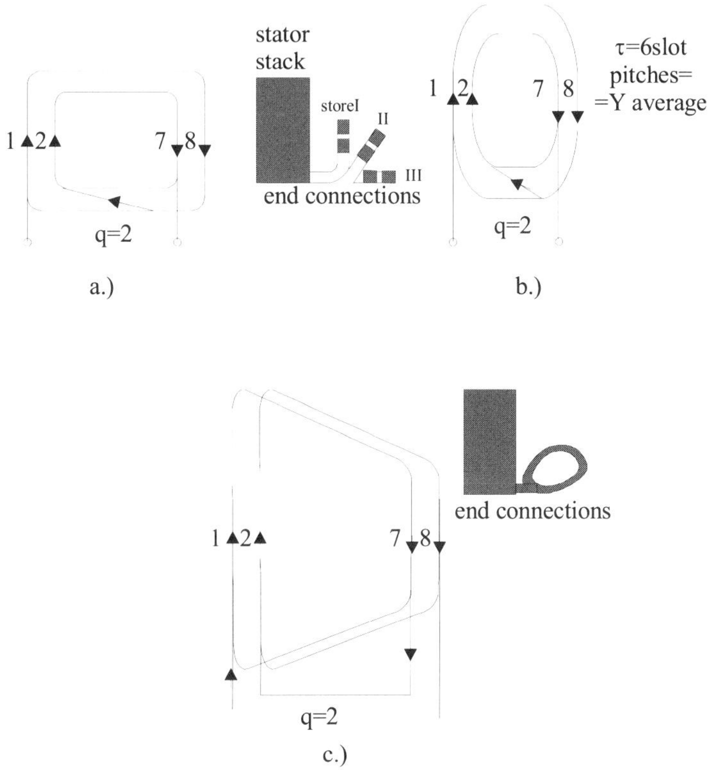

Figure 4.7 Full pitch coil groups/phase/pole–for q = 2–for single layer a.c. windings: a.) with concentrated rectangular shape coils and 2 (3) store end connections; b.) with concentrated rounded coils; c.) with chain shape coils.

A pure traveling stator mmf (4.13), with an open rotor winding and a constant airgap (slot opening effects are neglected), when the stator and iron core permeability is infinite, will produce a no-load ideal flux density in the airgap as

$$B_{g10}(x,t) = \frac{\mu_0 F_{1m}}{g} \cos\left(\frac{\pi}{\tau} x - \omega_1 t\right) \tag{4.25}$$

according to Biot – Savart law.

This flux density will self-induce sinusoidal emfs in the stator windings.

The emf induced in coil sides placed in neighboring slots are thus phase shifted by α_{es}

$$\alpha_{es} = \frac{2\pi p_1}{N_s} \tag{4.26}$$

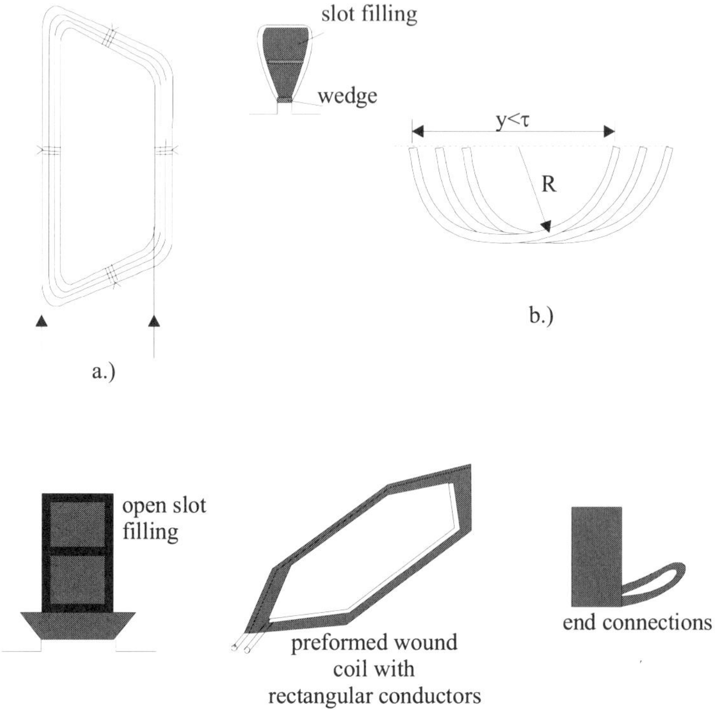

Figure 4.8 Typical coils for two-layer a.c. windings:
a. trapezoidal flexible coil (round wire);
b. rounded flexible coil (rounded wire);
c. preformed wound coil (of rectangular wire) for open slots.

The number of slots with emfs in phase, t, is

$$t = \text{greatest common divisor } (Ns, p_1) = \text{g.c.d.} (N_s, p_1) \le p_1 \tag{4.27}$$

Thus the number of slots with emfs of distinct phase is N_s/t. Finally the phase shift between neighboring distinct slot emfs α_{et} is

$$\alpha_{et} = \frac{2\pi t}{N_s} \tag{4.28}$$

If $\alpha_{es} = \alpha_{et}$, that is $t = p_1$, the counting of slots in the emf phasor star diagram is the real one in the machine.

Now consider the case of a single winding with $N_s = 24$, $2p_1 = 4$. In this case

$$\alpha_{es} = \frac{2\pi p_1}{N_s} = \frac{2\pi \cdot 2}{24} = \frac{\pi}{6} \tag{4.29}$$

$$t = \text{g.c.d.}\ (N_s, p_1) = \text{g.c.d.}(24,2) = 2 = p_1 \tag{4.30}$$

So the number of distinct emfs in slots is $N_s/t = 24/2 = 12$ and their phase shift $\alpha_{et} = \alpha_{es} = \pi/6$. So their counting (order) is the natural one (Figure. 4.9).

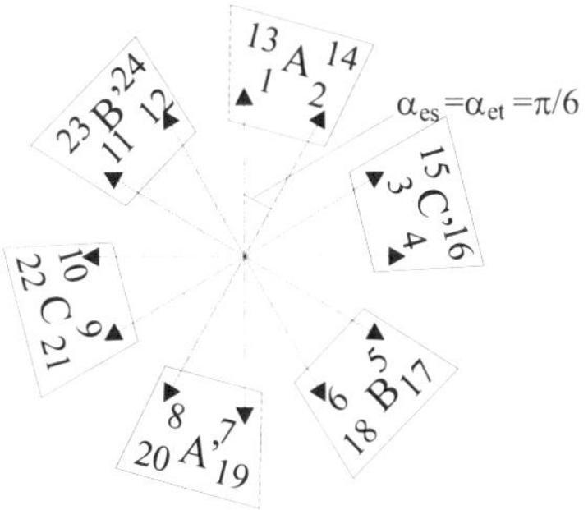

Figure 4.9 The star of slot emf phasors for a single-layer winding with $q = 2$, $2p_1 = 3$, $m = 3$, $N_s = 24$ slots.

The allocation of slots to phases to produce a symmetric winding is to be done as follows for

- single-layer windings
  - Built up the slot emf phasor star based on calculating α_{et}, α_{es}, N_s/t distinct arrows counting them in natural order after α_{es}.
  - Choose randomly $N_s/2m$ successive arrows to make up the inward current slots of phase A (Figure 4.9).
  - The outward current arrows of phase A are phase shifted by π radians with respect to the inward current ones.
  - By skipping $N_s/2m$ slots from phase A, we find the slots of phase B.
  - Skipping further $N_s/2m$ slots from phase B we find the slots of phase C.
- double-layer windings
  - Build up the slot emf phasor star as for single-layer windings.
  - Choose N_s/m arrows for each phase and divide them into two groups (one for inward current sides and one for outward current sides) such that they are as opposite as possible.
  - The same routine is repeated for the other phases providing a phase shift of $2\pi/3$ radians between phases.

It is well understood that the above rules are also valid for the case of fractional q. Fractional q windings are built only in two-layers and small q, to reduce the order of first slot harmonic.

➢ Placing the coils in slots

For single-layer, full pitch windings, the inward and outward side coil occupy entirely the allocated slots from left to right for each phase. There will be $N_s/2m$ coils/phase.

The chorded coils of double-layer windings, with a pitch y ($2\tau/3 \le y < \tau$ for integer q and single pole count windings) are placed from left to right for each phase, with one side in one layer and the other side in the second layer. They are connected observing the inward (A, B, C) and outward (A', B', C') directions of currents in their sides.

➢ Connecting the coils per phase

The $N_s/2m$ coils per phase for single-layer windings and the N_s/m coils per phase for double-layer windings are connected in series (or series/parallel) such that for the first layer the inward/outward directions are observed. With all coils/phase in series, we obtain a single current path (a = 1). We may obtain "a" current paths if the coils from $2p_1/a$ poles are connected in series and, then, the "a " chains in parallel.

Example 4.2. Let us design a single-layer winding with $2p_1 = 2$ poles, q = 4, m = 3 phases.

Solution

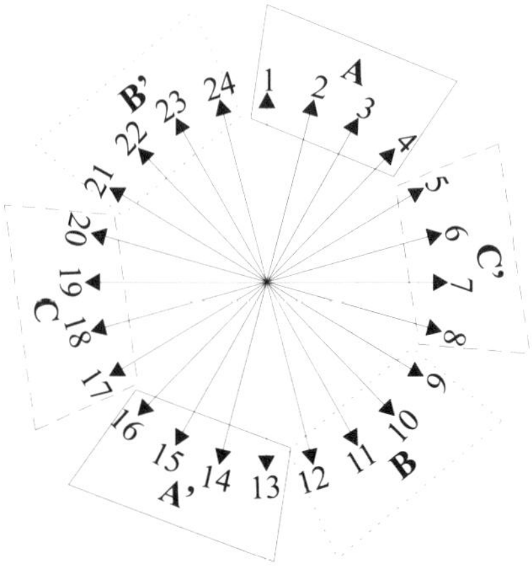

Figure 4.10 The star of slot emf phasors for a single-layer winding: q = 1, $2p_1 = 2$, m = 3, $N_s = 24$.

The angle α_{es} (4.26), t (4.27), α_{et} (4.28) are

$$N_s = 2p_1qm = 24;\ \alpha_{es} = \frac{2\pi P_1}{N_s} = \frac{2\pi \cdot 1}{24} = \frac{\pi}{12}$$

$$t = \text{g.c.d.}(N_s, P_1) = \text{g.c.d.}(24,1) = 1$$

$$\alpha_{et} = \frac{2\pi t}{N_s} = \frac{2\pi \cdot 1}{24} = \frac{\pi}{12}$$

Also the count of distinct arrows of slot emf star $N_s/t = 24/1 = 24$.

Consequently the number of arrows in the slot emf star is 24 and their order is the real (geometrical) one (1, 2, … 24)–Figure 4.10.

Making use of Figure 4.10, we may thus alocate the slots to phases as in Figure 4.11.

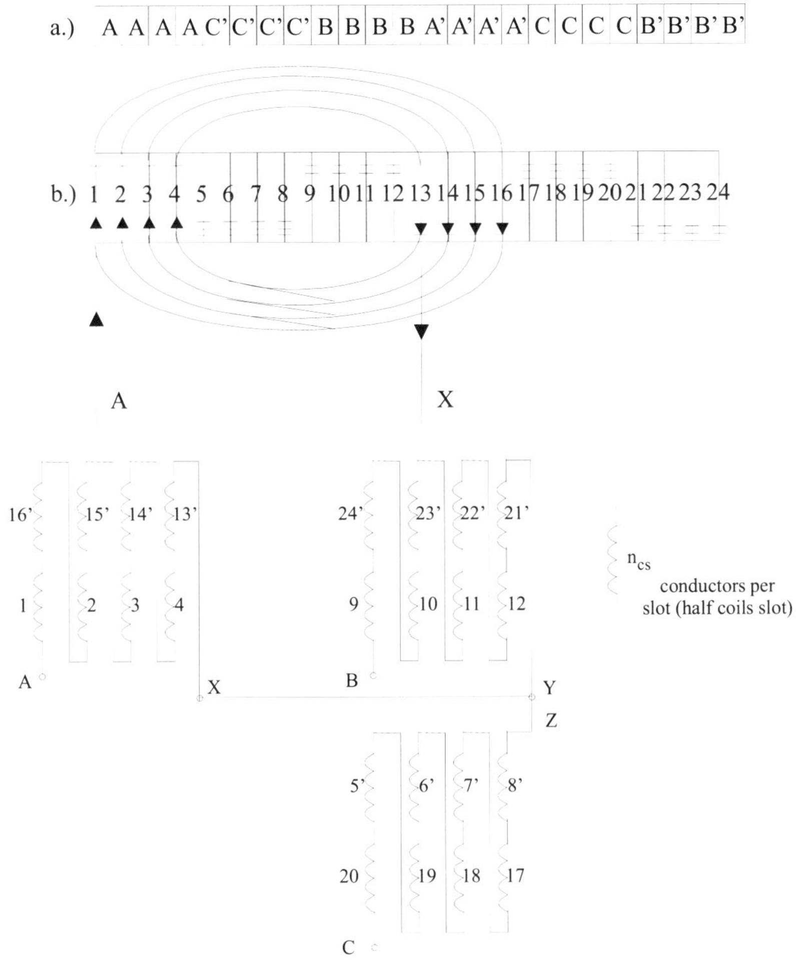

Figure 4.11 Single-layer winding layout
a.) slot/phase allocation; b.) rounded coils of phase A; c.) coils per phase

Example 4.3. Let us consider a double-layer three-phase winding with q = 3, $2p_1 = 4$, m = 3, ($Ns = 2p_1qm = 36$ slots), chorded coils $y/\tau = 7/9$ with a = 2 current paths.

Solution

Proceeding as explained above, we may calculate α_{es}, t, α_{et}:

$$\alpha_{es} = \frac{2\pi P_1}{N_s} = \frac{2\pi \cdot 2}{36} = \frac{\pi}{9}$$

$$t = g.c.d.(36,2) = 2$$

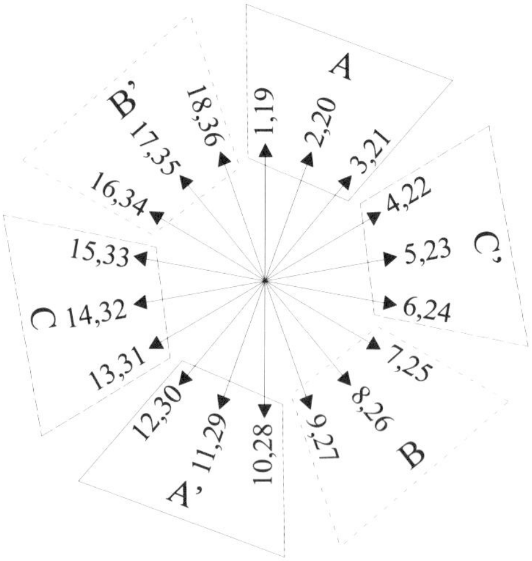

Figure 4.12 The star of slot emf phasors for a double-layer winding (one layer shown) with $2p_1 = 4$ poles,
q = 3 slots/pole/phase, m = 3, $N_s = 36$

$$\alpha_{et} = \frac{2\pi t}{N_s} = \frac{2\pi \cdot 2}{36} = \frac{\pi}{9} = \alpha_{es} \quad N_s / t = 36/2 = 18$$

There are 18 distinct arrows in the slot emf star as shown in Figure 4.12.

The winding layout is shown in Figure 4.13. We should notice the second layer slot allocation lagging by $\tau - y = 9 - 7 = 2$ slots, the first layer allocation.

Phase A produces 4 fully symmetric poles. Also, the current paths are fully symmetric. Equipotential points of two current paths U – U', V – V', W – W' could be connected to each other to handle circulating currents due to, say, rotor eccentricity.

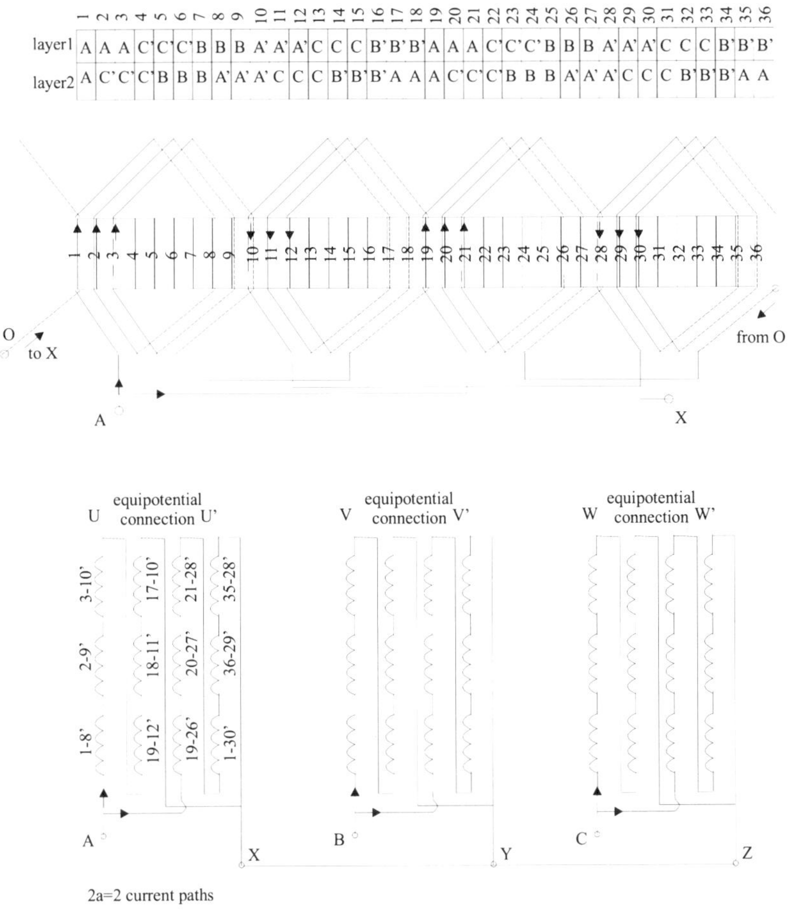

Figure 4.13 Double-layer winding: $2p_1 = 4$ poles, $q = 3$, $y/\tau = 7/9$, $N_s = 36$ slots, $a = 2$ current paths.

Having two current paths, the current in the coils is half the current at the terminals. Consequently, the wire gauge of conductors in the coils is smaller and thus the coils are more flexible and easier to handle.

Note that using wave coils is justified in single-bar coils to reduce the external leads to one by which the coils are connected to each other in series. Copper, labor, and space savings are the advantages of this solution.

4.7. BASIC FRACTIONAL q THREE-PHASE A.C. WINDINGS

Fractional q a.c. windings are not typical for induction motors due to their inherent pole asymmetry as slot/phase allocation under adjacent poles is not the same in contrast to integer q three-phase windings. However, with a small q ($q \leq$

3) to reduce the harmonics content of airgap flux density, by increasing the order of the first slot harmonic from $6q \pm 1$ for integer q to $6(ac + b) \pm 1$ for $q = (ca + b)/c$ = fractional two-layer such windings are favoured to single-layer versions. To set the rules to design such a standard winding–with identical coils–we proceed with an example.

Let us consider a small induction motor with $2p_1 = 8$ and $q = 3/2$, $m = 3$. The total number of slots $N_s = 2p_1qm = 2\cdot4\cdot3/2\cdot3 = 36$ slots. The coil span y is

$$y = \text{integer}(N_s/2p_1) = \text{integer}(36/8) = 4\text{slot pitches} \tag{4.31}$$

The parameters t, α_{es}, α_{et} are

$$t = \text{g.c.d.}(N_s, p_1) = \text{g.c.d.}(36,4) = 4 = p_1 \tag{4.33}$$

$$\alpha_{es} = \frac{2\pi p_1}{N_s} = \frac{\pi \cdot 8}{36} = \frac{2\pi}{9} = \alpha_{et} \tag{4.34}$$

The count of distinct arrows in the star of slot emf phasors is $N_s/t = 36/4 = 9$. This shows that the slot/phase allocation repeats itself after each pole pair (for an integer q it repeats after each pole). Thus mmf subharmonics, or fractional space harmonics, are still absent in this case of fractional q. This property holds for any $q = (2l + 1)/2$ for two-layer configurations.

The star of slot emf phasors has q arrows and the counting of them is the natural one ($\alpha_{es} = \alpha_{et}$) (Figure 4.14a).

A few remarks in Figure 4.14 are in order

- The actual value of q for each phase under neighboring poles is 2 and 1, respectively, to give an average of 3/2
- Due to the periodicity of two poles (2τ), the mmf distribution does not show fractional harmonics ($\nu < 1$)
- There are both odd and even harmonics, as the positive and negative polarities of mmf (Figure 4.14c) are not fully symmetric
- Due to a two pole periodicity we may have a = 1 (Figure 4.14d), or a = 2, 4
- The chording and distribution (spread) factors (K_{y1}, K_{q1}) for the fundamental may be determined from Figure 4.14e using simple phasor composition operations.

$$K_{y1} = \sin\left[\frac{\pi p_1}{N_s}\text{integer}(N_s/2p_1)\right] \tag{4.35}$$

$$K_{q1} = \frac{1 + 2\cos\left(\frac{\pi t}{N_s}\right)}{3} \tag{4.36}$$

This is a kind of general method valid both for integer and fractional q.

Extracting the fundamental and the space harmonics of the mmf distribution (Figure 4.14c) takes implicit care of these factors both for the fundamental and for the harmonics.

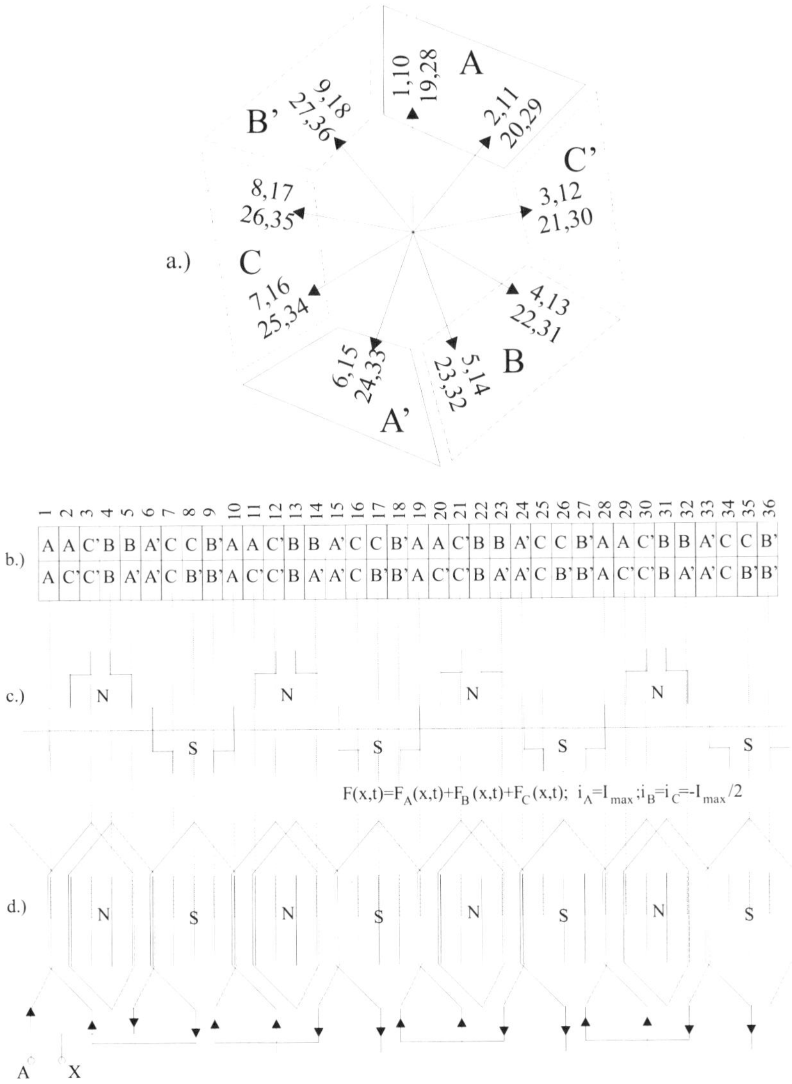

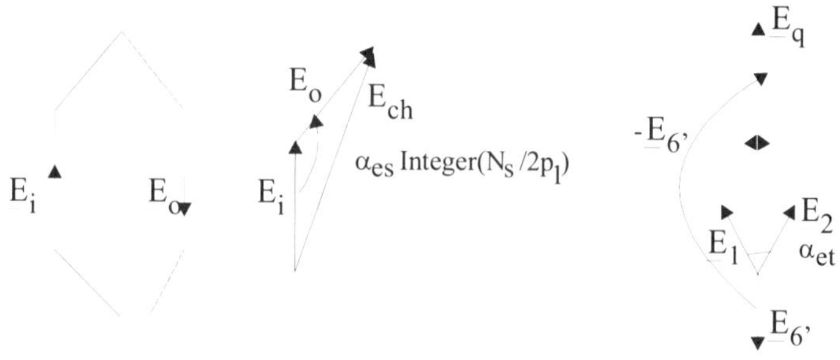

Figure 4.14 Fractionary q ($q = 3/2$, $2p_1 = 8$, $m = 3$, $N_s = 36$) winding
a.) emf star, b.) slot/phase allocation, c.) mmf,
d.) coils of phase A, e.) chording and spread factors

4.8. BASIC POLE-CHANGING THREE-PHASE A.C. WINDINGS

From (4.20) the speed of the mmf fundamental dx/dt is

$$\left(\frac{dx}{dt}\right)_{\nu=1} = 2\tau f_1 \tag{4.37}$$

The corresponding angular speed is

$$\Omega_1 = \frac{dx}{dt}\frac{2}{D} = \frac{2\pi f_1}{p_1}; \quad n_1 = \frac{f_1}{p_1} \tag{4.38}$$

The mmf fundamental wave travels at a speed $n_1 = f_1/p_1$. This is the ideal speed of the motor with a cage rotor.

Changing the speed may be accomplished either by changing the frequency (through a static power converter) or by changing the number of poles.

Changing the number of poles to produce a two-speed motor is a traditional method. Its appeal is still strong today due to low hardware costs where continuous speed variation is not required. In any case, the rotor should have a squirrel cage to accommodate both pole pitches. Even in variable speed drives with variable frequency static converters, when a very large constant power speed range (over 2(3) to 1) is required, such a solution should be considered to avoid a notable increase in motor weight (and cost).

Two – speed induction generators are also used for wind energy conversion to allow a notable speed variation to extract more energy from the wind speed.

There are two possibilities to produce a two-speed motor. The most obvious one is to place two distinct windings in the slots. The number of poles would be $2p_1 > 2p_2$. However the machine becomes very large and costly, while for the winding placed on the bottom of the slots the slot leakage inductance will be very large with all due consequences.

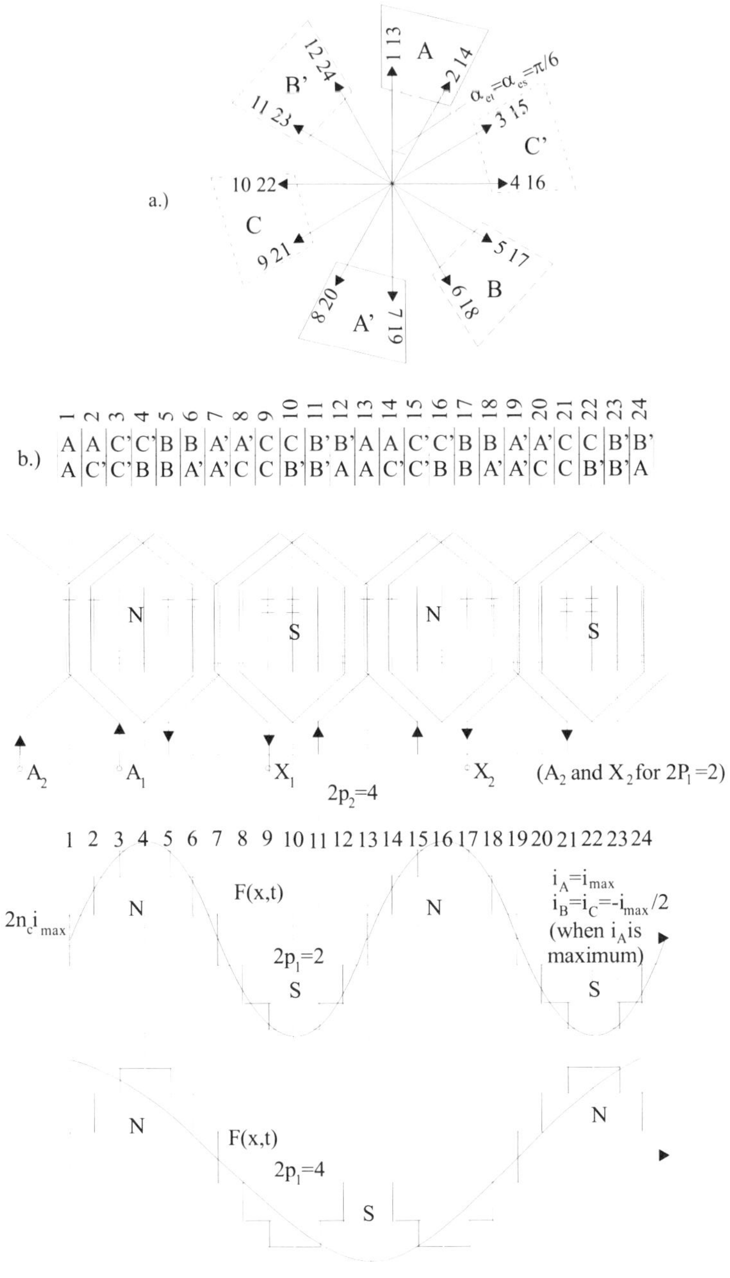

Figure 4.15 2/4 pole winding ($N_s = 24$)
a.) emf star, b.) slot/phase allocation, c.) coils of phase A, d.), e.) mmf for $2p_2 = 4$ and $2p_1 = 2$

Using a pole-changing winding seems thus a more practical solution. However standard pole-changing windings have been produced mainly for $2p_1/2p_2 = 1/2$ or $2p_1/2p_2 = 1/3$.

The most acclaimed winding has been invented by Dahlander and bears his name.

In essence the current direction (polarity) in half of a $2p_2$ pole winding is changed to produce only $2p_1 = 2p_2/2$ poles. The two-phase halves may be reconnected in series or parallel and Y or Δ connections of phases is applied. Thus, for a given line voltage and frequency supply, with various such connections, constant power or constant torque or a certain ratio of powers for the two speeds may be obtained.

Let us now proceed with an example and consider a two-layer three-phase winding with $q = 2$, $2p_2 = 4$, $m = 3$, $N_s = 24$slots, $y/\tau = 5/6$ and investigate the connection changes to switch it to a two pole ($2p_1 = 2$) machine.

The design of such a winding is shown on Figure 4.15. The variables are $t = g.c.d(N_s, p_2) = 2 = p_2$, $\alpha_{es} = \alpha_{et} = 2\pi p_2/N_s = \pi/6$, and $N_s/t = 12$. The star of slot emf phasors is shown in Figure 5.15a.

Figure 4.15c illustrates the fact that only the current direction in the section A2 --X2 of phase A is changed to produce a $2p_1 = 2$ pole winding. A similar operation is done for phases B and C.

A possible connection of phase halves called Δ/2Y is shown in Figure 4.16.

It may be demonstrated that for the Δ ($2p_2 = 4$)/2Y ($2p_1 = 2$) connection, the power obtained for the two speeds is about the same.

We also should notice that with chorded coils for $2p_2 = 4$ ($y/\tau = 5/6$), the mmf distribution for $2p_1 = 2$ has a rather small fundamental and is rich in harmonics.

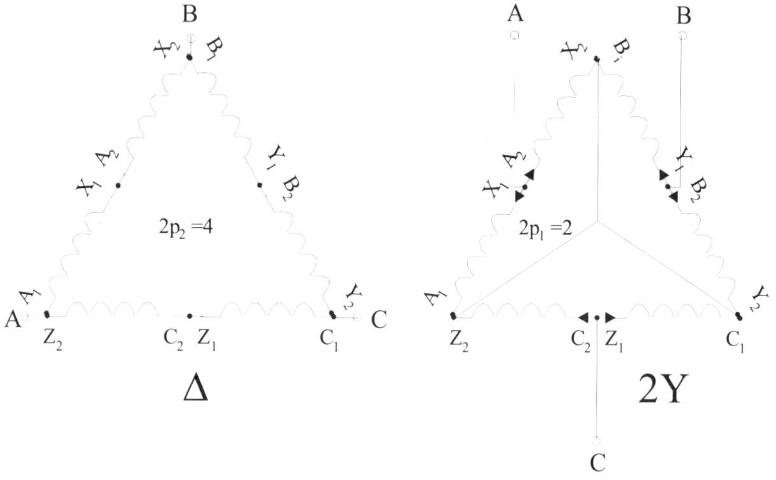

Figure 4.16 2/4 pole winding connection for constant power

In order to achieve about the same winding factor for both pole numbers, the coil span should be greater than the pole pitch for the large number of poles (y = 7, 8 slot pitches in our case). Even if close to each other, the two winding factors are, in general, below 0.85.

The machine is thus larger than usual for the same power and care must be exercised to maintain the acceptable noise level.

Various connections of phases may produce designs with constant torque (same rated torque for both speeds) or variable torque. For example, a Y – YY parallel connection for $2p_2/2p_1 = 2/1$ is producing a ratio of power $P_2/P_1 = 0.35 - 0.4$ as needed for fan driving. Also, in general, when switching the pole number we may need modify the phase sequence to keep the same direction of rotation.

One may check if this operation is necessary by representing the stator mmf for the two cases at two instants in time. If the positive maximum of the mmfs advances in time in opposite directions, then the phase sequence has to be changed.

4.9. TWO-PHASE A.C. WINDINGS

When only single phase supply is available, two-phase windings are used. One is called the main winding (M) and the other, connected in series with a capacitor, is called the auxiliary winding (Aux).

The two windings are displaced from each other, as shown earlier in this chapter, by 90^0 (electrical), and are symmetrized for a certain speed (slip) by choosing the correct value of the capacitance. Symmetrization for start (slip = 0) with a capacitance Cstart and again for rated slip with a capacitance C_{run} is typical (Cstart >> C_{run}).

When a single capacitor is used, as a compromise between acceptable starting and running, the two windings may be shifted in space by more than 90^0 (105^0 to 110^0) (Figure 4.17).

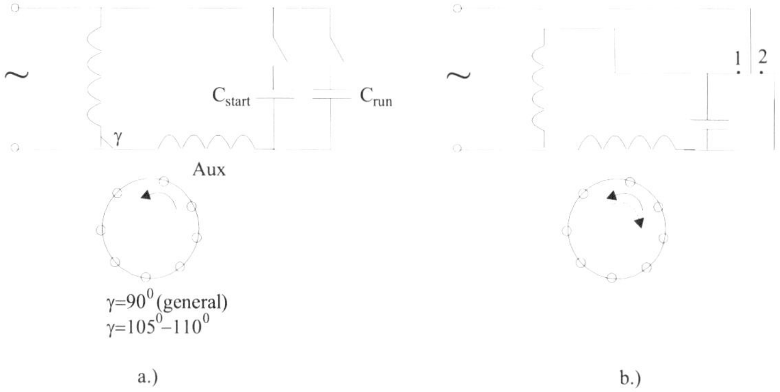

Figure 4.17 Two-phase induction motor
a.) for unidirectional motion,
b.) for bidirectional motion (1 - closed for forward motion; 2 - closed for backward motion)

Single capacitor configurations for good start are characterized by the disconnection of the auxiliary winding (and capacitor) after starting. In this case the machine is called capacitor start and the main winding occupies 66% of stator periphery. On the other hand, if bi-directional motion is required, the two windings should each occupy 50% of the stator periphery and should be identical.

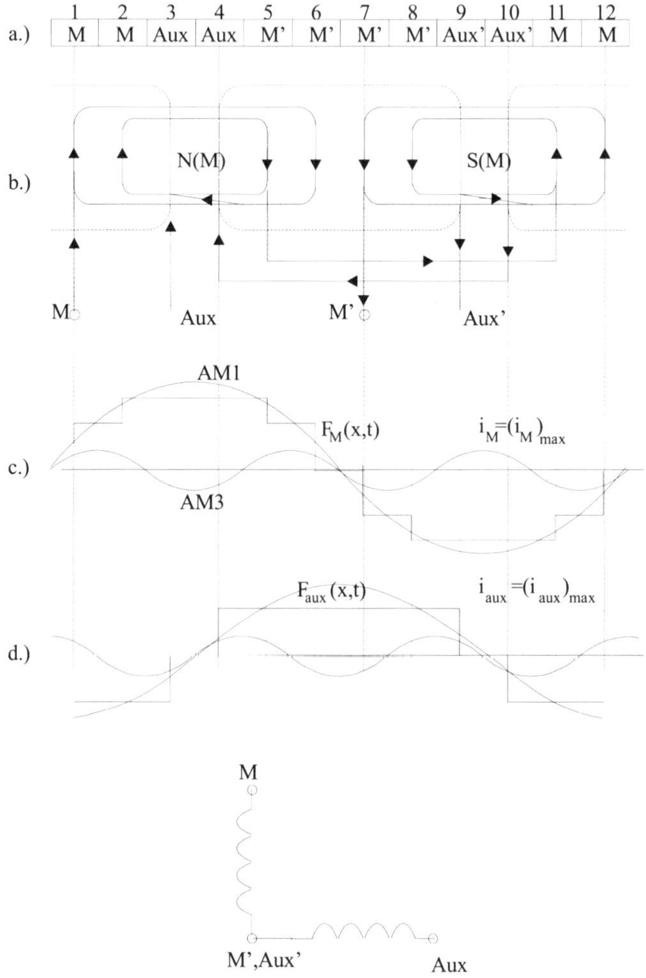

Figure 4.18 Capacitor start motor two-phase single-layer winding; $2p_1 = 2$ poles, $N_s = 12$ slots
a.) slot/phase allocation: M/Aux = 2/1, b.) coils per phase connections,
c.) mmf distribution for main phase (M), d.) mmf distribution for auxiliary phase (Aux)

Two-phase windings are used for low power (0.3 to 2 kW) and thus have rather low efficiency. As these motors are made in large numbers due to their use in home appliances, any improvement in the design may be implemented at competitive costs. For example, to reduce the mmf harmonics content, both phases may be placed in all (most) slots in different proportions. Notice that in two-phase windings multiple of three harmonics may exist and they may deteriorate performance notably.

Let us first consider the capacitor-start motor windings.

As the main winding (M) occupies 2/3 of all slots (Figure 4.18a, b), its mmf distribution (Figure 4.18c) is notably different from that of the auxiliary winding (Figure 4.18d). Also the distribution factors for the two phases (K_{q1M} and K_{q1Aux}) are expected to be different as $q_M = 4$ and $q_{Aux} = 2$.

For $N_s = 16$, $2p_1 = 2$, the above winding could be redesigned for reversible motion where both windings occupy the same number of slots. In that case the windings look like those shown in Figure 4.19.

The two windings have the same number of slots and same distribution factors and, have the same number of turns per coil and same wire gauge only for reversible motion when the capacitor is connected to either of the two phases. The quantity of copper is not the same since the end connections may have slightly different lengths for the two phases.

Double-layer windings, for this case ($q_M = q_{Aux}$ – slot/pole/phase), may be built as done for three-phase windings with identical chorded coils. In this case, even for different number of turns/coil and wire gauge–capacitor run motors– the quantity of copper is the same and the ratio of resistances and self inductances is proportional to the number of turns/coil squared.

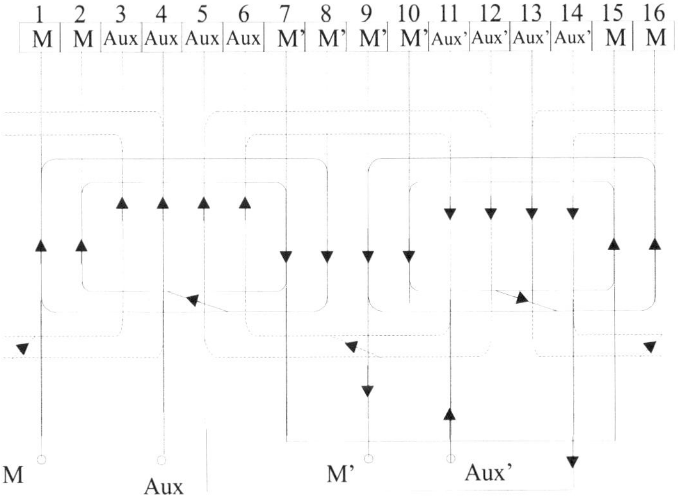

Figure 4.19 Reversible motion (capacitor run) two-phase winding with $2p_1 = 2$, $N_s = 16$, single-layer

As mentioned above and in Chapter 2, capacitor induction motors of high performance need almost sinusoidal mmf distribution to cancel the various bad space harmonic influences on performance.

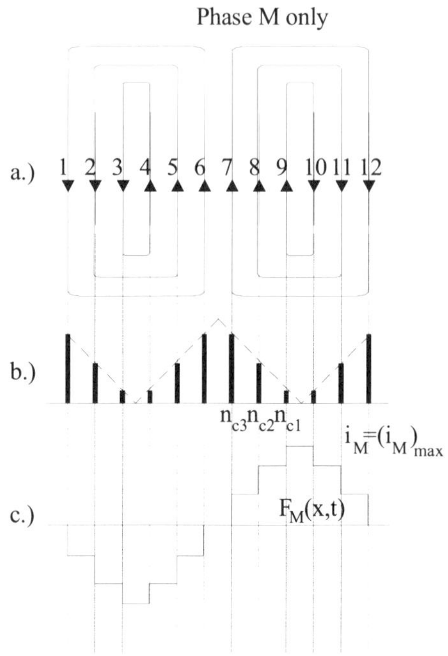

Figure 4.20. Quasi-sinusoidal winding (main winding M shown)
a.) coils in slot/one layer, b.) turns/coil/slot, c.) mmf distribution

One way to do it is to build almost sinusoidal windings; that is, all slots contain coils of both phases but in different proportions (number of turns) so as to obtain sinusoidal mmf. An example for the 2 pole 12 slot case is given in Figure 4.20.

Now if the two windings are equally strong, having the same amount of copper, the slots could be identical. If not, in the case of a capacitor-start, only auxiliary winding, part of the slots may have a smaller area. This way, the motor weight may be reduced. It is evident from Figure 4.20c that the mmf distribution is now much closer to a sinusoid.

Satisfactory results may be obtained for small power resistor started induction motors if the distribution of conductor/slot/phase runs linear rather than sinusoidal with a flat (open) zone of one third of a pole pitch. In this case, however, the number of slots N_s should be a multiple of $6p_1$.

An example of a two one-layer sinusoidal winding is shown in Figure 4.21, for $2p_1 = 2$, $N_s = 24$ slots.

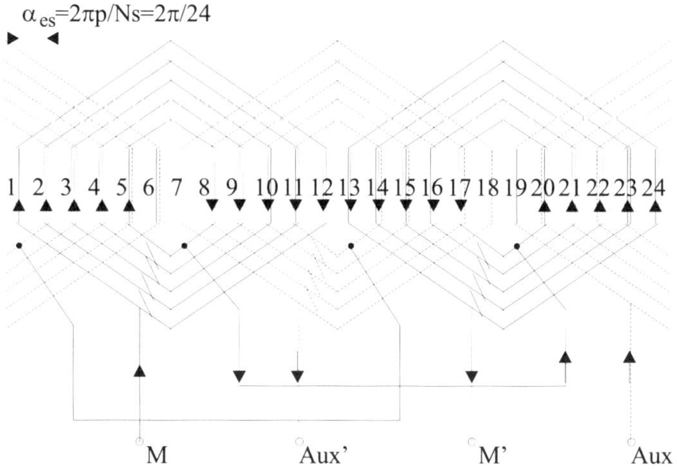

Figure 4.21 Two-pole two-phase winding with sinusoidal conductor count/slot/phase distribution

Similar to Figure 4.20b, the number of turns/coils for the main winding in slots 1 to 5 is

$$\begin{aligned}
n_{c1}^{m} &= K_M \cos\alpha_{es}/2 = n_{c12}^{m} = n_{c13}^{m} = n_{c24}^{m} = 0.991\cdot K_M \\
n_{c2}^{m} &= K_M \cos\frac{3}{2}\alpha_{es}/2 = n_{c11}^{m} = n_{c14}^{m} = n_{c23}^{m} = 0.924\cdot K_M \\
n_{c3}^{m} &= K_M \cos\frac{5}{2}\alpha_{es}/2 = n_{c10}^{m} = n_{c15}^{m} = n_{c22}^{m} = 0.7933\cdot K_M \\
n_{c4}^{m} &= K_M \cos\frac{7}{2}\alpha_{es}/2 = n_{c9}^{m} = n_{c16}^{m} = n_{c21}^{m} = 0.6087\cdot K_M \\
n_{c5}^{m} &= K_M \cos\frac{9}{2}\alpha_{es}/2 = n_{c8}^{m} = n_{c17}^{m} = n_{c20}^{m} = 0.38268\cdot K_M
\end{aligned} \tag{4.39}$$

A similar division of conductor counts is valid for the auxiliary winding.

$$\begin{aligned}
n_{c6}^{a} &= K_A \cos\alpha_{es}/2 = n_{c7}^{a} = n_{c18}^{a} = n_{c19}^{a} = 0.991\cdot K_A \\
n_{c5}^{a} &= K_A \cos\frac{3}{2}\alpha_{es}/2 = n_{c8}^{a} = n_{c17}^{a} = n_{c20}^{a} = 0.924\cdot K_A \\
n_{c4}^{a} &= K_A \cos\frac{5}{2}\alpha_{es}/2 = n_{c9}^{a} = n_{c16}^{a} = n_{c21}^{a} = 0.7933\cdot K_A \\
n_{c3}^{a} &= K_A \cos\frac{7}{2}\alpha_{es}/2 = n_{c10}^{a} = n_{c15}^{a} = n_{c22}^{a} = 0.6087\cdot K_A
\end{aligned} \tag{4.40}$$

$$n_{c2}^{a} = K_A \cos\frac{9}{2}\alpha_{es}/2 = n_{c11}^{a} = n_{c14}^{a} = n_{c23}^{a} = 0.38268\cdot K_A$$

4.10. POLE-CHANGING WITH SINGLE-PHASE SUPPLY INDUCTION MOTORS

A 2/1 = $2p_2/2p_1$ pole-changing winding for single-phase supply can be approached with a three-phase pole-changing winding. The Dahlander winding may be used (see Figure 4.16) as in Figure 4.22.

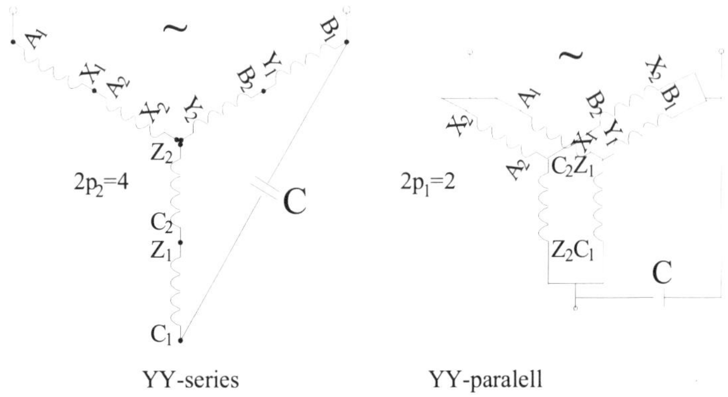

Figure 4.22 4/2 pole two-phase winding for single-phase supply made from a three-phase winding

4.11. SPECIAL TOPICS ON A.C. WINDINGS

Most of the windings treated so far in this chapter are in industrial use today. At the same time, new knowledge has surfaced recently. Here we review the most significant part of it.

- **A new general formula for mmf distribution**

Let us consider the slot opening angle β, n_{cK}–conductors/slot/phase, $i_K(t)$ current in slot K, θ_K–slot position per periphery.

A linear current density in a slot $I_K(\theta_K,t)$ may be defined as

$$\begin{cases} I_K(\theta,t) = \dfrac{2n_{cK} \cdot i_K(t)}{\beta D} & \text{for } \theta \in (\theta_K - \beta/2, \theta_K + \beta/2) \\ I_K(\theta,t) = 0 & \text{for } \theta \notin (\theta_K - \beta/2, \theta_K + \beta/2) \end{cases} \quad (4.41)$$

For a single slot, the linear current density may be decomposed in mechanical harmonics as

$$I_K(\theta,t) = \frac{2n_{cK} \cdot i_K(t)}{\pi D}\left[\frac{1}{2} + \sum_{\nu}^{\infty} K_\nu \cos(\theta - \theta_\nu)\right] \quad (4.42)$$

with

$$K_\nu = \frac{2}{\nu\beta}\sin\frac{\nu\beta}{2} \quad (4.43)$$

The number n_{cK} may be positive or negative depending on inward/outward sense of connections and it is zero if no conductors belong to that particular phase. By adding the contributions of all slots per phase A, we obtain its mmf [10,11].

$$F_A(\theta,t)=\frac{2}{\pi}W_1 i_A(t)\sum_{\nu}^{\infty}\frac{K_\nu K_{w\nu}}{\nu}\sin(\theta-\theta_{A\nu}) \tag{4.44}$$

with

$$K_{w\nu}=\frac{1}{2W_1}\sqrt{\left(\sum_{K=1}^{N_s}n_{cK}\cos\nu\theta_K\right)^2+\left(\sum_{K=1}^{N_s}n_{cK}\sin\nu\theta_K\right)^2} \tag{4.45}$$

$$\theta_{A\nu}=\frac{1}{\nu}\tan^{-1}\left[\left(\sum_{K=1}^{N_s}n_{cK}\sin\nu\theta_K\right)\Big/\left(\sum_{K=1}^{N_s}n_{cK}\cos\nu\theta_K\right)\right] \tag{4.46}$$

where W_1 is the total number of turns/phase. When there is more than one current path ($a > 1$), W_1 is replaced by W_{1a} (turns/phase/path) and only the slots/phase/path are considered.

$K_{w\nu}$ is the total winding factor: spread and chording factors multiplied. This formula is valid for the general case (integer q or fractional q windings, included). The effect of slot opening on the phase mmf is considered through the factor $K_\nu < 1$(which is equal to unity for zero slot opening – ideal case). In general, in an ideal case, the mmf harmonics amplitude is reduced by slot opening, but so is its fundamental.

- **Low space harmonic content windings**

Fractional pitch windings have been, in general, avoided for induction motor, especially due to low order (sub or fractional) harmonics and their consequences: parasitic torques, noise, vibration, and losses.

However, as basic resources for IM optimization are exhausted, new ways to reduce copper weight and losses are investigated. Among them, fractional windings with two pole symmetry ($q = (2l + 1)/2$) are investigated thoroughly [9,12]. They are characterised by the absence of sub (fractional) harmonics. The case of $q = 3/2$ is interesting low power motors for low-speed high-pole-count induction generators and motors. In essence, the number of turns of the coils in the two groups/phase belonging to neighboring poles (one with $q_1 = l$ and the other with $q_2 = l + 1$) varies so that designated harmonics could reduce or destroy. Also, using coils of a different number of conductors in various slots of a phase with standard windings (say $q = 2$) should lead to the cancellation of low order mmf harmonics and thus render the motors less noisy.

A reduction of copper weight and (or) an increase in efficiency may thus be obtained. The price is using nonidentical coils.

The general formula of the winding factor (4.45) may be simplified for the case of q = 3/2, 5/2, 7/2 … where two-pole symmetry is secured and thus the summation in (4.45) extends along only two poles.

With γ_i the coil pitch, the winding factor $K_{w\nu}$, for two poles, is [9]

$$K_{w\nu} = \sum_{i}^{q_1=l} n_{ci}'\sin\frac{\nu\gamma_i}{2} - (-1)^{\nu}\sum_{K}^{q_2=l+1} n_{cK}'\sin\frac{\nu\gamma_K}{2} \tag{4.47}$$

n_{ci}' and n_{cK}' are relative numbers of turns with respect to total turns/polepair/phase; therefore the coil pitches may be different under the two poles. Also, the number of such coils for the first pole is $q_1 = l$ and for the second is $q_2 = l + 1$ $(q = (2l + 1)/2)$.

As $q = (2l + 1)/2$, the coil pitch angles γ_i will refer to an odd number of slot pitches while γ_K refers to an even number of slot pitches. So it could be shown that the winding factor $K_{w\nu}$ of some odd and even harmonics pairs are equal to each other. For example, for q = 3/2, $K_{w4} = K_{w5}$, $K_{w2} = K_{w7}$, $K_{w1} = K_{w8}$. These relationships may be used when designing the windings (choosing γ_i, γ_K and n_{ci}', n_{cK}') to cancel some harmonics.

➢ **A quasisinusoidal two-layer winding [8]**

Consider a winding with 2l + 1 coils/phase/pole pairs. Consquently, q = (2l + 1)/2 slot/pole/phase.

There are two groups of coils per phase. One for the first pole containing K + 1 coils of span 3l + 1, 3l – 1 … and, for the second pole, K coils of span 3l, 3l – 2. The problem consists of building a winding in two-layers to cancel all harmonics except those in multiples of 3 (which will be inactive in star connection of phases anyway) and the 3(2l + 1) ± 1, or slot harmonics.

$$\sum K_{w\nu}(n_{ci}'+n_{cK}') = 0; \nu = 2,4,5,...(\text{in all } 2l \text{ values}) \tag{4.48}$$

with

$$\sum_{i=1}^{2l+1}(n_{ci}'+n_{cK}') = 1 \tag{4.49}$$

This linear system has a unique solution, which, in the order of increased number of turns, yields

$$n_{ci}' = 4\sin\left(\frac{\pi}{6(2l+1)}\right)\sin\frac{(2i-1)\pi}{6(2l+1)}; i = 1,2,3\ldots 2l+1 \tag{4.50}$$

For the fundamental mmf wave, the winding factor K_{w1} is

$$K_{w1} = (2l+1)\sqrt{3}\sin\frac{\pi}{6(2l+1)} \tag{4.51}$$

For l = 1, 2l +1 = 3 coils/phase/pole pair. In this case the relative number of turns per coil for phase A are (from 4.50): $n_{c1}' = 0.1206$, $n_{c2}' = 0.3473$, $n_{c3}' = 0.5321$.

As we can see, $n_{c1}' + n_{c3}' \cong 2\, n_{c2}'$ so the filling of all slots is rather uniform though not identical.

The winding layout is shown in Figure 4.23.

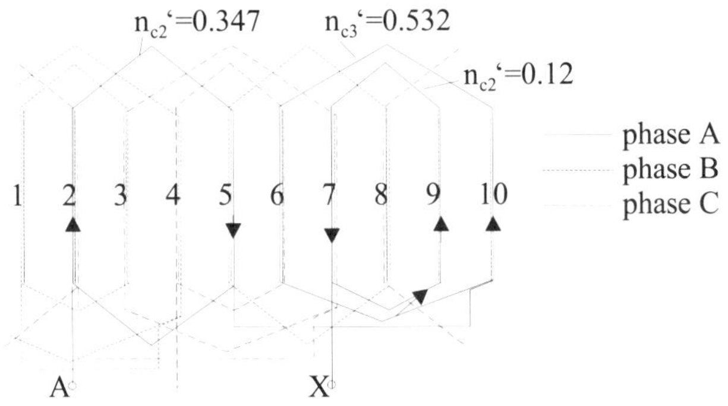

Figure 4.23 Sinusoidal winding with q = 3/2 in two-layers

The fundamental winding factor of this winding is $K_{w1} = 0.9023$, which is satisfactory (both distribution and chording factors are included). To further cancel the multiples of 3 (3l) harmonics, two/three layer windings based on the same principle have been successfully tested [9] though their fundamental winding factor is below 0.8.

The same methodology could be applied for q = integer, by using concentrated coils with various numbers of turns to reduce mmf harmonics. Very smooth operation has been obtained with such a 2.2 kW motor with q = 2, $2p_1 = 4$, $N_s = 24$ slots up to 6000 rpm. [12]

In general, for the same copper weight, such sinusoidal windings are claimed to produce a 20% increase in starting torque and 5 to 8dB noise reduction, at about the same efficiency.

- **Better pole-changing windings**

Pole amplitude modulation [3–5] and symmetrization [13–14] techniques have been introduced in the sixties and revisited in the nineties. [15] Like interspersing [10, pp.37–39], these methods have produced mixed results so far. On the other hand, the so called "3 equation principle" for better pole-changing winding has been presented in [7] for various pole count combinations. This is also a kind of symmetrization method but with a well defined methodology. The connections of phases is Δ/Δ and (3Y +2Y)/3Y for $2p_2$ and $2p_1$ pole counts, respectively. Higher fundamental winding factor, lower harmonics content, better winding and core utilisation, and simpler switching devices from $2p_2$ to $2p_1$ poles than with Dahlander connections are all merits of such a methodology. Two single-throw switches (soft starters) suffice (Figure 4.24) to change speed.

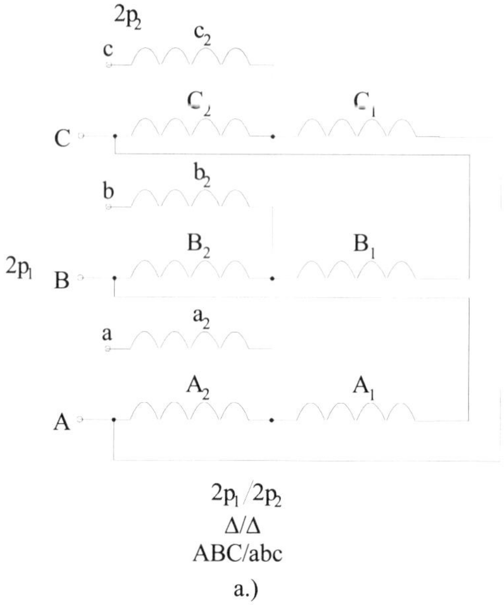

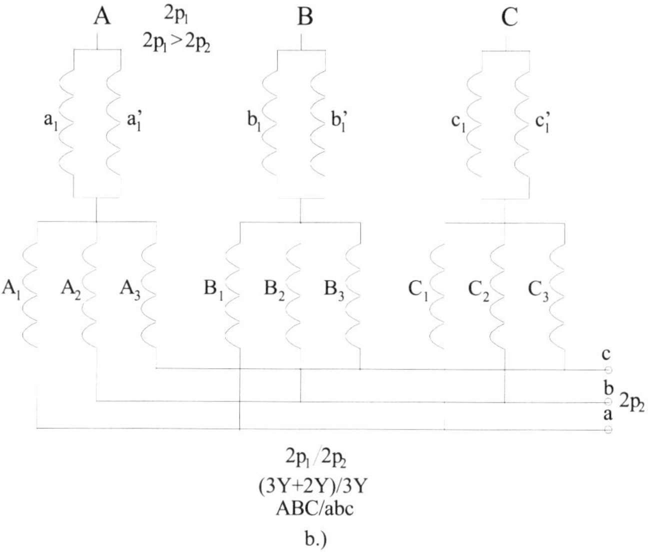

Figure 4.24 $2p_1/2p_2$ pole connections
a.) Δ/Δ; b.) (3Y + 2Y)/3Y

The principle consists of dividing the slot periphery into three equivalent parts by three slots counted n, $n + N_s/3$, and $n + 2N_s/3$. Now the electrical angle between these slots mmf is $2\pi/3p_i$ as expected ($p_i = p_1, p_2$). If $2p_i$ is not a multiple of three the coils in slots n, n + $N_s/3$ and n + $2N_s/3$ belong to all three-phases and the winding is to be symmetric. The Δ connection is used (Figure 4.24). In contrast, if $2p_i$ is a multiple of 3, the emfs in slots n, n + $N_s/3$, n + N_s + $2N_s/3$ are in phase and thus belong to the same phase and we may build three parallel branches (paths) per phase in a 3Y connection (Figure 4.24b). In general, n = 1, 2, … $N_s/3$.

As already meantioned, the Δ/Δ connection (Figure 4.24a) is typical for nonmultiple of 3 pole counts ($2p_1/2p_2 = 8/10$, for example).

Suppose every coil group $A_{1,2,3}$, $B_{1,2,3}$, $C_{1,2,3}$ is made of t_1 coils ($t_1 < N_s/6$). The coil group A_1 is made of t_1 coils in slots n_1, n_2, …, n_{t1} while groups B_1 and C_1 are displaced by $N_s/3$ and $2N_s/3$, respectively. The sections A_1, B_1, C_1, which belong to the three-phases, do not change phase when the number of poles is changed from $2p_1$ to $2p_2$.

On the other hand, coil groups A_2, B_2, C_2 are composed of $t_2 < N_s/6$ coils and thus coil group A_2 refers to slots n_1', n_2', …, n_{t2}' while coil groups B_2 and C_2 are displaced by $N_s/3$ and $2N_s/3$ with respect to group A_2. Again, sections A_2, B_2, C_2 are symmetric.

It may be shown that for $2p_1$ and $2p_2$ equal to $2p_1 = 6K_1 + 2(4)$ and $2p_2 = 6K_2 + 1$, respectively, the mmf waves for the two pole counts travel in the same direction.

On the other hand, if $2p_1 = 6K_1 + 2(4)$ and $2p_2 = 6K_2 + 4(2)$, the mmf waves for the two pole counts move in opposite directions. Swapping two phases is required to keep the same direction of motion for both pole counts (speeds).

In general, t_1 or t_2 are equal to $N_s/6$. For voltage adjustment for higher pole counts, $t_3 = N_s/3 - t_i$ slots/phase are left out and distributed later, also with symmetry in mind.

Table 4.1. (8/6 pole, Δ/Δ, N_s = 72 slots)

Coil group	Slot (coil) distribution principle	Slot (coil) number
A_1	$n(t_1 = 10)$	72,1,2,3,-10,-11,-65,-66,-67,-68
B_1	$n + N_s/3$	24,25,26,27,-34,-35,-17,-18,-19,-20
C_1	$n + 2N_s/3$	48,49,50,51,-58,-59,-41,-42,-43,-44
A_2	$n'(t_2 = 10)$	21,22,23,-29,-30,-31,-32,37,38,39
B_2	$n' + N_s/3$	45,46,47,-53,-54,-55,-56,61,62,63
C_2	$n' + 2N_s/3$	69,70,71,-5,-6,-7,-8,13,14,15
a_2	$n'' (t_3 = 4)$	-52,60,-9,16
b_2	$n'' + N_s/3$	-4,12,33,40
c_2	$n'' + 2N_s/3$	-28,26,-57,64

Adjusting groups of t_3 slots/phase should not produce emfs phase shifted by more than 30^0 with respect to groups $A_iB_iC_i$ to secure a high spread (distribution) factor.

An example of such an 8/6 pole Δ/Δ winding is shown in Table 4.1 [7]; the coil connections from table 4.1 are shown in Figure 4.25. [7]

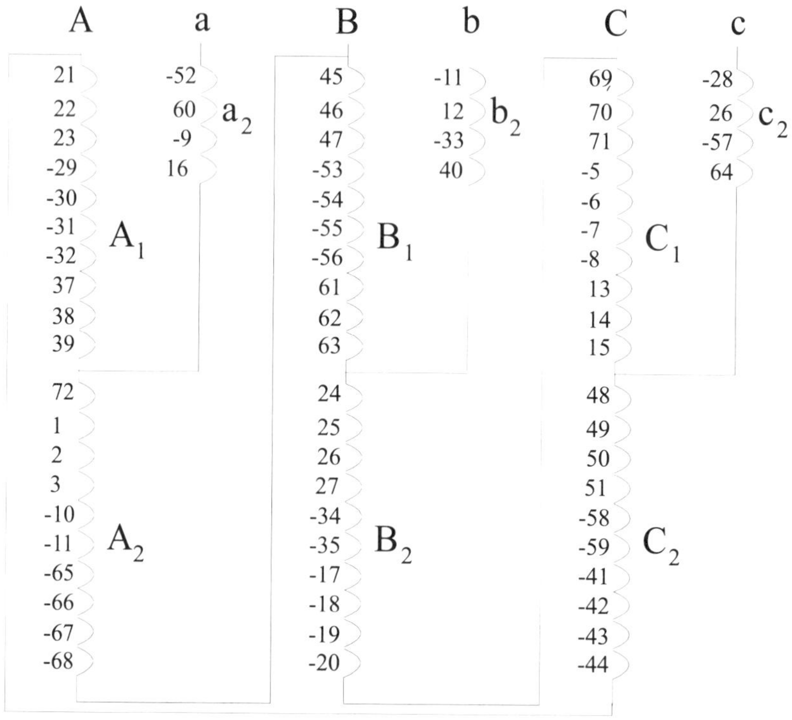

Figure 4.25 8/6 pole winding with 72 slots

Such a pole-changing winding seems suitable for fan or windmill applications as the power delivered is reduced to almost half when switching from 8 to 10 poles, that is, with a 20% reduction in speed. The procedure is similar for the (3Y + 2Y)/3Y connection. [6]

This type (Figure 4.24b) for $a_1 = a_1' = b_1 = b_1' = c_1 = c_1' = 0$ and pole count ratio of say $2p_2/2p_1 = 6/2$ has long been known and recently proposed for dual winding (doubly fed) stator nest – cage rotor induction machines [16].

Once the slot/phase allocation is completed the general formula for the winding factor (4.45) or the mmf step-shape distribution with harmonics decomposition may be used to determine the mmf fundamental and harmonics content for both pole counts. These results are part of any IM design process.

- **A pure mathematical approach to a.c. winding design**

As we have seen so far, a.c. winding design is more an art than a science. Intuition and symmetry serve as tools to this aim.

Since we have a general expression for the harmonics winding factor valid for all situations when we know the number of turns per coil and what current flows through those coils in each slot, we may develop an objective function to define an optimum performance.

Maximum airgap flux energy for the fundamental mmf, or minimum of some harmonics squared winding factors summation, or the maximum fundamental winding factor, may constitute practical objective functions. Now copper weight may be considered a constraint. So is the pole count $2p_1$. Some of the above objectives may become constraints. What then would the variables be for this problem? It seems that the slot/phase allocation and ultimately, the number of turns/coil are essential variables to play with, together with the number of slots.

In a pole-changing ($2p_2/2p_1$) winding, the optimization process has to be done twice. This way the problem becomes typical: non-linear, with constraints.

All optimization methods used now in electric machine design also could apparently be applied here.

The first attempt has been made in [7] by using the "simulated annealing method"–a kind of direct (random) search method, for windings with identical coils. The process starts with fixing the phase connection: Y or Δ and with a random allocation of slots to phases A, B, C. Then we take the instant when the current in phase A is maximum i_{Max} and $i_B = i_C = -i_{Max}/2$. Allowing for a given number of turns per coil n_c and 1 or 2 layers, we may start with given (random) mmfs in each phase.

Then we calculate the objective function for this initial situation and proceed to modify the slot/phase allocation either by changing the connection (beginning or end) of the coil or its position (see Figure 4.26). Various rules of change may be adopted here.

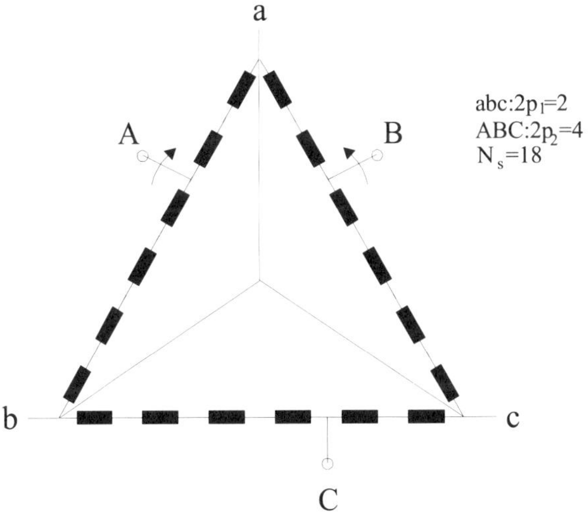

Figure 4.26 Initial (random) winding on the way to "rediscover" the Dahlander connection

The objective function is calculated again. The computation cycle is redone until sufficient convergence is obtained for a limited computation time.

In [7] by using such an approach the 4/2 pole Dahlander connection for an 18 slot IM is rediscovered after two hours CPU time on a large main frame computer. The computer time is still large, so the direct (random) search approach has to be refined or changed, but the method becomes feasible, especially for some particular applications.

4.12. THE MMF OF ROTOR WINDINGS

Induction machine rotors have either three-phase windings star-connected to three-phase slip rings or have squirrel cage type windings. The rotor three-phase windings are made of full pitch coils and one or two-layers located in semiclosed or semiopen slots. Their end connections are braced against centrifugal forces. Wound rotors are typical for induction machines in the hundreds and thousands of kW ratings and more recently in a few hundreds of MW as motor/generators in pumped storage hydro power plants.

The voltage rating of rotor windings is of the same order as that in the stator of the same machine. The mmf of such windings is similar to that of stator windings treated earlier in this chapter.

Squirrel-cage windings–single cage, double cage, deep bar, solid rotor, and nest cage types were introduced in Chapter 2. Here we deal with the mmf of a single cage winding. We consider a fully symmetrical cage (no broken bars or end rings) with straight slots (no skewing)–Figure 4.27a.

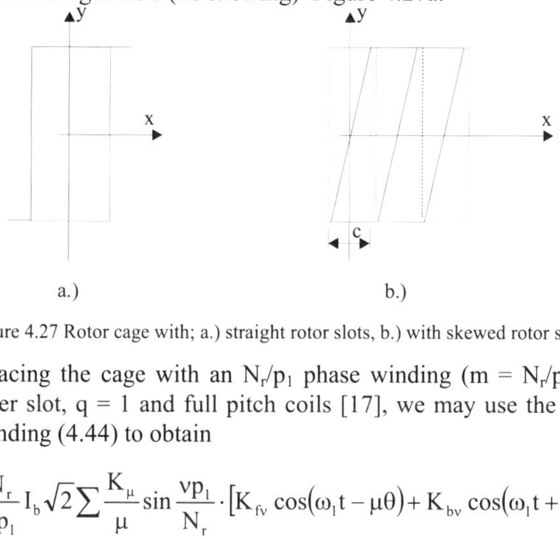

Figure 4.27 Rotor cage with; a.) straight rotor slots, b.) with skewed rotor slots

By replacing the cage with an N_r/p_1 phase winding ($m = N_r/p_1$) with one conductor per slot, $q = 1$ and full pitch coils [17], we may use the formula for the three winding (4.44) to obtain

$$F_2(\theta, t) = \frac{N_r}{\pi p_1} I_b \sqrt{2} \sum \frac{K_\mu}{\mu} \sin\frac{\nu p_1}{N_r} \cdot \left[K_{f\nu} \cos(\omega_1 t - \mu\theta) + K_{b\nu} \cos(\omega_1 t + \mu\theta)\right] \quad (4.52)$$

where I_b is the RMS bar current and

$$K_{f\nu,b\nu} = \frac{p_1}{N_s}\left[1 + \left(\frac{N_r}{p_1} - 1\right)\cos(\mu \pm p_1)\frac{2\pi}{N_r}\right] \quad (4.53)$$

Let us remember that μ is a spatial harmonic and θ is a geometrical angle: $\theta_e = p_1\theta$. Also, $\mu = p_1$ means the fundamental of mmf.

There are forward (f) and backward (b) harmonics. Only those of the order $\mu = KN_r + p_1$, have $K_{fv} = 1$ while for $\mu = KN_r - p_1$, they have $K_{bv} = 1$. All the other harmonics have $K_{fv,bv}$ as zero. But the above harmonics are, in fact, caused by the rotor slotting. The most important, are the first ones obtained for K = 1: $\mu_{fmin} = N_r - p_1$ and $\mu_{bmin} = N_r + p_1$.

For many sub-MW-power induction machines, the rotor slots are skewed, in general by about one stator slot pitch or more, to destroy the first stator-slot-caused stator mmf harmonic ($\nu = N_s - p_1$). Again ν is a spatial harmonic; that is $\nu_1 = p_1$ for the fundamental.

In this situation let us take only the case of the fundamental. The situation is as if the rotor position in (4.54) varies with y (Figure 4.27b).

$$F_2(\theta, t)_{\nu=p_1} \approx \frac{N_r}{\pi p_1} I_b\sqrt{2}\cdot\cos\left[\omega_1 t - \left(\theta_e - \frac{\pi y}{\tau}\right) - (\pi - \gamma_{1,2})\right];\ \theta_e = \frac{\pi x}{\tau} \quad (4.54)$$

$$\text{for } y \in [0.0, \pm c/2]$$

Now as ω_1, the stator frequency, was used in the rotor mmf, it means that (4.54) is written in a stator reference system,

$$\omega_1 = \omega_2 + \omega_r; \omega_r = \Omega_r p_1 \quad (4.56)$$

with ω_r as the rotor speed in electrical terms and ω_2 the frequency of the currents in the rotor. The angle $\gamma_{1,2}$ is a reference angle (at zero time) between the stator $F_1(\theta,t)$ and rotor $F_2(\theta,t)$ fundamental mmfs.

4.13. THE "SKEWING" MMF CONCEPT

In an induction motor on load, there are currents both in the stator and rotor. Let us consider only the fundamentals for the sake of simplicity. With $\theta_e = \pi x/\tau$, (4.13) becomes

$$F_1(\theta_e, t) = F_{1m}\cos(\omega_1 t - \theta_e) \quad (4.57)$$

with F1m from (4.14).

The resultant mmf is the sum of stator and rotor mmfs, $F_1(\theta_e,t)$, and $F_2(\theta_e,t)$. It is found by using (4.54) and (4.57).

$$F_{r1}(\theta_e, t) = F_{1m}\cos(\omega_1 t - \theta_e) + F_{2m}\cos\left(\omega_1 t - \theta_e - \frac{\pi y}{\tau} - (\pi - \gamma_{1,2})\right) \quad (4.58)$$

with
$$F_{1m} = \frac{3W_1 I\sqrt{2}K_{w1}}{\pi p_1}; \quad F_{2m} = \frac{N_r}{\pi p_1} I_b\sqrt{2} \quad (4.59)$$

The two mmf fundamental amplitudes, F_{1m} and F_{2m}, are almost equal to each other at standstill. In general,

$$\frac{F_{2m}}{F_{1m}} \approx \sqrt{1-\left(\frac{I_0}{I}\right)^2} \tag{4.60}$$

where I_0 is the no load current.

For rated current, with $I_0 = (0.7 \text{ to } 0.3)I_{Rated}$, F_{2m}/F_{1m} is about (0.7 to 0.95) at rated load. As at start (zero speed) $I_s = (4.5 \text{ to } 7)I_n$, it is now obvious that in this case,

$$\left(\frac{F_{2m}}{F_{1m}}\right)_{S=1} \approx 1.0 \tag{4.61}$$

The angle between the two mmfs $\gamma_{1,2}$ varies from a few degrees at standstill (S = 1) to 45^0 at peak (breakdown) slip S_K and then goes to zero at S = 0. The angle $\gamma_{1,2} > 0$ for motoring and $\gamma_{1,2} < 0$ for generating.

The resultant mmf $F_r(\theta_e,t)$, with (4.54) and (4.57), is

$$\begin{aligned} F_{r1}(\theta_e,t) &= \left[F_{1m}\cos(\omega_1 t-\theta_e) - F_{2m}\cos(\omega_1 t-\theta_e+\gamma_{1,2})\cos\frac{\pi y}{\tau}\right] + \\ &+ F_{2m}\sin(\omega_1 t-\theta_e+\gamma_{1,2})\cdot\sin\frac{\pi y}{\tau};\ y\in(-c/2,+c/2) \end{aligned} \tag{4.62}$$

The term outside the square brackets in (4.62) is zero if there is no skewing. Consequently the term inside the square brackets is the compensated or the magnetizing mmf which tends to be small as $\cos\pi y/\tau \cong 1$ and thus the skewing c = $(0.5 \text{ to } 2.0)\tau/mq$ is small and the angle $\gamma_{1,2} \in (\pi/4,-\pi/4)$ for all slip values (motoring and generating).

Therefore, we define F_{2skew} by

$$F_{2skew} = F_{2m}\sin(\omega_1 t-\theta_e+\gamma_{1,2})\sin\frac{\pi y}{\tau};\quad y\in(-c/2,+c/2) \tag{4.63}$$

the uncompensated rotor-produced mmf, which is likely to cause large airgap flux densities varying along the stator stack length at standstill (and large slip) when the rotor currents (and F2m) are large. Thus, heavy saturation levels are expected along main flux paths in the stator and rotor core at standstill as a form of leakage flux, which strongly influences the leakage inductances as shown later in this book.

4.14. SUMMARY

- A.C. windings design deals with assigning coils to slots and phases, establishing coil connections inside and between phases, and calculating the number of turns/coil and the wire gauge.

- The mmf of a winding means the spatial/time variation of ampere turns in slots along the stator periphery.
- A pure traveling mmf is the ideal; it may be approximately realized through intelligent placing of coils of phases in their slots.
- Traveling wave mmfs are capable producing ripple-free torque at steady state.
- The practical mmf wave has a spacial period of two pole pithces 2τ.
- There are p_1 electrical periods for one mechanical revolution.
- The speed of the traveling wave (fundamental) $n_1 = f_1/p_1$; f_1–current frequency.
- Two phases with step-wise mmfs, phase shifted by $\pi/2p_1$ mechanical degrees or m phases (3 in particular), phase shifted by π/mp_1 geometrical degrees, may produce a practical traveling mmf wave characterized by a large fundamental and small space harmonics for integer q. Harmonics of order $\nu = 5, 11, 17$ are reverse in motion, while $\nu = 7, 13, 19$ harmonics move forward at speed $n_\nu = f_\nu/(\nu p_1)$.
- Three-phase windings are built in one or two-layers in slots; the total number of coils equals half the number of stator slots N_s for single-layer configurations and is equal to N_s for two-layer windings.
- Full pitch and chorded coils are used; full pitch means π/p_1 mechanical radians and chorded coils means less than that; sometimes elongated coils are used; single-layer windings are built with full pitch coils.
- Windings for induction machines are built with integer and, rarely, fractional number of slots/pole/phase, q.
- The windings are characterized by their mmf fundamental amplitude (the higher the better) and the space harmonic contents, (the lower, the better); the winding factor $K_{w\nu}$ characterizes their performance.
- The star of slot emf phasors refers to the emf phasors in every slot conductor drawn with a common origin and based on the fact that the airgap field produced by the mmf fundamental is also a traveling wave; so, the emfs are sinusoidal in time.
- The star phasors are allocated to the three-phases to produce three resultant phasors 120^0 (electrical) apart; in partly symmetric windings, this angle is only close to 120^0.
- Pole-changing windings may be built by changing the direction of connections in half of each phase or by dividing each phase into a few sections (multiples of 3, in general) and switching them from one phase to another.
- Pole-changing is used to modify the speed as $n_1 = f_1/p_1$.
- New pole-changing windings need only two single throw switches while standard ones need more costly switches.
- Single phase supplies require two-phase windings–at least to start; a capacitor (or a resistor at very low power) in the auxiliary phase provides a traveling mmf for a certain slip (speed).

- Reducing harmonics content in mmfs may be achieved by varying the number of turns per coil in various slots and phases; this is standard in two-phase (single-phase supply) motors and rather new in three-phase motors, to reduce the noise level.
- Two phase (for single phase supply) pole-changing windings may be obtained from three-phase such windings by special connections.
- Rotor three-phase windings use full pitch coils in general.
- Rotor cage mmfs have a fundamental ($\mu = p_1$) and harmonics (in mechanical angles) $\mu = kn_r \pm p_1$, N_r – number of rotor slots.
- The rotor cage may be replaced by an N_r/p_1 phase winding with one conductor per slot.
- The rotor mmf fundamental amplitude F_{2m} varies with slip (speed) up to F_{1m} (of the stator) and so does the phase shift angle between them $\gamma_{1,2}$. The angle $\gamma_{1,2}$ varies in the interval $\gamma_{1,2} \in (0, \pm \pi/4)$.
- For skewed rotor slots a part of rotor mmf, variable along stator stack length (shaft direction), remains uncompensated; this mmf is called here the "skewing" mmf which may produce heavy saturation levels at low speed (and high rotor currents).
- The sum of the stator and rotor mmfs, which solely exist in absence of skewing constitutes the so called "magnetisation" mmf and produces the main (useful) flux in the machine.
- Even this magnetising mmf varies with slip (speed) – for constant voltage and frequency; however it decreases with slip (reduction in speed); this knowledge is to be used in later Chapters (5, 6, 8).

4.15. REFERENCES

1. H. Sequenz, The Windings of Electrical Machines, Vol.3., A.C. Machines, Springer Verlag, Vienna, 1950 (in German).
2. M.M. Liwschitz–Garik, Winding Alternating Current Machines, Van Nostrand Publications, 1950.
3. M. Bhattacharyya, Pole Amplitude Modulation Applied to Salient Pole Synchronous Machines, Ph.D. thesis, University of Bristol, U.K, 1966.
4. G.H. Rawcliffe, G.H. Burbidge, W. Fong, Induction Motor Speed Changing by Pole Amplitude Modulation, Proc IEE, Vol.105A, August, 1958.
5. B.J. Chalmers, Interspersed A.C. Windings, Proc. IEE, Vol.111, 1964, pp.1859.
6. C. Zhang, J. Yang, X. Chen, Y. Guo, A New Design Principle for Pole-Changing Winding–the Three equation Principle, EMPS Vol.22, No.2, 1994, pp.187 – 189.
7. M. Poloujadoff, J.C. Mipo, Designing 2 Pole and 2/4 Pole Windings by Simulated Annealing Method, EMPS, Vol26., No.10, 1998, pp.1059 – 1066.

8. M.V. Cistelecan, E. Demeter, A New Approach to the Three-phase Monoaxial Nonconventional Windings for A.C. Machines, Rec. of IEEE – IEMDC, 99, Seattle, U.S.A., 1999.
9. M.V. Cistelecan, E. Demeter, A Closer Approach of the Fractional Multilayer Three-phase A.C. Windings, Rec. of Electromotion – 1999, Patros, Greece, May 1999, Vol.1., pp.51 – 54.
10. B. Chalmers, A Williamson, A.C. Machines: Electromagnetics and Design, Research Studies Press Ltd., Somerset, England, 1991, pp.28.
11. D.M. Ionel, M.V. Cistelecan, T.J.E. Miller, M.I. McGilp, A New Analytical Method for the Computation of Air–gap Reactances in 3–phase Induction Motors, Rec. of IEEE – IAS – 1998 Annual Meeting, Vol.1.
12. M. Cistelecan, E. Demeter, New 3 Phase A.C. Windings with Low Spatial Harmonic Content, Rec. of Electromotion Symposium – 1997, Cluj – Napoca, Romania, pp.98 – 100.
13. W. Fong, Polyphase Symmetrization, U.K. Patent, 1964.
14. A.R.W. Broadway, Part–symmetrization of 3–phase Windings, Proc. IEE, Vol.122, July 1975, pp.125.
15. M. Bhattacharyya, G.K. Singh, Observations on the Design of Distributed Armature Windings by Poly–phase Symmetrization, EMPS Vol.19, No.3, 1991, pp.363 – 379.
16. R. Li, A. Wallace, R. Spee, Two–axis Model Development of Cage–rotor Brushless Doubly–fed Machines, IEEE Trans. Vol. EC – 6, No.3, 1991, pp.453 – 460.

Chapter 5

THE MAGNETIZATION CURVE AND INDUCTANCE

5.1 INTRODUCTION

As shown in Chapters 2 and 4, the induction machine configuration is quite complex. So far we elucidated the subject of windings and their mmfs. With windings in slots, the mmf has (in three-phase or two-phase symmetric windings) a dominant wave and harmonics. The presence of slot openings on both sides of the airgap is bound to amplify (influence, at least) the mmf step harmonics. Many of them will be attenuated by rotor-cage-induced currents. To further complicate the picture, the magnetic saturation of the stator (rotor) teeth and back irons (cores or yokes) also influence the airgap flux distribution producing new harmonics.

Finally, the rotor eccentricity (static and/or dynamic) introduces new harmonics in the airgap field distribution.

In general, both stator and rotor currents produce a resultant field in the machine airgap and iron parts.

However, with respect to fundamental torque-producing airgap flux density, the situation does not change notably from zero rotor currents to rated rotor currents (rated torque) in most induction machines, as experience shows.

Thus it is only natural and practical to investigate, first, the airgap field fundamental with uniform equivalent airgap (slotting accounted through correction factors) as influenced by the magnetic saturation of stator and rotor teeth and back cores, for zero rotor currents.

This situation occurs in practice with the wound rotor winding kept open at standstill or with the squirrel cage rotor machine fed with symmetrical a.c. voltages in the stator and driven at mmf wave fundamental speed ($n_1 = f_1/p_1$).

As in this case the pure travelling mmf wave runs at rotor speed, no induced voltages occur in the rotor bars. The mmf space harmonics (step harmonics due to the slot placement of coils, and slot opening harmonics etc.) produce some losses in the rotor core and windings. They do not notably influence the fundamental airgap flux density and, thus, for this investigation, they may be neglected, only to be revisited in Chapter 11.

To calculate the airgap flux density distribution in the airgap, for zero rotor currents, a rather precise approach is the FEM. With FEM, the slot openings could be easily accounted for; however, the computation time is prohibitive for routine calculations or optimization design algorithms.

In what follows, we first introduce the Carter coefficient K_c to account for the slotting (slot openings) and the equivalent stack length in presence of radial ventilation channels. Then, based on magnetic circuit and flux laws, we calculate the dependence of stator mmf per pole F_{1m} on airgap flux density accounting for magnetic saturation in the stator and rotor teeth and back cores, while accepting a pure sinusoidal distribution of both stator mmf F_{1m} and airgap flux density, B_{1g}.

The obtained dependence of $B_{1g}(F_{1m})$ is called the magnetization curve. Industrial experience shows that such standard methods, in modern, rather heavily saturated magnetic cores, produce notable errors in the magnetizing curves, at 100 to 130% rated voltage at ideal no load (zero rotor currents). The presence of heavy magnetic saturation effects such as airgap, teeth or back core flux density, flattening (or peaking), and the rough approximation of mmf calculations in the back irons are the main causes for these discrepancies.

Improved analytical methods have been proposed to produce satisfactory magnetization curves. One of them is presented here in extenso with some experimental validation.

Based on the magnetization curve, the magnetization inductance is defined and calculated.

Later the emf induced in the stator and rotor windings and the mutual stator/rotor inductances are calculated for the fundamental airgap flux density. This information prepares the ground to define the parameters of the equivalent circuit of the induction machine, that is, for the computation of performance for any voltage, frequency, and speed conditions.

5.2 EQUIVALENT AIRGAP TO ACCOUNT FOR SLOTTING

The actual flux path for zero rotor currents when current in phase A is maximum $i_A = I\sqrt{2}$ and $i_B = i_C = -I\sqrt{2}/2$, obtained through FEM, is shown in Figure 5.1. [4]

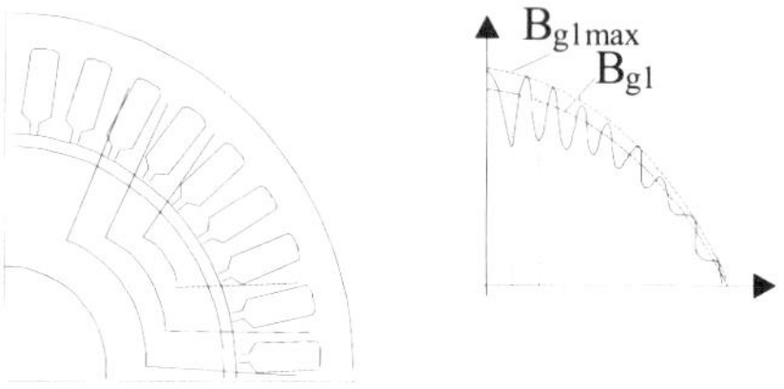

Figure 5.1 No-load flux plot by FEM when $i_B = i_C = -i_A/2$.

The corresponding radial airgap flux density is shown on Figure 5.1b. In the absence of slotting and stator mmf harmonics, the airgap field is sinusoidal, with an amplitude of B_{g1max}.

In the presence of slot openings, the fundamental of airgap flux density is B_{g1}. The ratio of the two amplitudes is called the Carter coefficient.

$$K_C = \frac{B_{g1\,max}}{B_{g1}} \tag{5.1}$$

When the magnetic airgap is not heavily saturated, K_C may also be written as the ratio between smooth and slotted airgap magnetic permeances or between a larger equivalent airgap g_e and the actual airgap g.

$$K_C = \frac{g_e}{g} \geq 1 \tag{5.2}$$

FEM allows for the calculation of Carter coefficient from (5.1) when it is applied to smooth and double-slotted structure (Figure 5.1).

On the other hand, easy to handle analytical expressions of K_C, based on conformal transformation or flux tube methods, have been traditionally used, in the absence of saturation, though. First, the airgap is split in the middle and the two slottings are treated separately. Although many other formulas have been proposed, we still present Carter's formula as it is one of the best.

$$K_{C1,2} = \frac{\tau_{s,r}}{\tau_{s,r} - \gamma_{1,2} \cdot g/2} \tag{5.3}$$

$\tau_{s,r}$–stator/rotor slot pitch, g–the actual airgap, and

$$\gamma_{1,2} = \frac{4}{\pi}\left[\frac{b_{os,r}}{g} / \tan\left(\frac{b_{os,r}}{g}\right) - \ln\sqrt{1+\left(\frac{b_{os,r}}{g}\right)^2}\right] \approx \frac{\left(2\frac{b_{os,r}}{g}\right)^2}{5+2\frac{b_{os,r}}{g}} \tag{5.4}$$

for $b_{os,r}/g >> 1$. In general, $b_{os,r} \approx (3 - 8)g$. Where $b_{os,r}$ is the stator(rotor) slot opening.

With a good approximation, the total Carter coefficient for double slotting is

$$K_C = K_{C1} \cdot K_{C2} \tag{5.5}$$

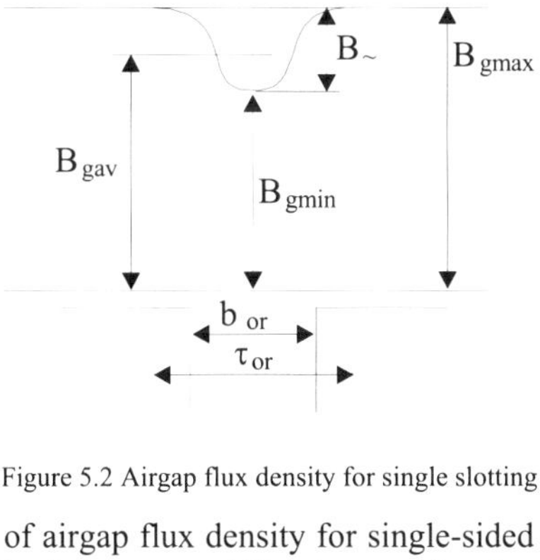

Figure 5.2 Airgap flux density for single slotting

The distribution of airgap flux density for single-sided slotting is shown on Figure 5.2. Again, the iron permeability is considered to be infinite. As the magnetic circuit becomes heavily saturated, some of the flux lines touch the slot bottom (Figure 5.3) and the Carter coefficient formula has to be changed. [2] In such cases, however, we think that using FEM is the best solution.

If we introduce the relation

$$B_{\sim} = B_{g\,max} - B_{g\,min} = 2\beta B_{g\,max} \tag{5.6}$$

the flux drop (Figure 5.2) due to slotting $\Delta\Phi$ is

$$\Delta\Phi_{s,r} = \sigma \frac{b_{os,r}}{2} B_{\sim s,r} \tag{5.7}$$

From [3],

$$\beta\sigma b_{os,r} = \frac{g}{2}\gamma_{1,2} \tag{5.8}$$

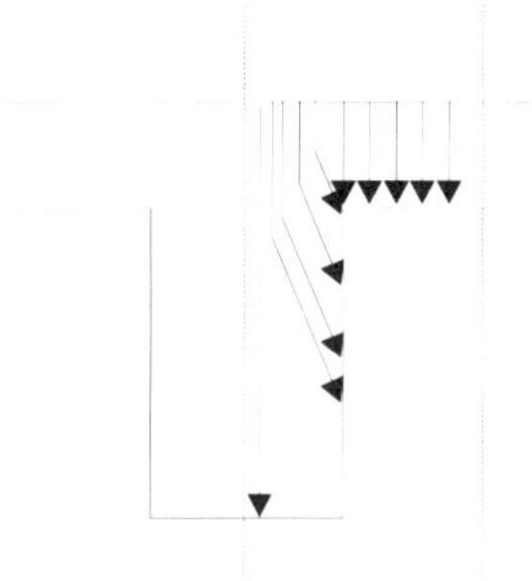

Figure 5.3 Flux lines in a saturated magnetic circuit

The two factors β and σ are shown on Figure 5.4 as obtained through conformal transformations. [3]

When single slotting is present, g/2 should be replaced by g.

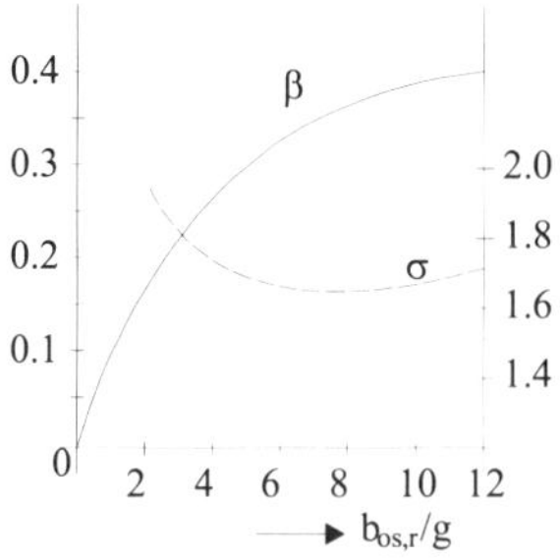

Figure 5.4 The factor β and σ as function of $b_{os,r}$ /(g/2)

Another slot-like situation occurs in long stacks when radial channels are placed for cooling purposes. This problem is approached next.

5.3 EFFECTIVE STACK LENGTH

Actual stator and rotor stacks are not equal in length to avoid notable axial forces, should any axial displacement of rotor occured. In general, the rotor stack is longer than the stator stack by a few airgaps (Figure 5.5).

$$l_r = l_s + (4-6)g \tag{5.9}$$

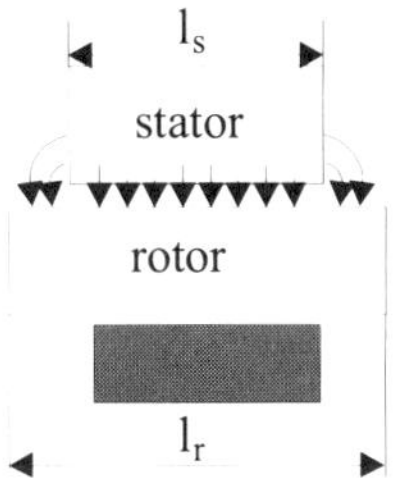

Figure 5.5. Single stack of stator and rotor

Flux fringing occurs at stator stack ends. This effect may be accounted for by apparently increasing the stator stack by (2 to 3)g,

$$l_{se} = l_s + (2 \div 3)g \tag{5.10}$$

The average stack length, l_{av}, is thus

$$l_{av} \approx \frac{l_s + l_r}{2} \approx l_{se} \tag{5.11}$$

As the stacks are made of radial laminations insulated axially from each other through an enamel, the magnetic length of the stack L_e is

$$L_e = l_{av} \cdot K_{Fe} \tag{5.12}$$

The stacking factor K_{Fe} (K_{Fe} = 0.9 – 0.95 for (0.35 – 0.5) mm thick laminations) takes into account the presence of nonmagnetic insulation between laminations.

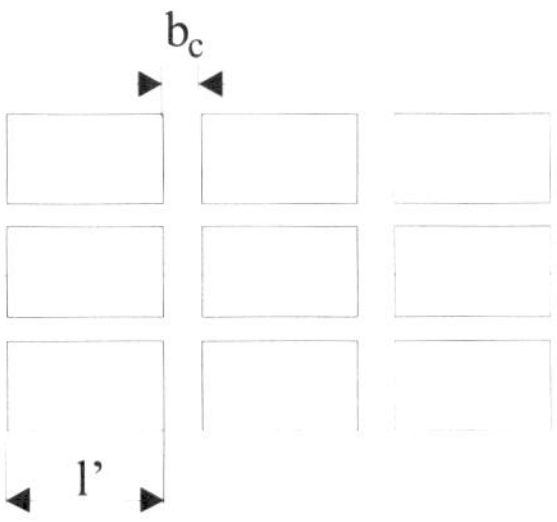

Figure 5.6 Multistack arrangement fcr radial cooling channels

When radial cooling channels (ducts) are used by dividing the stator into n elementary ones, the equivalent stator stack length L_e is (Figure 5.6)

$$L_e \approx (n+1)l' \cdot K_{Fe} + 2ng + 2g \tag{5.12}$$

with
$$b_c = 5 - 10\,mm; l' = 100 - 250\,mm \tag{5.13}$$

It should be noted that recently, with axial cooling, longer single stacks up to 500mm and more have been successfully built. Still, for induction motors in the MW power range, radial channels with radial cooling are in favor.

5.4 THE BASIC MAGNETIZATION CURVE

The dependence of airgap flux density fundamental B_{g1} on stator mmf fundamental amplitude F_{1m} for zero rotor currents is called the magnetization curve.

For mild levels of magnetic saturation, usually in general, purpose induction motors, the stator mmf fundamental produces a sinusoidal distribution of the flux density in the airgap (slotting is neglected). As shown later in this chapter by balancing the magnetic saturation of teeth and back cores, rather sinusoidal airgap flux density is maintained, even for very heavy saturation levels.

The basic magnetization curve ($F_{1m}(B_{g1})$ or $I_0(B_{g1})$ or I_o/I_n versus B_{g1}) is very important when designing an induction motor and notably influences the power factor and the core loss. Notice that I_0 and I_n are no load and full load stator phase currents and F_{1m0} is

$$F_{1m0} = \frac{3\sqrt{2}W_1 K_{w1} I_0}{\pi p_1} \tag{5.14}$$

The no load (zero rotor current) design airgap flux density is $B_{g1} = 0.6 - 0.8T$ for 50 (60) Hz induction motors and goes down to 0.4 to 0.6 T for (400 to 1000) Hz high speed induction motors, to keep core loss within limits.

On the other hand, for 50 (60) Hz motors, I_0/I_n (no-load current/rated current) decreases with motor power from 0.5 to 0.8 (in subkW power range) to 0.2 to 0.3 in the high power range, but it increases with the number of pole pairs.

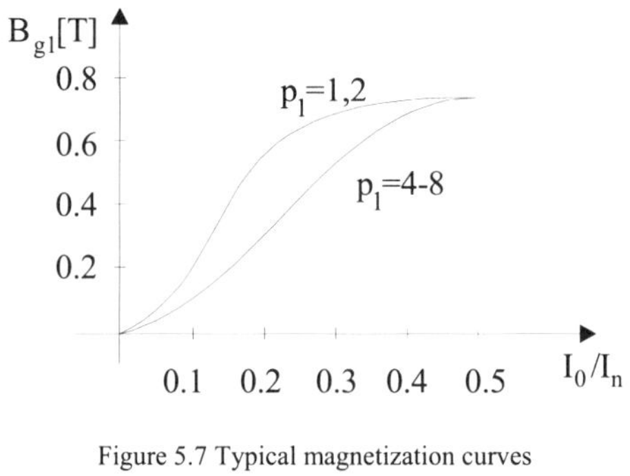

Figure 5.7 Typical magnetization curves

For low airgap flux densities, the no-load current tends to be smaller. A typical magnetization curve is shown in Figure 5.7 for motors in the kW power range at 50 (60) Hz.

Now that we do have a general impression on the magnetising (mag.) curve, let us present a few analytical methods to calculate it.

5.4.1 The magnetization curve via the basic magnetic circuit

We shall examine first the flux lines corresponding to maximum flux density in the airgap and assume a sinusoidal variation of the latter along the pole pitch (Figure 5.8a,b).

$$B_{g1}(\theta_e, t) = B_{g1m}\cos(p_1\theta - \omega_1 t);\ \theta_e = p_1\theta \tag{5.15}$$

For t = 0

$$B_g(\theta, 0) = B_{g1m}\cos p_1\theta \tag{5.16}$$

The stator (rotor) back iron flux density $B_{cs,r}$ is

$$B_{cs,r} = \frac{1}{2h_{cs,r}}\int_0^\theta B_{g1}(\theta, t)d\theta \cdot \frac{D}{2} \tag{5.17}$$

where $h_{cs,r}$ is the back core height in the stator (rotor). For the flux line in Figure 5.8a ($\theta = 0$ to π/p_1),

$$B_{cs1}(\theta, t) = \frac{1}{2}\frac{2}{\pi}\frac{\tau}{h_{cs}} \cdot B_{g1m}\sin(p_1\theta - \omega_1 t);\ \tau = \frac{\pi D}{2p_1} \tag{5.18}$$

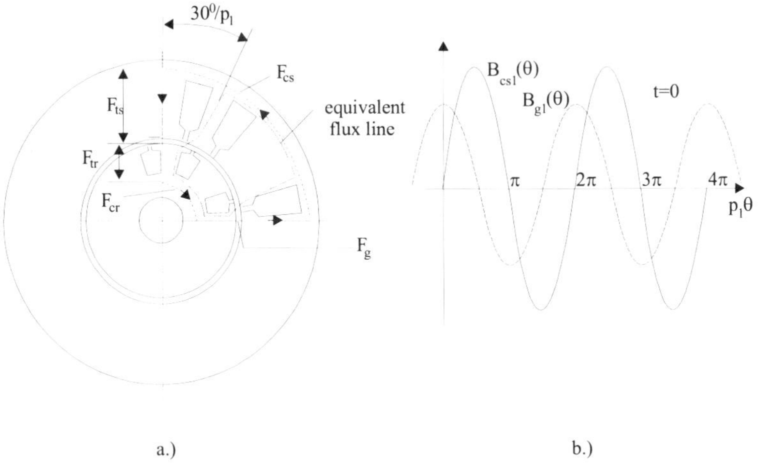

Figure 5.8 Flux path a.) and flux density types b.): ideal distribution in the airgap and stator core

Due to mmf and airgap flux density sinusoidal distribution along motor periphery, it is sufficient to analyse the mmf iron and airgap components F_{ts}, F_{tr} in teeth, F_g in the airgap, and F_{cs}, F_{cr} in the back cores. The total mmf is represented by F_{1m} (peak values).

$$2F_{1m} = 2F_g + 2F_{ts} + 2F_{tr} + F_{cs} + F_{cr} \tag{5.19}$$

Equation (5.19) reflects the application of the magnetic circuit (Ampere's) law along the flux line in Figure 5.8a.

In industry, to account for the flattening of the airgap flux density due to teeth saturation, B_{g1m} is replaced by the actual (designed) maximum flattened flux density B_{gm}, at an angle $\theta = 30°/p_1$, which makes the length of the flux lines in the back core 2/3 of their maximum length.

Then finally the calculated I_{1m} is multiplied by $2/\sqrt{3}$ (1/cos30°) to find the maximum mmf fundamental.

At $\theta_{er} = p_1\theta = 30°$, it is supposed that the flattened and sinusoidal flux density are equal to each other (Figure 5.9).

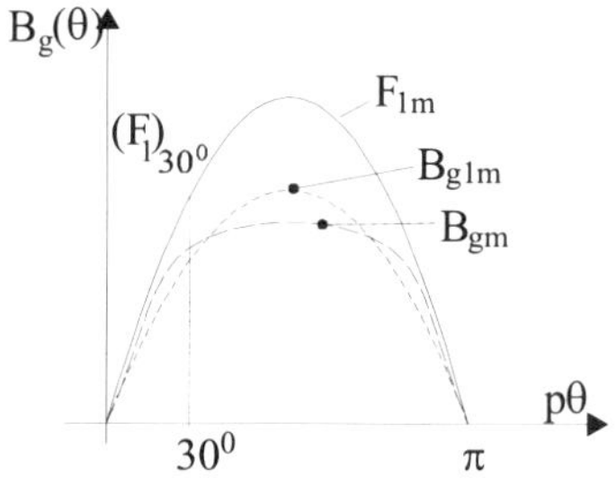

Figure 5.9 Sinusoidal and flat airgap flux density

We have to again write Ampere's law for this case (interior flux line in Figure 5.8a).

$$2(F_1)_{30^0} = 2F_g(B_{gm}) + 2F_{ts} + 2F_{tr} + F_{cs} + F_{cr} \tag{5.20}$$

and finally,

$$2F_{1m} = \frac{2(F_1)_{30^0}}{\cos 30^0} \tag{5.21}$$

For the sake of generality we will use (5.20) – (5.21), remembering that the length of average flux line in the back cores is 2/3 of its maximum.

Let us proceed directly with a numerical example by considering an induction motor with the geometry in Figure 5.10.

$$\begin{gathered} 2p_1 = 4; D = 0.1\text{m}; D_e = 0.176\text{m}; h_s = 0.025\text{m}; N_s = 24; \\ N_r = 18; b_{r1} = 1.2b_{tr1}; b_{s1} = 1.4b_{ts1}; g = 0.5\cdot 10^{-3}\text{m}; \\ h_r = 0.018\text{m}; D_{shaft} = 0.035\text{m}; B_{gm} = 0.7\text{T} \end{gathered} \tag{5.22}$$

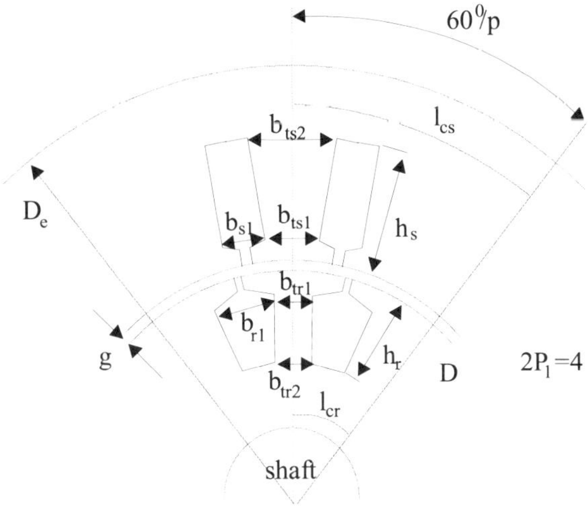

Figure 5.10 IM geometry for magnetization curve calculation

The B/H curve of the rotor and stator laminations is given in Table 5.1.

Table 5.1 B/H curve a typical IM lamination

B[T]	0.0	0.05	0.1	0.15	0.2	0.25	0.3	0.35	0.4	0.45	0.5	0.55	0.6
H[A/m]	0	22.8	35	45	49	57	65	70	76	83	90	98	206
B[T]	0.65	0.7	0.75	0.8	0.85	0.9	0.95	1	1.05	1.1	1.15	1.2	1.25
H[A/m]	115	124	135	148	177	198	198	220	237	273	310	356	417

1.3	1.35	1.4	1.45	1.5	1.55	1.6	1.65	1.7
482	585	760	1050	1340	1760	2460	3460	4800
1.75	1.8	1.85	1.9	1.95	2.0			
6160	8270	11170	15220	22000	34000			

Based on (5.20) – (5.21), Gauss law, and B/H curve in Table 5.1, let us calculate the value of F_{1m}.

To solve the problem in a rather simple way, we still assume a sinusoidal flux distribution in the back cores, based on the fundamental of the airgap flux density B_{g1m}.

$$B_{g1m} = \frac{B_{gm}}{\cos 30^0} = 0.7\frac{2}{\sqrt{3}} = 0.809\text{T} \tag{5.23}$$

The maximum stator and rotor back core flux densities are obtained from (5.18):

$$B_{csm} = \frac{1}{\pi} \cdot \frac{\pi D}{2p_1} \frac{1}{h_{cs}} B_{g1m} \tag{5.24}$$

$$B_{crm} = \frac{1}{\pi} \cdot \frac{\pi D}{2p_1} \frac{1}{h_{cr}} B_{g1m} \tag{5.25}$$

with $$h_{cs} = \frac{(D_e - D - 2h_s)}{2} = \frac{0.174 - 0.100 - 2 \cdot 0.025}{2} = 0.013m \tag{5.26}$$

$$h_{cr} = \frac{(D - 2g - D_{shaft} - 2h_r)}{2} = \frac{0.100 - 0.001 - 2 \cdot 0.018 - 0.036}{2} = 0.0135m$$

Now from (5.24) – (5.25),

$$B_{csm} = \frac{0.1 \cdot 0.809}{4 \cdot 0.013} = 1.555T \tag{5.28}$$

$$B_{crm} = \frac{0.1 \cdot 0.809}{4 \cdot 0.0135} = 1.498T \tag{5.29}$$

As the core flux density varies from the maximum value cosinusoidally, we may calculate an average value of three points, say B_{csm}, $B_{csm}\cos60^0$ and $B_{csm}\cos30^0$:

$$B_{csav} = B_{csm}\left(\frac{1 + 4\cos 60^0 + \cos 30^0}{6}\right) = 1.555 \cdot 0.8266 = 1.285T \tag{5.30}$$

$$B_{crav} = B_{crm}\left(\frac{1 + 4\cos 60^0 + \cos 30^0}{6}\right) = 1.498 \cdot 0.8266 = 1.238T \tag{5.31}$$

From Table 5.1 we obtain the magnetic fields corresponding to above flux densities. Finally, H_{csav} (1.285) = 460 A/m and H_{crav} (1.238) = 400 A/m.

Now the average length of flux lines in the two back irons are

$$l_{csav} \approx \frac{2}{3} \cdot \frac{\pi(D_e - h_{cs})}{2p_1} = \frac{2}{3}\pi \frac{(0.176 - 0.013)}{4} = 0.0853m \tag{5.32}$$

$$l_{crav} \approx \frac{2}{3} \cdot \frac{\pi(D_{shaft} + h_{cr})}{2p_1} = \frac{2}{3}\pi \frac{(0.36 + 0.135)}{4} = 0.02593m \tag{5.33}$$

Consequently, the back core mmfs are

$$F_{cs} = l_{csav} \cdot H_{csav} = 0.0853 \cdot 460 = 39.238 \text{Aturns} \tag{5.34}$$

$$F_{cr} = l_{crav} \cdot H_{crav} = 0.0259 \cdot 400 = 10.362 \text{Aturns} \tag{5.35}$$

The airgap mmf Fg is straightforward.

$$2F_g = 2g \cdot \frac{B_{gm}}{\mu_0} = 2 \cdot 0.5 \cdot 10^{-3} \cdot \frac{0.7}{1.256 \cdot 10^{-6}} = 557.324\text{Aturns} \tag{5.36}$$

Assuming that all the airgap flux per slot pitch traverses the stator and rotor teeth, we have

$$B_{gm} \cdot \frac{\pi D}{N_{s,r}} = \left(B_{ts,r}\right)_{av} \cdot \left(b_{ts,r}\right)_{av}; \quad b_{t_{s,r}} = \frac{\pi\left(D \pm h_{s,r}\right)}{N_{s,r}\left(1 + b_{s,r} / b_{ts,r}\right)} \tag{5.37}$$

Considering that the teeth flux is conserved (it is purely radial), we may calculate the flux density at the tooth bottom and top as we know the average tooth flux density for the average tooth width $(b_{ts,r})_{av}$. An average can be applied here again. For our case, let us consider $(B_{ts,r})_{av}$ all over the teeth height to obtain

$$\begin{aligned} \left(b_{ts}\right)_{av} &= \frac{\pi(0.1 + 0.025)}{(1 + 1.4) \cdot 24} = 6.81 \cdot 10^{-3}\text{m} \\ \left(b_{tr}\right)_{av} &= \frac{\pi(0.1 - 0.018)}{(1 + 1.2) \cdot 18} = 6.00 \cdot 10^{-3}\text{m} \end{aligned} \tag{5.38}$$

$$\begin{aligned} \left(B_{ts}\right)_{av} &= \frac{0.7 \cdot \pi \cdot 100}{24 \cdot 7.43 \cdot 10^{-3}} = 1.344\text{T} \\ \left(B_{tr}\right)_{av} &= \frac{0.7 \cdot \pi \cdot 100}{18 \cdot 6.56 \cdot 10^{-3}} = 1.878\text{T} \end{aligned} \tag{5.39}$$

From Table 5.1, the corresponding values of $H_{tsav}(B_{tsav})$ and H_{trav} (B_{trav}) are found to be H_{tsav} = 520 A/m, H_{trav} = 13,600 A/m. Now the teeth mmfs are

$$2F_{ts} \approx H_{tsav}\left(2h_s\right) = 520 \cdot (0.025 \cdot 2) = 26\text{Aturns} \tag{5.40}$$

$$2F_{tr} \approx H_{trav}\left(2h_r\right) = 13600 \cdot (0.018 \cdot 2) = 489.6\text{Aturns} \tag{5.41}$$

The total stator mmf $\left(F_1\right)_{30^0}$ is calculated from (5.20)

$$2\left(F_1\right)_{30^0} = 557.324 + 26 + 489.6 + 39.238 + 10.362 = 1122.56\text{Aturns} \tag{5.42}$$

The mmf amplitude F_{m10} (from 5.21) is

$$2F_{1m0} = \frac{2\left(F_1\right)_{30^0}}{\cos 30^0} = 1122.56 \cdot \frac{2}{\sqrt{3}} = 1297.764\text{Aturns} \tag{5.43}$$

Based on (5.14), the no-load current may be calculated with the number of turns/phase W_1 and the stator winding factor K_{w1} already known.

Varying as the value of B_{gm} desired the magnetization curve–$B_{gm}(F_{1m})$–is obtained.

Before leaving this subject let us remember the numerous approximations we operated with and define two partial and one equivalent saturation factor as K_{st}, K_{sc}, K_s.

$$K_{st} = 1 + \frac{2(F_{ts} + F_{tr})}{2F_g}; \quad K_{sc} = 1 + \frac{(F_{cs} + F_{cr})}{2F_g} \tag{5.44}$$

$$K_s = \frac{(F_1)_{30^0}}{2F_g} = K_{st} + K_{sc} - 1 \tag{5.45}$$

The total saturation factor K_s accounts for all iron mmfs as divided by the airgap mmf. Consequently, we may consider the presence of iron as an increased airgap g_{es}.

$$g_{es} = gK_cK_s; \quad K_s = K_s\left(\frac{I_0}{I_n}\right) \tag{5.46}$$

Let us notice that in our case,

$$K_{st} = 1 + \frac{26 + 489.6}{557.324} = 1.925; \; K_{ct} = 1 + \frac{39.23 + 10.362}{557.324} = 1.089$$
$$K_s = 1.925 + 1.089 - 1 = 2.103 \tag{5.47}$$

A few remarks are in order.

- The teeth saturation factor K_{st} is notable while the core saturation factor is low; so the tooth are much more saturated (especially in the rotor, in our case); as shown later in this chapter, this is consistent with the flattened airgap flux density.
- In a rather proper design, the teeth and core saturation factors K_{st} and K_{sc} are close to each other: $K_{st} \approx K_{sc}$; in this case both the airgap and core flux densities remain rather sinusoidal even if rather high levels of saturation are encountered.
- In 2 pole machines, however, K_{sc} tends to be higher than K_{st} as the back core height tends to be large (large pole pitch) and its reduction in size is required to reduce motor weight.
- In mildly saturated IMs, the total saturation factor is smaller than in our case: $K_s = 1.3 – 1.6$.

Based on the above theory, iterative methods, to obtain the airgap flux density distribution and its departure from a sinusoid (for a sinusoidal core flux density), have been recently introduced [2,4]. However, the radial flux density components in the back cores are still neglected.

5.4.2 Teeth defluxing by slots

So far we did assume that all the flux per slot pitch goes radially through the teeth. Especially with heavily saturated teeth, a good part of magnetic path passes through the slot itself. Thus, the tooth is slightly "discharged" of flux.

We may consider that the following are approximates:

$$B_t = B_{ti} - c_l B_g b_{s,r} / b_{ts,r}; \; c_l << 1.0 \tag{5.48}$$

$$B_{ti} = B_g \frac{b_{ts,r} + b_{s,r}}{b_{ts,r}} \tag{5.49}$$

The coefficient c_l is, in general, adopted from experience but it is strongly dependent on the flux density in the teeth B_{ti} and the slotting geometry (including slot depth [2]).

5.4.3 Third harmonic flux modulation due to saturation

As only inferred above, heavy saturation in stator (rotor) teeth and/or back cores tends to flatten or peak, respectively, the airgap flux distribution.

This proposition can be demonstrated by noting that the back core flux density $B_{cs,r}$ is related to airgap (implicitly teeth) flux density by the equation

$$B_{cs,r} = C_{s,r} \int_0^{\theta_{er}} B_{ts,r}(\theta_{er}) d\theta_{er}; \; \theta_{er} = p_1 \theta \tag{5.50}$$

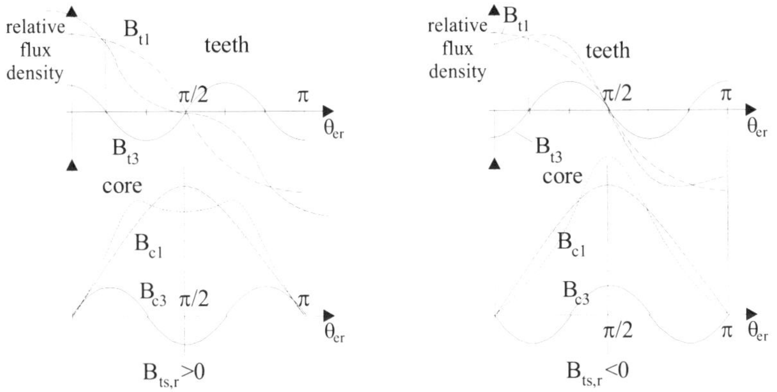

Figure 5.11 Tooth and core flux density distribution
a.) saturated back core ($B_{ts,r3} > 0$); b.) saturated teeth ($B_{ts,r3} < 0$)

Magnetic saturation in the teeth means flattening $B_{ts,r}(\theta)$ curve.

$$B_{ts,r}(\theta) = B_{ts,r1} \cos(\theta_{er} - \omega_1 t) + B_{ts,r3} \cos(3\theta_{er} - \omega_1 t) \tag{5.51}$$

Consequently, $B_{ts,r3} > 0$ means unsaturated teeth (peaked flux density, Figure 5.11a). With (5.51), equation (5.50) becomes

$$B_{cs,r} = C_{s,r}\left[B_{ts,r1}\sin(\theta_{er} - \omega_1 t) + \frac{B_{ts,r3}}{3}\sin(3\theta_{er} - \omega_1 t)\right] \tag{5.52}$$

Analyzing Figure 5.11, based on (5.51) – (5.52), leads to remarks such as

- Oversaturation of a domain (teeth or core) means flattened flux density in that domain (Figure 5.11b).
- In paragraph 5.4.1. we have considered flattened airgap flux density–that is also flattened tooth flux density–and thus oversaturated teeth is the case treated.
- The flattened flux density in the teeth (Figure 5.11b) leads to only a slightly peaked core flux density as the denominator 3 occurs in the second term of (5.52).
- On the contrary, a peaked teeth flux density (Figure 5.11a) leads to a flat core density. The back core is now oversaturated.
- We should also mention that the phase connection is important in third harmonic flux modulation. For sinusoidal voltage supply and delta connection, the third harmonic of flux (and its induced voltage) cannot exist, while it can for star connection. This phenomenon will also have consequences in the phase current waveforms for the two connections. Finally, the saturation produced third and other harmonics influence, notably the core loss in the machine. This aspect will be discussed in Chapter 11 dedicated to losses.

After describing some aspects of saturation – caused distribution modulation, let us present a more complete analytical nonlinear field model, which also allows for the calculation of actual spatial flux density distribution in the airgap, though with smoothed airgap.

5.4.4 The analytical iterative model (AIM)

Let us remind here that essentially only FEM [5] or extended magnetic circuit methods (EMCM) [6] are able to produce a rather fully realistic field distribution in the induction machine. However, they do so with large computation efforts and may be used for design refinements rather than for preliminary or direct optimization design algorithms.

A fast analytical iterative (nonlinear) model (AIM) [7] is introduced here for preliminary or optimization design uses.

The following assumptions are introduced: only the fundamental of m.m.f. distribution is considered; the stator and rotor currents are symmetric; - the IM cross-section is divided into five circular domains (Figure 5.12) with unique (but adjustable) magnetic permeabilities essentially distinct along radial (r) and tangential (θ) directions: μ_r, and μ_0; the magnetic vector potential A lays along the shaft direction and thus the model is two-dimensional; furthermore, the separation of variables is performed.

Magnetic potential, A, solution

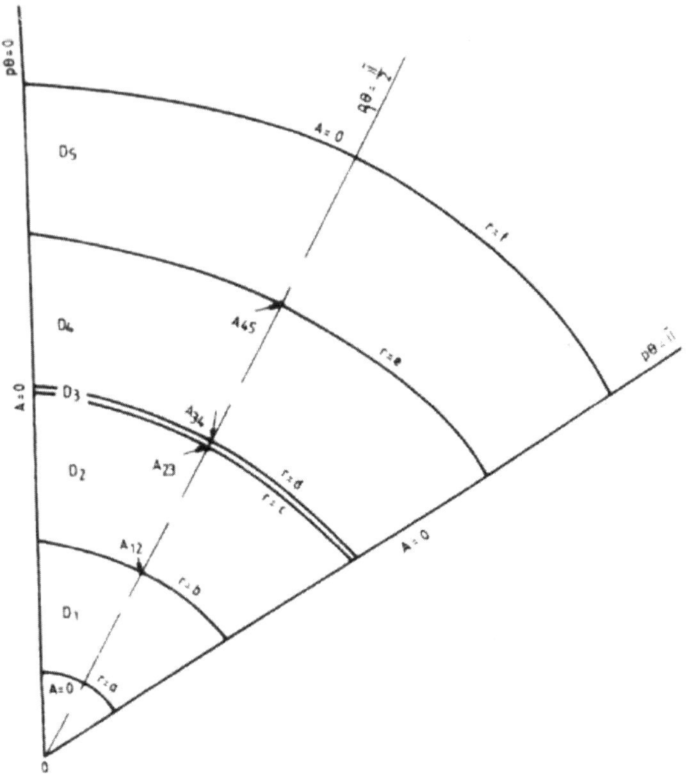

Figure 5.12 The IM cross-section divided into five domains

The Poisson equation in polar coordinates for magnetic potential A writes

$$\frac{1}{\mu_\theta}\left(\frac{1}{r}\frac{\partial A}{\partial r}+\frac{\partial^2 A}{\partial r^2}\right)+\frac{1}{\mu_r}\frac{1}{r^2}\frac{\partial^2 A}{\partial \theta^2}=-J \tag{5.53}$$

Separating the variables, we obtain

$$A(r,\theta) = R(r)\cdot T(\theta) \tag{5.54}$$

Now, for the domains with zero current (D_1, D_3, D_5) – J = O, Equation (5.53) with (5.54) yields

$$\frac{1}{R(r)}\left(r\frac{dR(r)}{dr}+r^2\frac{d^2R(r)}{dr^2}\right)=\lambda^2;\frac{\alpha^2}{T(\theta)}\frac{d^2T(\theta)}{d\theta^2}=-\lambda^2 \tag{5.55}$$

with $\alpha^2 = \mu_\theta/\mu_r$ and λ a constant.

Also a harmonic distribution along θ direction was assumed. From (5.55):

$$r^2\frac{d^2R(r)}{dr^2}+r\frac{dR(r)}{dr}-\lambda^2R=0 \tag{5.56}$$

and

$$\frac{d^2T(\theta)}{d\theta^2}+\frac{\lambda^2}{\alpha^2}T=0 \tag{5.57}$$

The solutions of (5.56) and (5.57) are of the form

$$R(r)=C_1r^{\lambda}+C_2r^{-\lambda} \tag{5.58}$$

$$T(\theta)=C_3\cos\left(\frac{\lambda}{\alpha}\theta\right)+C_4\sin\left(\frac{\lambda}{\alpha}\theta\right) \tag{5.59}$$

as long as $r \neq 0$.

Assuming further symmetric windings and currents, the magnetic potential is an aperiodic function and thus,

$$A(r,0)=0 \tag{5.60}$$

$$A\left(r,\frac{\pi}{p_1}\right)=0 \tag{5.61}$$

Consequently, from (5.59), (5.57) and (5.53), $A(r,\theta)$ is

$$A(r,\theta)=\left(g\cdot r^{p_1\alpha}+h\cdot r^{-p_1\alpha}\right)\sin(p_1\theta) \tag{5.62}$$

Now, if the domain contains a homogenous current density J,

$$J=J_m\sin(p_1\theta) \tag{5.63}$$

the particular solution $A_p(r,\theta)$ of (5.54) is:

$$A_p(r,\theta)=Kr^2\sin(P_1\theta) \tag{5.64}$$

with

$$K=-\frac{\mu_r\mu_\theta}{4\mu_r-p_1^{\,2}\mu_\theta}J_m \tag{5.65}$$

Finally, the general solution of A (5.53) is

$$A(r,\theta)=\left(g\cdot r^{p_1\alpha}+h\cdot r^{-p_1\alpha}+Kr^2\right)\sin(p_1\theta) \tag{5.66}$$

As (5.66) is valid for homogenous media, we have to homogenize the slotting domains D_2 and D_4, as the rotor and stator yokes (D_1, D_5) and the airgap (D_3) are homogenous.

Homogenizing the Slotting Domains.
The main practical slot geometries (Figure 5.13) are defined by equivalent center angles θ_s, and θ_t, for an equivalent (defined) radius r_{m4}, (for the stator) and r_{m2} (for the rotor). Assuming that the radial magnetic field H is constant along the circles r_{m2} and r_{m4}, the flux linkage equivalence between the homogenized and slotting areas yields

$$\mu_t H\theta_t r_x L_1 + \mu_0 H\theta_s r_x L_1 = \mu_r H(\theta_t + \theta_s) r_x L_1 \tag{5.67}$$

Consequently, the equivalent radial permeability μ_r, is

$$\mu_r = \frac{\mu_t\theta_t + \mu_0\theta_s}{\theta_t + \theta_s} \tag{5.68}$$

For the tangential field, the magnetic voltage relationship (along A, B, C trajectory on Figure 5.13), is

$$V_{mAC} = V_{mAB} + V_{mBC} \tag{5.69}$$

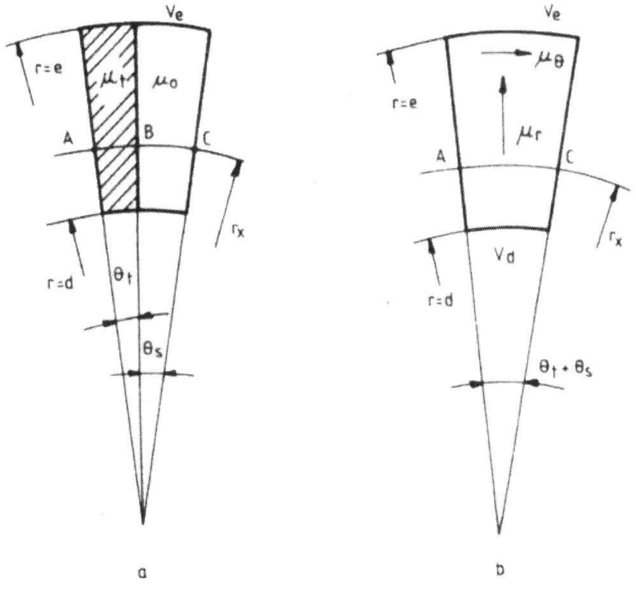

Figure 5.13 Stator and rotor slotting

With B_θ the same, we obtain

$$\frac{B_\theta}{\mu_\theta}(\theta_t + \theta_s)r_x = \frac{B_\theta}{\mu_t}\theta_t r_x + \frac{B_\theta}{\mu_0}\theta_s r_x \tag{5.70}$$

Consequently,

$$\mu_\theta = \frac{\mu_t\mu_0(\theta_t + \theta_s)}{\mu_0\theta_t + \mu_t\theta_s} \tag{5.71}$$

Thus the slotting domains are homogenized to be characterized by distinct permeabilities μ_r, and μ_θ along the radial and tangential directions, respectively.

We may now summarize the magnetic potential expressions for the five domains:

$$\begin{aligned}
&A_1(r,\theta) = \left(g_1 r^{p_1} + h_1 r^{-p_1}\right)\sin(p_1\theta);\ a' = \frac{a}{2} < r < b \\
&A_3(r,\theta) = \left(g_3 r^{p_1} + h_3 r^{-p_1}\right)\sin(p_1\theta);\ c < r < d \\
&A_5(r,\theta) = \left(g_5 r^{p_1} + h_5 r^{-p_1}\right)\sin(p_1\theta);\ e < r < f \\
&A_2(r,\theta) = \left(g_2 r^{p_1\alpha_2} + h_2 r^{-p_1\alpha_2} + k_2 r^2\right)\sin(p_1\theta);\ b < r < c \\
&A_4(r,\theta) = \left(g_4 r^{p_1\alpha_4} + h_4 r^{-p_1\alpha_4} + k_4 r^2\right)\sin(p_1\theta);\ d < r < e
\end{aligned} \tag{5.72}$$

with

$$\begin{aligned}
K_4 &= -\frac{\mu_{r4}\mu_{\theta 4}}{4\mu_{r4} - p_1{}^2\mu_{\theta 4}} J_{m4} \\
K_2 &= +\frac{\mu_{r2}\mu_{\theta 2}}{4\mu_{r2} - p_1{}^2\mu_{\theta 2}} J_{m2}
\end{aligned} \tag{5.73}$$

J_{m2} represents the equivalent demagnetising rotor equivalent current density which justifies the ⊕ sign in the second equation of (5.73). From geometrical considerations, J_{m2} and J_{m4} are related to the reactive stator and rotor phase currents I_{1s} and I_{2r}', by the expressions (for the three-phase motor),

$$\begin{aligned}
J_{m2} &= \frac{6\sqrt{2}}{\pi}\frac{1}{c^2 - b^2} W_1 K_{w1} I'_{2r} \\
J_{m4} &= \frac{6\sqrt{2}}{\pi}\frac{1}{e^2 - d^2} W_1 K_{w1} I_{1s}
\end{aligned} \tag{5.74}$$

The main pole-flux-linkage Ψ_{m1} is obtained through the line integral of A_3 around a pole contour Γ (L_1, the stack length):

$$\psi_{m1} = \oint_\Gamma A_3 \overline{dl} = 2L_1 A_3\left(d, \frac{\pi}{2P_1}\right) \tag{5.75}$$

Notice that $A_3(d, \pi/2p_1,) = -A_3(d, -\pi/2p_1,)$ because of symmetry. With A_3 from (5.72), ψ_{m1} becomes

$$\psi_{m1} = 2L_1\left(g_3 d^{p_1} + h_3 d^{-p_1}\right) \tag{5.76}$$

Finally, the e.m.f. E_1, (RMS value) is

$$E_1 = \pi\sqrt{2}f_1W_1K_{w1}\Psi_{m1} \tag{5.77}$$

Now from boundary conditions, the integration constants g_i and h_i are calculated as shown in the appendix of [7].

The computer program

To prepare the computer program, we have to specify a few very important details. First, instead of a (the shaft radius), the first domain starts at a'=a/2 to account for the shaft field for the case when the rotor laminations are placed directly on the shaft. Further, each domain is characterized by an equivalent (but adjustable) magnetic permeability. Here we define it. For the rotor and stator domains D_1, and D_5, the equivalent permeability would correspond to $r_{m1} = (a+b)/2$ and $r_{m5} = (e+f)/2$, and tangential flux density (and $\theta_0 = \pi/4$).

$$\begin{aligned} B_{1\theta}(r_{m1},\theta_0) &= -p_1\left[g_1\left(\frac{a+b}{2}\right)^{p_1-1} - h_1\left(\frac{a+b}{2}\right)^{-p_1-1}\right]\sin\frac{\pi}{4} \\ B_{5\theta}(r_{m5},\theta_0) &= -p_1\left[g_5\left(\frac{e+f}{2}\right)^{p_1-1} - h_5\left(\frac{e+f}{2}\right)^{-p_1-1}\right]\sin\frac{\pi}{4} \end{aligned} \tag{5.78}$$

For the slotting domains D_2 and D_4, the equivalent magnetic permeabilities correspond to the radiuses r_{m2} and r_{m4} (Figure 5.13) and the radial flux densities B_{2r}, and B_{4r}.

$$\begin{aligned} B_{2r}(r_{m2},\theta_0) &= p_1\left(g_2 r_{m2}^{\;p_1\alpha_2-1} + h_2 r_{m2}^{\;-p_1\alpha_2-1} + K_2 r_{m2}\right)\cos p_1\theta_0 \\ B_{4r}(r_{m4},\theta_0) &= p_1\left(g_4 r_{m4}^{\;p_1\alpha_4-1} + h_4 r_{m4}^{\;-p_1\alpha_4-1} + K_4 r_{m4}\right)\cos p_1\theta_0 \end{aligned} \tag{5.79}$$

with $\cos P_1\theta_0 = 0.9 \ldots 0.95$. Now to keep track of the actual saturation level, the actual tooth flux densities B_{4t}, and B_{2t}, corresponding to B_{2r}, H_{2r} and B_{4r}, H_{4r} are

$$\begin{aligned} B_{2t} &= \frac{\theta_{t1}+\theta_{s2}}{\theta_{t2}}B_{2r} - c_0\frac{\theta_{s2}}{\theta_{t2}}\mu_0 H_{2r} \\ B_{4t} &= \frac{\theta_{t4}+\theta_{s4}}{\theta_{t4}}B_{4r} - c_0\frac{\theta_{s4}}{\theta_{t4}}\mu_0 H_{4r} \end{aligned} \tag{5.80}$$

The empirical coefficient c_0 takes into account the tooth magnetic unloading due to the slot flux density contribution.

Finally, the computing algorithm is shown on Figure 5.14 and starts with initial equivalent permeabilities.

For the next cycle of computation, each permeability is changed according to

$$\mu^{(i+1)} = \mu^{(i)} + c_1\left(\mu^{(i+1)} - \mu^{(i)}\right) \tag{5.81}$$

It has been proved that $c_1 = 0.3$ is an adequate value, for a wide power range.

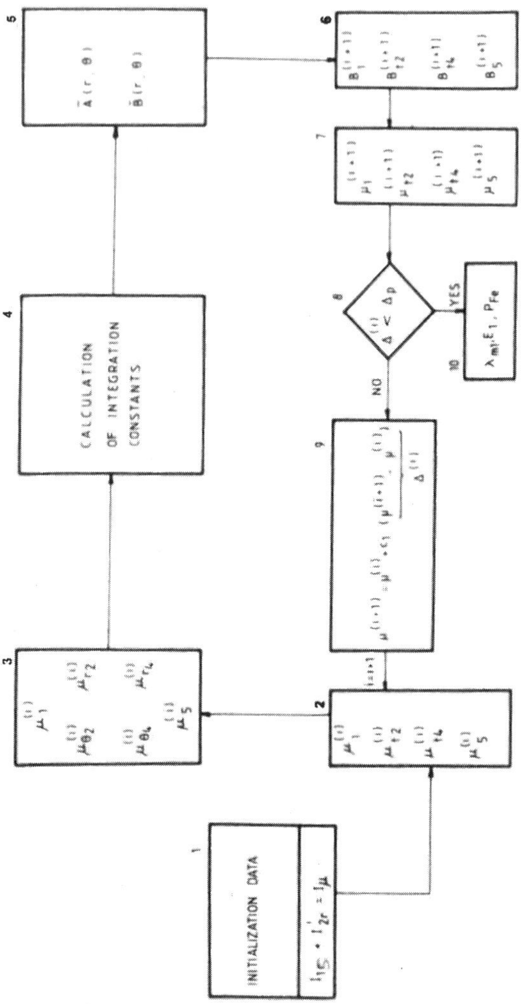

Figure 5.14 The computation algorithm

Model validation on no-load

The AIM has been applied to 12 different three-phase IMs from 0.75 kW to 15 kW (two-pole and four-pole motors), to calculate both the magnetization curve $I_{10} = f(E_1)$ and the core losses on no load for various voltage levels.

The magnetizing current is $I_\mu = I_{1r}$ and the no load active current I_{0A} is

$$I_{0A} = \frac{P_{Fe} + P_{mv} + P_{copper}}{3V_0} \tag{5.82}$$

The no-load current I_{10} is thus

$$I_{10} = \sqrt{I_\mu^{\ 2} + I_{0A}^{\ 2}} \tag{5.83}$$

and

$$E_1 \approx V_0 - \omega_1 L_{1\sigma} I_{10} \tag{5.84}$$

where $L_{1\sigma}$ is the stator leakage inductance (known). Complete expressions of leakage inductances are introduced in Chapter 6.

Figure 5.15 exhibits computation results obtained with the conventional nonlinear model, the proposed model, and experimental data.

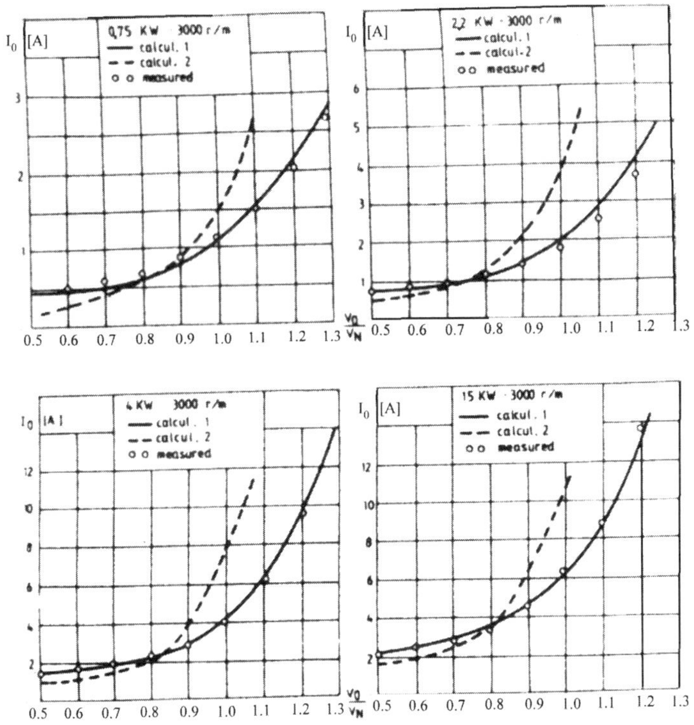

Figure 5.15 Magnetization characteristic validation on no load
1. conventional nonlinear model (paragraph 5.4.1); 2. AIM; 3. experiments

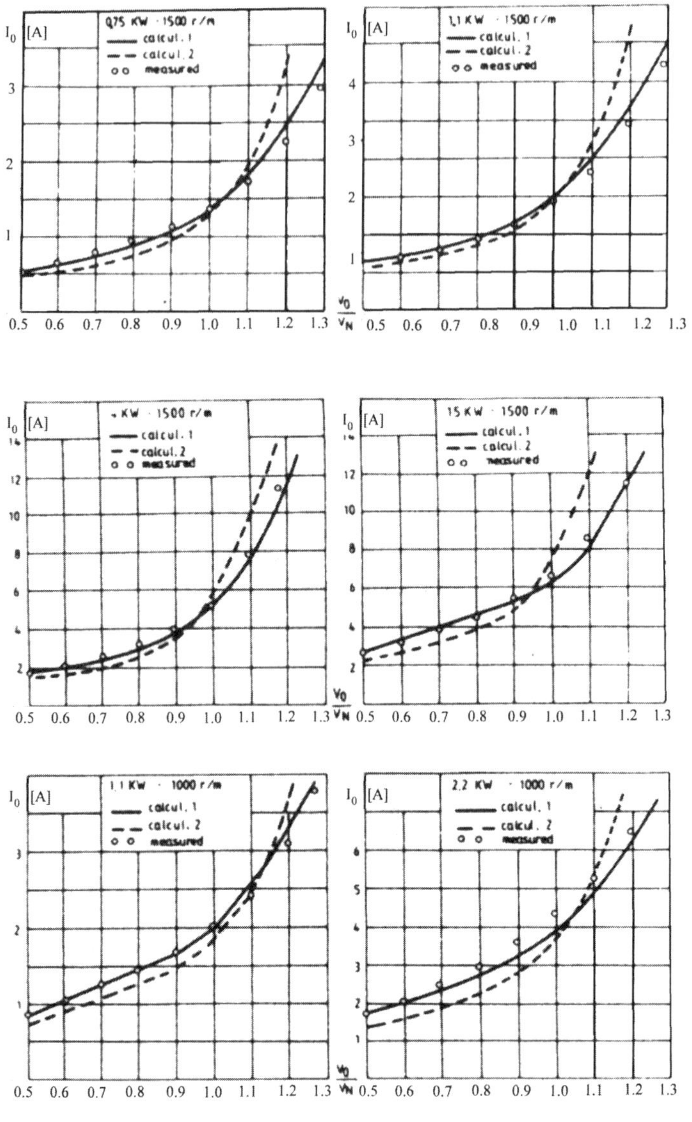

Figure 5.15 (continued)

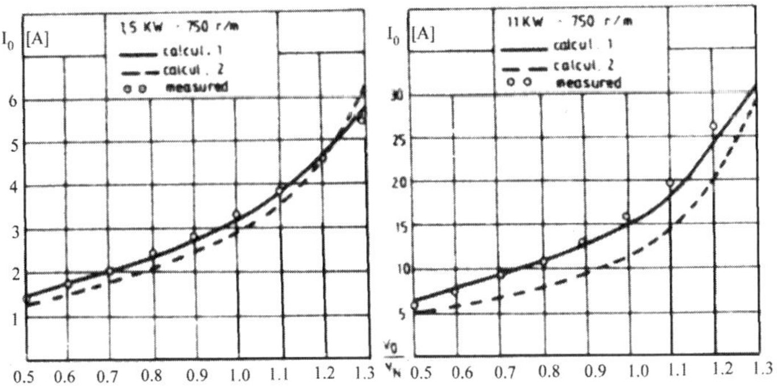

Figure 5.15 (continued)

It seems clear that AIM produces very good agreement with experiments even for high saturation levels.

A few remarks on AIM seem in order.

- The slotting is only globally accounted for by defining different tangential and radial permeabilities in the stator and rotor teeth.
- A single, but variable, permeability characterizes each of the machine domains (teeth and back cores).
- AIM is a bi-directional field approach and thus both radial and tangential flux density components are calculated.
- Provided the rotor equivalent current I_r (RMS value and phase shift with respect to stator current) is given, AIM allows calculation of the distribution of main flux in the machine on load.
- Heavy saturation levels ($V_0/V_n > 1$) are handled satisfactorily by AIM.
- By skewing the stack axially, the effect of skewing on main flux distribution can be handled.
- The computation effort is minimal (a few seconds per run on a contemporary PC).

So far AIM was used considering that the spatial field distribution is sinusoidal along stator bore. In reality, it may depart from this situation as shown in the previous paragraph. We may repeatedly use AIM to produce the actual spatial flux distribution or the airgap flux harmonics.

AIM may be used to calculate saturation-caused harmonics. The total mmf F_{1m} is still considered sinusoidal (Figure 5.16).

The maximum airgap flux density B_{g1m} (sinusoidal in nature) is considered known by using AIM for given stator (and eventually also rotor) current RMS values and phase shifts.

By repeatedly using the Ampere's law on contours such as those in Figure 5.17 at different position θ, we may find the actual distribution of airgap flux density, by admitting that the tangential flux density in the back core retains the sinusoidal distribution along θ. This assumption is not, in general, far from reality as was shown in paragraph 5.4.3.

Ampere's law on contour Γ (Figure 5.17) is

$$\oint_{\Gamma} \overline{H}\overline{dl} = \int_{-P_1\theta}^{P_1\theta} F_{1m} \sin p_1\theta d(p_1\theta) = 2F_{1m} \cos p_1\theta \tag{5.85}$$

The left side of (5.85) may be broken into various parts (Figure 5.17). It may easily be shown that, due to the absence of any mmf within contours ABB′B″A″A′A and DCC″D″,

$$\int_{ABB'A''B''A''} \overline{H}\overline{dl} = \int_{AA''} \overline{H}\overline{dl}; \quad \int_{CDD''C''} \overline{H}\overline{dl} = \int_{CC''} \overline{H}\overline{dl} \tag{5.86}$$

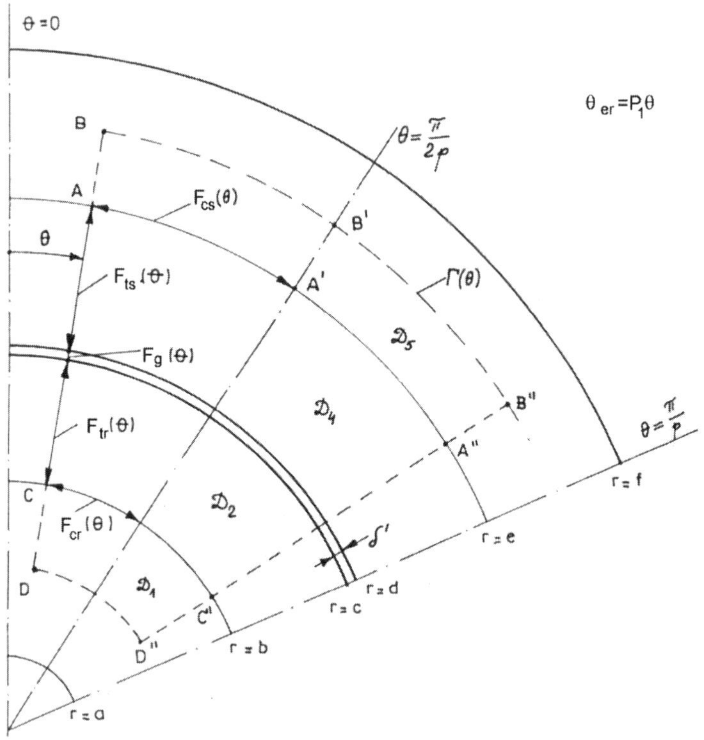

Figure 5.16 Ampere's law contours

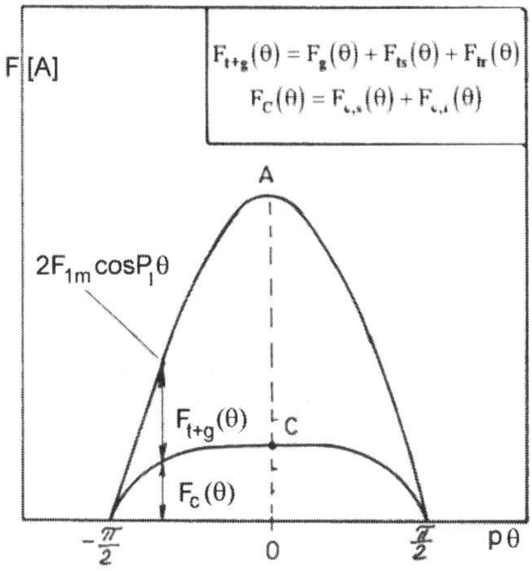

Figure 5.17 Mmf–total and back core component F

We may now divide the mmf components into two categories: those that depend directly on the airgap flux density $B_g(\theta)$ and those that do not (Figure 5.17).

$$2F_{t+g} = 2F_g(\theta) + 2F_{ts}(\theta) + 2F_{tr}(\theta) \tag{5.87}$$

$$F_c(\theta) = F_{cs}(\theta) + F_{cr}(\theta) \tag{5.88}$$

Also, the airgap mmf $F_g(\theta)$ is

$$F_g(\theta) = \frac{gK_c}{\mu_0} B_g(\theta) \tag{5.89}$$

Suppose we know the maximum value of the stator core flux density, for given F_{1m}, as obtained from AIM, Bcsm,

$$B_{cs}(\theta) = B_{csm} \sin p_1\theta \tag{5.90}$$

We may now calculate $F_{cs}(\theta)$ as

$$F_{cs}(\theta) = 2\int_{p_1\theta}^{\pi/2} [H_{cs}(B_{csm} \sin\alpha)] e d\alpha \tag{5.91}$$

The rotor core maximum flux density Bcrm is also known from AIM for the same mmf F1m. The rotor core mmf $F_{cr}(\theta)$ is thus

$$F_{cr}(\theta) = 2\int_{p_1\theta}^{\pi/2} [H_{cr}(B_{crm} \sin\alpha)] b d\alpha \tag{5.92}$$

From (5.88) $F_c(\theta)$ is obtained. Consequently,

$$F_{t+g} = 2F_{1m}\cos p\theta - F_{cr}(\theta) - F_{cs}(\theta) \tag{5.93}$$

Now in (5.87) we have only to express $F_{ts}(\theta)$ and $F_{tr}(\theta)$–tooth mmfs–as functions of airgap flux density $B_g(\theta)$. This is straightforward.

$$B_{ts}(\theta) = \frac{\beta_s}{\xi_s} B_g(\theta); \quad \xi_s = \left(\frac{\tau_s}{b_{ts}}\right)^{-1} \tag{5.94}$$

τ_s–stator slot pitch, b_{ts}–stator tooth, $\beta_s < 1$–ratio of real and apparent tooth flux density (due to the flux deviation through slot).

Similarly, for the rotor,

$$B_{tr}(\theta) = \frac{\beta_r}{\xi_1} B_g(\theta); \quad \xi_r = \left(\frac{\tau_r}{b_{tr}}\right)^{-1} \tag{5.95}$$

Now
$$F_{t+g}(\theta) = 2B_{g0}(\theta)\frac{g}{\mu_0} + 2h_s H_{ts}\left[\frac{\beta_s}{\xi_s} B_g(\theta)\right] + 2h_r H_{tr}\left[\frac{\beta_r}{\xi_r} B_g(\theta)\right] \tag{5.96}$$

In (5.96) $F_{t+g}(\theta)$ is known for given θ from (5.93). $B_{g0}(\theta)$ is assigned an initial value in (5.96) and then, based on the lamination B/H curve, H_{ts} and H_{tr} are calculated. Based on this, F_{t+g} is recalculated and compared to the value from (5.93). The calculation cycle is reiterated until sufficient convergence is reached.

The same operation is performed for 20 to 50 values of θ per half-pole to obtain smooth $B_g(\theta)$, $B_{ts}(\theta)$, $B_{tr}(\theta)$ curves per given stator (and rotor) mmfs (currents). Fourier analysis is used to calculate the harmonics contents. Sample results concerning the airgap flux density distribution for three designs of the same motor are given in Figures 5.18 though 5.21. [8]

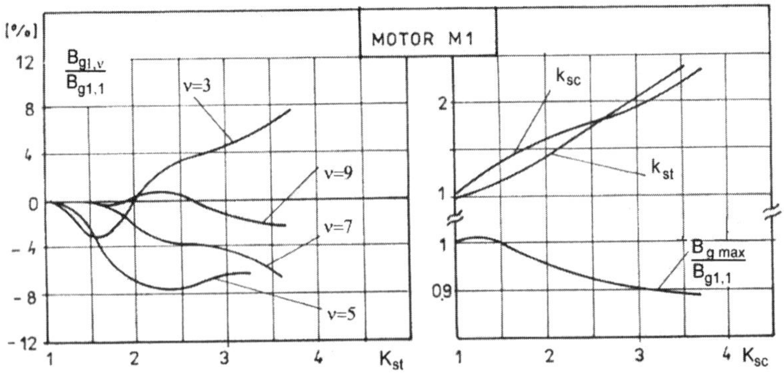

Figure 5.18 IM with equally saturated teeth and back cores

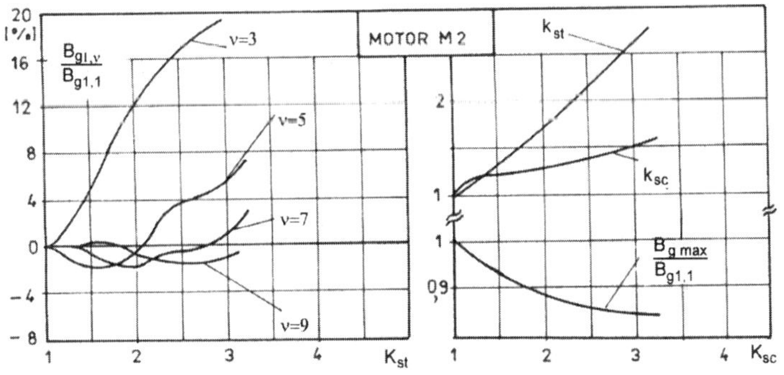

Figure 5.19 IM with saturated teeth ($K_{st} > K_{sc}$)

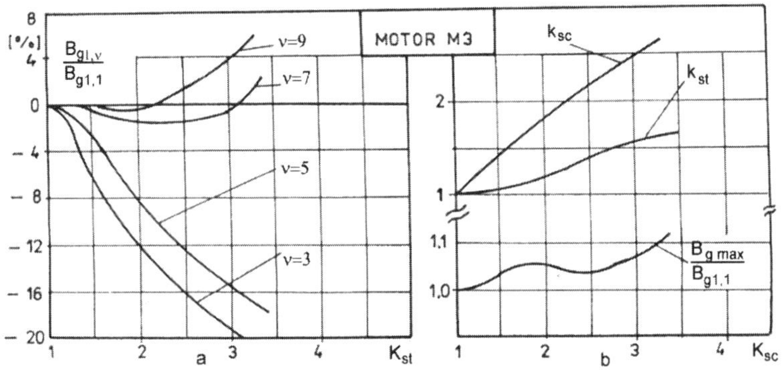

Figure 5.20 IM with saturated back cores
a.) airgap flux density b.) saturation factors K_{st}, K_{sc}, versus total saturation factor K_s

The geometrical data are

M1–optimized motor = equal saturation of teeth and cores: a = 0.015 m, b = 0.036 m, c = 0.0505 m, d = 0.051 m, e = 0.066 m, f = 0.091 m, $\xi_s = 0.54$, $\xi_r = 0.6$

M2–oversaturated stator teeth ($K_{st} > K_{sc}$): b = 0.04 m, e = 0.06 m, $\xi_s = 0.42$, $\xi_r = 0.48$

M3–oversaturated back cores ($K_{sc} > K_{st}$): b = 0.033 m, e = 0.069 m, $\xi_s = 0.6$, $\xi_r = 0.65$

The results on Figure 5.18 – 5.21 may be interpreted as follows:

- If the saturation level (factor) in the teeth and back cores is about the same (motor M1 on Figure 5.21), then even for large degrees of saturation (K_s = 2.6 on Figure 5.18b), the airgap flux harmonics are small (Figure 5.18a). This is why a design may be called optimized.

- If the teeth and back core saturation levels are notably different (oversaturated teeth is common), then the harmonics content of airgap flux density is rich. Flattened airgap flux distribution is obtained for oversaturated back cores (Figure 5.20).

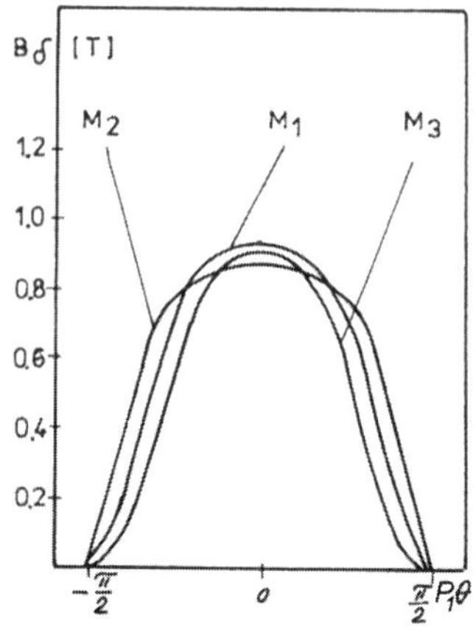

Figure 5.21 Airgap flux density distribution for three IM designs, M1, M2, M3.

- The above remarks seem to suggest that equal saturation level in teeth and back iron is to be provided to reduce airgap flux harmonics content. However, if core loss minimization is aimed at, this might not be the case, as discussed in Chapter 11 on core losses.

5.5 THE EMF IN AN A.C. WINDING

As already shown, the placement of coils in slots, the slot openings, and saturation cause airgap flux density harmonics (Figure 5.22).

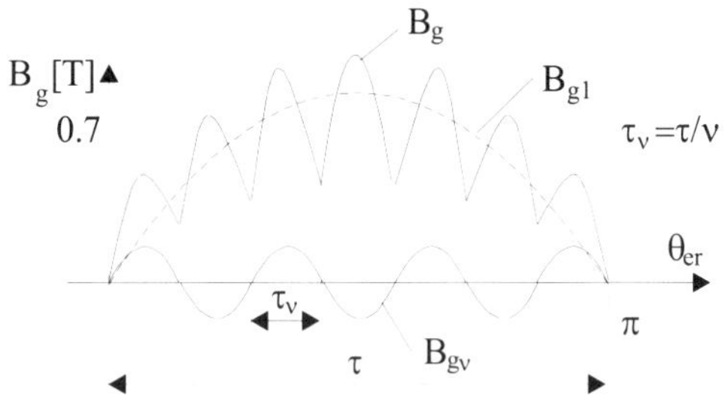

Figure 5.22 Airgap flux density and harmonics ν

The voltage (emf) induced in the stator and rotor phases (bars) is of paramount importance for the IM behavior. We will derive its general expression based on the flux per pole (Φ_ν) for the ν harmonic (ν = 1 for the fundamental here; it is an "electrical" harmonic).

$$\Phi_\nu = \frac{2}{\pi}\tau_\nu L_e B_{g\nu} \tag{5.97}$$

For low and medium power IMs, the rotor slots are skewed by a distance c (in meters here). Consequently, the phase angle of the airgap flux density along the inclined rotor (stator) slot varies continuously (Figure 5.23).

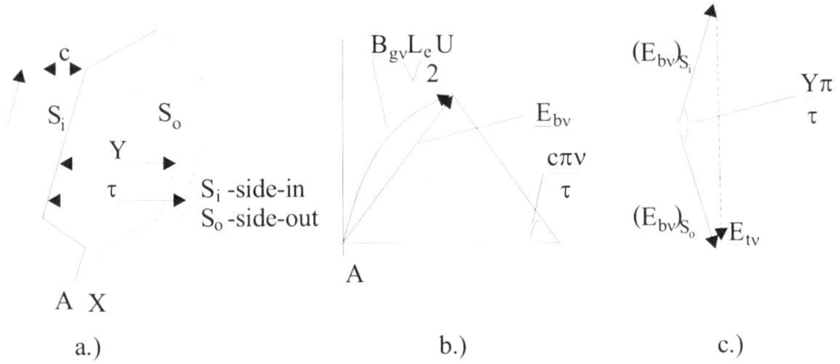

Figure 5.23 Chorded coil in skewed slots
a.) coil b.) emf per conductor c.) emf per turn

Consequently, the skewing factor $K_{skew\nu}$ is (Figure 5.23b)

$$K_{skew\nu} = \frac{\overline{AB}}{\overset{\frown}{AOB}} = \frac{\sin\frac{c}{\tau}\nu\frac{\pi}{2}}{\frac{c}{\tau}\nu\frac{\pi}{2}} \le 1 \tag{5.98}$$

The emf induced in a bar (conductor) placed in a skewed slot $E_{b\nu}$ (RMS) is

$$E_{b\nu} = \frac{1}{2\sqrt{2}}\omega_1 K_{skew\nu}\Phi_\nu \tag{5.99}$$

with ω_1–the primary frequency (time harmonics are not considered here).

As a turn may have chorded throw (Figure 2.23a), the emfs in the two turn sides S_i and S_o (in Figure 5.23c) are dephased by less than 180^0. A vector composition is required and thus $E_{t\nu}$ is

$$E_{t\nu} = 2K_{y\nu}E_{b\nu};\;\; K_{y\nu} = \sin\nu\frac{y}{2}\frac{\pi}{2} \le 1 \tag{5.100}$$

$K_{y\nu}$ is the chording factor.

There are n_c turns per coil, so the coil emf $E_{c\nu}$ is

$$E_{c\nu} = n_c E_{t\nu} \tag{5.101}$$

In all there are q neighbouring coils per pole per phase. Let us consider them in series and with their emfs phase shifted by $\alpha_{ecv} = 2\pi p_1 \nu / N_s$ (Figure 5.24). The resultant emf for the q coils (in series in our case) E_{cv} is

$$E_{cv} = qE_{cv}K_{qv} \tag{5.102}$$

K_{qv} is called the distribution factor. For an integer q (Figure 5.24) K_{qv} is

$$K_{qv} = \frac{\sin \nu\pi / 6}{q \sin(\nu\pi / 6q)} \leq 1 \tag{5.103}$$

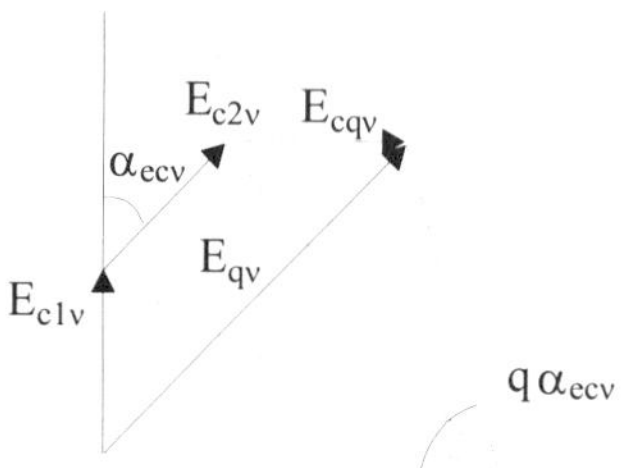

Figure 5.24 Distribution factor K_{qv}

The total number of q coil groups is p_1 (pole pairs) in single layer and $2p_1$ in double layer windings. Now we may introduce $a \leq p_1$ current paths in parallel.

The total number of turns/phase is $W_1 = p_1 q n_c$ for single layer and, $W_1 = 2p_1 q n_c$ for double layer windings, respectively. Per current path, we do have W_a turns.

$$W_a = \frac{W_1}{a} \tag{5.104}$$

Consequently the emf per current path E_a is

$$E_{av} = \pi\sqrt{2} f_1 W_a K_{qv} K_{yv} K_{cv} \Phi_v \tag{5.105}$$

For the stator phase emf, when the stator slots are straight, $K_{cv} = 1$.

For the emf induced in the rotor phase (bar) with inclined rotor slots, $K_{cv} \neq 1$ and is calculated with (5.98).

$$(E_{av})_{stator} = \pi\sqrt{2} f_1 W_a K^s{}_{qv} K^s{}_{yv} \Phi_v \tag{5.106}$$

$$(E_{bv})_{rotor} = \frac{\pi}{\sqrt{2}} f_1 K_{skewv} \Phi_v \tag{5.107}$$

$$(E_{av})_{rotor} = \pi\sqrt{2} f_1 W_a K^r{}_{qv} K^r{}_{yv} \Phi_v \tag{5.108}$$

The distribution K_{qv} and chording (K_{yv}) factors may differ in the wound rotor from stator.

Let us note that the distribution, chording, and skewing factors are identical to those derived for mmf harmonics. This is so because the source of airgap field distribution is the mmf distribution. The slot openings, or magnetic saturation influence on field harmonics is lumped into Φ_ν.

The harmonic attenuation or cancellation through chording or skewing is also evident in (5.102) – (5.104). Harmonics that may be reduced or cancelled by chording or skewing are of the order

$$\nu = Km \pm 1 \tag{5.109}$$

For one of them

$$\nu \frac{y}{\tau}\frac{\pi}{2} = K'\pi \quad \text{or} \quad \nu \frac{c}{\tau}\frac{\pi}{2} = K''\pi \tag{5.110}$$

These are called mmf step (belt) harmonics (5, 7, 11, 13, 17, 19). Third mmf harmonics do not exist for star connection. However, as shown earlier, magnetic saturation may cause 3rd order harmonics, even 5th, 7th, 11th order harmonics, which may increase or decrease the corresponding mmf step harmonics.

Even order harmonics may occur in two-phase windings or in three-phase windings with fractionary q. Finally, the slot harmonics orders ν_c are:

$$\nu_c = 2Kqm \pm 1 \tag{5.111}$$

For fractionary $q = a + b/c_1$,

$$\nu_c = 2(ac_1 + b)\left(\frac{K}{c_1}\right)m \pm 1 \tag{5.112}$$

We should notice that for all slot harmonics,

$$K_{q\nu} = \frac{\sin(2Kqm \pm 1)\pi/6}{q\sin[(2Kqm \pm 1)\pi/6q]} = K_{q1} \tag{5.113}$$

Slot harmonics are related to slot openings presence or to the corresponding airgap permeance modulation. All slot harmonics have the same distribution factor with the fundamental, they may not be destroyed or reduced. However, as their amplitude (in airgap flux density) decreases with ν_c increasing, the only thing we may do is to increase their lowest order. Here the fractionary windings come into play (5.112). With q increased from 2 to 5/2, the lowest slot integer harmonic is increased (see (5.111) – (5.112)) from $\nu_{cmin} = 11$ to $\nu_{cmin} = 29$! ($K/c_1 = 2/2$). As K is increased, the order ν of step (phase belt) harmonics (5.111) may become equal to ν_c.

Especially the first slot harmonics ($K = 1$, $K/c_1 = 1$) may thus interact and amplify (or attenuate) the effect of mmf step harmonics of the same order. Intricate aspects like this will be dealt with in Chapter 10.

5.6 THE MAGNETIZATION INDUCTANCE

Let us consider first a sinusoidal airgap flux density in the airgap, at no load (zero rotor currents). Making use of Ampere's law we get, for airgap flux amplitude,

$$B_{g1} \approx \frac{\mu_0 (F_1)_{phase}}{gK_cK_s}; \quad K_s > 1 \tag{5.114}$$

where F_1 is the phase mmf/pole amplitude,

$$(F_1)_{phase} = \frac{2W_1I_0\sqrt{2}K_{q1}K_{y1}}{\pi p_1} \tag{5.115}$$

For a three-phase machine on no load,

$$F_{1m} = \frac{3W_1I_0\sqrt{2}K_{q1}K_{y1}}{\pi p_1} \tag{5.116}$$

Now the flux per pole, for the fundamental Φ_1, is

$$\Phi_1 = \frac{2}{\pi}B_{g1}\tau L_e \tag{5.117}$$

So the airgap flux linkage per phase (a = 1) Ψ_{11h} is

$$\psi_{11h} = W_1K_{q1}K_{y1}\Phi_1 \tag{5.118}$$

The magnetization inductance of a phase, when all other phases on stator and rotor are open, L_{11m} is

$$L_{11m} = \frac{\psi_{11h}}{I_0\sqrt{2}} = \frac{4\mu_0(W_1K_{q1}K_{y1})^2}{\pi^2}\frac{L_e\tau}{p_1gK_cK_s} \tag{5.119}$$

K_s–total saturation factor.

In a similar manner, for a harmonic ν, we obtain

$$L_{11m\nu} = \frac{4\mu_0(W_1K_{q\nu}K_{y\nu})^2}{\pi^2}\frac{L_e\tau}{p_1gK_cK_{s\nu}}; \quad K_{s\nu} \approx (1-1.2) \tag{5.120}$$

The saturation coefficient $K_{s\nu}$ tends to be smaller than K_s, as the length of harmonics flux line in iron is usually shorter than for the fundamental.

Now, for three-phase supply we use the same rationale, but, in the (5.114) F_{1m} will replace $(F_1)_{phase}$.

$$L_{1m\nu} = \frac{6\mu_0(W_1K_{q\nu}K_{y\nu})^2}{\pi^2\nu^2}\frac{L_e\tau}{p_1gK_cK_{s\nu}} \tag{5.121}$$

When the no-load current (I_o/I_n) increases, the saturation factor increases, so the magnetization inductance L_{1m} decreases (Figure 5.25).

$$L_{1m} = \frac{6\mu_0 (W_1 K_{q1} K_{y1})^2}{\pi^2} \frac{L_e \tau}{p_1 g K_c K_s (I_0 / I_n)} \tag{5.122}$$

In general, a normal inductance, L_n, is defined as

$$L_n = \frac{V_n}{I_n \omega_{1n}} : l_{1m} = \frac{L_{1m}}{L_n} \tag{5.123}$$

V_n, I_n, ω_{1n} rated phase voltage, current, and frequency of the respective machine.

Well designed induction motors have l_{1m} (Figure 5.25) in the interval 2.5 to 4; in general, it increases with power level and decreases with large number of poles.

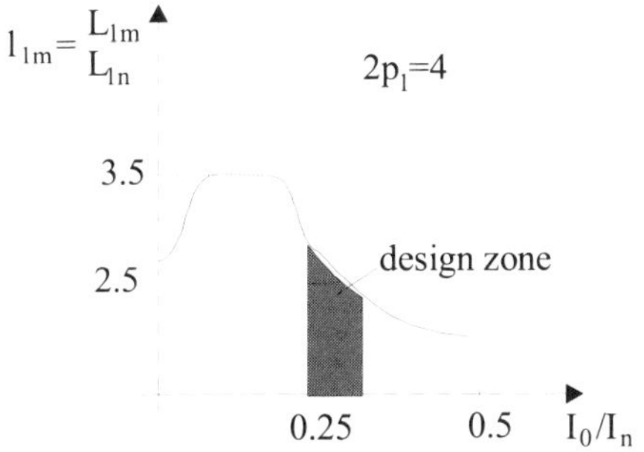

Figure 5.25 Magnetization inductance versus no load current I_0/I_n

It should be noticed that Figure 5.25 shows yet another way (scale) of representing the magnetization curve.

The magnetization inductance of the rotor L_{22m} has expressions similar to (5.119) – (5.120), but with corresponding number of turns phase, distribution, and chording factors.

However, for the mutual magnetising inductance between a stator and rotor phase L_{12m}, we obtain

$$L_{12mv} = \frac{4\mu_0}{\pi^2} \left(W_1 K_{qv}{}^s K_{yv}{}^s W_2 K_{qv}{}^r K_{yv}{}^r K_{cv}{}^r \right) \frac{L_e \tau}{p_1 g K_{sv}} \tag{5.124}$$

This time, the skewing factor is present (skewed rotor slots).

From (5.124) and (5.120),

$$\frac{L_{12mv}}{L_{11mv}} = \frac{W_2 K_{qv}{}^r K_{yv}{}^r K_{cv}{}^r}{W_1 K_{qv}{}^s K_{yv}{}^s} \tag{5.125}$$

It is now evident that for a rotor cage, $K_{qv}{}^{r} = K_{yv}{}^{r} = 1$ as each bar (slot) may be considered as a separate phase with $W_2 = ½$ turns.

The inductance L_{11m} or L_{22m} are also called main (magnetization) self inductances, while L_{1m} is called the cyclic (multiphase supply) magnetizing inductance. In general, for m phases,

$$L_{1m} = \frac{m}{2} L_{11m}; \qquad m = 3 \text{ for three phases} \tag{5.126}$$

Example 5.1. The magnetization inductance L_{1m} calculation

For the motor M1 geometry given in the previous paragraph, with double-layer winding and $N_s = 36$ slots, $N_r = 30$ slots, $2p_1 = 4$, $y/\tau = 8/9$, let us calculate L_{1m} and no-load phase current I_0 for a saturation factor $K_s = 2.6$ and $B_{g1} = 0.9T$, $K_{st} = K_{sr} = 1.8$, $K_c = 1.3$ (Figure 5.20, 5.21).

Stator (rotor) slot opening are: $b_{os} = 6g$, $b_{or} = 3g$, $W_1 = 120$turns/phase, $L_e = 0.12m$, and stator bore diameter $D_i/2 = 0.051m$.

Solution

- The Carter coefficient Kc ((5.4), (5.3)) is

$$K_{c1,2} = \frac{\tau_{s,r}}{\tau_{s,r} - \gamma_{1,2} \cdot g/2} = \begin{cases} 1.280 \\ 1.0837 \end{cases} \tag{5.127}$$

$$\begin{aligned} \tau_s &= \frac{\pi D_i}{N_s} = \frac{\pi \cdot 2 \cdot 0.051}{36} = 9.59 \cdot 10^{-3} m \\ \tau_r &= \frac{\pi(D_i - 2g)}{N_r} = \frac{\pi(2 \cdot 0.051 - 2 \cdot 0.0005)}{30} = 10.57 \cdot 10^{-3} m \end{aligned} \tag{5.128}$$

$$\begin{aligned} \gamma_1 &= \frac{(2b_{os}/g)^2}{5 + 2b_{os}/g} = \frac{12^2}{5+12} = 8.47 \\ \gamma_2 &= \frac{(2b_{or}/g)^2}{5 + 2b_{or}/g} = \frac{6^2}{5+6} = 3.27 \end{aligned} \tag{5.129}$$

$$K_c = K_{c1} \cdot K_{c2} = 1.280 \cdot 1.0837 = 1.327 \tag{5.130}$$

- The distribution and chording factors Kq1, Ky1, (5.100), (5.103) are

$$K_{q1} = \frac{\sin \pi/6}{3 \sin \pi/6 \cdot 3} = 0.9598; \; K_{y1} = \sin \frac{8}{9} \frac{\pi}{2} = 0.9848 \tag{5.131}$$

From (5.122),

$$L_{1m} = 6 \cdot \frac{1.25 \cdot 10^{-6}}{\pi^2} \cdot (300 \cdot 0.9598 \cdot 0.9848)^2 \cdot$$
$$\cdot \frac{9 \cdot 9.59 \cdot 10^{-3} \cdot 0.12}{2 \cdot 0.5 \cdot 10^{-3} \cdot 1.327 \cdot 2.6} = 0.1711\text{H} \tag{5.132}$$

- The no load current I_0 from (5.114) is

$$B_{g1} = \frac{\mu_0 F_{1m}}{gK_cK_s} = \frac{1.256 \cdot 10^{-3} \cdot 3\sqrt{2} \cdot 300 \cdot I_0 \cdot 0.9598 \cdot 0.9848}{3.14 \cdot 2 \cdot 0.5 \cdot 10^{-3} \cdot 1.327 \cdot 2.6} = 0.9\text{T} \tag{5.133}$$

So the no load phase current I_0 is

$$I_0 = 3.12\text{A} \tag{5.134}$$

Let us consider rated current $I_n = 2.5I_0$.

- The emf per phase for $f_1 = 60\text{Hz}$ is

$$E_1 = 2\pi f_1 L_{1m} I_0 = 2\pi 60 \cdot 0.1711 \cdot 3.12 = 201.17\text{V} \tag{5.135}$$

With a rated voltage/phase $V_1 = 210$ V, the relative value of L_{1m} is

$$l_{1m} = \frac{L_{1m}}{V_1/(\omega_1 I_n)} = \frac{0.1711}{210/(2\pi 60 \cdot 2.5 \cdot 3.12)} = 2.39 \tag{5.136}$$

It is obviously a small relative value, mainly due to the large value of saturation factor $K_s = 2.6$ (in the absence of saturation: $K_s = 1.0$).

In general, $E_1/V_1 \approx 0.9 - 0.98$; higher values correspond to larger power machines.

5.7 SUMMARY

- In absence of rotor currents, the stator winding currents produce an airgap flux density which contains a strong fundamental and spatial harmonics due to the placement of coils in slots (mmf harmonics), magnetic saturation, and slot opening presence. Essentially, FEM could produce a fully realistic solution to this problem.
- To simplify the study, a simplified analytical approach is conventionally used; only the mmf fundamental is considered.
- The effect of slotting is "removed" by increasing the actual airgap g to gK_c ($K_c > 1$); K_c–the Carter coefficient.
- The presence of eventual ventilating radial channels (ducts) is considered through a correction coefficient applied to the geometrical stack axial length.
- The dependence of peak airgap flux density on stator mmf amplitude (or stator current) for zero rotor currents, is called the magnetization curve.

- When the teeth are designed as the heavily saturated part, the airgap flux "flows" partially through the slots, thus "defluxing" to some the teeth extent.
- To account for heavily saturated teeth designs, the standard practice is to calculate the mmf $(F_1)_{30^0}$ required to produce the maximum (flat) airgap flux density and then increase its value to $F_{1m} = (F_1)_{30^0} / \cos 30^0$.
- A more elaborate two dimensional analytical iterative model (AIM), valid both for zero and nonzero rotor currents, is introduced to refine the results of the standard method. The slotting is considered indirectly through given but different radial and axial permeabilities in the slot zones. The results are validated on numerous low power motors, and sinusoidal airgap and core flux densities.
- Further on, AIM is used to calculate the actual airgap flux density distribution accounting for heavy magnetic saturation. For more saturated teeth, flat airgap flux density is obtained while, for heavier back core saturation, a peaked airgap and teeth flux density distribution is obtained.
- The presence of saturation harmonics is bound to influence the total core losses (as investigated in Chapter 11). It seems clear that if teeth and back iron cores are heavily but equally saturated, the flux density is still sinusoidal all along stator bore. The stator connection is also to be considered as for star connection, the stator no-load current is sinusoidal, and flux third harmonics may occur, while for the delta connection, the opposite is true.
- The expression of emf in a.c. windings exhibits the distribution, chording, and skewing factors already derived for mmfs in Chapter 4.
- The emf harmonics include mmf space (step) harmonics, saturation-caused space harmonics, and slot opening (airgap permeance variation) harmonics ν_c.
- Slot flux (emf) harmonics ν_c show a distribution factor $K_{q\nu}$ equal to that of the fundamental ($K_{q\nu} = K_{q1}$) so they cannot be destroyed. They may be attenuated by increasing the order of first slot harmonics $\nu_{cmin} = 2qm - 1$ with larger q slots per pole per phase or/and by increased airgap, or by fractional q.
- The magnetization inductance L_{1m} valid for the fundamental is the ratio of phase emf, to angular stator frequency to stator current. L_{1m} is decreasing with airgap, number of pole pairs, and saturation level (factor), but it is proportional to pole pitch and stack length and equivalent number of turns per phase squared.
- In relative values, l_{1m} increases with motor power and decreases with larger number of poles, in general.

5.8 REFERENCES

1. F.W. Carter, Airgap Induction, El. World and Engineering, 1901, p.884.
2. T.A. Lipo, Introduction to AC Machine Design, vol.1, pp.84, WEMPEC – University of Wisconsin, 1996.
3. B. Heller, V. Hamata, Harmonic Field Effects in Induction Machines, Elsevier Scientific, Amsterdam, 1977.
4. D.M. Ionel, M.V. Cistelecan, T.J.E. Miller, M.I. McGilp, A New Analytical Method for the Computation of Airgap Reactances in 3 Phase Induction Machines, Record of IEEE – IAS, Annual Meeting 1998, vol. 1, pp.65 – 72.
5. S.L. Ho, W.N. Fu, Review and Future Application of Finite Element Methods in Induction Motors, EMPS vol.26, 1998, pp.111 – 125.
6. V. Ostovic, Dynamics of Saturated Electric Machines, Springer Verlag, New York, 1989.
7. G. Madescu, I. Boldea, T.J.E Miller, An Analytical Iterative Model (AIM) for Induction Motor Design, Rec. of IEEE – IAS, Annual Meeting, 1996, vol.1., pp.566 – 573.
8. G. Madescu, "Contributions to The Modelling and Design of Induction Motors", Ph.D. Thesis, University Politehnica, Timisoara, Romania, 1995.

Chapter 6

LEAKAGE INDUCTANCES AND RESISTANCES

6.1. LEAKAGE FIELDS

Any magnetic field (H_i, B_i) zone within the IM is characterized by its stored magnetic energy (or coenergy) W_m.

$$W_{mi} = \frac{1}{2} \iiint_V \left(\overline{B} \cdot \overline{H}\right) dV = L_i \frac{I_i^2}{2} \tag{6.1}$$

Equation (6.1) is valid when, in that region, the magnetic field is produced by a single current source, so an inductance "translates" the field effects into circuit elements.

Besides the magnetic energy related to the magnetization field (investigated in Chapter 5), there are flux lines that encircle only the stator or only the rotor coils (Figure 6.1). They are characterized by some equivalent inductances called leakage inductances L_{sl}, L_{rl}.

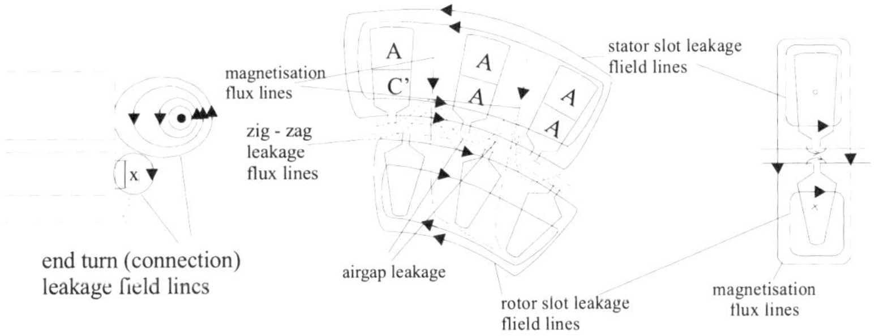

Figure 6.1 Leakage flux lines and components

There are leakage flux lines which cross the stator and, respectively, the rotor slots, end-turn flux lines, zig-zag flux lines, and airgap flux lines (Figure 6.1). In many cases, the differential leakage is included in the zig-zag leakage. Finally, the airgap flux space harmonics produce a stator emf as shown in Chapter 5, at power source frequency, so it should also be considered in the leakage category. Its torques occur at low speeds (high slips) and thus are not there at no–load operation.

6.2. DIFFERENTIAL LEAKAGE INDUCTANCES

As both the stator and rotor currents may produce space flux density harmonics in the airgap (only step mmf harmonics are considered here), there will be both a stator and a rotor differential inductance. For the stator, it is sufficient to add all $L_{1m\nu}$ harmonics, but the fundamental (5.122), to get L_{ds}.

$$L_{ds} = \frac{6\mu_0 \tau L_e W_1^2}{\pi^2 p_1 g K_c} \sum_{\nu \neq 1} \frac{K_{w\nu}^2}{\nu^2 K_{s\nu}}; \quad \nu = Km_1 \pm 1 \tag{6.2}$$

The ratio σ_d of L_{ds} to the magnetization inductance L_{1m} is

$$\sigma_{dS0} = \frac{L_{ds}}{L_{1m}} = \sum_{\nu \neq 1} \left(\frac{K_{ws\nu}^2}{\nu^2 K_{ws1}^2} \right) \cdot \frac{K_s}{K_{s\nu}} \tag{6.3}$$

where $K_{Ws\nu}$, K_{Ws1} are the stator winding factors for the harmonic ν and for the fundamental, respectively.

K_s and $K_{s\nu}$ are the saturation factors for the fundamental and for harmonics, respectively.

As the pole pitch of the harmonics is τ/ν, their fields do not reach the back cores and thus their saturation factor $K_{s\nu}$ is smaller then K_s. The higher ν, the closer $K_{s\nu}$ is to unity. In a first approximation,

$$K_{s\nu} \approx K_{st} < K_s \tag{6.4}$$

That is, the harmonics field is retained within the slot zones so the teeth saturation factor K_{st} may be used (K_s and K_{st} have been calculated in Chapter 5). A similar formula for the differential leakage factor can be defined for the rotor winding.

$$\sigma_{dr0} = \sum_{\mu \neq 1} \left(\frac{K_{wr\mu}^2}{\mu^2 K_{wr1}^2} \right) \cdot \frac{K_s}{K_{s\mu}} \tag{6.5}$$

As for the stator, the order μ of rotor harmonics is

$$\mu = K_2 m_2 \pm 1 \tag{6.6}$$

m_2–number of rotor phases; for a cage rotor $m_2 = Z_2/p_1$, also $K_{wr1} = K_{wr\mu} = 1$.

The infinite sums in (6.3) and (6.5) are not easy to handle. To avoid this, the airgap magnetic energy for these harmonics fields can be calculated. Using (6.1)

$$B_{g\nu} = \frac{\mu_0 F_{m\nu}}{g K_c K_{s\nu}} \tag{6.7}$$

We consider the step-wise distribution of mmf for maximum phase A current, (Figure 6.2), and thus

$$\sigma_{ds0} = \frac{\int_0^{2\pi} F^2(\theta)d\theta}{F_{1m}{}^2 2\pi} = \frac{1}{N_s}\frac{\sum_1^{N_s} F_j{}^2(\theta)}{F_{1m}{}^2} \quad (6.8)$$

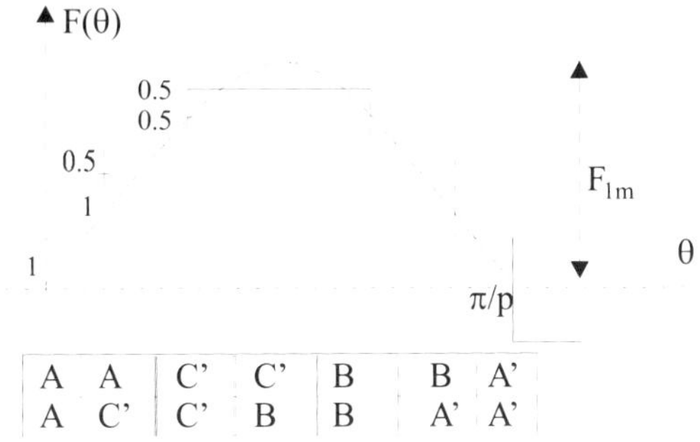

Figure 6.2 Step-wise mmf waveform (q = 2, y/τ = 5/6)

The final result for the case in Figure 6.2 is $\sigma_{dso} = 0.0285$.

This method may be used for any kind of winding once we know the number of turns per coil and its current in every slot.

For full-pitch coil three-phase windings [1],

$$\sigma_{ds} \approx \frac{5q^2+1}{12q^2}\cdot\frac{2\pi^2}{m_1{}^2 K_{w1}{}^2} - 1 \quad (6.9)$$

Also, for standard two-layer windings with chorded coils with chording length ε, σ_{ds} is [1]

$$\varepsilon = \tau - y = (1 - y/\tau)3q$$

$$\sigma_{ds} = \frac{2\pi^2}{m_1{}^2 K_{w1}{}^2}\cdot\frac{5q^2+1-\frac{3}{4}\left(1-\frac{y}{\tau}\right)\left[9q^2\left(1-\left(\frac{y}{\tau}\right)^2\right)+1\right]}{12q^2} - 1 \quad (6.10)$$

In a similar way, for the cage rotor with skewed slots,

$$\sigma_{r0} = \frac{1}{K'_{skew}{}^2 \eta_{r1}{}^2} - 1 \quad (6.11)$$

with

$$K'_{skew} = \frac{\sin\left(\alpha_{er}\cdot\frac{c}{\tau_r}\right)}{\alpha_{er}\cdot\frac{c}{\tau_r}};\ \eta_{r1} = \frac{\sin(\alpha_{er}/2)}{\alpha_{er}/2};\ \alpha_{er} = 2\pi\frac{p_1}{N_r} \quad (6.12)$$

The above expressions are valid for three-phase windings. For a single-phase winding, there are two distinct situations. At standstill, the a.c. field produced by one phase is decomposed into two equal traveling waves. They both produce a differential inductance and, thus, the total differential leakage inductance $(L_{ds1})_{s=1} = 2L_{ds}$.

On the other hand, at S = 0 (synchronism) basically the inverse (backward) field wave is almost zero and thus $(L_{ds1})_{S=0} \approx L_{ds}$.

The values of differential leakage factor σ_{ds} (for three- and two-phase machines) and σ_{dr}, as calculated from (6.9) and (6.10) are shown on Figures 6.3 and 6.4. [1]

A few remarks are in order.

- For q = 1, the differential leakage coefficient σ_{ds0} is about 10%, which means it is too large to be practical.
- The minimum value of σ_{ds0} is obtained for chorded coils with $y/\tau \approx 0.8$ for all q_s (slots/pole/phase).
- For same q, the differential leakage coefficient for two-phase windings is larger than for three-phase windings.
- Increasing the number of rotor slots is beneficial as it reduces σ_{dr0} (Figure 6.4).

Figures 6.3 do not contain the influence of magnetic saturation. In heavily saturated teeth IMs as evident in (6.3), $K_s/K_{st} > 1$, the value of σ_{ds} increases further.

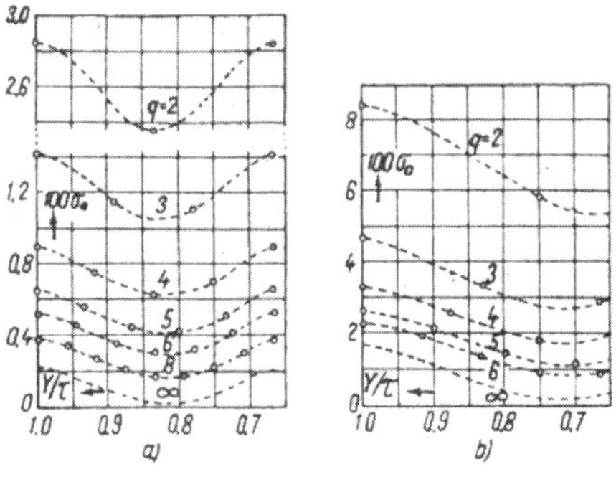

Figure 6.3 Stator differential leakage coefficient σ_{ds} for three phases a.) and two-phases b.) for various q_s

The stator differential leakage flux (inductance) is attenuated by the reaction of the rotor cage currents.

Coefficient Δ_d for the stator differential leakage is [1]

$$\Delta_d \approx 1 - \frac{1}{\sigma_{ds0}} \sum_{\nu \neq 1}^{2mq_1+1} \left(K'_{skew\nu}\, \eta_{r\nu} \frac{K_{w\nu}}{\nu K_{w1}} \right)^2 ; \quad \sigma_{ds} = \sigma_{ds0}\Delta_d \tag{6.13}$$

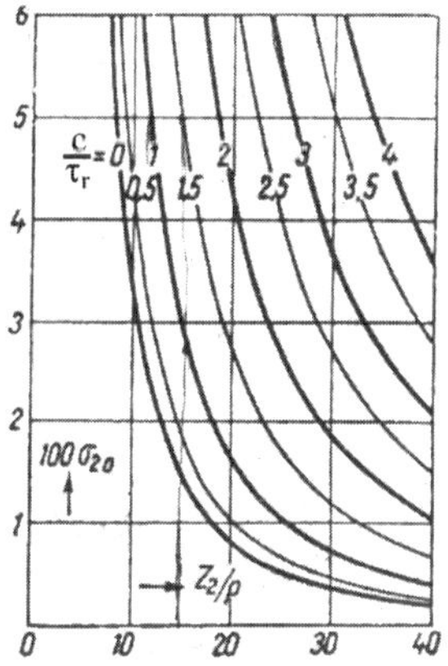

Figure 6.4 Rotor cage differential leakage coefficient σ_{dr} for various q_s and straight and single slot pitch skewing rotor slots ($c/\tau_r = 0,1,\ldots$)

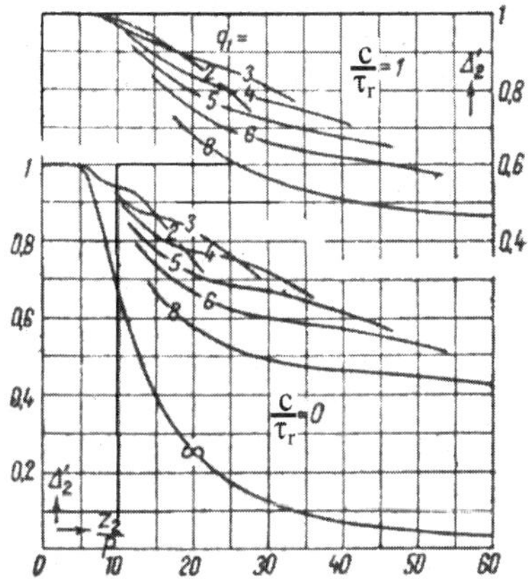

Figure 6.5. Differential leakage attenuation coefficient Δ_d for cage rotors with straight ($c/\tau_r = 0$) and skewed slots ($c/\tau_s = 1$)

As the stator winding induced harmonic currents do not attenuate the rotor differential leakage: $\sigma_{dr} = \sigma_{dr0}$.

A rather complete study of various factors influencing the differential leakage may be found in [Reference 2].

Example 6.1. For the IM in Example 5.1, with $q = 3$, $N_s = 36$, $2p_1 = 4$, $y/\tau = 8/9$, $K_{w1} = 0.965$, $K_s = 2.6$, $K_{st} = 1.8$, $N_r = 30$, stack length $L_e = 0.12$m, $L_{1m} = 0.1711$H, $W_1 = 300$ turns/phase, let us calculate the stator differential leakage inductance L_{ds} including the saturation and the attenuation coefficient Δ_d of rotor cage currents.

Solution

We will find first from Figure 6.3 (for q = 3, y/τ = 0.88) that $\sigma_{ds0} = 1.16\cdot10^{-2}$. Also from Figure 6.5 for c = 1τ_s (skewing), $Z_2/p_1 = 30/2 = 15$, $\Delta_d = 0.92$. Accounting for both saturation and attenuation coefficient Δ_d, the differential leakage stator coefficient K_{ds} is

$$K_{ds} = K_{ds0}\cdot\frac{K_s}{K_{st}}\Delta_d = 1.16\cdot10^{-2}\cdot\frac{2.6}{1.8}\cdot0.92 = 1.5415\cdot10^{-2} \qquad (6.14)$$

Now the differential leakage inductance L_{ds} is

$$L_{ds} = K_{ds}L_{1m} = 1.5415\cdot10^{-2}\cdot0.1711 = 0.2637\cdot10^{-3}H \qquad (6.15)$$

As seen from (6.13), due to the rather large q, the value of K_{ds} is rather small, but, as the number of rotor slots/pole pair is small, the attenuation factor Δ_d is large.

Values of q = 1,2 lead to large differential leakage inductances. The rotor cage differential leakage inductance (as reduced to the stator) L_{dr} is

$$L_{dr} = \sigma_{dr0}\frac{K_s}{K_{ts}}L_{1m} = 2.8\cdot10^{-2}\cdot\frac{2.6}{1.8}L_{1m} = 4.04\cdot10^{-2}L_{1m} \qquad (6.16)$$

σ_{dr0} is taken from Figure 6.4 for $Z_2/p_1 = 15$, $c/\tau_r = 1$, : $\sigma_{dr0} = 2.8\cdot10^{-2}$.

It is now evident that the rotor (reduced to stator) differential leakage inductance is, for this case, notable and greater than that of the stator.

6.3. RECTANDULAR SLOT LEAKAGE INDUCTANCE/SINGLE LAYER

The slot leakage flux distribution depends notably on slot geometry and less on teeth and back core saturation. It also depends on the current density distribution in the slot which may become nonuniform due to eddy currents (skin effect) induced in the conductors in slot by their a.c. leakage flux.

Let us consider the case of a rectangular stator slot where both saturation and skin effect are neglected (Figure 6.6).

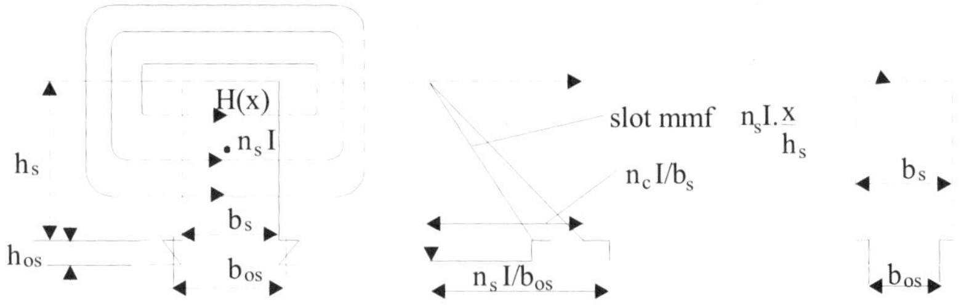

Figure 6.6 Rectangular slot leakage

Ampere's law on the contours in Figure 6.6 yields

$$H(x)b_s = \frac{n_s \cdot i \cdot x}{h_s}\ ;\ 0 \le x \le h_s$$
$$H(x)b_s = n_s \cdot i\ ;\ h_s \le x \le h_s + h_{0s} \tag{6.17}$$

The leakage inductance per slot, L_{sls}, is obtained from the magnetic energy formula per slot volume.

$$L_{sls} = \frac{2}{i^2} W_{ms} = \frac{2}{i^2} \cdot \frac{1}{2} \int_0^{h_s + h_{0s}} \mu_0 H(x)^2 dx \cdot L_e b_s = \mu_0 n_s^{\ 2} L_e \left[\frac{h_s}{3b_s} + \frac{h_{os}}{b_{os}} \right] \tag{6.18}$$

The term in square parenthesis is called the geometrical specific slot permeance.

$$\lambda_s = \frac{h_s}{3b_s} + \frac{h_{os}}{b_{os}} \approx 0.5 \div 2.5;\ h_{os} = (1 \div 3)10^{-3} m \tag{6.19}$$

It depends solely on the aspect of the slot. In general, the ratio $h_s/b_s < (5–6)$ to limit the slot leakage inductance to reasonable values.

The machine has N_s stator slots and N_s/m_1 of them belong to one phase. So the slot leakage inductance per phase L_{sl} is

$$L_{sl} = \frac{N_s}{m_1} L_{sls} = \frac{2p_1 q m_1}{m_1} L_{sls} = 2\mu_0 W_1^{\ 2} L_e \frac{\lambda_s}{p_1 q} \tag{6.20}$$

The wedge location has been replaced by a rectangular equivalent area on Figure 6.6. A more exact approach is also possible.

The ratio of slot leakage inductance L_{sl} to magnetizing inductance L_{1m} is (same number of turns/phase),

$$\frac{L_{sl}}{L_{1m}} = \frac{\pi^2}{3} \frac{g K_c K_s}{(K_{W1})^2} \frac{\tau}{q} \lambda_s \tag{6.21}$$

Suppose we keep a constant stator bore diameter D_i and increase two times the number of poles.

The pole pitch is thus reduced two times as $\tau = \pi D/2p_1$. If we keep the number of slots constant q will be reduced twice and, if the airgap and the winding factor are the same, the saturation stays low for the low number of poles. Consequently, L_{sl}/L_{1m} increases two times (as λ_s is doubled for same slot height).

Increasing q (and the number of slots/pole) is bound to reduce the slot leakage inductance (6.20) to the extent that λ_s does not increase by the same ratio. Our case here refers to a single-layer winding and rectangular slot.

Two-layer windings with chorded coils may be investigated the same way.

6.4. RECTANGULAR SLOT LEAKAGE INDUCTANCE/TWO LAYERS

We consider the coils are chorded (Figure 6.7).

Let us consider that both layers contribute a field in the slot and add the effects. The total magnetic energy in the slot volume is used to calculate the leakage inductance L_{sls}.

$$L_{sls} = \frac{2L_e}{i^2}\int_0^{h_{st}} \mu_0 [H_1(x) + H_2(x)]^2 dx \cdot b(x) \tag{6.22}$$

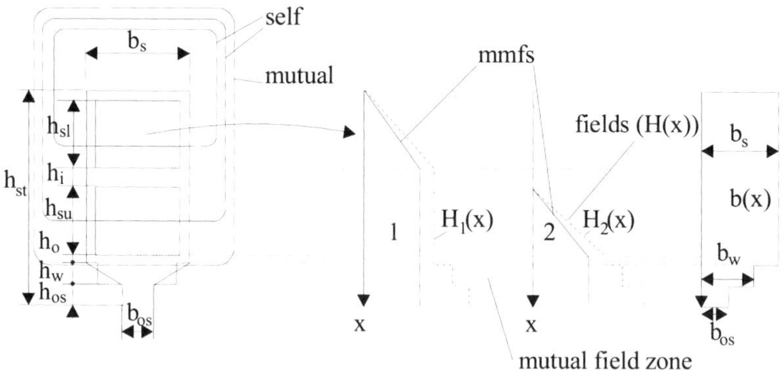

Figure 6.7 Two-layer rectangular semiclosed slots: leakage field

$$H_1(x) + H_2(x) = \begin{cases} \frac{n_{cl}I}{b_s}\cdot\frac{x}{h_{sl}}; \text{ for } 0 < x < h_{sl} \\ \frac{n_{cl}I}{b_s} \text{ for } h_{sl} < x < h_{sl} + h_i \\ \frac{n_{cl}I}{b_s} + \frac{n_{cu}I\cos\gamma_k}{b_s}\frac{(x - h_{sl} - h_i)}{h_{su}}; \\ \qquad \text{for } h_{sl} + h_i < x < h_{sl} + h_i + h_{su} \\ \frac{n_{cl}I}{b_i} + \frac{n_{cu}I\cos\gamma_k}{b_i}; \text{ for } x > h_{sl} + h_i + h_{su} \end{cases} \tag{6.23}$$

with $b_i = b_w$ or b_{os}

The phase shift between currents in lower and upper layer coils of slot K is γ_K and n_{cl}, n_{cu} are the number of turns of the two coils. Adding up the effect of all slots per phase (1/3 of total number of slots), the average slot leakage inductance per phase L_{sl} is obtained.

While (6.23) is valid for general windings with different number of turn/coil and different phases in same slots, we may obtain simplified solutions for identical coils in slots $n_{cl} = n_{cu} = n_c$.

$$\lambda_{sk} = \frac{L_{sl}}{\mu_0 (2n_c)^2 L_e} = \frac{1}{4}\left[\frac{\left(h_{sl} + h_{su}\cos^2\gamma_k\right)}{3b_s} + \frac{h_{su}}{b_s} + \frac{h_{su}\cos\gamma_k}{b_s} + \right.$$
$$\left. + \frac{h_i}{b_s} + (1+\cos\gamma)^2\left(\frac{h_o}{b_s} + \frac{h_w}{b_w} + \frac{h_{os}}{b_{os}}\right)\right] \quad (6.24)$$

Although (6.24) is quite general–for two-layer windings with equal coils in slots–the eventual different number of turns per coil can be lumped into cosγ as Kcosγ with $K = n_{cu}/n_{cl}$. In this latter case the factor 4 will be replaced by $(1 + K)^2$.

In integer and fractionary slot windings with random coil throws, (6.24) should prove expeditious. All phase slots contributions are added up.

Other realistic rectangular slot shapes for large power IMs (Figure 6.8) may also be handled via (6.24) with minor adaptations.

For full pitch coils ($\cos\gamma_K = 1.0$) symmetric winding ($h_{su} = h_{sl} = h_s'$) (6.24) becomes

$$\left(\lambda_{sk}\right)_{\substack{\gamma_k = 0 \\ h_{su} = h_{sl} = h_s'}} = \frac{2h_s'}{3b_s} + \frac{h_i}{4b_s} + \frac{h_0}{b_s} + \frac{h_w}{b_w} + \frac{h_{os}}{b_{os}} \quad (6.25)$$

Further on with $h_i = h_o = h_w = 0$ and $2h_s' = h_s$, we reobtain (6.19), as expected.

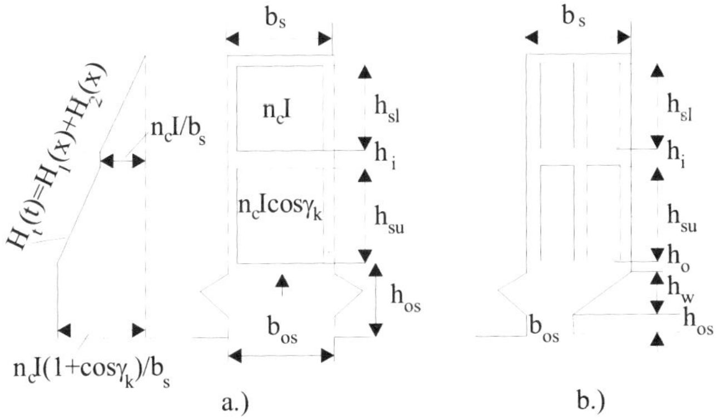

Figure 6.8 Typical high power IM stator slots

6.5. ROUNDED SHAPE SLOT LEAKAGE INDUCTANCE/TWO LAYERS

Although the integral in (6.2) does not have exact analytical solutions for slots with rounded corners, or purely circular slots (Figure 6.9), so typical to low-power IMs, some approximate solutions have become standard for design purposes:

- For slots a.) in Figure 6.9,

$$\lambda_{s,r} \approx \frac{2h_{s,r}K_1}{3(b_1+b_2)} + \left(\frac{h_{os,r}}{b_{os,r}} + \frac{h_o}{b_1} - \frac{b_{os,r}}{2b_1} + 0.785\right)K_2 \tag{6.26}$$

with

$$\begin{aligned} K_2 &\approx \frac{1+3\beta_y}{4}; \ \text{for} \frac{2}{3} \le \beta_y = \frac{y}{\tau} \le 1 \\ K_2 &\approx \frac{6\beta_y - 1}{4}; \ \text{for} \frac{1}{3} \le \beta_y \le \frac{2}{3} \\ K_2 &\approx \frac{3(2-\beta_y)+1}{4}; \ \text{for} \ 1 \le \beta_y \le 2 \\ K_1 &\approx \frac{1}{4} + \frac{3}{4}K_2 \end{aligned} \tag{6.27}$$

- For slots b.),

$$\lambda_{s,r} \approx \frac{2h_{s,r}K_1}{3(b_1+b_2)} + \left(\frac{h_{os,r}}{b_{os,r}} + \frac{h_o}{b_1} + \frac{3h_w}{b_1 + 2b_{os,r}}\right)K_2 \tag{6.28}$$

- For slots c.),

$$\lambda_r = 0.785 - \frac{b_{or}}{2b_1} + \frac{h_{or}}{b_{or}} \approx 0.66 + \frac{h_{or}}{b_{or}} \tag{6.29}$$

- For slots d.),

$$\lambda_r = \frac{h_r}{3b_1}\left(1 - \frac{\pi b_1^2}{8A_b}\right)^2 + 0.66 - \frac{h_{or}}{2b_1} + \frac{h_{or}}{b_{or}} \tag{6.30}$$

where A_b is the bar cross section.

If the slots in Figure 6.9c, d are closed ($h_o = 0$) (Figure 6.9e) the terms h_{or}/b_{or} in Equations (6.29, 6.30) may be replaced by a term dependent on the bar current which saturates the iron bridge.

$$\frac{h_{or}}{b_{or}} \rightarrow \approx 0.3 + 1.12h_{or}\frac{10^3}{I_b^2}; \ I_b > 5b_1 10^3; \ b_1 \ \text{in} \ [m] \tag{6.31}$$

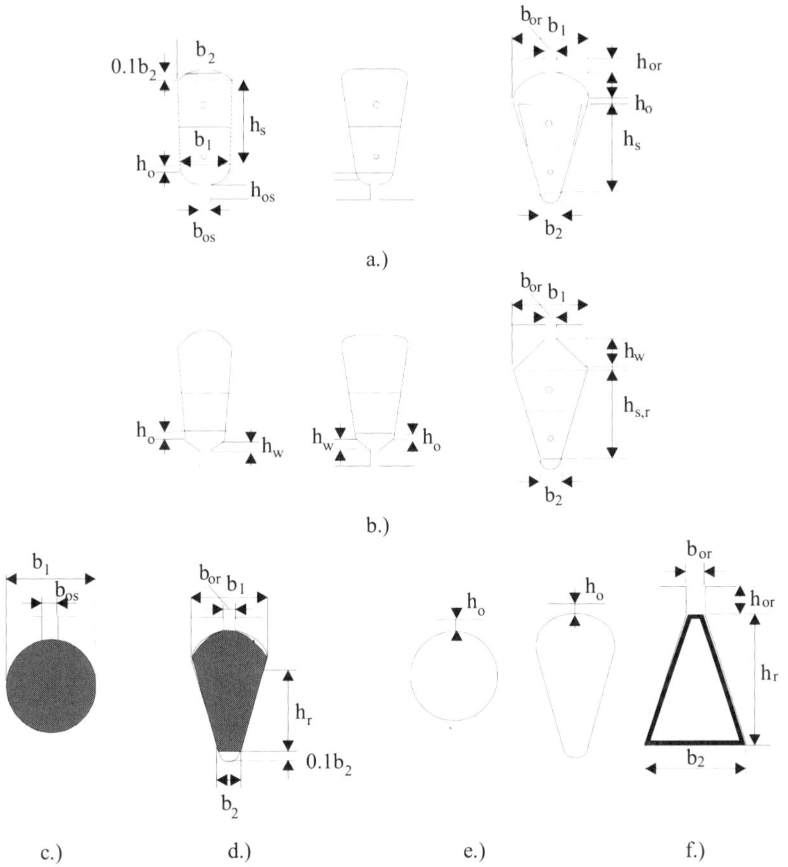

Figure 6.9 Rounded slots: oval, trapezoidal, and round

This is only an empirical approximation for saturation effects in closed rotor slots, potentially useful for very preliminary design purposes.

For the trapezoidal slot (Figure 6.9f), typical for deep rotor bars in high power IMs, by conformal transformations, the slot permeance is, approximately [3]

$$\lambda_r = \frac{1}{\pi}\left[\ln\frac{\left(\frac{b_2}{b_{or}}\right)^2 - 1}{4\frac{b_2}{b_{or}}} + \frac{\left(\frac{b_2}{b_{or}}\right)^2 + 1}{\frac{b_2}{b_{or}}}\cdot\ln\frac{\frac{b_2}{b_{or}} - 1}{\frac{b_2}{b_{or}} + 1}\right] + \frac{h_{or}}{b_{or}} \tag{6.32}$$

The term in square brackets may be used to calculate the geometrical permeance of any trapezoidal slot section (wedge section, for example).

Finally for stator (and rotors) with radial ventilation ducts (channels) additional slot leakage terms have to be added. [8]

For more complicated rotor cage slots used in high skin effect (low starting current, high starting torque) applications, where the skin effect is to be considered, pure analytical solutions are hardly feasible, although many are still in industrial use. Realistic computer-aided methods are given in Chapter 8.

6.6 ZIG-ZAG AIRGAP LEAKAGE INDUCTANCES

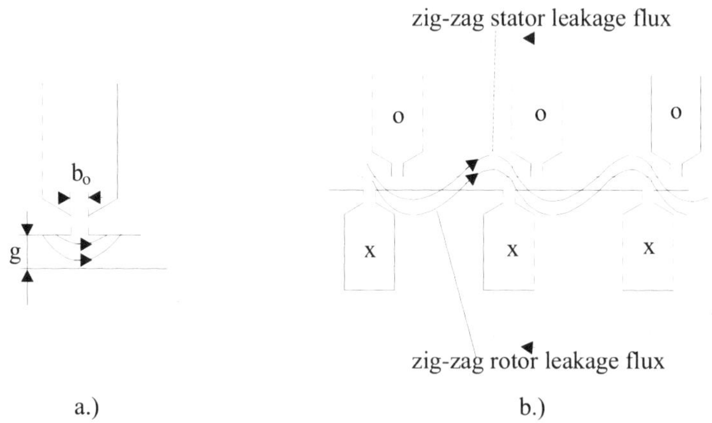

Figure 6.10 Airgap a.) and zig-zag b.) leakage fields

The airgap flux does not reach the other slotted structure (Figure 6.10a) while the zig-zag flux "snakes" out through the teeth around slot openings.

In general, they may be treated together either by conformal transformation or by FEM. From conformal transformations, the following approximation is given for the geometric permeance $\lambda_{zs,r}$ [3]

$$\lambda_{zs,r} \approx \frac{5gK_c / b_{os,r}}{5 + 4gK_c / b_{os,r}} \cdot \frac{3\beta_y + 1}{4} < 1.0; \ \beta_y = 1 \text{ for cage rotors} \tag{6.32}$$

The airgap zig-zag leakage inductance per phase in stator-rotor is

$$L_{zls} \approx 2\mu_0 \frac{W_1 L_e}{p_1 q} \lambda_{zs,r} \tag{6.33}$$

In [4], different formulas are given

$$L_{zls} = L_{1m} \cdot \frac{\pi^2 p_1^{\,2}}{12 N_s^{\,2}} \left[1 - \frac{a(1+a)(1-K')}{2K'}\right] \tag{6.34}$$

for the stator, and

$$L_{zlr} = L_{1m} \cdot \frac{\pi^2 p_1^{\ 2}}{12 N_s^{\ 2}} \left[\frac{N_s^{\ 2}}{N_r^{\ 2}} - \frac{a(1+a)(1-K')}{2K'} \right] \tag{6.35}$$

for the rotor with $K' = 1/K_c$, $a = b_{ts,r}/\tau_{s,r}$, $\tau_{s,r}$ = stator (rotor) slot pitch, $b_{ts,r}$–stator (rotor) tooth-top width.

It should be noticed that while expression (6.32) is dependent only on the airgap/slot opening, in (6.34) and (6.35) the airgap enters directly the denominator of L_{1m} (magnetization inductance) and, in general, (6.34) and (6.35) includes the number of slots of stator and rotor, N_s and N_r.

As the term in parenthesis is a very small number an error here will notably "contaminate" the results. On the other hand, iron saturation will influence the zig-zag flux path, but to a much lower extent than the magnetization flux as the airgap is crossed many times (Figure 6.10b). Finally, the influence of chorded coils is not included in (6.34) to (6.35). We suggest the use of an average of the two expressions (6.33) and (6.34 or 6.35).

In Chapter 7 we revisit this subject for heavy currents (at standstill) including the actual saturation in the tooth tops.

Example 6.2. Zig-zag leakage inductance

For the machine in Example 6.1, with $g = 0.5 \cdot 10^{-3}$m, $b_{os} = 6g$, $b_{or} = 3g$, $K_c = 1.32$, $L_{1m} = 0.1711$H, $p_1 = 2$, $N_s = 36$ stator slots, $N_r = 30$ rotor slots, stator bore $D_i = 0.102$m, $B_y = y/\tau = 8/9$ (chorded coils), and $W_1 = 300$ turns/phase, let us calculate the zig-zag leakage inductance both from (6.32 – 6.33) and (6.34 – 6.35).

Solution. Let us prepare first the values of $K' = 1/K_c = 1/1.32 = 0.7575$.

$$a_{s,r} = \frac{\tau_{s,r} - b_{os,r}}{\tau_{s,r}} =$$

$$= \begin{cases} 1 - \dfrac{b_{os} N_s}{\pi D_i} = 1 - \dfrac{6 \cdot 0.5 \cdot 10^{-3} \cdot 36}{\pi \cdot 0.102} = 0.6628 \\ 1 - \dfrac{b_{or} N_r}{\pi (D_i - 2g)} = 1 - \dfrac{3 \cdot 0.5 \cdot 10^{-3} \cdot 30}{\pi \cdot 0.101} = 0.858 \end{cases} \tag{6.37}$$

From (6.32),

$$\lambda_{Zs} = \frac{\dfrac{5 \cdot 0.5 \cdot 10^{-3} \cdot 1.32}{6 \cdot 0.1 \cdot 10^{-3}}}{5 + \dfrac{4 \cdot 0.1 \cdot 1.32}{6 \cdot 0.1}} = 0.187 \tag{6.38}$$

$$\lambda_{Zr} = \frac{\dfrac{5 \cdot 0.1 \cdot 10^{-3} \cdot 1.32}{3 \cdot 0.1 \cdot 10^{-3}}}{5 + \dfrac{4 \cdot 0.1 \cdot 1.32}{3 \cdot 0.1}} = 0.101 \tag{6.39}$$

The zig-zag inductances per phase $L_{zls,r}$ are calculated from (6.33).

$$L_{zls,r} = \begin{cases} \dfrac{2\cdot 1.256\cdot 10^{-6}\cdot 300^2\cdot 0.12}{2\cdot 3}\cdot 0.187 = 8.455\cdot 10^{-4}\,\mathrm{H} \\ \dfrac{2\cdot 1.256\cdot 10^{-6}\cdot 300^2\cdot 0.12}{2\cdot 3}\cdot 0.101 = 4.566\cdot 10^{-4}\,\mathrm{H} \end{cases} \tag{6.40}$$

Now from (6.34) – (6.35),

$$L_{zls} = 0.1711\cdot\frac{\pi^2 2^2}{12\cdot 36^2}\left[1-\frac{0.6628(1+0.6628)(1-0.7575)}{2\cdot 0.7575}\right] = 3.573\cdot 10^{-4}\,\mathrm{H} \tag{6.41}$$

$$L_{zlr} = 0.1711\cdot\frac{\pi^2 2^2}{12\cdot 36^2}\left[\left(\frac{36}{30}\right)^2-\frac{0.858(1+0.858)(1-0.7575)}{2\cdot 0.7575}\right] = 3.723\cdot 10^{-4}\,\mathrm{H} \tag{6.42}$$

All values are small in comparision to L_{1m} = $1711\cdot 10^{-4}$ H, but there are notable differences between the two methods. In addition it may be inferred that the zig-zag flux leakage also includes the differential leakage flux.

6.7. END-CONNECTION LEAKAGE INDUCTANCE

As seen in Figure 6.11, the three-dimensional character of end connection field makes the computation of its magnetic energy and its leakage inductance per phase a formidable task.

Analytical field solutions need bold simplifications. [5]. Biot-Savart inductance formula [6] and 3D FEM have all been also tried for particular cases.

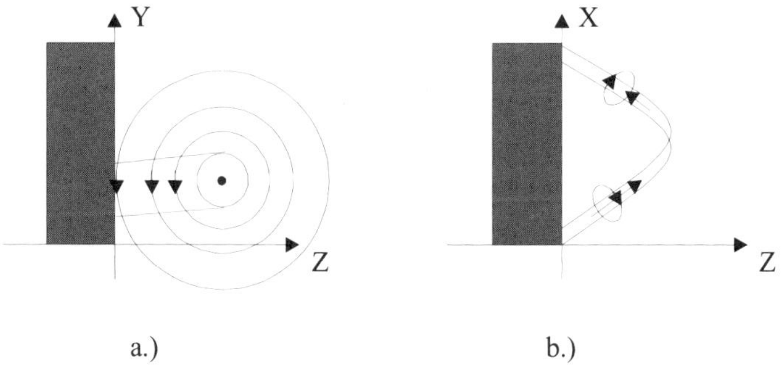

Figure 6.11 Three-dimensional end connection field

Some widely used expressions for the end connection geometrical permeances are as follows:

- Single-layer windings (with end turns in two "stores").

$$\lambda_{es,r} = 0.67 \cdot \frac{q_{s,r}}{L_e}\left(l_{es,r} - 0.64\tau\right) \tag{6.43}$$

- Single-layer windings (with end connections in three "stores").

$$\lambda_{es,r} = 0.47 \cdot \frac{q_{s,r}}{L_e}\left(l_{es,r} - 0.64\tau\right) \tag{6.44}$$

- Double-layer (or single-layer) chain windings.

$$\lambda_{es,r} = 0.34 \cdot \frac{q_{s,r}}{L_e}\left(l_{es,r} - 0.64y\right) \tag{6.45}$$

with $q_{s,r}$–slots/pole/phase in the stator/rotor, L_e–stack length, y–coil throw, $l_{es,r}$–end connection length per motor side.

For cage rotors,

- With end rings attached to the rotor stack.

$$\lambda_{ei} \approx \frac{2 \cdot 3 \cdot D_{ir}}{4N_r L_e \sin^2\left(\frac{\pi p_1}{N_r}\right)} l_g 4.7 \frac{D_{ir}}{a + 2b} \tag{6.46}$$

- With end rings distanced from the rotor st.ck:

$$\lambda_{ei} \approx \frac{2 \cdot 3 \cdot D_i}{4N_r L_e \sin^2\left(\frac{\pi p_1}{N_r}\right)} l_g 4.7 \frac{D_i}{2(a + b)} \tag{6.47}$$

with a and b the ring axial and radial dimensions, and D_{ir} the average end ring diameter.

6.8. SKEWING LEAKAGE INDUCTANCE

In Chapter 4 on windings, we did introduce the concept of skewing (uncompensated) rotor mmf, variable along axial length, which acts along the main flux path and produces a flux which may be considered of leakage character. Its magnetic energy in the airgap may be used to calculate the equivalent inductance.

This skewing inductance depends on the local level of saturation of rotor and stator teeth and cores, and acts simultaneously with the magnetization flux, which is phase shifted with an angle dependent on axial position and slip. We will revisit this complex problem in Chapter 8. In a first approximation, we can make use of the skewing factor K_{skew} and define $L_{skew,r}$ as

$$L_{skew,r} = \left(1 - K^2_{skew}\right)L_{1m} \tag{6.48}$$

$$K_{skew} = \frac{\sin \alpha_{skew}}{\alpha_{skew}}; \; \alpha_{skew} = \frac{c}{\tau}\frac{\pi}{2} \tag{6.49}$$

As this inductance "does not act" when the rotor current is zero, we feel it should all be added to the rotor.

Finally, the total leakage inductance of stator (rotor) is:

$$L_{ls} = L_{dls} + L_{zls} + L_{sls} + L_{els} = 2\mu_0 \frac{W_1 L_e}{pq} \sum \lambda_{si} \tag{6.50}$$

$$L_{lr} = L_{dlr} + L_{zlr} + L_{slr} + L_{elr} + L_{skew,r} \tag{6.51}$$

Although we discussed rotor leakage inductance as if basically reduced to the stator, this operation is due later in this chapter.

6.9. ROTOR BAR AND END RING EQUIVALENT LEAKAGE INDUCTANCE

The differential leakage L_{dlr}, zig-zag leakage $L_{zl,r}$ and $L_{skew,r}$ in (6.51) are already considered in stator terms (reduced to the stator). However, the terms L_{slr} and L_{elr} related to rotor slots (bar) leakage and end connection (end ring) leakage are not clarified enough.

To do so, we use the equivalent bar (slot) leakage inductance expression L_{be},

$$L_{be} = L_b + 2L_i \tag{6.52}$$

where L_b is the slot (bar) leakage inductance and L_i is the end ring (end connection) segment leakage inductance (Figure 6.13)

Using (6.18) for one slot with $n_c = 1$ conductors yields:

$$L_b = \mu_0 L_b \lambda_b \tag{6.53}$$

$$L_{ei} = \mu_0 L_i \lambda_{er} \tag{6.54}$$

where λ_b is the slot geometrical permeance, for a single layer in slot as calculated in paragraphs 6.4 and 6.5 for various slot shapes and λ_{er} is the geometrical permeance of end ring segment as calculated in (6.46) and (6.47).

Now all it remains to obtain L_{slr} and L_{elr} in (6.51), is to reduce L_{be} in (6.52) to the stator. This will be done in paragraph 6.13.

6.10. BASIC PHASE RESISTANCE

The stator resistance R_s is plainly

$$R_s = \rho_{Co} l_c \frac{W_a}{a} \frac{1}{A_{cos}} K_R \tag{6.55}$$

where ρ_{Co} is copper resistivity ($\rho_{Co} = 1.8 \cdot 10^{-8}\Omega$m at 25^{0}C), l_c the turn length:

$$l_c = 2L_e + 2b + 2l_{ec} \tag{6.56}$$

b–axial length of coil outside the core per coil side; l_{ec}–end connection length per stack side, L_e–stack length; A_{cos}–actual conductor area, a–number of current paths, W_a–number of turns per path. With a = 1, $W_a = W_1$ turns/phase. K_R is the ratio between the a.c. and d.c. resistance of the phase resistance.

$$K_R = \frac{R_{sac}}{R_{sdc}} \tag{6.57}$$

For 50 Hz in low and medium power motors, the conductor size is small with respect to the field penetration depth δ_{Co} in it.

$$\delta_{Co} = \sqrt{\frac{\rho_{Co}}{\mu_0 \pi f_1}} < d_{Co} \tag{6.58}$$

In other words, the skin effect is negligible. However, in high frequency (speed) special IMs, this effect may be considerable unless many thin conductors are transposed (as in litz wire). On the other hand, in large power motors, there are large cross sections (even above 60 mm^2) where a few elementary conductors are connected in parallel, even transposed in the end connection zone, to reduce the skin effect (Figure 6.12). For K_R expressions, check Chapter 9. For wound rotors, (6.55) is valid.

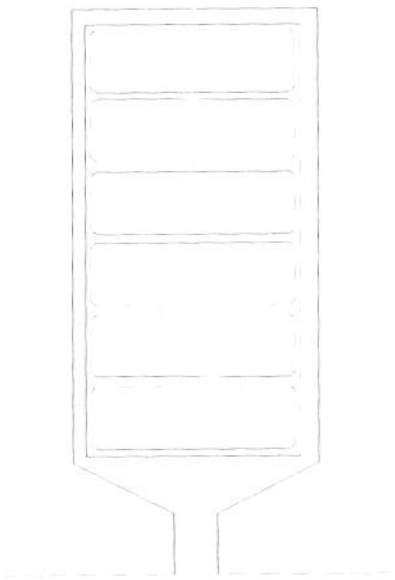

Figure 6.12 Multi-conductor single-turn coils for high power IMs

6.11. THE CAGE ROTOR RESISTANCE

The rotor cage geometry is shown in Figure 6.13.

Let us denote by R_b the bar resistance and by R_i the ring segment resistance:

$$R_b = \rho_b \frac{l_b}{A_b}; \quad R_i = \rho_i \frac{l_i}{A_i}; \quad A_i = a \cdot b \tag{6.59}$$

As the emf in rotor bars is basically sinusoidal, the phase shift α_{er} between neighbouring bars emf is

$$\alpha_{er} = \frac{2\pi p_1}{N_r} \tag{6.60}$$

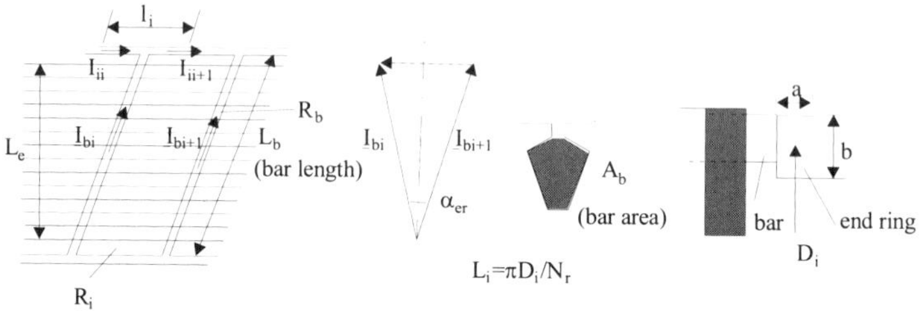

Figure 6.13. Rotor cage geometry

The current in a bar, I_{bi}, is the difference between currents in neighboring ring segments I_{ii} and I_{ii+1} (Figure 6.13).

$$I_{bi} = I_{ii+1} - I_{ii} = 2I_i \sin(\alpha_{er}/2) \tag{6.61}$$

The bar and ring segments may be lumped into an equivalent bar with a resistance R_{be}.

$$R_{be} I_b^2 = R_b I_b^2 + 2R_i I_i^2 \tag{6.62}$$

With (6.61), Equation (6.62), leads to

$$R_{be} = R_b + \frac{R_i}{2\sin^2\left(\frac{\pi p_1}{N_r}\right)} \tag{6.63}$$

When the number of rotor slot/pole pair is small or fractionary (6.49) becomes less reliable.

Example 6.3. Bar and ring resistance

For $N_r = 30$ slots per rotor $2p_1 = 4$, a bar current $I_b = 1000$ A with a current density $j_{cob} = 6$ A/mm^2 in the bar and $j_{coi} = 5$ A/mm^2 in the end ring, the average ring diameter $D_{ir} = 0.15$ m, $l_b = 0.14$ m, let us calculate the bar, ring cross section, the end ring current, and bar and equivalent bar resistance.

Solution

The bar cross-section A_b is

$$A_b = \frac{I_b}{j_{cob}} = 1000/6 = 166\text{mm}^2 \tag{6.64}$$

The current in the end ring I_i (from (6.61)) is:

$$I_i = \frac{I_b}{2\sin\alpha_{er}/2} = \frac{1000}{2\sin\frac{2\pi}{30}} = 2404.86A \tag{6.65}$$

In general, the ring current is greater than the bar current. The end ring cross-section A_i is

$$A_i = \frac{I_i}{j_{coi}} = \frac{2404.86}{5} = 480.97mm^2 \tag{6.66}$$

The end ring segment length,

$$L_i = \frac{\pi D_{ir}}{N_r} = \frac{\pi \cdot 0.150}{30} = 0.0157m \tag{6.67}$$

The end ring segment resistance R_i is

$$R_i = \rho_{Al}\frac{L_i}{A_i} = \frac{3\cdot 10^{-8}\cdot 0.0157}{480.97\cdot 10^{-6}} = 0.9\cdot 10^{-6}\Omega \tag{6.68}$$

The rotor bar resistance R_b is

$$R_b = \rho_{Al}\frac{L_b}{A_b} = \frac{3\cdot 10^{-8}\cdot 0.15}{166\cdot 10^{-6}} = 2.7\cdot 10^{-5}\Omega \tag{6.69}$$

Finally, (from (6.63)), the equivalent bar resistance R_{be} is

$$R_{be} = 2.7\cdot 10^{-5} + \frac{0.9\cdot 10^{-6}}{2\sin\frac{2\pi}{30}} = 3.804\cdot 10^{-5}\Omega \tag{6.70}$$

As we can see from (6.70), the contribution of the end ring to the equivalent bar resistance is around 30%. This is more than a rather typical proportion. So far we did consider that the distribution of the currents in the bars and end rings are uniform. However, there are a.c. currents in the rotor bars and end rings. Consequently, the distribution of the current in the bar is not uniform and depends essentially on the rotor frequency $f_2 = sf_1$.

Globally, the skin effect translates into resistance and slot leakage correction coefficients $K_R > 1$, $K_x < 1$ (see Chapter 9).

In general, the skin effect in the end rings is neglected. The bar resistance R_b in (6.69) and the slot permeance λ_b in (6.53) are modified to

$$R_b = \rho_{Al}\frac{L_b}{A_b}K_R \tag{6.71}$$

$$L_b = \mu_0 L_b \lambda_b K_x \tag{6.72}$$

6.12. SIMPLIFIED LEAKAGE SATURATION CORRECTIONS

Further on, for large values of currents (large slips), the stator (rotor) tooth tops tend to be saturated by the slot leakage flux (Figure 6.14).

Neglecting the magnetic saturation along the tooth height, we may consider that only the tooth top saturates due to the slot neck leakage flux produced by the entire slot mmf.

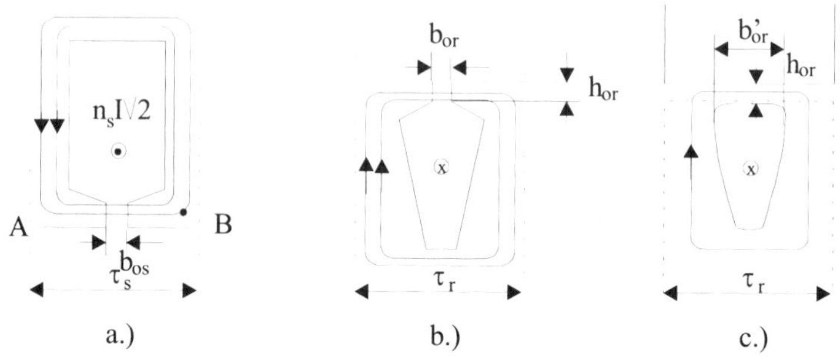

Figure 6.14 Slot neck leakage flux
a.) stator semiclosed slot b.) rotor semiclosed slot c.) rotor closed slot

A simple way to account for this leakage saturation that is used widely in industry consists of increasing the slot opening $b_{os,r}$ by the tooth top length t_t divided by the relative iron permeability μ_{rel} within it.

$$t_t = \tau_{s,r} - b_{os,r}$$
$$b'_{os,r} = b_{os,r} + \frac{t_t}{\mu_{rel}} \tag{6.73}$$

For the closed slot (Figure 6.14), we may use directly (6.31). To calculate μ_{rel}, we apply the Ampere's and flux laws.

$$B_t = \mu_0 H_0$$
$$n_s I\sqrt{2} = t_t H_t + H_0 b_{os,r} \tag{6.74}$$

$$\mu_0 n_s I\sqrt{2} = \mu_0 t_t H_t + B_t b_{os,r}; \; B_0 = B_t \tag{6.75}$$

We have to add the lamination magnetization curve, $B_t(H_t)$, to (6.75) and intersect them (Figure 6.15) to find B_t and H_t, and, finally,

$$\mu_{rel} = \frac{B_t}{\mu_0 H_t} \tag{6.76}$$

H_0–magnetic field in the slot neck.

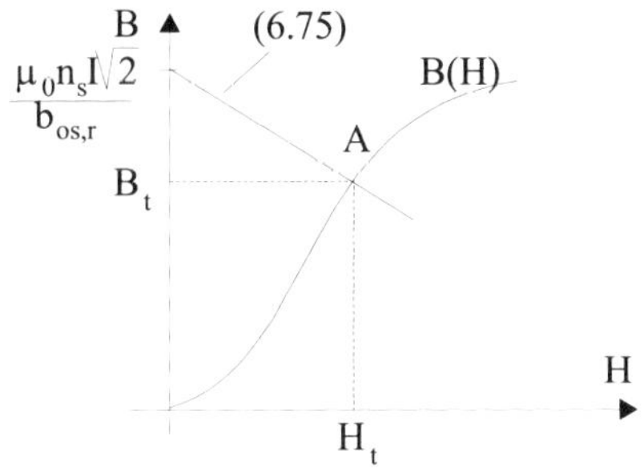

Figure 6.15 Tooth top flux density B_t

Given the slot mmf $n_s I\sqrt{2}$ (the current, in fact), we may calculate iteratively μ_{rel} (6.76) and, from (6.73), the corrected $b_{os,r}$ ($b'_{os,r}$) then to be used in geometrical slot permeance calculations.

As expected, for open slots, the slot leakage saturation is negligible. Simple as it may seem, a second iteration process is required to find the current, as, in general, a voltage type supply is used to feed the IM.

It may be inferred that both differential and skewing leakage are influenced by saturation, in the sense of reducing them. For the former, we already introduced the partial teeth saturation factor $K_{st} = K_{sd}$ to account for it (6.14). For the latter, in (6.48), we have assumed that the level of saturation is implicitly accounted for L_{1m} (that is, it is produced by the magnetization current in the machine). In reality, for large rotor currents, the skewing rotor mmf field, dependent on rotor current and axial position along stack, is quite different from that of the main flux path, at standstill.

However, to keep the formulas simple, Equation (6.73) has become rather standard for design purposes.

Zig-zag flux path tends also to be saturated at high currents. Notice that the zig-zag flux path occupies the teeth tops (Figure 6.10b). This aspect is conventionally neglected as the zig-zag leakage inductance tends to be a small part of total leakage inductance.

So far, we considered the leakage saturation through simple approximate correction factors, and treated skin effects only for rectangular coils. In Chapter 9 we will present comprehensive methods to treat these phenomena for slots of general shapes. Such slot shapes may be the result of design optimization methods based on various cost (objective) functions such as high starting torque, low starting current, large peak torque, etc.

6.13. REDUCING THE ROTOR TO STATOR

The main flux paths embrace both the stator and rotor slots passing through an airgap. So the two emfs per phase $\underline{E}_1$ (Chapter 5.10) and E_2 are

$$\underline{E}_1 = -j\pi\sqrt{2}W_1K_{q1}K_{y1}f_1\underline{\Phi}_1; \quad \Phi_1 = \frac{2}{\pi}\underline{B}_{g1}\tau L_e \tag{6.77}$$

$$\underline{E}_2 = -j\pi\sqrt{2}W_2K_{q2}K_{y2}K_{c2}f_1S\underline{\Phi}_1 \tag{6.77}$$

We have used phasors in (6.77) and (6.78) as E_1, E_2 are sinusoidal in time (only the fundamental is accounted for).

Also in (6.78), E_2 is already calculated as "seen from the stator side," in terms of frequency because only in this case Φ_1 is the same in both stator and rotor phases. In reality the flux in the rotor varies at Sf_1 frequency, while in the stator at f_1 frequency. The amplitude is the same and, with respect to each other, they are at stall.

Now we may proceed as for transformers by dividing the two equations to obtain

$$\frac{E_1}{E_2} = \frac{W_1K_{w1}}{W_2K_{w2}} = K_e = \frac{V_r'}{V_r} \tag{6.79}$$

K_e is the voltage reduction factor to stator.

Eventually the actual and the stator reduced rotor mmfs should be identical:

$$\frac{m_1\sqrt{2}K_{w1}I_r'W_1}{\pi p_1} = \frac{m_2\sqrt{2}K_{w2}I_rW_2}{\pi p_1} \tag{6.80}$$

From (6.80) we find the current reduction factor K_i.

$$K_i = \frac{I_r'}{I_r} = \frac{m_2K_{w2}W_2}{m_1K_{w1}W_1} \tag{6.81}$$

For the rotor resistance and reactance equivalence, we have to conserve the conductor losses and leakage field energy.

$$m_1R_r'I_r'^2 = m_2R_rI_r^2 \tag{6.82}$$

$$m_1\frac{L_{rl}'}{2}I_r'^2 = m_2\frac{L_{rl}}{2}I_r^2 \tag{6.83}$$

So,

$$R_r' = R_r\frac{m_2}{m_1}\frac{1}{K_i^2}; \quad L_{rl}' = L_{rl}\frac{m_2}{m_1}\frac{1}{K_i^2} \tag{6.84}$$

For cage rotors, we may still use (6.81) and (6.84) but with $m_2 = N_r$, $W_2 = ½$, and $K_{w2} = K_{skew}$ (skewing factor); that is, each bar (slot) represents a phase.

$$R_r' = R_{be}\frac{12K_{w1}^2W_1^2}{N_rK_{skew}^2} \tag{6.85}$$

$$L_{be}' = L_{be}\frac{12K_{w1}^2W_1^2}{N_rK_{skew}^2} \tag{6.86}$$

Notice that only the slot and ring leakage inductances in (6.86) have to be reduced to the stator in L_{rl} (6.81).

$$L_{rl}' = \left(L_{dlr} + L_{zlr} + L_{skew,r}\right) + L_{be}' \tag{6.87}$$

Example 6.4. Let us consider an IM with q = 3, p_1 = 2, (N_s = 36slots), W_1 = 300 turns/phase, K_{w1} = 0.965, N_r = 30 slots, one stator slot pitch rotor skewing (c/τ = 1/3q) (K_{skew} = 0.9954), and the rotor bar (slot) and end ring cross sections are rectangular (h_r/b_1 = 2/1), $h_{or} = 1.5\cdot10^{-3}$m, and $b_{or} = 1.5\cdot10^{-3}$m, L_e = 0.12m. Let us find the bar ring resistance and leakage inductance reduced to the stator.

Solution

From Example 6.3, we take D_{ir} = 0.15 m, Ai = a × b = 481 mm² for the end ring. We may assume a/b = ½ and, thus, a = $15.51\cdot10^{-3}$ m and b = 31 mm.

The bar cross-section in example 6.3 is A_b = 166 mm².

$$b_1 = \sqrt{\frac{A_b}{(h_r/b_1)}} = \sqrt{\frac{166}{2}} \approx 9.1\text{mm};\ h_r = 18.2\text{mm} \tag{6.88}$$

From example 6.3 we already know the equivalent bar resistance (6.70) R_{be} = $3.804\cdot10^{-5}$ Ω.

Also from (6.19), the rounded geometrical slot permenace λ_r is,

$$\lambda_r = \lambda_{bar} \approx \frac{h_r}{3b_1} + \frac{h_{or}}{b_{or}} = \frac{18.2}{3\cdot9.1} + \frac{1.5}{1.5} = 1.66 \tag{6.89}$$

From (6.46), the end ring segment geometrical permeance λ_{ei} may be found.

$$\lambda_{ei} = \frac{2.3D_{ir}}{4N_rL_e\sin^2\frac{\pi P_1}{N_r}}\lg\frac{4.7D_i}{a+2b} =$$

$$= \frac{2.3\cdot0.15}{4\cdot30\cdot0.12\cdot\sin^2\frac{\pi 2}{18}}\lg\frac{4.7\cdot0.15}{(1.5+2\cdot31)\cdot10^{-3}} = 0.1964 \tag{6.90}$$

with $L_i = \pi D_i/N_r = \pi\cdot150/30 = 15.7\cdot10^{-3}$m.

Now from (6.39) and (6.40), the bar and end ring leakage inductances are

$$L_b = \mu_0 l_b \lambda_{bar} = 1.256 \cdot 10^{-6} \cdot 0.14 \cdot 1.66 = 0.292 \cdot 10^{-6} H$$
$$L_{li} = \mu_0 l_i \lambda_{ei} = 1.256 \cdot 10^{-6} \cdot 0.0157 \cdot 1.964 = 0.3872 \cdot 10^{-8} H \quad (6.91)$$

The equivalent bar leakage inductance L_{be} (6.52) is written:

$$L_{be} = L_b + 2L_{ei} = 0.292 \cdot 10^{-6} + 2 \cdot 0.3872 \cdot 10^{-8} = 0.2997 \cdot 10^{-6} H \quad (6.92)$$

From (6.85) and (6.86) and we may now obtain the rotor slot (bar) and end ring equivalent resistance and leakage inductance reduced to the stator,

$$R_r' = R_{be} \frac{12 K_{w1}^2 W_1^2}{N_r K_{skew}^2} = \frac{3.804 \cdot 10^{-5} \cdot 12 \cdot 0.965^2 \cdot 300^2}{30 \cdot 0.99^2} = 1.28\Omega \quad (6.93)$$

$$L_{be}' = L_{be} \frac{12 K_{w1}^2 W_1^2}{N_r K_{skew}^2} = 0.2997 \cdot 10^{-6} \cdot 0.36 \cdot 10^{-5} = 1.008 \cdot 10^{-2} H \quad (6.94)$$

As a bonus, knowing from Example 6.3 that the magnetization inductance L_{1m} = 0.1711H, we may calculate the rotor skewing leakage inductance (from 6.48), which is already reduced to the stator because it is a fraction of L_{1m},

$$L_{skew,r} = \left(1 - K_{skew}^2\right) \cdot L_{1m} = \left(1 - 0.9954^2\right) \cdot 0.1711 = 1.57 \cdot 10^{-3} H \quad (6.95)$$

In this particular case the skewing leakage is more than 6 times smaller than the slot (bar) and end ring leakage inductance. Notice also that L_{be}'/L_{1m} = 0.059, a rather practical value.

6.14. SUMMARY

- Besides main path lines which embrace stator and rotor slots, and cross the airgap to define the magnetization inductance L_{1m}, there are leakage fields that encircle either the stator or the rotor conductors.
- The leakage fields are divided into differential leakage, zig-zag leakage, slot leakage, end-turn leakage, and skewing leakage. Their corresponding inductances are calculated from their stored magnetic energy.
- Step mmf harmonic fields through airgap induce emfs in the stator, while the space rotor mmf harmonics do the same. These space harmonics produced emfs have the supply frequency and this is why they are considered of leakage type. Magnetic saturation of stator and rotor teeth reduces the differential leakage.
- Stator differential leakage is minimum for y/τ = 0.8 coil throw and decreases with increasing q (slots/pole/phase).
- A quite general graphic-based procedure, valid for any practical winding, is used to calculate the differential leakage inductance.
- The slot leakage inductance is based on the definition of a geometrical permeance λ_s dependent on the aspect ratio. In general, $\lambda_s \in$ (0.5 to 2.5) to

keep the slot leakage inductance within reasonable limits. For a rectangular slot, some rather simple analytical expressions are obtained even for double-layer windings with chorded, unequal coils.

- The saturation of teeth tops due to high currents at large slips reduces λ_s and it is accounted for by a pertinent increase of slot opening which is dependent on stator (rotor) current (mmf).
- The zig-zag leakage flux lines of stator and rotor mmf snake through airgap around slot openings and close out through the two back cores. At high values of currents (large slip values), the zig-zag flux path, mainly the teeth tops, tends to saturate because the combined action of slot neck saturation and zig-zag mmf contribution.
- The rotor slot skewing leads to the existence of a skewing (uncompensated) rotor mmf which produces a leakage flux along the main paths but its maximum is phase shifted with respect to the magnetization mmf maximum and is dependent on slip and position along the stack length. As an approximation only, a simple analytical expression for an additional rotor-only skewing leakage inductance $L_{skew,r}$ is given.
- Leakage path saturation reduces the leakage inductance.
- The a.c. stator resistance is higher than the d.c. because of skin effect, accounted for by a correction coefficient K_R, calculated in Chapter 9. In most IMs, even at higher power but at 50 (60) Hz, the skin effect in the stator is negligible for the fundamental. Not so for current harmonics present in converter-fed IMs or in high-speed (high-frequency) IMs.
- The rotor bar resistance in squirrel cage motors is, in general, increased notably by skin effect, for rotor frequencies $f_2 = Sf_1 > (4 - 5)$Hz; $K_R > 1$.
- The skin effect also reduces the slot geometrical permeance ($K_x < 1$) and, finally, also the leakage inductance of the rotor.
- The rotor cage (or winding) has to be reduced to the stator to prepare the rotor resistance and leakage inductance for utilization in the equivalent circuit of IMs. The equivalent circuit is widely used for IM performance computation.
- Accounting for leakage saturation and skin effect in a comprehensive way for general shape slots (with deep bars or double rotor cage) is a subject revisited in Chapter 9.
- Now with Chapters 5 and 6 in place, we have all basic parameters– magnetization inductance L_{1m}, leakage inductances L_{sl}, L_{rl}' and phase resistances R_s, R_r'.
- With rotor parameters reduced to the stator, we are ready to approach the basic equivalent circuit as a vehicle for performance computation.

6.15. REFERENCES

1. R. Richter, Electric Machines – Vol.4. – Induction Machines, Verlag Birkhauser, Bassel/Stuttgart, 1954, (in German).
2. B. Heller, V. Hamata, Harmonic Field Effects in Induction Machines, Elsevier Scientific, Amsterdam, 1977.
3. I.B. Danilevici, V.V. Dombrovski, E.I. Kazovski, A.C. Machines Parameters", Science Publishers, St.Petersburg, 1965 (in Russian).
4. P.L. Alger, The Nature of Induction Machines, 2nd edition, Gordon & Breach, New York, 1970.
5. I.A. Voldek, The Computation of End Connection Leakage Inductances of A.C. Machine, Electriceskvo, nr.1, 1963 (in Russian).
6. S. Williamson, M.A. Mueller, Induction Motor End – Winding Leakage Reactance Calculation Using the Biot – Sawart Method, Taking Rotor Currents Into Account, Rec of ICEM 1990, MIT, vol.2., pp.480 – 484.
7. R.H. Engelmann, W.H. Middendorf, Handbook of Eelectric Motors, Marcel Dekker Inc., New York, 1995, Chapter 4.
8. A. Demenko, K. Oberretl, Calculation of Additional Slot Leakage Permeance in Electrical Machines due to Radial Ventilation Ducts, Compel vol.11. no.1, James & James Science Publishers Ltd, 1991, pp.93 – 96(c).

Chapter 7

STEADY STATE EQUIVALENT CIRCUIT AND PERFORMANCE

7.1 BASIC STEADY-STATE EQUIVALENT CIRCUIT

When the IM is fed in the stator from a three-phase balanced power source, the three-phase currents produce a traveling field in the airgap. This field produces emfs both in the stator and in the rotor windings: E_1 and E_{2s}. A symmetrical cage in the rotor may be reduced to an equivalent three-phase winding.

The frequency of E_{2s} is f_2.

$$f_2 = f_1 - np_1 = \left(\frac{f_1}{p_1} - n\right)p_1 = \frac{n_1 - n}{n_1} n_1 p_1 = Sf_1 \qquad (7.1)$$

This is so because the stator mmf travels around the airgap with a speed $n_1 = f_1/p_1$ while the rotor travels with a speed of n. Consequently, the relative speed of the mmf wave with respect to rotor conductors is $(n_1 - n)$ and thus the rotor frequency f_2 in (7.1) is obtained.

Now the emf in the short-circuited rotor "acts upon" the rotor resistance R_r and leakage inductance L_{rl}:

$$\underline{E}_{2s} = S\underline{E}_2 = (R_r + jS\omega_1 L_{rl})\underline{I}_r \qquad (7.2)$$

$$\frac{\underline{E}_{2s}}{S} = \underline{E}_2 = \left(\frac{R_r}{S} + j\omega_1 L_{rl}\right)\underline{I}_r \qquad (7.3)$$

If Equation (7.2) includes variables at rotor frequency $S\omega_1$, with the rotor in motion, Equation (7.3) refers to a circuit at stator frequency ω_1, that is with a "fictious" rotor at standstill.

Now after reducing $\underline{E}_2$, $\underline{I}_r$, R_r,and L_{rl} to the stator by the procedure shown in Chapter 6, Equation (7.3) yields

$$\underline{E}_2' = \underline{E}_1 = \left[R_r' + R_r'\left(\frac{1}{S} - 1\right) + j\omega_1 L_{rl}'\right]\underline{I}_r' \qquad (7.4)$$

$$\frac{E_2}{E_1} = \frac{K_{w2}W_2}{K_{w1}W_1} = K_E;\ K_{w2} = 1,\ W_2 = 1/2 \text{ for cage rotors}$$

$$\frac{\underline{I}_r}{\underline{I}_r'} = \frac{m_1 K_{w1} W_1}{m_2 K_{w2} W_2} = K_I;\ W_2 = 1/2, m_2 = N_r \text{ for cage rotors} \tag{7.5}$$

$$\frac{R_r}{\underline{R}_r'} = \frac{L_{rl}}{\underline{L}_{rl}'} = \frac{K_E}{K_I} \tag{7.6}$$

$$K_{w1} = K_{q1} \cdot K_{y1};\ K_{w2} = K_{q2} \cdot K_{y2} \cdot K_{skew} \tag{7.7}$$

W_1, W_2 are turns per phase (or per current path)
K_{w1}, K_{w2} are winding factors for the fundamental mmf waves
m_1, m_2 are the numbers of stator and rotor phases, N_r is the number of rotor slots

The stator phase equation is easily written:

$$-\underline{E}_1 = \underline{V}_s - \underline{I}_s(R_s + j\omega_1 L_{sl}) \tag{7.8}$$

because in addition to the emf, there is only the stator resistance and leakage inductance voltage drop.

Finally, as there is current (mmf) in the rotor, the emf E_1 is produced concurrently by the two mmfs ($\underline{I}_s$, $\underline{I}_r'$).

$$\underline{E}_1 = -\frac{d\psi_{1m}}{dt} = -j\omega_1 L_{1m}(\underline{I}_s + \underline{I}_r') \tag{7.9}$$

If the rotor is not short-circuited, Equation (7.4) becomes

$$\underline{E}_1 - \frac{\underline{V}_r'}{S} = \left[R_r' + R_r'\left(\frac{1}{S} - 1\right) + j\omega_1 L_{rl}'\right]\underline{I}_r' \tag{7.10}$$

The division of V_r (rotor applied voltage) by slip (S) comes into place as the derivation of (7.10) starts in (7.2) where

$$S\underline{E}_2 - \underline{V}_r = (R_r + jS\omega_1 L_{rl})\underline{I}_r \tag{7.11}$$

The rotor circuit is considered as a source, while the stator circuit is a sink. Now Equations (7.8) – (7.11) constitute the IM equations per phase reduced to the stator for the rotor circuit.

Notice that in these equations there is only one frequency, the stator frequency ω_1, which means that they refer to an equivalent rotor at standstill, but with an additional "virtual" rotor resistance per phase $R_r(1/S-1)$ dependent on slip (speed).

It is now evident that the active power in this additional resistance is in fact the electro-mechanical power of the actual motor

$$P_m = T_e \cdot 2\pi n = 3R_r'\left(\frac{1}{S} - 1\right)(\underline{I}_r')^2 \tag{7.12}$$

with
$$n = \frac{f_1}{p_1}(1 - S) \tag{7.13}$$

$$\frac{\omega_1}{p_1} T_e = \frac{3R_r'(I_r')^2}{S} = P_{elm} \tag{7.14}$$

P_{elm} is called the electromagnetic power, the active power which crosses the airgap, from stator to rotor for motoring and vice versa for generating.

Equation (7.14) provides a definition of slip which is very useful for design purposes:

$$S = \frac{3R_r'(I_r')^2}{P_{elm}} = \frac{P_{Cor}}{P_{elm}} \tag{7.15}$$

Equation (7.15) signifies that, for a given electromagnetic power P_{elm} (or torque, for given frequency), the slip is proportional to rotor winding losses.

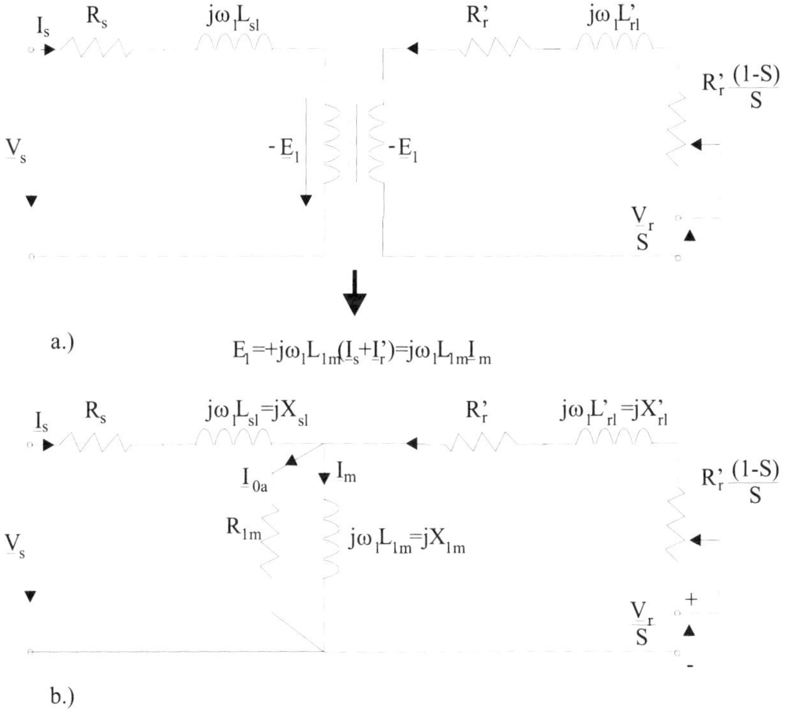

Figure 7.1 The equivalent circuit

Equations (7.8) – (7.11) lead progressively to the ideal equivalent circuit in Figure 7.1.

Still missing in Figure 7.1a are the parameters to account for core losses, additional losses (in the cores and windings due to harmonics), and the mechanical losses.

The additional losses P_{ad} will be left out and considered separately in Chapter 11 as they amount, in general, to up to 3% of rated power in well-designed IM.

The mechanical and fundamental core losses may be combined in a resistance R_{1m} in parallel with X_{1m} in Figure 7.1b, as at least core losses are produced by the main path flux (and magnetization current $\underline{I}_m$). R_{1m} may also be combined as a resistance in series with X_{1m}, for convenience in constant frequency IMs. For variable frequency IMs, however, the parallel resistance R_{1m} varies only slightly with frequency as the power in it (mainly core losses) is proportional to $E_1^2 = \omega_1^2 K_{w1}^2 \Phi_1^2$, which is consistent to eddy current core loss variation with frequency and flux squared.

R_{1m} may be calculated in the design stage or may be found through standard measurements.

$$R_{1m} = \frac{3E_1^2}{P_{iron}} = \frac{3X_{1m}^2 I_m^2}{P_{iron}}; \ I_{oa} << I_m \tag{7.16}$$

7.2 CLASSIFICATION OF OPERATION MODES

The electromagnetic (active) power crossing the airgap P_{elm} (7.14) is positive for $S > 0$ and negative for $S < 0$.

That is, for $S < 0$, the electromagnetic power flows from the rotor to the stator. After covering the stator losses, the rest of it is sent back to the power source. For $\omega_1 > 0$ (7.14) $S < 0$ means negative torque T_e. Also, $S < 0$ means $n > n_1 = f_1/p_1$. For $S > 1$ from the slip definition, $S = (n_1 - n)/n_1$, it means that either $n < 0$ and $n_1(f_1) > 0$ or $n > 0$ and $n_1(f_1) < 0$.

In both cases, as $S > 1$ ($S > 0$), the electromagnetic power $P_{elm} > 0$ and thus flows from the power source into the machine.

On the other hand, with $n > 0$, $n_1(\omega_1) < 0$, the torque T_e is negative; it is opposite to motion direction. That is braking. The same is true for $n < 0$ and $n_1(\omega_1) > 0$. In this case, the machine absorbs electric power through the stator and mechanical power from the shaft and transforms them into heat in the rotor circuit total resistances.

Now for $0 < S < 1$, $T_e > 0$, $0 < n < n_1$, $\omega_1 > 0$, the IM is motoring as the torque acts along the direction of motion.

The above reasoning is summarized in Table 7.1.

Positive $\omega_1(f_1)$ means positive sequence-forward mmf traveling wave. For negative $\omega_1(f_1)$, a companion table for reverse motion may be obtained.

Table 7.1 Operation modes ($f_1/p_1 > 0$)

S	- - - -	0	+ + + +	1	+ + + +
n	+ + + +	f_1/p_1	+ + + +	0	
T_e	0 - - - -	0	+ + + +	+ + + +	0
P_{elm}		0	+ + + +	+ + + +	0
Operation mode	Generator		Motor		Braking

7.3 IDEAL NO-LOAD OPERATION

The ideal no-load operation mode corresponds to zero rotor current. From (7.11), for $I_{r0} = 0$ we obtain

$$S_0\underline{E}_2 - \underline{V}_R = 0; \quad S_0 = \frac{\underline{V}_R}{\underline{E}_2} \tag{7.17}$$

The slip S_0 for ideal no-load depends on the value and phase of the rotor applied voltage $\underline{V}_R$. For $\underline{V}_R$ in phase with E_2: $S_0 > 0$ and, with them in opposite phase, $S_0 < 0$.

The conventional ideal no – load-synchronism, for the short-circuited rotor ($V_R = 0$) corresponds to $S_0 = 0$, $n_0 = f_1/p_1$. If the rotor windings (in a wound rotor) are supplied with a forward voltage sequence of adequate frequency $f_2 = Sf_1$ ($f_1 > 0$, $f_2 > 0$), subsynchronous operation (motoring and generating) may be obtained. If the rotor voltage sequence is changed, $f_2 = sf_1 < 0$ ($f_1 > 0$), supersynchronous operation may be obtained. This is the case of the doubly fed induction machine. For the time being we will deal, however, with the conventional ideal no-load (conventional synchronism) for which $S_0 = 0$.

The equivalent circuit degenerates into the one in Figure 7.2a (rotor circuit is open).

Building the phasor diagram (Figure 7.2b) starts with $\underline{I}_m$, continues with $jX_{1m}\underline{I}_m$, then $\underline{I}_{0a}$

$$\underline{I}_{oa} = \frac{jX_{1m}\underline{I}_m}{R_{1m}} \tag{7.18}$$

and

$$\underline{I}_{s0} = \underline{I}_{oa} + \underline{I}_m \tag{7.19}$$

Finally, the stator phase voltage $\underline{V}_s$ (Figure 7.2b) is

$$\underline{V}_s = jX_{1m}\underline{I}_m + R_s\underline{I}_{s0} + jX_{sl}\underline{I}_{s0} \tag{7.20}$$

The input (active) power P_{s0} is dissipated into electromagnetic loss, fundamental and harmonics stator core losses, and stator windings and space harmonics caused rotor core and cage losses. The driving motor covers the mechanical losses and whatever losses would occur in the rotor core and squirrel cage due to space harmonics fields and hysteresis.

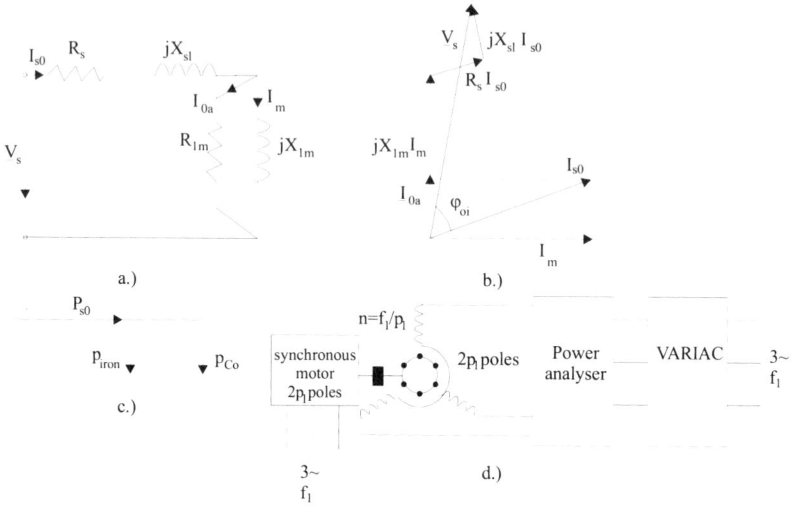

Figure 7.2 Ideal no-load operation ($V_R = 0$):
a.) equivalent circuit b.) phasor diagram c.) power balance d.) test rig

For the time being, when doing the measurements, we will consider only stator core and winding losses.

$$P_{s0} \approx 3R_{1m}I_{0a}^2 + 3R_sI_{s0}^2 = \left(3R_{1m}\frac{X_{1m}^2}{X_{1m}^2 + R_{1m}^2} + 3R_s\right)I_{s0}^2 = \\ = p_{iron} + 3R_sI_{s0}^2 \tag{7.21}$$

From d.c. measurements, we may determine the stator resistance R_s. Then, from (7.21), with P_{s0}, I_{s0} measured with a power analyzer, we may calculate the iron losses p_{iron} for given stator voltage V_s and frequency f_1.

We may repeat the testing for variable f_1 and V_1 (or variable V_1/f_1) to determine the core loss dependence on frequency and voltage.

The input reactive power is

$$Q_{s0} = \left(3X_{1m}\frac{R_{1m}^2}{X_{1m}^2 + R_{1m}^2} + 3X_{sl}\right)I_{s0}^2 \tag{7.22}$$

From (7.21)-(7.22), with R_s known, Q_{s0}, I_{s0}, P_{s0} measured, we may calculate only two out of the three unknowns (parameters): X_{1m}, R_{1m}, X_{sl}.

We know that $R_{1m} >> X_{1m} >> X_{sl}$. However, X_{sl} may be taken by the design value, or the value measured with the rotor out or half the stall rotor ($S = 1$) reactance X_{sc}, as shown later in this chapter.

Consequently, X_{1m} and R_{1m} may be found with good precision from the ideal no-load test (Figure 7.2d). Unfortunately, a synchronous motor with the same number of poles is needed to provide driving at synchronism. This is why the no-load motoring operation mode has become standard for industrial use, while the ideal no-load test is mainly used for prototyping.

Example 7.1 Ideal no-load parameters
An induction motor driven at synchronism ($n = n_1 = 1800$rpm, $f_1 = 60$Hz, $p_1 = 2$) is fed at rated voltage $V_1 = 240$ V (phase RMS) and draws a phase current $I_{s0} = 3$ A, the power analyzer measures $P_{s0} = 36$ W, $Q_{s0} = 700$ VAR, the stator resistance $R_s = 0.1\ \Omega$, $X_{sl} = 0.3\ \Omega$. Let us calculate the core loss p_{iron}, X_{1m}, R_{1m}.
Solution
From (7.21), the core loss p_{iron} is

$$p_{iron} = P_{s0} - 3R_s I_{s0}^2 = 36 - 3\cdot 0.1\cdot 3^2 = 33.3\text{W} \tag{7.23}$$

Now, from (7.21) and (7.22), we get

$$\frac{R_{1m}X_{1m}^2}{X_{1m}^2 + R_{1m}^2} = \frac{P_{s0} - 3R_s I_{s0}^2}{3I_{s0}^2} = \frac{36 - 3\cdot 0.1\cdot 3^2}{3\cdot 3^2} = \frac{33.3}{27} = 1.233\Omega \tag{7.24}$$

$$\frac{R_{1m}^2 X_{1m}}{X_{1m}^2 + R_{1m}^2} = \frac{Q_{s0} - 3X_{sl} I_{s0}^2}{3I_{s0}^2} = \frac{700 - 3\cdot 0.3\cdot 3^2}{3\cdot 3^2} = 25.626\Omega \tag{7.25}$$

Dividing (7.25) by (7.26) we get

$$\frac{R_{1m}}{X_{1m}} = \frac{25.626}{1.233} = 20.78 \tag{7.26}$$

From (7.25),

$$\frac{X_{1m}}{\left(\frac{X_{1m}}{R_{1m}}\right)^2 + 1} = 25.626; \quad \frac{X_{1m}}{(20.78)^2 + 1} = 25.626 \Rightarrow X_{1m} = 25.685 \tag{7.27}$$

R_{1m} is calculated from (7.26),

$$R_{1m} = X_{1m}\cdot 20.78 = 25.626\cdot 20.78 = 533.74\Omega \tag{7.28}$$

By doing experiments for various frequencies f_1 (and V_s/f_1 ratios), the slight dependence of R_{1m} on frequency f1 and on the magnetization current I_m may be proved.

As a bonus, the power factor $\cos\varphi_{0i}$ may be obtained as

$$\cos\varphi_{0i} = \cos\left(\tan^{-1}\left(\frac{Q_{s0}}{P_{s0}}\right)\right) = 0.05136 \tag{7.29}$$

The lower the power factor at ideal no-load, the lower the core loss in the machine (the winding losses are low in this case).

In general, when the machine is driven under load, the value of emf ($E_1 = X_{1m}I_m$) does not vary notably up to rated load and thus the core loss found from ideal no-load testing may be used for assessing performance during loading, through the loss segregation procedure. Note however that, on load, additional losses, produced by space field harmonics,occur. For a precise efficiency computation, these "stray load losses" have to be added to the core loss measured under ideal no-load or even for no-load motoring operation.

7.4 SHORT-CIRCUIT (ZERO SPEED) OPERATION

At start, the IM speed is zero (S = 1), but the electromagnetic torque is positive (Table 7.1), so when three-phase fed, the IM tends to start (rotate); to prevent this, the rotor has to be stalled.

First, we adapt the equivalent circuit by letting S = 1 and R_r' and X_{sl}, X_{rl}' be replaced by their values as affected by skin effect and magnetic saturation (mainly leakage saturation as shown in Chapter 6): $X_{slstart}$, R'_{rstart}, $X'_{rlstart}$ (Figure 7.3).

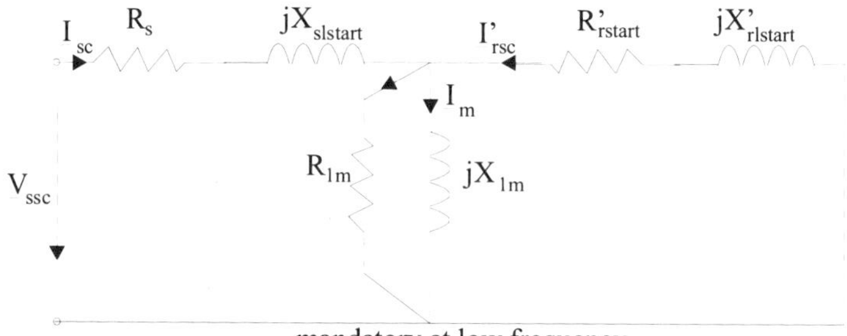

a.)

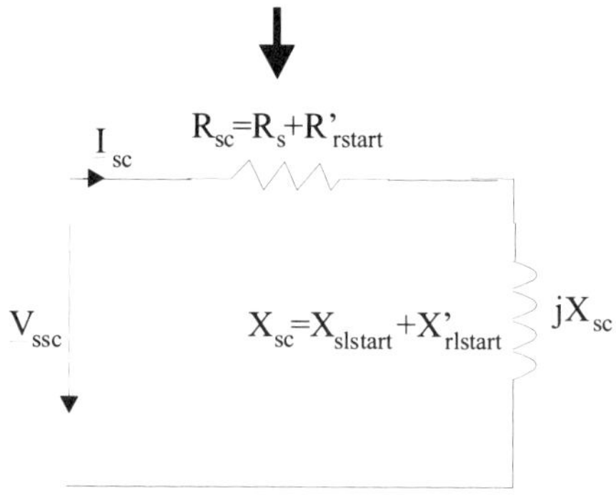

b.)

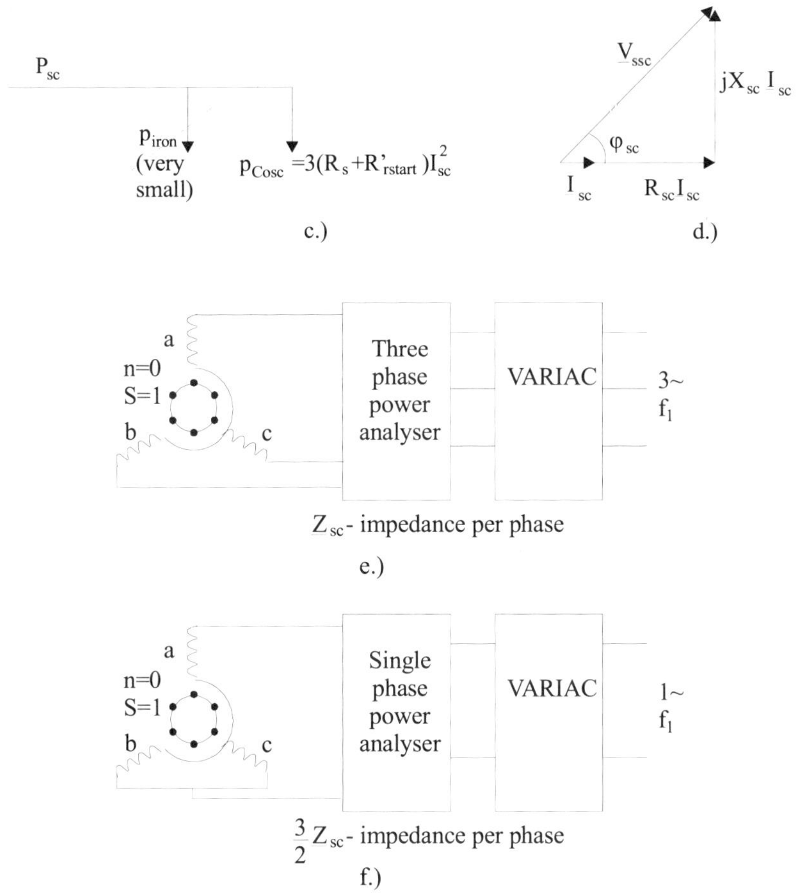

Figure 7.3 Short-circuit (zero speed) operation:
a.) complete equivalent circuit at S = 1, b.) simplified equivalent circuit S = 1, c.) power balance, d.) phasor diagram, e.) three-phase zero speed testing, f.) single-phase supply zero speed testing

For standard frequencies of 50(60) Hz, and above, $X_{1m} >> R'_{rstart}$. Also, $X_{1m} >> X'_{rlstart}$, so it is acceptable to neglect it in the classical short-circuit equivalent circuit (Figure 7.3b).

For low frequencies, however, this is not so; that is, $X_{1m} <> R'_{rstart}$, so the complete equivalent circuit (Figure 7.3a) is mandatory.

The power balance and the phasor diagram (for the simplified circuit of Figure 7.3b) are shown in Figure 7.3c and d. The test rigs for three-phase and single-phase supply testing are presented in Figure 7.3e and f.

It is evident that for single-phase supply, there is no starting torque as we have a non-traveling stator mmf aligned to phase a. Consequently, no rotor stalling is required.

The equivalent impedance is now (3/2)$\underline{Z}_{sc}$ because phase a is in series with phases b and c in parallel. The simplified equivalent circuit (Figure 7.3b) may be used consistently for supply frequencies above 50(60) Hz. For lower frequencies, the complete equivalent circuit has to be known. Still, the core loss may be neglected ($R_{1m} \approx \infty$) but, from ideal no-load test at various voltages, we have to have $L_{1m}(I_m)$ function. A rather cumbersome iterative (optimization) procedure is then required to find R'_{rstart}, $X'_{rlstart}$, and $X_{slstart}$ with only two equations from measurements of R_{sc}, V_{sc}/I_{sc}.

$$P_{sc} = 3R_s I_{sc}^2 + 3R'_{rstart} I'^2_{rsc} \tag{7.30}$$

$$\underline{Z}_{sc} = R_s + jX_{slstart} + \frac{jX_{1m}\left(R'_{rlstart} + jX'_{rlstart}\right)}{R'_{rlstart} + j\left(X_{1m} + X'_{rlstart}\right)} \tag{7.31}$$

This particular case, so typical with variable frequency IM supplies, will not be pursued further. For frequencies above 50(60) Hz the short-circuit impedance is simply

$$\underline{Z}_{sc} \approx R_{sc} + jX_{sc};\ R_{sc} = R_s + R'_{rlstart};\ X_{sc} = X_{slstart} + X'_{rlstart} \tag{7.32}$$

and with P_{sc}, V_{ssc}, I_{sc} measured, we may calculate

$$R_{sc} = \frac{P_{sc}}{3I_{sc}^2};\ X_{sc} = \sqrt{\left(\frac{V_{ssc}}{I_{sc}}\right)^2 - R_{sc}^2} \tag{7.33}$$

for three phase zero speed testing and

$$R_{sc} = \frac{2}{3}\frac{P_{sc\sim}}{I_{sc\sim}^2};X_{sc} = \frac{2}{3}\sqrt{\left(\frac{V_{ssc\sim}}{I_{sc\sim}}\right)^2 - \left(\frac{3}{2}R_{sc}\right)^2} \tag{7.34}$$

for single phase zero speed testing.

If the test is done at rated voltage, the starting current I_{start} $(I_{sc})_{Vsn}$ is much larger than the rated current,

$$\frac{I_{start}}{I_n} \approx 4.5 \div 8.0 \tag{7.35}$$

for cage rotors, larger for high efficiency motors, and

$$\frac{I_{start}}{I_n} \approx 10 \div 12 \tag{7.36}$$

for short-circuited rotor windings.

The starting torque T_{es} is:

$$T_{es} = \frac{3R'_{rstart} I_{start}^2}{\omega_1} p_1 \tag{7.37}$$

with

$$T_{es} = (0.7 \div 2.3) T_{en} \tag{7.38}$$

for cage rotors and

$$T_{es} = (0.1 \div 0.3) T_{en} \tag{7.39}$$

for short-circuited wound rotors.

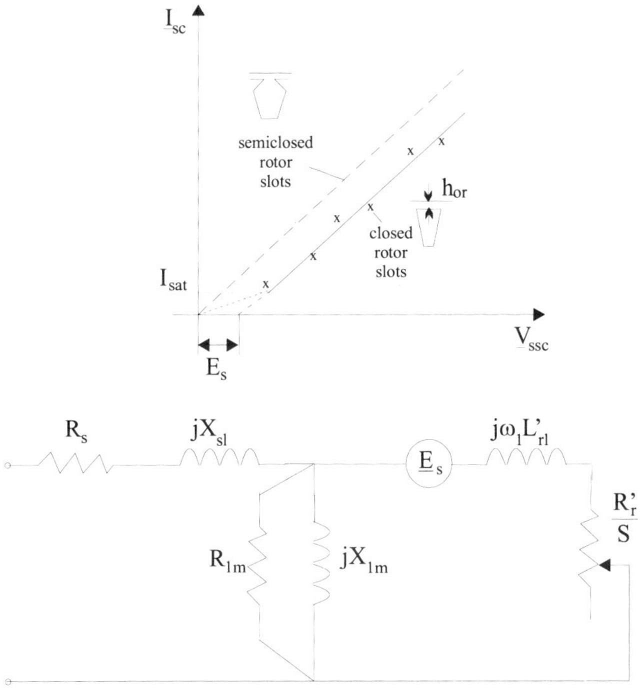

Figure 7.4 Stator voltage versus short-circuit current

Thorough testing of IM at zero speed generally completed up to rated current. Consequently, lower voltages are required, thus avoiding machine overheating. A quick test at rated voltage is done on prototypes or random IMs from, production line to measure the peak starting current and the starting torque. This is a must for heavy starting applications (nuclear power plant cooling pump-motors, for example) as both skin effect and leakage saturation notably modify the motor starting parameters: $X_{slstart}$, $X'_{rlstart}$, R'_{rstart}.

Also, closed slot rotors have their slot leakage flux path saturated at rotor currents notably less than rated current, so a low voltage test at zero speed should produce results as in Figure 7.4.

Intercepting the I_{sc}/V_{sc} curve with abscissa, we obtain, for the closed slot rotor, a non-zero emf E_s.[1] E_s is in the order of 6 to 12V for 220 V phase RMS, 50(60) Hz motors. This additional emf is sometimes introduced in the equivalent circuit together with a constant rotor leakage inductance to account for rotor slot–bridge saturation. E_s is 90^0 ahead of rotor current I_r' and is equal to

$$E_s \approx \frac{4}{\pi}\pi\sqrt{2}f_1(2W_1)\Phi_{bridge}K_{w1};\ \Phi_{bridge} = B_{sbridge}\cdot h_{or}\cdot L_e \tag{7.40}$$

$B_{sbridge}$ is the saturation flux density in the rotor slot bridges ($B_{sbridge}$ = (2 – 2.2)T). The bridge height is h_{or} = 0.3 to 1 mm depending on rotor peripheral speed. The smaller, the better.

A more complete investigation of combined skin and saturation effects on leakage inductances is to be found in Chapter 9 for both semiclosed and closed rotor slots.

Example 7.2 Parameters from zero speed testing

An induction motor with a cage semiclosed slot rotor has been tested at zero speed for V_{ssc} = 30 V (phase RMS, 60 Hz). The input power and phase current are: P_{sc} = 810 kW, I_{sc} = 30 A. The a.c. stator resistance $R_s = 0.1\Omega$. The rotor resistance, without skin effect, is good for running conditions, $R_r = 0.1\Omega$. Let us determine the short-circuit (and rotor) resistance and leakage reactance at zero speed, and the start-to-load rotor resistance ratio due to skin effect.

Solution

From (7.33) we may determine directly the values of short-circuit resistance and reactance, R_{sc} and X_{sc},

$$R_{sc} = \frac{P_{sc}}{3I_{sc}^2} = \frac{810}{3\cdot 30^2} = 0.3\Omega;$$
$$X_{sc} = \sqrt{\left(\frac{V_{ssc}}{I_{sc}}\right)^2 - R_{sc}^2} = \sqrt{\left(\frac{30}{30}\right)^2 - 0.3^2} = 0.954\Omega \tag{7.41}$$

The rotor resistance at start R'_{rstart} is

$$R'_{rstart} = R_{sc} - R_s = 0.3 - 0.1 = 0.2\Omega \tag{7.42}$$

So, the rotor resistance at start is two times that of full load conditions.

$$\frac{R'_{rstart}}{R'_r} = \frac{0.2}{0.1} = 2.0 \tag{7.43}$$

The skin effect is responsible for this increase.

Separating the two leakage reactances $X_{slstart}$ and $X'_{rlstart}$ from X_{sc} is hardly possible. In general, $X_{slstart}$ is affected at start by leakage saturation, if the stator slots are not open, while $X'_{rlstart}$ is affected by both leakage saturation and skin effect. However, at low voltage (at about rated current), the stator leakage and rotor leakage reactances are not affected by leakage saturation; only skin effect affects both R'_{rstart} and $X'_{rlstart}$. In this case it is common practice to consider

$$\frac{1}{2}X_{sc} \approx X'_{rlstart} \tag{7.44}$$

7.5 NO-LOAD MOTOR OPERATION

When no mechanical load is applied to the shaft, the IM works on no-load. In terms of energy conversion, the IM input power has to cover the core, winding, and mechanical losses. The IM has to develop some torque to cover the mechanical losses. So there are some currents in the rotor. However, they tend to be small and, consequently, they are usually neglected.

The equivalent circuit for this case is similar to the case of ideal no-load, but now the core loss resistance R_{1m} may be paralleled by a fictitious resistance R_{mec} which includes the effect of mechanical losses.

The measured values are P_0, I_0, and V_s. Voltages were tested varying from, in general, $1/3V_{sn}$ to $1.2V_{sn}$ through a Variac.

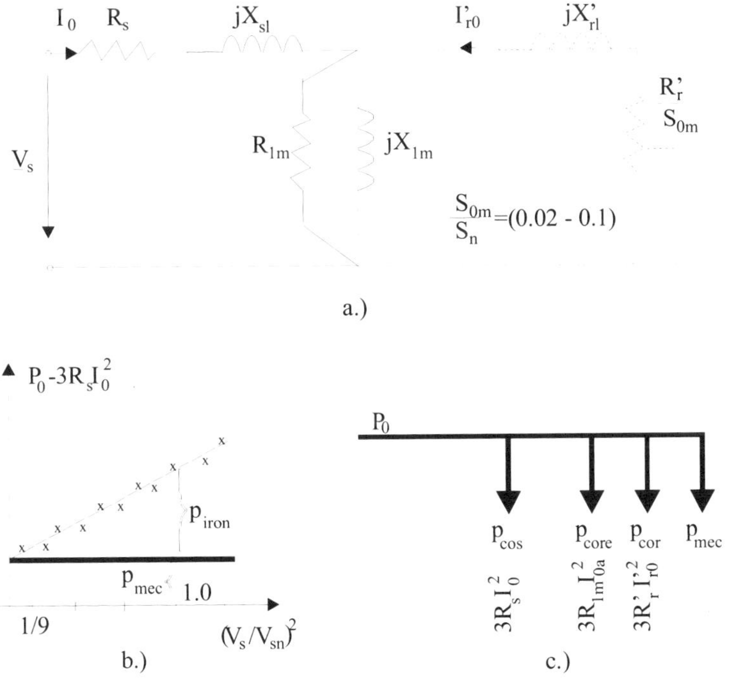

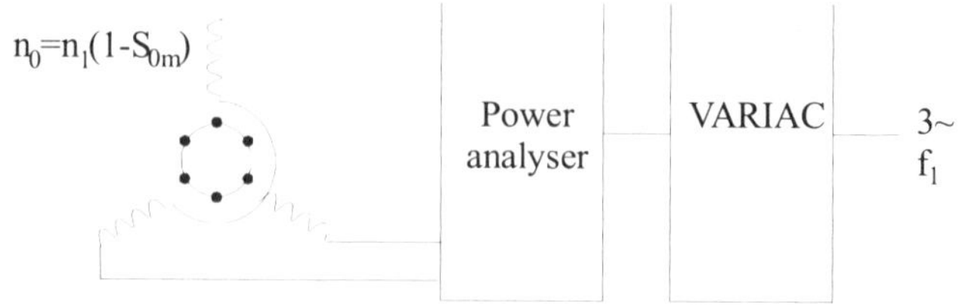

d.)

Figure 7.5 No-load motor operation
a.) equivalent circuit, b.) no-load loss segregation, c.) power balance, d.) test arrangement

As the core loss (eddy current loss, in fact) varies with $(\omega_1\Psi_1)^2$, that is approximately with V_s^2, we may end up with a straight line if we represent the function

$$P_0 - 3R_sI_0^2 = p_{iron} + p_{mec} = f\left(V_s^2\right) \tag{7.44}$$

The intersection of this line with the vertical axis represents the mechanical losses p_{mec} which are independent of voltage. But for all voltages the rotor speed has to remain constant and very close to synchronous speed.

Subsequently the core losses p_{iron} are found. We may compare the core losses from the ideal no-load and the no-load motoring operation modes. Now that we have p_{mec} and the rotor circuit is resistive ($R_r'/S_{0n} >> X_{rl}'$), we may calculate approximately the actual rotor current I_{r0}.

$$\frac{R_r'}{S_{0n}} I_{r0}' \approx V_{sn} \tag{7.45}$$

$$p_{mec} \approx \frac{3R_r' I_{r0}'^2}{S_{0n}}\left(1 - S_{0n}\right) \approx \frac{3R_r' I_{r0}'^2}{S_{0n}} \approx 3V_{sn}I_{r0}' \tag{7.46}$$

Now with R_r' known, S_{0n} may be determined. After a few such calculation cycles, convergence toward more precise values of I_{r0}' and S_{0n} is obtained.

Example 7.3. No-load motoring

An induction motor has been tested at no-load at two voltage levels: V_{sn} = 220V, P_0 = 300W, I_0 = 5A and, respectively, V_s' = 65V, P_0' = 100W, I_0' = 4A. With R_s = 0.1Ω, let us calculate the core and mechanical losses at rated voltage p_{iron}, p_{mec}. It is assumed that the core losses are proportional to voltage squared.

Solution

The power balance for the two cases (7.44) yields

$$P_0 - 3R_sI_0^2 = \left(p_{iron}\right)_n + p_{mec}$$

$$P_0' - 3R_s I_0'^2 = (p_{iron})_n \left(\frac{V_s'}{V_{sn}}\right)^2 + p_{mec} \quad (7.47)$$

$$300 - 3 \cdot 0.1 \cdot 5^2 = (p_{iron})_n + p_{mec} = 292.5$$

$$100 - 3 \cdot 0.1 \cdot 4^2 = (p_{iron})_n \left(\frac{65}{220}\right)^2 + p_{mec} = 95.2$$

From (7.47),

$$\begin{aligned} p_{iron} &= \frac{292.5 - 95.25}{1 - 0.0873} = 216.17\text{W} \\ p_{mec} &= 292.5 - 216.17 = 76.328\text{W} \end{aligned} \quad (7.48)$$

Now, as a bonus, from (7.46) we may calculate approximately the no-load current I_{r0}'.

$$I_{r0}' \approx \frac{p_{mec}}{3V_{sn}} = \frac{76.328}{3 \cdot 220} = 0.1156\text{A} \quad (7.49)$$

It should be noted the rotor current is much smaller than the no-load stator current I_0 = 5A! During the no-load motoring tests, especially for rated voltage and above, due to teeth or/and back core saturation, there is a notable third flux and emf harmonic for the star connection. However, in this case, the third harmonic in current does not occur. The 5th, 7th saturation harmonics occur in the current for the star connection.

For the delta connection, the emf (and flux) third saturation harmonic does not occur. It occurs only in the phase currents (Figure 7.6).

As expected, the no-load current includes various harmonics (due to mmf and slot openings).

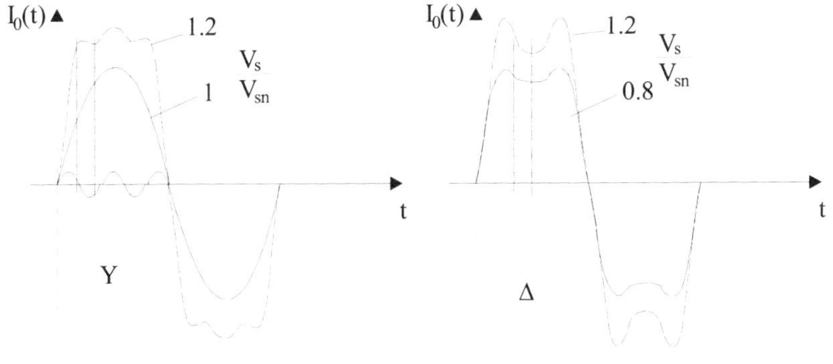

Figure 7.6 No-load currents, a.) star connection, b.) delta connection

They are, in general, smaller for larger q slot/pole/phase chorded coils and skewing of rotor slots. More details in Chapter 10. In general the current harmonics content decreases with increasing load.

7.6 THE MOTOR MODE OF OPERATION

The motor mode of operation occurs when the motor drives a mechanical load (a pump, compressor, drive-train, machine tool, electrical generator, etc.). For motoring, $T_e > 0$ for $0 < n < f_1/p_1$ and $T_e < 0$ for $0 > n > -f_1/p_1$. So the electromagnetic torque acts along the direction of motion. In general, the slip is $0 < S < 1$ (see paragraph 7.2). This time the complete equivalent circuit is used (Figure 7.1).

The power balance for motoring is shown in Figure 7.7.

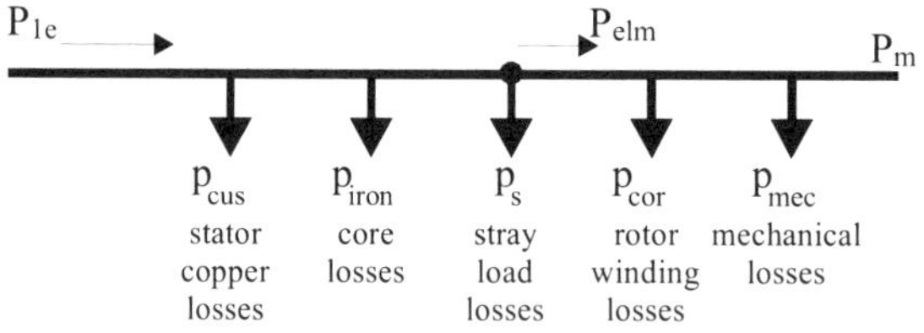

Figure 7.7 Power balance for motoring operation

The motor efficiency η is

$$\eta = \frac{\text{shaft power}}{\text{input electric power}} = \frac{P_m}{P_{1e}} = \frac{P_m}{P_m + p_{Cos} + p_{iron} + p_{Cor} + p_s + p_{mec}} \tag{7.50}$$

The rated power speed n is

$$n = \frac{f_1}{P_1}(1 - S) \tag{7.51}$$

The rated slip is $S_n = (0.08 - 0.006)$, larger for lower power motors.

The stray load losses p_s refer to additional core and winding losses due to space (and eventual time) field and voltage time harmonics. They occur both in the stator and in the rotor. In general, due to difficulties in computation, the stray load losses are still assigned a constant value in some standards (0.5 or 1% of rated power). More on stray losses in Chapter 11.

The slip definition (7.15) is a bit confusing as P_{elm} is defined as active power crossing the airgap. As the stray load losses occur both in the stator and rotor, part of them should be counted in the stator. Consequently, a more realistic definition of slip S (from 7.15) is

$$S = \frac{p_{cor}}{P_{elm} - p_s} = \frac{p_{cor}}{P_{1e} - p_{Cos} - p_{iron} - p_s} \tag{7.52}$$

As slip frequency (rotor current fundamental frequency) $f_{2n} = Sf_{1n}$ it means that in general for $f_1 = 60(50)$ Hz, $f_{2n} = 4.8(4)$ to 0.36(0.3) Hz.

For high speed (frequency) IMs, the value of f_2 is much higher. For example, for $f_{1n} = 300$ Hz (18,000 rpm, $2p_1 = 2$) $f_{2n} = 4 - 8$ Hz, while for $f_{1n} =$ 1,200 Hz it may reach values in the interval of 16 – 32 Hz. So, for high frequency (speed) IMs, some skin effect is present even at rated speed (slip). Not so, in general, in 60(50) Hz low power motors.

7.7 GENERATING TO POWER GRID

As shown in paragraph 7.2, with $S < 0$ the electromagnetic power travels from rotor to stator ($P_{eml} < 0$) and thus, after covering the stator losses, the rest of it is sent back to the power grid. As expected, the machine has to be driven at the shaft at a speed $n > f_1/p_1$ as the electromagnetic torque T_e (and P_{elm}) is negative (Figure 7.8).

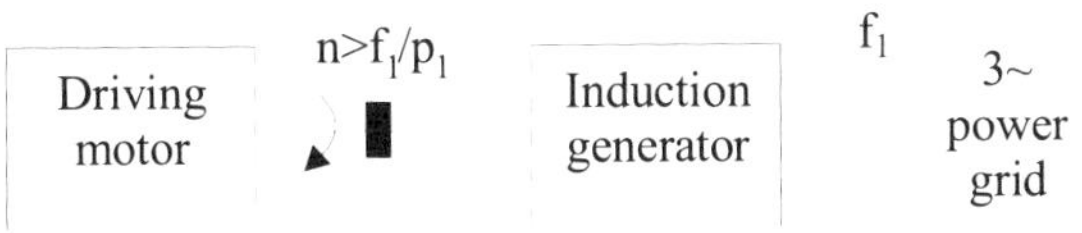

Figure 7.8 Induction generator at power grid

The driving motor could be a wind turbine, a Diesel motor, a hydraulic turbine etc. or an electric motor (in laboratory tests).

The power grid is considered stiff (constant voltage and frequency) but, sometimes, in remote areas, it may also be rather weak.

To calculate the performance, we may use again the complete equivalent circuit (Figure 7.1) with $S < 0$. It may be easily proved that the equivalent resistance R_e and reactance X_e, as seen from the power grid, vary with slip as shown on Figure 7.9.

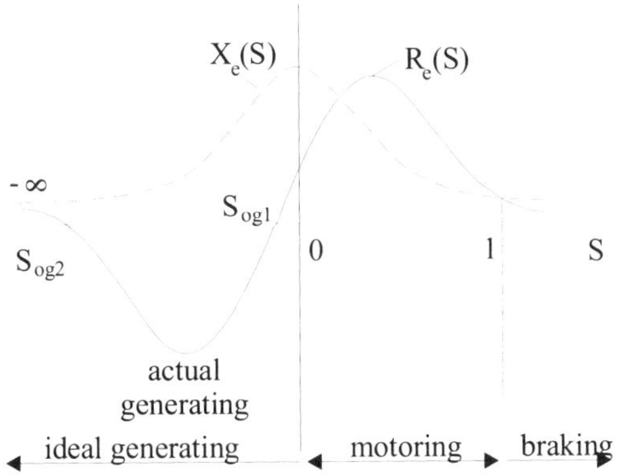

Figure 7.9 Equivalent IM resistance R_e and reactance X_e versus slip S

It should be noted that the equivalent reactance remains positive at all slips. Consequently, the IM draws reactive power in any conditions. This is necessary for a short-circuited rotor. If the IM is doubly fed as shown in Chapter 19, the situation changes as reactive power may be infused in the machine through the rotor slip rings by proper rotor voltage phasing, at $f_2 = Sf_1$.

Between S_{0g1} and S_{0g2} (both negative), Figure 7.9, the equivalent resistance R_e of IM is negative. This means it delivers active power to the power grid.

The power balance is, in a way, opposite to that for motoring (Figure 7.10).

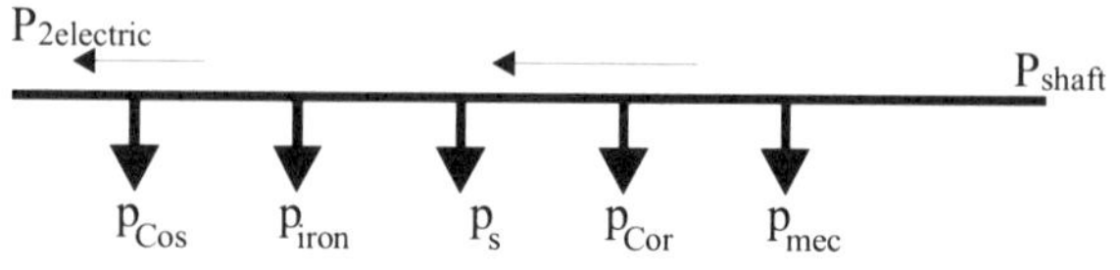

Figure 7.10. Power balance for generating

The efficiency η_g is now

$$\eta = \frac{\text{electric power output}}{\text{shaft power input}} = \frac{P_{2electric}}{P_{1shaft}} = \frac{P_{2electric}}{P_{2electric} + p_{Cos} + p_{iron} + p_{Cor} + p_s + p_{mec}} \quad (7.53)$$

Above the speed
$$n_{max} = \frac{f_1}{P_1}\left(1 - S_{0g2}\right); \; S_{0g2} < 0, \quad (7.54)$$

as evident in Figure 7.9, the IM remains in the generator mode but all the electric power produced is dissipated as loss in the machine itself.

Induction generators are used more and more for industrial generation to produce part of the plant energy at convenient timing and costs. However, as it still draws reactive power, "sources" of reactive power (synchronous generators, or synchronous capacitors or capacitors) are also required to regulate the voltage in the power grid.

Example 7.4 Generator at power grid

A squirrel cage IM with the parameters $R_s = R_r' = 0.6\ \Omega$, $X_{sl} = X_{rl}' = 2\ \Omega$, $X_{1ms} = 60\ \Omega$, and $R_{1ms} = 3\ \Omega$ (the equivalent series resistance to cover the core losses, instead of a parallel one)–Figure 7.11 – works as a generator at the power grid. Let us find the two slip values S_{0g1} and S_{0g2} between which it delivers power to the grid.

Solution

The switch from parallel to series connection in the magnetization branch, used here for convenience, is widely used.

The condition to find the values of S_{0g1} and S_{0g2} for which, in fact, (Figure 7.9) the delivered power is zero is

$$R_e\left(S_{0g}\right) = 0.0 \quad (7.55)$$

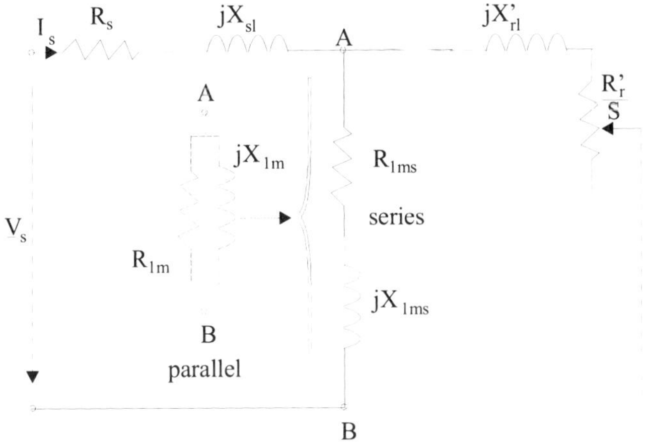

Figure 7.11 Equivalent circuit with series magnetization branch

Equation (7.55) translates into

$$\begin{gathered}(R_{1ms}+R_s)\left(\frac{R_r'}{S_{0g}}\right)^2+\frac{R_r'}{S_{0g}}\left(R_{1ms}{}^2+X_{1ms}{}^2+2R_{1ms}R_s\right)+\\ +R_{1ms}X'_{rl}{}^2+R_{1ms}{}^2R_s+R_s\left(X_{1ms}+X'_{rl}\right)^2=0\end{gathered} \quad (7.56)$$

In numbers:

$$\begin{gathered}3.6\cdot 0.6^2\left(\frac{1}{S_{0g}}\right)^2+0.6\left(3^2+60^2+2\cdot 3\cdot 0.6\right)\frac{1}{S_{0g}}+\\ +3\cdot 2^2+3^2\cdot 0.6+0.6(60+2)^2=0\end{gathered} \quad (7.57)$$

with the solutions $S_{og1} = -0.33\cdot 10^{-3}$ and $S_{0g2} = -0.3877$.

Now with f_1 = 60 Hz and with $2p_1$ = 4 poles, the corresponding speeds (in rpm) are

$$\begin{aligned}n_{og1}&=\frac{f_1}{P_1}\left(1-S_{0g1}\right)\cdot 60=\frac{60}{2}\left(1-\left(-0.33\cdot 10^{-3}\right)\right)\cdot 60=1800.594\text{rpm}\\ n_{og2}&=\frac{f_1}{P_1}\left(1-S_{0g2}\right)\cdot 60=\frac{60}{2}\left(1-\left(-0.3877\right)\right)\cdot 60=2497.86\text{rpm}\end{aligned} \quad (7.58)$$

7.8 AUTONOMOUS INDUCTION GENERATOR MODE

As shown in paragraph 7.7, to become a generator, the IM needs to be driven above no-load ideal speed n_1 ($n_1 = f_1/p_1$ with short-circuited rotor) and to be provided with reactive power to produce and maintain the magnetic field in the machine.

As known, this reactive power may be "produced" with synchronous condensers (or capacitors)–Figure 7.12.

The capacitors are Δ connected to reduce their capacitance as they are supplied by line voltage. Now the voltage V_s and frequency f_1 of the IG on no-load and on load depend essentially on machine parameters, capacitors C_Δ, and speed n. Still $n > f_1/p_1$.

Let us explore the principle of IG capacitor excitation on no-load. The machine is driven at a speed n.

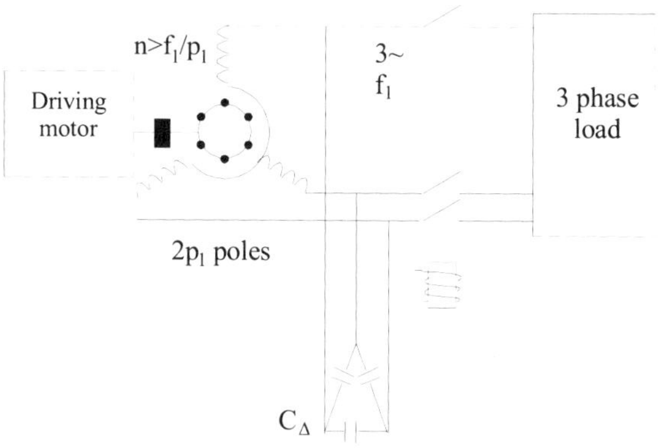

Figure 7.12 Autonomous induction generator (IG) with capacitor magnetization

The d.c. remanent magnetization in the rotor, if any (if none, d.c. magnetization may be provided as a few d.c. current surges through the stator with one phase in series with the other two in parallel), produces an a.c. emf in the stator phases. Then three-phase emfs of frequency $f_1 = p_1 \cdot n$ cause currents to flow in the stator phases and capacitors. Their phase angle is such that they are producing an airgap field that always increases the remanent field; then again, this field produces a higher stator emf and so on until the machine settles at a certain voltage V_s and frequency $f_1 \approx p_1 n$. Changing the speed will change both the frequency f_1 and the no-load voltage V_{s0}. The same effect is produced when the capacitors C_Δ are changed.

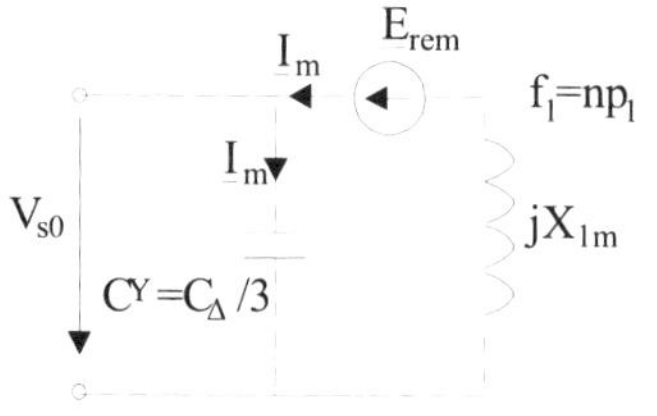

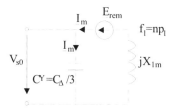

Figure 7.13 Ideal no-load IG per phase equivalent circuit with capacitor excitation

A quasiquantitative analysis of this selfexcitation process may be produced by neglecting the resistances and leakage reactances in the machine. The equivalent circuit degenerates into that in Figure 7.13.

The presence of rotor remanent flux density (from prior action) is depicted by the resultant small emf E_{rem} (E_{rem} = 2 to 4 V) whose frequency is $f_1 = np_1$. The frequency f_1 is essentially imposed by speed.

The machine equation becomes simply

$$V_{s0} = jX_{1m}\underline{I}_m + \underline{E}_{rem} = -j\frac{1}{\omega_1 C_Y}\underline{I}_m = V_{s0}(I_m) \tag{7.59}$$

As we know, the magnetization characteristic (curve)–$V_{s0}(I_m)$–is, in general, nonlinear due to magnetic saturation (Chapter 5) and may be obtained through the ideal no-load test at $f_1 = p_1n$. On the other hand, the capacitor voltage depends linearly on capacitor current. The capacitor current on no-load is, however, equal to the motor stator current. Graphically, Equation (7.59) is depicted on Figure 7.14.

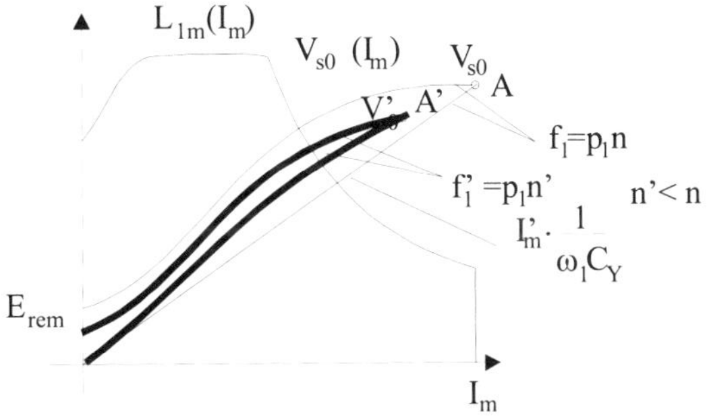

Figure 7.14 Capacitor selfexcitation of IG on no-load

Point A represents the no-load voltage V_{s0} for given speed, n, and capacitor C_Y. If the selfexcitation process is performed at a lower speed n′ (n′ < n), a lower no-load voltage (point A′), V_{s0}', at a lower frequency $f_1' \approx p_1n'$ is obtained.

Changing (reducing) the capacitor C_Y produces similar effects. The selfexcitation process requires, as seen in Figure 7.14, the presence of remanent magnetization ($E_{rem} \neq 0$) and of magnetic saturation to be successful, that is, to produce a clear intersection of the two curves.

When the magnetization curve $V_{1m}(I_m)$ is available, we may use the complete equivalent circuit (Figure 7.1) with a parallel capacitor C_Y over the terminals to explore the load characteristics. For a given capacitor bank and speed n, the output voltage V_s versus load current I_s depends on the load power factor (Figure 7.15).

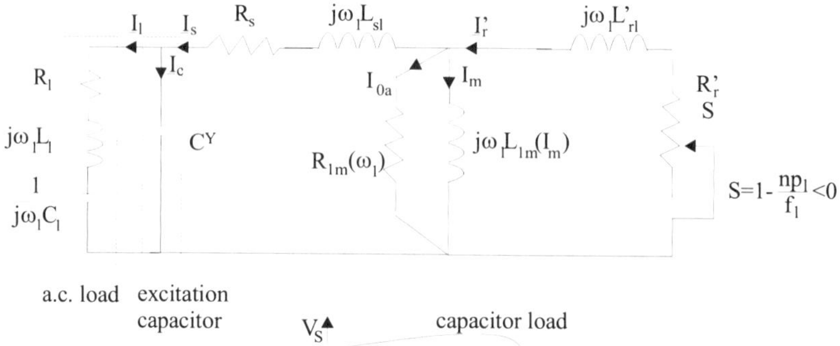

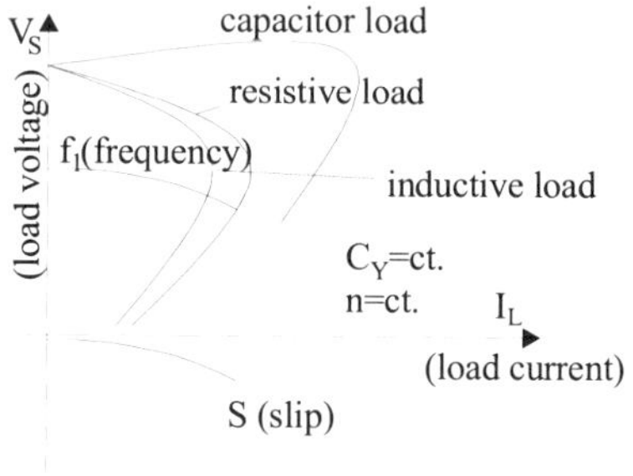

Figure 7.15 Autonomous induction generator on load
a.) equivalent circuit b.) load curves

The load curves in Figure 7.15 may be obtained directly by solving the equivalent circuit in Figure 7.15a for load current, for given load impedance, speed n, and slip S. However, the necessary nonlinearity of magnetization curve $L_{1m}(I_m)$, Figure 7.14, imposes an iterative procedure to solve the equivalent circuit. This is now at hand with existing application software such Matlab, etc.

Above a certain load, the machine voltage drops gradually to zero as there will be a deficit of capacitor energy to produce machine magnetization. Point A on the magnetization curve will drop gradually to zero.

As the load increases, the slip (negative) increases and, for given speed n, the frequency decreases. So the IG can produce power above a certain level of magnetic saturation and above a certain speed for given capacitors. The voltage and frequency decrease notably with load.

A variable capacitor would keep the voltage constant with load. Still, the frequency, by principle, at constant speed, will decrease with load.

Only simultaneous capacitor and speed control may produce constant voltage and frequency for variable load. More on autonomous IGs in Chapter 19.

7.9 THE ELECTROMAGNETIC TORQUE

By electromagnetic torque, T_e, we mean, the torque produced by the fundamental airgap flux density in interaction with the fundamental rotor current.

In paragraph 7.1, we have already derived the expression of T_e (7.14) for the singly fed IM. By singly fed IM, we understand the short-circuited rotor or the wound rotor with a passive impedance at its terminals. For the general case, an $R_l L_l C_l$ impedance could be connected to the slip rings of a wound rotor (Figure 7.16).

Even for this case, the electromagnetic torque T_e may be calculated (7.14) where instead of rotor resistance R_r', the total series rotor resistance $R_r' + R_l'$ is introduced.

Notice that the rotor circuit and thus the R_l, C_l, L_l have been reduced to primary, becoming R_l', C_l', L_l'.

Both rotor and stator circuit blocks in Figure 7.16 are characterized by the stator frequency ω_1.

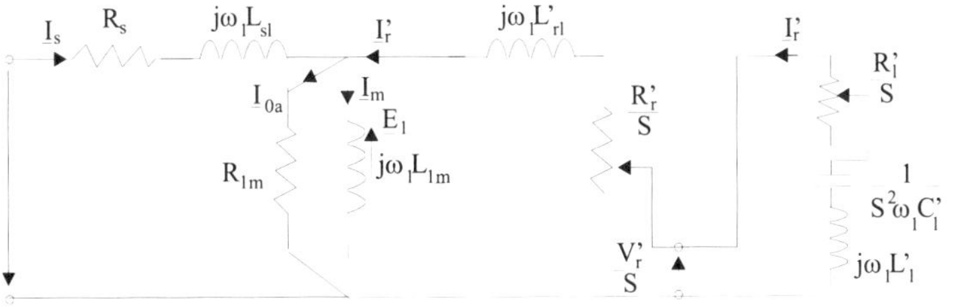

Figure 7.16 Singly – fed IM with additional rotor impedance

Again, Figure 7.16 refers to a fictitious IM at standstill which behaves as the real machine but delivers active power in a resistance $(R_r' + R_l')(1 - S)/S$ dependent on slip, instead of producing mechanical (shaft) power.

$$T_e = \frac{3R_{re}'}{S} I_r'^2 \frac{p_1}{\omega_1}; \; R_{re}' = R_r' + R_l' \tag{7.60}$$

In industry the wound rotor slip rings may be connected through brushes to a three-phase variable resistance or a diode rectifier, or a d.c. – d.c. static converter and a single constant resistance, or a semi (or fully) controlled rectifier and a constant single resistance for starting and (or) limited range speed control as shown later in this paragraph. In essence, however, such devices have models that can be reduced to the situation in Figure 7.16. To further explore the torque, we will consider the rotor with a total resistance R_{re}' but without any additional inductance and capacitance connected at the brushes (for the wound rotor).

From Figure 7.16, with $V_r' = 0$ and R_r' replaced by R_{re}', we can easily calculate the rotor and stator currents $\underline{I}_r'$ and $\underline{I}_s$ as

$$\underline{I}_r' = -\frac{\underline{I}_1 \cdot \underline{Z}_{1m}}{\frac{R_{re}'}{S} + jX_{rl}' + \underline{Z}_{1m}} \tag{7.61}$$

$$\underline{I}_s = \frac{\underline{V}_s}{R_s + jX_{sl} + \dfrac{\underline{Z}_{1m}\left(\dfrac{R_{re}'}{S} + jX_{rl}'\right)}{\dfrac{R_{re}'}{S} + jX_{rl}' + \underline{Z}_{1m}}} \tag{7.62}$$

$$\underline{I}_{0a} + \underline{I}_m = \underline{I}_s + \underline{I}_r' = \underline{I}_{0s} \tag{7.63}$$

$$\underline{Z}_{1m} = \frac{R_{1m} jX_{1m}}{R_{1m} + jX_{1m}} = R_{1ml} + jX_{1ml} \tag{7.64}$$

With constant parameters, we may approximate

$$\frac{\underline{Z}_{1m} + jX_{sl}}{\underline{Z}_{1m}} \approx \frac{X_{1ml} + X_{sl}}{X_{1ml}} = C_1 \approx (1.02 - 1.08) \tag{7.65}$$

In this case

$$\underline{I}_s \approx \frac{\underline{V}_s}{R_s + \dfrac{C_1 R_{re}'}{S} + j(X_{sl} + C_1 X_{rl}')} \tag{7.66}$$

Substituting I_s from (7.66) into (7.61) and then I_r' from (7.61) into (7.60), T_e becomes

$$T_e = \frac{3V_s^2 P_1}{\omega_1} \frac{\dfrac{R_{rl}'}{S}}{\left(R_s + \dfrac{C_1 R_{re}'}{S}\right)^2 + (X_{sl} + C_1 X_{rl}')^2} \tag{7.67}$$

As we can see, T_e is a function of slip S. Its peak values are obtained for

$$\frac{\partial T_e}{\partial S} = 0 \rightarrow S_k = \frac{\pm C_1 R_{re}'}{\sqrt{R_s^2 + (X_{sl} + C_1 X_{rl}')^2}} \tag{7.68}$$

Finally, the peak (breakdown) torque T_{ek} is obtained from the critical slip S_k in (7.67)

$$T_{ek} = (T_e)_{sk} = \frac{3P_1}{\omega_1} \frac{V_s^2}{2C_1\left[R_s \pm \sqrt{R_s^2 + (X_{sl} + C_1 X_{rl}')}\right]} \tag{7.69}$$

The whole torque/slip (or speed) curve is shown in Figure 7.17.

Concerning the T_e versus S curve, a few remarks are in order

- The peak (breakdown) torque T_{ek} value is independent of rotor equivalent (cumulated) resistance, but the critical slip S_k at which it occurs is proportional to it.

- When the stator resistance R_s is notable (low power, subkW motors), the generator breakdown torque is slightly larger than the motor breakdown torque. With R_s neglected, the two are equal to each other.

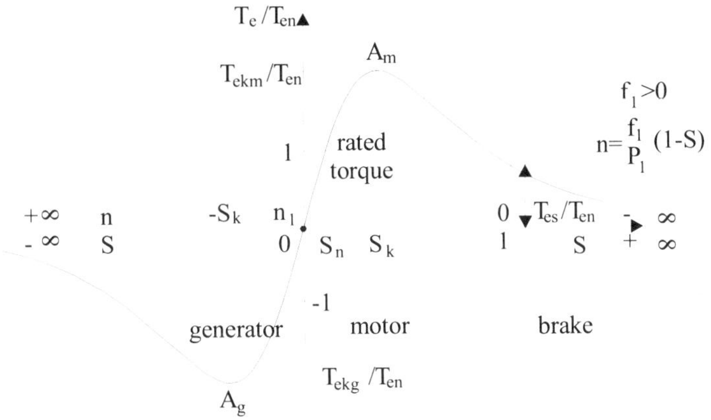

Figure 7.17 The electromagnetic torque T_e (in relative units) versus slip (speed)

- With R_s neglected (large power machines above 10 kW) S_k, from (7.68) and T_{ek} from (7.69), become

$$S_k \approx \frac{\pm C_1 R_{re}'}{(X_{sl} + C_1 X_{rl}')} \approx \frac{\pm C_1 R_{re}'}{\omega_1 (L_{sl} + C_1 L_{rl}')} \approx \frac{\pm R_{re}'}{\omega_1 L_{sc}} \tag{7.70}$$

$$T_{ek} \approx \pm 3 p_1 \left(\frac{V_s}{\omega_1}\right)^2 \frac{1}{2C_1 (L_{sl} + C_1 L_{rl}')} \approx 3 p_1 \left(\frac{V_s}{\omega_1}\right)^2 \frac{1}{2L_{sc}} \tag{7.71}$$

- In general, as long as $I_s R_s / V_s < 0.05$, we may safely approximate the breakdown torque to Equation (7.71).
- The critical slip speed in (7.70) $S_k \omega_1 = \pm R_{re}'/L_{sc}$ is dependent on rotor resistance (acumulated) and on the total leakage (short-circuit) inductance. Care must be exercised to calculate L_{sl} and L_{rl}' for the actual conditions at breakdown torque, where skin and leakage saturation effects are notable.
- The breakdown torque in (7.71) is proportional to voltage per frequency squared and inversely proportional to equivalent leakage inductance L_{se}. Notice that (7.70) and (7.71) are not valid for low values of voltage and frequency when $V_s \le 0.05 I_s R_s$.
- When designing an IM for high breakdown torque, a low short-circuit inductance configuration is needed.
- In (7.70) and (7.71), the final form, $C_1 = 1$, which, in fact, means that $L_{1m} \approx \infty$ or the iron is not saturated. For deeply saturated IMs, $C_1 \neq 1$ and the first form of T_{ek} in (7.71) is to be used.

- Stable operation under steady state is obtained when torque decreases with speed:

$$\frac{\partial T_e}{\partial n} < 0 \text{ or } \frac{\partial T_e}{\partial S} > 0 \tag{7.72}$$

- With $R_s = 0$, $C_1 = 1$, the relative torque T_e/T_{ek} is

$$\frac{T_e}{T_{ek}} \approx \frac{2}{\frac{S}{S_k} + \frac{S_k}{S}} \tag{7.73}$$

This is known as simplified Kloss's formula.

- The margin of stability is often defined as the ratio between breakdown torque and rated torque.

$$T_{ek}/(T_e)_{S_n} > 1.6; \ (T_e)_{S_n} - \text{rated torque} \tag{7.74}$$

- Values of $T_{ek}/(T_e)_{S_n}$ larger than 2.3–2.4 are not easy to obtain unless very low short-circuit inductance designs are applied. For loads with unexpected, frequent, large load bursts, or load torque time pulsations (compressor loads), large relative breakdown torque values are recommended.
- On the other hand, the starting torque depends essentially on rotor resistance and on the short-circuit impedance Z_{sc} for given voltage and frequency:

$$T_{es} = \frac{3p_1}{\omega_1} V_s^2 \frac{R_{re}'}{(R_s + C_1 R_{re}')^2 + (X_{sl} + C_1 X_{rl}')^2} \approx \frac{3p_1}{\omega_1} V_s^2 \frac{R_{re}'}{|\underline{Z}_{sc}|^2} \tag{7.75}$$

Cage rotor IMs have a relative starting torque $T_{es}/T_{en} = 0.7$ to 1.1 while deep bar or double cage rotor (as shown in the next chapter) have large starting torque but smaller breakdown torque and larger critical slip S_k because their short-circuit inductance (reactance) is larger.

Slip ring rotors (with the phases short-circuited over the slip rings) have much lower starting torque $T_{es}/T_{en} < 0.3$ in general, larger peak torque at lower critical slip S_k.

- The current/slip curve for a typical single cage and, respectively, wound or short-circuited rotor are shown on Figure 7.18.

Due to negative slip the stator current increases even more with absolute slip value in the generator than in the motor operation mode.

Also, while starting currents of (5 to 7)I_n (as in squirrel cage rotors) are accepted for direct starting values in excess of 10, typical for wound short-circuited rotor are not acceptable. A wound rotor IM should be started either with a rotor resistance added or with a rotor power supply to absorb part of rotor losses and reduce rotor currents.

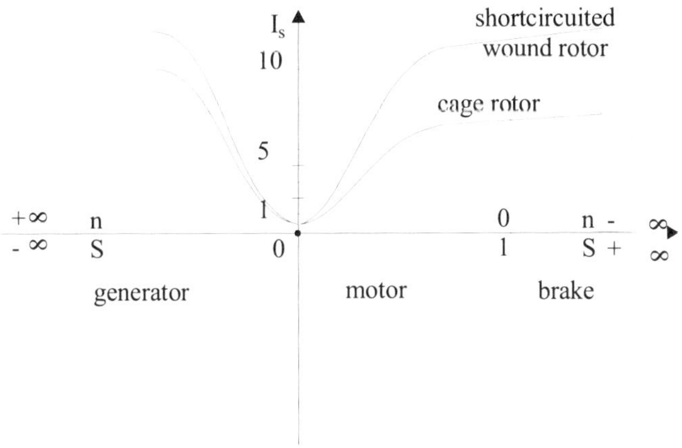

Figure 7.18 Current versus slip (V_s = ct, f_1 = ct)

On the other hand, high efficiency motors favor today for energy saving are characterized by lower stator and rotor resistances under load, but also at start. Thus higher starting currents are expected: I_s/I_n = 6.5 to 8. To avoid large voltage sags at start, the local power grids have to be strengthened (by additional transformers in parallel) to allow for large starting currents.

These additional hardware costs have to be added to high efficiency motor costs and then compared to the better efficiency motor energy savings and maintenance, and cost reductions. Also, high efficiency motors tend to have slightly lower starting torque (Figure 7.19).

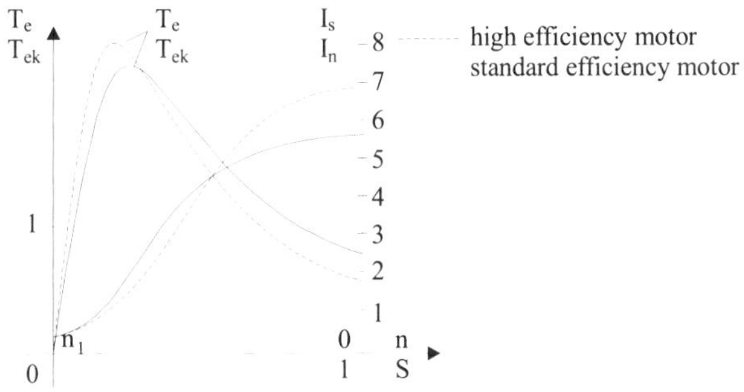

Figure 7.19 Torque and stator current versus slip for motoring

More on these aspects are to be found in chapters on design methodologies (Chapters 14 through 18).

7.10 EFFICIENCY AND POWER FACTOR

The efficiency η has already been defined, both for motoring and generating, earlier in this chapter. The power factor $\cos\varphi_1$ may be defined in relation to the equivalent circuit as

$$\cos\varphi_1 = \frac{|R_e|}{Z_e} \tag{7.76}$$

Figure 7.9 shows that Re is negative for the actual generating mode. This explains why in (7.76) the absolute value of R_e is used. The power factor, related to reactive power flow, has the same formula for both motoring and generating, while the efficiency needs separate formulas (7.50) and (7.53).

The dependence of efficiency and power factor on load (slip or speed) is essential when the machine is used for variable load (Figure 7.20).

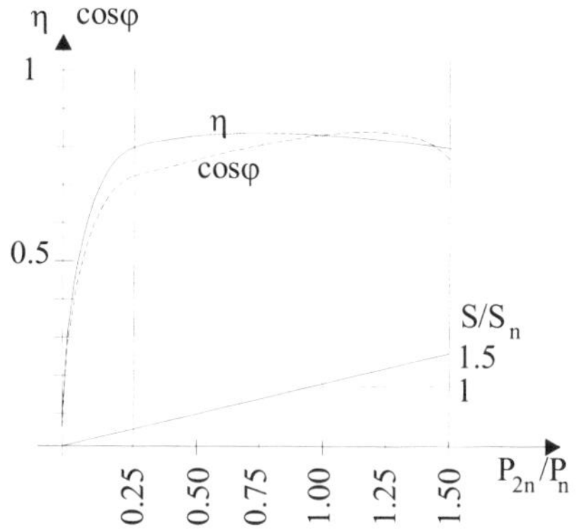

Figure 7.20 IM efficiency, η, power factor, cosφ, and slip, S/S_n versus relative power load at the shaft: P_{2m}/P_n

Such curves may be calculated strictly based on the equivalent circuit (Figure 7.1) with known (constant or variable) parameters, and given values of slip. Once the currents are known all losses and input power may be calculated. The mechanical losses may be lumped into the core loss resistance R_{1m} or, if not, they have to be given from calculations or tests performed in advance. For applications with highly portable loads rated from 0.25 – to 1.5 the design of the motor has to preserve a high efficiency.

A good power factor will reduce VAR compensation hardware ratings and costs, as an industrial plant average power factor should not be lagging below 0.95 to avoid VAR penalties.

Example 7.5. Motor performance

An induction motor with deep bars is characterized by rated power P_n = 20 kW, supply line voltage V_{sl} = 380 V (star connection), f_1 = 50 Hz, η_n = 0.92, p_{mec} = 0.005 P_n, p_{iron} = 0.015 P_n, p_s = 0.005 P_n, p_{cosn} = 0.03 P_n, $\cos\varphi_n$ = 0.9, p_1 = 2, starting current $I_{sc} = 5.2I_n$, power factor $\cos\varphi_{sc} = 0.4$, and no-load current $I_{on} = 0.3I_n$.

Let us calculate

a.) Rotor cage losses p_{corn}, electromagnetic power P_{elm}, slip, S_n, speed, n_n, rated current I_n and rotor resistance R_r', for rated load

b.) Stator resistance R_s and rotor resistance at start R_{rs}'

c.) Electromagnetic torque for rated power and at start

d.) Breakdown torque

Solution

a.) Based on (7.50), we notice that, at rated load, we lack the rotor cage losses, p_{corn} from all losses.

$$p_{corn} = \frac{P_n}{\eta_n} - \left(p_{\cos n} + p_{Sn} + p_{iron} + p_{iron}\right) - P_n = \tag{7.77}$$

$$= \frac{20000}{0.92} - (0.03 + 0.005 + 0.015 + 0.015) \cdot 20000 - 20000 = 639.13\text{W}$$

On the other hand, the rated current In comes directly from the power input expression

$$I_n = \frac{P_n}{\eta_n \sqrt{3} V_{sl} \cos\varphi_n} = \frac{20000}{0.92 \cdot \sqrt{3} \cdot 380 \cdot 0.9} = 36.742\text{A} \tag{7.78}$$

The slip expression (7.52) yields

$$S_n = \frac{p_{corn}}{\frac{P_n}{\eta_n} - p_{\cos n} - p_{Sn} - p_{iron}} = \tag{7.79}$$

$$= \frac{639.13}{\frac{20000}{0.92} - (0.03 + 0.005 + 0.015)20000} = 0.0308$$

The rated speed n_n is

$$n_n = \frac{f_1}{P_1}\left(1 - S_n\right) = \frac{50}{2}(1 - 0.0308) = 24.23\text{rps} = 1453.8\text{rpm} \tag{7.80}$$

The electromagnetic power P_{elm} is

$$P_{elm} = \frac{p_{corn}}{S_n} = \frac{639.13}{0.0308} = 20739\text{W} \tag{7.81}$$

To easily find the rotor resistance, we need the rotor current. At rated slip, the rotor circuit is dominated by the resistance R_r'/S_n and thus the no-load (or magnetizing current) I_{0n} is about 90^0 behind the rotor resistive current I_{rn}'. Thus, the rated current I_n is

$$I_n = \sqrt{I_{0n}^2 + I_{rn}'^2} \tag{7.82}$$

$$I_{rn}' = \sqrt{I_n^2 - I_{0n}^2} = \sqrt{36.742^2 - (0.3 \cdot 36.672)^2} = 35.05A \tag{7.83}$$

Now the rotor resistance at load R_r' is

$$R_r' = \frac{p_{corn}}{3I_{rn}'^2} = \frac{639.13}{3 \cdot 35.05^2} = 0.1734\Omega \tag{7.84}$$

b.) The stator resistance R_s comes from

$$R_s = \frac{p_{cos\,n}}{3 \cdot I_n^2} = \frac{0.03 \cdot 20000}{3 \cdot 36.742^2} = 0.148\Omega \tag{7.85}$$

The starting rotor resistance R_{rs}' is obtained based on starting data:

$$R_{rs}' = \frac{V_{sl}}{\sqrt{3}} \frac{\cos \varphi_{sc}}{I_{sc}} - R_s = \frac{380}{\sqrt{3}} \frac{0.4}{5.2 \cdot 36.742} - 0.148 = 0.31186\Omega \tag{7.86}$$

Notice that
$$K_r = \frac{R_{rs}'}{R_r'} = \frac{0.31186}{0.1734} = 1.758 \tag{7.87}$$

This is mainly due to skin effect.

c.) The rated electromagnetic torque T_{en} is

$$T_{en} = \frac{P_{elm}}{\omega_1} p_1 = 20739 \cdot \frac{2}{2\pi 50} = 132.095\text{Nm} \tag{7.88}$$

The starting torque T_{es} is

$$T_{es} = \frac{3R_{rs}' I_{sc}^2}{\omega_1} p_1 = \frac{3 \cdot 0.31186 \cdot (5.2 \cdot 36.742)^2}{2\pi 50} = 108.76\text{Nm} \tag{7.89}$$
$$T_{es} / T_{en} = 108.76 / 132.095 = 0.833$$

d.) For the breakdown torque, even with approximate formula (7.71), the short-circuit reactance is needed. From starting data,

$$X_{sc} = \frac{V_{sl}}{\sqrt{3}} \frac{\sin \varphi_{sc}}{I_{sc}} = \frac{380}{\sqrt{3}} \frac{\sqrt{1 - 0.4^2}}{5.2 \cdot 36.672} = 1.0537\Omega \tag{7.90}$$

The rotor leakage reactance $X_{rls} = X_{sc} - X_{sl} = 1.0537 - 0.65 = 0.4037\ \Omega$. Due to skin effect and leakage saturation, this reactance is smaller than the one "acting" at rated load and even at breakdown torque conditions. Knowing that $K_r = 1.758$ (7.87), resistance correction coefficient value due to skin effect, we could find $K_x \approx 0.9$ from Chapter 8, for a rectangular slot.

With leakage saturation effects neglected the rotor leakage reactance at critical slip S_k (or lower) is

$$X_{rl}' = K_x^{-1} X_{rls}' = \frac{1}{0.9} \cdot 0.4 = 0.444\Omega \tag{7.91}$$

Now, with C1 ≈ 1 and Rs ≈ 0, from (7.71) the breakdown torque T_{ek} is

$$\begin{aligned} T_{ek} &\approx 3\left(\frac{V_{sl}}{\sqrt{3}}\right)^2 \frac{p_1}{\omega_1} \frac{1}{2(X_{sl}+X_{rl}')} = \\ &= 3\left(\frac{380}{\sqrt{3}}\right)^2 \frac{2}{2\pi 50} \frac{1}{2(0.65+0.444)} = 422.68\text{Nm} \end{aligned} \tag{7.92}$$

The ratio T_{ek}/T_{en} is

$$T_{ek}/T_{en} = \frac{422.68}{132.095} = 3.20$$

This is an unusually large value facilitated by a low starting current and a high power factor at start; that is, a low leakage reactance X_{sc} was considered.

7.11 PHASOR DIAGRAMS: STANDARD AND NEW

The IM equations under steady state are, again,

$$\begin{cases} \underline{I}_s(R_s + jX_{sl}) - \underline{V}_s = \underline{E}_s \\ \underline{I}_r'\left(\frac{R_r'}{S} + jX_{rl}'\right) + \frac{\underline{V}_r'}{S} = \underline{E}_s \\ \underline{E}_s = -\underline{Z}_{1m}(\underline{I}_s + \underline{I}_r') = -\underline{Z}_{1m}\underline{I}_{0s} \\ \underline{Z}_{1m} = \frac{R_{1m} \cdot jX_{1m}}{R_{1m} + jX_{1m}} = R_{1ms} + jX_{1ms} \end{cases}$$

$$R_{1m} >> X_{1m};\ R_{1ms} << X_{1ms};\ R_{1ms} << R_{1m};\ X_{1ms} \approx X_{1m} \tag{7.93}$$

They can be and have been illustrated in phasor diagrams. Such diagrams for motoring (S > 0) and generating (S < 0) are shown in Figure 7.21a and b, for $\underline{V}_r' = 0$ (short-circuited rotor).

As expected, for the short-circuited rotor, the reactive power remains positive for generating ($Q_1 > 0$) as only the active stator current changes sign from motoring to generating.

$$\begin{aligned} P_1 &= 3\underline{V}_s\underline{I}_s \cos\varphi_1 <> 0 \\ Q_1 &= 3\underline{V}_s\underline{I} \sin\varphi_1 > 0 \end{aligned} \tag{7.94}$$

Today the phasor diagrams are used mainly to explain IM action, while, before the occurrence of computers, they served as a graphical method to determine steady-state performance without solving the equivalent circuit.

However, rather recently the advance of high performance variable speed a.c. drives has led to new phasor (or vector) diagrams in orthogonal axis models of an ac machine. Such vector diagrams may be accompanied by similar phasor diagrams valid for each phase of a.c. machines. To simplify the derivation let us neglect the core loss ($R_{1m} \approx \infty$ or $R_{1ms} = 0$). We start by defining three flux linkages per phase: stator flux linkage $\underline{\Psi}_s$, rotor flux linkage $\underline{\Psi}_r'$, and airgap (magnetizing) $\underline{\Psi}_m$.

$$\begin{aligned} \underline{\psi}_s &= L_{sl}\underline{I}_s + \underline{\psi}_m \\ \underline{\psi}_r' &= L_{rl}'\underline{I}_r' + \underline{\psi}_m \\ \underline{\psi}_m &= L_{1m}\left(\underline{I}_s + \underline{I}_r'\right) = L_{1m}\underline{I}_m \end{aligned} \tag{7.95}$$

$$\underline{E}_s = -j\omega_1 L_{1m}\underline{I}_m = -j\omega_1\underline{\psi}_m \tag{7.96}$$

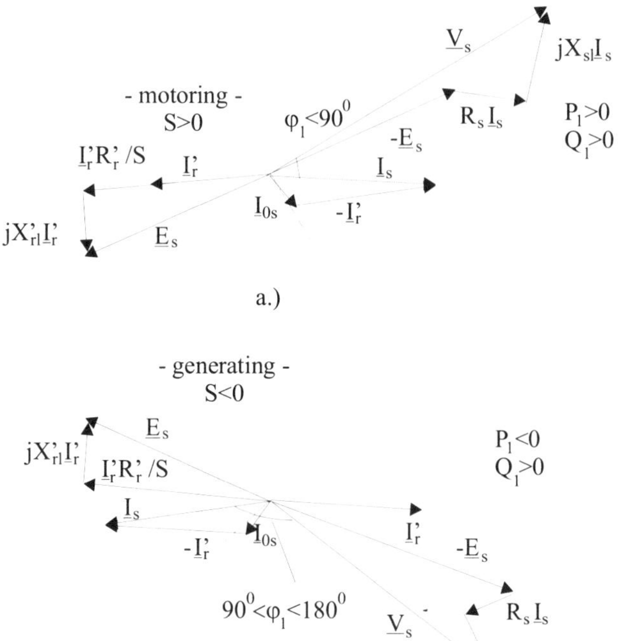

Figure 7.21 Standard phasor diagrams, a.) for motoring, b.) for generating

The stator and rotor equations in (7.93) become

$$\underline{I}_s R_s - \underline{V}_s = -j\omega_1 \underline{\psi}_s \tag{7.97}$$

$$\underline{I}_r' \frac{R_r'}{S} + \frac{\underline{V}_r'}{S} = -j\omega_1 \underline{\psi}_r' \tag{7.98}$$

For zero rotor voltage (short-circuited rotor), Equation (7.98) becomes:

$$\underline{I}_r' \frac{R_r'}{S} = -j\omega_1 \underline{\psi}_r'; \ \ \underline{V}_r' = 0 \tag{7.99}$$

Equation (7.99) shows that at steady state, for $V_r' = 0$, the rotor current I_r' and rotor flux $\underline{\Psi}_r'$ are phase shifted by 90^0.

Consequently, the torque T_e expression (7.60) becomes:

$$T_e = \frac{3R_r' I_r'^2}{S} \frac{p_1}{\omega_1} = 3p_1 \psi_r' I_r' \tag{7.100}$$

Equation (7.100) may be considered the basis for modern (vector) IM control. For constant rotor flux per phase amplitude (RMS), the torque is proportional to rotor phase current amplitude (RMS). With these new variables, the phasor diagrams based on Equations (7.97) and (7.98) are shown in Figure 7.22.

Such phasor diagrams could be instrumental in computing IM performance when fed from a static power converter (variable voltage, variable frequency).

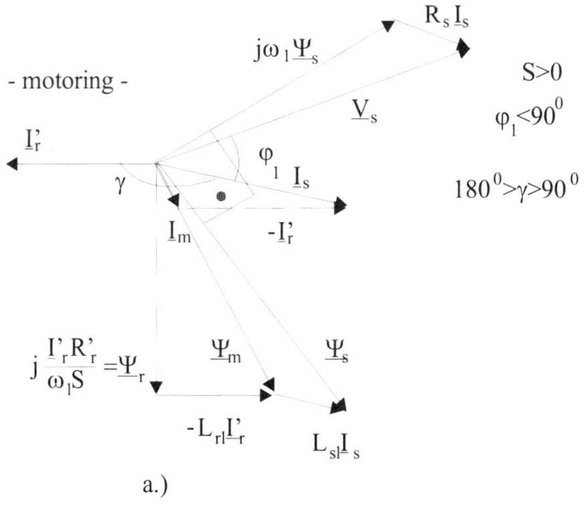

a.)

Figure 7.22 Phasor diagram with stator, rotor and magnetization flux linkages, $\underline{\Psi}_r'$, $\underline{\Psi}_m$, $\underline{\Psi}_s$ a.) motoring b.) generating (continued)

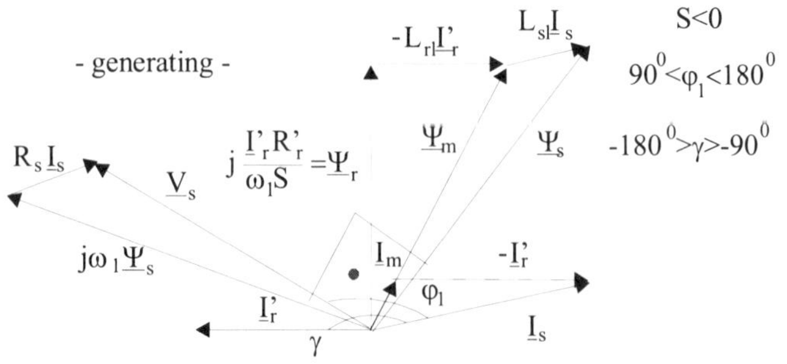

b.)

Figure 7.22 (continued)

The torque expression may also be obtained from (7.97) by taking the real part of it after multiplication by I_s^*.

$$\mathrm{Re}\left[3\underline{I}_s\underline{I}_s^* R_s + 3j\omega_1\underline{\psi}_s\underline{I}_s^*\right] = 3\,\mathrm{Re}\left(\underline{V}_s\underline{I}_s^*\right) \tag{7.101}$$

The second term in (7.101), with core loss neglected, is, in fact, the electromagnetic power P_{elm}

$$P_{elm} = 3\omega_1\,\mathrm{Imag}\left(\underline{\psi}_s\underline{I}_s^*\right) = T_e\left(\omega_1/p_1\right) \tag{7.102}$$

$$T_e = 3p_1\,\mathrm{Imag}\left(\underline{\psi}_s\underline{I}_s^*\right) = 3p_1L_{1m}\,\mathrm{Imag}\left(\underline{I}_r'\underline{I}_s^*\right) \tag{7.103}$$

or

$$T_e = 3p_1L_{1m}\underline{I}_r'\underline{I}_s\sin\gamma;\ \ |\gamma| > 90^0 \tag{7.104}$$

In (7.104), γ is the angle between the rotor and stator phase current phasors with $\gamma > 0$ for motoring and $\gamma < 0$ for generating.

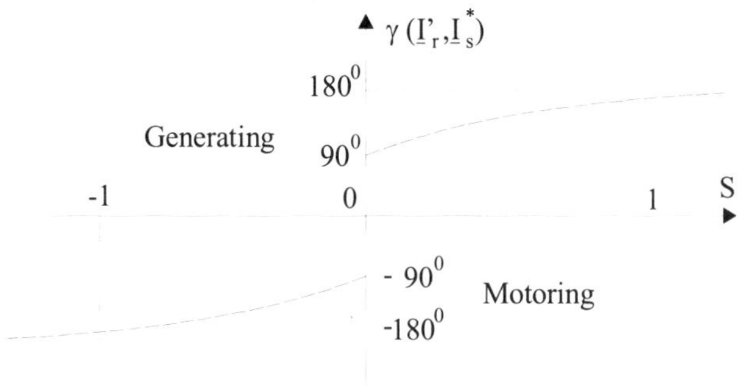

Figure 7.23 Angle γ ($\underline{I}_r'$, $\underline{I}_s^*$) versus slip S

Now, from the standard equivalent circuit (Figure 7.1 with $R_{1m} \approx \infty$),

$$\underline{I}_r' = -\underline{I}_s \frac{j\omega_1 L_{1m}}{\frac{R_r'}{S} + j\omega_1 (L_{1m} + L_{rl}')} \tag{7.105}$$

So the angle γ between $\underline{I}_r'$ and $\underline{I}_s^*$ depends on slip S and frequency ω_1 and motor parameters (Figure 7.23). It should be noted that the angle γ is close to $\pm 180^0$ for $S = \pm 1$ and close to $\pm 90^0$ toward $|S| = 0$.

7.12 ALTERNATIVE EQUIVALENT CIRCUITS

Alternative equivalent circuits for IMs abound in the literature. Here we will deal with some that have become widely used in IM drives. In essence [2], a new rotor current is introduced.

$$\underline{I}_{ra}' = \frac{\underline{I}_r'}{a} \tag{7.106}$$

Using this new variable in (7.95) through (7.98), we easily obtain:

$$\begin{aligned}
&\underline{I}_s R_s + j\omega_1 [L_s - aL_{1m}] \underline{I}_s - \underline{V}_s = \underline{E}_a \\
&\underline{E}_a = -j\omega_1 \underline{\psi}_{ma};\ \underline{\psi}_{ma} = aL_{1m}(\underline{I}_s + \underline{I}_{ra}');\ \underline{I}_{ma} = \underline{I}_s + \frac{\underline{I}_r'}{a} \\
&\underline{I}_{ra}' R_r' a^2 + j\omega_1 (aL_r' - L_{1m}) a \underline{I}_{ra}' + \frac{V_r' a}{S} = \underline{E}_a \\
&L_s = L_{sl} + L_{1m};\ L_r' - L_{rl}' + L_{1m}
\end{aligned} \tag{7.107}$$

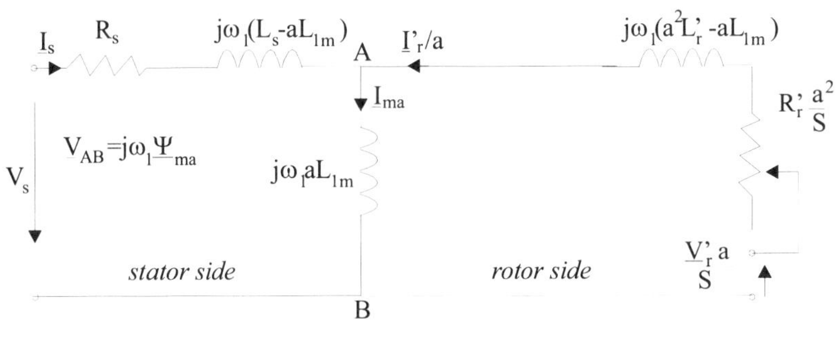

Figure 7.24 General equivalent circuit (core loss neglected)

For a = 1, we reobtain, as expected, Equations (7.95) through (7.98). L_s, L_r' represent the total stator and rotor inductances per phase when the IM is three-phase fed.

Now the general equivalent circuit of (7.107), similar to that of Figure 7.1, is shown in Figure 7.24. For a = 1, the standard equivalent circuit (Figure 7.1) is obtained. If $a = L_s/L_{1m} \approx 1.02 - 1.08$, the whole inductance (reactance) is occurring in the rotor side of equivalent circuit and $\underline{\Psi}_{ma} = \underline{\Psi}_s$ (Figure 7.25a). The equivalent circuit directly evidentiates the stator flux.

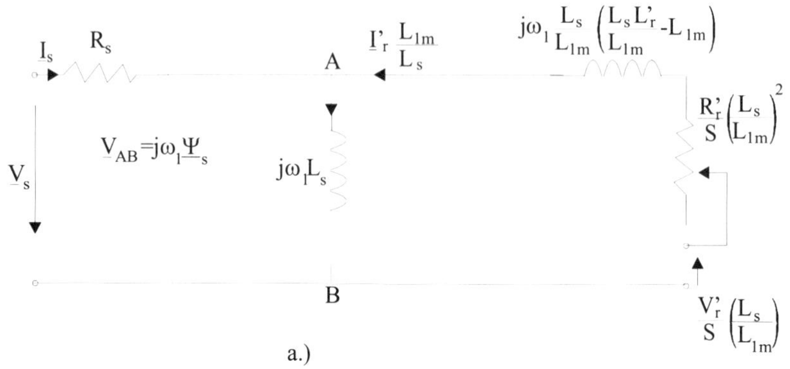

a.)

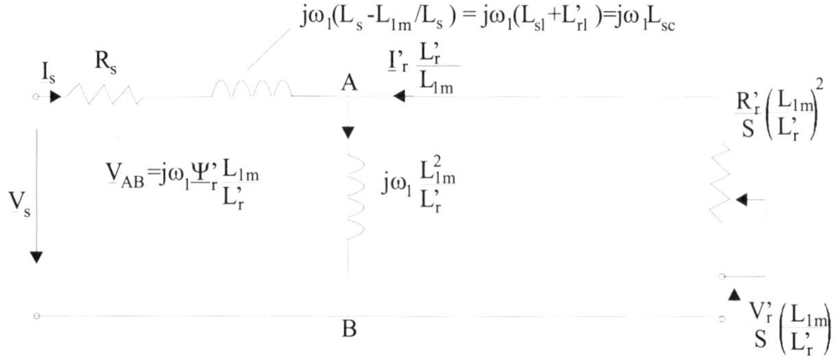

b.)

Figure 7.25 Equivalent circuit
a.) with stator flux shown b.) with rotor flux shown

For $a = L_{1m}/L_r' \approx 0.93 - 0.97$, $\underline{\psi}_{ma} = \frac{L_{1m}}{L_r'}\psi_r'$, the equivalent circuit is said to evidentiate the rotor flux.

Advanced IM control is performed at constant stator, Ψ_s, rotor Ψ_r, or magnetization flux, Ψ_m amplitudes. That is voltage and frequency, V_s, f_1, are changed to satisfy such conditions.

Consequently circuits in Figure 7.25 should be instrumental in calculating IM steady-state performance under such flux linkage constraints.

Alternative equivalent circuits to include leakage and main saturation and skin effect are to be treated in Chapter 8.

Example 7.6. Flux linkage calculations

A cage-rotor induction motor has the following parameters: $R_s = 0.148\Omega$, $X_{sl} = X_{rl}' = 0.5\Omega$, $X_m = 20\Omega$, $2p_1 = 4$poles, $V_s = 220$V/phase (RMS)–star connection, $f_1 = 50$Hz, rated current $I_{sn} = 36$A, $\cos\varphi_n = 0.9$. At start, the current $I_{start} = 5.8I_n$, $\cos\varphi_{start} = 0.3$.

Let us calculate the stator, rotor and magnetization flux linkages Ψ_s, Ψ_r'. Ψ_m at rated speed and at standstill (core loss is neglected). With S = 0.02, calculate the rotor resistance at full load.

Solution

To calculate the RMS values of Ψ_s, Ψ_r'. Ψ_m we have to determine, in fact, V_{AB} on Figure 7.24, for $a_s = L_s/L_{1m}$, $a_r = L_{1m}/L_r'$ and $a_m = 1$, respectively.

$$\psi_s\sqrt{2} = \frac{\sqrt{2}\left|\underline{V}_s - \underline{I}_s R_r\right|}{\omega_1} \tag{7.108}$$

$$\psi_r'\sqrt{2} = \sqrt{2}\frac{\left|\underline{V}_s - \underline{I}_s R_r - \underline{I}_s j\omega_1\left(L_{sl} + L_{rl}'\right)\right|}{\omega_1\left(\frac{L_{1m}}{L_r'}\right)} \tag{7.109}$$

$$\psi_m\sqrt{2} = \frac{\sqrt{2}\left|\underline{V}_s - \underline{I}_s R_r - \underline{I}_s j\omega_1 L_{sl}\right|}{\omega_1} \tag{7.110}$$

$$\underline{I}_s = I_s\left(\cos\varphi - j\sin\varphi\right) \tag{7.111}$$

These operations have to be done twice, once for $I_s = I_n$, $\cos\varphi_n$ and once for $I_s = I_{start}$, $\cos\varphi_{start}$.

For $I_n = 36$A, $\cos\varphi_n = 0.9$,

$$\begin{aligned}\psi_s\sqrt{2} &= \frac{\sqrt{2}\left|220 - 36\cdot(0.9 - j4359)0.148\right|}{2\pi 50} = 0.9654\text{Wb}\\ \psi_r'\sqrt{2} &= \frac{\sqrt{2}\left|220 - 36\cdot(0.9 - j4359)(0.148 + j1)\right|}{2\pi 50\left(\frac{20}{20.5}\right)} = 0.9314\text{Wb}\end{aligned} \tag{7.112}$$

$$\psi_m\sqrt{2} = \frac{\sqrt{2}\left|220 - 36\cdot(0.9 - j4359)(0.148 + j1)\right|}{2\pi 50} = 0.9369\text{Wb} \tag{7.113}$$

The same formulas are applied at zero speed (S = 1), with starting current and power factor,

$$\left(\psi_s\sqrt{2}\right)_{start} = \frac{\sqrt{2}\left|220-36\cdot 5.8\cdot(0.3-j0.888)\cdot 0.148\right|}{2\pi 50} = 0.9523\text{Wb}$$

$$\left(\psi_r'\sqrt{2}\right)_{start} = \frac{\sqrt{2}\left|220-36\cdot 5.8\cdot(0.3-j0.888)(0.148+j1)\right|}{2\pi 50\left(\frac{20}{20.5}\right)} = 0.1998\text{Wb} \quad (7.114)$$

$$\left(\psi_m\sqrt{2}\right)_{start} = \frac{\sqrt{2}\left|220-36\cdot 5.8\cdot(0.3-j0.888)(0.148+j1)\right|}{2\pi 50} = 0.5306\text{Wb}$$

The above results show that while at full load the stator, magnetization and rotor flux linkage amplitudes do not differ much, they do so at standstill. In particular, the magnetization flux linkage is reduced at start to 55–65% of its value at full load, so the main magnetic circuit of IMs is not saturated at standstill.

7.13 UNBALANCED SUPPLY VOLTAGES

Steady state performance with unbalanced supply voltages may be treated by the method of symmetrical components. The three-wire supply voltages $\underline{V}_a$, $\underline{V}_b$, $\underline{V}_c$ are decomposed into forward and backward components:

$$\begin{aligned}\underline{V}_{af} &= \frac{1}{3}\left(\underline{V}_a + a\underline{V}_b + a^2\underline{V}_c\right); \quad a = e^{j\frac{2\pi}{3}} \\ \underline{V}_{ab} &= \frac{1}{3}\left(\underline{V}_a + a^2\underline{V}_b + a\underline{V}_c\right)\end{aligned} \quad (7.115)$$

$$\underline{V}_{bf} = a^2\underline{V}_{af};\ \underline{V}_{cf} = a\underline{V}_{af};\ \underline{V}_{bb} = a\underline{V}_{ab};\ \underline{V}_{cb} = a^2\underline{V}_{ab}; \quad (7.116)$$

Note also that the slip for the forward component is $S_f = S$, while for the backward component, S_b is

$$S_b = \frac{-\left(\frac{f_1}{p_1}\right) - n}{-\left(\frac{f_1}{p_1}\right)} = 2 - S \quad (7.117)$$

So, in fact, we obtain two equivalent circuits (Figure 7.26) as

$$\underline{V}_a = \underline{V}_{af} + \underline{V}_{ab} \quad (7.118)$$

$$\underline{I}_a = \underline{I}_{af} + \underline{I}_{ab} \quad (7.119)$$

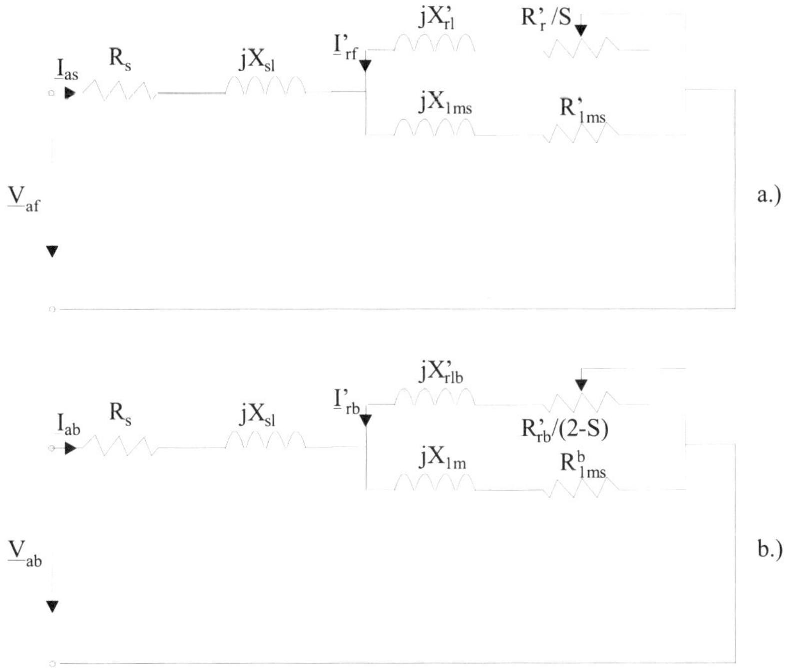

Figure 7.26 Forward and backward component equivalent circuits per phase

For S = 0.01 – 0.05 and f_1 = 60(50) Hz, the rotor frequency for the backward components is $f_2 = (2 - S)f_1 \approx 100(120)$ Hz. Consequently, the rotor parameters R_r' and L_{rl}' are notably influenced by the skin effect, so

$$R_{rb}' > R_r'; \quad X_{rlb}' < X_{rl}' \tag{7.120}$$

Also for the backward component, the core losses are notably less than those of, forward component. The torque expression contains two components:

$$T_e = \frac{3R_r'(I_{rf}')^2}{S}\frac{p_1}{\omega_1} + \frac{3R_{rb}'(I_{rb}')^2}{(2-S)}\frac{p_1}{(-\omega_1)} \tag{7.121}$$

With phase voltages given as amplitudes and phase shifts, all steady-state performance may be calculated with the symmetrical component method.

The voltage imbalance index $V_{imbalance}$ (in %) may be defined as

$$V_{imbalance} = \frac{\Delta V_{max}}{V_{ave}}100\%; \quad \Delta V_{max} = V_{max} - V_{min}; \quad V_{ave} = \frac{(V_a + V_b + V_c)}{3} \tag{7.122}$$

V_{max} = maximum phase voltage; V_{min} = phase with minimum voltage.

An alternative definition would be

$$V_{unb}[\%] = \frac{V_{ab}}{V_{af}}100 \tag{7.123}$$

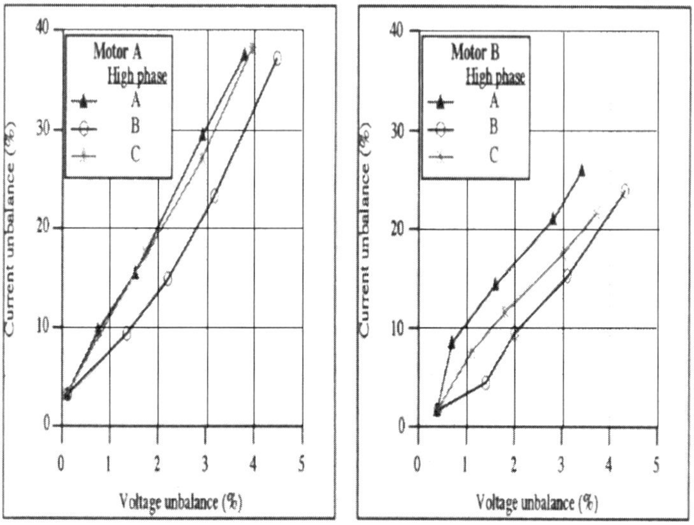

Figure 7.27 Current imbalance versus voltage imbalance [2]
a.) cost optimized motor (motor A)
b.) premium motor (motor B)

Due to very different values of slip and rotor parameters for forward and backward components, a small voltage imbalance is likely to produce a rather large current imbalance. This is illustrated in Figure 7.27.

Also, for rated slip (power), the presence of backward (braking) torque and its losses leads to a lower efficiency η.

$$\eta = \frac{T_e \frac{\omega_1}{P_1}(1-S)}{3\,\mathrm{Re}\left[\underline{V}_{af}\underline{I}_{af}^* + \underline{V}_{ab}\underline{I}_{ab}^*\right]} \tag{7.124}$$

Apparently cost-optimised motors–with larger rotor skin effect in general–are more sensitive to voltage imbalances than premium motors (Figure 7.28). [2]

As losses in the IM increase with voltage imbalance, machine derating is to be applied to maintain rated motor temperature. NEMA 14.35 standards recommend an IM derating with voltage imbalance as shown in Figure 7.29. [3]

Voltage imbalance also produces, as expected, $2f_1$ frequency vibration. As small voltage imbalance can easily occur in local power grids, care must be exercised in monitoring it and the motor currents and temperature.

An extreme case of voltage imbalance is one phase open.

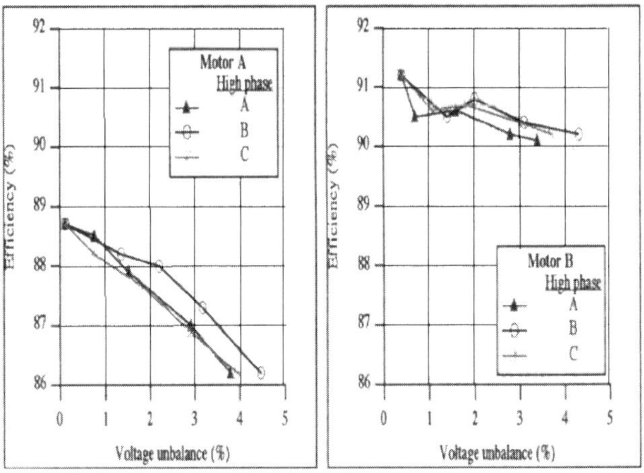

Figure 7.28. Efficiency versus voltage imbalance
a.) cost optimised motor
b.) premium motor

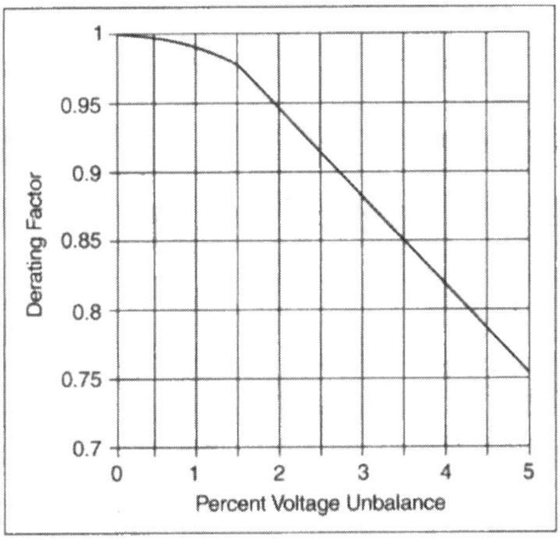

Figure 7.29. Derating versus voltage imbalance (NEMA Figure 14.1)

7.14 ONE STATOR PHASE IS OPEN

Let us consider the IM under steady state with one stator phase open (Figure 7.30a).

This time it is easy to calculate first the symmetrical current components $\underline{I}_{af}$, $\underline{I}_{ab}$, $\underline{I}_{ao}$.

$$\underline{I}_{af} = \frac{1}{3}\left(\underline{I}_a + a\underline{I}_b + a^2\underline{I}_c\right) = \frac{a-a^2}{3}\underline{I}_b = \frac{j\underline{I}_b}{\sqrt{3}}$$
$$\underline{I}_{ab} = \frac{1}{3}\left(\underline{I}_a + a^2\underline{I}_b + a\underline{I}_c\right) = -\frac{a-a^2}{3}\underline{I}_b = -\underline{I}_{af} \tag{7.125}$$

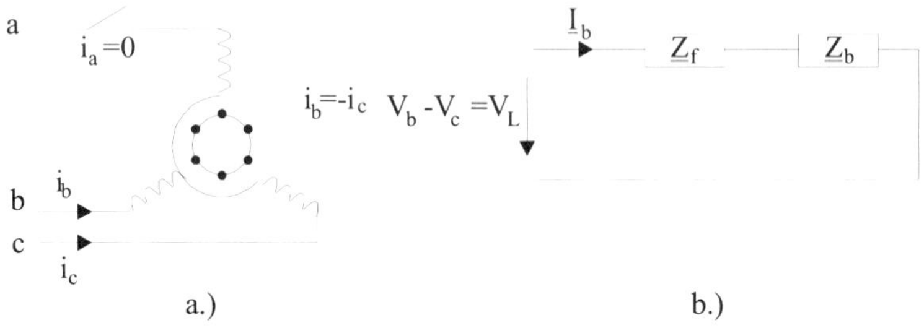

Figure 7.30 One stator phase is open
a.) phase a is open
b.) equivalent single phase circuit

Let us replace the equivalent circuits in Figure 7.26 by the forward and backward impedances Z_f, Z_b.

$$\underline{V}_{af} = \underline{Z}_f\underline{I}_{af};\ \ \underline{V}_{ab} = \underline{Z}_b\underline{I}_{ab} \tag{7.126}$$

Similar relations are valid for $\underline{V}_{bf} = a^2\underline{V}_{af}$, $\underline{V}_{bb} = a\underline{V}_{ab}$, $\underline{V}_{cf} = a\underline{V}_{af}$, $\underline{V}_{cb} = a^2\underline{V}_{ab}$:

$$\begin{aligned}\underline{V}_b - \underline{V}_c &= \underline{V}_{bf} + \underline{V}_{bb} - \underline{V}_{cf} - \underline{V}_{cb} = a^2\underline{Z}_f\underline{I}_{af} + a\underline{Z}_b\underline{I}_{ab} - a\underline{Z}_f\underline{I}_{af} - a^2\underline{Z}_b\underline{I}_{ab} = \\ &= \left(a^2 - a\right)\underline{I}_{af}\left(\underline{Z}_f + \underline{Z}_b\right)\end{aligned} \tag{7.127}$$

With (7.125),

$$\underline{V}_b - \underline{V}_c = \underline{I}_b\left(\underline{Z}_f + \underline{Z}_b\right) \tag{7.128}$$

The electromagnetic torque still retains Equation (7.121), but (7.128) allows a handy computation of current in phase b and then from (7.115) through (7.126), I_{af} and I_{ab} are calculated.

At standstill (S = 1) $\underline{Z}_f = \underline{Z}_b = \underline{Z}_{sc}$ (Figure 7.26) and thus the short-circuit current I_{sc1} is

$$\underline{I}_{sc1} = \frac{\underline{V}_L}{2\underline{Z}_{sc}} = \frac{\sqrt{3}}{2}\frac{\underline{V}_{phase}}{\underline{Z}_{sc}} = \frac{\sqrt{3}}{2}\underline{I}_{sc3} \tag{7.129}$$

The short-circuit current I_{sc1} with one phase open is thus $\sqrt{3}/2$ times smaller than for balanced supply. There is no danger from this point of view at start. However, as $\left(\underline{Z}_f\right)_{S=1} = \left(\underline{Z}_b\right)_{S=1}$, $I_{rf}' = I_{rb}'$ and thus the forward and backward torque components in (7.121) are cancelling each other. The IM will not start with one stator phase open.

The forward and backward torque/slip curves differ by the synchronous speed: ω_1/p_1 and $-\omega_1/p_1$, respectively, and consequently, by the slips S and 2 – S, respectively. So they are anti-symmetric (Figure 7.31).

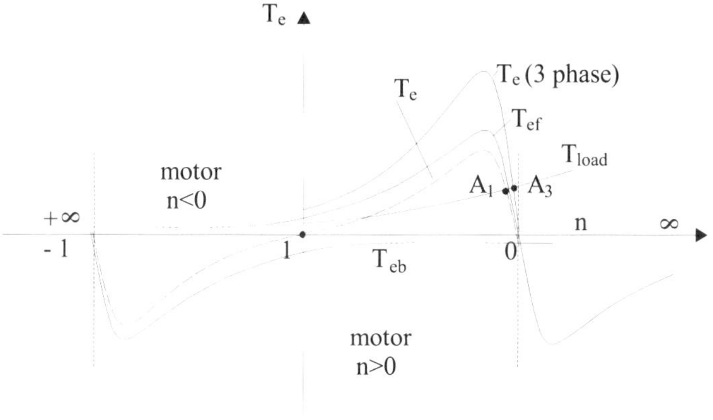

Figure 7.31 Torque components when one stator phase is open

If the IM is supplied by three-phase balanced voltages and works in point A_3 and one phase is open, the steady state operation point moves to A_1.

So, the slip increases (speed decreases). The torque slightly decreases. Now the current in phase b is expected to increase. Also, the power factor decreases because the backward equivalent circuit (for small S) is strongly inductive and the slip increases notably.

It is therefore not practical to let a fully loaded three-phase IM operate a long time with one phase open.

Example 7.7 One stator phase is open

Let us consider a three-phase induction motor with the data $V_L = 220V$, $f_1 = 60$ Hz, $2p_1 = 4$, star stator connection $R_s = R_r' = 1\ \Omega$, $X_{sl} = X_{rl}' = 2.5\ \Omega$, $X_{1m} = 75\ \Omega$, $R_{1m} = \infty$ (no core losses). The motor is operating at S = 0.03 when one stator phase is opened. Calculate: the stator current, torque, and power factor before one phase is opened and the slip, current, and power factor for same torque with one phase open.

Solution

We make use of Figure 7.26a.

With balanced voltages, only Z_f counts.

$$\underline{V}_a = \underline{Z}_f \underline{I}_a;\ \ C_1 = \frac{X_{sl} + X_{1m}}{X_{1m}} = \frac{75 + 2.5}{75} = 1.033 \tag{7.130}$$

$$I_a = I_b = I_c \approx \frac{\frac{V_L}{\sqrt{3}}}{\sqrt{\left(R_s + C_1\frac{R_r'}{S}\right)^2 + \left(X_{sl} + C_1 X_{rl}'\right)^2}} = \tag{7.131}$$

$$= \frac{\frac{220}{\sqrt{3}}}{\sqrt{\left(1 + 1.033\frac{1}{0.03}\right)^2 + \left(2.5 + 1.033 \cdot 2.5\right)^2}} = 3.55\text{A} \ ; \ \cos\varphi = 0.989!$$

The very high value of cosφ indicates that (7.131) does not approximate the phase of stator current while it correctly calculates the amplitude. Also,

$$\underline{I}_r' = \underline{I}_s \left| \frac{jX_{1m}}{\frac{R_r'}{S} + j\left(X_{1m} + X_{rl}'\right)} \right| = 3.55 \cdot \frac{75}{\sqrt{\left(\frac{1}{0.03}\right)^2 + \left(75 + 2.5\right)^2}} = 3.199\text{A} \tag{7.132}$$

From (7.121),

$$T_{e3} = \frac{3R_r'}{S} I_r'^2 \frac{p_1}{\omega_1} = 3 \cdot \frac{1}{0.03} \cdot 3.199^2 \frac{2}{2\pi 60} = 5.4338\text{Nm} \tag{7.133}$$

It is not very simple to calculate the slip S for which the IM with one phase open will produce the torque of (7.133).

To circumvent this difficulty we might give the slip a few values higher than S = 0.03 and plot torque T_{e1} versus slip until $T_{e1} = T_{e3}$.

Here we take only one slip value, say, S = 0.05 and calculate first the current I_b from (7.128), then $I_{af} = -I_{ab}$ from (7.125), and, finally, the torque T_{e1} from (7.121).

$$I_b = \frac{V_L}{\left|\underline{Z}_f + \underline{Z}_b\right|} \approx \frac{V_L}{\left| 2\left(R_s + jX_{sl}\right) + \frac{\frac{R_r'}{S} jX_{1m}}{\frac{R_r'}{S} + j\left(X_{1m} + X_{rl}'\right)} + \frac{R_r'}{2-S} + jX_{rl}' \right|} = \tag{7.134}$$

$$= \frac{220}{\left| 2\left(1 + j2.5\right) + \frac{\frac{1}{0.05} j75}{\frac{1}{0.05} + j\left(75 + 2.5\right)} + \frac{1}{2-S} + j2.5 \right|} = \frac{220}{23.709} = 9.2789\text{A}$$

$$\cos\varphi_1 = \frac{20.34}{23.709} = 0.858$$

From (7.125),

$$I_{af} = I_{ab} = \frac{I_b}{\sqrt{3}} = \frac{9.2789}{5.73} = 5.3635A \tag{7.135}$$

Figure (7.26) yields

$$\underline{I}_{rf}' = I_{af}\left|\frac{jX_{1m}}{\frac{R_r'}{S} + j(X_{1m} + X_{rl}')}\right| = \left|\frac{5.3635 \cdot j \cdot 7.5}{\frac{1}{0.05} + j(75 + 2.5)}\right| = 5.0258A \tag{7.136}$$

$$I_{rb}' \approx I_{ab} = 5.3635A$$

Now, from (7.121), the torque T_{e1} is

$$T_{e1} = \frac{3 \cdot 2 \cdot 1}{2\pi 60}\left[\frac{5.0258^2}{0.05} - \frac{5.3635^2}{2 - 0.05}\right] = 7.809\text{Nm} \tag{7.137}$$

In this particular case, the influence of a backward component on torque has been small. A great deal of torque is obtained at S = 0.05, but at a notably large current and smaller power factor. The low values of leakage reactances and the large value of magnetization reactance explain, in part, the impractically high power factor calculation. For more correct calculations, the complete equivalent circuit is to be solved. The phase b current is I_b = 9.2789A for phase a open at S = 0.05 and only 3.199A for balanced voltages at S = 0.03.

7.15 UNBALANCED ROTOR WINDINGS

Wound rotors may be provided with external three-phase resistances which may not be balanced. Also, broken bars in the rotor cage lead to unbalanced rotor windings. This latter case will be treated in Chapter 13 on transients.

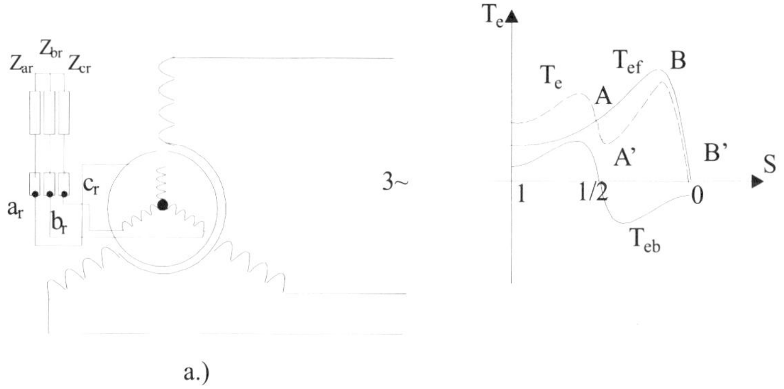

Figure 7.32 Induction motor with unbalanced wound rotor winding a.) and torque / speed curve b.)

However, the wound rotor unbalanced windings (Figure 7.32) may be treated here with the method of symmetrical components.

We may start decomposing the rotor phase currents $\underline{I}_{ar}$, $\underline{I}_{br}$ and $\underline{I}_{cr}$ into symmetrical components:

$$\begin{aligned}\underline{I}_{ar}{}^{f} &= \frac{1}{3}\left(\underline{I}_{ar} + a\underline{I}_{br} + a^2\underline{I}_{cr}\right)\\ \underline{I}_{ar}{}^{b} &= \frac{1}{3}\left(\underline{I}_{ar} + a^2\underline{I}_{br} + a\underline{I}_{cr}\right)\\ \underline{I}_{ar}{}^{0} &= \frac{1}{3}\left(\underline{I}_{ar} + \underline{I}_{br} + \underline{I}_{cr}\right) = 0\end{aligned} \tag{7.138}$$

Also, we do have

$$\underline{V}_{ar} = -\underline{Z}_{ar}\underline{I}_{ar};\ \underline{V}_{br} = -\underline{Z}_{br}\underline{I}_{br};\ \underline{V}_{cr} = -\underline{Z}_{cr}\underline{I}_{cr} \tag{7.139}$$

In a first approximation, all rotor currents have the frequency $f_2 = Sf_1$ at steady state. The forward mmf, produced by $\underline{I}_{ar}{}^{f}$, $\underline{I}_{br}{}^{f}$, $\underline{I}_{cr}{}^{f}$, interacts as usual with the stator winding and its equations are

$$\begin{aligned}&\underline{I}_r{}'^{f} R_r' - \underline{V}_r{}'^{f} = -jS\omega_1\underline{\psi}_r{}'^{f};\ \underline{\psi}_r{}'^{f} = L_r'\underline{I}_r{}'^{f} + L_{1m}\underline{I}_s{}^{f}\\ &\underline{I}_s{}^{f}R_s - \underline{V}_s = -j\omega_1\underline{\psi}_s{}^{f};\ \underline{\psi}_s{}^{f} = L_s\underline{I}_s{}^{f} + L_{1m}\underline{I}_r{}'^{f}\end{aligned} \tag{7.140}$$

The backward mmf component of rotor currents rotates with respect to the stator at the speed n_1'.

$$n_1' = n - S\frac{f_1}{p_1} = \frac{f_1}{p_1}(1-2S) \tag{7.141}$$

So it induces stator emfs of a frequency $f_1' = f_1(1 - 2S)$. The stator may be considered short-circuited for these emfs as the power grid impedance is very small in relative terms. The equations for the backward component are

$$\begin{aligned}&\underline{I}_r{}'^{b} R_r' - \underline{V}_r{}'^{b} = -jS\omega_1\underline{\psi}_r{}'^{b};\ \underline{\psi}_r{}'^{b} = L_r'\underline{I}_r{}'^{b} + L_{1m}\underline{I}_s{}^{b}\\ &\underline{I}_s{}^{b}R_s = -j(1-2S)\omega_1\underline{\psi}_s{}^{b};\ \underline{\psi}_s{}^{b} = L_s\underline{I}_s{}^{b} + L_{1m}\underline{I}_r{}'^{b}\end{aligned} \tag{7.142}$$

For given slip, rotor external impedances $\underline{Z}_{ar}$, $\underline{Z}_{br}$, $\underline{Z}_{cr}$, motor parameters stator voltage and frequency, Equation (7.139) and their counterparts for rotor voltages, (7.140) through (7.142) to determine $\underline{I}_r{}^{f}$, $\underline{I}_r{}^{b}$, $\underline{I}_s{}^{f}$, $\underline{I}_s{}^{b}$, $\underline{V}_r{}'^{f}$, $\underline{V}_r{}'^{b}$. Note that (7.140) and (7.142) are, per-phase basis and are thus valid for all three phases in the rotor and stator.

The torque expression is

$$T_e = 3P_1L_{1m}\left[\mathrm{Imag}\left(\underline{I}_s{}^{f}\underline{I}_r{}'^{f*}\right) + \mathrm{Imag}\left(\underline{I}_s{}^{b}\underline{I}_r{}'^{b*}\right)\right] = T_{ef} + T_{eb} \tag{7.143}$$

The backward component torque is positive (motoring) for $1 - 2S < 0$ or $S > ½$ and negative (braking) for $S < ½$. At start, the backward component torque is motoring. Also for $S = ½$ $\underline{I}_s^b = 0$ and, thus, the backward torque is zero. This backward torque is also called monoaxial or Geörge's torque.

The torque/speed curve obtained is given in Figure 7.33. A stable zone, AA′ around $S = ½$ occurs, and the motor may remain "hanged" around half ideal no-load speed.

The larger the stator resistance R_s, the larger the backward torque component is. Consequently, the saddle in the torque-speed curve will be more visible in low power machines where the stator resistance is larger in relative values $r_s = R_s I_n / V_n$. Moreover the frequency f_1' of stator current is close to f_1 and thus visible stator backward current pulsations occur. This may be used to determine the slip S as $f' - f_1 = 2Sf_1$. Low frequency and noise at $2Sf_1$ may be encountered in such cases.

7.16 ONE ROTOR PHASE IS OPEN

An extreme case of unbalanced rotor winding occurs when one rotor phase is open (Figure 7.33). Qualitatively the phenomenon occurs as described for the general case in paragraph 7.13. After reducing the rotor variables to stator, we have:

$$\underline{I}_{ar}'^{f} = -\underline{I}_{ar}'^{b} = -\frac{j}{\sqrt{3}}\underline{I}_{br}' \tag{7.144}$$

$$\underline{V}_r'^{f} = \underline{V}_r'^{b} = \frac{1}{3}\left(\underline{V}_{ar}' - \underline{V}_{br}'\right) \tag{7.145}$$

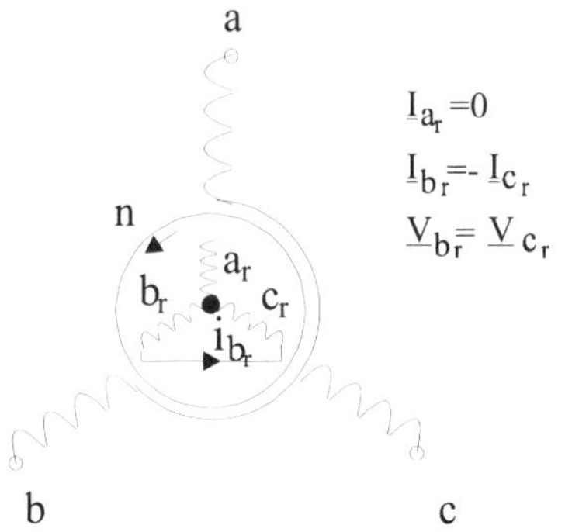

Figure 7.33 One rotor phase is open

Now, with (7.140) and (7.142), the unknowns are $\underline{I}_s^f$, $\underline{I}_s^b$, $\underline{I}_r'^f$, $\underline{I}_r'^b$, and, $\underline{V}_r'^f$.

This time we may get (from (7.142) with (7.144 and 7.145)):

$$\underline{I}_s^{\,b} = \frac{-j\omega_1(1-2S)L_{1m}\underline{I}_r^{\,\prime b}}{R_s + j\omega_1(1-2S)L_s} \tag{7.146}$$

It is again clear that with $R_s = 0$, the stator backward current is

$$\left(\underline{I}_s^{\,b}\right)_{R_s=0} = \frac{-L_{1m}}{L_s}\underline{I}_r^{\,\prime b} \tag{7.147}$$

Consequently the backward torque component in (7.143) becomes zero.

7.17 WHEN VOLTAGE VARIES AROUND RATED VALUE

It is common practice that, in industry, the local power grid voltage varies around rated values as influenced by the connection (disconnection) of other loads and of capacitors used for power factor correction. Higher supply voltages, even when balanced, notably influence the induction machine under various load levels.

A ±10% voltage variation around the rated value is, in general, accepted in many standards. However, the IMs have to be designed to incur such voltage variations without excessive temperature rise at rated power.

In essence, if the voltage increases above the rated value, the core losses p_{iron} in the machine increase notably as

$$p_{iron} \approx \left(C_h f_1 + C_e f_1^{\,2}\right)V^2 \tag{7.148}$$

In the same time, with the core notably more saturated, L_{1m} decreases accordingly.

The rated torque is to be obtained at lower slips. The power factor decreases so the stator current tends to increase while, for lower slip, the rotor current does not change much. So the efficiency, not only the power factor, decreases for rated load.

The temperature tends to rise. In the design process, a temperature reserve is to be allowed for nonrated voltage safe operation.

On the other hand, if low load operation at higher than rated voltage occurs, the high core losses mean excessive power loss. Motor disconnection instead of no-load operation is a practical solution in such cases.

If the voltage decreases below rated value, the rated torque is obtained at higher slip and, consequently, higher stator and rotor currents occur. The winding losses increase while the core losses decrease, because the voltage (and magnetizing flux linkage) decreases.

The efficiency should decrease slowly, while the power factor might also decrease slightly or remain the same as with rated voltage.

A too big voltage reduction leads, however, to excessive losses and lower efficiency at rated load. With partial load, lower voltage might be beneficial in terms of both efficiency and power factor, as the core plus winding losses decrease toward a minimum.

When IMs are designed for two frequencies – 50 or 60Hz – care is exercised to meet the over temperature limitation for the most difficult situation (lower frequency).

7.18 SUMMARY

- The relative difference between mmf speed $n_1 = f_1/p_1$ and the rotor speed n is called slip, $S = 1\text{-}np_1/f_1$.
- The frequency of the emf induced by the stator mmf field in the rotor is $f_2 = Sf_1$.
- The rated slip S_n, corresponding to rated n_n, $S_n = 0.08 - 0.01$; larger values correspond to smaller power motors (under 1 kW).
- At zero slip (S = 0), for short-circuited rotor, the rotor current and the torque is zero; this is called the ideal no-load mode $n = n_1 = f_1/p_1$.
- When the rotor windings are fed by balanced three-phase voltages V_r' of frequency $f_2 = Sf_1$, the zero rotor current and zero torque is obtained at a slip $S_0 = V_r'/E_1$, where E_1 is the stator phase emf. S_0 may be positive (subsynchronous operation) or negative (supersynchronous operation) depending on the phase angle between $\underline{V}_r'$ and $\underline{E}_1$.
- The active power traveling through an airgap (related to Pointing's vector) is called the electromagnetic power P_{elm}.
- The electromagnetic torque, for a short-circuited rotor (or in presence of a passive, additional, rotor impedance) is

$$T_e = P_{elm}\frac{P_1}{\omega_1}$$

- At ideal no-load speed (zero torque), the value of P_{elm} is zero, so the machine driven at that speed absorbs power in the stator to cover the winding and core loss.
- Motor no-load mode is when there is no-load torque at the shaft. The input active power now covers the core and stator winding losses and the mechanical losses.
- The induction motor operates as a motor when $(0 < S < 1)$.
- For generator mode $S < 0$ and for braking $S > 1$.
- In all operation modes, the singly fed IM motor "absorbs" reactive power for magnetization.
- Autonomous generating mode may be obtained with capacitors at terminals to produce the reactive power for magnetization.
- At zero speed, the torque is nonzero with stator balanced voltages. The starting torque $T_{es} = (0.5 - 2.2)T_{en}$. T_{en} – rated torque; starting current at rated voltage is $I_{start} = (5 - 7(8))I_n$; I_n – rated current. Higher values of starting current correspond to high efficiency motors.
- At high currents ($S >> S_n$), the slot leakage flux path saturates and the short-circuit (zero speed) inductance decreases by 10 to 40%. In the same time the rotor frequency $f_2 = Sf_1 = f_1$ and the skin effect causes a reduction

of rotor slot leakage inductance and a higher increase in rotor resistance. Lower starting current and larger starting torque are thus obtained.

- The closed-slot rotor leakage inductance saturates well below rated current. Still, it remains higher than with half-closed rotor slots.
- No-load and short-circuit (zero speed) tests may be used to determine the IM parameters–resistances and reactances.
- The electromagnetic torque T_e versus slip curve has two breakdown points: one for motoring and one for generating. The breakdown torque is inversely proportional to short-circuit leakage reactance X_{sc}, but independent of rotor resistance.
- Efficiency and power factor also have peak values both for motoring and generating at distinct slips. The rather flat maxima allow for a large plateau of good efficiency for loads from 25 to 150%.
- Adequate phasor diagrams evidentiating stator, rotor, and airgap (magnetization) flux linkages (per phase) show that at steady-state, the rotor flux linkage and rotor current per phase are time-phase shifted by 90^0. It is $+90^0$ for motor and -90^0 for generating. If the rotor flux amplitude may also be maintained constant during transients, the 90^0 phase shift of rotor flux linkage and rotor currents would also stand for transients. Independent rotor flux and torque control may thus be obtained. This is the essence of vector control.
- Unbalanced stator voltages cause large imbalances in stator currents. Derating is required for sustained stator voltage imbalance.
- Higher than rated balanced voltages cause lower efficiency and power factor for rated power. In general, IMs are designed (thermally) to stand ±10% voltage variation around the rated value.
- Unbalanced rotor windings cause an additional stator current at the frequency $f_1(1 - 2S)$, besides the one at f_1. Also an additional rotor-initiated backward torque which is zero at $S = ½$ ($n = f_1/2p_1$), Geörge's torque, is produced. A saddle in the torque versus slip occurs around $S = ½$ for IMs with relatively large stator resistance (low-power motors). The machine may be "hanged" (stuck) around half the ideal no-load speed.
- In this chapter, we dealt only with the single-cage rotor IM and the fundamental mmf and airgap field performance during steady state for constant voltage and frequency. The next chapter treats starting and speed control methods.

7.19 REFERENCES

1. P.T. Lagonotte, H.Al. Miah, N. Poloujadoff, Modelling and Identification of Parameters of Saturated Induction Machine Operating under Motor and Generator Conditions, EMPS: Vol.27, No.2, 1999, pp.107 – 121.
2. J. Kneck, D.A. Casada, P.J. Otday, A Comparision of Two Energy Efficient Motors, IEEE Trans Vol.EC – 13, No.2., 1998, pp.140 – 147.
3. A.H. Bonnett, G.C. Soukup, NEMA Motor-Generator Standards for Three Phase Induction Motors, IEEE – IA Magazine, Vol.5, No.3, 1999, pp.49 – 63.

Chapter 8

STARTING AND SPEED CONTROL METHODS

Starting refers to speed, current, and torque variations in an induction motor when fed directly or indirectly from a rather constant voltage and frequency local power grid.

A "stiff" local power grid would mean rather constant voltage even with large starting currents in the induction motors with direct full-voltage starting (5.5 to 5.6 times rated current is expected at zero speed at steady state). Full-starting torque is produced in this case and starting over notable loads is possible.

A large design KVA in the local power grid, which means a large KVA power transformer, is required in this case. For starting under heavy loads, such a large design KVA power grid is mandatory.

On the other hand, for low load starting, less stiff local power grids are acceptable. Voltage decreases due to large starting currents will lead to a starting torque, which decreases with voltage squared. As many local power grids are less stiff, for low starting loads, it means to reduce the starting currents, although in most situations even larger starting torque reduction is inherent for cage rotor induction machines.

For wound-rotor induction machines, additional hardware connected to the rotor brushes may provide even larger starting torque while starting currents arc reduced. In what follows, various starting methods and their characteristics are presented. Speed control means speed variation with given constant or variable load torque. Speed control can be performed by either open loop (feed forward) or close loop (feedback). In this chapter, we will introduce the main methods for speed control and the corresponding steady state characteristics.

Transients related to starting and speed control are treated in Chapter 13. Close loop speed control methods are beyond the scope of this book as they are fully covered by literature presented by References 1 and 2

8.1 STARTING OF CAGE-ROTOR INDUCTION MOTORS

Starting of cage-rotor induction motors may be performed by:

- Direct connection to power grid
- Low voltage auto-transformer
- Star-delta switch connection
- Additional resistance (reactance) in the stator
- Soft starting (through static variacs)

8.1.1 Direct starting

Direct connection of cage-rotor induction motors to the power grid is used when the local power grid is off when rather large starting torques are required.

Typical variations of stator current and torque with slip (speed) are shown in Figure 8.1.

For single cage induction motors, the rotor resistance and leakage inductance are widely influenced by skin effect and leakage saturation. At start, the current is reduced and the torque is increased due to skin effect and leakage saturation.

In deep-bar or double-cage rotor induction motors, the skin effect is more influential as shown in Chapter 9. When the load torque is notable from zero speed on (> 0.5 T_{er}) or the inertia is large (J_{total} > 3J_{motor}), the starting process is slower and the machine may be considered to advance from steady state to steady state until full-load speed is reached in a few seconds to minutes (in large motors).

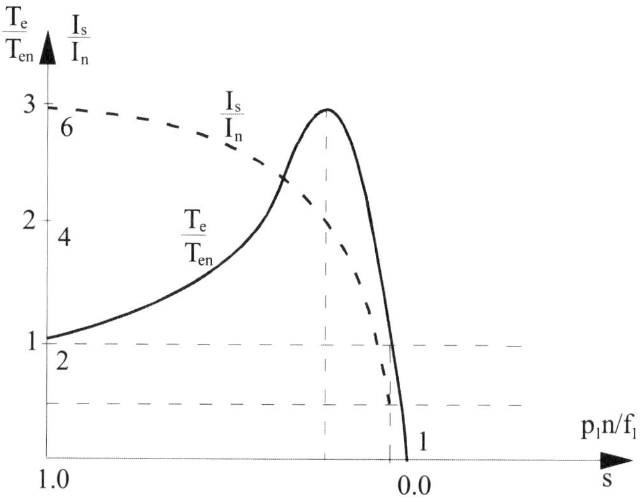

Figure 8.1 Current and torque versus slip (speed) in a single induction motions

If the induction motor remains at stall for a long time, the rotor and stator temperatures become too large, so there is a maximum stall time for each machine design.

On the other hand, for frequent start applications, it is important to count the rotor acceleration energy.

Let us consider applications with load torque proportional to squared speed (fans, ventilators). In this case we may, for the time being, neglect the load torque presence during acceleration. Also, a large inertia is considered and thus the steady state torque/speed curve is used.

$$\frac{J}{p_1}\cdot\frac{d\omega_r}{dt} \approx T_e(\omega_r);\ \omega_r = \omega_1(1-S) \tag{8.1}$$

The rotor winding loss p_{cor} is

$$p_{cor} = S \cdot P_{elm} = S \cdot T_e\left(\frac{\omega_1}{P}\right) \tag{8.2}$$

with T_e from (8.1), the rotor winding losses W_{cos} are

$$W_{cor} = \int_0^{t_s}\left(S \cdot T_e \cdot \frac{\omega_1}{p_1}\right)dt = \int_0^{t_r}\frac{J}{p_1}\cdot\frac{d\omega_r}{dt}\cdot\frac{\omega_1}{p_1}\cdot S dt \approx -\frac{J}{p_1^2}\int_1^0 \omega_1^2 \cdot S dS = \\ = +\frac{J}{2}\left(\frac{\omega_1}{p_1}\right)^2; \; S_{initial} = 1.0; \; S_{final} = 0.0 \tag{8.3}$$

On the other hand, the stator winding losses during motor acceleration under no load W_{cos} are:

$$W_{cos} = 3\int_0^{t_s} I_s^2(S) R_s dt \approx 3\int I_r'^2 \cdot R_r' \cdot \frac{R_s}{R_r'} dt \approx W_{cor} \cdot \frac{R_s}{R_r'} \tag{8.4}$$

Consequently, the total winding energy losses W_{co} are

$$W_{co} = W_{cos} + W_{cor} = \frac{J}{2}\left(\frac{\omega_1}{p_1}\right)^2\left(1 + \frac{R_s}{R_r'}\right) \tag{8.5}$$

A few remarks are in order.

- The rotor winding losses during rotor acceleration under no load are equal to the rotor kinetic energy at ideal no-load speed
- Equation (8.5) tends to suggest that for given stator resistance R_s, a larger rotor resistance (due to skin effect) is beneficial
- The temperature transients during such frequent starts are crucial for the motor design, with (8.5) as a basic evaluation equation for total winding losses
- The larger the rotor attached inertia J, the larger the total winding losses during no load acceleration.

Returning to the starting current issue, it should be mentioned that even in a rather stiff local power grid a voltage drop occurs during motor acceleration, as the current is rather large. A 10% voltage reduction in such cases is usual.

On the other hand, with an oversized reactive power compensation capacitor, the local power grid voltage may increase up to 20% during low energy consumption hours.

Such a large voltage has an effect on voltage during motor starting in the sense of increasing both the starting current and torque.

Example 8.1 Voltage reduction during starting

The local grid power transformer in a small company has the data $S_n = 700$ KVA, secondary line voltage $V_{L2} = 440$ V (star connection), short circuit voltage $V_{SC} = 4\%$, $\cos\varphi_{SC} = 0.3$. An induction motor is planned to be installed for direct starting. The IM power $P_n = 100$ kW, $V_L = 440$ V (star connection), rated efficiency $\eta_n = 95\%$, $\cos\varphi_n = 0.92$, starting current $\frac{I_{start}}{I_n} = \frac{6.5}{1}$, and $\cos\varphi_{start} = 0.3$.

Calculate the transformer short circuit impedance, the motor starting current at rated voltage, and impedance at start. Finally determine the voltage drop at start, and the actual starting current in the motor.

Solution

First we have to calculate the rated transformer current in the secondary I_{2n}

$$I_{2n} = \frac{S_n}{\sqrt{3}\cdot V_{L2}} = \frac{700\times10^3}{\sqrt{3}\cdot 440} = 919.6\text{A} \tag{8.6}$$

The short circuit voltage V_{SC} corresponds to rated current

$$V_{SC2} = 0.04\cdot\frac{V_{L2}}{\sqrt{3}} = I_{2n}\left|\underline{Z}_{SC2}\right| \tag{8.8}$$

$$\left|\underline{Z}_{SC2}\right| = \frac{0.04\cdot 440}{\sqrt{3}\cdot 919.6} = 11.0628\times10^{-3}\Omega \tag{8.8}$$

$$R_{SC} = \left|\underline{Z}_{SC}\right|\cos\varphi_{SC} = 11.0628\cdot10^{-3}\cdot0.3 = 3.3188\cdot10^{-3}\Omega \tag{8.9}$$

$$X_{SC} = \left|\underline{Z}_{SC}\right|\sin\varphi_{SC} = 11.0628\cdot10^{-3}\cdot\sqrt{1-0.3^2} = 10.5532\cdot10^{-3}\Omega$$

For the rated voltage, the motor rated current I_{sn} is

$$I_{sn} = \frac{P_n}{\sqrt{3}\eta_n\cos\varphi_n V_L} = \frac{100\times10^3}{\sqrt{3}\cdot0.95\cdot0.92\cdot440} = 150.3\text{A} \tag{8.10}$$

The starting current is

$$I_{start} = 6.5\times150.3 = 977\ \text{A} \tag{8.11}$$

Now the starting motor impedance $Z_{start} = R_{start} + jX_{start}$ is

$$R_{start} = \frac{V_L}{\sqrt{3}I_{start}}\cos\varphi_{start} = \frac{440}{\sqrt{3}\cdot977}\cdot0.3 = 78.097\times10^{-3}\Omega \tag{8.12}$$

$$X_{start} = \frac{V_L}{\sqrt{3}I_{start}}\sin\varphi_{start} = \frac{440}{\sqrt{3}\cdot977}\cdot\sqrt{1-0.3^2} = 0.24833\Omega \tag{8.13}$$

Now the actual starting current in the motor/transformer I'_{start} is

$$I'_{start} = \frac{V_L}{\sqrt{3}\left|R_{SC} + R_{start} + j(X_{SC} + X_{start})\right|} =$$

$$= \frac{440}{10^{-3}\sqrt{3}\left|3.3188 + 78.097 + j(10.5542 + 248.33)\right|} \tag{8.14}$$

$$== 937[0.3 + j0.954]$$

The voltage at motor terminal is:

$$\frac{V'_L}{V_L} = \frac{I'_{start}}{I_{start}} = \frac{937}{977} = 0.959 \tag{8.15}$$

Consequently the voltage reduction $\frac{\Delta V_L}{V_L}$ is

$$\frac{\Delta V_L}{V_L} = \frac{V_L - V'_L}{V_L} = 1.0 - 0.959 = 0.04094 \tag{8.16}$$

A 4.1% voltage reduction qualifies the local power grid as stiff for the starting of the newly purchased motor. The starting current is reduced from 977 A to 937 A, while the starting torque is reduced $\left(\frac{V'_L}{V_L}\right)^2$ times. That is, $0.959^2 \approx$ 0.9197 times. A smaller KVA transformer and/or a larger shortcircuit transformer voltage would result in a larger voltage reduction during motor starting, which may jeopardize a safe start under heavy load.

Notice that high efficiency motors have larger starting currents so the voltage reduction becomes more severe. Larger transformer KVA is required in such cases.

8.1.2 Autotransformer starting

Although the induction motor power for direct starting has increased lately, for large induction motors (MW range) and not so stiff local power grids voltage, reductions below 10% are not acceptable for the rest of the loads and, thus, starting current reduction is mandatory. Unfortunately, in a cage rotor motor, this means a starting torque reduction, so reducing the stator voltage will reduce the stator current K_i times but the torque K_i^2.

$$K_i = \frac{V_L}{V'_L} = \frac{I_L}{I'_L} = \sqrt{\frac{T_e}{T'_e}}; \quad I_s = \frac{\frac{V_L}{\sqrt{3}}}{\left|\underline{Z}_e(s)\right|} \tag{8.17}$$

because the current is proportional to voltage and the torque with voltage squared.

Autotransformer voltage reduction is adequate only with light high starting loads (fan, ventilator, pump loads).

A typical arrangement with three-phase power switches is shown on Figure 8.2.

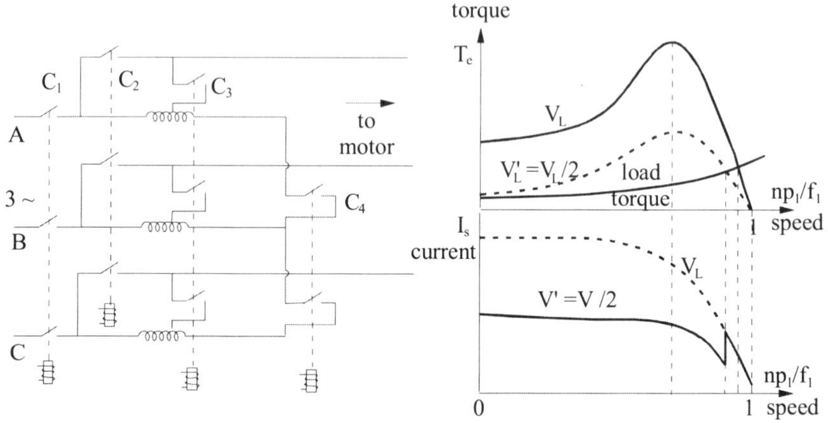

Figure.8.2 Autotransformer starting

Before starting, C_4 and C_3 are closed. Finally, the general switch C_1 is closed and thus the induction motor is fed through the autotransformer, at the voltage

$$V_L^{'}\left(\frac{V_L^{'}}{V_L}\cong 0.5, 0.65, 0.8\right)$$

To avoid large switching current transients when the transformer is bypassed, and to connect the motor to the power grid directly, first C_4 is opened and C_2 is closed with C_3 still closed. Finally, C_3 is opened. The transition should occur after the motor accelerated to almost final speed or after a given time interval for start, based on experience at the commissioning place. Autotransformers are preferred due to their smaller size, especially with $\frac{V_L^{'}}{V_L}=0.5$ when the size is almost halved.

8.1.3 Wye-delta starting

In induction motors that are designed to operate with delta stator connection it is possible, during starting, to reduce the phase voltage by switching to wye connection (Figure 8.3).

During wye connection, the phase voltage V_s becomes

$$V_S=\frac{V_L}{\sqrt{3}} \tag{8.18}$$

so the phase current, for same slip, I_{sY}, is reduced $\sqrt{3}$ times

$$I_{s\lambda} = \frac{I_{s\Delta}}{\sqrt{3}} \tag{8.19}$$

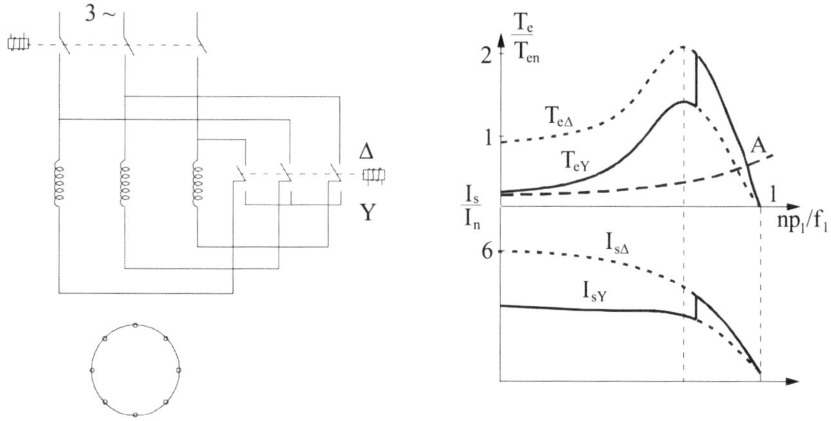

Figure 8.3 Wye-delta starting

Now the line current in Δ connection $I_{l\Delta}$ is

$$I_{l\Delta} = \sqrt{3} I_{S\Delta} = 3 I_{sY} \tag{8.20}$$

so the line current is three times smaller for wye connection. The torque is proportional to phase voltage squared

$$\frac{T_{e\lambda}}{T_{e\Delta}} = \left(\frac{V_{sY}}{V_L}\right)^2 = \frac{1}{3} \tag{8.21}$$

therefore, the wye-delta starting is equivalent to an $\frac{\sqrt{3}}{1}$ reduction of phase voltage and a 3 to 1 reduction in torque. Only low load torque (at low speeds) and mildly frequent start applications are adequate for this method.

A double-throw three-phase power switch is required and notable transients are expected to occur when switching from wye to delta connection takes place.

An educated guess in starting time is used to figure when switching is to take place.

The series resistance and series reactance starting methods behave in a similar way as voltage reduction in terms of current and torque. However, they are not easy to build especially in high voltage (2.3 kV, 4 kV, 6 kV) motors. At the end of the starting process they have to be shortcircuited. With the advance of softstarters, such methods are used less and less frequently.

8.1.4 Softstarting

We use here the term softstarting for the method of a.c. voltage reduction through a.c. voltage controllers called softstarters.

In numerous applications such as fans, pumps, or conveyors, softstarters are now common practice when speed control is not required.

Two basic symmetric softstarter configurations are shown in Figure 8.4. They use thyristors and enjoy natural commutation (from the power grid). Consequently, their cost is reasonably low to be competitive.

In small (sub kW) power motors, the antiparallel thyristor group may be replaced by a triac to cut costs.

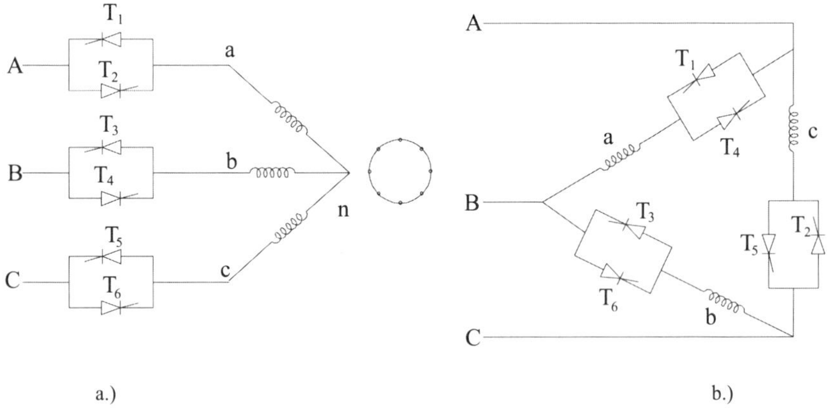

Figure 8.4 Softstarters for three-phase induction motors: a.) wye connection, b.) delta connection

Connection a.) in Figure 8.4 may also be used with delta connection and this is why it became a standard in the marketplace.

Connection b.) in Figure 8.4 reduces the current rating of the thyristor by $\sqrt{3}$ in comparison with connection a.). However, the voltage rating is basically the same as the line voltage, and corresponds to a faulty condition when thyristors in one phase remain on while all other phases would be off.

To apply connection b.), both phase ends should be available, which is not the case in many applications.

The 6 thyristors in Figure 8.4 b.) are turned on in the order T_1, T_2, T_3, T_4, T_5, T_6 every 60°. The firing angle α is measured with respect to zero crossing of V_{an} (Figure 8.5). The motor power factor angle is φ_{10}.

The stator current is continuous if $\alpha < \varphi_1$ and discontinuous (Figure 8.5) if $\alpha > \varphi_1$.

As the motor power factor angle varies with speed (Figure 8.5), care must be exercised to keep $\alpha > \varphi_1$ as a current (and voltage) reduction for starting is required.

So, besides voltage, current fundamentals, and torque reductions, the soft starters produce notable voltage and current low-order harmonics. Those harmonics pollute the power grid and produce additional losses in the motor. This is the main reason why softstarters do not produce notable energy savings when used to improve performance at low loads by voltage reduction. [3]

However, for light load starting, they are acceptable as the start currents are reduced. The acceleration into speed time is increased, but it may be programmed (Figure 8.6).

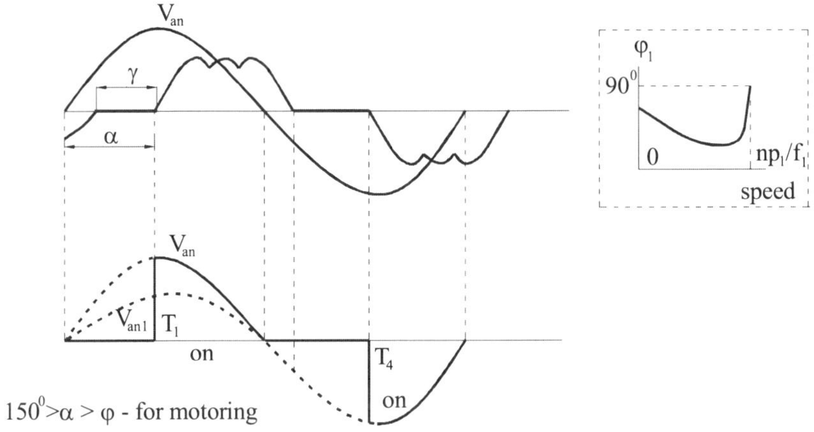

Figure 8.5 Softstarter phase voltage and current

During starting, either the stator current or the torque may be controlled. After the start ends, the soft starter will be isolated and a bypass power switch takes over. In some implementations only a short-circuiting (bypass) power switch is closed after the start ends and the softstarter command circuits are disengaged (Figure 8.7). Dynamic braking is also performed with softstarters. The starting current may be reduced to twice rated current or more.

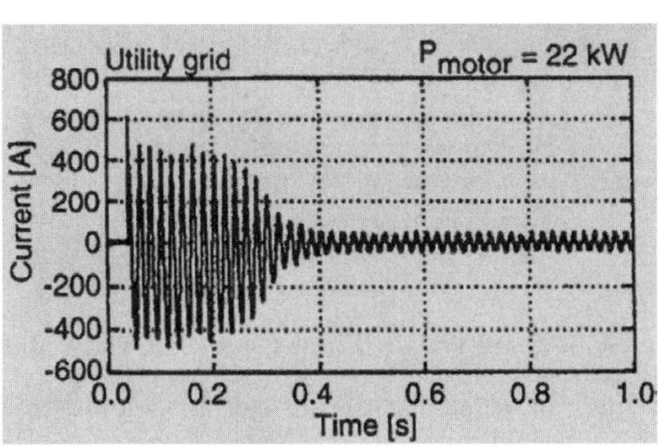

Figure 8.6 Start under no load of a 22 kW motor
a) direct starting; b) softstarting (continued)

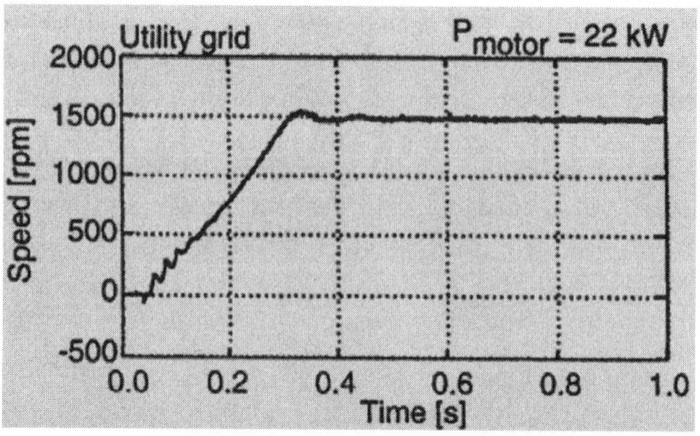

a.)

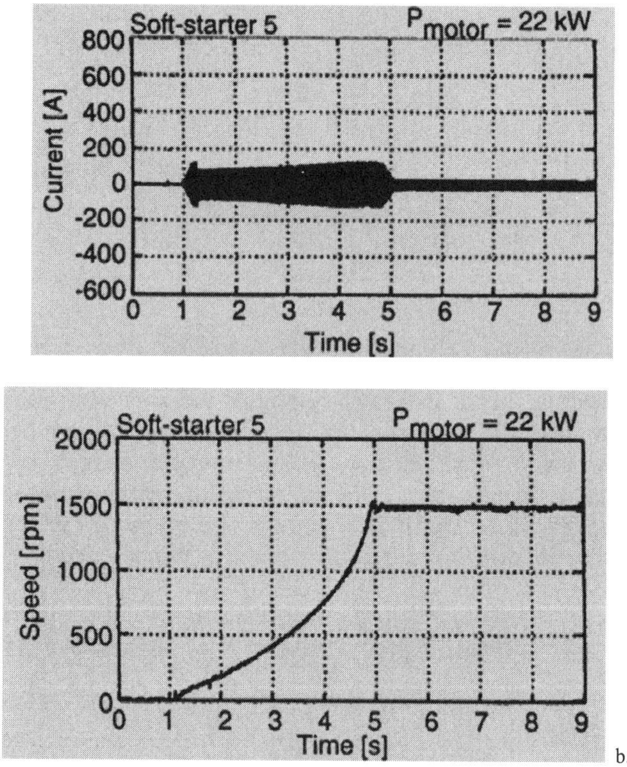

b.)

Figure 8.6 (continued)

8.2 STARTING OF WOUND-ROTOR INDUCTION MOTORS

A wound-rotor induction motor is built for heavy load frequent starts and (or) limited speed control motoring or generating at large power loads (above a few hundred kW).

Here we insist on the classical starting method used for such motors: variable resistance in the rotor circuit (Figure 8.8). As discussed in the previous chapter, the torque/speed curve changes with additional rotor resistance in terms of critical slip S_k, while the peak (breakdown) torque remains unchanged (Figure 8.8.b.).

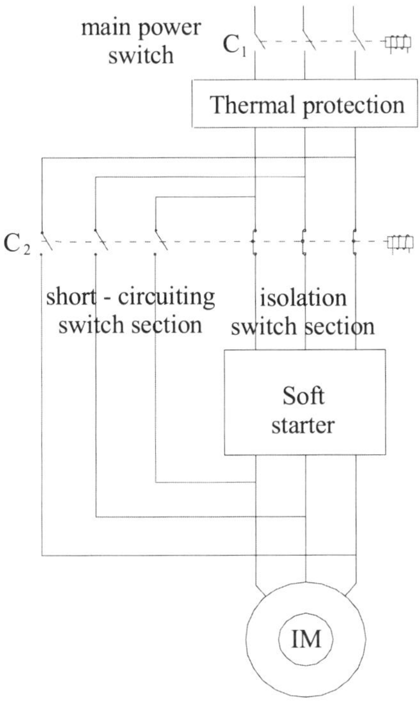

Figure 8.7 Softstarter with isolation and short-circuiting power switch C_2

$$S_K = \frac{C_1(R'_r + R'_{ad})}{\sqrt{R_s^2 + (X_{sl} + C_1 X'_{rl})^2}}$$
$$T_{ek} = \frac{3p_1}{2C_1}\frac{V_s^2}{\omega_1} \cdot \frac{1}{R_1 \pm \sqrt{R_s^2 + (X_{sl} + C_1 X'_{rl})^2}} \tag{8.22}$$

As expected, the stator current, for given slip, also decreases with R'_{ad} increasing (Figure 8.8 c.). It is possible to start ($S'' = 1$) with peak torque by providing $S_K'' = 1.0$. When R'_{ad} increases, the power factor, especially at high slips, improves the torque/current ratio. The additional losses in R'_{ar} make this method poor in terms of energy conversion for starting or sustained low-speed operation.

However, the peak torque at start is an extraordinary feature for heavy starts and this feature has made the method popular for driving elevators or overhead cranes.

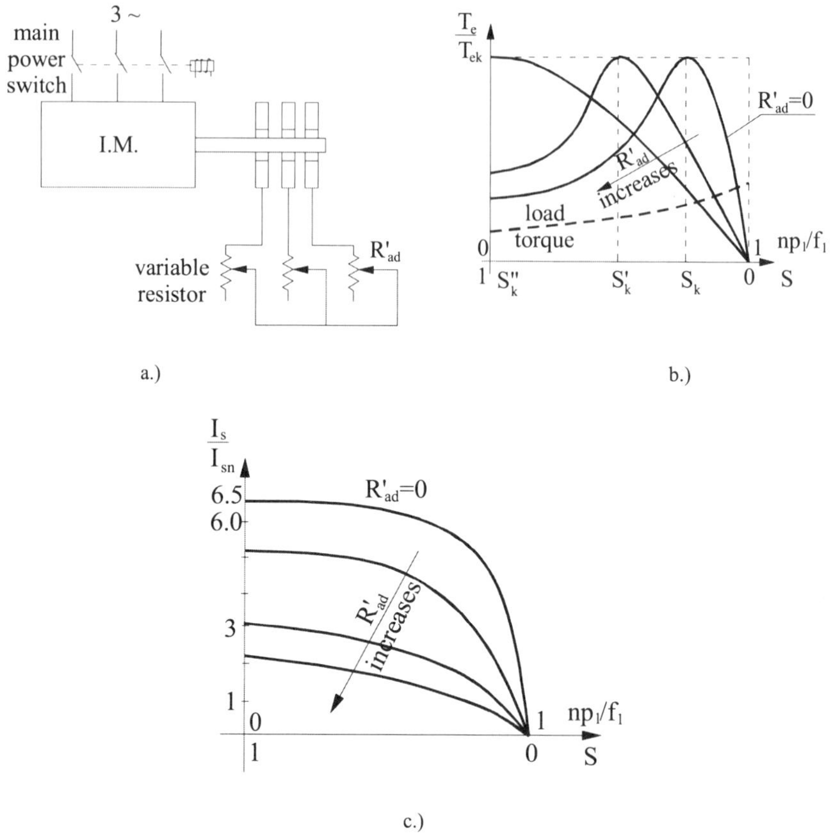

Figure 8.8 Starting through additional rotor resistance R'_{ad}
a.) general scheme, b.) torque/speed, c.) current/speed

There a few ways to implement the variable rotor resistance method as shown in Figure 8.9 a,b,c.

The half-controlled rectifier and the diode-rectifier-static switch methods (Figure 8.8 a,b) allow for continuous stator (or rotor) current close loop control during starting. Also, only a fix resistance is needed.

The diode rectifier-static-switch method implies a better power factor and lower stator current harmonics but is slightly more expensive.

A low cost solution is shown on Figure 8.9 c, where a three-phase pair of constant resistances and inductances lead to an equivalent resistance R_{oe} (Figure 8.9 c), which increases when slip increases (or speed decreases).

The equivalent reactance of the circuit decreases with slip increases. In this case, there is no way to intervene in controlling the stator current unless the inductance L_0 is varied through a d.c. coil which controlls the magnetic saturation in the coil laminated magnetic core.

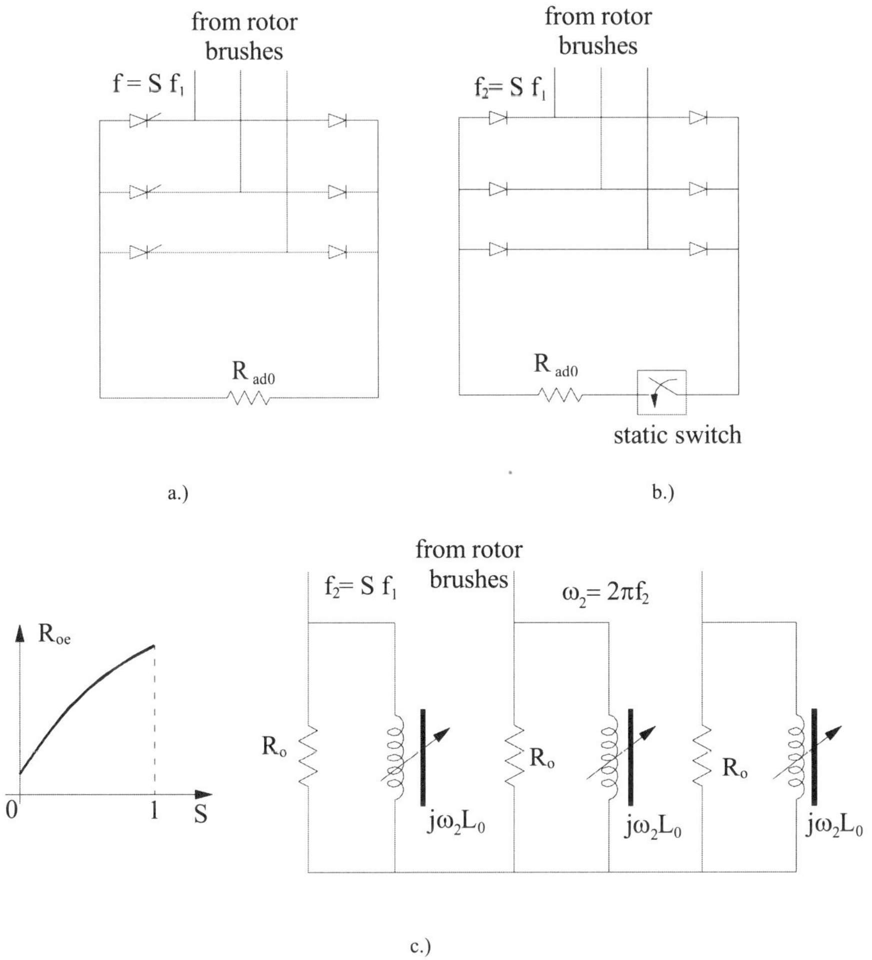

Figure 8.9 Practical implementations of additional rotor resistance
a.) with half-controlled rectifier; b.) with diode rectifier and static switch; c.) with self-adjustable resistance

8.3 SPEED CONTROL METHODS FOR CAGE-ROTOR INDUCTION MOTORS

Speed control means speed variation under open loop (feedforward) or close loop conditions. We will discuss here only the principles and steady-state characteristics of speed control methods.

For cage-rotor induction motors, all speed control methods have to act on the stator windings as only they are available.

To start, here is the speed/slip relationship.

$$n = \frac{f_1}{p_1}(1 - S) \tag{8.23}$$

Equation (8.23) suggests that the speed may be modified through

- Slip S variation: through voltage reduction
- Pole number $2p_1$ change: through pole changing windings
- Frequency f_1 control: through frequency converters

8.3.1 The voltage reduction method

When reducing the stator phase (line) voltage through an autotransformer or an a.c. voltage controller as inferred from (8.22), the critical slip s_K remains constant, but the peak (breakdown) torque varies with voltage V_s squared (Figure 8.10).

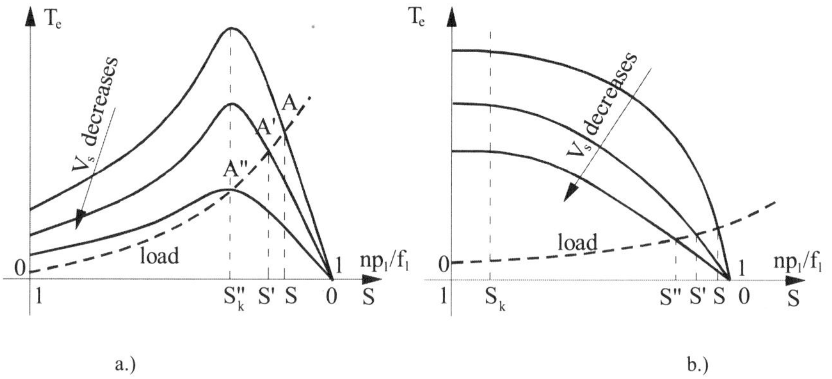

Figure 8.10 Torque versus speed for different voltage levels V_s
a.) standard motor: S_K=0.04 – 0.10
b.) high rotor resistance (solid iron) rotor motor S_K > 0.7 – 0.8

$$T_e = f(S) \cdot V_s^2 \tag{8.24}$$

The speed may be varied downward until the critical slip speed n_K is reached.

$$n_K = \frac{f_1}{p_1}(1 - S_K) \tag{8.25}$$

In standard motors, this means an ideal speed control range of

$$(n_1\text{-}n_k)/n_1 = S_K < 0.1 = 10\% \tag{8.26}$$

On the contrary, in high rotor resistance rotor motors, such as solid rotor motors where the critical slip is high, the speed control range may be as high as 100%.

However, in all cases, when increasing the slip, the rotor losses increase accordingly, so the wider the speed control range, the poorer the energy conversion ratio.

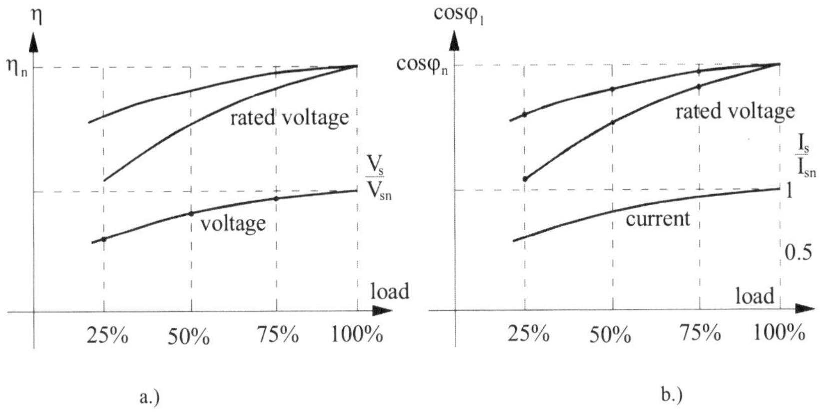

Figure 8.11 Performance versus load for reduced voltage
a.) voltage V_s/V_{sn} and efficiency η
b.) cos φ_1 and stator current

Finally, a.c. voltage controllers (soft starters) have been proposed to reduce voltage, when the load torque decreases, to reduce the flux level in the machine and thus reduce the core losses and increase the power factor and efficiency while stator current also decreases.

The slip remains around the rated value. Figure 8.11. show a qualitative illustration of the above claims.

The improvement in performance at light loads, through voltage reduction, tends to be larger in low power motors (less than 10 kW) and lower for larger power levels. [3] In fact, above 50% load the overall efficiency decreases due to significant soft starter losses.

For motor designs dedicated to long light load operation periods, the efficiency decreases only 3 to 4% from 100% to 25% load and, thus, reduced voltage by soft starters does not produce significant performance improvements. In view of the above, voltage reduction has very limited potential for speed control.

8.3.2 The pole-changing method.

Changing the number of poles, $2p_1$, changes the ideal no-load speed $n_1 = f_1/p_1$ accordingly. In Chapter 4, we discussed pole-changing windings and their connections of phases to produce constant power or constant torque for the two different pole numbers $2p_2$ and $2p_1$ (Figure 8.12).

The IM has to be sized carefully for the largest torque conditions and with careful checking for performance for both $2p_1$ and $2p_2$ poles.

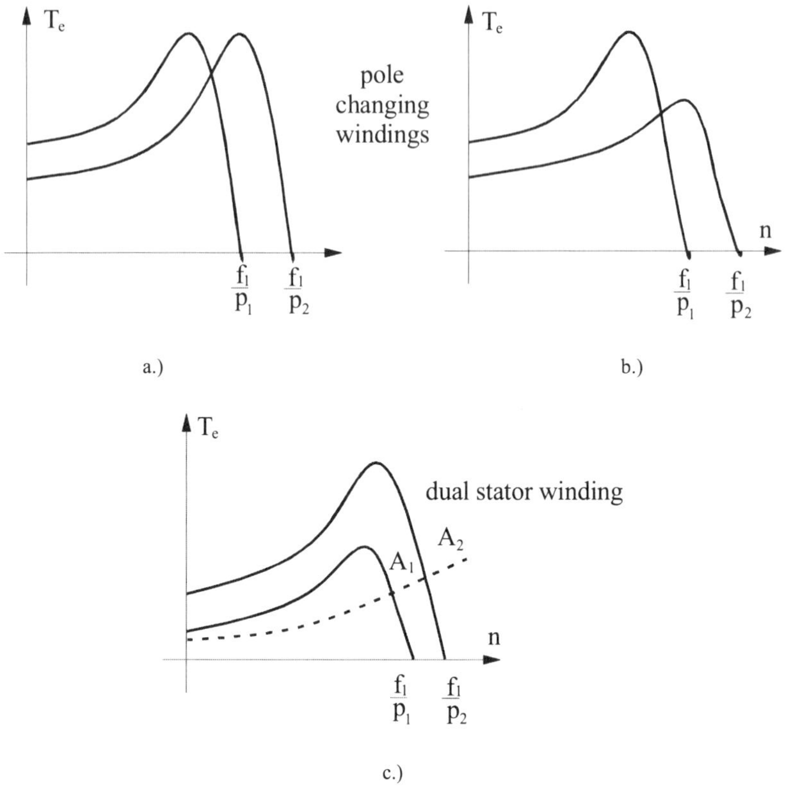

Figure 8.12 Pole-changing torque/speed curves
a.) constant torque; b.) constant power; c.) dual winding

Switching from $2p_2$ to $2p_1$ and back in standard pole-changing (p_2/p_1 = 2 Dahlander) windings implies complicated electromechanical power switches.

Better performance, new pole-changing windings (Chapter 4) that require only 2 single throw power switches have been proposed recently.

For applications where the speed ratio is 3/2, 4/3, 6/4, etc. and the power drops dramatically for the lower speed (wind generators), dual windings may be used. The smaller power winding will occupy only a small part of slot area. Again, only two power switches are required-the second one of notably smaller power rating.

Pole-changing windings are also useful for wide speed range $\omega_{max}/\omega_b > 3$ power induction motor drives (spindle drives or electric (or hybrid) automobile electric propulsion). This solution is a way to reduce motor size for $\omega_{max}/\omega_b > 3$.

8.4 VARIABLE FREQUENCY METHODS

When changing frequency f_1, the ideal no-load speed $n_1=f_1/p_1$ changes and so does the motor speed for given slip.

Frequency static converters are capable of producing variable voltage and frequency, V_s, f_1. A coordination of V_s with f_1 is required.

Such a coordination may be "driven" by an optimization criterion or by flux linkage control in the machine to secure fast torque response.

The various voltage-frequency relationships may be classified into 4 main categories:

- V/f scalar control
- Rotor flux vector control
- Stator flux vector control
- Direct torque and flux control

Historically, the V/f scalar control was first introduced and is used widely today for open loop speed control in driving fans, pumps, etc., which have the load torque dependent on speed squared, or more. The method is rather simple, but the torque response tends to be slow.

For high torque response performance, separate flux and torque control much like in a d.c. machine, is recommended. This is called vector control.

Either rotor flux or stator flux control is performed. In essence, the stator current is decomposed into two components. One is flux producing while the other one is torque producing. This time the current or voltage phase and amplitude and frequency are continuously controlled. Direct torque and flux control (DTFC) [2] shows similar performance.

Any torque/speed curve could thus be obtained as long as voltage and current limitations are met. Also very quick torque response, as required in servodrives, is typical for vector control.

All these technologies are now enjoying very dynamic markets worldwide.

8.4.1 V/f scalar control characteristics

The frequency converter, which supplies the motor produces sinusoidal symmetrical voltages whose frequency is ramped for starting. Their amplitude is related to frequency by a certain relationship of the form

$$V = V_0 + K_0(f_1) \cdot f_1 \tag{8.27}$$

V_0 is called the voltage boost destined to cover the stator resistance voltage at low frequency (speed).

Rather simple $K_0(f_1)$ functions are implemented into digitally controlled variable frequency converters (Figure 8.13).

As seen in Figure 8.13 a), a slip frequency compensator may be added to reduce speed drop with load (Figure 8.14).

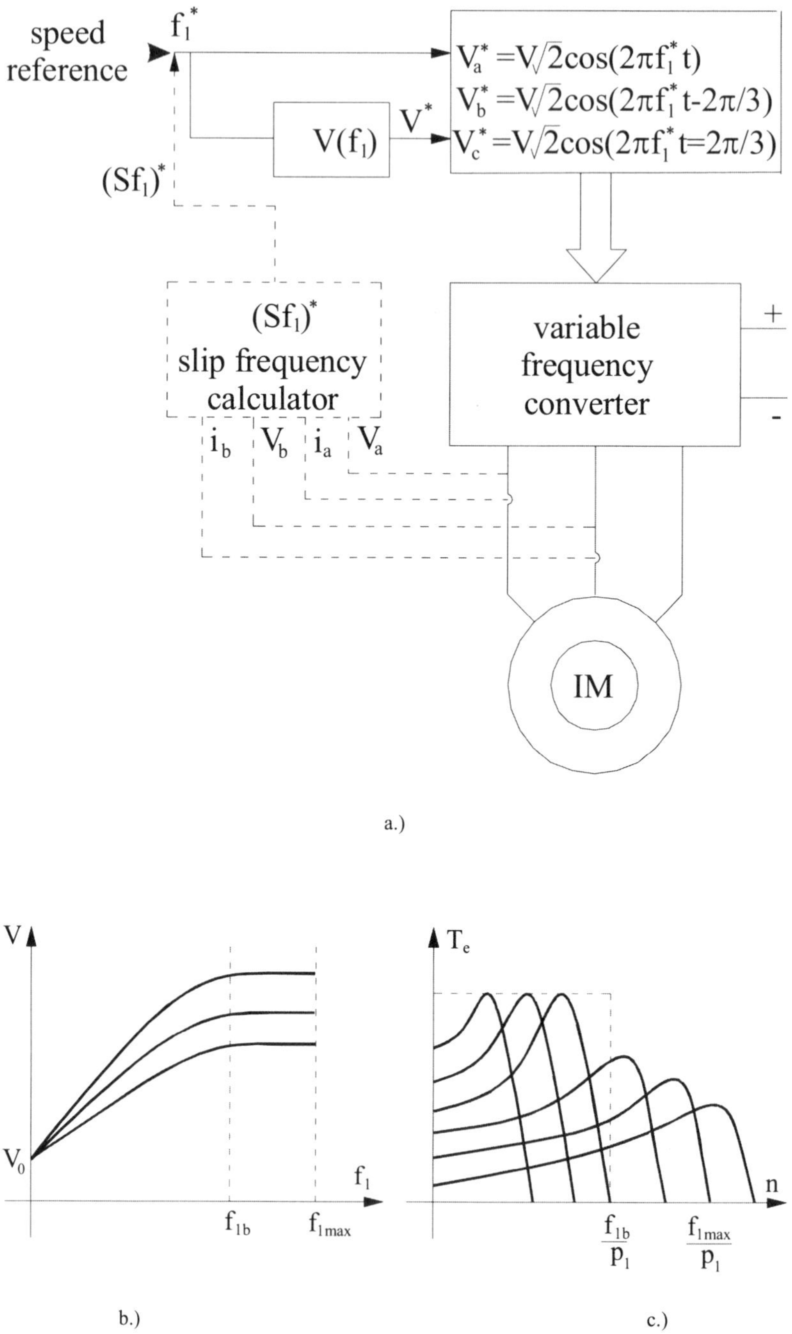

Figure 8.13 V/f speed control
a) structural diagram b) V/f relationship c) torque/speed curve

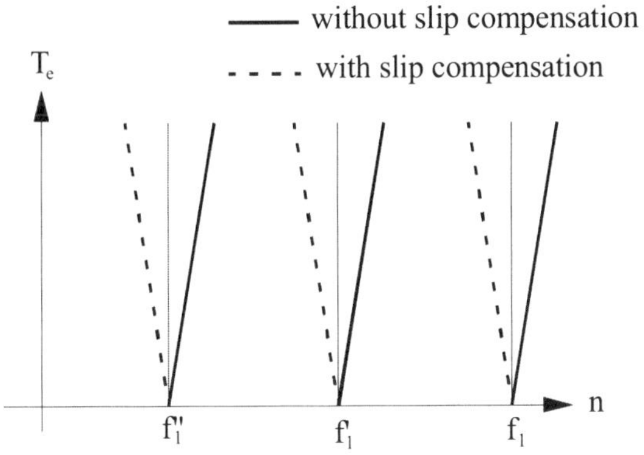

Figure 8.14 Torque/speed curves with slip compensation

Safe operation is provided above f_{1min} = 3 Hz as torque response is rather slow (above 20 milliseconds).

Example 8.2 V/f speed control

An induction motor has the following design data: $P_n = 10$ kW, $V_{Ln} = 380$ V (Y), $f_{1b} = 50$ Hz, $\eta_n = 0.92$, $\cos\varphi_n = 0.9$, $2p_1 = 4$, $I_{start}/I_n = 6/1$, $I_0/I_{sn} = 0.3$, $p_{mec} = 0.015P_n$, $p_{add} = 0.015P_n$; core losses are neglected and $R_s = R'_r$ and $X_{sl} = X'_{rl}$. Such data are known from the manufacturer. Let us calculate: rated current, motor parameters R_s, X_{sl}, X_{1m}, critical slip S_K and breakdown torque T_{eK} at f_{1b}, and f_{1max} for rated voltage; voltage for critical slip S_K and minimum frequency $f_{1min} = 3$ Hz to provide rated breakdown torque.

Find the voltage boost V_0 for linear V/f dependence up to base speed.

Solution

$$f_{1b}=50\text{Hz}$$

Based on the efficiency expression

$$\eta_n = \frac{P_n}{\sqrt{3}V_L I_{sn} \cos\varphi_n} \tag{8.28}$$

The rated current

$$I_{sn} = \frac{10000}{\sqrt{3}\cdot 0.92\cdot 380\cdot 0.9} = 18.37\text{A}$$

The rotor and stator winding losses $p_{cos} + p_{cor}$ are

$$p_{cos} + p_{cor} = \frac{P_n}{\eta_n} - P_n - p_{mec} - p_{add} \tag{8.29}$$

$$p_{cos} + p_{cor} = 10000\left(\frac{1}{0.92} - 1 - 0.015 - 0.015\right) = 569.56\text{W}$$

Now:

$$p_{cos} + p_{cor} = 3R_s I_{sn}^2 + 3R_r I'^2_{rn} \tag{8.30}$$

$$I'_{rn} \approx \sqrt{I_{sn}^2 - I_{0n}^2} = 18.37 \cdot \sqrt{1 - 0.3^2} = 17.52\text{A} \tag{8.31}$$

From (8.30) and (8.31)

$$R_s = R_r = \frac{569.56}{3 \cdot (18.37^2 + 17.52^2)} = 0.4316\Omega$$

Neglecting the skin effect,

$$X_{sc} = X_{sl} + X'_{rl} \approx \sqrt{\left(\frac{V_{sn}}{I_{start}}\right)^2 - (R_s + R'_r)^2} = \sqrt{\left(\frac{380/\sqrt{3}}{6 \cdot 18.37}\right)^2 - 0.8632^2} = 1.80\Omega \tag{8.32}$$

Therefore,

$$X_{sl} = X'_{rl} = 0.9\Omega$$

The critical slip S_K (8.22) and breakdown torque T_{eK} (8.22) are, for $f_l = f_{lb} = 50$ Hz,

$$(S_K)_{50Hz} = \frac{R'_r}{\sqrt{R_s^2 + X_{sc}^2}} = \frac{0.4316}{\sqrt{0.4316^2 + 1.8^2}} = 0.233 \tag{8.33}$$

$$\begin{aligned}(T_{eK})_{50Hz} &= \frac{3p_1}{2}\frac{(V_{Ln}/\sqrt{3})^2}{2\pi f_{1b}} \cdot \frac{1}{R_s + \sqrt{R_s^2 + X_{sc}^2}} = \\ &= 3 \cdot \frac{2}{2} \cdot \frac{220^2}{2\pi 50} \cdot 0.4378 = 202.44\text{Nm}\end{aligned} \tag{8.34}$$

for $f_{1max} = 100$ Hz

$$(s_K)_{100Hz} = \frac{R'_r}{\sqrt{R_s^2 + X_{sc}^2 \cdot \left(\frac{f_{1\,max}}{f_{1b}}\right)^2}} = \frac{0.4316}{\sqrt{0.4316^2 + 1.8^2 \cdot \left(\frac{100}{50}\right)^2}} = 0.119 \tag{8.35}$$

$$\left(T_{eK}\right)_{100Hz} = \frac{3p_1}{2}\frac{\left(V_{Ln}/\sqrt{3}\right)^2}{2\pi f_{1\,max}} \cdot \frac{1}{R_s + \sqrt{R_s^2 + \left(X_{sc}\cdot\frac{f_{1\,max}}{f_{1b}}\right)^2}} = \tag{8.36}$$

$$= 3\cdot\frac{2}{2}\cdot\frac{220^2}{2\pi 100}\cdot\frac{1}{0.4316+\sqrt{0.4316^2+1.8^2\left(\frac{100}{50}\right)^2}} = 113.97\text{Nm}$$

$$\left(S_K\right)_{f1\,min} = \frac{R'_r}{\sqrt{R_s^2+\left(X_{sc}\cdot\frac{f_{1\,min}}{f_{1b}}\right)^2}} = \frac{0.4316}{\sqrt{0.4316^2+\left(1.8\cdot\frac{3}{50}\right)^2}} = 0.97! \tag{8.37}$$

From (8.36), written for f_{1min} and $(T_{eK})_{50Hz}$, the stator voltage $(V_s)_{3Hz}$ is

$$\left(V_s\right)_{3Hz} = \sqrt{\frac{2(T_{eK})_{50Hz}\cdot 2\pi f_{1\,min}\left(\sqrt{R_s^2+\left(X_{sc}\cdot\frac{f_{1\,min}}{f_{1b}}\right)^2}+R_s\right)}{3\cdot p_1}} \tag{8.38}$$

$$= \sqrt{\frac{2\cdot 202.44\cdot 2\pi\cdot 3}{3\cdot 2}\left(\sqrt{0.4316^2+\left(1.8\cdot\frac{3}{50}\right)^2}+0.4316\right)} = 33.38\text{V}$$

And, according to (8.27), the voltage increases linearly with frequency up to base (rated) frequency $f_{1b} = 50$ Hz,

$$\begin{aligned} 33.38 &= V_0 + K_0\cdot 3 \\ 220 &= V_0 + K_0\cdot 50 \end{aligned} \tag{8.39}$$

$$K_0 = 3.97\text{V/Hz}; V_0 = 21.47\text{V}$$

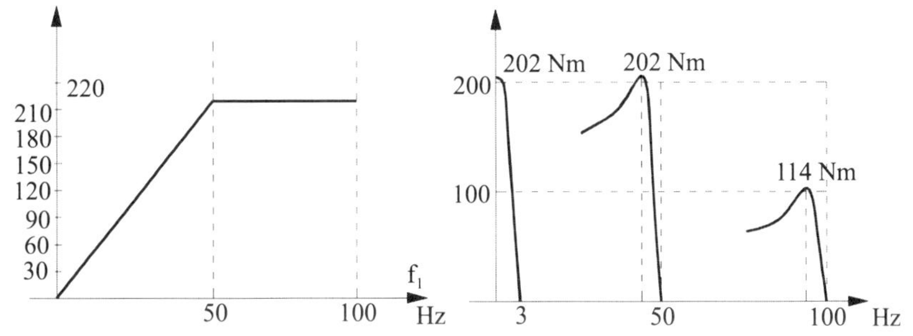

Figure 8.15 V/f and peak torque for V/f control

The results are synthesized on Figure 8.15. Toward minimum frequency f_{1min} = 3 Hz, the neglect of magnetizing reactance branch in the critical slip calculation may produce notable errors.

A check of stator current and torque at $(S_K)_{3Hz} = 0.97$, $V_{sn} = 33.38$ V, f_{1min} = 3 Hz is recommended. This aspect suggests that assigning a value for voltage boost V_0 is a sensitive issue.

Any variation of parameters due to temperature (resistances) and magnetic saturation (inductances) may lead to serious stability problems. This is the main reason why V/f control method, though simple, is to be used only with light load start applications.

8.4.2 Rotor flux vector control

As already mentioned, when firm starting or stable low speed performance is required, vector control is needed. To start, let us reconsider the IM equations for steady state (Chapter 7, paragraph 7.10, Equations (7.97) and (7.98)).

$$\begin{aligned} &\underline{I}_s R_s - \underline{V}_s = -j\omega_1 \underline{\Psi}_s \\ &\underline{I}'_r R'_r = -j\omega_1 S \underline{\Psi}_r \end{aligned} \quad \text{for } V'_r = 0 \text{ (cage rotor)} \tag{8.40}$$

Also from (7.95),

$$\underline{\Psi}_s = \frac{L_{1m}}{L'_r} \cdot \underline{\Psi}'_r + L_{sc}\underline{I}_s ; \underline{I}'_r = \frac{\underline{\Psi}'_r}{L'_r} - \frac{L_{1m}}{L'_r} \cdot \underline{I}_s \tag{8.41}$$

The torque (7.100) is,

$$T_e = 3p_1 \Psi'_r I'_r \tag{8.42}$$

It is evident in (8.40) that for steady state in a cage rotor IM, the rotor flux and current per equivalent phase are phase-shifted by $\pi/2$. This explains the torque formula (8.42) which is very similar to the case of a d.c. motor with separate excitation.

Separate (decoupled) rotor flux control represents the original vector control method. [4]

Now from (8.41) and (8.40),

$$\underline{I}_s = \frac{\underline{\Psi}'_r}{L_{1m}} + jS\omega_1 \frac{L'_r}{R'_r} \cdot \frac{\underline{\Psi}'_r}{L_{1m}} ; T_r = L'_r / R'_r \tag{8.43}$$

or, with Ψ'_r along real axis,

$$\underline{I}_s = I_M + jI_T ; I_M = \Psi_r / L_{1m} ; I_T = jS\omega_1 T_r \cdot I_M \tag{8.44}$$

Equations (8.43) and (8.44) show that the stator current may be decomposed into two separate components, one, I_M in phase with rotor flux $\underline{\Psi}_r$ called flux current, and the other, shifted ahead 90^0, I_T, called the torque current.

With (8.43) and (8.44), the torque equation (8.42) may be progressively written as

$$T_e = 3p_1 \frac{L^2_{1m}}{L'_r} \cdot I_M \cdot I_T = \frac{3p_1 \Psi'^2_r S\omega_1}{R'_r} \tag{8.45}$$

Consequently, for constant rotor flux, the torque/speed curve represents a straight line for a separately excited d.c. motor (Figure 8.16).

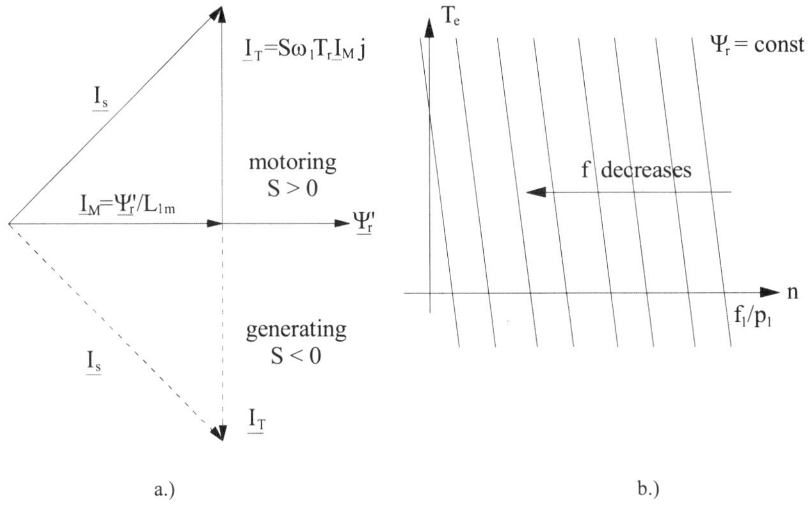

Figure 8.16 Rotor flux vector control: a) stator current components; b) torque versus speed curves for variable frequency f_1 at constant rotor flux

As expected, keeping the rotor flux amplitude constant is feasible until the voltage ceiling in the frequency converter is reached. This happens above the base frequency f_{1b}. Above f_{1b}, Ψ_r has to be decreased, as the voltage is constant. Consequently, a kind of flux weakening occurs as in d.c. motors.

The IM torque-speed curve degenerates into V/f torque/speed curves above base speed.

As long as the rotor flux transients are kept at zero, even during machine transients, the torque expression (8.45) and rotor Equation (8.40) remain valid. This explains the fastest torque response claims with rotor flux vector control. The bonus is that for constant rotor flux the mathematics to handle the control is the simplest. This explains enormous commercial success.

A basic structural diagram for a rotor flux vector control is shown on Figure 8.17

The rotor flux and torque reference values are used to calculate the flux and torque current components I_M and I_T as amplitudes. Then the slip frequency ($S\omega_1$) is calculated and added to the measured (or calculated on line) speed value ω_r to produce the primary reference frequency ω_1*. Its integral is the angle θ_1 of rotor flux position. With I_M, I_T, θ_1 the three phase reference currents i_a*, i_b*, i_c* are calculated.

Then a.c. current controllers are used to produce a pulse width modulation (PWM) strategy in the frequency converter to copy the reference currents. There are some delays in this "copying" process but they are small, so fast response in torque is provided.

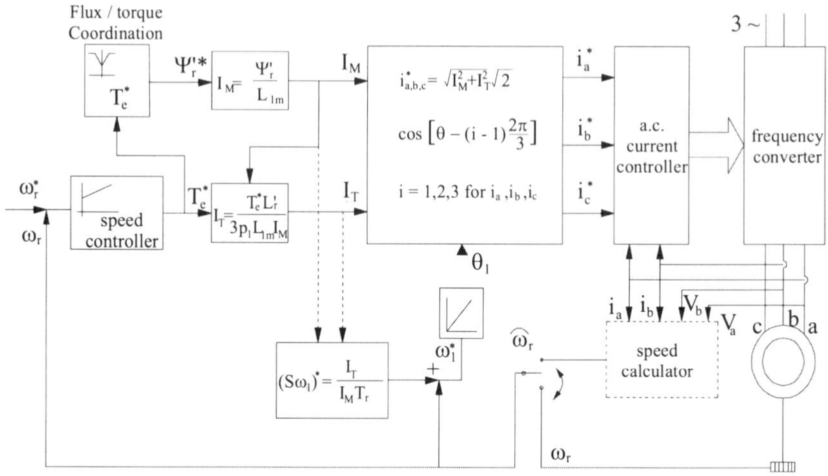

Figure 8.17 Basic rotor flux vector control system

Three remarks are in order.

- to produce regenerative braking it is sufficient to reduce the reference speed ω_r* below ω_r; I_T will become negative and so will be the torque
- The calculation of slip frequency is heavily dependent on rotor resistance (T_r) variation with temperature.
- For low speed, good performance, the rotor resistance has to be corrected on line.

Example 8.3 Rotor flux vector speed control

For the induction motor in example 8.2 with the data $2p_1 = 4$, $R_s = R_r = 0.4316\ \Omega$, $L_{sl} = L'_{rl} = \frac{X_{sc}}{2 \cdot 2\pi \cdot f_{1b}} = \frac{1.8}{2 \cdot 2\pi \cdot 50} = 2.866 \cdot 10^{-3} H$, and $L_{1m} \approx \frac{V_s}{I_{0m} \cdot 2\pi \cdot f_{1b}} - L_{sl} = \frac{220}{18.37 \cdot 0.3 \cdot 314} - 2.866 \times 10^{-3} = 0.12427H$, the rotor flux magnetizing current $I_M = 6A$ and the torque current $I_T = 20A$.

For speed n = 600 rpm, calculate the torque, rotor flux Ψ_r, stator flux slip frequency $S\omega_1$, frequency ω_1, and voltage required.

Solution

The rotor flux Ψ'_r (8.44) is

$$\Psi'_r = L_{1m} \cdot I_M = 0.12427 \cdot 6 = 0.74256 Wb$$

Equation (8.45) yields

$$T_e = 3p_1 \frac{L_{1m}^2}{L'_r} I_M \cdot I_T = 3 \cdot 2 \cdot \frac{0.12427^2}{2.866 \times 10^{-3} + 0.12427} \cdot 6 \times 20 = 87.1626\text{Nm}$$

The slip frequency is calculated from (8.44)

$$S\omega_1 = \frac{1}{T_r} \frac{I_T}{I_M} = \frac{20}{6} \cdot \frac{0.4316}{0.12417 + 2.866 \times 10^{-3}} = 11.30\text{rad/s}$$

Now the frequency ω_1 is

$$\omega_1 = 2\pi n \cdot p_1 + S\omega_1 = 2\pi \cdot \frac{600}{60} \cdot 2 + 11.30 = 136.9\text{rad/s}$$

To calculate the voltage, we have to use Equations. (8.40) and (8.41) progressively.

$$\underline{\Psi}_s = \frac{L_{1m}}{L'_r} \cdot L_{1m} I_M + L_{sc} \cdot (I_M + jI_T) = L_s I_M + jL_{sc} I_T \tag{8.46}$$

So the stator flux has two components, one produced by I_M through the no-load inductance $L_s = L_{sl} + L_{1m}$ and the other produced by the torque current through the shortcircuit inductance L_{sc}.

$$\underline{\Psi}_s = (2.866 \times 10^{-3} + 0.12427) \cdot 6 + j \cdot 2 \cdot 2.866 \times 10^{-3} \cdot 20 = 0.7628 + j \cdot 0.11464 \tag{8.47}$$

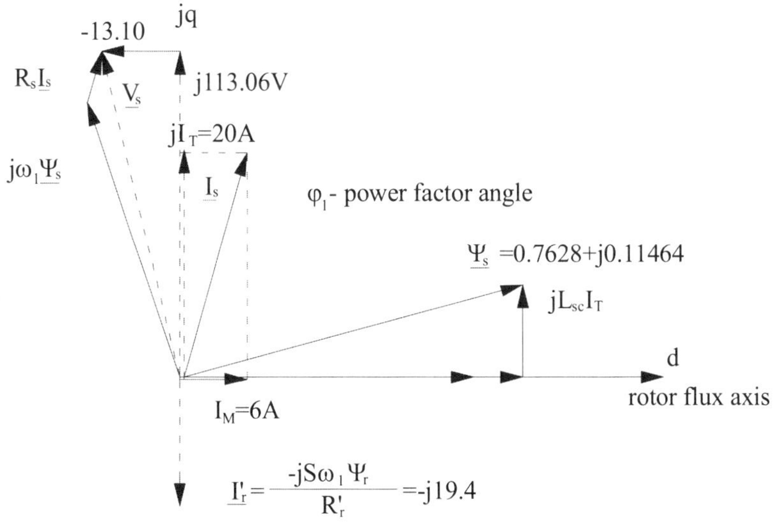

Figure 8.18 Phasor diagram with rotor and stator fluxes shown

The results in terms of fluxes and voltage are summarized in Figure 8.18

For negative torque (regenerative braking) only I_T becomes negative and so does the slip frequency $S\omega_1$.

Now from (8.40):

$$\underline{V}_s = j\omega_1\underline{\Psi}_s + \underline{I}_s R_s = j\cdot 136.9\cdot(0.7628 + j\cdot 0.11464) + (6 + j\cdot 20)\cdot 0.4316 = \\ = -13.10 + j\cdot 113.06$$

$$V_s = \sqrt{13.10^2 + 113.06^2} = 113.81\ \text{V}$$

Let us further exploit the torque expression T_e which may be alternative to (8.42).

$$T_e = 3p_1\,\text{Re}(j\underline{\Psi}_s\underline{I}_s{}^*) = 3p_1(L_s - L_{sc})I_M \cdot I_T \tag{8.48}$$

Also the rotor current (8.38) is

$$I'_r = S\omega_1 \frac{L_{1m}I_M}{R'_r} \tag{8.49}$$

Torque Equation (8.48) may be interpreted as pertaining to a reluctance synchronous motor with constant high magnetic saliency as $L_s/L_{sc} > 10 - 20$ in general.

This is true only for constant rotor flux.

The apparent magnetic saliency is created by the rotor current $\underline{I}'_r$ which is opposite to jI_T in the stator (Figure 8.18) to kill the flux in the rotor along axis q. The situation is similar to the short-circuited secondary effect on the transformer equivalent inductance.

Stator flux vector control may be treated in a similar way. For detailed information on advanced IM drives see References [5,6].

8.5 SPEED CONTROL METHODS FOR WOUND ROTOR IMs

When the rotor of IMs is provided with a wound rotor, speed control is performed by

- Adding a variable resistor to the rotor circuit
- Connecting a power converter to the rotor to introduce or extract power from the rotor. Let us call this procedure additional voltage to the rotor.

As the method of additional rotor resistance has been presented for starting, it will suffice to stress here that only for limited speed control range (say 10%) and medium power such a method is still used in some special applications. In the configuration of self-adjustable resistance, it may also be used for 10 to 15% speed variation range in generators for windpower conversion systems.

In general, for this method, a unidirectional or a bidirectional power flow frequency converter is connected to the rotor brushes while the stator winding is fed directly, through a power switch, to the power grid (Figure 8.19).

Large power motors (2-3 MW to 15 MW) or very large power generators/motors, for pump-back hydropower plants (in the hundreds of MW/unit), are typical for this method.

A half-controlled rectifier-current source inverter has constituted for a long while one of the most convenient ways to extract energy from the rotor while the speed decreases 75 to 80% of its rated value. Such a drive is called a slip recovery drive and it may work only below ideal no-load speed (S > 0).

8.5.1 Additional voltage to the rotor (the doubly-fed machine)

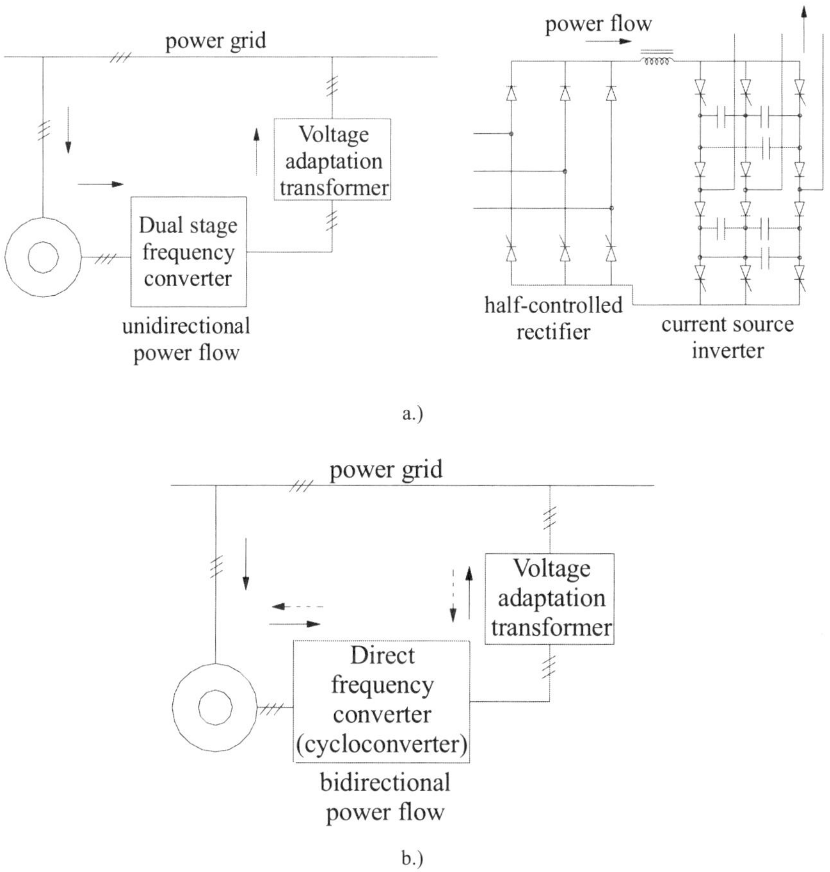

Figure 8.19 Additional voltage to the rotor
a.) with unidirectional power flow; b.) with bidirectional power flow

On the other hand, with bidirectional power flow, when the phase sequence of the voltages in the rotor may be changed, the machine may work both as a motor or as a generator, both below and above ideal no-load speed f_1/p_1.

With such direct frequency converters, it is possible to keep the rotor slip frequency f_2 constant and adjust the rotor voltage phase angle δ with respect to stator voltage after reduction to stator frequency.

The speed is constant under steady state and the machine behaves like a synchronous machine with the torque dependent on the power angle δ ($S = f_2/f_1 = ct$).

On the other hand, the frequency f_2 may be varied with speed such that

$$f_2 = f_1 - np_1;\ \ s = f_2 / f_1 - \text{variable} \tag{8.50}$$

In this case, the phase angle δ may be kept constant.

The rotor equation is

$$\underline{I}'_r \frac{R'_r}{s} + \underline{V}'_r = -jS\omega_1 \underline{\Psi}'_r = S\underline{E}'_r \tag{8.51}$$

For $\underline{V}_r$ in phase (or in phase oposition for $V'_r < 0$) with $\underline{E}'_r$

$$\underline{I}'_r R'_r = S\underline{E}'_r - \underline{V}'_r \tag{8.52}$$

For zero rotor current, the torque is zero. This corresponds to the ideal no-load speed (slip, S_0)

$$s_0 E'_r - V'_r = 0; s_0 = \frac{V'_r}{E'_r} \tag{8.53}$$

So, may be positive if $V'_r > 0$, that is $\underline{V}'_r$ and $\underline{E}'_r$ are in phase while it is negative for $V'_r<0$ or V'_r or $\underline{V}'_r$ and $\underline{E}'_r$ in phase opposition.

For constant values of V'_r (+ or -), the torque speed curves obtained by solving the equivalent circuit may be calculated as

$$T_e = \frac{P_{elm} p_1}{S\omega_1} = \frac{3p_1}{S\omega_1}\left[I'^2_r R'_r + V'_r I'_r \cos\theta'_r\right] \tag{8.54}$$

θ'_r is the angle between $\underline{V}'_r$ and I'_r. In our case, $\theta_r = 0$ and V'_r is positive or negative, and given.

It is seen (Figure 8.20) that with such a converter, the machine works well as a motor above synchronous conventional speed f_1 / p_1 and as a generator below f_1 / p. This is true for voltage control. A different behavior is obtained for constant rotor current control. [7]

However, under subsynchronous motor and generator modes, all are feasible with adequate control.

Example 8.4 Doubly - fed IM

An IM with wound rotor has the data: $V_{sn} = 220$ V/phase, $f_1 = 50$ Hz, $R_s = R'_s = 0.01\ \Omega$, $X_{sl} = X'_{rl} = 0.03\ \Omega$, $X_{1m} = 2.5\ \Omega$, $2p_1 = 2$, $V_{rn} = 300$ V/phase, $S_n = 0.01$. Calculate

- the stator current and torque at $s_n = 0.01$ for the short-circuited rotor;
- for same stator current at n =1200 rpm calculate the rotor voltage V_r in phase with stator voltage V_{sn} (Figure 8.21) and the power extracted from the rotor by frequency converter.

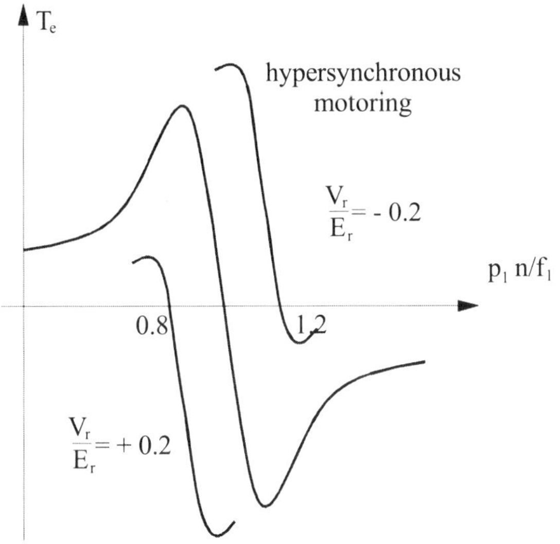

Figure 8.20 Torque speed curves for $V_r/E_r = \pm 0.2$, $\theta'_r = 0$

Solution

The stator current (see chapter 7) is

$$\left(I'_s\right)_{V'_r=0} \approx \frac{V_s}{\sqrt{\left(R_s + C_1\frac{R'_r}{s}\right)^2 + \left(X_{ls} + C_1 X'_{ri}\right)^2}} =$$

$$= \frac{220}{\sqrt{\left(0.01 + 1.012\frac{0.01}{0.01}\right)^2 + \left(0.03 + 1.012 \cdot 0.03\right)^2}} = 217.17\text{A}$$

with

$$C_1 = \frac{X_{ls}}{X_{lm}} + 1 = 1 + \frac{0.03}{2.5} = 1.012 \quad (8.55)$$

The actual rotor current I_r per phase is

$$I_r = I'_r \cdot \frac{V_s}{V_{rn}} = 217.17 \cdot \frac{220}{300} \approx 152\text{A} \quad (8.56)$$

The torque T_e is

$$T_e = \frac{3R'_r I'^2_r}{S\omega_1} \cdot p_1 = \frac{3 \cdot 0.01 \cdot 217.172}{0.01 \cdot 2\pi \cdot 50} = 450.60\text{Nm} \quad (8.57)$$

If $\underline{V}'_r$ is in phase with $\underline{V}_{sn}$, then

$$\left(I'_r\right)_{S'} \approx I_s = \frac{V_s + V'_r / S'}{\sqrt{\left(R_s + C_1 \frac{R'_r}{S'}\right)^2 + \left(X_{sl} + C_1 X'_{rl}\right)^2}} =$$

$$= \frac{220 + V'_r / 0.2}{\sqrt{\left(0.01 + 1.012 \frac{0.01}{0.2}\right)^2 + \left(0.03 + 1.012 \cdot 0.03\right)^2}} = 217.17 \quad (8.58)$$

$$S' = 1 - \frac{np_1}{f_1} = 1 - \frac{1200}{60} \cdot \frac{2}{50} = 0.2; f_2 = S'f_1 = 0.2 \cdot 50 = 10\text{Hz}$$

So $V'_r = -40.285$V

In this case the angle between $\underline{V}'_r$ and $\underline{I}'_r$ corresponds to the angle + $\underline{V}'_s$ and - $\underline{I}'_s$ (from 8.58)

$$\theta'_r(\underline{V}'_r, \underline{I}'_r) \approx -45^0 + 180^0 = 135^0 \quad (8.59)$$

Now the torque (8.54) is

$$T_e = \frac{3 \cdot 2}{0.2 \cdot 2\pi \cdot 50}\left[0.01 \cdot 217.17^2 + (-40.28)(217.17)(-0.707)\right] = 635.94\text{Nm} \quad (8.60)$$

Note that the new torque is larger than for S = 0.01 with the short – circuited rotor.

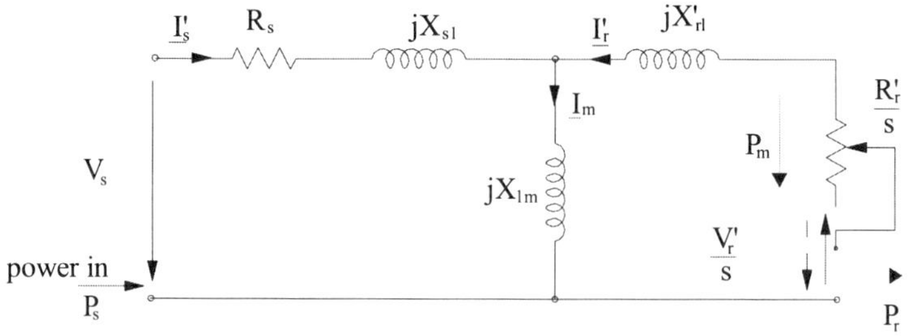

Figure 8.21 Equivalent circuit with additional rotor voltage $\underline{V}'_r$

The input active power P_s, electric power out of rotor P_r and mechanical power P_m are

$$\begin{aligned} P_s &= 3V_s I_s \cos\varphi_s = 3 \cdot 220 \times 217.17 \times 0.707 = 101.335 \times 10^3\,\text{W} \\ P_r &= 3V_r I_r \cos\theta_1 = 3 \cdot 40.38 \times 217.17 \cdot 0.707 \approx 18.6 \times 10^3\,\text{W} \\ P_m &= T_e \cdot 2\pi n = 635.94 \cdot 2\pi \cdot 1200/60 = 79.874 \times 10^3\,\text{W} \end{aligned} \quad (8.61)$$

The efficiency at 1200 rpm η is

$$\eta = \frac{P_m}{P_s - P_r} = \frac{79.874}{101.335 - 18.6} = 0.9654 \tag{8.62}$$

Note that core, mechanical, and stray load losses have been neglected.

$$-j\omega_1 \underline{\Psi}_r = -\left(\frac{L'_r}{L_{1m}} \cdot \underline{\Psi}_s - L_{sc}\underline{I}_s\right) j\omega_1 = -\frac{L'_r}{L_{1m}}\left[\left(V_s - R_s\underline{I}_s\right) - jX_{sc}\underline{I}_s\right] = \underline{E}'_r \tag{8.63}$$

$$\underline{E}'_r = -\frac{(2.5+0.03)}{2.5}\left[220 - (217.17 - j217.17)0.707(0.01 + j0.06036)\right] = -(209.49 - j7.74) \tag{8.64}$$

Now $\underline{I}'_r$ is recalculated from (8.51)

$$\underline{I}'_r = \frac{S\underline{E}'_r - \underline{V}'_r}{R'_r} = \frac{-0.2(211.767 - j7.74) - (-40.301)}{0.01} = (-158.88 + j154.8) = -\underline{I}_s \tag{8.65}$$

Both components of $\underline{I}'_r$ are close to those of (8.58).

When the rotor slip frequency f_2 is constant, the speed is constant so only the rotor voltage sequence, amplitude, and phase may be modified.

The phase sequence information is contained into S_2 sign:

$$S_2 = f_2 / f_1 >< 0 \tag{8.66}$$

We may calculate the stator current $\underline{I}_s$ and the rotor current I'_r (reduced to the stator) from the equivalent circuit (Figure 8.21) is:

$$\underline{I}_s = \frac{\underline{V}_s(R'_r/s + jX'_r) - jX_{1m}\underline{V}'_r/s}{(R_s + jX_s)(R'_r/s + jX'_r) + X_{1m}^2}; \underline{X}_s = X_{sl} + X_{1m} \tag{8.67}$$

$$\underline{I}'_r = \frac{\underline{V}'_r(R_s + jX_s)/s - jX_{1m}\underline{V}_s}{(R_s + jX_s)(R'_r/s + jX'_r) + X_{1m}^2}; X'_r = X'_{lr} + X_{1m} \tag{8.68}$$

The stator and rotor power input powers are

$$P_s = 3\,\mathrm{Re}\left(\underline{V}_s\underline{I}_s{}^*\right); T_e \approx \left(P_s - 3R_sI_s^2\right)\frac{p_1}{\omega_1}$$
$$Q_s = 3\,\mathrm{Imag}\left(\underline{V}_s\underline{I}_s{}^*\right) \tag{8.69}$$

Similarly the powers out of rotor are

$$P_r = 3\,\mathrm{Re}\left(\underline{V}'_r\underline{I}'_r{}^*\right)$$
$$Q_r = 3\,\mathrm{Imag}\left(\underline{V}'_r\underline{I}'_r{}^*\right) \tag{8.70}$$

With S = const, such a control method can work at constant rotor voltage V'_r but with variable phase γ (V_s, V'_r). Alternatively, it may operate at constant stator or rotor current [8,9]. Finally, vector control is also feasible. [6] Such schemes are currently proposed for powers up to 300 MW in pump-storage power plants where, both for pumping and for generating, a 10 to 25% of speed control improves the hydroturbine-pump output.

The main advantage is the limited rating of the rotor-side frequency converter $S_r = S_{max} \cdot S_n$. As long as $S_{max} < 0.2 - 0.15$, the savings in converter costs are very good. Notice that resistive starting is required in such cases.

For medium and lower power limited speed control, dual stator winding stator nest cage rotor induction motors have been proposed. The two windings have different pole numbers p_1, p_2, and the rotor has a pole count $p_r = p_1+p_2$. One winding is fed from the power grid at the frequency f_1 and the other at frequency f_2 from a limited power frequency converter. The machine speed n is

$$n = \frac{f_1 \pm f_2}{p_1 + p_2} \tag{8.71}$$

The smaller f_2, the smaller the rating of the corresponding frequency converter. The behavior is typical to that of a synchronous machine when f_2 is const. Low speed applications such as wind generators or some pump applications with low initial cost for very limited speed control range (less than 20%) might be suitable for those configurations with rather large rotor losses.

8.6 SUMMARY

- Starting methods are related to IMs fed from the industrial power grid.
- With direct starting and stiff local power grids, the starting current is in the interval of 580 to 650% (even higher for high efficiency motors).
- For direct starting at no mechanical load, the rotor winding energy losses during machine acceleration equals the rotor-attached kinetic energy at no load.
- For direct frequent starting at no load, and same stator, rotors with higher rotor resistance lead to lower total energy input for acceleration.
- To avoid notable voltage sags during direct starting of a newly installed IM, the local transformer KVA has to be oversized accordingly.
- For light load, starting voltage reduction methods are used, as they reduce the line currents. However, the torque/current ratio is also reduced.
- Voltage reduction is performed through an autotransformer, wye/delta connection, or through softstarters.
- Softstarters are now built to about 1MW motors at reasonable costs as they are made with thyristors. Input current harmonics and motor additional losses are the main drawbacks of softstarters. However, they are continually being improved and are expected to be common practice in light load starting applications (pumps, fans) where speed

control is not required but soft (low current), slow but controlled, starting are required.

- Rotor resistive starting of wound rotor IMs is traditional. The method produces up to peak torque at start, but at the expense of very large additional losses.
- Self adjustable resistance-reactance paralleled pairs may also be used for the scope to cut the cost of controls.
- Speed control in cage rotor IMs may be approached by voltage amplitude control, for a very limited range (up to 10 to 15%, in general).
- Pole changing windings in the stator can produce two speed motors and are used in some applications, especially for low power levels.
- Dual stator windings and cage rotors can also produce two speed operations efficiently if the power level for one speed is much smaller than for the other.
- Coordinated frequency/voltage speed control represents the modern solution to adjustable speed drives.
- V/f scalar control is characterized by an almost linear dependence of voltage amplitude on frequency; a voltage boost $V_0 = (V)_{f1} = 0$ is required to produce sufficient torque at lowest frequency $f_1 \approx 3$ Hz. Slip feedforward compensation is added. Still slow transient response is obtained. For pumps, fans, etc., such a method is adequate. This explains its important market share.
- Rotor flux vector control keeps the rotor flux amplitude constant and requires the motor to produce two stator current components: a flux current I_M and a torque current I_T, 90^0 apart. These two components are decoupled to be controlled separately. Fast dynamics, quick torque availability, stable, high performance drives are built and sold based on rotor flux vector control or on other related forms of decoupled flux and torque control.
- Linear torque/speed curves, ideal for control, are obtained.
- Flux, torque coordination through various optimization criteria could be applied to cut energy losses or widen the torque-speed range.
- For constant rotor flux, the IM behaves like a high saliency reluctance synchronous motor. However, the apparent saliency is produced by the rotor currents phase- shifted by 90^0 with respect to rotor flux for cage rotors. As for the reluctance synchronous motor, a maximum power factor for the loss, less motor can be defined as

$$\cos\varphi_{max} = \frac{1 - L_{sc}/L_s}{1 + L_{sc}/L_s}; L_s/L_{sc} > 15 \div 20 \tag{8.72}$$

  L_s – no-load inductance; L_{sc} – short-circuit inductance.

- Wound rotor speed control is to be approached through frequency converters connected to the rotor brushes. The rotor converter rating is low. Considerable converter cost saving is obtained with limited speed control so characteristic to such drives. With adequate frequency converter, motoring and generating over or under conventional synchronous speed ($n_1 = f_1/p_1$) is possible. High and very high power motor/generator systems are main applications as separated active and reactive power control for limited speed variation at constant frequency is feasible. In fact, the largest electric motor with no-load starting has been built for such purposes (for pump storage power plants).

- Details on control of electric drives with IMs are to be found in the rich literature dedicated to this very dynamic field of engineering.

8.7 REFERENCES

1. W. Leonhard, Control of electric drives, 2nd edition, Springer Verlag, 1995.
2. I.Boldea and S. A. Nasar, Electric drives, CRC Press, 1998.
3. F. Blaabjerg et. al, Can Softstarters Save Energy, IEEE – IA Magazine, Sept/Oct.1997, pp. 56 – 66.
4. F. Blaschke, The Principle of Field Orientation As Applied to the New Transvector Closed Loop Control System For Rotating Field Machines, Siemens Review, vol. 34, 1972, pp. 217 – 220 (in German).
5. D. W. Novotny and T. A. Lipo, Vector control and dynamics of a.c. drives, Clarendon Press, Oxford, 1996.
6. Boldea and S. A. Nasar, Vector Control of A.C. Drives, CRC Press, 1992.
7. Masenoudi and M. Poloujadoff, Steady State Operation of Doubly Fed Synchronous Machines Under Voltage and Current Control, EMPS, vol.27, 1999.
8. L. Schreir, M. Chomat, J. Bendl, Working Regions of Adjustable Speeds Unit With Doubly Fed Machines, Record of IEEE – IEMDC – 99, Seattle, USA, pp. 457 – 459.
9. L. Schreir, M. Chomat, J. Bendl, Analysis of Working Regions of Doubly Fed Generators, Record of ICEM – 1998, vol. 3, pp. 1892 – 1897.

Chapter 9

SKIN AND ON – LOAD SATURATION EFFECTS

9.1. INTRODUCTION

So far we have considered that resistances, leakage and magnetization inductances are invariable with load.

In reality, the magnetization current I_m varies only slightly from no-load to full load (from zero slip to rated slip $S_n \approx 0.01 - 0.06$), so the magnetization inductance L_{1m} varies little in such conditions.

However, as the slip increases toward standstill, the stator current increases up to (5.5 – 6.5) times rated current at stall (S = 1).

In the same time, as the slip increases, even with constant resistances and leakage inductances, the magnetization current I_m decreases.

So the magnetization current decreases while the stator current increases when the slip increases (Figure 9.1).

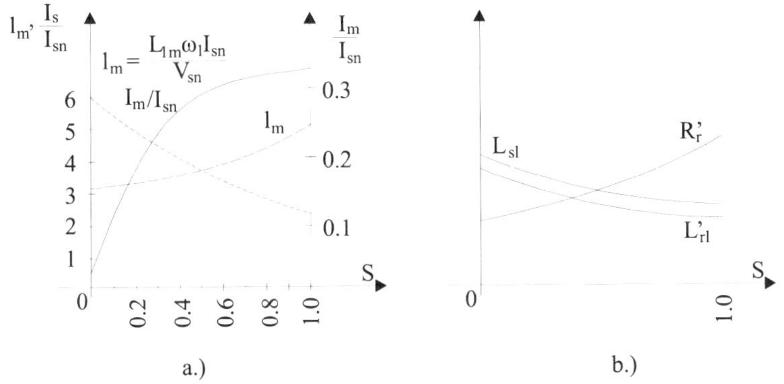

Figure 9.1 Stator I_s/I_{sn} and magnetization I_m current, magnetization inductance (l_m) in p.u. a.), leakage inductance and rotor resistance versus slip b.)

When the rotor (stator) current increases with slip, the leakage magnetic field path in iron tends to saturate. With open slots on stator, this phenomenon is limited, but, with semiopen or semiclosed slots, the slot leakage flux path saturates the tooth tops both in the stator and rotor (Figure 9.2) above (2–3) times rated current.

Also, the differential leakage inductance which is related to main flux path is affected by the tooth top saturation caused by the circumpherential flux produced by slot leakage flux lines (Figure 9.2). As the space harmonics flux paths are contained within τ/π from the airgap, only the teeth saturation affects them.

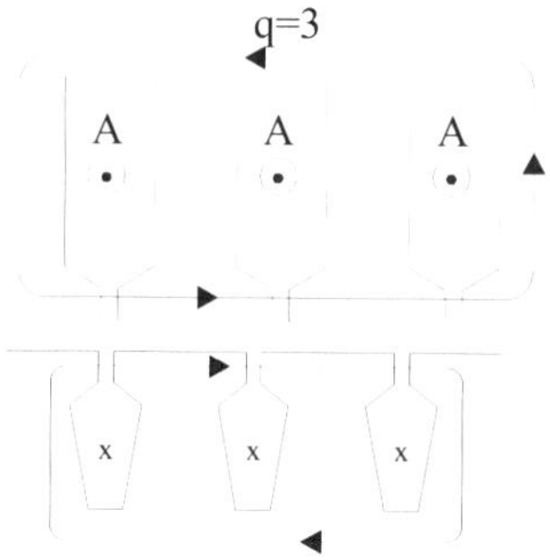

Figure 9.2 Slot leakage flux paths

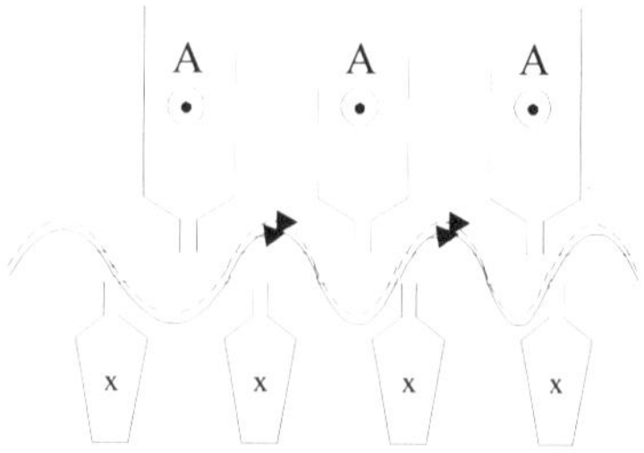

Figure 9.3 Zig-zag flux lines

Further on, for large values of stator (and rotor) currents, the zig-zag flux becomes important and contributes notably to teeth top magnetic saturation in addition to slot leakage flux contribution.

Rotor slot skewing is also known to produce variable main flux path saturation along the stack length together with the magnetization current. However the flux densities from the two contributions are phase shifted by an angle which varies and increases towards 90^0 at standstill. The skewing contribution to the main flux path saturation increases with slip and dominates the picture for $S > S_k$ as the magnetization flux density, in fact, decreases with slip so that at standstill it is usually 55 to 65% of its rated value.

A few remarks are in order.

- The magnetization saturation level in the core decreases with slip, such that at standstill only 55 – 65% of rated airgap flux remains.
- The slot leakage flux tends to increase with slip (current) and saturates the tooth top unless the slots are open.
- Zig – zag circumpherential flux and skewing accentuate the magnetic saturation of teeth top and of entire main flux path, respectively, for high currents (above 2 to 3 times rated current).
- The differential leakage inductance is also reduced when stator (and rotor) current increases as slot, zig-zag, and skewing leakage flux effects increase.
- As the stator (rotor) current increases the main (magnetising) inductance and leakage inductances are simultaneously influenced by saturation. So leakage and main path saturation are not independent of each other. This is why we use the term: on-load saturation.

As expected, accounting for these complex phenomena simultaneously is not an easy tractable mathematical endeavour. Finite element or even refined analytical methods may be suitable. Such methods are presented in this chapter after more crude approximations ready for preliminary design are given.

Besides magnetic saturation, skin (frequency) effect influences both the resistances and slot leakage inductances. Again, a simultaneous treatment of both aspects may be practically done only through FEM.

On the other hand, if slot leakage saturation occurs only on the teeth top and the teeth, additional saturation due to skewing does not influence the flux lines distribution within the slot, the two phenomena can be treated separately.

Experience shows that such an approximation is feasible. Skin effect is treated separately for the slot body occupied by a conductor. Its influence on equivalent resistance and slot body leakage geometrical permeance is accounted for by two correction coefficients, K_R and K_X. The slot neck geometry is corrected for leakage saturation.

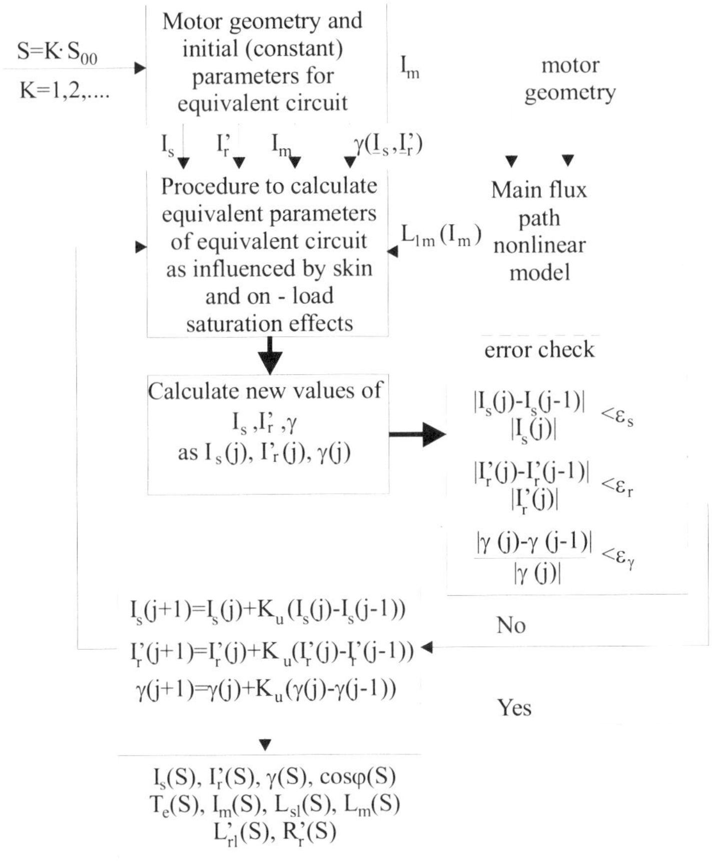

Figure 9.4 Iterative algorithm to calculate IM performance and parameters as influenced by skin and on-load saturation effects.

Finally, the on load saturation effects are treated iteratively for given slip values to find, from the equivalent circuit with variable parameters, the steady state performance. The above approach may be summarized as in Figure 9.4.

The procedure starts with the equivalent circuit with constant parameters and calculates initial values of stator and rotor currents I_s, I_r' and their phase shift angle γ. Now that we described the whole picture, let us return to its different facets and start with skin effect.

9.2. THE SKIN EFFECT

As already mentioned, skin effects are related to the flux and current density distribution in a conductor (or a group of conductors) flowed by a.c. currents and surrounded by a magnetic core with some airgaps.

Easy to use analytical solutions have been found essentially only for rectangular slots, but adaptation for related shapes has also become traditional.

More general slots with notable skin effect (of general shape) have been so far treated through equivalent multiple circuits after slicing the conductor(s) in slots in a few elements.

A refined slicing of conductor into many sections may be solved only numerically, but within a short computation time. Finally, FEM may also be used to account for skin effect. First, we will summarize some standard results for rectangular slots.

9.2.1. Single conductor in rectangular slot

Rectangular slots are typical for the stator of large IMs and for wound rotors of the same motors. Trapezoidal (and rounded) slots are typical for low power motors.

The case of a single conductor in slot is (Figure 9.5) typical to single (standard) cage rotors and is commonplace in the literature. The main results are given here.

The correction coefficients for resistance and slot leakage inductance K_R and K_X are

$$K_R = \xi\frac{(\sinh 2\xi + \sin 2\xi)}{(\cosh 2\xi - \cos 2\xi)} = \frac{R_{ac}}{R_{dc}}; \quad K_X = \frac{3}{2\xi}\frac{(\sinh 2\xi - \sin 2\xi)}{(\cosh 2\xi - \cos 2\xi)} = \frac{(L_{sls})_{ac}}{(L_{sls})_{dc}} \tag{9.1}$$

with

$$\xi = \beta h_s = \frac{h_s}{\delta_{Al}}; \beta = \frac{1}{\delta_{Al}} = \sqrt{\frac{S\omega_1\mu_0\sigma_{Al}}{2}\frac{b_c}{b_s}}; \sigma_{Al} - \text{electrical conductivity} \tag{9.2}$$

The slip S signifies that in this case the rotor (or secondary) of the IM is considered.

Figure 9.5 depicts K_R and K_x as functions of ξ, which, in fact, represents the ratio between the conductor height and the field penetration depth δ_{Al} in the conductor for given frequency $S\omega_1$. With one conductor in the slot, the skin effects, as reflected in K_R and K_x, increase with the slot (conductor) height, h_s, for given slip frequency $S\omega_1$.

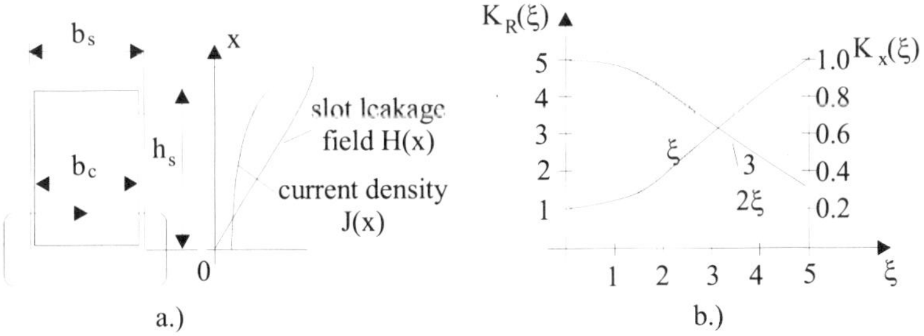

Figure 9.5 Rectangular slot
a.) slot field (H(x)) and current density (J(x)) distributions
b.) resistance K_R and slot leakage inductance K_X skin effect correction factors

This rotor resistance increase, accompanied by slot leakage inductance (reactance) decrease, leads to both a lower starting current and a higher starting torque.

This is how the deep bar cage rotor has evolved. To increase further the skin effects, and thus increase starting torque for even lower starting current (I_{start} = (4.5–5)I_{rated}), the double cage rotor was introduced by the turn of this century already by Dolivo – Dobrovolski and later by Boucherot.

The advent of power electronics, however, has led to low frequency starts and thus, up to peak torque at start, may be obtained with (2.5–3) times rated current. Skin effect in this case is not needed. Reducing skin effect in large induction motors with cage rotors lead to particular slot shapes adequate for variable frequency supply.

9.2.2. Multiple conductors in rectangular slots: series connection

Multiple conductors are placed in the stator slots, or in the rotor slots of wound rotors (Figure 9.6).

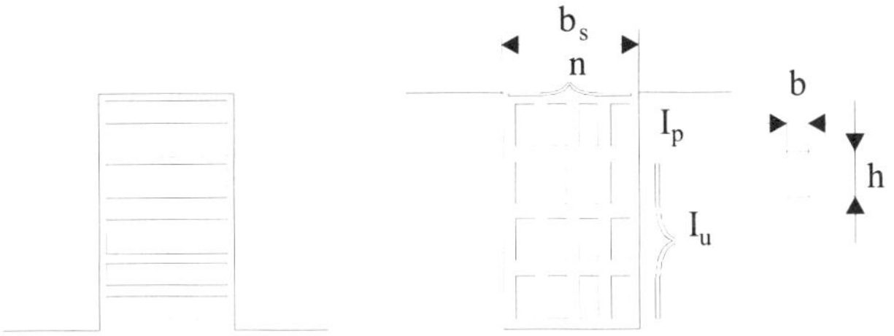

Figure 9.6 Multiple conductors in rectangular slots

According to Emde and R.Richter [1,2] who continued the classic work of Field [3], the resistance correction coefficient K_{RP} for the pth layer in slot (Figure 9.6) with current I_p, when total current below p^{th} layer is I_u, is

$$K_{RP} = \varphi(\xi) + \frac{I_u(I_u \cos\gamma + I_p)}{I_p^2}\psi(\xi) \tag{9.3}$$

$$\varphi(\xi) = \xi\frac{(\sinh 2\xi + \sin 2\xi)}{(\cosh 2\xi - \cos 2\xi)}; \quad \psi(\xi) = 2\xi\frac{(\sinh\xi - \sin\xi)}{(\cosh\xi + \cos\xi)} \tag{9.4}$$

$$\xi = \beta_n h; \quad \beta_n = \sqrt{\frac{S\omega_1\mu_0\sigma_{Al}}{2}\frac{nb}{b_s}}$$

There are n conductors in each layer and γ is the angle between $\underline{I}_p$ and $\underline{I}_u$ phasors.

In two-layer windings with chorded coils, there are slots where the current in all conductors is the same and some in which two phases are located and thus the currents are different (or there is a phase shift $\gamma = 60^0$).

For the case of $\gamma = 0$ with $I_u = I_p(p - 1)$ Equation (9.3) becomes

$$K_{RP} = \varphi(\xi) + (p^2 - p)\psi(\xi) \tag{9.5}$$

This shows that the skin effect is not the same in all layers. The average value of K_{RP} for m layers,

$$K_{Rm} = \frac{1}{m}\sum_1^m K_{RP}(p) = \varphi(\xi) + \frac{m^2 - 1}{3}\psi(\xi) > 1 \tag{9.6}$$

Based on [4], for $\gamma \neq 0$ in (9.6) $(m^2-1)/3$ is replaced by

$$\frac{m^2(5 + 3\cos\gamma)}{24} - \frac{1}{3} \tag{9.6'}$$

A similar expression is obtained for the slot-body leakage inductance correction K_x [4].

$$K_{xm} = \varphi'(\xi) + \frac{(m^2 - 1)\psi'(\xi)}{m^2} < 1 \tag{9.7}$$

$$\varphi'(\xi) = \frac{3}{2\xi}\frac{(\sinh 2\xi - \sin 2\xi)}{(\cosh 2\xi - \cos 2\xi)} \tag{9.8}$$

$$\psi'(\xi) = \frac{(\sinh\xi + \sin\xi)}{\xi(\cosh\xi + \cos\xi)} \tag{9.9}$$

Please note that the first terms in K_{Rm} and K_{xm} are identical to K_R and K_x of (9.1) valid for a single conductor in slot. As expected, K_{Rm} and K_{xm} degenerate into K_R and K_x for one layer (conductor) per slot. The helping functions φ, ψ, φ', ψ' are quite general (Figure 9.7).

For a given slot geometry, increasing the number of conductor layers in slot reduces their height $h = h_s/m$ and thus reduces ξ, which ultimately reduces $\psi(\xi)$ in (9.6). On the other hand, increasing the number of layers, the second term in (9.6) tends to increase.

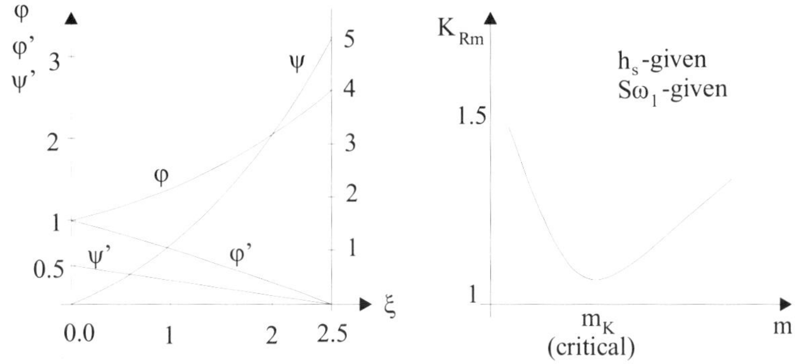

Figure 9.7 Helping functions φ, Ψ, φ', Ψ' versus ξ

It is thus evident that there is a critical conductor height h_c for which the resistance correction coefficient is minimum. Reducing the conductor height below h_c does not produce a smaller K_{Rm}.

In large power or in high speed (frequency), small/medium power machines this problem of critical conductor height is of great importance to minimize the additional (a.c.) losses in the windings.

A value of $K_{Rm} \approx (1.1 - 1.2)$ is in most cases, acceptable. At power grid frequency (50 – 60 Hz), the stator skin effect resistance correction coefficient is very small (close to 1.0) as long as power is smaller than a few hundred kW.

Inverter-fed IMs, however, show high frequency time harmonics for which K_{Rm} may be notable and has to be accounted for.

Example 9.1. Derivation of resistance and reactance corrections

Let us calculate the magnetic field H(x) and current density J(x) in the slot of an IM with m identical conductors (layers) in series making a single layer winding.

Solution

To solve the problem we use the field equation in complex numbers for the slot space where only along slot depth (OX) the magnetic field and current density vary.

$$\frac{\partial^2 \underline{H}(x)}{\partial x^2} = j\frac{b}{b_s}\omega\mu_0\sigma_{Co}\underline{H}(x) \tag{9.10}$$

The solution of (9.10) is

$$\underline{H}(x) = \underline{C}_1 e^{-(1+j)\beta x} + \underline{C}_2 e^{+(1+j)\beta x};\ \beta = \sqrt{\frac{b}{b_s}\frac{\omega_1\mu_0\sigma}{2}} \tag{9.11}$$

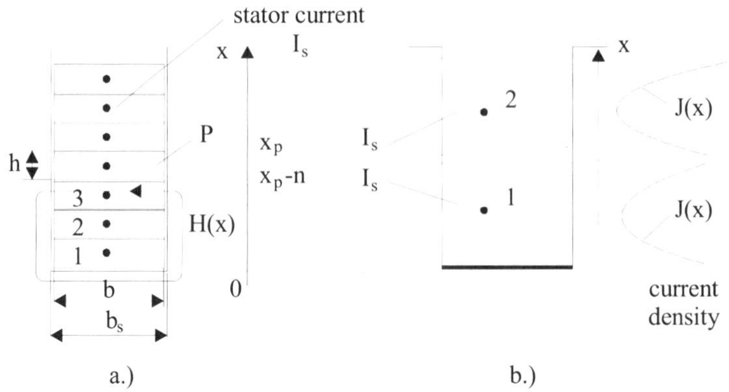

Figure 9.8 Stator slot with single coil with m layers (conductors in series) a.) and two conductors in series b.)

The boundary conditions are

$$\begin{aligned} &\underline{H}(x_p)\cdot b_s = \underline{I}_s p; \quad x = x_p; \; x_p = ph \\ &\underline{H}(x_p - h)\cdot b_s = \underline{I}_s(p-1); \quad x = x_p - h \end{aligned} \tag{9.12}$$

From (9.11) and (9.12), we get the expressions of the constants $\underline{C}_1$ and $\underline{C}_2$

$$\begin{aligned} \underline{C}_1 &= \frac{\underline{I}_s}{2b_s \sinh[(1+j)\beta h]}\left[(p-1)e^{(1+j)\beta x_p} - pe^{(1+j)\beta(x_p-h)}\right] \\ \underline{C}_2 &= \frac{\underline{I}_s}{2b_s \sinh[(1+j)\beta h]}\left[-(p-1)e^{-(1+j)\beta x_p} + pe^{-(1+j)\beta(x_p-h)}\right] \end{aligned} \tag{9.13}$$

The current density $\underline{J}(x)$ is

$$\underline{J}(x) = -\frac{b_s}{b}\frac{\partial \underline{H}(x)}{\partial x} = -\frac{b_s}{b}\beta(1+j)\left[\underline{C}_1 e^{-(1+j)\beta x} - \underline{C}_2 e^{(1+j)\beta x}\right] \tag{9.14}$$

For m = 2 conductors in series per slot, the current density distribution (9.14) is as shown qualitatively in Figure 9.8.

The active and reactive powers in the pth conductor $\underline{S}_p$ is calculated using the Poyting vector [4].

$$\underline{S}_{a.c.} = P_{a.c.} + jQ_{a.c.} = \frac{b_s L}{\sigma_{Co}}\left[\left(\frac{\underline{J}}{2}\frac{\underline{H}^*}{2}\right)_{x=x_p-h} - \left(\frac{\underline{J}}{2}\frac{\underline{H}^*}{2}\right)_{x=x_p}\right] \tag{9.15}$$

Denoting by R_{pa} and X_{pa} the a.c. resistance and reactance of conductor p, we may write

$$P_{ac} = R_{ac} I_s^2 \quad Q_{ac} = X_{ac} I_s^2 \tag{9.16}$$

The d.c. resistance R_{dc} and reactance X_{dc} of conductor p,

$$R_{dc} = \frac{1}{\sigma_{Co}}\frac{L}{hb}; \; X_{dc} = \omega\mu_0\frac{b_s L}{3h}; \; L\text{ - stack length} \tag{9.17}$$

The ratios between a.c. and d.c. parameters K_{Rp} and K_{xp} are

$$K_{Rp} = \frac{R_{ac}}{R_{dc}}; \quad K_{xp} = \frac{X_{ac}}{X_{dc}} \tag{9.18}$$

Making use of (9.11) and (9.14) leads to the expressions of K_{Rp} and K_{xp} represented by (9.5) and (9.6).

9.2.3. Multiple conductors in slot: parallel connection

Conductors are connected in parallel to handle the phase current, In such a case, besides the skin effect correction K_{Rm}, as described in paragraph 9.3.2 for series connection, circulating currents will flow between them. Additional losses are produced this way.

When multiple round conductors in parallel are used, their diameter is less than 2.5(3) mm and thus, at least for 50(60) Hz machines, the skin effect may be neglected altogether. In contrast, for medium and large power machines, with rectangular shape conductors (Figure 9.9), the skin effect influence has at least to be verified. In this case also, the circulating current influence is to be considered.

A simplified solution to this problem [5] is obtained by neglecting, for the time being, the skin effect of individual conductors (layers), that is by assuming a linear leakage flux density distribution along the slot height. Also the inter-turn insulation thickness is neglected.

At the junction between elementary conductors (strands), the average a.c. magnetic flux density $B_{ave} \approx B_m/4$ (Figure 9.11a). The a.c. flux through the cross section of a strand Φ_{ac} is

$$\Phi_{ac} = B_{ave} h l_{stack} \tag{9.19}$$

The d.c. resistance of a strand R_{dc} is

$$R_{ac} \approx R_{dc} = \frac{1}{\sigma_{Co}}\frac{l_{turn}}{bh} \tag{9.20}$$

Now the voltage induced in a strand turn E_{ac} is

$$E_{ac} = \omega\Phi_{ac} \tag{9.21}$$

So the current in a strand I_{st}, with the leakage inductance of the strand neglected, is:

$$I_{st} = E_{ac} / R_{ac} \tag{9.22}$$

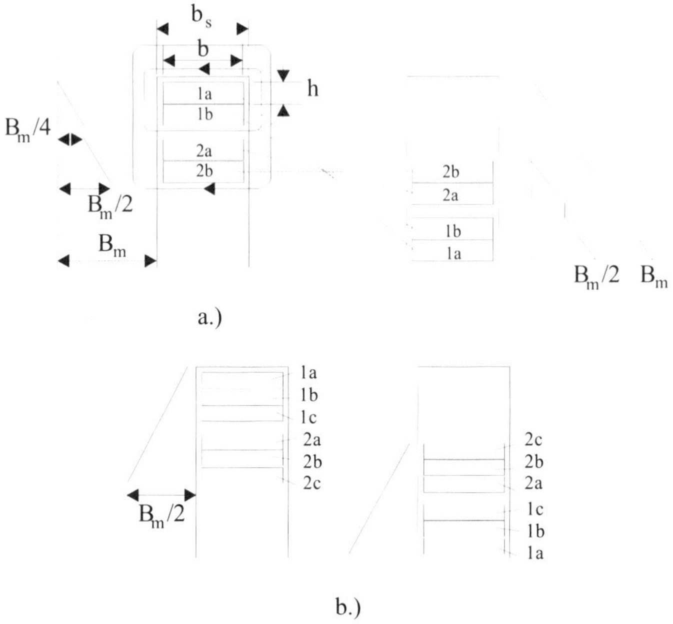

Figure 9.9 Slot leakage flux density for coil sides: two turn coils
a.) two elementary conductors in parallel (strands) b.) three elementary conductors in parallel

The loss in a strand P_{strand} is

$$P_{strand} = \frac{E^2{}_{ac}}{R_{ac}} = \frac{\omega^2 B^2{}_{ave} h^2 l^2_{stack}}{\dfrac{1}{\sigma_{Co}} \dfrac{l_{turn}}{bh}} \tag{9.23}$$

As seen from Figure 9.9a, the average flux density B_{ave} is

$$B_{ave} = \frac{B_m}{4} = \frac{\mu_0 n_{coil} I_{phase}(1+\cos\gamma)}{4b_s} \tag{9.24}$$

I_{phase} is the phase current and γ is the angle between the currents in the upper and lower coils. Also, n_{coil} is the number of turns per coil (in our case n_{coil} = 2,3).

The usual d.c. loss in a strand with current (two vertical strands / coil) is

$$P_{dc} = R_{dc}\left(\frac{I_{phase}}{2}\right)^2 \tag{9.25}$$

We may translate the circulating new effect into a resistance additional coefficient, K_{Rad}.

$$K_{Rad} \approx \frac{P_{strand}}{P_{dc}} = \omega^2 \mu_0{}^2 \sigma_{Co}^2 \frac{b^2 h^4}{b_s{}^2} \left(\frac{l_{stack}}{l_{turn}}\right)^2 \frac{n_{coil}^2 (1+\cos\gamma)^2}{4} \quad (9.26)$$

Expression (9.26) is strictly valid for two vertical strands in parallel. However as B_{ave} seems to be the same for other number of strands/turn, Equation (9.26) should be valid in general.

Adding the skin effect coefficient K_{Rm} as already defined to the one due to circulating current between elementary conductors in parallel, we get the total skin effect coefficient $K_{R||}$.

$$K_{R||} = K_{Rm} \frac{l_{stack}}{l_{turn}} + K_{Rad} \quad (9.26')$$

Even with large power IMs, $K_{R||}$ should be less than 1.25 to 1.3 with K_{Rad} < 0.1 for a proper design.

Example 9.2. Skin effect in multiple vertical conductors in slot

Let us consider a rather large induction motor with 2 coils, each made of 4 elementary conductors in series, respectively, and, of two turns, each of them made of two vertical strands (conductors in parallel) per slot in the stator. The size of the elementary conductor is h·b = 5·20 [mm·mm] and the slot width b_s = 22 mm; the insulation thickness along slot height is neglected. The frequency f_1 = 60 Hz. Let us determine the skin effect in the stack zone for the two cases, if l_{stack}/l_{turn} = 0.5.

Solution

As the elementary conductor is the same in both cases, the first skin effect resistance correction coefficient K_{Rm} may be computed first from (9.6) with ξ from (9.4),

$$\xi = \beta_n h; \ h = 8\text{mm}$$

$$\beta_n = \sqrt{\frac{\omega_1 \mu_0 \sigma_{Co}}{2} \frac{b}{b_s}} = \sqrt{\frac{2\pi 60 \cdot 1.256 \cdot 10^{-6}}{2 \cdot 1.8 \cdot 10^{-8}} \frac{20}{22}} = 109.32\text{m}^{-1}$$

$$\xi = 109.32 \cdot 5 \cdot 10^{-3} = 0.5466$$

The helping functions φ(ξ) and ψ(ξ) are (from (9.7)): φ(ξ) = 1.015, ψ(ξ) = 0.04. Now with m = 8 layers in slot K_{Rm} (9.6) is

$$K_{Rm} = 1.015 + \frac{8^2 - 1}{3} 0.04 = 1.99$$

Now, for the parallel conductors (2 in parallel), the additional resistance correction coefficient K_{Rad} (9.26) for circulating currents is

$$K_{Rad} = \left(1.256 \cdot 10^{-6}\right)^2 (2\pi 60)^2 \frac{1}{\left(1.8 \cdot 10^{-8}\right)^2} \left(\frac{20}{22}\right)^2 \cdot$$

$$\cdot \frac{\left(5 \cdot 10^{-3}\right)^4 \cdot 0.5^2 \cdot 2^2 \cdot (1+1)^2}{4} = 0.3918!$$

The coefficient K_{Rad} refers to the whole conductor (turn) length, that is, it includes the end-turn part of it. K_{Rm} is too large, to be practical.

9.2.4. The skin effect in the end turns

There is a part of stator and rotor windings that is located outside the lamination stack, mainly in air: the end turns or endrings.

The skin effect for conductors in air is less pronounced than in their portions in slots.

As the machine power or frequency increases, this kind of skin effect is to be considered. In Reference [6] the resistance correction coefficient K_R for a single round conductor (d_{Co}) is also a function of β in the form (Figure 9.10).

$$\xi = d_{Co}\sqrt{\frac{\omega_1 \mu_0 \sigma_{Co}}{2}} \tag{9.27}$$

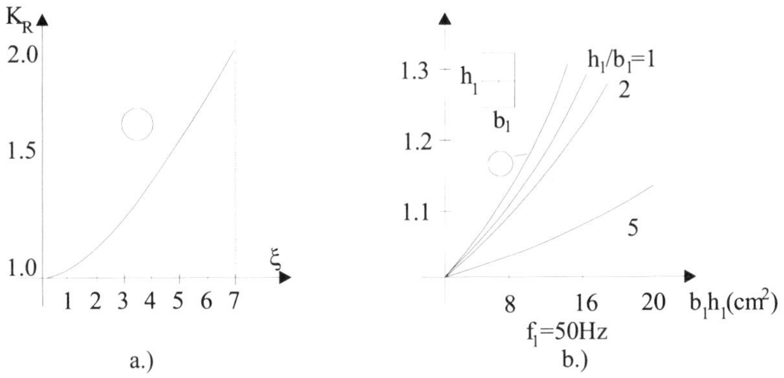

Figure 9.10 Skin effect correction factor K_R for a round conductor in air:
a.) circular b.) rectangular

On the other hand, a rectangular conductor in air [7] presents the resistance correction coefficient (Figure 9.10) based on the assumption that there are magnetic field lines that follow the conductor periphery.

In general, there are m layers of round or rectangular conductors on top of each other (Figure 9.11).

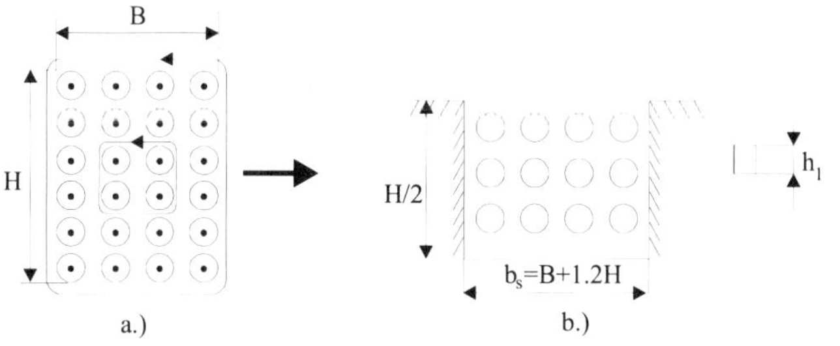

Figure 9.11 Four layer coil in air a.) and its upper part placed in an equivalent (fictious) slot

Now the value of ξ is

$$\begin{aligned} \xi &= d_{Co}\sqrt{\frac{\omega_1\mu_0\sigma}{2}\frac{B}{B+1.2H}} \quad \text{for round conductors} \\ \xi &= h_1\sqrt{\frac{\omega_1\mu_0\sigma}{2}\frac{B}{B+1.2H}} \quad \text{for rectangular conductors} \end{aligned} \tag{9.28}$$

As the skin effect is to be reduced, ξ should be made smaller than 1.0 by design. And, in this case, for rectangular conductors displaced in m layers [2], the correction coefficient K_{Rme} is

$$K_{Rme} = 1 + \frac{\left(m^2 - 0.8\right)}{36}\xi^4 \tag{9.29}$$

For a bundle of Z round conductors [24] K_{Rme} is

$$K_{Rme} = 1 + 0.005 \cdot Z \cdot (d/cm)^4 (f/50Hz)^2 \tag{9.29'}$$

The skin effect in the endrings of rotors may be treated as a single rectangular conductor in air. For small induction machines, however, the skin effect in the endrings may be neglected. In large IMs, a more complete solution is needed. This aspect will be treated later in this chapter.

For the IM in example 9.2, with m = 4, $\xi = 0.5466$, the skin effect in the end turns K_{Rme} (9.29) is

$$K_{Rme} = 1 + \frac{4^2 - 0.8}{36}0.5466^2 = 1.0377!$$

As expected, $K_{Rme} << K_{Rm}$ corresponding to the conductors in slot. The total skin effect resistance correction coefficient K_{Rt} is

$$K_{Rt} = \frac{K_{Rm} l_{stack} + K_{Rme}\left(l_{coil} - l_{stack}\right)}{l_{coil}} + K_{Rad} \tag{9.30}$$

For the case of example 9.2,

$$K_{Rt} = \frac{1.99 + 1.0377\left(1 - \frac{1}{1.5}\right)}{1.5} + 0.3918 = 1.5572 + 0.3918 = 1.949$$

for 2 conductors in parallel and K_{Rt} = 1.5572, for all conductors in series.

9.3. SKIN EFFECTS BY THE MULTILAYER APPROACH

For slots of more general shape, adopted to exploit the beneficial effects of rotor cages, a simplified solution is obtained by dividing the rotor bar into n layers of height h_t and width b_j (Figure 9.12). The method originates in [1].

For the pth layer Faraday's law yields

$$R_p \underline{I}_p - R_{p+1}\underline{I}_{p+1} = -jS\omega_1 \Delta\underline{\Phi}_p \tag{9.31}$$

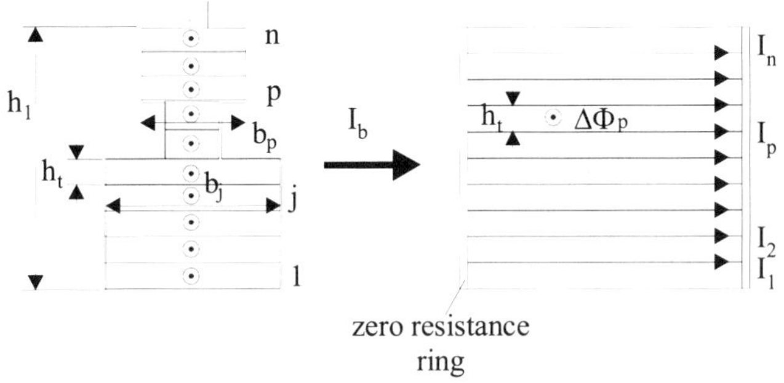

Figure 9.12 More general shape rotor bars

$$R_p = \frac{1}{\sigma_{Al}}\frac{l_{stack}}{b_p h_t};\; R_{p+1} = \frac{1}{\sigma_{Al}}\frac{l_{stack}}{b_{p+1} h_t};\; \Delta\Phi_p = \frac{\mu_0 l_{stack} h_t}{b_p}\sum_{j=1}^{p} I_j \tag{9.32}$$

R_p and R_{p+1} represent the resistances of pth and $(p+1)^{th}$ layer and L_p the inductance of pth layer.

$$L_p = \frac{\mu_0 l_{stack} h_t}{b_p} \tag{9.33}$$

With (9.33), Equation (9.31) becomes

$$I_{p+1} = \frac{R_p}{R_{p+1}} I_p + j\frac{S\omega_1 L_p}{R_{p+1}}\sum_{j=1}^{p} I_j \tag{9.34}$$

Let us consider p = 1,2 in (9.34)

$$\underline{I}_2 = \frac{R_1}{R_2}\underline{I}_1 + j\frac{S\omega_1 L_1}{R_2}\underline{I}_1 \tag{9.35}$$

$$\underline{I}_3 = \frac{R_2}{R_3}\underline{I}_2 + j\frac{S\omega_1 L_2}{R_3}\left(\underline{I}_1 + \underline{I}_2\right) \tag{9.36}$$

If we assign a value to I_1 in relation to total current I_b, say,

$$\left(\underline{I}_1\right)_{initial} = \frac{I_b}{n}, \tag{9.37}$$

we may use Equations (9.34) through (9.36) to determine the current in all layers. Finally,

$$\left(I_b\right)' = \left|\sum_{j=1}^{n} I_j\right| \tag{9.38}$$

As expected, I_b and I_b' will be different. Consequently, the currents in all layers will be multiplied by I_b/I_b' to obtain their real values. On the other hand, Equations (9.35) – (9.36) lead to the equivalent circuit in Figure 9.12.

Once the layer currents $\underline{I}_1, \ldots \underline{I}_n$ are known, the total losses in the bar are

$$P_{ac} = \sum_{j=1}^{n} \left|I_j\right|^2 R_j \tag{9.39}$$

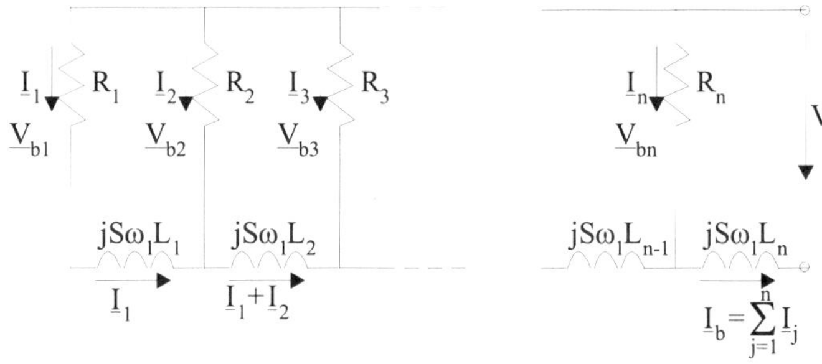

Figure 9.13 Equivalent circuit for skin effect evaluation

In a similar manner, the magnetic energy in the slot W_{mac} is

$$W_{mac} = \frac{1}{2}\sum_{j=1}^{n} L_j \left|\sum_{1}^{j} I_k\right|^2 \tag{9.40}$$

The d.c. power loss in the slot (for given total bar current) P_{dc} is

$$P_{dc} = \sum I_{jdc}^2 R_j; \quad I_{jdc} = \frac{I_b'}{A_{bar}} h_t b_j \tag{9.41}$$

Also the d.c. magnetic energy in the slot

$$W_{mdc} = \frac{1}{2}\sum_{j=1}^{n} L_j \left| \sum_{1}^{j} I_{kdc} \right|^2 \tag{9.42}$$

Now the skin effect resistance and inductance correction coefficients K_R, K_x are

$$K_R = \frac{P_{ac}}{P_{dc}} = \frac{\sum_{j=1}^{n} I_j^2 R_j}{\sum_{j=1}^{n} I_{jdc}^2 R_j} \tag{9.43}$$

$$K_x = \frac{\sum_{j=1}^{n} L_j \left| \sum_{k=1}^{j} I_k \right|^2}{\sum_{j=1}^{n} L_j \left| \sum_{k=1}^{j} I_{kdc} \right|^2} \tag{9.44}$$

Example 9.3. Let us consider a deep bar of the shape in Figure 9.12 with the dimensions as in Figure 9.14.

Let us divide the bar into only 6 layers, each 5 mm high (h_t = 5 mm) and calculate the skin effects for S = 1 and f_1 = 60 Hz.

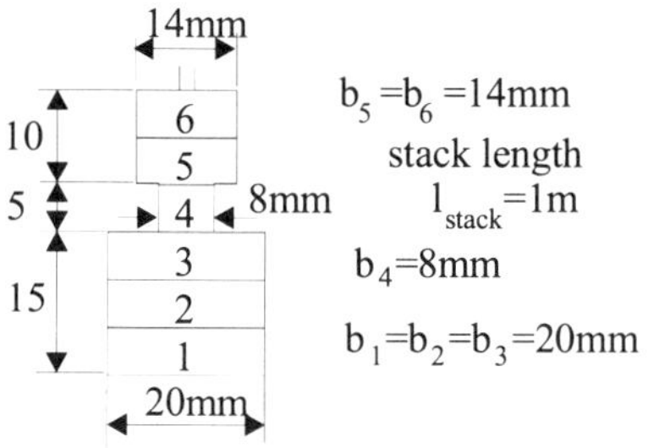

Figure 9.14 Deep bar geometry

Solution

From Figure 9.14, the layer resistances and inductances are (9.32 – 9.33).

$$R_1 = R_2 = R_3 = \frac{1}{\sigma_{Al}} \frac{l_{stack}}{b_1 h_t} = \frac{1}{3 \cdot 10^7} \frac{1}{20 \cdot 10^{-3} \cdot 5 \cdot 10^{-3}} = 0.333 \cdot 10^{-3} \Omega$$

$$R_4 = \frac{1}{\sigma_{Al}}\frac{l_{stack}}{b_4 h_t} = \frac{1}{3\cdot10^7}\frac{1}{8\cdot10^{-3}\cdot5\cdot10^{-3}} = 0.833\cdot10^{-3}\Omega$$

$$R_5 = R_6 = \frac{1}{\sigma_{Al}}\frac{l_{stack}}{b_5 h_t} = \frac{1}{3\cdot10^7}\frac{1}{14\cdot10^{-3}\cdot5\cdot10^{-3}} = 0.476\cdot10^{-3}\Omega$$

From (9.33)

$$L_1 = L_2 = L_3 = \frac{\mu_0 l_{stack} h_t}{b_1} = 1.256\cdot10^{-6}\cdot1\cdot\frac{0.005}{0.020} = 0.314\cdot10^{-6}\mathrm{H}$$

$$L_4 = \frac{\mu_0 l_{stack} h_t}{b_5} = 1.256\cdot10^{-6}\cdot\frac{0.005}{0.008} = 0.785\cdot10^{-6}\mathrm{H}$$

$$L_5 = L_6 = \frac{\mu_0 l_{stack} h_t}{b_5} = 1.256\cdot10^{-6}\cdot\frac{0.005}{0.014} = 0.44857\cdot10^{-6}\mathrm{H}$$

Let us now consider that the bar current is $I_b = 3600\mathrm{A}$ and $I_1 = I_b/n = I_b/6 = 600\mathrm{A}$. Now $\underline{I}_2$ (in the second layer from slot bottom) is

$$\underline{I}_2 = \frac{R_1}{R_2}\underline{I}_1 + j\frac{S\omega_1 L_1}{R_2}\underline{I}_1 = 600 + j\cdot1\cdot2\pi60\cdot\frac{0.314\cdot10^{-6}}{0.333\cdot10^{-3}}\cdot600 = 600 + j213.18\mathrm{A}$$

$$|\underline{I}_2| = 634.74\mathrm{A}$$

$$\underline{I}_3 = \frac{R_2}{R_3}\underline{I}_2 + j\frac{S\omega_1 L_2}{R_3}\left(\underline{I}_1 + \underline{I}_2\right) = 600 + j213.18 +$$

$$+ j\cdot1\cdot2\pi60\cdot\frac{0.314\cdot10^{-6}}{0.333\cdot10^{-3}}\cdot\left(600 + 600 + j213.18\right) = 524.25 + j640\mathrm{A}$$

$$|\underline{I}_3| = 827.3\mathrm{A}$$

$$\underline{I}_4 = \frac{R_3}{R_4}\underline{I}_3 + j\frac{S\omega_1 L_3}{R_4}\left(\underline{I}_1 + \underline{I}_2 + \underline{I}_3\right) = \frac{0.333\cdot10^{-3}}{0.833\cdot10^{-3}}\left(524.25 + j426.26\right) +$$

$$+ j\cdot1\cdot2\pi60\cdot\frac{0.314\cdot10^{-6}}{0.833\cdot10^{-3}}\cdot\left(600 + 600 + j213.18 + 524 + j640\right)$$
$$= 55.5 + j490.2\mathrm{A}$$

$$|\underline{I}_4| = 493.27\mathrm{A}$$

$$\underline{I}_5 = \frac{R_4}{R_5}\underline{I}_4 + j\frac{S\omega_1 L_4}{R_5}\left(\underline{I}_1 + \underline{I}_2 + \underline{I}_3 + \underline{I}_4\right) = \frac{0.833\cdot 10^{-3}}{0.476\cdot 10^{-3}}(55.5 + j490.2) +$$

$$+ j\cdot 1\cdot 2\pi 60\cdot \frac{0.785\cdot 10^{-6}}{0.4485\cdot 10^{-3}}\cdot (1779.75 + j1342)$$

$$= -712.088 + j2030.4A$$

$$\left|\underline{I}_5\right| = 2151.2A$$

$$\underline{I}_6 = \frac{R_5}{R_6}\underline{I}_5 + j\frac{S\omega_1 L_5}{R_6}\left(\underline{I}_1 + \underline{I}_2 + \underline{I}_3 + \underline{I}_4 + \underline{I}_5\right) =$$

$$= \frac{0.476\cdot 10^{-3}}{0.476\cdot 10^{-3}}(-712.088 + j2030.4) +$$

$$+ j\cdot 1\cdot 2\pi 60\cdot \frac{0.74857\cdot 10^{-6}}{0.476\cdot 10^{-3}}\cdot (-712.088 + j2030.4 + 1724.25 + j853)$$

$$= -1735.75 + j2389.4A$$

$$\left|\underline{I}_6\right| = 2953.31A$$

Now the total current

$$\underline{I}_b' = \underline{I}_1 + \underline{I}_2 + \underline{I}_3 + \underline{I}_4 + \underline{I}_5 + \underline{I}_6 = -712.088 + j2030.4 - 1735.75 + j2389 =$$

$$= -2447.75 + j4419.4A$$

$$I_b' \approx 5050A$$

The a.c. power in the bar is

$$P_{ac} = \sum_{j=1}^{n} I_j^2 R_j = 0.333\cdot 10^{-3}\left(600^2 + 636.74^2 + 827.30^2\right) +$$

$$+ 0.833\cdot 10^{-3}\cdot 493.27^2 + 0.476\cdot 10^{-3}\left(2151.2^2 + 2955.31^2\right) =$$

$$= 482 + 202.68 + 6360 = 7044.68W$$

The d.c. current distribution in the 6 layer is uniform, therefore

$$I_{1dc} = I_{2dc} = I_{3dc} = \frac{\underline{I}_b'}{A_{bar}} b_1 h_t = \frac{5050\cdot 20.5}{(14.10 + 8.5 + 15.20)} = 1052.08A$$

$$I_{4dc} = \frac{\underline{I}_b'}{A_{bar}} b_4 h_t = \frac{5050\cdot 8.5}{480} = 420.83A$$

$$I_{5dc} = I_{6dc} = \frac{I_b'}{A_{bar}} b_5 h_t = \frac{5050 \cdot 14.5}{480} = 736.458A$$

$$P_{dc} = \sum_{j=1}^{n} I_{jdc}^2 R_j = 0.333 \cdot 10^{-3} \cdot 3 \cdot 1052.08^2 +$$

$$+0.833 \cdot 10^{-3} \cdot 420.83^2 + 0.476 \cdot 10^{-3} \cdot 2 \cdot 736.458^2 = 1768.81W$$

and the skin effect resistance correction factor K_R is

$$K_R = \frac{P_{ac}}{P_{dc}} = \frac{7044.68}{1768.81} = 3.9827!$$

The magnetic energy ratio K_x is

$$K_x = \frac{A}{B}$$

$$A = L_1\left(|\underline{I}_1|^2 + |\underline{I}_1 + \underline{I}_2|^2 + |\underline{I}_1 + \underline{I}_2 + \underline{I}_3|^2\right) + L_4|\underline{I}_1 + \underline{I}_2 + \underline{I}_3|^2 +$$
$$+ L_5|\underline{I}_1 + \underline{I}_2 + \underline{I}_3 + \underline{I}_4 + \underline{I}_5|^2 + L_5|\underline{I}_1 + \underline{I}_2 + \underline{I}_3 + \underline{I}_4 + \underline{I}_5 + \underline{I}_6|^2$$

$$B = L_1\left[I_{1dc}^2 + (I_{1dc} + I_{2dc})^2 + (I_{1dc} + I_{2dc} + I_{3dc})^2\right] +$$
$$+ L_4(I_{1dc} + I_{2dc} + I_{3dc} + I_{4dc})^2 + L_5(I_b' - I_6)^2 + L_5 I_b'^2$$

$$A = 0.314 \cdot 10^{-6}\left(600^2 + 1.217^2 10^6 + 1.9237^2 10^6\right) + 0.785 \cdot 10^{-6} 2.229^2 10^6 +$$
$$+ 0.44857 \cdot 10^{-6}\left(3.0554^2 10^6 + 5.050^2 10^6\right)$$

$$B = 0.314 \cdot 10^{-6}\left(1.052^2 10^6 + 2.104^2 10^6 + 3.156^2 10^6\right) +$$
$$+ 0.785 \cdot 10^{-6} 3.576^2 10^6 + 0.44857 \cdot 10^{-6}\left(4.3128^2 10^6 + 5.05^2 10^6\right)$$

$$K_x = \frac{21.267}{34.685} = 0.613!$$

The inductance coefficient refers only to the slot body (filled with conductor) and not to the slot neck, if any.

A few remarks are in order.

- The distribution of current in the various layers is nonuniform when the skin effect occurs.
- Not only the amplitude, but the phase angle of bar current in various layers varies due to skin effect (Figure 9.14).
- At S = 1 (f_1 = 60 Hz) most of the current occurs in the upper part of the slot.

- The equivalent circuit model can be easily put into computer form once the layers geometry–h_t (height) and b_j (width)–are given. For various practical slots special subroutines may provide b_j, h_t when the number of layers is given.
- To treat a double cage by this method, we have only to consider zero the current in the empty slot layers between the upper and lower cage (Figure 9.15).

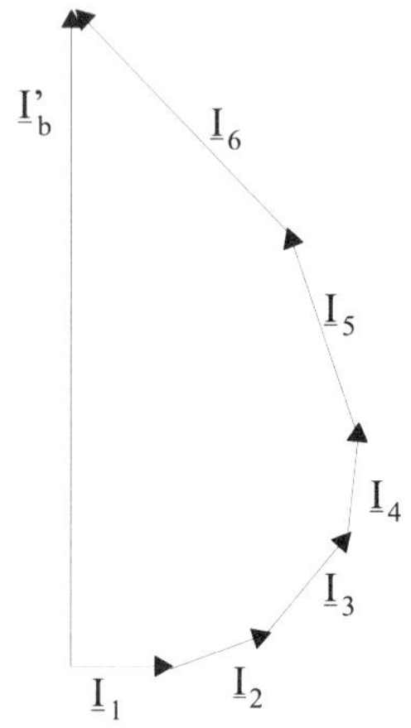

Figure 9.14 Layer currents and the bar current with skin effect

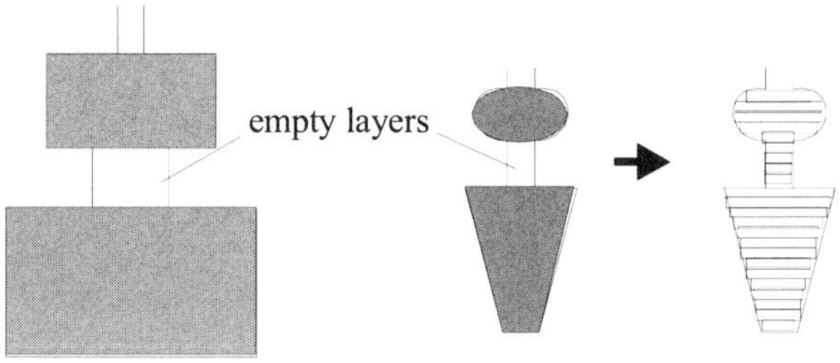

Figure 9.15 Treating skin effect with equivalent circuit (or multilayer) method

Now that both K_R and K_x are known, the bar resistance and slot body leakage geometrical specific permeance λ_{sbody} is modified to account for skin effect.

$$(\lambda_{sbody})_{ac} = (\lambda_{sbody})_{dc} K_x \tag{9.45}$$

From d.c. magnetic energy W_{mdc} (9.42), we write

$$L_{dc} = \frac{1}{2}\frac{W_{mdc}}{I_b'^2} = \mu_0 l_{stack} (\lambda_{sbody})_{dc} \tag{9.46}$$

The slot neck geometrical specific permeance is still to be added to account for the respective slot leakage flux. This slot neck geometrical specific permeance is to be corrected for leakage flux saturation discussed later in this chapter.

9.4. SKIN EFFECT IN THE END RINGS VIA THE MULTILAYER APPROACH

As the end rings are placed in air, although rather close to the motor laminated stack, the skin effect in them is routinely neglected. However, there are applications where the value of slip goes above unity (S = up to 3.0 in standard elevator drives) or the slip frequency is large as in high frequency (high speed) motors to be started at rated frequency (400 Hz in avionics).

For such cases, the multilayer approach may be extended to end rings. To do so we introduce radial and circumpherential layers in the end rings (Figure 9.16) as shown in Reference [7].

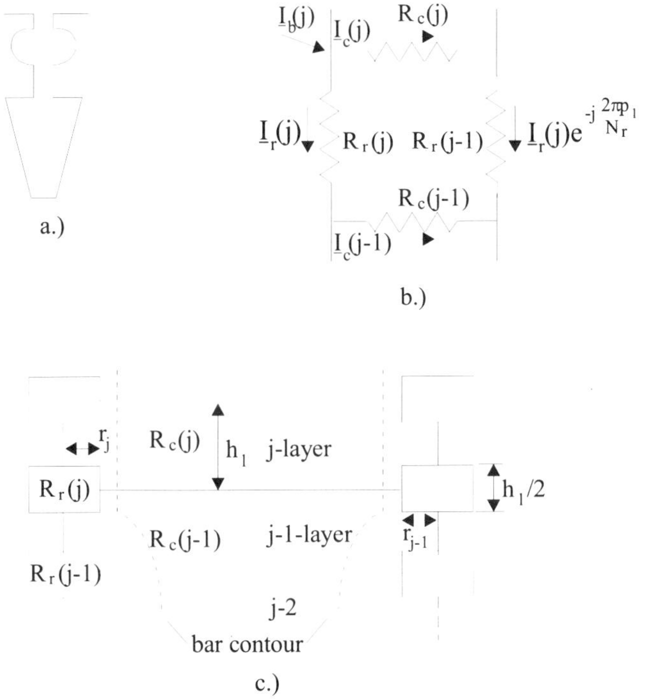

Figure 9.16 Bar-end ring transition
a.) slot cross – section b.) radial and circumpherential layer end ring currents
c.) geometry of radial and circumpherential end ring layers

In all layers, the current density is considered uniform. It means that their radial dimension has to be less than the depth of field penetration in aluminum.

The currents in neighboring slots are considered phase-shifted by $2\pi p_1/N_r$ radians (N_r–number of rotor slots).

The relationship between bar and end ring layer currents (Figure 9.16) is

$$I_{b(j)} + \underline{I}_{r(j+1)} + \underline{I}_{r(j)} e^{j\frac{2\pi p_1}{N_r}} = \underline{I}_{r(j)} + \underline{I}_{c(j)} \tag{9.47}$$

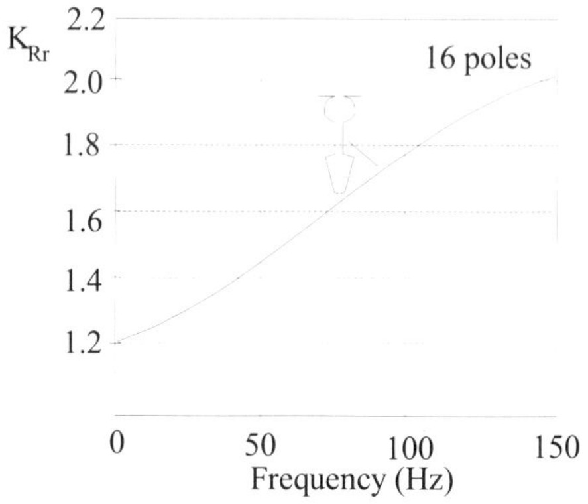

Figure 9.17 End ring skin effect resistance coefficient K_{Rr}

$$R_{c(j)}\underline{I}_{c(j)} + R_{r(j)}\underline{I}_{r(j)} e^{-j\frac{2\pi p_1}{N_r}} = R_{c(j-1)}\underline{I}_{c(j-1)} + R_{r(j)}\underline{I}_{r(j)} \tag{9.48}$$

The circumpherential extension of the radial layer r_j is assigned a value at start. Now if we add the equations for the bar layer currents, we may solve the system of equations. As long as the radial currents increase, γ_j is increased in the next iteration cycle until sufficient convergence is met. Some results, after [2], are given in Figure 9.17.

As the slot total height is rather large (above 25 mm), the end ring skin effect is rather large, especially for rotor frequencies above 50(60) Hz. In fact, a notable part of this resistance rise is due to the radial ring currents which tend to distribute the bar currents, gathered toward the slot opening, into most of end ring cross section.

9.5. THE DOUBLE CAGE BEHAVES LIKE A DEEP BAR CAGE

In some applications, very high starting torque–$T_{start}/T_{rated} \geq 2.0$–is required. In such cases, a double cage is used. It has been proved that it behaves like a deep bar cage, but it produces even higher starting torque at lower starting current. For the case when skin effect can be neglected in both cages, let us consider a double cage as configured in Figure 9.18. [8]

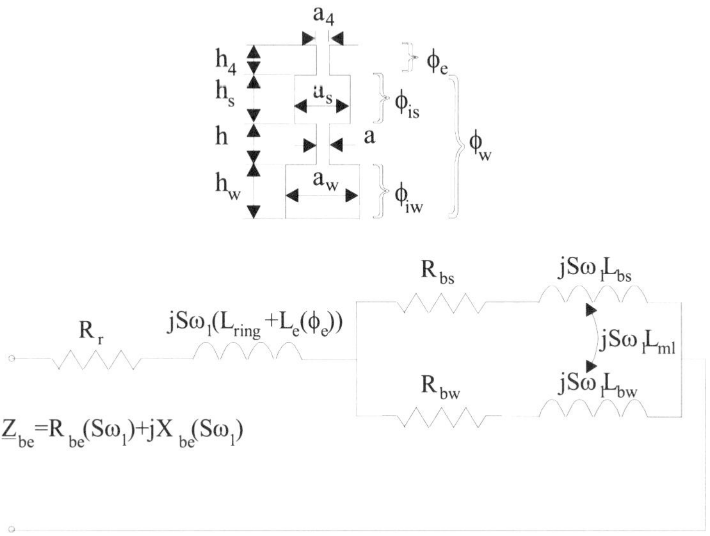

Figure 9.18 Double cage rectangular-shape geometry a.) and equivalent circuit b.)

The equivalent single bar circuit is given in Figure 9.18b. For the common ring of the two cages

$$R_r = R_{ring},\ R_{bs} = R_{bs\ upper\ bar},\ R_{bw} = R_{bw\ lower\ bar} \tag{9.49}$$

For separate rings

$$R_r = 0,\ R_{bs} = R_{bs} + R^e_{rings},\ R_{bw} = R_{bw} + R^e_{ringw} \tag{9.50}$$

The ring segments are included into the bar resistance after approximate reduction as shown in Chapter 6. The value of L_{ring} is the common ring inductance or is zero for separate rings. Also for both cases, $L_e(\Phi_e)$ refers to the slot neck flux.

$$L_e(\Phi_e) = \mu_0 l_{stack} \frac{h_4}{a_4} \tag{9.51}$$

We may add into L_e the differential leakage inductance of the rotor.

The start (upper) and work (lower) cage inductances L_{bs} and L_{bw} include the end ring inductances only for separate rings. Otherwise, the bar inductances are

$$\begin{aligned} L_{bs} &= \mu_0 l_{stack} \frac{h_s}{3a_s} \\ L_{bw} &= \mu_0 l_{stack}\left(\frac{h_w}{3a_w} + \frac{h}{a} + \frac{h_s}{a_s}\right) \end{aligned} \tag{9.52}$$

There is also a flux common to the two cages represented by the flux in the starting cage. [3]

$$L_{ml} = \mu_0 l_{stack} \frac{h_s}{2a_s} \tag{9.53}$$

In general, L_{ml} is neglected though it is not a problem to consider in solving the equivalent circuit in Figure 9.18. It is evident (Figure 9.18a) that the starting (upper) cage has a large resistance (R_{bs}) and a small slot leakage inductance L_{bs}, while for the working cage the opposite is true.

Consequently, at high slip frequency, the rotor current resides mainly in the upper (starting) cage while, at low slip frequency, the current flows mainly into the working (lower) cage. Thus both R_{be} and X_{be} vary with slip frequency as they do in a deep bar single cage (Figure 9.19).

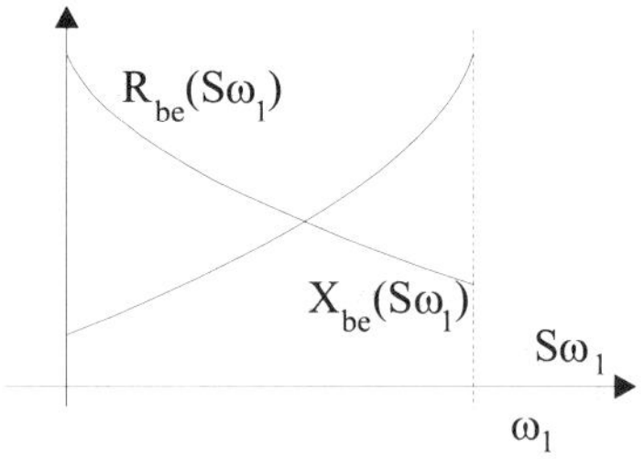

Figure 9.19 Equivalent parameters of double cage versus slip frequency

9.6. LEAKAGE FLUX PATH SATURATION–A SIMPLIFIED APPROACH

Leakage flux path saturation occurs mainly in the slot necks zone for semiclosed slots for currents above 2 to 3 times rated current or in the rotor slot iron bridges for closed slots even well below the rated current (Figure 9.20).

Consequently,

$$\begin{aligned} a_{os}' &= a_{os} + \frac{(\tau_{ss} - a_{os})}{\mu_s}\mu_0; \quad \tau_{ss} = \frac{\pi D_i}{N_s} \\ a_{or}' &= a_{or} + \frac{(\tau_{sr} - a_{or})}{\mu_r}\mu_0; \quad \tau_{sr} = \frac{\pi (D_i - 2g)}{N_s} \\ a_{or}' &= \frac{b_{or}\mu_b}{\mu_{br}} \end{aligned} \tag{9.54}$$

The slot neck geometrical permeances will be changed to: a'_{os}/h_{os}, a'_{or}/h_{or}, or a'_{or}/h_{or} dependent on stator (rotor) current.

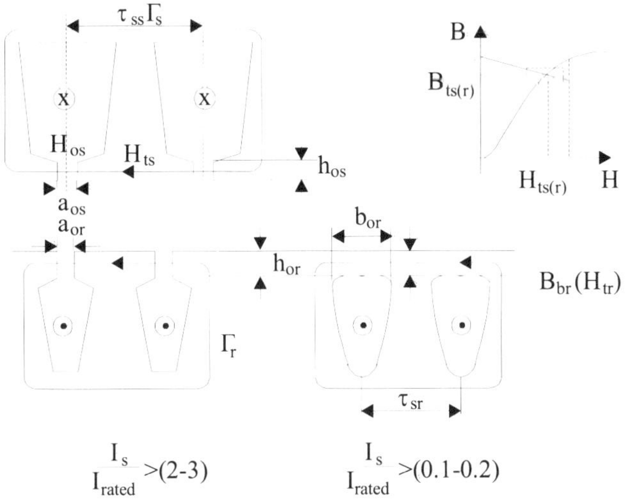

Figure 9.20 Leakage flux path saturation conditions

With n_s the number of turns (conductors) per slot, and I_s and I_b the stator and rotor currents, the Ampere's law on Γ_s, Γ_r trajectories in Figure 9.20 yields

$$\begin{aligned} &H_{ts}(\tau_{ss} - a_{os}) + H_{os}a_{os} \approx n_s I_s \sqrt{2};\ \mu_s H_{ts} = B_{ts} = \mu_0 H_{os} \\ &H_{tr}(\tau_{sr} - a_{or}) + H_{or}a_{or} \approx I_b \sqrt{2};\ \mu_r H_{tr} = B_{tr} = \mu_0 H_{or} \\ &H_{tr}b_{or} = I_b\sqrt{2};\ \mu_{br} = \frac{B_{tb}}{H_{tb}} \end{aligned} \tag{9.55}$$

The relationship between the equivalent rotor current I_r' (reduced to the stator) and I_b is (Chapter 8)

$$I_b = I_r' \frac{6p_1qn_sK_{w1}}{N_r} = I_r' n_s K_{w1} \frac{N_s}{N_r} \tag{9.56}$$

N_s–number of stator slots; K_{w1}–stator winding factor.

When the stator and rotor currents I_s and I_r' are assigned pertinent values, iteratively, using the lamination magnetization curves, Equations (9.55) may be solved (Figure 9.20) to find the iron permeabilities of teeth tops or of closed rotor slot bridges. Finally, from (9.54), the corrected slot openings are found.

With these values, the stator and rotor parameters (resistances and leakage inductances) as influenced by the skin effect (in the slot body zone) and by the leakage saturation (in the slot neck permeance) are recalculated. Continuing with these values, from the equivalent circuit, new values of stator and rotor currents I_s, I_r' are calculated for given stator and voltage, frequency and slip.

The iteration cycles continue until sufficient convergence is obtained for stator current.

Example 9.4. An induction motor has semiclosed rectangular slots whose geometry is shown in Figure 9.21.

The current density design for rated current is $j_{Co} = 6.5$ A/mm^2 and the slot fill factor $K_{fill} = 0.45$. The starting current is 6 times rated current. Let us calculate the slot leakage specific permeance at rated current and at start.

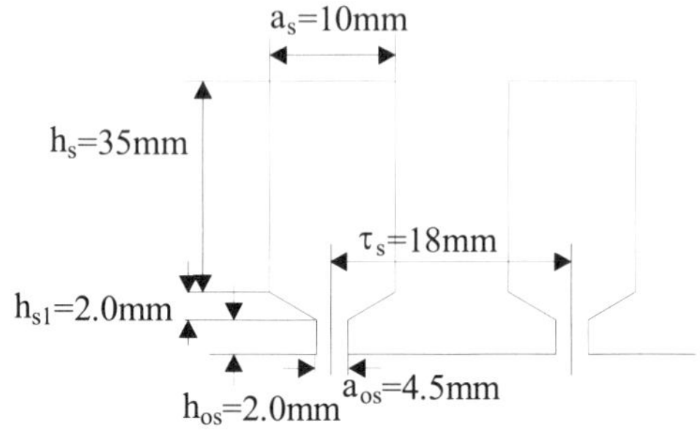

Figure 9.21 Slot geometry

Solution

The slot mmf n_sI_s is obtained from

$$n_sI_s = (h_sa_s)K_{fill}j_{Co} = 10 \cdot 35 \cdot 0.45 \cdot 6.5 = 1023.75 \text{Ampereturns}$$

Making use of (9.54),

$$\left[H_{ts}(18-4.5)+\frac{B_{ts}}{\mu_0}4.5\right] \cdot 10^{-3} = 1023.75\sqrt{2}$$

$$B_{ts} = \left(1023.75\sqrt{2} - H_{ts} \cdot 13.5 \cdot 10^{-3}\right)\frac{1.256 \cdot 10^{-6}}{4.5 \cdot 10^{-3}} = 0.4029 - H_{ts}3\mu_0$$

It is very clear that for rated current $B_{ts} < 0.4029$, so the tooth heads are not saturated. In (9.54), $a_{os}' = a_{os}$ and the slot specific geometrical permeance λ_s is

$$\lambda_s = \frac{h_s}{3a_s} + \frac{2h_{s1}}{a_{os}+a_s} + \frac{h_{os}}{a_{os}} = \frac{35}{3 \cdot 10} + \frac{2 \cdot 2.5}{4.5+10} + \frac{2}{4.5} = 1.9559$$

For the starting conditions,

$$\left[H_{tstart}(18-4.5) \cdot 10^{-3} + \frac{B_{ts}}{\mu_0}4.5\right] \cdot 10^{-3} = 6142.5\sqrt{2}$$

$$B_{tstart} = 2.41 - H_{tstart}3\mu_0$$

Now we need the lamination magnetization curve. If we do so we get $B_{tstart} \approx 2.16$T, $H_{tstart} \approx 70{,}000$ A/m. The iron permeability μ_s is

$$\frac{\mu_s}{\mu_0} = \frac{B_{tstart}}{H_{tstart}}\mu_0^{-1} = \frac{2.16}{70000 \cdot 1.256 \cdot 10^{-6}} = 24.56$$

Consequently, $a_{os}' = a_{os} + \frac{(\tau_s - a_{os})}{\mu_s}\mu_0 = 4.5 + \frac{(18-4.5)}{24.56} = 5.05mm$

The slot geometric specific permeance is, at start,

$$\lambda_s' = \frac{h_s}{3a_s} + \frac{2h_{sl}}{a_{os}'+a_s} + \frac{h_{os}}{a_{os}'} = \frac{35}{3 \cdot 10} + \frac{2 \cdot 2.5}{5.05+10} + \frac{2}{5.05} = 1.895$$

The leakage saturation of tooth heads at start is not very important in our case. In reality, it could be more important for semiclosed stator (rotor) slots.

Example 9.5. Let us consider a rotor bar with the geometry of Figure 9.21. The bar current at rated current density $j_{Alr} = 6.0$ A/mm^2 is

$$I_{br} = j_{Alr} h_r \frac{(a_r + b_r)}{2} = 25\frac{(11+5)}{2} \cdot 6 = 1200A$$

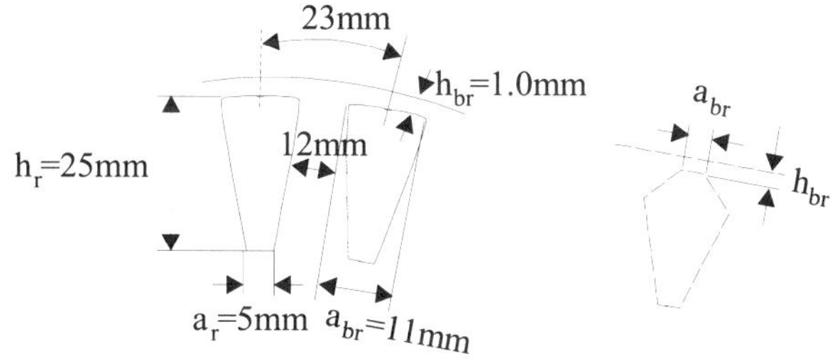

Figure 9.22. Closed rotor slots

Calculate the rotor bridge flux density for 10%, 20%, 30%, 50%, 100% of rated current and the corresponding slot geometrical specific permeance.

Solution

Let us first introduce a typical magnetization curve.

B(T)	0.05	0.1	0.15	0.2	0.25	0.3	0.35	0.4	0.45	0.5
H(A/m)	22.8	35	45	49	57	65	70	76	83	90
B(T)	0.55	0.6	0.65	0.7	0.75	0.8	0.85	0.9	0.95	1
H(A/m)	98	106	115	124	135	148	162	177	198	220

B(T)	1.05	1.1	1.15	1.2	1.25	1.3	1.35	1.4	1.45	1.5
H(A/m)	237	273	310	356	417	482	585	760	1050	1340
B(T)	1.55	1.6	1.65	1.7	1.75	1.8	1.85	1.9	1.95	2.0
H(A/m)	1760	2460	3460	4800	6160	8270	11170	15220	22000	34000

Neglecting the tooth top saturation and including only the bridge zone, the Ampere's law yields

$$H_{br}a_{br} = I_b; \quad H_{br} = \frac{I_b}{a_{br}}; \quad a_{br} = 11\text{mm}$$

For various levels of bar current, H_{br} and B_{br} become

I_b (A)	120	240	360	600	1200
H_{br}	10909	21818	32727	54545	109090
B_{br}	1.84	1.95	1.98	2.1	2.2
$\mu_{rel} = \frac{\mu_{br}}{\mu_0}$	134.28	71.16	48.165	30.653	16.056
$a_{or}' = \frac{a_{br}}{\mu_{rel}}$ (mm)	0.0819	0.1546	0.2284	0.3588	0.6851
$\lambda_{slot} = \frac{2h_r}{3(a_r + a_{br})} + \frac{h_{br}}{a_{or}'}$	14.293	75.1	5.42	3.8287	2.5013

Note that at low current levels the slot geometrical specific permeance is unusually high. This is because the iron bridge is long (a_{br} = 11mm); so H_{br} is small at small currents.

Also, the iron bridge is 1mm thick in our case, while 0.5 to 0.6 mm would produce better results (lower λ_{slot}). The slot height is quite large (h_r = 25 mm) so significant skin effect is expected at high slips. We investigated here only the currents below the rated one so $Sf_1 < 3$ Hz in general. Should the skin effect occur, it would have reduced the first term in λ_{slot} (slot body specific geometrical permeance) by the factor K_x.

As a conclusion to this numerical example, we may infer that a thinner bridge with a smaller length would saturate sooner (at smaller currents below the rated one), producing finally a lower slot leakage permeance λ_{slot}. The advantages of closed slots are related to lower stray load losses and lower noise.

9.7. LEAKAGE SATURATION AND SKIN EFFECTS–A COMPREHENSIVE ANALYTICAL APPROACH

Magnetic saturation in induction machines occurs in the main path, at low slip frequencies and moderate currents unless closed rotor slots are used when their iron bridges saturate the leakage path.

In contrast, for high slip frequencies and high currents, the leakage flux paths saturate as the main flux decreases with slip frequency for constant stator voltage and frequency.

The presence of slot openings, slot skewing, and winding distribution in slots, make the problem of accounting for magnetic saturation, in presence of rotor skin effects, when the induction machine is voltage fed, a very difficult task.

Ultimately a 3D finite element approach would solve the problem, but for prohibitive computation time.

Less computation, intensive, analytical solutions have been gradually introduced [9–15]. However, only the starting conditions have been treated (with $I_s = I_r'$ known), with skin effect neglected. Main flux path saturation level, skewing, zig-zag leakage flux have all to be considered for a more precise assessment of induction motor parameters for large currents.

In what follows, an attempt is made to allow for magnetic saturation in the main and leakage path (with skin effect accounted for by appropriate correction coefficients K_R, K_x) for given slip, frequency, voltage, motor geometry, no-load curve, by iteratively recalculating the stator magnetization and rotor currents until sufficient convergence is obtained. So, in fact, the saturation and skin effects are explored from no-load to motor stall conditions.

The computation algorithm presupposes that the motor geometry and the no-load curve as $L_m(I_m)$, magnetising inductance versus no-load current, is known based on analytical or FEM modeling or from experiments.

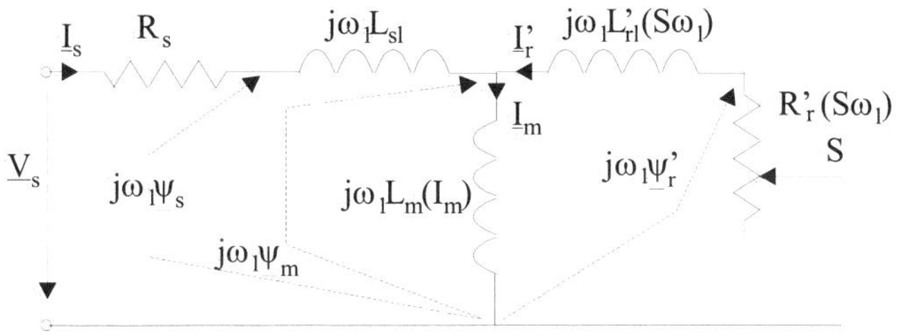

Figure 9.23 Equivalent circuit with main flux path saturation and skin effects accounted for

The leakage inductances are initially considered unsaturated (as derived in Chapter 6) with their standard expressions. The influence of skin effect on slot–body specific geometric permeance and on resistances is already accounted for in these parameters.

Consequently, given the slip value, the equivalent circuit with the variable parameter $L_m(I_m)$, Figure 9.23, may be solved.

The phasor diagram corresponding to the equivalent circuit in Figure 9.23 is shown in Figure 9.24.

From the equivalent circuit, after a few iterations, based on the $L_m(I_m)$ saturation function, we can calculate I_s, φ_s, I_r', φ_r, I_m, φ_m.

As expected, at higher slip frequency ($S\omega_1$) and high currents, the main flux ψ_m decreases for constant voltage V_s and stator frequency ω_1. At start ($S = 1$, rated voltage and frequency), due to the large voltage drop over ($R_s + j\,\omega_1 L_{sl}$), the main flux Ψ_m decreases to 50 to 60% of its value at no-load.

This ψ_m decreasing with slip and current rise led to the idea of neglecting the main flux saturation level when leakage flux path saturation was approached. However, at a more intrusive view, in the teeth top main flux (Ψ_m), slot neck leakage flux (Φ_{st1}, Φ_{st2}), zig-zag leakage flux (Φ_{z1}, Φ_{z2}), the so-called skewing flux (Φ_{sk1}, Φ_{sk2}), and the differential leakage flux (Φ_{d1}, Φ_{d2}) all meet to produce the resultant flux and saturation level (Figure 9.25). It is true that these fluxes have, at a certain point of time, their maxima at different positions along rotor periphery.

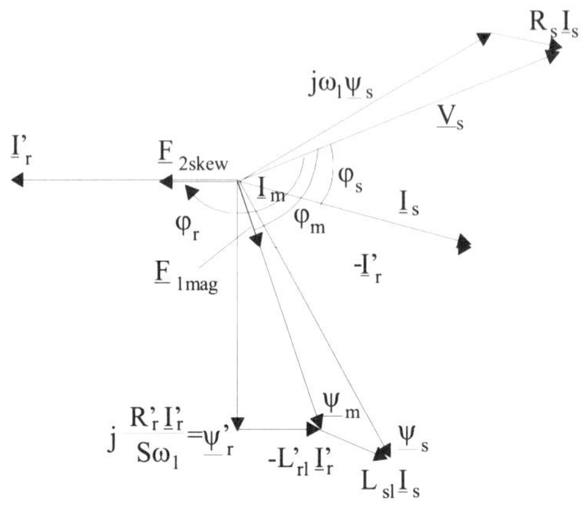

Figure 9.24 The phasor diagram

In essence, the main flux phase angle is φ_m, while Φ_{st1}, Φ_{z1}, Φ_{d1}, Φ_{st2}, Φ_{z2}, Φ_{d2}, have the phase angles φ_s, φ_r, while Φ_{sk1}, Φ_{sk2} are related to rotor current uncompensated mmf and thus is linked to φ_r again. As slip varies, not only I_s, I_r', I_m vary but also their phase angles: especially φ_s and φ_r' with φ_m about the same.

The differential leakage fluxes (Φ_{d1}, Φ_{d2}) are not shown in Figure 9.25 though they also act within airgap and teeth top zones, AB and DE.

To make the problem easy and amenable to computation, it is assumed that when saturation occurs in some part of the machine, the ratios between various flux contributions to total flux remain as they were before saturation.

The case of closed rotor slots is to be treated separately. Let us now define the initial expressions of various fluxes per tooth and their geometrical specific leakage permeances.

The main flux per tooth Φ_{m1} is

$$\Phi_{m1} = \frac{\psi_m \cdot 2p_1}{W_1 K_{w1} N_s}; \quad \psi_m = L_m(I_m) I_m \sqrt{2} \tag{9.57}$$

On the other hand, noticing that the stator slot mmf $AT_{1m} = n_s I_m \sqrt{2}$, the main flux per stator tooth is

$$\Phi_{m1} = AT_{1m} \mu_0 l_{stack} \lambda_m; \quad \Phi_{m2} = \Phi_{m1} \frac{N_s}{N_r} \tag{9.58}$$

where λ_m is the equivalent main flux geometrical specific permeance.

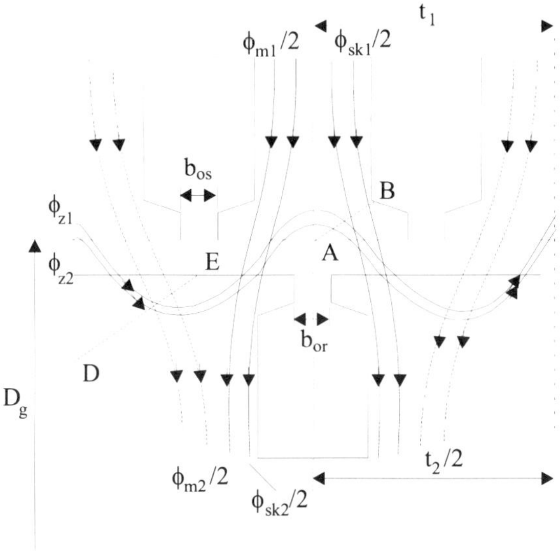

Figure 9.25 Computational flux lines

In a similar manner, the zig-zag fluxes Φ_{z1} and Φ_{z2} are

$$\begin{aligned}\Phi_{z1} &= AT_1\mu_0 l_{stack}\lambda_{z1}\\ \Phi_{z2} &= AT_2\mu_0 l_{stack}\lambda_{z2}\\ AT_1 = n_s I_s\sqrt{2};&\quad AT_2 = n_r I'_r\sqrt{2};\end{aligned} \tag{9.59}$$

From (9.56) and (9.59) n_r is

$$n_r = \frac{I_b\sqrt{2}}{I'_r\sqrt{2}} = n_s K_{w1}\frac{N_s}{N_r} \tag{9.60}$$

The unsaturated zig-zag specific permeances λ_{z1} and λ_{z2} are [5]

$$\lambda_{z1} = \frac{K_{w1}{}^2\pi D_g}{12N_s g}\left(1.2a_1 - 0.2\right) \tag{9.61}$$

$$\lambda_{z2} = \frac{K_{w1}{}^2\pi D_g}{12N_r g}\left(1.2a_2 - 0.2\right) \tag{9.62}$$

$$a_1 = \frac{t_1\left(5g + b_{os}\right) - b_{or}{}^2}{t_1\left(5g + b_{os}\right)} \tag{9.63}$$

$$a_2 = \frac{t_2(5g + b_{or}) - b_{os}^2}{t_2(5g + b_{or})} \quad (9.64)$$

The slot neck fluxes Φ_{st1} and Φ_{st2} are straightforward.

$$\Phi_{st1} = AT_1 \mu_0 l_{stack} \lambda_{st1}; \quad \lambda_{st1} = \frac{h_{os}}{b_{os}} \quad (9.65)$$

$$\Phi_{st2} = AT_2 \mu_0 l_{stack} \lambda_{st2}; \quad \lambda_{st2} = \frac{h_{or}}{b_{or}} \quad (9.66)$$

The differential leakage, specific permeance $\lambda_{d1,2}$ may be derived from a comparison between magnetization harmonic inductances $L_{m\nu}$ and the leakage inductance expressions L_{sl} and L_{rl}' (Chapter 7).

$$L_{m\nu} = \frac{6\mu_0}{\pi^2} \frac{(\tau/\nu) l_{stack} (W_1 K_{w\nu})^2}{(p_1 \nu) g K_c (1 + K_{steeth})} \quad (9.67)$$

$$L_{sl} = 2\mu_0 \frac{W_1^2}{p_1 q}(\lambda_{ss} + \lambda_{z1} + \lambda_{endcon1} + \lambda_{st1} + \lambda_{d1}) \quad (9.68)$$

$$L_{sr} = 2\mu_0 \frac{W_1^2}{p_1 q}(\lambda_{sr} + \lambda_{z2} + \lambda_{endcon2} + \lambda_{d2} + \lambda_{st2} + \lambda_{sk2}) \quad (9.69)$$

$$\lambda_{d1} = \sum_{\nu>1}^{\infty} 3 \frac{K_{w\nu}^2}{g K_c \nu^2} \frac{q_1 \tau}{(1 + K_{steeth})} \approx \lambda_{m1}(\tau_{d1} - \Delta\tau_{d1})$$

$$\tau_{d1} = \pi^2 \frac{(10 q_1^2 + 2)}{27} \sin^2\left(\frac{30^0}{q_1}\right) - 1; \quad \Delta\tau_{d1} = K_{z1} \frac{1}{9 q_1^2}$$

$$\lambda_{d2} = \lambda_{m1} \frac{N_s}{N_r}(\tau_{d2} - \Delta\tau_{d2}); \quad \Delta\tau_{d2} = K_{z2}\left(\frac{2p_1}{N_r}\right)^2 \quad (9.70)$$

$$\tau_{d2} = \frac{\left(\frac{\pi p_1}{N_r}\right)^2}{\sin^2\left(\frac{\pi p_1}{N_r}\right)} \frac{\left(\frac{\pi p_1 \cdot skew}{N_r}\right)^2}{\sin^2\left(\frac{\pi p_1 \cdot skew}{N_r}\right)} - 1$$

The coefficient K_{z1} and K_{z2} are found from Figure 9.26.

The slot-specific geometric permeances λ_{ss} and λ_{sr}, for rectangular slots, with rotor skin effect, are

$$\lambda_{ss} = \frac{h_s}{3b_s} + \lambda_{st1} \tag{9.71}$$

$$\lambda_{sr} = \frac{h_r}{3b_r} \cdot K_X + \lambda_{st2}; \quad K_x - \text{skin effect correction} \tag{9.72}$$

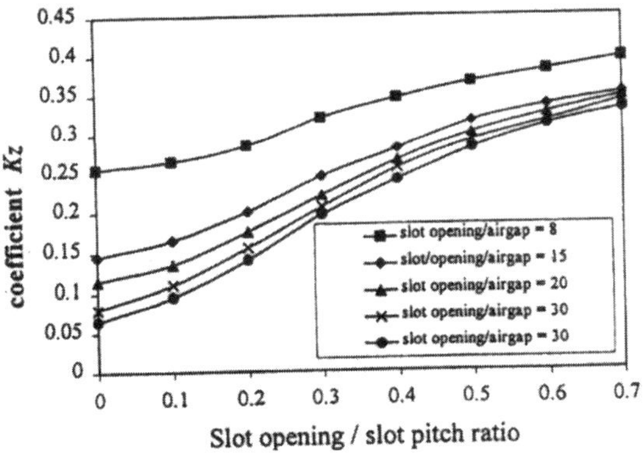

Figure 9.26 Correction coefficient for differential leakage damping

The most important "ingredient" in leakage inductance saturation is, for large values of stator and rotor currents, the skewing (uncompensated) mmf.

9.7.1. The skewing mmf

The rotor (stator) slot skewing, performed to reduce the first slot harmonic torque (in general) has also some side effects such as: a slight reduction in the magnetization inductance L_m accompanied by an additional leakage inductance component, L'_{rskew}. We decide to "attach" this new leakage inductance to the rotor as it "disappears" at zero rotor current.

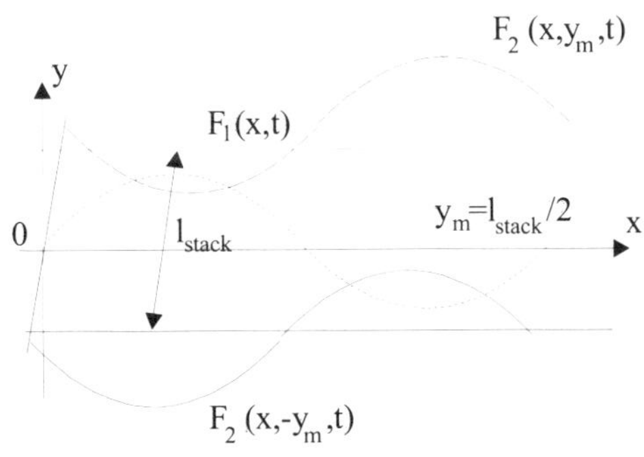

Figure 9.27. Phasing of stator and rotor mmfs for skewed rotor slots

To account for L_m alteration due to skewing, we use the standard skewing factor K_{skew}^{conv}.

$$K_{skew}^{conv} = \frac{\sin\left(\frac{\pi}{2} skew / m_1 q_1\right)}{\frac{\pi}{2} skew / m_1 q_1}; \quad K_{skew}^{conv} \approx 0.98 - 1.0 \tag{9.73}$$

and include a rotor current linear dependence to make sure that the influence disappears at zero rotor current,

$$K_{skew} = 1 - \left(1 - K_{skew}^{conv}\right)\frac{I_r'}{I_s} \tag{9.74}$$

skew–is the skewing in stator slot pitch counts.

To account for the skewing effect, we first consider the stator and rotor mmf fundamentals for a skewed rotor.

As the current in the bar is the same all along the stack length (with interbar currents neglected), the rotor mmf phasing varies along OY, beside OX and time (Figure 9.27) so the resulting mmf fundamental is

$$F(x, y, t) = F_{1m} \sin\left(\frac{\pi}{\tau} x - 2\pi f_1 t\right) + \\ + F_{2m} \sin\left(\frac{\pi}{\tau} x - 2\pi f_1 t - (\varphi_s - \varphi_r) - \frac{\pi}{m_1 q_1} skew \frac{y}{l_{stack}}\right) \\ -\frac{l_{stack}}{2} \le y \le \frac{l_{stack}}{2} \tag{9.75}$$

We may consider this as made of two terms, the magnetization mmf F_{1mag} and the skewing mmf F_{2skew}.

$$F(x, y, t) = F_{1mag} + F_{2skew} \tag{9.76}$$

$$F_{1mag} = F_{1m} \sin\left(\frac{\pi}{\tau} x - 2\pi f_1 t\right) + \\ + F_{2m} \cos\left(\frac{\pi}{m_1 q_1} skew \frac{y}{l_{stack}}\right) \cdot \sin\left(\frac{\pi}{\tau} x - 2\pi f_1 t - (\varphi_s - \varphi_r)\right) \tag{9.77}$$

$$F_{2skew} = -F_{2m} \sin\left(\frac{\pi}{m_1 q_1} skew \frac{y}{l_{stack}}\right) \cdot \cos\left(\frac{\pi}{\tau} x - 2\pi f_1 t - (\varphi_s - \varphi_r)\right) \tag{9.78}$$

F_{1mag} is only slightly influenced by skewing (9.77) as also reflected in (9.73).

On the other hand, the skewing (uncompensated) mmf varies with rotor current (F_{2m}) and with the axial position y. It tends to be small even at rated current in amplitude (9.78) but it gets rather high at large rotor currents. Notice also that the phase angle of F_{1mag} (φ_m) and F_{2skew} (φ_r) are different by slightly more than 90^0 (Figure 9.24), above rated current,

$$\varphi_{skew} = \varphi_r - \varphi_m \tag{9.79}$$

We may now define a skewing magnetization current I_{mskew}:

$$I_{mskew} = \pm I_m \frac{AT_2}{AT_{1m}} \sin(\pi 2p_1 SKW_1 / N_s) \tag{9.80}$$

Example 9.6. The skewing mmf

Let us consider an induction motor with the following data: no-load rated current: I_m = 30% of rated current; starting current. 600% of rated current; the number of poles: $2p_1 = 4$, skew = 1 stator slot pitch; number of stator slots N_s = 36. Assessment of the maximum skewing magnetization current I_{mskew}/I_m for starting conditions is required.

Solution

Making use of (9.80), the skewing magnetization current for SKW_1 = ±skew.

$$\frac{I_{mskew}}{I_m} = \pm \frac{AT_2}{AT_{1m}} \sin\left(\frac{\pi 2p_1(\text{skew}/2)}{N_s}\right) = \pm\left(\frac{I_r'}{I_m}\right)\sin\left(\frac{\pi \cdot 4 \cdot 1/2}{36}\right) \approx$$

$$\approx \pm\left(\frac{I_{start}}{I_m}\right) \cdot 0.15643 = \pm\frac{600\%}{30\%} \cdot 0.15643 = \pm 3.1286!$$

This shows that, at start, the maximum flux density in the airgap produced by the skewing magnetization current in the extreme axial segments will be maximum and much higher than the main flux density in the airgap, which is dominant below rated current when I_{mskew} becomes negligible (Figure 9.28).

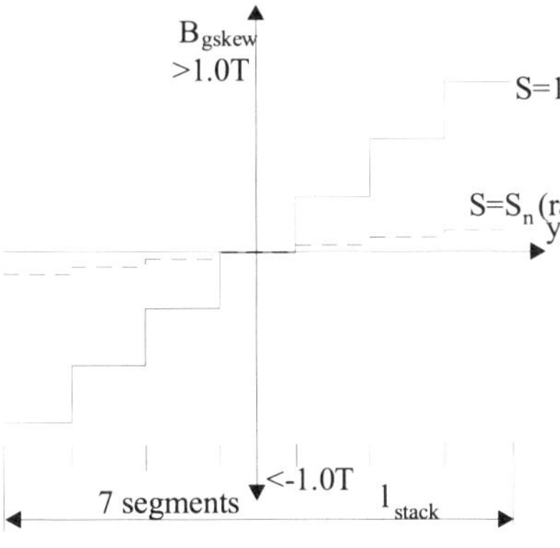

Figure 9.28 Skewing magnetization current produced airgap flux density

Airgap flux densities in excess of 1.0 T are to be expected at start and thus heavy saturation all over magnetic circuit should occur.

Let us remember that in these conditions the magnetization current is reduced such that the airgap flux density of main flux is (0.6 – 0.65) of its rated value, and its maximum is phase shifted with respect to the skewing flux by more than 90^0 (Figure 9.24). Consequently at start (high currents), the total airgap flux density is dominated by the skewing mmf contribution! This phenomenon has been documented recently through FEM [16].

Dividing the stator stack in (N_{seg} + 1) axial segments, we have:

$$SKW_1 = 0.0, \frac{skew}{N_{seg}}, \frac{skew}{2} \tag{9.81}$$

Now, based on I_{mskew}, from the magnetization curve ($L_m(I_m)$), we may calculate the magnetization inductance for skewing magnetization current $L_m(I_{mskew})$. Finally we may define a specific geometric permeance for skewing λ_{sk1}.

$$\lambda_{sk1} = \frac{L_m(I_{mskew})}{N_s K_{w1} W_1 n_s \mu_0 l_{stack}} \tag{9.82}$$

W_1–turns/phase.

λ_{sk1} depends on the segment considered (6 to 10 segments are enough) and on the level of rotor current.

Now the skewing flux/tooth in the stator and rotor is

$$\Phi_{sk1} = AT_2 \sin\left(\frac{\pi 2 p_1 SKW1}{N_s}\right)\mu_0 l_{stack}\lambda_{sk1} \tag{9.83}$$

$$\Phi_{sk2} = \Phi_{sk1}\frac{N_s}{N_r} \tag{9.84}$$

In (9.83), the value of Φ_{sk1} is calculated as if the value of segmented skewing is valid all along the stack length. This means that all calculations will be made n_{seg} times and then average values will be used to calculate the final values of leakage inductances.

Now if the skewing flux occurs both in the stator and in the rotor, when the leakage inductance is calculated, the rotor one will include the skewing permeance λ_{sk2}.

$$\lambda_{sk2} = \lambda_{sk1}\frac{N_s}{N_r} \tag{9.85}$$

9.7.2. Flux in the cross section marked by AB (Figure 9.25)

The total flux through AB, Φ_{AB} is

$$\Phi_{AB} = \frac{\Phi_{m1}}{2} + \frac{\Phi_{sk1}}{2} + \Phi_{z1} + \Phi_{z2} + \Phi_{st1} \tag{9.86}$$

Let us denote

$$C_{m1} = \frac{\Phi_{m1}}{2\Phi_{AB}} \qquad C_{sk1} = \frac{\Phi_{sk1}}{2\Phi_{AB}} \qquad C_{z1} = \frac{\Phi_{z1}}{\Phi_{AB}} \tag{9.87}$$

$$C_{st1} = \frac{\Phi_{st1}}{\Phi_{AB}} \qquad C_{z2} = \frac{\Phi_{z2}}{\Phi_{AB}} \tag{9.88}$$

These ratios, calculated before slot neck saturation is considered, are also taken as valid for a total leakage saturation condition.

There are two different situations: one with the stator tooth saturating first and the second with the saturated rotor teeth.

9.7.3. The stator tooth top saturates first

Making use of Ampere's and flux laws, the slot neck zone is considered (Figure 9.29).

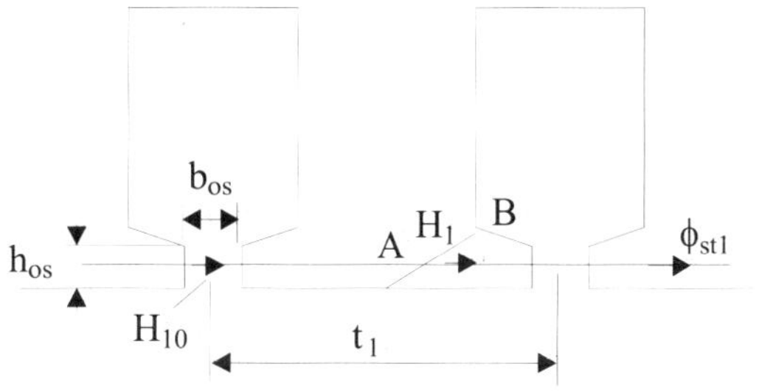

Figure 9.29. Stator slotting geometry

$$(t_1 - b_{os})H_1 + b_{os}H_{10} = AT_1 \tag{9.89}$$

The total flux through AB is

$$\Phi_{ABsat} = AB \cdot l_{stack} \cdot B_1 \tag{9.90}$$

On the other end, with C_{st1} known from unsaturated conditions, the slot neck flux Φ_{st1} is

$$\Phi_{st1} = C_{st1}\Phi_{ABsat} = \mu_0 H_{10} h_{os} l_{stack} \tag{9.91}$$

From (9.89) – (9.91), we finally obtain

$$B_1 = \mu_0 h_{os} \frac{|AT_1 - t_1' H_1|}{b_{os} C_{st1} AB}; \quad t_1' = t_1 - b_{os} \tag{9.92}$$

Equation (9.92) corroborated with the lamination $B_1(H_1)$ magnetization curve leads to B1 and H1 (Figure 9.30).

An iterative procedure as in Figure 9.30 may be used to solve (9.92) for B_1 and H_1. Finally, from (9.90), the saturated value of $(\Phi_{AB})_{sat}$ is obtained.

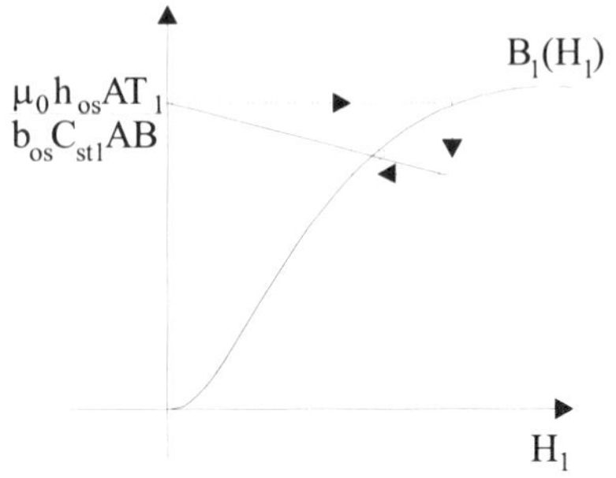

Figure 9.30 Tooth top flux density

Further on, new values of various teeth flux components are obtained.

$$\left(\frac{\Phi_{m1}}{2}\right)_{sat} = C_{m1}\left(\Phi_{AB}\right)_{sat}$$
$$\left(\Phi_{sk1}\right)_{sat} = C_{sk1}\left(\Phi_{AB}\right)_{sat}$$
$$\left(\Phi_{z1}\right)_{sat} = C_{z1}\left(\Phi_{AB}\right)_{sat} \qquad (9.93)$$
$$\left(\Phi_{st1}\right)_{sat} = C_{st1}\left(\Phi_{AB}\right)_{sat}$$

The new (saturated) values of stator leakage permeance components are

$$\left(\lambda_{st1}\right)_{sat} = \frac{\left(\Phi_{st1}\right)_{sat}}{AT_1\mu_0 l_{stack}}$$
$$\left(\lambda_{z1}\right)_{sat} = \frac{\left(\Phi_{z1}\right)_{sat}}{AT_1\mu_0 l_{stack}} \qquad (9.94)$$

The end connection and differential geometrical permeances λ_{endcon} and λ_{d1} maintain, in (9.68), their unsaturated values, while $(\lambda_{st1})_{sat}$ and $(\lambda_{z1})_{sat}$ (via (9.94)) enter (9.68) and thus the saturated value of stator leakage inductance L_{sl} for given stator and rotor currents and slip is obtained.

We have to continue with the rotor. There may be two different situations.

9.7.4. Unsaturated rotor tooth top

In this case the saturation flux from the stator tooth is "translated" into the rotor by the ratio of slot numbers N_s/N_r.

$$\left(\lambda_{st2}\right)_{sat} = \frac{\left(\Phi_{st1}\right)_{sat}\dfrac{N_s}{N_r}}{AT_1\mu_0 l_{stack}}$$
$$\left(\lambda_{z2}\right)_{sat} = \frac{\left(\Phi_{z1}\right)_{sat}\dfrac{N_s}{N_r}}{AT_1\mu_0 l_{stack}} \quad (9.95)$$
$$\left(\lambda_{sk2}\right)_{sat} = \lambda_{sk1}\frac{N_s}{N_r}$$

In Equation (9.95) it is recognised that the skewing flux "saturation" is not much influenced by the tooth top flux zone saturation as it represents a small length flux path in comparison with the main flux path available for skewing flux.

By now, with (9.69), the saturated rotor leakage inductance L_{rl}' may be calculated. With these new, saturated, leakage inductances, new stator, magnetization, and rotor currents are calculated.

The process is repeated until sufficient convergence is obtained with respect to the two leakage inductances. A new slip is then chosen and the computation cycle is repeated.

9.7.5. Saturated rotor tooth tip

In this case, after the AB zone was explored as above the DE cross section is investigated (Figure 9.31). Again the Ampere's law across tooth tip and slot neck yields

$$\left(t_2 - b_{or}\right)H_2 + b_{or}H_{20} = AT_2 \quad (9.96)$$

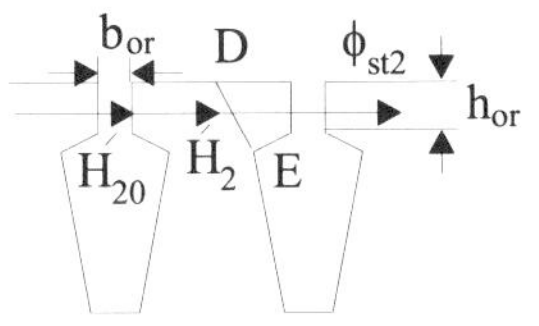

Figure 9.31 Rotor slot neck flux Φ_{st2}

The flux law results in

$$\Phi_{DE} = \Phi_1 + \left(\Phi_{st2}\right)_{sat};\ B_2 = \mu_0 H_{20} \quad (9.97)$$

where
$$\Phi_1 = \Phi_{ABsat}\frac{N_s}{N_r} \tag{9.98}$$

Also
$$(\Phi_{st2})_{sat} = DE \cdot l_{stack} \cdot B_2(H_2); \quad t_2' = t_2 - b_{or} \tag{9.99}$$

$$(\Phi_{st2})_{sat} = \mu_0 H_{20} h_{os} l_{stack} = \mu_0 h_{or}\frac{(AT_2 - t_2'H_2)}{b_{or}} l_{stack} \tag{9.100}$$

Finally,
$$B_2(H_2) = \frac{1}{l_{stack}DE}\left[\Phi_{ABsat}\frac{N_s}{N_r} + \mu_0 h_{or}\frac{(AT_2 - t_2'H_2)}{b_{or}} l_{stack}\right] \tag{9.101}$$

Equation (9.101), together with the rotor lamination magnetization curve, $B_2(H_2)$, is solved iteratively (as illustrated in Figure 9.30 for the stator).

Then,

$$(\lambda_{st2})_{sat} = \frac{(\Phi_{st2})_{sat}\lambda_{st2}}{\Phi_{st2}}; \quad \lambda_{st2} = \frac{h_{or}}{b_{or}} \tag{9.102}$$

$$\Phi_{st2} = \mu_0 AT_2 l_{stack}\lambda_{st2} \tag{9.103}$$

The other specific geometrical permeances in the rotor leakage inductance $(\lambda_{z2})_{sat}$ and $(\lambda_{sk2})_{sat}$ are calculated as in (9.95). Now we calculate the saturated value of L_{rl}' in (9.69). Then, from the equivalent circuit, new values of currents are calculated and a new iteration is initiated until sufficient convergence is reached.

9.7.6. The case of closed rotor slots

Closed rotor slots (Figure 9.32) are used to reduce starting current, noise, vibration, and windage friction losses though the breakdown and starting torque are smaller and so is the rated power factor.

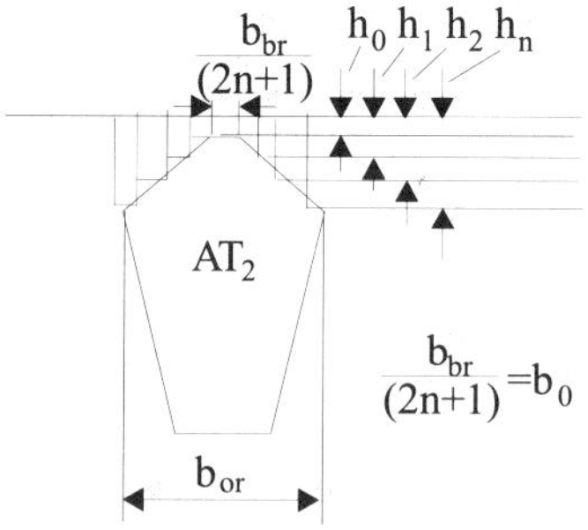

Figure 9.32 Closed rotor slot bridge discretisation

The slot iron bridge region is discretized in 2n + 1 zones of constant (but distinct) permeability μ_j. A value of n = (4 – 6) suffices. The Ampere's and flux laws for this region provide the following equations:

$$b_0\left(H_0 + 2\sum_{1}^{n} H_j\right) = AT_2 \tag{9.104}$$

$$B_0 h_0 = B_1 h_1 = B_2 h_2 = \ldots = B_n h_n \tag{9.105}$$

Given an initial value for B_0, for given AT_2, B_1, …, B_n are determined from (9.105). From the lamination magnetization curve, H_0, H_1, …, H_n are found. A new value of AT_2 is calculated, then compared with the given one and a new B_0 is chosen. The iteration cycle is revisited until sufficient convergence is obtained.

Finally, the slot bridge geometrical specific permeance $(\lambda_{st2})_{sat}$ is

$$(\lambda_{st2})_{sat} = \frac{1}{\dfrac{b_0}{h_0}\dfrac{B_0}{\mu_0 H_0} + \dfrac{2b_0}{h_1}\dfrac{B_1}{\mu_0 H_1} + \ldots + \dfrac{2b_0}{h_n}\dfrac{B_n}{\mu_0 H_n}} \tag{9.106}$$

All the other permeances are calculated as in previous paragraphs.

9.7.7. The algorithm

The algorithm for the computation of permeance in the presence of magnetic saturation and skin effects, as unfolded gradually so far, can be synthesized as following:

- Load initial data (geometry, winding data, name-plate data etc.)
- Load magnetization file: B(H)
- Load magnetization inductance function $L_m(I_m)$ – from FEM, design or tests
- Compute differential leakage specific permeances
- for S = S_{min}, S_{step}, S_{max}

```
    for SKW1 = 0:skew/nseg:skew
        compute unsaturated specific permeances
        while magnetization current error < 2%
            compute induction motor model using nonlinear Lm(Im)
    function
        end;
        Save unsaturated parameters
        1. if case = closed rotor slots;
            compute closed rotor case;
        else
            while H1 error < 2%;
                compute stator saturation case;
            end;
            compute unsaturated rotor case;
```

```
            if case = saturated rotor;
                while H2 error < 2%;
                    compute saturated rotor case
                end;
            end.
            Compute saturated parameters; compare them with those of
            previous cycle and go to 1. Return to 1 until sufficient
            convergence in Lsl and Lrl' is reached.
        end.
    end (axial segments cycle)
end (slip cycle)
    Save saturated parameters
    for SKW1 = 0:skew/nseg:skew;
        Compute saturated parameters average values; plot the IM
        parameters (segment number, slip);
        Compute IM equivalent circuit using an average value for each
        parameter (for the nseg, axial segments);
        Compute stator, rotor, magnetization current, torque, power factor
    versus slip;
        Plot the results
end.
```

Applying this algorithm for S = 1 an 1.1 kW, 2 pole, three-phase IM with skewed rotor, the results shown on Figure 9.33 have been obtained [18]. Heavy saturation occurs at high voltage level. The model seems to produce satisfactory results.

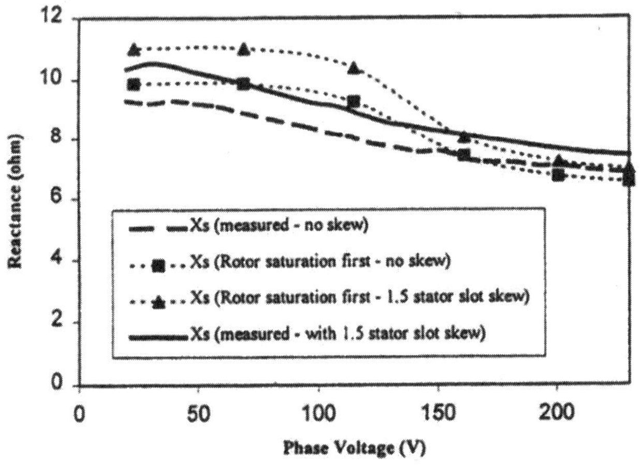

Figure 9.33 Leakage reactance for a skewed rotor versus voltage at stall

9.8. THE FEM APPROACH

The FEM has so far proved successful in predicting the main path flux saturation or the magnetization curve $L_m(I_m)$. It has also been used to calculate the general shape deep bar [17] and double cage bar [19] resistance and leakage inductance for given currents and slip frequency (Figure 9.34).

The essence of the procedure is to take into consideration only a slot sector and apply symmetry rules (Figure 9.34b).

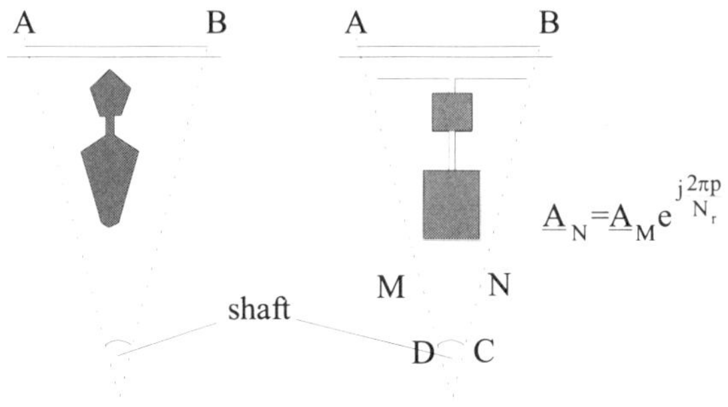

a.)

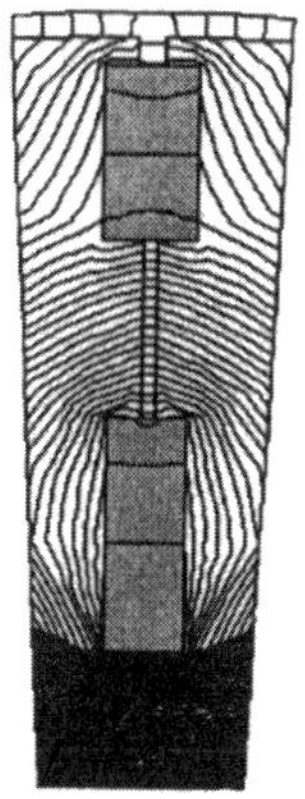

b.)

Figure 9.34 a.) Deep bar and double-cage rotor geometry; b.) FEM flux line

The stator is considered slotless and an infinitely thin stator current mmf, placed along AB, produces the main flux while the stator core has infinite permeability. Moreover, when the stator (and rotor) mmf have time harmonics, for inverter-fed situations, they may be included, but the saturation background is "provided" by the fundamental to simplify the computation process.

Still the zig-zag and skewing flux contributions to load saturation (in the teeth tips) are not accounted for.

The entire machine, with slotting on both sides of the airgap and the rotor divided into a few axial slices to account for skewing, may be approached by modified 2D-FEM. But the computation process has to be repeated until sufficient convergence in saturation then in stator and rotor resistances and leakage inductances is obtained for given voltage, frequency, and slip.

This enterprise requires a great amount of computation time, but it may be produced more quickly as PCs are getting stronger every day.

9.9. PERFORMANCE OF INDUCTION MOTORS WITH SKIN EFFECT

As already documented in previous paragraphs and in Chapter 7, the induction machine steady state performance is illustrated by torque, current, efficiency, and power factor versus slip for given voltage and frequency.

These characteristics may be influenced many ways. Among them the influences of magnetic saturation and skin effects are paramount.

It is the locked-rotor (starting) torque, breakdown torque, pull-up torque, starting current, rated slip, rated efficiency, and rated power factor that interest both manufacturers and users.

In an effort to put some order into this pursuit, NEMA has defined 5 designs for induction machines with cage rotors. They are distinguishable basically by the torque/speed curve (Figure 9.35).

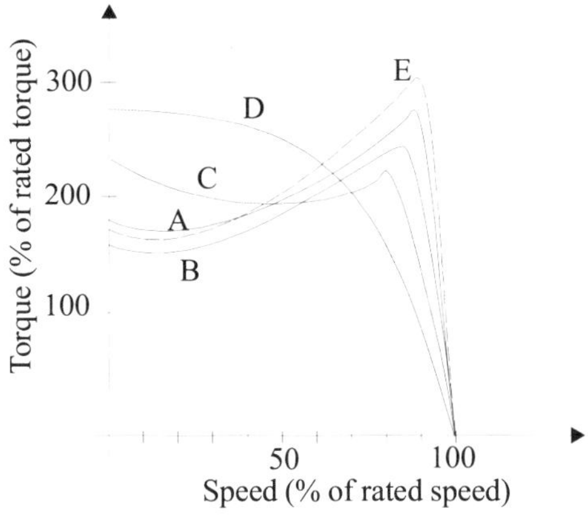

Figure 9.35 Torque/speed curve designs

Design A.

The starting in-rush current is larger than for general use design B motors, However, starting torque, pull-up and breakdown torques are larger than for design B.

While design A is allowed for larger starting current for larger starting torque, it also larger. These characteristics are obtained with a lower leakage inductance, mainly in the rotor and notable skin effect. Mildly deep rotor bars will do it in general.

Design B.

Design B motors are designed for given maximum lock – rotor current and minimum breakdown, locked-rotor and pull-up torques to make sure the typical load torque/speed curve is exceeded for all slips. They are called general purpose induction motors and their rated slip varies in general from 0.5 to 3 (maximum 5)%, depending on power and speed.

A normal cage with moderate skin effect is likely to produce design B characteristics.

Design C.

Design C motors exhibit a very large locked-rotor torque (larger than 200%, in general) at the expense of lower breakdown torque and larger rated slip–lower efficiency and power factor–than design A & B. Also lower starting currents are typical (I_{start} < 550%). Applications with very high break-away torque such as conveyors are typical for Design C.

Rather deep bars (of rather complex shape) or double cages are required on the rotor to produce C designs.

Design D.

Design D induction motors are characterized by a high breakdown slip, and large rated slip at very high starting torque and lower starting current (I_{start} < 450%).

As they are below the torque/speed curve D (Figure 9.35) it means that the motor will accelerate a given load torque/speed plant the fastest. Also the total energy losses during acceleration will be the lowest (see Chapter 8, the high rotor resistance case).

The heat to be evacuated over the load acceleration process will be less stringent. Solid rotors made of iron without or with axial slits or slots with copper bars are candidates for design D motors. They are designated to applications where frequent accelerations are more important than running at rated load. Punch presses are typical applications for D designs as well as very high speed applications.

Design E.

Design E induction motors are characterized as high efficiency motors and their efficiency is 1 to 4% higher than the B designs for the same power and speed. This superior performance is paid for by larger volume (and initial cost) and higher starting current (up to 30%, in general).

Lower current density, lower skin effect, and lower leakage inductances are typical for design E motors. It has been demonstrated that even a 1% increase in efficiency in a kW range motor has the payback in energy savings, in a rather loaded machine, of less than 3 years. As the machine life is in the range of more than 10 to 15 years, it pays off to invest more in design E (larger in volume) induction motor in many applications.

The increase in the starting current, however, will impose stronger local power grids which will tend to increase slightly the above-mentioned 3 year payback time for E designs.

9.10. SUMMARY

- As the slip (speed) varies for constant voltage and frequency, the IM parameters: magnetization inductance (L_m), leakage inductances (L_{sl}, L_{rl}'), and rotor resistance R_r' vary.
- In general, the magnetization flux decreases with slip to 55 to 65% of its full load value at stall.
- As the slip frequency (slip) increases, the rotor cage resistance and slot body leakage inductance vary due to skin effect: the first one increases and the second one decreases.
- Moreover at high slips (and currents), the leakage flux path saturate to produce a reduction of stator and rotor leakage inductance.
- In rotors with skewed slots, there is an uncompensated skewing rotor mmf which varies from zero in the axial center of the stack to maximum positive and negative values at stack axial ends. This mmf acts as a kind of independent magnetization current whose maximum is phase shifted by an angle slightly over 90^0 unless the rotor current is notably smaller than the rated value.
- At stall, the skewing magnetization current I_{mskew} may reach values 3 times the rated magnetization current of the motor in the axial zones close to the axial stack end.

  As the main flux is smaller than usual at stall, the level of saturation seems to be decided mainly by the skewing magnetization current.
- By slicing the stack axialy and using the machine magnetization curve $L_m(I_m)$, the value of $L_m(I_{mskew})$ in each segment is calculated. Severe saturation of the core occurs in the marginal axial segments. The skewing introduces an additional notable component in the rotor leakage inductance. It reduces slightly the magnetization inductance.
- By including the main flux, slot neck flux, zig-zag flux, differential leakage flux, and skewing flux per tooth in the stator and rotor, for each axial segment of stack, the stator and rotor saturated leakage inductances are calculated for given stator, rotor, and magnetization current and slip frequency.
- The skin effect is treated separately for stator windings with multiple conductors in slots and for rotor cages with one bar per slot. Standard correction coefficients for bar resistance and leakage inductance are derived. For the general shape rotor bar (including the double cage), the multiple layer approach is extended both for the bar and for the end ring.
- The end ring skin effect is notable, especially for $S \geq 1$ in applications such as freight elevators, etc.

- Skin and leakage saturation effects in the rotor cage are beneficial for increasing the starting torque with less starting current. However, the slot depth tends to be high which leads to a rather high rotor leakage inductance and thus a smaller breakdown torque and a larger rated slip. Consequently, the efficiency and power factor at full load are slightly lower.
- Skin effect is to be avoided in inverter-fed induction motors as it only increases the winding losses. However, it appears for current time harmonics anyway. Leakage saturation may occur only in servodrives with large transient (or short duration) torque requirements where the currents are large.
- In closed slot cage rotor motors, the rotor slot iron bridge saturates at values of rotor current notably less than the rated value.
- Skin and leakage saturation effects may not be neglected in line-start motors when assessing the starting, pull-up or even breakdown-torque. The larger the motor power, the more severe is this phenomenon.
- Skin and leakage saturation effects may be controlled to advantage by slot geometry optimization. This way, the 5 NEMA designs (A, B, C, D, E) have been born.
- While skin and leakage (on load) saturation effects can be treated with modified 2D and 3D FEMs [19, 21], the computation time is still prohibitive for routine calculations. Rather sophisticated analytical tools should be used first with FEM and then applied later for final refinements.

9.11. REFERENCES

1. Emde, About Current Redistribution, EUM, 1922, pp.301 (in German).
2. R. Richter, Electrical Machines", Vol.1., pp.233 – 243, Verlag Birkhäuser, Basel, 1951 (in German).
3. A.B. Field, Eddy Currents in Conductors Placed in Sslots, AIEE, 1905, pp.659.
4. I.B. Danilevici, V.V. Dombrovski, E.I. Kazovski, A.c. Machine Parameters, Science Publishers, St. Petersburg, 1965 (in Russian).
5. P.L. Cochran, Polyphase Induction Motors, pp.295 – 312, Marcel Dekker Inc. New York, 1989.
6. K. Vogt, Electrical Machines, pp.315, 4th edition, Verlag Berlin 1988 (in German).
7. J. Martinez – Roman, L. Serrano – Iribarnegaray, Torque Speed Characteristic of Elevator Deep Rotor Bar and Double Stator Winding Asynchronous Machines. Modelling and Measurement, Rec. of ICEM – 1998, Istanbul, Turkey, Sept, 1998, Vol.1, pp.1314 – 1319.
8. R. Richter, Electric Machines, Vol.IV, "Induction Machines", Ch.J, Verlag Birkhäuser Bassel (Stuttgart, 1954 in German).
9. P.D. Agarwal, P.L. Alger, Saturation Factors for Leakage Reactance of Induction Motors, AIEE Trans. (1961), pp. 1037-1042.
10. G. Angst, Saturation Factors for Leakage Reactance of Induction Motors with Skewed Rotors, AEEE Trans. 1963, No. 10 (October), pp. 716-722.

11. B. J. Chalmers, R. Dodgson, Saturated Leakage Reactances of Cage Induction Motors, Proc. EEE, Vol. 116, No. 8, august 1969, pp. 1395-1404.
12. G. J. Rogers and D.S. Benaragama, An Induction Motor Model with Deep-bar Effect and Leakage Inductance Saturation, AfE. Vol. 60, 1978, pp. 193-201.
13. M. Akbaba, S.O. Fakhleo, Saturation Effects in Three-phase Induction Motors, EMPS. Vol.12, 1987, pp. 179-193.
14. Y. Ning, C. Zhong, K. Shao, Saturation Effects in Three-phase Induction Motors", Record of ICEM, 1994, Paris, France, D1 section;
15. P. Lagonotte, H. Al Miah, M. Poloujadoff, Modelling and Identification of Parameters under Motor and Generator Conditions, EMPS Vol. 27, No. 2, 1999, pp. 107-121.
16. B-Il Kown, B-T Kim, Ch-S Jun, Analysis of Axially Nonuniform Distribution in 3 Phase Induction Motor Considering Skew Effect, IEEE Trans Vol. – MAG – 35, No.3, 1999, pp.1298 – 1301.
17. S. Williamson, M.J. Robinson, Calculation of Cage Induction Motor Equivalent Circuit Oarameters Using Finite Elements, Proc.IEE – EPA – 138, 1991, pp.264 – 276.
18. I. Boldea, D.G. Dorel, C.B. Rasmussen, T.J.E. Miller, Leakage Reactance Saturation in Induction Motors, Rec. of ICEM – 2000, Vol.1., pp.203 – 207.
19. S. Williamson, D.G. Gersh, Finite Element Calculation of Double Cage Rotor Equivalent Circuit Parameters, IEEE Trans. Vol.EC – 11, No.1., 1996, pp.41 – 48.
20. M. Sfaxi, F. Bouillault, M. Gabsi, J.F. Rialland, Numerical Method for Determination the Stray Losses of Conductors in Armature Slots, Rec. of ICEM – 1998, Istanbul, Turkey, Vol.3, pp.1815 – 1820.
21. P. Zhou, J. Gilmore, Z. Badic, Z.J. Cendes, Finite Element Analysis of Induction Motors Based on Computing Detailed Equivalent Circuit Parameters, IEEE Trans. Vol. MAG – 34, No.5, 1998, pp.3499 – 3502.

Chapter 10

AIRGAP FIELD SPACE HARMONICS, PARASITIC TORQUES, RADIAL FORCES, AND NOISE

The airgap field distribution in induction machines is influenced by stator and rotor mmf distributions and by magnetic saturation in the stator and rotor teeth and yokes (back cores).

Previous chapters introduced the mmf harmonics but were restricted to the fundamentals. Slot openings were considered but only in a global way, through an apparent increase of airgap by the Carter coefficient.

We first considered magnetic saturation of the main flux path through its influence on the airgap flux density fundamental. Later on a more advanced model was introduced (AIM) to calculate the airgap flux density harmonics due to magnetic saturation of main flux path (especially the third harmonic).

However, as shown later in this chapter, slot leakage saturation, rotor static, and dynamic eccentricity together with slot openings and mmf step harmonics produce a multitude of airgap flux density space harmonics. Their consequences are parasitic torque, radial uncompensated forces, and harmonics core and winding losses. The harmonic losses will be treated in the next chapter.

In what follows we will use gradually complex analytical tools to reveal various airgap flux density harmonics and their parasitic torques and forces. Such treatment is very intuitive but is merely qualitative and leads to rules for a good design. Only FEM–2D and 3D–could depict the extraordinary involved nature of airgap flux distribution in IMs under various factors of influence, to a good precision, but at the expense of much larger computing time and in an intuitiveless way. For refined investigation, FEM is, however, “the way”.

10.1. STATOR MMF PRODUCED AIRGAP FLUX HARMONICS

As already shown in Chapter 4 (Equation 4.27), the stator mmf per-pole-stepped waveform may be decomposed in harmonics as

$$
\begin{aligned}
F_1(x,t) = \frac{3W_1I_1\sqrt{2}}{\pi p_1}\Bigg[& K_{w1}\cos\left(\frac{\pi}{\tau}x-\omega_1 t\right)+\frac{K_{w5}}{5}\cos\left(\frac{5\pi}{\tau}x+\omega_1 t\right)+ \\
& +\frac{K_{w7}}{7}\cos\left(\frac{7\pi}{\tau}x-\omega_1 t\right)+\frac{K_{w11}}{11}\cos\left(\frac{11\pi}{\tau}x+\omega_1 t\right)+\frac{K_{w13}}{13}\cos\left(\frac{13\pi}{\tau}x-\omega_1 t\right)\Bigg]
\end{aligned}
\quad (10.1)
$$

where $K_{w\nu}$ is the winding factor for the νth harmonic,

$$K_{w\nu} = K_{q\nu}K_{y\nu};\ K_{q\nu} = \frac{\sin\frac{\nu\pi}{6}}{q\sin\frac{\nu\pi}{6q}};\ K_{y\nu} = \sin\frac{\nu\pi y}{2\tau} \tag{10.2}$$

In the absence of slotting, but allowing for it globally through Carter's coefficient and for magnetic circuit saturation by an equivalent saturation factor $K_{s\nu}$, the airgap field distribution is

$$\begin{aligned} B_{g1}(\theta,t) = \frac{3\mu_0 W_1 I_1\sqrt{2}}{\pi p_1 g K_c}\Bigg[&\frac{K_{w1}}{1+K_{s1}}\cos(\theta-\omega_1 t) + \frac{K_{w5}}{5(1+K_{s5})}\cos(5\theta+\omega_1 t) + \\ &+ \frac{K_{w7}}{7(1+K_{s7})}\cos(7\theta-\omega_1 t) + \frac{K_{w11}}{11(1+K_{s11})}\cos(11\theta+\omega_1 t) + \\ &+ \frac{K_{w13}}{13(1+K_{s13})}\cos(13\theta-\omega_1 t)\Bigg];\quad \theta = \frac{\pi}{\tau}x \end{aligned} \tag{10.3}$$

In general, the magnetic field path length in iron is shorter as the harmonics order gets higher (or its wavelength gets smaller). $K_{s\nu}$ is expected to decrease with ν increasing.

Also, as already shown in Chapter 4, but easy to check through (10.2) for all harmonics of the order ν,

$$\nu = C_1\frac{N_s}{p_1} \pm 1 \tag{10.4}$$

the distribution factor is the same as for the fundamental.

For three-phase symmetrical windings (with integer q slots/pole/phase), even order harmonics are zero and multiples of three harmonics are zero for star connection of phases. So, in fact,

$$\nu = 6C_1 \pm 1 \tag{10.5}$$

As shown in Chapter 4 (Equations 4.17 – 4.19) harmonics of 5th, 11th, 17th, … order travel backwards and those of 7th, 13th, 19th, … order travel forwards – see Equation (10.1).

The synchronous speed of these harmonics ω_ν is

$$\omega_\nu = \frac{d\theta_\nu}{dt} = \frac{\omega_1}{\nu} \tag{10.6}$$

In a similar way, we may calculate the mmf and the airgap field flux density of a cage winding.

10.2. AIRGAP FIELD OF A SQUIRREL CAGE WINDING

A symmetric (healthy) squirrel cage winding may be replaced by an equivalent multiphase winding with N_r phases, ½ turns/phase and unity winding factor. In this case, its airgap flux density is

$$B_{g2}(\theta, t) = \frac{N_r \mu_0 I_1 \sqrt{2}}{\pi p_1 g K_c} \sum_{\mu=1}^{\infty} \frac{\cos(\mu\theta \mp \omega_1 t - \varphi_{12v})}{\mu(1 + K_{sv})} = \frac{\mu_0 F_2(t)}{g K_c} \tag{10.7}$$

The harmonics order which produces nonzero mmf amplitudes follows from the applications of expressions of band factors K_{BI} and K_{BII} for $m = N_r$:

$$\mu = C_2 \frac{N_r}{p_1} \pm 1 \tag{10.8}$$

Now we have to consider that, in reality, both stator and rotor mmfs contribute to the magnetic field in the airgap and, if saturation occurs, superposition of effects is not allowed. So either a single saturation coefficient is used (say $K_{sv} = K_{s1}$ for the fundamental) or saturation is neglected ($K_{sv} = 0$).

We have already shown in Chapter 9 that the rotor slot skewing leads to variation of airgap flux density along the axial direction due to uncompensated skewing rotor mmf. While we investigated this latter aspect for the fundamental of mmf, it also applies for the harmonics. Such remarks show that the above analytical results should be considered merely as qualitative.

10.3. AIRGAP CONDUCTANCE HARMONICS

Let us first remember that even the step harmonics of the mmf are due to the placement of windings in infinitely thin slots. However, the slot openings introduce a kind of variation of airgap with position. Consequently, the airgap conductance, considered as only the inverse of the airgap, is

$$\frac{1}{g(\theta)} = f(\theta) \tag{10.9}$$

Therefore the airgap change $\Delta(\theta)$ is

$$\Delta(\theta) = \frac{1}{f(\theta)} - g \tag{10.10}$$

With stator and rotor slotting,

$$g(\theta) = g + \Delta_1(\theta) + \Delta_2(\theta - \theta_r) = \frac{1}{f_1(\theta)} + \frac{1}{f_2(\theta - \theta_r)} - g \tag{10.11}$$

$\Delta_1(\theta)$ and $\Delta_2(\theta - \theta_r)$ represent the influence of stator and rotor slot openings alone on the airgap function.

As $f_1(\theta)$ and $f_2(\theta - \theta_r)$ are periodic functions whose period is the stator (rotor) slot pitch, they may be decomposed in harmonics:

$$f_1(\theta) = a_0 - \sum_{\nu=1}^{\infty} a_\nu \cos \nu N_s \theta$$
$$f_2(\theta - \theta_r) = b_0 - \sum_{\nu=1}^{\infty} b_\nu \cos \nu N_r (\theta - \theta_r) \quad (10.12)$$

Now, if we use the conformal transformation for airgap field distribution in presence of infinitely deep, separate slots–essentially Carter's method–we obtain [1]

$$a_\nu, b_\nu = \frac{\beta}{g} F_\nu\left(\frac{b_{os,r}}{t_{s,r}}\right) \quad (10.13)$$

$$F_\nu\left(\frac{b_{os,r}}{t_{s,r}}\right) = \frac{1}{\nu}\frac{4}{\pi}\left[0.5 + \frac{\left(\frac{\nu b_{os,r}}{t_{s,r}}\right)^2}{0.18 - 2\left(\frac{\nu b_{os,r}}{t_{s,r}}\right)^2}\right]\sin\left(1.6\frac{\pi \nu b_{os,r}}{t_{s,r}}\right) \quad (10.14)$$

β is [2]

Table 10.1 $\beta(b_{os,r}/g)$

$b_{os,r}/g$	0	0.5	1.0	1.5	2.0	3.0	4.0	5.0
β	0.0	0.0149	0.0528	0.1	0.1464	0.2226	0.2764	0.3143
$b_{os,r}/g$	6.0	7.0	8.0	10.0	12.0	40.0	∞	
β	0.3419	0.3626	0.3787	0.4019	0.4179	0.4750	0.5	

Equations (10.13 – 10.14) are valid for $\nu = 1, 2, \ldots$. On the other hand, as expected,

$$a_0 \approx \frac{1}{K_{c1} g} \qquad b_0 \approx \frac{1}{K_{c2} g} \quad (10.15)$$

where K_{c1} and K_{c2} are Carter's coefficients for the stator and rotor slotting, respectively, acting separately.

Finally, with a good approximation, the inversed airgap function $1/g(\theta,\theta_r)$ is

$$\lambda_g(\theta,\theta_r) = \frac{1}{g(\theta,\theta_r)} \approx \frac{1}{g}\left\{\frac{1}{K_{c1}K_{c2}} - \frac{a_1}{K_{c2}}\cos N_s \frac{\theta}{p_1} - \frac{b_1}{K_{c1}}\cos N_r \frac{(\theta - \theta_r)}{p_1} + \right.$$
$$\left. + a_1 b_1\left[\cos\left(\frac{(N_s + N_r)}{p_1}\theta - \frac{N_r}{p_1}\theta_r\right) + \cos\left(\frac{(N_s - N_r)}{p_1}\theta + \frac{N_r}{p_1}\theta_r\right)\right]\right\} \quad (10.16)$$

As expected, the average value of $\lambda_g(\theta,\theta_r)$ is

$$\left(\lambda_g(\theta,\theta_r)\right)_{average} = \frac{1}{K_{c1}K_{c2}g} = \frac{1}{K_c g} \quad (10.17)$$

θ, θ_r – electrical angles.

We should notice that the inversed airgap function (or airgap conductance) λ_g has harmonics related directly to the number of stator and rotor slots and their geometry.

10.4. LEAKAGE SATURATION INFLUENCE ON AIRGAP CONDUCTANCE

As discussed in Chapter 9, for semiopen or semiclosed stator (rotor) slots at high currents in the rotor (and stator), the teeth heads get saturated. To account for this, the slot openings are increased. Considering a sinusoidal stator mmf and only stator slotting, the slot opening increased by leakage saturation b_{os}' varies with position, being maximum when the mmf is maximum (Figure 10.1).

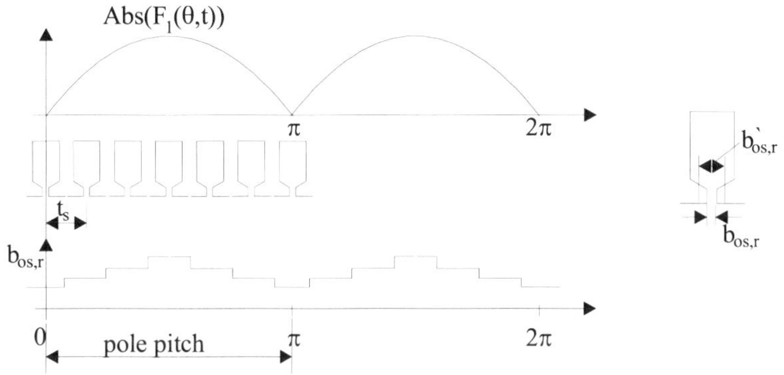

Figure 10.1 Slot opening $b_{os,r}$ variation due to slot leakage saturation

We might extract the fundamental of $b_{os,r}(\theta)$ function:

$$b'_{os,r} \approx b^0_{os,r} - b''_{os,r}\cos(2\theta - \omega_1 t); \quad \theta\text{ - electric angle} \quad (10.18)$$

The leakage slot saturation introduces a $2p_1$ pole pair harmonic in the airgap permeance. This is translated into a variation of a_0, b_0.

$$a_0, b_0 = a_0^0, b_0^0 - a_0', b_0' \sin(2\theta - \omega_1 t) \quad (10.19)$$

a_o^o, b_o^o are the new average values of $a_0(\theta)$ and $b_0(\theta)$ with leakage saturation accounted for.

This harmonic, however, travels at synchronous speed of the fundamental wave of mmf. Its influence is notable only at high currents.

Example 10.1. Airgap conductance harmonics

Let us consider only the stator slotting with $b_{os}/t_s = 0.2$ and $b_{os}/g = 4$ which is quite a practical value, and calculate the airgap conductance harmonics.

Solution

Equations (10.13 and 10.14) are to be used with β from Table 10.1, β(4) = 0.2764,

with
$$K_{cl} \approx \frac{t_s}{t_s - 1.6\beta b_{os}} = \frac{1}{1 - 1.6 \cdot 0.2764 \cdot 0.2} = 1.097$$

$$a_0 = \frac{1}{g}\frac{1}{K_{cl}} = \frac{1}{g}\frac{1}{1.097} = \frac{0.91155}{g}$$

$$a_1 = \frac{1}{g}0.2764\frac{4}{\pi}\left[0.5 + \frac{(1 \cdot 0.2)^2}{0.78 - 2 \cdot (0.2)^2}\right]\sin(1.6\pi \cdot 1 \cdot 0.2) = \frac{1}{g}0.1542$$

$$a_2 = \frac{1}{g}0.2764\frac{4}{2\pi}\left[0.5 + \frac{(2 \cdot 0.2)^2}{0.78 - 2 \cdot (2 \cdot 0.2)^2}\right]\sin(1.6\pi \cdot 2 \cdot 0.2) = \frac{1}{g}0.04886$$

The airgap conductance $\lambda_1(\theta)$ is

$$\lambda_1(\theta) = \frac{1}{g(\theta)} = \frac{1}{g}\left[0.91155 - 0.1542\cos(N_s\theta) - 0.04886\cos(2N_s\theta)\right]$$

If for the νth harmonics the "sine" term in (10.14) is zero, so is the νth airgap conductance harmonic. This happens if

$$1.6\pi\nu\frac{b_{os,r}}{t_{s,r}} = \pi; \quad \nu\frac{b_{os,r}}{t_{s,r}} = 0.625$$

Only the first 1, 2 harmonics are considered, so only with open slots may the above condition be approximately met.

10.5. MAIN FLUX SATURATION INFLUENCE ON AIRGAP CONDUCTANCE

In Chapter 9 an iterative analytical model (AIM) to calculate the main flux distribution accounting for magnetic saturation was introduced. Finally, we showed the way to use AIM to derive the airgap flux harmonics due to main flux path (teeth or yokes) saturation. A third harmonic was particularly visible in the airgap flux density.

As expected, this is the result of a virtual second harmonic in the airgap conductance interacting with the mmf fundamental, but it is the resultant (magnetizing) mmf and not only stator (or rotor) mmfs alone.

The two second order airgap conductance harmonics due to leakage slot flux path saturation and main flux path saturation are phase shifted as their originating mmfs are by the angle $\varphi_m - \varphi_1$ for the stator and $\varphi_m - \varphi_2$ for the rotor.

φ_1, φ_2, φ_m are the stator, rotor, magnetization mmf phase shift angle with respect to phase A axis.

$$\lambda_s(\theta_s) = \frac{1}{gK_c}\left[\frac{1}{1+K_{s1}} + \frac{1}{1+K_{s2}}\sin(2\theta - \omega_1 t - (\varphi_m - \varphi_1))\right] \quad (10.20)$$

The saturation coefficients K_{s1} and K_{s2} result from the main airgap field distribution decomposition into first and third harmonics.

There should not be much influence (amplification) between the two second order saturation-caused airgap conductance harmonics unless the rotor is skewed and its rotor currents are large (> 3 to 4 times rated current), when both the skewing rotor mmf and rotor slot mmf are responsible for large main and leakage flux levels, especially toward the axial ends of stator stack.

10.6. THE HARMONICS-RICH AIRGAP FLUX DENSITY

It has been shown [1] that, in general, the airgap flux density $B_g(\theta,t)$ is

$$B_g(\theta, t) = \mu_0 \lambda_{1,2}(\theta) F_\nu \cos(\nu\theta \mp \omega t - \varphi_\nu) \quad (10.21)$$

where F_ν is the amplitude of the mmf harmonic considered, and $\lambda(\theta)$ is the inversed airgap (airgap conductance) function. $\lambda(\theta)$ may be considered as containing harmonics due to slot openings, leakage, or main flux path saturation. However, using superposition in the presence of magnetic saturation is not correct in principle, so mere qualitative results are expected by such a method.

10.7. THE ECCENTRICITY INFLUENCE ON AIRGAP MAGNETIC CONDUCTANCE

In rotary machines, the rotor is hardly ever located symmetrically in the airgap either due to the rotor (stator) unroundedness, bearing eccentric support, or shaft bending.

An one-sided magnetic force (uncompensated magnetic pull) is the main result of such a situation. This force tends to increase further the eccentricity, produce vibrations, noise, and increase the critical rotor speed.

When the rotor is positioned off center to the stator bore, according to Figure 10.2, the airgap at angle θ_m is

$$g(\theta_m) = R_s - R_r - e\cos\theta_m = g - e\cos\theta_m \quad (10.22)$$

where g is the average airgap (with zero eccentricity: e = 0.0).

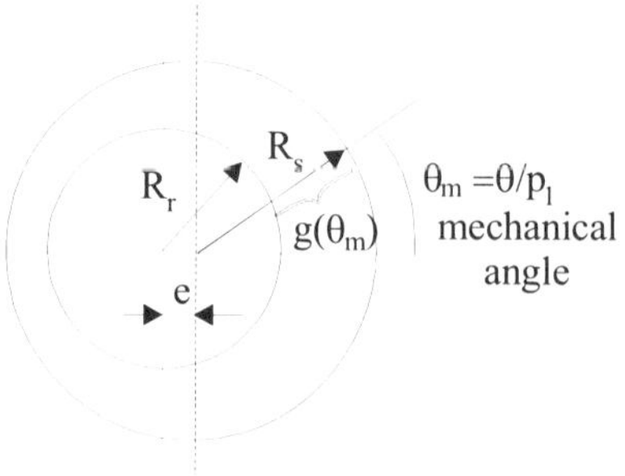

Figure 10.2 Rotor eccentricity R_s – stator radius, R_r – rotor radius

The airgap magnetic conductance $\lambda(\theta_m)$ is

$$\lambda(\theta_m) = \frac{1}{g(\theta_m)} = \frac{1}{g(1-\varepsilon\cos\theta_m)}; \quad \varepsilon = \frac{e}{g} \tag{10.23}$$

Now (10.23) may be easily decomposed into harmonics to obtain

$$\lambda(\theta_m) = \frac{1}{g}(c_0 + c_1\cos\theta_m + \ldots) \tag{10.24}$$

with

$$c_0 = \frac{1}{\sqrt{1-\varepsilon^2}}; \quad c_1 = \frac{2(c_0 - 1)}{\varepsilon} \tag{10.25}$$

Only the first geometrical harmonic (notice that the period here is the entire circumpherence: $\theta_m = \theta/p_1$) is hereby considered.

The eccentricity is static if the angle θ_m is a constant, that is if the rotor revolves around its axis but this axis is shifted with respect to stator axis by e.

In contrast, the eccentricity is dynamic if θ_m is dependent on rotor motion.

$$\theta_m = \theta_m - \frac{\omega_r t}{p_1} = \theta_m - \frac{\omega_1(1-S)t}{p_1} \tag{10.26}$$

It corresponds to the case when the axis of rotor revolution coincides with the stator axis but the rotor axis of symmetry is shifted.

Now using Equation (10.21) to calculate the airgap flux density produced by the mmf fundamental as influenced only by the rotor eccentricity, static and dynamic, we obtain

$$B_g(\theta,t) = \mu_0 F_1\cos(\theta-\omega_1 t)\frac{1}{g}\left[c_0 + c_1\cos\left(\frac{\theta}{p_1} - \varphi_s\right) + c_1'\cos\left(\frac{\theta-\omega_1(1-S)t}{p_1} - \varphi_d\right)\right]$$

(10.27)

As seen from (10.27) for $p_1 = 1$ (2 pole machines), the eccentricity produces two homopolar flux densities, $B_{gh}(t)$,

$$B_{gh}(t)=\frac{\left[\frac{\mu_0 F_1 c_1}{g}\cos(\omega_1 t-\varphi_s)+\frac{\mu_0 F_1 c_1'}{g}\cos(S\omega_1 t-\varphi_d)\right]}{1+c_h} \qquad (10.28)$$

These homopolar components close their flux lines axially through the stator frame then radially through the end frame, bearings, and axially through the shaft (Figure 10.3).

The factor c_h accounts for the magnetic reluctance of axial path and end frames and may have a strong influence on the homopolar flux. For nonmagnetic frames and (or) insulated bearings c_h is large, while for magnetic steel frames, c_h is smaller. Anyway, c_h should be much larger than unity at least for the static eccentricity component because its depth of penetration (at ω_1), in the frame, bearings, and shaft, is small. For the dynamic component, c_h is expected to be smaller as the depth of penetration in iron (at $S\omega_1$) is larger.

The d.c. homopolar flux may produce a.c. voltage along the shaft length and, consequently, shaft and bearing currents, thus contributing to bearing deterioration.

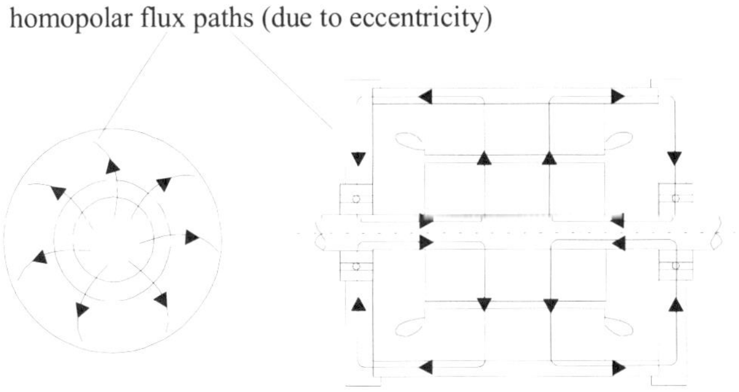

Figure 10.3 Homopolar flux due to rotor eccentricity

10.8. INTERACTIONS OF MMF (OR STEP) HARMONICS AND AIRGAP MAGNETIC CONDUCTANCE HARMONICS

It is now evident that various airgap flux density harmonics may be calculated using (10.21) with the airgap magnetic conductance $\lambda_{1,2}(\theta)$ either from (10.16) to account for slot openings, with a_0, b_0 from (10.19) for slot leakage saturation, or with $\lambda_s(\theta_s)$ from (10.20) for main flux path saturation, or $\lambda(\theta_m)$ from (10.24) for eccentricity.

$$B_g(x,t)=\frac{\mu_0}{g}\Big[\underbrace{F_1(\theta,t)+F_2(\theta,t)}_{(10.1)}\Big]\left[\underbrace{\lambda_{1,2}(\theta)}_{(10.16)\text{with}(10.19)}+\underbrace{\lambda_s(\theta)}_{(10.20)}+\underbrace{\lambda\left(\frac{\theta}{p_1}\right)}_{(10.24)}\right] \quad (10.29)$$

As expected, there will be a very large number of airgap flux density harmonics and its complete exhibition and analysis is beyond our scope here.

However, we noticed that slot openings produce harmonics whose order is a multiple of the number of slots (10.12). The stator (rotor) mmfs may produce harmonics of the same order either as sourced in the mmf or from the interaction with the first airgap magnetic conductance harmonic (10.16).

Let us consider an example where only the stator slot opening first harmonic is considered in (10.16).

$$B_{g_v}(\theta,t)\approx\frac{\mu_0F_{1v}}{g}\cos(v\theta'-\omega_1t)\left[\frac{1}{K_{c1}}+a_1\cos\frac{N_s\theta}{p_1}\right] \quad (10.30)$$

with

$$\theta'=\theta+\frac{\pi}{N_s}p_1 \quad (10.31)$$

θ' takes care of the fact that the axis of airgap magnetic conductance falls in a slot axis for coil chordings of 0, 2, 4 slot pitches. For odd slot pitch coil chordings, $\theta' = \theta$. The first step (mmf) harmonic which might be considered has the order

$$v=c_1\frac{N_s}{p_1}\pm1=\frac{N_s}{p_1}\pm1 \quad (10.32)$$

Writing (10.30) into the form

$$B_{g_v}(\theta,t)=\frac{\mu_0F_{1v}}{gK_{c1}}\cos\left[v\left(\theta+\frac{\pi p_1}{N_s}\right)-\omega_1t\right]+$$
$$+\frac{a_1\mu_0F_{1v}}{2}\left[\cos\left[v\left(\theta+\frac{\pi p_1}{N_s}\right)+\frac{N_s\theta}{p_1}-\omega_1t\right]+\cos\left[v\left(\theta+\frac{\pi p_1}{N_s}\right)-\frac{N_s\theta}{p_1}-\omega_1t\right]\right] \quad (10.33)$$

For $v=\frac{N_s}{P_1}+1$, the argument of the third term of (10.33) becomes

$$\left[\left(\frac{N_s}{p_1}+1\right)\left(\theta+\frac{\pi p_1}{N_s}\right)-\pi-\omega_1t\right]$$

But this way, it becomes the opposite of the first term argument, so the first step, mmf harmonic $\frac{N_s}{p_1}+1$, and the first slot opening harmonics subtract each other. The opposite is true for $\nu = \frac{N_s}{p_1} - 1$ when they are added. So the slot openings may amplify or attenuate the effect of the step harmonics of order $\nu = \frac{N_s}{p_1} \mp 1$, respectively.

Other effects such as differential leakage fields affected by slot openings have been investigated in Chapter 6 when the differential leakage inductance has been calculated. Also we have not yet discussed the currents induced by the flux harmonics in the rotor and in the stator conductors.

In what follows, some attention will be paid to the main effects of airgap flux and mmf harmonics: parasitic torque and radial forces.

10.9. PARASITIC TORQUES

Not long after the cage-rotor induction motors reached industrial use, it was discovered that a small change in the number of stator or rotor slots prevented the motor to start from any rotor position or the motor became too noisy to be usable. After Georges (1896), Punga (1912), Krondl, Lund, Heller, Alger, and Jordan presented detailed theories about additional (parasitic) asynchronous torques, Dreyfus (1924) derived the conditions for the manifestation of parasitic synchronous torques.

10.9.1. When do asynchronous parasitic torques occur?

Asynchronous parasitic torques occur when a harmonic ν of the stator mmf (or its airgap flux density) produces in the rotor cage currents whose mmf harmonic has the same order ν.

The synchronous speed of these harmonics $\omega_{1\nu}$ (in electrical terms) is

$$\omega_{1\nu} = \frac{\omega_1}{\nu} \tag{10.34}$$

The stator mmf harmonics have orders like: $\nu = -5, +7, -11, +13, -17, +19, \ldots$, in general, $6c_1 \pm 1$, while the stator slotting introduces harmonics of the order $\frac{c_1 N_s}{p_1} \pm 1$. The higher the harmonic order, the lower its mmf amplitude. The slip S_ν of the νth harmonic is

$$S_\nu = \frac{\omega_{1\nu} - \omega_r}{\omega_{1\nu}} = 1 - \nu(1 - S) \tag{10.35}$$

For synchronism $S_\nu = 0$ and, thus, the slip for the synchronism of harmonic ν is

$$S = 1 - \frac{1}{\nu} \tag{10.36}$$

For the first mmf harmonic ($\nu = -5$), the synchronism occurs at

$$S_5 = 1 - \frac{1}{(-5)} = \frac{6}{5} = 1.2 \qquad (10.37)$$

For the seventh harmonic ($\nu = +7$)

$$S_7 = 1 - \frac{1}{(+7)} = \frac{6}{7} \qquad (10.38)$$

All the other mmf harmonics have their synchronism at slips

$$S_5 > S_\nu > S_7 \qquad (10.39)$$

In a first approximation, the slip synchronism for all harmonics is $S \approx 1$. That is close to motor stall, only, the asynchronous parasitic torques occur.

As the same stator current is at the origin of both the stator mmf fundamental and harmonics, the steady state equivalent circuit may be extended in series to include the asynchronous parasitic torques (Figure 10.4).

The mmf harmonics, whose order is lower than the first slot harmonic $\nu_{s\min} = \frac{N_s}{p_1} \pm 1$, are called phase belt harmonics. Their order is:–5, +7, …. They are all indicated in Figure 10.4 and considered to be mmf harmonics.

One problem is to define the parameters in the equivalent circuit. First of all the magnetizing inductances X_{m5}, X_{m7}, … are in fact the "components" of the up to now called differential leakage inductance.

$$L_{m\nu} = \frac{6\mu_0 \tau L}{\pi^2 p_1 g K_c (1 + K_{st})} \left(\frac{W_1 K_{w\nu}}{\nu} \right)^2 \qquad (10.40)$$

The saturation coefficient K_{st} in (10.40) refers to the teeth zone only as the harmonics wavelength is smaller than that of the fundamental and therefore the flux paths close within the airgap and the stator and rotor teeth/slot zone.

The slip $S_\nu \approx 1$ and thus the slip frequency for the harmonics $S_\nu \omega_1 \approx \omega_1$. Consequently, the rotor cage manifests a notable skin effect towards harmonics, much as for short-circuit conditions. Consequently, $R'_{r\nu} \approx (R'_r)_{start}$, $L'_{rl\nu} \approx (L'_{rl})_{start}$.

As these harmonics act around $S = 1$, their torque can be calculated as in a machine with given current $I_s \approx I_{start}$.

$$T_{e\nu} \approx \frac{3R'_{rstart}}{S_\nu \omega_1} I_{start}^2 \frac{L_{m\nu}^2}{\left[(L'_{rl})_{start} + L_{m\nu} \right]^2 + \left(\frac{R'_{rstart}}{S_\nu \omega_1} \right)^2} \qquad (10.41)$$

In (10.41), the starting current I_{start} has been calculated from the complete equivalent circuit where, in fact, all magnetization harmonic inductances $L_{m\nu}$

have been lumped into $X_{sl} = \omega_1 L_{sl}$ as a differential inductance as if $S_v = \infty$. The error is not large.

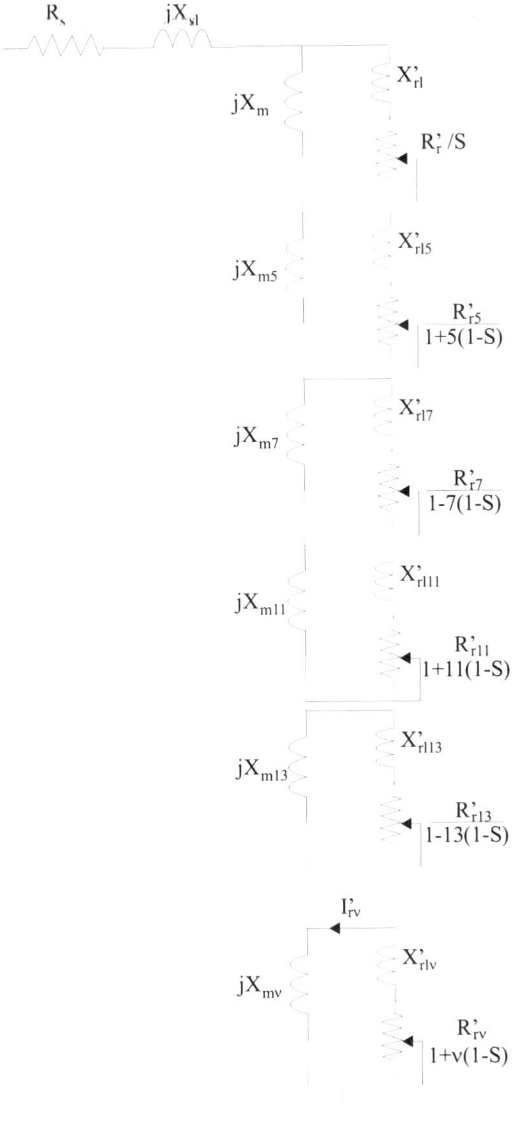

Figure 10.4 Equivalent circuit including asynchronous parasitic torques due to mmf harmonics (phase band and slot driven)

Now (10.41) reflects a torque/speed curve similar to that of the fundamental (Figure 10.5). The difference lies in the rather high current, but factor K_{wv}/v is rather small and overcompensates this situation, leading to a small harmonic torque in well-designed machines.

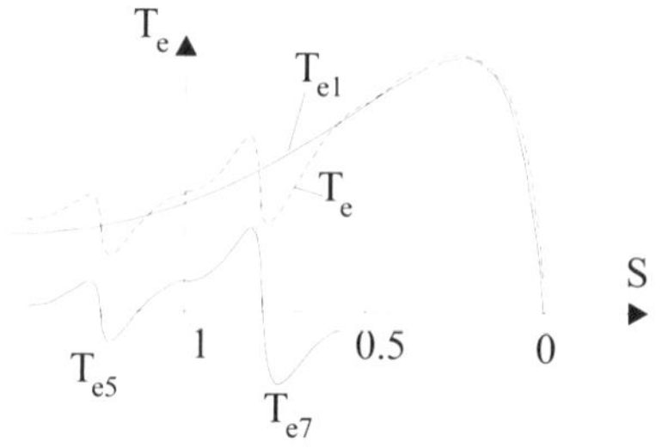

Figure 10.5 Torque/slip curve with 5th and 7th harmonic asynchronous parasitic torques

We may dwell a little on (10.41) noticing that the inductance L_{mv} is of the same order (or smaller) than $(L'_{rl})_{start}$ for some harmonics. In these cases, the coupling between stator and rotor windings is small and so is the parasitic torque.

Only if $(L'_{rl})_{start} < L_{mv}$ a notable parasitic torque is expected. On the other hand, to reduce the first parasitic asynchronous torques T_{e5} or T_{e7}, chording of stator coils is used to make $K_{w5} = 0$.

$$K_{w5} = \frac{\sin\frac{5\pi}{6}}{q\sin\left(\frac{5\pi}{6q}\right)}\sin\frac{5\pi y}{2\tau} \approx 0; \quad \frac{y}{\tau} = \frac{4}{5} \tag{10.42}$$

As $y/\tau = 4/5$ is not feasible for all values of q, the closest values are y/τ = 5/6, 7/9, 10/12, and 12/15. As can be seen for q = 5, the ratio 12/15 fulfils exactly condition (10.42). In a similar way the 7th harmonic may be cancelled instead, if so desired.

The first slot harmonics torque ($\nu_{smin} = \frac{N_s}{p_1} \pm 1$) may be cancelled by using skewing.

$$K_{w\nu_{smin}} = \frac{\sin\frac{\nu_{smin}\pi}{6}}{q\sin\left(\frac{\nu_{smin}\pi}{6q}\right)}\sin\frac{\nu_{smin}\pi y}{2\tau} \cdot \frac{\sin \nu_{smin}\frac{c}{\tau}\frac{\pi}{2}}{\nu_{smin}\frac{c}{\tau}\frac{\pi}{2}} \approx 0 \tag{10.43}$$

From (10.43),

$$\nu_{smin}\frac{c}{\tau}\frac{\pi}{2} = K\pi; \quad \frac{c}{\tau} = \frac{2K}{\nu_{smin}} = \frac{2K}{\frac{N_s}{p_1} \pm 1} \tag{10.44}$$

In general, the skewing c/τ corresponds to 0.5 to 2 slot pitches of the part (stator or rotor) with less slots.

For $N_s = 36$, $p_1 = 2$, $m = 3$, $q = N_0/2p_1m = 36/(2\cdot2\cdot3) = 3$: $c/\tau = 2/17(19) \approx 1/9$. This, in fact, means a skewing of one slot pitch as a pole has $q\cdot m = 3\cdot3 = 9$ slot pitches.

10.9.2. Synchronous parasitic torques

Synchronous torques occur through the interaction of two harmonics of the same order (one of the stator and one of the rotor) but originating from different stator harmonics. This is not the case, in general, for wound rotor whose harmonics do not adapt to stator ones, but produces about the same spectrum of harmonics for the same number of phases.

For cage-rotors, however, the situation is different. Let us first calculate the synchronous speed of harmonic ν_1 of the stator.

$$n_{1\nu_1} = \frac{f_1}{p_1\nu_1} \tag{10.45}$$

The relationship between a stator harmonic ν and the rotor harmonic ν' produced by it in the rotor is

$$\nu - \nu' = c_2\frac{N_r}{p_1} \tag{10.46}$$

This is easy to accept as it suggests that the difference in the number of periods of the two harmonics is a multiple of the number of rotor slots N_r.

Now the speed of rotor harmonics ν' produced by the stator harmonic ν with respect to rotor $n_{2\nu\nu'}$ is

$$n_{2\nu\nu'} = p\frac{S_\nu f_1}{p_1\nu'} = \frac{n_1}{\nu'}\left[1 - \frac{\nu}{1-S}\right] \tag{10.47}$$

With respect to stator, $n_{1\nu\nu'}$ is

$$n_{1\nu\nu'} = n + n_{2\nu\nu'} = \frac{f_1}{p_1\nu'}\left[1 + (\nu - \nu')(1 - S)\right] \tag{10.48}$$

With (10.46), Equation (10.48) becomes

$$n_{1\nu\nu'} = \frac{f_1}{p_1\nu'}\left[1 + c_2\frac{N_r}{p_1}(1 - S)\right] \tag{10.49}$$

Synchronous parasitic torques occur when two harmonic field speeds $n_{1\nu_1}$ and $n_{1\nu\nu'}$ are equal to each other.

$$n_{1\nu_1} = n_{1\nu\nu'} \tag{10.50}$$

or

$$\frac{1}{\nu_1} = \frac{1}{\nu'}\left[1 + c_2 \frac{N_r}{p_1}(1 - S)\right] \tag{10.51}$$

It should be noticed that ν_1 and ν' may be either positive (forward) or negative (backward) waves. There are two cases when such torques occur.

$$\begin{aligned} \nu_1 &= +\nu' \\ \nu_1 &= -\nu' \end{aligned} \tag{10.52}$$

When $\nu_1 = +\nu'$, it is mandatory (from (10.51)) to have S = 1 or zero speed. On the other hand, when $\nu_1 = -\nu'$,

$$(S)_{\nu_1 = +\nu'} = 1; \ \text{standstill} \tag{10.53}$$

$$(S)_{\nu_1 = -\nu'} = 1 + \frac{2p_1}{c_2 N_r} <> 1 \tag{10.54}$$

As seen from (10.54), synchronous torques at nonzero speeds, defined by the slip $(S)_{\nu_1 = -\nu'}$, occur close to standstill as $N_r >> 2p_1$.

Example 10.2. Synchronous torques

For the unusual case of a three-phase IM with N_s = 36 stator slots and N_r = 16 rotor slots, $2p_1$ = 4 poles, frequency f_1 = 60Hz, let us calculate the speed of the first synchronous parasitic torques as a result of the interaction of phase belt and step (slot) harmonics.

Solution

First, the stator and rotor slotting-caused harmonics have the orders

$$\nu_{1s} = c_1 \frac{N_s}{p_1} + 1; \quad \nu_s' = c_2 \frac{N_r}{p_1} + 1; \ c_1 c_2 <> 0 \tag{10.55}$$

The first stator slot harmonics ν_{1s} occur for $c_1 = 1$.

$$\nu_{1f} = +1\frac{36}{2} + 1 = 19; \quad \nu_{1b} = -1\frac{36}{2} + 1 = -17 \tag{10.56}$$

On the other hand, a harmonic in the rotor $\nu_f' = +17 = -\nu_{1b}$ is obtained from (10.55) for $c_2 = 2$.

$$\nu_f' = +2\frac{16}{2} + 1 = +17 \tag{10.57}$$

The slip S for which the synchronous torque occurs (10.54),

$$S_{(17)} = 1 + \frac{2 \cdot 2}{2 \cdot 16} = \frac{9}{8}; \ n_{17} = \frac{f_1}{P_1}\left(1 - S_{(17)}\right) = \frac{60}{2}\left(1 - \frac{9}{8}\right) \cdot 60 = -255\text{rpm} \tag{10.58}$$

Now if we consider the first (phase belt) stator mmf harmonic ($6c_1 \pm 1$), $\nu_f =$ +7 and ν_b = -5, from (10.55) we discover that for $c_2 = -1$.

$$\nu_b' = -1\frac{16}{2} + 1 = -7 \tag{10.59}$$

So the first rotor slot harmonic interacts with the 7th order (phase belt) stator mmf harmonic to produce a synchronous torque at the slip.

$$S_{(7)} = 1 + \frac{2 \cdot 2}{(-1) \cdot 16} = \frac{3}{4};\ n_7 = \frac{f_1}{P_1}\left(1 - S_{(7)}\right) = \frac{60}{2}\left(1 - \frac{3}{4}\right) \cdot 60 = 450\text{rpm} \tag{10.60}$$

These two torques occur as superposed on the torque/speed curve, as discontinuities at corresponding speeds and at any other speeds their average values are zero (Figure 10.6).

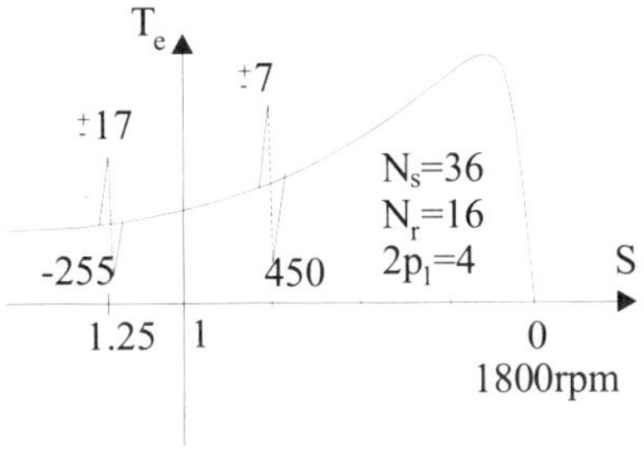

Figure 10.6 Torque/speed curve with parasitic synchronous torques

To avoid synchronous torques, the conditions (10.52) have to be avoided for harmonics orders related to the first rotor slot harmonics (at least) and stator mmf phase belt harmonics.

$$6c_1 + 1 \neq c_2\frac{N_r}{p_1} + 1$$
$$c_1 = \pm 1, \pm 2 \text{ and } c_2 = \pm 1, \pm 2 \tag{10.61}$$

For stator and rotor slot harmonics

$$c_1\frac{N_s}{p_1} + 1 \neq c_2\frac{N_r}{p_1} + 1 \tag{10.62}$$

Large synchronous torque due to slot/slot harmonics may occur for $c_1 = c_2 =$ ±1 when

$$N_s - N_r = 0;\ \ \nu_1 = \nu_2' \tag{10.63}$$

$$N_s - N_r = 2p_1;\ \ \nu_1 = -\nu_2' \tag{10.64}$$

When $N_s = N_r$, the synchronous torque occurs at zero speed and makes the motor less likely to start without hesitation from any rotor position. This situation has to be avoided in any case, so $N_s \neq N_r$.

Also from (10.61), it follows that (c_1, $c_2 > 0$), for zero speed,

$$c_2 N_r \neq 6 c_1 p_1 \tag{10.65}$$

Condition (10.64) refers to first slot harmonic synchronous torques occurring at nonzero speed:

$$n_{\nu_1 = \nu_2'} = \frac{f_1}{p_1}(1 - S) = \frac{f_1}{p_1}\frac{2p_1}{c_2 N_r} = \frac{-2f_1}{c_2 N_r} \tag{10.66}$$

In our case, $c_2 = \pm 1$. Equation (10.66) can also be written in stator harmonic terms as

$$n_{\nu_1 = -\nu_2'} = \frac{f_1}{p_1 \nu_1} = \frac{f_1}{p_1}\frac{1}{\left(c_1 \frac{Z_1}{p_1} + 1\right)} = \frac{f_1}{c_1 Z_1 + p_1} \tag{10.67}$$

with $c_1 = \pm 1$.

The synchronous torque occurs at the speed $n_{\nu_1 = -\nu_2'}$ if there is a geometrical phase shift angle γ_{geom} between stator and rotor mmf harmonics ν_1 and $-\nu_2'$, respectively. The torque should have the expression $T_{s\nu_1}$.

$$T_{s\nu_1} = \left(T_{s\nu_1}\right)_{max} \sin\left(c_1 Z_1 + p_1\right) \cdot \gamma_{geom} \tag{10.68}$$

The presence of synchronous torques may cause motor locking at the respective speed. This latter event may appear more frequently with IMs with descending torque/speed curve (Design C, D), Figure 10.7.

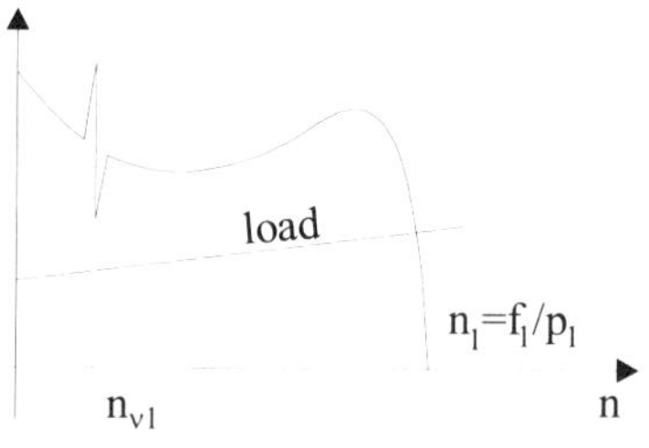

Figure 10.7 Synchronous torque on descending torque/speed curve

So far we considered mmf harmonics and slot harmonics as sources of synchronous torques. However, there are other sources for them such as the airgap magnetic conductance harmonics due to slot leakage saturation.

10.9.3. Leakage saturation influence on synchronous torques

The airgap field produced by the first rotor mmf slot harmonic $\nu' = +\frac{N_r}{p_1} + 1$ due to the airgap magnetic conduction fluctuation caused by leakage saturation (second term in (10.19)) is

$$B_{g\nu'} = \mu_0 F_{2\nu_m} \cdot \cos\left[\left(\pm\frac{N_r}{p_1} + 1\right)\theta \mp \left(\omega_1 \pm N_r \Omega_r\right)t\right] \cdot \frac{a_0'}{g}\sin(2\theta - \omega_1 t) \quad (10.69)$$

If B_{gv} contains harmonic (term) multiplied by $\cos(\nu\theta \mp \omega_1 t)$ for any n = $\pm 6c_1 + 1$ that corresponds to the stator mmf harmonics, a synchronous torque condition is again obtained. Decomposing (10.69) into two terms we find that only if

$$\frac{N_r}{p_1} \pm 3 \neq (6c_1 \pm 1) \quad (10.70)$$

such torques are avoided. Condition (10.70) becomes

$$N_r \neq 2p_1(3c_1 \mp 1); \quad c_1 = 1,2,3,... \quad (10.71)$$

The largest such torque occurs in interaction with the first stator slot mmf harmonic $\nu_1 = \frac{N_s}{p_1} \pm 1$ unless

$$\frac{N_s}{p_1} \pm 1 \neq \frac{N_r}{p_1} \pm 3 \quad (10.72)$$

So,

$$N_r \neq N_s \mp 4p_1 \quad (10.73)$$

As an evidence of the influence of synchronous torque at standstill with rotor bars skewed by one stator slot pitch, some experimental results are given on Table 10.2. [1]

Table 10.2. Starting torque (maximum and minimum value)

Motor power (kW) $f_1 = 50$Hz $2p_1 = 2$ poles	N_s	N_r	Starting torque (Nm)	
			max	min
0.55	24	18	1.788	0
0.85	24	22	0.75	0.635
1.4	24	26	0.847	0.776
2	24	28	2	1.7647
3	24	30	5.929	2.306

As Table 10.2 shows the skewing of rotor bars cannot prevent the occurrence of synchronous torque and a small change in the number of rotor slots drastically changes the average starting torque. The starting torque variation (with rotor position) is a clear indication of synchronous torque presence at stall.

The worst situation occurs with $N_r = 18 = 6c_1p_1$ (10.65). The speed for this synchronous torque comes from (10.69) as

$$\pm 3\omega_1 + N_r\Omega_r = \pm\omega_1 \tag{10.74}$$

or

$$\Omega_r = \pm\frac{4\omega}{N_r}; \; n = \frac{4f_1}{N_r} \tag{10.75}$$

Calculating the synchronous torques (even at zero speed) analytically is a difficult task. It may be performed by calculating the stator mmf harmonics airgap field interaction with the pertinent rotor current mmf harmonic.

After [1], the maximum tangential force $f_{\theta max}$ per unit rotor area due to synchronous torques at standstill should be

$$f_{\theta \max} \approx \frac{\mu_0\left(n_s I_s \sqrt{2}\right)^2 d}{4\gamma_{slot}} \tag{10.76}$$

where $n_s I_s \sqrt{2}$ is the peak slot mmf, γ_{slot} is the stator slot angle, and d is the greatest common divisor of N_s and N_r.

From this point of view, the smaller d, the better. The number d = 1 if Ns and Nr are prime numbers and it is maximum for $N_s = N_r = d$.

With $N_s = 6p_1q_1$, for q integer, that is an even number with N_r an odd number, with d = 1 the smallest variation with rotor position (safest start) of starting torque is obtained.

However, with N_r an odd number, the IM tends to be noisy, so most IMs have an even number of rotor slots, but with d = 2 if possible.

10.9.4. The secondary armature reaction

So far in discussing the synchronous torques, we considered only the reaction of rotor currents on the fields of the fundamental stator current. However, in some cases, harmonic rotor currents can induce in the stator windings additional currents of a frequency different from the power grid frequency. These additional stator currents may influence the IM starting properties and add to the losses in an IM. [4,5] These currents are related to the term of secondary armature reaction. [4]

With delta stator connection, the secondary armature reaction currents in the primary windings find a good path to flow, especially for multiple of 3 order harmonics. Parallel paths in the stator winding also favor this phenomenon by its circulating currents.

For wound (slip ring) rotor IMs, the field harmonics are not influenced by delta connection or by the parallel path. [6] Let us consider the first step (slot) stator harmonic $\frac{N_s}{p_1} \pm 1$ whose current $I_{1\left(\frac{N_s}{p_1}\pm 1\right)}$ produces a rotor current $I_{2\left(\frac{N_s}{p_1}\pm 1\right)}$.

$$I_{2\left(\frac{N_s}{p_1}\pm 1\right)} \approx \frac{I_{1\left(\frac{N_s}{p_1}\pm 1\right)}}{1+\tau_{d\left(\frac{N_s}{p_1}\pm 1\right)}} \quad (10.77)$$

τ_d is the differential leakage coefficient for the rotor cage. [1]

$$\tau_{d\left(\frac{N_s}{p_1}\pm 1\right)} = \left[\frac{\pi(N_s \pm p_1)}{N_r}\right]^2 \cdot \frac{1}{\sin^2\left[\frac{\pi(N_s \pm p_1)}{N_r}\right]} - 1 \quad (10.78)$$

As known, N_s and N_r are numbers rather close to each other, so $\tau_{d\left(\frac{N_s}{p_1}\pm 1\right)}$ might be a large number and thus the rotor harmonic current is small. A strong secondary armature reaction may change this situation drastically. To demonstrate this, let us notice that $I_{2\left(\frac{N_s}{p_1}\pm 1\right)}$ can now create harmonics of order ν' (10.46), for $c_2 = 1$ and $\nu = \left(\frac{N_s}{p_1} \pm 1\right)$:

$$\nu' = \frac{N_r}{p_1} \pm \left(\frac{N_s}{p_1} \pm 1\right) \quad (10.79)$$

The ν' rotor harmonic mmf $F_{\nu'}$ has the amplitude:

$$F_{\nu'} = F_{2\left(\frac{N_s}{p_1}\pm 1\right)} \cdot \frac{N_s \pm p_1}{N_r \pm N_s \pm p_1} \quad (10.80)$$

If these harmonics induce notable currents in the stator windings, they will, in fact, reduce the differential leakage coefficient from $\tau_{d\left(\frac{N_s}{p_1}\pm 1\right)}$ to $\tau'_{d\left(\frac{N_s}{p_1}\pm 1\right)}$. [1]

$$\tau'_{d\left(\frac{N_s}{p_1}\pm 1\right)} \approx \tau_{d\left(\frac{N_s}{p_1}\pm 1\right)} - \left(\frac{N_s \pm p_1}{N_r - (N_s \pm p_1)}\right)^2 \quad (10.81)$$

Now the rotor current $I_{2\left(\frac{N_s}{p_1}\pm1\right)}$ of (10.77) will become $I'_{2\left(\frac{N_s}{p_1}\pm1\right)}$.

$$I'_{2\left(\frac{N_s}{p_1}\pm1\right)} = \frac{I_{1\left(\frac{N_s}{p_1}\pm1\right)}}{\tau'_{d\left(\frac{N_s}{p_1}\pm1\right)}} \qquad (10.82)$$

The increase of rotor current $I'_{2\left(\frac{N_s}{p_1}\pm1\right)}$ is expected to increase both asynchronous and synchronous torques related to this harmonic.

In a delta connection, such a situation takes place. For a delta connection, the rotor current (mmf) harmonics

$$\nu' = \frac{N_r}{p_1} - \left(\frac{N_s}{p_1} \pm 1\right) = 3c \qquad c = 1,3,5,... \qquad (10.83)$$

can flow freely because their induced voltages in the three stator phases are in phase (multiple of 3 harmonics). They are likely to produce notable secondary armature reaction.

To avoid this undesirable phenomenon,

$$|N_s - N_r| \neq (3c \mp 1)p_1 = 2p_1, 4p_1, ... \qquad (10.84)$$

An induction motor with $2p_1 = 6$, $N_s = 36$, $N_r = 28$ slots does not satisfy (10.84), so it is prone to a less favourable torque-speed curve in Δ connection than it is in star connection.

In parallel path stator windings, circulating harmonics current may be responsible for notable secondary armature reaction with star or delta winding connection.

These circulating currents may be avoided if all parallel paths of the stator winding are at any moment at the same position with respect to rotor slotting.

However, for an even number of current paths, the stator current harmonics of the order $\frac{N_r}{p_1} - \frac{N_s}{p_1} - 1$ or $\frac{N_r}{p_1} - \frac{N_s}{p_1} + 1$ may close within the winding if the two numbers are multiples of each other. The simplest case occurs if

$$|N_s - N_r| = 3p_1 \qquad (10.85)$$

When (10.85) is fulfilled, care must be exercised in using parallel path stator windings.

We should also notice again that the magnetic saturation of stator and rotor teeth lead to a further reduction of differential leakage by 40 to 70%.

10.9.5. Notable differences between theoretical and experimental torque/speed curves

By now the industry has accumulated enormous data on the torque/speed curves of IM of all power ranges.

In general, it was noticed that there are notable differences between theory and tests especially in the braking regime (S > 1). A substantial rise of torque in the braking regime is noticed. In contrast, a smaller reduction of torque in the large slip motoring regime appears frequently.

A similar effect is produced by transverse (lamination) rotor currents between skewed rotor bars.

However, in this latter case, there should be no difference between slip ring and cage rotor, which is not the case in practice.

Finally these discrepancies between theory and practice occur even for insulated copper bar cage rotors where the transverse (lamination) rotor currents cannot occur (Figure 10.8).

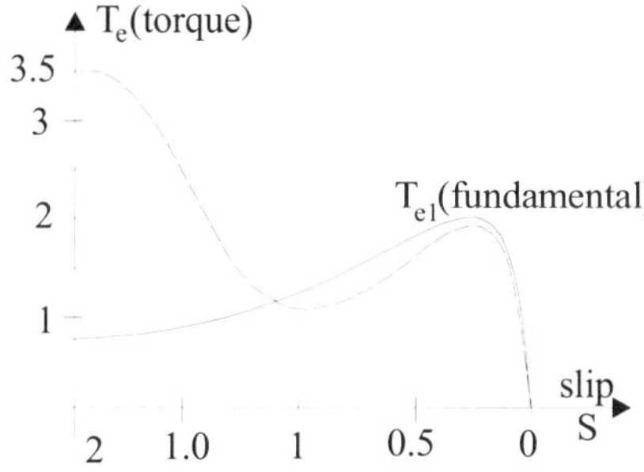

Figure 10.8 Torque-slip curve (relative values)
for $N_s = 36$, $N_r = 28$, $2p_1 = 4$, skewed rotor (one slot pitch), insulated copper bars

What may cause such a large departure of experiments from theory? Although a general, simple, answer to this question is still not available, it seems intuitive to assume that the airgap magnetic conductance harmonics due to slot openings and the tooth head saturation due leakage path saturation are the main causes of spectacular braking torque augmentation and notable reduction high slip motoring torque in.

This is true when the contact resistance of rotor bars (practically insulated rotor bars) is high and, thus, the transverse (lamination) currents in skewed rotors are small.

Slot openings have been shown to increase considerably the parasitic torques. (Remember we showed earlier in this chapter that the first slot (step) stator mmf harmonic $\nu_{s\min} = \frac{N_s}{p_1} - 1$ is amplified by the influence of the first airgap magnetic conductance–due to slot openings – (10.33)).

Now the leakage path saturation produced airgap magnetic conductance (10.11) may add further to amplify the parasitic torques.

For skewed rotors, at high slips, where parasitic torques occur, main path flux saturation due to the uncompensated (skewing) rotor mmf may lead to a fundamental torque increase in the braking region (due to higher current, for lower leakage inductance) and to the amplification of parasitic torque through main path saturation-caused airgap conductance harmonics (10.20). And, finally, rotor eccentricity may introduce its input to this braking torque escalation.

We have to stress again that leakage flux path saturation and skin effects alone, also cause important increases in rotor resistance, decreases in motor reactance, and thus produce also more fundamental torque.

In fact, the extreme case on Figure 10.8 with a steady increase of torque with slip (for S > 1) seems to be a mixture of such intricated contributions. However, the minimum torque around zero speed shows the importance of parasitic torques as the skin effect and leakage saturation depend on $\sqrt{S}$; so they change slower with slip than Figure 10.7 suggests.

10.9.6. A case study: $N_s/N_r = 36/28$, $2p_1 = 4$, $y/\tau = 1$ and 7/9; m = 3 [7]

As there is no general answer, we will summarize the case of a three-phase induction motor with $P_n = 11$ kW, $N_s/N_r = 36/28$, $2p_1 = 4$, $y/\tau = 9/9$ and 7/9 (skewed rotor by one stator slot pitch).

For this machine, and $y/\tau = 9/9$, with straight slots, condition (10.71) is not satisfied as

$$N_r = 28 = 2p_1(3c \mp 1) = 2 \cdot 2(3c + 1); \text{ for } c = 2 \tag{10.86}$$

This means that a synchronous torque occurs at the speed

$$n = \frac{4f_1}{N_r} = \frac{4 \cdot 50 \cdot 60}{28} = 215\text{rpm} \tag{10.87}$$

Other synchronous, but smaller, torques may be detected. A very large synchronous torque has been measured at this speed (more than 14 Nm, larger than the peak torque for 27% voltage). This leads to a negative total torque at 215 rpm; successful acceleration to full speed is not always guaranteed. It is also shown that as the voltage increases and saturation takes place, the synchronous torque does not increase significantly over this value.

Also, a motor of 70 kW, $N_s/N_r = 48/42$, $y/\tau = 10/12$ with Δ connection and 4 parallel current paths has been investigated. As expected, a strong synchronous torque has been spotted at standstill because condition (10.65) is not met as

$$N_r = 6\frac{c_1}{c_2}p_1 = 6\frac{7}{2} \cdot 2 = 42 \tag{10.88}$$

The synchronous torque at stall is amplified by Δ connection with 4 paths in parallel stator winding as $N_s - N_r = 48 - 42 = 6 = 3p_1 = 3 \cdot 2$.

The situation is so severe that, for 44% voltage and 30% load torque, the motor remains blocked at zero speed.

Strong variations of motor torque with position have been measured and calculated satisfactorily with the finite difference method. Saturation consideration seems mandatory for realistic torque calculations during starting transients.

It has also been shown that the amplification of parasitic torque due to slot openings is higher for $N_r > N_s$ because the differential leakage coefficient $\tau_{d\left(\frac{N_s}{p_1}\pm 1\right)}$ is smaller in this case, allowing for large rotor mmf harmonics. From the point of view of parasitic torques, the rotor with $N_r < N_s$ is recommended and may be used without skewing for series-connected stator windings.

For $N_r > N_s$ skewing is recommended unless magnetic wedges in the stator slots are used.

10.9.7. Evaluation of parasitic torques by tests (after [1])

To assist potential designers of IMs, here is a characterization of torque/speed curve for quite a few stator /rotor slot numbers of practical interest.

<u>For N_s = 24 stator slots</u>, f_1 = 50 Hz, $2p_1$ = 4

- N_r = 10; small synchronous torques at n = 150, 300, 600 rpm and a large one at n = –300 rpm
- N_r = 16; unsuitable (high synchronous torque at n = 375 rpm)
- N_r = 18; very suitable (low asynchronous torque dips at n = 155 rpm and at –137 rpm)
- N_r = 19; unsuitable because of high noise, although only small asynchronous torque dips occur at n = 155 rpm and n = –137 rpm
- N_r = 20; greatest synchronous torque of all situations investigated here at n = –300 rpm, unsuitable for reversible operation.
- N_r = 22; slight synchronous torques at n = 690, 600, ±550, 214, -136 rpm
- N_r = 24; unsuitable because of high synchronous torque at n = 0 rpm
- N_r = 26; larger asynchronous torque deeps at n = 115, –136 rpm and synchronous one at n = -300 rpm
- N_r = 27; high noise
- N_r = 28; high synchronous torque dip at n = 214 rpm and slight torque dips at n = 60, –300, –430 rpm. Asynchronous torque dips at n = 115, –136 rpm. With skewing, the situation gets better, but at the expense of 20% reduction in fundamental torque.
- N_r = 30; typical asynchronous torque dips at n = 115, –136 rpm, small synchronous ones at n = 600, ±400, ±300, 214, ±200, –65, –100, –128 rpm. Practically noiseless.
- N_r = 31; very noisy.

<u>Number of slots N_s = 36</u>, f_1 = 50Hz, $2p_1$ = 4

This time, the asynchronous torques are small. At n = 79 to 90 rpm, they become noticeable for higher number of rotor slots (44, 48).

- $N_r = 16$; large synchronous torque at n = –188, +376 rpm
- $N_r = 20$; synchronous torque at n = –376 rpm
- $N_r = 24$; large synchronous torque at n = 0 rpm, impractical
- $N_r = 28$; large synchronous torque at n = –214 rpm, –150, –430 rpm
- $N_r = 31$; good torque curve but very noisy
- $N_r = 32$; large synchronous torque at n = –188 rpm and a smaller one at n = –375 rpm;
- $N_r = 36$; impractical, motor locks at zero speed;
- $N_r = 44$; large synchronous torque at n = –136 rpm, smaller ones at n = 275, –40, –300 rpm
- $N_r = 48$; moderate synchronous torque dip at zero speed

Number of slots $N_s = 48$, $f_1 = 50$ Hz, $2p_1 = 4$

- $N_r = 16$; large synchronous torque at n = –375, a smaller one at n=-188 rpm
- $N_r = 20$; large synchronous torque at n = –300 rpm, small ones at n = 600, 375, 150, 60, –65, –136, –700 rpm
- $N_r = 22$; large synchronous torque dip at n = –136 rpm due to the second rotor slip harmonic ($c_2 = 2$, $n = \frac{2f_1}{c_2 N_r} = \frac{-2\cdot 50\cdot 60}{2\cdot 22} = -136\,\text{rpm}$)
- $N_r = 28$; high synchronous torque at n = 214 rpm, smaller ones at n = 375, 150, –65, –105, -136, –214rpm
- $N_r = 32$; high synchronous torque at n = –188 rpm, small ones at n = 375, 188, –90, 90, –300 rpm
- $N_r = 36$; large torque dip at zero speed, not good start
- $N_r = 44$; very large synchronous torque at n = –136 rpm, smaller ones at n = 275, 214, 60, –300 rpm
- $N_r = 48$; motor locks at zero speed ($N_s = N_r = 48!$), after warming up, however, it starts!

10.10. RADIAL FORCES AND ELECTROMAGNETIC NOISE

Tangential forces in IMs are producing torques which smoothly drive the IM (or cause torsional vibrations) while radial forces are causing vibration and noise.

This kind of vibration and noise is produced by electromagnetic forces which vary along the rotor periphery and with time. Using Maxwell tensor, after neglecting the tangential field component in the airgap (as it is much lower than the radial one), the radial stress on the stator is

$$p_r = \frac{B^2(\theta_m, t)}{2\mu_0}; \; \theta = \theta_m p_1 \tag{10.89}$$

where $B(\theta,t)$ is the resultant airgap flux density.

The expression of $B(\theta_m,t)$ is, as derived earlier in this chapter,

$$B(\theta_m, t) = \frac{\mu_0}{g}\left(F_1(\theta_m, t) + F_2(\theta_m, t)\right) \cdot \lambda(\theta_m, t) \tag{10.90}$$

with $F_1(\theta_m,t)$ and $F_2(\theta_m,t)$ the stator and rotor mmf and $\lambda(\theta_m,t)$ the airgap magnetic conductance as influenced by stator and rotor slot openings, leakage, and main path saturation and eccentricity.

$$\lambda(\theta, t) = \lambda_0 + \sum_{\sigma=1}^{\infty} \lambda_\sigma \cos(\sigma\theta_m + \varphi_\sigma) + \sum_{\rho=1}^{\infty} \lambda_\rho \cos\left(\rho\left(\theta_m - \frac{\omega_r t}{p_1}\right) + \varphi_\rho\right) + \ldots \tag{10.91}$$

$$F_1(\theta_m, t) = \sum F_{1\nu} \cos(\nu\theta_m \pm \omega_\nu t) \tag{10.92}$$

$$F_2(\theta_m, t) = \sum F_{2\mu} \cos(\mu\theta_m \pm \omega_\mu t + \varphi_\mu) \tag{10.93}$$

In (10.89 through 10.93) we have used the mechanical angle θ_m as the permeance harmonics are directly related to it, and not the electrical angle $\theta = p_1\theta_m$. Consequently, now the fundamental, with p_1 pole pairs, will be $\nu_1 = p_1$ and not $\nu = 1$.

Making use of (10.92 and 10.93) in (10.89) we obtain terms of the form

$$p_r = A_r \cos(\gamma\theta_m - \Omega_r t) \tag{10.94}$$

$\gamma = 0,1,2,\ldots$. The frequency of these forces is $f_r = \frac{\Omega_r}{2p_1}$, which is not be confused to the rotor speed ω_r/p_1.

If the number of pole pairs (γ) is small, stator vibrations may occur due to such forces. As inferred from (10.90), there are terms from same harmonic squared and interaction terms where 1(2) mmf harmonics interact with an airgap magnetic conductance. Such combinations are likely to yield low orders for γ.

For $\gamma = 0$, the stress does not vary along rotor periphery but varies in time producing variable radial pressure on the stator (and rotor).

For $\gamma = 1$ (10.94) yields a single-sided pull on the rotor which circulates with the angular velocity of Ω_r. If Ω_r corresponds to a mechanical resonance of the machine, notable vibration will occur.

Case $\gamma = 1$ results from the interaction of two mechanical harmonics whose order differs by unity.

For $\gamma = 2, 3, 4, \ldots$ stator deflections occur. In general, all terms p_r with $\gamma \le 4$ should be avoided, by adequately choosing the stator and rotor slot number pairs N_s and N_r.

However, as mechanical resonance of stators (without and with frames attached) decreases with motor power increase, it means that a certain N_s, N_r combination may lead to low noise in low power motors but to larger noise in large power motors.

Now as in (10.90), the mmf step harmonics and the airgap magnetic conductance occur, we will treat their influence separately to yield some favourable N_s, N_r combinations.

10.10.1. Constant airgap (no slotting, no eccentricity)

In this case $\lambda(\theta_m,t) = \lambda_0$ and a typical p_r term occurs from two different stator and rotor mmf harmonics.

$$(p_r)_{\gamma=\nu-\mu} = \frac{\mu_0 F_{1\nu} F_{2\mu}}{2g^2}\lambda_0^{\,2}\cos\left[(\gamma \mp \mu)\theta_m \pm (\omega_\nu + \omega_\mu) - \varphi_\mu\right] \qquad (10.95)$$

In this case $\gamma = \nu - \mu$, that is the difference between stator and rotor mmf mechanical harmonics order.

Now if we are considering the first slot (step) mmf harmonics ν, μ,

$$\begin{aligned} \nu &= N_s \pm p_1 \\ \mu &= N_r \pm p_1 \end{aligned} \qquad (10.96)$$

It follows that to avoid such forces with $\gamma = 0, 1, 3, 4$,

$$\begin{aligned} |N_s - N_r| &\neq 0,1,2,3,4 \\ |N_s - N_r| &\neq 2p_1, 2p_1 \pm 1, 2p_1 \pm 2, 2p_1 \pm 3, 2p_1 \pm 4 \end{aligned} \qquad (10.97)$$

The time variation of all stator space harmonics has the frequency $\omega_\nu = \omega_1 = 2\pi f_1$, while for the rotor,

$$\omega_\mu = \omega_1 \mp cN_r\omega_r = \omega\left[1 \mp c\frac{N_r}{p_1}(1-S)\right] \qquad (10.98)$$

Notice that ω_r is the rotor mechanical speed here.

Consequently the frequency of such forces is $\omega_\nu \pm \omega_\mu$.

$$\omega_\xi = \omega_\nu \pm \omega_\mu = \begin{cases} \omega_1 c\dfrac{N_r}{p_1}(1-S) \\ 2\omega_1 + \omega_1 c\dfrac{N_r}{p_1}(1-S) \end{cases} \qquad (10.99)$$

In general, only frequencies ω_ξ low enough (c = 1) can cause notable stator vibrations.

The frequency ω_ξ is twice the power grid frequency or zero at zero speed (S = 1) and increases with speed.

10.10.2. Influence of stator/rotor slot openings, airgap deflection and saturation

As shown earlier in this chapter, the slot openings introduce harmonics in the airgap magnetic conductance which have the same order as the mmf slot (step) harmonics.

Zero order ($\gamma = 0$) components in radial stress p_r may be obtained when two harmonics of the same order, but different frequency, interact. One springs from the mmf harmonics and one from the airgap magnetic conductance harmonics. To avoid such a situation simply,

$$\begin{matrix} N_s \pm p_1 \neq N_r \\ N_r \pm p_1 \neq N_s \end{matrix} \quad \text{or} \quad |N_s - N_r| \neq \pm p_1 \tag{10.100}$$

Also, to avoid first order stresses for $\gamma = 2$,

$$2(N_r - N_s \mp p_1) \neq 2 \tag{10.101}$$

$$|N_s - N_r| \neq \pm p_1 \pm 1 \tag{10.102}$$

The frequency of such stresses is still calculated from (10.100) as

$$\omega_\xi = \left[\omega_1 \pm \omega_1 \frac{N_r}{p_1}(1 - S)\right]2 \tag{10.103}$$

The factor 2 above comes from the fact that this stress component comes from a term squared so all the arguments (of space and time) are doubled.

It has been shown that airgap deformation itself, due to such forces, may affect the radial stresses of low order.

In the most serious case, the airgap magnetic conductance becomes

$$\begin{aligned} \lambda(\theta_m, t) = \frac{1}{g(\theta_m, t)} &= [a_0 + a_1 \cos N_s\theta_m + a_2 \cos N_r(\theta_m - \omega_r t)] \cdot \\ &\cdot [1 + B\cos(K\theta_m - \Omega t)] \end{aligned} \tag{10.104}$$

The last factor is related to airgap deflection. After decomposition of $\lambda(\theta_m,t)$ in simple sin(cos) terms for $K = p_1$, strong airgap magnetic conductance terms may interact with the mmf harmonics of orders p_1, $N_s \pm p_1$, $N_r \pm p_1$ again. So we have to have

$$\begin{aligned} &(N_s \pm p_1) - (N_r \pm p_1) \neq p_1 \\ &\text{or } |N_s - N_r| \neq p_1 \\ &\text{and } |N_s - N_r| \neq 3p_1 \end{aligned} \tag{10.115}$$

In a similar way, slot leakage flux path saturation, already presented earlier in this chapter with its contribution to the airgap magnetic conductance, yields conditions as above to avoid low order radial forces.

10.10.3. Influence of rotor eccentricity on noise

We have shown earlier in this chapter that rotor eccentricity–static and dynamic–introduces a two-pole geometrical harmonics in the airgap magnetic conductance. This leads to a $p_1 \pm 1$ order flux density strong components (10.90) in interaction with the mmf fundamentals. These forces vary at twice stator frequency $2f_1$, which is rather low for resonance conditions. In interaction with another harmonic of same order but different frequency, zero order ($\gamma = 0$) vibrations may occur again. Consider the interference with first mmf slot harmonics $N_s \pm p_1$, $N_r \pm p_1$

$$\begin{aligned} &(N_s \pm p_1) - (N_r \pm p_1) \neq p_1 \pm 1 \\ \text{or } &|N_s - N_r| \neq 3p_1 \pm 1 \\ &|N_s - N_r| \neq p_1 \pm 1 \end{aligned} \tag{10.106}$$

These criteria, however, have been found earlier in this paragraph. The dynamic eccentricity has been also shown to produce first order forces ($\gamma = 1$) from the interaction of $p_1 \pm 1$ airgap magnetic conductance harmonic with the fundamental mmf (p_1) as their frequencies is different. Noise is thus produced by the dynamic eccentricity.

10.10.4. Parallel stator windings

Parallel path stator windings are used for large power, low voltage machines (now favored in power electronics adjustable drives). As expected, if consecutive poles are making current paths and some magnetic asymmetry occurs, a two-pole geometric harmonic occurs. This harmonic in the airgap magnetic conductance can produce, at least in interaction with the fundamental mmf, radial stresses with the order $\gamma = p_1 \pm 1$ which might cause vibration and noise.

On the contrary, with two parallel paths, if north poles make one path and south poles the other current path, with two layer full pitch windings, no asymmetry will occur. However, in single layer winding, this is not the case and stator harmonics of low order are produced ($\nu = N_s/2c$). They will never lead to low order radial stress, because, in general, N_s and N_r are not far from each other in actual motors.

Parallel windings with chorded double layer windings, however, due to circulating current, lead to additional mmf harmonics and new rules to eliminate noise, [1]

$$\left|N_s - N_r\right| = \left|\frac{p_1}{a} \pm p_1 \pm 1\right|; \quad a\text{ - odd}$$
$$\left|N_s - N_r\right| = \left|\frac{2p_1}{a} \pm p_1 \pm 1\right|; \quad a\text{ - even} \tag{10.107}$$

where a is the number of current paths.

Finally, for large power machines, the first slot (step) harmonic order $N_s \pm p_1$ is too large so its phase belt harmonics $n = (6c_1 \pm 1)p_1$ are to be considered in (10.117), [1]

$$\left|N_s \pm \frac{p_1}{a} - (6c_1 \pm 1)p_1\right| \neq 1; \quad a > 2, \text{odd}$$
$$\left|N_r \pm \frac{2p_1}{a} - (6c_1 \pm 1)p_1\right| \neq 1; \quad a > 2, \text{even} \tag{10.108}$$
$$\left|N_r \pm 2p_1 - (6c_1 \pm 1)p_1\right| \neq 1; \quad a = 2$$

with c_1 a small number $c_1 = 1,2,\ldots$.

Symmetric (integer q) chorded windings with parallel paths introduce, due to circulating current, a $2p_1$ pole mmf harmonic which produces effects that may be avoided by fulfilling (10.108).

10.10.5. Slip-ring induction motors

The mmf harmonics in the rotor mmf are now $\nu = c_2N_r \pm p_1$ only. Also, for integer q_2 (the practical case), the number of slots is $N_r = 2\cdot3\cdot p_1\cdot q_2$. This time the situation of $N_s = N_r$ is to be avoided; the other conditions are automatically fulfilled due to this constraint on N_r. Using the same rationale as for the cage rotor, the radial stress order, for constant airgap, is

$$r = \nu = \mu = (6c_1 \pm 1)p_1 - (6c_2 \pm 1)p_1 \tag{10.109}$$

Therefore γ may only be an even number.

Now as soon as eccentricity occurs, the absence of armature reaction leads to higher radial uncompensated forces than the cage rotors. The one for $\gamma = p_1 \pm 1$–that in interaction with the mmf fundamental, which produces a radial force at $2f_1$ frequency–may be objectionable because it is large (no armature reaction) and produces vibrations anyway.

So far, the radial forces have been analyzed in the absence of magnetic saturation and during steady state. The complexity of this subject makes experiments in this domain mandatory.

Thorough experiments [8] for a few motors with cage and wound rotors confirm the above analysis in general, and reveal the role of magnetic saturation of teeth and yokes in reducing these forces (above 70 to 80% voltage). The radial forces due to eccentricity level off or even decrease.

During steady state, the unbalanced magnetic pull due to eccentricity has been experimentally proved much higher for the wound rotor than for the cage rotor! During starting transients, both are, however, only slightly higher than for the wound rotor under steady state. Parallel windings have been proven to reduce the uncompensated magnetic pull force ($\gamma = 0$).

It seems that even with 20% eccentricity, UMP may surpass the cage rotor mass during transients and the wound rotor mass even under steady state. [8]

The effect of radial forces on the IM depends not only on their amplitude and frequency but also on the mechanical resonance frequencies of the stator, so the situation is different for low medium power and large power motors.

10.10.6 Mechanical resonance stator frequencies

The stator ring resonance frequency for r = 0 (first order) radial forces F_0 is [1]

$$F_0 = \frac{8 \cdot 10^2}{R_a} \sqrt{\frac{G_{yoke}}{G_{yoke} + G_{teeth}}} \tag{10.110}$$

with R_a–average stator yoke radius, G_{yoke} and G_{teeth} stator yoke and teeth weight.

For $r \neq 0$,

$$F_r = F_0 \frac{1}{2\sqrt{3}} \frac{h_{yoke}}{R_a} \frac{r(r^2 - 1)}{\sqrt{r^2 + 1}} K_r \tag{10.111}$$

h_{yoke}–yoke radial thickness.

$K_r = 1$ for $h_{yoke}/R_a > 0.1$ and $0.66 < K_r < 1$ for $h_{yoke}/R_a < 0.1$.

The actual machine has a rather elastic frame fixture to ground and a less than rigid attachment of stator laminations stack to the frame.

This leads to more than one resonance frequency for a given radial force of order r. A rigid framing and fixture will reduce this resonance "multiplication" effect.

To avoid noise, we have to avoid radial force frequencies to be equal to resonance frequencies. For the step harmonics of mmfs contribution (10.100) applies.

$$\text{for } r = 0, \text{at } S = 0 \text{ (zero slip)},$$
$$\frac{F_0}{f_1} \neq c\frac{N_r}{p_1} \quad \text{or} \quad \frac{F_0}{f_1} \neq c\frac{N_r}{p_1} \pm 2; \quad c = 1,2 \tag{10.122}$$

We may proceed, for r = 1, 2, 3, by using F_r instead of F_0 in (10.112) and with c = 1.

In a similar way, we should proceed for the frequency of other radial forces (derived in previous paragraphs).

In small power machines, such conditions are easy to meet as the stator resonance frequencies are higher than most of radial stress frequencies. So radial forces of small pole pairs r = 0, 1, 2 are to be checked. For large power motors radial forces with r = 3, 4 are more important to account for.

The subject of airgap flux distribution, torque pulsations (or parasitic torques), and radial forces has been approached recently by FEM [9,10] with good results, but at the price of prohibitive computation time.

Consequently, it seems that intuitive studies, based on harmonics analysis, are to be used in the preliminary design study and, once the number of stator/rotor slots are defined, FEM is to be used for precision calculation of parasitic torques and of radial forces.

10.11. SUMMARY

- The fundamental component of resultant airgap flux density distribution in interaction with the fundamental of stator (or rotor) mmf produces the fundamental (working) electromagnetic torque in the IM.
- The placement of windings in slots (even in infinitely thin slots) leads to a stepped-like waveform of stator (rotor) mmfs which exhibit space harmonics besides the fundamental wave (with p_1 pole pairs). These are called mmf harmonics (or step harmonics); their lower orders are called phase-belt harmonics ν = 5, 7, 11, 13, … to the first "slot" harmonic order $\nu_{s\min} = \frac{N_s}{p_1} \pm 1$. In general, $\nu = (6c_1 \pm 1)$.
- A wound rotor mmf has its own harmonic content $\mu = (6c_2 \pm 1)$.
- A cage rotor, however, adapts its mmf harmonic content (μ) to that of the stator such that

$$\nu - \mu = \frac{c_2 N_r}{p_1}$$

- ν and μ are electric harmonics, so ν = μ = 1 means the fundamental (working) wave.
- The mechanical harmonics ν_m, μ_m are obtained if we multiply ν and μ by the number of pole pairs,

$$\nu_m = p_1\nu; \quad \mu_m = p_1\mu;$$

- In the study of parasitic torques, it seems better to use electric harmonics, while, for radial forces, mechanical harmonics are favored. As the literature uses both concepts, we thought it appropriate to use them both in a single chapter.
- The second source of airgap field distribution is the magnetic specific airgap conductance (inversed airgap function) $\lambda_{1,2}(\theta_m,t)$ variation with stator or/and rotor position. In fact, quite a few phenomena are accounted for by decomposing $\lambda_{1,2}(\theta_m,t)$ into a continuous component and various harmonics. They are outlined below.

- The slot openings, both on the stator and rotor, introduce harmonics in $\lambda_{1,2}(\theta_m,t)$ of the orders c_1N_s/p_1, c_2N_r/p_1 and $(c_1N_s \pm c_2N_r)p_1$ with $c_1 = c_2 = 1$ for the first harmonics.
- At high currents, the teeth heads saturate due to large leakage fluxes through slot necks. This effect is similar to a fictitious increase of slot openings, variable with slot position with respect to stator (or rotor) mmf maximum. A $\lambda_{1,2}(\theta_m,t)$ second order harmonic ($4p_1$ poles) traveling at the frequency of the fundamental occurs mainly due to this aspect.
- The main flux path saturation produces a similar effect, but its second order harmonic ($4p_1$ poles) is in phase with the magnetization mmf and not with stator or rotor mmfs separately!
- The second permeance harmonic also leads to a third harmonic in the airgap flux density.
- The rotor eccentricity, static and dynamic, produces mainly a two-pole harmonic in the airgap conductance. For a two-pole machine, in interaction with the fundamental mmf, a homopolar flux density is produced. This flux is closing axially through the frame, bearings, and shaft, producing an a.c. shaft voltage and bearing current of frequency f_1 (for static eccentricity) and Sf_1 (for the dynamic eccentricity).
- The interaction between mmf and airgap magnetic conductance harmonics is producing a multitude of harmonics in the airgap flux distribution.
- Stator and rotor mmf harmonics of same order and same stator origin produce asynchronous parasitic torques. Practically all stator mmf harmonics produce asynchronous torques in cage rotors as the latter adapts to the stator mmf harmonics. The no-load speed of asynchronous torque is $\omega_r = \omega_1/\nu$.
- As $\nu \geq 5$ in symmetric (integer q) stator windings, the slip S for all asynchronous torques is close to unity. So they all occur around standstill.
- The rotor cage, as a short-circuited multiphase winding, may attenuate asynchronous torques.
- Chording is used to reduce the first asynchronous parasitic torques (for $\nu = -5$ or $\nu = +7$); skewing is used to cancel the first slot mmf harmonic $\nu = 6q_1 \pm 1$. (q_1–slots/pole/phase).
- Synchronous torques are produced at some constant specific speeds where two harmonics of same order $\nu_1 = \pm\nu'$ but of different origin interact.
- Synchronous torques may occur at standstill if $\nu_1 = +\nu'$ and at

$$S = 1 + \frac{2p_1}{c_2 N_r}; \ -1 \geq C_2 \geq 1$$

  for $\nu_1 = -\nu'$.
- Various airgap magnetic conductance harmonics and stator and rotor mmf harmonics may interact in a cage rotor IM many ways to produce synchronous torques. Many stator/rotor slot number N_s/N_r combinations are to be avoided to eliminate most important synchronous torque. The main

benefit is that the machine will not lock into such speed and will accelerate quickly to the pertinent load speed.

- The harmonic current induced in the rotor cage by various sources may induce certain current harmonics in the stator, especially for Δ connection or for parallel stator windings (a >1).
- This phenomenon, called secondary armature reaction, reduces the differential leakage coefficient τ_d of the first slot harmonic $\nu_{smin} = N_s/p_1 \pm 1$ and thus, in fact, increases its corresponding (originating) rotor current. Such an augmentation may lead to the amplification of some synchronous torques.
- The stator harmonics currents circulating between phases (in Δ connection) or in between current paths (in parallel windings), whose order is multiples of three may be avoided by forbidding some N_s, N_r combinations $(Ns-Nr) \neq 2p_1, 4p_1, \ldots$.
- Also, if the stator current paths are in the same position with respect to rotor slotting, no such circulating currents occur.
- Notable differences, between the linear theory and tests have been encountered with large torque amplifications in the braking region (S > 1). The main cause of this phenomenon seems to be magnetic saturation.
- Slot opening presence also tends to amplify synchronous torques. This tendency is smaller if $N_r < N_s$; even straight rotor slots (no skewing) may be adopted. Attention to noise for no skewing!
- As a result of numerous investigations, theoretical and experimental, clear recommendations of safe stator/rotor slot combinations are now given in some design books. Attention is to be paid to the fact that, when noise is concerned, low power machines and large power machines behave differently.
- Radial forces are somehow easier to calculate directly by Maxwell's stress method from various airgap flux harmonics.
- Radial stress (force per unit stator area) p_r is a wave with a certain order r and a certain electrical frequency Ω_r. Only r = 0, 1, 2, 3, 4 cases are important.
- Investigating again the numerous combination contributions to the mechanical stress components coming from the mmfs and airgap magnetic conductance harmonics, new stator/rotor slot number N_s/N_r restrictions are developed.
- Slip ring IMs behave differently as they have clear cut stator and rotor mmf harmonics. Also damping of some radial stress component through induced rotor current is absent in wound rotors.
- Especially radial forces due to rotor eccentricity are much larger in slip ring rotor than in cage rotor with identical stators, during steady state. The eccentricity radial stress during starting transients are however about the same.

- The circulating current of parallel winding might, in this case, reduce some radial stresses. By increasing the rotor current, they reduce the resultant flux density in the airgap which, squared, produces the radial stress.
- The effect of radial stress in terms of stator vibration amplitude manifests itself predominantly with zero, first, and second order stress harmonics (r = 0, 1, 2).
- For large power machines, the mechanical resonance frequency is smaller and thus the higher order radial stress r = 3, 4 are to be avoided if low noise machines are to be built.
- For precision calculation of parasitic torques and radial forces, after preliminary design rules have been applied, FEM is to be used, though at the price of still very large computation time.
- Vibration and noise is a field in itself with a rich literature and standardization. [11,12]

10.12. REFERENCES

1. B. Heller, V. Hamata, Harmonics Effects in Induction Motors, Chapter 6, Elsevier, 1977.
2. R. Richter, Electric Machines, Second edition, Vol.1, pp.173, Verlag Birkhäuser, Basel, 1951 (in German).
3. K. Vogt, Electric Machines–Design of Rotary Electric Motors, Chapter 10, VEB Verlag Technik, Berlin, 1988. (in German)
4. K. Oberretl, New Knowledge on Parasitic Torques in Cage Rotor Induction Motors, Buletin Oerlikon 348, 1962, pp.130 – 155.
5. K. Oberretl, The Theory of Harmonic Fields of Induction Motors Considering the Influence of Rotor Currents on Additional Stator Harmonic Currents in Windings with Parallel Paths, Archiv für Electrotechnik, Vol.49, 1965, pp.343 – 364 (in German).
6. K. Oberretl, Field Harmonics Theory of Slip Ring Motor Taking Multiple Reaction into Account, Proc of IEE, No.8, 1970, pp.1667 – 1674.
7. K. Oberretl, Parasitic Synchronous and Pendulation Torques in Induction Machines; the Influence of Transients and Saturation, Part II – III, Archiv für Elek. Vol.77, 1994, pp.1 – 11, pp.277 – 288 (in German).
8. D.G. Dorrel, Experimental Behaviour of Unbalanced Magnetic Pull in 3-phase Induction Motors with Eccentric Rotors and the Relationship with Teeth Saturation, IEEE Trans Vol. EC – 14, No.3, 1999, pp.304 – 309.
9. J.F. Bangura, N.A. Dermerdash, Simulation of Inverter-fed Induction Motor Drives with Pulse-width-modulation by a Time-stepping FEM Model–Flux Linkage-based Space Model, IEEE Trans, Vol. EC – 14, No.3, 1999, pp.518 – 525.
10. A. Arkkio, O. Lingrea, Unbalanced Magnetic Pull in a High Speed Induction Motor with an Eccentric Rotor, Record of ICEM – 1994, Paris, France, Vol.1, pp.53 – 58.
11. S.J. Yang, A.J. Ellison, Machinery Noise Measurement, Clarendon Press, Oxford, 1985.

12. P.L. Timar, Noise and Vibration of Electrical Machines, Technical Publishers, Budapest, Hungary, 1986.

Chapter 11

LOSSES IN INDUCTION MACHINES

Losses in induction machines occur in windings, magnetic cores, besides mechanical friction and windage losses. They determine the efficiency of energy conversion in the machine and the cooling system that is required to keep the temperatures under control.

In the design stages, it is natural to try to calculate the various types of losses as precisely as possible. After the machine is manufactured, the losses have to be determined by tests. Loss segregation has become a standard method to determine the various components of losses, because such an approach does not require shaft-loading the machine. Consequently, the labor and energy costs for testings are low.

On the other hand, when prototyping or for more demanding applications, it is required to validate the design calculations and the loss segregation method. The input-output method has become standard for the scope. It is argued that, for high efficiency machines, measuring of the input and output P_{in}, P_{out} to determine losses Σp on load

$$\Sigma p = P_{in} - P_{out} \tag{11.1}$$

requires high precision measurements. This is true, as for a 90% efficiency machine a 1% error in P_{in} and P_{out} leads to a 10% error in the losses.

However, by now, less than (0.1 to 0.2)% error in power measurements is available so this objection is reduced considerably.

On the other hand, shaft-loading the IM requires a dynamometer, takes time, and energy. Still, as soon as 1912 [1] it was noticed that there was a notable difference between the total losses determined from the loss segregation method (no-load + short – circuit tests) and from direct load tests. This difference is called "stray load losses." The dispute on the origin of "stray load losses" and how to measure them correctly is still on today after numerous attempts made so far. [2 - 8]

To reconcile such differences, standards have been proposed. Even today, only in the USA (IEEE Standard 112B) the combined loss segregation and input-output tests are used to calculate aposteriori the "stray load losses" for each motor type and thus guarantee the efficiency.

In most other standards, the "stray load losses" are assigned 0.5 or 1% of rated power despite the fact that all measurements suggest much higher values.

The use of static power converters to feed IMs for variable speed drives complicates the situation even more, as the voltage time harmonics are producing additional winding and core losses in the IM.

Faced with such a situation, we decided to retain only the components of losses which proved notable (greater than (3 to 5%) of total losses) and explore their computation one by one by analytical methods.

Further on numerical, finite element, loss calculation results are given.

11.1. LOSS CLASSIFICATIONS

The first classification of losses, based on their location in the IM, includes:

- Winding losses – stator and rotor
- Core losses – stator and rotor
- Friction & windage losses – rotor

Electromagnetic losses include only winding and core losses.

A classification of electromagnetic losses by origin would include

- Fundamental losses
    - Fundamental winding losses (in the stator and rotor)
    - Fundamental core losses (in the stator)
- Space harmonics losses
    - Space harmonics winding losses (in the rotor)
    - Space harmonic core losses (stator and rotor)
- Time harmonic losses
    - Time harmonics winding losses (stator and rotor)
    - Time harmonic core losses (stator and rotor)

Time harmonics are to be considered only when the IM is static converter fed, and thus the voltage time harmonics content depends on the type of the converter and the pulse width modulation used with it. The space harmonics in the stator (rotor) mmf and in the airgap field are related directly to mmf space harmonics, airgap permeance harmonics due to slot openings, leakage, or main path saturation.

All these harmonics produce additional (stray) core and winding losses called

- Surface core losses (mainly on the rotor)
- Tooth flux pulsation core losses (in the stator and rotor teeth)
- Tooth flux pulsation cage current losses (in the rotor cage)

Load, coil chording, and the rotor bar-tooth contact electric resistance (interbar currents) influence all these stray losses.

No-load tests include all the above components, but at zero fundamental rotor current. These components will be calculated first; then corrections will be applied to compute some components on load.

11.2. FUNDAMENTAL ELECTROMAGNETIC LOSSES

Fundamental electromagnetic losses refer to core loss due to space fundamental airgap flux density– essentially stator based–and time fundamental conductor losses in the stator and in the rotor cage or winding.

The fundamental core losses contain the hysteresis and eddy current losses in the stator teeth and core,

$$P_{Fe1} \approx C_h f_1 \left[\left(\frac{B_{1ts}}{1} \right)^n G_{teeth} + \left(\frac{B_{1cs}}{1} \right)^n G_{core} \right] + \\ + C_e f_1^2 \left[\left(\frac{B_{1ts}}{1} \right)^2 G_{teeth} + \left(\frac{B_{1cs}}{1} \right)^2 G_{core} \right] \quad (11.2)$$

where C_h [W/kg], C_e [W/kg] n = (1.7 – 2.0) are material coefficients for hysteresis and eddy currents dependent on the lamination hysteresis cycle shape, the electrical resistivity, and lamination thickness; G_{teeth} and G_{core}, the teeth and back core weights, and B_{1ts}, B_{1cs} the teeth and core fundamental flux density values.

At any instant in time, the flux density is different in different locations and, in some regions around tooth bottom, the flux density changes direction, that is, it becomes rotating.

Hysteresis losses are known to be different with alternative and rotating, respectively, fields. In rotating fields, hysteresis losses peak at around 1.4 to 1.6 T, while they increase steadily for alternative fields.

Moreover, the mechanical machining of stator bore (when stamping is used to produce slots) is known to increase core losses by, sometimes, 40 to 60%.

The above remarks show that the calculation of fundamental core losses is not a trivial task. Even when FEM is used to obtain the flux distribution, formulas like (11.2) are used to calculate core losses in each element, so some of the errors listed above still hold. The winding (conductor) fundamental losses are

$$P_{co} = 3R_s I_{1s}^2 + 3R_r I_{1r}^2 \quad (11.3)$$

The stator and rotor resistances R_s and R_r' are dependent on skin effect. In this sense $R_s(f_1)$ and $R_r'(Sf_1)$ depend on f_1 and S. The depth of field penetration in copper $\delta_{Co}(f_1)$ is

$$\delta_{Co}(f_1) = \sqrt{\frac{2}{\mu_0 2\pi f_1 \sigma_{Co}}} = \sqrt{\frac{2}{1.256 \cdot 10^{-6} 2\pi 60 \left(\frac{f_1}{60} \right)}} = 0.94 \cdot 10^{-2} \sqrt{\frac{60}{f_1}} \text{m} \quad (11.4)$$

If either the elementary conductor height d_{Co} is large or the fundamental frequency f_1 is large, whenever

$$\delta_{Co} > \frac{d_{Co}}{2} \quad (11.5)$$

the skin effect is to be considered. As the stator has many layers of conductors in slot even for $\delta_{Co} \approx d_{Co}/2$, there may be some skin effect (resistance increase). This phenomenon was treated in detail in Chapter 9.

In a similar way, the situation stands for the wound rotor at large values of slip S. The rotor cage is a particular case for one conductor per slot. Again,

Chapter 9 treated this phenomenon in detail. For rated slip, however, skin effect in the rotor cage is, in general, negligible.

In skewed cage rotors with uninsulated bars (cast aluminum bars), there are interbar currents (Figure 11.1a).

The treatment of various space harmonics losses at no-load follows.

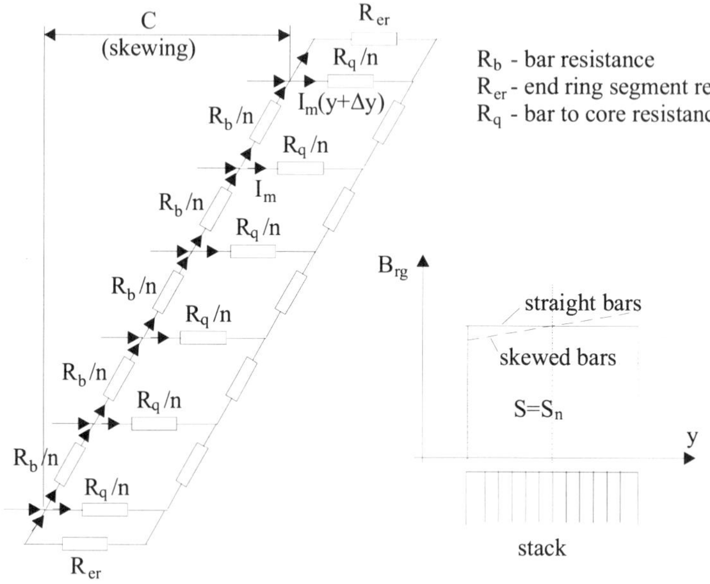

Figure 11.1 Interbar currents ($I_m(Y)$) in a skewed cage rotor a.) and the resultant airgap flux density along stack length b.)

Depending on the relative value of the transverse (contact) resistance R_q and skewing c, the influence of interbar currents on fundamental rotor conductor losses will be notable or small.

The interbar currents influence also depends on the fundamental frequency as for $f_1 = 500$ Hz, for example, the skin effect notably increases the rotor cage resistance even at rated slip.

On the other hand, skewing leaves an uncompensated rotor mmf (under load) which modifies the airgap flux density distribution along stack length (Figure 11.1b).

As the flux density squared enters the core loss formula, it is obvious that the total fundamental core loss will change due to skewing.

As skewing and interbar currents also influence the space harmonics losses, we will treat their influence once for all harmonics. The fundamental will then become a particular case.

Fundamental core losses seem impossible to segregate in a special test which would hold the right flux distribution and frequency. However a standstill test at rated rotor frequency $f_2 = Sf_1$ would, in fact, yield the actual value of $R_r'(S_nf_1)$. The same test at various frequencies would produce precise results on conductor fundamental losses in the rotor. The stator fundamental conductor losses might be segregated from a standstill a.c. single-phase test with all phases in series at rated frequency as, in this case, the core fundamental and additional losses may be neglected by comparison.

11.3. NO-LOAD SPACE HARMONICS (STRAY NO-LOAD) LOSSES IN NONSKEWED IMs

Let us remember that airgap field space harmonics produce on no-load in nonskewed IMs the following types of losses:

- Surface core losses (rotor and stator)
- Tooth flux pulsation core losses (rotor and stator)
- Tooth flux pulsation cage losses (rotor)

The interbar currents produced by the space harmonics are negligible in nonskewed machines if the rotor end ring (bar) resistance is very small (R_{cr}/R_b < 0.15).

11.3.1. No-load surface core losses

As already documented in Chapter 10, dedicated to airgap field harmonics, the stator mmf space harmonics (due to the very placement of coils in slots) as well the slot openings produce airgap flux density harmonics. Further on, main flux path heavy saturation may create third flux harmonics in the airgap.

It has been shown in Chapter 10 that the mmf harmonics and the first slot opening harmonics with a number of pole pairs $\nu_s = N_s \pm p_1$ are attenuating and augmenting each other, respectively, in the airgap flux density.

For these mmf harmonics, the winding factor is equal to that of the fundamental. This is why they are the most important, especially in windings with chorded coils where the 5th, 7th, 11th, and 17th stator mmf harmonics are reduced considerably.

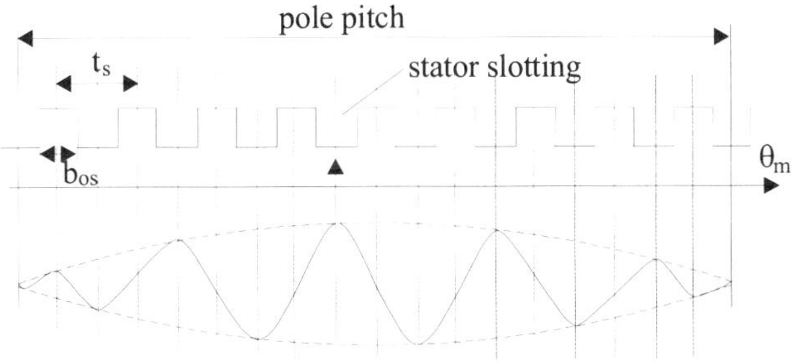

Figure 11.2 First slot opening (airgap permeance) airgap flux density harmonics

Let us now consider the fundamental stator mmf airgap field as modulated by stator slotting openings (Figure 11.2).

$$B_{N_s \pm p_1}(\theta_m, t) = \frac{\mu_0 F_{1m} a_1}{a_0} \cos(\omega_1 t - p_1\theta_m) \cos N_s \theta_m \qquad (11.6)$$

This represents two waves, one with $N_s + p_1$ pole pairs and one with $N_s - p_1$ pole pairs, that is, exactly, the first airgap magnetic conductance harmonic.

Let us consider that the rotor slot openings are small or that closed rotor slots are used. In this case, the rotor surface is flat.

The rotor laminations have a certain electrical conductivity, but they are axially insulated from each other by an adequate coating.

For frequencies of 1,200 Hz, characteristic to $N_s \pm p_1$ harmonics, the depth of field penetration in silicon steal for $\mu_{Fe} = 200\ \mu_0$ is (from 11.4) $\delta \approx 0.4$ mm, for 0.6 mm thick laminations. This means that the skin effect is not significant.

Therefore we may neglect the rotor lamination-induced current reaction to the stator field. That is, the airgap field harmonics (11.6) penetrate the rotor without being disturbed by induced rotor surface eddy currents. The eddy currents along the axial direction are neglected.

But now let us consider a general harmonic of stator mmf produced field B_{vg}.

$$B_{vg} = B_v \cos\left(\frac{v\pi x_m}{p_1 \tau} - S_v \omega t\right) = B_v \cos\left(\frac{v x_m}{R} - S_v \omega t\right) \qquad (11.7)$$

R–rotor radius; τ–pole pitch of the fundamental.

The slip for the vth mmf harmonic, S_v, is

$$(S_v)_{S=0} = \left(1 - \frac{v}{p_1}(1 - S)\right)_{S=0} = 1 - \frac{v}{p_1} \qquad (11.8)$$

For constant iron permeability μ, the field equations in the rotor iron are

$$\text{rot}B_v = 0,\ \text{div}B_v = 0,\ \ B_z = 0 \qquad (11.9)$$

This leads to

$$\frac{\partial^2 B_x}{\partial x^2} + \frac{\partial^2 B_x}{\partial y^2} = 0 \qquad \frac{\partial^2 B_y}{\partial x^2} + \frac{\partial^2 B_y}{\partial y^2} = 0 \qquad (11.10)$$

In iron these flux density components decrease along y (inside the rotor) direction

$$B_y = f_y(y)\cos\left(\frac{v x_m}{R} - S_v \omega t\right) \qquad (11.11)$$

According to $\frac{\partial B_x}{\partial x} + \frac{\partial B_y}{\partial y} = 0$.

$$B_x = \left(\frac{\frac{\partial f_y(y)}{\partial y}}{\frac{\nu}{R}} \right) \sin\left(\frac{\nu x_m}{R} - S_\nu \omega t \right) \tag{11.12}$$

From (11.10),

$$\frac{\partial^2 f_y(y)}{\partial y^2} - \left(\frac{\nu}{R} \right)^2 f_y(y) = 0 \tag{11.13}$$

or

$$f_y(y) = B_\nu e^{\frac{\nu y}{R}} \tag{11.14}$$

Equation (11.14) retains one term because y < 0 inside the rotor and $f_y(y)$ should reach zero for $y \approx -\infty$.

The resultant flux density amplitude in the rotor iron $B_{r\nu}$ is

$$B_{r\nu} = B_\nu e^{\frac{\nu y}{R}} \tag{11.15}$$

But the Faraday law yields

$$\mathrm{rot}\,\overline{E} = \mathrm{rot}\left(\frac{1}{\sigma_{Fe}} \overline{J} \right) = -\frac{\partial \overline{B}}{dt} \tag{11.16}$$

In our case, the flux density in iron has two components B_x and B_y, so

$$\begin{aligned} \frac{\partial J_y}{\partial z} &= -\sigma_{Fe} \frac{\partial B_x}{\partial t} \\ \frac{\partial J_x}{\partial z} &= -\sigma_{Fe} \frac{\partial B_y}{\partial t} \end{aligned} \tag{11.17}$$

The induced current components J_x and J_y are thus

$$\begin{aligned} J_x &= -\sigma_{Fe} S_\nu \omega B_\nu e^{\frac{\nu y}{R}} z \cos\left(\frac{\nu x_m}{R} - S_\nu \omega t \right) \\ J_y &= -\sigma_{Fe} S_\nu \omega B_\nu e^{\frac{\nu y}{R}} z \sin\left(\frac{\nu x_m}{R} - S_\nu \omega t \right) \end{aligned} \tag{11.18}$$

The resultant current density amplitude $J_{r\nu}$ is

$$J_{r\nu} = \sqrt{J_x^2 + J_y^2} = \sigma_{Fe} S_\nu \omega B_\nu e^{\frac{\nu y}{R}} z \tag{11.19}$$

The losses per one lamination (thickness d_{Fe}) is

$$P_{lamv} = \frac{1}{\sigma_{Fe}} \int_{-d_{Fe}/2}^{d_{Fe}/2} \int_{y=0}^{y=-\infty} \int_{x=0}^{2\pi R} J_{rv}^2 dxdydz \tag{11.20}$$

For the complete rotor of axial length l_{stack},

$$P_{0v} = \frac{\sigma_{Fe}}{24} B_v^2 (S_v \omega)^2 d_{Fe}^2 \frac{R}{v} 2\pi R l_{stack} \tag{11.21}$$

Now, if we consider the first and second airgap magnetic conductance harmonic (inversed airgap function) with $a_{1,2}$ (Chapter 10),

$$a_{1,2} = \frac{\beta}{g} F_{1,2}\left(\frac{b_{os}}{t_s}\right) \tag{11.22}$$

$\beta(b_{os,r}/g)$ and $F_{1,2}$ $(b_{os,r}/t_{s,r})$ are to be found from Table 10.1 and (10.14) in Chapter 10 and

$$B_v = B_{g1} \frac{a_{1,2}}{a_0}; \; a_0 = \frac{1}{K_c g} \tag{11.23}$$

where B_{g1} is the fundamental airgap flux density with

$$v = N_s \pm p_1 \quad \text{and} \quad S_v = 1 - \frac{(N_s \pm p_1)}{p_1} \approx \frac{N_s}{p_1} \tag{11.24}$$

The no-load rotor surface losses P_{ov} are thus

$$(P_{0v})_{N_s \pm P_1} \approx 2\frac{\sigma_{Fe}}{24} B_{g1}^2 \left(\frac{N_s}{p_1}\right)^2 (2\pi f_1)^2 d_{Fe}^2 \frac{Rp_1}{N_s} 2\pi R l_{stack} \left[\frac{a_1^2 + 2a_2^2}{a_0^2}\right] \tag{11.25}$$

Example 11.1.

Let us consider an induction machine with open stator slots and 2R = 0.38 m, N_s = 48 slots, t_s = 25 mm, b_{os} = 14 mm, g = 1.2 mm, d_{Fe} = 0.5 mm, B_{g1} = 0.69 T, $2p_1$ = 4, n_0 = 1500 rpm, N_r = 72, t_r = 16.6 mm, b_{or} = 6 mm, $\sigma_{Fe} = 10^8/45\ (\Omega m)^{-1}$. To determine the rotor surface losses per unit area, we first have to determine a_1, a_2, a_0 from (11.22). For b_{os}/g = 14/1.2 = 11.66 from table 10.1 β = 0.4. Also, from Equation (10.14), $F_1(14/25) = 1.02$, $F_2(0.56) = 0.10$.

Also, $a_0 = \frac{1}{K_{c1,2} g}$; K_c (from Equations (5.3 – 5.5)) is $K_{c1,2}$ = 1.85.

Now the rotor surface losses can be calculated from (11.25).

$$\frac{P_{0v}}{2\pi R l_{stack}} = \frac{2 \cdot 2.22 \cdot 10^6}{24} \cdot 0.69^2 \cdot \left(\frac{48}{2}\right)(2\pi 60)^2 \cdot 0.5^2 \cdot 10^{-6} \cdot 0.19 \cdot$$
$$\cdot \left[\frac{(0.4 \cdot 1.02)^2 + 2(0.4 \cdot 0.1)^2}{1.85^2}\right] = 8245.68 W/m^2$$

As expected, with semiclosed slots, a_1 and a_2 become much smaller; also, the Carter coefficient decreases. Consequently, the rotor surface losses will be much smaller. Also, increasing the airgap has the same effect. However, at the price of larger no-load current and lower power factor of the machine. The stator surface losses produced by the rotor slotting may be calculated in a similar way by replacing $F_1(b_{os}/t_s)$, $F_2(b_{os}/t_s)$, $\beta(b_{os}/g)$ with $\beta(b_{or}/g)$, $F_1(b_{or}/t_s)$, $F_2(b_{or}/t_s)$, and N_s/p_1 with N_r/p_1.

As the rotor slots are semiclosed, $b_{or} << b_{os}$, the stator surface losses are notably smaller than those of the rotor, They are, in general, neglected.

11.3.2. No-load tooth flux pulsation losses

As already documented in the previous paragraph, the stator (and rotor) slot openings produce variation in the airgap flux density distribution (Figure 11.3).

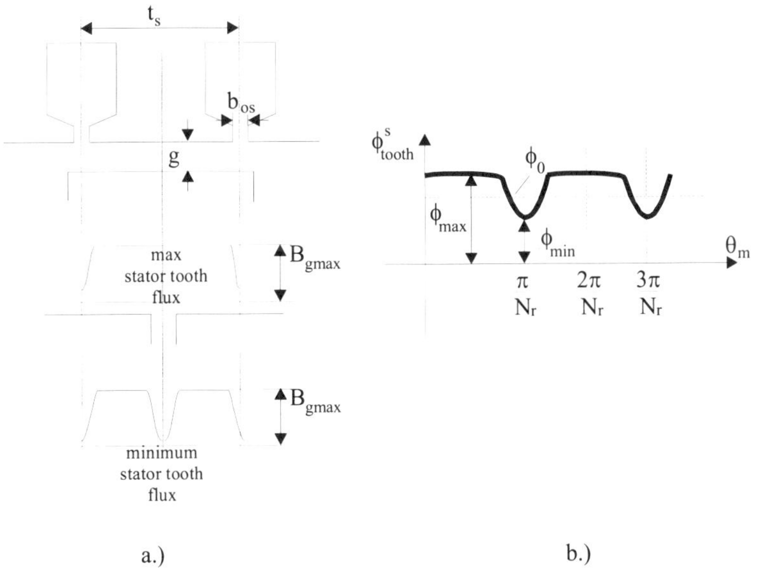

Figure 11.3 Airgap flux density as influenced by rotor and stator slotting a.) and stator tooth flux versus rotor position b.)

In essence, the total flux in a stator and rotor tooth varies with rotor position due to stator and rotor slot openings only in the case where the number of stator and rotor slots are different from each other. This is, however, the case, as $N_s \neq N_r$ at least to avoid large synchronous parasitic torques at zero speed (as demonstrated in Chapter 10).

The stator tooth flux pulsates due to rotor slot openings with the frequency $f_{PS} = N_r f \frac{(1-S)}{p_1} = N_r \frac{f}{p_1}$, for S = 0 (no-load). The flux variation coefficient K_ϕ is

$$K_\phi = \frac{\phi_{max} - \phi_{min}}{2\phi_0} \tag{11.26}$$

The coefficient K_ϕ, as derived when the Carter coefficient was calculated in [6]

$$K_\phi = \frac{\gamma_2 g}{2t_s} \tag{11.27}$$

with γ_2 as

$$\gamma_2 = \frac{\left(\frac{b_{or}}{g}\right)^2}{5 + \frac{b_{or}}{g}} \tag{11.28}$$

Now, by denoting the average flux density in a stator tooth with B_{ots}, the flux density pulsation B_{p1} in the stator tooth is

$$B_{Ps} = B_{ots}\frac{\gamma_2 g}{2t_s} \tag{11.29}$$

$$B_{ots} = B_{g0}\frac{t_s}{t_s - b_{os}} \tag{11.30}$$

B_{g0}–airgap flux density fundamental.

We are now in the classical case of an iron region with an a.c. magnetic flux density B_{Ps}, at frequency $f_{Ps} = N_r f/p_1$. As $N_r f/p_1$ is a rather high frequency, eddy current losses prevail, so

$$P_{Ps0} = C_{ep}\left(\frac{B_{Ps}}{1}\right)^2\left(\frac{f_{Ps}}{50}\right)^2 G_{steeth};\ G_{steeth} - \text{stator teeth weight} \tag{11.29}$$

In (11.31), C_{ep} represents the core losses at 1T and 50Hz. It could have been for 1T and 60 Hz as well.

Intuitively, the magnetic saturation of main flux path places the pulsation flux on a local hysteresis loop with lower (differential) permeability, so saturation is expected to reduce the flux pulsation in the teeth. However, Equation (11.31) proves satisfactory even in the presence of saturation. Similar tooth flux pulsation core losses occur in the rotor due to stator slotting.

Similar formulas as above are valid.

$$P_{Pr0} = C_{ep}\left(\frac{B_{Pr}}{1}\right)^2\left(\frac{f_{Pr}}{50}\right)^2 G_{teeth};\ f_{pr} = N_s\frac{f}{p_1}$$

$$B_{Pr} = B_{otr}\frac{\gamma_1 g}{2t_r}; \quad B_{otr} = B_{g0}\frac{t_r}{t_r - b_{or}} \tag{11.32}$$

$$\gamma_1 = \frac{\left(\frac{b_{os}}{g}\right)^2}{5 + \frac{b_{os}}{g}}$$

As expected, C_{ep} – power losses per Kg at 1T and 50 Hz slightly changes when the frequency increases, as it does at $N_s\frac{f}{p_1}$ or $N_r\frac{f}{p_1}$. $C_{ep} = C_e K$, with K = 1.1 – 1.2 as an empirical coefficient.

Example 11.2

For the motor in Example 10.1, let us calculate the tooth flux pulsation losses per Kg of teeth in the stator and rotor.

Solution

In essence, we have to determine the flux density pulsations B_{Ps}, B_{Pr}, then f_{Ps}, f_{Pr} and apply Equation (11.31).

From (11.26) and (11.32),

$$\gamma_2 = \frac{\left(\frac{b_{or}}{g}\right)^2}{5 + \frac{b_{or}}{g}} = \frac{\left(\frac{6}{1.2}\right)^2}{5 + \frac{6}{1.2}} = 2.5$$

$$\gamma_1 = \frac{\left(\frac{b_{os}}{g}\right)^2}{5 + \frac{b_{os}}{g}} = \frac{\left(\frac{14}{1.2}\right)^2}{5 + \frac{14}{1.2}} = 8.166$$

$$B_{Ps} = B_{g0}\frac{t_s}{t_s - b_{or}}\frac{\gamma_2 g}{2t_s} = \frac{0.69 \cdot 2.5 \cdot 1.2}{2(25 - 14)} = 0.094\text{T}$$

$$B_{Pr} = B_{g0}\frac{t_r}{t_r - b_{or}}\frac{\gamma_1 g}{2t_r} = \frac{0.69 \cdot 8.166 \cdot 1.2}{2(16.6 - 6)} = 0.3189\text{T}$$

With C_{ep} = 3.6 W/Kg at 1T and 60 Hz,

$$P_{Ps} = C_{ep}\left(\frac{B_{Ps}}{1}\right)^2\left(\frac{f_{Pr}}{50}\right)^2 = 3.6\left(\frac{0.094}{1}\right)^2\left(\frac{72 \cdot 60}{2 \cdot 50}\right)^2 = 59.3643\text{W}/\text{Kg}$$

$$P_{Pr} = C_{ep}\left(\frac{B_{Pr}}{1}\right)^2\left(\frac{f_{Ps}}{50}\right)^2 = 3.6\left(\frac{0.3189}{1}\right)^2\left(\frac{48\cdot 60}{2\cdot 50}\right)^2 = 303.66W/Kg$$

The open slots in the stator produce large rotor tooth flux pulsation no-load specific losses (P_{Pr}).

The values just obtained, even for straight rotor (and stator) slots, are too large.

Intuitively we feel that at least the rotor tooth flux pulsations will be notably reduced by the currents induced in the cage by them. At the expense of no-load circulating cage-current losses, the rotor flux pulsations are reduced.

Not so for the stator unless parallel windings are used where circulating currents would play the same role as circulating cage currents.

When such a correction is done, the value of B_{Pr} is reduced by a subunitary coefficient, [4]

$$B'_{Pr} = B_{Pr}\cdot K_{tk}\cdot \sin\left(\frac{Kt_r p_1}{2R}\right) = K_{dk}B_{Pr};\ K = \frac{N_s}{p_1} \tag{11.33}$$

$$K_{tk} = 1 - \frac{L_{mk} + L_{2dk}}{L_{mk} + L_{2dk} + L_{slot} + L_{mk}\left(\frac{1}{K^2_{skewk}} - 1\right)} \tag{11.34}$$

where L_{mk} is the magnetizing inductance for harmonic K, L_{2dk} the differential leakage inductance of K_{th} harmonic, K_{skewk} the skewing factor for harmonic K, and L_{slot} – rotor slot leakage inductance.

$$L_{mk} \approx L_m \frac{1}{K^2} \tag{11.35}$$

$$L_{2dk} \approx K_{dlk}\cdot \tau_{2dk}\cdot L_{mk} \tag{11.36}$$

$$K_{dlk} = \tanh\left(\frac{d_{Fe}}{2\delta_{Fe}}\right) \Big/ \frac{d_{Fe}}{2\delta_{Fe}};\ d_{Fe} - \text{lamination thickness;}$$

$$\delta_{Fe} = \sqrt{\frac{2}{\mu_{Fediff}S_k\omega\sigma_{Fe}}} \tag{11.37}$$

where μ_{Fediff} is the the iron differential permeability for given saturation level of fundamental.

$$S_k\omega = 2\pi f_k = 2\pi\frac{N_s}{p_1} \tag{11.38}$$

$$\tau_{2dk} = \left(\frac{\frac{\pi p_1 K}{R}}{\sin\left(\frac{\pi p_1 K}{R}\right)} \right)^2 - 1 \tag{11.39}$$

$$K_{skewk} = \frac{\sin\left(\frac{K p_1 c}{R}\right)}{\frac{K p_1 c}{R}} \tag{11.40}$$

Skewing is reducing the cage circulating current reaction which increases the value of K_{tk} and, consequently, the teeth flux pulsation core losses stay undamped.

Typical values for K_{dk}–the total damping factor due to cage circulating currents–would be in the range K_{dk} = 1 to 0.05, depending on rotor number slots, skewing, etc.

A skewing of one stator slot pitch is said to reduce the circulating cage reaction to almost zero ($K_{dk} \approx 1$), so the rotor tooth flux pulsation losses stay high. For straight rotor slots with $K_{dk} \approx 0.1 - 0.2$, the rotor tooth flux pulsation core losses are reduced to small relative values, but still have to be checked.

11.3.3. No-load tooth flux pulsation cage losses

Now that the rotor tooth flux pulsation, as attenuated by the corresponding induced bar currents, is known, we may consider the rotor bar mesh in Figure 11.4.

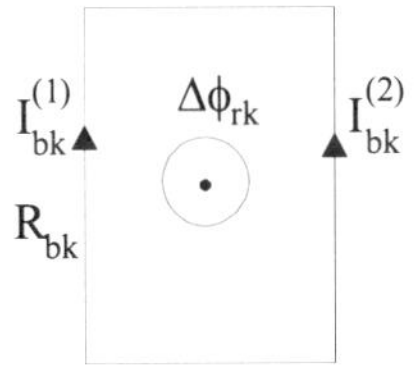

Figure 11.4 Rotor mesh with two bars

$$\Delta\phi_{rk} = K_{dk} B_{prk} (t_r - b_{or}) l_{stack} \tag{11.41}$$

In our case, $K = \frac{N_s}{p_1}$. For this harmonic, the rotor bar resistance is increased due to skin effect by K_{Rk} ($f_k = \frac{N_s}{p_1} f_1$).

The skin effect in the end ring is much smaller, so the ring resistance may be neglected. If we also neglect the leakage reactance in comparison with rotor bar resistance, the equation of the mesh circuit in Figure 11.4 becomes

$$2\pi fK\Delta\phi_{rk} = R_b K_{Rk}\left(I_{bk}^{(1)} - I_{bk}^{(2)}\right) \tag{11.42}$$

The currents I_{bk} in the neighbouring bars (1) and (2) are phase shifted by an angle $2\pi\frac{N_s}{N_r}$, so

$$I_{bk}^{(1)} - I_{bk}^{(2)} = 2I_{bk}\sin\pi\frac{N_s}{N_r} \tag{11.43}$$

From (11.41) and (11.42),

$$I_{bk} = \frac{\Delta\phi_{rk}2\pi fK}{2\left(\sin\frac{\pi N_s}{N_r}\right)R_b K_{Rk}} \tag{11.44}$$

Finally, for all bars, the cage losses for harmonic K are

$$P_{ocagek} = N_r R_b K_{Rk}\left(\frac{I_{bk}}{\sqrt{2}}\right)^2 \tag{11.45}$$

or

$$P_{ocagek} \approx \frac{N_r}{8}\frac{(\Delta\phi_{rk})^2(2\pi f)^2\left(\frac{N_s}{p_1}\right)^2}{\sin^2\left(\pi\frac{N_s}{N_r}\right)R_b K_{Rk}} \tag{11.46}$$

Expression (11.46) is valid for straight rotor slots. In a skewed rotor, the rotor tooth flux pulsation is reduced (per stack length) and, thus, the corresponding no-load cage losses are also reduced. They should tend to zero for one stator slot pitch skewing.

Example 11.3. Consider the motor in Example 11.1, 11.2 with the stack length $l_{stack} = 0.4$ m, rotor bar class section $A_{bar} = 250$ mm^2, ($N_s = 48$, $N_r = 72$, $2p_1 = 4$). The rotor bar skin effect coefficient for $f_k = \frac{N_s}{p_1}f = \frac{48}{2}\cdot 60 = 1440\text{Hz}$ is $K_{Rk} =$ 15. Let us calculate the cage losses due to rotor tooth flux pulsations if the attenuating factor of $K_{dk} = 0.05$.

Solution

With $K_{dk} = 0.05$ and $B_{pr} = 0.3189$, from Example 11.2 we may calculate $\Delta\phi_k$ (from 11.41).

$$\Delta\phi_{rk} = K_{dk} B_{prk} (t_r - b_{or}) l_{stack} = 0.05 \cdot 0.3189 \cdot (16.6 - 6) \cdot 10^{-3} \cdot 0.4 = \\ = 0.067 \cdot 10^{-3}\,\text{Wb}$$

The d.c. bar resistance R_b is

$$R_b \approx \frac{l_{stack}}{A_{bar}} \frac{1}{\sigma_{Al}} = \frac{0.4}{250 \cdot 10^{-6}} \frac{1}{3.0 \cdot 10^{7}} = 0.533 \cdot 10^{-4}\,\Omega$$

So the bar current $(I_{bk})_{K=\frac{N_r}{p_1}}$ is (11.43)

$$I_{bk} = \frac{0.067 \cdot 10^{-3} \cdot 2\pi 60 \cdot \dfrac{48}{2}}{2 \sin \pi \dfrac{48}{72} \cdot 0.533 \cdot 10^{-4} \cdot 15} = 443\text{A}$$

This is indeed a large value, but let us remember that the rotor bars are many ($N_r >> N_s$) and the stator slots are open, with a small airgap.

Now the total rotor flux pulsation losses P_{ocagek} are (11.45)

$$P_{0cagek} = 72 \cdot \left(\frac{443}{\sqrt{2}}\right)^2 \cdot 0.533 \cdot 10^{-4} \cdot 15 = 5682.2\text{W}$$

Let us compare this with the potential power of the IM under consideration ($f_x = 3 \cdot 10^4 \text{N/cm}^2$ – specific force),

$$P_n \approx f_x \cdot 2\pi R \cdot L \cdot R \cdot \frac{2\pi f}{p_1} = 3 \cdot 10^4 \cdot 2\pi \cdot 0.19^2 \cdot 0.4 \cdot \frac{2\pi 50}{2} \approx 427\text{kW}$$

In such a case, the no-load cage losses would represent more than 1% of all power.

11.4. LOAD SPACE HARMONICS (STRAY LOAD) LOSSES IN NONSKEWED IMs

Again, slot opening and mmf space harmonics act together to produce space harmonics load (stray load) losses in the presence of larger stator and rotor currents.

In general, the no-load stray losses are augmented by load, component by component. The mmf space harmonics of the stator, for integral-slot windings, is

$$\nu = (6c \pm 1) p_1 \tag{11.47}$$

The slot opening (airgap magnetic conductance) harmonics are

$$\nu_s = c_1 N_s \pm p_1 \tag{11.48}$$

As expected, they overlap. The first slot harmonics ($c_1 = 1$) $\nu_{smin} = N_s \pm p_1$ are most important as their winding factor is the same as for the fundamental.

The mmf harmonic F_ν amplitude is

$$F_\nu = F_1 \frac{K_{w\nu}}{K_{w1}} \frac{p_1}{\nu} \tag{11.49}$$

where K_{w1} and $K_{w\nu}$ are the winding factors of the fundamental and of harmonic ν, respectively.

If the number of slots per pole and phase (q) is not very large, the first slot (opening) harmonics $N_s \pm p_1$ produce the largest field in the airgap.

Only for full – pitch windings may the phase belt mmf harmonics (ν = 5, 7, 11, 13) produce airgap fields worthy of consideration because their winding factors are not small enough (as they are in chorded coil windings). However, even in such cases, the losses produced by the phase belt mmf harmonics may be neglected by comparison with the first slot opening harmonics $N_s \pm p_1$. [6]

Now, for these first slot harmonics, it has been shown in Chapter 10 that their mmf companion harmonics of same order produce an increase for $N_s - p_1$ and a decrease for $N_s + p_1$, [6]

$$\xi_{(N_s - p_1)0} = \left(1 + \frac{N_s - p_1}{p_1} \frac{a_1}{2a_0}\right) \frac{1}{K_c} > 1 \tag{11.50}$$

$$\xi_{(N_s + p_1)0} = \left(1 - \frac{N_s + p_1}{p_1} \frac{a_1}{2a_0}\right) \frac{1}{K_c} < 1$$

where a_1, a_0, K_s have been defined earlier in this chapter for convenience.

For load conditions, the amplification factors $\xi_{(N_s - p_1)0}$ and $\xi_{(N_s + p_1)0}$ are to be replaced by [6]

$$\begin{aligned} \xi_{(N_s - p_1)n} &= \frac{1}{K_c} \sqrt{1 + 2(\sin\varphi_n) \frac{I_0}{I_n} \frac{N_s - p_1}{p_1} \frac{a_1}{2a_0} + \left(\frac{I_0}{I_n} \frac{N_s - p_1}{p_1} \frac{a_1}{2a_0}\right)^2} \\ \xi_{(N_s + p_1)n} &= \frac{1}{K_c} \sqrt{1 - 2(\sin\varphi_n) \frac{I_0}{I_n} \frac{N_s + p_1}{p_1} \frac{a_1}{2a_0} + \left(\frac{I_0}{I_n} \frac{N_s + p_1}{p_1} \frac{a_1}{2a_0}\right)^2} \end{aligned} \tag{11.51}$$

I_0 is the no-load current, I_n the current under load, and φ_n the power factor angle on load.

From now on using the above amplification factors, we will calculate the correction coefficients for load to multiply the various no-load stray losses and thus find the load stray losses.

Based on the fact that losses are proportional to harmonic flux densities squared, for $N_s \pm P_1$ slot harmonics we obtain

- <u>Rotor surface losses on load</u>

  The ratio between load P_{0n} and no-load P_{00} surface losses C_{loads} is

$$P_{0n} = P_{00} \cdot C_{loads}; \quad C_{loads} > 1 \tag{11.52}$$

$$C_{loads} = \left(\frac{I_n}{I_0}\right)^2 \frac{\left[\left(\frac{p_1}{N_s - p_1}\right)^2 \xi^2{}_{(N_s - p_1)n} + \left(\frac{p_1}{N_s + p_1}\right)^2 \xi^2{}_{(N_s + p_1)n}\right]}{\left[\left(\frac{p_1}{N_s - p_1}\right)^2 \xi^2{}_{(N_s - p_1)0} + \left(\frac{p_1}{N_s + p_1}\right)^2 \xi^2{}_{(N_s + p_1)0}\right]} \tag{11.53}$$

- Tooth flux pulsation load core losses

For both stator and rotor, the no-load surface P_{sp0}, P_{rp0} are to be augmented by amplification coefficients similar to C_{load},

$$P_{spn} = P_{sp0} \cdot C_{loadr}$$
$$P_{rpn} = P_{rp0} \cdot C_{loads} \tag{11.54}$$

where C_{loadr} is calculated with N_r instead of N_s.

- Tooth flux pulsation cage losses on load

For chorded coil windings the same reasoning as above is used.

$$P_{ncage} = P_{0cage} \cdot C_{loadr} \tag{11.55}$$

For full pitch windings the 5th and 7th (phase belt) mmf harmonics losses are to be added.

The cage equivalent current $I_{r\nu}$ produced by a harmonic ν is

$$I_{r\nu} \sim \frac{F_\nu \nu}{1 + \tau_\nu} = F_1 \frac{K_{w\nu}}{K_{w1}} \frac{p_1}{1 + \tau_\nu} \tag{11.56}$$

τ_ν is the leakage (approximately differential leakage) coefficient for harmonic ν ($\tau_\nu \approx \tau_{d\nu}$).

The cage losses,

$$P_{r\nu} \sim I_{r\nu}{}^2 R_b K_{R\nu} \tag{11.57}$$

$K_{R\nu}$ is, again, the skin effect coefficient for frequency f_ν.

$$f_\nu = S_\nu f_1; \quad S_\nu = 1 - \frac{\nu}{p_1}(1 - S) \tag{11.58}$$

The differential leakage coefficient for the rotor (Chapter 6),

$$\tau_{d\nu} = \left(\frac{\pi\nu}{N_r}\right)^2 \frac{1}{\sin^2\left(\frac{\pi\nu}{N_r}\right)^2} - 1 \tag{11.59}$$

Equation (11.57) may be applied both for 5th and 7th ($5p_1$, $7p_1$), phase belt, mmf harmonics and for the first slot opening harmonics $N_s \pm p_1$.

Adding these four terms for load conditions [6] yields

$$P_{ncage} \approx P_{0cage} \cdot C_{loadr}\left[1 + \frac{2\left(\frac{K_{w5}}{K_{w1}}\right)^2 \frac{\left(1+\tau_{dN_s}\right)^2}{\left(1+\tau_{d6}\right)^2}\sqrt{\frac{6p_1}{N_s}}}{\xi^2_{(N_s-P_1)n} + \xi^2_{(N_s+P_1)n}}\right] \tag{11.60}$$

$$\left(1+\tau_{d6}\right) = \frac{\left(\frac{6\pi p_1}{N_r}\right)^2}{\sin^2\left(\frac{6\pi p_1}{N_r}\right)}; \quad \left(1+\tau_{dN_s}\right) = \frac{\left(\frac{\pi N_s}{N_r}\right)^2}{\sin^2\left(\frac{\pi N_s}{N_r}\right)} \tag{11.61}$$

Magnetic saturation may reduce the second term in (11.60) by as much as 60 to 80%. We may use (11.61) even for chorded coil windings, but, as K_{w5} is almost zero, the factor in parenthesis is almost reduced to unity.

Example 11.4. Let us calculate the stray load amplification with respect to no-load for a motor with nonskewed insulated bars and full pitch stator winding; open stator slots, $a_0 = 0.67/g$, $a_1 = 0.43/g$, $I_0/I_n = 0.4$, $\cos\varphi_n = 0.86$, $N_s = 48$, $N_r = 40$, $2p_1 = 4$, $\beta = 0.41$, $K_c = 1.8$.

Solution

We have to first calculate from (11.50) the amplification factors for the airgap field $N_s \pm P_1$ harmonics by the same order mmf harmonics.

$$\xi_{(N_s-p_1)0} = \frac{1}{K_c}\left(1 + \frac{N_s - p_1}{p_1}\frac{a_1}{2a_0}\right) = \frac{1}{K_c}\left(1 + \frac{(48-2)}{2}\cdot\frac{0.43}{2\cdot 0.67}\right) = \frac{8.4}{K_c}$$

$$\xi_{(N_s+p_1)0} = \frac{1}{K_c}\left(1 - \frac{N_s + p_1}{p_1}\frac{a_1}{2a_0}\right) = \frac{1}{K_c}\left(1 - \frac{(48-2)}{2}\cdot\frac{0.43}{2\cdot 0.67}\right) = \frac{-7}{K_c}$$

Now, the same factors under load are found from (11.51).

$$\xi_{(N_s-p_1)n} = \frac{1}{K_c}\sqrt{1 + 2\cdot 0.53\cdot 0.4\cdot\frac{48-2}{2}\cdot\frac{0.43}{2\cdot 0.67} + \left(0.4\cdot\frac{48-2}{2}\cdot\frac{0.43}{2\cdot 0.67}\right)^2} = \frac{3.6}{K_c}$$

$$\xi_{(N_s+p_1)n} = \frac{1}{K_c}\sqrt{1 - 2\cdot 0.53\cdot 0.4\cdot\frac{48+2}{2}\cdot\frac{0.43}{2\cdot 0.67} + \left(0.4\cdot\frac{48+2}{2}\cdot\frac{0.43}{2\cdot 0.67}\right)^2} = \frac{2.8}{K_c}$$

The stray loss load amplification factor C_{loads} is (11.53)

$$C_{loads} \approx \left(\frac{I_n}{I_0}\right)^2 \left[\frac{\xi^2{}_{(N_s-p_1)n} + \xi^2{}_{(N_s+p_1)n}}{\xi^2{}_{(N_s-p_1)0} + \xi^2{}_{(N_s+p_1)0}}\right] = \left(\frac{1}{0.4}\right)^2 \left[\frac{3.60^2 + 2.80^2}{8.4^2 + 7^2}\right] = 1.0873!$$

So the rotor surface and the rotor tooth flux pulsation (stray) load losses are increased at full load only by 8.73% with respect to no-load.

Let us now explore what happens to the no-load rotor tooth pulsation (stray) cage losses under load. This time we have to use (11.60),

$$\frac{P_{ncage}}{P_{0cage}} = C_{load} \cdot \left[1 + \frac{2K_s\left(\frac{K_{w5}}{K_{w1}}\right)^2 \frac{\left(1+\tau_{dN_s}\right)^2}{\left(1+\tau_{d6}\right)^2}\sqrt{\frac{6p_1}{N_s}}}{\xi^2{}_{(N_s-p_1)n} + \xi^2{}_{(N_s+p_1)n}}\right]$$

with K_s a saturation factor ($K_s = 0.2$), $K_{w5} = 0.2$, $K_{w1} = 0.965$. Also τ_{d48} and τ_{d6} from (11.61) are $\tau_{d48} = 42$, $\tau_{d6} = 0.37$. Finally,

$$\frac{P_{ncage}}{P_{0cage}} = 1.0873 \cdot \left[1 + \frac{2\cdot 0.2\left(\frac{0.2}{0.965}\right)^2 \frac{(1+42)^2}{(1+0.37)^2}\sqrt{\frac{6\cdot 2}{48}}}{3.60^2 + 2.80^2}\right] = 1.43!$$

As expected, the load stray losses in the insulated nonskewed cage are notably larger than for no-load conditions.

11.5. FLUX PULSATION (STRAY) LOSSES IN SKEWED INSULATED BARS

When the rotor slots are skewed and the rotor bars are insulated from rotor core, no interbar currents flow between neighboring bars through the iron core.

In this case, the cage no-load stray losses P_{0cage} due to first slot (opening) harmonics $N_s \pm p_1$ are corrected as [6]

$$P_{0cages} = \frac{P_{0cage}}{2}\left[\left(\frac{\sin\frac{\pi c}{t_s N_s}(N_s + p_1)}{\frac{\pi c}{t_s N_s}(N_s + p_1)}\right)^2 + \left(\frac{\sin\frac{\pi c}{t_s N_s}(N_s - p_1)}{\frac{\pi c}{t_s N_s}(N_s - p_1)}\right)^2\right] \quad (11.62)$$

where c/t_s is the skewing in stator slot pitch t_s units.

When skewing equals one stator slot pitch ($c/t_s = 1$), Equation (11.62) becomes

$$\left(P_{0cages}\right)_{c/t_s=1} \approx P_{0cage} \cdot \left(\frac{p_1}{N_s}\right)^2 \quad (11.63)$$

Consequently, in general, skewing reduces the stray losses in the rotor insulated (on no-load and on load) as a bonus from (11.60). For one stator pitch skewing, this reduction is spectacular.

Two things we have to remember here.

- When stray cage losses are almost zero, the rotor flux pulsation core losses are not attenuated and are likely to be large for skewed rotors with insulated bars.
- Insulated bars are made generally of brass or copper.

For the vast majority of small and medium power induction machines, cast aluminum uninsulated bar cages are used. Interbar current losses are expected and they tend to be augmented by skewing.

We will treat this problem separately.

11.6. INTERBAR CURRENT LOSSES IN NONINSULATED SKEWED ROTOR CAGES

For cage rotor IMs with skewed slots (bars) and noninsulated bars, transverse or cross-path or interbar additional currents occur through rotor iron core between adjacent bars.

Measurements suggest that the cross-path or transverse impedance Z_d is, in fact, a resistance R_d up to at least f = 1 kHz. Also, the contact resistance between the rotor bar and rotor teeth is much larger than the cross-path iron core resistance. This resistance tends to increase with the frequency of the harmonic considered and it depends on the manufacturing technology of the cast aluminum cage rotor. To have a reliable value of R_d, measurements are mandatory.

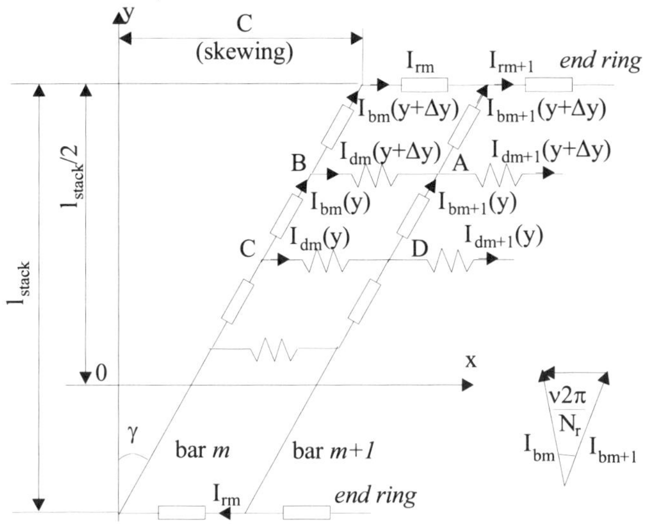

Figure 11.5 Bar and interbar currents in a rotor cage with skewed slots

To calculate the interbar currents (and losses), a few analytical procedures have been put forward [9, 6, 10]. While [9, 10] do not refer especially to the first slot opening harmonics, [6] ignores the end ring resistance.

In what follows, we present a generalization of [9, 11] to include the first slot opening harmonics and the end ring resistance.

Let us consider the rotor stack divided into many segments with the interbar currents lumped into definite resistances (Figure 11.5).

The skewing angle γ is

$$\tan\gamma = \frac{c}{l_{stack}} \tag{11.64}$$

What airgap field harmonics are likely to produce interbar currents? In principle, all space harmonics, including the fundamental, do so. Additionally, time harmonics have a similar effect. However, for chorded coil stator windings, the first slot harmonic $N_s - p_1$ is augmented by the same stator mmf harmonic; $N_s + p_1$ harmonic is attenuated in a similar way as shown in previous paragraphs.

Only for full-pitch stator windings, the losses produced by the first two 5th and 7th phase belt stator mmf harmonics are to be added. (For chorded coil windings, their winding factors, and consequently their amplitudes, are negligible).

At 50(60) Hz apparently the fundamental component losses in the cross-path resistance for skewed noninsulated bar rotor cages may be neglected. However, this is not so in high (speed) frequency IMs (above 300Hz fundamental frequency). Also, inverter fed IMs show current and flux time harmonics, so additional interbar current losses due to the space fundamental and time harmonics are to be considered. Later in this chapter, we will return to this subject.

For the time being, let us consider only one stator frequency f with a general space harmonic ν with its slip S_ν.

Thus, even the case of time harmonics can be dealt with one by one changing only f, but for $\nu = p_1$ (pole pairs), the fundamental.

The relationship between adjacent bar and cross-path resistance currents on Figure 11.5 are

$$\begin{aligned}
I_{bm}(y) - I_{bm+1}(y) &= 2je^{-\frac{j\pi\nu}{N_r}} I_{bm}(y)\sin\frac{\pi\nu}{N_r} \\
I_{dm}(y) - I_{dm+1}(y) &= 2je^{-\frac{j\pi\nu}{N_r}} I_{dm}(y)\sin\frac{\pi\nu}{N_r} \\
I_{rm} - I_{rm+1} &= 2je^{-\frac{j\pi\nu}{N_r}} I_{rm}\sin\frac{\pi\nu}{N_r}
\end{aligned} \tag{11.65}$$

Kirchoff's first law in node A yields

$$I_{dm}(y+\Delta y) - I_{dm+1}(y+\Delta y) \approx \frac{I_{bm}(y+\Delta y) - I_{bm+1}(y)}{\Delta y} \tag{11.66}$$

Note that the cross path currents I_{dm}, I_{dm+1} refer to unity stack length.

With (11.65) in (11.66), we obtain

$$I_{dm}(y) \approx -j\frac{e^{-\frac{j\pi\nu}{N_r}}}{2\sin\frac{\pi\nu}{N_r}} \cdot \frac{d}{dy} I_{bm}(y) \tag{11.67}$$

Also, for the ring current I_{rm},

$$I_{rm} - I_{rm+1} = \left(I_{bm+1}\right)_{y=\frac{l_{stack}}{2}} \tag{11.68}$$

with (11.65), (11.68) becomes

$$I_{rm} = j\frac{e^{-\frac{j\pi\nu}{N_r}}}{2\sin\frac{\pi\nu}{N_r}} \left(I_{bm}\right)_{y=\frac{l_{stack}}{2}} \tag{11.69}$$

The second Kirchoff's law along the closed path ABCD yields

$$R_d l_{stack}\left(I_{dm}(y+\Delta y) - I_{dm}(y)\right) + \left(R_{b\nu} + jS_\nu X_{0\nu}\left(1+\tau_{d\nu}\right)\right) \cdot$$
$$\cdot \left(I_{bm}(y) - I_{bm+1}(y)\right) \cdot \frac{\Delta y}{l_{stack}} = E_m^\nu \cdot e^{-\frac{j\gamma\nu y}{l_{stack}}} \cdot \frac{\Delta y}{l_{stack}} \tag{11.70}$$

E_m^ν represents the stator current produced induced voltage in the contour ABCD. $X_{0\nu}$ represents the airgap reactance for harmonic ν seen from the rotor side; $\tau_{d\nu}$ is the differential leakage coefficient for the rotor (11.59).

Making use of (11.67) into (11.70) yields

$$R_{de} l_{stack}^2 \frac{d^2 I_{bm}(y)}{dy^2} - Z_{be} I_{bm}(y) = E_m^\nu e^{-j\frac{\gamma\nu y}{l_{stack}}} \tag{11.71}$$

The boundary conditions at $y = \pm l_{stack}/2$ are

$$\left(I_{rm}\right)_{y=\pm l_{stack}/2} \cdot Z_{er} = \mp\left(I_{dm}\right)_{y=\pm l_{stack}/2} R_d l_{stack};\quad R_{de} = \frac{R_d}{4\sin^2\frac{\nu\pi}{N_r}} \tag{11.72}$$

Z_{er} represents the impedance of end ring segment between adjacent bars.

$$\underline{Z}_{er} = R_{erv} + jX_{erv}; \quad Z_{ere} = \frac{\underline{Z}_{er}}{4\sin^2 \frac{\nu\pi}{N_r}}; \quad \underline{Z}_{be} = R_{be} + jX_{be} \tag{11.73}$$

R_{erv}, X_{erv} are the resistance and reactance of the end ring segment. Z_{be} is bar equivalent impedance.

Solving the differential Equation (11.71) with (11.72) yields [10]

$$I_{bm}(y) = \frac{-E_{mv}}{\left(\underline{Z}_{be} + \nu^2\gamma^2 R_{de} l_{stack}\right)} \cdot \left[\left(\cos\frac{\gamma\nu y}{l_{stack}} - j\sin\frac{\gamma\nu y}{l_{stack}}\right) - A\cosh\frac{y}{l_{stack}}\sqrt{\frac{Z_{be}}{R_{de}}} - B\sinh\frac{y}{l_{stack}}\sqrt{\frac{Z_{be}}{R_{de}}}\right] \tag{11.74}$$

The formula of the cross-path current $I_{dm}(y)$ is obtained from (11.67) with (11.74).

Also,

$$A = \frac{Z_{ere}\cos\left(\frac{\nu\gamma}{2}\right) - \nu\gamma R_{de}\sin\left(\frac{\nu\gamma}{2}\right)}{Z_{ere}\cosh\left(\frac{1}{2}\sqrt{\frac{Z_{be}}{R_{de}}}\right) + R_{de}\sqrt{\frac{Z_{be}}{R_{de}}}\sinh\left(\frac{1}{2}\sqrt{\frac{Z_{be}}{R_{de}}}\right)} \tag{11.75}$$

$$B = -j\frac{Z_{ere}\sin\left(\frac{\nu\gamma}{2}\right) - \nu\gamma R_{de}\cos\left(\frac{\nu\gamma}{2}\right)}{Z_{ere}\sinh\left(\frac{1}{2}\sqrt{\frac{Z_{be}}{R_{de}}}\right) + R_{de}\sqrt{\frac{Z_{be}}{R_{de}}}\cosh\left(\frac{1}{2}\sqrt{\frac{Z_{be}}{R_{de}}}\right)} \tag{11.76}$$

When the skewing is zero ($\gamma = 0$), $B = 0$ but A is not zero and, thus, the bar current in (11.74) still varies along y (stack length) and therefore some cross-path currents still occur. However, as the end ring impedance Z_{ere} tends to be small with respect to R_{de}, the cross-path currents (and losses) are small.

However, only for zero skewing and zero end ring impedance, the interbar current (losses) are zero (bar currents are independent of y). Also, as expected, for infinite transverse resistance (Rde ~ ∞), again the bar current does not depend on y in terms of amplitude. We may now calculate the sum of the bar losses and interbar (transverse) losses together, P^d_{cagev}.

$$P^d_{cagev} = N_r\left[\int_{-\frac{l_{stack}}{2}}^{\frac{l_{stack}}{2}} \left|I_{bm}{}^2(y)\right|\frac{R_b}{l_{stack}}dy + \int_{-\frac{l_{stack}}{2}}^{\frac{l_{stack}}{2}} R_d l_{stack}\left|I_{dm}(y)\right|^2 dy + 2R_{er}\left|I_{rm}\right|^2\right] \tag{11.77}$$

Although (11.77) looks rather cumbersome, it may be worked out rather comfortably with I_{bm} (bar current) from (11.74), I_{dm} from (11.67), and I_{rm} from (11.69).

We still need the expressions of emfs E_m^{ν} which would refer to the entire stack length for a straight rotor bar pair.

$$E_m^{\nu} = \frac{V_{r\nu}\left(B_{\nu} - B_{\nu}e^{-j\frac{2\nu\pi}{N_r}}\right)l_{stack}}{\sqrt{2}}; \text{ (RMS value)} \tag{11.78}$$

$V_{r\nu}$ is the field harmonic ν speed with respect to the rotor cage and B_{ν} the νth airgap harmonic field. For the case of a chorded coil stator winding, only the first stator slot (opening) harmonics $\nu = N_s \pm p_1$ are to be considered.

For this case, accounting for the first airgap magnetic conductance harmonic, the value of B_{ν} is

$$B_{N_s} = B_{g1}\frac{a_1}{2a_0}; \; N_s = (N_s \pm p_1) \mp p_1 \tag{11.79}$$

a_1 and a_0 from (11.22 and 11.23).

The speed $V_{r\nu}$ is

$$V_{r\nu} = R\frac{2\pi f}{p_1 60}\left(1 \pm \frac{p_1}{\nu}\right) \tag{11.80}$$

with

$$\nu = N_s \pm p_1 \qquad V_{r\nu} \approx R\frac{2\pi f}{p_1 60} = V_{N_s} \tag{11.81}$$

The airgap reactance for the $\nu = N_s \pm p_1$ harmonics, $X_{0\nu}$, is

$$X_{0N_s} \approx X_{0p}\frac{p_1}{N_s} \tag{11.82}$$

X_{0p}, the airgap reactance for the fundamental, as seen from the rotor bar,

$$X_{0p} = \frac{X_m}{\alpha^2}; \; \alpha = \frac{4 \cdot 3 \cdot W_1^2 K_{w1}^2}{N_r} \tag{11.83}$$

where X_m is the main (airgap) reactance for the fundamental (reduced to the stator),

$$X_m = \frac{6\mu_0\omega_1}{\pi^2}\frac{W_1^2 K_{w1}^2 \tau l_{stack}}{p_1 g K_c (1 + K_s)} \tag{11.84}$$

(For the derivation of (11.83), see Chapter 5.) W_1–turns/phase, K_{w1}–winding factor for the fundamental, K_s–saturation factor, g–airgap, p_1–pole pairs, τ–pole pitch of stator winding.

Now that we have all data to calculate the cage losses and the transverse losses for skewed uninsulated bars, making use of a computer programming environment like Matlab etc., the problem could be solved rather comfortably in terms of numbers.

Still we can hardly draw any design rules this way. Let us simplify the solution (11.74) by considering that the end ring impedance is zero ($Z_{er} = 0$) and that the contact (transverse) resistance is small.

- <u>Small transverse resistance</u>

Let us consider $R_q = R_{dl} \cdot l_{stack}$, the contact transverse resistance per unit rotor length and that R_q is so small that

$$R_q \left(\frac{\nu\gamma}{R}\right)^2 << 4\sin^2\left(\frac{\nu\pi}{N_r}\right)\frac{X_{0\nu}(1+\tau_{d\nu})}{l_{stack}} \tag{11.85}$$

For this case in [6], the following solution has been obtained:

$$I_{bm}(y) = \frac{-jE_m{}^{\nu} e^{-j\frac{\nu\gamma y}{l_{stack}}}}{X_{0\nu}(1+\tau_{d\nu})}; \quad I_{dm} = \frac{-jE_m{}^{\nu} e^{-j\frac{\nu\gamma y}{l_{stack}}}}{2\sin\left(\frac{\nu\pi}{N_r}\right) X_{0\nu}(1+\tau_{d\nu})} \gamma \frac{l_{stack}}{n}\frac{1}{R} \tag{11.86}$$

n–the number of stack axial segments.

The transverse losses P_{d0} are

$$P_{d0} \approx \frac{R_q}{l_{stack}}\frac{N_r{}^3}{4\pi^2}\left(\frac{E_m{}^{\nu}}{X_{0\nu}}\right)^2 \cdot \frac{1}{1+\tau_{d\nu}} \cdot \left(\frac{c}{R}\right)^2 \tag{11.87}$$

The bar losses P_{b0} are:

$$P_{b0} = N_r R_{be} \frac{\left(E_m{}^{\nu}\right)^2}{X^2{}_{0\nu}(1+\tau_{d\nu})^2} \tag{11.88}$$

Equations (11.87 and 11.88) lead to remarks such as

- For zero end ring resistance and low transverse resistance R_d, the transverse (interbar) rotor losses P_d are proportional to R_d and to the skewing squared.
- The bar losses are not influenced by skewing (11.88), so skewing is not effective in this case and it shall not be used.
- <u>Large transverse resistance</u>

In this case, the opposite of (11.85) is true. The final expressions of transverse and cage losses P_{d0}, P_{b0} are

$$P_{q0} \approx \frac{2\pi^2}{N_r}\frac{\left(\frac{E_m{}^{\nu}}{l_{stack}}\right)^2}{\left(\frac{c}{R}\right)^2} l_{stack}{}^3 \cdot \frac{1}{(1+\tau_{d\nu})} \cdot \frac{\left(1-\frac{1}{3}\sin^2\left(\frac{\nu l_{stack}}{2R}\right)\right)}{\frac{R_q}{2}} \tag{11.89}$$

$$P_{b0} = N_r R_{be} \left(\frac{E_m^{\ \nu}}{X_{0\nu}\left(1+\tau_{d\nu}\right)} \right)^2 K_{skew\nu}^2; \tag{11.90}$$

$$K_{skew\nu} = \frac{\sin\left(\frac{\gamma\nu l_{stack}}{2R}\right)}{\frac{\gamma\nu l_{stack}}{2R}}$$

The situation is now different.

- The transverse losses are proportional to the third power of stack length, to the square of skewing factor, and inversely proportional to transverse resistance per unit length $R_q(\Omega m)$ ($R_d \cdot l_{stack} = R_q$).
- The rotor bar losses P_b are proportional with skewing squared and, thus, by proper skewing for the first slot (opening) harmonics $\nu \approx N_s$, this harmonic bar losses are practically zero; skewing is effective. Also, with long stacks, only chorded windings are practical.
- With long stacks, large transverse resistivities have to be avoided by good contact between bars and tooth walls or by insulated bars. Plotting the transverse losses by the two formulas, (11.87) and (11.89), versus transverse resistance per unit length R_q shows a maximum (Figure 11.6).

So there is a critical transverse resistance R_{q0} for which the transverse losses are maximum. Such conditions are to be avoided.

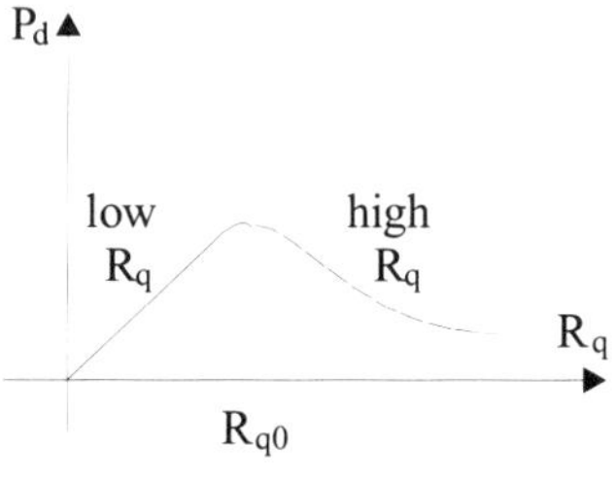

Figure 11.6 Transverse losses versus transverse resistance R_q

In [6] the critical value of R_q, R_{q0}, is

$$R_{q0} \approx \frac{X_{0\nu} l_{stack} 4\pi^2 \sqrt{\left[1 - \frac{1}{3}\sin^2\left(\frac{\nu l_{stack}}{2R}\right)\right]}}{N_r^{\ 2}\left(\frac{c}{R}\nu\right)^2} \tag{11.91}$$

X_{on}, R_{be}, Z_{er}, R_d are in Ω and E_{mv} in volt.

Increasing the stack length tends to increase the critical transverse resistivity R_q. R_q is inversely proportional with skewing. Also, it seems that $(P_b)_{R_{q0}}$, or maximum transverse rotor losses, do depend on skewing.

11.7. NO-LOAD ROTOR SKEWED NONINSULATED CAGE LOSSES

For small or high transverse resistivity (R_q) expressions (11.87 – 11.90), use the computation of transverse and cage no-load stray losses with E_m^{ν} from (11.78) and $\nu = N_s \pm p_1$. It has been shown that for this case, with $\nu \approx N_s$, the transverse losses have a minimum of $0.9 < N_r/N_s < 1.1$.

It means that for low contact (transverse) resistivity, the rotor and stator slot numbers should not be too far away from each other. Especially for $N_r/N_s < 1$, small transverse rotor losses are expected.

11.8. LOAD ROTOR SKEWED NONINSULATED CAGE LOSSES

While under no-load, the airgap permeance first harmonic was important as it acted upon the airgap flux distribution; under load, it is the stator mmf harmonics of same order $N_s \pm p_1$ that are important as the stator current increases with load. This is true for chorded pitch stator windings when the 5th and 7th pole pair phase belt harmonics are small.

The airgap flux density B_ν will now be coming from a different source:

$$B_\nu = \frac{p_1}{\nu}\frac{K_{w\nu}}{K_{w1}}\frac{\mu_0 F_{1p}}{gK_c(1+K_s)} \tag{11.92}$$

F_{1p} is the stator mmf fundamental amplitude.

For full pitch windings however, the 5th and 7th (phase belt) harmonics are to be considered and the conditions of low or high transverse resistivity (11.85) has to be verified for them as well with $\nu = 5p_1, 7p_1$.

Transverse cage losses for both smaller or higher transverse resistivity R_q are inversely proportional to differential leakage coefficient $\tau_{d\nu}$ which, for $\nu = N_s$, is

$$\tau_{dN_s} = \frac{\left(\frac{\pi N_s}{N_r}\right)^2}{\sin^2\left(\frac{\pi N_s}{N_r}\right)^2} - 1 \tag{11.93}$$

It may be shown that τ_{dN_s} increases when $N_s/N_r > 1$.

Building IMs with $N_s/N_r > 1$ seems very good to reduce the transverse cage losses with skewed noninsulated rotor bars. For such designs, it may be adequate to even use non-skewed rotor slots when the additional transverse losses are almost zero.

Care must be exercised to check if the parasitic torques are small enough to secure safe starts. We should also notice that skewing leads to a small attenuation of tooth flux pulsation core losses by the rotor cage currents.

In general, the full load transverse cage loss P_{dn} is related to its value under no-load P_{d0} by the load multiplication factor C_{loads} (11.53), for chorded pitch stator windings.

$$P_{dn} = P_{d0}C_{loads} \tag{11.94}$$

For a full pitch (single layer) winding C_{loads} has to be changed to add the 5th and 7th phase belt harmonics in a similar way as in (11.60 and 11.61).

$$P_{dn} = P_{d0}C_{loads}\left[1+\frac{2\left(\frac{K_{w5}}{K_{w1}}\right)^2\frac{\left(1+\tau_{dN_s}\right)}{\left(1+\tau_{d6}\right)}}{\xi^2_{(N_s-P_1)n}+\xi^2_{(N_s+P_1)n}}\right] \tag{11.95}$$

with $\xi_{(N_s-p_1)n}$ and $\xi_{(N_s+p_1)n}$ from (11.51).

Example 11.5. For the motor with the data in Example 11.4, let us determine the P_{dn}/P_{d0} (load to no-load transverse rotor losses).

Solution

We are to use (11.95).

From Example 11.4, $\xi_{(N_s-p_1)n} = 3.6/K_c$, $K_c = 1.8$, $K_{w5} = 0.2$, $K_{w1} = 0.965$, $C_{loads} = 1.0873$, $\xi_{(N_s+p_1)n} = 2.8/K_c$, $1+\tau_{dN_s} = 43$, $1+\tau_{d6} = 1.37$.

We have now all data to calculate

$$\frac{P_{dn}}{P_{d0}} = 1.0873\left[1+\frac{2\left(\frac{0.2}{0.965}\right)^2\frac{43}{1.37}}{\left(\frac{3.6}{1.8}\right)^2+\left(\frac{2.8}{1.8}\right)^2}\right] = 1.54!$$

11.9. RULES TO REDUCE FULL LOAD STRAY (SPACE HARMONICS) LOSSES

So far the rotor surface core losses, rotor and stator tooth flux pulsation core losses, and space harmonic cage losses have been included in stray load losses. They all have been calculated for motor no-load, then corrected for load conditions by adequate amplification factors.

Insulated and noninsulated, skewed and nonskewed bar rotors have been investigated. Chorded pitch and full pitch windings cause differences in terms of stray losses. Other components such as end-connection leakage flux produced losses, which occur in the windings surroundings, have been left out as their study by analytical methods is almost impractical. In high power machines, such losses are to be considered.

Based on analysis, such as the one corroborated above, with those of [12], we line up a few rules to reduce full load stray losses

- Large number of slots per pole and phase, q, if possible, to increase the first phase belt and first slot (opening) harmonics.
- Insulated or large transverse resistance cage bars in long stack skewed rotors to reduce transverse cage losses.
- Skewing is not adequate for low transverse resistance as is does not reduce the stray cage losses while the transverse cage losses are large. Check the tooth flux pulsation core losses in skewed rotors.
- $N_r < N_s$ to reduce the differential leakage coefficient of the first slot (opening) harmonics $N_s \pm p_1$, and thus reduce the transverse cage losses.
- For $N_r < N_s$, skewing may be eliminated after the parasitic torques are checked and found small enough. For q = 1, 2, skewing is mandatory.
- Chorded coil windings with $y/\tau \approx 5/6$ are adequate as they reduce the first phase belt harmonics.
- With full pitch winding, use large numbers of slot/pitch/phase whenever possible.
- Skewing seems efficient for noninsulated rotor bars with high transverse resistance as it reduces the transverse rotor losses.
- With delta connection $(N_s-N_r) \neq 2p_1, 4p_1, 8p_1$.
- With parallel path winding, the circulating stator currents induced by the bar current mmf harmonics have to be avoided by observing a certain symmetry.
- Use small stator and rotor slot openings, if possible, to reduce the first slot opening flux density harmonics and their losses. Check the starting and peak torque as they tend to decrease due to slot leakage inductance increase.
- Use magnetic wedges for open slots, but check for the additional eddy current losses in them and secure their mechanical ruggedness.
- Increase the airgap, but maintain good power factor and efficiency.
- For one conductor per layer in slot and open slots, check that the slot opening $b_{os,r}$ to elementary conductor $d_{s,r}$ are $b_{os}/d_s \leq 1$, $b_{or}/d_r \leq 3$. This way, the stray load conductor losses by space harmonics fields are reduced.
- Use sharp tools and annealed lamination sheets (especially for low power motors), to reduce surface core losses.
- Return rotor surface to prevent laminations to be short-circuited and thus reduce rotor surface core losses.
- As storing the motor with cast aluminum (noninsulated) cage rotors after fabrication leads to a marked increase of rotor bar to slot wall contact electrical resistivity, a reduction of stray losses of (40 to 60)% may be obtained after 6 months of storage.

11.10. HIGH FREQUENCY TIME HARMONICS LOSSES

High frequency time harmonics in the supply voltage of IMs may occur either because the IM itself is fed from a PWM static power converter for variable speed or because, in the local power grid, some other power electronic devices produce voltage time harmonics at IM terminals.

For voltage-source static power converters, the time harmonics frequency content and distribution depends on the PWM strategy and the carrier ratio c_r ($2c_r$ switchings per period). For high performance symmetric regular sampled, asymmetric regular sampled, and optimal regular sampled PWM strategies, the main voltage harmonics are $(c_r \pm 2)f_1$ and $(2c_r \pm 1)f_1$, respectively, [13], Figure 11.7.

It seems that accounting for time harmonics losses at frequencies close to carrier frequency suffices. As of today, the carrier ratio c_r varies from 20 to more than 200. Smaller values relate to larger powers.

Switching frequencies up to 20 kHz are typical for low power induction motors fed from IGBT voltage-source converters. Exploring the conductor and core losses up to such large frequencies becomes necessary.

High carrier frequencies tend to reduce the current harmonics and thus reduce the conductor losses associated with them, but the higher frequency flux harmonics may lead to larger core loss. On the other hand, the commutation losses in the PWM converter increase with carrier frequency.

The optimum carrier frequency depends on the motor and the PWM converter itself. The 20 kHz is typical for hard switched PWM converters. Higher frequencies are practical for soft switched (resonance) converters. We will explore, first rather qualitatively and then quantitatively, the high frequency time harmonics losses in the stator and rotor conductors and cores.

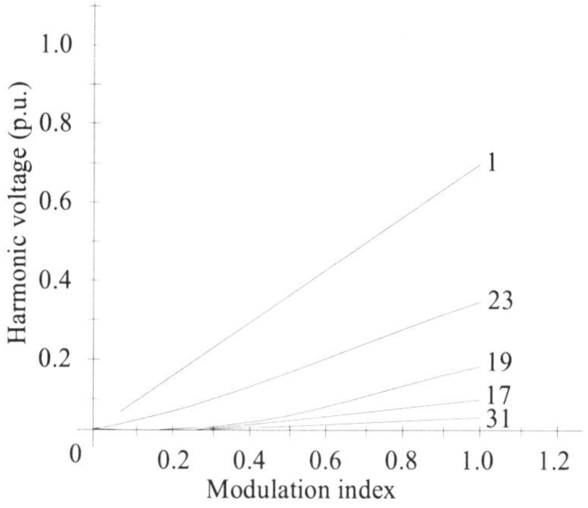

a.)

Figure 11.7 Harmonic voltages with modulation index for cr = 21
a.) symmetric regular sampled PWM b.) asymmetric regular sampled PWM (continued)

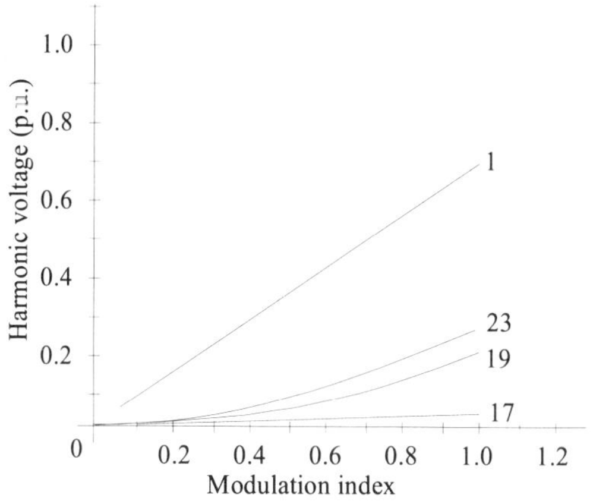

b.)

Figure 11.7 (continued)

At high carrier frequency, the skin effect, both in the conductors and iron cores, may not be neglected.

11.10.1. Conductor losses

The variation of resistance R and leakage L_l inductance for conductors in slots with frequency, as studied in Chapter 9, is at first rapid, being proportional to f^2. As the frequency increases further, the field penetration depth gets smaller than the conductor height and the rate of change of R and L_l decreases to become proportional to $f^{1/2}$ (Figure 11.8).

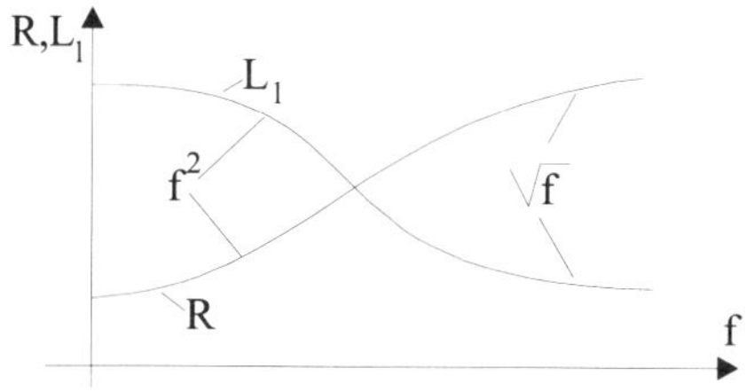

Figure 11.8 R and L_l variation with frequency

For the end connections (or end rings), the skin effect is less pronounced, but it may be calculated with similar formulas providing virtual larger slots are defined (Chapter 9, Figure 9.10).

For high frequencies, the equivalent circuit of the IM may be simplified by eliminating the magnetization branch (Figure 11.9).

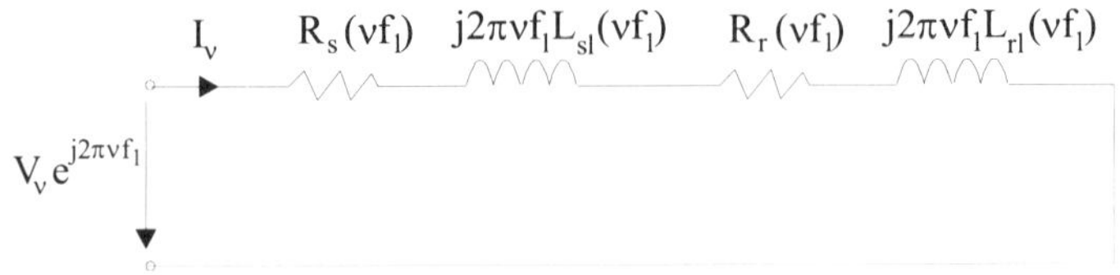

Figure 11.9 Equivalent circuit for time harmonic ν

Notice that the slip $S_v \approx 1$.

In general, the reactances prevail at high frequencies,

$$I_v \approx \frac{V_v}{2\pi f_v L_\sigma(f_v)}; \quad L_\sigma(f_v) = L_{sl}(f_v) + L_{rl}(f_v) \tag{11.95}$$

The conductor losses are then

$$P_{con} = 3I_v^2\left(R_s(f_v) + R_r(f_v)\right) \approx \frac{3V_v^2}{\left(2\pi f_v L_\sigma(f_v)\right)^2}\left(R_s(f_v) + R_r(f_v)\right) \tag{11.96}$$

For a given (fixed) value of the current harmonic I_v, the conductor losses will increase steadily with frequency. This case would be typical for current control.

For voltage control, when voltage harmonics V_v are given, however, (11.96) shows that it is possible to have conductor losses increasing with frequency if the decrease of leakage inductance $L_\sigma(f_v)$ with f_v and the increase factors of $\left(R_s(f_v) + R_r(f_v)\right)$ with f_v are less than proportional to f_v^2.

Measurements have shown that the leakage inductance decrease to 0.5 to 0.3 of the rated frequency (60 Hz) value at 20 kHz.

So, in general,

$$L_\sigma(f_v) \approx K_L f_v^{-0.16} \tag{11.97}$$

For high frequencies R_r variation with frequency is in the $f_v^{0.5}$ range, approximately, so the rotor conductor losses are

$$P_{conr} \approx \frac{3V_v^2}{\left(2\pi f_v K_L\right)^2 f_v^{-0.32}} K_R f_v^{0.5} \sim \frac{V_h^2}{f_v^{1.18}} \tag{11.98}$$

The rotor conductor losses drop notably as the time harmonic frequency increases.

The situation in the stator is different as there are many conductors in every slot (at least in small power induction machines).

So the skin effect for R_s will remain f_v dependent in the initial stages ($K_{Rs} = C_{Rs} \cdot f_v$),

$$P_{cons} \sim V_v^2 f_v^{0.32} \tag{11.99}$$

The stator conductor losses tend to increase slightly with frequency. For large power IMs (MW range) the stator conductor skin effect is stronger and the situation comes closer to that of the rotor: the stator conductor losses slightly decrease at higher frequencies. We should also mention that the carrier frequency in large power is only 1 to 3 kHz.

For low and medium power motors, as the carrier frequency reaches high levels (20 kHz or more), the skin effect in the stator conductors enters the $f_v^{0.5}$ domain and the stator conductor losses, for given harmonic voltage, behave like the rotor cage losses (decrease slightly with frequency (11.98)). This situation occurs when the penetration depth becomes smaller than conductor height.

11.10.2. Core losses

Predicting the core loss at high frequencies is difficult because the flux penetration depth in lamination becomes comparable with (or smaller than) the lamination thickness.

The leakage flux paths may then prevail and thus the reaction of core eddy currents may set up significant reaction fields.

The field penetration depth in laminations δ_{Fe} is

$$\delta_{Fe} = \sqrt{\frac{1}{\pi f_v \sigma_{Fe} \mu_{Fe}}} \tag{11.100}$$

σ_{Fe}–iron electrical conductivity; μ_{Fe}–iron magnetic permeability.

For f = 60 Hz, $\sigma_{Fe} = 2\cdot 10^6\,(\Omega m)^{-1}$, $\mu_{Fe} = 800\,\mu_0$, $\delta_{Fe} = 1.63$ mm, while at 20 kHz $\delta_{Fe} = 0.062$ mm. In contrast for copper $(\delta_{Co})_{60Hz} = 9.31$ mm, $(\delta_{Co})_{20kHz} = 0.51$ mm and in aluminum $(\delta_{Al})_{60Hz} = 13.4$ mm, $(\delta_{Al})_{20kHz} = 0.73$ mm.

The penetration depth at $800\mu_0$ and 20 kHz, $\delta_{Fe} = 0.062$ mm, shows the importance of skin effect in laminations. To explore the dependence of core losses on frequency, let us distinguish three cases.

- Case 1 – No lamination skin effect: $\delta_{Fe} >> d$

This case corresponds to low frequency time harmonics. Both hysteresis and eddy current losses are to be considered.

$$P_{Fe} = \left(K_{hl}' B_v^{\,n} f_v + K_{el}' B_v^{\,2} f_v^{\,2}\right) A_l l \tag{11.101}$$

where B_v is the harmonic flux density:

$$B_v = \frac{\phi_v}{A_l} \approx \frac{V_v}{2\pi f_v A_l} \tag{11.102}$$

A_l is the effective area of the leakage flux path and l is its length.

With (11.102), (11.101) becomes

$$P_{Fe} = \left(K_{hl} V_v^{\,n} f_v^{\,1-n} + K_{el} V_v\right) A_l l \tag{11.103}$$

Since n > 1, the hysteresis losses decrease with frequency while the eddy current losses stay constant. P_{Fe} is almost constant in these conditions.

- <u>Case 2 – Slight lamination skin effect, $\delta_{Fe} \approx d$</u>

When $\delta_{Fe} \approx d$, the frequency f_v is already high and thus $\delta_{Al} < d_{Al}$ and a severe skin effect in the rotor slot occurs. Consequently, the rotor leakage flux is concentrated close to the rotor surface. The "volume" where the core losses occur in the rotor decreases. In general then, the core losses tend to decrease slowly and level out at high frequencies.

- <u>Case 3 – Strong lamination skin effect, $\delta_{Fe} < d$</u>

With large enough frequencies, the lamination skin depth $\delta_{Fe} < d$ and thus the magnetic field is confined to a skin depth layer around the stator slot walls and on the rotor surface (Figure 11.10).

The conventional picture of rotor leakage flux paths around the rotor slot bottom is not valid in this case.

The area of leakage flux is now, for the stator, $A_l = l_f\,\delta_{Fe}$, with l_f the length of the meander zone around the stator slot.

$$l_f = (2h_s + b_s)N_s \tag{11.104}$$

Now the flux density B_v is

$$B_v = \frac{K_v V_v}{f_v l_f \delta_{Fe}} \sim K f_v^{-\frac{1}{2}} \tag{11.105}$$

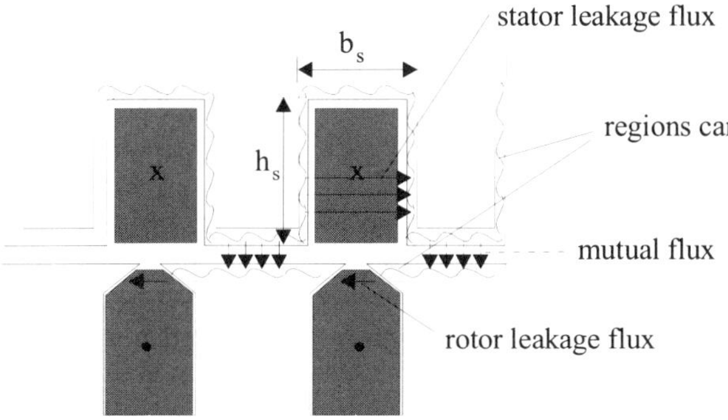

Figure 11.10 Leakage flux paths at high frequency

Consequently, the core losses (with hysteresis losses neglected) are

$$P_{Fee} = K_e B_v^2 f_v^2 l_f l_{stack} \delta_{Fe} \sim K_1 V_v^2 f_v^{\frac{1}{2}} \tag{11.106}$$

A slow steady state growth of core losses at high frequencies is thus expected.

11.10.3. Total time harmonics losses

As the discussion above indicates, for a given harmonic voltage, above certain frequency, the conductor losses tend to decrease as $f_v^{-1.2}$ (11.98) while the core losses (11.106) increase.

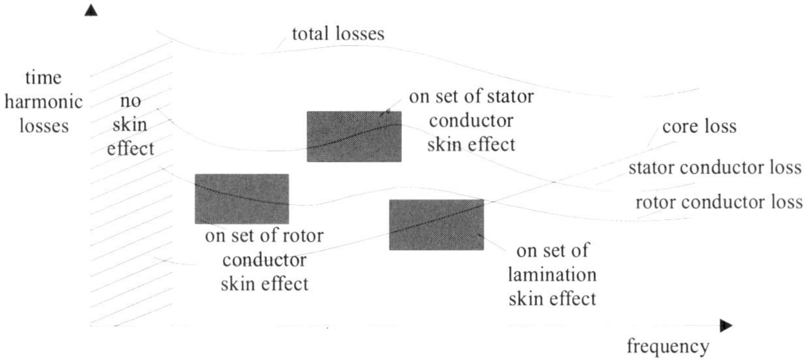

Figure 11.11 Time harmonic IM losses with frequency (constant time harmonic voltage)

Consequently, the total time harmonics may level out above a certain frequency (Figure 11.11). Even a minimum may be observed. At what frequency such a minimum occurs depends on the machine design, power, and PWM frequency spectrum.

11.11. COMPUTATION OF TIME HARMONICS CONDUCTOR LOSSES

As Equation (11.96) shows, to compute the time harmonics conductor losses, the variation of stator and rotor leakage inductances and resistances with frequency due to skin effect is needed. This problem has been treated in detail, for a single, unassigned, frequency in Chapter 9, both for the stator and rotor.

Here we simplify the correction factors in Chapter 9 to make them easier to use and interpret.

In essence, with different skin effect in the slot and end-connection (end ring) zones, the stator and resistance $R_s(f_v)$ is

$$R_s(f_v) = R_{sdc}\left[K_{Rss}(f_v)\frac{l_{stack}}{l_{coil}} + K_{Rse}(f_v)\frac{l_{endcon}}{l_{coil}}\right] \tag{11.107}$$

l_{coil}–coil length, l_{endcon}–coil end connection length.

$K_{Rss} > 1$ is the resistance skin effect correction coefficient for the slot zone and K_{Rse} corresponds to the end connection. A similar expression is valid for the stator phase leakage inductance.

$$L_{sl}(f_v) = L_{sldc}\left[K_{Xlss}(f_v)\frac{l_{stack}}{l_{coil}} + K_{Xlse}(f_v)\frac{l_{endcon}}{l_{coil}}\right] \quad (11.108)$$

K_{Xlss} and K_{Xlse} are leakage inductance correction factors.

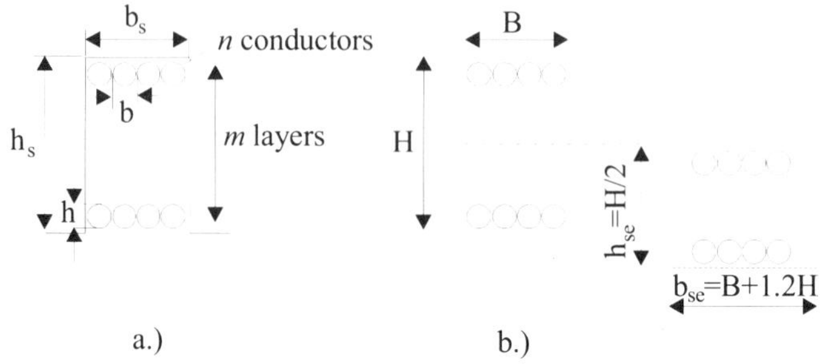

Figure 11.12 Stator slots a.) and end connection b.) geometries

Now with m layers of conductors per slot from Chapter 9, Equation (9.6),

$$K_{Rss} = \varphi(\xi) + \frac{m^2 - 1}{3}\psi(\xi) > 1 \quad (11.109)$$

$$\xi = \beta_n h, \quad \beta_n = \sqrt{\pi f_v \mu_0 \sigma_{Co}\frac{nb}{b_s}}$$

h – conductor height, n – conductors per layer, b – conductor width (or diameter), b_s – slot width. Equation (11.109) is strictly valid for rectangular slots (Figure 11.12). More evolved methods, as described in Chapter 9, may be used for general shape slots.

The same formula (11.109) may be used for end connections coefficient K_{Rse} but with b_{se} instead of b_s, h_{se} instead of h_s,and m/2 instead of m.

In a similar way, the reactance correction coefficient K_{Xss} is (Equation (9.7))

$$K_{Xss} = \frac{\varphi'(\xi) + (m^2 - 1)\psi'(\xi)}{m^2} < 1 \quad (11.110)$$

Also, from (9.4) and (9.8), (9.9)

$$\varphi(\xi) = \xi\frac{(\sinh(2\xi) + \sin(2\xi))}{(\cosh(2\xi) - \cos(2\xi))}; \quad \psi(\xi) = 2\xi\frac{(\sinh(\xi) - \sin(\xi))}{(\cosh(\xi) + \cos(\xi))} \quad (11.111)$$

$$\varphi'(\xi) = \frac{3}{2\xi}\frac{(\sinh(2\xi) - \sin(2\xi))}{(\cosh(2\xi) - \cos(2\xi))}; \quad \psi'(\xi) = \frac{(\sinh(\xi) + \sin(\xi))}{\xi(\cosh(\xi) + \cos(\xi))} \quad (11.112)$$

As the number of layers in the virtual slot of end connection is m/2 and the slot width is much larger than for the actual slot $b_{se} - b_s \approx 1.2h_s \approx (4 - 6)b_s$, the skin effect coefficients for the end connection zone are smaller than for the slot zone but still, at high frequencies, they are to be considered.

For the cage rotor resistance R_r and leakage inductance L_{rl}, expressions similar to (11.107 – 11.108) are valid.

$$R_r(f_v) = R_{bdc}K_{Rb}(f_v) + R_{erdc}K_{Re}(f_v) \tag{11.113}$$

$$L_{rl}(f_v) = L_{bdc}K_{Xb}(f_v) + L_{erdc}K_{Xe}(f_v) \tag{11.114}$$

Again, expressions (11.109) and (11.110) are valid, but with m = 1, with $n_b = b_s$, and σ_{Al} in (11.109).

Also, in (11.110), ξ is to be calculated with $h_{se} = h$, $b_{se} = b_s + K_h h$. In other words, the end ring has been assimilated to an end connection, but the factor K_h should be 1.2 for a distant ring and less than that for an end ring placed close to the rotor stack.

Example 11.6. Time harmonics conductor losses. An 11 kW, 6 pole, $f_1 = 50$ Hz, 220 V induction motor is fed from a PWM converter at carrier frequency $f_v = 20$ kHz. For this frequency, the voltage $V_v = 126$ V (see [14]).

For the rotor: $\xi = 18.38$, $K_{Xb} = 0.0816$, $K_{Rb} = 18.38$, $R_r'(f_v) = 2.2\ \Omega$ (end ring resistance neglected), $X'_{rl} = 2\pi f_v \cdot L_{rl}(f_v) = 47.9\ \Omega$, and for the stator: $\xi = 3.38$, $K_{Rss} = K_{Rse} = 79.4$, $R_s(f_v) = 11.9\ \Omega$, $K_{Xss} = K_{Xse} = 0.3$, $X_{sl} = 2\pi f_v \cdot L_{rl}(f_v) = 63.3\ \Omega$

The harmonic current I_v is

$$I_v = \frac{V_v}{\sqrt{(R_r'+R_s)^2 + (X_{rl} + X_{sl})^2}} = \frac{126}{\sqrt{(2.2+11.9)^2 + (47.9+63.3)^2}} \approx 1.12\text{A}$$

Now the conductor losses at 20 kHz are

$$(P_{con})_{20kHz} = 3(R'_r + R_s)I_v^{\ 2} = 3(2.2+11.9)1.12^2 = 53.06\text{W}$$

We considered above $V_v \approx V_1$,which is not the case, in general, although in some PWM strategies at some modulation factor values, such a situation may occur.

11.12. TIME HARMONICS INTERBAR ROTOR CURRENT LOSSES

Earlier in this chapter we dealt with space harmonic induced losses in skewed noninsulated bar rotor cages. This phenomenon occurs also due to time harmonics. At the standard fundamental frequency (50 or 60 Hz) the additional transverse rotor losses due to interbar currents are negligible. In high frequency motors ($f_1 > 300$ Hz), these losses count. [11] Also, time harmonics are likely to produce transverse losses for skewed noninsulated bar rotor cages.

To calculate such losses, we may use the theory developed in Paragraph 11.6, with S = 1, $\nu = 1$ but for given frequency f_v and voltage V_v.

However, to facilitate the computation process, we adopt the final results of Reference [11] which calculates the rotor resistance and leakage reactance including the skin effect and the transverse (interbar currents) losses,

$$R_r^*(f_v) = \left(R_{bdc}K_{Rb}(f_v) + R_{erdc}K_{Re}(f_v)\right)\cdot|\alpha_0|\cos\gamma_0 + \\ + 2\pi f_v\left(L_{bdc}K_{Xb}(f_v) + L_{erdc}K_{XRe}(f_v)\right)\cdot|\alpha_0|\sin\gamma_0 + X_m(f_v)\cdot\frac{\sin\gamma_0}{\eta^2\left|\dot{K}\right|} \tag{11.115}$$

$$X_{rl}^*(f_v) = 2\pi f_v\left(L_{bdc}K_{Xb}(f_v) + L_{erdc}K_{XRe}(f_v)\right)\cdot|\alpha_0|\cos\gamma_0 - \\ - \left(R_{bdc}K_{Rb}(f_v) + R_{erdc}K_{Re}(f_v)\right)\cdot|\alpha_0|\sin\gamma_0 + X_m(f_v)\cdot\left(\frac{\cos\gamma_0}{\eta^2\left|\dot{K}\right|} - 1\right) \tag{11.116}$$

with

$$\alpha_0 = \frac{4\cdot 3\cdot 2\pi f_v K_{w1}{}^2}{N_r\cdot\dot{K}};\quad \dot{K} = \left|\dot{K}\right|e^{j\gamma_0};\quad \eta = \frac{\sin\left(\frac{\pi}{N_r}\right)}{\frac{\pi}{N_r}} \tag{11.117}$$

$$\dot{K} = \frac{Z_{be} + R_{ere}}{Z_{be} + R_{de}\cdot(\gamma^2)}\cdot\left[1 - (A+B)\frac{\sinh\frac{1}{2}\left(\sqrt{\frac{Z_{be}}{R_{de}}} + j\gamma\right)}{\left(\sqrt{\frac{Z_{be}}{R_{de}}} + j\gamma\right)} - \\ - (A-B)\frac{\sinh\frac{1}{2}\left(\sqrt{\frac{Z_{be}}{R_{de}}} + j\gamma\right)}{\left(\sqrt{\frac{Z_{be}}{R_{de}}} + j\gamma\right)}\right] \tag{11.118}$$

where γ is the skewing angle.

A and B are defined in (11.75) and (11.76). Z_{be} and R_{de} are defined in (11.73) and refer to rotor bar impedance and equivalent rotor end ring and bar tooth wall resistance R_{de} is reduced to the rotor bar.

$X_m(f_v)$ is the airgap reactance for frequency f_v and fundamental pole pitch,

$$X_m(f_v) - X_m(f_1) \cdot \frac{f_v}{f_1} \qquad (11.119)$$

In [11], for a 500 Hz fundamental, high speed, 2.2 kW motor, the variation of rectangular bar rotor resistance $R_r^*(R_{de})$ for various frequencies was found to have a maximum whose position depends only slightly on frequency (Figure 11.13a).

The rotor leakage reactance $X_{rl}^*(R_{de})$ levels out for all frequencies at high bar-tooth wall contact resistance values (Figure 11.13b).

We should note that Equations (11.115) and (11.116) and Figure 11.13 refer both to skin effect and transverse losses.

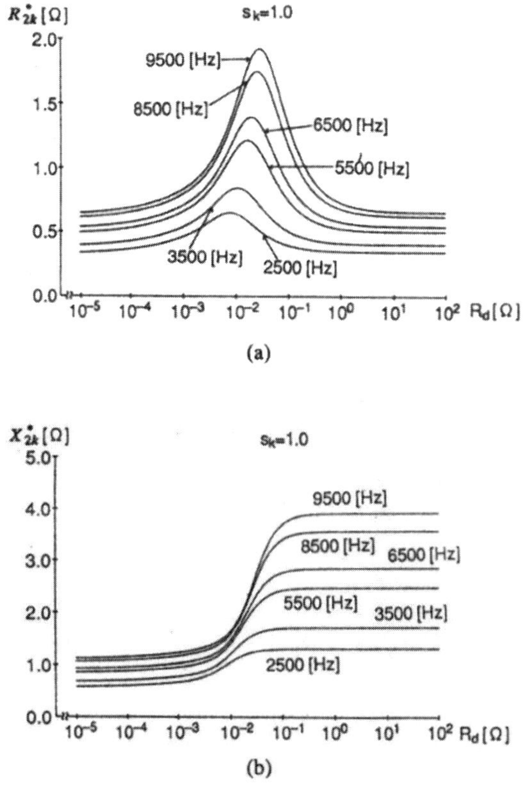

Figure 11.13 Rotor cage resistance and leakage reactance versus bar-tooth wall contact resistance R_{de}, for various time harmonics

As the skin effect at, say, 9500 Hz is very strong and only the upper part of the rotor bar is active, the value of transverse resistance Rde is to be increased in the ratio

$$R_{de} = R_{dedc}\frac{h}{\delta_{Al}(f_v)} \quad (11.120)$$

This is as if we would read values of R_r^*, X_{rl}^* in Figure 11.13 calculated for $(R_{de})_{dc}$ at the abscissa R_{de} of (11.120).

Also, as X_{rl}^* increases with R_{de}, the harmonic current and conductor losses tend to decrease for large R_{de}.

Increasing the frequency tends to push the transverse resistance R_{de} to larger values, beyond the critical value and thus lower interbar currents and losses are to be obtained.

However, such a condition has to be verified up to carrier frequency so that a moderate influence of interbar currents is secured.

11.13. COMPUTATION OF TIME HARMONICS CORE LOSSES

The computation of core losses for high frequency in IMs is a very difficult task. However, for the case when $\delta_{Fe} < d$ (large skin effect) and all the flux paths are located around stator and rotor slots and on rotor surface in a thin layer (due to airgap flux pulsation caused by stator slot openings), such an attempt may be made easily through analytical methods.

Core losses may also occur due to axial fluxes, in the stack-end laminations because of end connection high frequency leakage flux. So we have three types of losses here.

a. Slot wall time harmonic core losses
b. Airgap flux pulsation (zig-zag) time harmonic core losses
c. End connection leakage flux time harmonic core losses

11.13.1. Slot wall core losses

Due to the strong skin effect, the stator and rotor currents are crowded toward the slot opening in each conductor layer (Figure 11.14).

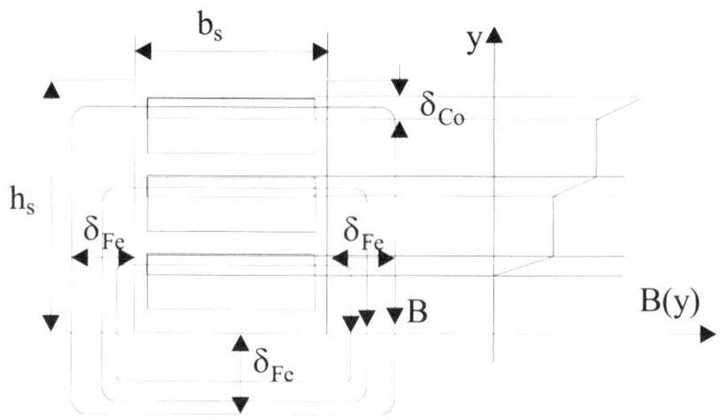

a.)

Figure 11.14 Stator a.) and rotor b.) slot leakage flux path and flux density distribution (continued)

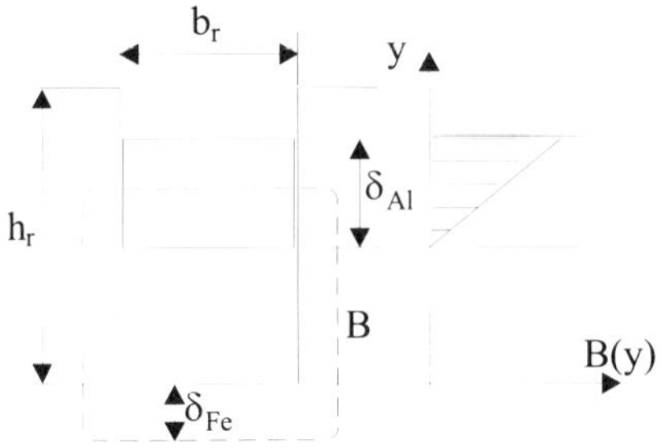

Figure 11.14 (continued)

The stator average leakage flux ϕ_{vs} is

$$\phi_{vs} = \frac{1}{W_1}\left[\frac{L_{slslot}(f_v)I_v}{\frac{N_s}{3}}\right]\cdot\frac{1}{\left(\frac{h_s}{\xi}\right)}[Wb/m] \tag{11.121}$$

where W_1 – turns per phase
$N_s/3$ – slots per stator phase
h_s/ξ – the total height of m skin depths
m conductor layers in slot

From [14] the surface core losses, with skin effect considered for the stator, are

$$P_{sw} = \frac{\phi_{vs}^{\ 2}}{\delta_{Fe}^3\sigma_{Fe}\mu_{Fe}^2}\cdot\frac{\left(\sinh\left(\frac{2d}{\delta_{Fe}}\right)-\sin\left(\frac{2d}{\delta_{Fe}}\right)\right)}{\left(\cosh\left(\frac{2d}{\delta_{Fe}}\right)-\cos\left(\frac{2d}{\delta_{Fe}}\right)\right)}\cdot K_{er}N_s l_{stack}(2h_s+b_s)\delta_{Fe} \tag{11.122}$$

K_{er} is the reaction-field factor: $K_{er} = \varphi'$ (see (11.112)) for

$$\xi = \frac{2d}{\delta_{Fe}} \tag{11.123}$$

In a similar way, we may proceed for the rotor slot, but here m = 1 (conductors per slot). The rotor average slot leakage flux ϕ_{vr} is

$$\phi_{vr} = \frac{1}{W_2}\left[\frac{L_{rlslot}(f_v)I_v}{\frac{N_r}{m_2}}\right]\cdot\frac{1}{\left(\frac{h_r}{\xi}\right)}[Wb/m] \tag{11.124}$$

W_2–turns per rotor phase, m_2–rotor equivalent phases. For a cage rotor we may consider $W_2 = W_1$ and $m_2 = 3$ as $L_{rlslot}(\nu)$ is already reduced to the stator.

The rotor slot wall losses P_{rw} are

$$P_{rw} = \frac{{\phi_{vr}}^2}{\delta_{Fe}^3 \sigma_{Fe} \mu_{Fe}^2} \cdot \frac{\left(\sinh\left(\frac{2d}{\delta_{Fe}}\right) - \sin\left(\frac{2d}{\delta_{Fe}}\right)\right)}{\left(\cosh\left(\frac{2d}{\delta_{Fe}}\right) - \cos\left(\frac{2d}{\delta_{Fe}}\right)\right)} \cdot K_{er} N_r l_{stack} (2h_r + b_r)\delta_{Fe} \quad (11.125)$$

11.13.2. Zig-zag rotor surface losses

The time harmonic airgap flux density pulsation on the rotor tooth heads, similar to zig-zag flux, produces rotor surface losses.

The maximum airgap flux density pulsation $B_{\nu K}$, due to stator slot opening is

$$B_{\nu K} = \frac{\pi}{4} \frac{\beta}{2-\beta} B_{\nu g} \quad (11.126)$$

The factor β (b_{os}/g) has been defined in Chapter 5 (Figure 5.4) or Table 10.1. It varies from zero to 0.41 for b_{os}/g from zero to 12. The frequency of $B_{\nu K}$ is $f_\nu(N_s/p_1 \pm 1)$. $B_{\nu g}$ is the f_ν frequency airgap space fundamental flux density.

To calculate $B_{\nu g}$, we have to consider that all the airgap flux is converted into leakage rotor flux. So they are equal to each other.

$$\left(\frac{2}{\pi} B_{\nu g} \tau l_{stack}\right) \frac{W_1 K_{w1}}{\sqrt{2}} = I_\nu L_{r\nu} \quad (11.127)$$

Based on (11.126 and 11.127) in Reference [14], the following result has been obtained for zig-zag rotor surface losses:

$$P_{zr} = 2(\pi D \delta_{Fe}) \frac{1}{8} \frac{\pi^2}{16} \left(\frac{\beta}{2-\beta}\right)^2 {B_{\nu g}}^2 l_{stack}^2 \cdot \frac{K_{er} C_{\nu 1r}}{\delta_{Fe}^3 \sigma_{Fe} \mu_{Fe}^2} \cdot \frac{\left(\sinh\left(\frac{2d}{\delta_{Fe}}\right) - \sin\left(\frac{2d}{\delta_{Fe}}\right)\right)}{\left(\cosh\left(\frac{2d}{\delta_{Fe}}\right) - \cos\left(\frac{2d}{\delta_{Fe}}\right)\right)} \quad (11.128)$$

$$C_{\nu 1r} = \left(\frac{\sin\left(\frac{K\pi K_2}{2R'}\right)}{\left(\frac{K\pi K_2}{2R'}\right)}\right)^2 + \left(\frac{\sin\left(\frac{K\pi K_2}{R'}\right)}{\left(\frac{K\pi K_2}{R'}\right)}\right)^2 - \\ - \frac{\sin\left(\frac{K\pi K_2}{2R'}\right)}{\left(\frac{K\pi K_2}{2R'}\right)} \cdot \frac{\sin\left(\frac{K\pi K_2}{R'}\right)}{\left(\frac{K\pi K_2}{R'}\right)} \cdot \cos\frac{K\pi K_2}{2R'} << 1 \quad (11.129)$$

$$K = \frac{N_s}{p_1} \pm 1; \ K_2 = \frac{(t_r - b_{or})}{t_r}; \ R' = \frac{N_r}{2p_1} \tag{11.130}$$

where t_r–rotor slot pitch; b_{or}–rotor slot opening.

Zig-zag rotor surface losses tend to be negligible, at least in small power motors. End connection leakage core losses proved negligible in small motors. In high power motors, only 3D FEM are able to produce trustworthy results.

In fact, it seems practical to calculate only the slot wall time harmonics core losses in the stator and rotor. For the motor, in Example 11.6 in [14], it was found that $P_{sw} + P_{rw}' = 6.8 + 7.43 = 14.23$W. This is about 4 times less than conductor losses at 20 kHz.

11.14. LOSS COMPUTATION BY FEM

The FEM, even in its 2D version, allows for the computation of magnetic field distribution once the stator and rotor currents are known. Under no-load, only the stator current waveform fundamental is required. It is thus possible to calculate the additional currents induced in the rotor cage by conjugating field distribution with circuit equations.

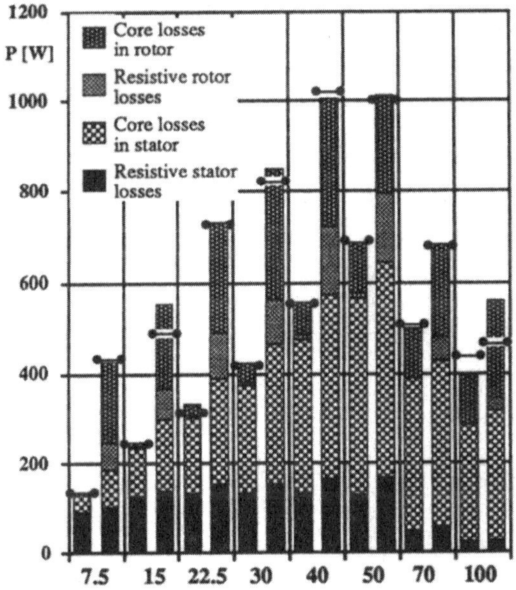

Figure 11.15 No-load losses (37 kW IM) at 7.5 Hz, …, 100 Hz fundamental frequency.
Left column for sinusoidal voltage and right column for PWM voltage.
Measured losses marked by cross bars [16]

The core loss may be calculated only from the distribution of field in the machine with zero electrical conductivity in the iron core. This is a strong approximation, especially at high time harmonic frequencies (see previous paragraphs). In References [15, 16] such a computation approach is followed for both sinusoidal and PWM voltages. Sample results of losses for a 37 kW motor on no-load are shown on Figure 11.15. [16]

More important, the flux density radial (B_r) and tangential (B_t) flux density hodographs in 3 points a, b, c (Figure 11.16c) are shown on Figure 11.16a, b. [16] This is proof that the rotor slot opening (point a) and tooth (point b) experience a.c. field while point c (slot bottom) experiences a quasi-traveling field. Although such knowledge is standard, a quantitative proof is presented here.

When the motor is under load, for sinusoidal voltage supply, FEM has been also used to calculate the losses for a skewed bar cage rotor machine. Insulated bars have been used for the computation. This time, again, but justified, the iron skin effect was neglected as time harmonics do not exist. Semiempirical loss formulas, as used with analytical models, are still used with FEM.

The influence of skewing in the rotor is considered separately, and then by using the coupling field-circuit FEM. The stack was sliced into 5 to 8 axial segments properly shifted to consider the skewing effects. [17]

The already documented axial variation of airgap and core flux density due to skewing changes the balance of losses by increasing the stator fundamental and especially the rotor (stray) core losses and decreasing the rotor stray (additional) cage losses.

In low power induction motors, where conductor losses dominate, skewing tends to reduce total losses on load, while for large machines where core loss is relatively more important, the total losses on load tend to increase slightly due to skewing. [17]

The network – field coupled time – stepping finite element 2D model with axial stack segmentation of skewed-rotor cage IMs has been used to include also the interbar currents [18].

The fact that such complex problems can be solved by quasi 2D–FEM today is encouraging as rather reasonable computation times are required, However, all the effects are mixed and no easy way to derive design hints seems in sight.

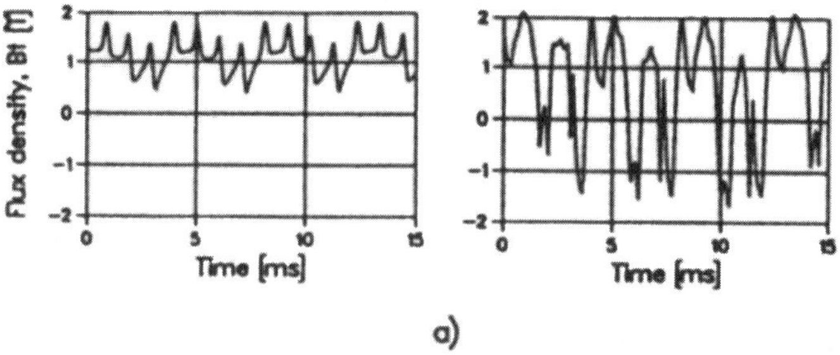

Figure 11.16 No-load flux density hodograph in points a, b, c, at 40 Hz, left – sinusoidal voltage, right – PWM voltage (continued)

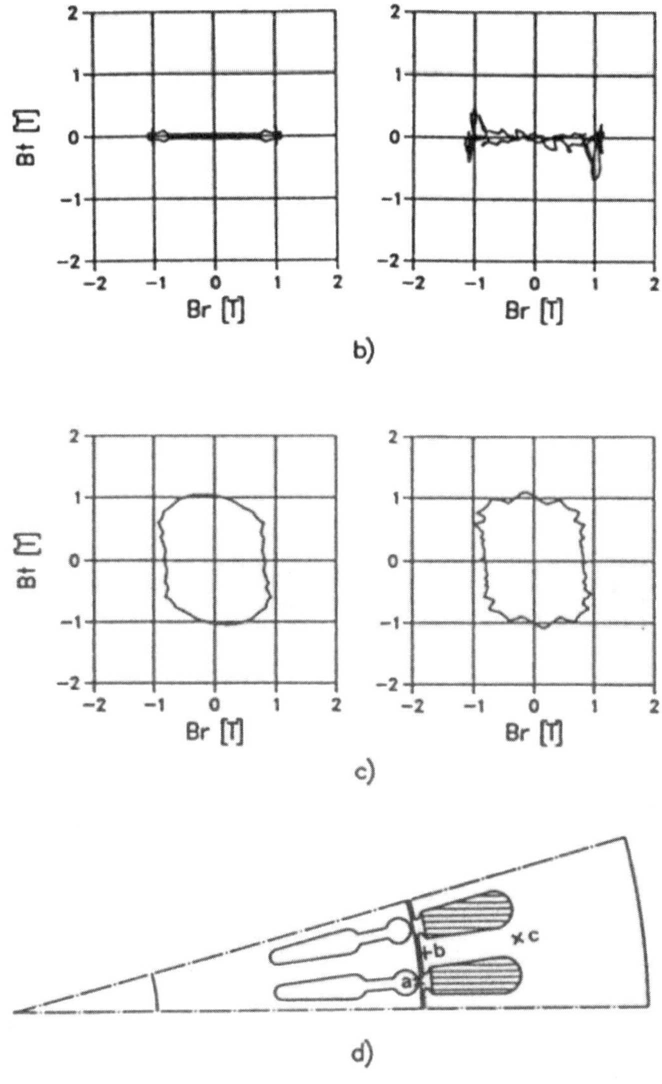

Figure 11.16 (continued)

For other apparently simpler problems, however, as the usage of magnetic wedges for stator slots in large machines, FEM is extremely useful. [19] A notable reduction of rotor stray core losses is obtained. Consequently, higher efficiency for open stator slots is expected.

A similar situation occurs when the hysteresis losses in a cage IM are calculated by investigating the IM from rated to zero positive and negative slip. [20]

A pertinent review of FEM usage in the computation of IM performance (and losses) is given in Reference [21].

11.15. SUMMARY

- Conductor, core, frictional, and windage losses occur in IMs.
- Conductor and core losses constitute electromagnetic losses.
- Electromagnetic losses may be divided into fundamental and harmonic (space and/or time harmonics) losses.
- Fundamental conductor losses depend on skin effect, temperature, and the machine power and specific design.
- Fundamental core losses depend on the airgap flux density, yoke and teeth flux densities, and the supply voltage frequency.
- In a real IM, fundamental electromagnetic losses hardly exist separately.
- Nonfundamental electromagnetic losses are due to space or/and time harmonics.
- The additional (stray) losses, besides the fundamental, are caused by space airgap flux density harmonics or by voltage time harmonics (when PWM converters are used to feed the IM).
- The airgap flux density space harmonics are due to mmf space harmonics, airgap magnetic conductance space harmonics due to slot opening, and leakage or main flux path magnetic saturation.
- No-load stray losses are space harmonics losses at no-load. They are mainly: surface core losses, tooth flux pulsation core losses and tooth flux pulsation cage losses.
- For phase-belt (mmf first) harmonics, 5, 7, 11, 13 and up to the first slot opening harmonic ($N_s/p_1 \pm 1$), the skin depth in laminations is much larger than lamination thickness. So the reaction of core eddy currents on airgap harmonic field is neglected., This way, the rotor (stator) surface core losses are calculated.
- As the number of stator and rotor slots are different from each other $N_s \neq N_r$, the stator and rotor tooth flux pulsates with N_r and N_s pole pairs, respectively, per revolution. Consequently, tooth flux pulsation core losses occur both in the stator and rotor teeth.

  However, such tooth flux pulsations and losses are attenuated by the corresponding currents induced in the rotor cage.
- For the rotor slot skewing by one stator slot pitch, the reaction of cage currents to rotor tooth flux pulsation is almost zero and thus large rotor tooth flux pulsation core losses persist (undamped).
- For straight slot rotors, the rotor tooth flux pulsation produces, as stated above, no-load tooth flux pulsation cage losses. They are not to be neglected, especially in large power motors.

- All space harmonics losses under load are called stray load losses.
- Under load, the no-load stray losses are "amplified". A load amplification factor C_{load} is defined and used to correct the stator and rotor tooth flux pulsation core losses. A distinct load correction factor is used for calculating rotor tooth flux pulsation cage losses. For full pitch stator winding when calculating stray losses, besides the first slot harmonics $N_s \pm p_1$, the first phase belt harmonics (5, 7) are contributing to losses. In skewed rotors, the load stray cage losses are also corrected by a special factor.
- Cast aluminum cages are used with low to medium power induction machines. Their bars are noninsulated from slot walls and thus a transverse (cross-path) resistance between bars through iron occurs. In this case, interbar current losses occur especially for skewed rotors. Thus, despite of the fact that with proper skewing, the stray cage losses are reduced, notable interbar current losses may occur making the skewing less effective.
- There is a critical transverse resistance for which the interbar current losses are maximum. This condition is to be avoided by proper design. Interbar current losses occur both on no-load and on load.
- A number of rules to reduce stray losses are presented. The most interesting is $N_s > N_r$ where even straight rotor slots may be used after safe starting is secured.
- With the advent of power electronics, supply voltage time harmonics occur. With PWM static power converters, around carrier frequency, the highest harmonics occur unless random PWM is used. Frequencies up to 20 kHz occur this way.
- Both stator and rotor conductor losses, due to these voltage time harmonics, are heavily influenced by the skin effect (frequency). In general, as the frequency rises over 2 to 3kHz, for given time harmonic voltage, the conductor losses decrease slightly with frequency while core losses increase with frequency.
- The computation of core losses at high time frequencies (up to 20 kHz) is made accounting for the skin depth in iron δ_{Fe} as all field occurs around slots and on the rotor surface in a thin layer (δ_{Fe} – thick). These slot wall and rotor surface core losses are calculated. Only slot wall core losses are not negligible and they represent about 20 to 30% of all time harmonic losses.

  FEM has been applied recently to calculate all no-load or load losses, thus including implicitly the space harmonics losses. Still the core losses are determined with analytical expressions where the local flux densities variation in time is considered. For field distribution computation, the laminated core electrical conductivity is considered zero. So the computation of time harmonic high frequencies (20 kHz) core losses including the iron skin depth is not available yet with FEM. The errors vary from 5 to 30%.

- However, the effect of skewing, interbar currents, magnetic wedges, and relative number of slots N_s/N_r has been successfully investigated by field-coupled circuit 2D FEMs for reasonable amounts if computation time.
- New progress with 3D FEM is expected in the near future.
- Measurements of losses will be dealt with separately in the chapter dedicated to IM testing.

11.16. REFERENCES

1. E.M. Olin, Determination of Power Efficiency of Rotating Electric Machines: Summation of Losses Versus Input-Output Method, AIEE Trans. Vol. 31, part.2, 1912, pp.1695 – 1719.
2. L. Dreyfus, The Additional Core Losses in A.C. Synchronous Machines, Elektrotechnik und Maschinenbau, Vol.45, 1927, pp.737 – 756 (in German).
3. P.L. Alger, G. Angst, E.J. Davies, Stray Load Losses in Polyphase Induction Machines, AIEE Trans, Vol78, 1957, pp.349 – 357.
4. N. Christofieds, Origin of Load Losses in Induction Machines with Cast Aluminum Rotors, Proc. IEE, Vol.112, 1965, pp.2317 – 2332.
5. A. Odok, Stray Load Losses and Stray Torques in Induction Machines, AIEE Trans Vol.77, part 2, 1958, pp.43 – 53.
6. B. Heller, V. Hamata, Harmonic Field Effects in Induction Machines, (Elsevier Scientific, 1977).
7. A.A. Jimoh, S.R.D. Findlay, M. Poloujadoff, Stray Losses in Induction Machines, part 1 and 2, IEEE Trans Vol.PAS – 104, No.6, 1985, pp.1500 – 1512.
8. C.N. Glen, Stray Load Losses in Induction Motors: A Challenge to Academia, Record of EMD – 1977, IEE Publication, No.444, pp.180 – 184.
9. A.M. Odok, Stray Load Losses and Stray Torques in Induction Machines, AIEE Trans. Vol.77, No.4, 1958, pp.43 – 53.
10. R. Woppler, A Contribution to the Design of Induction Motors with Uninsulated-Cage Rotors, A fur E, Vol.50, No.4, 1966, pp.248 – 252 (in German).
11. K. Matsuse, T. Hayashida, I. Miki, H. Kubota, Y. Yoshida, Effect of Crosspath Resistance Between Adjacent Rotor Bars on Performance of Inverter-fed High Speed Induction Motor, IEEE Trans. Vol.30, No.3., 1994, pp.621 – 627.
12. K. Oberretl, 13 Rules for Minimum Stray Losses in Induction Machines, Bull – Oerlikon, 1969, No.389/390, pp.2-12.
13. J. Singh, Harmonic Analysis and Loss Comparision of Microcomputer-based PWM Strategies for Induction Motor Drive, EMPS Journal, Vol.27, No.10, 1999, pp.1129 – 1140.
14. D.W. Novotny, S.A. Nasar, High Frequency Losses in Induction Motors, Part II, Contract no. MAG 3-940, Final Report, University of Wisconsin, ECE Dept, 1991.

15. G. Bertotti et al., An Improved Estimation of Core Losses in Rotating Electrical Machines, IEEE Trans. Vol.MAG-27, 1991, pp.5007 – 5009.
16. A. Arkkio, A. Miemenmaa, Estimation of Losses in Cage Induction Motors Using FEM, Record of ICEM – 1992, Vol.1, pp.317 – 321.
17. C.I. McClay, S. Williamson, The Variation of Cage Motor Losses with Skew, Record of IEEE – IAS – 1998, Vol.1, pp.79 – 86.
18. S.L. Ho, H.L. Li, W.N. Fu, Inclusion of Interbar Currents in the Network Field-coupled Time-stepping FEM of Skewed Rotor Induction Motors, IEEE Trans. Vol.MAG – 35, No.5, 1999, pp.4218 – 4225.
19. T.J. Flack, S. Williamson, On The Possible Case of Magnetic Slot Wedges to Reduce Iron Losses in Cage Motors, Record of ICEM – 1998, Vol.1, pp.417 – 422.

Chapter 12

THERMAL MODELING AND COOLING

12.1. INTRODUCTION

Besides electromagnetic, mechanical and thermal designs are equally important.

Thermal modeling of an electric machine is in fact more nonlinear than electromagnetic modeling. Any electric machine design is highly thermally constrained.

The heat transfer in an induction motor depends on the level and location of losses, machine geometry, and the method of cooling.

Electric machines work in environments with temperatures varied, say from -20^0C to 50^0C, or from 20^0 to 100^0 in special applications.

The thermal design should make sure that the motor windings temperatures do not exceed the limit for the pertinent insulation class, in the worst situation. Heat removal and the temperature distribution within the induction motor are the two major objectives of thermal design. Finding the highest winding temperature spots is crucial to insulation (and machine) working life.

The maximum winding temperatures in relation to insulation classes shown in Table 12.1.

Table 12.1. Insulation classes

Insulation class	Typical winding temperature limit [^{0}C]
Class A	105
Class B	130
Class F	155
Class H	180

Practice has shown that increasing the winding temperature over the insulation class limit reduces the insulation life L versus its value L_0 at the insulation class temperature (Figure 12.1).

$$\mathrm{Log}\,L \approx a + \frac{b}{T} \qquad (12.1)$$

It is very important to set the maximum winding temperature as a design constraint. The highest temperature spot is usually located in the stator end connections. The rotor cage bars experience a larger temperature, but they are not, in general, insulated from the rotor core. If they are, the maximum (insulation class dependent) rotor cage temperature also has to be observed.

The thermal modeling depends essentially on the cooling approach.

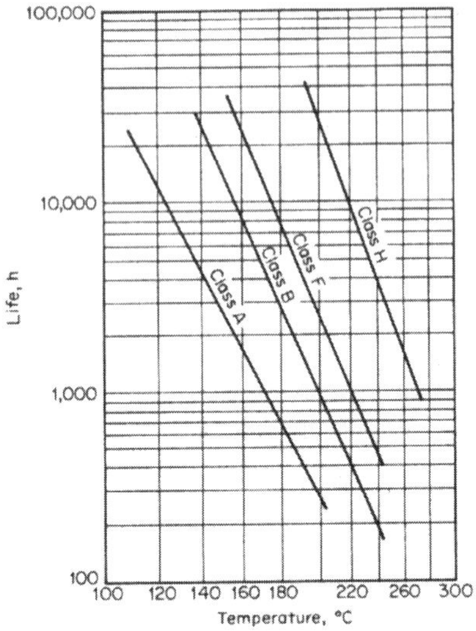

Figure 12.1 Insulation life versus temperature rise

12.2. SOME AIR COOLING METHODS FOR IMs

For induction motors, there are four main classes of cooling systems

- Totally enclosed design with natural (zero air speed) ventilation (TENV)
- Drip-proof axial internal cooling
- Drip-proof radial internal cooling
- Drip-proof radial-axial cooling

In general, fan air-cooling is typical for induction motors. Only for very large powers is a second heat exchange medium (forced air or liquid) used in the stator to transfer the heat to the ambient.

TENV induction motors are typical for special servos to be mounted on machine tools etc., where limited space is available. It is also common for some static power converter-fed IMs, that operate at large loads for extended periods of time at low speeds to have an external ventilator running at constant speed to maintain high cooling in all conditions.

The totally enclosed motor cooling system with external ventilator only (Figure 12.2b) has been extended lately to hundreds of kW by using finned stator frames.

Radial and radial-axial cooling systems (Figure 12.2c, d) are in favor for medium and large powers.

However, axial cooling with internal ventilator and rotor, stator axial channels in the core, and special rotor slots seem to gain ground for very large power as it allows lower rotor diameter and, finally, greater efficiency is obtained, especially with two pole motors (Figure 12.3). [2]

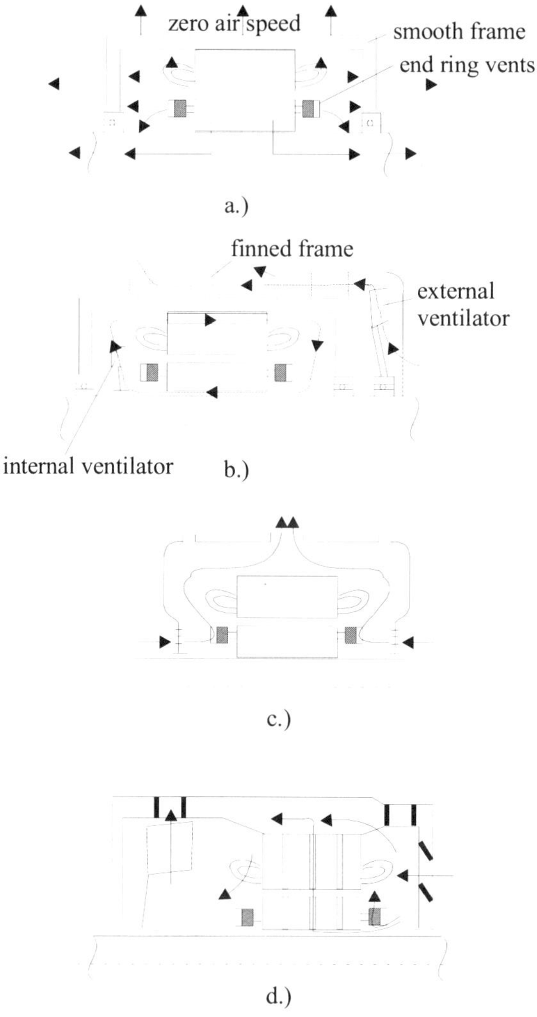

Figure 12.2 Cooling methods for induction machines
a.) totally enclosed naturally ventilated (TENV);
b.) totally enclosed motor with internal and external ventilator
c.) radially cooled IM d.) radial – axial cooling system

The rotor slots are provided with axial channels to facilitate a kind of direct cooling.

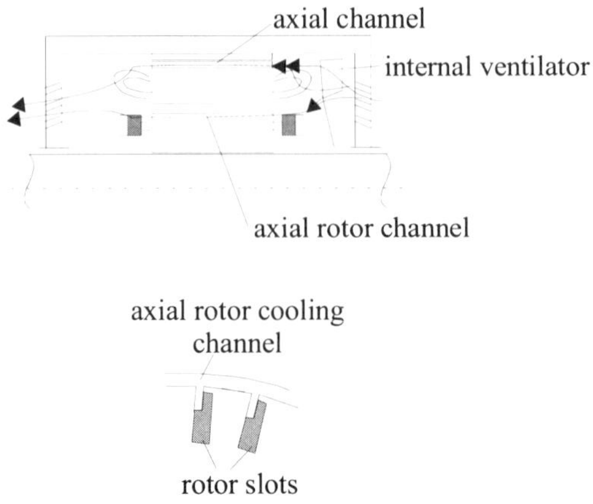

Figure. 12.3 Axial cooling of large IMs

The rather complex (anisotropic) structure of the IM for all cooling systems presented in Figures 12.2 and 12.3 suggests that the thermal modeling has to be rather difficult to build.

There are thermal circuit models and distributed (FEM) models. Thermal circuit models are similar to electric circuits and they may be used both for thermal steady state and transients. They are less precise but easy to handle and require a smaller computation effort. In contrast, distributed (FEM) models are more precise but require large amounts of computation time.

We will define first the elements of thermal circuits based on the three basic methods of heat transfer: conduction, convection and radiation.

12.3. CONDUCTION HEAT TRANSFER

Heat transfer is related to thermal energy flow from a heat source to a heat sink.

In electric (induction) machines, the thermal energy flows from the windings in slots to laminated core teeth through the conductor insulation and slot line insulation.

On the other hand, part of the thermal energy in the end-connection windings is transferred through thermal conduction through the conductors axially toward the winding part in slots. A similar heat flow through thermal conduction takes place in the rotor cage and end rings.

There is also thermal conduction from the stator core to the frame through the back core iron region and from rotor cage to rotor core, respectively, to shaft and axially along the shaft. Part of the conduction heat now flows through the slot insulation to core to be directed axially through the laminated core. The presence of lamination insulation layers will make the thermal conduction along the axial direction more difficult. In long stack IMs, axial temperature differentials of a few degrees (less than 10^0C in general), (Figure 12.4), occur.

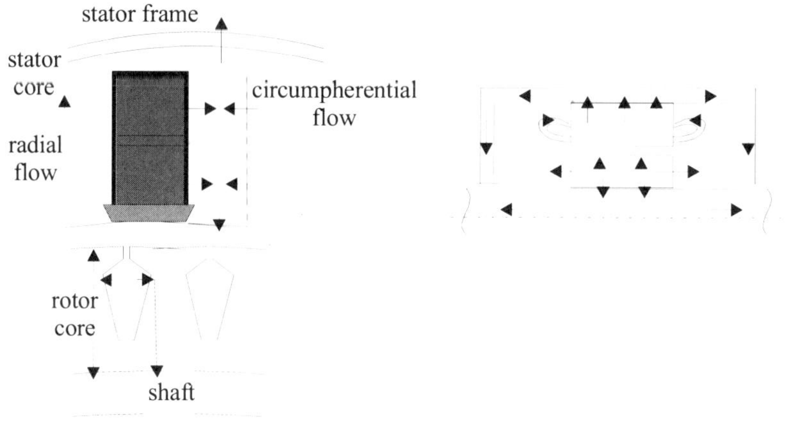

Figure 12.4 Heat conduction flow routs in the IM

So, to a first approximation, the axial heat flow may be neglected.

Second, after accounting for conduction heat flow from windings in slots to the core teeth, the machine circumferential symmetry makes possible the neglecting of circumferential temperature variation.

So we end up with a one-dimensional temperature variation, along the radial direction. For this crude approximation defining thermal conduction, convection, and radiation, and of the equivalent circuit becomes a rather simple task.

The Fourier's law of conduction may be written, for steady state, as

$$\nabla(-K\Delta\theta) = q \tag{12.2}$$

where q is heat generation rate per unit volume (W/m^3); K is thermal conductivity (W/m, ^{0}C) and θ is local temperature.

For one-dimensional heat conduction, Equation (12.2), with constant thermal conductivity K, becomes:

$$-K\frac{\partial^2\theta}{\partial x^2} = q \tag{12.3}$$

A basic heat conduction element (Figure 12.5) shows that power Q transported along distance l of cross section A is

$$Q \approx q \cdot l \cdot A \tag{12.4}$$

with q, A – constant along distance l.

The thermal conduction resistance R_{con} may be defined as similar to electrical resistance.

$$R_{con} = \frac{l}{KA}\left[{}^0C / W\right] \tag{12.5}$$

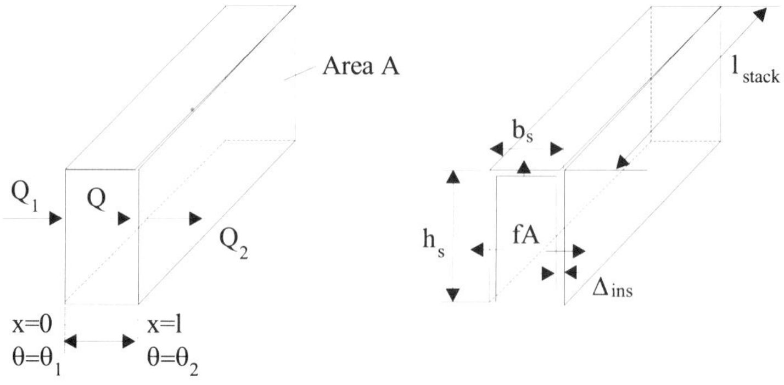

Figure 12.5 One dimensional heat conduction

Temperature takes the place of voltage and power (losses) replaces the electrical current.

For a short l, the Fourier's law in differential form yields

$$f \approx -K\frac{\Delta\theta}{\Delta x};\ f - \text{heat flow density}\ \left[W/m^2\right] \tag{12.6}$$

If the heat source is in a thin layer,

$$f = \frac{p_{cos}}{A} \tag{12.7}$$

p_{cos} in watts is the electric power producing losses and A the cross-section area.

For the heat conduction through slot insulation Δ_{ins} (total, including all conductor insulation layers from the slot middle (Figure 12.5)), the conduction area A is

$$A = (2h_s + b_s)l_{stack}N_s;\ \ N_s - \text{slots / stator} \tag{12.8}$$

The temperature differential between winding in slots and the core teeth $\Delta\theta_{Co}$ is

$$\Delta\theta_{Cos} = p_{cos}R_{con};\ R_{con} = \frac{\Delta_{ins}}{AK} \tag{12.9}$$

In well-designed IMs, $\Delta\Theta_{cos} < 10^0C$ with notably smaller values for small power induction motors.

The improvement of insulation materials in terms of thermal conductivity and in thickness reduction have been decisive factors in reducing the slot insulation conductor temperature differential. Thermal conductivity varies with temperature and is constant only to a first approximation. Typical values are given in Table 12.2. The low axial thermal conductivity of the laminated cores is evident.

Table 12.2. Thermal conductivity

Material	Thermal conductivity (W/m^0C)	Specific heat coefficient C_s ($J/Kg/^0C$)
Copper	383	380
Aluminum Carbonsteel	204 45	900
Motor grade steel	23	500
Si steel lamination – Radial; – Axial	 20 – 30 2.0	490
Micasheet	0.43	-
Varnished cambric	2.0	-
Press board Normex	0.13	-

12.4. CONVECTION HEAT TRANSFER

Convection heat transfer takes place between the surface of a solid body (the stator frame) and a fluid (air, for example) by the movement of the fluid.

The temperature of a fluid (air) in contact with a hotter solid body rises and sets a fluid circulation and thus heat transfer to the fluid occurs.

The heat flow out of a body by convection is

$$q_{conv} = hA\Delta\theta \tag{12.10}$$

where A is the solid body area in contact with the fluid; $\Delta\theta$ is the temperature differential between the solid body and bulk of the fluid, and h is the convection heat coefficient ($W/m^{2}\cdot{}^0C$).

The convection heat transfer coefficient depends on the velocity of the fluid, fluid properties (viscosity, density, thermal conductivity), the solid body geometry, and orientation. For free convention (zero forced air speed and smooth solid body surface [2])

$$\begin{aligned} h_{Co} &\approx 2.158(\Delta\theta)^{0.25} \quad \text{vertical up - } W/\left(m^{2\,0}C\right) \\ h_{Co} &\approx 0.496(\Delta\theta)^{0.25} \quad \text{vertical down - } W/\left(m^{2\,0}C\right) \\ h_{Co} &\approx 0.67(\Delta\theta)^{0.25} \quad \text{horizontal - } W/\left(m^{2\,0}C\right) \end{aligned} \tag{12.11}$$

where $\Delta\theta$ is the temperature differential between the solid body and the fluid.

For $\Delta\theta = 20^0C$ (stator frame $\theta_1 = 60^0C$, ambient temperature $\theta_2 = 40^0C$) and vertical – up surface

$$h_{Co} = 2.158(60-40)^{0.25} = 4.5W/\left(m^{2\,0}C\right)$$

When air is blown with a speed U along the solid surfaces, the convection heat transfer coefficient h_c is

$$h_C{}^0(u) = h_{Co}\left(1 + K\sqrt{U}\right) \tag{12.12}$$

with K = 1.3 for perfect air blown surface; K = 1.0 for the winding end connection surface, K = 0.8 for the active surface of rotor, K = 0.5 for the external stator frame.

Alternatively,

$$h_C(u) = 1.77\frac{U^{0.75}}{L^{0.25}} \quad W/\left(m^{2\,0}C\right) \text{ for } U > 5m/s \tag{12.13}$$

U in m/s and L is the length of surface in m.

For a closed air blowed surface – inside the machine:

$$h_C^{\ c}(U) = h_{Co}\left(1 + K\sqrt{U}\right)\left(1 - a/2\right);\quad a = \frac{\theta_{air}}{\theta_a} \tag{12.14}$$

θ_{air}–local air heating; θ_a–heating (temperature) of solid surface.

In general, θ_{air} = 35 – 40^{0}C while θ_a varies with machine insulation class. So, in general, a < 1.

For convection heat transfer coefficient in axial channels of length, L (12.13) is to be used.

In radial cooling channels, $h_c^{\ c}(U)$ does not depend on the channel's length, but only on speed.

$$h_c^{\ c}(U) \approx 23.11U^{0.75}\left(W/m^{2\,0}C\right) \tag{12.15}$$

12.5. HEAT TRANSFER BY RADIATION

Between two bodies at different temperatures there is a heat transfer by radiation. One body radiates heat and the other absorbs heat. Bodies which do not reflect heat, but absorb it, are called black bodies.

Energy radiated from a body with emissivity ε to black surroundings is

$$q_{rad} = \sigma\varepsilon A\left(\theta_1^4 - \theta_2^4\right) = \sigma\varepsilon A\left(\theta_1 + \theta_2\right)\left(\theta_1^2 + \theta_2^2\right)\left(\theta_1 - \theta_2\right) \tag{12.16}$$

σ–Boltzmann's constant: σ = $5.67\cdot10^{-8}$ W/(m^{2}K^4); ε – emissivity; for a black painted body ε = 0.9; A–radiation area.

In general, for IMs, the radiated energy is much smaller than the energy transferred by convection except for totally enclosed natural ventilation (TENV) or for class F(H) motor with very hot frame (120 to 150°C).

For the case when θ_2 = 40° and θ_1 = 80°C, 90°C, 100°C, ε = 0.9, h_{rad} = 7.67, 8.01, and 8.52 W/(m^2 °C).

For TENV with h_{Co} = 4.56 W/(m^2, °C) (convection) the radiation is superior to convection and thus it cannot be neglected. The total (equivalent) convection coefficient

$h_{(c+r)0} = h_{Co} + h_{rad} \geq 12$ W/(m^2, °C).

The convection and radiation combined coefficients $h_{(c+r)0} \approx 14.2$W/(m^2, °C) for steel unsmoothed frames, $h_{(c+r)0}$ = 16.7W/(m^2, °C) for steel smoothed frames,

$h_{(c+r)0}$ = 13.3W/(m^2, °C) for copper/aluminum or lacquered or impregnated copper windings. In practice, for design purposes, this value of h_{Co}, which enters Equations (12.12 through 12.14), is, in fact, $h_{(c+r)0}$, the combined convection radiation coefficient.

It is well understood that the heat transfer is three dimensional and as K, h_c and h_{rad} are not constants, the heat flow, even under thermal steady state, is a very complex problem. Before advancing to more complex aspects of heat flow, let us work out a simple example.

Example 12.1. One – dimensional simplified heat transfer

In an induction motor with p_{Co1} = 500 W, p_{Co2} = 400 W, p_{iron} = 300 W, the stator slot perimeter $2h_s + b_s$ = (2.25 + 8) mm, 36 stator slots, stack length: l_{stack} = 0.15 m, an external frame diameter D_e = 0.30 m, finned area frame (4 to 1 area increase by fins), frame length 0.30m, let us calculate the winding in slots temperature and the frame temperature, if the air temperature increase around the machine is 10°C over the ambient temperature of 30°C and the slot insulation total thickness is 0.8 mm. The ventilator is used and the end connection/coil length is 0.4.

Solution

First, the temperature differential of the windings in slots has to be calculated. We assume here that all rotor heat losses crosses the airgap and it flows through the stator core toward the stator frame.

In this case, the stator winding in slot temperature differential is (12.3)

$$\Delta\theta_{cos} = \frac{\Delta_{ins} p_{Co1}\left(1 - \frac{l_{endcon}}{l_{coil}}\right)}{K_{ins} N_s (2h_s + b_s) l_{stack}} = \frac{0.8 \cdot 10^{-3} \cdot 500 \cdot 0.6}{2.0 \cdot 0.36 \cdot 0.058 \cdot 0.15} = 3.83\ ^0\mathrm{C}$$

Now we consider that stator winding in slot losses, rotor cage losses, and stator core losses produce heat that flows radially through stator core by conduction without temperature differential (infinite conduction!).

Then all these losses are transferred to ambient through the motor frame through combined free convection and radiation.

$$\theta_{core} - \theta_{air} = \frac{q_{total}}{h_{(c+r)0} A_{frame}} = \frac{(500 + 400 + 300)}{14.2 \cdot \pi \cdot 0.30 \cdot 0.3 \cdot (4/1)} = 74.758\ ^0\mathrm{C}$$

with θ_{air} = 40°, $\theta_{ambient}$ = 30°, the frame (core) temperature θ_{core} = 40 + 74.758 = 114.758°C and the winding in slots temperature $\theta_{cos} = \theta_{core} + \Delta\theta_{cos}$ = 114.758 + 3.83 = 118.58°C. In such TENV induction machines, the unventilated stator winding end turns are likely to experience the highest temperature spot. However, it is not at all simple to calculate the end connection temperature distribution.

12.6. HEAT TRANSPORT (THERMAL TRANSIENTS) IN A HOMOGENOUS BODY

Although the IM is not a homogenous body, let us consider the case of a homogenous body – where temperature is the same all over.

The temperature of such a body varies in time if the heat produced inside, by losses in the induction motor, is applied at a certain point in time–as after starting the motor. The heat balance equation is

$$\underset{\substack{\text{losses}\\\text{per unit}\\\text{time in W}}}{P_{loss}} = \underset{\substack{\text{heat accumulation}\\\text{in the body}}}{Mc_t \frac{d(T - T_0)}{dt}} + \underset{\substack{\text{heat transfer from the body}\\\text{through convection, conduction, radiation}}}{A\,h_{\substack{cond\\(conv)}}(T - T_0)} \tag{12.17}$$

M–body mass (in Kg), c_t–specific heat coefficient (J/(Kg·^{0}C))
A–area of heat transfer from (to) the body
h–heat transfer coefficient

Denoting by

$$C_t = Mc_t \text{ and } R_{\substack{conv\\(rad)}} = \frac{1}{Ah}; \left(R_{cond} = \frac{1}{KA}\right) \tag{12.18}$$

equation (12.17) becomes

$$P_{loss} = C_t \frac{d(T - T_0)}{dt} + \frac{(T - T_0)}{R_t} \tag{12.19}$$

This is similar to a R_t, C_t parallel electric circuit fed from a current source P_{loss} with a voltage $T - T_0$ (Figure 12.6).

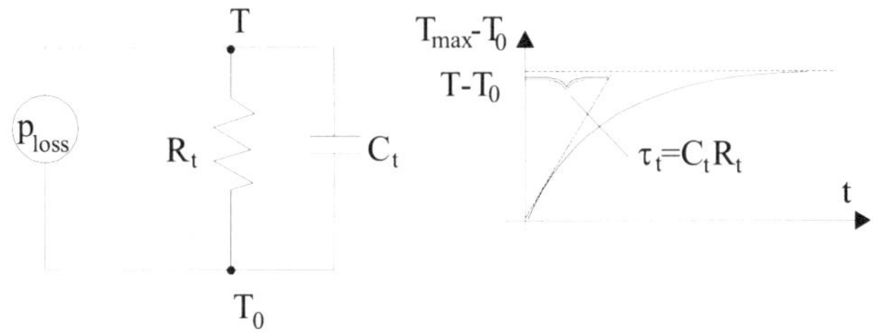

Figure 12.6 Equivalent thermal circuit

For steady state, C_t does not enter Equation (12.17) and the equivalent circuit (Figure 12.6).

The solution of this electric circuit is evident.

$$T = (T_{max} - T_0)\left(1 - e^{-\frac{t}{\tau_t}}\right) + T_0 e^{-\frac{t}{\tau_t}} \tag{12.20}$$

The thermal time constant $\tau_t = C_t R_t$ is very important as it limits the machine working time with a certain level of losses and given cooling conditions. Intermittent operation, however, allows for more losses (more power) for the same given maximum temperature, T_{max}.

The thermal time constant increases with machine size and effectivity of the cooling system. A TENV motor is expected to have a smaller thermal time constant than a constant speed ventilator-cooled configuration.

12.7. INDUCTION MOTOR THERMAL TRANSIENTS AT STALL

The IM at stall is characterized by very large conductor losses. Core loss may be neglected by comparison. If the motor remains at stall the temperature of the windings and cores increases in time. There is a maximum winding temperature limit T_{max}^{copper} given by the insulation class, (155^0C for class F) which should not be surpassed. This is to maintain a reasonable working life for conductor insulation. The machine is designed for lower winding temperatures at full continuous load.

To simplify the problem, let us consider two extreme cases, one with long end connection stator winding and the other a long stack and short end connections.

For the first case we may neglect the heat transfer by conduction to the winding in slots portion. Also, if the motor is totally enclosed, the heat transfer through free convection to the air inside the machine is rather small (because this air gets hot easily). In fact, all the heat produced in the end connection (p_{Coend}) serves to increase end winding temperature.

$$\frac{\Delta\theta_{endcon}}{\Delta t} \approx \frac{p_{Coend}}{C_{endcon}};\ C_{endcon} = M_{endcon} c_{tcopper} \tag{12.21}$$

with p_{Coend} = 1000 W, M_{endcon} = 1 Kg, $c_{tcopper}$ = 380 J/Kg/^{0}C, the winding would heat up 115°C (from 40 to 155°C) in a time interval Δt.

$$(\Delta t)_{40\to 155^0 C} = \frac{115\cdot 1\cdot 380}{1000} = 43.7\text{seconds} \tag{12.22}$$

Now if the machine is already hot at, say, 100^0C, $\Delta\theta_{endcon} = 155^0 - 100^0 = 55^0$C. So the time allowed to keep the machine at stall is reduced to

$$(\Delta t)_{100\to 155^0 C} = \frac{55\cdot 1\cdot 380}{1000} = 20.9\,\text{seconds}$$

The equivalent thermal circuit for this oversimplified case is shown on Figure 12.7a.

On the contrary, for long stacks, only the winding losses in slots are considered. However, this time some heat accumulated in the core and the same heat is transferred through thermal conduction through insulation from slot conductors to core.

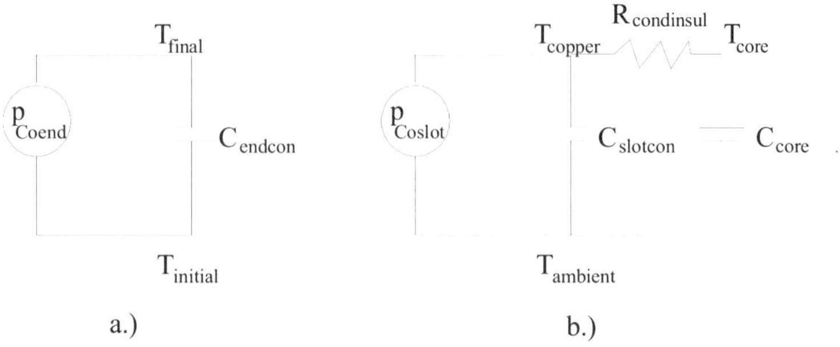

Figure 12.7 Simplified thermal equivalent circuits for stator winding temperature rise at stall a.) long end connections; b.) long stacks

With p_{Coslot} = 1000 W, $M_{slotcopper}$ = 1 Kg, $C_{slotcopper}$ = 380 J/Kg/^{0}C, insulation thickness 0.3 mm K_{ins} = 6 W/m/^{0}C, c_{tcore} = 490 J/Kg/^{0}C, slot height h_s = 20 mm, slot width b_s = 8 mm, slot number: N_s = 36, M_{core} = 5 Kg, stack length l_{stack} = 0.1 m,

$$R_{condinsul} = \frac{\Delta_{ins}}{(2h_s + b_s)l_{stack} N_s K_{ins}} = \frac{0.3 \cdot 10^{-3}}{(2 \cdot 20 + 8) \cdot 10^{-3} \cdot 0.1 \cdot 36 \cdot 2} = 8.68 \cdot 10^{-4} \left({}^0 C / m\right) \tag{12.23}$$

$$\begin{aligned} C_{slotcon} &= M_{slotcon} c_{tcopper} = 1 \cdot 380 = 380 J/{}^0 C \\ C_{core} &= M_{core} c_{tcore} = 5 \cdot 490 = 2450 J/{}^0 C \end{aligned} \tag{12.24}$$

The temperature rise in the copper and core versus time (solving the circuit of Figure 12.7b) is

$$\begin{aligned} T_{copper} - T_{ambient} &= P_{Coslot} \left[\frac{t}{C_{slotcon} + C_{core}} + \frac{\tau_t^{\,2}}{C_{slotcon} \tau_{con}} \left(1 - e^{-\frac{t}{\tau_t}} \right) \right] \\ T_{core} - T_{ambient} &= \frac{P_{Coslot}}{C_{slotcon} + C_{core}} \left[t - \tau_t \left(1 - e^{-\frac{t}{\tau_t}} \right) \right] \end{aligned} \tag{12.25}$$

with

$$\tau_{con} = C_{slotcon} R_{condinsul}; \ \tau_t = \frac{C_{slotcon} \cdot C_{core}}{C_{slotcon} + C_{core}} R_{condinsul} \tag{12.26}$$

As expected, the copper temperature rise is larger than core temperature rise. Also, the core accumulates a good part of the winding-produced heat, so the time after which the conductor insulation temperature limit (155^{0}C for class F) is reached at stall is larger than for the end connection windings.

The thermal time constant τ_t is

$$\tau_t = \frac{380 \cdot 2450}{2450 + 380} \cdot 8.68 \cdot 10^{-4} = 0.2855\text{seconds} \tag{11.27}$$

The second term in (12.25) dies out quickly so, in fact, only the first, linear term counts. As $C_{core} >> C_{slotcon}$, the time to reach the winding insulation temperature limit is increased a few times: for $T_{ambient} = 40^0C$ and $T_{copper} = 155^0C$ from (12.25).

$$(\Delta t)_{40^0 C \to 155^0 C} = \frac{(155-40)(380+2450)}{1000} = 325.45\text{seconds}$$

Consequently, longer stack motors seem advantageous if they are to be used frequently at or near stall at high currents (torques).

12.8. INTERMITTENT OPERATION

Intermittent operation with IMs occurs both in line-start constant frequency and voltage, and in variable speed drives (variable frequency and voltage).

In most line-start applications, as the voltage and frequency stay constant, the magnetization current I_m is constant. Also, the rotor circuit is dominated by the rotor resistance term (R_r/S) and thus the rotor current I_r is 90^0 ahead of I_m and the torque may be written as

$$T_e \approx 3p_1 L_m I_m I_r = 3p_1 L_m I_m \sqrt{I_s^2 - I_m^2} \tag{12.28}$$

The torque is proportional to the rotor current, and the stator and rotor winding losses and core losses are related to torque by the expression

$$P_{dis} = p_{core} + p_{Costator} + p_{Corotor} \approx 3R_s I_s^2 + 3R_r I_r^2 + p_{core} = \\ = 3R_s\left(I_m^2 + \left(\frac{T_e}{3p_1 L_m I_m}\right)^2\right) + 3R_r\left(\frac{T_e}{3p_1 L_m I_m}\right)^2 + \frac{3(\omega_1 L_m I_m)^2}{R_{m||}} \tag{12.29}$$

For fractional power (sub kW) or low speed ($2p_1 = 10, 12$), motors I_m (magnetization current) may reach 70 to 80% of rated current I_{sn} and thus (12.29) remains a rather complicated expression of torque, with I_n = const.

For medium and large power (and $2p_1 = 2, 4, 6$) IMs, in general $I_m < 30\% I_{sn}$ and I_m may be neglected in (12.29), which becomes

$$P_{dis} \approx (p_{core})_{const} + 3(R_s + R_r)\left(\frac{T_e}{3p_1 L_m I_m}\right)^2 \tag{12.30}$$

Electromagnetic losses are proportional to torque squared. For variable speed drives with IMs, the magnetization current is reduced with torque reduction to cut down (minimize) core and winding losses together.

Thus, (12.29) may be used to obtain $\partial P_{dis}/\partial I_m = 0$ and obtain $I_m(T_e)$ and, again, from (12.29), $P_{dis}(T_e)$. Qualitatively for the two cases, the electromagnetic loss variation with torque is shown on Figure 12.8.

As expected, for an on-off sequence (t_{ON}, t_{OFF}), more than rated (continuous duty) losses are acceptable during on time. Therefore, motor overloading is permitted. For constant magnetization current, however, as the losses are proportional to torque squared, the overloading is not very large but still similar to the case of PM motors [3], though magnetization losses $3R_sI_m^2$ are additional for the IM.

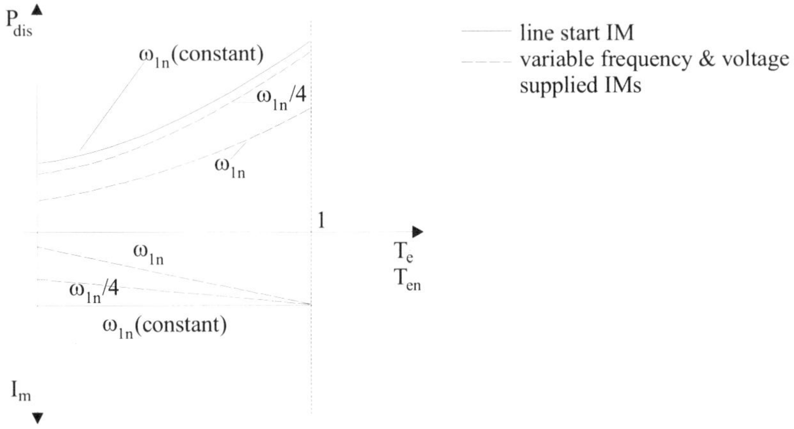

Figure 12.8 Electromagnetic losses P_{dis} and magnetisation current I_m versus torque

The duty cycle d may be defined as

$$d = \frac{t_{ON}}{t_{ON} + t_{OFF}} \tag{12.31}$$

Complete use of the machine in intermittent operation is made if, at the end of ON time, the rated temperature of windings is reached. Evidently the average losses during ON time P_{dis} may surpass the rated losses P_{disn}, for continuous steady state operation. By how much depends both on the t_{ON} value and on the machine equivalent thermal time constant τ_t.

$$\tau_t = R_t C_t \tag{12.32}$$

C_t–thermal capacity of winding (J/^{0}C); R_t–thermal resistance between windings and the surroundings (^{0}C/W). The value of τ_t depends on machine geometry, rated power and speed, and on the cooling system, and may run from tens of seconds to tens of minutes or even several hours.

12.9. TEMPERATURE RISE (T_{ON}) AND FALL (T_{OFF}) TIMES

The loss (dissipated power) P_{dis} may be considered approximately proportional to load squared.

$$\frac{P_{dis}}{P_{disn}} \approx \left(\frac{T_e}{T_{en}}\right)^2 = K_{load}^2 \tag{12.33}$$

The temperature rise, for an equivalent homogeneous body, during t_{ON} time is (12.19),

$$T - T_0 = RP_{dis}\left(1 - e^{-\frac{t_{ON}}{\tau_t}}\right) + (T_c - T_0)e^{-\frac{t_{ON}}{\tau_t}} \tag{12.34}$$

where T_c is the initial temperature and T_0 the ambient temperature ($T_{max} - T_0 = R \cdot P_{dis}$), with P_{disn} ($K_{load} = 1$), $T(t_{ON}) = T_{rated}$. Replacing in (12.34) P_{dis} by

$$P_{dis} = P_{disn} \cdot K_{load}^2 = \frac{(T_{rated} - T_0)}{R} \cdot K_{load}^2 \tag{12.35}$$

with $T_{max} = T_{rated}$,

$$(T_{rated} - T_0)\left[1 - K_{load}^2\left(1 - e^{-\frac{t_{ON}}{\tau_t}}\right)\right] = (T_c - T_0)e^{-\frac{t_{ON}}{\tau_t}} \tag{12.36}$$

Equation (12.36) shows the dependence of t_{ON} time, to reach the rated winding temperature, from an initial temperature T_c for a given overload factor K_{load}. As expected t_{ON} time decreases with the rise of initial winding temperature T_c.

During t_{OFF} time, the losses are zero, and the initial temperature is $T_c = T_r$. So with $K_{load} = 0$ and $T_c = T_r$, (12.36) becomes

$$T - T_0 = (T_r - T_0) \cdot e^{-\frac{t_{OFF}}{\tau_t}} \tag{12.37}$$

For steady state intermittent operation, however, the temperature at the end of OFF time is equal, again, to T_c.

$$T_c - T_0 = (T_r - T_0) \cdot e^{-\frac{t_{OFF}}{\tau_t}} \tag{12.38}$$

For given initial (low) T_c, final (high) T_{rated} temperatures, load factor K_{load}, and thermal time constant τ_t, Equations (12.36) and (12.38) allow for the computation of t_{ON} and t_{OFF} times.

Now, introducing the duty cycle $d = \frac{t_{ON}}{t_{ON} + t_{OFF}}$ to eliminate t_{OFF}, from (12.36) and (12.38) we obtain

$$K_{load} = \sqrt{\frac{1 - e^{-\frac{t_{ON}}{d\tau_t}}}{1 - e^{-\frac{t_{ON}}{\tau_t}}}} \tag{12.39}$$

It is to be noted that using (12.36)–(12.38) is most practical when $t_{ON} < \tau_t$ as it is known that the temperature stabilizes after 3 to $4\tau_t$.

For example, with $t_{ON} = 0.2\ \tau_t$ and d = 25%, $K_{load} = 1.743$, τ_t = minutes, it follows that t_{ON} = 15 minutes and t_{OFF} = 45 minutes.

For very short on-off cycles $((t_{ON} + t_{OFF}) < 0.2\ \tau_t)$, we may use Taylor's formula to simplify (12.39) to

$$K_{load} = \sqrt{\frac{1}{d}} \tag{12.40}$$

For short cycles, when the machine is overloaded as in (12.40), the medium loss will be the rated one.

For a single pulse, we may use d = 0 in (12.39) to obtain

$$K_{load} = \sqrt{\frac{1}{1 - e^{-\frac{t_{ON}}{\tau_t}}}} \tag{12.41}$$

As expected for one pulse, K_{load} allowed to reach rated temperature for given t_{ON} is larger than for repeated cycles.

With same start and end of the cooling period temperature T_c, the t_{ON} and t_{OFF} times are again obtained from (12.39) and (12.38), respectively, even for a single cycle (heat up, cool down).

$$t_{OFF} = -\tau_t \ln\left[K_{load}^2 - \left(K_{load}^2 - 1\right)e^{\frac{t_{ON}}{\tau_t}}\right] \tag{12.42}$$

For given K_{load} from (12.41), we may calculate t_{ON}/τ, while from (12.42) t_{OFF} time, for a single steady state cycle, T_c to T_r to T_c temperature excursion ($T_r > T_c$) is obtained. It is also feasible to set t_{OFF} and, for given K_{load}, to determine from (12.42), t_{ON}.

Rather simple formulas as presented in this chapter, may serve well in predicting the thermal transients for given overload and intermittent operation.

After this almost oversimplified picture of IM thermal modeling, let us advance one more step by building more realistic thermal equivalent circuits.

12.9 MORE REALISTIC THERMAL EQUIVALENT CIRCUITS FOR IMs

Let us consider the overall heating of the stator (or rotor) winding with radial channels. The air speed and temperatures inside the motor are taken as known. (The ventilator design is a separate problem which, produces the airflow rate and temperatures of air as its output, for given losses in the machine and its geometry.)

A half longitudinal cross section is shown in Figure 12.9a for the stator and in Figure 12.9b for the rotor.

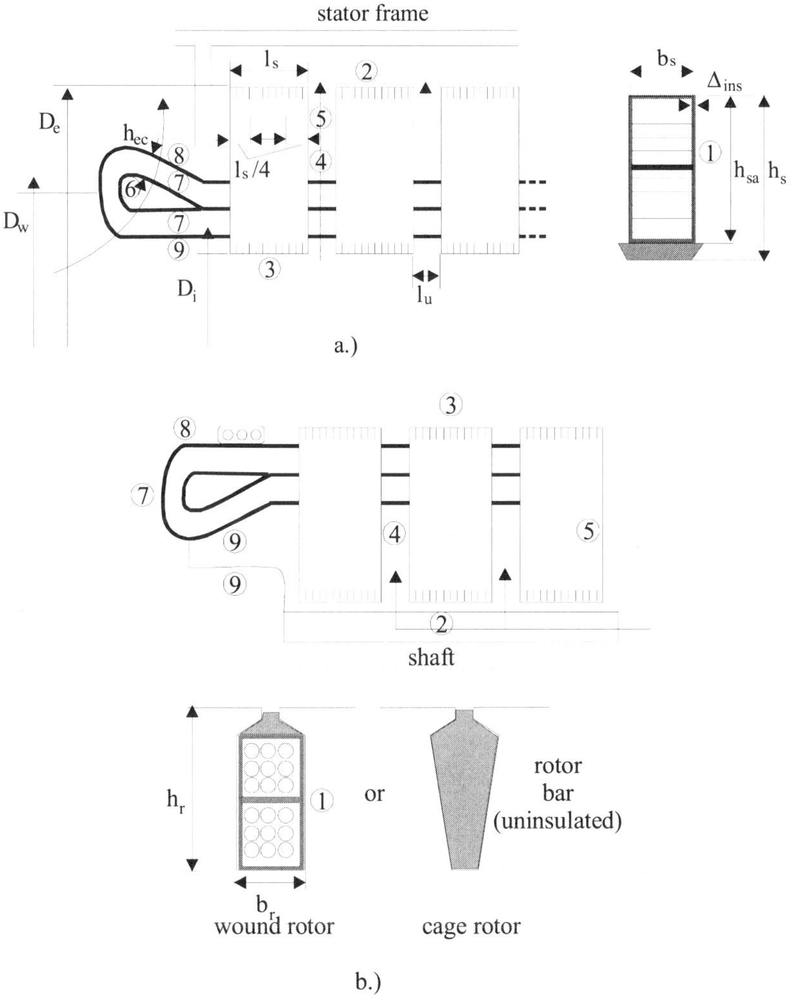

Figure 12.9 IM with radial ventilating channels
a.) stator winding, b.) rotor winding

The objective here is to set a more realistic equivalent thermal circuit and explicitate the various thermal resistances R_{t1}, … R_{t9}.

To do so a few assumptions are made.

- The winding end connection losses do not contribute to the stator (rotor) stack heating
- The end-connection and in-slot winding temperature, respectively, do not vary axially or radially
- The core heat center is placed $l_s/4$ away from elementary stack radial channel

The equivalent circuit with thermal resistances is shown in Figure 12.10.

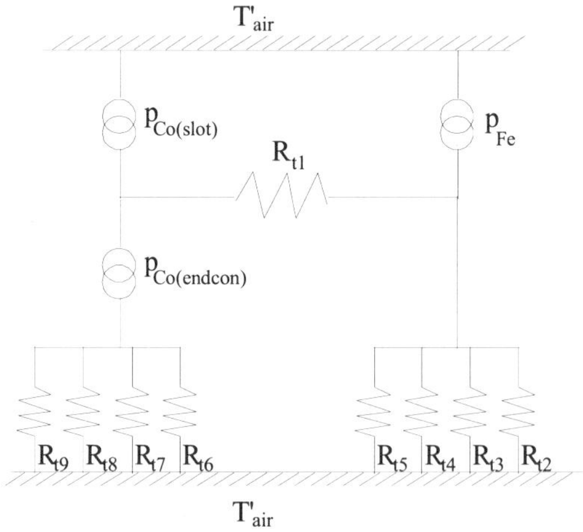

Figure 12.10 Equivalent thermal circuit for stator or rotor windings

T'_{air} – air temperature from the entrance into the machine to the winding surface.

In Figure 12.10,

$P_{Co(slot)}$ – winding losses for the part situated in slots

$P_{Co(endcon)}$ – end connection winding losses

P_{Fe} – iron losses

R_{t1} – slot insulation thermal resistance from windings in slots to core

R_{t2} – thermal resistance from core to air which is exterior (interior) to it – stator core to frame; rotor core to shaft

R_{t3} – thermal resistance from core to airgap cooling air

R_{t4} – thermal resistance from the iron core to the air in the ventilation channels

R_{t5} – thermal resistance from core to air in ventilation channel

R_{t6} – thermal resistance towards the air inside the end connections (it is ∞ for round conductor coils)

R_{t7} – thermal resistance from the frontal side of end connections to the air between neighbouring coils

R_{t8} – thermal resistance from end connections to the air above them

R_{t9} – thermal resistance from end connections to the air below them

Approximate formulas for $R_{t1} - R_{t9}$ are

$$R_{t1} = \frac{\Delta_{ins}}{K_{ins} A_1}; \quad A_1 \approx (2h_s + b_s) n_s l_s N_s \tag{12.43}$$

n_s – number of elementary stacks ($n_s - 1$ – radial channels)
l_s – elementary stack length; N_s – stator slot number
K_{ins} – slot insulation heat transfer coefficient

$$R_{t2} = \frac{h_{cs(r)}}{K_{Fe} A_2} + \frac{1}{h_{c2} A_2}; \quad A_2 = (\pi D_e - b_f n_f) n_s l_s \tag{12.44}$$

b_f – fins width, n_f – fins number for the stator. For the rotor, $b_f = 0$ and D_e is replaced by rotor core interior diameter, $h_{cs(r)}$ – back core radial thickness, K_{core} – core radial thermal conductivity and h_{c2} – thermal convection coefficient (all parameters in IS units). Note that h_{c3} is influenced by the air speed as in (12.14).

$$R_{t3} = \frac{h_{sa}}{K_{Fe} A_3} + \frac{1}{h_{c3} A_3}; \quad A_3 = (\pi D_i - b_f N_s) n_s l_s \tag{12.45}$$

h_{c3} is the convection thermal coefficient as influenced by the air speed in the airgap in presence of radial channels (Equation 12.14).

$$R_{t4} = \frac{\Delta_{inscon}}{K_{copper} A_4} + \frac{1}{h_{c4} A_4}; \quad A_4 = N_s (2h_{s0} + b_s)(n_s - 1) l_v \tag{12.46}$$

l_v – axial length of ventilation channel

$$R_{t5} = \frac{2(n_s - 1)(l_s / 4)}{K_{Felong} A_{5s(r)}} + \frac{1}{h_{c5} A_{5s(r)}}$$

$$A_{5s} = 2n_s \left(\frac{\pi}{4} \left(D_e^2 - (D_i + 2h_s)^2 \right) + N_s b_{ts} h_s \right) \tag{12.47}$$

$$A_{5r} = 2n_s \left(\frac{\pi}{4} \left((D_i - 2h_r)^2 - D_{ir}^2 \right) + N_r b_{tr} h_s \right) \tag{12.48}$$

N_r – rotor slots, $b_{ts(r)}$ – stator (rotor) tooth width, h_{c5} – convection thermal coefficient as influenced by the air speed in the radial channels.

$$R_{t6} = \frac{\Delta_{inscon}}{K_{copper} A_6} + \frac{1}{h_{c6} A_6}$$

$$R_{t7} = \frac{\Delta_{inscon}}{K_{copper} A_7} + \frac{1}{h_{c7} A_7}$$

$$R_{t8} = \frac{\Delta_{inscon}}{K_{copper} A_8} + \frac{1}{h_{c8} A_8}$$
$$R_{t9} = \frac{\Delta_{inscon}}{K_{copper} A_9} + \frac{1}{h_{c9} A_9} \quad (12.49)$$

$R_{t6} - R_{t9}$ refer to winding end connection heat transfer by thermal conduction through the electrical insulation and, by convention, through the circulating air in the machine. Areas of heat transfer $A_6 - A_9$ depend heavily on the coils shape and their arrangement as end connections in the stator (or rotor).

For round wire coils with insulation between phases, the situation is even more complicated as the heat flow through the end connections toward their interior or circumferentially may be neglected ($R_6 = R_7 = \infty$).

As the air temperature inside the machine was considered uniform, the stator and rotor equivalent thermal circuits as in Figure 12.10 may be treated rather independently ($p_{Fe} = 0$ in the rotor, in general). In the case where there is one stack (no radial channels), the above expressions are still valid with $n_s = 1$ and, thus, all heat transfer resistances related to radial channels are ∞ ($R_4 = R_5 = \infty$).

12.10. A DETAILED THERMAL EQUIVALENT CIRCUIT FOR TRANSIENTS

The ultimate detailed thermal equivalent circuit of the IM should account for the three dimensional character of heat flow in the machine.

Although this may be done, a two dimensional model is used. However we may break the motor axially into a few segments and "thermally" connect these segments together.

To account for thermal transients, the thermal equivalent circuit should contain thermal resistances R_{ti}(^{0}C/W) and capacitors C_{ti}(J/^{0}C) and heat sources (W) (Figure 12.11).

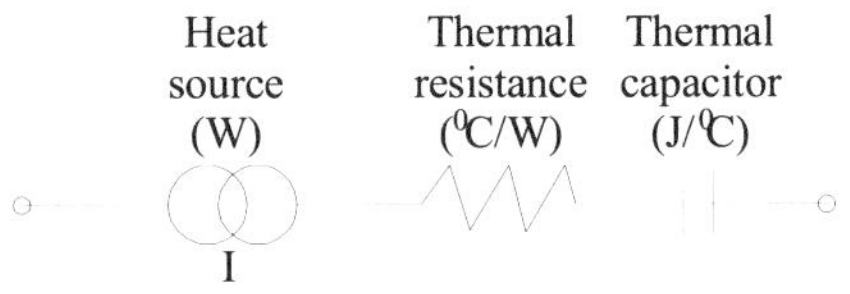

Figure 12.11 Thermal circuit elements with units

A detailed thermal equivalent circuit–in the radial plane–emerges from the more realistic thermal circuit of Figure 12.9 by dividing the heat sources into more components (Figure 12.12).

The stator conductor losses are divided into their in-slot and overhang (end-connection) components. The same thing could be done for the rotor (especially for wound rotors). Also, no heat transport through conduction from end connections to the coils section in slot is considered in Figure 12.12, as the axial heat flow is neglected.

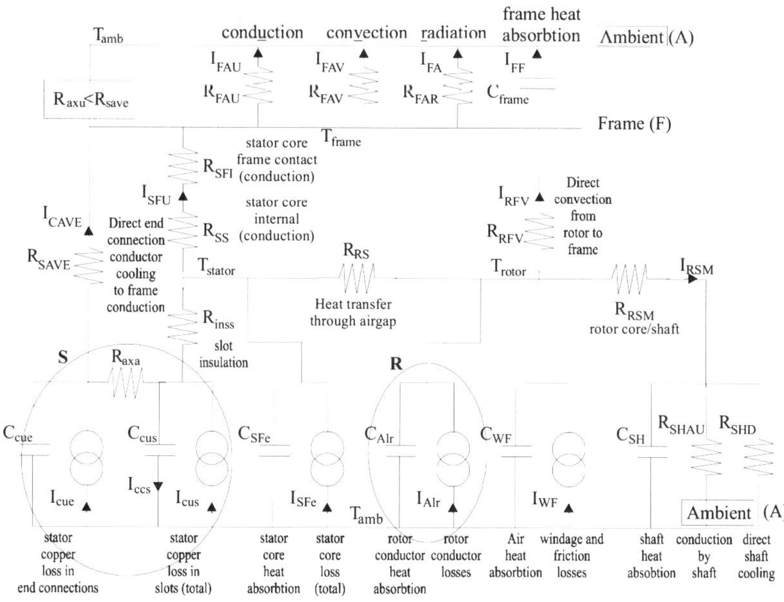

Figure 12.12 A detailed thermal equivalent circuit for IMs

Due to the machine pole symmetry, the model in Figure 12.12 is in fact one dimensional, that is the temperatures vary only radially. A kind of similar model is to be found in Reference [3] for PM brushless motors and in Reference 4 for induction motors.

In Reference [5], a thermal model for three-dimensional heat flow is presented.

It is possible to augment the model with heat transfer along circumferential direction and along axial dimension to obtain a rather complete thermal equivalent circuit with hundreds of nodes.

12.11. THERMAL EQUIVALENT CIRCUIT IDENTIFICATION

As shown in paragraph 12.9, the various thermal resistances R_{ti} (or conductances $G_{ti} = 1/R_{ti}$) and thermal capacitances (C_{ti}) may be approximately calculated through analytical formulas.

As the thermal conductivities, K_i, convection and radiation heat transfer coefficients h_i are dependent on various geometrical factors, cooling system and average local temperature, and at least for thermal transients, their identification through tests would be beneficial.

Once the thermal equivalent circuit structure is settled (Figure 12.12, for example), with various temperatures as unknowns, its state-space matrix equation system is

$$\frac{dX}{dt} = AX + BP \tag{12.50}$$

$$\begin{aligned} X &= [T_1.....T_j.....T_n]^t \text{ - temperature matrix} \\ P &= [P_1.....P_j.....P_m]^t \text{ - power loss matrix} \end{aligned} \tag{12.51}$$

P_j have to be known from the electromagnetic model. A and B are coefficient matrixes built with R_{ti}, C_{ti}.

Ideally n temperature sensors to measure T_1, ... T_n versus time would be needed. If it is not feasible to install so many, the model is to be simplified so that all temperatures in the model are measured.

Having experimental values of $T_i(t)$, system (12.50) may be used to determine, by an optimization method, the parameters of the equivalent circuit. In essence, the squared error between calculated and measured (after filtering) temperatures is to be minimum over the entire time span. In Reference 6 such a method is used and the results look good.

As some of the thermal parameters may be calculated, the method can be used to identify them from the losses and then check the heat division from its center.

For example, it may be found that for low power IMs at rated speed, 65% of the rotor cage losses is evacuated through airgap, 20% to the internal frame, and 15% by shaft bearing.

Also the heat produced in the stator end connection windings for such motors is divided as: 20% to the internal frame and end shields (brackets) by convection and the rest of 80% to the stator core by conduction (axially).

Consequently, based on such results, the detailed equivalent circuit should contain a conduction resistance branch from end connections to stator core as 80% of the heat goes through it (R_{axa} in Figure 12.12).

As an example of an ingenious procedure to measure the R_i, C_i parameters, or the loss distribution, we notice here the case of turning off the IM and measuring the temperature decrease in location of interest versus time.

From (12.18) in the steady state conditions:

$$p_{loss} = \frac{1}{R_t}(T_0 - T_a) \tag{12.52}$$

The temperature derivative at t = 0 (from 12.18), when the heat input is turned off, is

$$\left(\frac{dT}{dt}\right)_{t=0} = -\frac{1}{R_t C_t}(T_0 - T_a) \tag{12.53}$$

Finally

$$p_{loss} = -C_t\left(\frac{dT}{dt}\right)_{t\to 0} \tag{12.54}$$

Measuring the temperature gradient at the moment when the motor is turned off, with C_t known, allows for the calculation of local power loss in the machine just before the machine was turned off. [7,8]

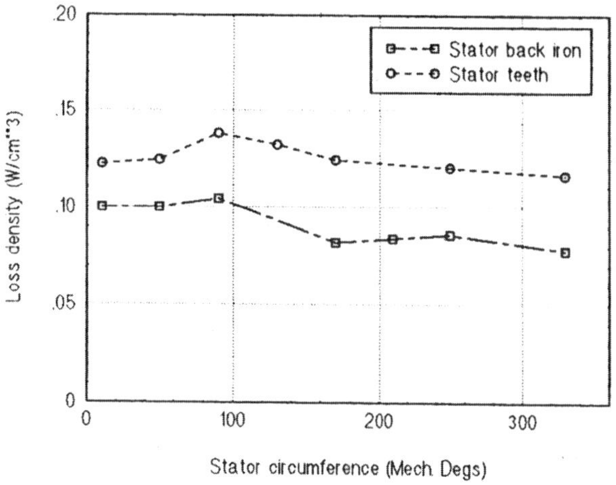

a.)

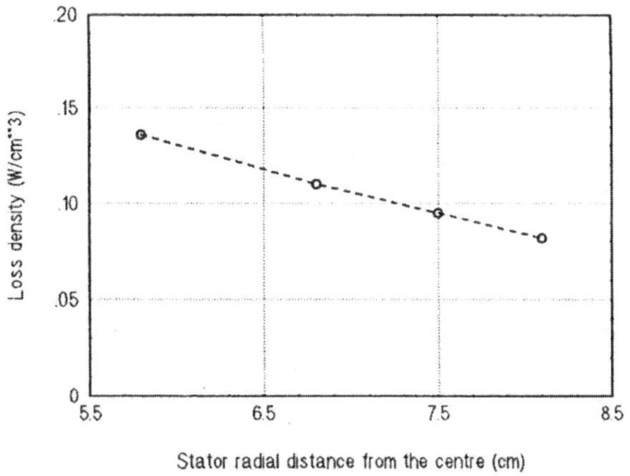

b.)

Figure 12.13 Iron loss density distribution (W/cm^3)
a.) along circumpherential direction, b.) along radial direction, c.) along axial direction (continued)

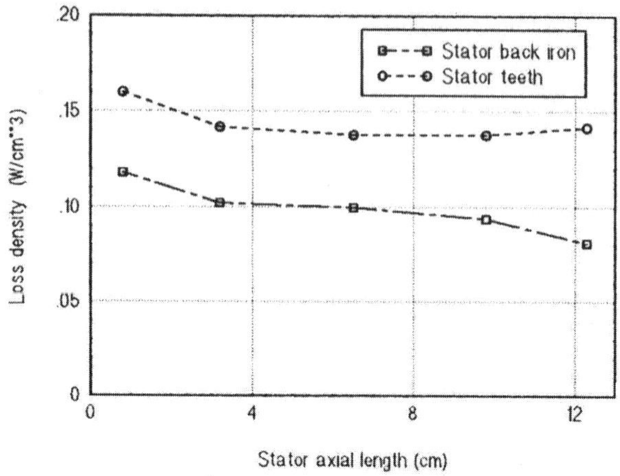

c.)

Figure 12.13 (continued)

This way the radial or axial variation of losses (especially in the core) may be obtained, provided that small enough temperature sensors are placed in key locations. The winding average temperature is measured after turn-off by the d.c. voltage/current method of stator winding resistance.

$$R_s(T) = (R_s)_{20^0C}\left[1 + \frac{1}{273}(T - T_{amb})\right] \tag{12.55}$$

Typical results are shown in Figure 12.13. [8] For the 4 pole IM in question, the temperature variation along circumferential and axial directions in the back core is small, but it is notable in the stator teeth. A notable decrease of loss density with radial distance is present, as expected, (Figure 12.13c).

12.12. THERMAL ANALYSIS THROUGH FEM

In theory, the three-dimensional FEM alone could lead to a fully realistic temperature distribution for a machine of any power and size, provided the localization of heat sources and their levels are known.

The differential equation for heat flow is

$$\nabla(K\Delta T) + p_{loss} = \gamma c \frac{\partial T}{\partial t} \tag{12.56}$$

with K – local thermal conductivity (W/m^0C); γ – local density (Kg/m^3), p_{loss} – losses per unit volume (W/m^3); and c – specific heat coefficient ($J/^0CKg$). The coefficients K, γ, c vary throughout the machine.

Two types of boundary conditions are usually present:

- Dirichlet conditions:

$$T(x_b, y_b, z_b, t) = T^* \tag{12.57}$$

The ambient temperature is such a boundary around the machine.

- Newman conditions:

$$q - K_x \frac{\partial T}{\partial x} n_x - K_y \frac{\partial T}{\partial y} n_y - K_z \frac{\partial T}{\partial z} n_z = 0 \tag{12.58}$$

n_x, n_y, n_z are the x, y, z components of unit vector rectangular to the respective boundary surface; q – the heat flow through the surface.

As a 3D FEM would require large amounts of computation time, 2D FEM models have been built to study the temperature distribution either in the radial cross section or in the axial cross section.

The radial cross section has a geometrical symmetry as seen in Figure 12.14.

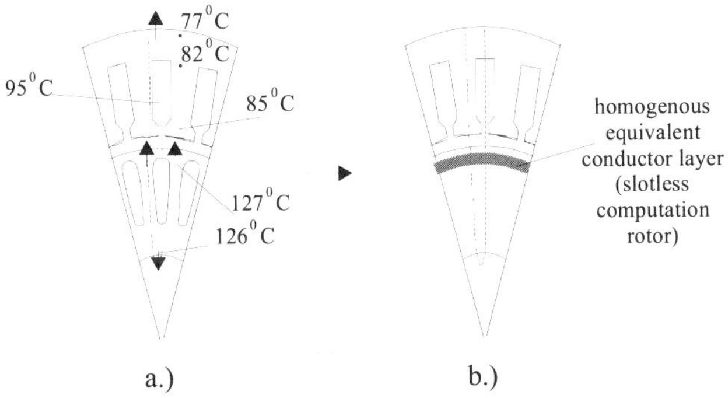

Figure 12.14 Computation sector a.) and its restructuring to avoid motion influence b.)

To avoid the rotation influence on the modeling in the airgap zone, the rotor sector is replaced by a motion-independent computation sector (Figure 12.14b). [9]

After the temperature on the rotor surface is calculated and considered "frozen," the actual rotor sector is modeled.

It was found that the stator temperature varies radially and axially in a visible manner, while the temperature gradient is much smaller in the rotor (Figure 12.14a). [9] Similar results with 2D FEM are presented in Reference [10].

3D FEM may be used for a more complete IM thermal modeling, treated independently or together with the electromagnetic model for load (or speed) transients; however the computation time becomes large.

Still, due to the dependence of thermal resistances and capacitors and loss distribution on many "imponderable" factors, experimental methods (such as advanced calorimetric ones) are needed to validate theoretical results, especially when new machine configurations or design specifications are encountered.

12.13. SUMMARY

- Any IM design is thermally limited according to the conductor insulation class mainly: A(105^{0}C), B(130^{0}C), F(155^{0}C), H(180^{0}C)
- The insulation life decreases very rapidly with temperature increase over the rated value (for extra 10^{0}C the life may be halved).
- Cooling methods, used to extract the heat from the machine and thus limit the temperature, are paramount in IM design.
- From totally enclosed, naturally ventilated (TENV) induction motors through drip-proof axial internal, radial internal to radial-axial cooling, the sophistication of cooling methods and their heat removal capacity increases steadily.
- Heat transfer takes place through thermal conduction, convection, and radiation. For induction machine design, convection and radiation heat transfer coefficients are usually lumped together. Radiation is notable for TENV configurations.
- The heat transfer steady state and transients may be approached either by equivalent thermal circuits (similar to R–C electric circuits) or by numerical methods (FEM).
- Thermal resistances R_{ti} (^{0}C/W) are introduced for conduction, convection, and radiation heat transfer and thermal capacitances C_{ti} (J/^{0}C) stand for heat absorption in the various parts of the motor.
- The heat sources p_{lossi}(W) represent current sources in the equivalent circuit.
- Thermal resistances decrease with cooling air speed.
- The windings have the highest temperature spots, in general.
- It has been shown that, at least in small power IMs, about 80% of the heat (loss) in the stator coils overhangs is transmitted through conduction to the winding part in slots and to the core.
- It has also been shown that 60 to 70% of rotor losses are transferred through the airgap to the stator core.
- The axial variation of temperature depends on the relative stack length, method of cooling, and machine power. However, for powers up to hundreds of kW, unistack motors with axial external cooling and finned frames have recently become popular for general purpose designs.
- Information like this should be instrumental into developing adequate simplified equivalent thermal circuits. FEM–2D and 3D, are very instrumental in getting information for designing practical equivalent circuits.
- Due to the thermal modeling complexity, theory and tests should go hand in hand in the future.

12.14. REFERENCES

1. S.A. Nasar, Handbook of Electric Machines, McGraw-Hill Inc., Chapter 12 (by A.J.Spisak).
2. W.H. McAdams, Heat Transmission, McGraw-Hill, New York, 1942, 2nd edition.
3. J. Hondershot, T.J.E. Miller, PM Brushless Motor Designs, OUP, 1995, Chapter 15.
4. G. Bellenda, L. Ferraris, A. Tenconi, A New Simplified Thermal Model for Induction Motors for EVs Applications, Record of 7th International Conference on EMD, IEE Conf, Publ. 412, 195, Durham, UK, pp.11 – 15.
5. W. Guyglewics – Kacerka, J. Mukosiej, Thermal Analysis of Induction Motor with Axial and Radial Cooling Ducts, Record of ICEM – 1992, Vol.3., pp.971 – 975.
6. G. Champenois, D. Roye, D.S. Zhu, Electrical and Thermal Performance Predictions in Inverter-fed Squirrel Cage Induction Motor Drives, EMPS – Vol.22, No.3, 1994, pp.355 – 369.
7. M. Benamrouche et al., Determination of Iron Loss Distribution in Inverter-fed Induction Motors, EMPS Vol.25, No.6, 1997, pp.649 – 660.
8. H. Benamrouche et.al., Determination of Iron and Stray Load Losses in Induction Motors Using a Thermometric Method, EMPS Vol.26, No.1, 1998, pp.3 – 12.
9. J. Roger, G. Jinenez, The Finite Element Application to the Study of the Temperature Distribution Inside Electric Rotating Machines, Record of ICEM – 1992, Vol.3., pp.976 – 980.
10. D. Sarkar, Approximate Analysis of Temperature Rise in an Induction Motor During Dynamic Braking, EMPS Vol.26, No.6, 1998, p.585 – 599.

Chapter 13

INDUCTION MACHINE TRANSIENTS

13.1. INTRODUCTION

Induction machines undergo transients when voltage, current, and (or) speed undergo changes. Turning on or off the power grid leads to starting transients an induction motor.

Reconnecting an induction machine after a short-lived power fault (zero current) is yet another transient. Bus switching for large power induction machines feeding urgent loads also qualifies as large deviation transients.

Sudden short-circuits, at the terminals of large induction motors lead to very large peak currents and torques. On the other hand more and more induction motors are used in variable speed drives with fast electromagnetic and mechanical transients.

So, modeling transients is required for power-grid-fed (constant voltage and frequency) and for PWM converter-fed IM drives control.

Modeling the transients of induction machines may be carried out through circuit models or coupled field/circuit models (through FEM). We will deal first with phase-coordinate abc model with inductance matrix exhibiting terms dependent on rotor position.

Subsequently, the space phasor (d–q) model is derived. Both single and double rotor circuit models are dealt with. Saturation is also included in the space-phasor (d–q) model. The abc–dq model is then derived and applied, as it is adequate for nonsymmetrical voltage supplies and PWM converter-fed IMs.

Reduced order d–q models are used to simplify the study of transients for low and large motors, respectively.

Modeling transients with the computation of cage bar and end-ring currents is required when cage and/or end-ring faults occur. Finally the FEM coupled field circuit approach is dealt with.

Autonomous generator transients are left out as they are treated in the chapter dedicated to induction generators.

13.2. THE PHASE COORDINATE MODEL

The induction machine may be viewed as a system of electric and magnetic circuits which are coupled magnetically and/or electrically.

An assembly of resistances, self inductances, and mutual inductances is thus obtained. Let us first deal with the inductance matrix.

A symmetrical (healthy) cage may be replaced by a wound three-phase rotor. [2] Consequently, the IM is represented by six circuits, (phases) (Figure 14.1). Each of them is characterized by a self inductance and 5 mutual inductances.

The stator and rotor phase self inductances do not depend on rotor position if slot openings are neglected. Also, mutual inductances between stator phases and rotor phases, respectively, do not depend on rotor position. A sinusoidal distribution of windings is assumed. Finally, stator/rotor phase mutual inductances depend on rotor position ($\theta_{er} = p_1\theta_r$).

The induction matrix, $L_{abca_rb_rc_r}(\theta_{er})$ is

$$\left[L_{abca_rb_rc_r}(\theta_{er})\right] = \begin{bmatrix} L_{aa} & L_{ab} & L_{ac} & L_{aa_r} & L_{ab_r} & L_{ac_r} \\ L_{ab} & L_{bb} & L_{bc} & L_{ba_r} & L_{bb_r} & L_{bc_r} \\ L_{ac} & L_{bc} & L_{cc} & L_{ca_r} & L_{cb_r} & L_{cc_r} \\ L_{aa_r} & L_{ba_r} & L_{ca_r} & L_{a_ra_r} & L_{a_rb_r} & L_{a_rc_r} \\ L_{ab_r} & L_{bb_r} & L_{cb_r} & L_{a_rb_r} & L_{b_rb_r} & L_{b_rc_r} \\ L_{ac_r} & L_{bc_r} & L_{cc_r} & L_{a_rc_r} & L_{b_rc_r} & L_{c_rc_r} \end{bmatrix} \tag{13.1}$$

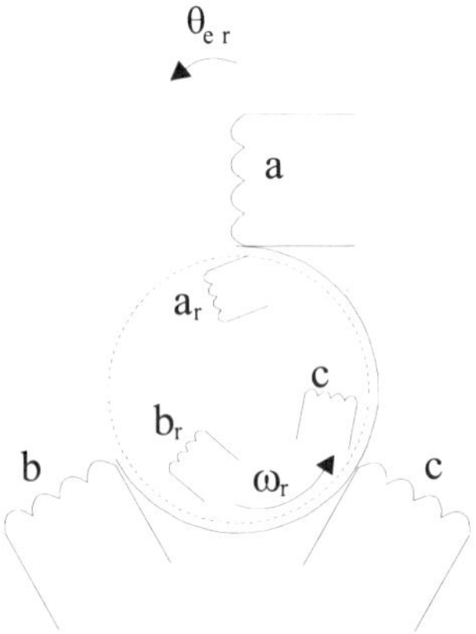

Figure 13.1 Three-phase IM with equivalent wound rotor

with

$$L_{aa} = L_{bb} = L_{cc} = L_{ls} + L_{ms}; \quad L_{ab} = L_{ac} = L_{bc} = -L_{ms}/2;$$

$$L_{aa_r} = L_{bb_r} = L_{cc_r} = L_{srm}\cos\theta_{er}; \quad L_{a_ra_r} = L_{b_rb_r} = L_{c_rc_r} = L_{lr}^r + L_{mr}^r;$$

$$L_{c_ra} = L_{a_rb} = L_{b_rc} = L_{srm}\cos\left(\theta_{er} - \frac{2\pi}{3}\right); \quad L_{a_rb_r} = L_{a_rc_r} = L_{b_rc_r} = -L_{mr}^r/2; \tag{13.2}$$

$$L_{c_rb} = L_{b_ra} = L_{a_rc} = L_{srm}\cos\left(\theta_{er} + \frac{2\pi}{3}\right)$$

Assuming a sinusoidal distribution of windings, it may be easily shown that

$$L_{srm} = \sqrt{L_{ms} \cdot L_{mr}^r} \tag{13.3}$$

Reducing the rotor to stator is useful especially for cage rotor IMs, no access to rotor variables is available.

In this case, the mutual inductance becomes equal to self inductance $L_{srm} \rightarrow L_{sm}$ and the rotor self inductance equal to the stator self inductance $L_{mr}^{r} \rightarrow L_{sm}$.

To conserve the fluxes and losses with stator reduced variables,

$$\frac{i_{ar}}{i_{ar}^{r}} = \frac{i_{br}}{i_{br}^{r}} = \frac{i_{cr}}{i_{cr}^{r}} = \frac{L_{srm}}{L_{sm}} = K_{rs} \tag{13.4}$$

$$\frac{V_{ar}}{V_{ar}^{r}} = \frac{V_{br}}{V_{br}^{r}} = \frac{V_{cr}}{V_{cr}^{r}} = \frac{i_{ar}^{r}}{i_{ar}} = \frac{i_{br}^{r}}{i_{br}} = \frac{i_{cr}^{r}}{i_{cr}} = \frac{1}{K_{rs}} \tag{13.5}$$

$$\frac{R_r}{R_r^{r}} = \frac{L_{lr}}{L_{lr}^{r}} = \frac{1}{K_{rs}^{2}} \tag{13.6}$$

The expressions of rotor resistance R_r, leakage inductance L_{lr}, both reduced to the stator for both cage and wound rotors are given in Chapter 6.

The same is true for R_s, L_{ls} of the stator. The magnetization self inductance L_{sm} has been calculated in Chapter 5.

Now the matrix form of phase coordinate (variable) model is

$$\begin{gathered}
[V] = [R][i] + \frac{d}{dt}[\Psi] \\
[V] = \left[V_a, V_b, V_c, V_{a_r}, V_{b_r}, V_{c_r}\right]^T \\
[i] = \left[i_a, i_b, i_c, i_{a_r}, i_{b_r}, i_{c_r}\right]^T \\
[R] = \mathrm{Diag}\left[R_s, R_s, R_s, R_r, R_r, R_r\right]
\end{gathered} \tag{13.7}$$

$$[\Psi] = \left[L_{abca_rb_rc_r}(\theta_{er})\right][i] \tag{13.8}$$

$$\left[L_{abca_rb_rc_r}(\theta_{er})\right] =$$

$$= \begin{bmatrix}
L_{ls}+L_{sm} & -L_{sm}/2 & -L_{sm}/2 & L_{sm}\cos\theta_{er} & L_{srm}\cos\left(\theta_{er}+\frac{2\pi}{3}\right) & L_{srm}\cos\left(\theta_{er}-\frac{2\pi}{3}\right) \\
-L_{sm}/2 & L_{ls}+L_{sm} & -L_{sm}/2 & L_{srm}\cos\left(\theta_{er}-\frac{2\pi}{3}\right) & L_{sm}\cos\theta_{er} & L_{srm}\cos\left(\theta_{er}+\frac{2\pi}{3}\right) \\
-L_{sm}/2 & -L_{sm}/2 & L_{ls}+L_{sm} & L_{srm}\cos\left(\theta_{er}+\frac{2\pi}{3}\right) & L_{srm}\cos\left(\theta_{er}-\frac{2\pi}{3}\right) & L_{sm}\cos\theta_{er} \\
L_{sm}\cos\theta_{er} & L_{srm}\cos\left(\theta_{er}-\frac{2\pi}{3}\right) & L_{srm}\cos\left(\theta_{er}+\frac{2\pi}{3}\right) & L_{ls}+L_{sm} & -L_{sm}/2 & -L_{sm}/2 \\
L_{srm}\cos\left(\theta_{er}+\frac{2\pi}{3}\right) & L_{sm}\cos\theta_{er} & L_{srm}\cos\left(\theta_{er}-\frac{2\pi}{3}\right) & -L_{sm}/2 & L_{ls}+L_{sm} & -L_{sm}/2 \\
L_{srm}\cos\left(\theta_{er}-\frac{2\pi}{3}\right) & L_{srm}\cos\left(\theta_{er}+\frac{2\pi}{3}\right) & L_{sm}\cos\theta_{er} & -L_{sm}/2 & -L_{sm}/2 & L_{ls}+L_{sm}
\end{bmatrix} \tag{13.9}$$

With (13.8), (13.7) becomes

$$[V]=[R][i]+\left([L]+\left[\frac{\partial L}{\partial i}\right][i]\right)\frac{d[i]}{dt}+\frac{d[L]}{d\theta_{er}}[i]\frac{d\theta_{er}}{dt} \quad (13.10)$$

Multiplying (13.10) by $[i]^T$ we get

$$[i]^T[V]=[i]^T R[i]+\frac{d}{dt}\left(\frac{1}{2}[L][i][i]^T\right)+\frac{1}{2}[i]^T\frac{d}{d\theta_{er}}[L][i]\omega_r \quad (13.11)$$

The first term represents the winding losses, the second, the stored magnetic energy variation, and the third, the electromagnetic power P_e.

$$P_e=T_e\frac{\omega_r}{p_1}=\frac{1}{2}[i]^T\frac{d[L]}{d\theta_{er}}[i]\omega_r \quad (13.12)$$

The electromagnetic torque T_e is

$$T_e=\frac{1}{2}p_1[i]^T\frac{d[L]}{d\theta_{er}}[i] \quad (13.13)$$

The motion equation is

$$\frac{J}{p_1}\frac{d\omega_r}{dt}=T_e-T_{load}; \quad \frac{d\theta_{er}}{dt}=\omega_r \quad (13.14)$$

An 8th order nonlinear model with time-variable coefficients (inductances) has been obtained, even with core loss neglected.

Numerical methods are required to solve it, but the computation time is prohibitive. Consequently, the phase coordinate model is to be used only for special cases as the inductance and resistance matrix may be assigned any values and rotor position dependencies.

The complex or space variable model is now introduced to get rid of rotor position dependence of parameters.

13.3. THE COMPLEX VARIABLE MODEL

Let us use the following notations:

$$a=e^{j\frac{2\pi}{3}}; \quad \cos\frac{2\pi}{3}=\mathrm{Re}[a]; \quad \cos\frac{4\pi}{3}=\mathrm{Re}[a^2]$$
$$\cos\left(\theta_{er}+\frac{2\pi}{3}\right)=\mathrm{Re}\left[ae^{j\theta_{er}}\right]; \quad \cos\left(\theta_{er}+\frac{4\pi}{3}\right)=\mathrm{Re}\left[a^2e^{j\theta_{er}}\right] \quad (13.15)$$

Based on the inductance matrix, expression (13.9), the stator phase a and rotor phase a_r flux linkages Ψ_a and Ψ_{ar} are

$$\Psi_a=L_{ls}i_a+L_{ms}\mathrm{Re}\left[i_a+ai_b+a^2i_c\right]+L_{ms}\mathrm{Re}\left[\left(i_{a_r}+ai_{b_r}+a^2i_{c_r}\right)e^{j\theta_{er}}\right] \quad (13.16)$$

$$\Psi_{a_r} = L_{lr} i_{a_r} + L_{ms} \mathrm{Re}\left[i_{a_r} + a i_{b_r} + a^2 i_{c_r}\right] + L_{ms} \mathrm{Re}\left[\left(i_a + a i_b + a^2 i_c\right) e^{-j\theta_{er}}\right] \quad (13.17)$$

We may now introduce the following complex variables as space phasors: [1]

$$\bar{i}_s^s = \frac{2}{3}\left(i_a + a i_b + a^2 i_c\right) \quad (13.18)$$

$$\bar{i}_r^r = \frac{2}{3}\left(i_{a_r} + a i_{b_r} + a^2 i_{c_r}\right) \quad (13.19)$$

Also,

$$\mathrm{Re}\left(\bar{i}_s^s\right) = i_a - \frac{1}{3}\left(i_a + i_b + i_c\right) \quad (13.20)$$

$$\mathrm{Re}\left(\bar{i}_r^r\right) = i_{a_r} - \frac{1}{3}\left(i_{a_r} + i_{b_r} + i_{c_r}\right) \quad (13.21)$$

In symmetric steady-state and transient regimes,

$$i_a + i_b + i_c = i_{a_r} + i_{b_r} + i_{c_r} = 0 \quad (13.22)$$

With the above definitions, Ψ_a and Ψ_{ar} become

$$\Psi_a = L_{ls} \mathrm{Re}\left(\bar{i}_s^s\right) + L_m \mathrm{Re}\left(\bar{i}_s^s + \bar{i}_r^r e^{j\theta_{er}}\right); \quad L_m = \frac{3}{2} L_{ms} \quad (13.23)$$

$$\Psi_{a_r} = L_{lr} \mathrm{Re}\left(\bar{i}_r^r\right) + L_m \mathrm{Re}\left(\bar{i}_r^r + \bar{i}_s^s e^{-j\theta_{er}}\right) \quad (13.24)$$

Similar expressions may be derived for phases b_r and c_r. After adding them together, using the complex variable definitions (13.18) and (13.19) for flux linkages and voltages, also, we obtain

$$\begin{aligned} \bar{V}_s^s &= R_s \bar{i}_s^s + \frac{d\bar{\Psi}_s^s}{dt}; \quad \bar{\Psi}_s^s = L_s \bar{i}_s^s + L_m \bar{i}_r^r e^{j\theta_{er}}; \\ \bar{V}_r^r &= R_r \bar{i}_r^r + \frac{d\bar{\Psi}_r^r}{dt}; \quad \bar{\Psi}_r^r = L_r \bar{i}_r^r + L_m \bar{i}_s^s e^{-j\theta_{er}} \end{aligned} \quad (13.25)$$

where

$$L_s = L_{sl} + L_m; \quad L_r = L_{rl} + L_m \quad (13.26)$$

$$\bar{V}_s^s = \frac{2}{3}\left(V_a + aV_b + a^2 V_c\right); \quad \bar{V}_r^r = \frac{2}{3}\left(V_{a_r} + aV_{b_r} + a^2 V_{c_r}\right) \quad (13.27)$$

In the above equations, stator variables are still given in stator coordinates and rotor variables in rotor coordinates.

Making use of a rotation of complex variables by the general angle θ_b in the stator and $\theta_b - \theta_{er}$ in the rotor, we obtain all variables in a unique reference rotating at electrical speed ω_b,

$$\omega_b = \frac{d\theta_b}{dt} \tag{13.28}$$

$$\begin{aligned} \overline{\Psi}_s^s &= \overline{\Psi}_s^b e^{j\theta_b}; \quad \overline{i}_s^s = \overline{i}_s^b e^{j\theta_b}; \quad \overline{V}_s^s = \overline{V}_s^b e^{j\theta_b}; \\ \overline{\Psi}_r^r &= \overline{\Psi}_r^b e^{j(\theta_b - \theta_{er})}; \quad \overline{i}_r^r = \overline{i}_r^b e^{j(\theta_b - \theta_{er})}; \quad \overline{V}_r^r = \overline{V}_r^b e^{j(\theta_b - \theta_{er})} \end{aligned} \tag{13.29}$$

With these new variables Equations (13.25) become

$$\begin{aligned} \overline{V}_s &= R_s \overline{i}_s + \frac{d\overline{\Psi}_s}{dt} + j\omega_b \overline{\Psi}_s; \quad \overline{\Psi}_s = L_s \overline{i}_s + L_m \overline{i}_r \\ \overline{V}_r &= R_r \overline{i}_r + \frac{d\overline{\Psi}_r}{dt} + j(\omega_b - \omega_r)\overline{\Psi}_r; \quad \overline{\Psi}_r = L_r \overline{i}_r + L_m \overline{i}_s \end{aligned} \tag{13.30}$$

For convenience, the superscript b was dropped in (13.30). The electromagnetic torque is related to motion-induced voltage in (13.30).

$$T_e = \frac{3}{2} \cdot p_1 \cdot \mathrm{Re}\left(j \cdot \overline{\psi}_s \cdot \overline{i}_s^*\right) = -\frac{3}{2} \cdot p_1 \cdot \mathrm{Re}\left(j \cdot \overline{\psi}_r \cdot \overline{i}_r^*\right) \tag{13.31}$$

Adding the equations of motion, the complete complex variable (space-phasor) model of IM is obtained.

$$\frac{J}{p_1}\frac{d\omega_r}{dt} = T_e - T_{load}; \quad \frac{d\theta_{er}}{dt} = \omega_r \tag{13.32}$$

The complex variables may be decomposed in plane along two orthogonal d and q axes rotating at speed ω_b to obtain the d–q (Park) model. [2]

$$\begin{aligned} \overline{V}_s &= V_d + j \cdot V_q; \quad \overline{i}_s = i_d + j \cdot i_q; \quad \overline{\Psi}_s = \Psi_d + j \cdot \Psi_q \\ \overline{V}_r &= V_{dr} + j \cdot V_{qr}; \quad \overline{i}_r = i_{dr} + j \cdot i_{qr}; \quad \overline{\Psi}_r = \Psi_{dr} + j \cdot \Psi_{qr} \end{aligned} \tag{13.33}$$

With (13.33), the voltage Equations (13.30) become

$$
\begin{aligned}
\frac{d\Psi_d}{dt} &= V_d - R_s \cdot i_d + \omega_b \cdot \Psi_q \\
\frac{d\Psi_q}{dt} &= V_q - R_s \cdot i_q - \omega_b \cdot \Psi_d \\
\frac{d\Psi_{dr}}{dt} &= V_{dr} - R_r \cdot i_{dr} + (\omega_b - \omega_r) \cdot \Psi_{qr} \\
\frac{d\Psi_{qr}}{dt} &= V_{qr} - R_r \cdot i_{qr} - (\omega_b - \omega_r) \cdot \Psi_{dr} \\
T_e &= \frac{3}{2} p_1 (\Psi_d i_q - \Psi_q i_d) = \frac{3}{2} p_1 L_m (i_q i_{dr} - i_d i_{qr})
\end{aligned}
\tag{13.34}
$$

Also from (13.27) with (13.19), the Park transformation for stator $P(\theta_b)$ is derived.

$$
\begin{bmatrix} V_d \\ V_q \\ V_0 \end{bmatrix} = [P(\theta_b)] \cdot \begin{bmatrix} V_a \\ V_b \\ V_c \end{bmatrix}
\tag{13.35}
$$

$$
[P(\theta_b)] = \frac{2}{3} \cdot \begin{bmatrix} \cos(-\theta_b) & \cos\left(-\theta_b + \frac{2\pi}{3}\right) & \cos\left(-\theta_b - \frac{2\pi}{3}\right) \\ \sin(-\theta_b) & \sin\left(-\theta_b + \frac{2\pi}{3}\right) & \sin\left(-\theta_b - \frac{2\pi}{3}\right) \\ \frac{1}{2} & \frac{1}{2} & \frac{1}{2} \end{bmatrix}
\tag{13.36}
$$

The inverse Park transformation is

$$
[P(\theta_b)]^{-1} = \frac{3}{2} \cdot [P(\theta_b)]^T
\tag{13.37}
$$

A similar transformation is valid for the rotor but with $\theta_b - \theta_{er}$ instead of θ_b.

It may be easily proved that the homopolar (real) variables V_0, i_0, V_{0r}, i_{0r}, Ψ_0, Ψ_{0r} do not interface in energy conversion

$$
\begin{aligned}
\frac{d\Psi_0}{dt} &= V_0 - R_s \cdot i_0; \quad \Psi_0 \approx L_{0s} \cdot i_0 \\
\frac{d\Psi_{0r}}{dt} &= V_{0r} - R_r \cdot i_{0r}; \quad \Psi_{0r} \approx L_{0r} \cdot i_0
\end{aligned}
\tag{13.38}
$$

L_{0s} and L_{0r} are the homopolar inductances of stator and rotor. Their values are equal or lower (for chorded coil windings) to the respective leakage inductances L_{ls} and L_{lr}.

A few remarks on the complex variable (space phasor) and d–q models are in order.

- Both models include, in the form presented here, only the space fundamental of mmfs and airgap flux distributions.
- Both models exhibit inductances independent of rotor position.
- The complex variable (space phasor) model is credited with a reduction in the number of equations with respect to the d–q model but it operates with complex variables.
- When solving the state space equations, only the d–q model, with real variables, benefits the existing commercial software (Mathematica, Matlab–Simulink, Spice, etc.).
- Both models are very practical in treating the transients and control of symmetrical IMs fed from symmetrical voltage power grids or from PWM converters.
- Easy incorporation of magnetic saturation and rotor skin effect are two additional assets of complex variable and d–q models. The airgap flux density retains a sinusoidal distribution along the circumferential direction.
- Besides the complex variable which enjoys widespread usage, other models (variable transformations), that deal especially with asymmetric supply or asymmetric machine cases have been introduced (for a summary see Reference. [3, 4])

13.4. STEADY-STATE BY THE COMPLEX VARIABLE MODEL

By IM steady-state we mean constant speed and load. For a machine fed from a sinusoidal voltage symmetrical power grid, the phase voltages at IM terminals are

$$V_{a,b,c} = V\sqrt{2}\cdot\cos\left(\omega_1 t - (i-1)\cdot\frac{2\pi}{3}\right);\quad i = 1,2,3 \tag{13.39}$$

The voltage space phasor $\overline{V_s}^b$ in random coordinates (from (13.27)) is

$$\overline{V_s}^b = \frac{2}{3}\left(V_a(t) + aV_b(t) + a^2V_c(t)\right)e^{-j\theta_{er}} \tag{13.40}$$

From (13.39) and (13.40),

$$\overline{V_s}^b = V\sqrt{2}\left[\cos(\omega_1 t - \theta_b) + j\sin(\omega_1 t - \theta_b)\right] \tag{13.41}$$

Only for steady-state,

$$\theta_b = \omega_b t + \theta_0 \tag{13.42}$$

Consequently,

$$\overline{V_s}^b = V\sqrt{2}e^{j[(\omega_1 - \omega_b)t + \theta_0]} \tag{13.43}$$

For steady-state, the current in the space phasor model follows the voltage frequency: $(\omega_1 - \omega_b)$. Steady-state in the state space equations means replacing d/dt with $j(\omega_1 - \omega_b)$.

Using this observation makes Equations (13.30) become

$$\overline{V}_{s0} = R_s \bar{i}_{s0} + j\omega_1 \overline{\Psi}_{s0}; \quad \overline{\Psi}_{s0} = L_{sl} \bar{i}_{s0} + \overline{\Psi}_{m0};$$
$$\overline{\Psi}_{r0} = L_{rl} \bar{i}_{r0} + \overline{\Psi}_{m0}; \quad \bar{i}_{m0} = \bar{i}_{s0} + \bar{i}_{r0} \tag{13.44}$$
$$\overline{V}_{r0} = R_r \bar{i}_{r0} + jS\omega_1 \overline{\Psi}_{r0}; \quad S = \frac{(\omega_1 - \omega_r)}{\omega_r}; \quad \overline{\Psi}_{m0} = L_m \bar{i}_{m0}$$

So the form of space phasor model voltage equations under the steady-state is the same irrespective of the speed of the reference system ω_b.

When ω_b, only the frequency changes of voltages, currents, flux linkages in the space phasor model varies as it is $\omega_1 - \omega_b$.

No wonder this is so, as only Equations (13.44) exhibit the total emf, which should be independent of reference system speed ω_b. S is the slip, a variable well known by now.

Notice that for $\omega_b = \omega_1$ (synchronous coordinates), $d/dt = (\omega_1 - \omega_b) = 0$. Consequently, for synchronous coordinates the steady-state means d.c. variables.

The space phasor diagram of (13.44) is shown in Figure 13.2 for a cage rotor IM.

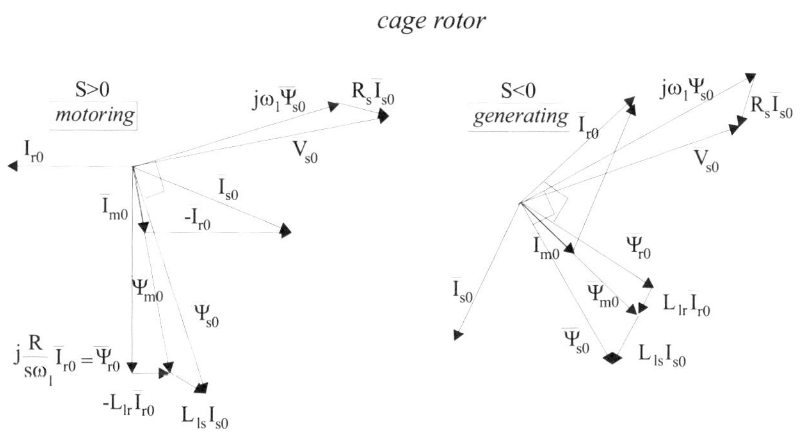

Figure 13.2 Space phasor diagram for steady-state

From the stator space Equations (13.44) the torque (13.31) becomes

$$T_e = \frac{3}{2} p_1 \left(j\overline{\Psi}_{r0} \bar{i}^*_{r0} \right) = \frac{3}{2} P_1 \Psi_{r0} i_{r0} \tag{13.45}$$

Also, from (13.44),

$$\bar{i}_{r0} = -jS\omega_1 \frac{\overline{\Psi}_{r0}}{R_r} \tag{13.46}$$

With (13.46), alternatively, the torque is

$$T_e = \frac{3}{2} p_1 \frac{\Psi_{r0}^2}{R_r} S\omega_1 \tag{13.47}$$

Solving for Ψ_{r0} in Equations (13.44) lead to the standard torque formula.

$$T_e \approx \frac{3p_1}{\omega_1} \frac{V_s^{\ 2} \frac{R_r}{S}}{\left(R_s + C_1 \frac{R_r}{S}\right)^2 + \omega_1^{\ 2}\left(L_{ls} + C_1 L_{lr}\right)^2}; \quad C_1 = 1 + \frac{L_{ls}}{L_m} \tag{13.48}$$

Expression (13.47) shows that, for constant rotor flux space-phasor amplitude, the torque varies linearly with speed as it does in a separately excited d.c. motor. So all steady-state performance may be calculated using the space-phasor model as well.

13.5. EQUIVALENT CIRCUITS FOR DRIVES

Equations (13.30) lead to a general equivalent circuit good for transients, especially in variable speed drives (Figure 13.3).

$$\begin{aligned} \overline{V}_s &= R_s \bar{i}_s + (p + j\omega_b) L_{sl} \bar{i}_s + (p + j\omega_b)\overline{\Psi}_m \\ \overline{V}_r &= R_r \bar{i}_r + (p + j(\omega_b - \omega_r)) L_{rl} \bar{i}_r + (p + j(\omega_b - \omega_r))\overline{\Psi}_m \end{aligned} \tag{13.49}$$

The reference system speed ω_b may be random, but three particular values have met with rather wide acceptance.

- Stator coordinates: $\omega_b = 0$; for steady-state: $p \rightarrow j\omega_1$
- Rotor coordinates: $\omega_b = \omega_r$; for steady-state: $p \rightarrow jS\omega_1$
- Synchronous coordinates: $\omega_b = \omega_1$; for steady-state: $p \rightarrow 0$

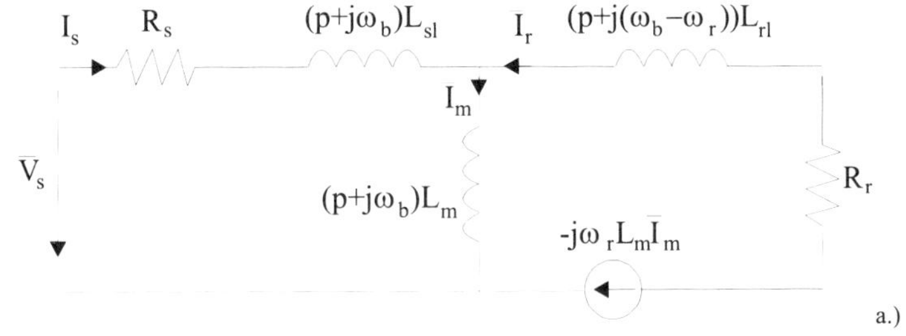

Figure 13.3 The general equivalent circuit a.) and for steady-state b.) (continued)

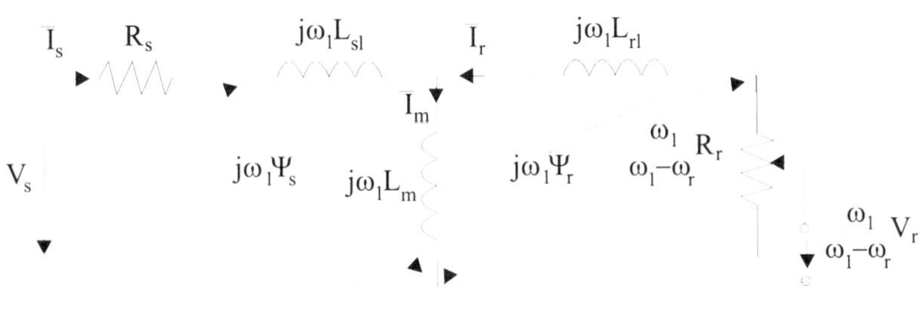

b.)

Figure 13.3 (continued)

Also, for steady-state in variable speed drives, the steady-state circuit, the same for all values of ω_b, comes from Figure 13.3a with $p \to j(\omega_1 - \omega_b)$ and Figure 13.3b.

Figure 13.3b shows, in fact, the standard T equivalent circuit of IM for steady-state, but in space phasors and not in phase phasors.

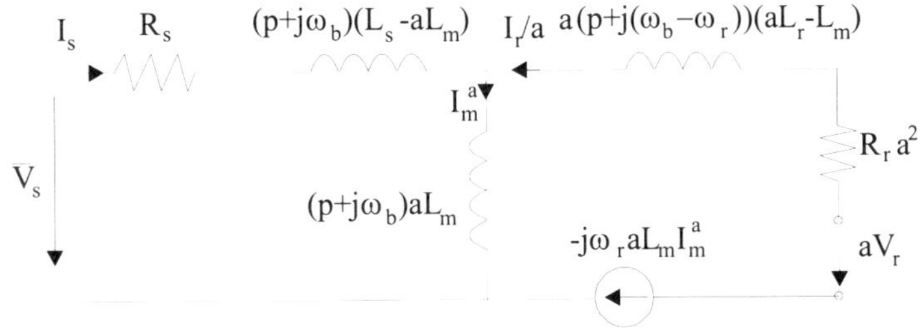

Figure 13.4 Generalized equivalent circuit

A general method to "arrange" the leakage inductances L_{sl} and L_{rl} in various positions in the equivalent circuit consists of a change of variables.

$$\bar{i}_r^a = \bar{i}_r / a; \quad \overline{\Psi}_{ma} = aL_m\left(\bar{i}_s + \bar{i}_r^a\right) \tag{13.50}$$

Making use of this change of variables in (13.49) yields

$$\begin{aligned} \overline{V}_s &= R_s \bar{i}_s + (p + j\omega_b)(L_s - aL_m)\bar{i}_s + (p + j\omega_b)\overline{\Psi}_m^a \\ a\overline{V}_r &= a^2 R_r \bar{i}_r^a + (p + j(\omega_b - \omega_r))a(aL_{rl} - L_m)\bar{i}_r^a + (p + j(\omega_b - \omega_r))\overline{\Psi}_m^a \end{aligned} \tag{13.51}$$

An equivalent circuit may be developed based on (13.51), Figure 13.4.

The generalized equivalent circuit in Figure 13.4 warrants the following comments:

- For a = 1, the general equivalent circuit of Figure 13.4 is reobtained and $a\overline{\Psi}_m^a = \overline{\Psi}_m$: the main flux

- For $a = L_m/L_r$ the inductance term in the "rotor section" "disappears", being moved to the primary section and

$$a\overline{\Psi}_m^a = \frac{L_m}{L_r} L_m \left(\overline{i}_s + \overline{i}_r \frac{L_r}{L_m} \right) = \frac{L_m}{L_r} \overline{\Psi}_r \tag{13.52}$$

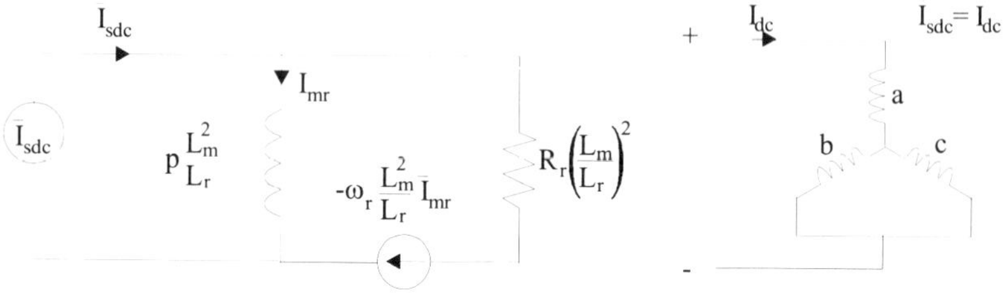

Figure 13.5 Equivalent circuit for dc braking

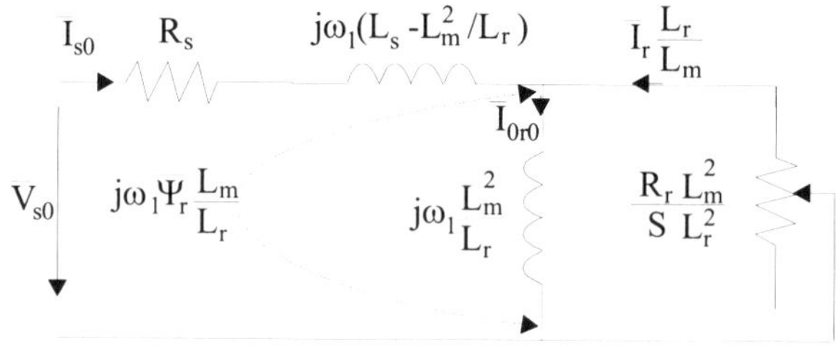

a.)

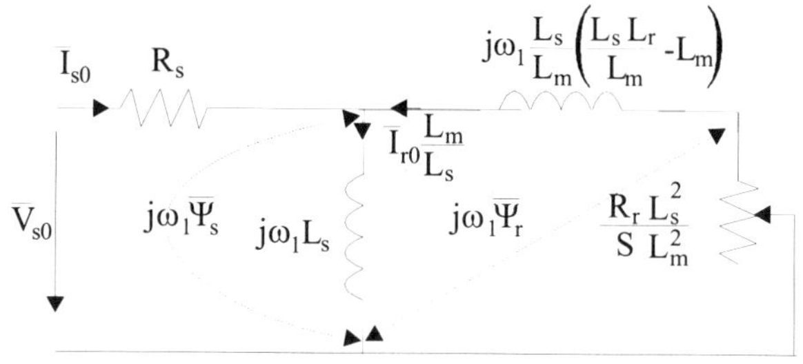

b.)

Figure 13.6 Steady-state equivalent circuits
a.) rotor flux oriented b.) stator flux oriented

- For $a = L_s/L_m$, the leakage inductance term is lumped into the "rotor section" and

$$a\overline{\Psi}_m^a = \frac{L_s}{L_m} L_m \left(\bar{i}_s + \bar{i}_r \frac{L_m}{L_s} \right) = \overline{\Psi}_s \tag{13.53}$$

- This type of equivalent circuit is adequate for stator flux orientation control
- For d.c. braking, the stator is fed with d.c. The method is used for variable speed drives. The model for this regime is obtained from Figure 13.4 by using a d.c. current source, $\omega_b = 0$ (stator coordinates, $V_r = 0$, $a = L_m/L_r$)

The result is shown in Figure 13.5.

For steady-state, the equivalent circuits for $a_r = L_m/L_r$ and $a_r = L_s/L_m$ and V_r=0 are shown in Figure 13.6.

Example 13.1. The constant rotor flux torque/speed curve.

Let us consider an induction motor with a single rotor cage and constant parameters: $R_s = R_r = 1\ \Omega$, $L_{sl} = L_{rl} = 5$ mH, $L_m = 200$ mH, $\Psi_{r0}' = 1$ Wb, $S = 0.2$, $\omega_1 = 2\pi 6$ rad/s, $p_1 = 2$, $V_r = 0$. Find the torque, rotor current, stator current, stator and main flux and voltage for this situation. Draw the corresponding space phasor diagram.

Solution

We are going to use the equivalent circuit in Figure 13.6a and the rotor current and torque expressions (13.46 and 13.47):

$$T_e = \frac{3}{2} p_1 \frac{\Psi_{r0}^2}{R_r} S\omega_1 = \frac{3}{2} 2 \frac{1^2}{1} 0.2 \cdot 2\pi 6 = 22.608\text{Nm}$$

$$I_{r0} = -S\omega_1 \frac{\Psi_{r0}}{R_r} = -0.2 \cdot 2\pi 6 \frac{1}{1} = -7.536\text{A}$$

The rotor current is placed along real axis in the negative direction. The rotor flux magnetization current $\bar{I}_{r0}$ (Figure 13.6a) is

$$\bar{I}_{0r0} = -j \frac{(-I_{r0}) \frac{R_r}{S} \frac{L_m}{L_r}}{\omega_1 \frac{L_m^2}{L_r}} = -j \frac{(-I_{r0}) R_r}{S\omega_1 L_m} = -j \frac{(+7.536) \cdot 1}{0.2 \cdot 2\pi 6 \cdot 0.2} = -j5\text{A}$$

The stator current

$$\bar{I}_{s0} = -\bar{I}_{r0} \frac{L_r}{L_m} + \bar{I}_{0r0} = 7.536 \frac{0.205}{0.2} - j5 = 7.7244 - j5.0\text{A}$$

The stator flux $\overline{\Psi}_s$ is

$$\overline{\Psi}_{s0} = L_s \bar{I}_{s0} + L_m \bar{I}_{r0} = 0.205(7.7244 - j5.0) + 0.2(-7.536) = 0.076 - j1.025$$

The airgap flux $\overline{\Psi}_m$ is

$$\overline{\Psi}_{m0} = L_m\left(\bar{I}_{s0} + \bar{I}_{r0}\right) = 0.2(7.7244 - j5.0 - 7.536) = 0.03768 - j1.0$$

The rotor flux is

$$\overline{\Psi}_{r0} = \overline{\Psi}_{m0} + L_{rl}\bar{I}_{r0} = 0.03768 - j1.0 + 0.005(-7.536) = -j1.0 \text{ (as expected)}$$

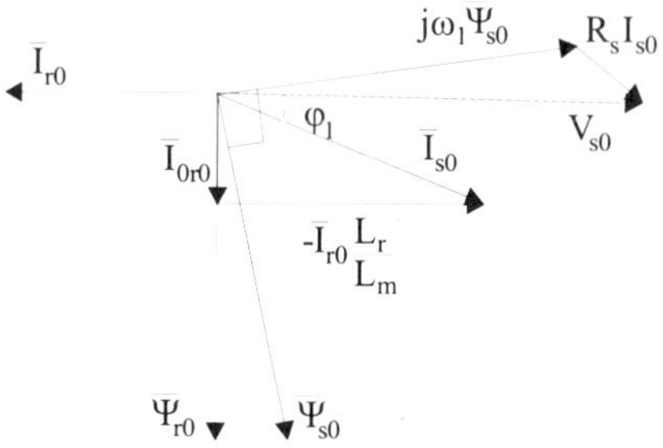

Figure A The space phasor diagram

The voltage V_{so} is:

$$V_{s0} = j\omega_1\overline{\Psi}_{s0} + R_s\bar{I}_{s0} = j2\pi6(0.076 - j1.025) + 1(7.7244 - j5.0) = 46.346 - j2.136$$

The corresponding space phasor diagram is shown in Figure A.

13.6. ELECTRICAL TRANSIENTS WITH FLUX LINKAGES AS VARIABLES

Equations (13.30) may be transformed by changing the variables through currents elimination.

$$\bar{i}_s = \sigma^{-1}\left[\frac{\overline{\Psi}_s}{L_s} - \overline{\Psi}_r\frac{L_m}{L_sL_r}\right]; \quad \sigma = 1 - \frac{L_m^{\ 2}}{L_sL_r}$$
$$\bar{i}_r = \sigma^{-1}\left[\frac{\overline{\Psi}_r}{L_r} - \overline{\Psi}_s\frac{L_m}{L_sL_r}\right] \tag{13.54}$$

$$\tau_s'\frac{d\overline{\Psi}_s}{dt} + (1 + j\cdot\omega_b\cdot\tau_s')\cdot\overline{\Psi}_s = \tau_s'\overline{V}_s + K_r\overline{\Psi}_r$$
$$\tau_r'\frac{d\overline{\Psi}_r}{dt} + (1 + j\cdot(\omega_b - \omega_r)\cdot\tau_r')\cdot\overline{\Psi}_r = \tau_r'\overline{V}_r + K_s\overline{\Psi}_s \tag{13.55}$$

with

$$K_s = \frac{L_m}{L_s}; \quad K_r = \frac{L_m}{L_r}$$
$$\tau_s' = \tau_s \cdot \sigma; \quad \tau_r' = \tau_r \cdot \sigma \tag{13.56}$$
$$\tau_s = \frac{L_s}{R_s}; \quad \tau_r = \frac{L_r}{R_r}$$

By electrical transients we mean constant speed transients. So both ω_b and ω_r are considered known. The inputs are the two voltage space phasors $\overline{V}_s$ and $\overline{V}_r$ and the outputs are the two flux linkage space phasors, $\overline{\Psi}_s$ and $\overline{\Psi}_r$.

The structural diagram of Equations (13.55) is shown in Figure 13.7.

The transient behavior of stator and rotor flux linkages as complex variables, at constant speed ω_b and ω_r, for standard step or sinusoidal voltages $\overline{V}_s$, $\overline{V}_r$ signals has analytical solutions. Finally, the torque transients also have analytical solutions as

$$T_e = \frac{3}{2} p_1 \operatorname{Re}\left(j\Psi_s i_s^*\right) \tag{13.57}$$

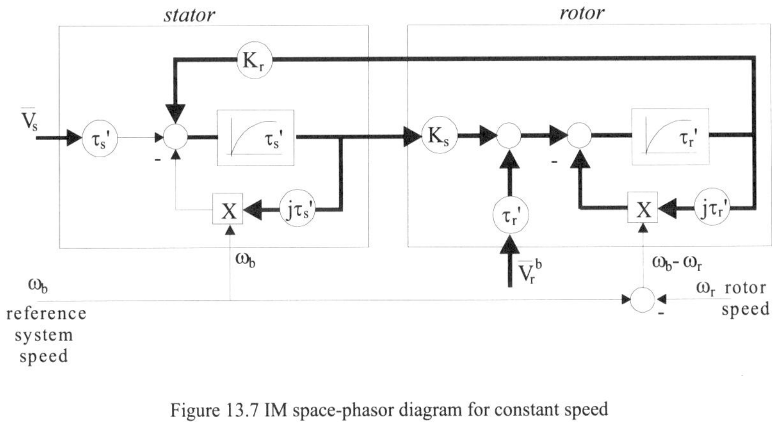

Figure 13.7 IM space-phasor diagram for constant speed

The two complex eigenvalues of (13.55) are obtained from

$$\begin{vmatrix} \tau_s'p + 1 + j\omega_b\tau_s' & -K_r \\ -K_s & \tau_r'p + 1 + j(\omega_b - \omega_r)\tau_r' \end{vmatrix} = 0 \tag{13.58}$$

As expected, the eigenvalues $p_{1,2}$ depend on the speed of the motor and on the speed of the reference system ω_b.

Equation (13.58) may be put in the form

$$\begin{aligned} &p^2\tau_s'\tau_r' + p\left[(\tau_s' + \tau_r') + j\tau_s'\tau_r'(2\omega_b - \omega_r)\right] - K_sK_r + \\ &+ (1 + j\cdot\omega_b\cdot\tau_s')(1 + j\cdot(\omega_b - \omega_r)\cdot\tau_r') = 0 \end{aligned} \tag{13.58'}$$

In essence, in-rush current and torque (at zero speed), for example, has a rather straightforward solution through the knowledge of eigenvalues, with $\omega_r = 0 = \omega_b$.

$$p^2\tau_s'\tau_r' + p(\tau_s' + \tau_r') - K_s K_r + 1 = 0$$

$$(p_{1,2})_{\omega_r=\omega_b=0} = \frac{-(\tau_s'+\tau_r') \pm \sqrt{(\tau_s'+\tau_r')^2 - 4\tau_s'\tau_r'(-K_s K_r + 1)}}{2\tau_s'\tau_r'} \quad (13.59)$$

The same capability of yielding analytical solutions for transients is claimed by the spiral vector theory. [5]

For constant amplitude stator or rotor flux conditions

$$\left(\frac{d\overline{\Psi}_s}{dt} = j(\omega_1 - \omega_b)\overline{\Psi}_s \text{ or } \frac{d\overline{\Psi}_r}{dt} = j(\omega_1 - \omega_b)\overline{\Psi}_r;\ \omega_1 - \text{stator frequency}\right)$$

Equations (13.55) are left with only one complex eigenvalue. Even a simpler analytical solution for electrical transients is feasible for constant stator and rotor flux conditions, so typical in fast response modern vector control drives.

Also, at least at zero speed, the eigenvalue with voltage supply is about the same for stator or for rotor supply.

The same equations may be expressed with $\bar{i}_s$ and $\overline{\Psi}_r$ as variables, by simply putting $a = L_m/L_r$ and by eliminating $\bar{i}_r$ from (13.51) with (13.50).

13.7. INCLUDING MAGNETIC SATURATION IN THE SPACE PHASOR MODEL

To incorporate magnetic saturation easily into the space phasor model (13.49), we separate the leakage saturation from main flux path saturation with pertinent functions obtained *a priori* from tests or from field solutions (FEM).

$$L_{sl} = L_{sl}(|\bar{i}_s|);\ L_{lr} = L_{lr}(|\bar{i}_r|);\ \overline{\Psi}_m = L_m(|\bar{i}_m|)\bar{i}_m \quad (13.60)$$

$$|\bar{i}_m| = |\bar{i}_s + \bar{i}_r| \quad (13.61)$$

Let us consider the reference system oriented along the main flux $\overline{\Psi}_m$, that is, $\overline{\Psi}_m = \Psi_m$, and eliminate the rotor current, maintaining Ψ_m and $\bar{i}_s$ as variables.

$$\begin{aligned}\overline{V}_s &= \left[R_s + (p + j\omega_b)L_{sl}(|\bar{i}_s|)\right]\cdot \bar{i}_s + (p + j\omega_b)\cdot\overline{\Psi}_m\\ \overline{V}_r &= \left[R_r + (p + j(\omega_b - \omega_r))L_{lr}(|\bar{i}_m - \bar{i}_s|)\right]\cdot(\bar{i}_m - \bar{i}_s) + (p + j(\omega_b - \omega_r))\cdot\overline{\Psi}_m\end{aligned} \quad (13.62)$$

$$i_m = \frac{\Psi_m}{L_m(i_m)} \tag{13.63}$$

We may add the equation of motion,

$$\frac{Jp\omega_r}{P_1} = \frac{3}{2}p_1 \, Real\left(j\overline{\Psi}_m \overline{i}_s^{\,*}\right) - T_{load} \tag{13.64}$$

Provided the magnetization curves $L_{sl}(i_s)$, $L_{lr}(i_r)$ are known, Equations (13.62)–(13.64) may be solved only by numerical methods after splitting Equations (13.62) along d and q axis.

For steady-state, however, it suffices that in the equivalent circuits, Lm is made a function of i_m, L_{sl} of i_{s0} and $L_{lr}(i_{ro})$ (Figure 13.8). This is only in synchronous coordinates where steady-state means dc variables.

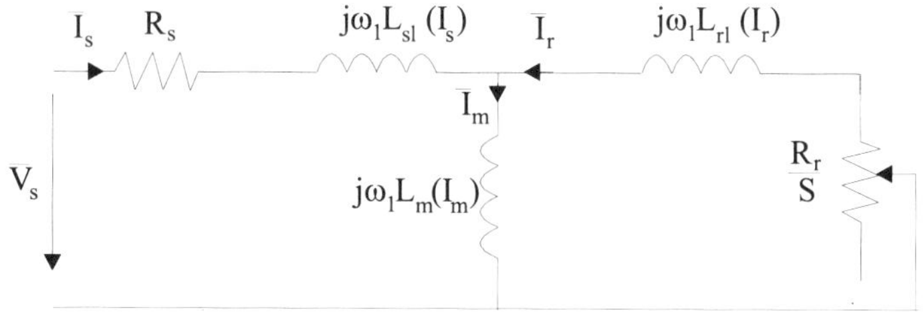

Figure 13.8 Standard equivalent circuit for steady-state leakage and main flux path saturation

In reality, both in the stator and rotor, the magnetic fields are a.c. at frequency ω_1 and $S\omega_1$. So, in fact, in Figure 13.8, the transient (a.c.), inductances should be used.

$$\begin{aligned}
L_{ls}^{t}(i_s) &= L_{ls}(i_s) + \frac{\partial L_{ls}}{\partial i_s} i_s < L_{ls}(i_s) \\
L_{lr}^{t}(i_r) &= L_{lr}(i_r) + \frac{\partial L_{lr}}{\partial i_r} i_r < L_{lr}(i_r) \\
L_{m}^{t}(i_m) &= L_{m}(i_m) + \frac{\partial L_{m}}{\partial i_m} i_m < L_{m}(i_m)
\end{aligned} \tag{13.65}$$

Typical curves are shown in Figure 13.9.

As the transient (a.c.) inductances are even smaller than the normal (d.c.) inductances, the machine behavior at high currents is expected to show further increased currents.

Furthermore, as shown in Chapter 6, the leakage flux circumferential flux lines at high currents influence the main (radial) flux and contribute to the resultant flux in the machine core. The saturation in the stator is given by the stator flux Ψ_s and in the rotor by the rotor flux for high levels of currents.

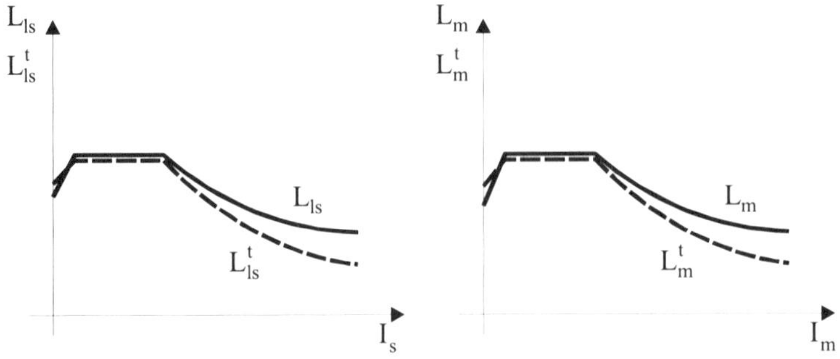

Figure 13.9 Leakage and main flux inductances versus current

So, for large currents, it seems more appropriate to use the equivalent circuit with stator and rotor flux shown (Figure 13.10). However, two new variable inductances, L_{si} and L_{ri}, are added. L_g refers only to the airgap only. Finally the stator and rotor leakage inductances are related only to end connections and slot volume: L_{ls}^e, L_{lr}^e. [6]

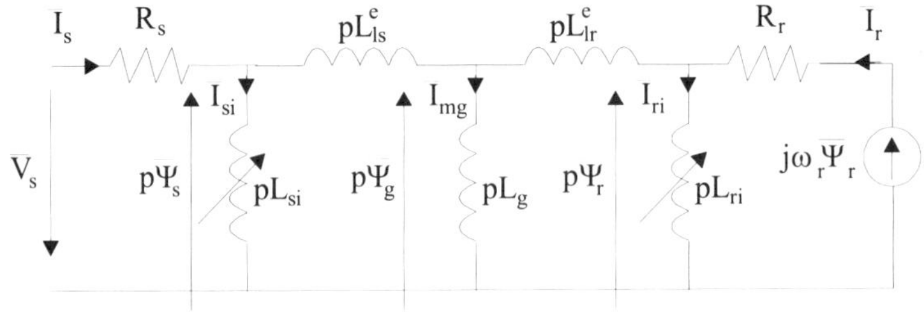

Figure 13.10 Space-phasor equivalent circuit with stator and rotor core saturation included (stator coordinates)

As expected, the functions $\Psi_s(i_{si}) = L_{si}i_{si}$ and $\Psi_r(i_{ri}) = L_{ri}i_{ri}$ have to be known together with L_{ls}^e, L_{lr}^e, R_s, R_r, L_g which are hereby considered constant.

In the presence of skin effect, L_{lr}^e and R_r are functions of slip frequency $\omega_{sr} = \omega_1 - \omega_r$. Furthermore we should notice that it is not easy to measure all parameters in the equivalent circuit on Figure 13.10. It is, however, possible to calculate them either by sophisticated analytical or through FEM models. Depending on the machine design and load, the relative importance of the two variable inductances L_{si} and L_{ri} may be notable or negligible.

Consequently, the equivalent circuit may be simplified. For example, for heavy loads the rotor saturation may be ignored and thus only L_{si} remains. Then L_{si} and L_g in parallel may be lumped into an equivalent variable inductance L_{ms} and L_{ls}^e and L_{lr}^e into the total constant leakage inductance of the machine L_l. Then the equivalent circuit of Figure 13.10 degenerates into the one of Figure 13.11.

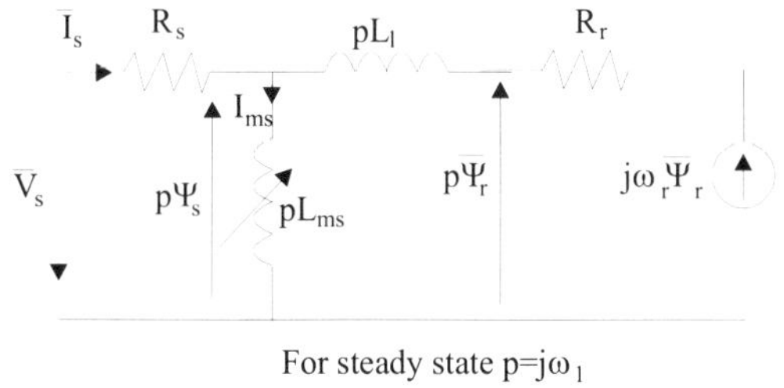

Figure 13.11 Space-phasor equivalent circuit with stator saturation included (stator coodinates)

When high currents occur, during transients in electric drives, the equivalent circuit of Figure 13.11 indicates severe saturation while the conventional circuit (Figure 13.8) indicates moderate saturation of the main path flux and some saturation of the stator leakage path.

So when high current (torque) transients occur, the real machine, due to the L_{ms} reduction, produces torque performance quite different from the predictions by the conventional equivalent circuit (Figure 13.8), up to 2 to 2.5 p.u. current, however, the differences between the two models are negligible.

In IMs designed for extreme saturation conditions (minimum weight), models like that in Figure 13.10 have to be used, unless FEM is applied.

13.8. SATURATION AND CORE LOSS INCLUSION INTO THE STATE-SPACE MODEL

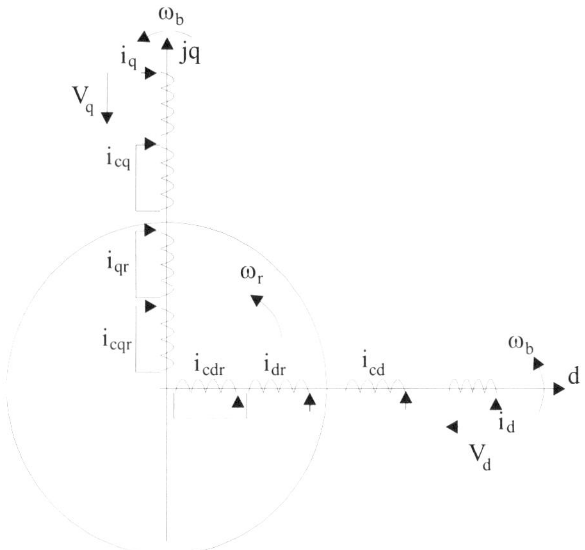

Figure 13.12 d–q model with stator and core loss windings

To include the core loss in the space phasor model of IMs, we assume that the core losses occur, both in the stator and rotor, into equivalent orthogonal windings: $c_d - c_q$, $c_{dr} - c_{qr}$ (Figure 13.12).

Alternatively, when rotor core loss is neglected (cage rotor IMs), the $c_{dr} - c_{qr}$ windings account for the skin effect via the double-cage equivalence principle.

Let us consider also that the core loss windings are coupled to the other windings only by the main flux path.

The space phasor equations are thus straightforward (by addition).

$$
\begin{aligned}
&R_s \bar{i}_s - \bar{V}_s = -\frac{d\bar{\Psi}_s}{dt} - j\omega_b \bar{\Psi}_s;\ \bar{\Psi}_s = \bar{\Psi}_m + L_{ls}\bar{i}_s \\
&R_{cs} \bar{i}_{cs} = -\frac{d\bar{\Psi}_{cs}}{dt} - j\omega_b \bar{\Psi}_{cs};\ \bar{\Psi}_{cs} = \bar{\Psi}_m + L_{lcs}\bar{i}_{cs} \\
&R_r \bar{i}_r - \bar{V}_r = -\frac{d\bar{\Psi}_r}{dt} - j(\omega_b - \omega_r)\bar{\Psi}_r;\ \bar{\Psi}_r = \bar{\Psi}_m + L_{lr}\bar{i}_r \\
&R_{cr} \bar{i}_{cr} = -\frac{d\bar{\Psi}_{cr}}{dt} - j(\omega_b - \omega_r)\bar{\Psi}_{cr};\ \bar{\Psi}_{cr} = \bar{\Psi}_m + L_{lcr}\bar{i}_{cr}
\end{aligned}
\tag{13.66}
$$

$$
\bar{i}_m = \left(\bar{i}_s + \bar{i}_{cs} + \bar{i}_r + \bar{i}_{cr}\right);\ \bar{\Psi}_m = L_m(i_m)\bar{i}_m \tag{13.67}
$$

The airgap torque now contains two components, one given by the stator current and the other one (braking) given by the stator core losses.

$$
T_e = \frac{3}{2} p_1 \operatorname{Re}\left[j\left(\bar{\Psi}_s \bar{i}_s^* + \bar{\Psi}_{cs} \bar{i}_{cs}^*\right)\right] = \frac{3}{2} p_1 \operatorname{Re}\left[j\bar{\Psi}_m\left(\bar{i}_s^* + \bar{i}_{cs}^*\right)\right] \tag{13.68}
$$

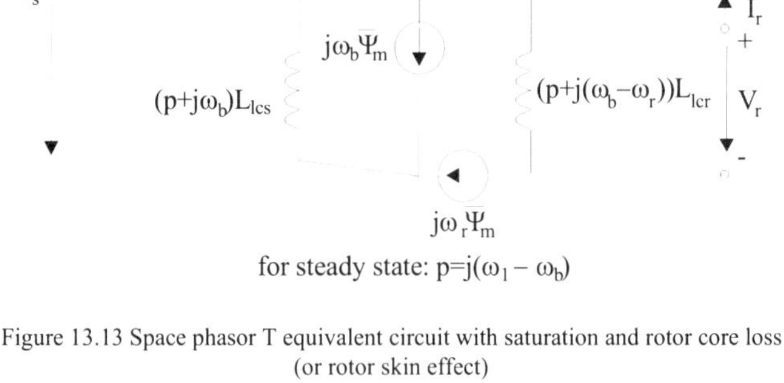

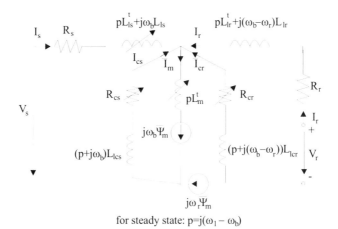

Figure 13.13 Space phasor T equivalent circuit with saturation and rotor core loss (or rotor skin effect)

Equations (13.66) and (13.67) lead to a fairly general equivalent circuit when we introduce the transient magnetization inductance L_{mt} (13.65) and consider separately main flux and leakage paths saturation (Figure 13.13).

The apparently involved equivalent circuit in Figure 13.13 is fairly general and may be applied for many practical cases such as

- The reference system may be attached to the stator, $\omega_b = 0$ (for cage rotor and large transients), to the rotor (for the doubly fed IM), or to stator frequency $\omega_b = \omega_1$ for electric drives transients.
- To simplify the solving of Equations (13.66) and (13.67) the reference system may be attached to the main flux space phasor: $\overline{\Psi}_m = \Psi_m(i_m)$ and eventually using i_m, i_{cs}, i_{cr}, i_r, ω_r as variables with $\bar{i}_s$ as a dummy variable. [7,8] As expected, d–q decomposition is required.
- For a cage rotor with skin effect (medium and large power motors fed from the power grid directly), we should simply make $V_r = 0$ and consider the rotor core loss winding (with its equations) as the second (fictitious) rotor cage.

Example 13.2. Saturation and core losses

Simulation results are presented in what follows. The motor constant parameters are: $R_s = 3.41\Omega$, $R_r = 1.89\Omega$, $L_{sl} = 1.14\cdot10^{-2}$H, $L_{rl} = 0.9076\cdot10^{-2}$H, $J = 6.25\cdot10^{-3}$Kgm^2. The magnetization curve $\Psi_m(i_m)$ is shown in Figure 13.14 together with core loss in the stator at 50 Hz.

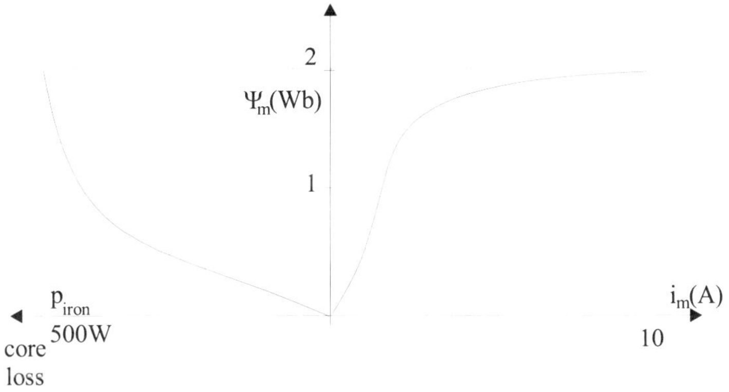

Figure 13.14 Magnetization curve and stator core loss

The stator core loss resistance R_{cs}

$$R_{cs} = \frac{3}{2}\frac{\omega_1^2\Psi_m^2}{P_{iron}} \approx \frac{3}{2}\frac{(2\pi 50)^2\cdot 2^2}{500} = 1183.15\Omega$$

Considering that $L_{lcs}\cdot\omega_1 = R_{cs}$; $L_{lcs} = 1183.15/(2\pi 50) = 3.768$ H.

For steady-state and synchronous coordinates (d/dt = 0), Equations (13.66) become

$$\begin{aligned} R_s \bar{i}_{s0} - \overline{V}_{s0} &= -j\omega_1 \overline{\Psi}_{s0}; \quad V_{s0} = 380\sqrt{2}e^{j\delta_0} \text{ - V d.c.} \\ R_{cs} \bar{i}_{cs0} &= -j\omega_1 \overline{\Psi}_{cs0}; \quad \bar{i}_{m0} = \bar{i}_{s0} + \bar{i}_{cs0} + \bar{i}_{r0} \\ R_r \bar{i}_{r0} &= -jS\omega_1 \overline{\Psi}_{r0} \end{aligned} \tag{13.69}$$

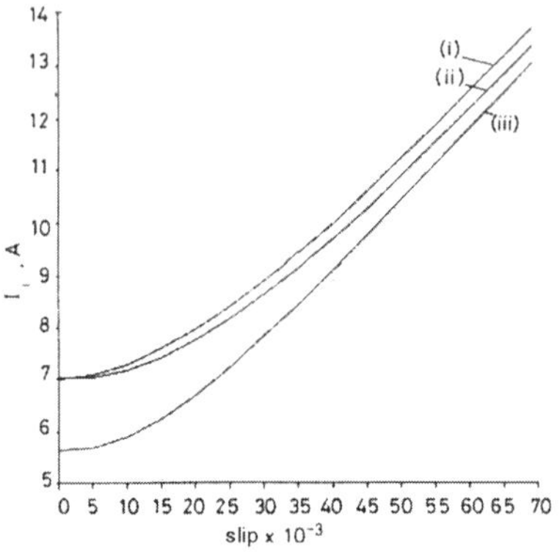

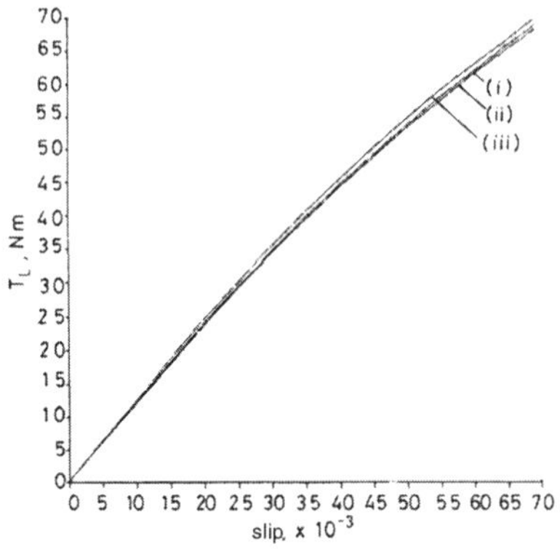

Figure 13.15 Steady-state phase current and torque versus slip
(i) saturation and core loss considered (ii) saturation considered, core loss neglected
(iii) no saturation, no core loss

For given values of slip S, the values of stator current $i_{s0}/\sqrt{2}$ (RMS/phase) and the electromagnetic torque are calculated for steady-state making use of (13.69) and the flux/current relationships (13.66) and (13.67) and Figure 13.14.

The results are given in Figure 13.15a and b. There are notable differences in stator current due to saturation. The differences in torque are rather small. Efficiency–power factor product and the magnetization current versus slip are shown in Figure 13.16a and b.

Again, saturation and core loss play an important role in reducing the efficiency–power factor although core loss itself tends to increase the power factor while it tends to reduce the efficiency.

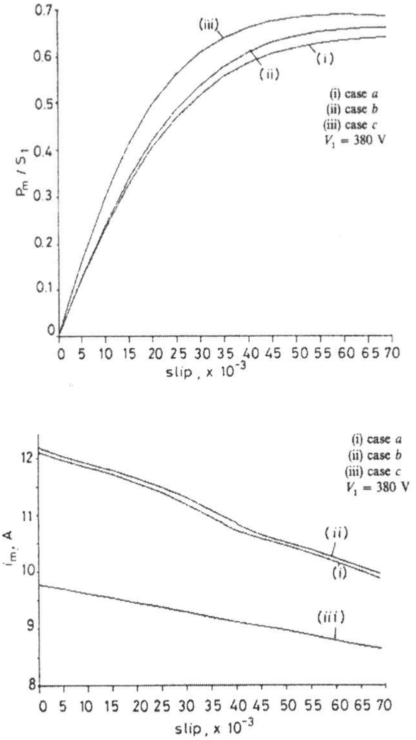

Figure 13.16 Efficiency–power factor product and magnetization current versus slip

The reduction of magnetization current i_m with slip is small but still worth considering.

Two transients have been investigated by solving (13.66) and (13.67) and the motion equations through the Runge–Kutta–Gill method.

1. Sudden 40% reduction of supply voltage from 380 V per phase (RMS), steady-state constant load corresponding to S = 0.03.
2. Disconnection of the loaded motor at S = 0.03 for 10 ms and reconnection (at $\delta_0 = 0$, though any phasing could be used). The load is constant again.

The transients for transient 1 are shown in Figure 13.17a, b, c.

Some influence of saturation and core loss occurs in the early stages of the transients, but in general the influence of them is moderate because at reduced voltage, saturation influence diminishes considerably.

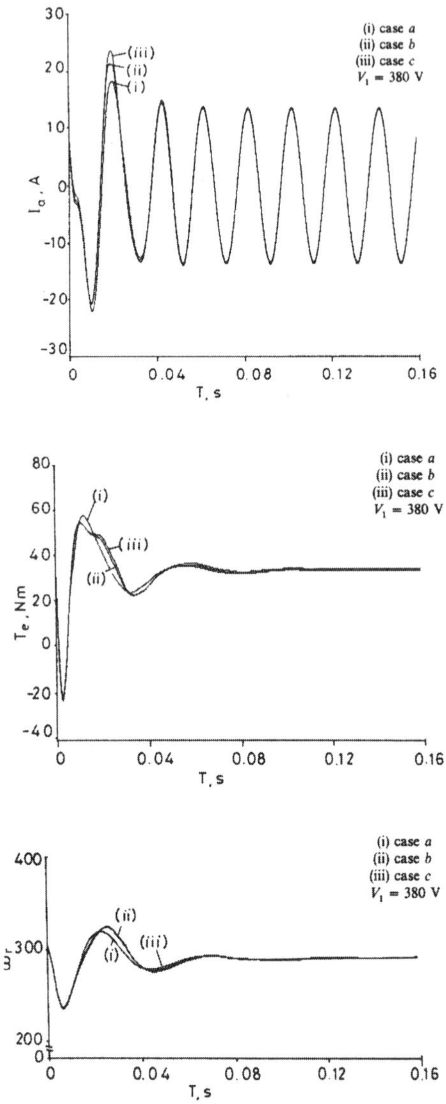

Figure 13.17 Sudden 40% voltage reduction at S = 0.03 constant load
a.) stator phase current, b.) torque, c.) speed

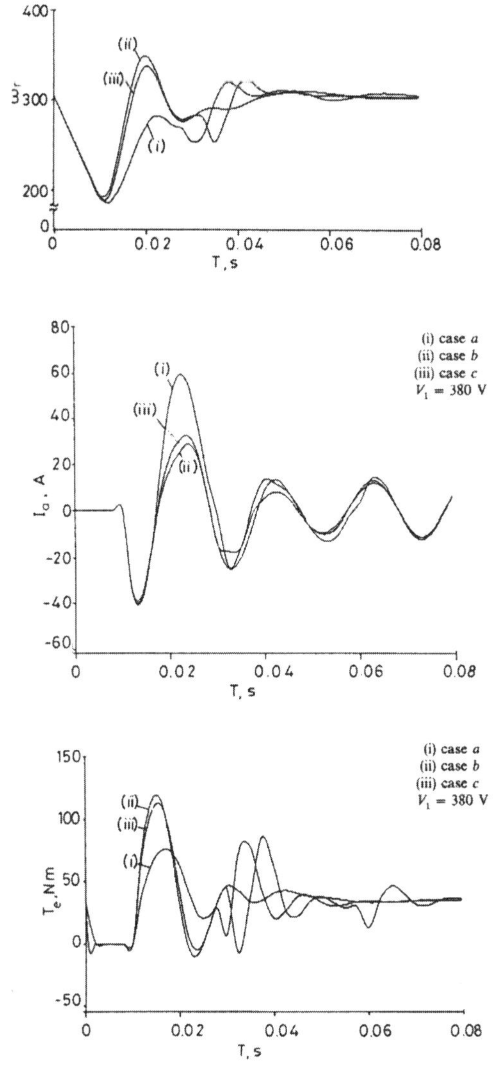

Figure 13.18 Disconnection–reconnection transients at high voltage: V_{phase} = 380V (RMS)

The second transient, occurring at high voltage (and saturation level), causes more important influences of saturation (and core loss) as shown in Figure 13.18a, b, c.

The saturation and core loss lead to higher current peaks, but apparently for low torque and speed transients.

In high-performance variable-speed drives, high levels of saturation may be inflicted to increase the torque transient capabilities of IM. Vector control detuning occurs and it has to be corrected. Also, in very precise torque control drives, such second order phenomena are to be considered. Skin effect also influences the transients of IMs and the model on Figure 13.13 can handle it for cage rotor IMs directly as shown earlier in this paragraph.

13.9. REDUCED ORDER MODELS

The rather involved d–q (space phasor) model with skin effect and saturation may be used directly to investigate the induction machine transients. More practical (simpler) solutions have also been proposed. They are called reduced order models. [9,10,11]

The complete model has, for a single cage, a fifth order in d–q writing and a third order in complex variable writing ($\overline{\Psi}_s$, $\overline{\Psi}_r$, ω_r).

In general, the speed transients are considered slow, especially with inertia or high loads, while stator flux $\overline{\Psi}_s$ and rotor flux $\overline{\Psi}_r$ transients are much faster for voltage-source supply.

The intuitive way to obtain reduced models is to ignore fast transients, of the stator, $\frac{d\overline{\Psi}_s}{dt} = 0$ (in synchronous coordinates) or stator and rotor transients $\left(\frac{d\overline{\Psi}_s}{dt} = \frac{d\overline{\Psi}_r}{dt} = 0\right)$ (in synchronous coordinates) and up date the speed by solving the motion equation by numerical methods.

This way, third order $\left(\frac{d\overline{\Psi}_s}{dt} = 0\right)$ or first order $\left(\frac{d\overline{\Psi}_s}{dt} = \frac{d\overline{\Psi}_r}{dt} = 0\right)$ models are obtained. The problem is that the results of such order obscure reductions inevitably fast transients. Electric (supply frequency) torque transients (due to rotor flux transients) during starting, may still be visible in such models, but for too long a time interval during starting in comparison with the reality.

So there are two questions about model order reduction.

- What kind of transients are to be used
- What torque transients have to be preserved

In addition to this, small and large power IMs seem to require different reduced orders to produce practical results in simulating a group of IMs when the computation time is critical.

13.9.1. Neglecting stator transients

In this case, in synchronous coordinates, the stator flux derivative is considered zero (Equations (13.55)).

$$0 + (1 + j\omega_1\tau_s')\overline{\Psi}_s = \tau_s'\overline{V}_s + K_r\overline{\Psi}_r$$

$$\tau_r'\frac{d\overline{\Psi}_r}{dt} + (1 + j(\omega_1 - \omega_r))\overline{\Psi}_r = \tau_r'\overline{V}_r + K_s\overline{\Psi}_s$$

$$T_e = -\frac{3}{2}p_1\,\mathrm{Re}(j\overline{\Psi}_r\bar{i}_r^*);\ \bar{i}_r^* = \sigma^{-1}\left(\frac{\overline{\Psi}_r}{L_r} - \overline{\Psi}_s\frac{L_m}{L_sL_r}\right) \tag{13.70}$$

$$\frac{J}{p_1}\frac{d\omega_r}{dt} = T_e - T_{load}$$

With two algebraic equations, there are three differential ones (in real d–q variables).

For cage rotor IMs ,

$$\overline{\Psi}_s = \frac{\tau_s'\overline{V}_s + K_r\overline{\Psi}_r}{1 + j\omega_1\tau_s'} \tag{13.71}$$

$$\frac{d\overline{\Psi}_r}{dt} = \left[-(1 + j(\omega_1 - \omega_r)) + \frac{K_sK_r}{1 + j\omega_1\tau_s'}\right]\frac{\overline{\Psi}_r}{\tau_r'} + \frac{\tau_s'\cdot\overline{V}_s\cdot K_s}{(1 + j\omega_1\tau_s')\tau_r'} \tag{13.72}$$

$$T_e = \frac{3}{2}p_1\,\mathrm{Im}\left[\sigma^{-1}\overline{\Psi}_r\overline{\Psi}_s^*\frac{L_m}{L_sL_r}\right] = \frac{3}{2}p_1\frac{\sigma^{-1}L_m}{L_sL_r}\mathrm{Im}\left[\overline{\Psi}_r\frac{\tau_s'\overline{V}_s^* + K_r\overline{\Psi}_r^*}{1 - j\omega_1\tau_s'}\right]$$

$$\frac{d\omega_r}{dt} = \frac{p_1}{J}(T_e - T_{load}) \tag{13.73}$$

Fast (at stator frequency) transient torque pulsations are absent in this third-order model (Figure 13.8b).

The complete model and third-order model for a motor with L_s = 0.05H, L_r = 0.05H, L_m = 0.0474H, R_s = 0.29Ω, R_r = 0.38Ω, ω_1 = 100πrad/s, V_s = $220\sqrt{2}$ V, p_1 = 2, J = 0.5 Kgm^2 yields, for no-load, starting transient results as shown in Figure 13.19. [11]

It is to be noted that steady-state torque at high slips (Figure 13.19a) falls into the middle of torque pulsations, which explains why calculating no-load starting time with steady-state torque produces good results.

A second-order system may be obtained by considering only the amplitude transients of $\overline{\Psi}_r$ in (13.73) [11], but the results are not good enough in the sense that the torque fast (grid frequency) transients are present during start up at higher speed than for the full model.

Modified second-order models have been proposed to improve the precision of torque results [11] (Figure 13.19c), but the results are highly dependent on motor parameters and load during starting. So, in fact, for starting transients when the torque pulsations (peaks) are required with precision, model reduction may be exercised with extreme care.

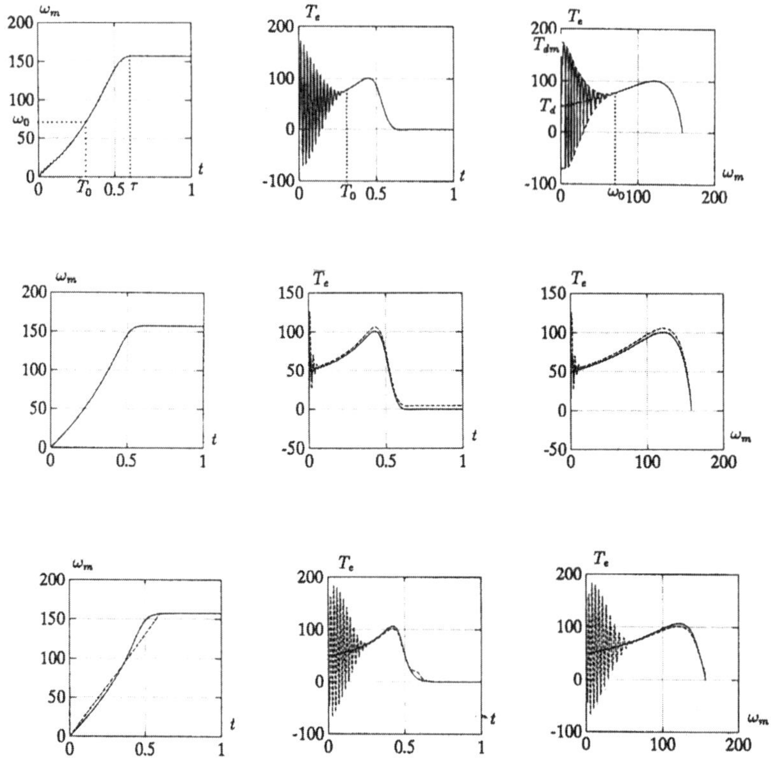

Figure 13.19 Starting transients [11]
a.) complete model (5th order),
b.) third model (stator transients neglected), c.) modified second order

The presence of leakage saturation makes the above results more of academic interest.

13.9.2. Considering leakage saturation

As shown in Chapter 9, the leakage inductance ($L_{sc} = L_{sl} + L_{rl} = L_l$) decreases with current. This reduction is accentuated when the rotor has closed slots and more moderate when both, stator and rotor, have semiclosed slots (Figure 13.20).

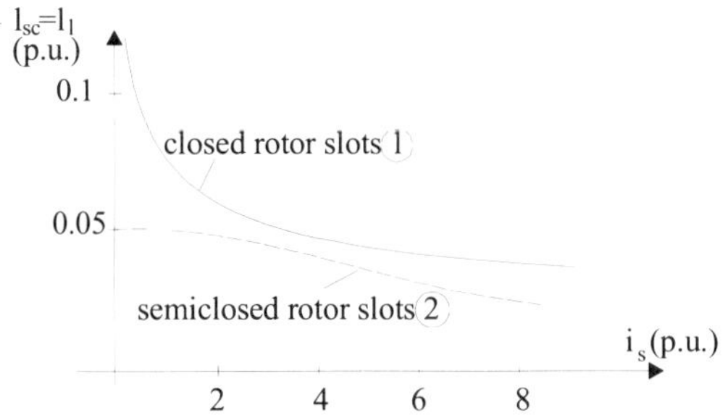

Figure 13.20 Typical leakage inductance versus stator current for semiclosed stator slots

The calculated (or measured) curves of Figure 13.20 may be fitted with analytical expressions of RMS rotor current such as [12]

$$X_l = \omega_1 L_l = K_1 + K_2 \left(\frac{I_r}{\sqrt{2}}\right)^{-K} \text{ for curve 1} \tag{13.74}$$

$$X_l = \omega_1 L_l = K_3 - K_4 \tanh\left(K_3 I_r^3 - K_6\right) + K_7 \cos\left(K_8 I_r\right) \text{ for curve 2} \tag{13.75}$$

During starting on no-load the stator current envelope $I_{sp}(t)$ varies approximately as [12]

$$I_{sp} \approx \frac{Q_1 I_{rK}}{\sqrt{3\left(1 + \sqrt{\frac{t}{t_{K0}}} + \left(\frac{t}{t_{K0}}\right)^2\right)}} \tag{13.76}$$

Q_1 reflects the effect of point on wave connection when nonsymmetrical switching of supply is considered to reduce torque transients (peaks) [13]

$$Q_1 \approx \sqrt{2 + \left(\frac{X_l(S_K)}{R_s + R_r}\right) \cdot \left(\cos^2\alpha + \sin^2(\alpha + \beta)\right)} \tag{13.77}$$

α–point on wave connection of first phase to phase (b-c) voltage.
β–delay angle in connecting the third phase.
(It has been shown that minimum current (and torque) transients are obtained for $\alpha = \pi/2$ and $\beta = \pi/2$. [14])

As the steady-state torque gives about the same starting time as the transient one, we may use it in the motion equation (for no-load).

$$\frac{J}{p_1}\frac{d\omega_r}{dt} = \frac{3V_{ph}^2 R_r}{\omega_1 p_1} \cdot \frac{1}{\left(R_s + \frac{R_r}{S}\right)^2 S + \omega_1 L_l^2 S} = T_e \tag{13.78}$$

We may analytically integrate this equation with respect to the time up to the breakdown point: S_K slip, noting that $\frac{d\omega_r}{dt} = -\omega_1 \frac{dS}{dt}$.

$$t_{K0} = \frac{J\omega_1^{\ 2}}{3V_{ph}^{\ 2}R_r^{\ 2}} \cdot \left[\frac{1}{2}\left(R_s^{\ 2} + (\omega_1 L_1(S_{K0}))^2\right)\left(1 - S_{K0}^2\right) + \right. \\ \left. + 2R_s R_r\left(1 - S_{K0}\right) + R_r^{\ 2}\ln\left(\frac{1}{S_{K0}}\right)\right] \tag{13.79}$$

with

$$S_{K0} = \frac{R_r}{\sqrt{R_s^{\ 2} + (\omega_1 L_1(S_{K0}))^2}} \tag{13.80}$$

On the other hand, the base current I_{rK} is calculated at standstill.

$$I_{rK} = \frac{V_{ph}}{\sqrt{(R_s + R_r)^2 + (\omega L_1(1))^2}} \tag{13.81}$$

Introducing L_l as function of I_r from (13.74) or (13.75) into (13.81) we may solve it numerically to find I_{rK}.

Now we may calculate the time variation of I_{sp} (the current envelope) versus time during no-load starting. The current envelope compares favourably with complete model results and with test results. [12]

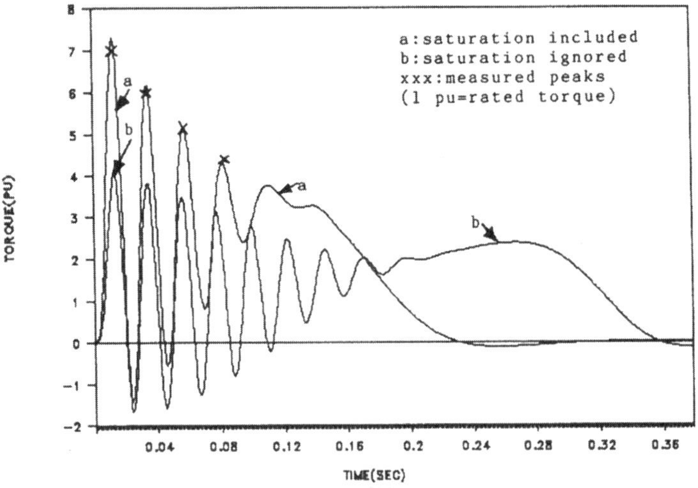

Figure 13.21 Torque transients during no-load starting

As the current envelope varies with time, so does the leakage inductance $L_l(I_{sp})$ and, finally, the torque value in (13.78) is calculated as a function of time through its peak values, since from Equation (13.78) we may calculate slip (S) versus time.

Sample results in Figure 13.21 [12] show that the approximation above is really practical and that neglecting the leakage saturation leads to an underestimation of torque peaks by more than 40%!

13.9.3. Large machines: torsional torque

Starting of large induction machines causes not only heavy starting currents but also high torsional torque due to the large masses of the motor and load which may exhibit a natural torsional frequency in the range of electric torque oscillatory components.

Sudden voltage switching (through a transformer or star/delta), low leakage time constants τ_s' and τ_r', and a long starting interval cause further problems in starting large inertia loads in large IMs. The layout of such a system is shown in Figure 13.22.

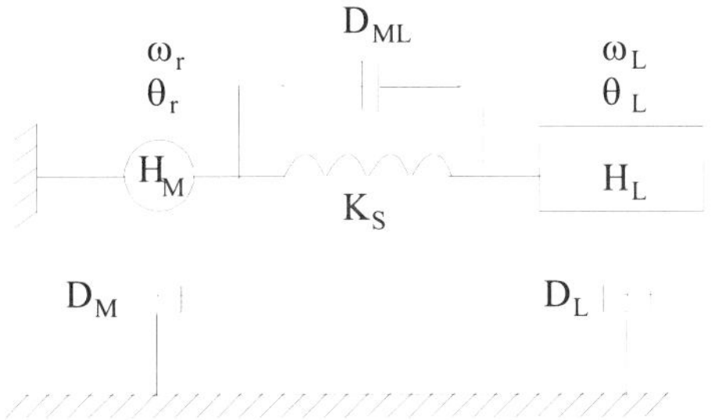

Figure 13.22 Large IM with inertia load

The mechanical equations of the rotating masses in Figure 13.22, with elastic couplings, is straightforward. [15]

$$\begin{bmatrix} \dot{\omega}_m \\ \dot{\theta}_m \\ \dot{\omega}_L \\ \dot{\theta}_L \end{bmatrix} = \begin{bmatrix} -\frac{(D_M + D_{ML})}{2H_M} & -\frac{K_S\omega_0}{2H_M} & \frac{D_{ML}}{2H_M} & \frac{K_S\omega_0}{2H_M} \\ 1 & 0 & 0 & 0 \\ -\frac{D_{ML}}{2H_M} & \frac{K_S\omega_0}{2H_L} & -\frac{(D_L + D_{HL})}{2H_L} & -\frac{K_S\omega_0}{2H_L} \\ 0 & 0 & 1 & 0 \end{bmatrix} \cdot \begin{bmatrix} \omega_m \\ \theta_m \\ \omega_L \\ \theta_L \end{bmatrix} \quad (13.82)$$

ω_0 – base speed, H_M, H_L – inertia in seconds, D_{HL}, D_{ML}, D_M in p.u./rad/s, K_s in p.u./rad.

The rotor d–q model Equations (13.34), with single rotor cage, in p.u. (relative units) is, after a few manipulations,

$$\begin{vmatrix} x_s & 0 & x_m & 0 \\ 0 & x_s & 0 & x_m \\ x_m & 0 & x_r & 0 \\ 0 & x_m & 0 & x_r \end{vmatrix} \cdot \frac{1}{\omega_0}\frac{d}{dt}\begin{vmatrix} i_q \\ i_d \\ i_{qr} \\ i_{dr} \end{vmatrix} =$$

$$\begin{vmatrix} -r_s & -\frac{\omega_m x_s}{\omega_0} & 0 & -\frac{\omega_m x_m}{\omega_0} \\ \frac{\omega_m x_s}{\omega_0} & -r_s & \frac{\omega_m x_m}{\omega_0} & 0 \\ 0 & -Sx_m & -r_r & -Sx_r \\ 0 & 0 & Sx_r & -r_r \end{vmatrix} \begin{vmatrix} i_q \\ i_d \\ i_{qr} \\ i_{dr} \end{vmatrix} + \begin{vmatrix} 1 & 0 \\ 0 & 1 \\ 0 & 0 \\ 0 & 0 \end{vmatrix}\begin{vmatrix} V_d \\ V_q \\ 0 \\ 0 \end{vmatrix} \tag{13.83}$$

where r_s, r_r, x_s, x_r, x_m are the p.u. values of stator (rotor) resistances and stator, rotor and magnetization reactances

$$r_s = \frac{R_s}{X_n};\ x_s = \frac{\omega_0 L_s}{X_n};\ X_n = \frac{V_{nph}}{I_{nph}} \tag{13.84}$$

$$H = \frac{J\left(\frac{\omega_0}{p_1}\right)^2}{S_n}\ \text{(seconds)} \tag{13.85}$$

$$V_d + jV_q = \frac{2}{3}\left(V_a + V_b e^{j\frac{2\pi}{3}} + V_c e^{-j\frac{2\pi}{3}}\right)e^{-j\omega_0 t} \tag{13.86}$$

The torque is
$$t_e = x_m\left(i_{dr}i_q - i_{qr}i_d\right) \tag{13.87}$$

Equations (13.82) and (13.83) may be solved by numerical methods such as Runge–Kutta–Gill. For the data [15] r_s = 0.0453 p.u., x_s = 2.1195 p.u., r_r = 0.0272 p.u., x_r = 2.0742 p.u., x_m = 2.042 p.u., H_M = 0.3 seconds, H_L = 0.74 seconds, $D_L = D_M = 0$, D_{ML} = 0.002 p.u./(rad/s), K_S = 30 p.u./rad, T_L = 0.0, ω_0 = 377 rad/s, the electromagnetic torque (t_e), the shaft torque (t_{sh}), and the speed ω_r are shown in Figure 13.23. [15]

The current transients include a stator frequency current component in the rotor, a slip frequency current in the rotor, and d.c. decaying components both in the stator and rotor. As expected, their interaction produces four torque components: unidirectional torque (by the rotating field components), at supply frequency ω_0, slip frequency S ω_0, and at speed frequency ω_m.

The three a.c. components may interact with the mechanical part whose natural frequency f_m is [15]

$$f_m = \frac{1}{2\pi}\sqrt{\frac{K_S(H_M + H_L)}{2H_M H_L}} = 26\text{Hz} \tag{13.88}$$

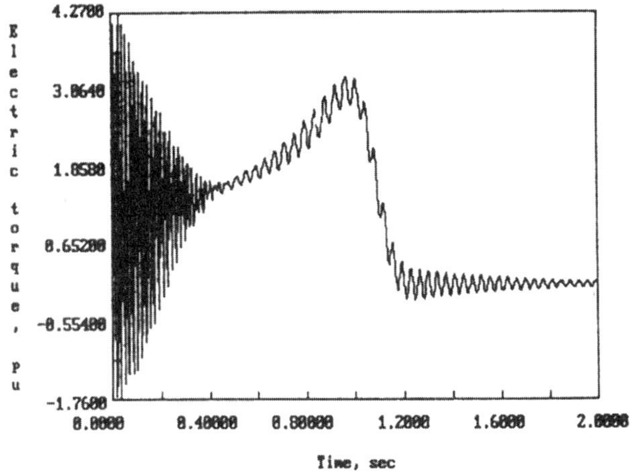

a- Electromagnetic torque

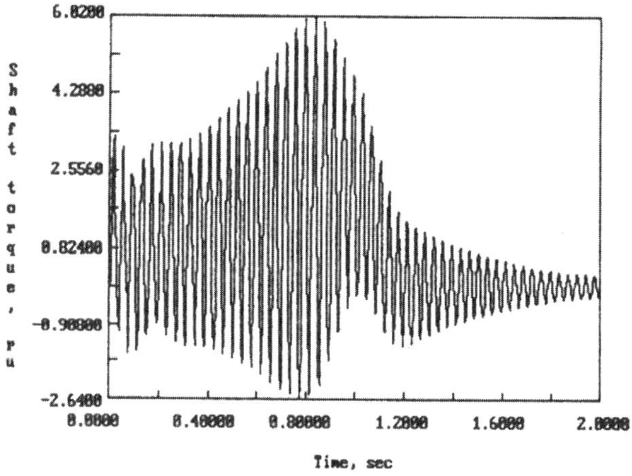

b- Shaft torque

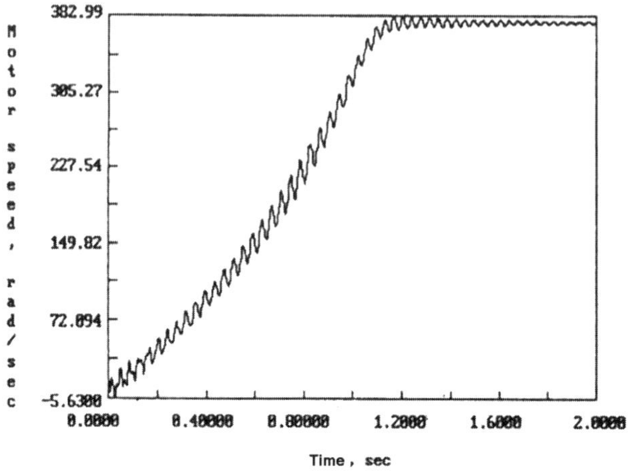

c– Motor speed

Figure 13.23 Starting transients of a large motor with elastic coupling of inertial load

The torsional torque in Figure 13.23b (much higher than the electromagnetic torque) may be attributed to the interaction of the slip frequency component of electromagnetic torque, which starts at 60 Hz (S = 1) and reaches 26 Hz as the speed increases (S·60 Hz = 26Hz, $\omega_m = (1 - S)\cdot\omega_0$ = 215rad/s). The speed frequency component (ω_m) of electromagnetic torque is active when ω_m = 162 rad/s ($\omega_m = 2\pi f_m$), but it turns out to be small.

A good start would imply the avoidance of torsional torques to prevent shaft damage.

For example, in the star/delta connection starting, the switching from star to delta may be delayed until the speed reaches 80% of rated speed to avoid the torsional torques occurring at 215 rad/s and reducing the current and torque peaks below it.

The total starting time with star/delta is larger than for direct (full voltage) starting, but shaft damage is prevented.

Using an autotransformer with 40%, 60%, 75%, 100% voltage steps further reduces the shaft torque but at the price of even larger starting time.

Also, care must be exercised to avoid switching voltage steps near ω_m = 215 rad/s (in our case).

13.10. THE SUDDEN SHORT-CIRCUIT AT TERMINALS

Sudden short-circuit at the terminals of large induction motors produces severe problems in the corresponding local power grid. [16]

Large induction machines with cage rotors are skin-effect influenced so the double cage representation is required. The space-phasor model for double-cage IMs is obtained from (13.66) with $\overline{V}_r = 0$ and eliminating the stator core loss equation, with the rotor core loss equation now representing the second rotor cage. In rotor coordinates,

$$\begin{aligned}
&\overline{V}_s = R_s \overline{i}_s + p\overline{\Psi}_s + j\omega_r \overline{\Psi}_s; \quad \overline{\Psi}_s = \overline{\Psi}_m + L_{sl}\overline{i}_s; \quad \overline{\Psi}_m = L_m \overline{i}_m \\
&0 = R_{r1}\overline{i}_{r1} + p\overline{\Psi}_{r1}; \quad \overline{i}_m = \overline{i}_s + \overline{i}_{r1} + \overline{i}_{r1}; \quad \overline{\Psi}_{r1} = \overline{\Psi}_m + L_{rl1}\overline{i}_{r1} \\
&0 = R_{r2}\overline{i}_{r2} + p\overline{\Psi}_{r2}; \quad \overline{\Psi}_{r2} = \overline{\Psi}_m + L_{rl2}\overline{i}_{r2}
\end{aligned} \tag{13.89}$$

To simplify the solution, we may consider that the speed remains constant (eventually with initial slip $S_0 = 0$).

Eliminating $\overline{i}_{r1}$ and $\overline{i}_{r2}$ from (13.89) yields

$$\overline{\Psi}_s(p) = L(p)\overline{i}_s(p) \tag{13.90}$$

The operational inductance L(p) is

$$L(p) = L_s \frac{(1+T'p)(1+T''p)}{(1+T_0'p)(1+T_0''p)} \tag{13.91}$$

The transient and subtransient inductances L_s', L_s'' are defined as for synchronous machines.

$$\begin{aligned}
&L_s = \lim_{s\to 0} L(p) \\
&L_s' = \lim_{p\to\infty}\left(L(p)\right)_{T''=T_0''=0} = L_s \frac{T'}{T_0'} < L_s
\end{aligned} \tag{13.92}$$

$$L_s'' = \lim_{p\to\infty} L(p) = L_s \frac{T'T''}{T_0'T_0''} << L_s \tag{13.93}$$

Using Laplace transformation (for $\omega_r = \omega_1 = ct$) makes the solution much easier to obtain. First, the initial (d.c.) stator current at no-load in rotor (synchronous) coordinates ($\omega_r = \omega_1$) $i_s^0(t)$ is

$$i_s^0 = \frac{V_s e^{j\gamma}}{j\omega_1 L_s}; \quad V_s = V\sqrt{2}; R_s \approx 0 \tag{13.94}$$

$$V_{a,b,c} = V\sqrt{2}\cos\left(\omega_1 t + \gamma - (i-1)\frac{2\pi}{3}\right) \tag{13.95}$$

To short-circuit the machine model, $-V_s e^{j\gamma}$ must be applied to the terminals and thus Equation (13.89) with (13.90) yields (in Laplace form):

$$-\frac{V_s e^{j\gamma}}{p} = \left[R_s + (p + j\omega_1)L(p)\right] i_s(p) \tag{13.96}$$

Denoting
$$\alpha = \frac{R_s}{L(p)} \approx \frac{R_s}{L_s''} = ct \tag{13.97}$$

the current $i_s(p)$ becomes

$$i_s(s) = \frac{-V_s e^{j\gamma}}{pL(p)(p + j\omega_1 + \alpha)} \tag{13.98}$$

with L(p) of (13.91) written in the form

$$\frac{1}{L(p)} = \frac{1}{L_s} + \frac{\left(\frac{1}{L_s'} - \frac{1}{L_s}\right)p}{\frac{1}{T'} + p} + \frac{\left(\frac{1}{L_s''} - \frac{1}{L_s'}\right)p}{\frac{1}{T''} + p} \tag{13.99}$$

$i_s(p)$ may be put into an easy form to solve.

Finally, $i_s(t)$ is

$$\bar{i}_s(t) = -\frac{V_s}{j\omega_1} e^{j\gamma}\left[\frac{1}{L_s}\left(1 - e^{-(\alpha + j\omega_1)t}\right) + \left(\frac{1}{L_s'} - \frac{1}{L_s}\right)\left(e^{-\frac{t}{T'}} - e^{-(\alpha + j\omega_1)t}\right)\right.$$
$$\left. + \left(\frac{1}{L_s''} - \frac{1}{L_s'}\right)\left(e^{-\frac{t}{T''}} - e^{-(\alpha + j\omega_1)t}\right)\right] \tag{13.100}$$

Now the current in phase a, $i_a(t)$, is

$$i_a(t) = \mathrm{Re}\left[\left(i_s(t) + i_s^0(t)\right)e^{j\omega_1 t}\right] =$$
$$V\sqrt{2}\left[\frac{1}{\omega_1 L_s''} e^{-\alpha t}\sin\gamma - \left[\left(\frac{1}{\omega_1 L_s'} - \frac{1}{\omega_1 L_s}\right)e^{-\frac{t}{T'}} + \left(\frac{1}{\omega_1 L_s''} - \frac{1}{\omega_1 L_s'}\right)e^{-\frac{t}{T''}}\right]\sin(\omega_1 t + \gamma)\right] \tag{13.101}$$

A few remarks are in order.

- The subtransient component, related to α, T", L_s" reflects the decay of the leakage flux towards the rotor surface (upper cage)
- The transient component, related to T', Ls', reflects the decay of leakage flux pertaining to the lower cage
- The final value of short-circuit current is zero

The torque expression first requires the stator flux solution.

$$T_e = \frac{3}{2} p_1 \left(\overline{\Psi}_s \bar{i}_s^* \right) \quad (13.102)$$

The initial value of stator flux (Rs ≈ 0) is

$$\overline{\Psi}_s^0 = \frac{V_s e^{j\gamma}}{j\omega_1} = L_s i_s^0 \quad (13.103)$$

with i_s^0 from (13.94).

After short-circuit,

$$\Psi_s(p) = \frac{-V_s e^{j\gamma}}{p[p + j\omega_1 + \alpha]} = L(p) i_s(p) \quad (13.104)$$

with $i_s(p)$ from (13.98).

So,

$$\overline{\Psi}_s(t) = \frac{-V_s e^{j\gamma}}{\alpha + j\omega_1}\left(1 - e^{-(\alpha + j\omega_1)t}\right) \quad (13.105)$$

We may neglect α and add $\overline{\Psi}_s(t)$ with $\overline{\Psi}_s^0$ to obtain $\overline{\Psi}_{st}(t)$.

$$\overline{\Psi}_{st}(t) = \frac{V_s e^{j\gamma}}{j\omega_1} e^{-(\alpha + j\omega_1)t} \quad (13.106)$$

Now, from (13.100), (13.105), and (13.102), the torque is

$$T_e = 3V^2 \frac{p_1}{\omega_1}\left[\left(\frac{1}{\omega_1 L_s'} - \frac{1}{\omega_1 L_s}\right) e^{-\frac{t}{T'}} + \left(\frac{1}{\omega_1 L_s''} - \frac{1}{\omega_1 L_s'}\right) e^{-\frac{t}{T''}}\right] \sin(\omega_1 t) \quad (13.107)$$

Neglecting the damping components in (13.107), the peak torque is

$$T_{epeak} \approx -\frac{3p_1 V^2}{\omega_1} \frac{1}{\omega_1 L_s''} \quad (13.108)$$

This result is analogous to that of peak torque at start up.

Typical parameters for a 30 kW motors are [17, 18]

T' = 50.70ms, T" = 3.222 ms, T_0' = 590 ms, T_0" = 5.244 ms, 1/α = 20.35ms, L_m = 30.65 mH, L_{sl} = 1 mH (L_s = 31.65 mH), R_s = 0.0868 Ω

Analyzing, after acquisition, the sudden short-circuit current, the time constants (T_0', T_0", T', T") may be determined by curve-fitting methods.

Using power electronics, the IM may be separated very quickly from the power grid and then short-circuited at various initial voltages and frequency. [17] Such tests may be an alternative to standstill frequency response tests in determining the machine parameters.

Finally, we should note that the torque and current peaks at sudden short-circuit are not (in general) larger than for direct connection to the grid at any speed. It should be stressed that when direct connection to the grid is applied at various initial speeds, the torque and current peaks are about the same.

Reconnection, after a short turn-off period before the speed and the rotor current have decreased, notably produces larger transients if the residual voltage at stator terminals is in phase opposition with the supply voltage at the reconnection instant.

The question arises: which is the most severe transient? So far apparently there is no definite answer to this question, but the most severe transient up to now is reported in [18].

13.11. MOST SEVERE TRANSIENTS (SO FAR)

We inferred above that reconnection, after a short supply interruption, could produce current and torque peaks higher than for direct starting or sudden short-circuit .

Whatever cause would prolong the existence of d.c. decaying currents in the stator (rotor) at high levels would also produce severe current and torque oscillations due to interaction between the a.c. source frequency current component caused by supply reconnection and the large d.c. decaying currents already in existence before reconnection.

One more reason for severe torque peaks would be the phasing between the residual voltage at machine terminals at the moment of supply reconnection. If the two add up, higher transients occur.

The case may be [13] of an IM supplied through a power transformer whose primary is turned off after starting the motor, at time t_d with the secondary connected to the motor, that is left for an interval Δt, before supply reconnection takes place (Figure 13.24).

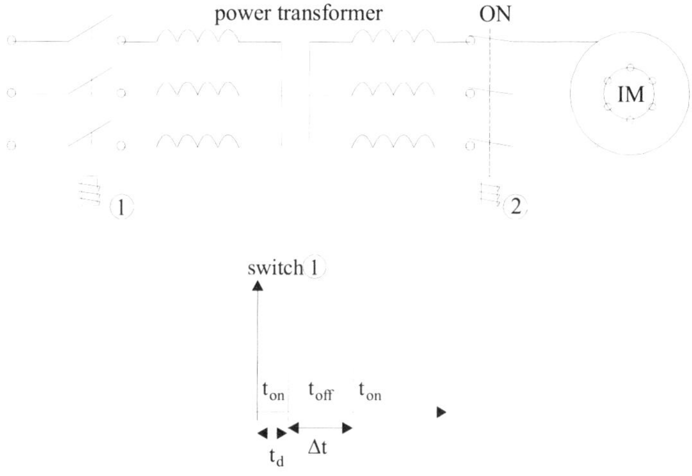

Figure 13.24 Arrangement for most severe transients

During the transformer primary turn off, the total (no-load) impedance of the transformer (r_{ts}, x_{t0s}), as seen from the secondary side, is connected in series with the IM stator resistance and leakage reactance r_s, x_{ls}, given in relative (or absolute) units.

To investigate the above transients, we simply use the d–q model in p.u. in (13.83) with the simple motion equation,

$$H\frac{d\omega_r}{dt} = t_e - t_L \tag{13.109}$$

During the Δt interval, the model is changed only in the sense of replacing r_s with $r_s + r_{ts}$ and x_{ls} with $x_{ls} + x_{tos}$. When the transformer primary is reconnected, r_{ts} and x_{tos} are eliminated.

Digital simulations have been run for a 7.5 and a 500 HP IM.

The parameters of the two motors [18] are shown in Table 13.1.

Table 13.1. Machine parameters

Size HP	V_L line	rpm	T_n Nm	I_n A	R_s Ω	X_{ls} Ω	X_m Ω	R_r Ω	X_{lr} Ω	J Kgm^2	f_1 Hz
7.5	440	1440	37.2	9.53	0.974	2.463	68.7	1.213	2.463	0.042	50
500	2300	1773	1980	93.6	0.262	1.206	54.04	0.187	1.206	11.06	60

For the 7.5 HP machine, results, as shown in Figure 13.25a, b, c have been obtained. For the 500 HP machine, Figure 13.26 shows only 12 p.u. peak torque while 26 p.u. peak torque is noticeable for the 7.5 HP machine.

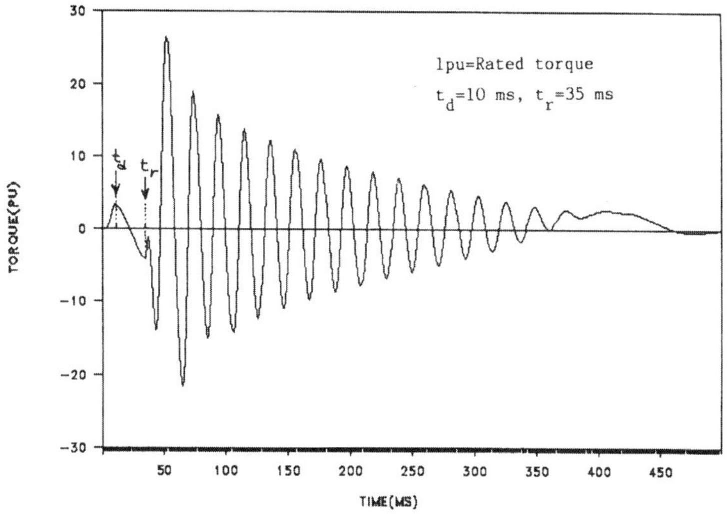

Torque-time pattern-7.5 HP machine

a.)

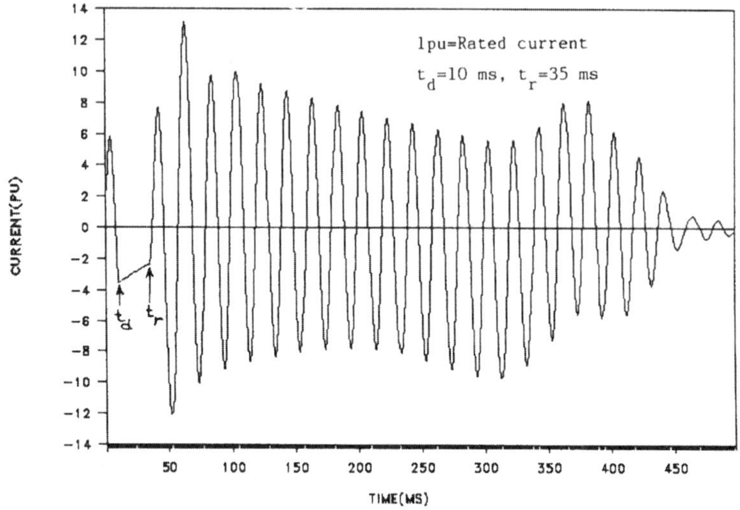

Current–time pattern–7.5 HP machine

b.)

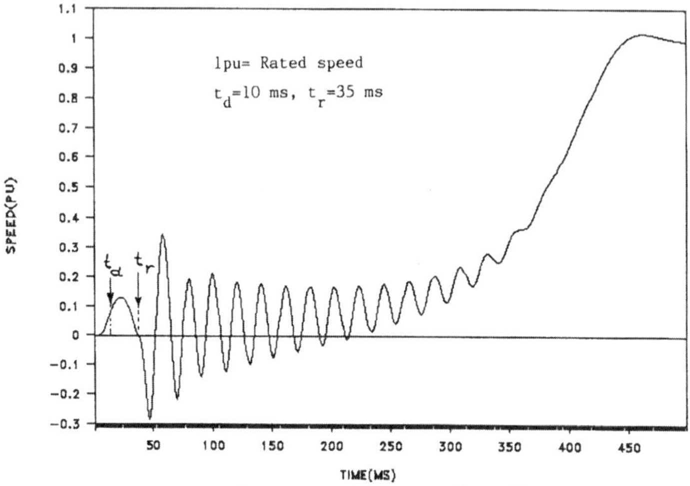

Speed - time pattern - 7.5HP machine

c.)

Figure 13.25 Transformer primary turn-off and reconnection transients (7.5HP)

The influence of turn-off moment t_d (after starting) and of the transformer primary turn-off period Δt on peak torque is shown on Figure 13.27 for the 7.5 HP machines.

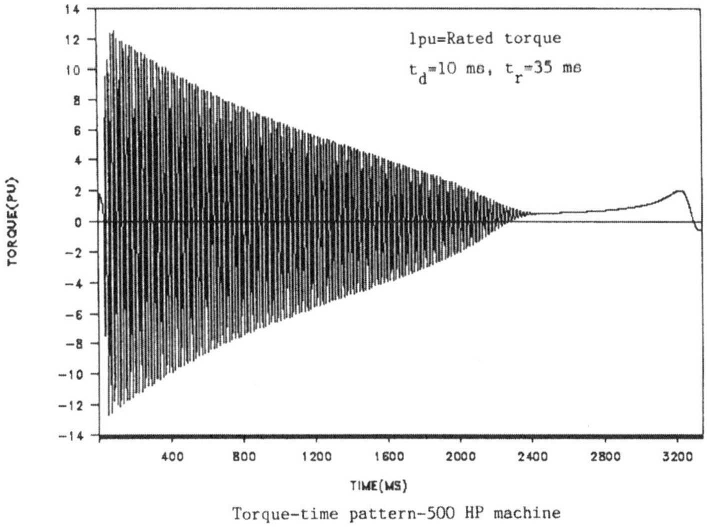

Figure 13.26. Transformer primary turn off and reconnection transients (500HP)

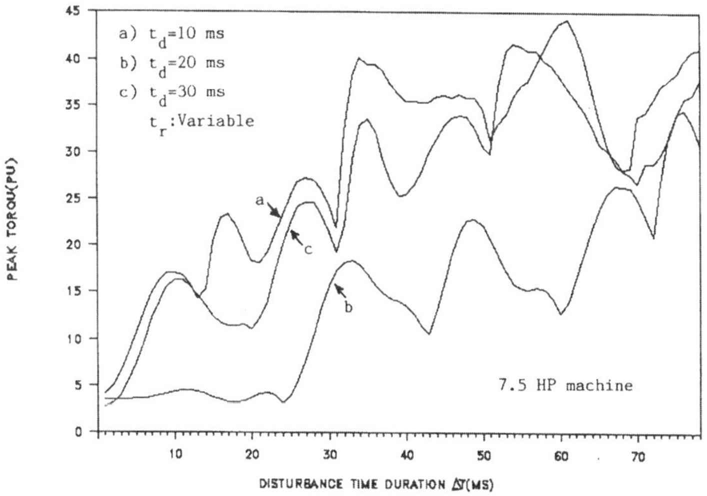

Figure 13.27 Peak torque versus perturbation duration (7.5HP machine)

Figure 13.27 shows very high torque peaks. They depend on the turn-off time after starting t_d. There is a period difference between t_d values for curves a and c and their behavior is similar. This shows that the residual voltage phasing at the reconnection instant is important. If it adds to the supply voltage, the torque peaks increase. When it subtracts, the peak torques decrease. The periodic behaviour with Δt is a proof of this aspect.

Also, up to a point, when the perturbation time Δt increases, the d.c. (nonperiodic) current components are high as the combined stator and secondary transformer (subtransient) and IM time constant is large.

This explains the larger value of peak torque with Δt increasing. Finally, the speed shows oscillations (even to negative values, for low inertia (power)) and the machine total starting time is notably increased.

Unusually high peak torques and currents are expected with this kind of transients for a rather long time, especially for low power machines while, even for large power IMs, they are at least 20% higher than for sudden short-circuit .

13.12. THE ABC–DQ MODEL FOR PWM INVERTER FED IMs

Time domain modeling of PWM inverter fed IMs, both for steady-state and transients, symmetrical and fault conditions, may be approached through the so called abc–dq model (Figure 13.28).

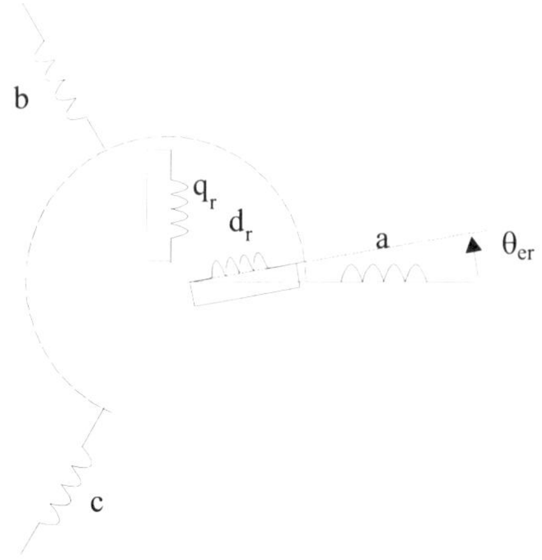

Figure 13.28 The abc – dq model

The derivation of the abc–dq model in stator coordinates may be obtained starting again from, abc–$a_rb_rc_r$ model with Park transformation of only rotor variables:

$$\left|P_{abcdq0}\right| = \begin{vmatrix} [1] & [0] \\ [0] & [P_{3\times3}(\theta_b - \theta_{er})] \end{vmatrix} \tag{13.110}$$

For stator coordinates, $\theta_b = 0$.

Instead of phase voltages, the line voltages may be used:
V_{ab}, V_{bc}, V_{ca} instead of V_a, V_b, V_c,

$$V_{ab} = V_a - V_b;\ V_{bc} = V_b - V_c;\ V_{ca} = V_c - V_a \tag{13.111}$$

Making use of the 6 × 6 inductance matrix $\left|L_{abca_1b_1c_1}(\theta_{er})\right|$ of (13.1), the abc–dq model equations become

$$[V] = [R][I] + \omega_r G[I] + L\frac{d}{dt}[I] \tag{13.112}$$

$$\begin{aligned}[R] &= \mathrm{Diag}[R_s, R_s, R_s, R_r, R_r, R_r] \\ [V] &= [V_{ab}, V_{bc}, V_{ca}, 0, 0] \\ [I] &= [i_a, i_b, i_c, i_{dr}, i_{qr}]\end{aligned} \tag{13.113}$$

$$[G] = \begin{bmatrix} 0 & 0 & 0 & 0 & 0 \\ 0 & 0 & 0 & 0 & 0 \\ 0 & 0 & 0 & 0 & 0 \\ 0 & \frac{L_m}{\sqrt{2}} & -\frac{L_m}{\sqrt{2}} & 0 & L_r \\ -\sqrt{\frac{2}{3}}L_m & \frac{L_m}{\sqrt{6}} & \frac{L_m}{\sqrt{6}} & -L_r & 0 \end{bmatrix} \tag{13.114}$$

$$[L] = \begin{bmatrix} L_s & -\frac{L_m}{3} & -\frac{L_m}{3} & L_m\sqrt{\frac{2}{3}} & 0 \\ -\frac{L_m}{3} & L_s & -\frac{L_m}{3} & -\frac{L_m}{\sqrt{6}} & \frac{L_m}{\sqrt{2}} \\ -\frac{L_m}{3} & -\frac{L_m}{3} & L_s & -\frac{L_m}{\sqrt{6}} & \frac{L_m}{\sqrt{2}} \\ \sqrt{\frac{2}{3}}L_m & -\frac{L_m}{\sqrt{6}} & -\frac{L_m}{\sqrt{6}} & L_r & 0 \\ 0 & \frac{L_m}{\sqrt{2}} & -\frac{L_m}{\sqrt{2}} & 0 & L_r \end{bmatrix} \tag{13.115}$$

The parameters L_s, L_r, L_m refer to phase inductances,

$$L_s = L_m + L_{ls};\ L_r = L_m + L_{lr} \tag{13.116}$$

The torque equation becomes

$$T_e = p_1 [I]^T [G][I] \tag{13.117}$$

Finally, the motion equation is

$$\frac{J}{p_1}\frac{d\omega_r}{dt} = T_e - T_{load} - B\omega_r \tag{13.118}$$

For PWM excitation and mechanical steady-state (ω_r = const.) based on $V_{ab}(t)$, $V_{bc}(t)$, $V_{ca}(t)$ functions, as imposed by the corresponding PWM strategy, the currents may be found from (13.112) under the form

$$\left[\dot{I}\right] = -\left[L\right]^{-1}\left[\left[R\right] + \omega_r\left[G\right]\right]\left[I\right] + \left[L\right]^{-1}\left[V\right] \tag{13.119}$$

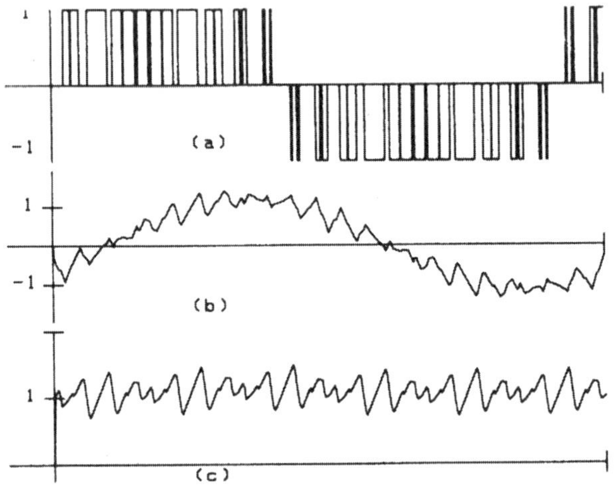

Figure 13.29 Computed waveforms for sinusoidal PWM scheme (F_{nc} = 21), f = 60 Hz
a.) line to line voltage V_{ab}, b.) line current i_a, c.) developed torque

Initial values of currents are to be given for steady-state sinusoidal supply and the same fundamental.

Although a computer program is required, the computation time is rather low as the coefficients in (13.119) are constant.

Results obtained for an IM with $R_s = 0.2\Omega$, $R_r = 0.3\Omega$, $X_m = 16\Omega$, and $X_r = X_s = 16.55\Omega$, fed with a fundamental frequency f_n = 60 Hz, at a switching frequency f_c = 1260 Hz and sinusoidal PWM, are shown in Figure 13.29. [19]

When the speed varies, the PWM switching patterns may change and thus produce current and torque transients. A smooth transition is needed for good performance. The abc–dq model, including the motion equation, also serves this purpose (for example, switching from harmonic elimination PWM (N = 5) to square wave, Figure 13.30 [19].)

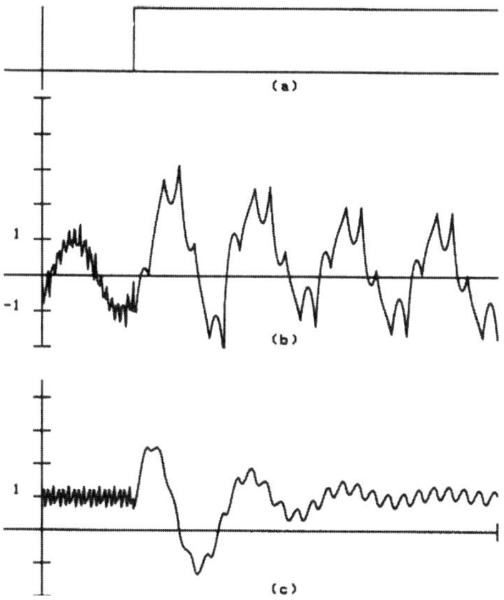

Figure 13.30 Computed transient response for a change from harmonic (N = 5) elimination PWM to square wave excitation at 60 Hz
a.) voltage, b.) current, c.) torque

The current waveform changes to the classic six-pulse voltage response after notable current and torque transients.

a.) *Fault conditions*

Open line conditions or single phasing due to loss of gating signals of a pair of inverter switches are typical faults. Let us consider delta connection when line a is opened (Figure 13.31a). The relationships between the system currents [i_{ab}, i_{bc}, i_{ca}, i_d, i_q] and the "fault" currents i_{bac}, i_{bc}, i_d, i_q are (Figure 13.31a).

$$\begin{vmatrix} i_{ab} \\ i_{bc} \\ i_{ca} \\ i_d \\ i_q \end{vmatrix} = |C| \begin{vmatrix} i_{bac} \\ i_{bc} \\ i_d \\ i_q \end{vmatrix}; \ [I_n] = [C][i_0] \quad (13.120)$$

$$[C] = \begin{vmatrix} -1 & 0 & 0 & 0 \\ 0 & 1 & 0 & 0 \\ -1 & 0 & 0 & 0 \\ 0 & 0 & 1 & 0 \\ 0 & 0 & 0 & 1 \end{vmatrix} \quad (13.121)$$

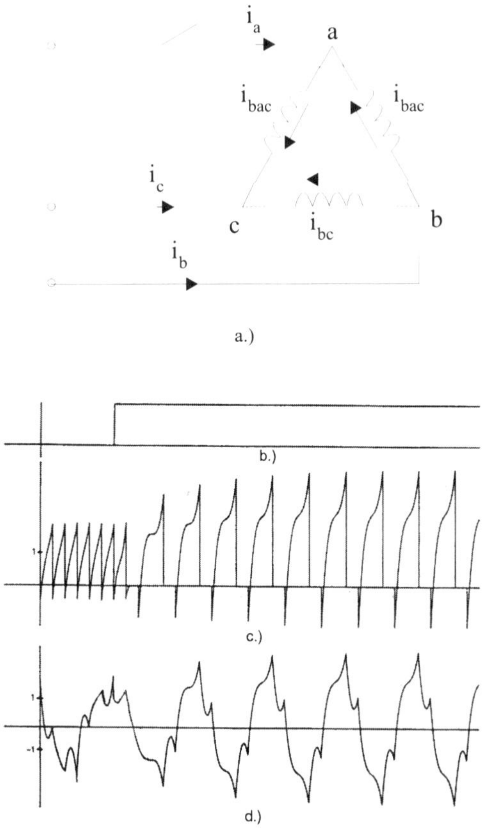

Figure 13.31 Computed transient response for a single phasing fault on line 'a' at 60 Hz
a.) open line condition b.) instant when single phasing occurs
c.) dc link current d.) line current i_s

The voltage relationship is straightforward as power has to be conserved:

$$\begin{bmatrix} V_{bac} \\ V_{bc} \\ V_d \\ V_q \end{bmatrix} = [C]^T \begin{bmatrix} V_{ab} \\ V_{bc} \\ V_{ca} \\ V_d \\ V_q \end{bmatrix}; \quad [V_n] = [C]^T [V_0] \tag{13.122}$$

Now we may replace the voltage matrix in (13.112) to obtain

$$[V_n] = [C]^T [R][C][I_n] + [C]^T \omega_r [G][C][I_n] + [C]^T [L][C] \frac{d[I_n]}{dt} \tag{13.123}$$

All we need to be able to solve Equation (13.123) are the initial conditions.

In reality, (Figure 13.31a) at time t = 0, the currents i_{ba} and i_{ca} are forced to be equal to I_{bac} when open line occurs. To avoid a discontinuity in the stored energy, the flux conservation law is applied to (13.112).

We impose $i_{ab} = i_{ca} = -i_{bac}$ and the conservation of flux will give the variation of rotor current to comply with flux conservation. Typical results for such a transient [19] are shown in Figure 13.31b,c,d. Severe d.c. line pulsations and line b current transients are visible.

Other fault conditions may be treated in a similar way.

13.13. FIRST ORDER MODELS OF IMs FOR STEADY-STATE STABILITY IN POWER SYSTEMS

The quasistatic (first order) or slip model of IMs with neglected stator and rotor transients may be obtained from the general model (13.70) with $\frac{d\overline{\Psi}_s}{dt} = \frac{d\overline{\Psi}_r}{dt} = 0$ in synchronous coordinates.

$$\begin{gathered}(1 + j\omega_1\tau_s')\overline{\Psi}_s = \tau_s' V_s + K_r\overline{\Psi}_r \\ (1 + j(\omega_1 - \omega_r))\overline{\Psi}_r = K_s\overline{\Psi}_s \\ T_e = -\frac{3}{2}p_1 \operatorname{Re}\left(j\overline{\Psi}_r \bar{i}_r^*\right); \; \bar{i}_r = \sigma^{-1}\left(\frac{\overline{\Psi}_r}{L_r} - \overline{\Psi}_s\frac{L_m}{L_s L_r}\right) \\ \frac{J}{p_1}\frac{d\omega_r}{dt} = T_e - T_{load}\end{gathered} \tag{13.124}$$

It has been noted that this overly simplified speed model becomes erroneous for large induction machines.

Near rated conditions, the structural dynamics of low and large power induction machines differ. The dominant behavior of low power IMs is a first order speed model while the large power IMs is characterized by a first order voltage model. [20, 21]

Starting again with the third order model of (13.70), we may write the rotor flux space phasor as

$$\overline{\Psi}_r = \Psi_r e^{j\delta} \tag{13.125}$$

This way we may separate the real and imaginary part of rotor equation of (13.70) as

$$\begin{gathered}\frac{\tau_r'}{K_r}\frac{d\Psi_r}{dt}K_r = \frac{x}{x'}\Psi_r K_r + \frac{x - x'}{x'}V\cos\delta; \; K_r \approx K_s = \frac{L_m}{L_s} = \frac{L_m}{L_r} \\ \frac{d\delta}{dt} = \omega_r - \omega_1 - \frac{x - x'}{x'}K_r\frac{V\sin\delta}{\tau_r'\Psi_r K_r}\end{gathered}$$

$$H'\frac{d\omega_r}{dt} = -V\frac{\Psi_r}{x'}K_r \sin\delta - t_{load} \tag{13.125'}$$

where V (voltage), Ψ_r – rotor flux amplitude, x – no-load reactance, x' – short-circuit (transient) reactance in relative units (p.u.) and τ_r', ω_1, ω_r in absolute units and H' in seconds^2. The stator resistance is neglected.

In relative units, $\Psi_r K_r$ may be replaced by the concept of the voltage behind the transient reactance x', E'.

$$\begin{aligned}
\frac{\tau_r'}{K_r}\frac{dE'}{dt} &= \frac{x}{x'}E' + \frac{x-x'}{x'}V\cos\delta \\
\frac{d\delta}{dt} &= \omega_r - \omega_1 - \frac{x-x'}{x'}K_r\frac{V\sin\delta}{\tau_r'E'} \\
H'\frac{d\omega_r}{dt} &= -V\frac{E'}{x'}\sin\delta - t_{load}
\end{aligned} \tag{13.126}$$

- The concept of E' is derived from the similar synchronous motor concept
- The angle δ is the rotor flux vector angle with respect to the synchronous reference system
- The steady-state torque is obtained from (13.126) with d/dt = 0

$$E' = \frac{x-x'}{x}V\cos\delta \tag{13.127}$$

$$\omega_r - \omega_1 = \frac{x-x'}{x'}K_r\frac{V\sin\delta}{\tau_r'E'} < 0 \text{ for motoring} \tag{13.128}$$

$$t_e = -V\frac{E'}{x'}\sin\delta = -\left(\frac{1}{x'} - \frac{1}{x}\right)\frac{V^2}{2}\sin 2\delta > 0 \text{ for motoring} \tag{13.129}$$

for positive torque δ<0 and for peak torque $\delta_K = \pi/4$ as in a fictitious reluctance synchronous motor with x_d = x (no-load reactance) and x_q = x' (transient reactance).

For low power IMs, E' and δ are the fast variables and $\omega_r - \omega_1$, the slow variable.

Consequently, we may replace δ and E' from the steady-state Equations (13.127) and (13.128) in the motion equation (13.126).

$$x'H'\frac{d(\omega_r - \omega_1)}{dt} = -V^2\frac{x-x'}{x}\frac{(\omega_r - \omega_1)T'}{1+T'(\omega_r - \omega_1)^2} - x't_{load};\ T' = \frac{\tau_r'}{K_r} \tag{13.130}$$

Also, from (13.127) and (13.128),

$$E' = \frac{(x - x')V}{x\sqrt{1 + T'^2(\omega_r - \omega_1)^2}}; \ \delta \approx \tan^{-1} T'(\omega_r - \omega_1) \tag{13.131}$$

In contrast, for large power induction machines, the slow variable is E' and the fast variables are δ and $\omega_r - \omega_1$. With $d\omega_r/dt = 0$ and $d\delta/dt = 0$ in (13.126),

$$\delta = \sin^{-1}\left(\frac{-x' t_{load}}{VE'}\right); \ (\omega_r - \omega_1) = 0 \tag{13.132}$$

It is evident that $\omega_r = \omega_1 \neq 0$.
A nontrivial first order approximation of $\omega_r - \omega_1$ is [21]

$$\omega_r - \omega_1 \approx \frac{x' t_{load}}{T'\sqrt{(VE')^2 - (x' t_{load})^2}} \tag{13.133}$$

Substituting δ of (13.132) in (13.126) yields

$$T'\frac{dE'}{dt} = -\frac{x}{x'}E' + \frac{x - x'}{x'}V\sqrt{1 - \left(\frac{x' t_{load}}{VE'}\right)^2} \tag{13.134}$$

Equations (13.132) through (13.134) represent the modified first order model for large power induction machines. Good results with these first order models are reported in [20] for 50 HP and 500 HP, respectively, in comparison with the third order model (Figure 13.32).

Still better results for large IMs are claimed in [21] with a heuristic first order voltage model based on sensitivity studies.

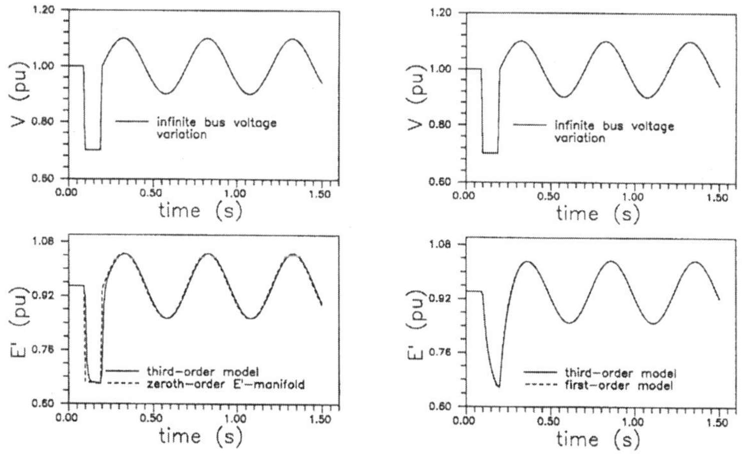

Figure 13.32 Third and first order model transients a.) 50 HP

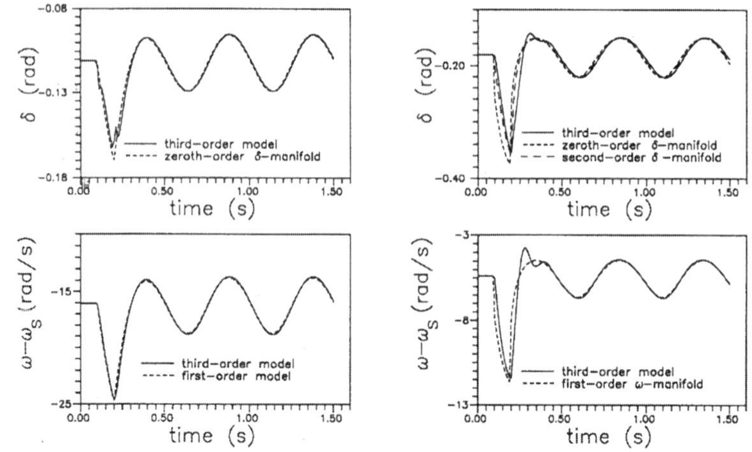

Figure 13.32 (continued) Third and first order model transients b.) 500 HP, IM

13.14. MULTIMACHINE TRANSIENTS

A good part of industrial loads is represented by induction motors. The power range of induction motors in industry spans from a few kW to more than 10 MW per unit. A local (industrial) transformer feeds a group of induction motors through pertinent switchgears which may undergo randomly load perturbations, direct turn on starting or turn off. Alternatively, bus transfer schemes are providing energy-critical duty loads to thermal nuclear power plants, etc.

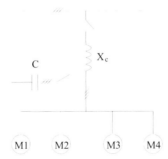

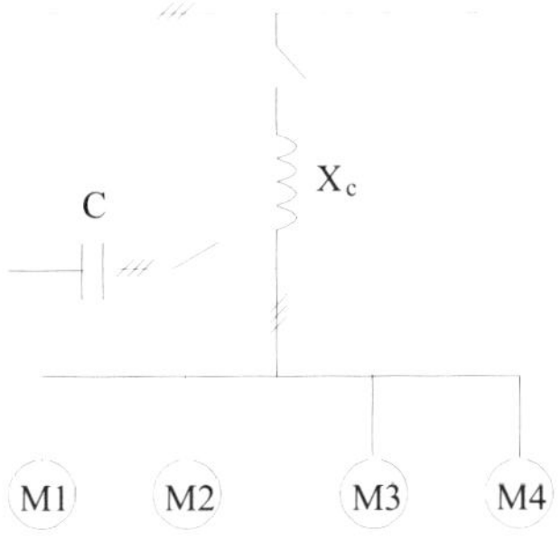

Figure 13.33 A group of induction motors

During the time interval between turn off from one power grid until the turn on to the emergency power grid, the group of induction motors, with a series reactance common feeder or with a parallel capacitor bank (used for power factor correction), exhibits a residual voltage, which slowly dies down until reconnection takes place (Figure 13.33). During this period, the residual rotor currents in the various IMs of the group are producing stator emfs and, depending on their relative phasing and amplitude (the speed or inertia), some of them will act as generators and some as motors until their mechanical and magnetic energies dry out.

The obvious choice to deal with multimachine transients is to use the complete space phasor model with flux linkages $\overline{\Psi}_s$ and $\overline{\Psi}_r$ and ω_r as variables ((13.54 through 13.56) and (13.64)).

Adding up to the 5th order model of all n IMs (5n equations), the equations related to the terminal capacitor (and the initial conditions before the transient is initiated) seems the natural way to solve any multimachine transient. A unique synchronous reference system may be chosen to speed up the solution.

Furthermore, to set the initial conditions, say, when the group is turned off, we may write that total stator current $\overline{i}_{st}(0+)=0$. With a capacitor at the terminal, the total current is zero.

$$\sum_{j=1}^{n}\overline{i}_{sj}(0+)+\overline{i}_c=0;\ \frac{d\overline{V}_s}{dt}=\frac{\overline{i}_c}{C} \tag{13.135}$$

with $\overline{i}_s$ from (13.54),

$$\sum_{j=1}^{n}\overline{i}_{sj}=\sum_{j=1}^{n}\sigma_j^{-1}\left(\frac{\overline{\Psi}_{sj}}{L_{sj}}-\frac{\overline{\Psi}_{rj}L_{mj}}{L_{sj}L_{rj}}\right) \tag{13.136}$$

Differentiating Equations (13.135) with (13.136) with respect to time will produce an equation containing a relationship between the stator and rotor flux time derivatives. This way, the flux conservation law is met.

We take the expressions of these derivatives from the space phasor model of each machine and find the only unknown, $\left(\overline{V}_s\right)_{t=0+}$, as the values of all variables $\overline{\Psi}_{sj}$, $\overline{\Psi}_{rj}$, ω_{rj}, before the transient is initiated, are known.

Now, if we neglect the stator transients and make use of the third order model, we may use the stator equations of any machine in the group to calculate the residual voltage $\overline{V}_s(t)$ as long as (13.135) with (13.136) is fulfilled.

This way no iteration is required and only solving each third model of each motor is, in fact, done. [22, 23] Unfortunately, although the third order model forecasts the residual voltage $V_s(t)$ almost as well as the full (5th order) model, the test results show a sudden drop in this voltage in time. This sharply contrasts with the theoretical results.

A possible explanation for this is the influence of magnetic saturation which maintains the “selfexcitation” process of some of the machines with larger inertia for some time, after which a fast deexcitation of them by the others, which act as motors, occurs.

Apparently magnetic saturation has to be accounted for to get any meaningful results related to residual voltage transient with or without terminal capacitors.

Simpler first order models have been introduced earlier in this chapter. They have been used to study load variation transients once the flux in the machine is close to its rated value. [24, 25]

Furthermore, multimachine transients (such as limited time bus fault, etc.) with unbalanced power grid voltages have also been treated, based on the dq (space phasor) model–complete or reduced. However, in all these cases, no strong experimental validation has been brought up so far.

So we feel that model reduction, or even saturation and skin effect,, has to be pursued with extreme care when various multimachine transients are investigated.

13.15. SUBSYNCHRONOUS RESONANCE (SSR)

Subsynchronous series resonance may occur when the induction motors are fed from a power source that shows a series capacitance (Figure 13.34). It may lead to severe oscillations in speed, torque, and current. It appears, in general, during machine starting or after a power supply fault.

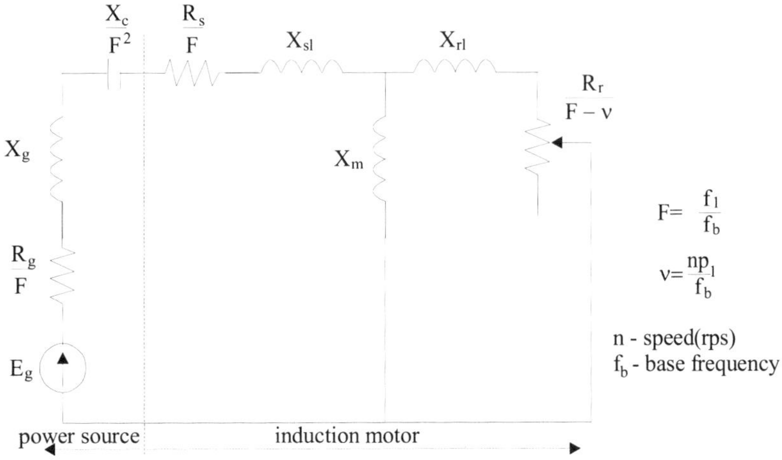

Figure 13.34 Equivalent circuit with power source parameters, R_g, X_g, X_c in series

f_b is the base frequency, f_1 is the current frequency, n is the rotor speed. The reactances are written at base frequency.

It is self evident that the space phasor model may be used directly to solve this problem completely, provided a new variable (capacitor voltage V_c) is introduced.

$$\frac{d\overline{V}_c}{dt} = \frac{1}{C}\overline{i}_s \quad (13.137)$$

The total solution contains a forced solution and a free solution. The complete mathematical model is complex when magnetic saturation is considered. It simplifies greatly if the magnetization reactance X_m is considered constant (though at a saturated value). In contrast to the induction capacitor-excited generator, the free solution here produces conditions that lead to machine heavy saturation.

In such conditions we may treat separately the resonance conditions for the free solution.

To do so we lump the power source impedance R_g, X_g into stator parameters to get

$$R_e = R_s + R_g;\ X_e = X_g + X_{sl} \quad (13.138)$$

and eliminate the generator (power source) emf E_g (Figure 13.35).

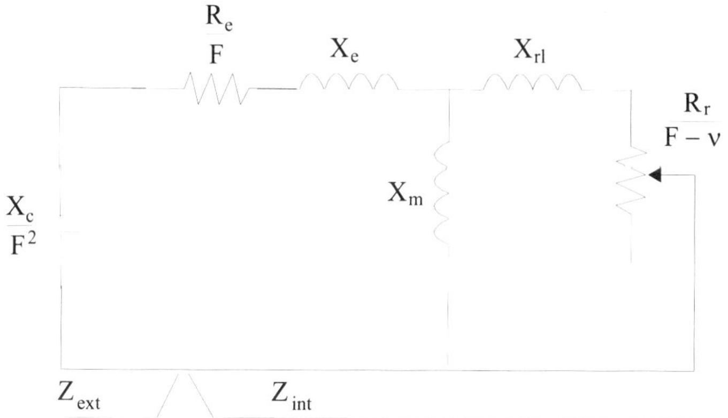

Figure 13.35 Equivalent circuit for free solution

To simplify the analysis, let us use a graphical solution. The internal impedance locus is, as known, a circle with the center 0 at $0\left(\frac{R_e}{F}, a\right)$. [28]

$$a \approx X_e + \frac{X_e^2}{2(X_e + X_m)} \quad (13.139)$$

and the radius r

$$r \approx \frac{X_e^2}{2(X_e + X_m)} \quad (13.140)$$

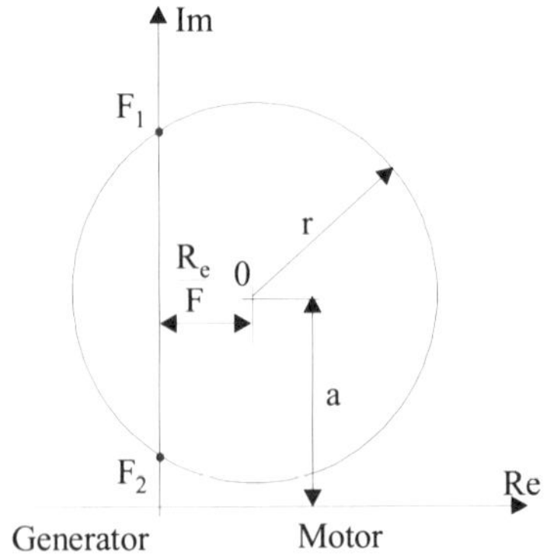

Figure 13.36 Impedance locus

On the other hand, the external (the capacitor, in fact) impedance is represented by the imaginary axis (Figure 13.36). The intersection of the circle with the imaginary axis, for a given capacitor, takes place in two points F_1 and F_2 which correspond to reference conditions. F_1 corresponds to the lower slip and is the usual self excitation point of induction generators.

The p.u. values of frequencies F_1 and F_2, for a given series capacitor X_c and resistor, R_e, reactance X_e, are

$$F_1 \approx \sqrt{\frac{X_e + X_m}{X_c}}$$
$$F_2 \approx \sqrt{\frac{X_e + \frac{X_m X_{rl}}{X_m + X_{rl}}}{X_c}} \tag{13.141}$$

Alternatively, for a given resonance frequency and capacitor, we may obtain the resistance R_e to cause subsynchronous resonance. With $F = F_1$ given, we may use the impedance locus to obtain R_e.

$$\frac{X_c}{F_1^2} = a + \sqrt{r^2 - \left(\frac{R_e}{F_1}\right)^2} \tag{13.142}$$

Also, from the internal impedance at $R_e(\underline{Z}_{int}) = 0$ equal to X_c/F_1^2, the value of speed (slip: F–ν) may be obtained.

$$\text{slip} = F_1 - \nu \approx \pm\sqrt{\frac{R_r\left(X_m - X_e - \frac{X_c}{F_1^2}\right)}{(X_m + X_{rl})\left[\left(\frac{X_c}{F_1^2} - X_e\right)(X_m + X_{rl}) - X_m X_{rl}\right]}} \tag{13.143}$$

In general, we may use Equation (13.143) with both ± to calculate the frequencies for which resonance occurs.

$$F^4\left(a^2 - r^2\right) + F^2\left(R_e^{\ 2} - 2aX_c\right) + X_c^{\ 2} = 0 \tag{13.144}$$

The maximum value of R_e, which still produces resonance, is obtained when the discriminant of (13.144) becomes zero.

$$R_{e\max} = \sqrt{2X_c\left(\sqrt{a^2 - r^2} + a\right)} = KX_c \tag{13.145}$$

The corresponding frequency,

$$F(R_{e\max}) = \sqrt{\frac{2X_c^{\ 2}}{2aX_c - R_{e\max}^{\ 2}}} \tag{13.146}$$

This way the range of possible source impedance that may produce subsynchronous resonance (SSR) has been defined.

Preventing SSR may be done either by increasing R_e above R_{emax} (which means notable losses) or by connecting a resistor R_{ad} in parallel with the capacitor (Figure 13.37), which means less losses.

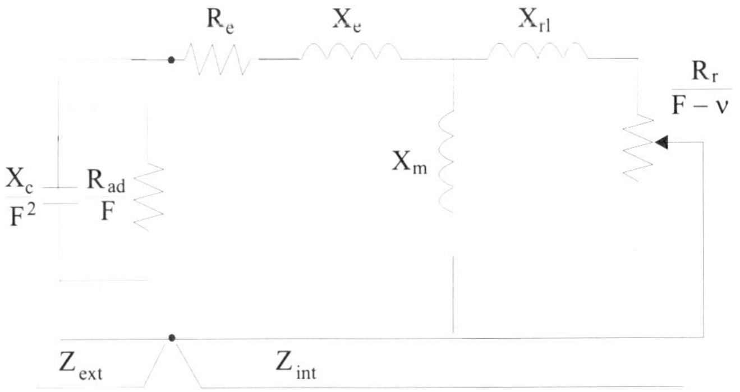

Figure 13.37 Additional resistance R_{ad} for preventing SSR

A graphical solution is at hand as $\underline{Z}_{int}$ locus is already known (the circle). The impedance locus of the $R_{ad}||X_c$ circuit is a half circle with the diameter R_{ad}/F and the center along the horizontal negative axis (Figure 13.38).

The SSR is eliminated if the two circles, at same frequency, do not intersect each other. Analytically, it is possible to solve the equation:

$$\underline{Z}_{ext} + \underline{Z}_{int} = 0 \tag{13.146'}$$

again, with $R_{ad}||X_c$ as $\underline{Z}_{ext}$.

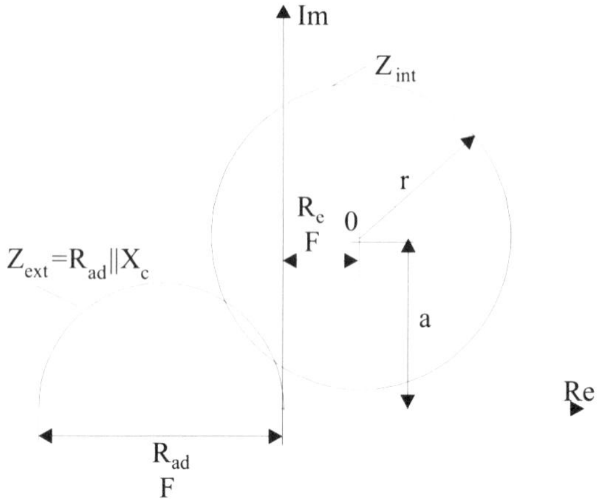

Figure 13.38 Impedances locus for SSR prevention

A 6th order polynomial equation is obtained. The situation with a double real root and 4 complex roots corresponds to the two circles being tangent. Only one SSR frequency occurs. This case defines the critical value of R_{ad} for SSR.

In general, R_{ad} (critical) increases with X_c and so does the corresponding SSR frequency. [27, 28]

Using a series capacitor for starting medium power IMs to reduce temporary supply voltage sags and starting time is a typical situation when the prevention SSR is necessary.

13.16. THE m/N_r ACTUAL WINDING MODELING FOR TRANSIENTS

By m/N_r winding model, we mean the induction motor with m stator windings and N_r actual (rotor loop) windings. [29]

In the general case, the stator windings may exhibit faults such as local short-circuits, or some open coils, when the self and mutual inductances of stator windings have special expressions that may be defined using the winding function method. [30, 31, 32]

While such a complex methodology may be justifiable in large machine preventive fault diagnostics, for a symmetrical stator the definition of inductances is simpler.

The rotor cage is modeled through all its N_r loops (bars). One more equation for the end ring is added. So the machine model is characterized by (m + N_r + 1 + 1) equations (the last one is the motion equation). Rotor bar and end ring faults so common in IMs may thus be modeled through the m/N_r actual winding model.

The trouble is that it is not easy to measure all inductances and resistances entering the model. In fact, relying on analysis (or field) computation is the preferred choice.

The stator phase equations (in stator coordinates) are

$$V_a = R_s i_a + \frac{d\Psi_a}{dt}$$
$$V_b = R_s i_b + \frac{d\Psi_b}{dt} \qquad (13.147)$$
$$V_c = R_s i_c + \frac{d\Psi_c}{dt}$$

The rotor cage structure and unknowns are shown in Figure 13.39.

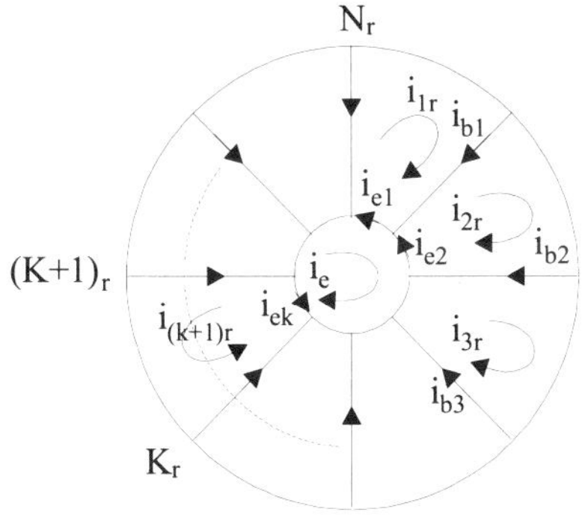

Figure 13.39 Rotor cage with rotor loop currents

The K_r^{th} rotor loop equation (in rotor coordinates) is

$$0 = 2\left(R_b + \frac{R_e}{N_r}\right)i_{kr} + \frac{d\Psi_{kr}}{dt} - R_b\left(i_{(k-1)r} + i_{(k+1)r}\right) \qquad (13.148)$$

Also, for one of the end rings,

$$0 = R_e i_e + L_e \frac{di_e}{dt} - \sum_1^{N_r}\left(\frac{R_e}{N_r} i_{kr} + L_e \frac{di_{kr}}{dt}\right) \qquad (13.149)$$

As expected, for a healthy cage, the end ring loop cage current $i_e = 0$.

The relationships between the bar loop and end ring currents i_b, i_e are given by Kirchoff's law,

$$i_{bk} = i_{kr} - i_{(k+1)r};\ \ i_{ek} = i_{kr} - i_e \qquad (13.150)$$

This explains why only $N_r + 1$ independent rotor current variables remain.

The self and mutual stator phase inductances are given through their standard expressions (see Chapter 5).

$$L_{aa} = L_{bb} = L_{cc} = L_{ls} + L_{ms}$$
$$L_{ab} = L_{bc} = L_{ca} = -\frac{1}{2}L_{ms} \quad (13.151)$$
$$L_{ms} = \frac{4\mu_0 (W_1 K_{w1})^2 \tau L}{\pi^2 K_c p_1 g(1+K_s)}$$

The saturation coefficient K_s is considered constant. L_{ls} – leakage inductance, τ – pole pitch, L – stack length, p_1 – pole pairs, g – airgap, W_1 – turns/phase, K_{w1} – winding factor. The self inductance of the rotor loop calculated on the base of rotor loop area,

$$L_{K_r,K_r} = \frac{2\mu_0 (N_r - 1) p_1 \tau L}{N_r^2 K_c g(1+K_s)} \quad (13.152)$$

The mutual inductance between rotor loops is

$$L_{K_r,(K+j)_r} = -\frac{2\mu_0 p_1 \tau L}{N_r^2 K_c g(1+K_s)} \quad (13.153)$$

This is a kind of average small value, as in reality, the coupling between various rotor loops varies notably with the distance between them.

The stator–rotor loops mutual inductances L_{aKr}, L_{bKr}, L_{cKr} are

$$L_{aK_r}(\theta_r) = L_{sr}\cos[\theta_{er} + (K_r - 1)\alpha]$$
$$L_{bK_r}(\theta_r) = L_{sr}\cos\left[\theta_{er} + (K_r - 1)\alpha - \frac{2\pi}{3}\right] \quad (13.154)$$
$$L_{cK_r}(\theta_r) = L_{sr}\cos\left[\theta_{er} + (K_r - 1)\alpha + \frac{2\pi}{3}\right]$$

$$\alpha = p_1\frac{2\pi}{N_r};\ \theta_{er} = p_1\theta_r;\ p_1 - \text{pole pairs} \quad (13.155)$$

$$L_{sr} = \frac{-(W_1 K_{w1})\mu_0 2P_1 \tau L}{4P_1 g K_c (1+K_s)}\sin\frac{\alpha}{2} \quad (13.156)$$

The m + N_r + 1 electrical equations may be written in matrix form as

$$[V] = [R][i] + [L'(\theta_{er})]\left[\frac{di}{dt}\right] + [G][i] \quad (13.157)$$

$$[V] = [V_a, V_b, V_c, 0, 0, \ldots 0]^T$$
$$[I] = [i_a, i_b, i_c, i_{1r}, i_{2r}, \ldots i_{N_r}, i_e]^T \quad (13.158)$$

$$L'(\theta_{er}) = \begin{bmatrix} L_{aa}(\theta_{er}) & L_{ab}(\theta_{er}) & & & & & & L_{aN_r}(\theta_{er}) & 0 \\ L_{ab}(\theta_{er}) & L_{bb}(\theta_{er}) & & & & & & L_{bN_r}(\theta_{er}) & 0 \\ L_{ac}(\theta_{er}) & L_{bc}(\theta_{er}) & & & & & & L_{cN_r}(\theta_{er}) & 0 \\ \cdots & \cdots & & & & & & \cdots & -\dfrac{L_e}{N_r} \\ \cdots & \cdots & & & & & & \cdots & \cdots \\ L_{a3r}(\theta_{er}) & L_{b3r}(\theta_{er}) & & & & & & L_{3rN_r}(\theta_{er}) & -\dfrac{L_e}{N_r} \\ \cdots & \cdots & & & & & & \cdots & \cdots \\ L_{aN_r}(\theta_{er}) & L_{bN_r}(\theta_{er}) & & & & & & L_{N_rN_r}(\theta_{er}) & -\dfrac{L_e}{N_r} \\ 0 & 0 & 0 & -\dfrac{L_e}{N_r} & -\dfrac{L_e}{N_r} & -\dfrac{L_e}{N_r} & -\dfrac{L_e}{N_r} & -\dfrac{L_e}{N_r} & +L_e \end{bmatrix}$$

(13.159)

$$[G] = \begin{bmatrix} & a & b & c & 1r & & & & N_r & \\ a & \dfrac{dL_{aa}(\theta_{er})}{d\theta_{er}} & & & & & & & \dfrac{dL_{aN_r}(\theta_{er})}{d\theta_{er}} & 0 \\ b & & & & & & & & & 0 \\ c & & & & & & & & & 0 \\ 1r & \cdots & & & & & & & \cdots & 0 \\ 2r & \cdots & & & & & & & \cdots & 0 \\ 3r & & & & & & & & & 0 \\ \cdots & \cdots & & & & & & & \cdots & \cdots \\ N_r & \dfrac{dL_{aN_r}(\theta_{er})}{d\theta_{er}} & & & & & & & \dfrac{dL_{N_rN_r}(\theta_{er})}{d\theta_{er}} & 0 \\ & 0 & 0 & 0 & 0 & 0 & 0 & 0 & 0 & 0 \end{bmatrix} \qquad (13.160, 161)$$

$$[R] = \begin{bmatrix} R_s & 0 & 0 & 0 & & 0 & & 0 & 0 \\ 0 & R_s & 0 & 0 & & & & & 0 \\ 0 & 0 & R_s & 0 & & & & & 0 \\ 0 & 0 & 0 & 2\left(R_b + \dfrac{R_e}{N_r}\right) & -R_b & 0 & 0 & -R_b & -\dfrac{R_e}{N_r} \\ 0 & 0 & 0 & -R_b & 2\left(R_b + \dfrac{R_e}{N_r}\right) & -R_b & & 0 & -\dfrac{R_e}{N_r} \\ \cdots & \cdots & \cdots & 0 & -R_b & \cdots & \cdots & & -\dfrac{R_e}{N_r} \\ \cdots & \cdots & \cdots & \cdots & & & \cdots & -R_b & \cdots \\ 0 & 0 & 0 & 0 & 0 & 0 & -R_b & 2\left(R_b + \dfrac{R_e}{N_r}\right) & -\dfrac{R_e}{N_r} \\ 0 & 0 & 0 & -\dfrac{R_e}{N_r} & -\dfrac{R_e}{N_r} & -\dfrac{R_e}{N_r} & -\dfrac{R_e}{N_r} & -\dfrac{R_e}{N_r} & R_e \end{bmatrix}$$

The motion equations are

$$\frac{d\omega_r}{dt} = \frac{p_1}{J}\left(T_e - T_{load}\right); \ \frac{d\theta_{er}}{dt} = \omega_r \tag{13.162}$$

The torque T_e is

$$T_e = p_1 [I]^T \frac{\partial[L(\theta_{er})]}{d\theta_{er}}[I] \tag{13.163}$$

Finally,

$$\begin{aligned} T_e = p_1 L_{sr}\Bigg[\left(i_a - \frac{1}{2}i_b - \frac{1}{2}i_c\right)\sum_1^{N_r} i_{Kr}\sin[\theta_{er} + (K_r - 1)\alpha] + \\ + \frac{\sqrt{3}}{2}(i_b - i_c)\sum_1^{N_r} i_{Kr}\cos[\theta_{er} + (K_r - 1)\alpha]\Bigg] \end{aligned} \tag{13.164}$$

A few remarks are in order.

- The $(3 + N_r + 1)$ order system has quite a few coefficients (inductances) depending on rotor position θ_{er}.
- When solving such a system a matrix inversion is required for each integration step when using a Runge–Kutta–Gill or the like method to solve it.
- The case of healthy cage can be handled by a generalized space vector (phasor) approach when the coefficients dependence on rotor position is eliminated.
- The case of faulty bars is handled by increasing the resistance of that (those) bar (R_b) a few orders of magnitude.
- When an end ring segment is broken, the corresponding R_e/N_r term in the resistance matrix is increased a few orders of magnitude (both along the diagonal and along its vertical and horizontal direction in the last row and column, respectively.

In Reference [33], a motor with the following data was investigated:

R_s = 10 Ω, $R_b = R_e$ = 155 μΩ, L_{ls} = 0.035 mH, L_{ms} = 378 mH, W_1 = 340 turns/phase, K_{w1} = 1, N_r = 30 rotor slots, p_1 = 2 pole pairs, $L_e = L_b$ = 0.1 μH, L_{sr} = 0.873 mH, τ = 62.8 mm (pole pitch), L = 66 mm (stack length), g = 0.375 mm, K_c = 1, K_s = 0, J = $5.4\cdot10^{-3}$ Kgm^2, f = 50 Hz, V_L = 380 V, P_n = 736 W, I_0 = 2.1 A.

Numerical results for bar 2 broken ($R_{b2} = 200R_b$) with the machine under load T_L = 3.5 Nm are shown in (Figure 13.40).

For the ring segment 3 broken ($R_{e3} = 10^3\ R_e/N_r$), the speed and torque are given in Figure 13.41 [33].

- Small pulsations in speed and torque are caused by a single bar or end ring segment faults.

- A few broken bars would produce notable torque and speed pulsations. Also, a $2Sf_1$ frequency component will show up, as expected, in the stator phase currents.

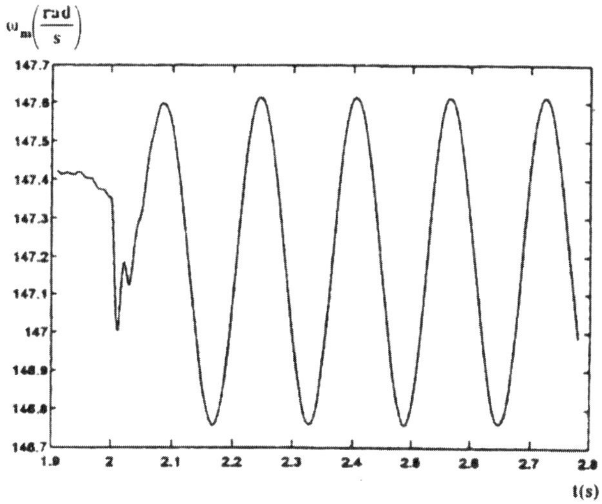

Mechanical angular velocity ω_m. Bar 2 is broken

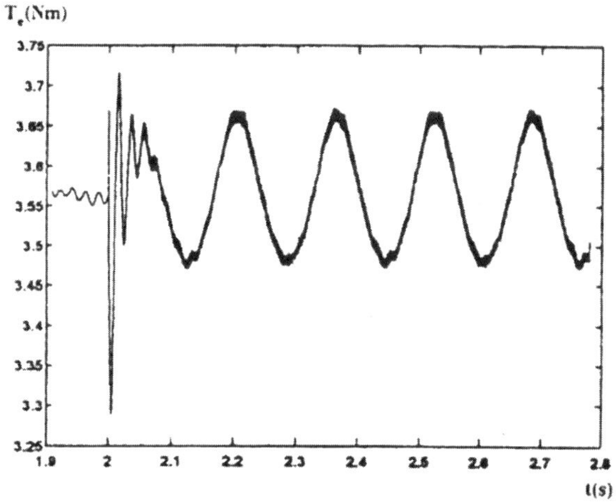

Electromagnetic torque T_e. Bar 2 is broken

Figure 13.40 Bar 2 broken at steady-state (T_L = 3.5Nm)
a.) speed, b.) torque, c.) broken bar current (continued)

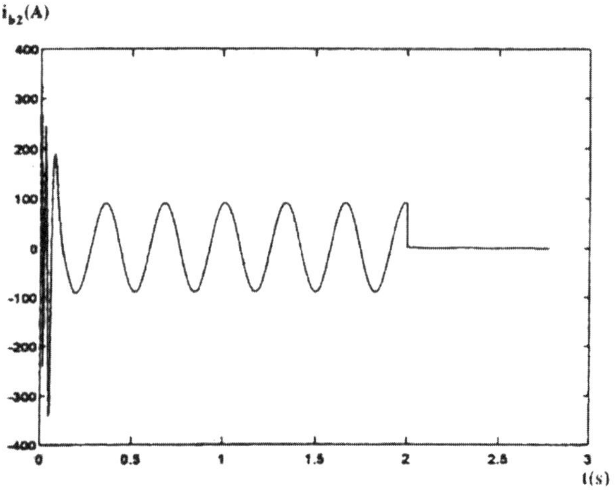

Current of broken bar. Bar 2 is broken

Figure 13.40 (continued)

The assumption that the interbar resistance is infinitely large (no interbar currents or insulated bars) is hardly true in reality. In presence of interbar currents, the effect of 1 to 2 broken bars tends to be further diminished. [34]

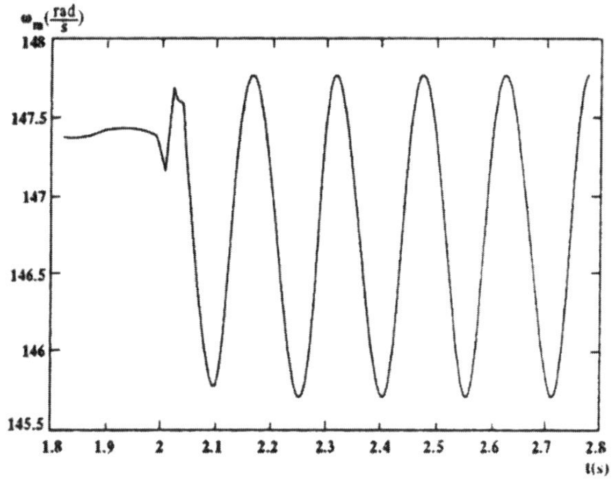

Mechanical angular velocity. Ring-section 3 is broken

Figure 13.41 End ring 3 broken at steady-state
a.) speed, b.) torque (continued)

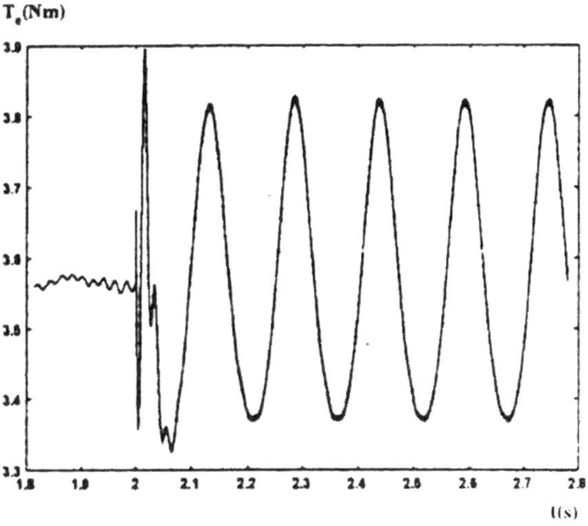

Electromagnetic torque. Ring-section 3. is broken

Figure 13.41 (continued)

The transients of IM have also been approached by FEM, specifically by time stepping coupled FEM circuit (or state space) models. [35, 36] The computation time is still large, but the results are getting better and better as new contributions are made. The main advantage of this method is the possibility to account for most detailed topological aspects and non-linear material properties. Moreover, coupling the electromagnetic to thermal and mechanical modeling seems feasible in the near future.

13.17. SUMMARY

- Induction machines undergo transients when the amplitude and frequency of electric variables and (or) the speed vary in time.
- Direct starting, after turn-off, sudden mechanical loading, sudden short-circuit , reconnection after a short supply fault, behavior during short intervals of supply voltage reduction, performance when PWM converter-fed, are all typical examples of IM transients.
- The investigation of transients may be approached directly by circuit models or by coupled FEM circuit models of different degrees of complexity.
- The phase coordinate model is the "natural" circuit model, but its stator–rotor mutual inductances depend on rotor position. The order of the model is 8 for a single three-phase winding in the rotor. The symmetrical rotor

cage may be replaced by one, two, n, three-phase windings to cater for skin (frequency) effects.

- Solving such a model for transients requires a notable computation effort on a contemporary PC. Dedicated numerical software (such as Matlab) has quite a few numerical methods to solve systems of nonlinear equations.
- To eliminate the parameter (inductance) dependence on rotor position, the space phasor (complex variable) or d–q model is used. The order of the system (in d–q, real, variables) is not reduced, but its solution is much easier to obtain through numerical methods.
- In complex variables, with zero homopolar components, only two electrical and one mechanical equations remain for a single cage rotor.
- With stator and rotor flux space phasors $\overline{\Psi}_s$ and $\overline{\Psi}_r$ as variables and constant speed, the machine model exhibits only two complex eigen values. Their expressions depend on the speed of the reference system.
- As expected, below breakdown torque speed, the real part of the complex eigen values tends to be positive suggesting unstable operation.
- The model has two transient electrical time constants τ_s', τ_r', one of the stator and one of the rotor. So, for voltage supply, the stator and rotor flux transients are similar. Not so for current supply (only the rotor equation is used) when the rotor flux transients are slow (marked by the rotor time constant $\tau_r = L_r/R_r$).
- Including magnetic saturation in the complex variable model is easy for main flux path if the reference system is oriented to main flux space phasor $\overline{\Psi}_m = \Psi_m$, as $\Psi_m = L_m(i_m)i_m$.
- The leakage saturation may be accounted for separately by considering the leakage inductance dependence on the respective (stator or rotor) current.
- The standard equivalent circuit may be ammended to reflect the leakage and main inductances, L_{sl}, L_{rl}, L_m, dependence on respective currents i_s, i_r, i_m. However, as both the stator and rotor cores experience a.c. fields, the transient values of these inductances should be used to get better results.
- For large stator (and rotor) currents such as in high performance drives, the main and leakage flux contribute to the saturation of teeth tops. In such cases, the airgap inductance L_g is separated as constant and one stator and one rotor total core inductance L_{si}, L_{ri}, variable with respective currents, are responsible for the magnetic saturation influence in the machine.
- Further on, the core loss may be added to the complex variable model by orthogonal short-circuited windings in the stator (and rotor). The ones in the rotor may be alternatively used as a second rotor cage.
- Core losses influence slightly the efficiency torque and power factor during steady-state. Only in the first few miliseconds the core loss "leakage" time constant influences the IM transients behavior. The slight detuning in field orientation controlled drives, due to core loss, is to be considered mainly when precise (sub 1% error) torque control is required.

- Reduced order models of IMs are used in the study of power system steady-state or transients, as the number of motors is large.
- Neglecting the stator transients (stator leakage time constant τ_s') is the obvious choice in obtaining a third order model (with $\overline{\Psi}_r$ and ω_r as state variables). It has to be used only in synchronous coordinates (where steady-state means dc). However, as the supply frequency torque oscillations are eliminated, such a simplification is not to be used in calculating starting transients.
- The supply frequency oscillations in torque during starting are such that the average transient torque is close to the steady-state torque. This is why calculating the starting time of a motor via the steady-state circuit produces reasonable results.
- Considering leakage saturation in investigating starting transients is paramount in calculating correctly the peak torque and current values.
- Large induction machines are coupled to plants through a kind of elastic coupling which, together with the inertia of motor and load, may lead to torsional frequencies which are equal to those of transient torque components. Large torsional torques occur at the load shaft in such cases. Avoiding such situations is the practical choice.
- The sudden short-circuit at the terminals of large power IMs represents an important liability for not-so-strong local power grids. The torque peaks are not, however, severely larger than those occuring during starting. The sudden short-circuit test may be used to determine the IM parameters in real saturation and frequency effects conditions.
- More severe transients than direct starting, reconnection on residual voltage, or sudden short-circuit have been found. The most severe case so far occurs when the primary of transformer feeding an IM is turned off for a short interval (tens of milliseconds), very soon after direct starting. The secondary of the transformer (now with primary on no-load) introduces a large stator time constant which keeps alive the d.c. decaying stator current components. This way, very large torque peaks occur. They vary from 26 – 40 p.u. in a 7.5 HP to 12 to 14 p.u. in a 500 HP machine.
- To treat unsymmetrical stator voltages (connections) so typical with PWM converter fed drives, the abc–dq model seems the right choice. Fault conditions may be treated rather easily as the line voltages are the inputs to the system (stator coordinates).
- For steady-state stability in power systems, first order models are recommended. The obvious choice of neglecting both stator and rotor electrical transients in synchronous coordinates is not necessarily the best one. For low power IMs, the first order system thus obtained produces acceptably good results in predicting the mechanical (speed) transients. The subtransient voltage (rotor flux) E' and its angle δ represent the fast variables.
- For large power motors, a modified first order model is better. This new model reflects the subtransient voltage E' (rotor flux, in fact) transients,

with speed ω_r and angle δ as the fast transients calculated from algebraic equations.

- In industry, a number of IMs of various power levels are connected to a local power grid through a power bus that contains a series reactance and a parallel capacitor (to increase power factor).
  Connection and disconnection transients are very important in designing the local power grid. Complete d–q models, with saturation included, proved to be necessary to simulate residual voltage (turn-off) of a few IMs when some of them act as motors and some as generators until the mechanical and magnetic energy in the system die down rather abruptly. The parallel capacitor delays the residual voltage attenuation, as expected.
- Series capacitors are used in some power grids to reduce the voltage sags during large motor starting. In such cases, subsynchronous resonance (SSR) conditions might occur.
- In such a situation, high current and torque peaks occur at certain speed. The maximum value of power bus plus stator resistance R_e for which SSR occurs, shows how to avoid SSR. A better solution (in terms of losses) to avoid SSR is to put a resistance in parallel with the series capacitor. In general the critical resistance R_{ad} (in parallel) for which SSR might occur increases with the series capacitance and so does the SSR frequency.
- Rotor bar and end ring segment faults occur frequently. To investigate them, a detailed modeling of the rotor cage loops is required. Detailed circuit models to this end are available. In general, such fault introduces mild torque and speed pulsations and $(1\text{-}2S)f_1$ pulsations in the stator current. Information like this may be instrumental in IM diagnosis and monitoring. Interbar currents tend to attenuate the occurring asymmetry.
- Finite element coupled circuit models have been developed in the last 10 years to deal with the IM transients. Still when skin and saturation effects, skewing and the rotor motion, and the IM structural details are all considered, the computation time is still prohibitive. World-wide aggressive R & D FEM efforts should render such complete (3D) models feasible in the near future (at least for prototype design refinements with thermal and mechanical models linked together).

13.18. REFERENCES

1. K.P. Kovacs, I. Racz, Transient Regimes of A.C. Machines, Springer Verlag, 1985 (original edition in German, 1959).
2. R.H. Park, Two Reaction Theory of Synchronous Machines: Generalised Method Analysis, AIEE Trans. Vol.48, 1929, pp.716 – 730.
3. I. Boldea, S.A. Nasar, Dynamics of Electric Machines, MacMillan Publ. Comp., 1986.
4. C.V. Jones, The Unified Theory of Electric Machines, Butterworth, London, 1979.
5. S. Yamamura, A.c. Motors for High Performance Applications, Marcel Dekker, New York, 1986.

6. G.R. Slemon, Modeling of Induction Machines for Electric Drives, IEEE Trans. Vol. IA – 25, No.6., 1989, pp. 1126 – 1131.
7. I. Boldea, S. Nasar, Unified Treatment of Core Losses and Saturation in the Orthogonal Axis Model of Electric Machines, Proc. IEE, Vol.134, Part B. No.6, 1987, pp.355 – 363.
8. P. Vas, Simulation of Saturated Double-cage Induction Machines, EMPS Journal, Vol.25, No.3, April 1997, pp.271 – 285.
9. P.C. Krause, F. Nazari, T.L. Skvarenina, D.W. Olive, The Theory of Neglecting Stator Transients, IEEE Trans. Vol. PAS – 98, No.1, 1979.
10. G.G. Richards, O.T. Tan, Simplified Models for Induction Machine under Balanced and Unbalanced Conditions, IEEE Trans. Vol.IA – 17, No.1, 1981, pp.15 – 21.
11. N. Derbel, B.A. Kamon, M. Poloujadoff, On the Order Reduction of Induction Machine During Start up, IEEE Trans. Vol. EC – 10, No.4, 1995, pp.655 – 660.
12. M. Akbaba, Incorporating the Saturated Leakage Reactance of Induction Motors into Transients Computations, Record of ICEM – 1990, MIT, Vol.3, pp.994 – 999.
13. R.D. Slater, W.S. Wood, Constant Speed Solution Applied to the Evaluation of Induction Motor Torque Peaks, Proc. IEE, Vol.114 (10), 1967, pp.1429 – 1435.
14. I. Bendl, L. Schreier, Torque and Constant Stress of a Three-phase Induction Motor Duc to Nonsimultaneous Switch on, EMPS Journal Vol.21, No.5, 1993, pp.591 – 603.
15. A.A. Shaltout, Analysis of Torsional Torques in Starting of Large Squirrel Cage Induction Machines, IEEE Trans.vol.EC – 9, No.1, 1994, pp.135 – 141.
16. S.A. Nasar, I. Boldea, Electric Machines: Dynamics and Control, CRC Press, 1993.
17. N. Retiere, M. Ivanes, D. Diallo, P.J. Chrozan, Modeling and Estudy of 3 Phase Short-circuits of a Double-cage Induction Machine, EMPS Journal, Vol.27, No.4, 1999, pp.343 – 362.
18. M. Akbaba, A Phenomenon That Causes Most Severe Transients in Three-phase Induction Motors, EMPS Journal, Vol.18, No.2, 1990, pp.149 – 162.
19. P.N. Enjeti, J.F. Lindsay, Steady-state and Transient Behaviour of PWM Inverter-fed Induction Motors, EMPS Journal, Vol.16, No.1, 1989, pp.1 – 13.
20. S. Ahmed-Zaid, M. Taleb, Structural Modeling of Small and Large Induction Machines Using Integral Manifolds, IEEE Trans. Vol. EC – 6, 1991, pp.529 – 533.
21. M. Taleb, S. Ahmed-Zaid, W.W. Price, Induction Machine Models Near Voltage Collapse, EMPS Journal Vol.25, No.1, 1997, pp.15 – 28.
22. S.C. Srivastava, K.N. Srivastava, G.H. Murty, Transient Residual Voltage Analysis During Isolated Operation of a Group of Induction Motor Loads, EMPS Journal Vol.22, No.2, 1994, pp.289 – 309.

23. I.A. M.Abdel-Halim, M.A. Al-Ahmar, M.Z. El-Sherif, Transient Performance of a Group of Induction Motors with Terminal Capacitors Following Supply Disconnection, IBID Vol.26, No.3, 1998, pp.235 – 247.
24. S. Srihoran, L.H. Tan, H.M. Ting, Reduced Transient Model of a Group of Induction Motors, IEEE Trans Vol.EC – 8, No.4, 1993, pp.769 – 777.
25. R.G. Harley et al., Induction Motor Model for the Study of Transient Stability in Both Balanced and Unbalanced Multimachine Networks, IBID Vol.EC – 7, No.1, 1992, pp.209 – 215.
26. D.J.N. Limebeer, R.G. Harley, Subsynchronous Resonance of Single-cage Induction Motors, Proc IEE Vol.128B, No.1, 1981, pp.33 – 42.
27. D.J.N. Limebeer, R.G. Harley, Subsynchronous Resonance of Deep Bar Induction Motors, Proc IEE, Vol.128B, No.1, 1981, pp.43 – 51.
28. P.G. Casielles, L. Zaranza, J. Sanz, Subsynchronous Resonance in Self-excited Induction Machines–Analysis and Evaluation, Record of ICEM 1990, MIT, Vol.3, pp.971 – 975.
29. S.A. Nasar, Electromechanical Energy Conversion in nm–winding Double Cylindrical Structures in Presence of Space Harmonics, IEEE Trans Vol.PAS – 87, No.4, 1968, pp.1099 – 1106.
30. H.A. Tolyat, T.A. Lipo, Feasibility Study of a Converter Optimised Induction Motor, Electric Power Research Institute, EPRI Final Report 2624 – 02, Jan, 1989.
31. H.A. Tolyat, T.A. Lipo, Transient Analysis of Cage Induction Machines under Stator, Rotor , Bar and End Ring Faults, IEEE Tran. Vol.EC – 10, No.2, 1995, pp.241 – 247.
32. H.A. Tolyat, M.S. Arefeen, A.G. Parlos, A Method for Dynamic Simulation of Airgap Eccentricity in Induction Machines, IEEE Trans. Vol. IA-32, No.4, 1996, pp.910 – 918.
33. St. Manolas, J.A. Tegopoulos, Analysis of Squirrel Cage Induction Motors with Broken Bars and Endrings, Record of IEEE – IEMDC – 1997, pp.TD2 – 1.1 – 1.3.
34. I. Kerszenbaum, C.F. Landy, The Existance of Large Interbar Current in Three-phase Cage Motors with Rotor Bar and/or End Ring Faults, IEEE Trans. Vol.PAS – 103, July, 1984, pp.1854 – 1861.
35. S.L. Ho, W.N. Fu, Review and Future Application of Finite Element Methods in Induction Machines, EMPS Journal, Vol.26, No.1, 1998, pp.111 – 125.
36. J.F. Bangura, H.A. Demerdash, Performance Characterisation of Torque Ripple Reduction in Induction Motor Adjustable Speed Drives Using Time Stepping Coupled F.E. State Space Techniques, Part 1+2, Record of IEEE – IAS 1998, Annual Meeting Vol.1, pp.218 – 236.

Chapter 14

MOTOR SPECIFICATIONS AND DESIGN PRINCIPLES

14.1 INTRODUCTION

Induction motors are used to drive loads in various industries for powers from less than 100W to 10MW and more per unit. Speeds encountered go up to tens of thousands of rpm.

There are two distinct ways to supply an induction motor to drive a load.

- Constant voltage and frequency (constant V/f) – power grid connection
- Variable voltage and frequency – PWM static converter connection

The load is represented by its shaft torque–speed curve (envelope).

There are a few basic types of loads. Some require only constant speed (constant V/f supply) and others request variable speed (variable V/f supply).

In principle, the design specifications of the induction motor for constant and variable speed, respectively, are different from each other. Also, an existing motor, that was designed for constant V/f supply may, at some point in time, be supplied from variable V/f supply for variable speed.

It is thus necessary to lay out the specifications for constant and variable V/f supply and check if the existing motor is the right choice for variable speed. Selecting an induction motor for the two cases requires special care.

Design principles are common to both constant and variable speed. However, for the latter case, because the specifications are different, with machine design constraints, or geometrical aspects (rotor slot geometry, for example) lead to different final configurations. That is, induction motors designed for PWM static converter supplies are different.

It seems that in the near future more and more IMs will be designed and fabricated for variable speed applications.

14.2 TYPICAL LOAD SHAFT TORQUE/SPEED ENVELOPES

Load shaft torque/speed envelopes may be placed in the first quadrant or in 2, 3, or 4 quadrants (Figure 14.1a, b).

Constant V/f fed induction motors may be used only for single quadrant load torque/speed curves.

In modern applications (high performance machine tools, robots, elevators), multiquadrant operation is required. In such cases only variable V/f (PWM static converter) fed IMs are adequate.

Even in single quadrant applications, variable speed may be required (from point A to point B in Figure 14.1a) to reduce energy consumption for lower speeds, by supplying the IM through a PWM static converter at variable V/f (Figure 14.2).

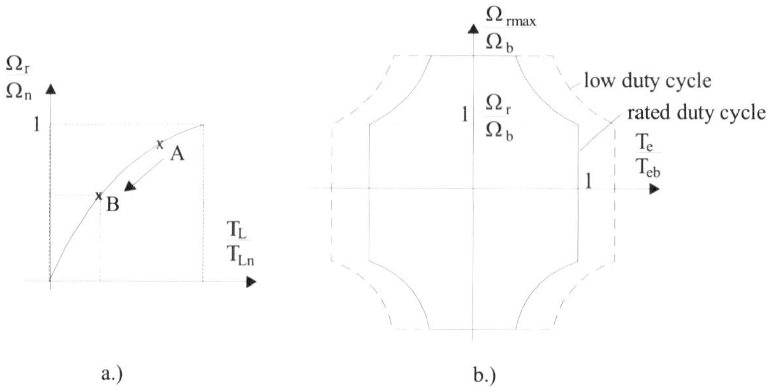

Figure 14.1 Single (a) and multiquadrant (b) load speed/torque envelopes

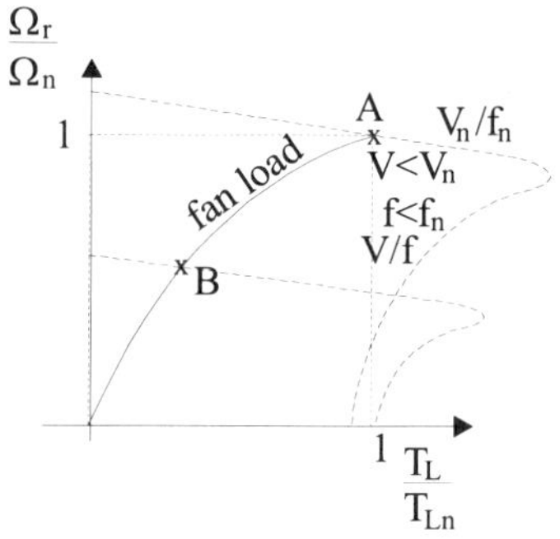

Figure 14.2 Variable V/f for variable speed in single quadrant operation

The load torque/speed curves may be classified into 3 main categories

- Squared torque: (centrifugal pumps, fans, mixers, etc.)

$$T_L = T_{Ln}\left(\frac{\Omega_r}{\Omega_n}\right)^2 \tag{14.1}$$

- Constant torque: (conveyors, rollertables, elevators, extruders, cement kilns, etc.)

$$T_L = T_{Ln} = \text{constant} \tag{14.2}$$

- Constant power

$$\begin{aligned} T &= T_{Lb} \ \text{ for } \ \Omega_r \le \Omega_b \\ T &= T_{Lb}\frac{\Omega_b}{\Omega_r} \ \text{ for } \ \Omega_r > \Omega_b \end{aligned} \tag{14.3}$$

A generic view of the torque/speed envelopes for the three basic loads is shown in Figure 14.3.

The load torque/speed curves of Figure 14.3 show a marked diversity and, especially, the power/speed curves indicate that the induction motor capability to meet them depends on the motor torque/speed envelope and on the temperature rise for the rated load duty-cycle.

There are two main limitations concerning the torque/speed envelope deliverable by the induction motor. The first one is the mechanical characteristic of the induction machine itself and the second is the temperature rise.

For a general purpose design induction motor, when used with variable V/f supply, the torque/speed envelope for continuous duty cycle is shown in Figure 14.4 for self ventilation (ventilator on shaft) and separate ventilator (constant speed ventilator) ,respectively.

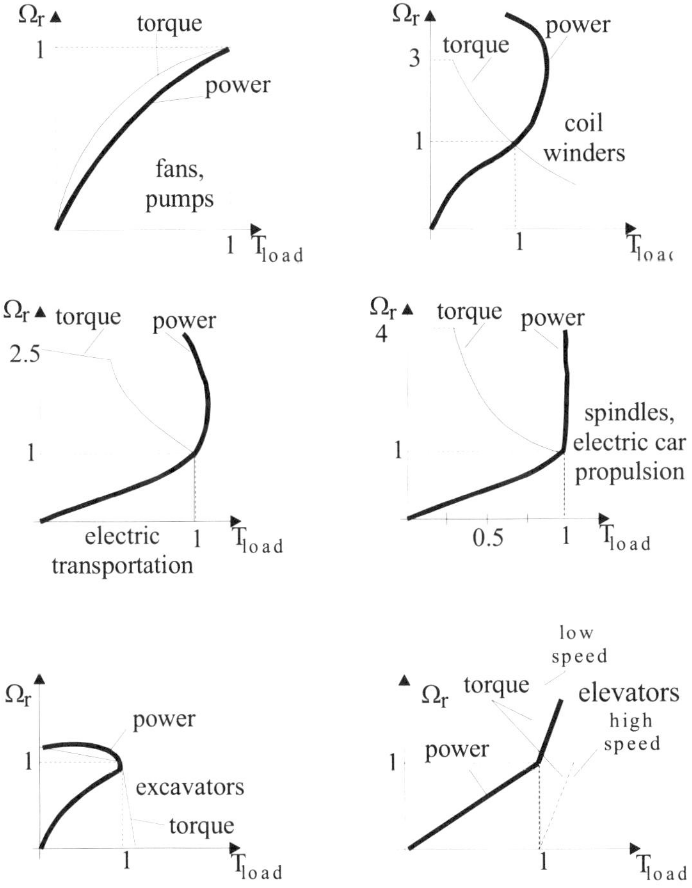

Figure 14.3 Typical load speed/torque curves (first quadrant shown)

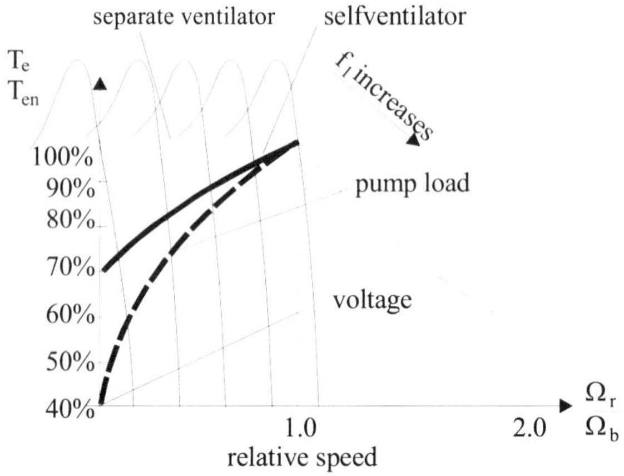

Figure 14.4 Standard induction motor torque/speed envelope for variable V/f supply

Sustained operation at large torque levels and low speed is admitted only with separate (constant speed) ventilator cooling. The decrease of torque with speed reduction is caused by temperature constraints.

As seen from Figure 14.4, the quadratic torque load (pumps, ventilators torque/speed curve) falls below the motor torque/speed envelope under rated speed (torque). For such applications only self ventilated IM design are required.

Not so for servodrives (machine tools, etc) where sustained operation at low speed and rated torque is necessary.

A standard motor capable of producing the extended speed/torque of Figure 14.4 has to be fed through a variable V/f source (a PWM static converter) whose voltage and frequency has to vary with speed as in Figure 14.5.

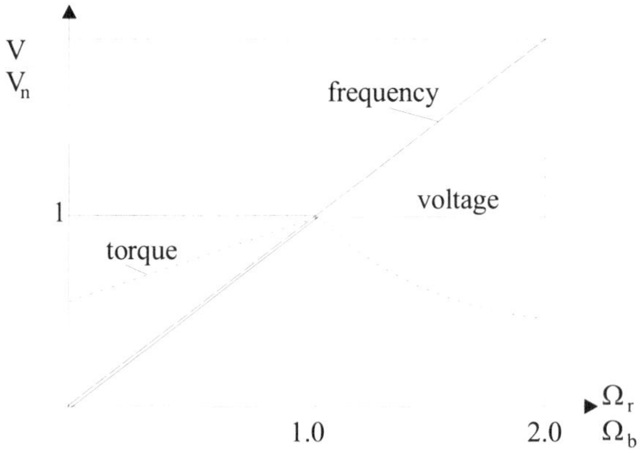

Figure 14.5 Voltage and frequency versus speed

The voltage ceiling of the inverter is reached at base speed Ω_b. Above Ω_b, constant voltage is applied for increasing frequency. How to manage the IM flux linkage (rotor flux) to yield the maximum speed/torque envelope is a key point in designing an IM for variable speed.

14.3 DERATING

Derating is required when an induction motor designed for sinusoidal voltage and constant frequency is supplied from a power grid that has a notable voltage harmonic content due to increasing use of PWM static converters for other motors or due to its supply from similar static power converters. In both cases the time harmonic content of motor input voltages is the cause of additional winding and core losses (as shown in Chapter 11). Such additional losses for rated power (and speed) would mean higher than rated temperature rise of windings and frame. To maintain the rated design temperature rise, the motor rating has to be reduced.

The rise of switching frequency in recent years for PWM static power converters for low and medium power IMs has led to a significant reduction of voltage time harmonic content at motor terminals. Consequently, the derating has been reduced. NEMA 30.01.2 suggests derating the induction motor as a function of harmonic voltage factor (HVF), Figure 14.6.

Reducing the HVF via power filters (active or passive) becomes a priority as the variable speed drives extension becomes more and more important.

In a similar way, when IMs designed for sinewave power source are fed from IGBT PWM voltage source inverters, typical for induction motors now up to 2MW (as of today), a certain derating is required as additional winding and core losses due to voltage harmonics occur.

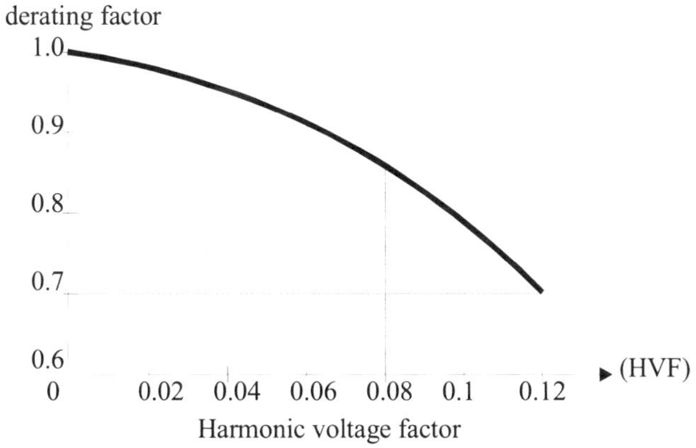

Figure 14.6 Derating for harmonic content of standard motors operating on sinewave power with harmonic content

This derating is not yet standardized, but it should be more important when power increases as the switching frequency decreases. A value of 10% derating for such a situation is now common practice.

When using an IM fed from a sinewave power source with line voltage V_L through a PWM converter, the motor terminal voltage is somewhat reduced with respect to V_L due to various voltage drops in the rectifier and inverter power switches, etc.

The reduction factor is 5 to 10% depending on the PWM strategy in the converter.

14.4 VOLTAGE AND FREQUENCY VARIATION

When matching an induction motor to a load, a certain supply voltage reduction has to be allowed for which the motor is still capable to produce rated power for a small temperature rise over rated value. A value of voltage variation of ±10% of rated value at rated frequency is considered appropriate (NEMA 12.44).

Also, a ±5% frequency variation at rated voltage is considered acceptable. A combined 10% sum of absolute values, with a frequency variation of less than 5%, has to be also handled successfully. As expected in such conditions, the motor rated speed efficiency and power factor for rated power will be slightly different from rated label values.

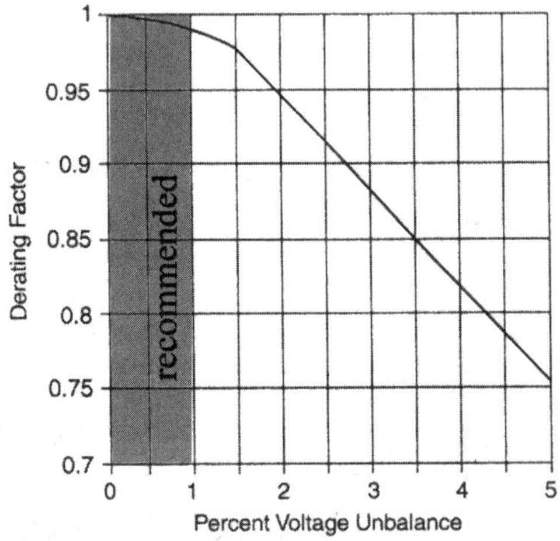

Figure 14.7 Derating due to voltage imbalance in %

Through the negative sequence voltage imbalanced voltages may produce, additional winding stator and rotor losses. In general, a 1% imbalance in voltages would produce a 6 – 10% imbalance in phase currents.

The additional winding losses occurring this way would cause notable temperature increases unless the IM is derated (NEMA Figure 14.1) Figure

14.7. A limit of 1% in voltage imbalance is recommended for medium and large power motors.

14.5 INDUCTION MOTOR SPECIFICATIONS FOR CONSTANT V/f

Key information pertaining to motor performance, construction, and operating conditions is provided for end users' consideration when specifying induction motors.

National (NEMA in U.S.A. [1]) and international (IEC in Europe) standards deal with such issues to provide harmonization between manufacturers and users worldwide.

Table 14.1. summarizes most important headings and the corresponding NEMA section.

Table 14.1. NEMA standards for 3 phase IMs (with cage rotors)

Heading	NEMA section
Nameplate markings	NEMA MG – 1 10.40
Terminal markings	NEMA MG – 1 2.60
NEMA size starters	
NEMA enclosure types	
Frame dimensions	NEMA MG – 1 11
Frame assignments	NEMA MG – 1 10
Full load current	NEC Table 430 – 150
Voltage	NEMA MG – 1 12.44, 14.35
Impact of voltage, frequency variation	
Code letter	NEMA MG – 1 10.37
Starting	NEMA MG – 1 12.44, 54
Design letter and torque	NEMA MG – 1 12
Winding temperature	NEMA MG – 1 12.43
Motor efficiency	NEMA MG – 12 – 10
Vibration	NEMA MG – 17
Testing	NEMA MG – 112, 55, 20, 49 / IEEE-112B
Harmonics	NEMA MG – 1 30
Inverter applications	NEMA MG – 1, 30, 31

Among these numerous specifications, that show the complexity of IM design, nameplate markings are of utmost importance.

The following data are given on the nameplate:

a. Designation of manufacturer's motor type and frame
b. kW (HP) output
c. Time rating
d. Maximum ambient temperature
e. Insulation system
f. RPM at rated load
g. Frequency
h. Number of phases
i. Rated load amperes
j. Line voltage

Table 14.2. 460V, 4 pole, open frame design B and E performance NEMA defined performance

Appendix II: 460 V, 4 Pole, Open Frame, Design B and E Motors—Comparison of NEMA Defined Performance

HP	Full Load Amperes (FLA) per NEC Table 430-150	Maximum Locked Rotor Amperes (LRA) per NEMA MG 1, Tables 12.35 and 12.35A		Ratio: LRA/FLA		Nominal Full Load Efficiencies (%) per NEMA MG 1, Tables 12-10 and 12-11 (open 4 pole)		Efficiency Ratio: Design E/ Design B	Minimum Locked Rotor Torque (%) per NEMA MG 1, Tables 12-2 and 12.38.4		Minimum Breakdown Torque (%) per NEMA MG 1, Tables 12.39.1 and 12.39.3		Minimum Pull-up Torque (%) per NEMA MG 1, Tables 12.40.1 and 12.40.3	
	Designs B and E	Design B	Design E	Design B	Design E	Design B*	Design E		Design B	Design E	Design B	Design E	Design B	Design E
3	4.8	32	37	6.7	7.7	86.5	89.5	1.04	215	180	250	200	150	120
5	7.6	46	61	6.1	8.0	87.5	90.2	1.03	185	170	225	200	130	120
7 ½	11	64	92	5.8	8.4	88.5	91.0	1.03	175	160	215	200	120	110
10	14	81	113	5.8	8.1	89.5	91.7	1.03	165	160	200	200	115	110
15	21	116	169	5.5	8.0	91.0	92.4	1.02	160	150	200	200	110	110
20	27	145	225	5.4	8.3	91.0	93.0	1.02	150	150	200	200	105	110
25	34	183	281	5.4	8.3	91.7	93.6	1.02	150	140	200	190	105	100
30	40	218	337	5.5	8.4	92.4	94.1	1.02	150	140	200	190	105	100
40	52	290	412	5.6	7.9	93.0	94.5	1.02	140	130	200	190	100	100
50	65	363	515	5.6	8.0	93.0	95.4	1.03	140	130	200	190	100	100
60	77	435	618	5.6	7.5	93.6	95.4	1.02	140	120	200	180	100	90
75	96	543	723	5.7	7.6	94.1	95.4	1.01	140	120	200	180	100	90
100	124	725	937	5.8	7.5	94.1	95.4	1.01	125	110	200	180	100	80
125	156	908	1171	5.8	7.8	94.5	95.4	1.01	110	110	200	180	100	80
150	180	1085	1405	6.0	7.8	95.0	95.8	1.01	110	100	200	170	100	80
200	240	1450	1873	6.0	7.8	95.0	95.8	1.01	100	100	200	170	90	80
250	302	1825	2344	6.0	7.8	95.4	96.2	1.01	80	90	175	170	75	70
300	361	2200	2809	6.1	7.8	95.4	96.2	1.01	80	90	175	170	75	70
350	414	2550	3277	6.2	7.9	95.4	96.5	1.01	80	75	175	160	75	60
400	477	2900	3745	6.1	7.9	95.4	96.5	1.01	80	75	175	160	75	60
450	515	3250	4214	6.3	8.2	95.8	96.8	1.01	80	75	175	160	75	60
500	590	3625	4682	6.1	7.9	95.8	96.8	1.01	80	75	175	160	75	60

* - Applies to induction motors labeled "Premium Efficiency" or "Energy Efficient"

k. Locked-rotor amperes or code letter for locked-rotor kVA per HP for motor ½ HP or more
l. Design letter (A, B, C, D, E)
m. Nominal efficiency
n. Service factor load if other than 1.0
o. Service factor amperes when service factor exceeds 1.15
p. Over-temperature protection followed by a type number, when over-temperature device is used
q. Information on dual voltage/frequency operation conditions

Rated power factor does not appear on NEMA nameplates, but is does so according to most European standards.

Efficiency is perhaps the most important specification of an electric motor as the cost of energy per year even in an 1 kW motor is notably higher than the initial motor cost. Also, a 1% increase in efficiency saves energy whose costs in 3 to 4 years cover the initial extra motor costs.

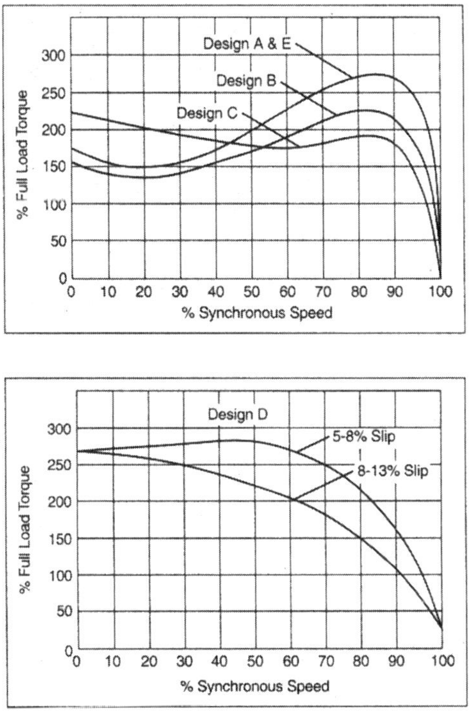

Figure 14.8. NEMA designs A, B, C, E (a) and D (b) torque/speed curves

Standard and high efficiency IM classes have been defined and standardized by now worldwide. As expected, high efficiency (class E) induction motors have higher efficiency than standard motors but their size, initial cost, and locked-rotor current are higher. This latter aspect places an additional burden on the local power grid when feeding the motor upon direct starting. If softstarting or inverter operation is used, the higher starting current does not have any effect on the local power grid rating. NEMA defines specific efficiency levels for design B and E (high efficiency) IMs (Table 14.2).

On the other hand, EU established three classes EFF1, EFF2, EFF3 of efficiencies, giving the manufacturers an incentive to qualify for the higher classes.

The torque/speed curves reveal, for constant V/f fed IMs, additional specifications such as starting, pull-up, and breaking torque for the five classes (letters: A, B, C, D, E design) of induction motors (Figure 14.8).

The performance characteristics of the A, B, C, D, E designs are summarized in Table 14.3 from NEMA Table 2.1 with their typical applications.

Table 14.3. Motor designs (after NEMA Table 2.1)

Classification	Locked rotor torque (% rated load torque)	Breakdown torque (% rated load torque)	Locked rotor current (% rated load current)	Slip %	Typical applications	Rel. η
Design B Normal locked rotor torque and normal locked rotor current	70 – 275*	175 – 300*	600 - 700	0.5 - 5	Fans, blowers, centrifugal pumps and compressors, motor – generator sets, etc., where starting torque requirements are relatively low	Medium or high
Design C High locked rotor torque and normal locked rotor current	200 – 250*	190 – 225*	600 - 700	1 - 5	Conveyors, crushers, stirring machines, agitators, reciprocating pumps and compressors, etc., where starting under load is required	Medium
Design D High locked rotor torque and high slip	275	275	600 – 700		High peak loads with or without fly wheels, such as punch presses, shears, elevators, extractors, winches, hoists, oil – well pumping and wire – drawing machines	Medium
Design E IEC 34-12 Design N locked rotor torques and currents	75 – 190*	160 – 200*	800 – 1000	0.5 - 3	Fans, blowers, centrifugal pumps and compressors, motor – generator sets, etc. where starting torque requirements are relatively low	High

Note – Design A motor performance characteristics are similar to those for Design B except that the locked rotor starting current is higher than the values shown in the table above
* Higher values are for motors having lower horsepower ratings

14.6 MATCHING IMs TO VARIABLE SPEED/TORQUE LOADS

IMs are, in general, designed for 60(50) Hz; when used for variable speed with variable V/f supply, they operate at variable frequency. Below the rated frequency, the machine is capable of full flux linkage, while above that, flux weakening occurs.

For given load speed and load torque with variable V/f supply, we may use IMs with $2p_1 = 2, 4, 6$. Each of them, however, works at a different (distinct) frequency.

Figures 14.9 show the case of quadratic torque (pump) load with the speed range of 0 to 2000 rpm, load of 150 kW at 2000 rpm, 400 V, 50 Hz (network). Two different motors are used: one of 2 poles and one of 4 poles.

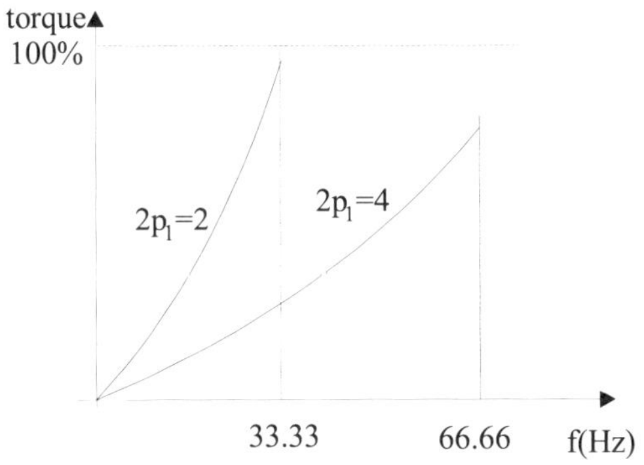

Figure 14.9 Torque versus motor frequency (and speed) pump load

At 2000 rpm the 2 pole IM works at 33.33 Hz with full flux, while the 4 pole IM operates at 66.66 Hz in the flux-weakening zone. Which of the two motors is used is decided by the motor costs. Note however, that the absolute torque (in Nm) of the motor has to be the same in both cases.

For a constant torque (extruder) load with the speed range of 300 – 1100 rpm, 50kW at 1200 rpm, network: 400 V, 50 Hz, two motors compete. One, of 4 pole, will work at 40 Hz and one, of 6 pole, operating at 60 Hz (Figure 14.10).

Again, both motors can satisfy the specifications for the entire speed range as the load torque is below the available motor torque. Again the torque in Nm is the same for both motors and the choice between the two motors is decided by motor costs and total losses.

While starting torque and current are severe design constraints for IMs designed for constant V/f supply, they are not for variable V/f supply.

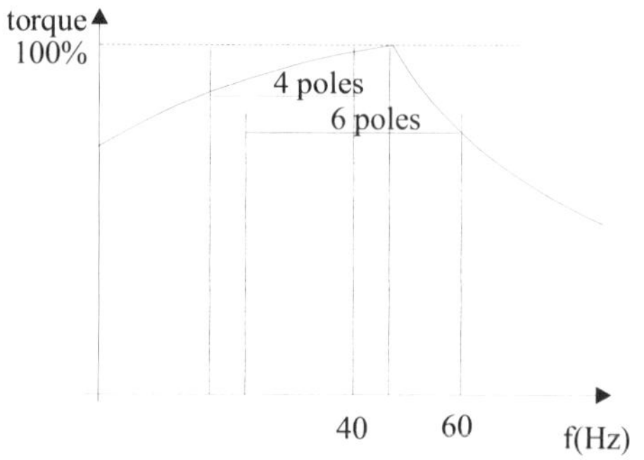

Figure 14.10 Torque versus motor frequency (and speed) constant torque load

Skin effect is important for constant V/f supply as it reduces the starting current and increases the starting torque. In contrast to this, for variable V/f suply, skin effect is to be reduced, especially for high performance speed control systems.

Breakdown torque may become a much more important design factor for variable V/f supply, when a large speed zone for constant power is required. A spindle drive or an electric car drive may require more than 4-to-1 constant power range (Figure 14.11).

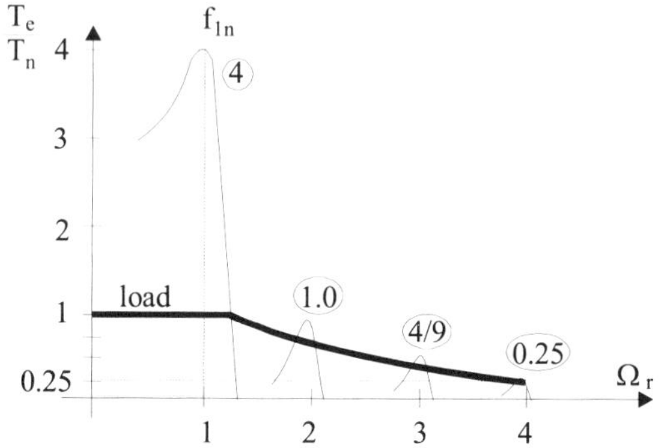

Figure 14.11 Induction motor torque/speed curves for various values of frequency and a 4/1 constant power speed range

The peak torque of IM is approximately

$$T_{ek} \approx 3\left(\frac{V_{phn}}{2\pi f_{1n}}\right)^2 \left(\frac{f_{1n}}{f_1}\right)^2 \frac{p_1}{2L_{sc}} = T_{ekf_{1n}}\left(\frac{f_{1n}}{f_1}\right)^2 \quad (14.4)$$

The peak torque for constant (rated) voltage is inversely proportional to frequency squared. To produce a 4/1 constant power speed range, the peak torque has to be 4 times the rated torque. Only in this case, the motor may produce at $f_{1max} = 4f_{1n}$, 25% of rated torque.

Consequently, if the load maximum torque is equal to the rated torque, then at $4f_{1n}$ the rated power is still produced.

In reality, a breakdown torque of 400% is hardly practical. However, efforts to reduce the short-circuit leakage inductance (L_{sc}) have led up to 300% breakdown torque.

So there are two solutions to provide the required load torque/speed envelope: increase the motor rating (size) and costs or increase the flux (voltage) level in the machine by switching from star to delta connection (or by reducing the number of turns per phase by switching off part of the stator coils).

The above rationale was intended to suggest some basic factors that guide the IM design.

Relating the specifications to a dedicated machine geometry is the object of design (or dimensioning). This enterprise might be as well be called sizing the IM.

Because there are many geometrical parameters and their relationships to specifications (performance) are in general nonlinear, the design process is so complicated that it is still a combination of art and science, based solidly on existing experience (motors) with tested (proven) performance. In the process of designing an induction motor, we will define a few design factors, features, and sizing principles.

14.7 DESIGN FACTORS

Factors that influence notably the induction machine design are as follows:

- *Costs*

  Costs in most cases, are the overriding consideration in IM design. But how do we define costs? It maybe the costs of active materials with or without the fabrication costs. Fabrication costs depend on machine size, materials available or not in stock, manufacturing technologies, and man power costs.

  The costs of capitalized losses per entire motor active life surpass quite a few times the initial motor costs. So loss reduction (through higher efficiency or via variable V/f supply) pays off generously. This explains the rapid extension of variable speed drives with IMs worldwide.

  Finally, maintenance costs are also important but not predominant. We may now define the global costs of an IM as

$$\begin{aligned}\text{Global costs} &= \text{material costs} + \text{fabrication and selling costs} + \\ &+ \text{losses capitalized costs} + \text{maintenance costs}\end{aligned} \quad (14.5.)$$

  Global costs are also a fundamental issue when we have to choose between repairing an old motor or replacing it with a new motor (with higher efficiency and corresponding initial costs).

- *Material limitations*

  The main materials used in IM fabrication are magnetic-steel laminations, copper and aluminum for windings, and insulation materials for windings in slots.

  Their costs are commensurate with performance. Progress in magnetic and insulation materials has been continuous. Such new improved materials drastically affect the IM design (geometry), performance (efficiency), and costs.

  Flux density, B(T), losses (W/kg) in magnetic materials, current density J (A/mm^2) in conductors and dielectric rigidity E (V/m) and thermal conductivity of insulation materials are key factors in IM design.

- *Standard specifications*

  IM materials (lamination thickness, conductor diameter), performance indexes (efficiency, power factor, starting torque, starting current, breakdown torque), temperature by insulation class, frame sizes, shaft height, cooling types, service classes, protection classes, etc. are specified in national (or international) standards (NEMA, IEEE, IEC, EU, etc.) to facilitate globalization in using induction motors for various applications. They limit, to some extent, the designer's options, but provide solutions that are widely accepted and economically sound.

- *Special factors*

  In special applications, special specifications–such as minimum weight and maximum reliability in aircraft applications–become the main concern. Transportation applications require ease of maintaining, high reliability, and good efficiency. Circulating water home pumps require low noise, highly reliable, induction motors.

  Large compressors have large inertia rotors and thus motor heating during frequent starts is severe. Consequently, maximum starting torque/current becomes the objective function.

14.8 DESIGN FEATURES

The major issues in designing an IM may be divided into 5 area: electrical, dielectric, magnetic, thermal and mechanical.

- *Electrical design*

  To supply the IM, the supply voltage, frequency, and number of phases are specified. From this data and the minimum power factor and a target efficiency, the phase connection (start or delta), winding type, number of poles, slot numbers and winding factors are calculated. Current densities (or current sheets) are imposed.

- *Magnetic design*

  Based on output coefficients, power, speed, number of poles, type of cooling, and the rotor diameter is calculated. Then, based on a specific current loading (in A/m) and airgap flux density, the stack length is determined.

Fixing the flux densities in various parts of the magnetic circuit with given current densities and slot mmfs, the slot sizing, core height, and external stator diameter D_{out} are all calculated. Choosing D_{out}, which is standardized, the stack length is modified until the initial current density in the slot is secured.

It is evident that sizing the stator and rotor core may be done many ways based on various criteria.

- *Insulation design*

  Insulation material and its thickness, be it slot/core insulation, conductor insulation, end connection insulation, or terminal leads insulation depends on machine voltage insulation class and the environment in which the motor operates.

  There are low voltage 400V/50Hz, 230V/60Hz, 460V/60Hz 690V/60Hz or less or high voltage machines (2.3kV/60Hz, 4kV/50Hz, 6kV/50Hz). When PWM converter fed IMs are used, care must be exercised in reducing the voltage stress on the first 20% of phase coils or to enforce their insulation or to use random wound coils.

- *Thermal design*

  Extracting the heat caused by losses from the IM is imperative to keep the windings, core, and frame temperatures within safe limits. Depending on application or power level, various types of cooling are used. Air cooling is predominant but stator water cooling in the stator of high speed IMs (above 10,000 rpm) is frequently used. Calculating the loss and temperature distribution and the cooling system represents the thermal design.

- *Mechanical design*

  Mechanical design refers to critical rotating speed, noise, and vibration modes, mechanical stress in the shaft, and its deformation displacement, bearings design, inertia calculation, and forces on the winding end coils during most severe current transients.

We mentioned here the output coefficient as an experience, proven theoretical approach to a tentative internal stator (stator bore) diameter calculation. The standard output coefficient is $D_{is}^2 \cdot L$, where D_{is} is the stator bore diameter and L, the stack length.

Besides elaborating on $D_{is}^2 \cdot L$, we introduce here the rotor tangential stress σ_{tan} (in N/cm^2), that is, the tangential force at rotor surface at rated and peak torque.

This specific force criterion may be used also for linear motors. It turns out that σ_{tan} varies from 0.2 to 0.3 N/cm^2 for hundred watt IMs to less than 3 to 4 N/cm^2 for large IMs. Not so for the output coefficient $D_{is}^2 \cdot L$, which is related to rotor volume and thus increases steadily with torque (and power).

14.9 THE OUTPUT COEFFICIENT DESIGN CONCEPT

To calculate the relationship between the $D_{is}^2 \cdot L$ and the machine power and performance, we start by calculating the airgap apparent power S_g,

$$S_{gap} = 3E_1 I_{1n} \tag{14.6}$$

where E_1 is the airgap emf per phase and I_{1n} rated current (RMS values).

Based on the phasor diagram with zero stator resistance ($R_s = 0$), Figure 14.12.

$$\underline{I}_{1n} R_s - \underline{V}_{1n} = \underline{E}_1 - jX_{1s}\underline{I}_{1n} \tag{14.7}$$

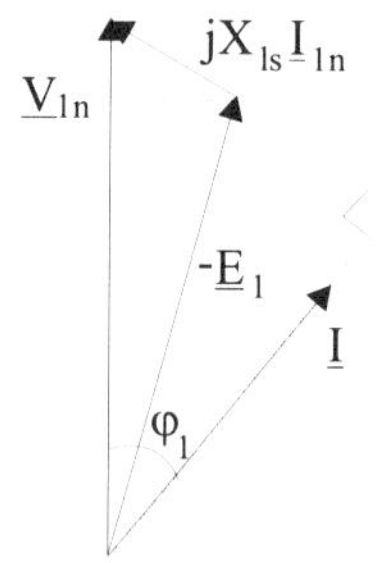

Figure 14.12 Simplified phasor diagram

Or

$$K_E = \frac{E_1}{V_{1n}} \approx 1 - x_{1s} \cdot \sin \varphi_1 \tag{14.8}$$

with

$$x_{1s} = \frac{X_{1s} I_{1n}}{V_{1n}} \tag{14.9}$$

The p.u. value of stator leakage reactance increases with pole pairs p_1 and so does $\sin\varphi_1$ (power factor decreases when p_1 increases).

$$K_E \approx 0.98 - 0.005 \cdot p_1 \tag{14.10}$$

Also, the input apparent power S_{1n} is

$$S_{1n} = 3V_{1n} I_{1n} = \frac{P_n}{\eta_n \cos\varphi_{1n}} \tag{14.11}$$

where P_n is the rated output power and η_n and $\cos\varphi_{1n}$ are the assigned values of rated efficiency and power factor based on past experience.

Typical values of efficiency have been given in Table 14.3 for Design B and E (NEMA). Each manufacturer has its own set of data.

Efficiency increases with power and decreases with the number of poles. Efficiency of wound rotor IMs is slightly larger than that of cage rotor IMs of same power and speed because the rotor windings are made of copper and the total additional load (stray) losses are lower.

As efficiency is defined with stray losses p_{stray} of 0.5 to1.0% of rated power in Europe (still!) and with the latter (p_{stray}) measured in direct load tests in the U.S.A., differences in actual losses (in IMs of same power and nameplate efficiency) of even more than 20% may be encountered when motors fabricated in Europe are compared with those made in the U.S.A.

Anyway, the assigned value of efficiency is only a starting point for design as iterations are performed until the best performance is obtained.

The power factor also increases with power and decreases with the number of pole pairs with values slightly smaller than corresponding efficiency for existing motors. More data on initial efficiency and power factor data will be given in subsequent chapters on design methodologies.

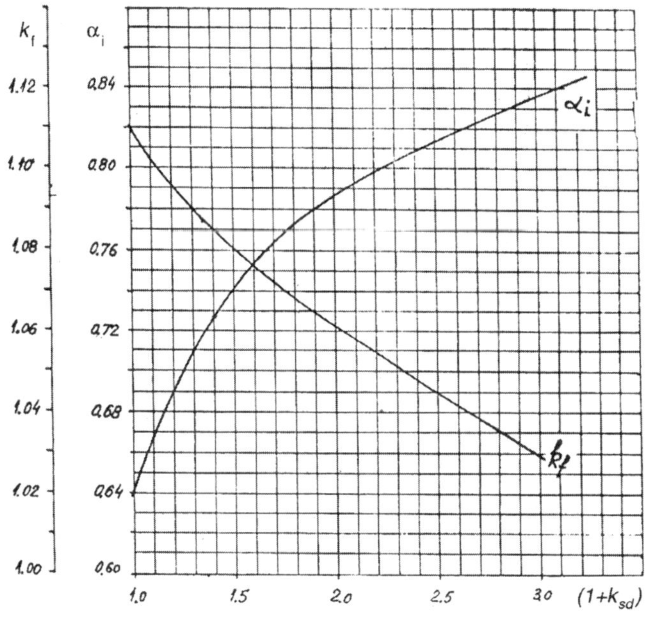

Figure 14.13 Form factor K_f and flux density shape factor α_i versus teeth saturation

The emf E_1 may be written as a function of airgap pole flux ϕ,

$$E_1 = 4f_1K_fW_1K_{w1}\phi \tag{14.12}$$

where f_1 is frequency, $1.11 > K_f > 1.02$ form factor (dependent on teeth saturation) (Figure 14.13), W_1 is turns per phase, and K_{w1} is winding factor, ϕ pole flux.

$$\phi = \alpha_i \tau L B_g \tag{14.13}$$

where α_i is flux density shape factor dependent on the magnetic saturation coefficient of teeth (Figure 14.13) and B_g is flux density in the airgap. The pole pitch τ is

$$\tau = \frac{\pi D_{is}}{2p_1}; \; n_1 = \frac{f_1}{p_1} \tag{14.14}$$

Finally, S_{gap} is

$$S_{gap} = K_f \alpha_i K_{w1} \pi^2 D_{is}^2 L \frac{n_1}{60} A_1 B_g \tag{14.15}$$

with A_1 the specific stator current load A_1 (A/m),

$$A_1 = \frac{6W_1 I_{1n}}{\pi D_{is}} \tag{14.16}$$

We might separate the volume utilization factor C_0 (Esson's constant) as

$$C_0 = K_f \alpha_i K_{w1} \pi^2 A_1 B_g = \frac{60 S_{gap}}{D_{is}^2 L n_1} \tag{14.17}$$

C_0 is not a constant as both the values of A_1(A/m) and airgap flux density (B_g) increase with machine torque and with the number of pole pairs.

The $D_{is}^2 \cdot L$ output coefficient may be calculated from (14.17) with S_{gap} from (14.6) and (14.11).

$$D_{is}^2 L = \frac{1}{C_0} \frac{60}{n_1} \frac{K_E P_n}{\eta_n \cos\varphi_{1n}} \tag{14.18}$$

Typical values of C_0 as a function S_{gap} with pole pairs p_1 as parameter for low power IMs is given in Figure 14.14.

The $D_{is}^2 \cdot L$ (internal) output constant (proportional to rotor core volume) is, in fact, almost proportional to machine rated shaft torque. Torque production apparently requires less volume as the pole pairs number p_1 increases, C_0 increases with p_1 (Figure 14.14).

It is standard to assign a value λ to the stack length to pole pitch ratio

$$\lambda = \frac{L}{\tau} = \frac{2Lp_1}{\pi D_{is}}; \; 0.6 < \lambda < 3.0 \tag{14.19}$$

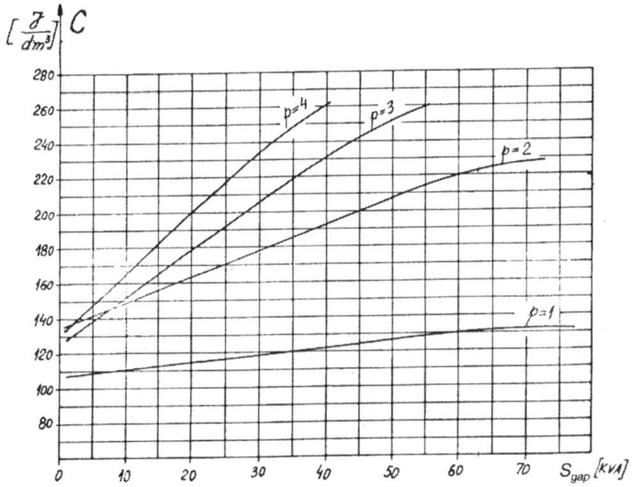

Figure 14.14 Esson's "constant" C_0 versus S_{gap} (airgap apparent power)

The stator bore diameter may now be calculated from (14.18) with (14.19).

$$D_{is} = \sqrt[3]{\frac{2p_1}{\pi\lambda}\frac{1}{C_0}\frac{p_1}{f_1}\frac{K_E P_n}{\eta_n \cos\varphi_{1n}}} \tag{14.20}$$

This is a standard design formula. However it does not say enough on the machine total volume (weight). Moreover, in many designs, the stator external (frame internal) diameters are standardized.

A similar (external) output coefficient $D_{is}^2 \cdot L$ may be derived if we first adopt a design current density $J_{con}(A/m^2)$ and consider the slot fill factor (with conductors), $K_{fill} = 0.4$ to 0.6,
given together with the tooth and stator back iron flux densities B_{ts} and B_{cs}.

With the airgap flux and tooth flux densities B_g and B_{ts} considered known, the stator slot height h_s is approximately

$$h_s = \frac{6W_1I_n}{\frac{B_g}{B_{ts}}j_{con}K_{fill}}\frac{1}{\pi D_{is}} = \frac{A_1}{\frac{B_g}{B_{ts}}j_{con}K_{fill}} \tag{14.21}$$

Now the core radial height h_{cs} is

$$h_{cs} = \frac{\phi}{2LB_{cs}} = \frac{\alpha_i}{2}\left(\frac{\pi D_{is}}{2p_1}\right)\frac{B_g}{B_{cs}} \tag{14.22}$$

The outer stator diameter D_{out} is

$$D_{out} = D_{is} + 2(h_s + h_{cs}) \quad (14.23)$$

We may replace D_{is} from (14.23) in $D_{is}^2 \cdot L$ with h_s and h_{cs} from (14.21) and (14.22).

$$D_{is}^2 L = D_{out}^2 L \cdot f_0(D_{is}) \quad (14.24)$$

$$f_0(D_{is}) = \frac{1}{\left[1 + \frac{2(h_s + h_{cs})}{D_{is}}\right]^2} \quad (14.25)$$

And, finally,

$$f_0(D_{is}) = \frac{1}{\left[1 + \frac{2A_1 B_{ts}}{j_{con} K_{fill} B_g D_{is}} + \frac{\alpha_i}{2} \frac{\pi}{p_1} \frac{B_g}{B_{cs}}\right]^2} \quad (14.26)$$

From (14.24),

$$D_{out}^2 L = \frac{D_{is}^2 L}{f_0(D_{is})} = \frac{1}{C_0 f_0(D_{is})} \frac{p_1}{f_1} \frac{K_E P_n}{\eta_n \cos\varphi_{1n}} \quad (14.27)$$

As

$$L = \lambda \frac{\pi D_{is}}{2p_1} \quad (14.28)$$

$$D_{out} = \sqrt{\frac{2p_1^2}{\pi \lambda C_0 f_1} \frac{K_E P_n}{\eta_n \cos\varphi_{1n}} \frac{1}{D_{is} f_0(D_{is})}} \quad (14.29)$$

Although (14.29) through the function $[D_{is} f_0(D_{is})]^{-1}$ suggests that a minimum D_{out} may be obtained for given λ, B_g/B_{co}, B_g/B_t, j_{con}, and A_1, it seems to us more practical to use (14.29) to find the outer stator diameter D_{out} after the stator bore diameter was obtained from (14.20). Now if this value is not a standard one and a standard frame is a must, the aspect ratio λ is modified until D_{out} matches a standardized value.

The specific current loading A_1 depends on pole pitch τ and number of poles on D_{is}, once a certain cooling system (current density) is adopted.

In general, it increases with D_{is} from values of less than 10^3A/m for $D_{is} = 4 \cdot 10^{-2}$m to 45,000 A/m for D_{is} = 0.4 m and $2p_1$ = 2 poles. Smaller values are common for larger number of poles and same stator bore diameter.

On the other hand, the design current density j_{con} varies in the interval $j_{con} = (3.5 - 8.0) \cdot 10^6$A/m^2 for axial or axial-radial air cooling. Higher values are designated to high speed IMs (lower pole pair numbers p_1) or for liquid cooling. While A_1 varies along such a large span and the slot height h_s to slot width b_s

ratio is limited to K_{aspect} (3 – 6), to limit the slot leakage inductance, using A_1 may be avoided by calculating slot height h_s as

$$h_s = K_{aspect} b_s = K_{aspect} \frac{\pi D_{is}}{N_s}\left(1 - \frac{b_t}{\tau_{slot}}\right) = K_{aspect} \frac{\pi D_{is}}{N_s}\left(1 - \frac{B_g}{B_{ts}}\right) \quad (14.30)$$

Higher values of aspect ratios are typical to larger motors.

This way, $D_{out}^2 L$ is

$$D_{out}^2 L = \frac{1}{C_0}\frac{P_1}{f_1}\frac{K_E P_n}{\eta_n \cos\varphi_{1n}}\left[1 + \frac{2K_{aspect}\pi}{N_s}\left(1 - \frac{B_g}{B_{ts}}\right) + \frac{\alpha_i}{2}\frac{\pi}{p_1}\frac{B_g}{B_{cs}}\right]^2 \quad (14.31)$$

Also,

$$\frac{D_{out}}{D_{is}} \approx 1 + \frac{2K_{aspect}\pi}{N_s}\left(1 - \frac{B_g}{B_{ts}}\right) + \frac{\alpha_i}{2}\frac{\pi}{p_1}\frac{B_g}{B_{cs}} \quad (14.32)$$

To start, we may calculate D_{is}/D_{out} as a function of only pole pairs p_1 if B_g/B_{ts} = ct and B_g/B_{cs} = ct, with K_{aspect} and N_s (slots/stator) also assigned corresponding values (Table 14.4).

Table 14.4 Outer to inner stator diameter ratios

$2p_1$	2	4	6	8	≥ 10
$\frac{D_{out}}{D_{is}}$	1.65 – 1.69	1.46 – 1.49	1.37 – 1.40	1.27 – 1.30	1.24 – 1.26

The stack aspect ratio λ is assigned an initial value in a rather large interval: 0.6 to 3.

In general, longer stacks, allowing for a smaller stator bore diameter (for given torque) lead to shorter stator winding end connections, lower winding losses, and lower inertia, but the temperature rise along the stack length may become important. An optimal value of λ is highly dependent on IM design specifications and the objective function taken into consideration. There are applications with space shape constraints that prevent using a long motor.

Example 14.1 Output coefficient

Let us consider a 55 kW, 50 Hz, 400 V, $2p_1 = 4$ induction motor whose assigned (initial) rated efficiency and power factor are $\eta_n = 0.92$, $\cos\varphi_n = 0.92$.

Let us determine the stator internal and external diameters D_{out} and D_{is} for $\lambda = L/\tau = 1.5$.

Solution

The emf coefficient K_E (14.10) is: $K_E = 0.98 - 0.005 \cdot 2 = 0.97$

The airgap apparent power S_{gap} (14.3) is

$$S_{gap} = 3K_E V_1 I_{1n} = K_E \frac{P_n}{\cos\varphi_n \eta_n} = \frac{0.97 \cdot 55 \cdot 10^3}{0.92 \cdot 0.92} = 63.03 \cdot 10^3 VA$$

Esson's constant C_0 is obtained from Figure 14.14 for $p_1 = 2$ and $S_{gap} = 63.03 \cdot 10^3 VA$: $C_0 = 222 \cdot 10^3 J/m^3$.

For an airgap flux density $B_g = 0.8T$, $K_{w1} = 0.955$, $\alpha_i = 0.74$, $K_f = 1.08$ (teeth saturation coefficient $1 + K_{st} = 1.5$, Figure 4.13). The specific current loading A_1 is (14.17).

$$A_1 = \frac{C_0}{K_f \alpha_i K_{w1} \pi^2 B_g} = \frac{222 \cdot 10^3}{1.08 \cdot 0.74 \cdot 0.955 \cdot 0.8\pi^2} = 36.876 \cdot 10^3 A/m$$

with $\lambda = 1.5$ from (14.20) stator internal diameter D_{is} is obtained.

$$D_{is} = \sqrt[3]{\frac{2 \cdot 2}{2 \cdot 1.5} \cdot \frac{1}{222 \cdot 10^3} \cdot \frac{2}{50} \cdot 63.03 \cdot 10^3} = 0.2477m$$

The stack length L (14.19) is

$$L = \lambda \frac{\pi D_{is}}{2p_1} = \frac{1.5 \cdot \pi \cdot 0.2477}{2 \cdot 2} = 0.2917m$$

with $j_{con} = 6 \cdot 10^6 A/m$, $K_{fill} = 0.5$, $B_{ts} = B_{cs} = 1.6$ T, the stator slot height h_s is (14.21).

$$h_s = \frac{36.876 \cdot 10^3}{\frac{0.8}{1.6} \cdot 6 \cdot 10^6 \cdot 0.5} = 24.584 \cdot 10^{-3} m$$

The back iron height h_{cs} (14.22) is

$$h_{cs} = \frac{\alpha_i}{2} \frac{\pi D_{is}}{2p_1} \frac{B_g}{B_{cs}} = \frac{0.74 \cdot \pi \cdot 0.2477}{2 \cdot 2 \cdot 2} \cdot \frac{0.8}{1.6} \approx 36 \cdot 10^{-3} m$$

The external stator diameter D_{out} becomes

$$D_{out} = D_{is} + 2(h_{cs} + h_s) = 0.2477 + 2(0.024584 + 0.036) = 0.3688m$$

With $N_s = 48$ slots/stator and a slot aspect ratio $K_{aspect} = 3.03$, the value of slot height h_s (14.30) is

$$h_s = K_{aspect} \frac{\pi D_{is}}{N_s}\left(1 - \frac{B_g}{B_{ts}}\right) = 3.03\pi \frac{0.2477}{48}\left(1 - \frac{0.8}{1.6}\right) = 0.0246m$$

About the same value of h_s as above has been obtained. It is interesting to calculate the approximate value of the specific tangential force σ_{tan}.

$$\sigma_{tan} \approx \frac{P_n}{\pi D_{is}\left(\frac{D_{is}}{2}\right)L}\frac{p_1}{2\pi f_1} = \frac{55\cdot 10^3}{\frac{\pi}{2}\cdot 0.2917\cdot 0.2477^2}\cdot\frac{2}{2\pi 50} =$$

$$= 1.246\cdot 10^4\,N/m^2 = 1.246N/cm^2$$

This is not a high value and the slot aspect ratio $K_{aspect} = h_s/b_s = 3.03$ is a clear indication of this situation.

Apparently the machine stator internal diameter may be reduced by increasing A_1 (in fact, C_0 is Esson's constant). For the same λ, the stack length will be reduced, while the stator external diameter will also be slightly reduced (the back iron height h_{cs} decreases and the slot height increases).

Given the simplicity of the above analytical approach further speculations on better (eventually optimized) designs are considered inappropriate here.

14.10 THE ROTOR TANGENTIAL STRESS DESIGN CONCEPT

The rotor tangential stress $\sigma_{tan}(N/m^2)$ may be calculated from the motor torque T_e.

$$\sigma_{tan} = \frac{T_{en}\cdot 2}{(\pi D_{is}L)\cdot D_{is}}\left(N/m^2\right) \tag{14.33}$$

The electromagnetic torque T_{en} is approximately

$$T_{en} \approx \frac{p_1 P_n\left(1+\frac{p_{mec}}{P_n}\right)}{2\pi f_1(1-S_n)} \tag{14.34}$$

P_n is the rated motor power; S_n = rated slip.

The rated slip is less than 2 to 3% for most induction motors and the mechanical losses are around 1% of rated power.

$$T_{en} \approx \frac{p_1 P_n\cdot 1.01}{2\pi f_1 0.98} = 0.1641\cdot P_n\frac{p_1}{f_1} \tag{14.35}$$

Choosing σ_{tan} in the interval 0.2 to 5 N/cm^2 or 2,000 to 50,000 N/m^2, we may use (14.33) directly with $\lambda = \frac{2p_1L}{\pi D_{is}}$ to determine the internal stator diameter.

$$D_{is} = \sqrt[3]{\frac{4p_1}{\pi^2\lambda\sigma_{tan}}\left(0.1641\cdot P_n\cdot\frac{p_1}{f_1}\right)} \tag{14.36}$$

No apparent need occurs to adopt at this stage efficiency and power factor values for rated load.

We may now adopt the no-load value of airgap flux density B_{g0},

$$B_{g0} = \frac{\mu_0 3\sqrt{2} W_1 K_{w1} I_0}{\pi p_1 K_c g (1 + K_s)} \tag{14.37}$$

where the no load current I_0 and the number of turns/phase are unknown and the airgap g, Carter's coefficient K_c, and saturation factor K_s are assigned pertinent values.

$$\begin{aligned} g &\approx \left(0.1 + 0.02\sqrt[3]{P_n}\right)\cdot 10^{-3}[m]; \text{ for } p_1 = 1 \\ g &\approx \left(0.1 + 0.012\sqrt[3]{P_n}\right)\cdot 10^{-3}[m]; \text{ for } p_1 \geq 2 \end{aligned} \tag{14.38}$$

Typical values of airgap are 0.35, 0.4, 0.45, 0.5, 0.55 … mm, etc. Also, $K_c \approx (1.15 - 1.35)$ for semiclosed slots and $K_c = 1.5 - 1.7$ for open stator slots (large power induction motors). The saturation factor is typically $K_s = 0.3 - 0.5$ for $p_1 \geq 2$ and larger for $2p_1 = 2$.

The airgap flux density B_g is

$$\begin{aligned} B_g &= (0.5 - 0.7)T \quad \text{for } 2p_1 = 2 \\ B_g &= (0.65 - 0.75)T \quad \text{for } 2p_1 = 4 \\ B_g &= (0.7 - 0.8)T \quad \text{for } 2p_1 = 6 \\ B_g &= (0.75 - 0.85)T \quad \text{for } 2p_1 = 8 \end{aligned} \tag{14.39}$$

The larger values correspond to larger motors.

The product, W_1I_0, is thus obtained from (14.37). The number of turns W_1 may be calculated from the emf E_1 (14.12 and (14.13).

$$W_1 = \frac{E_1}{4f_1K_fK_{w1}\phi} = \frac{E_1 2p_1}{4f_1K_fK_{w1}\alpha_i\pi D_{is}LB_g} \tag{14.40}$$

with W_1I_0 and W_1 known, the no load (magnetization) current I_0 may be obtained. The airgap active power P_{gap} is

$$P_{gap} = T_{en}\frac{2\pi f_1}{p_1} = 3K_EV_1I_T \tag{14.41}$$

where I_T is the stator current torque component (in phase with E_1). With I_T determined from (14.41), we may now calculate the stator rated current I_{1n}.

$$I_{1n} \approx \sqrt{I_0^2 + I_T^2} \tag{14.42}$$

The rotor bar current (for a cage rotor) I_b is

$$I_b \approx \frac{2mW_1K_{w1}I_T}{N_r} \tag{14.43}$$

N_r – number of rotor slots, m – number of stator phases.

We may now check the product $\eta_n\cos\varphi_{1n}$.

$$\eta_n \cos\varphi_n = \frac{P_n}{3V_l I_{ln}} < 1 \qquad (14.44)$$

The linear current loading A_l may be also checked,

$$A_l = \frac{2mW_l I_{ln}}{\pi D_{is}} \qquad (14.45)$$

and eventually compared with data from existing similar motors.

With all these data available, the sizing of stator and rotor slots and their windings is feasible. Then the machine reactances and resistances and the steady-state performance may be calculated. Knowing the motor geometry and the loss breakdown, the thermal aspects (design) may be approached. Finally, if the temperature rise or other performance are not satisfactory, the design process is repeated.

Given the complexity of such an enterprise, some coherent methodologies are in order. They will be developed in subsequent chapters.

Example 14.2 Tangential stress

Let us consider the motor data of Example 14.1, adopt $\sigma_{tan} = 1.5 \cdot 10^4 N/m^2$, and determine the values of D_{is}, L, W_l, I_0, I_{ln}, $\eta_n \cos\varphi_n$.

Solution

With $p_l = 2$, $P_n = 55$ kW, $f_l = 50$ Hz, $\lambda = 1.5$, from (14.36),

$$D_{is} = \sqrt[3]{\frac{4 \cdot 2 \cdot 0.1641 \cdot 55 \cdot 10^3 \cdot 2}{\pi^2 1.5 \cdot 1.5 \cdot 10^4 \cdot 50}} = 0.2352m$$

The stack length L is

$$L = \lambda \frac{\pi D_{is}}{2p_l} = 1.5 \frac{\pi \cdot 0.2352}{2 \cdot 2} = 0.277m$$

with $B_g = 0.8$, $K_f = 1.08$, $\alpha_i = 0.74$, $K_{wl} = 0.955$, $K_E = 0.97$, and from (14.40), the number of turns per phase W_l is

$$W_l = \frac{0.97 \cdot \left(\frac{400}{\sqrt{3}}\right) \cdot 2 \cdot 2}{4 \cdot 50 \cdot 1.08 \cdot 0.955 \cdot 0.74 \cdot \pi \cdot 0.2352 \cdot 0.277 \cdot 0.8} = 36 turns/phase$$

The rated electromagnetic torque T_{en} (14.35) is

$$T_{en} = 0.1641 \cdot P_n \frac{p_l}{f_l} = 0.1641 \cdot 55 \cdot 10^3 \cdot \frac{2}{50} = 361.02Nm$$

Now, from (14.41), the torque current component I_T is

$$I_T = \frac{T_{en} \cdot 2\pi f_1}{3K_E V_1 p_1} = \frac{361.02 \cdot 2\pi 50}{3 \cdot 0.97 \cdot \frac{400}{\sqrt{3}} \cdot 2} = 84.24A$$

The magnetization current I_0 is obtained from (14.37).

$$I_0 = \frac{B_{g0}\pi p_1 K_c g(1+K_s)}{\mu_0 3\sqrt{2} W_1 K_{w1}} = \frac{0.8 \cdot \pi \cdot 2 \cdot 1.25 \cdot 0.55 \cdot (1+0.5) \cdot 10^{-3}}{1.256 \cdot 10^{-6} \cdot 36 \cdot 0.955 \cdot 3\sqrt{2}} = 28.36A$$

The airgap g (14.38) is

$$g = \left(0.1 + 0.012\sqrt[3]{55000}\right) \cdot 10^{-3} = 0.55 \cdot 10^{-3} m$$

The stator rated current I_{1n} is

$$I_{1n} = \sqrt{I_0^2 + I_T^2} = \sqrt{28.36^2 + 84.24^2} = 88.887A$$

$$\eta_n \cos\varphi_{1n} = \frac{P_n}{3V_1 I_{1n}} = \frac{55 \cdot 10^3}{3 \cdot \frac{400}{\sqrt{3}} \cdot 88.887} = 0.894 < 1$$

This corresponds to a rather high (say) $\eta_n = \cos\varphi_{1n} = 0.9455$.

Note that these values appear at design starting, before all the losses in the machine have been assessed. They provide a design start without Esson's (output) constant which changed continuously over the last decade as material quality and cooling systems improved steadily.

14.11 SUMMARY

- Mechanical loads are characterised by torque/speed curves.
- Single quadrant and multiquadrant load torque/speed curves are typical.
- Constant V/f supply IMs are suitable only for constant speed single quadrant loads.
- For single and multiquadrant variable speed loads, variable V/f supply IMs are required. They result is energy savings commensurable with speed control range.
- Three load torque/speed curves are typical: quadratic torque/speed (pumps), constant torque (elevators), and constant power (machine tool, spindles, traction, etc.).
- The standard IM design torque/speed envelope, to match the load, includes two regions: below and above base speed Ω_b. For base speed full voltage, full torque, is delivered at rated service cycle and rated temperature rise.
- With self ventilation the machine overtemperature leads to torque reduction with speed reduction. For constant torque below base speed, separate ventilation is required.

- Above base speed–constant voltage and increasing frequency, the torque available decreases and so does the flux linkage in the machine.
- A 2/1 constant power speed range (from Ω_b to 2 Ω_b) is typical with standard IM designs at constant voltage.
- When an induction motor designed for sine wave power is faced with a notable harmonic content in the power grid due to presence of power electronic equipment nearby, it has to be derated. In general a harmonic voltage factor (HVF) of less than 3% is considered harmless (Figure 14.6).
- A standard sine wave IM, when fed from a PWM voltage source inverter due to the additional (time harmonics) core and winding losses, has to be derated. A derating of 10% is considered acceptable with today's IGBT converters.
- Further on, the presence of a static power converter leads to a 5% voltage reduction at motor terminals with respect to the power grid voltage.
- Finally, an additional derating occurs due to unbalanced power grid voltage. The derating is significant for voltage imbalance above 2% (Figure 14.7).
- Induction motor specifications for constant V/f motors are lined up in pertinent standards. Nameplate markings refer to a miryad of specifications for the user's convenience.
- Efficiency is the most important nameplate marking as the cost of losses per year is about 30 - 40% of initial motor costs.
- Standard and high efficiency motors are now available. NEMA and EU regulations refer to high efficiency thresholds (Table 14.2 and 14.3).
- Designs A, B, C, D, E reveals through their torque speed curves, the starting, pull-up, and breakdown torques which are important factors in most constant V/f supply IMs.
- Matching a constant V/f IM to a load refers to equality of load and motor torque at rated speed and lower load torque below rated speed.
- For variable speed drive, two pole pairs count motors at two different frequencies, one below base speed (flux zone) and one above base speed (flux weakening, constant voltage zone) may be used.
- For constant power large speed ranges $\Omega_{max}/\Omega_b > 3$, very large breakdown torque designs are required (above 300%). Alternatively, the voltage per phase is increased above base speed by star/delta connection or a larger torque (larger size) IM is chosen.
- Design of an IM means sizing the motor for given specifications of power supply parameters and load torque/speed envelope.
- Main design factors are: costs of active materials, fabrication, and selling, capitalised costs, maintenance costs, material limitations (magnetic, electric, dielectric, thermal, mechanical), and special application specifications.
- The IM design features 5 issues.
    - Electric design
    - Dielectric design
    - Magnetic design

- Thermal design
- Mechanical design

- IM sizing is both a science and an art based on prior experience.
- D_{is}^2L output coefficient design concept has gained widespread acceptance due to Esson's output constant and, with efficiency and power factor known, the stator bore diameter D_{is}, may be calculated for given power, speed, and stack length L per pitch τ ratio λ given.
- Further on, with given stator winding current density, airgap, stator teeth, and back core flux densities B_g, B_{ts}, B_{cs}, the outer stator diameter is obtained. Based on this data, the stator/rotor slot sizing, wire gauge, machine parameters, performance, losses, and temperatures may be approached. Such a complex enterprise requires coherent methodologies, to be developed in subsequent chapters.
- The rotor tangential stress $\sigma_{tan} = (0.2 - 4)N/cm^2$ is defined above as a more general design concept valid for both rotary and linear induction motors. This way there is no need to assign initial values to efficiency and power factor to perform the complete design (sizing) process.
- More on design principles in References [4–6].

14.12 REFERENCES

1. A.H. Bonett, G.C. Soukup, NEMA Motor-generator Standards for Three-phase Induction Motors, IEE – IAS Magazine, May – June 1999, pp.49 – 63.
2. B. De Vault, D. Heckenkamp, T. King, Selection of Shortcircuit Protection and Control for Design E Motors, IBID pp.26 – 37.
3. J.C. Andreas, Energy-efficient Electric Motors, Marcel Dekker Inc., New York, 1982.
4. E. Levi, Polyphase Motors–A Direct Approach to Their Design, Wiley Interscience, 1985.
5. B.J. Chalmers, A Williamson, A.C. Machines–Electromagnetics and Design, Research studies Press LTD, 1991, John Wiley & Sons Inc.
6. E.S. Hamdi, Design of Small Electrical Machines, John Wiley & Sons Ltd, Chichester, England, 1993.

Chapter 15

IM DESIGN BELOW 100 KW AND CONSTANT V AND f

15.1. INTRODUCTION

The power of 100 kW is traditionally considered the border between a small and medium power induction machine. In general, sub 100 kW motors use a single stator and rotor stack (no radial cooling channels) and a finned frame washed by air from a ventilator externally mounted at the shaft end (Figure 15.1). It has an aluminum cast cage rotor and, in general, random wound stator coils made of round magnetic wire with 1 to 6 elementary conductors (diameter $\leq$ 2.5mm) in parallel and 1 to 3 current paths in parallel, depending on the number of pole pairs. The number of pole pairs $2p_1 = 1, 2, 3, \ldots 6$.

Figure 15.1 Low power 3 phase IM with cage rotor

Induction motors with power below 100 kW constitute a sizable portion of the world electric motor markets. Their design for standard or high efficiency is a nature mixture of art and science, at least in the preoptimization stage. Design optimization will be dealt with separately in a dedicated chapter.

For the most part, IM design methodologies are proprietory.

Here we present what may constitute a sample of such methodologies. For further information, see also [1].

15.2. DESIGN SPECIFICATIONS BY EXAMPLE

Standard design specifications are

- Rated power: $P_n[W] = 5.5kW$
- Synchronous speed: $n_1[rpm] = 1800$
- Line supply voltage: $V_1[V] = 460V$
- Supply frequency: $f_1[Hz] = 60$
- Number of phases $m = 3$
- Phase connections: star
- Targeted power factor: $\cos\varphi_n = 0.83$
- Targeted efficiency: $\eta_n = 0.895$ (high efficiency motor)
- p.u. locked rotor torque: $t_{LR} = 1.75$
- p.u. locked rotor current: $i_{LR} = 6$
- p.u. breakdown torque: $t_{bK} = 2.5$
- Insulation class: F; temperature rise: class B
- Protection degree: IP55 – IC411
- Service factor load: 1.0
- Environment conditions: standard (no derating)
- Configuration (vertical or horizontal shaft etc.): horizontal shaft

15.3. THE ALGORITHM

The main steps in IM design are shown in Figure 15.2. The design process may start with (1) design specs and assigned values of flux densities and current densities and (2) calculate in the stator bore diameter D_{is}, stack length, stator slots, and stator outer diameter D_{out}, after stator and rotor currents are found. The rotor slots, back iron height, and cage sizing follows.

All dimensions are adjusted in (3) to standardized values (stator outer diameter, stator winding wire gauge, etc.). Then in (4), the actual magnetic and electric loadings (current and flux densities) are verified.

If the results on magnetic saturation coefficient $(1 + K_{st})$ of stator and rotor tooth are not equal to assigned values, the design restarts (1) with adjusted values of tooth flux densities until sufficient convergence is obtained in $1 + K_{st}$.

Once this loop is surpassed, stages (5) to (8) are traveled by computing the magnetization current I_0 (5); equivalent circuit parameters are calculated in (6), losses, rated slip S_n, and efficiency are determined in (7) and then power factor, locked rotor current and torque, breakdown torque, and temperature rise are assessed in (8).

In (9) all this performance is checked and if found unsatisfactory, the whole process is restarted in (1) with new values of flux densities and/or current densities and stack aspect ratio $\lambda = L/\tau$ (τ – pole pitch).

The decision in (9) may be made based on an optimization method which might result in going back to (1) or directly to (3) when the chosen construction and geometrical data are altered according to an optimization method (deterministic or evolutionary) as shown in Chapter 18.

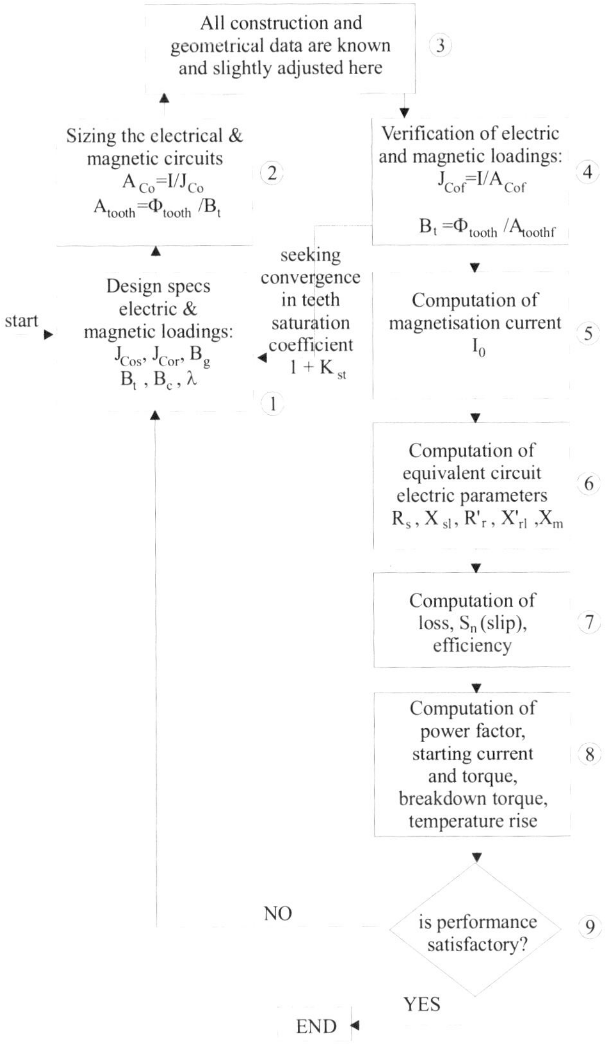

Figure 15.2 The design algorithm

So, IM design is basically an iterative procedure whose output–the resultant machine to be built–depends on the objective function(s) to be minimized and on the corroborating constraints related to temperature rise, starting current (torque), breakdown torque, etc.

The objective function may be active materials or costs or $(\text{efficiency})^{-1}$ or global costs or a weighted combination of them.

Before treating the optimization stage in Chapter 18, let us perform here a practical design.

15.4. MAIN DIMENSIONS OF STATOR CORE

Here we are going to use the widely accepted D_{is}^2L output constant concept detailed in the previous chapter. For completely new designs, the rotor tangential stress concept may be used.

Based on this, the stator bore diameter D_{is} (14.15) is

$$D_{is} = \sqrt[3]{\frac{2p\ p_1}{\pi\lambda}\frac{p_1}{f_1}\frac{S_{gap}}{C_0}};\quad K_E = 0.98 - 0.005p_1 = 0.97 \qquad (15.1)$$

with
$$S_{gap} = \frac{K_E P_n}{\eta_n \cos\varphi_{1n}};\quad \lambda = L\left(\frac{2p_1}{\pi D_{is}}\right) = \frac{L}{\tau} \qquad (15.2)$$

From past experience, λ is given in Table 15.1.

Table 15.1. Stack aspect ratio λ

$2p_1$	2	4	6	8
λ	0.6 – 1.0	1.2 – 1.8	1.6 – 2.2	2 -3

From (15.2), the apparent airgap power S_{gap} is

$$S_{gap} = \frac{0.97\cdot 5.5\cdot 10^3}{0.895\cdot 0.83} = 7181.8\text{VA}$$

C_o is extracted from Figure 14.14 for S_{gap} = 7181.8VA, C_0 = $147\cdot10^3$J/m^3 and $\lambda = 1.5$, $f_1 = 60$Hz, $p_1 = 2$. So D_{is} from (15.1) is

$$D_{is} = \sqrt[3]{\frac{2\cdot 2\cdot 2}{\pi\cdot 15\cdot 60}\cdot\frac{7181.8}{147\cdot 10^3}} = 0.1116\text{m}$$

The stack length L (from 15.2) is

$$L = \frac{1.5\pi\cdot 0.1116}{2\cdot 2} = 0.1315\text{m}$$

The pole pitch

$$\tau = \frac{\pi\cdot 0.1116}{2\cdot 2} = 0.0876\text{m}$$

The number of stator slots per pole 3q may be 3·2 = 6 or 3·3 = 9. For q = 3, the slot pitch τ_s will be around

$$\tau_s = \frac{\tau}{3q} = \frac{0.0876}{3\cdot 3} = 9.734\cdot 10^{-3}\text{m} \qquad (15.3)$$

In general the larger q gives better performance (space field harmonics and losses are smaller).

The slot width at airgap is to be around 5 to 5.3 mm with a tooth of 4.7 to 4.4 mm which is mechanically feasible.

From past experience (or from optimal lamination concept, developed later in this chapter), the ratio of the internal to external stator diameter D_{is}/D_{out}, bellow 100 kW for standard motors is given in Table 15.2.

Table 15.2. Inner/outer stator diameter ratio

$2p_1$	2	4	6	8
$\frac{D_{is}}{D_{out}}$	0.54 – 0.58	0.61 – 0.63	0.68 – 0.71	0.72 – 0.74

With $2p_1 = 4$, we choose $\frac{D_{is}}{D_{out}} = K_D = 0.62$ and thus

$$D_{out} = \frac{D_{is}}{K_D} = \frac{0.1116}{0.62} = 0.18\text{m} \tag{15.4}$$

Let us suppose that this value is normalized. The airgap value has also been introduced in Chapter 14 as

$$\begin{aligned} g &= \left(0.1 + 0.02 \cdot \sqrt[3]{P_n}\right) \cdot 10^{-3}\text{m} \quad \text{for } 2p_1 = 2 \\ g &= \left(0.1 + 0.012 \cdot \sqrt[3]{P_n}\right) \cdot 10^{-3}\text{m} \quad \text{for } 2p_1 \geq 2 \end{aligned} \tag{15.5}$$

In our case,

$$g = \left(0.1 + 0.012 \cdot \sqrt[3]{5500}\right) \cdot 10^{-3} = 0.3111 \cdot 10^{-3} \approx 0.35 \cdot 10^{-3}\text{m}$$

As known, too small airgap would produces large space airgap field harmonics and additional losses while a too large one would reduce the power factor and efficiency.

15.5. THE STATOR WINDING

Induction motor windings have been presented in Chapter 4. Based on such knowledge, we choose the number of stator slots N_s.

$$N_s = 2p_1 qm = 2 \cdot 2 \cdot 3 \cdot 3 = 36 \tag{15.6}$$

A two layer winding with chorded coils: $y/\tau = 7/9$ is chosen as $7/9 = 0.777$ is close to 0.8, which would reduce the first (5th order) stator mmf space harmonic.

The electrical angle between emfs in neighboring slots α_{ec} is

$$\alpha_{ec} = \frac{2\pi p_1}{N_s} = \frac{2\pi 2}{36} = \frac{\pi}{9} \tag{15.7}$$

The largest common divisor of N_s and p_1 (36, 2) is $t = p_1 = 2$ and thus the number of distinct stator slot emfs $N_s/t = 36/2 = 18$. The star of emf phasors has 18 arrows (Figure 15.3a) and the distribution of phases in slots of Figure 15.3b.

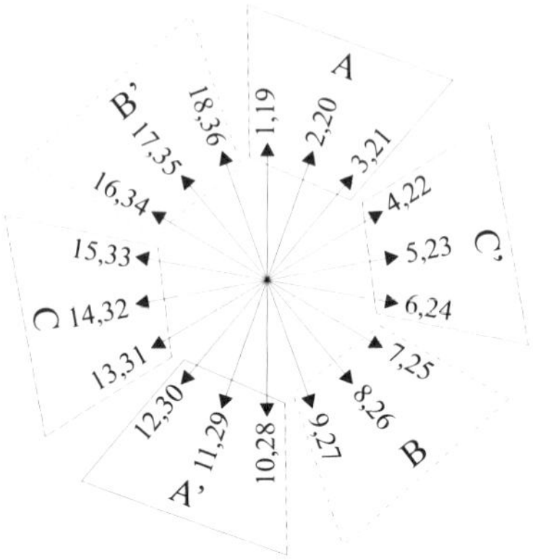

1	2	3	4	5	6	7	8	9	10	11	12	13	14	15	16	17	18
A	A	A	C'	C'	C'	B	B	B	A'	A'	A'	C	C	C	B'	B'	B'
A	C'	C'	C'	B	B	B	A'	A'	A'	C	C	C	B'	B'	B'	A	A

19	20	21	22	23	24	25	26	27	28	29	30	31	32	33	34	35	36
A	A	A	C'	C'	C'	B	B	B	A'	A'	A'	C	C	C	B'	B'	B'
A	C'	C'	C'	B	B	B	A'	A'	A'	C	C	C	B'	B'	B'	A	A

Figure 15.3. A 36 slots, $2p_1 = 4$ poles, 2 layer, chorded coils ($y/\tau = 7/9$) three phase winding

The zone factor K_{q1} is

$$K_{q1} = \frac{\sin\frac{\pi}{6}}{q\sin\left(\frac{\pi}{6q}\right)} = \frac{0.5}{3\sin\left(\frac{\pi}{18}\right)} = 0.9598 \tag{15.8}$$

The chording factor K_{y1} is

$$K_{y1} = \sin\frac{\pi}{2}\frac{y}{\tau} = \sin\frac{\pi}{2}\frac{7}{9} = 0.9397 \tag{15.9}$$

So, the stator winding factor K_{w1} becomes

$$K_{w1} = K_{q1}K_{y1} = 0.9598 \cdot 0.9397 = 0.9019$$

The number of turns per phase is based on the pole flux ϕ,

$$\phi = \alpha_i \tau L B_g \tag{15.10}$$

The airgap flux density is recommended in the intervals

$$\begin{aligned} B_g &= (0.5 - 0.75)T \quad \text{for } 2p_1 = 2 \\ B_g &= (0.65 - 0.78)T \quad \text{for } 2p_1 = 4 \\ B_g &= (0.7 - 0.82)T \quad \text{for } 2p_1 = 6 \\ B_g &= (0.75 - 0.85)T \quad \text{for } 2p_1 = 8 \end{aligned} \tag{15.11}$$

The pole spanning coefficient α_i (Chapter 14, Figure 14.3) depends on the tooth saturation factor $1 + K_{st}$.

Let us consider $1 + K_{st} = 1.4$ with $\alpha_i = 0.729$, $K_f = 1.085$. Now from (15.10) with $B_g = 0.7T$:

$$\phi = 0.729 \cdot 0.0876 \cdot 0.1315 \cdot 0.7 = 5.878 \cdot 10^{-3}\,\text{Wb}$$

The number of turns per phase W_1 (from Chapter 14, (14.9)) is:

$$W_1 = \frac{K_E V_{1ph}}{4K_f K_{w1} f_1 \phi} = \frac{0.97 \cdot \left(\dfrac{460}{\sqrt{3}}\right)}{4 \cdot 1.085 \cdot 0.902 \cdot 60 \cdot 5.878 \cdot 10^{-3}} = 186.8\,\text{turns/phase} \tag{15.12}$$

The number of conductors per slot n_s is

$$n_s = \frac{a_1 W_1}{p_1 q} \tag{15.13}$$

where a_1 is the number of current paths in parallel.

In our case, $a_1 = 1$ and

$$n_s = \frac{1 \cdot 186.8}{2 \cdot 3} = 31.33 \tag{15.14}$$

It should be an even number as there are two distinct coils per slot in a double layer winding, $n_s = 30$. Consequently, $W_1 = p_1 q n_s = 2 \cdot 3 \cdot 30 = 180$.

Going back to (15.12), we have to recalculate the actual airgap flux density B_g.

$$B_g = 0.7 \cdot \frac{186.8}{180} = 0.726T \tag{15.15}$$

The rated current I_{1n} is

$$I_{1n} = \frac{P_n}{\eta_n \cos\varphi_n \sqrt{3} V_1} = \frac{5500}{0.895 \cdot 0.83 \cdot 1.73 \cdot 460} = 9.303\text{A} \tag{15.16}$$

As high efficiency is required and, in general, at this power level and speed, winding losses are predominant from the recommended current densities.

$$\begin{aligned} J_{cos} &= (4 \ldots 7)\text{A/mm}^2 \quad \text{for } 2p_1 = 2,4 \\ J_{cos} &= (5 \ldots 8)\text{A/mm}^2 \quad \text{for } 2p_1 = 6,8 \end{aligned} \tag{15.17}$$

we choose $J_{cos} = 4.5A/mm^2$.

The magnetic wire cross section A_{Co} is

$$A_{Co} = \frac{I_{1n}}{J_{cos}a_1} = \frac{9.303}{4.5 \cdot 1} = 2.06733mm^2 \quad (15.18)$$

Table 15.3. Standardized magnetic wire diameter

Rated diameter [mm]	Insulated diameter [mm]
0.3	0.327
0.32	0.348
0.33	0.359
0.35	0.3795
0.38	0.4105
0.40	0.4315
0.42	0.4625
0.45	0.4835
0.48	0.515
0.50	0.536
0.53	0.567
0.55	0.5875
0.58	0.6185
0.60	0.639
0.63	0.6705
0.65	0.691
0.67	0.7145
0.70	0.742
0.71	0.7525
0.75	0.749
0.80	0.8455
0.85	0.897
0.90	0.948
0.95	1.0
1.0	1.051
1.05	1.102
1.10	1.153
1.12	1.173
1.15	1.2035
1.18	1.2345
1.20	1.305
1.25	1.305
1.30	1.356
1.32	1.3765
1.35	1.407
1.40	1.4575
1.45	1.508
1.5	1.559

With the wire gauge diameter d_{Co}

$$d_{Co} = \sqrt{\frac{4A_{Co}}{\pi}} = \sqrt{\frac{4 \cdot 2.06733}{\pi}} = 1.622mm \quad (15.19)$$

In general, if $d_{Co} > 1.3$ mm in low power IMs, we may use a few conductors in parallel a_p.

$$d_{Co}' = \sqrt{\frac{4A_{Co}}{\pi a_p}} = \sqrt{\frac{4\cdot 2.06733}{\pi\cdot 2}} = 1.15\text{mm} \qquad (15.20)$$

Now we have to choose a standardized bare wire diameter from Table 15.3.

The value of 1.15 mm is standardized, so each coil is made of 15 turns and each turn contains 2 elementary conductors in parallel (diameter $d_{Co}' = 1.15$ mm).

If the number of conductors in parallel $a_p > 4$, the number of current paths in parallel has to be increased. If, even in this case, a solution is not found, use is made of rectangular cross section magnetic wire.

15.6. STATOR SLOT SIZING

As we know by now, the number of turns per slot n_s and the number of conductors in parallel a_p with the wire diameter d_{Co}', we may calculate the useful slot area A_{su} provided we adopt a slot fill factor K_{fill}. For round wire, $K_{fill} \approx 0.35$ to 0.4 below 10 kW and 0.4 to 0.44 above 10 kW.

$$A_{su} = \frac{\pi d_{Co}'^2 a_p n_s}{4K_{fill}} = \frac{\pi\cdot 1.15^2\cdot 2\cdot 30}{4\cdot 0.40} = 155.7\text{mm}^2 \qquad (15.21)$$

For the case in point, trapezoidal or rounded semiclosed shape is recommended (Figure 15.4).

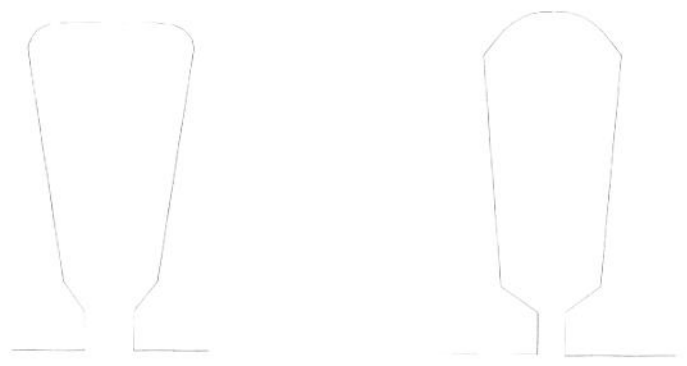

Figure 15.4 Recommended stator slot shapes

For such slot shapes, the stator tooth is rectangular (Figure 15.5). The variables b_{os}, h_{os}, h_w are assigned values from past experience: $b_{os} = 2$ to 3 mm $\leq 8g$, $h_{os} = (0.5$ to $1.0)$ mm, wedge height $h_w = 1$ to 4 mm.

The stator slot pitch τ_s (from 15.3) is $\tau_s = 9.734$ mm.

Assuming that all the airgap flux passes through the stator teeth:

$$B_g \tau_s L \approx B_{ts} b_{ts} L K_{Fe} \qquad (15.22)$$

$K_{Fe} \approx 0.96$ for 0.5 mm thick lamination constitutes the influence of lamination insulation thickness.

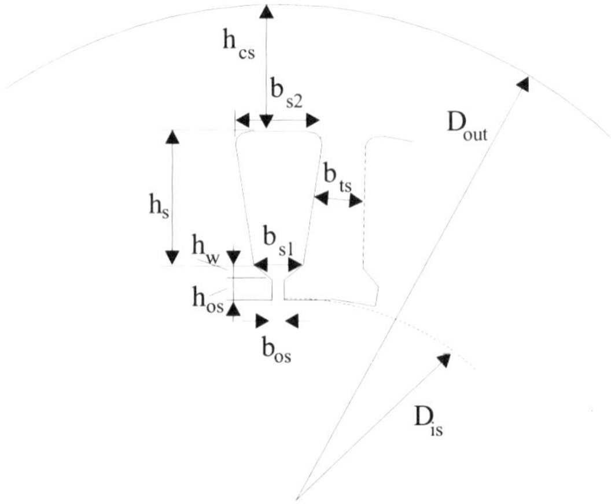

Figure 15.5 Stator slot geometry

With $B_{ts} = 1.5 - 1.65$ T, ($B_{ts} = 1.55$ T), from (15.22) the tooth width b_{ts} may be determined.

$$b_{ts} = \frac{0.726 \cdot 9.734 \cdot 10^{-3}}{1.55 \cdot 0.96} = 4.75 \cdot 10^{-3}\,\mathrm{m}$$

From technological limitations, the tooth width should not be under $3.5 \cdot 10^{-3}$m.

With $b_{os} = 2.2 \cdot 10^{-3}$m, $h_{os} = 1 \cdot 10^{-3}$m, $h_w = 1.5 \cdot 10^{-3}$m, the slot lower width b_{s1} is

$$b_{s1} = \frac{\pi\left(D_{is} + 2h_{os} + 2h_w\right)}{N_s} - b_{ts} = \\ = \frac{\pi(111.6 + 2 \cdot 1 + 2 \cdot 1.5) \cdot 10^{-3}}{36} - 4.75 \cdot 10^{-3} = 5.42 \cdot 10^{-3}\,\mathrm{m} \tag{15.23}$$

The useful area of slot A_{su} may be expressed as:

$$A_{su} = h_s \frac{\left(b_{s1} + b_{s2}\right)}{2} \tag{15.24}$$

Also,

$$b_{s2} \approx b_{s1} + 2h_s \tan\frac{\pi}{N_s} \tag{15.25}$$

From these two equations, the unknowns b_{s2} and h_s may be found.

$$b_{s2}^{\ 2} - b_{s1}^{\ 2} = 4A_{su} \tan\frac{\pi}{N_s} \tag{15.26}$$

$$b_{s2} = \sqrt{4A_{su}\tan\frac{\pi}{N_s} + b_{s1}^2} = 10^{-3}\sqrt{4\cdot 155.72\tan\frac{\pi}{36} + 5.42^2} \approx 9.16\cdot 10^{-3}\text{m} \quad (15.27)$$

The slot useful height h_s (15.24) writes

$$h_s = \frac{2A_{su}}{b_{s1} + b_{s2}} = \frac{2\cdot 155.72}{5.42 + 9.16}\cdot 10^{-3} = 21.36\cdot 10^{-3}\text{m} \quad (15.28)$$

Now we proceed in calculating the teeth saturation factor $1 + K_{st}$ by assuming that stator and rotor tooth produce same effects in this respect.

$$1 + K_{st} = 1 + \frac{F_{mts} + F_{mtr}}{F_{mg}} \quad (15.29)$$

The airgap mmf F_{mg} is

$$F_{mg} \approx 1.2\cdot g\cdot\frac{B_g}{\mu_0} = 1.2\cdot 0.35\cdot 10^{-3}\cdot\frac{0.726}{1.256\cdot 10^{-6}} = 242.77\text{Aturns}$$

with B_{ts} = 1.55T, from the magnetization curve table (Table 15.4), H_{ts} = 1760 A/m. Consequently, the stator tooth mmf F_{mts} is

$$F_{mts} = H_{ts}(h_s + h_{os} + h_w) = 1760(21.36 + 1 + 1.5)\cdot 10^{-3} = 41.99\text{Aturns} \quad (15.30)$$

Table 15.4. Lamination magnetization curve $B_m(H_m)$

B[T]	H[A/m]	B[T]	H[A/m]
0.05	22.8	1.05	237
0.1	35	1.1	273
0.15	45	1.15	310
0.2	49	1.2	356
0.25	57	1.25	417
0.3	65	1.3	482
0.35	70	1.35	585
0.4	76	1.4	760
0.45	83	1.45	1050
0.5	90	1.5	1340
0.55	98	1.55	1760
0.6	106	1.6	2460
0.65	115	1.65	3460
0.7	124	1.7	4800
0.75	135	1.75	6160
0.8	148	1.8	8270
0.85	162	1.85	11170
0.9	177	1.9	15220
0.95	198	1.95	22000
1.0	220	2.0	34000

From (15.29), we may calculate the value of rotor tooth mmf F_{mtr} which corresponds to $1 + K_{st}$ = 1.4.

$$F_{mtr} = K_{st}F_{mg} - F_{mts} = 0.4 - \frac{41.99}{242.77} = 55.11\text{Aturns} \quad (15.31)$$

As this value is only slightly larger than that of stator tooth, we may go on with the design process.

However, if $F_{mtr} << F_{mts}$ (or negative) in (15.31) it would mean that for given $1 + K_{st}$, a smaller value of flux density B_g is required.

Consequently, the whole design procedure has to be brought back to Equation (15.10). The iterative procedure is closed for now when $F_{mtr} \approx F_{mts}$.

As the outer diameter of stator has been calculated in (15.4) at $D_{out} = 0.18$m, the stator back iron height h_{cs} becomes

$$b_{cs} = \frac{D_{out} - (D_{is} + 2(h_{os} + h_w + h_s))}{2} = \\ = \frac{180 - (111.6 + 2(21.36 + 1.5 + 1))}{2} = 10.34\text{mm} \quad (15.32)$$

The back core flux density B_{cs} has to be verified here with $\phi = 5.878 \cdot 10^{-3}$Wb (from 15.10).

$$B_{cs} = \frac{\phi}{2Lh_{cs}} = \frac{5.878 \cdot 10^{-3}}{2 \cdot 0.1315 \cdot 10.34 \cdot 10^{-3}} = 2.16\text{T!!} \quad (15.33)$$

Evidently B_{cs} is too large. There are three main ways to solve this problem. One is to simply increase the stator outer diameter until $B_{cs} \approx 1.4$ to 1.7 T. The second solution consists in going back to the design start (Equation 15.1) and introducing a larger stack aspect ratio λ which eventually would result in a smaller D_{is}, and, finally, a larger back iron height b_{cs} and thus a lower B_{cs}. The third solution is to increase current density and thus reduce slot height h_s. However, if high efficiency is the target, such a solution is to be used cautiously.

Here we decide to modify the stator outer diameter to $D_{out}' = 0.190$ m and thus obtain

$$B_{cs} = 2.16 \frac{b_{cs}}{b_{cs} + \frac{0.190 - 0.180}{2}} = \frac{2.16 \cdot 10.34 \cdot 10^{-3}}{(10.34 + 5) \cdot 10^{-3}} = 1.456\text{T}$$

This is considered a reasonable value.

From now on, the outer stator diameter will be $D_{out}' = 0.190$ m.

15.7. ROTOR SLOTS

For cage rotors, as shown in Chapters 10 and 11, care must be exercised in choosing the correspondence between the stator and rotor numbers of slots to reduce parasitic torque, additional losses, radial forces, noise, and vibration. Based on past experience (Chapters 10 and 11 backs this up with pertinent explanations), the most adequate number of stator and rotor slot combinations are given in Table 15.5.

Table 15.5. Stator / rotor slot numbers

$2p_1$	N_s	N_r – skewed rotor slots
2	24	18, 20, 22, 28, 30, ,33,34
	36	25,27,28,29,30,43
	48	30,37,39,40,41
4	24	16,18,20,30,33,34,35,36
	36	28,30,32,34,45,48
	48	36,40,44,57,59
	72	42,48,54,56,60,61,62,68,76
6	36	20,22,28,44,47,49
	54	34,36,38,40,44,46
	72	44,46,50,60,61,62,82,83
8	48	26,30,34,35,36,38,58
	72	42,46,48,50,52,56,60
12	72	69,75,80
	90	86,87,93,94

For our case, let us choose $N_s \neq N_r = 36/28$.

As the starting current is rather large–high efficiency is targeted–the skin effect is not very pronounced. Also, as the locked rotor torque is large, the leakage inductance will not be large. Consequently, from the four typical slot shapes of Figure 15.6, that of Figure 15.6c is adopted.

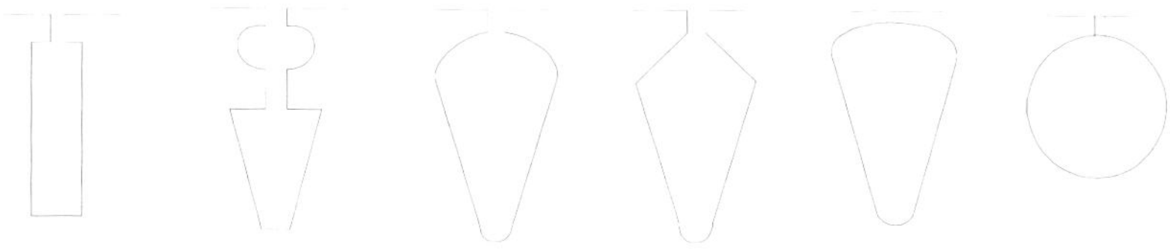

a.) b.) c.) d.) e.) f.)

Figure 15.6 Typical rotor cage slots

First, we need the value of rated rotor bar current I_b,

$$I_b = K_I \frac{2mW_1K_{w1}}{N_r} I_{1n} \tag{15.34}$$

with $K_I = 1$, the rotor and stator mmf would have equal magnitudes. In reality, the stator mmf is slightly larger.

$$K_I \approx 0.8 \cdot \cos\varphi_{1n} + 0.2 = 0.8 \cdot 0.83 + 0.2 = 0.864 \tag{15.35}$$

From (15.34), the bar current I_b is

$$I_b = \frac{0.864 \cdot 2 \cdot 3 \cdot 180 \cdot 0.9019 \cdot 9.303}{28} = 279.6A$$

For high efficiency, the current density in the rotor bar j_b = 3.42 A/mm^2 The rotor slot area A_b is

$$A_b = \frac{I_b}{J_b} = \frac{279.6}{3.42 \cdot 10^6} = 81.65 \cdot 10^{-6} m^2 \quad (15.36)$$

The end ring current I_{er} is

$$I_{er} = \frac{I_b}{2\sin\frac{\pi p_1}{N_r}} = \frac{279.6}{2\sin\left(\frac{2\pi}{28}\right)} = 628.255A \quad (15.37)$$

The current density in the end ring J_{er} = (0.75 – 0.8)J_b. The higher values correspond to end rings attached to the rotor stack as part of the heat is transferred directly to rotor core.

With $J_{er} = 0.75 \cdot J_b = 0.75 \cdot 342 \cdot 10^6 = 2.55 \cdot 10^6 A/m^2$, the end ring cross section, A_{er}, is

$$A_{er} = \frac{I_{er}}{J_{er}} = \frac{628.255}{2.565 \cdot 10^6} = 245 \cdot 10^{-6} m^2 \quad (15.38)$$

We may now proceed to rotor slot sizing based on the variables defined on Figure 15.7.

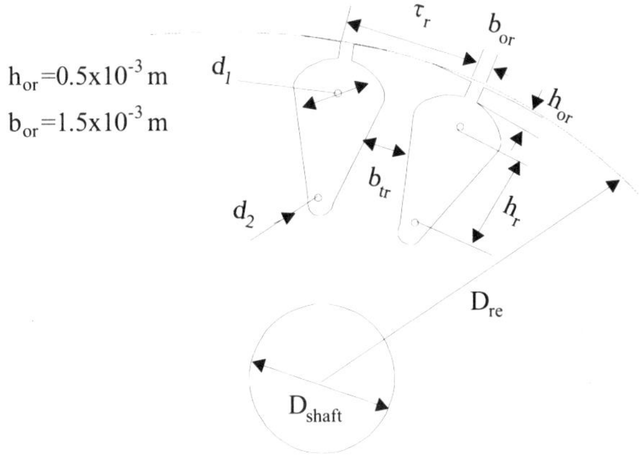

Figure 15.7 Rotor slot geometry

The rotor slot pitch τ_r is

$$\tau_r = \frac{\pi(D_{is} - 2g)}{N_r} = \frac{\pi(111.6 - 0.7) \cdot 10^{-3}}{28} = 12.436 \cdot 10^{-3} m \quad (15.39)$$

With the rotor tooth flux density $B_{tr} = 1.60T$, the tooth width b_{tr} is

$$b_{tr} \approx \frac{B_g}{K_{Fe} B_{tr}} \cdot \tau_r = \frac{0.726}{0.96 \cdot 1.6} \cdot 12.436 \cdot 10^{-3} = 5.88 \cdot 10^{-3} m \quad (15.40)$$

The diameter d_1 is obtained from

$$\frac{\pi(D_{re} - 2h_{or} - d_1)}{N_r} = d_1 + b_{tr} \quad (15.41)$$

$$d_1 = \frac{\pi(D_{re} - 2h_{or}) - N_r b_{tr}}{\pi + N_r} = \quad (15.42)$$

$$= \frac{\pi(111.6 - 0.7 - 1) - 28 \cdot 5.88 \cdot 10^{-3}}{\pi + 28} = 5.70 \cdot 10^{-3} m; \quad h_{or} = 0.5 \cdot 10^{-3} m$$

To completely define the rotor slot geometry, we use the slot area equations,

$$A_b = \frac{\pi}{8}\left(d_1^2 + d_2^2\right) + \frac{(d_1 + d_2)h_r}{2} \quad (15.43)$$

$$d_1 - d_2 = 2h_r \tan\frac{\pi}{N_r} \quad (15.44)$$

Solving (15.43) and (15.44), we will obtain d_2 and h_r (with $d_1 = 5.70 \cdot 10^{-3}m$, $A_b = 81.65 \cdot 10^{-6} m^2$) as $d_2 = 1.2 \cdot 10^{-3}m$ and $h_r = 20 \cdot 10^{-3}m$.

Now we have to verify the rotor teeth mmf F_{mtr} for $B_{tr} = 1.6T$, $H_{tr} = 2460A/m$ (Table 15.4).

$$F_{mtr} = H_{tr}\left(h_r + h_{or} + \frac{(d_1 + d_2)}{2}\right) = 2460\left(20 + 0.5 + \frac{(1.2 + 5.70)}{2}\right) \cdot 10^{-3} = 60.134 \text{Aturns} \quad (15.45)$$

This is rather close to the value of $V_{mtr} = 55.11$ Aturns of (15.39). The design is acceptable so far.

If V_{mtr} had been too large, we might have reduced the flux density, thus increasing tooth width b_{tr} and the bar current density. Increasing the slot height is not practical as already $d_2 = 1.2 \cdot 10^{-3}m$. This bar current density increase could reduce the efficiency below the target value. We may alternatively increase $1 + K_{st}$, and redo the design from (15.10).

When the power factor constraint is not too tight, this is a good solution. To maintain same efficiency, the stator bore diameter has to be increased. So the design should restart from Equation (15.1). The process ends when V_{mtr} is within bounds.

When V_{mtr} is too small, we may increase B_{tr} and return to (15.40) until sufficient convergence is obtained. The required rotor back core may be calculated after allowing for a given flux density B_{cr} = 1.4 – 1.7 T. With B_{cr} = 1.65 T, the rotor back core height h_{cr} is

$$h_{cr} = \frac{\phi}{2}\frac{1}{L \cdot B_{cr}} = \frac{5.878 \cdot 10^{-3}}{2 \cdot 0.1315 \cdot 1.65} = 13.55 \cdot 10^{-3}\text{m} \tag{15.46}$$

The maximum diameter of the shaft D_{shaft} is

$$(D_{shaft})_{max} \le D_{is} - 2g - 2\left(h_{or} + \frac{d_1 + d_2}{2} + h_r + h_{cr}\right) =$$
$$= \left(111.6 - 2 \cdot 0.35 - 2\left(1.5 + \frac{(1.2 + 5.69)}{2} + 20 + 13.55\right)\right) \cdot 10^{-3} \approx 35 \cdot 10^{-3}\text{m} \tag{15.47}$$

The shaft diameter corresponds to the rated torque and is given in tables based on mechanical design and past experience. The rated torque is approximately

$$T_{en} = \frac{P_n}{2\pi \frac{f_1}{p_1}(1 - S_n)} \approx \frac{5.5 \cdot 10^{-3}}{2\pi \frac{60}{2}(1 - 0.02)} = 33.56\text{Nm} \tag{15.48}$$

For the case in point, the $36 \cdot 10^{-3}$m left for the shaft diameter suffices. The end ring cross section is shown in Figure 15.8.

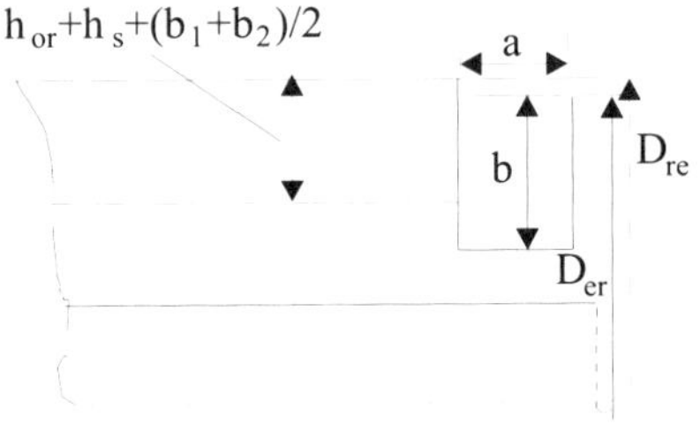

Figure 15.8 End ring cross - section

In general, $D_{re} - D_{er} = (3 - 4) \cdot 10^{-3}$m.
Also,

$$b = (1.0 - 1.2)\left(h_r + h_{or} + \frac{(b_1 + b_2)}{2}\right) \tag{15.49}$$

For

$$b = 1.0 \cdot \left(1 + 20 + \frac{(5.69 + 1.2)}{2}\right) = 24.445 \cdot 10^{-3}\text{m} \tag{15.50}$$

the value of a is

$$a = \frac{A_{er}}{b} = \frac{245 \cdot 10^{-6}}{24.445 \cdot 10^{-3}} = 10.02 \cdot 10^{-3} \text{m} \tag{15.51}$$

15.8. THE MAGNETIZATION CURRENT

The magnetization mmf F_{1m} is

$$F_{1m} = 2\left(K_c g \frac{B_g}{\mu_0} + F_{mts} + F_{mtr} + F_{mcs} + F_{mcr}\right) \tag{15.52}$$

So far we have considered $K_c = 1.2$ in F_{mg} (15.29). We do know all of the variables to calculate Carter's coefficient K_c.

$$\gamma_1 = \frac{b_{os}^{\ 2}}{5g + b_{os}} = \frac{2.2^2 \cdot 10^{-3}}{5 \cdot 0.35 + 2.2} = 1.2253 \cdot 10^{-3} \text{m} \tag{15.53}$$

$$\gamma_2 = \frac{b_{or}^{\ 2}}{5g + b_{or}} = \frac{1.5^2 \cdot 10^{-3}}{5 \cdot 0.35 + 1.5} = 0.692 \cdot 10^{-3} \text{m} \tag{15.54}$$

$$K_{c1} = \frac{\tau_s}{\tau_s - \gamma_1} = \frac{9.734}{9.734 - 1.2253} = 1.144 \tag{15.55}$$

$$K_{c2} = \frac{\tau_r}{\tau_r - \gamma_2} = \frac{12.436}{12.436 - 0.692} = 1.059 \tag{15.56}$$

The total Carter coefficient K_c is

$$K_c = K_{c1} K_{c2} = 1.144 \cdot 1.059 = 1.2115 \tag{15.57}$$

This is very close to the assigned value of 1.2, so it holds. With $F_{mts} = 42$ Aturns (15.30) and $F_{mtr} = 60.134$ Aturns (15.41) as definitive values (for $(1 + K_{st}) = 1.4$), we still have to calculate the back core mmfs F_{mcs} and F_{mcr}.

$$F_{mcs} = C_{cs} \frac{\pi(D_{out} - h_{cs})}{2p_1} H_{cs}(B_{cs}) \tag{15.58}$$

$$F_{mcr} = C_{cr} \frac{\pi(D_{shaft} + h_{cr})}{2p_1} H_{cr}(B_{cr}) \tag{15.59}$$

C_{cs} and C_{cr} are subunitary empirical coefficients that define an average length of flux path in the back core. They may be calculated by AIM methodology, Chapter 5.

$$C_{cs,r} \approx 0.88 \cdot e^{-0.4B_{cs,r}^2} \tag{15.60}$$

with B_{cs} = 1.456T and B_{cr} = 1.6T from Table 15.4 H_{cs} = 1050 A/m, H_{cr} = 2460 A/m. From (15.58) and (15.59),

$$F_{mcs} = 0.88 \cdot e^{-0.4 \cdot 1.456^2} \frac{\pi(190 - 15.34) \cdot 10^{-3}}{2 \cdot 2} 1050 = 54.22 \text{Aturns}$$

$$F_{mcr} = 0.88 \cdot e^{-0.4 \cdot 1.6^2} \frac{\pi(36 + 13.55) \cdot 10^{-3}}{2 \cdot 2} 2460 = 23.04 \text{Aturns}$$

Finally, from (15.52) and (15.29),

$$F_{1m} = 2 \cdot (242.77 + 42 + 60.134 + 54.22 + 23.04) = 844.328$$

The total saturation factor K_s comes from

$$K_s = \frac{F_{1m}}{2F_{mg}} - 1 = \frac{844.328}{2 \cdot 242.77} - 1 = 1.739 \tag{15.61}$$

The magnetization current I_μ is

$$I_\mu = \frac{\pi p_1 (F_{1m}/2)}{3\sqrt{2} W_1 K_{w1}} = \frac{\pi \cdot 2 \cdot 844.328}{6\sqrt{2} \cdot 180 \cdot 0.9019} = 3.86\text{A} \tag{15.62}$$

The relative (p.u.) value of I_μ is

$$i_\mu = \frac{I_\mu}{I_{1n}} = \frac{3.86}{9.303} = 0.415 = 41.5\% \tag{15.62'}$$

15.9. RESISTANCES AND INDUCTANCES

The resistances and inductances refer to the equivalent circuit (Figure 15.9).

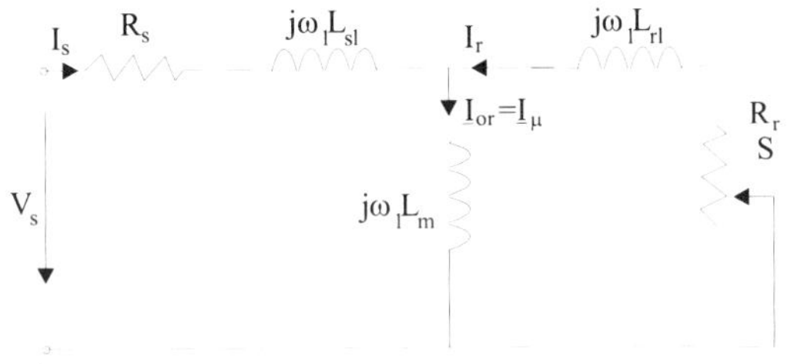

Figure 15.9 The T equivalent circuit (core losses not evident)

The stator phase resistance,

$$R_s = \rho_{Co}\frac{l_c W_1}{A_{co} a_1} \tag{15.63}$$

The coil length l_s includes the active part 2L and the end connection part $2l_{end}$

$$l_c = 2(L + l_{end}) \tag{15.64}$$

The end connection length depends on the coil span y, number of poles, shape of coils, and number of layers in the winding.

In general, manufacturing companies developed empirical formulas such as

$$\begin{aligned} l_{end} &= 2y - 0.04 \text{ m for } 2p_1 = 2 \\ l_{end} &= 2y - 0.02 \text{ m for } 2p_1 = 4 \\ l_{end} &= \frac{\pi}{2}y + 0.018 \text{ m for } 2p_1 = 6 \\ l_{end} &= 2.2y - 0.012 \text{ m for } 2p_1 = 8 \end{aligned} \tag{15.65}$$

$\frac{y}{\tau} = \beta$ with β the chording factor. In general,

$$\frac{2}{3} \le \beta \le 1 \tag{15.66}$$

In our case for $\frac{y}{\tau} = \frac{7}{9}$, we do have

$$y = \frac{7}{9}\tau = \frac{7}{9}\cdot 0.0876 = 0.06813\text{m} \tag{15.67}$$

And from (15.65) for $2p_1 = 4$,

$$l_{end} = 2y - 0.02 = 2\cdot 0.06813 - 0.02 = 0.11626\text{m} \tag{15.68}$$

The copper resistivity at 20^0C and 115^0C is $(\rho_{Co})_{20^0C} = 1.78\cdot 10^{-8}\Omega\text{m}$ and $(\rho_{Co})_{115^0C} = 1.37(\rho_{Co})_{20^0C}$. We do not know yet the rated stator temperature but the high efficiency target indicates that the winding temperature should not be too large even if the insulation class is F. We use here $(\rho_{Co})_{80^0C}$.

$$(\rho_{Co})_{80^0C} = (\rho_{Co})_{20^0C}\left(1 + \frac{1}{273}(80 - 20)\right) = 2.1712\cdot 10^{-8}\Omega\text{m} \tag{15.69}$$

From (15.63): $R_s = 2.1712\cdot 10^{-8}\cdot\frac{2(0.1315 + 0.11626)\cdot 180}{2.06733\cdot 10^{-6}} = 0.93675\Omega$

The rotor bar/end ring segment equivalent resistance R_{be} is

$$R_{be} = \rho_{Al}\left[\frac{L}{A_b}K_R + \frac{l_{er}}{2A_{er}\sin^2\left(\frac{\pi P_1}{N_r}\right)}\right] \quad (15.70)$$

The cast aluminium resistivity at 20^{0}C $(\rho_{Al})_{20^0C} = 3.1\cdot10^{-8}\Omega m$ and the end ring segment length l_{er} is

$$l_{er} = \frac{\pi(D_{er}-b)}{N_r} = \frac{\pi(111.6-2\cdot0.35-6-24.445)\cdot10^{-3}}{28} = 9.022\cdot10^{-3}m \quad (15.71)$$

K_r, the skin effect resistance coefficient for the bar (Chapter 9, Equation 9.1), is approximately (as for a rectangular bar)

$$K_R = \xi\frac{(\sinh 2\xi + \sin 2\xi)}{(\cosh 2\xi - \cos 2\xi)} \approx \xi \quad (15.72)$$

$$\xi = \beta_s h_r\sqrt{S};\ \ \beta_s = \sqrt{\frac{\omega_1\mu_0}{2\rho_{Al}}} = \sqrt{\frac{2\pi60\cdot1.25\cdot10^{-6}}{2\cdot3.1\cdot10^{-8}}} = 87(m)^{-1} \quad (15.73)$$

For h_r = 20·10^{-3}m and S = 1, ξ = 87·20·10^{-3}·1 = 1.74. K_R ≈ 1.74. From (15.70) the value of R_{be} is

$$R_{be\,80^0}^{S=1} = 3.1\cdot10^{-8}\left(1+\frac{1}{273}(80-20)\right)\left(\frac{0.1315\cdot1.73}{81.65\cdot10^{-6}} + \frac{9.022\cdot10^{-3}}{2\cdot245\cdot10^{-6}\sin^2\left(\frac{2\pi}{28}\right)}\right) =$$

$$= 1.194\cdot10^{-4}\Omega$$

The rotor cage resistance reduced to the stator R_r' is

$$(R_r')_{S=1} = \frac{4m_1}{N_r}(W_1K_{w1})^2R_{be\,80^0} =$$
$$= \frac{4\cdot3}{28}(180\cdot0.9019)^2\cdot1.194\cdot10^{-4} = 1.1295\Omega \quad (15.74)$$

The stator phase leakage reactance X_{sl} is

$$X_{sl} = 2\mu_0\omega_1 L\frac{W_1^2}{p_1 q}(\lambda_s + \lambda_{ds} + \lambda_{ec});\ \beta = \frac{y}{\tau} \quad (15.75)$$

λ_s, λ_d, λ_{ec} are the slot differential and end ring connection coefficients:

$$\lambda_s = \left[\frac{2}{3}\frac{h_s}{(b_{s1}+b_2)} + \frac{2h_w}{(b_{os}+b_{s1})} + \frac{h_{os}}{b_{os}}\right]\left(\frac{1+3\beta}{4}\right) =$$
$$= \left[\frac{2}{3}\frac{21.36}{(5.42+9.16)} + \frac{2\cdot 1.5}{(2.2+5.42)} + \frac{1}{2.2}\right]\left(\frac{1+3\cdot(7/9)}{4}\right) = 1.523 \quad (15.76)$$

An expression of λ_{ds} has been developed in Chapter 9, Equation (9.85). An alternative approximation is given here.

$$\lambda_{ds} \approx \frac{0.9\tau_s q^2 K_{w1}{}^2 C_s \gamma_{dc}}{K_c g(1+K_{st})} \quad (15.77)$$

$$C_s = 1 - 0.033\frac{b_{os}{}^2}{g\tau_s}$$

$$\begin{aligned}
\gamma_{ds} &= (0.11\sin\varphi_1 + 0.28)\cdot 10^{-2};\ \text{for } q = 8\\
\gamma_{ds} &= (0.11\sin\varphi_1 + 0.41)\cdot 10^{-2};\ \text{for } q = 6\\
\gamma_{ds} &= (0.14\sin\varphi_1 + 0.76)\cdot 10^{-2};\ \text{for } q = 4\\
\gamma_{ds} &= (0.18\sin\varphi_1 + 1.24)\cdot 10^{-2};\ \text{for } q = 3\\
\gamma_{ds} &= (0.25\sin\varphi_1 + 2.6)\cdot 10^{-2};\ \text{for } q = 2\\
\gamma_{dc} &= 9.5\cdot 10^{-2};\ \text{for } q = 1\\
\varphi_1 &= \pi(6\beta - 5.5)
\end{aligned} \quad (15.78)$$

For $\beta = 7/9$ and $q = 3$, γ_{ds} (from (15.78)) is

$$\gamma_{ds} = \left[0.18\sin\left(\pi\left(\frac{6\cdot 7}{9} - 5.5\right)\right) + 1.24\right]\cdot 10^{-2} = 1.15\cdot 10^{-2}$$

$$C_s = 1 - 0.033\frac{2.2^2}{0.35\cdot 9.734} = 0.953$$

From (15.77),

$$\lambda_{ds} = \frac{0.9\cdot 9.734\cdot 10^{-3}\cdot 3^2\cdot 0.9019^2\cdot 0.953\cdot 1.15\cdot 10^{-2}}{1.21\cdot 0.35\cdot 10^{-3}(1+0.4)} = 1.18$$

For two-layer windings, the end connection specific geometric permeance coefficient λ_{ec} is

$$\lambda_{ec} = 0.34\frac{q}{L}(l_{end} - 0.64\cdot\beta\cdot\tau) =$$

$$= 0.34 \cdot \frac{3}{0.1315}\left(0.11626 - 0.64 \cdot \frac{7}{9} \cdot 0.0876\right) = 0.5274 \quad (15.79)$$

From (15.75) the stator phase reactance X_{sl} is

$$X_{sl} = 2 \cdot 1.126 \cdot 10^{-6} \cdot 2\pi 60 \cdot 0.1315 \cdot \frac{180^2}{2 \cdot 3}(1.523 + 0.18 + 0.5274) = 2.17\Omega \quad (15.80)$$

The equivalent rotor bar leakage reactance X_{be} is

$$X_{be} = 2\pi f_1 \mu_0 L(\lambda_r K_X + \lambda_{dr} + \lambda_{er}) \quad (15.80')$$

where λ_r are the rotor slot, differential, and end ring permeance coefficient. For the rounded slot of Figure 15.7 (see Chapter 6),

$$\lambda_r \approx 0.66 + \frac{2h_r}{3(d_1 + d_2)} + \frac{h_{or}}{b_{or}} = 0.66 + \frac{2 \cdot 20}{3(5.7 + 1.2)} + \frac{0.5}{1.5} = 2.922 \quad (15.81)$$

The value of λ_{dr} is (Chapter 6)

$$\lambda_{dr} = \frac{0.9\tau_r \gamma_{dr}}{K_c g}\left(\frac{N_r}{6p_1}\right)^2; \; \gamma_{dr} = 9\left(\frac{6p_1}{N_r}\right)^2 10^{-2} \quad (15.82)$$

$$\gamma_{dr} = 9\left(\frac{6 \cdot 2}{28}\right)^2 10^{-2} = 1.653 \cdot 10^{-2}$$

$$\lambda_{dr} = \frac{0.9 \cdot 12.436}{1.21 \cdot 0.35} \cdot 1.653 \cdot 10^{-2}\left(\frac{28}{6 \cdot 2}\right)^2 = 2.378$$

$$\lambda_{er} = \frac{2.3(D_{er} - b)}{N_r \cdot L \cdot 4\sin^2\left(\frac{\pi P_1}{N_r}\right)} \log \frac{4 \cdot 7(D_{er} - b)}{b + 2a} =$$

$$= \frac{2.3 \cdot 80.455}{28 \cdot 131.5 \cdot 4\sin^2\left(\frac{\pi \cdot 2}{28}\right)} \log\left(\frac{4.7 \cdot 80.455}{24.445 + 20}\right) = 0.2255 \quad (15.83)$$

The skin effect coefficient for the leakage reactance K_x is, for $\xi = 1.74$,

$$K_x \approx \frac{3}{2\xi}\frac{(\sinh(2\xi) - \sin(2\xi))}{(\cosh(2\xi) - \cos(2\xi))} \approx \frac{3}{2\xi} = 0.862 \quad (15.84)$$

From (15.80), X_{be} is

$$X_{be} = 2\pi 60 \cdot 1.256 \cdot 10^{-6} \cdot 0.1315(2.922 \cdot 0.862 + 2.378 + 0.2255) = 3.1877 \cdot 10^{-4}\Omega$$

The rotor leakage reactance X_{rl} becomes

$$X_{rl} = 4m\frac{(W_1K_{w1})^2}{N_r}X_{be} - 12\cdot\frac{(180\cdot 0.9019)^2}{28}3.1387\cdot 10^{-4} = 3.6506\Omega \quad (15.85)$$

For zero speed (S = 1), both stator and rotor leakage reactances are reduced due to leakage flux path saturation. This aspect has been treated in detail in Chapter 9. For the power levels of interest here, with semiclosed stator and rotor slots:

$$\begin{aligned}(X_{sl})_{sat}^{S=1} &= X_{sl}(0.7-0.8) \approx 2.17\cdot 0.75 = 1.625\Omega\\ (X_{rl})_{sat}^{S=1} &= X_{rl}(0.6-0.7) \approx 3.938\cdot 0.65 = 2.56\Omega\end{aligned} \quad (15.86)$$

For rated slip (speed), both skin and leakage saturation effects have to be eliminated ($K_R = K_x = 1$).

From (15.70), $R_{be\,80^0}$ is

$$\left(R_{be\,80^0}\right)_{S_n} = 3.1\cdot 10^{-8}\left(1+\frac{1}{273}(80-20)\right)\left[\frac{0.1315\cdot 1}{81.65\cdot 10^{-6}} + \frac{9.022\cdot 10^{-3}}{2.245\cdot 10^{-6}\sin^2\left(\frac{\pi\cdot 2}{28}\right)}\right] =$$

$$= 0.7495\cdot 10^{-3}\Omega$$

So the rotor resistance $(R_r')_{S_n}$ is

$$(R_r')_{S_n} = (R_r')_{S=1}\cdot\frac{R_{be80^0}^{S=S_n}}{R_{be80^0}^{S=1}} = 1.1295\cdot\frac{0.7495\cdot 10^{-4}}{1.194\cdot 10^{-4}} = 0.709\Omega \quad (15.87)$$

In a similar way, the equivalent rotor leakage reactance at rated slip S_n, $X_{rl}^{S=S_n} = 3.938\,\Omega$.

The magnetization X_m is

$$X_m = \sqrt{\left(\frac{V_{ph}}{I_\mu}\right)^2 - R_s^{\,2}} - X_{sl} = \sqrt{\left(\frac{460}{3.86\sqrt{3}}\right)^2 - 0.93675^2} - 2.17 = 66.70\Omega \quad (15.88)$$

Skewing effect on reactances

In general, the rotor slots are skewed. A skewing C of one stator slot pitch τ_s is typical ($c = \tau_s$).

The change in parameters due to skewing is discussed in detail in Chapter 9. Here the approximations are used.

$$X_m = X_m K_{skew} \quad (15.89)$$

$$K_{skew} = \frac{\sin\frac{\pi}{2}\frac{c}{\tau}}{\frac{\pi}{2}\frac{c}{\tau}} = \frac{\sin\frac{\pi}{2}\frac{\tau_s}{\tau}}{\frac{\pi}{2}\frac{\tau_s}{\tau}} = \frac{\sin\frac{\pi}{2}\frac{1}{3q}}{\frac{\pi}{2}\frac{1}{3q}} = \frac{\sin\frac{\pi}{18}}{\frac{\pi}{18}} = 0.9954 \qquad (15.90)$$

Now with (15.88) and (15.89),

$$X_m = 66.70 \cdot 0.9954 = 66.3955\Omega$$

Also, as suggested in Chapter 9, the rotor leakage inductance (reactance) is augmented by a new term X'_{rlskew}.

$$X'_{rlskew} = X_m\left(1 - K_{skew}^2\right) = 66.70\left(1 - 0.9954^2\right) = 0.6055\Omega \qquad (15.91)$$

So, the final values of rotor leakage reactance at $S = 1$ and $S = S_n$, respectively, are

$$\left(X_{rl}\right)_{skew}^{S=1} = \left(X_{rl}\right)_{sat}^{S=1} + X_{rlskew} = 2.56 + 0.6055 = 3.165\Omega \qquad (15.92)$$

$$\left(X_{rl}\right)_{skew}^{S=S_n} = X_{rl} + X_{rlskew} = 3.6506 + 0.6055 = 4.256\Omega \qquad (15.93)$$

15.10. LOSSES AND EFFICIENCY

The efficiency is defined as output per input power:

$$\eta = \frac{P_{out}}{P_{in}} = \frac{P_{out}}{P_{in} + \sum losses} \qquad (15.94)$$

The loss components are

$$\sum losses = p_{Co} + p_{Al} + p_{iron} + p_{mv} + p_{stray} \qquad (15.95)$$

p_{Co} represents the stator winding losses,

$$p_{Co} = 3R_s I_{1n}^2 = 3 \cdot 0.93675 \cdot 9.303^2 = 243.215W \qquad (15.96)$$

p_{Al} refers to rotor cage losses (at $S = S_n$).

$$p_{Al} = 3\left(R_r\right)_{S_n} I_{rn}^2 = 3R_r K_I^2 I_{1n}^2 \qquad (15.97)$$

With (15.87) and (15.35), we get

$$p_{Al} = 3 \cdot 0.709 \cdot 0.864^2 \cdot 9.303^2 = 137.417W$$

The mechanical/ventilation losses are considered as $p_{mv} = 0.03P_n$ for $p_1 = 1$, $0.012P_n$ for $p_1 = 2$, and $0.008P_n$ for $p_1 = 3,4$.

The stray losses p_{stray} have been dealt with in detail in Chapter 11. Here their standard value $p_{stray} = 0.01P_n$ is considered.

The core loss p_{iron} is made of fundamental p^1_{iron} and additional (harmonics) p^h_{iron} iron loss.

The fundamental core losses occur only in the teeth and back iron (p_{t1}, p_{y1}) of the stator as the rotor (slip) frequency is low ($f_2 < (3-4)$Hz).

The stator teeth fundamental losses (see Chapter 11) are:

$$p_{t1} \approx K_t p_{10}\left(\frac{f_1}{50}\right)^{1.3} B_{ts}^{1.7} G_{t1} \tag{15.98}$$

where p_{10} is the specific losses in W/Kg at 1.0 Tesla and 50 Hz (p_{10} = (2 – 3)W/Kg; it is a catalog data for the lamination manufacture). K_t = (1.6 – 1.8) accounts for core loss augmentation due to mechanical machining (stamping value depends on the quality of the material, sharpening of the cutting tools, etc.).

G_{t1} is the stator tooth weight,

$$\begin{gathered} G_{t1} = \gamma_{iron} \cdot N_s \cdot b_{ts} \cdot (h_s + h_w + h_{os}) \cdot L \cdot K_{Fe} = \\ = 7800 \cdot 36 \cdot 4.75 \cdot 10^{-3} \cdot (21.36 + 1.5 + 1) \cdot 10^{-3} \cdot 0.1315 \cdot 0.95 = 3.975\text{Kg} \end{gathered} \tag{15.99}$$

With B_{ts} = 1.55 T and f_1 = 60 Hz, from (15.98), p_{t1} is

$$p_{t1} = 1.7 \cdot 2 \cdot \left(\frac{60}{50}\right)^{1.3} \cdot 1.55^{1.7} \cdot 3.975 = 36.08\text{W}$$

In a similar way, the stator back iron (yoke) fundamental losses p_{y1} is

$$p_{y1} = K_y p_{10}\left(\frac{f_1}{50}\right)^{1.3} B_{cs}^{1.7} G_{y1} \tag{15.100}$$

Again, K_y = 1.6 – 1.9 takes care of the influence of mechanical machining and the yoke weight G_{y1} is

$$\begin{gathered} G_{y1} = \gamma_{iron} \frac{\pi}{4}\left[D_{out}^{\ 2} - (D_{out} - 2h_{cs})^2\right] \cdot L \cdot K_{Fe} = \\ = 7800\frac{\pi}{4}\left[0.19^2 - \left(0.19 - 2 \cdot 15.34 \cdot 10^{-3}\right)^2\right] \cdot 0.1315 \cdot 0.95 = 8.275\text{Kg} \end{gathered} \tag{15.101}$$

with K_y = 1.6, B_{cs} = 1.6T, p_{y1} from (15.100) is

$$p_{y1} = 1.6 \cdot 2 \cdot \left(\frac{60}{50}\right)^{1.3} \cdot 1.6^{1.7} \cdot 8.275 = 74.62\text{W}$$

So, the fundamental iron losses p^1_{iron} is

$$p^1_{iron} = p_{t1} + p_{y1} = 36.08 + 74.62 = 110.70\text{W} \tag{15.102}$$

The tooth flux pulsation core loss constitutes the main components of stray losses (Chapter 11) [1].

$$p_{iron}^{s} \approx 0.5\cdot 10^{-4}\left[\left(N_r\frac{f_1}{p_1}K_{ps}B_{ps}\right)^2 G_{ts} + \left(N_s\frac{f_1}{p_1}K_{pr}B_{pr}\right)^2 G_{tr}\right] \quad (15.103)$$

$$K_{ps} \approx \frac{1}{2.2 - B_{ts}} = \frac{1}{2.2 - 1.55} = 1.5385$$
$$K_{pr} = \frac{1}{2.2 - B_{tr}} = \frac{1}{2.2 - 1.6} = 1.666 \quad (15.104)$$

$$B_{ps} \approx (K_{c2} - 1)B_g = (1.059 - 1.0)\cdot 0.726 = 0.0428T \quad (15.105)$$

$$B_{pr} \approx (K_{c1} - 1)B_g = (1.144 - 1.0)\cdot 0.726 = 0.1045T \quad (15.106)$$

The rotor teeth weight G_{tr} is

$$G_{tr} = \gamma_{iron}\cdot L\cdot K_{Fe}\cdot N_r\cdot\left(h_r + \frac{d_1 + d_2}{2}\right)\cdot b_{tr} =$$
$$= 7800\cdot 0.1315\cdot 0.95\cdot 28\cdot\left(20 + \frac{5.7 + 1.2}{2}\right)\cdot 10^{-3}\cdot 5.88\cdot 10^{-3} = 3.710Kg \quad (15.107)$$

Now, from (15.103),

$$p_{iron}^{s} = 0.5\cdot 10^{-4}\left[\left(28\cdot\frac{60}{2}\cdot 1.538\cdot 0.0429\right)^2 3.975 + \left(36\cdot\frac{60}{2}\cdot 1.666\cdot 0.1045\right)^2 3.710\right] =$$
$$= 7.2W$$

The total core loss p_{iron} is

$$p_{iron} = p_{iron}^{1} + p_{iron}^{s} = 110.70 + 7.2 = 117.90W \quad (15.108)$$

Total losses (15.95) is

$$\sum losses = 243.215 + 137.417 + 117.90 + 2.2\cdot 10^{-2}\cdot 5500 = 609.6W$$

The efficiency η_n (from 15.87) becomes

$$\eta_n = \frac{5500}{5500 + 609.6} = 0.9002!$$

The targeted efficiency was 0.895, so the design holds. If the efficiency had been smaller than the target value, the design should have returned to square one by adopting a larger stator bore diameter D_{is}, then smaller current densities, etc. A larger size machine would have been obtained, in general.

15.11. OPERATION CHARACTERISTICS

The operation characteristics are defined here as active no load current I_{0a}, rated slip S_n, rated torque T_n, breakdown slip and torque S_k, T_{bk}, current Is and power factor versus slip, starting current, and torque I_{LR}, T_{LR}.

The no load active current is given by the no load losses

$$I_{0a} = \frac{p_{iron} + p_{mv} + 3 \cdot I_\mu^2 \cdot R_s}{3V_{ph}} = \frac{117.9 + 56.1 + 3 \cdot 3.86^2 \cdot 0.936}{3\left(460/\sqrt{3}\right)} = 0.271\text{A} \quad (15.109)$$

The rated slip S_n is

$$S_n = \frac{p_{Al}}{P_N + p_{Al} + p_{mv} + p_{stray}} = \frac{137.417}{5500 + 137.417 + 56.1 + 27.5} = 0.024! \quad (15.110)$$

The rated shaft torque T_n is

$$T_n = \frac{P_n}{2\pi \frac{f_1}{p_1}\left(1 - S_n\right)} = \frac{5500}{2\pi \frac{60}{2}\left(1 - 0.024\right)} = 29.91\text{Nm} \quad (15.111)$$

The approximate expressions of torque versus slip (Chapter 7) are

$$T_e = \frac{3p_1}{\omega_1} \frac{V_{ph}^2 \frac{R_r}{S}}{\left(R_s + C_m \frac{R_r}{S}\right)^2 + \left(X_{sl} + C_m X_{rl}\right)^2} \quad (15.112)$$

with

$$C_m \approx 1 + \frac{X_{sl}}{X_m} = 1 + \frac{2.17}{66.4} = 1.0327 \quad (15.113)$$

From (15.112), the breakdown torque T_{bk} is

$$T_{bk} = \frac{3p_1}{2\omega_1} \frac{V_{ph}^2}{\left[R_s + \sqrt{R_s^2 + \left(X_{sl} + C_1 X_{rl}\right)^2}\right]} = \frac{3 \cdot 2}{2 \cdot 2\pi 60} \frac{\left(460/\sqrt{3}\right)^2}{\left[0.936 + \sqrt{0.936^2 + \left(2.17 + 4.256\right)^2}\right]} = 75.48\text{Nm} \quad (15.114)$$

The starting current I_{LR} is

$$I_{LR} = \frac{V_{ph}}{\sqrt{\left(R_s + R_r^{S=1}\right)^2 + \left(X_{sl}^{S=1} + X_{rl}^{S=1}\right)^2}} = \frac{460/\sqrt{3}}{\sqrt{(0.936+1.1295)^2 + (1.6275+3.165)^2}} = 54.44\text{A} \quad (15.115)$$

The starting torque T_{LR} may now be computed as

$$T_{LR} = \frac{3R_r^{S=1}{I_{LR}}^2}{\omega_1} P_1 = \frac{3 \cdot 1.1295 \cdot 54.44^2 \cdot 2}{2\pi 60} = 53.305\text{Nm} \quad (15.116)$$

In the specifications, the rated power factor $\cos\varphi_{1n}$, T_{bk}/T_{en}, T_{LR}/T_{en} and I_{LR}/I_{1n} are given. So we have to check the final design values of these constraints.

$$\cos\varphi_{1n} = \frac{P_n}{3V_{ph}I_{1n}\eta_n} = \frac{5500}{3\frac{460}{\sqrt{3}}9.303 \cdot 0.90} = 0.825 \approx 0.83 \quad (15.117)$$

$$t_{bk} = \frac{T_{bk}}{T_{en}} = \frac{75.48}{29.91} = 2.523 \approx 2.5 \quad (15.118)$$

$$t_{LR} = \frac{T_{LR}}{T_{en}} = \frac{53.305}{29.91} = 1.7828 \approx 1.75 \quad (15.119)$$

$$i_{LR} = \frac{I_{LR}}{I_{1n}} = \frac{54.44}{9.303} = 5.85 \approx 6.0 \quad (15.120)$$

Apparently the design does not need any iterations. This is pure coincidence "combined" with standard specifications. Higher breakdown or starting ratios t_{bk}, t_{LR} would, for example, need lower rotor leakage inductance and higher rotor resistance as influenced by skin effect. A larger stator bore diameter is again required. In general, it is not easy to make a few changes to get the desired operation characteristics. Here the optimization design methods come into play.

15.12. TEMPERATURE RISE

Any electromagnetic design has to be thermally valid (Chapter 12). Only a coarse verification of temperature rise is given here.

First the temperature differential between the conductors in slots and the slot wall $\Delta\theta_{co}$ is calculated.

$$\Delta\theta_{co} \approx \frac{P_{co}}{\alpha_{cond} A_{ls}} \quad (15.121)$$

Then, the frame temperature rise $\Delta\theta_{frame}$ with respect to ambient air is determined.

$$\Delta\theta_{frame} = \frac{\sum losses}{\alpha_{cond} A_{frame}} \tag{15.122}$$

For IMs with selfventilators placed outside the motor (below 100kW).

$$\alpha_{conv}\left(W/m^2K\right) = \begin{cases} 60; \text{ for } 2p_1 = 2; \\ 50; \text{ for } 2p_1 = 4; \\ 40; \text{ for } 2p_1 = 6; \\ 32; \text{ for } 2p_1 = 8; \end{cases} \tag{15.123}$$

More precise values are given in Chapter 12.

The slot insulation conductivity plus its thickness lumped into α_{cond},

$$\alpha_{cond} = \frac{\lambda_{ins}}{h_{ins}} = \frac{0.25}{0.3 \cdot 10^{-3}} = 833 W/m^2K \tag{15.124}$$

λ_{ins} is the insulation thermal conductivity in (W/m^{0}K) and h_{ins} total insulation thickness from the slot middle to teeth wall.

The stator slot lateral area A_{ls} is

$$A_{ls} \approx (2h_s + b_{s2}) \cdot L \cdot N_s = (2 \cdot 21.36 + 9.16) \cdot 10^{-3} \cdot 0.1315 \cdot 36 = 0.2456 m^2 \tag{15.125}$$

The frame area S_{frame} (including the finns area) is

$$A_{frame} = \pi D_{out}(L + \tau) \cdot K_{fin} = \pi \cdot 190(0.1315 + 0.0876) \cdot 3.0 = 0.392 m^2 \tag{15.126}$$

Now, from (15.121)

$$\Delta\theta_{co} = \frac{234.215}{833 \cdot 0.2456} = 1.18^0 C$$

and from (15.122),

$$\Delta\theta_{frame} = \frac{609.6}{50 \cdot 0.392} = 31.10^0 C$$

Suppose that the ambient temperature is $\theta_{amb} = 40^0$C.

In this case, the winding temperature is

$$\theta_{Co} = \theta_{amb} + \Delta\theta_{Co} + \Delta\theta_{frame} = 40 + 1.18 + 31.10 = 72.28^0 C < 80^0 C \tag{15.127}$$

For this particular design, the K_{fin} = 3.0, which represents the frame area multiplication by fins, provided to increase heat transfer, may be somewhat reduced, especially if the ambient temperature would be 20^{0}C as usual.

15.13. SUMMARY

- 100 kW is, in general, considered the border between low and medium power induction machines.
- Low power IMs use a single stator and rotor stack and a finned frame cooled by an external ventilator mounted on the motor shaft.
- Low power IMs use a rotor cage made of cast aluminum and a random wound coil stator winding made of round (in general) magnetic wire with 1 to 6 elementary conductors in parallel and 1 to 3 current paths in parallel.
- The number of poles is $2p_1 = 2, 4, 6, 8, 10$, in general.
- In designing an IM, we understand sizing of the IM; the breakdown torque, starting torque, and current are essential among design specifications.
- The design algorithm contains 9 main stages: design specs, sizing the electric and magnetic circuits, adjustment of sizing data, verification of electric and magnetic loading (with eventual return to step one), computation of magnetization current, of equivalent circuit parameters, of losses, rated slip and efficiency, of power factor, temperature rise, and performance check with eventual returns to step one with adjusted electric and magnetic loadings.
- With values assigned to efficiency and power factor, based on past experience (Esson's output coefficient), the stator bore diameter D_{is} is calculated after assigning a value (dependent on and increasing with the number of poles) for λ = stack length/pole pitch = 0.6 – 3.
- The airgap should be small to increase the power factor, but not too small to avoid too high stray losses.
- Double layer, chorded coils stator windings are used in most cases.
- The essence of slot design is the design current density, which depends on the IM design letter A, B, C, D, E, type of cooling, power level, and number of poles. Values in the range of 4 to 8 A/mm^2 are typical below 100 kW.
- Semiclosed trapezoidal (rounded) stator and rotor slot with rectangular teeth are most adequate below 100 kW.
- A key factor in the design is the teeth saturation factor $1 + K_{st}$ which is assigned an initial value that has to be met after a few design iterations.
- There , are special stator/rotor slot number combinations to avoid synchronous parasitic torques, radial forces, noise, and vibration. They are given as Table 15.5.
- An essential design variable is the ratio K_D = outer/inner stator diameter. Its recommended value intervals increase with the number of poles, based on past experience.
- When the lamination cross section is designed to produce the maximum airgap flux density for a given rated current density in the stator [3], the ratio K_D is

$2p_1$	2	4	6	8
K_D	1.58	0.65	0.69	0.72

Using values around these would most probably yield practical designs.

- Care must be exercised to correct the rotor resistance and leakage reactances for skin effect and leakage magnetic saturation at zero speed (S = 1). Otherwise, both starting torque and starting current contain notable errors.
- No electromagnetic design is practical unless the temperature rise corresponds to the winding insulation class. Detailed thermal design methodologies are presented in Chapter 12.
- Design optimization methods will be introduced in Chapter 18.

15.14. REFERENCES

1. K. Vogt, Electrical Machines. Design of Rotary Electric Machines, Fourth edition (in German), Chapter 16, VEB Verlag Technik Berlin, 1988.
2. G. Madescu, I. Boldea, T.J.E. Miller, An Analytical Iterative Model (AIM) for Induction Motor Design", Record of IEEE – IAS – 1996 Annual Meeting, Vol.1, pp.566 – 573.
3. G. Madescu, I. Boldea, T.J.E. Miller, The Optimal Lamination Approach (OLA) for Induction Motor Design, IEEE Trans Vol.IA – 34, No.2, 1998, pp.1 – 8.

Chapter 16

INDUCTION MOTOR DESIGN ABOVE 100KW AND CONSTANT V/f

16.1 INTRODUCTION

Induction motors above 100 kW are built for low voltage (480 V/50 Hz, 460 V/60 Hz, 690V/50Hz) or higher voltages, 2.4 kV to 6 kV and 12 kV in special cases.

The advent of power electronic converters, especially those using IGBTs, caused the increase of power/unit limit for low voltage IMs, 400V/50Hz to 690V/60Hz, to more than 2 MW. Although we are interested here in constant V/f fed IMs, this trend has to be observed.

High voltage, for given power, means lower cross section easier to wind stator windings. It also means lower cross section feeding cables. However, it means thicker insulation in slots, etc. and thus a low slot-fill factor; and a slightly larger size machine. Also, a high voltage power switch tends to be costly. Insulated coils are used. Radial – axial cooling is typical, so radial ventilation channels are provided. In contrast, low voltage IMs above 100 kW are easy to build, especially with round conductor coils (a few conductors in parallel with copper diameter below 3.0 mm) and, as power goes up, with more than one current path, $a_1 > 1$. This is feasible when the number of poles increases with power: for $2p_1 = 6, 8, 10, 12$. If $2p_1 = 2, 4$ as power goes up, the current goes up and preformed coils made of stranded rectangular conductors, eventually with 1 to 2 turns/coil only, are required. Rigid coils are used and slot insulation is provided.

Axial cooling, finned-frame, unistack configuration low-voltage IMs have been recently introduced up to 2.2 MW for low voltages (690V/60Hz and less).

Most IMs are built with cage rotors but, for heavy starting or limited speed-control applications, wound rotors are used.

To cover most of these practical cases, we will unfold a design methodology treating the case of the same machine with: high voltage stator and a low voltage stator, and deep bar cage rotor, double cage rotor, and wound rotor, respectively.

The electromagnetic design algorithm is similar to that applied below 100 kW. However the slot shape and stator coil shape, insulation arrangements, and parameters expressions accounting for saturation and skin effect are slightly, or more, different with the three types of rotors.

Knowledge in Chapters 9 and 11 on skin and saturation effects, respectively, and for stray losses is directly applied throughout the design algorithm.

The deep bar and double-cage rotors will be designed based on fulfilment of breakdown torque and starting torque and current, to reduce drastically the number of iterations required. Even when optimization design is completed, the latter will be much less time consuming, as the "initial" design is meeting approximately the main constraints. Unusually high breakdown/rated torque ratios ($t_{be} = T_{bk}/T_{en} > 2.5$) are to be approached with open stator slots and larger l_i/τ ratios to obtain low stator leakage inductance values.

$$T_{bk} \approx \frac{3p_1}{2}\left(\frac{V_{ph}}{\omega_1}\right)^2 \frac{1}{L_{sc}}; \quad L_{sc} = L_{ls} + L_{lr} \tag{16.1}$$

where L_{sl} is the stator leakage and L_{lr} is the rotor leakage inductance at breakdown torque. It may be argued that, in reality, the current at breakdown torque is rather large ($I_k/I_{1n} \geq T_{bk}/T_{en}$) and thus both leakage flux paths saturate notably and, consequently, both leakage inductances are somewhat reduced by 10 to 15%. While this is true, it only means that ignoring the phenomenon in (16.1) will yield conservative (safe) results.

The starting torque T_{LR} and current I_{LR} are

$$T_{LR} \approx \frac{3(R_r)_{S=1} K^2_{istart} I_{LR}^2 p_1}{\omega_1} \tag{16.2}$$

$$I_{LR} \approx \frac{V_{1ph}}{\sqrt{\left(R_s + (R_r)_{S=1}\right)^2 + \omega_1^2\left((L_{sl})_{S=1} + (L_{rl})_{S=1}\right)^2}} \tag{16.3}$$

In general, $K_{istart} = 0.9 - 0.975$ for powers above 100 kW. Once the stator design, based on rated performance requirements, is done, with R_s and L_{sl} known, Equations (16.1) through (16.3) yield unique values for $(R_r)_{S=1}$, $(L_{rl})_{S=1}^{sat}$ and $(L_{rl})_{S=S_n}$. For a targeted efficiency with the stator design done and core loss calculated, the rotor resistance at rated power (slip) may be calculated approximately,

$$(R_r)_{S=S_n} = \left(\frac{P_n}{\eta_n} - 3R_s I_{1n}^2 - p_{iron} - p_{stray} - p_{mec}\right)\frac{1}{3(K_i I_{1n})^2} \tag{16.4}$$

with

$$K_i = \frac{(I_r)_{S=S_n}}{I_{1n}} \approx 0.8\cos\varphi_{1n} + 0.2; \quad I_{1n} = \frac{P_n}{\sqrt{3}V_{1n}\cos\varphi_{1n}\eta_n} \tag{16.5}$$

We may assume that rotor bar resistance and leakage inductance at S = 1 represent 0.80 to 0.95 of their values calculated from (16.1 through 16.4).

$$(R_{be})_{S=1} = (0.85-0.95)\frac{(R_r)_{S=1}}{K_{bs}};\; K_{bs} = \frac{4m(W_1K_{W1})^2}{N_r} \tag{16.6}$$

$$(L_{be})_{S=1} = (0.75-0.80)\frac{(L_{rl})_{S=1}^{sat}}{K_{bs}} \tag{16.7}$$

Their values for rated slip are

$$(R_{be})_{S=S_n} = (0.7-0.85)\frac{(R_r)_{S=S_n}}{K_{bs}} \tag{16.8}$$

$$(L_{be})_{S=S_n} = (0.8-0.85)\frac{(L_{rl})_{S=S_n}}{K_{bs}} \tag{16.9}$$

With rectangular semiclosed rotor slots, the skin effect K_R and K_x coefficients are

$$K_R = \frac{(R_{be})_{S=1}}{(R_{be})_{S=S_n}} \approx \xi = \beta_{Skin}h_r;\; \beta_{Skin} = \sqrt{\frac{\pi f_1 \mu_0}{\rho_{Al}}} \tag{16.10}$$

$$\frac{(L_{be})_{S=1}^{unsat}}{(L_{be})_{S=S_n}} \approx \frac{\frac{h_r}{3b_r}K_x + \frac{h_{or}}{b'_{or}}}{\frac{h_r}{3b_r} + \frac{h_{or}}{b_{or}}} \tag{16.11}$$

Apparently, by assigning a value for h_{or}/b_{or}, Equation (16.11) allows us to calculate b_r because

$$K_x \approx \frac{3}{2\beta_{skin}h_r} \tag{16.12}$$

Now the bar cross section for given rotor current density j_{AL}, ($A_b = h_r \cdot b_r$) is

$$A_b = \frac{I_b}{j_{Al}} = \frac{K_i I_{1n}}{K_{bi} j_{AL}};\; K_{bi} = \frac{2m_1 W_1 K_{w1}}{N_r} \tag{16.13}$$

If A_b from (16.13) is too far away from $h_r \cdot b_r$, a more complex than rectangular slot shape is to be looked for to satisfy the values of K_R and K_X calculated from (16.10 and 16.11).

It should be noted that the rotor leakage inductance has also a differential component which has not been considered in (16.9) and (16.11).

Consequently, the above rationale is merely a basis for a closer-to-target rotor design from the point of view of breakdown, starting torques, and starting current.

A similar approach may be taken for the double cage rotor, but to separate the effects of the two cages, the starting and rated power conditions are taken to design the starting and working cage, respectively.

16.2 HIGH VOLTAGE STATOR DESIGN

To save space, the design methodology will be unfolded simultaneously with a numerical example of an IM with the following specifications:

- $P_n = 736$ kW (1000HP)
- Targeted efficiency: 0.96
- $V_{1n} = 4$ kV (Δ)
- $f_1 = 60$ Hz, $2p_1 = 4$ poles, m = 3phases;

Service: Si_1 continuous, insulation class F, temperature rise for class B (maximum 80 K).

The rotor will be designed separately for three cases: deep bar cage, double cage, and wound rotor configurations.

➢ Main stator dimensions

As we are going to again use Esson's constant (Chapter 14), we need the apparent airgap power S_{gap}.

$$S_{gap} = 3EI_{1n} = 3K_E V_{1ph} I_{1n} \tag{16.14}$$

with
$$K_E = 0.98 - 0.005 \cdot p_1 = 0.98 - 0.005 \cdot 2 = 0.97. \tag{16.15}$$

The rated current I_{1n} is

$$I_{1n} = \frac{P_n}{\sqrt{3} V_{1n} \cos\varphi_n \eta_n} \tag{16.16}$$

To find I_{1n}, we need to assign target values to rated efficiency η_n and power factor $\cos\varphi_n$, based on past experience and design objectives.

Although the design literature uses graphs of η_n, $\cos\varphi_n$ versus power and number of pole pairs p_1, continuous progress in materials and technologies makes the η_n graphs quickly obsolete. However, the power factor data tend to be less dependent on material properties and more dependent on airgap/pole pitch ratio and on the leakage/magnetization inductance ratio (L_{sc}/L_m) as

$$(\cos\varphi)_{\substack{max \\ zero\ loss}} \approx \frac{1 - \frac{L_{sc}}{L_m}}{1 + \frac{L_{sc}}{L_m}} \tag{16.17}$$

Because L_{sc}/L_m ratio increases with the number of poles, the power factor decreases with the number of poles increasing. Also, as the power goes up, the ratio L_{sc}/L_m goes down, for given $2p_1$ and $\cos\varphi_n$ increases with power.

Furthermore, for high breakdown torque, L_{sc} has to be small as the maximum power factor increases. Adopting a rated power factor is not easy. Data of Figure 16.1 are to be taken as purely orientative.

Corroborating (16.1) with (16.17), for given breakdown torque, the maximum ideal power factor $(\cos\varphi)_{max}$ may be obtained.

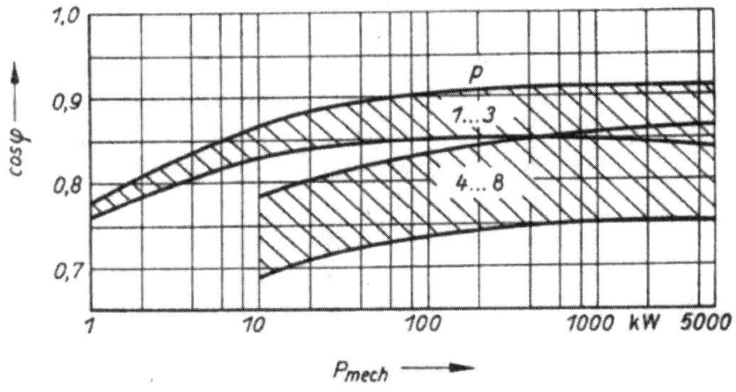

Figure 16.1 Typical power factor of cage rotor IMs

For our case $\cos\varphi_n = 0.92 - 0.93$.

Rated efficiency may be purely assigned a desired, though realistic, value. Higher values are typical for high efficiency motors. However, for $2p_1 < 8$, and $P_n > 100$ kW the efficiency is above 0.9 and goes up to more than 0.95 for $P_n > 2000$ kW. For high efficiency motors, efficiency at 2000 kW goes as high as 0.98 with recent designs.

With $\eta_n = 0.96$ and $\cos\varphi_n = 0.92$, the rated phase current I_{1nf} (16.16) is

$$I_{1nf} = \frac{736 \cdot 10^{-3}}{3 \cdot 4 \cdot 10^{3} \cdot 0.92 \cdot 0.96} = \frac{120.42}{\sqrt{3}} \mathrm{A}$$

From (16.14), the airgap apparent power S_{gap} becomes

$$S_{gap} = \sqrt{3} \cdot 0.97 \cdot 4000 \cdot 120.42 = 808.307 \cdot 10^{3} \mathrm{VA}$$

➢ Stator main dimensions

The stator bore diameter D_{is} may be determined from Equation (15.1) of Chapter 15, making use of Esson's constant,

$$D_{is} = \sqrt[3]{\frac{2p_1}{\pi\lambda_1} \frac{p_1}{f_1} \frac{S_{gap}}{C_0}} \tag{16.18}$$

From Figure 14.14 (Chapter 14), $C_0 = 265 \cdot 10^{3} \mathrm{J/m^3}$, $\lambda = 1.1$ = stack length/pole pitch (Table 15.1, Chapter 15) with (16.18), D_{is} is

$$D_{is} = \sqrt[3]{\frac{2\cdot 2}{\pi \cdot 1.1}\frac{2}{60}\frac{808.307\cdot 10^3}{265\cdot 10^3}} = 0.4\text{m}$$

The airgap is chosen at $g = 1.5\cdot 10^{-3}$ m as a compromise between mechanical constraints and limitation of surface and tooth flux pulsation core losses.

The stack length l_i is

$$l_i = \lambda\tau = \lambda\cdot\frac{\pi D_{is}}{2p_1} = 1.1\frac{\pi\cdot 0.49}{2\cdot 2} = 0.423\text{m} \quad (16.19)$$

➢ Core construction

Traditionally the core is divided between a few elementary ones with radial ventilation channels between. Such a configuration is typical for radial-axial cooling (Figure 16.2). [1]

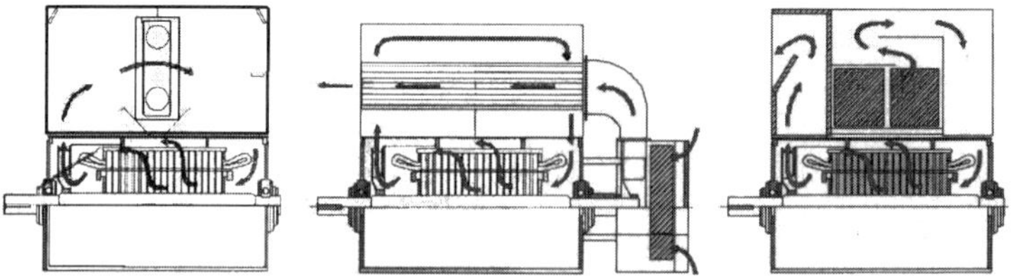

Figure 16.2 Divided core with radial-axial air cooling (source ABB)

Recently the unistack core concept, rather standard for low power (below 100 kW), has been extended up to more than 2000 kW both for high and low voltage stator IMs. In this case axial aircooling of the finned motor frame is provided by a ventilator on the motor shaft, outside bearings (Figure 16.3). [2]

As both concepts are in use and as, in Chapter 15, the unistack case has been considered, the divided stack configuration will be considered here for a high voltage stator case.

The outer/inner stator diameter ratio intervals have been recommended in Chapter 15, Table 15.2. For $2p_1 = 4$, let us consider $K_D = 0.63$.

Consequently, the outer stator diameter D_{out} is

$$D_{out} = \frac{D_{is}}{K_D} = \frac{0.49}{0.63} = 0.777\text{m} \approx 780\text{mm} \quad (16.20)$$

Figure 16.3 Unistack with axial air cooling (source, ABB)

The airgap flux density is taken as B_g = 0.8 T. From Equation (14.14) (Chapter 14), C_0 is

$$C_o = K_B\alpha_i K_{w1}\pi A_1 B_g 2p_1 \tag{16.21}$$

Assuming a tooth saturation factor $(1 + K_{st}) = 1.25$, from Figure 14.13 Chapter 14, $K_B = 1.1$, $\alpha_i = 0.69$. The winding factor is given a value $K_{w1} \approx 0.925$. With Bg = 0.8T, $2p_1 = 4$, and $C_0 = 265\cdot103 J/m^3$, the stator rated current sheet A_1 is

$$A_1 = \frac{265\cdot10^3}{1.1\cdot0.69\cdot0.925\cdot\pi\cdot0.8\cdot4} = 37.565\cdot10^3 \text{Aturns/m}$$

This is a moderate value.

The pole flux ϕ is

$$\phi = \alpha_i\tau l_i B_g; \quad \tau = \frac{\pi D_{is}}{2p_1} \tag{16.22}$$

$$\tau = \frac{\pi\cdot0.49}{2\cdot2} = 0.314\text{m}; \quad \phi = 0.69\cdot0.314\cdot0.423\cdot0.8 = 0.0733\text{Wb}$$

The number of turns per phase W_1 (a_1 = 1 current paths) is

$$W_1 = \frac{K_E V_{ph}}{4K_B f_1 K_{w1}\phi} = \frac{0.97\cdot4000}{4\cdot1.1\cdot60\cdot0.925\cdot0.0733} = 207.8 \tag{16.23}$$

The number of conductors per slot n_s is written as

$$n_s = \frac{2m_1a_1W_1}{N_s} \tag{16.24}$$

The number of stator slots, N_s, for $2p_1 = 4$ and $q = 6$, becomes

$$N_s = 2p_1q_1m_1 = 2\cdot 2\cdot 6\cdot 3 = 72 \tag{16.25}$$

So

$$n_s = \frac{2\cdot 3\cdot 1\cdot 207.8}{72} = 17.31$$

We choose $n_s = 18$ conductors/slot, but we have to decrease the ideal stack length l_i to

$$l_i = l_i\cdot\frac{17.31}{18} = 0.423\cdot\frac{17.31}{18} \approx 0.406\text{m}$$

The flux per pole $\phi = \phi\cdot\frac{18}{17.31} = 0.07049\text{W}$

The airgap flux density remains unchanged ($B_g = 0.8$ T).

As the ideal stack length l_i is final (provided the teeth saturation factor K_{st} is confirmed later on), the former may be divided into a few parts.

Let us consider $n_{ch} = 6$ radial channels, each 10^{-2}m wide ($b_{ch} = 10^{-2}$m). Due to axial flux fringing its equivalent width $b_{ch}' \approx 0.75b_{ch} = 7.5\cdot 10^{-3}$m ($g = 1.5$ mm). So the total geometrical length L_{geo} is

$$L_{geo} = l_i + n_{ch}b_{ch}' = 0.406 + 6\cdot 0.0075 = 0.451\text{m} \tag{16.26}$$

On the other hand, the length of each elementary stack is

$$l_s = \frac{L_{geo} - n_{ch}b_{ch}}{n_{ch}+1} = \frac{0.451 - 6\cdot 0.01}{6+1} \approx 0.056\text{m} \tag{16.27}$$

As lamination are 0.5 mm thick, the number of laminations required to make l_s is easy to match. So there are 7 stacks each 56 mm long (axially).

➢ The stator winding

For high voltage IMs, the winding is made of form-wound (rigid) coils. The slots are open in the stator so that the coils may be introduced in slots after prefabrication (Figure 16.4). The number of slots per pole/phase q_1 is to be chosen rather large as the slots are open and the airgap is only $g = 1.5\cdot 10^{-3}$m.

The stator slot pitch τ_s is

$$\tau_s = \frac{\pi D_{is}}{N_s} = \frac{\pi\cdot 0.49}{72} = 0.02137\text{m} \tag{16.28}$$

The coil throw is taken as $y/\tau = 15/18 = 5/6$ ($q_1 = 6$). There are 18 slots per pole to reduce drastically the 5th mmf space harmonic.

-- MICADUR - Compact Industry Insulation System (MCI)

-- voidless insulation construction

-- tight fit in the slots

-- free of partial discharges

-- high thermal conductivity

-- good moisture resistance

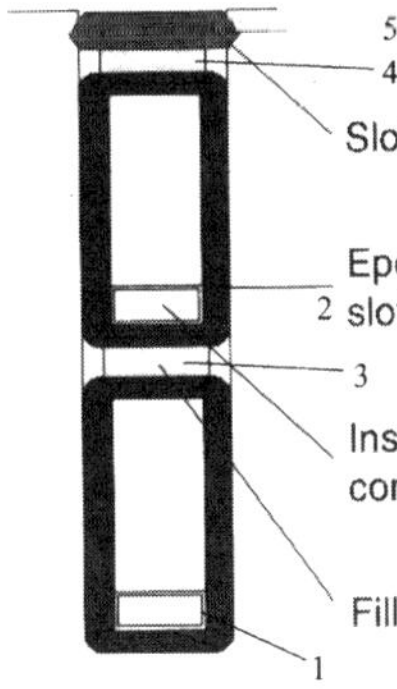

Figure 16.4 Open stator slot for high voltage winding with form-wound (rigid) coils

The winding factor K_{w1} is

$$K_{w1} = \frac{\sin\frac{\pi}{6}}{6\cdot\sin\frac{\pi}{6\cdot 6}}\sin\frac{\pi}{2}\frac{5}{6} = 0.9235$$

The winding is fully symmetric with N_s/m_1a_1 = 24 (integer), $2p_1/a_1$ = 4/1 (integer). Also, t = g.c.d(N_s,p_1) = p_1 = 2, and N_s/m_1t = 72/(3·2) = 12 (integer).

The conductor cross section A_{Co} is (delta connection)

$$A_{Co} = \frac{I_{1nf}}{a_1 J_{Co}};\ a_1 = 1,\ J_{Co} = 6.3\text{A}/\text{mm}^2;\ I_{1nf} = 69.36 \tag{16.29}$$

$$A_{Co} = \frac{120.42}{1\cdot 6.3\sqrt{3}} = 11.048\text{mm}^2 = a_c \cdot b_c$$

A rectangular cross section conductor will be used. The rectangular slot width b_s is

$$b_s = \tau_s \cdot (0.36 \div 0.5) = 0.021375 \cdot (0.36 \div 0.5) = 7.7 \div 10.7\text{mm} \tag{16.30}$$

Before choosing the slot width, it is useful to discuss the various insulation layers (Table 16.1).

The available conductor width in slot a_c is

$$a_c = b_s - b_{ins} = 10.0 - 4.4 = 5.6\text{mm} \tag{16.31}$$

Table 16.1 Stator slot insulation at 4kV

Figure 16.4	Denomination	thickness (mm)	
		tangential	radial
1	conductor insulation (both sides)	$1\cdot 04 = 0.4$	$18\cdot 0.4 = 7.2$
2	epoxy mica coil and slot insulation	4	$4\cdot 2 = 8.0$
3	interlayer insulation	-	$2\cdot 1 = 2$
4	wedge	-	$1\cdot 4 = 4$
	Total	$b_{ins} = 4.4$	$h_{ins} = 21.2$

This is a standardised value and it was considered when adopting $b_s = 10$ mm (16.30). From (16.19), the conductor height b_c becomes

$$b_c = \frac{A_{Co}}{a_c} = \frac{11.048}{5.6} \approx 2\text{mm} \tag{16.32}$$

So the conductor size is 2×5.6 mm×mm.

The slot height h_s is written as

$$h_s = h_{ins} + n_s b_c = 21.2 + 18\cdot 2 = 57.2\text{mm} \tag{16.33}$$

Now the back iron radial thickness h_{cs} is

$$h_{cs} = \frac{D_{out} - D_{is}}{2} - h_s = \frac{780 - 490}{2} - 57.2 = 87.8\text{mm} \tag{16.34}$$

The back iron flux density B_{cs} is

$$B_{cs} = \frac{\phi}{2l_i h_{cs}} = \frac{0.07049}{2\cdot 0.406\cdot 0.0878} = 0.988\text{T} \tag{16.35}$$

This value is too small so we may reduce the outer diameter to a lower value: $D_{out} = 730$ mm; the back core flux density will now be close to 1.4T.

The maximum tooth flux density B_{tmax} is:

$$B_{t\max} = \frac{\tau_s B_g}{\tau_s - b_s} = \frac{21.37\cdot 0.8}{21.37 - 10} = 1.5\text{T} \tag{16.36}$$

This is acceptable though even higher values (up to 1.8 T) are used as the tooth gets wider and the average tooth flux density will be notably lower than B_{tmax}.

The stator design is now complete but it is not definitive. After the rotor is designed, performance is computed. Design iterations may be required, at least to converge K_{st} (teeth saturation factor), if not for observing various constraints (related to performance or temperature rise).

16.3 LOW VOLTAGE STATOR DESIGN

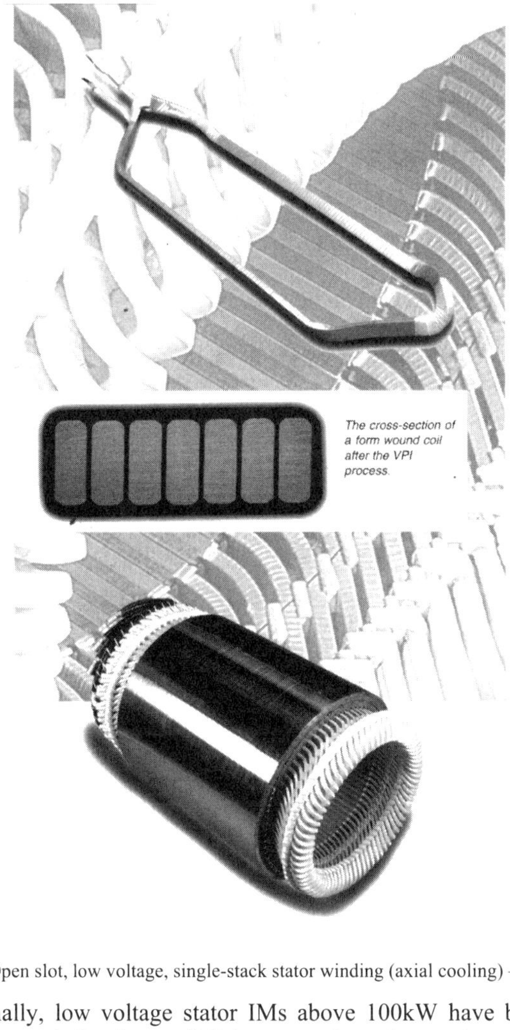
The cross-section of
a form wound coil
after the VPI
process.

Figure 16.5 Open slot, low voltage, single-stack stator winding (axial cooling) – (source, ABB)

Traditionally, low voltage stator IMs above 100kW have been built with round conductors (a few in parallel) in cases where the number of poles is large so that many current paths in parallel are feasible ($a_1 = p_1$).

Recent extension of variable speed IM drives tends to lead to the conclusion that low voltage IMs up to 2000 kW and more at 690V/60Hz (660V, 50Hz) or at (460V/50Hz, 400V/50Hz) are to be designed for constant V and f, but having in view the possibility of being used in general purpose variable speed drives with voltage source PWM IGBT converter supplies. To this end, the machine insulation is enforced by using almost exclusively form-wound coils and open stator slots as for high voltage IMs. Also insulated bearings are used to reduce bearing stray currents from PWM converters (at switching frequency).

Low voltage PWM converters are a costly advantage. Also, a single stator stack is used (Figure 16.5).

The form-wound (rigid) coils (Figure 16.5) have a small number of turns (conductors) and a kind of crude transposition occurs in the end-connections zone to reduce the skin effect. For large powers and $2p_1 = 2, 4$, even 2 – 3 elementary conductors in parallel and $a_1 = 2, 4$ current paths may be used to keep the elementary conductors within a size with limited skin effect (Chapter 9).

In any case, skin effect calculations are required as the power goes up and the conductor cross section follows path. For a few elementary conductors or current path in parallel, additional (circulating current) losses occur as detailed in Chapter 9 (paragraphs 9.2 and 9.3).

Aside from these small differences, the stator design follows the same path as high voltage stators.

This is why it will not be further treated here.

16.4 DEEP BAR CAGE ROTOR DESIGN

We will now resume the design methodology in paragraph 16.2 with the deep bar cage rotor design. More design specifications are needed for the deep bar cage.

$$\frac{\text{breakdown torque}}{\text{rated torque}} = \frac{T_{bk}}{T_{en}} = 2.7$$

$$\frac{\text{starting current}}{\text{rated current}} = \frac{I_{LR}}{I_n} \leq 6.1$$

$$\frac{\text{starting torque}}{\text{rated torque}} = \frac{T_{LR}}{T_{en}} = 1.2$$

The above data are merely an example.

As shown in paragraph 16.1, in order to size the deep bar cage, stator leakage reactance X_{sl} is required. As the stator design is done, X_{sl} may be calculated.

➢ Stator leakage reactance X_{sl}

As documented in Chapter 9, the stator leakage reactance X_{sl} may be written as

$$X_{sl} = 15.8\left(\frac{f_1}{100}\right)\left(\frac{W_1}{100}\right)^2 \frac{l_i}{p_1 q_1}\sum\lambda_{is} \tag{16.37}$$

where $\sum\lambda_{is}$ is the sum of the leakage slot (λ_{ss}), differential (λ_{ds}), and end connection (λ_{fs}) geometrical permeance coefficients.

$$\lambda_{ss} = \frac{(h_{s1} - h_{s3})}{3b_s} K_\beta + \frac{h_{s2}}{b_s} K_\beta + \frac{h_{s3}}{4b_s}$$
$$K_\beta = \frac{1 + 3\beta}{4}; \ \beta = \frac{y}{\tau} = \frac{5}{6} \quad (16.38)$$

In our case, (see Table 16.1 and Figure 16.6).

$$h_{s1} = n_s \cdot b_c + n_s \cdot 0.4 + 2 \cdot 2 + 1 = 18 \cdot 2 + 18 \cdot 0.4 + 2 \cdot 2 + 1 = 48.2\text{mm} \quad (16.39)$$

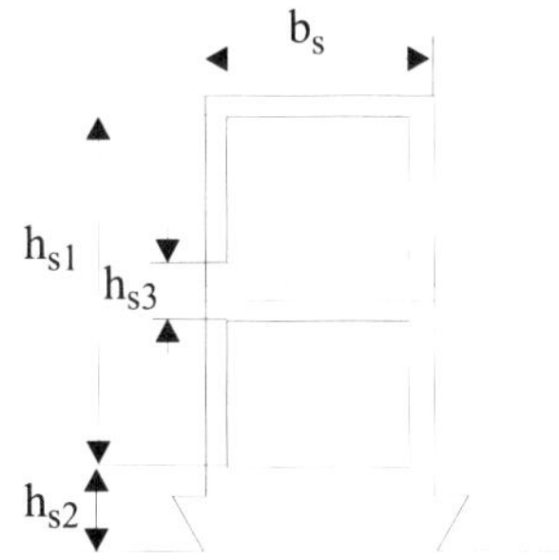

Figure 16.6 Stator slot geometry

Also, $h_{s3} = 2{\cdot}2 + 1 = 5$ mm, $h_{s2} = 1{\cdot}2 + 4 = 6$ mm. From (16.37),

$$\lambda_{ss} = \left(\frac{48.2 - 5}{3 \cdot 10} + \frac{6}{10}\right)\left(\frac{1 + 3\frac{5}{6}}{4}\right) + \frac{5}{4 \cdot 10} = 1.91$$

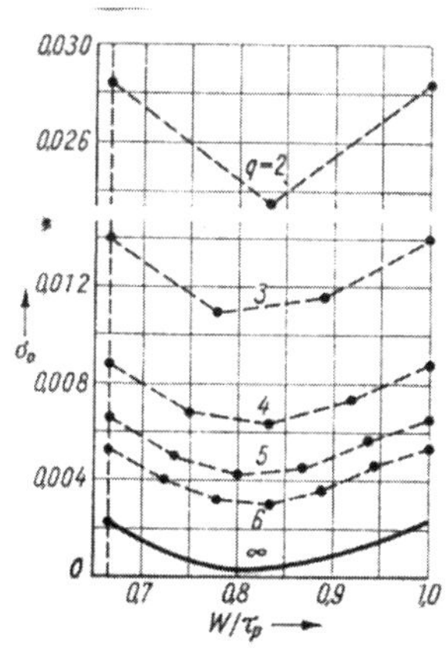

Figure 16.7 Differential leakage coefficient.

The differential geometrical permeance coefficient λ_{ds} is calculated as in Chapter 6.

$$\lambda_{ds} = \frac{0.9 \cdot \tau_s (q_1 K_{w1})^2 K_{01} \sigma_{d1}}{K_c g} \tag{16.40}$$

with
$$K_{01} \approx 1 - 0.033\frac{b_s^2}{g\tau_s} = 1 - 0.033\frac{10^2}{1.5 \cdot 21.37} = 0.8975 \tag{16.41}$$

σ_{d1} is the ratio between the differential leakage and the main inductance, which is a function of coil chording (in slot pitch units) and q_s (slot/pole/phase) – Figure 16.7: $\sigma_{d1} = 0.3 \cdot 10^{-2}$.

The Carter coefficient K_c (as in Chapter 15, Equations 15.53–15.56) is

$$K_c = K_{c1} K_{c2} \tag{16.42}$$

K_{c2} is not known yet but, as the rotor slot is semiclosed, $K_{c2} < 1.1$ with $K_{c1} >> K_{c2}$ due to the fact that the stator has open slots,

$$\gamma_1 = \frac{b_s^2}{5g + b_s} = \frac{10^2}{5 \cdot 1.5 + 10} = 5.714; \; K_{c1} = \frac{\tau_s}{\tau_s - \gamma_1} = \frac{21.37}{21.37 - 5.71} = 1.365 \tag{16.43}$$

Consequently, $K_c \approx 1.365 \cdot 1.1 = 1.50$.

From (16.40),

$$\lambda_{ds} = \frac{0.9 \cdot 21.37 (6 \cdot 0.923)^2 0.895 \cdot 0.3 \cdot 10^{-2}}{1.5 \cdot 1.5} = 0.6335$$

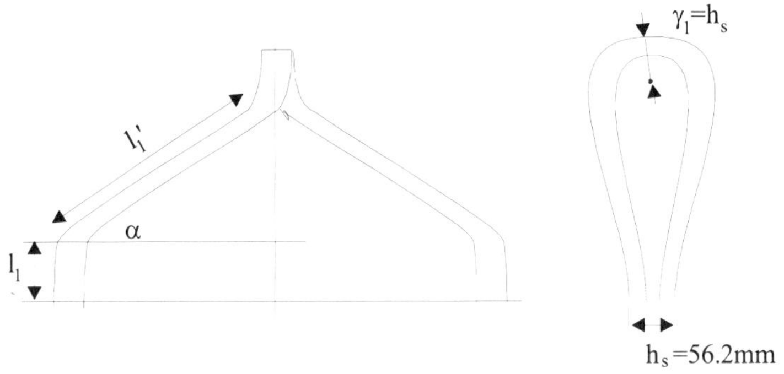

Figure 16.8 Stator end connection coil geometry

The end-connection permeance coefficient λ_{fs} is

$$\lambda_{fs} = 0.34 \frac{q_1}{l_i}(l_{fs} - 0.64\beta\tau) \tag{16.44}$$

l_{fs} is the end connection length (per one side of stator) and may be calculated based on the end connection geometry in Figure 16.8.

$$l_{fs} \approx 2(l_l + l_l') + \pi\gamma_l = 2\left(l_l + \frac{\beta\tau}{\sin\alpha}\right) + \pi h_s =$$
$$= 2\left(0.015 + \frac{5}{6}\frac{1}{2}\frac{0.314}{\sin 40^0}\right) + \pi \cdot 0.0562 = 0.548\text{m} \tag{16.45}$$

So, from (16.44),

$$\lambda_{ts} = 0.34\frac{6}{0.406}\left(0.548 - 0.64\frac{5}{6}0.314\right) = 1.912$$

Finally, from (16.37), the stator leakage reactance X_{ls} (unaffected by leakage saturation) is

$$X_{ls} = 15.8\left(\frac{60}{100}\right)\left(\frac{2\cdot 6\cdot 18}{100}\right)^2\frac{0.406}{2\cdot 6}(1.91 + 0.6335 + 1.912) = 6.667\Omega$$

As the stator slots are open, leakage flux saturation does not occur even for S = 1 (standstill), at rated voltage. The leakage inductance of the field in the radial channels has been neglected.

The stator resistance R_s is

$$R_s = K_R\rho_{Co80^0}\frac{W_1 2}{A_{Co}}(L_{geo} + l_{fs}) =$$
$$= 1\cdot 1.8\cdot 10^{-3}\left(1 + \frac{80-20}{272}\right)\cdot 12\cdot 18\cdot\frac{2\cdot(0.451 + 0.548)}{11.048\cdot 10^{-6}} = 0.8576\Omega \tag{16.46}$$

Although the rotor resistance at rated slip may be approximated from (16.4), it is easier to compare it to stator resistance,

$$(R_r)_{S=S_n} = (0.7 \div 0.8)R_s = 0.8\cdot 0.8576 = 0.686\Omega \tag{16.47}$$

for aluminium bar cage rotors and high efficiency motors. The ratio of 0.7 – 0.8 in (16.47) is only orientative to produce a practical design start. Copper bars may be used when very high efficiency is targeted for single-stack axially-ventilated configurations.

- The rotor leakage inductance L_{rl} may be computed from the breakdown torque's expression (16.1)

$$L_{rl} = \frac{3p_1}{2T_{LR}}\left(\frac{V_{1ph}}{\omega_1}\right)^2 - L_{sl};\; L_{sl} = \frac{X_{sl}}{\omega_1} = \frac{6.667}{2\pi 60} = 0.0177\text{H} \tag{16.48}$$

with

$$T_{LR} = t_{LR} T_{en} = t_{LR} \frac{P_n}{\omega_1} p_1 \quad (16.49)$$

Now, L_{rl} is

$$L_{rl} = \frac{3p_1 V_{1ph}^{\;2}}{2t_{LR} P_n \omega_1} - L_{sl} = \frac{3 \cdot 4000^2}{2 \cdot 2.7 \cdot 736 \cdot 10^3 \cdot 2\pi 60} - 0.0177 = 0.01435\text{H} \quad (16.50)$$

From starting current and torque expressions (16.2 and 16.3),

$$(R_r)_{S=1} = \frac{\omega_1 t_{LR} P_n \left(\frac{\omega_1}{p_1}\right)^{-1}}{3p_1 K_{istart} I_{LRphase}^{\;2}} = \frac{1.2 \cdot 736 \cdot 10^3}{3 \cdot 0.975^2 \cdot \left(5.6\frac{120}{\sqrt{3}}\right)^2} = 2.0538\Omega \quad (16.51)$$

$$(L_{rl})_{S=1} = \frac{1}{\omega_1}\sqrt{\left(\frac{V_{1ph}}{I_{LRphase}}\right)^2 - \left(R_s + (R_r)_{S=1}\right)^2} - (L_{sl})_{S=1} =$$

$$= \frac{1}{2\pi 60}\left[\sqrt{\left(\frac{4000}{5.6\frac{120}{\sqrt{3}}}\right)^2 - (1.083 + 2.0537)^2} - 6.667\right] = 8.436 \cdot 10^{-3}\text{H} \quad (16.52)$$

Note that due to skin effect $(R_r)_{S=1} = 2.914 (R_s)_{S=S_n}$ and to both leakage saturation and skin effect, the rotor leakage inductance at stall is $(L_{rl})_{S=1} = 0.5878 (L_{rl})_{S=S_n}$.

Making use of (16.8) and (16.10), the skin effect resistance ratio K_R is

$$K_R = \frac{(R_r)_{S=1}}{(R_r)_{S=S_n}} \cdot \frac{0.95}{0.8} = 2.999 \cdot \frac{0.95}{0.8} = 3.449 \quad (16.53)$$

The deep rotor bars are typically rectangular, but other shapes are also feasible. A modern aluminum rectangular bar insulated from core by a resin layer is shown in Figure 16.9. For a rectangular bar, the expression of K_R (when skin effect is notable) is (16.10).

$$K_R \approx \beta_{skin} h_r \quad (16.54)$$

with $$\beta_{skin} = \sqrt{\frac{\pi f_1 \mu_0}{\rho_{Al}}} = \sqrt{\frac{\pi \cdot 60 \cdot 1.256 \cdot 10^{-6}}{3.1 \cdot 10^{-8}}} = 87 m^{-1} \quad (16.55)$$

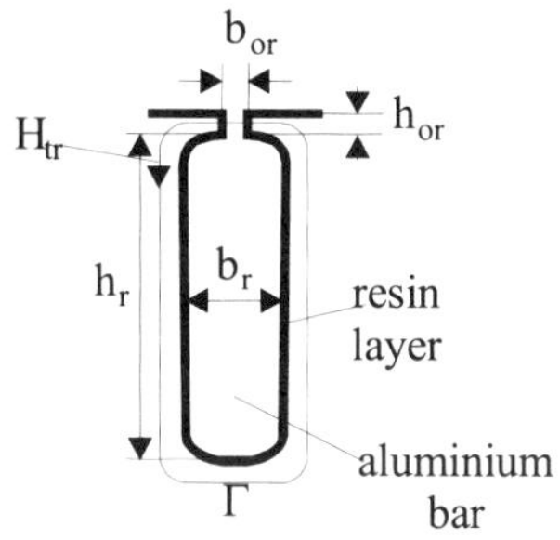

Figure 16.9 Insulated aluminum bar

The rotor bar height h_r is

$$h_r = \frac{3.449}{87} = 3.964 \cdot 10^{-2} m$$

From (16.11),

$$\frac{(L_{rl})_{S=1}}{(L_{rl})_{S=S_n}} \frac{0.75}{0.85} = \frac{\frac{h_r}{3b_r} K_x + \frac{h_{or}}{b'_{or}}}{\frac{h_r}{3b_r} + \frac{h_{or}}{b_{or}}} = 0.5878 \cdot \frac{0.75}{0.85} = 0.51864 \quad (16.56)$$

$$K_x \approx \frac{3}{2\beta_{skin} h_r} = \frac{3}{2 \cdot 87 \cdot 3.977 \cdot 10^{-2}} = 0.43 \quad (16.57)$$

We have to choose the rotor slot neck $h_{or} = 1.0 \cdot 10^{-3} m$ for mechanical reasons. A value of b_{or} has to be chosen, say, $b_{or} = 2 \cdot 10^{-3} m$. Now we have to check the saturation of the slot neck at start which modifies b_{or} into b_{or}' in (16.56). We use the approximate approach developed in Chapter 9 (paragraph 9.8).

First the bar current at start is

$$I_{bstart} = I_n \left(\frac{I_{start}}{I_n} \right) \cdot 0.95 \cdot K_{bs} \quad (16.58)$$

with $N_r = 64$ and straight rotor slots.

K_{bs} is the ratio between the reduced-to-stator and actual bar current.

$$K_{bs} = \frac{2mW_1K_{w1}}{N_r} = \frac{2\cdot 3\cdot (12\cdot 18)\cdot 0.923}{64} = 18.69075 \quad (16.59)$$

$$I_{bstart} = 5.6\cdot 0.95\cdot \frac{120}{\sqrt{3}}\cdot 18.69 = 6896.9A \quad (16.60)$$

Making use of Ampere's law in the Γ contour in Figure 16.9 yields

$$\frac{B_{tr}}{\mu_{rel}\mu_0}\left[\tau_r - b_{or} + b_{os}\mu_{rel}(H_{tr})\right] = I_{bstart}\sqrt{2} \quad (16.61)$$

Iteratively, making use of the lamination magnetization curve (Table 15.4, Chapter 15), with the rotor slot pitch τ_r,

$$\tau_r = \frac{\pi(D_{is} - 2g)}{N_r} = \frac{\pi(0.49 - 2\cdot 1.5\cdot 10^{-3})}{64} = 23.893\cdot 10^{-3}m \quad (16.62)$$

the solution of (16.61) is $B_{tr} = 2.29$ T and $\mu_{rel} = 12$!

The new value of slot opening b_{or}', to account for tooth top saturation at S = 1, is

$$b_{or}' = b_{or} + \frac{\tau_r - b_{or}}{\mu_{rel}} = 2\cdot 10^{-3} + \frac{(23.893 - 2)\cdot 10^{-3}}{12} = 3.8244\cdot 10^{-3}m \quad (16.63)$$

Now, with $N_r = 64$ slots, the minimum rotor slot pitch (at slot bottom) is

$$\tau_{r\,min} = \frac{\pi(D_{is} - 2g - 2h_r)}{N_r} = \frac{\pi(49 - 2\cdot 1.5 - 2\cdot 39.64)\cdot 10^{-3}}{64} = 20\cdot 10^{-3}m \quad (16.64)$$

With the maximum rotor tooth flux density $B_{tmax} = 1.7T$, the maximum slot width b_{rmax} is

$$b_{r\,max} = \tau_r - \frac{B_g}{B_{t\,max}}\tau_r = 20\cdot 10^{-3}\left(1 - \frac{0.8}{1.7}\right) = 10.58mm \quad (16.65)$$

The rated bar current density j_{Al} is

$$j_{Al} = \frac{I_b}{h_r b_r} = \frac{K_i I_n K_{bs}}{h_r b_r} = \frac{0.936\cdot 120\cdot 18.69}{39.64\cdot 10\sqrt{3}} = 3.061A/mm^2 \quad (16.66)$$

with
$$K_i = 0.8\cos\varphi_n + 0.2 = 0.8\cdot 0.92 + 0.2 = 0.936 \quad (16.67)$$

We may now verify (16.56).

$$0.51864 \geq \frac{3\cdot\frac{39.64}{10}\cdot 0.43+\frac{1}{3.824}}{\frac{39.64}{3\cdot 10}\cdot 1+\frac{1}{2}} = \frac{0.829}{1.821} = 0.455!$$

The fact that approximately the large cage dimensions of $h_r = 39.64$ m and $b_r = 10$ mm with $b_{or} = 2$ mm, $h_{or} = 1$ mm fulfilled the starting current, starting and breakdown torques, for a rotor bar rated current density of only 3.06A/mm^2, means that the design leaves room for further reduction of slot width.

We may now proceed, based on the rated bar current $I_b = 1213.38$A (16.66), to the detailed design of the rotor slot (bar), end ring, and rotor back iron.

Then the teeth saturation coefficient K_{st} is calculated. If notably different from the initial value, the stator design may be redone from the beginning until acceptable convergence is obtained. Further on, the magnetization current equivalent circuit parameters, losses, rated efficiency and power factor, rated slip, torque, breakdown torque, starting torque, and starting current are all calculated.

Most of these calculations are to be done with the same expressions as in Chapter 15, which is why we do not repeat them here.

16.5 DOUBLE CAGE ROTOR DESIGN

When a higher starting torque for lower starting current and high efficiency are all required, the double cage rotor comes into play. However, the breakdown torque and the power factor tend to be slightly lower as the rotor cage leakage inductance at load is larger.

The main constraints are

$$\frac{T_{bk}}{T_{en}} = t_{ek} > 2.0$$

$$\frac{I_{LR}}{I_{1n}} < 5.35$$

$$\frac{T_{LR}}{T_{en}} = t_{LR} \geq 1.5$$

Typical geometries of double cage rotors are shown in Figure 16.10.

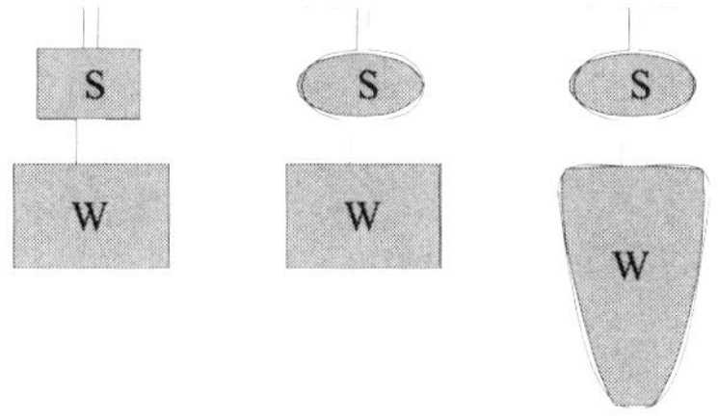

Figure 16.10 Typical rotor slot geometries for double cage rotors.

The upper cage is the starting cage as most of rotor current flows through it at start, mainly because the working cage leakage reactance is high, so its current at primary (f_1) frequency is small.

In contrast, at rated slip ($S < 1.5\cdot10^{-2}$), most of the rotor current flows through the working (lower) cage as its resistance is smaller and its reactance is smaller than the starting cage resistance.

In principle, it is fair to say that there is always current in both cages, but, at high rotor frequency ($f_2 = f_1$), the upper cage is more important, while at rated rotor frequency ($f_2 = S_nf_1$), the lower cage takes more current.

The end ring may be common to both cages, but, when frequent starts are considered, separate end rings are preferred because of thermal expansion. (Figure 16.11). It is also possible to make the upper cage of brass and the lower cage of copper.

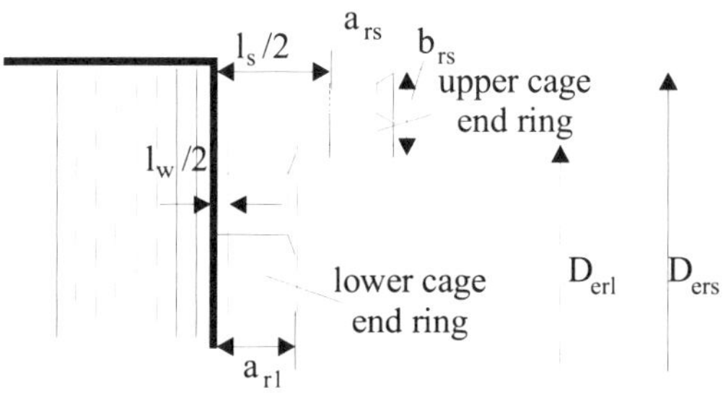

Figure 16.11 Separate end rings

The equivalent circuit for the double cage has been introduced in Chapter 9 (paragraph 9.7) and is inserted here only for easy reference (Figure 16.12).

For the common end ring case, R_{ring} = R_{ring}, ring reduced equivalent resistance) reduced to bar resistance), and R_{bs} and R_{bw} are the upper and lower bar resistances.

For separate end rings, $R_{ring} = 0$, but the rings resistances $R_{bs} \rightarrow R_{bs} + R_{rings}$, $R_{bw} \rightarrow R_{bw} + R_{ringw}$. L_{lr} contains the differential leakage inductance and only for common end ring, the inductance of the latter.

Also,
$$L_{bs} \approx \mu_0\left(l_{geo} + l_s\right)\frac{h_{rs}}{b_{rs}};\; L_e = \mu_0 l_{geo}\frac{h_{or}}{b_{or}} \tag{16.68}$$

$$L_{bw} \approx \mu_0\left(l_{geo} + l_w\right)\left(\frac{h_{rw}}{3b_{rw}} + \frac{h_n}{a_n} + \frac{h_{rs}}{b_{rs}}\right)$$

The mutual leakage inductance L_{ml} is

$$L_{ml} \approx \mu_0 l_{geo}\frac{h_{rs}}{2b_{rs}} \tag{16.69}$$

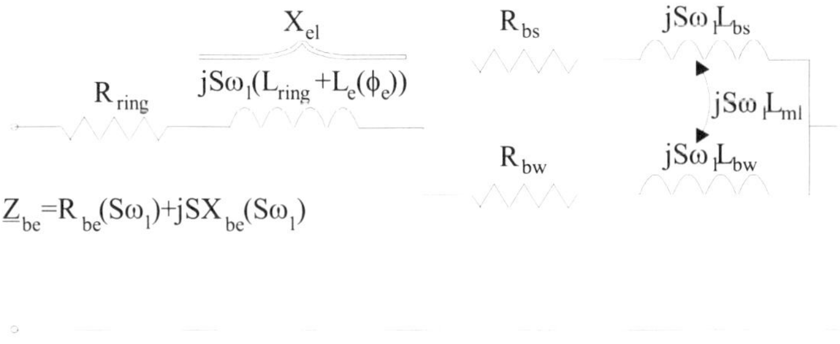

a.)

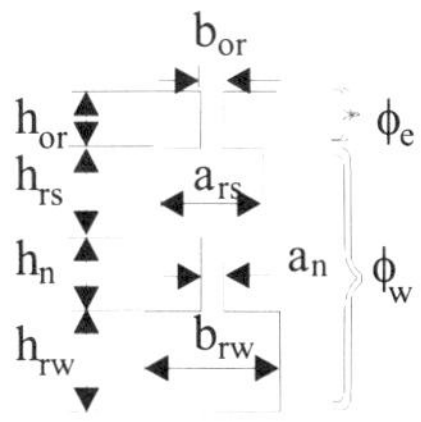

b.)

Figure 16.12 Equivalent circuit of 16.11 double cage a.) and slot geometrical parameters b.)

In reality, instead of l_{geo}, we should use l_i but in this case the leakage inductance of the rotor bar field in the radial channels is to be considered. The two phenomena are lumped into l_{geo} (the geometrical stack length).

The lengths of bars outside the stack are l_s and l_w, respectively.

First we approach the starting cage, made of brass (in our case) with a resistivity $\rho_{brass} = 4\rho_{Co} = 4 \cdot 2.19 \cdot 10^{-8} = 8.76 \cdot 10^{-8}$ (Ωm). We do this based on the fact that, at start, only the starting (upper) cage works.

$$T_{LR} = t_{LR} T_{en} \approx t_{LR} \frac{P_n}{\omega_1} p_1 = 3 \frac{p_1}{\omega_1} (R_r)_{S=1} 0.95 I_{LR}^2 \tag{16.70}$$

$$I_{LR} = 5.35 I_{1n} = 5.35 \frac{120}{\sqrt{3}} = 371.1\text{A} \tag{16.71}$$

From (16.70), the rotor resistance ($(R_r)_{S=1}$) is

$$(R_r)_{S=1} = \frac{t_{LR} P_n}{3 \cdot 0.95 \cdot I_{LR}^2} = \frac{1.5 \cdot 736 \cdot 10^3}{3 \cdot 0.95 \cdot 371.1^2} = 2.8128\Omega \tag{16.72}$$

From the equivalent circuit at start, the rotor leakage inductance at S = 1, $(L_{rl})_{S=1}$, is

$$(L_{rl})_{S=1} = \left(\sqrt{\left(\frac{V_{ph}}{I_{LR}}\right)^2 - \left(R_s + (R_r)_{S=1}\right)^2} - L_{ls}\right)\frac{1}{\omega_1} =$$

$$= \left(\sqrt{\left(\frac{4000}{371.1}\right)^2 - (0.8576 + 2.8128)^2} - 6.667\right)\cdot\frac{1}{2\pi 120} = 9.174\cdot 10^{-3}\,\mathrm{H} \quad (16.73)$$

If it were not for the rotor working cage large leakage in parallel with the starting cage, $(R_r)_{S=1}$ and $(L_{rl})_{S=1}$ would simply refer to the starting cage whose design would then be straightforward.

To produce realistic results, we first have to design the working cage.

To do so we again assume that the copper working cage (made of copper) resistance referred to the stator is

$$(R_r)_{S=S_n} = 0.8R_s = 0.8\cdot 0.8576 = 0.6861\Omega \quad (16.74)$$

This is the same as for the aluminum deep bar cage, though it is copper this time. The reason is to limit the slot area and depth in the rotor.

- Working cage sizing

The working cage bar approximate resistance R_{be} is

$$R_{be} \approx (0.7 \div 0.8)\frac{(R_r)_{S=S_n}}{K_{bs}} = \frac{0.75\cdot 0.6861}{7452.67} = 0.6905\cdot 10^{-4}\Omega \quad (16.75)$$

From (16.6), K_{bs} is

$$K_{bs} = \frac{4m(W_1K_{w1})^2}{N_r} = \frac{4\cdot 3\cdot(12\cdot 18\cdot 0.923)^2}{64} = 7452.67 \quad (16.76)$$

The working cage cross section A_{bw} is

$$A_{bw} = \frac{\rho_{Co}(l_{geo} + l_w)}{R_{bl}} = 2.2\cdot 10^{-8}\frac{0.451 + 0.01}{0.6905\cdot 10^{-4}} = 1.468\cdot 10^{-4}\mathrm{m}^2 \quad (16.77)$$

The rated bar current I_b (already calculated from (16.66)) is I_b = 1213.38A and thus the rated current density in the copper bar j_{Cob} is

$$j_{Cob} = \frac{I_b}{A_{bw}} = \frac{1213.38}{1.468\cdot 10^{-4}} = 8.26\cdot 10^6\,\mathrm{A/m^2} \quad (16.78)$$

This is a value close to the maximum value acceptable for radial-axial air cooling.

A profiled bar $1\cdot10^{-2}$m wide and $1.5\cdot10^{-2}$m high is used: $b_{rl} = 1\cdot10^{-2}$m, $h_{rl} = 1.5\cdot10^{-2}$ m.

The current density in the end ring has to be smaller than in the bar (about 0.7 to 0.8) and thus the working (copper) end ring cross section A_{rw} is

$$A_{rw} = \frac{A_{bw}}{0.75\cdot 2\sin\frac{\pi p_1}{N_r}} = \frac{1.468\cdot10^{-4}}{1.5\sin\frac{\pi\cdot 2}{64}} = 9.9864\cdot10^{-4}\,m^2 \quad (16.79)$$

So (from Figure 16.11),

$$a_{rl}b_{rl} = 9.9846\cdot10^{-4}\,m^2 \quad (16.80)$$

We may choose $a_{rl} = 2\cdot10^{-2}$m and $b_{rl} = 5.0\cdot10^{-2}$m.

The slot neck dimensions will be considered as (Figure 16.12b) $b_{or} = 2.5\cdot10^{-3}$m, $h_{or} = 3.2\cdot10^{-3}$m (larger than in the former case) as we can afford a large slot neck permeance coefficient h_{or}/b_{or} because the working cage slot leakage permeance coefficient is already large.

Even for the deep cage, we could afford a larger h_{or} and b_{or} to stand the large mechanical centrifugal stresses occurring during full operation speed.

We go on to calculate the rotor differential geometrical permeance coefficient for the cage (Chapter 15, Equation (15.82)).

$$\lambda_{dr} = \frac{0.9\cdot\tau_r\gamma_{dr}}{K_c g}\left(\frac{N_r}{6p_1}\right)^2; \quad \gamma_{dr} = 9\left(\frac{6p_1}{N_r}\right)^2\cdot10^{-2} \quad (16.81)$$

$$\gamma_{dr} = 9\left(\frac{6\cdot2}{64}\right)^2\cdot10^{-2} = 0.3164\cdot10^{-2}$$

$$\lambda_{dr} = \frac{0.9\cdot23.893}{1.5\cdot1.5}\cdot0.3164\cdot10^{-2}\cdot\left(\frac{64}{12}\right)^2 = 0.86$$

The saturated value of λ_{drl} takes into account the influence of tooth saturation coefficient K_{st} assumed to be $K_{st} = 0.25$.

$$\lambda_{drs} = \frac{\lambda_{drl}}{1+K_{st}} = \frac{0.96}{1.25} = 0.688 \quad (16.82)$$

The working end ring specific geometric permeance λ_{erl} is (Chapter 15, Equation (15.83)):

$$\lambda_{erl} = \frac{2.3\cdot(D_{erl}-b_{rl})}{N_r l_{geo} 4\sin^2\left(\frac{\pi p_1}{N_r}\right)}\log\frac{4.7\cdot(D_{erl}-b_{rl})}{b_{rl}+2a_{rl}} \approx$$

$$\approx \frac{2.3\cdot\left(0.49-3\cdot10^{-3}-4\cdot10^{-2}-5\cdot10^{-2}\right)}{64\cdot0.451\cdot4\sin^2\left(\frac{\pi\cdot2}{64}\right)}\log\frac{397\cdot10^{-3}}{(50+2\cdot20)\cdot10^{-3}} = 0.530 \quad (16.83)$$

The working end ring leakage reactance X_{rlel} (Chapter 11, Equation 15.85) is

$$X_{rlel} \approx \mu_0 2\pi f_1 l_{geo}\cdot10^{-8}\lambda_{erl} = 7.85\cdot60\cdot0.451\cdot10^{-6}\cdot0.530 = 1.1258\cdot10^{-4}\Omega \quad (16.84)$$

The common reactance (Figure 16.12) is made of the differential and slot neck components, unsaturated, for rated conditions and saturated at S = 1.

$$\begin{aligned}(X_{el})_{S=S_n} &= 2\pi f_1 l_{geo}\mu_0\left(\lambda_{dr}+\frac{h_{or}}{b_{or}}\right)\\ (X_{el})_{S=1} &= 2\pi f_1 l_{geo}\mu_0\left(\lambda_{drs}+\frac{h_{or}}{b'_{or}}\right)\end{aligned} \quad (16.85)$$

The influence of tooth top saturation at start is considered as before (in 16.60–16.63).

With $b_{or}' = 1.4b_{or}$ we obtain

$$(X_{el})_{S=S_n} = 2\pi60\cdot0.451\cdot\left(0.86+\frac{2.5}{3.2}\right)\cdot1.256\cdot10^{-6} = 3.503\cdot10^{-4}\Omega$$

$$(X_{el})_{S=1} = 2\pi60\cdot0.451\cdot\left(0.688+\frac{2.5}{3.5+1.4}\right)\cdot1.256\cdot10^{-6} = 2.6595\cdot10^{-4}\Omega$$

Now we may calculate the approximate values of starting cage resistance from the value of the equivalent resistance at start R_{start}.

$$R_{start} = \frac{(R_r)_{S=1}}{K_{bs}} = \frac{2.8128}{7452.67} = 3.7742\cdot10^{-4}\Omega \quad (16.86)$$

From Figure 16.12a, the starting cage resistance R_{bes} is

$$R_{bes} \approx \frac{R_{start}^2+X_{el}^2}{R_{start}} = \frac{\left(3.7742\cdot10^{-4}\right)^2+\left(2.6595\cdot10^{-4}\right)^2}{\left(3.7742\cdot10^{-4}\right)} = 5.648\cdot10^{-4}\Omega \quad (16.87)$$

For the working cage reactance X_{rlw}, we also have

$$X_{rlw} = \frac{R_{start}^2+X_{el}^2}{X_{el}} = \frac{\left(3.7742\cdot10^{-4}\right)^2+\left(2.6595\cdot10^{-4}\right)^2}{2.6595\cdot10^{-4}} = 8.0156\cdot10^{-4}\Omega \quad (16.88)$$

The presence of common leakage X_{el} makes starting cage resistance R_{bes} larger than the equivalent starting resistance R_{start}. The difference is notable and affects the sizing of the starting cage.

The value of R_{bes} includes the influence of starting end ring. The starting cage bar resistance R_b is approximately

$$R_{bs} \approx R_{bes}(0.9 \div 0.95) = 0.9 \cdot 5.648 \cdot 10^{-4} = 5.083 \cdot 10^{-4}\Omega \qquad (16.89)$$

The cross section of the starting bar A_{bs} is

$$A_{bs} = \rho_{brass}\frac{l_{geo}+l_s}{R_{bs}} = 4 \cdot 2.2 \cdot 10^{-8}\frac{0.451+0.05}{5.083 \cdot 10^{-4}} = 0.8673 \cdot 10^{-4} m^2 \quad (16.90)$$

The utilization of brass has reduced drastically the starting cage bar cross section. The length of starting cage bar was prolonged by $l_s = 5 \cdot 10^{-2}$m (Figure 16.11) as only the working end ring axial length $a_{rl} = 2 \cdot 10^{-2}$m on each side of the stack.

We may adopt a rectangular bar again (Figure 16.12b) with $b_{rs} = 1 \cdot 10^{-2}$m and $h_{rs} = 0.86 \cdot 10^{-2}$m.

The end ring cross section A_{rs} (as in 16.79) is

$$A_{rs} = \frac{A_{bs}}{0.75 \cdot 2 \cdot \sin\frac{\pi p_1}{N_r}} = \frac{0.8673 \cdot 10^{-4}}{1.5 \cdot \sin\frac{\pi \cdot 2}{64}} = 5.898 \cdot 10^{-4}\Omega \qquad (16.91)$$

The dimensions of the starting cage ring are chosen to be $a_{rs} \times b_{rs}$ (Figure 16.11) = $2 \cdot 10^{-2} \times 3 \cdot 10^{-2}$m×m.

Now we may calculate more precisely the starting cage bar equivalent resistance, R_{bes}, and of the working cage, R_{bew},

$$R_{bes} = R_{bs} + \frac{R_{rs}}{2\sin^2\left(\frac{\pi p_1}{N_r}\right)}; \; R_{rs} = \rho_{brass}\frac{l_{rs}}{A_{rs}}$$
$$R_{bew} = R_{bw} + \frac{R_{rw}}{2\sin^2\left(\frac{\pi p_1}{N_r}\right)}; \; R_{rw} = \rho_{Co}\frac{l_{rw}}{A_{rw}} \qquad (16.92)$$

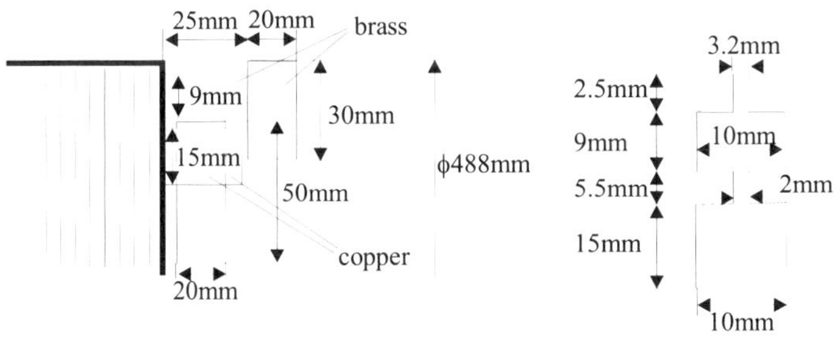

Figure 16.13 Double cage design geometry

The size of end rings is visible in Figure 16.13.

It is now straightforward to calculate R_{bes} and R_{bew} based on l_{rs} and l_{rw}, the end ring segments length of Figure 16.13.

The only unknowns are (Figure 16.13) the stator middle neck dimensions a_n and h_n.

From the value of the working cage reactance (X_{rlw} 16.88), if we subtract the working end ring reactance (X_{rlel}, (16.84)), we are left with the working cage slot reactance in rotor terms.

$$X_{rlslot} = X_{rlw} - X_{rlel}$$
$$X_{rlslot} = 8.0156 \cdot 10^{-4} - 1.1258 \cdot 10^{-4} = 6.8898 \cdot 10^{-4}\Omega \quad (16.93)$$

The slot geometrical permeance coefficient for the working cage λ_{rl} is

$$\lambda_{rl} = \frac{X_{rlslot}}{2\pi f_1 \mu_o l_{geo}} = \frac{6.8898 \cdot 10^{-4}}{2\pi 60 \cdot 1.256 \cdot 10^{-6} \cdot 0.451} = 3.243! \quad (16.94)$$

Let us remember that λ_{rl} refers only to working cage body and the middle neck interval,

$$\lambda_{rl} = \frac{h_{rw}}{3b_{rw}} + \frac{h_n}{a_n} \quad (16.95)$$

Finally,

$$\frac{h_n}{a_n} = 3.243 - \frac{15}{3 \cdot 10} = 2.7435$$

with $$a_n = 2 \cdot 10^{-3}m; \ \ h_n = 5.5 \cdot 10^{-3}m$$

As for the deep bar cage rotor, we were able to determine all working and starting cage dimensions so that they meet (approximately) the starting torque, current, and rated efficiency and power factor.

Again, all the parameters and performance may now be calculated as for the design in Chapter 15.

Such an approach would drastically reduce the number of design iterations until satisfactory performance is obtained with all constraints observed. Also having a workable initial sizing helps to approach the optimization design stage if so desired.

16.6 WOUND ROTOR DESIGN

For the stator, as in previous paragraphs, we approach the wound rotor design methodology. The rotor winding has diametrical coils and is placed in two layers. As the stator slots are open, to limit the airgap flux pulsations (and, consequently, the tooth flux pulsation additional core losses), the rotor slots are to be half – open (Figure 16.13). This leads to the solution with wave-shape half-preformed coils made of a single bar with one or more (2) conductors in parallel.

The half-shaped uniform coils are introduced frontally in the slots and then the coil unfinished terminals are bent as required to form wave coils which suppose less material to connect the coils into phases.

The number of rotor slots N_r is now chosen to yield an integer q_2 (slots/pole/pitch) which is smaller than $q_1 = 6$. Choosing $q_2 = 5$ the value of N_r is

$$N_r = 2p_1 m q_2 = 2 \cdot 2 \cdot 3 \cdot 5 = 60 \tag{16.96}$$

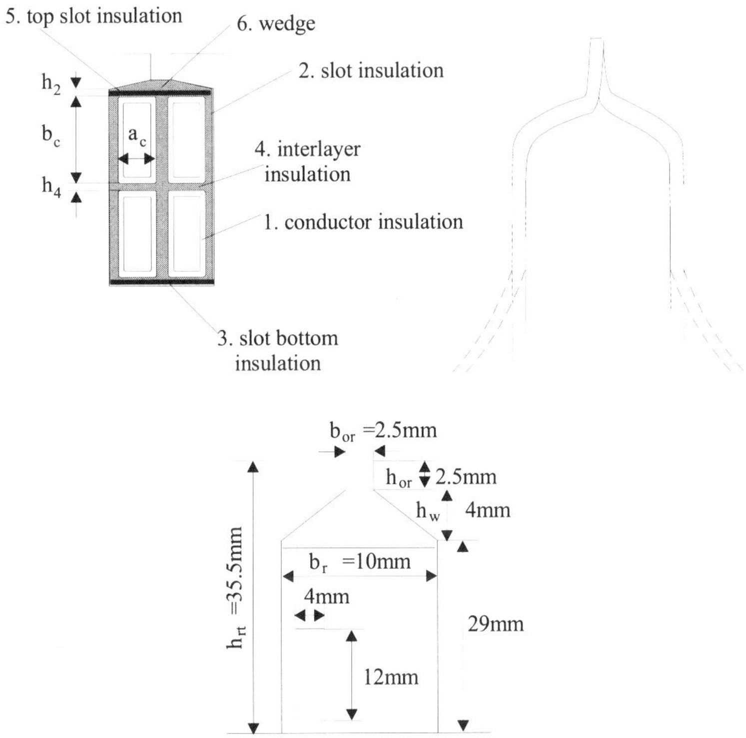

Figure 16.14 Wound rotor slot a.) and half–preformed wave coils b.)

A smaller number of rotor than stator slots leads to low tooth flux pulsation core losses (Chapter 11).

The winding factor is now the distribution factor.

$$K_{w2} = \frac{\sin\frac{\pi}{6}}{q_2 \sin\frac{\pi}{6q_2}} = \frac{0.5}{5\sin\left(\frac{\pi}{6.5}\right)} = 0.9566 \tag{16.97}$$

The rotor tooth pitch τ_r is

$$\tau_r = \frac{\pi(D_{is} - 2g)}{N_r} = \frac{\pi(490 - 2\cdot 1.5)\cdot 10^{-3}}{60} = 25.486\cdot 10^{-3}\text{m} \qquad (16.98)$$

As we decided to use wave coils, there will be one conductor (turn) per coil and thus $n_r = 2$ conductor/slot.

The number of turns per phase, with a_2 current path in parallel, W_2, is (a_2 = 1).

$$W_2 = \frac{N_r n_r}{2m_1 a_2} = \frac{60\cdot 2}{2\cdot 3\cdot 1} = 20\text{turns/phase} \qquad (16.99)$$

It is now straightforward to calculate the emf E_2 in the rotor phase based on $E_1 = K_E V_{ph}$ in the stator.

$$E_2 = E_1\frac{W_2 K_{w2}}{W_1 K_{w1}} = \frac{0.97\cdot 4000\cdot 20\cdot 0.9566}{12\cdot 18\cdot 0.923} = 372.33\text{V} \qquad (16.100)$$

Usually star connection is used and the inter – slip – ring (line) rotor voltage V_{21} is

$$V_{21} = E_2\sqrt{3} = 372.33\cdot 1.73 = 644\text{V} \qquad (16.101)$$

We know by now the rated stator current and the ratio $K_i = 0.936$ (for $\cos\varphi_n = 0.92$) between rotor and stator current (16.67). Consequently, the rotor phase rated current is

$$I_{2n} = K_i\frac{W_1 K_{w1}}{W_2 K_{w2}} I_n = 0.936\cdot\frac{12\cdot 18\cdot 0.923}{20\cdot 0.9566}\cdot\frac{120}{\sqrt{3}} = 676.56\text{A} \qquad (16.102)$$

The conductor area may be calculated after adopting the rated current density: $j_{Cor} > j_{Co} = 6.3$ A/mm², by (5 to 15)%. We choose here $j_{Cor} = 7$ A/mm².

$$A_{Cor} = \frac{I_{2n}}{j_{Cor}} = \frac{676.56}{7\cdot 10^6} = 96.65\cdot 10^{-6}\text{m}^2 \qquad (16.103)$$

As the voltage is rather low, the conductor insulation is only 0.5 mm thick/side; also 0.5mm foil slot insulation per side and an interlayer foil insulation of 0.5mm suffice.

The slot width b_r is

$$b_r < \tau_r\left(1 - \frac{B_g}{B_{tr\,min}}\right) = 28.486\cdot 10^{-3}\left(1 - \frac{0.8}{1.4}\right) = 10.92\cdot 10^{-3}\text{m} \qquad (16.104)$$

Let us choose $b_r = 10.5\cdot 10^{-3}$m.

Considering the insulation system width (both side) of $2\cdot10^{-3}$m, the conductor allowable width $a_{cond} \approx 8.5\cdot10^{-3}$m. We choose 2 vertical conductors, so each has $a_c = 4\cdot10^{-3}$m. The conductors height b_c is

$$b_c = \frac{A_{Cor}}{2a_c} = \frac{96.65}{2\cdot 4.0} \approx 12\cdot10^{-3}\,\text{m} \tag{16.105}$$

Rectangular conductor cross sections are standardized, so the sizes found in (16.105) may have to be slightly corrected.

Now the final slot size (with $1.5\cdot10^{-3}$m total insulation height (radially)) is shown in Figure 16.14.

The maximum rotor tooth flux density B_{trmax} is

$$B_{tr\max} = \left(\left(\pi(D_{is} - 2g) - 2h_{rt}\right) - N_r b_r\right)^{-1} \pi(D_{is} - 2g)B_g = \\ = \frac{\pi(490-3)\cdot 0.8}{\left[\pi(490-3-2\cdot 35.5) - 60\cdot 10.5\right]} = 1.809\text{T} \tag{16.106}$$

This is still an acceptable value.

The key aspects of wound rotor designs are now solved. However, this time we will pursue the whole design methodology to prove the performance.

For the wound rotor, the efficiency, power factor ($\cos\varphi_n = 0.92$), and the breakdown p.u. torque $t_{bk} = 2.5$ (in our case) are the key performance parameters.

The starting performance is not so important as such a machine is to use either power electronics or a variable resistance for limited range speed control and for starting, respectively.

- The rotor back iron height

The rotor back iron has to flow half the pole flux, for a given maximum flux density $B_{cr} = (1.4 - 1.6)$ T. The back core height h_{cr} is

$$h_{cr} = \frac{B_g}{B_{cr}} \frac{\tau}{\pi} \frac{1}{K_{Fe}} = \frac{0.8}{1.4} \frac{0.314}{\pi\cdot 0.98} = 0.06\text{m} \tag{16.107}$$

The interior lamination diameter D_{ir} is

$$D_{ir} = D_{is} - 2g - 2h_{rt} - 2h_{cs} = (490-3-71-120)\cdot 10^{-3} = 0.296\text{m} \tag{16.108}$$

So there is enough radial room to accommodate the shaft.

16.7 IM WITH WOUND ROTOR-PERFORMANCE COMPUTATION

We start the performance computation with the magnetization current to continue with wound rotor parameters (stator parameters have already been computed); then losses and efficiency follow. The computation ends with the no-load current, short-circuit current, breakdown slip, rated slip, breakdown p.u. torque, and rated power factor.

➢ Magnetization mmfs

The mmf per pole contains the airgap F_g stator and rotor teeth (F_{st}, F_{rt}) stator and rotor back core (F_{cs}, F_{cr}) components (Figure 16.15).

$$F_{1m} = F_g + (F_{st} + F_{rt}) + \frac{F_{cs}}{2} + \frac{F_{cr}}{2} \tag{16.109}$$

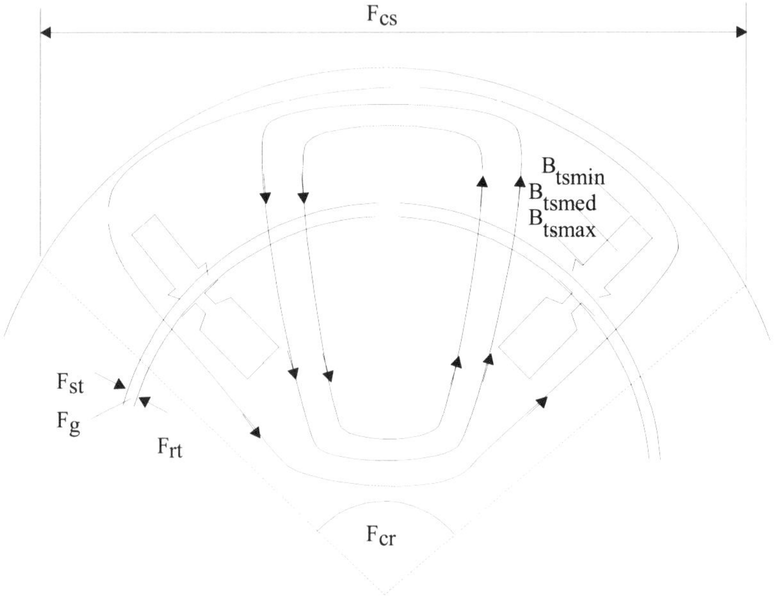

Figure 16.14 Main flux path and its mmfs

➢ The airgap F_g is

$$F_g = K_c g \frac{B_g}{\mu_0} \tag{16.110}$$

The Carter's coefficient K_c is

$$K_c = K_{c1} K_{c2}$$

with
$$K_{c1,2} = \frac{\tau_{s,r}}{\tau_{s,r} - \gamma_{s,r} g}; \quad \gamma_{s,r} = \frac{\left(\frac{b_{os,r}}{g}\right)^2}{5 + \frac{b_{os,r}}{g}} \tag{16.111}$$

K_{c1} has been already calculated in (16.43): K_{c1} = 1.365. With $b_{or} = 5 \cdot 10^{-3}$m, $g = 1.5 \cdot 10^{-3}$m and $\tau_r = 25.486 \cdot 10^{-3}$m, γ_r is

$$\gamma_r = \frac{\left(\frac{5}{1.5}\right)^2}{5+\frac{5}{1.5}} = 1.333; \quad K_{c2} = \frac{25.486}{25.486 - 1.5\cdot 1.333} = 1.0852$$

Finally, $K_c = 1.365 \cdot 1.0852 = 1.4812 < 1.5$ (as assigned from start).
From (16.110), F_g is

$$F_g = 1.485 \cdot 1.5 \cdot 10^{-3} \frac{0.8}{1.256 \cdot 10^{-6}} = 1418\text{Aturns}$$

- The stator teeth mmf

With the maximum stator tooth flux density $B_{tmax} = 1.5$ T, the minimum value B_{tsmin} occurs at slot bottom diameter.

$$\begin{aligned} B_{ts\min} &= \left(\pi(D_{is} + 2h_{st}) - N_s b_s\right)^{-1} B_g \pi D_{is} = \\ &= \left(\pi(490 + 2\cdot 57.2) - 72\cdot 10\right)^{-1} \pi \cdot 490 \cdot 0.8 = 1.045\text{T} \end{aligned} \tag{16.112}$$

The average value B_{tsmed} is

$$B_{tsmed} = \frac{\pi D_{is} B_g}{\pi(D_{is} + h_{st}) - N_s b_s} = \frac{\pi \cdot 490 \cdot 0.8}{\pi(490 + 57.2) - 72 \cdot 10} = 1.233\text{T} \tag{16.113}$$

For the three flux densities, the lamination magnetization curve (Chapter 15, Table 15.4) yields:

$$H_{ts\max} = 1340\text{A/m}, \quad H_{tsmed} = 400\text{A/m}, \quad H_{ts\min} = 230\text{A/m}$$

The average value of H_{ts} is

$$\begin{aligned} H_{ts} &= \frac{1}{6}\left(H_{ts\max} + 4H_{tsmed} + H_{ts\min}\right) = \\ &= \frac{1}{6}(1340 + 4\cdot 400 + 230) = 528.33\text{A/m} \end{aligned} \tag{16.114}$$

The stator tooth mmf F_{ts} becomes

$$F_{ts} = h_{st} H_{ts} = 57.2 \cdot 10^{-3} \cdot 528.33 = 30.22\text{Aturns} \tag{16.115}$$

- Rotor tooth mmf (F_{tr}) computation

It is done as for the stator. The average tooth flux density B_{trmed} is

$$B_{trmed} = \frac{\pi D_{is} B_g}{\pi(D_{is} - h_{rt}) - N_r b_r} = \frac{\pi \cdot 490 \cdot 0.8}{\pi(490 - 35.52) - 60 \cdot 10.5} = 1.544\text{T} \tag{16.116}$$

The corresponding iron magnetic fields for $B_{trmin} = 1.4$ T, $B_{trmed} = 1.544$ T, $B_{trmax} = 1.809$T are

$$H_{tr\,max} = 9000A/m, \quad H_{trmed} = 1750A/m, \quad H_{tr\,min} = 760A/m$$

The average H_{tr} is

$$H_{tr} = \frac{1}{6}\left(H_{tr\,max} + 4H_{trmed} + H_{tr\,min}\right) = 2793.33A/m \qquad (16.117)$$

The rotor tooth mmf F_{tr} is

$$F_{tr} = h_{rt}H_{tr} = 35.5 \cdot 10^{-3} \cdot 2793.33 = 99.163\text{Aturns} \qquad (16.118)$$

We are now in position to check the teeth saturation coefficient 1 + K_{st} (adopted as 1.25).

$$1 + K_{st} = 1 + \frac{F_{ts} + F_{tr}}{F_g} = 1 + \frac{30.22 + 99.163}{1418} = 1.0912 < 1.25 \qquad (16.119)$$

The value of 1 + K_{st} is a bit too small indicating that the stator teeth are weakly saturated. Slightly wider stator slots would have been feasible. In case the stator winding losses are much larger than stator core loss, increasing the slot width by 10% could lead to higher conductor loss reduction. Another way to take advantage of this situation is illustrated in [4], Figure 16.16, where axial cooling is used, and the rotor diameter is reduced due to better cooling. For a 2 pole 1.9 MW, 6600 V machine, the reduction of mechanical loss led to 1% efficiency rise (from 0.96 to 0.97).

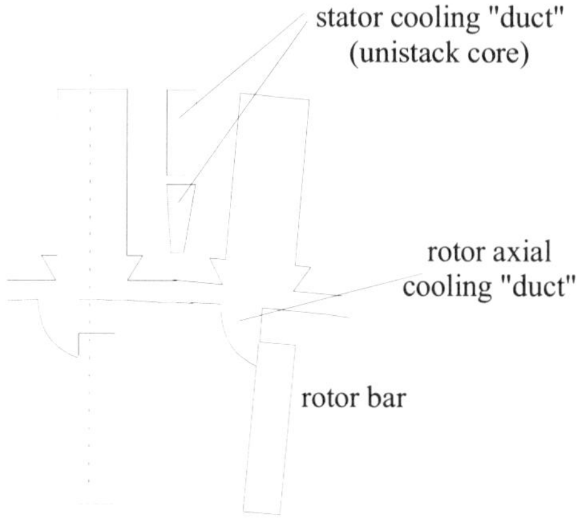

Figure 16.16 Tooth axial ventilation design (high voltage stator)

The stator back core mmf F_{cs} is

$$F_{cs} = l_{cs}H_{cs} \qquad (16.120)$$

As both B_{cs} and l_{cs} vary within a pole pitch span, a correct value of F_{cs} would warrant FEM calculations.

We take here B_{csmax} as reference.

$$B_{cs\max} = \frac{B_g \tau}{\pi h_{cs}} = \frac{0.8 \cdot 0.314}{\pi\left(\frac{0.730 - 0.6044}{2}\right)} = 1.274\text{T} \tag{16.121}$$

$$H_{cs\max} = 450\text{A/m}$$

The mean length l_{cs} is

$$l_{cs} = \frac{\pi(D_{out} - h_{cs})}{2P_1} K_{cs} = \frac{\pi(0.730 - 628 \cdot 10^{-3})}{2 \cdot 2} 0.4 = 0.2095\text{m} \tag{16.122}$$

K_{cs} is an amplification factor which essentially depends on the flux density level in the back iron.

Again B_{c1max} is a bit too small. Consequently, if the temperature rise allows, the outer stator diameter may be reduced, perhaps to 0.700 m (from 0.73 m).

$$F_{cs} = l_{cs} H_{c1\max} = 0.2095 \cdot 450 = 94.27\text{Aturns}$$

- Rotor back iron mmf F_{cr} (as for the stator)

$$F_{cr} = H_{cr} l_{cr} = 760 \cdot 0.1118 = 84.95\text{Aturns} \tag{16.123}$$

$$H_{cr}(B_{cr}) = H_{cr}(1.4) = 760\text{A/m}$$

$$l_{cr} = \frac{\pi(D_{ir} + h_{cr})}{2P_1} K_{cr} = \frac{\pi(0.296 - 0.06)}{2 \cdot 2} 0.4 = 0.1118\text{m} \tag{16.124}$$

From (16.109), the total mmf per pole F_{1m} is

$$F_{1m} = 1418 + 30.22 + 99.163 + \frac{94.27 + 84.95}{2} = 1636.99$$

The mmf per pole F_{1m} is

$$F_{1m} = \frac{3\sqrt{2} I_\mu W_1 K_{w1}}{p_1 \pi} \tag{16.125}$$

So, the magnetization current I_μ is

$$I_\mu = \frac{F_{1m} p_1 \pi}{3\sqrt{2} W_1 K_{w1}} = \frac{1636.99 \cdot 2 \cdot \pi}{3\sqrt{2} \cdot 12 \cdot 18 \cdot 0.923} = 12.19\text{A} \tag{16.126}$$

The ratio between the magnetization and rated phase current I_μ/I_{1n},

$$\frac{I_\mu}{I_{1n}} = \frac{12.19}{\frac{120}{\sqrt{3}}} = 0.175$$

Even at this power level and $2p_1 = 4$, because the ratio $\tau/g = 314/1.5 = 209.33$ (pole-pitch/airgap) is rather large and the saturation level is low, the magnetization current is lower than 20% of rated current. The machine has slightly more iron than needed or the airgap may be increased from $1.5\cdot10^{-3}$m to $(1.8–2)\cdot10^{-3}$m. As a bonus the additional surface and tooth flux pulsation core losses will be reduced. We may simply reduce the outer diameter to notably saturate the stator back iron. The gain will be less volume and weight.

➢ The rotor winding parameters

As the stator phase resistance R_s and leakage reactance (X_{ls}) have already been computed ($R_s = 0.8576\Omega$, $X_{ls} = 6.667\Omega$), only the rotor winding parameters R_r and X_{lr} have to be calculated.

The computation of R_r, X_{lr} requires the calculation of rotor coil end connection length based on its geometry (Figure 16.16).

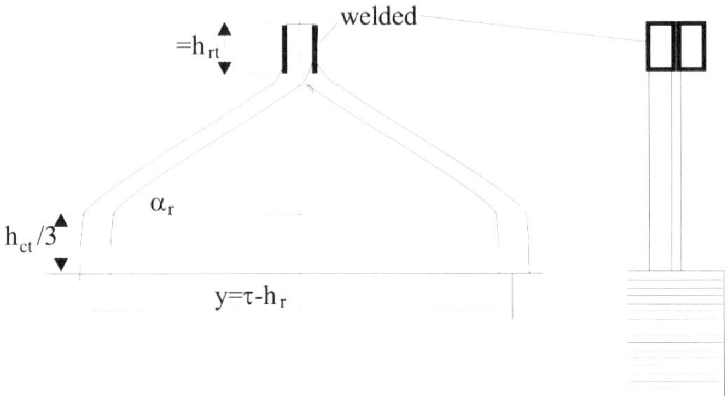

Figure 16.17 Rotor half-formed coil winding

The end connection length l_{fr} per rotor side is

$$l_{fr} = \frac{y}{\cos\alpha_r} + \left(\frac{h_{rt}}{3} + h_{rt}\right)\cdot 2 = \frac{\left(0.314 - 35.5\cdot10^{-3}\cdot\frac{\pi}{4}\right)}{0.707} + \tag{16.127}$$

$$+\frac{4}{3}\cdot 2\cdot 35.5\cdot 10^{-3} = 0.5\text{m}$$

So the phase resistance R_r^r (before reduction to stator) is

$$R_r^{\,r} = K_R \rho_{Co80^0} \frac{2(l_{fr} + l_{geo})W_2}{A_{cor}} = 1.0 \cdot 2.2 \cdot 10^{-8} \frac{2(0.5 + 0.451) \cdot 20}{96 \cdot 10^{-6}} = 0.87 \cdot 10^{-2}\Omega \quad (16.128)$$

➢ The rotor slot leakage geometrical permeance coefficient λ_{sr} is (Figure 16.14)

$$\lambda_{s,r} \approx \frac{2b_c}{3b_r} + \frac{h_4}{4b_r} + \frac{h_2}{b_s} + \frac{h_{or}}{b_{or}} + \frac{2h_w}{(b_{or} + b_r)} = \frac{2 \cdot 12}{3 \cdot 10.5} + \frac{1.2}{4 \cdot 10.5} + \frac{1.8}{10.5} + \frac{2.5}{5} + \frac{2 \cdot 4}{5 + 10.5} = 1.978 \quad (16.129)$$

The rotor differential leakage permeance coefficient λ_{dr} is (as in [3])

$$\lambda_{dr} = \frac{0.9 \cdot \tau_r (q_2 K_{w2})^2 K_{or} \sigma_{dr}}{K_c g} \quad (16.130)$$

$$K_{or} = 1.0 - 0.033 \frac{b_{or}^{\,2}}{g\tau_r} = 1.0 - 0.033 \frac{5^2}{1.5 \cdot 24.86} = 0.9778 \quad (16.131)$$

σ_{dr} is the differential leakage coefficient as defined in Figure 6.3, Chapter 6.3. For $q_2 = 5$ and diametrical coils $\sigma_{dr} = 0.64 \cdot 10^{-2}$.

$$\lambda_{dr} = 0.9 \frac{24.86}{1.8 \cdot 1.5} (5 \cdot 0.9566)^2 \cdot 0.9778 \cdot 0.64 \cdot 10^{-2} = 1.4428$$

The end connection permeance coefficient λ_{fr} (defined as for the stator) is

$$\lambda_{fr} = 0.34 \frac{q_2}{l_{ir}} (l_{fr} - 0.64\tau) = 0.34 \frac{5}{0.406} (0.5 - 0.64 \cdot 0.314) = 1.252 \quad (16.132)$$

The rotor phase leakage reactance X_{lr} (before reduction to stator) is

$$X_{lr}^{\,r} = 4\pi f_1 \mu_0 W_2^{\,2} \frac{l_i}{P_1 q_2} \sum \lambda_{ri} = 4 \cdot 3.14 \cdot 60 \cdot 1.256 \cdot 10^{-6} \cdot 20^2 \frac{0.406}{2 \cdot 5} (1.978 + 1.4428 + 1.252) = 7.1828 \cdot 10^{-2}\Omega \quad (16.133)$$

After reduction to stator, the rotor resistance and leakage reactances $R_r^{\,r}$, $X_{lr}^{\,r}$ become R_r and X_{lr}.

$$R_r = R_r^{r}\left(\frac{W_1K_{w1}}{W_2K_{w2}}\right)^2 = 0.87\cdot 10^{-2}\cdot\left(\frac{12\cdot 18\cdot 0.923}{20\cdot 0.9566}\right)^2 = 0.94\Omega$$

$$X_{lr} = X_{lr}^{r}\left(\frac{W_1K_{w1}}{W_2K_{w2}}\right)^2 = 7.1828\cdot 10^{-2}\cdot 108.6 = 7.8\Omega \qquad (16.134)$$

The magnetization reactance X_m is

$$X_m \approx \frac{(V_{ph} - I_\mu X_{ls})}{I_\mu} = \frac{(4000 - 12.19\cdot 6.667)}{12.19} = 326.56\Omega \qquad (16.135)$$

The p.u. parameters are

$$X_n = \frac{V_{ph}}{I_{1n}} = \frac{4000}{\frac{120}{\sqrt{3}}} = 57.66\Omega$$

$$r_s = \frac{R_s}{X_n} = \frac{0.8756}{57.66} = 0.01487\text{p.u.}$$

$$r_r = \frac{R_r}{X_n} = \frac{0.94}{57.66} = 0.0163\text{p.u.} \qquad (16.136)$$

$$x_{ls} = \frac{X_{ls}}{X_n} = \frac{6.667}{57.66} = 0.1156\text{p.u.}$$

$$x_{lr} = \frac{X_{lr}}{X_n} = \frac{7.8}{57.66} = 0.1352\text{p.u.}$$

$$x_m = \frac{X_m}{X_n} = \frac{326.56}{57.66} = 5.66\text{p.u.} > (3.2 - 4)\text{p.u.}$$

Again, x_m is visibly higher than usual but, as explained earlier, this may be the starting point for further reduction in machine size. A simple reduction of the outer stator diameter to saturate the stator back iron would bring x_m into the standard interval of (3.5 to 4)p.u. for 4 pole IMs of 700 kW power range with wound rotors. An increase in airgap will produce similar effects, with additional core loss reduction.

We leave x_m as it is in (16.136) as a reminder that the IM design is an iterative procedure.

➢ Losses and efficiency

The losses in induction machines may be classified as

- Conductor (winding or electric) losses
- Core losses
- Mechanical / ventilation losses

The stator winding rated losses p_{cos} are

$$p_{cos} = 3R_sI_{1n}^{\ 2} = 3\cdot 0.8576\cdot 69.36^2 = 12.378\cdot 10^3\,W \qquad (16.137)$$

For the rotor winding:

$$p_{cor} = 3R_rI_{1n}^{\ 2} = 3R_rK_i^{\ 2}I_{1n}^{\ 2} = 3\cdot 0.94\cdot(0.936\cdot 69.36)^2 = 11.885\cdot 10^3\,W \quad (16.138)$$

Considering a voltage drop, V_{ss} = 0.75 V in the slip ring and brushes, the slip-ring brush losses p_{sr} (with I_{2n} from (16.102)) is

$$p_{sr} = 3V_{sr}I_{2n} = 3\cdot 0.75\cdot 676.56 = 1.522\cdot 10^3\,W$$

Additional winding losses due to skin and proximity effects (Chapter 11) may be neglected for this case due to proper conductor sizing.

The core losses have three components (Chapter 11).

- Fundamental core losses
- Additional no load core losses
- Additional load (stray) core losses

The fundamental core losses are calculated by empirical formulas (FEM may also be used) as in Chapter 15.

The stator teeth fundamental core losses P_{iront} is (15.98)

$$P_{iront} = K_t p_{\frac{10}{50}}\left(\frac{f_1}{50}\right)^{1.3} B_{ts}^{1.7} G_{ts} \qquad (16.139)$$

K_t = 1.6 – 1.8, takes into account the mechanical machining influence on core losses; p_{10}(1T, 50 Hz) = (2 – 3) W/Kg for 0.5 mm thick laminations.

The value of B_t varies along the tooth height so the average value $B_{ts}(H_{ts})$, Equation (16.114) is B_{ts}(528 A/m) = 1.32 T.

The stator teeth weight G_{ts} is

$$G_{ts} = \left[\frac{\pi}{4}\left((D_{is} + 2h_{st})^2 - D_{is}^{\ 2}\right) - h_{ts}b_sN_s\right]l_i\gamma_{iron} =$$

$$= \left[\frac{\pi}{4}\left((0.490 + 2\cdot 56.2\cdot 10^{-3})^2 - 0.490^2\right) - 56.2\cdot 10^{-3}\cdot 10\cdot 10^{-3}\cdot 72\right]\cdot \qquad (16.140)$$

$$\cdot 0.406\cdot 7800 = 175.87\,Kg$$

From (16.139),

$$P_{iront} = 1.8\cdot 2.4\left(\frac{60}{50}\right)^{1.3} 1.32^{1.7}\cdot 175.87 = 1.543\cdot 10^3\,W$$

Similarly, the stator back iron (yoke) fundamental losses $p_{iron}^{\ y}$ are

$$P_{iron}^{\ y} = K_y p_{\frac{10}{50}}\left(\frac{f_1}{50}\right)^{1.3} B_{cs}^{1.7} G_{ys} \qquad (16.141)$$

$K_y = 1.2 - 1.4$ takes care of the influence of mechanical machining on yoke losses (for lower power machines this coefficient is larger).

The stator yoke weight G_{ys} is

$$G_{ys} = \frac{\pi\left(D_{out}^{2} - \left(D_{is} + 2h_{st}\right)^2\right)}{4} l_i \gamma_{iron} =$$
$$= \frac{\pi}{4}\left(0.73^2 - \left(0.49 + 2\cdot 0.0562\right)^2\right)\cdot 0.406\cdot 7800 = 422.6\text{Kg} \tag{16.142}$$

$$P_{iron}^{y} = 1.35\cdot 2.4\cdot\left(\frac{60}{50}\right)^{1.7}\cdot 1.274^{1.7}\cdot 422.6 = 2.619\cdot 10^3\,\text{W}$$

Stator and rotor surface core losses, due to slot openings (Chapter 11), are approximately

$$P_{siron}^{s} = 2p_1\tau\frac{\left(\tau_s - b_{os}\right)}{\tau_s} l_i K_{Fe} p_{siron}^{s} \tag{16.143}$$

with the specific surface stator core losses (in W/m^2) p_{siron}^{s} as

$$p_{siron}^{s} \approx 5\cdot 10^5 B_{os}^{\ 2}\tau_r^{\ 2} K_0\left(\frac{N_r 60 f_1}{p_1\cdot 10000}\right)^{1.5} \tag{16.144}$$

B_{os} is the airgap flux pulsation due to rotor slot openings

$$B_{os} = \beta_s K_c B_g = 0.2\cdot 1.48\cdot 0.8 = 0.2368\text{T} \tag{16.145}$$

β_s comes from Figure 5.4 (Chapter 5). With $b_{or}/g = 5/1.5$, $\beta_s \approx 0.2$. K_0 again depends on mechanical factors: $K_0 = 1.5$, in general.

$$p_{siron}^{s} = 5\cdot 10^5\cdot 0.2368^2\cdot 0.02486^2\cdot 1.5\cdot\left(\frac{60\cdot 60\cdot 60}{2\cdot 10000}\right)^{1.5} = 922.49\text{W}/\text{m}^2$$

From (16.143):

$$P_{siron}^{s} = 2\cdot 2\cdot 0.314\frac{\left(21.37 - 10\right)}{21.37}\cdot 0.406\cdot 0.95\cdot 922.49 = 0.237\cdot 10^3\,\text{W}$$

As the rotor slot opening b_{or} per gap g ratio is only 5/1.5, the stator surface core losses are expected to be small.

The rotor surface core losses p_{siron}^{r} are

$$P_{siron}^{r} = 2p_1\frac{\left(\tau_r - b_{or}\right)}{\tau_s}\tau l_i K_{Fe} p_{siron}^{r} \tag{16.146}$$

with $$p^r_{siron} = 5\cdot10^5 B_{or}^2 \tau_s^2 K_0 \left(\frac{N_r 60 f_1}{p_1\cdot10000}\right)^{1.5};\quad B_{or} = \beta_r K_c B_g$$

The coefficient β_r for $b_{os}/g = 10/1.5$ is, Figure 5.4, Chapter 5, $\beta_r = 0.36$. Consequently,

$$B_{or} = 0.36\cdot1.48\cdot0.8 = 0.426T$$

$$p^r_{siron} = 5\cdot10^5\cdot0.426^2\cdot0.0214^2\cdot2.0\cdot\left(\frac{72\cdot60\cdot60}{2\cdot10000}\right)^{1.5} = 3877W/m^2$$

$K_0 = 2.0$ as the rotor is machined to a rated airgap.

$$P^r_{siron} = 2\cdot2\frac{(24.86-5)}{24.86}\cdot0.314\cdot0.406\cdot0.95\cdot3877 = 1.5\cdot10^3 W$$

The tooth flux pulsation core losses are still to be determined.

$$P_{puls} = P^s_{puls} + P^r_{puls} \approx K_0^{'} 0.5\cdot10^{-4}\left[\left(N_r\frac{f_1}{p_1}\beta_{ps}\right)^2 G_{ts} + \left(N_s\frac{f_1}{p_1}\beta_{pr}\right)^2 G_{tr}\right] \quad (16.147)$$

$$\beta_{ps} \approx (K_{c2}-1)B_g = (1.085-1)\cdot0.8 = 0.068T$$
$$\beta_{pr} \approx (K_{c1}-1)B_g = (1.36-1)\cdot0.8 = 0.288T$$

$K_0' > 1$ is a technological factor to account for pulsation loss increases due to saturation and type of lamination manufacturing technology (cold or hot stripping). We take here $K_0' = 2.0$.

The rotor teeth weight G_{tr} is

$$G_{tr} = \left[\frac{\pi\left((D_{is}-2g)^2 - (D_{is}-2g-2h_{rt})^2\right)}{4} - N_r h_{rt} b_r\right] l_i \gamma_{iron} =$$

$$= \left[\frac{\pi\left((0.49-3\cdot10^{-3})^2 - (0.487-2\cdot0.0355)^2\right)}{4} - 60\cdot0.0355\cdot105\cdot10^{-2}\right]\cdot \quad (16.148)$$

$$\cdot0.406\cdot7800 = 88.55Kg$$

From (16.147),

$$P_{puls} = 2\cdot0.5\cdot10^{-4}\left[\left(60\cdot\frac{60}{2}\cdot0.068\right)^2\cdot175.87 + \left(72\cdot\frac{60}{2}\cdot0.288\right)^2\cdot88.55\right] =$$

$$= 3.684\cdot10^3 W$$

The mechanical losses due to friction (brush-slip ring friction and ventilator on-shaft mechanical input) depend essentially on the type of cooling machine design.

We take them all here as being $P_{mec} = 0.01P_n$. For two-pole or high-speed machines, $P_{mec} > 0.01P_n$ and should be, in general, calculated as precisely as feasible in a thorough mechano thermal design.

- The machine rated efficiency η_n is

$$\eta_n = \frac{P_n}{P_n + p_{cos} + p_{cor} + P_{sr} + P^{t}_{iron} + P^{y}_{iron} + P^{s}_{siron} + P^{r}_{siron} + P_{puls} + p_{mec}} =$$

$$= \frac{736 \cdot 10^3}{10^3(736 + 12.378 + 11.885 + 1.522 + 1.543 + 2.619 + 1.237 + 3.877 + 3.684 + 7.36)} = $$

$$= 0.943 \quad (16.149)$$

As the efficiency is notably below the target value (0.96), efforts to improve the design are in order. Additional losses are to be reduced. Here is where optimization comes into play.

- The rated slip S_n (with short-circuited slip rings) may be calculated as

$$S_n \approx \frac{R_r K_i I_{1n}}{K_e V_{ph}} = \frac{0.94 \cdot 0.936 \cdot 69.36}{0.97 \cdot 4000} = 0.0157 \quad (16.150)$$

Alternatively, accounting for all additional core losses in the stator,

$$S_n = \frac{p_{cor}}{P_n + p_{cor} + p_{mec}} = \frac{11.885 \cdot 10^3}{10^3 \cdot (736 + 11.885 + 7.36)} = 0.015736 \quad (16.151)$$

The rather large value of rated slip is an indication of not so high efficiency. A reduction of current density both in the stator and rotor windings seems appropriate to further increase efficiency.

The rather low teeth flux densities allow for slot area increases to cause winding loss reductions.

- The breakdown torque

The approximate formula for breakdown torque is

$$T_{bk} \approx \frac{3p_1}{2\omega_1} \frac{V_{ph}^{2}}{X_{ls} + X_{lr}} = \frac{3 \cdot 2}{2 \cdot 2\pi \cdot 60} \frac{4000^2}{(6.67 + 7.8)} = 8.8035 \cdot 10^3\,\text{Nm} \quad (16.152)$$

The rated torque T_{en} is

$$T_{en} = \frac{p_1 P_n}{\omega_1 (1 - S_n)} = \frac{2 \cdot 736 \cdot 10^3}{2\pi \cdot 60 \cdot (1 - 0.015)} = 3.966 \cdot 10^3\,\text{Nm} \quad (16.153)$$

The breakdown torque ratio t_{ek} is

$$t_{ek} = \frac{T_{bk}}{T_{en}} = \frac{8.8035}{3.966} \approx 2.2 \quad (16.154)$$

This is an acceptable value.

16.8 SUMMARY

- Above 100 kW, induction motors are built for low voltage (≤690 V) and for high voltage (between 1 KV and 12 KV).
- Low voltage IMs are designed and built with up to more than 2 MW power per unit, in the eventuality of their use with PWM converter supplies.
- For constant voltage and frequency supply, besides efficiency and initial costs, the starting torque and current and the peak torque are important design factors.
- The stator design is based on targeted efficiency and power factor, while rotor design is based on the starting performance and breakdown torque.
- Deep bar cage rotors are used when moderate p.u. starting torque is required ($t_{LR} < 1.2$). The bar depth and width are found based on breakdown torque, starting torque, and starting current constraints. This methodology reduces the number of design iterations drastically.
- The stator design is dependent on the cooling system: axial cooling (unistack core configuration) and radial-axial cooling (multistack core with radial ventilation channels, respectively).
- For high voltage stator design, open rectangular slots with form-wound chorded (rigid) coils, built into two-layer windings, are used. Practically only rectangular cross section conductors are used.
- The rather large total insulation thickness makes the stator slot fill in high voltage windings rather low, with slot aspect ratios $h_{st}/b_s > 5$. Large slot leakage reactance values are expected.
- In low voltage stator designs, above 100 kW, semiclosed-slot, round-wire coils are used only up to 200 to 250 kW for $2p_1 = 2, 4$; also, for larger powers and large numbers of poles ($2p_1 = 8, 10$), with quite a few current paths in parallel ($a_1 = p_1$) may be used.
- For low voltage stator designs, prepared also for PWM converter use, only rectangular shape conductors in form-wound coils are used to secure good insulation as required for PWM converter supplies. To reduce skin effect, the conductors are divided into a few elementary conductors in parallel. Stranding is performed twice, at the end connection zone (Chapter 9).
- Deep bar design handles large breakdown torques, $t_{bk} \approx 2.5 – 2.7$ and more, with moderate starting torques $t_{LR} < 1.2$ and starting currents $i_{LR} < 6.1$. Based on breakdown torque, starting torque, and current values, with stator already designed (and thus stator resistance and leakage reactance values known), the rotor resistance and leakage reactance values (with and without skin effect and leakage saturation consideration) are calculated. This leads

directly to the adequate rectangular deep bar geometry. If other bar geometries are used, the skin effect coefficients are calculated by the refined multilayer approach (Chapter 9).

- Double cage rotor designs lead to lower starting currents $i_{LR} < 5.5$, larger starting torque $t_{LR} > (1.4$ to $1.5)$ and good efficiency but moderate power factor.
  A similar methodology, based on starting and rated power performance leads to the starting and working cage sizing, respectively. Thus, a first solution meeting the main constraints is obtained. Refinements should follow eventually through design optimization methods.
- For limited variable speed range, or for heavy and frequent starting (up to breakdown torque at start), wound rotor designs are used.
- The wound-rotor two-layer windings are, in general, low voltage type, with half-formed diametrical wave uniform coils in semiopen slots (to reduce surface and tooth flux pulsation core loss). A few elementary conductors in parallel may be needed to handle the large rotor currents as the power goes up.
- The wound rotor design is similar to stator design, being based only on rated targeted efficiency (and power factor) and assigned rotor current density.
- This chapter presents design methodologies through a single motor (low and high voltage stator 736 kW (1,000 HP) design with deep bar cage, double cage, and wound rotor, respectively). Only for the wound rotor, all parameters (losses, efficiency, rated slip, breakdown torque, and rated power factor) are calculated. Ways to improve the resultant performance are indicated. Temperature rise calculations may be done as in Chapter 15, but with more complex mathematics. If temperature rise is not met (insulation class F and temperature rise class B) and it is too high, a new design with a larger geometry or lower current densities or (and) flux densities has to be tried.
- Chapters 14, 15, and 16 laid out the machine models (analysis) and a way to size the machine for given performance and constraints (preliminary synthesis).
  The mathematics in these chapters may be included in a design optimization software. Design optimization will be treated in a separate chapter. FEM is to be used for refinements.
- The above methodologies are by no means unique to the scope, as IM design is both a science and an art.

16.9 REFERENCES

1. ABB Industry Oy, Machines, Product Information, AMA Machines, PIF 2b000-999/364.
2. ABB Industry Oy, Machines, Product Information, HXR Machines, PIF 2a000-999/364.

3. P.L. Alger, Induction Machines. Their Behaviour and Uses, Second edition, 1995, Gordon & Breach Publishers.
4. C.M. Glew, G.D. LeFlem, J.D. Walker, The Customer Is Always Right–Improved Performance of High Speed Induction Motors, Record of "Electrical Machines and Drives", 11 – 13, Sept, 1995, Conference Publication No.412, IEE (London) 1995, pp.37 – 40.
5. A. Demenkco, K. Oberretl, Calculation of Additional Slot Leakage Permeance in Electrical Machines Due to Radial Ventilation Ducts, COMPEL – The Journal for Computation and Mathematics in E.E.E., Vol.11, No.1, pp.93 – 96.
6. E. Levi, Polyphase Motors–A Direct Approach to Their Design, Wiley Interscience, 1985.

Chapter 17

INDUCTION MACHINE DESIGN FOR VARIABLE SPEED

17.1 INTRODUCTION

Variable speed drives with induction motors are by now a mature technology with strong and dynamic markets for applications in all industries. Based on the load torque/speed envelope, three main types of applications may be distinguished:

- Servodrives: no constant power speed range
- General drives: moderate constant power speed range ($\omega_{max}/\omega_b \leq 2$)
- Constant power drives: large constant power speed range ($\omega_{max}/\omega_b \geq 2$)

Servodrives for robots, machine tools, are characterized, in general, by constant torque versus speed up to base speed ω_b. The base speed ω_b is the speed for which the motor can produce (for continuous service) for rated voltage and rated temperature rise, the rated (base) power P_b, and rated torque T_{eb}.

Servodrives are characterized by fast torque and speed response and thus for short time, during transients, the motor has to provide a much higher torque T_{ek} than T_{eb}. The higher the better for speed response quickness. Also servodrives are characterized by sustained very low speed and up to rated (base) torque operation, for speed or position control. In such conditions, low torque pulsations and limited temperature rise are imperative.

Temperature rise has to be limited to avoid both winding insulation failure and mechanical deformation of the shaft which would introduce errors in position control.

In general, servodrives have a constant speed (separate shaft) power, grid fed, ventilator attached to the IM at the non-driving end. The finned stator frame is thus axially cooled through the ventilator's action. Alternatively, liquid cooling of the stator may be provided.

Even from such a brief introduction, it becomes clear that the design performance indexes of IMs for servodrives need special treatment. However, fast torque and speed response and low torque pulsations are paramount. Efficiency and power factor are second order performance indexes as the inverter KVA rating is designed for the low duration peak torque (speed) transients requirements.

General drives, which cover the bulk of variable speed applications, are represented by fans, pumps, compressors, etc.

General drives are characterized by a limited speed control range, in general, from $0.1\omega_b$ to $2\omega_b$. Above base speed ω_b constant power is provided. A limited constant power speed range $\omega_{max}/\omega_b = 2.0$ is sufficient for most cases. Above base speed, the voltage stays constant.

Based on the stator voltage circuit equation at steady state,

$$\overline{V}_s = \overline{I}_s R_s + j\omega_1 \overline{\Psi}_s \tag{17.1}$$

with Rs ≈ 0

$$\Psi_s \approx \frac{V_s}{\omega_1} \tag{17.2}$$

Above base speed (ω_b), the frequency ω_1 increases for constant voltage. Consequently, the stator flux level decreases. Flux weakening occurs. We might say that general drives have a 2/1-flux weakening speed range.

As expected, there is some torque reserve for fast acceleration and braking at any speed. About 150% to 200% overloading is typical.

General drives use IMs with on-the-shaft ventilators. More sophisticated radial-axial cooling systems with a second cooling agent in the stator may be used.

General drives may use high efficiency IM designs as in this case efficiency is important.

Made with class F insulated preformed coils and insulated bearings for powers above 100 kW and up to 2000 kW, and at low voltage (maximum 690 V), such motors are used in both constant and variable speed applications. While designing IMs on purpose for general variable speed drives is possible, it may seem more practical to have a single design both for constant and variable speed: the high efficiency induction motor.

Constant power variable speed applications, such as spindles or hybrid (or electric) car propulsion, generator systems, the main objective is a large flux weakening speed range $\omega_{max}/\omega_b > 2$, in general more than 3–4 , even 6–7 in special cases. Designing an IM for a wide constant power speed range is very challenging because the breakdown torque T_{bk} is in p.u. limited: $t_{bk} < 3$ in general.

$$T_{eK} \approx \frac{3p_1}{2}\left(\frac{V_{ph}}{\omega_1}\right)^2 \cdot \frac{1}{L_{sc}} \tag{17.3}$$

Increasing the breakdown torque as the base speed (frequency) increases could be done by

- Decreasing the pole number $2p_1$
- Increasing the phase voltage
- Decreasing the leakage inductance L_{sc} (by increased motor size, winding tapping, phase connection changing, special slot (winding) designs to reduce L_{sc})
  Each of these solutions has impact on both IM and static power converter costs. The global cost of the drive and the capitalized cost of its losses are solid criteria for appropriate designs. Such applications are most challenging. Yet another category of variable speed applications is represented by super-high speed drives.

- For fast machine tools, vacuum pumps etc., speeds which imply fundamental frequencies above 300 Hz (say above 18,000 rpm) are considered here for up to 100 kW powers and above 150 Hz (9000 rpm) for higher powers.

As the peripheral speed goes up, above (60–80) m/s, the mechanical constraints become predominant and thus novel rotor configurations become necessary. Solid rotors with copper bars are among the solutions considered in such applications. Also, as the size of IM increases with torque, high-speed machines tend to be small (in volume/power) and thus heat removal becomes a problem. In many cases forced liquid cooling of the stator is mandatory.

Despite worldwide efforts in the last decade, the design of IMs for variable speed, by analytical and numerical methods, did not crystallize in widely accepted methodologies.

What follows should be considered a small step towards such a daring goal.

As basically the design expressions and algorithms developed for constant V/f (speed) are applicable to variable speed design, we will concentrate only on what distinguishes this latter enterprise.

- In the end, a rather detailed design example is presented. Among the main issues in IM design for variable speed, we treat here
- Power and voltage derating
- Reducing skin effect
- Reducing torque pulsations
- Increasing efficiency
- Approaches to leakage inductance reduction
- Design for wide constant power wide speed range
- Design for variable very high speed

17.2 POWER AND VOLTAGE DERATING

An induction motor is only a part of a variable speed drive assembly (Figure 17.1).

As such, the IM is fed from the power electronics converter (PEC) directly, but indirectly, in most cases, from the industrial power grid.

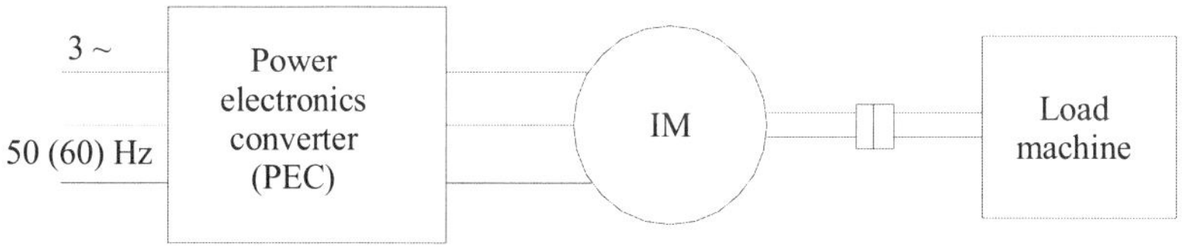

Figure 17.1 Induction machine in a variable speed drive system

There are a few cases where the PEC is fed from a dc source (battery).

The PEC inflicts on the motor voltage harmonics (in a voltage source type) or current harmonics (in a current source type). In this way, voltage and current harmonics, whose frequency and amplitude are dependent on the PWM (control) strategy and power level, are imposed on the induction motor.

Additionally, high frequency common voltage mode currents may occur in the stator phases in high frequency PWM voltage source converter IM drives. All modern PECs have incorporated filtering methods to reduce the additional current and voltage (flux) harmonics in the IMs as they produce additional losses.

Analytical and even finite element methods have been proposed to cater to these time harmonics losses (see Chapter 11). Still, these additional time harmonics core and winding losses depend not only on machine geometry and materials, but also on PWM, switching frequency and load level. [1,2]

On top of this, for given power grid voltage, the maximum fundamental voltage at motor terminals depends on the type of PEC (single or double stage, type of power electronics switches (PES)) and PWM (control) strategy.

Though each type of PEC has its own harmonics and voltage drop signature, the general rule is that lately both these indexes have decreased. The matrix converter is a notable exception in the sense that its voltage derating (drop) is larger (up to 20%) in general.

Voltage derating – less than 10%, in general 5%–means that the motor design is performed at a rated voltage V_m which is smaller than the a.c. power grid voltage V_g:

$$V_m = V_g(1 - v_{derat}); v_{der} < 0.1 \tag{17.4}$$

Power derating comes into play in the design when we choose the value of Esson's constant C_0 (W/m^3), as defined by past experience for sinusoidal power supply, and reduce it to $C_0^{'}$ for variable V/f supply:

$$C_0^{'} = C_0(1 - p_{derat}) \; ; p_{derat} \approx (0.08 - 0.12) \tag{17.5}$$

It may be argued that this way of handling the PEC-supplied IM design is quite empirical. True, but this is done only to initiate the design (sizing) process. After the sizing is finished, the voltage drops in the PEC and the time harmonics core and winding losses may be calculated (see Chapter 11). Then design refinements are done. Alternatively, if prototyping is feasible, test results are used to validate (or correct) the loss computation methodologies.

There are two main cases: one when the motor exists, as designed for sinusoidal power supply, and the other when a new motor is to be designed for the scope.

The derating concepts serve both these cases in the same way.

However, the power derating concept is of little use where no solid past experience exists, such as in wide constant power speed range drives or in super-high speed drives. In such cases, the tangential specific force (N/cm^2), Chapter 14, with limited current sheet (or current density) and flux densities, seem to be the right guidelines for practical solutions. Finally, the temperature rise and performance (constraints) checks may lead to design iterations. As already mentioned in Chapter 14, the rated (base) tangential specific force (σ_t) for sinusoidal power supply is

$$\sigma_t^{sin} \approx (0.3 - 4.0) N / cm^2 \qquad (17.6)$$

Derating now may be applied to σ_t^{sin} to get σ_t^{PEC}

$$\sigma_t^{PEC} = \sigma_t^{sin}(1 - p_{derat}) \qquad (17.7)$$

for same rated (base) torque and speed.

The value of σ_t^{PEC} increases with rated (base) torque and decreases with base speed.

17.3 REDUCING THE SKIN EFFECT IN WINDINGS

In variable speed drives, variable V and f are used. Starting torque and current constraints are not relevant in designing the IM. However, for fast torque (speed) response during variable frequency and voltage starting or loading or for constant power wide speed range applications, the breakdown torque has to be large.

Unfortunately, increasing the breakdown torque without enlarging the machine geometry is not an easy task.

On the other hand, rotor skin effect that limits the starting current and produces larger starting torque, based on a larger rotor resistance is no longer necessary.

Reducing skin effect is now mandatory to reduce additional time harmonics winding losses.

Skin effect in winding losses depends on frequency, conductor size, and position in slots. First, the rotor and stator skin effect at fundamental frequency is to be reduced. Second, the rotor and stator skin effect has to be checked and limited at PEC switching frequency. The amplitude of currents is larger for the fundamental than for time harmonics. Still the time harmonics conductor losses at large switching frequencies are notable. In super-high speed IMs the fundamental frequency is already large, (300-3(5)000) Hz. In this case the fundamental frequency skin effects are to be severely checked and kept under control for any practical design as the slip frequency may reach tenth of Hz (up to 50-60 Hz).

As the skin effect tends to be larger in the rotor cage we will start with this problem.

Rotor bar skin effect reduction

The skin effect is a direct function of the parameter:

$$\zeta = h\sqrt{\frac{S\omega_1\sigma_{cor}\mu_0}{2}} \qquad (17.8)$$

The slot shape also counts. But once the slot is rectangular or circular, only the slot diameter, and respectively, the slot height counts.

Rounded trapezoidal slots may also be used to secure constant tooth flux density and further reduce the skin effects (Figure 17.2).

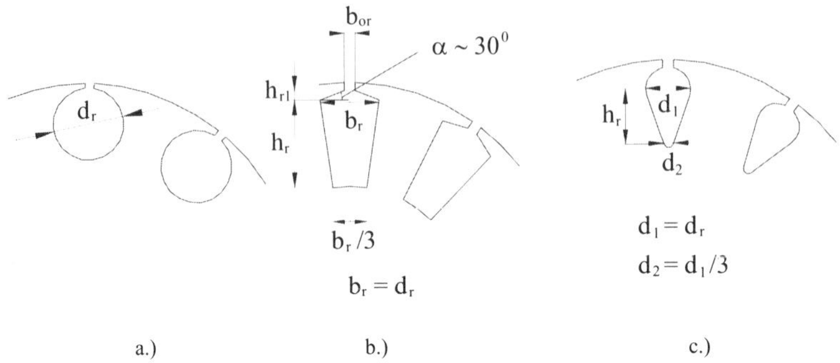

Figure 17.2 Rotor bar slots with low skin effect
a.) round shape b.) rectangular c.) pear-shape

For given rotor slot (bar) area A_b (Figure17.2 a,b,c), we have

$$\begin{aligned} A_b &= \frac{\pi d_r^2}{4} \\ &= h_r \cdot \frac{2d_r}{3} + \frac{h_{r1}}{2}\left(b_{or} + d_r\right) \\ &= h_r \frac{2}{3} d_1 + \frac{\pi 5}{36} d_r^2 \end{aligned} \tag{17.9}$$

For the rectangular slot, the skin effect coefficients K_r and K_x have the standard formulas

$$\begin{aligned} K_R &= \zeta \frac{\sinh 2\zeta + \sin 2\zeta}{\cosh 2\zeta - \cos 2\zeta} \\ K_X &= \frac{3}{2\zeta} \frac{\sinh 2\zeta - \sin 2\zeta}{\cosh 2\zeta - \cos 2\zeta} \end{aligned} \tag{17.10}$$

In contrast, for round or trapezoidal-round slots, the multiple-layer approach of Chapter 9, has to be used.

A few remarks are in order:

- As expected, for given geometry and slip frequency, skin effects are more important in copper than in aluminum bars
- For given rotor slot area, the round bar has limited use
- As the bar area (bar current or motor torque) increases, the maximum slip frequency $f_r = Sf_1$ for which $K_R < 1.1$ diminishes
- Peak slip frequency f_{srk} varies from 2 Hz to 10 Hz
- The smaller values correspond to larger (MW) machines and larger values to subKW machines designed for base frequencies of 50 (60) Hz. For f_{srK}, $K_R < 1.1$ has to be fulfilled if rotor additional losses are to

be limited. Consequently, the maximum slot depth depends heavily on motor peak torque requirements

- For super-high speed machines, f_{srk} may reach even 50 (60) Hz, so extreme care in designing the rotor bars is to be exercised (in the sense of severe limitation of slot depth, if possible)
- Maintaining reduced skin effect at f_{srK} means, apparently, less deep slots and thus, for given stator bore diameter, longer lamination stacks. As shown in the next paragraph this leads to slightly lower leakage inductances, and thus to larger breakdown torque. That is, a beneficial effect.
- When the rotor skin effect for f_{srK} may not be limited by reducing the slot depth, we have to go so far as to suggest the usage of a wound rotor with shortcircuited phases and mechanically enforced end-connections against centrifugal forces
- To reduce the skin effect in the end rings, they should not be placed very close to the laminated stack, though their heat transmission and mechanical resilience is a bit compromised
- Using copper instead of aluminium leads to a notable reduction of rotor bar resistance for same bar cross-section though the skin effect is larger. A smaller copper bar cross-section is allowed, for same resistance as aluminum, but for less deep slots and thus smaller slot leakage inductance. Again, larger breakdown torque may be obtained. The extracost of copper may prove well worth while due to lower losses in the machine.
- As the skin effect is maintained low, the slot-body geometrical specific permeance λ_{sr} for the three cases mentioned earlier (Figure 17.1) is:

$$\begin{aligned} &\lambda_{sr}^{round} \approx 0.666 \\ &\lambda_{sr}^{rect} \approx \frac{h_r}{3b_r} + \frac{2}{3\sqrt{3}} \\ &b_r = d_r \\ &\lambda_{sr}^{trap} \approx \frac{h_r}{2d_r} + 0.4 \end{aligned} \tag{17.11}$$

Equations (17.11) suggest that, in order to provide for identical slot geometrical specific permeance λ_{sr}, $h_r/b_r \leq 1.5$ for the rectangular slot and $h_r/d_r < 0.5$ for the trapezoidal slot. As the round part of slot area is not negligible, this might be feasible ($h_r/d_r \approx \pi/8 < 0.5$), especially for low torque machines.

Also for the rectangular slot with $b_r = d_r$, $h_r = (\pi/4)\ d_r << 1.5$, so the rectangular slot may produce

$$\lambda_{st}^{root}(\pi/4) = \pi/12 + 2/3\sqrt{3} = 0.67 \approx \lambda_{sr}^{round} \tag{17.11'}$$

In reality, as the rated torque gets larger, the round bar is difficult to adopt as it would lead to a too small number of rotor slots or too a larger rotor diameter. In general, a slot aspect ratio $h_r/b_r \leq 3$ may be considered acceptable for many practical cases.

- The skin effect in the stator windings, at least for fundamental frequencies less than 100(120) Hz is negligible in well designed IMs for all power levels. For large powers, elementary rectangular cross section conductors in parallel are used. They are eventually stranded in the end-connection zone. The skin effect and circulating current additional losses have to be limited in large motors
- In super-high speed IMs, for fundamental frequencies above 300 Hz (up to 3 kHz or more), stator skin effect has to be carefully investigated and suppressed by additional methods such as Litz wire, or even by using thin wall pipe conductors with direct liquid cooling when needed
- Skin-effect stator and rotor winding losses at PWM inverter carrier frequency are to be calculated as shown in Chapter 11, paragraph 11.12.

17.4 TORQUE PULSATIONS REDUCTION

Torque pulsations are produced both by airgap flux density space harmonics in interaction with stator (rotor) m.m.f. space harmonics and by voltage (current) time harmonics produced by the power electronics converter (PEC) which supplies the IM to produce variable speed.

As torque time harmonics pulsations depend mainly on the PEC type and power level we will not treat them here. The space harmonic torque pulsations are produced by the so called parasitic torques (see Chapter 10). They are of two categories: asynchronous and synchronous and depend on the number of rotor and stator slots, slot opening/airgap ratios and airgap/pole pitch ratio, and the degree of saturation of stator (rotor) core. They all however occur at rather large values of slip: $S > 0.7$ in general.

This fact seems to suggest that for pump/fan type applications, where the minimum speed hardly goes below 30% base speed, the parasitic torques occur only during starting.

Even so, they should be considered, and the same rules apply, in choosing stator rotor slot number combinations, as for constant V and f design (Chapter 15, table 15.5).

- As shown in Chapter 15, slot openings tend to amplify the parasitic synchronous torques for $N_r > N_s$ (N_r – rotor slot count, N_s – stator slot count). Consequently $N_r < N_s$ appears to be a general design rule for variables V and f, even without rotor slot skewing (for series connected stator windings).
- Adequate stator coil throw chording (5/6) will reduce drastically asynchronous parasitic torque.
- Carefully chosen slot openings to mitigate between low parasitic torques and acceptable slot leakage inductances are also essential.
- Parasitic torque reduction is all the more important in servodrive applications with sustained low (even very low) speed operation. In

such cases, additional measures such as skewed resin insulated rotor bars and eventually closed rotor slots and semiclosed stator slots are necessary. FEM investigation of parasitic torques may become necessary to secure sound results.

17.5 INCREASING EFFICIENCY

Increasing efficiency is related to loss reduction. There are fundamental core and winding losses and additional ones due to space and time harmonics.

Lower values of current and flux densities lead to a larger but more efficient motor. This is why high efficiency motors are considered suitable for variables V and f.

Additional core losses and winding losses have been treated in detail in Chapter 11.

Here we only point out that the rules to reduce additional losses, presented in Chapter 11 still hold. They are reproduced here and extended for convenience and further discussion.

- Large number of slots/pole/phase in order to increase the order of the first slot space harmonic
- Insulated or uninsulated high bar-slot wall contact resistance rotor bars in long stack skewed rotors, to reduce interbar current losses
- Skewing is not adequate for low bar-slot wall contact resistance as it does not reduce the harmonics (stray) cage losses while it does increase interbar current losses
- $0.8N_s < N_r < N_s$ – to reduce the differential leakage coefficient of the first slot harmonics ($N_s \pm p_1$), and thus reduce the interbar current losses
- For $N_r < N_s$ skewing may be altogether eliminated after parasitic torque levels are checked. For q = 1,2 skewing seems mandatory
- Usage of thin special laminations (0.1 mm) above f_{1n} = 300Hz is recommended to reduce core loss in super-high speed IM drives
- Chorded coils ($y/\tau \approx 5/6$) reduce the asynchronous parasitic torque produced by the first phase belt harmonic ($\upsilon = 5$)
- With delta connection of stator phases: $(N_s - N_r) \neq 2p_1, 4p_1, 8p_1$.
- With parallel paths stator windings, the stator interpath circulating currents produced by rotor bar current m.m.f. harmonics have to be avoided by observing certain symmetry of stator winding paths
- Small stator (rotor) slot openings lead to smaller surface and tooth flux pulsation additional core losses but they tend to increase the leakage inductances and thus reduce the breakdown torque
- Carefully increase the airgap to reduce additional core and cage losses without compromising too much the power factor and efficiency
- Use sharp tools and annealed laminations to reduce surface core losses
- Return core losses rotor surface to prevent rotor lamination shortcircuits which would lead to increased rotor surface core losses

- Use only recommended N_s, N_r combinations and check for parasitic torque and stray load levels
- To reduce the time and space harmonics losses in the rotor cage, U shape bridge rotor slots have been proposed (Figure 17.3). [3]

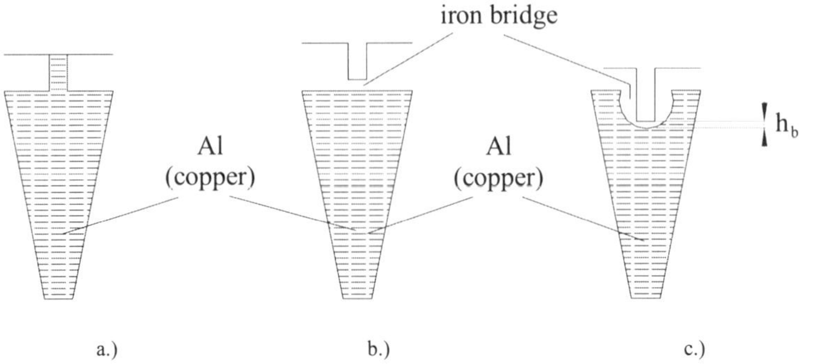

Figure 17.3 Rotor slot designs
a) conventional b) straight bridge closed slot c) u-bridge close slot

In essence in conventional rotor slots, the airgap flux density harmonics induce voltages which produce eddy currents in the aluminium situated in the slot necks. By providing a slit in the rotor laminations (Figure 17.3b, 17.3c), the rotor conductor is moved further away from the airgap and thus the additional cage losses are reduced.

However, this advantage comes with three secondary effects.

First, the eddy currents in the aluminium cage close to airgap damp the airgap flux density variation on the rotor surface and in the rotor tooth. This, in turn, limits the rotor core surface and tooth flux pulsation core losses.

In our new situation, it no longer occurs. Skewed rotor slots seem appropriate to keep the rotor surface and tooth flux pulsation core losses under control.

Second, the iron bridge height h_b above the slot, even when saturated, leads to a notable additional slot leakage geometrical permeance coefficient: λ_b.

Consequently, the value of L_{sc} is slightly increased, leading to a breakdown torque reduction.

Third, the mechanical resilience of the rotor structure is somewhat reduced which might prevent the usage of this solution to super-high speed IMs.

17.6 INCREASING THE BREAKDOWN TORQUE

As already inferred, a large breakdown torque is desirable either for high transient torque reserve or for widening the constant power speed range. Increasing the breakdown torque boils down to **leakage inductance decreasing**, when the base speed and stator voltage are given (17.3).

The total leakage inductance of the IM contains a few terms as shown in Chapter 6.

The stator and rotor leakage inductances are (Chapter 6)

$$L_{sl} = 2\mu_0 L_i \cdot n_s^2 p_1 q_1 \left(\lambda_{ss} + \lambda_{zs} + \lambda_{ds} + \lambda_{end}\right) \tag{17.12}$$

n_s – conductors slot

$$L_{rl} = 4m_1 \frac{\left(W_1 K_{W1}\right)^2}{N_r} \times 2\mu_0 L_i \left(\lambda_b + \lambda_{er} + \lambda_{zr} + \lambda_{dr} + \lambda_{skew}\right) \tag{17.13}$$

m_1 – number of phases
L_i – stack length
q_1 – slots/pole/phase
λ_{ss} – stator slot permeance coefficient
λ_{zs} – stator zig-zag permeance coefficient
λ_{ds} – stator differential permeance coefficient
λ_{end} – stator differential permeance coefficient
λ_b – rotor slot permeance coefficient
λ_{er} – end ring permeance coefficient
λ_{zr} – rotor zig-zag permeance coefficient
λ_{dr} – rotor differential permeance coefficient
λ_{skew} – rotor skew leakage coefficient

The two general expressions are valid for open and semi-closed slots. For closed rotor slots λ_b has the slot iron bridge term as rotor current dependent.

With so many terms, out of which very few may be neglected in any realistic analysis, it becomes clear that an easy sensitivity analysis of L_{sl} and L_{rl} to various machine geometrical variables is not easy to wage.

The main geometrical variables which influence $L_{sc} = L_{sl}+L_{rl}$ are

- Pole number: $2p_1$
- Stack length/pole pitch ratio: L_i/τ
- Slot/tooth width ratio: b_{s1r}/b_{ts1r}
- Stator bore diameter
- Stator slots/pole/phase q_1
- Rotor slots/pole pair N_r/p_1
- Stator (rotor) slot aspect ratio h_{ss1r}/b_{s1r}
- Airgap flux density level B_g
- Stator (rotor) base torque (design) current density

Simplified sensitivity analysis [4] of j_{cos}, j_{Al} to t_{sc} (or t_{bk} – breakdown torque in p.u.) have revealed that 2,4,6 poles are the main pole counts to consider except for very low direct speed drives – conveyor drives – where even $2p_1 = 12$ is to be considered, only to eliminate the mechanical transmission.

Globally, when efficiency, breakdown torque, and motor volume are all considered, the 4 pole motor seems most desirable.

Reducing the number of poles to $2p_1 = 2$, for given speed $n_1 = f_1/p_1$ (rps) means lower frequency, but for the same stator bore diameter, it means larger pole pitch and longer end connections and thus, larger λ_{end} in (17.12).

Now with q_1 larger, for $2p_1 = 2$ the differential leakage coefficient λ_{ds} is reduced. So, unless the stack length is not small at design start (pancake shape), the stator leakage inductance (17.12) decreases by a ratio of between 1 and 2 when the pole count increases from 2 to 4 poles, for the same current and flux density. Considering the two rotors identical, with the same stack length, L_{rl} in (17.13) would not be much different. This leads us to the peak torque formula (17.3).

$$T_{ek} \approx \frac{3p_1}{2}\frac{(\Psi_{sph})^2}{2L_{sc}}; \Psi_{sph} = \frac{V_{ph}}{\omega_1} \tag{17.14}$$

For the same number of stator slots for $2p_1 = 2$ and 4, conductors per slot n_s, same airgap flux density, and same stator bore diameter, the phase flux linkage ratio in the two cases is

$$\frac{(\Psi_{sph})_{2p_1=2}}{(\Psi_{sph})_{2p_1=4}} = \frac{2}{1} \tag{17.15}$$

as the frequency is doubled for $2p_1 = 4$, in comparison with $2p_1 = 2$, for the same no load speed (f_1/p_1).

Consequently,

$$\frac{(L_{sc})_{2p_1=2}}{(L_{sc})_{2p_1=4}} = 1.5 - 1.8 \tag{17.16}$$

Thus, for the same bore diameter, stack length, and slot geometry,

$$\frac{(T_{ek})_{2p_1=2}}{(T_{ek})_{2p_1=4}} \approx \frac{2}{1.5 - 1.8} \tag{17.17}$$

From this simplified analysis, we may draw the conclusion that the $2p_1 = 2$ pole motor is better. If we consider the power factor and efficiency, we might end up with a notably better solution.

For super-high speed motors, $2p_1 = 2$ seems a good choice ($f_{1n} > 300$ Hz).

For a given total stator slot area (same current density and turns/phase), and the same stator bore diameter, increasing q_1 (number of slot/pole/phase) does not essentially influence L_{sl} (17.12) – the stator slot leakage and end connection leakage inductance components – as $n_s p_1 q_1 = W_1 = ct$ and slot depth remains constant while the slot width decreases to the extent q_1 increases, and so does λ_{end}

$$\lambda_{end} \approx \frac{0.34}{L_i}(l_{end} - 0.64 \cdot y) \cdot q_1 \tag{17.18}$$

with l_{end} = end connection length; y – coil throw; L_i – stack length.

However, λ_{ds} decreases and apparently so does λ_{zs}. In general, with larger q_1, the total stator leakage inductance will decrease slightly. In addition, the stray losses have been proved to decrease with q_1 increasing.

A similar rationale is valid for the rotor leakage inductance L_{rl} (17.13) where the number of rotor slots increases. It is well understood that the condition $0.8N_s < N_r < N_s$ is to be observed.

A safe way to reduce the leakage reactance is to reduce the slot aspect ratio $h_{ss,r}/b_{ss,r} < 3.0\text{-}3.5$. For given current density this would lead to lower q_1 (or N_s) for a larger bore diameter, that is, a larger machine volume.

However, if the design current density is allowed to increase (sacrificing to some extent the efficiency) with a better cooling system, the slot aspect ratio could be kept low to reduce the leakage inductance L_{sc}.

A low leakage (transient) inductance L_{sc} is also required for current source inverter IM drives. [4]

So far, we have considered same current and flux densities, stator bore diameter, stack length, but the stator and yoke radial height for $2p_1 = 2$ is doubled with respect to the 4 pole machine.

$$\frac{\left(h_{cs,r}\right)_{2p_1=2}}{\left(h_{cs,r}\right)_{2p_1=4}} \approx \frac{(1.5-2)}{1} \tag{17.19}$$

Even if we oversaturated the stator and rotor yokes, and more for the two pole machine, the outer stator diameter will still be larger in the latter. It is true that this leads to a larger heat exchange area with the environment, but still the machine size is larger.

So, when the machine size is crucial, $2p_1 = 4$ [5] even $2p_1 = 6$ is chosen (urban transportation traction motors).

Whether to use long or short stack motors is another choice to make in the pursuit of smaller leakage inductance L_{sc}. Long stator stacks may allow smaller stator bore diameters, smaller pole pitches and thus smaller stator end connections.

Slightly smaller L_{sc} values are expected. However, a lower stator (rotor) diameter does imply deeper slots for the same current density. An increase in slot leakage occurs.

Finally, increasing the stack length leads to limited breakdown torque increase.

When low inertia is needed, the stack length is increased while the stator bore diameter is reduced. The efficiency will vary little, but the power factor will likely decrease. Consequently, the PEC KVA rating has to be slightly increased. The KVA ratings for two pole machines with the same external stator diameter and stack length, torque and speed, is smaller than for a 4 pole machine because of higher power factor. So when the inverter KVA is to be limited, the 2 pole machine might prevail.

A further way to decrease the stator leakage inductance may be to use four layers (instead of two) and chorded coils to produce some cancelling of mutual leakage fluxes between them. The technological complication seems to render such approaches as less than practical.

17.7 WIDE CONSTANT POWER SPEED RANGE VIA VOLTAGE MANAGEMENT

Constant power speed range varies for many applications from 2 to 1 to 5(6) to 1 or more.

The obvious way to handle such requirements is to use an IM capable to produce, at base speed ω_b, a breakdown torque T_{bk}:

$$\frac{T_{bk}}{T_{en}} = \frac{\omega_{max}}{\omega_b} = C_\omega \tag{17.20}$$

• A larger motor

In general, IMs may not develop a peak to rated torque higher than 2.5 (3) to 1 (Figure 17.4a).

In the case when a large constant power speed range C_ω is required, it is only intuitive to use a larger IM (Figure 17.4b).

Adopting a larger motor, to have enough torque reserve up to maximum speed for constant power may be done either with an IM with $2p_1 = 2,4$ of higher rating or a larger number of pole motor with the same power. While such a solution is obvious for wide constant power speed range ($C_\omega > 2.0 - 3.0$), it is not always acceptable as the machine size is notably increased.

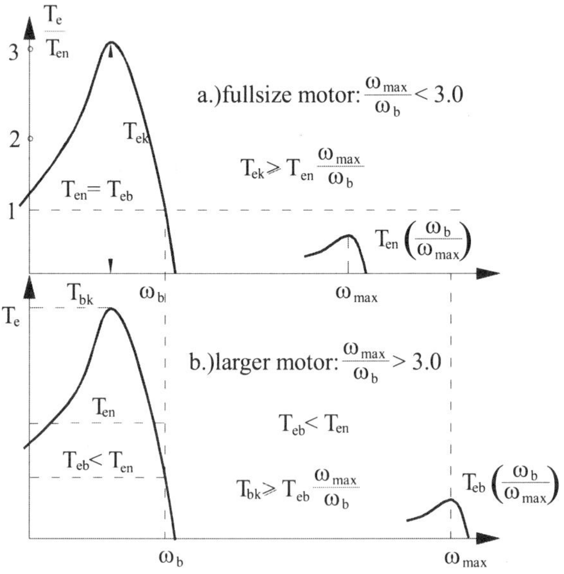

Figure 17.4 Torque-speed envelope for constant power
a) full size motor b) larger motor

• Higher voltage/phase

The typical torque/speed, voltage/speed, and current/speed envelopes for moderate constant power speed range are shown on Figure 17.5.

The voltage is considered constant above base speed. The slip frequency f_{sr} is rather constant up to base speed and then increases up to maximum speed. Its maximum value f_{srmax} should be less or equal to the critical value (that corresponding to breakdown torque)

$$f_{sr} \leq f_{sr\max} = \frac{R_r}{2\pi L_{sc}} \tag{17.21}$$

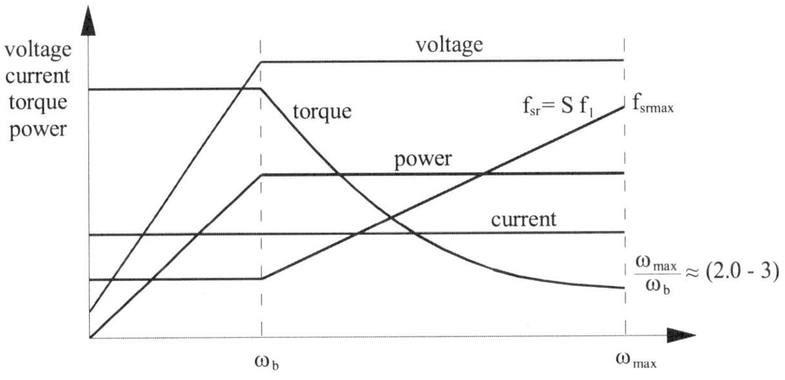

Figure 17.5 Torque, voltage, current versus frequency

$$\left(T_{bk}\right)_{\omega\max} = \frac{3}{2} p_1 \left(\frac{V_{ph}}{\omega_{r\max}}\right)^2 \cdot \frac{1}{L_{sc}} \geq T_{eb} \cdot \frac{\omega_b}{\omega_{max}} \tag{17.22}$$

If

$$\left(T_{bk}\right)_{\omega\max} < T_{eb} \frac{\omega_b}{\omega_{max}}, \tag{17.23}$$

it means that the motor has to be oversized to produce enough torque at maximum speed.

If a torque (power reserve) is to be secured for fast transients, a larger torque motor is required.

Alternatively, the phase voltage may be increased during the entire constant power range (Figure 17.6)

To provide a certain constant overloading over the entire speed range, the phase voltage has to increase over the entire constant power speed range. This means that for base speed ω_b, the motor will be designed for lower voltage and thus larger current. The inverter current loading (and costs) will be increased.

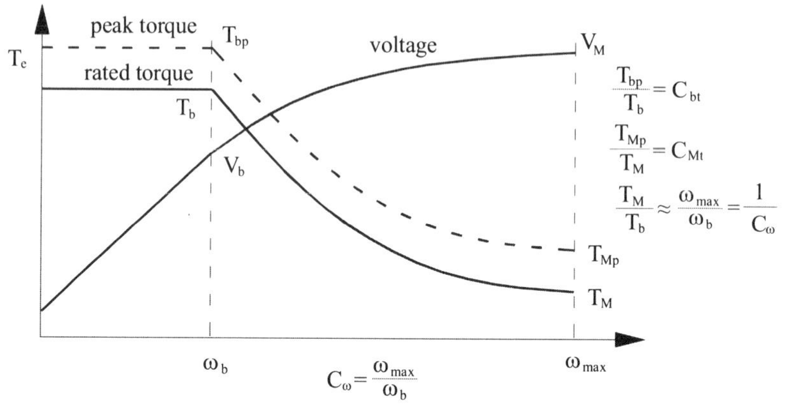

Figure 17.6 Raising voltage for the constant power speed range

Let us calculate the torque T_e, and stator current I_s from the basic equivalent circuit.

$$T_e \approx \frac{3p_1}{\omega_1} \cdot \frac{V_{ph}^2 \cdot R_r/s}{\left(R_s + C_1 \frac{R_r}{s}\right)^2 + \omega_1^2 L_{sc}^{'2}}; \tag{17.24}$$

$$L_{sc}^{'} = L_{sl} + C_1 L_{rl}; C_1 = 1 + \frac{L_{sl}}{L_m} \approx 1.02 - 1.08$$

$$I_s \approx V_{ph} \sqrt{\left(\frac{1}{\omega_1 L_m}\right)^2 + \frac{1}{\left(R_s + C_1 \frac{R_r}{s}\right)^2 + \omega_1^2 L_{sc}^{'2}}} \tag{17.25}$$

At maximum speed, the machine has to develop the peak torque T_{Mp} at maximum voltage V_M.

Now if for both T_{bp} and T_{Mp}, the breakdown torque conditions are met,

$$T_{bP} = \frac{3p_1}{2}\left(\frac{V_b}{\omega_b}\right)^2 \cdot \frac{1}{2L_{sc}},$$
$$T_{Mp} = \frac{3p_1}{2}\left(\frac{V_M}{\omega_{max}}\right)^2 \cdot \frac{1}{2L_{sc}} \tag{17.26}$$

the voltage ratios V_M/V_b is (as in [6])

$$\left(\frac{V_M}{V_b}\right)^2 = \left(\frac{\omega_{max}}{\omega_b}\right)^2 \frac{T_{Mp}}{T_{bp}} = \frac{\omega_{max}}{\omega_b} \cdot \frac{\left(\frac{T_{Mp}}{T_M}\right)}{\left(\frac{T_{bp}}{T_b}\right)} = C_\omega \frac{C_{Mt}}{C_{bt}} \tag{17.27}$$

For $\omega_{max}/\omega_b = 4$, $C_{Mt} = 1.5$, $C_{bt} = 2.5$, we find from (17.17) that $V_M/V_b = 1.55$. So the current rating of the inverter (motor) has to be increased by 55%.

Such a solution looks extravagant in terms of converter extra-costs, but it does not suppose IM overrating (in terms of power). Only a special design for lower voltage V_b at base speed ω_b is required.

For

$$S = S_{K,b,m} = \frac{C_1 R_r}{\sqrt{R_s^2 + \omega_{b,M}^2 L_{sc}^{'2}}} \tag{17.28}$$

and $\omega_1 = \omega_b$ and, respectively, $\omega_1 = \omega_M$ from (17.25), the corresponding base and peak current values I_b, I_M may be obtained.

These currents are to be used in the motor (I_b) and converter $\left(I_M\sqrt{2}\right)$ design.

Now, based on the above rationale, the various cases, from constant overloading $C_{MT} = C_{bt}$ to zero overloading at maximum speed $C_{MT} = 1$, may be treated in terms of finding V_M/V_b and I_M/I_b ratios and thus prepare for motor and converter design (or selection). [6]

As already mentioned, reducing the slot leakage inductance components through lower slot aspect ratios may help to increase the peak torque by as much as (50–60%) in some cases, when forced air cooling with a constant speed fan is used. [6]

• High constant power speed range: $\omega_{max}/\omega_b > 4$

For a large constant power speed range, the above methods are not generally sufficient. Changing the phase voltage by switching motor phase connection from star to delta (Y to Δ) leads to a sudden increase in phase voltage by $\sqrt{3}$. A notably larger constant power speed range is obtained this way (Figure 17.7).

The Y to Δ connection switching has to be done through a magnetic switch with the PEC control temporarily inhibited and then so smoothly reconnected that no notable torque transients occurred. This voltage change is done "on the fly."

An even larger constant power speed range extension may be obtained by using a winding tap (Figure 17.8) to switch from a large to a small number of turns per phase (from W_1 to W'_1, $W_1/W'_1 = C_{W1} > 1$).

A double single-throw magnetic switch suffices to switch from high (W_1) to low number of turns (W_1/C_{W1}) for high speed.

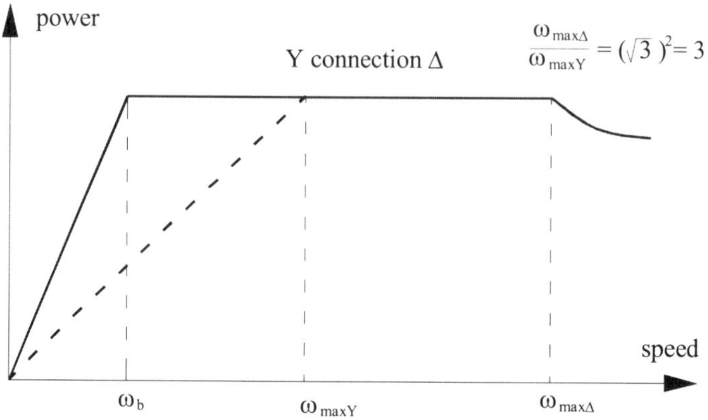

Figure 17.7 Extended constant power speed range by Y to Δ stator connection switching

Once the winding is designed with the same current density for both winding sections, the switching from W_1 to W_1/C_{W1} turns leads to stator and rotor resistance reduction by $C_{W1} > 1$ times and a total leakage inductance L_{sc} reduction by C^2_{W1}.

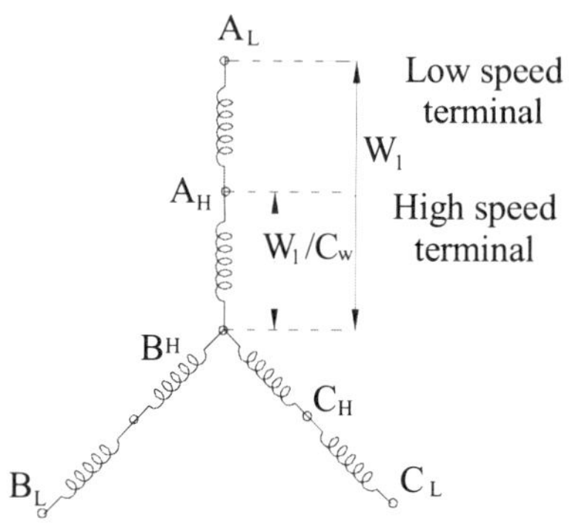

Figure 17.8 Tapped stator winding

In terms of peak (breakdown torque), it means a strong increase by C^2_{W1}. The constant power speed range is extended by

$$\frac{\omega_{max\,H}}{\omega_{max\,L}} = C^2_{W_1} \tag{17.29}$$

The result is similar to that obtained with Y to Δ connection switching but this time C_{W1} may be made larger than $\sqrt{3}$ and thus a larger extension of constant power speed range is obtained without oversizing the motor. The price to be paid is the magnetic (or thyristor-soft) switch. The current ratios for the two winding situations for the peak torque are:

$$\frac{I_{s\max H}}{I_{s\max L}} \approx C_{W1} \tag{17.30}$$

The leakage inductance L_{sc} decreases with C^2_W while the frequency increases C_W times. Consequently, the impedance decreases, at peak torque conditions, C_W times. This is why the maximum current increases C_{W1} times at peak speed ω_{maxH} with respect to its value at ω_{maxL}. Again, the inverter rating has to be increased accordingly while the motor cooling must be adequate for these high demands in winding losses. The core losses are much smaller but they are generally smaller than copper losses unless super-high speed (or large power) is considered. As for how to build the stator winding, to remain symmetric in the high speed connection, it seems feasible to use two unequal coils per layer and make two windings, with W_1/C_{W1}, and $W_1(1 - C_{W1}^{-1})$ turns per phase, respectively, and connect them in series. The slot filling factor will be slightly reduced but, with the lower turns coil on top of each slot layer, a further reduction of slot leakage inductance for high speed connection is obtained as a bonus.

• **Inverter pole switching**

It is basically possible to reduce the number of poles in the ratio of 2 to 1 through using two twin half-rating inverters and thus increase the constant power speed range by a factor of two. [7]

No important oversizing of the converter seems needed. In applications when PEC redundancy for better feasibility is a must, such a solution may prove adequate.

It is also possible to use a single PEC and two simple three phase soft switches (with thyristors) to connect it to the two three phase terminals corresponding to two different pole count windings (Figure 17.9)

Such windings with good winding factors are presented in Chapter 4.

As can be seen from Figure 17.9, the current remains rather constant over the extended constant power speed range, as does the converter rating. Apart from the two magnet (or static) switches, only a modest additional motor over-cost due to dual pole count winding is to be considered the price to pay for speed range extension at constant power. For cost sensitive applications, this solution might be practical.

17.8 DESIGN FOR HIGH AND SUPER-HIGH SPEED APPLICATIONS

As previously mentioned, super-high speeds range start at about 18 krpm (300 Hz in 2 pole IMs). Between 3 krpm (50 Hz) and 18 krpm (300 Hz) the interval of high speeds is located (Figure 17.10)

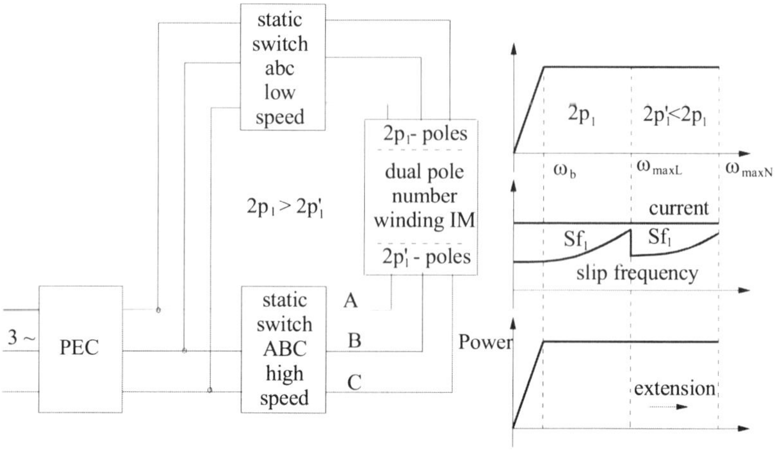

Figure 17.9 Dual static switch for dual pole count winding IM

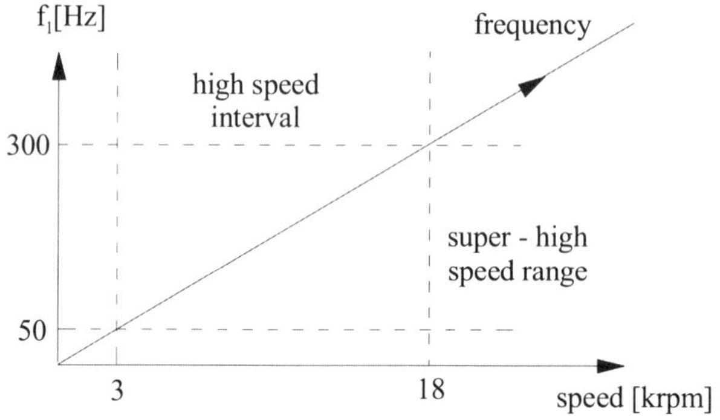

Figure 17.10 High and super-high speed division

Below 300 Hz, where standard PWM-PEC are available, up to tens of kW (recently up to MW/unit) but above 50 Hz, the design is very much similar to that for speeds below 50 Hz. Two pole motors are favored, with lower flux densities to limit core losses, forced stator cooling, and carefully limited mechanical losses. Laminations and conductors remain of the standard type with close watch on skin effect containment.

Above 300 Hz, in the super-high speed applications, the increase in mechanical stresses, mechanical and core loss, and in skin effects have triggered specialized worldwide research efforts. We will characterize the most representative of them in what follows.

17.8.1 Electromagnetic limitations

The Esson's constant C_0 still holds (Chapter 14, Equation (14.14)),

$$C_0 \approx \frac{\pi^2}{\sqrt{2}} K_{W1} A_1 B_g \tag{17.31}$$

The linear current density A_1 (Aturns/m) is limited to $(5 - 20)\times 10^3$A/m for most super-high speed IMs, and in general increases with stator bore diameter D_{is}.

The stator bore diameter D_{is} is related to motor power P_n as (Chapter 14, Equation (14.15))

$$D_{is} = \sqrt{\frac{K_E P_n}{\eta \cos\varphi} \cdot \frac{p_1}{f_1} \cdot \frac{1}{C_0} \cdot \frac{1}{L}}; K_E \approx 0.98 - p_1 \cdot 5 \cdot 10^{-3} \tag{17.32}$$

f_1 – frequency, p_1 – pole pairs, L – stack length, η – efficiency, K_E – e.m.f. coefficient.

As the no load speed $n_1 = f_1/p$ is large, for given stack length L, the stator bore diameter D_{is} may be obtained for given C_0, from (7.32).

For super-high speed motors, even with thin laminations (0.1 mm thick or less), the airgap flux density is lowered to $B_g < (0.5 - 0.6)$T to cut the core losses which tend to be large as frequency increases. Longer motors (L – larger) tend to require, as expected, low stator bore diameters D_{is}.

17.8.2 Rotor cooling limitations

Due to the uni-stack rotor construction, made either from solid iron or from laminations, the heat due to rotor losses – total conductor, core and air friction losses – has to be transferred to the cooling agent almost entirely through the rotor surface. Thus, the rotor diameter $D_{er} \approx D_{is}$ is limited by

$$D_{is} \geq \frac{rotor\_losses}{\pi L \cdot \alpha \Delta\theta_{ra}} \tag{17.33}$$

α – heat transfer factor W/m^2 ^{0}K, $\Delta\theta_{ra}$ – rotor to internal air temperature differential.

Among rotor losses, besides fundamental cage and core losses, space and time harmonics losses in the core and rotor cage are notable.

17.8.3 Rotor mechanical strength

The centrifugal stress on the rotor increases with speed squared. At the laminated rotor bore $D_{er} \approx D_{is}$, the critical stress σ_K(N/m^2) should not be surpassed. [6]

$$D_{is} < \sqrt{\frac{16\sigma_K}{\gamma\omega_r^2(3+\mu)} - \frac{(1-\mu)}{3+\mu} \cdot D_{int}^2} \tag{17.34}$$

γ – rotor material density (kg/m^3), ω_r – rotor angular speed (rad/s), μ – Poisson's ratio, D_{int}(m) – interior diameter of rotor laminations ($D_{int} = 0$ for solid rotor).

For 100 kW IMs with $L/D_{is} = 3$ (long stack motors) and $C_0 = 60\times10^3$ J/m^3, the limiting curves given by (17.32) – (17.34) are shown in Figure 17.12. [8]

Results in Figure 17.11 lead to remarks such that

- To increase the speed, improved (eventually liquid) rotor cooling may be required.
- When speed and the Esson's constant increase, the rotor losses per rotor volume increase and thus thermal limitations become the main problem.
- The centrifugal stress in the laminated rotor restricts the speed range (for 100 kW).

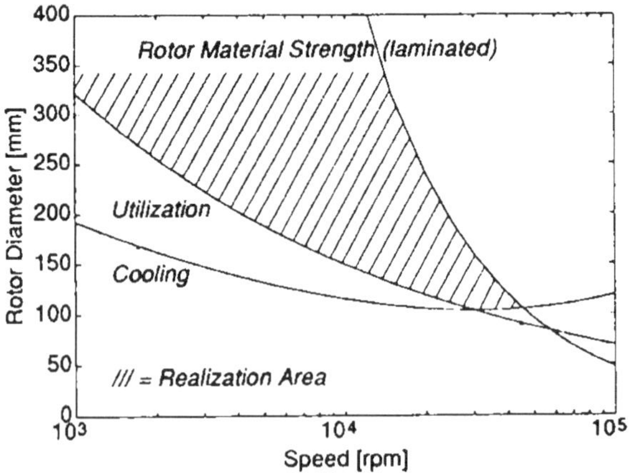

Figure 17.11 Stator bore diameter limiting curves for 100 kW machines at very high speed

17.8.4 The solid iron rotor

As the laminated rotor shows marked limitations, extending the speed range for given power leads, inevitably, to the solid rotor configuration. The absence of the central hole and the solid structure produce a more rugged configuration.

The solid iron rotor is also better in terms of heat transmission, as it allows for good axial heat exchange by thermal conductivity.

Unfortunately, the smooth solid iron rotor is characterized by a large equivalent resistance R_s and a rather limited magnetization inductance. This is so because the depth of field penetration in the rotor δ_{iron} is [9]

$$\delta_i \approx \left| \mathrm{Re}\left(\sqrt{\frac{1}{(\pi/\tau)^2 + j2\pi f_1 S \frac{\sigma_{iron}}{K_T} \cdot \mu_0 \cdot \mu_{rel}}} \right) \right| \tag{17.35}$$

τ – pole pitch, μ_{rel} – relative iron permeability, K_T – conductivity correction factor to be explained later in this paragraph:

$$a - (\pi / \tau)^2$$
$$b = 2\pi f_1 S \frac{\sigma_{iron}}{K_T} \cdot \mu_0 \mu_{rel} \tag{17.36}$$

$$\delta_i = \frac{\sqrt{a + \sqrt{a^2 + b^2}}}{\sqrt{2(a^2 + b^2)}} \tag{17.37}$$

For $\tau = 0.1$m, $\mu_{rel} = 18$, $Sf_1 = 5$ Hz, $\sigma_{iron} = 10^{-1}\sigma_{co} = 5\times10^6$ $(\Omega m)^{-1}$, from (17.36 – 17.37), $\delta i = 1.316\times10^{-2}$m.

To provide reasonable airgap flux density $B_g \approx 0.5 - 0.6$T the field penetration depth in iron δ_i for a highly saturated iron, with $\mu_{rel} = 36$ (for $Sf_1 = 5$ Hz) and $B_{iron} = 2.3$T ($H_{iron} = 101733.6$ A/m), must be

$$\delta_{iron} \geq \frac{B_g \cdot \tau / \pi}{B_{iron}} = \frac{0.55}{\pi \times 2.3} \times \tau = 7.6 \cdot 10^{-2} \times \tau \ [m] \tag{17.38}$$

For $\delta_{iron} = 0.01316$ m, the pole pitch τ should be $\tau < 0.173$ m. As the pole pitch was supposed to be $\tau = 0.1$m, the iron permeability may be higher.

For a $2p_1 = 2$ pole machine, typical for super-high speed IMs, the stator bore diameter D_{is},

$$D_{is} < \frac{2p_1\tau}{\pi} = \frac{2\times1\times0.173}{\pi} = 0.110 \ \text{m!} \tag{17.39}$$

This is a severe limitation in terms of stator bore diameter (see Figure 17.12). For higher diameters D_{is}, smaller relative permeabilities have to be allowed for. The consequence of the heavily saturated rotor iron is a large magnetization m.m.f. contribution which in relative values to airgap m.m.f. is

$$\frac{F'_{iron}}{F_{airgap}} \approx \frac{\frac{1}{3}\tau \times H_{iron} \cdot \mu_0}{B_g \cdot 2g} \approx \frac{\frac{1}{3}0.1\times101733\times1.256\times10^{-6}}{0.55\times2\times1.0\times10^{-3}} = 3.872! \tag{17.40}$$

A rather low power factor is expected. The rotor leakage reactance to resistance ratio may be approximately considered as 1.0 (for constant, even low permeability)

$$\frac{\omega_1 L_{rl}}{R_r} \approx 1 \tag{17.41}$$

From [8] the rotor resistance reduced to the stator is

$$R_r = \frac{6L \cdot K_T}{\delta_i \tau p_1 \sigma_{iron}} \cdot (K_{W1} W_1)^2 \quad (17.42)$$

In (17.42), the rotor length was considered equal to stator stack length and K_T is a transverse edge coefficient which accounts for the fact that the rotor current paths have a circumferential component. (Figure 17.12)

From [8],

$$K_T \approx \frac{1}{1 - \frac{\tanh(\pi L / 2\tau \cdot n)}{(\pi L / 2\tau \cdot K)}} > 1 \quad (17.43)$$

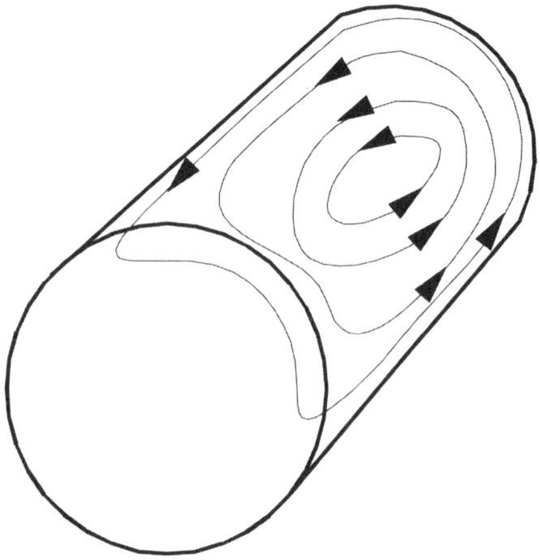

Figure 17.12 Smooth solid rotor induced current paths

The longer the stack length/pole pitch ratio L/τ, the smaller is this coefficient.

Here n = 1. While K_T reduces the equivalent apparent conductivity of iron, it does increase the rotor resistivity as $\sqrt{K_T}$, lowering the torque for given slip frequency Sf_1.

This is why the rotor structure may be slitted with the rotor longer than the stack to get lower transverse coefficient. With copper end rings, we may consider $K_T = 1$ in (17.43) and in (17.35), (since the end rings resistance is much smaller than the rotor iron resistance (Figure 17.13)).

How deep the rotor slits should be is a matter of optimal design as the main flux paths have to close below their bottom. So definitely $h_{slit} > \delta_i$. Notably larger output has been demonstrated for given stator with slitted rotor and copper end rings. [10]

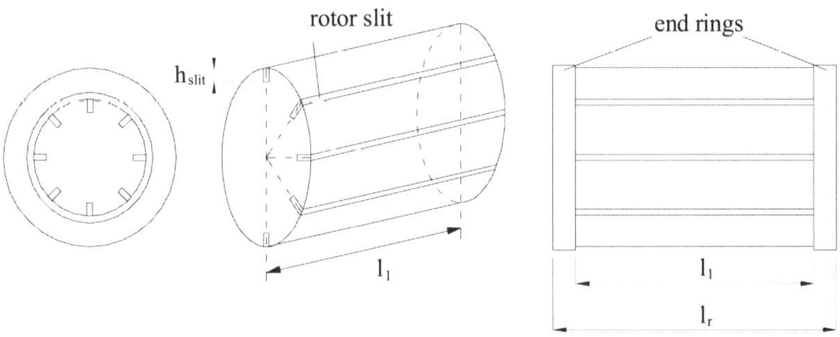

Figure 17.13 Radially slitted rotor with end rings

The radial slots may be made wider and filled with copper bars (Figure 17.14a); copper end rings are added. As expected, still larger output for better efficiency and power factor has been obtained. [10]

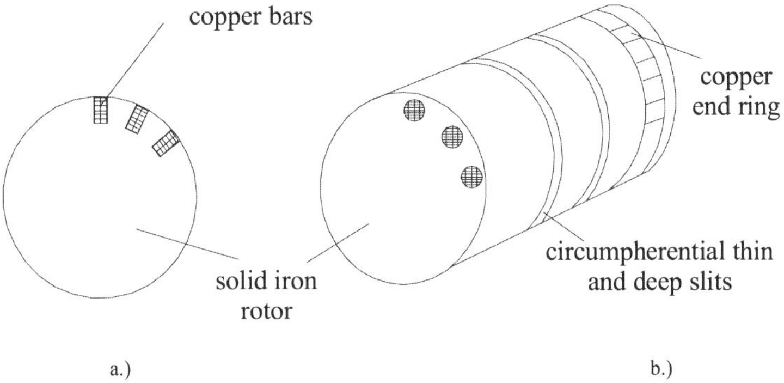

Figure 17.14 Solid iron rotor with copper bars
a.) copper bars in radial open slots (slits)
b.) copper bars in closed slots

Another solution would be closed rotor slots with copper bars and circumferential deep slits in the rotor (Figure 17.14b).

This time, before the copper bars are inserted in slots the rotor is provided with thin and deep circumferential slits such that to increase the transverse edge effect coefficient K_T (17.43) where n is the number of stack axial sections produced by slits.

This time they serve to destroy the solid iron eddy currents as the copper bars produce more efficiently torque, that is for lower rotor resistance. The value of n may be determined such that K_T is so large that depth of penetration $\delta_{iron} \geq 2d_{co}$ (d_{co} – copper bar diameter). Three to six such circumferential slits generally suffice.

The closed rotor slots serve to keep the copper bar tight while the slitted iron serves as a rugged rotor and better core and the space harmonics losses in the rotor iron are kept reasonably small.

A thorough analysis of the solid rotor with or without copper cage with and without radial or circumferential slits may require a full fledged 3D FEM approach. This is time consuming and thus full scale careful testing may not be avoided.

17.8.5 21 kW, 47,000 rpm, 94% efficiency with laminated rotor [11]

As seen in Figure 17.12, at 50,000 rpm, a stator bore diameter D_{is} of up to 80 mm is acceptable from the rotor material strength point of view.

Reference [5] reports a good performance 21 kW, 47,000 rpm (f_1 = 800 Hz) IM with a laminated rotor which is very carefully designed and tested for mechanical and thermal compliance.

High silicon 3.1% laminations (0.36 mm thick) were specially thermally treated to reduce core loss at 800 Hz to less than 30 W/kg at 1T, and to avoid brittleness, while still holding a yield strength above 500 MPa. Closed rotor slots are used and the rotor lamination stress in the slot bridge zone was carefully investigated by FEM.

Al 25 (85% copper conductivity) was used for the rotor bars, while the more rugged Al 60 (75% copper conductivity) was used to make the end rings. The airgap was g = 1.27 mm for a rotor diameter of 51 mm, and a stack length L = 102 mm and $2p_1$ = 2 poles. The stator/rotor slot numbers are N_s/N_r = 24/17, rated voltage 420 V, and current 40 A. A rather thick shaft is provided (30 mm in diameter) for mechanical reasons. However, in this case part of it is used as rotor yoke for the main flux. As the rated slip frequency Sf_1 = 5 Hz the flux penetrates enough in the shaft to make the latter a good part of the rotor yoke. The motor is used as a direct drive for a high speed centrifugal compressor system. Shaft balancing is essential. Cooling of the motor is done in the stator by using a liquid cold refrigerant. The refrigerant also flows through the end connections and through the airgap zone in the gaseous form. To cut the pressure drop, a rather large airgap (g = 1.27 mm) was adopted. The motor is fed from an IGBT voltage source PWM converter switched at 15 kHz.

The solution proves to be the least expensive by a notable margin when compared with PM-brushless and switched reluctance motor drives of equivalent performance.

This is a clear example how, by carefully pushing forward existing technologies new boundaries are reached. [11,12]

17.9 SAMPLE DESIGN APPROACH FOR WIDE CONSTANT POWER SPEED RANGE

Design specifications are
Base power: P_n = 7.5 kW
Base speed: n_b = 500 rpm
Maximum speed n_{max} = 6000 rpm
Constant power speed range: 500 rpm to 6000 rpm
Power supply voltage: 400 V, 50 Hz, 3 phase

Solution characterization

The design specifications indicate an unusually large constant power speed range n_{max}/n_b = 12:1. Among the solutions for this case, it seems that a combination of winding connection switching from Y to Δ and high peak torque design with lower than rated voltage at base speed might do it.

The Y to Δ connection produces an increase of constant power speed zone by

$$\frac{n_{max\,\Delta}}{n_{max\,Y}} = \left[\frac{(V_{ph})_\Delta}{(V_{ph})_Y}\right]^2 = \frac{3}{1} \tag{17.44}$$

It remains a 4/1 ratio constant power speed range for the Y (low speed) connection of windings.

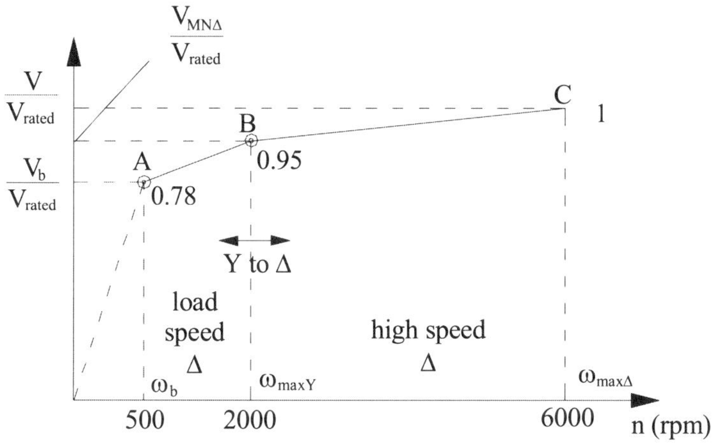

Figure 17.15 Line voltage envelope versus speed

As the peak/rated torque seldom goes above 2.5-2.7, the voltage at base speed has to be reduced notably.

To leave some torque reserve up to maximum speed, even in the Δ (high speed) connection, a voltage reserve has to be left.

First, from (17.27), we may calculate the $V_{MY/\Delta}/V_{rated}$ from the high speed zone,

$$\left(\frac{V_{MY/\Delta}}{V_{rated}}\right)^{-2} = \frac{\omega_{max\,\Delta}}{\omega_{max\,Y}} \cdot \frac{C_{Mt}}{C_{bt}} \tag{17.45}$$

C_{MT} is the overload capacity at maximum speed. Let us consider $(C_{MT}) = 1$ for $\omega_{max\Delta}$(6000 rpm). C_{bT} is the ratio between peak torque and base torque, a matter of motor design. Let us take $C_{bT} = 2.7$. In this case from (17.45)(point B),

$$\left(\frac{V_{MY/\Delta}}{V_{rated}}\right) = \sqrt{\frac{2.7}{3\times 1}} = 0.95 \tag{17.46}$$

Now for point A in Figure 17.15, we may apply the same formula,

$$\left(\frac{V_b}{V_{rated}}\right)^{-2} = \left(\frac{V_{mY/\Delta}}{V_{rated}}\right)^{-2} \cdot \frac{\omega_{max\,Y}}{\omega_b} \cdot \frac{C_{Mt}}{C_{bt}} \tag{17.46'}$$

As for point A V < V_{rated}, there is a torque reserve, so we may consider again $C_{Mt} = 1.0$

$$\frac{V_b}{V_{rated}} = 0.95\sqrt{\frac{1}{4}\cdot\frac{2.7}{1.0}} = 0.78 \tag{17.47}$$

Admitting a voltage derating of 5% ($V_{derat} = 0.05$) due to the PEC, the base line voltage at motor terminals, which should produce rated power at $n_b = 500$ rpm, is

$$(V_b)_{line} = V_{line} \cdot \frac{V_b}{V_{rated}} \cdot (1 - V_{derat}) = 400 \times 0.78(1 - 0.05) = 296\text{V} \tag{17.48}$$

For constant power factor and efficiency, reducing the voltage for a motor does increase the current in proportion:

$$\frac{I_b'}{I_b} \approx \frac{V_{rated}}{V_b} = \frac{400}{296} = 1.35 \tag{17.49}$$

As this overloading is not needed during steady state, it follows that the power electronics converter does not need to be overrated as it allows 50% current overload for short duration by design.

Now the largest torque T_b is needed at base speed (500 rpm), with lowest voltage $V_{bline} = 296$ V.

$$T_b = \frac{P_b}{2\pi \cdot \frac{n_b}{60}} = \frac{7500 \times 60}{2\pi \cdot 500} = 143.3\text{Nm} \tag{17.50}$$

The needed peak (breakdown) torque T_{bk} is

$$T_{bK} = C_{bT} \times T_b = 2.7 \times 143.3 = 387\text{Nm} \tag{17.51}$$

The electromagnetic design has to be done for base power, base speed, base voltage with the peak torque ratio constraint.

The thermal analysis has to consider the operation times at different speeds. Also the core (fundamental and additional) and mechanical losses increase with speed, while the winding losses also slightly increase due to power factor decreasing with speed and due to additional winding loss increase with frequency.

Basically, from now on the electromagnetic design is similar to that with constant V/f (Chapter 15,16). However, care must be exercised to additional losses due to skin effects, space and time harmonics, and their influence on temperature rise and efficiency.

17.10 SUMMARY

- Variable speed drives with induction motors may be classified by the speed range for constant output power.
- Servodrives lack any constant power speed range and the base torque is available for fast acceleration and deceleration provided the inverter is oversized to handle the higher currents typical for breakdown torque conditions.
- General drives have a moderate constant power speed range of $\omega_{max}/\omega_b < 2.5$, for constant (ceiling) voltage from the PEC. Increases in the ratio ω_{max}/ω_b are obtainable basically only by reducing the total leakage inductance of the machine. Various parameters such as stack length, bore diameter, or slot aspect ratio may be used to increase the "natural" constant power speed range up to $\omega_{max}/\omega_b \leq T_{bk}/T_b < (2.2 - 2.7)$. No overrating of the PEC or of the IM is required.
- Constant power drives are characterized by $\omega_{max}/\omega_b > T_{bK}/T_{eb}$. There are quite a few ways to get around this challenging task. Among them is designing the motor at base speed for a lower voltage and allowing the voltage to increase steadily and reach its peak value at maximum speed.
- The constant power speed zone may be extended notably this way but at the price of higher rated (base) current in the motor and PEC. While the motor does not have to be oversized, the PEC does. This method must be used with cautious.
- Using a winding tap, and some static (or magnetic) switch, is another very efficient approach to extend the constant power speed zone because the leakage inductance is reduced by the number of turns reduction ratio for high speeds. A much higher peak torque is obtained. This explains the widening of constant power speed range. The current raise is proportional to the turns reduction ratio. So PEC oversizing is inevitable. Also, only a part of the winding is used at high speeds.
- Winding connection switching from Y to Δ is capable of increasing the speed range at constant power by as much as 3 times without overrating the motor or the converter. A dedicated magnetic (or static) switch with PEC short time shutdown is required during the connection switching.
- Pole changing from higher to lower also extends the constant power speed zone by p_1/p_1' times. For $p_1/p_1' = 2/1$, a twin half rated power PEC may be used to do the switching of pole count. Again, no oversizing of motor or total converter power is required. Alternatively, a single full power PEC may be used to supply a motor with dual pole

count winding terminals by using two conventional static (or magnetic) three phase power switches.

- The design for variable speed has to reduce the skin effect which has a strong impact on rotor cage sizing.
- The starting torque and current constraints typical for constant V/f design do no longer exist.
- High efficiency IMs with insulated bearings designed for constant V/f may be (in general) safely supplied by PEC to produce variable speed drives.
- Super-high speed drives (centrifugal compressors, vacuum pumps, spindles) are considered here for speeds above 18,000 rpm (or 300 Hz fundamental frequency). This classification reflects the actual standard PWM–PECs performance limitations
- For super-high speeds, the torque density is reduced and thus the rotor thermal and mechanical limitations become even more evident. Centrifugal force – produced mechanical stress might impose, at certain peripheral rotor speed, the usage of a solid iron rotor.
- While careful design of laminated rotor IMs has reached 21 kW at 47,000 rpm (800 Hz), for larger powers and equivalent or higher peripheral speed, solid iron rotors may be mandatory.
- To reduce iron equivalent conductivity and thus increase field penetration depth, the solid iron rotor with circumpherential deep (and thin) slits and copper bars in slots seems a practical solution.
- The most challenging design requirements occur for wide constant power speed range applications (ω_{max}/ω_b > peak_torque/base_torque). The electromagnetic design is performed for base speed and base power. The thermal design has to consider the loss evolution with increasing speed.
- Once the design specifications and the most demanding performance are identified, the design (sizing) of IM for variable speed may follow a similar path as for constant V/f operation (Chapter 14–16).

17.11 REFERENCES

1. J. Singh, Harmonic Analysis and Loss Comparison of Microcomputer Based PWM Strategies for Induction Motor Drive" EMPS vol. 27, no.10, 1999, pp. 1129-1139.

2. A. Boglietti, P. Ferraris, M. Lazzari, M. Pastorelli, Change in Iron Losses with the Switching Frequency in Soft Magnetic Materials Supplied by PWM Inverter, IEEE – Trans vol. MAG – 31, no.6, 1995, pp. 4250-4255.

3. H. P. Nee, Rotor Slot Design of Inverter-Fed Induction Motors, Record of 1995 EMD International Conference, IEEE Conf. Public. No. 412, pp.52-56.

4. K. N. Pavithran, R. Pavimelalagan, G. Sridhara, J. Holtz, Optimum Design of an Induction Motor for Operation with Current Source Inverters" Proc. IEEE, vol. 134, Pt. B, no. 1, 1987, pp.1-7.

5. J. L. Oldenkamp and S. C. Peak, Selection and Design of an Inverter Driven Induction Motor for a Traction Drive Application, IEEE Trans, vol. IA-21, no. 1, 1985, pp. 285-295.

6. A. Bogllietti, P. Ferraris, M. Lazzari, F. Profumo, A New Design Criterion for Spindle Drive Induction Motors Controlled by Field Oriented Technique, EMPS vol. 21, no. 2, 1993, pp. 171-182.

7. M. Osama and T. A. Lipo, A New Inverter Control Scheme for Induction Motor Drives Requiring Wide Speed Range, Record of IEEE-IAS-1995-Annual Meeting vol. 1, pp. 350-355.

8. G. Pasquarella and K. Reichert, Development of Solid Rotors for a High Speed Induction Machine with Magnetic Bearings, Record of ICEM-1990, at MIT, vol. 2, pp. 464-469.

9. I. Boldea and S. A Nasar, Linear Motion Electromagnetic Systems, book, John Wiley, 1985, pp. 88-91.

10. J. Huppunen and Juha Pirhönen, Choosing the Main Dimensions of a Medium Speed (<30,000rpm) solid rotor induction motor, Record of ICEM-1998, vol. 1, pp. 296-301.

11. W. L. Soong, G. B. Kliman, R. N. Johnson, R. White, J. Miller, Novel High Speed Induction Motor for a Commercial Centrifugal Compressor, Record of ICEM-1998, vol. 1, pp. 296-301.

12. A. Boglietti, P. Ferraris, M. Lazzari, F. Profumo, About the Design of Very High Frequency Induction Motors for Spindle Applications, EMPS vol. 25, no. 4, 1997, pp. 387-409.

Chapter 18

OPTIMIZATION DESIGN

18.1. INTRODUCTION

As we have seen in previous chapters, the design of an induction motor means to determine the IM geometry and all data required for manufacturing so as to satisfy a vector of performance variables together with a set of constraints.

As induction machines are now a mature technology, there is a wealth of practical knowledge, validated in industry, on the relationship between performance constraints and the physical aspects of the induction machine itself.

Also, mathematical modelling of induction machines by circuit, field or hybrid models provides formulas of performance and constraint variables as functions of design variables.

The path from given design variables to performance and constraints is called *analysis*, while the reverse path is called *synthesis*.

Optimization design refers to ways of doing efficiently synthesis by repeated analysis such that some single (or multiple) objective (performance) function is maximized (minimized) while all constraints (or part of them) are fulfilled (Figure 18.1).

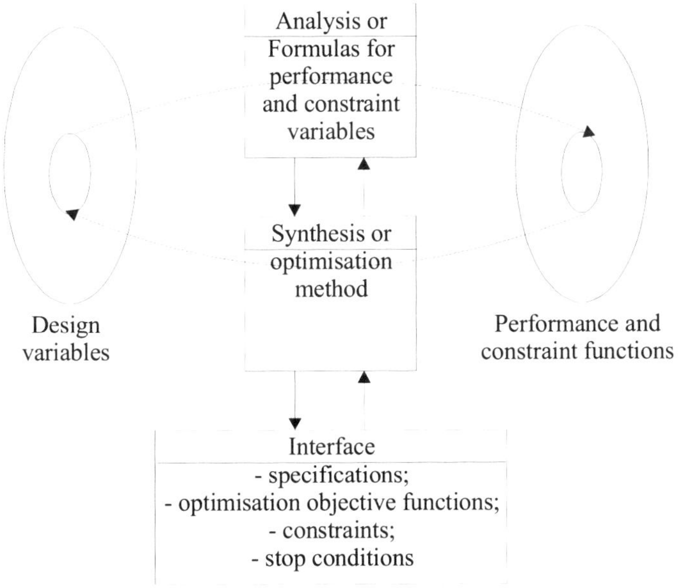

Figure 18.1 Optimization design process

Typical single objective (optimization) functions for induction machines are:

- Efficiency: η
- Cost of active materials: c_{am}
- Motor weight: w_m
- Global cost (c_{am} + cost of manufacturing and selling + loss capitalized cost + maintenance cost)

While single objective function optimization is rather common, multiobjective optimization methods have been recently introduced [1]

The IM is a rather complex artifact and thus there are many design variables that describe it completely. A typical design variable set (vector) of limited length is given here.

- Number of conductors per stator slot
- Stator wire gauge
- Stator core (stack) length
- Stator bore diameter
- Stator outer diameter
- Stator slot height
- Airgap length
- Rotor slot height
- Rotor slot width
- Rotor cage end – ring width

The number of design variables may be increased or reduced depending on the number of adopted constraint functions. Typical constraint functions are

- Starting/rated current
- Starting/rated torque
- Breakdown/rated torque
- Rated power factor
- Rated stator temperature
- Stator slot filling factor
- Rated stator current density
- Rated rotor current density
- Stator and rotor tooth flux density
- Stator and rotor back iron flux density

The performance and constraint functions may change attributes in the sense that any of them may switch roles. With efficiency as the only objective function, the other possible objective functions may become constraints.

Also, breakdown torque may become an objective function, for some special applications, such as variable speed drives. It may be even possible to turn one (or more) design variables into a constraint. For example, the stator outer diameter or even the entire stator lamination may be fixed to cut manufacturing costs.

The constraints may be equalities or inequalities. Equality constraints are easy to handle when their assigned value is used directly in the analysis and thus the number of design variables is reduced.

Not so with an equality constraint such as starting torque/rated torque, or starting current/rated current as they are calculated making use, in general, of all design variables.

Inequality constraints are somewhat easier to handle as they are not so tight restrictions.

The optimization design main issue is the computation time (effort) until convergence towards a global optimum is reached.

The problem is that, with such a complex nonlinear model with lots of restrictions (constraints), the optimization design method may, in some cases, converge too slowly or not converge at all.

Another implicit problem with convergence is that the objective function may have multiple maxima (minima) and the optimization method gets trapped in a local rather than the global optimum (Figure 18.2).

It is only intuitive that, in order to reduce the computation time and increase the probability of reaching a global optimum, the search in the subspace of design variables has to be thorough.

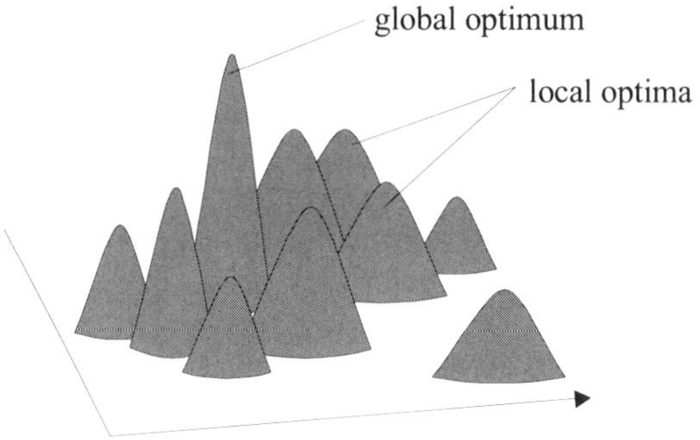

Figure 18.2 Multiple maxima objective function for 2 design variables

This process gets simplified if the number of design variables is reduced. This may be done by intelligently using the constraints in the process. In other words, the analysis model has to be wisely manipulated to reduce the number of variables.

It is also possible to start the optimization design with a few different sets of design variable vectors, within their existence domain. If the final objective function value is the same for the same final design variables and constraint violation rate, then the optimization method is able to find the global optimum.

But there is no guarantee that such a happy ending will take place for other IM with different specifications, investigated with same optimization method.

These challenges have led to numerous optimization method proposals for the design of electrical machines, IMs in particular.

18.2. ESSENTIAL OPTIMIZATION DESIGN METHODS

Most optimization design techniques employ nonlinear programing (NLP) methods. A typical form for a uni-objective NLP problem can be expressed in the form

$$\text{minimize}\, F(x) \tag{18.1}$$

$$\begin{gathered}\text{subject to:}\\ g_j(x) = 0;\ \ j = 1...m_e\end{gathered} \tag{18.2}$$

$$g_j(x) \geq 0;\ \ j = m_e + 1,...,m \tag{18.3}$$

$$X_{low} \leq X \leq X_{high} \tag{18.4}$$

where

$$X = \{x_1, x_1, ..., x_n\} \tag{18.5}$$

is the design variable vector, F(x) is the objective function and $g_j(x)$ are the equality and inequality constraints. The design variable vector X is bounded by lower (X_{low}) and upper (X_{high}) limits.

The nonlinear programming (NLP) problems may be solved by direct methods (DM) and indirect methods (IDM). The DM deals directly with the constraints problem into a simpler, unconstrained problem by integrating the constraints into an augmented objective function.

Among the direct methods, the complex method [2] stands out as an extension of the simplex method. [3] It is basically a stochastic approach. From the numerous indirect methods, we mention first the sequential quadratic programming (SQP). [4,5] In essence, the optimum is sought by successively solving quadratic programming (QP) subproblems which are produced by quadratic approximations of the Lagrangian function.

The QP is used to find the search direction as part of a line search procedure. Under the name of "augmented Lagrangian multiplies method" (ALMM) [6], it has been adopted for inequality constraints. Objective function and constraints gradients must be calculated.

The Hooke–Jeeves [7,8] direct search method may be applied in conjunction with SUMT (segmented unconstrained minimization technique) [8] or without it. No gradients are required. Given the large number of design variables, the problem nonlinearity, the multitude of constraints has ruled out many other general optimization techniques such as grid search, mapping linearization, simulated annealing, when optimization design of the IM is concerned.

Among the stochastic (evolutionary) practical methods for IM optimization design the genetic algorithms (GA) method [9] and the Montecarlo approach [10] have gained most attention.

Finally, a fuzzy artificial experience-based approach to optimization design of double cage IM is mentioned here. [11]

Evolutionary methods start with a few vectors of design variables (initial population) and use genetics inspired operations such as selection (reproduction) crossover and mutation to approach the highest fitness chromosomes by the survival of the fittest principle.

Such optimization approaches tend to find the global optimum but for a larger computation time (slower convergence). They do not need the computation of the gradients of the fitness function and constraints. Nor do they require an already good initial design variable set as most nongradient deterministic methods do.

No single optimization method has gained absolute dominance so far and stochastic and deterministic methods have complimentary merits. So it seems that the combination of the two is the way of the future. First, the GA is used to yield in a few generations a rough global optimization. After that, ALMM or Hooke–Jeeves methods may be used to secure faster convergence and larger precision in constraints meeting.

The direct method called the complex (random search) method is also claimed to produce good results. [12] A feasible initial set of design variables is necessary, but no penalty (wall) functions are required as the stochastic search principle is used. The method is less probable to land on a local optimum due to the random search approach applied.

18.3. THE AUGMENTED LAGRANGIAN MULTIPLIER METHOD (ALMM)

To account for constraints, in ALMM, the augmented objective function L(x,r,h) takes the form

$$L(X,r,h) = F(X) + r\sum_{i=1}^{m} \min\left[0, g_i(X) + h_{(i)}/r\right]^2 \tag{18.6}$$

where X is the design variable vector, $g_i(s)$ is the constraint vector (18.2)–(18.3), $h_{(i)}$ is the multiplier vector having components for all m constraints; r is the penalty factor with an adjustable value along the optimization cycle.

An initial value of design variables set (vector X_0) and of penalty factor r are required. The initial values of the multiplier vector h_0 components are all considered as zero.

As the process advances, r is increased.

$$r_{k+1} = C \cdot r_k; \quad C = 2 \div 4 \tag{18.7}$$

Also a large initial value of the maximum constraint error is set. With these initial settings, based on an optimization method, a new vector of design variables X_k which minimizes L(X,r,h) is found. A maximum constraint error δ_k is found for the most negative constraint function $g_i(X)$.

$$\delta_k = \max_{1\le i\le m}\left|\min[0, g_i(X_k)]\right| \tag{18.8}$$

The large value of δ_0 was chosen such that $\delta_1 < \delta_0$. With the same value of the multiplier, $h_{(i)k}$ is set to

$$h_{(i)k} = \min_{1\le i\le m}\left[0, r_k g_i(X_k) + h_{(i)k-1}\right] \tag{18.9}$$

to obtain $h_{i(1)}$. The minimization process is then repeated.

The multiplier vector is reset as long as the iterative process yields a 4/1 reduction of the error δ_k. If δ_k fails to decrease the penalty factor, r_k is increased. It is claimed that ALMM converges well and that even an infeasible initial X_0 is acceptable. Several starting (initial) X_0 sets are to be used to check the reaching of the global (and not a local) optimum.

18.4. SEQUENTIAL UNCONSTRAINED MINIMIZATION

In general, the induction motor design contains not only real but also integer (slot number, conductor/coil) variables. The problem can be treated as a multivariable nonlinear programming problem if the integer variables are taken as continuously variable quantities. At the end of the optimization process, they are rounded off to their closest integer feasible values. Sequential quadratic programming (SQP) is a gradient method. [4, 5] In SQP, SQ subproblems are successively solved based on quadratic approximations of Lagrangian function. Thus, a search direction (for one variable) is found as part of the line search procedure. SQP has some distinctive merits.

- It does not require a feasible initial design variable vector
- Analytical expressions for the gradients of the objective functions or constraints are not needed. The quadratic approximations of the Lagrangian function along each variable direction provides for easy gradient calculations.

To terminate the optimization process there are quite a few procedures:

- Limited changes in the objective function with successive iterations;
- Maximum acceptable constraint violation;
- Limited change in design variables with successive iterations;
- A given maximum number of iterations is specified.

One or more of them may in fact be applied to terminate the optimization process.

The objective function is also augmented to include the constraints as

$$f'(X) = f(X) + \gamma\sum_{i=1}^{m}\left| < g_i(X) > \right|^2 \tag{18.10}$$

where γ is again the penalty factor and

$$< g_i(X) >= \begin{cases} g_i(X) \text{ if } g_i(X) \geq 0 \\ 0 \text{ if } g_i(X) < 0 \end{cases} \tag{18.11}$$

As in (18.7), the penalty factor increases when the iterative process advances.

The minimizing point of f'(X) may be found by using the univariate method of minimizing steps. [13] The design variables change in each iteration as:

$$X_{j+1} = X_j + \alpha_j S_j \tag{18.12}$$

where S_j are unit vectors with one nonzero element. $S_1 = (1, 0, \ldots, 0)$; $S_2 = (0, 1, \ldots, 0)$ etc.

The coefficient α_j is chosen such that

$$f'(X_{j+1}) < f(X_j) \tag{18.13}$$

To find the best α, we may use a quadratic equation in each point.

$$f'(X + \alpha) = H(\alpha) = a + b\alpha + c\alpha^2 \tag{18.14}$$

$H(\alpha)$ is calculated for three values of α,

$$\alpha_1 = 0, \quad \alpha_2 = d, \quad \alpha_3 = 2d, \quad (d \text{ is arbitrary})$$

$$\begin{aligned} H(0) &= t_1 = a \\ H(d) &= t_2 = a + bd + cd^2 \\ H(2d) &= t_3 = a + 2bd + 4cd^2 \end{aligned} \tag{18.15}$$

From (18.15), a, b, c are calculated. But from

$$\frac{\partial H}{\partial \alpha} = 0; \quad \alpha_{opt} = \frac{-b}{2c} \tag{18.16}$$

$$\alpha_{opt} = \frac{4t_2 - 3t_1 - t_3}{4t_2 - 3t_3 - 2t_1} d \tag{18.17}$$

To be sure that the extreme is a minimum,

$$\left| \frac{\partial^2 H}{\partial \alpha^2} \right| = c > 0; \quad t_3 + t_1 > 2d \tag{18.18}$$

These simple calculations have to be done for each iteration and along each design variable direction.

18.5. A MODIFIED HOOKE–JEEVES METHOD

A direct search method may be used in conjunction with the pattern search of Hooke-Jeeves. [7] Pattern search relies on evaluating the objective function for a sequence of points (within the feasible region). By comparisons, the optimum value is chosen. A point in a pattern search is accepted as a new point if the objective function has a better value than in the previous point.

Let us denote

$$X^{(k-1)} - \text{previous base point}$$
$$X^{(k)} - \text{current base exploratory point}$$
$$X^{(k+1)} - \text{pattern point (after the pattern move)}$$

The process includes exploratory and pattern moves. In an exploratory move, for a given step size (which may vary during the search), the exploration starts from $X^{(k-1)}$ along each coordinate (variable) direction.

Both positive and negative directions are explored. From these three points, the best $X^{(k)}$ is chosen. When all n variables (coordinates) are explored, the exploratory move is completed. The resulting point is called the current base point $X^{(k)}$.

A pattern move refers to a move along the direction from the previous to the current base point. A new pattern point is calculated

$$X^{(k+1)} = X^{(k)} + a\left(X^{(k)} - X^{(k-1)}\right) \tag{18.19}$$

a is an accelerating factor.

A second pattern move is initiated.

$$X^{(k+2)} = X^{(k+1)} + a\left(X^{(k+1)} - X^{(k)}\right) \tag{18.20}$$

The success of this second pattern move $X^{(k+2)}$ is checked. If the result of this pattern move is better than that of point $X^{(k+1)}$, then $X^{(k+2)}$ is accepted as the new base point. If not, then $X^{(k+1)}$ constitutes the new current base point.

A new exploratory-pattern cycle begins but with a smaller step search and the process stops when the step size becomes sufficiently small.

The search algorithm may be summarized as

Step 1: Define the starting point $X^{(k-1)}$ in the feasible region and start with a large step size;

Step 2: Perform exploratory moves in all coordinates to find the current base point $X^{(k)}$;

Step 3: Perform a pattern move: $X^{(k+1)} = X^{(k)} + a\left(X^{(k)} - X^{(k-1)}\right)$ with $a < 1$;

Step 4: Set $X^{(k-1)} = X^{(k)}$;

Step 5: Perform tests to check if an improvement took place. Is $X^{(k+1)}$ a better point?

If "YES", set $X^{(k)} = X^{(k+1)}$ and go to step 3.

If "NO", continue;

Step 6: Is the current step size the smallest?

If "YES", stop with $X^{(k)}$ as the optimal vector of variables.

If "NO", reduce the step size and go to step 2.

To account for the constraints, the augmented objective function f'(X), (8.10 – 8.11) – is used. This way the optimization problem becomes an unconstraint one. In all nonevolutionary methods presented so far, it is necessary to do a few runs for different initial variable vectors to make sure that a global optimum is obtained. It is necessary to have a feasible initial variable vector. This requires some experience from the designer. Comparisons between the above methods reveal that the sequential unconstrained minimization method (Han & Powell) is a very powerful but time consuming tool while the modified Hooke–Jeeves method is much less time consuming. [14, 15, 16]

18.6. GENETIC ALGORITHMS

Genetic algorithms (GA) are computational models which emulate biological evolutionary theories to solve optimization problems. The design variables are grouped in finite length strings called chromosomes. GA maps the problem to a set of strings (chromosomes) called population. An initial population is adopted by way of a number of chromosomes. Each string (chromosome) may constitute a potential solution to the optimization problem.

The string (chromosome) can be constituted with an orderly alignment of binary or real coded variables of the system. The chromosome–the set of design variables – is composed of genes which may take a number of values called alleles. The choice of the coding type, binary or real, depends on the number and type of variables (real or integer) and the required precision. Each design variable (gene) is allowed a range of feasible values called search space. In GA, the objective function is called fitness value. Each string (chromosome) of population of generation i, is characterised by a fitness value.

The GA manipulates upon the population of strings in each generation to help the fittest survive and thus, in a limited number of generations, obtain the optimal solution (string or set of design variables). This genetic manipulation involves copying the fittest string (elitism) and swapping genes in some other strings of variables (genes).

Simplicity of operation and the power of effect are the essential merits of GA. On top of that, they do not need any calculation of gradients (of fitness function) and provide more probably the global rather than a local optimum.

They do so because they start with a random population–a number of strings of variables–and not only with a single set of variables as nonevolutionary methods do.

However their convergence tends to be slow and their precision is moderate. Handling the constraints may be done as for nonevolutionary methods through an augmented fitness function.

Finally, multi-objective optimization may be handled mainly by defining a comprehensive fitness function incorporating as linear combinations (for example) the individual fitness functions.

Though the original GAs make use of binary coding of variables, real coded variables seem more practical for induction motor optimization as most variables are continuous. Also, in a hybrid optimization method, mixing GAs with a nonevolutionary method for better convergence, precision, and less computation time, requires real coded variables.

For simplicity, we will refer here to binary coding of variables. That is, we describe first a basic GA algorithm.

A simple GA uses three genetic operations:

- Reproduction (evolution and selection)
- Crossover
- Mutation

18.6.1. Reproduction (evolution and selection)

Reproduction is a process in which individual strings (chromosomes) are copied into a new generation according to their fitness (or scaled fitness) value. Again, the fitness function is the objective function (value).

Strings with higher fitness value have a higher probability of contributing one or more offsprings in the new generation. As expected, the reproduction rate of strings may be established, many ways.

A typical method emulates the *biased roulette wheel* where each string has a roulette slot size proportional to its fitness value.

Let us consider as an example 4 five binary digit numbers whose fitness value is the decimal number value (Table 18.1).

Table 18.1.

String number	String	Fitness value	% of total fitness value
1	01000	64	5.5
2	01101	469	14.4
3	10011	361	30.9
4	11000	576	49.2
	Total	1170	100

The percentage in Table 18.1 may be used to draw the corresponding biased roulette wheel (Figure 18.2).

Each time a new offspring is required, a simple spin of the biased roulette produces the reproduction candidate. Once a string has been selected for reproduction, an exact replica is made and introduced into the mating pool for the purpose of creating a new population (generation) of strings with better performance.

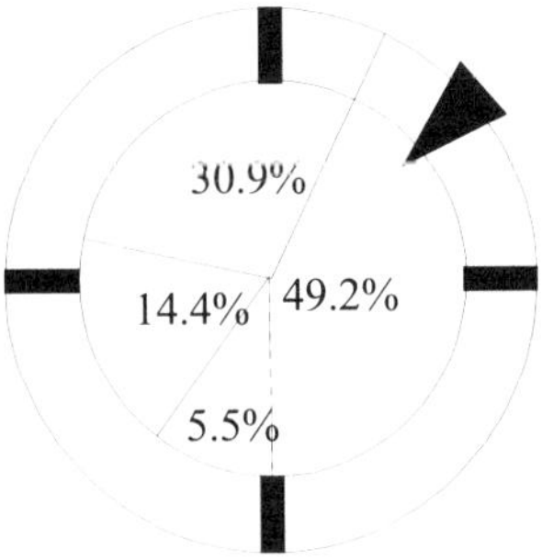

Figure 18.3 The biased roulette wheel

The biased roulette rule of reproduction might not be fair enough in reproducing strings with very high fitness value. This is why other methods of selection may be used. The selection by the arrangement method for example, takes into consideration the diversity of individuals (strings) in a population (generation). First, the m individuals are arranged in the decreasing order of their fitness in m rows.

Then the probability of selection ρ_i as offspring of one individual situated in row i is [1]

$$\rho_i = \frac{[\phi - (r_i - 1)(2\phi - 2)/(m-1)]}{m} \tag{18.21}$$

ϕ – pressure of selection $\phi = (1 - 2)$;
m – population size (number of strings);
r_i – row of ith individual (there are m rows);
ρ_i – probability of selection of ith row (individual).

Figure 18.4 shows the average number of offspring versus the row of individuals. The pressure of selection ϕ is the average number of offspring of best individual. For the worst, it will be $2 - \phi$. As expected, an integer number of offspring is adopted.

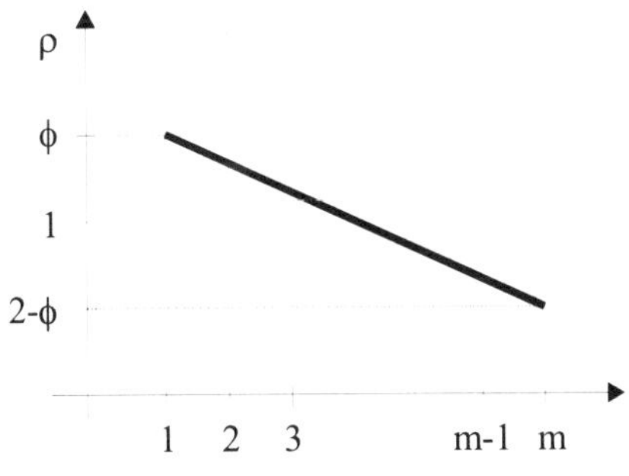

Figure 18.4 Selection by arrangement

By the pressure of selection ϕ value, the survival chance of best individuals may be increased as desired.

18.6.2. Crossover

After the reproduction process is done *crossover* may proceed. A simple crossover contains two steps:

- Choosing randomly two individuals (strings) for mating
- Mating by selection of a crossover point K along the chromosome length l. Two new strings are created by swapping all characters in positions K + 1 to l (Figure 18.5).

Besides simple (random) crossover, at the other end of scale, completely continued crossover may be used. Let us consider two individuals of the tth generation A(t) and B(t), whose genes are real variables $a_1, \ldots, a_n$ and $b_1, \ldots, b_n$,

$$A(t) = [a_1, a_2, \ldots, a_n] \tag{18.22}$$

$$B(t) = [b_1, b_2, \ldots, b_n] \tag{18.23}$$

The two new offspring A(t+1) and B(t+1) may be produced by linear combination of parents A(t) and B(t).

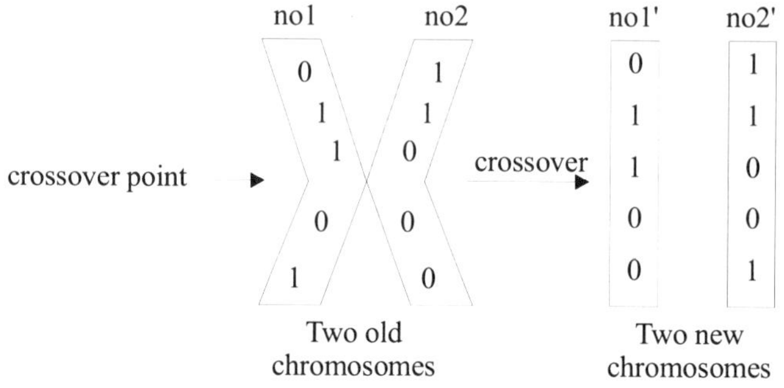

Figure 18.5 Simple crossover operation

$$A(t+1) = [\rho_1 a_1 + (1-\rho_1) b_1, \ldots, \rho_n a_n + (1-\rho_n) b_n] \tag{18.24}$$

$$B(t+1) = [\rho_1 b_1 + (1-\rho_1) a_1, \ldots, \rho_n b_n + (1-\rho_n) a_n] \tag{18.25}$$

where $\rho_1, \ldots, \rho_n \in [0,1]$ are random values (uniform probability distribution).

18.6.3. Mutation

Mutation is required because reproduction and crossover tend to become overzealous and thus lose some potentially good "genetic material".

In simple GAs, mutation is a low probability (occasional) random alteration of one or more genes in one or more offspring chromosomes. In binary coding

GAs, a zero is changed to a 1. The mutation plays a secondary role and its frequency is really low.

Table 18.2 gives a summary of simple GAs.

Table 18.2. GA stages

Stage	Chromosomes	Fitness value
Initial population: (size 3, with 8 genes and its fitness values)	P_1: 11010110 P_2: 10010111 P_3: 01001001	$F(P_1) = 6\%$ $F(P_2) = 60\%$ $F(P_3) = 30\%$
Reproduction: based on fitness value a number of chromosomes survive. There will be two P_2 and one P_3.	P_2: 10010111 P_2: 10010111 ←|→ P_3: 01001001	no need to recalculate fitness as P_2 and P_3 will mate
Crossover: some portion of P_2 and P_3 will be swapped	P_2: 10010111 P_2': 10010001 P_3': 01001111	no need of it here
Mutation: some binary genes of some chromosomes are inverted	P_2: 10010111 P_2'': 10011001 P_3': 01001111	A new generation of individuals has been formed; a new cycle starts

In real coded GAs, the mutation changes the parameters of selected individuals by a random change in predefined zones. If A(t) is an individual of tth generation, subject to mutation, each gene will undergo important changes in first generations. Gradually, the rate of change diminishes. For the t_{th} generation, let us define two numbers (p) and (r) which are randomly used with equal probabilities.

$$\begin{aligned} &p = +1 \quad - \text{positive alteration} \\ &p = -1 \quad - \text{negative alteration} \\ &r \in [0,1] \ - \text{uniform distribution} \end{aligned} \tag{18.26}$$

The factor r selected for uniform distribution determines the amplitude of change. The mutated (K^{th}) parameter (gene) is given by [18]

$$\begin{aligned} &a_k' = a_k + (a_{kmax} - a_k)\left(1 - r^{\left(1-\frac{t}{T}\right)^5}\right); \ \text{for } p = +1 \\ &a_k' = a_k - (a_k - a_{kmin})\left(1 - r^{\left(1-\frac{t}{T}\right)^5}\right); \ \text{for } p = -1 \end{aligned} \tag{18.27}$$

a_{kmax} and a_{kmin} are the maximum and minimum feasible values of a_k parameter (gene).

T is the generation index when mutation is cancelled. Figure 18.6 shows the mutation relative amplitude for various generations (t/T varies) as a function of the random number r, based on (18.27).

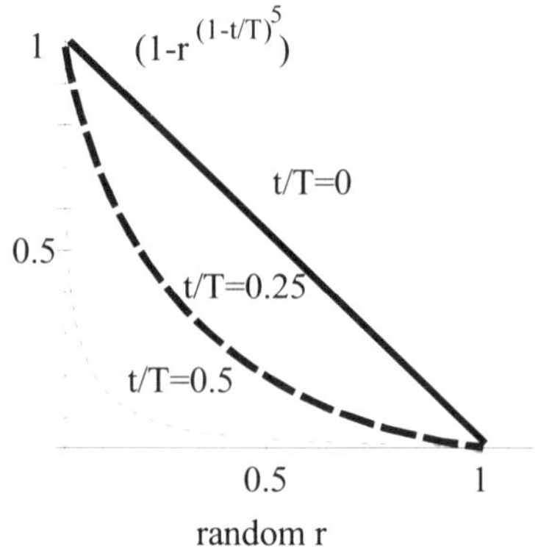

Figure 18.6 Mutation amplitude for various generations (t/T)

18.6.4. GA performance indices

As expected, GAs are supposed to produce a global optimum with a limited number (%) of all feasible chromosomes (variable sets) searched. It is not easy to assess the minimum number of generations (T) or the population size (m) required to secure a sound optimal selection. The problem is still open but a few attempts to solve it have been made.

A schema, as defined by Holland, is a similarity template describing a subset of strings with similarities at certain string positions.

If to a binary alphabet {0,1} we add a new symbol * which may be indifferent, a ternary alphabet is created {1,0,*}.

A subset of 4 members of a schema like *111* is shown in (18.28).

$$\{11111, 11110, 01111, 01110\} \tag{18.28}$$

The length of the string $n_l = 5$ and thus the number of possible schemata is 3^5. In general, if the ordinality of the alphabet is K (K = 2 in our case), there are $(K + 1)^n$ schemata.

The exact number of unique schemata in a given population is not countable because we do not know all the strings in this particular population. But a bound for this is feasible.

A particular binary string contains 2^{n_l} schemata as each position may take its value or * thus a population of size m contains between 2^{n_l} and $m \cdot 2^{n_l}$ schemata. How many of them are usefully processed in a GA? To answer this question, let us introduce two concepts.

- The order o of a schema H (δ(H)): the number of fixed positions in the template. For 011**1**, the order o(H) is 4.
- The length of a schema H, (δ(H)): the distance between the first and the last specified string position. In 011**1**, δ(H) = 6 – 1 = 5.

The key issue now is to estimate the effect of reproduction, crossover and mutation on the number of schemata processed in a simple GA.

It has been shown [9] that short, low order, above-average fitness schemata receive exponentially increasing chances of survival in subsequent generations. This is the fundamental theorem of GAs.

Despite the disruption of long, high-order schemata by crossover and mutation, the GA implicitly process a large number of schemata (%n_l^3) while in fact they explicitly process a relatively small number of strings (m in each population). So, despite the processing n_l of structures, each generation GAs process approximately n_l^3 schemata in parallel with no memory of bookkeeping. This property is called *implicit parallelism* of computation power in GAs.

To be efficient, GAs must find solutions by exploring only a few combinations in the variable search space. Those solutions must be found with a better probability of success than with random search algorithms. The traditional performance criterion of GA is the evolution of average fitness of the population through subsequent generations.

It is almost evident that such a criterion is not complete as it may concentrate the point around a local, rather than global optimum.

A more complete performance criterion could be the ratio between the number of success runs per total given number of runs (1000 for example, [19], P_{ag} (A)):

$$P_{ag}(A) = \frac{\text{number of success runs}}{1000} \tag{18.29}$$

For a pure random search algorithm, it may be proved that the theoretical probability to find p optima in a searching space of M possible combinations, after making A% calls of the objective functions, is [19]

$$P_{rand}(A) = \sum_{i=1}^{K} \frac{p}{M}\left(1 - \frac{p}{M}\right)^{i-1}; \quad A = \frac{K}{M} \cdot 1000 \tag{18.30}$$

The initial population size M in GA is generally M > 50.

We should note that when the searching space is discretised the number of optima may be different than the case of continuous searching space.

In general, the crossover probability P_c = 0.7 and the mutation probability $P_m \leq 0.005$ in practical GAs. For a good GA algorithm, searching less than A = 10% of the total searching space should produce a probability of success $P_{ag}(A) > P_{rand}(A)$ with $P_{ag}(A) > 0.8$ for A = 10% and $P_{ag}(A) > 0.98$ for A = 20%.

Special selection mechanisms, different from the biased roulette wheel selection are required. Stochastic remainder techniques [17] assign offsprings to strings based on the integer part of the expected number of offspring. Even such solutions are not capable to produce performance as stated above.

However if the elitist strategy is used to make sure that the best strings (chromosomes) survive intact in the next generation, together with selection by stochastic remainder technique and real coding, the probability of success may be drastically increased. [19]

Recently the GAs and some deterministic optimization methods have been compared in IM design [20–23]. Mixed conclusions have resulted. They are summarized below.

18.7. SUMMARY

- GAs offer some advantages over deterministic methods.
- GAs reduce the risk of being trapped in local optima.
- GAs do not need a good starting point as they start with a population of possible solutions. However a good initial population reduces the computation time.
- GAs are more time consuming though using real coding, elitist strategy and stochastic remainder selection techniques increases the probability of success to more than 90% with only 10% of searching space investigated.
- GAs do not need to calculate the gradient of the fitness function; the constraints may be introduced in an augmented fitness function as penalty function as done with deterministic methods. Many deterministic methods do not require computation of gradients.
- GAs show in general lower precision and slower convergence than deterministic methods.
- Other refinements such as scaling the fitness function might help in the selection process of GAs and thus reduce the computation time for convergence.
- It seems that hybrid approaches which start with GAs (with real coding), to avoid trapping in local optima and continue with deterministic (even gradient) methods for fast and accurate convergence, might become the way of the future.
- If a good initial design is available, optimization methods produce only incremental improvements; however, for new designs (high speed IMs for example), they may prove indispensable.
- Once optimization design is performed based on nonlinear analytical IM models, with frequency effects approximately considered, the final solution (design) performance may be calculated more precisely by FEM as explained in previous chapters.
- For IM with involved rotor slot configurations – deep bars, closed slots, double-cage – it is also possible to leave the stator unchanged after optimization and change only the main rotor slot dimensions (variables) and explore by FEM their effect on performance and constraints (starting torque, starting current) until a practical optimum is reached.
- Any optimization approach may be suited to match the FEM. In [14], a successful attempt with Fuzzy Logic optimiser and FEM assistance is presented for a double-cage IM. Only a few tens of 2DFEM runs are used and thus the computation time remains practical.

- Notable advances in matching FEM with optimization design methods are expected as the processing power of PCs is continuously increasing.

18.8. REFERENCES

1. U. Sinha, A Design and Optimization Assistant for Induction Motors and Generators, Ph.D. Thesis, MIT, June 1998.
2. M.Box, A New Method of Constrained Optimization and a Comparison With Other Methods, Computer Journal, Vol.8, 1965, pp.42–52.
3. J. Helder, R. Mead, A Simplex Method for Function Minimization, Computer Journal, Vol.7, 1964, pp.308 – 313.
4. S.P. Han, A Globally Convergent Method for Nonlinear Programming, Journal of optimization theory and applications", Vol22., 1977, pp.297 – 309.
5. M.J.D. Powell, A Fast Algorithm for Nonlinearly Constrained Optimization Calculation, Numerical Analysis, Editor: D.A.Watson, Lecture notes in mathematics, Springer Verlag, 630, 1978, pp.144 – 157.
6. R. Rockafeller, Augmented Lagrange Multiplier Functions and Duality in Convex Programming, SIAM, J.Control, Vol.12, 1994, pp.268 – 285.
7. R. Hooke, T.A. Jeeves, Direct Search Solution of Numerical and Statistical Problem, Journal of ACM, Vol.8, 1961, pp.212.
8. R. Ramarathnam, B.G. Desa, V.S. Rao, A Comparative Study of Minimization Techniques for Optimization of IM Design, IEEE Trans. Vol.PAS – 92, 1973, pp.1448 – 1454.
9. D.E. Goldberg, Genetic Algorithms in Search Optimization and Machine Learning, Addison Wesley Longman Inc., 1989.
10. U. Sinha, A Design Assistant for Induction Motors, S.M.Thesis, Department of Mechanical Engineering, MIT, August, 1993.
11. N. Bianchi, S. Bolognani, M. Zigliotto, Optimised Design of a Double Cage Induction Motor by Fuzzy Artificial Experience and Finite Element Analysis, Record of ICEM – 1998, Vol.1/3.
12. Q. Changtoo, Wu Yaguang, Optimization Design of Electrical Machines by Random Search Approach, Record of ICEM – 1994, Paris, session D16, pp.225 – 229.
13. A.V. Fiacco, G.P. McCormick, Nonlinear Programming Sequential Unconstrained Minimization Techniques, John Wiley, 1969.
14. J. Appelbaum, E.F. Fuchs, J.C. White, I.A. Kahn, Optimization of Three Phase Induction Motor Design, Part I + II, IEEE Trans. Vol. EC – 2, No.3, 1987, pp.407 – 422.
15. Ch. Li, A. Rahman, Three–phase Induction Motor Design Optimization Using The Modified Hooke–Jeeves Method, EMPS Vol.18, No.1, 1990, pp.1 – 12.
16. C. Singh, D. Sarkas, Practical Considerations in the Optimization of Induction Motor Design, Proc.IEE, Vol.B – 149, No.4, 1992, pp.365 – 373.
17. J.E. Baker, Adaptive Selection Methods of Genetic Algorithms, Proc. of the first International Conference on Genetic Algorithms, 1998, pp.101 – 111.

18. C.Z. Janikov, Z. Michaleewiez, An Experimental Comparison of Binary and Floating Point Representation in Genetic Algorithms, Proc. of Fourth International Conference on Genetic Algorithms, 1991, pp.31 – 36.
19. F. Wurtz, M. Richomme, J. Bigeon, J.C. Sabonnadiere, A few Results for Using Genetic Algorithms in the Design of Electrical Machines, IEEE Trans., Vol.MAG – 33, No.2, 1997, pp.1892 – 1895.
20. M. Srinivas, L.M. Patniak, Genetic Algorithms: A Survey, in Computer, an IEEE review, June 1994, pp.17 – 26.
21. S. Hamarat, K. Leblebicioglu, H.B. Ertan, Comparison of Deterministic and Nondeterministic Optimization Algorithms for Design Optimization of Electrical Machines, Record of ICEM – 1998, pp.1477 – 1481.
22. Ö. Göl, J.P. Wieczorek, A Comparison of Deterministic and Stochastic Optimization Methods in Induction Motor Design, IBID, pp.1472 – 1476.
23. S.H. Shahalami, S. Saadate, Genetic Algorithm Approach in the Identification of Squirrel Cage Induction Motor's Parameters, Record of ICEM – 1998, pp.908 – 913.

Chapter 19

THREE PHASE INDUCTION GENERATORS

19.1 INTRODUCTION

In Chapter 7 we alluded to the induction generator mode both in stand alone (capacitor excited) and grid-connected situations. In essence for the cage-rotor generator mode, the slip S is negative ($S < 0$). As the IM with a cage rotor is not capable of producing, reactive power, the energy for the machine magnetization has to be provided from an external means, either from the power grid or from constant (or electronically controlled) capacitors.

The generator mode is currently used for braking advanced PWM converter fed drives for industrial and traction purposes. Induction generators-grid connected or isolated (capacitor excited)-are used for constant or variable speed and constant or variable voltage/frequency, in small hydro power plants, wind energy systems, emergency power supplies, etc. [1]. Both cage and wound rotor configurations are in use. For a summary of these possibilities, see Table 19.1. Cogeneration of electric power in industry at the grid (constant voltage and frequency) for low range variable speed and motor/generator operation in pump-back hydropower plant are all typical applications for wound rotor IMs.

In what follows, we will treat the main performance issues of IGs first in stand alone configurations and then at power grid. Though changes in performance owing to variable speed are a key issue here, we will not deal with power electronics or control issues. We choose to do so as IG systems now constitute a mature technology whose in-depth (useful) treatment could be the subject matter of an entire book.

Table 19.1 IG configurations

IG type	speed		grid	isolated	frequency		voltage	
	constant	variable	connected		constant	variable	constant	variable
wound rotor	-	*	*	-	*	-	*	-
cage rotor	*	*	*	*	*	*	*	*

-impractical; * practical;

Typical configurations of IGs are shown in Figure 19.1-19.4.

Figure 19.1 portrays a WR-IG whose rotor is connected through a bi-directional power flow converter and transformer to the power grid which has constant voltage and frequency.

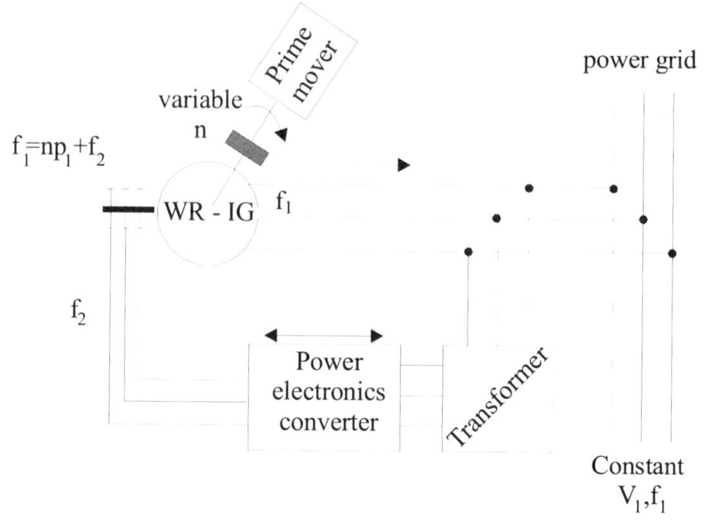

Figure 19.1 Advanced (bi-directional rotor power flow) wound rotor IG (WR - IG) at power grid

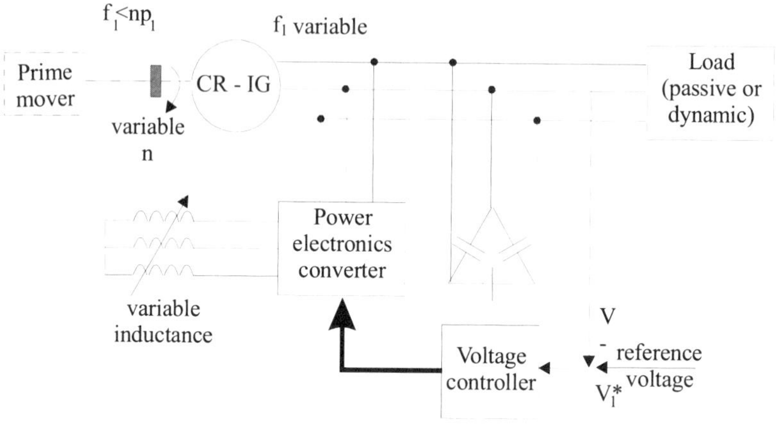

Figure 19.2 Isolated cage-rotor IG (CR - IG) with constant voltage V_1 and variable frequency output f_1 for variable speed

For limited prime mover speed variation (X%), the power converter rating is limited to X% of IG rated power. Consequently reasonably lower costs are encountered.

Figure 19.2 shows an isolated cage-rotor IG (CR - IG) with variable frequency but constant voltage for variable prime mover speed. The power electronics converter in Figure 19.2 has limited rating and simplified control as it acts as a Variac to control (reduce) the capacitance reactive power flow into the IG for constant voltage control.

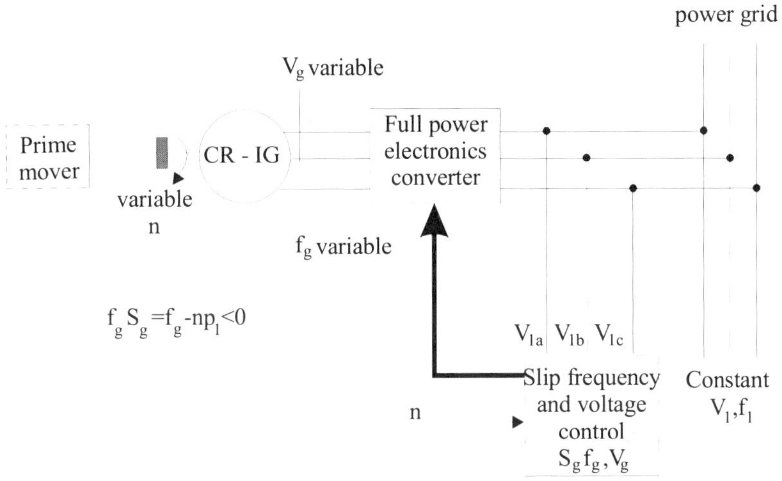

Figure 19.3 Power grid connected CR-IG variable speed constant V_1 & f_1

The a.c.-a.c. power electronics converter (PEC) in Figure 19.3 makes the transition from the variable voltage and frequency V_g, f_g of the generator to the constant voltage and frequency of the power grid. In the process, the same PEC transfers reactive power from the power grid to provide the magnetization of the cage-rotor induction generator (CR - IG).

The PEC is rated at full power and thus adds to the total costs of the equipment while the CR-IG is rugged and costs less.

Handling limited variable speed prime movers (wind or constant head small hydraulic turbines) to extract most of the available primary energy may be done by simpler methods such as pole changing windings for CR-IG (Figure 19.4a) or even a parallel connected R_{ad}/L_{ad} circuit in the rotor of the WR-IG for grid connected IGs (Figure 19.4b).

On the other hand, for less frequency sensitive loads voltage regulation for variable speed variable frequency but constant voltage, long shunt capacitor connections or saturable load interfacing transformers may be used in conjunction with CR-IGs.

Solutions like those shown in Figure 19.4 are characterized by low costs. But the power flow control and voltage regulation (for isolated systems) are only moderate. For low/medium power applications with limited primer mover speed variation they are adequate.

The system configurations presented in Figures 19.1-19.4 are meant to show the multitude of solutions that are feasible and have been proposed. Some of them are extensively used in wind power and small hydropower systems.

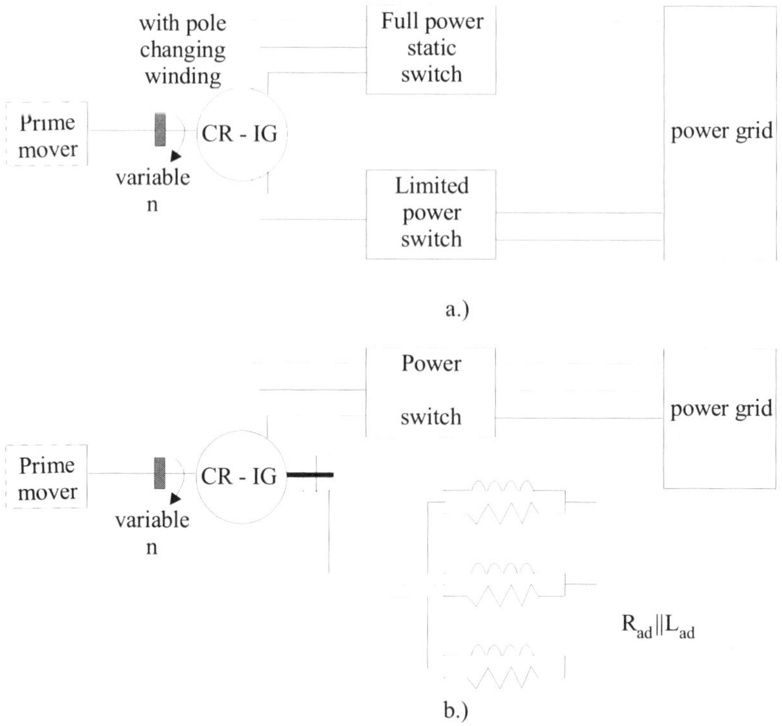

Figure 19.4 Simplified IG systems
a.) CR-IG with pole changing winding and two static power switches for grid connection
b.) WR-IG with parallel R_{ad} || L_{ad} in the rotor for grid connection system

In the following we will focus on IG behavior in such schemes rather than on the systems themselves. We will first investigate in depth the self excitation and load performance steady state and transients of CR-IG with capacitors for isolated systems.

Then the WR-IG with bi–directional power capability PEC in the rotor for grid connected systems will be dealt with in some detail in terms of stability limits and performance.

19.2 SELF-EXCITED INDUCTION GENERATOR (SEIG) MODELING

By SEIG, we mean a cage-rotor IG with capacitor excitation (Figure 19.5a). The standard equivalent circuit of SEIG on a per phase bases is shown in Figure 19.5b.

First, with the switch S open, the machine is driven by the prima mover. As the SEIG picks up speed slowly, the no load terminal voltage increases and settles at a certain value. This is the self-excitation process, which has been known from 1930s. [2]

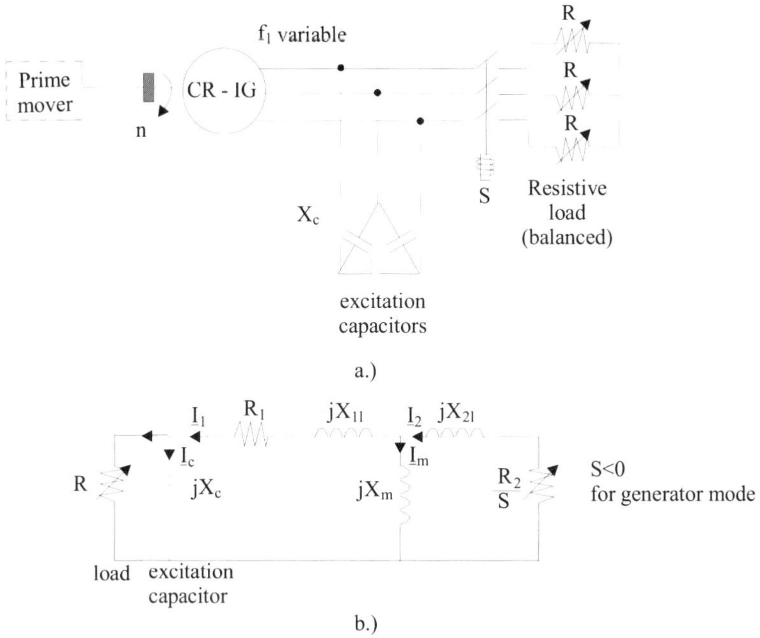

Figure 19.5 Selfexcited induction generator (SEIG) with resistive load
a.) general scheme; b.) standard equivalent circuit

In essence, first the residual magnetism in the rotor laminated core (from previous IG operation) produces by motion an e.m.f. in the stator windings. Its frequency is $f_{10} = np_1$. This e.m.f. is applied to the machine terminals and produces in the RLC circuit of each phase a magnetization current which produces an airgap field. This field adds to the remnant field of the rotor to produce a higher e.m.f.. The process goes on until an equilibrium is reached for a given speed n and a given capacitance C at a voltage level V_0. However this process is stable as long as the machine is saturated $X_m(I_m)$ is a nonlinear function (as shown already in Chapter 7).

In a very simplified form, with X_{1l}, X_{2l}, R_1, R_2-neglected, the equivalent circuit for no load degenerates to X_m in parallel with a capacitor and a small e.m.f. determined by the remnant rotor field (Figure 19.6a).

The mandatory nonlinear $L_m(I_m)$ relationship is evident from Figure 19.6b where the final no load voltage V_{10} occurs at the intersection of the no load characteristic (to be found by the standard no load test, or by design)-with the capacitor voltage straight line. Also the necessary presence of remnant rotor field (E_{rem}) is self evident.

Once the SEIG is loaded, the terminal voltage changes depending on speed, SEIG parameters, the nature of the load and its level.

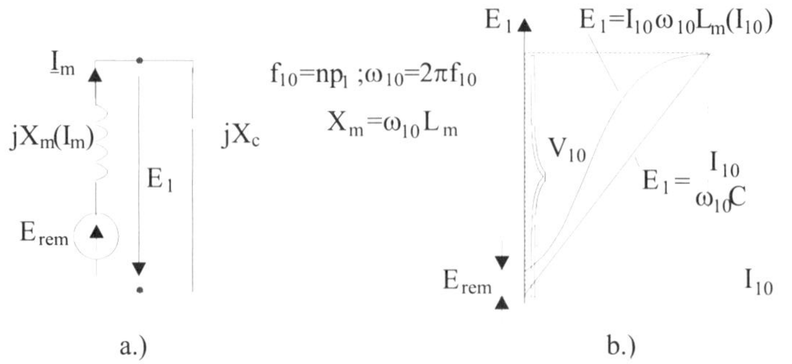

Figure 19.6 Oversimplified circuit for explaining self-excitation
a.) the circuit, b.) the characteristics

The occurrence of load implies rotor currents, that is a non zero slip, $S \neq 0$. Even if the speed of the prime mover is kept constant, the frequency f_1 varies with load

$$f_1 = \frac{np_1}{1+|S|}; \quad f_{10} = (f_1)_{S=0} = np_1 \tag{19.1}$$

Calculating the variation of voltage V_1, frequency f_1, stator current I_1, power factor, efficiency, with speed n, load and capacitor C, means, in fact, to determine the steady state performance of SEIG. In this arrangement, V_1 and f_1 are fundamental unknowns.

It is also feasible to have the capacitance C and frequency f_1 as the main unknowns for given speed, load and output voltage V_1. Apparently the problem is simple use the standard circuit of Figure 19.5b with given $L_m(I_m)$ function. The next paragraph deals extensively with this issue.

19.3 STEADY STATE PERFORMANCE OF SEIG

Various analytical methods (models) have been developed in order to predict the steady state performance of SEIG.

Among them, two are predominant

- the impedance model;
- the admittance model;

The impedance model is based on the single phase equivalent circuit shown in Figure 19.5a, which is expressed in per unit terms
f-frequency f_1/rated frequency f_{1b}, $f = f_1/f_{1b}$;
v-speed/synchronous speed for f_{1b} $v = np_1/f_{1b}$.

The final form of the circuit is shown in Figure 19.7.

The R, L character of the load, the presence of core loss resistance R_{core} (which may also vary slightly with frequency f), the nonlinearity of $X_m(I_m)$ dependence with the unknowns X_m (the real output is V_1) and f makes the solving of this model possible only through a numerical procedure. Once X_m

and f are calculated, the entire circuit model may be solved in a straightforward manner.

Fifth or fourth order polynomial equations in f or X_m are obtained from the conditions that the real and imaginary parts of the equivalent impedance are zero. A wealth of literature on this subject is available [3–5]. Recently a fairly general solution of the impedance model, based on the optimisation approach has been introduced. [6]

However, the high order of system nonlinearity prevents an easy understanding of performance sensitivity to various parameters. In search of a simpler solution, the admittance model has been proposed. [7]

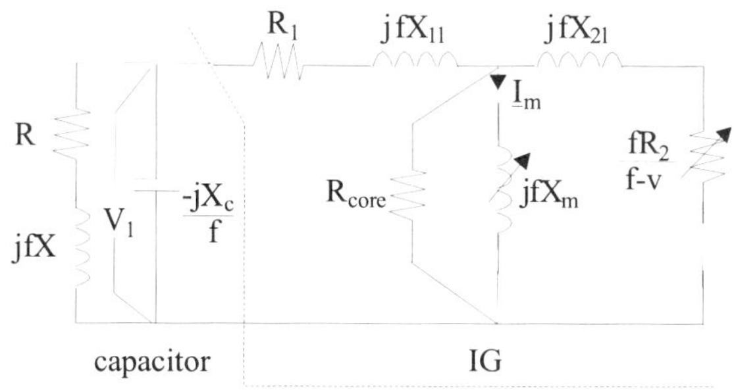

Figure 19.7 The impedance model of SEIG
f-p.u. frequency, v = p.u. speed.

While the X_m equation is simple, calculating f involves a complex procedure. In [8, 9] admittance models that lead to quadratic equations for the unknowns are obtained for balanced resistive load without additional simplifying assumptions.

19.4 THE SECOND ORDER SLIP EQUATION MODEL FOR STEADY STATE

The second order equation model may be obtained from the standard circuit shown in Figure 19.8.

The slip is negative for generator mode

$$S = \frac{(f - v)}{f} \tag{19.2}$$

The airgap voltage E_a for frequency f (in p.u.)

$$\underline{E}_a = f\underline{E}_1 = \underline{I}_2\left(\frac{R_2}{S} + jX_{21}\right) \tag{19.3}$$

E_1-airgap voltage at rated frequency.

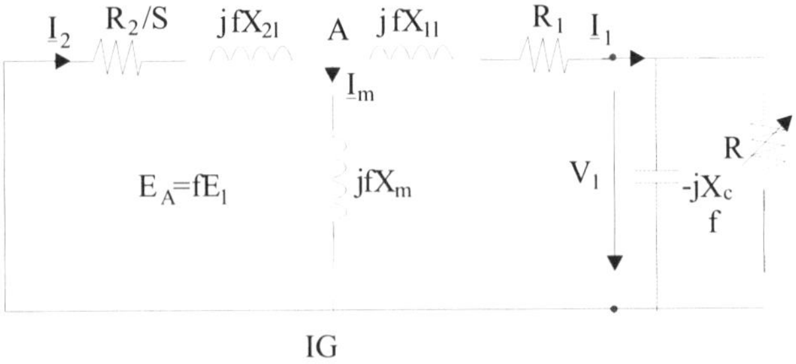

Figure 19.8. Equivalent circuit of SEIG with slip S and frequency f (p.u.) shown

For simplicity, the core loss resistance is neglected while the load is purely resistive. The parallel capacitor-load resistance circuit may be transformed into a series one as

$$R_L - jX_L = \frac{R\left(-j\frac{X_C}{f}\right)}{R - j\frac{X_C}{f}} = \frac{R}{1+\frac{R^2 X_C^{\ 2}}{f^2}} - \frac{j\frac{X_C}{f}}{1+\frac{X_C^{\ 2}}{f^2 R^2}} \qquad (19.4)$$

Now we may lump R_l and R_L into $R_{lL} = R_l + R_L$ and fX_{ll} and X_L into $X_{lL} = fX_{ll} - X_L$ to obtain the simplified equivalent circuit in Figure 19.9.

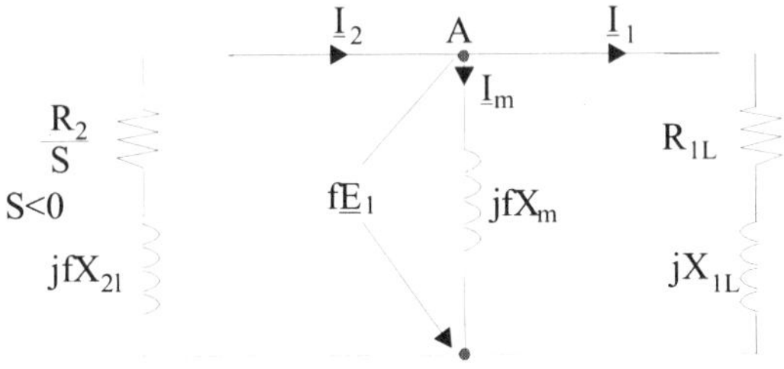

Figure 19.9 A simple nodal form of SEIG equivalent circuit

Notice that frequency f will be given and the new unknowns are S and X_m while $E_l(I_m)$ or $X_m(E_l)$ come from the no load curve of IG. Also the load resistance R, the capacitance C, and the values of R_2, X_{2l}, R_l, X_{ll} are given. Consequently, with f known and S calculated, the speed v will be computed as

$$v = f(1-S) \qquad (19.5)$$

If the speed is known, a simpler iterative procedure is required to change f until the desired v is obtained.

We should mention that, with given f, the presence of any type of load does not complicate the problem rather than the expressions of R_L and X_L in (19.4). An induction motor load is such a typical dynamic load.

For self-excitation, the summation of currents in node A (Figure 19.9) must be zero (implicitly E1 ≠ 0)

$$0 = -\underline{I}_2 + \underline{I}_1 + \underline{I}_m; \tag{19.6}$$

$$f\underline{E}_1\left(\frac{1}{jfX_m} + \frac{1}{R_{1L} + jX_{1L}} + \frac{S}{R_2 + jSfX_{21}}\right) = 0 \tag{19.7}$$

The same result is obtained in [9] after introducing a voltage source in the rotor.

The real and imaginary parts of (19.7) must be zero

$$\frac{R_{1L}}{R_{1L}^2 + X_{1L}^2} + \frac{SR_2}{R_2^2 + S^2f^2X_{21}^2} = 0 \tag{19.8}$$

$$\frac{1}{fX_m} - \frac{X_{1L}}{R_{1L}^2 + X_{1L}^2} + \frac{SfX_{21}}{R_2^2 + S^2f^2X_{21}^2} = 0 \tag{19.9}$$

For a given f load and IG parameters, (19.8) has the slip S as the only unknown

$$aS^2 + bS + c = 0 \tag{19.10}$$

with

$$a = f^2X_{21}^2R_{1L};\ b = +R_2\left(R_{1L}^2 + X_{1L}^2\right);\ c = R_{1L}R_2^2 \tag{19.11}$$

Equation (19.10) has two solutions but only the smaller one (S_1) refers to a real generator mode. The larger one refers to a braking regime (all the power is consumed in the machine losses).

$$S_{1,2} = \frac{-b \pm \sqrt{b^2 - 4ac}}{2a} < 0 \tag{19.12}$$

Complex solutions $S_{1,2}$ imply that self excitation cannot take place.

Once S_1 is known, the corresponding speed v, for a given frequency f, capacitance and load, is calculated from (19.5). When the speed is given, f is changed until the desired speed v is obtained.

Now, with S, f, etc. known, the only unknown in (19.9) is X_m, which is given by

$$X_m = \frac{R_2\left(R_{1L}^2 + X_{1L}^2\right)}{-\left(SfX_{2L}R_{1L} + R_2X_{1L}\right)f};\ S < 0 \tag{19.13}$$

With X_m determined, E_1 may be directly obtained from the no load curve at the rated frequency (Figure 19.10).

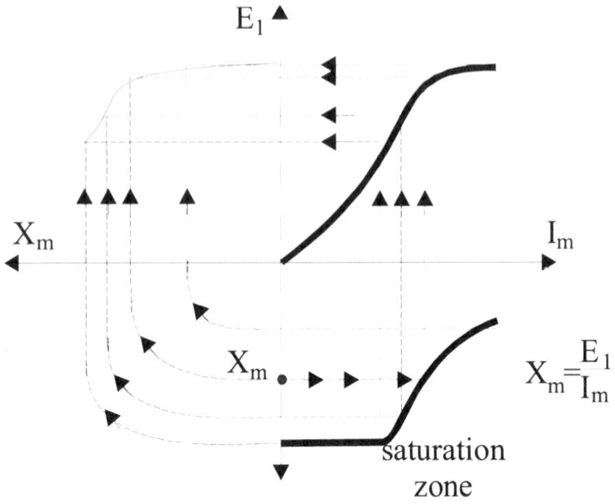

Figure 19.10 No-load curve of IG at rated frequency

As $\underline{E}_1$ and f, S, X_m are known, from the parallel equivalent circuit of Figure 19.8, we may simply calculate $\underline{I}_2$ and $\underline{I}_1$, as I_m comes directly from the magnetization curve (Figure 19.10)

$$\underline{I}_2 = \frac{-fE_1}{\frac{R_2}{S} + jfX_{2l}} \tag{19.14}$$

$$\underline{I}_1 = \frac{fE_1}{R_{lL} + jX_{lL}}; \; X_{lL} < 0 \tag{19.15}$$

It is now simple to construct the terminal voltage phasor $\underline{V}_1$ as

$$\underline{V}_1 = \underline{I}_1\left[(R_{lL} - R_1) + j(X_{lL} - X_{ll})\right] \tag{19.16}$$

or

$$\underline{V}_1 = fE_1 - (R_1 + jfX_{ll})\underline{I}_1 \tag{19.17}$$

Capacitor and load currents, I_C and I_L are, respectively,

$$\underline{I}_C = +\underline{V}_1 jf / X_C; \; \underline{I}_L = \underline{I}_1 - \underline{I}_C \tag{19.18}$$

Equations (19.14)-(19.18) are illustrated on the phasor diagram of Figure 19.11.

As the direction of the stator current, $\underline{I}_1$, and voltage, $\underline{V}_1$, have been chosen for the generator, φ_1-the power factor angle shows the current ahead of voltage. This is a clear sign that the machine is magnetised from outside.

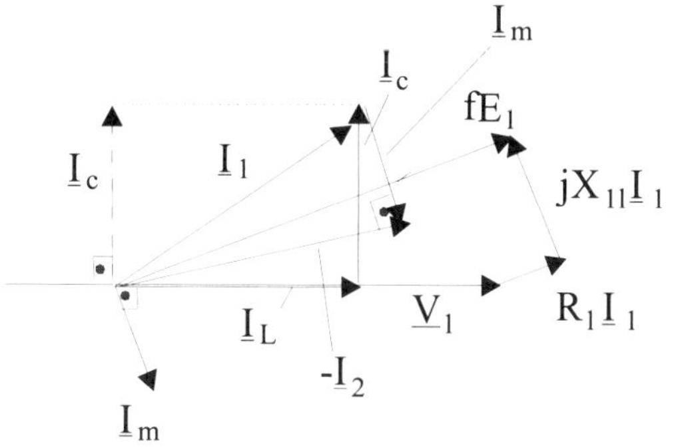

Figure 19.11. SEIG phasor diagram under resistive load

As expected, for a resistive load, the terminal voltage $\underline{V}_1$ and the load current $\underline{I}_L$ end up in phase (Figure 19.11). We may approximately calculate the core losses, p_{core}, as

$$p_{core} = \frac{3(fE_1)^2}{R_{core}} \tag{19.19}$$

with R_{core} determined from no load tests at the rated frequency. In reality R_{core} varies slightly with frequency and a no load test for different frequencies would give the $R_{core}(f)$ functional.

The SEIG efficiency η_g is

$$\eta_g = \frac{3V_1I_L\cos\varphi_L}{3V_1I_L\cos\varphi_L + 3R_1I_1^2 + 3R_2I_2^2 + p_{core} + p_{mec}} \tag{19.20}$$

Mechanical losses are determined separately from the standard no load tests at variable voltage and, eventually, frequency.

As we can see the direct steady state problem with f_1 and capacitance given (and the load) machine parameters and no load characteristic given and S and X_m as unknowns, may be solved in a straightforward manner. In the case when the speed v, instead of f, is fixed, the problem may be solved the same way through a few iterations by changing f in (19.10) until the value of S<0 will produce the required speed v = f(1-S).

The solution to (19.10) leads to a few solution existence conditions

$$fX_{1l} < X_L; \quad X_L = \frac{\frac{X_C}{f}}{1 + \frac{X_C^2}{f^2R^2}} \tag{19.21}$$

where X_L refers to the capacitor in parallel with the load impedance.

Without capacitive predominance, the voltage collapses. With given X_C (capacitor), (19.21) reduces to a load resistance R condition given by

$$R \geq X_C\sqrt{\frac{X_{1l}}{X_C - f^2 X_{1l}}} \quad (19.22)$$

The load resistance R has to be a real number in (19.22)

$$X_C \geq f^2 X_{1l} \quad (19.23)$$

Thus, (19.22) sets the value of the minimum load resistance R_{min} for a given capacitance, frequency f and stator leakage reactance X_{1l}. The smaller the X_{1l}, the lower the R_{min}, that is, the larger the maximum load.

Also the slip equation (19.10) solutions must be real such that

$$b^2 - 4ac = R_2^2\left(R_{1L}^2 + X_{1L}^2\right)^2 - 4f^2 X_{21}^2 R_{1L}^2 R_2^2 > 0 \quad (19.24)$$

or

$$2fX_{21} \leq \frac{\left(R_{1L}^2 + X_{1L}^2\right)}{R_{1L}} \quad (19.25)$$

A low rotor leakage reactance X_{2l} seems beneficial from this point of view.

The maximum possible value of the slip S_{max} corresponds to b2-4ac = 0, and from (19.12) with (19.25), it becomes

$$S_{max} = -\frac{b}{2a} = -\frac{R_2}{fX_{21}} \quad (19.26)$$

For this limiting value of the slip (19.21) is not satisfied and more. So the composite capacitor-load reactance is not capacitive any more. Consequently, the voltage collapses at S_{max}. However, the elegant expression (19.26), which corresponds essentially to an ideal maximum load power, represents a design constraint condition. We may remember that S_{max} of (19.26) corresponds to the peak torque slip of constant airgap flux vector controlled induction machine [10].

Example 19.1 The slip S and X_m problem solution.

Let us consider an IM with the data P_n = 7.36 kW, f_{1b} = 60 Hz, $2p_1$ = 2 poles, star-connection, V_{1n} = 300 V stator resistance R_1 = 0.148 Ω, rotor resistance R_2 = 0.144 Ω, stator leakage reactance (at 60Hz) X_{1l} = 0.42 Ω, the rotor leakage reactance X_{2l} = 0.25 Ω and the no load curve $E(X_m)$ at 60 Hz (Figure 19.10) may be broken into a few straight line segments. [9]

$$E_1 = \begin{cases} 0; & X_m \geq 24.0\Omega \\ 304.1 - 7.67X_m; & 19 \leq X_m \leq 24 \\ 220 - 3.16X_m; & 16 \leq X_m \leq 19 \\ 206 - 2.17X_m; & X_m < 16 \end{cases} \quad (19.27)$$

Let us determine the voltage and speed for $f_g = 43.33$ Hz and for $f_g = 58.33$ Hz and zero load ($R = \infty$).

Solution

We start with a given frequency for no load $R = \infty$ in (19.4), with $f \approx 43.33/60 \approx 0.722$, $R_L = 0$, $X_L = X_C/f$, $R_{1L} = R_1 + R = R_1 = 0.148\Omega$.

$$X_{1L} = fX_{1l} - \frac{X_c}{f} = 0.722 \cdot 0.42 - \frac{1}{2\pi 60 \cdot 0.722 \cdot 180 \cdot 10^{-6}} \approx -20.11\Omega$$

Now from (19.11) a, b, c are

$$a = f^2 X_{21}^{\ 2} R_{1L} = 0.722^2 \cdot 0.25^2 \cdot 0.148 = 4.821 \cdot 10^{-2}$$
$$b = R_2\left(R_{1L}^{\ 2} + X_{1L}^{\ 2}\right) = 0.144 \cdot \left(0.148^2 + 33.42^2\right) = 160.833$$
$$c = R_{1L} R_2^{\ 2} = 0.148 \cdot 0.144^2 = 3.069 \cdot 10^{-3}$$

From (19.12)

$$S_1 = \frac{-b + \sqrt{b^2 - 4ac}}{2a}$$

For 4ac << b, we may approximate S_1 by

$$S_1 = \frac{-2ac}{2ab} = -\frac{c}{b} = \frac{3.069 \cdot 10^{-3}}{2 \cdot 160.33} \approx 1 \cdot 10^{-5} \approx 0$$

This small value was expected.

From (19.13) the magnetization reactance is

$$X_m = \frac{R_2\left(R_{1L}^{\ 2} + X_{1L}^{\ 2}\right)}{-R_2 X_{1L} f} \approx \frac{-X_{1L}}{f} = \frac{20.11}{0.722} = 27.864 > 24.0\Omega$$

with zero rotor current X_m is expected to be fully compensated by the capacitance reactance minus the stator leakage reactance. The e.m.f. value E_1, for $X_m = 27.864 > 24$ in (19.27), is $E_1 = 0$, that is, at a low frequency (speed), with a capacitance $C = 180\mu F$ (per phase) the machine does not self excite even on no load!

For $f_g = 58.33$Hz ($f = 0.9722$), however, and $R = \infty$, X_{1L} becomes

$$X_{1L} = fX_{1l} - \frac{X_C}{f} = 0.9722 \cdot 0.42 - \frac{1}{2\pi 60 \cdot 0.9722 \cdot 180 \cdot 10^{-6}} = -14.75\Omega$$

In this case $$X_m = -\frac{X_{1L}}{f} = -\frac{14.75}{0.9722} = 15.179\Omega$$

Now from (19.27)

$$E_1 = 206 - 2.27 \cdot 15.179 = 171.54V$$

The phase current I_1 (equal, in this case, to capacitor current) is (from (19.15))

$$\underline{I}_1 = \frac{fE_1}{R_{1L} + jX_{1L}} = \frac{0.9722 \cdot 171.54}{0.148 - j14.75} = 1.13 \cdot 10^{-2} + j11.306$$

The terminal voltage $\underline{V}_1$ (from 19.17) is

$$\underline{V}_1 = fE_1 - (R_1 + jfX_{1l})\underline{I}_1 =$$
$$= 0.9722 \cdot 171.54 - (0.148 + j0.9722 \cdot 0.42)(1.13 \cdot 10^{-2} + j11.306) =$$
$$= 171.422 - j1.67$$

Due to capacitor current $(V_1) > fE_1$ as expected. Note also that with S = 0, $I_2' = 0$.

When the machine is loaded (R = 40Ω), as expected, at f = 43.33Hz and same capacitor, the machine again will not self-excite (it did not under no load).

However, for f = 58.33Hz it may self-excite.

With R = 40Ω and $X_C = \frac{1}{2\pi 60 \cdot 180 \cdot 10^{-6}} = 14.744\Omega$ from (19.4)

$$R_L = \frac{R}{1 + \frac{R^2 f^2}{X_C^2}} = \frac{40}{1 + \frac{40^2 \cdot 0.9722^2}{14.744^2}} = 5.027\Omega$$

$$X_L = \frac{\frac{X_C}{f}}{1 + \frac{X_C^2}{R^2 f^2}} = \frac{\frac{14.744}{0.9722}}{1 + \frac{14.744^2}{0.9722^2 \cdot 40^2}} = 13.26\Omega$$

So,
$$R_{1L} = R_1 + R_l = 0.148 + 5.027 = 5.175\Omega$$
$$X_{1L} = fX_{1l} - X_L = 0.9722 \cdot 0.42 - 13.26 = -12.851\Omega$$

The coefficients a, b, c are obtained from (19.11)

$$a = f^2 X_{21}^2 R_{1L} = 0.9722^2 \cdot 0.25^2 \cdot 5.175 = 0.3057$$
$$b = R_2 (R_{1L}^2 + jX_{1L}^2) = 0.144(5.475^2 + 12.551^2) = 27.6377$$
$$c = R_{1L} R_2^2 = 5.175 \cdot 0.25^2 = 0.323$$

Still $b^2 >> 4ac$, so S_1 is

$$S_1 = \frac{-c}{b} = \frac{-0.323}{27.6377} = -1.17 \cdot 10^{-2}$$

Now the speed n is

$$n = f_g(1 - S_1) = 58.33 \cdot (1 + 1.17 \cdot 10^{-2}) = 59.126\text{rps} = 3540.75\text{rpm}$$

The magnetizing reactance X_m is (19.13)

$$X_m = \frac{R_2(R_{1L}{}^2 + X_{1L}{}^2)}{-(SfX_{21}R_{1L} + R_2X_{1L})f} =$$

$$= \frac{0.144(5.175^2 + 12.851^2)}{-(-1.17 \cdot 10^{-2} \cdot 0.9722 \cdot 0.25 \cdot 5.175 + 0.144 \cdot (-12.851)) \cdot 0.9722} = 15.241\Omega$$

The e.m.f. E_1 is thus (19.27)

$$E_1 = 206 - 2.27 \cdot 15.241 = 171.40\text{V}$$

The stator current I_1 (19.15) writes

$$\underline{I}_1 = \frac{fE_1}{R_{1L} + jX_{1L}} = \frac{0.9722 \cdot 171.54}{5.175 - j12.851} = 4.493 + j11.157$$

The terminal phase voltage $\underline{V}_1$ (19.16) is

$$\underline{V}_1 = fE_1 - (R_1 + jfX_{11})\underline{I}_1 =$$
$$= 0.9722 \cdot 171.40 - (0.148 + j0.9722 \cdot 0.148)(4.493 + j11.157) =$$
$$= 166.729 - j2.29; \quad V_1 = 166.74\text{V}$$

Notice that the terminal voltage decreased from no load to R = 40Ω load, from 171.54V to 166.74V, while the speed required for the same frequency f_g = 58.33Hz had to be increased from 3500 rpm to 3540.75 rpm.

The load is still rather small as the output power P_2 is

$$P_2 = 3R_LI_1{}^2 = 3 \cdot 5.027 \cdot (4.493^2 + 11.157^2) = 2.181\text{kW} << 7.36\text{kW}$$

If the speed v is given, the computation of slip from (19.12) is done starting with a few frequencies lower than the speed until the required speed is obtained.

Once computerized, the rather straightforward computation procedure presented here allows for all performance computation, for given frequency f (or speed), capacitance, load and machine parameters. Any load can be handled directly putting it in form of a series impedance.

The slip and capacitor problem occurs when the voltage V_1, speed and load are given. To retain simplicity in the computation process, the same algorithm as above may be used repeatedly for a few values of frequency and then for capacitance (above the no load value at same voltage) until the required voltage and speed are obtained.

The iterative procedure converges rapidly as voltage in general increases with capacitance C.

19.5 STEADY STATE CHARACTERISTICS OF SEIG FOR GIVEN SPEED AND CAPACITOR

The main steady state characteristics of SEIGs are to be obtained at constant (given) speed though a prime mover, such as a constant speed small hydroturbines, does not have constant speed if unregulated.

They are

- Voltage/current characteristic –for given speed and load power factor
- Voltage versus power-for given speed and load power factor
- Frequency (slip) versus load power for given speed and power factor

All these curves may be obtained by using the second order slip equation model.

Such qualitative characteristics are shown in Figure 19.11a, b, c.

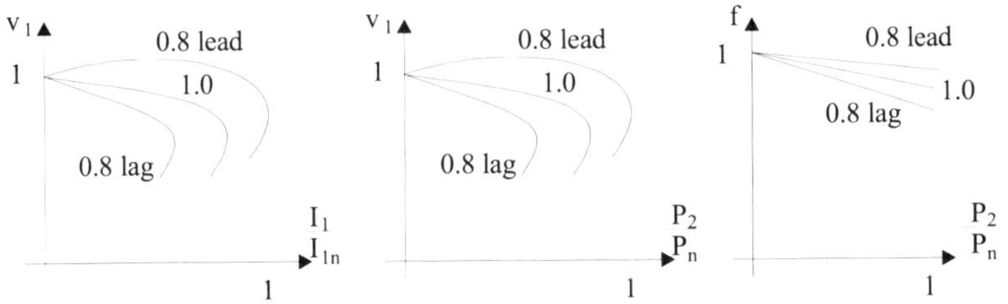

Figure 19.12 Steady state curves for given speed and capacitor
a.) voltage/current, b.) voltage/power, c.) frequency/power

The slight increase and then decrease in output voltage with power (and current) for leading power factor is expected. For lagging power factor, the voltage drops rapidly and thus the critical slip is achieved at lower power levels. The frequency is influenced little by the power factor for given load. On the other hand, for low value of slips, S_1, from (19.12), becomes

$$S_1 = \frac{-c}{b} = \frac{-R_{1L}R_2}{R_{1L}^2 + X_{1L}^2} \tag{19.28}$$

The larger X_{1L}, the smaller S_1 will be. Note that X_{1L} includes the self excitation capacitance in parallel with the load.

The characteristics notably change if the prime mover speed is not regulated. The same methodology as above may be used, with the speed versus power curve as given, to obtain results as shown in [1].

For constant head hydroturbines, the speed decreases with power and thus the voltage regulation is even more pronounced. Variable capacitance is needed to limit the voltage regulation.

19.6 PARAMETER SENSITIVITY IN SEIG ANALYSIS

By parameter sensitivity to performance we mean the influence of IG resistances R_1, R_2 and reactances X_{1l}, X_{2l}, X_m and of parallel or series-parallel excitation capacitor on performance.

Parameter sensitivity studies of SEIG may be performed for unregulated (variable speed) [1] or regulated (constant speed) prime movers. [6] The main conclusions of such studies are as follows

<u>For constant speed</u>

- The no load voltage increases with capacitance
- The terminal voltage and maximum output power increase notably with capacitance
- For constant voltage the needed excitation capacitance increases with power
- At no load the magnetisation reactance decreases with the supply voltage increase
- In the stable load area, voltage increases with IG leakage reactance and decreases with rotor resistance

<u>For variable speed (constant head hydroturbine)</u>

- Along a rather large (40%) load resistance variation range, the output power varies only a little
- Within the range of rather constant power the voltage drops sharply with load, more so with inductive-resistive loads
- Up to peak load power the frequency and speed decrease with power. They tend to increase after that which is to be translated into stable operation from this point of view
- The maximum power depends approximately on $C^{2.2}$ [3] and the corresponding voltage on $C^{0.5}$ [11]
- For no load the minimum capacitance requirement is inversely proportional to speed squared
- As expected, under load the minimum required capacitance depends on load impedance, load power factor and speed
- When the capacitance is too small it produces negligible capacitive current for magnetisation and thus the SEIG cuts off. At the other end, with too large a capacitance, the rotor impedance of the generator causes deexcitation and the voltage collapses again. In between there should be an optimal value of capacitance C for which both output power and efficiency are high
- When feeding an IM load the SEIG's power rating should be 2 to 3 times higher than that of the IM and about twice the capacitance needed for a 0.8

lag power factor passive load. These requirements are mainly imposed by the IM transients during starting. [12]

- The long shunt connection decreases the voltage regulation and increases the maximum load accordingly. Avoiding low frequency oscillations makes subsynchronous resonance important. [13]
- Saturable reactances (or transformers) have been proposed also to reduce voltage regulation at low costs. [14]

19.7 POLE CHANGING SEIGS

Pole changing winding stator SEIGs have been proposed for isolated generator systems with variable speed operation when an up to 3(4) to 1 ratio is required. To keep the flux density in the airgap about the same for both pole numbers, the numbers of poles have to be carefully chosen.

Adequate numbers are 4/6, 6/8, 8/10, etc. The connection of phases for the two pole numbers is also important for same airgap flux density. Parallel-star and series-star for 4/6 pole combination was found adequate for the scope. [8]

The pole switching should occur near (but before) the drop-out speed for the low pole number.

The voltage drop (for single capacitor) with load, for both numbers of poles, is about the same. Operating at lower number of poles (higher speed), the power is doubled with respect to the case of lower number of poles. Also the efficiency is higher, as expected, at higher speed.

For constant voltage, when the speed is doubled, the capacitance and VAR demands in both 6 and 4 pole configurations drop to $1/7^{th}$ and $1/3^{rd}$, When the load increases, the capacitance and VAR demands increase almost two fold. [5]

High winding factor pole changing windings with simple (two three-phase ends) pole switching power switch configurations are required (see Chapter 4). SEIGs with dual windings (of different power ratings) are also commercially available.

19.8 UNBALANCED STEADY STATE OPERATION OF SEIG

Failure of one capacitor or unbalanced load impedances in a three phase SEIG lead to unbalanced operation.

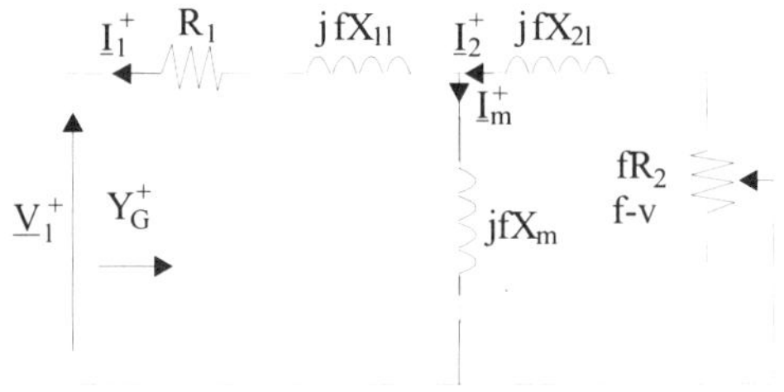

a.)

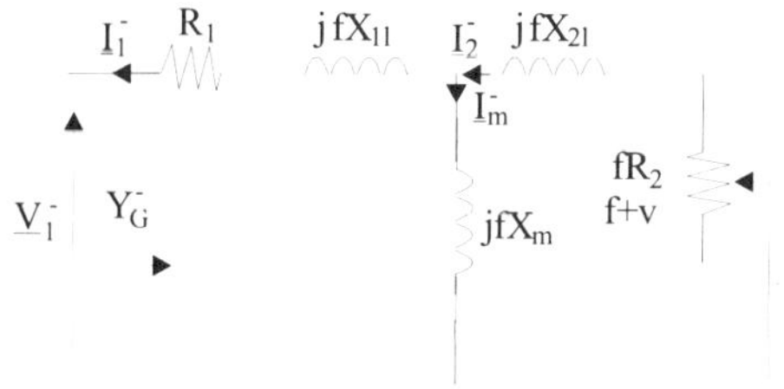

b.)

Figure 19.13 +/- sequence equivalent circuits of SEIG

A general approach to unbalanced operation is provided here by using the symmetrical component method. [15] Both the SEIG and the load may be either *delta* or *star* connected.

The equivalent circuit of IG for the positive and negative sequence, with frequency f and speed v in p.u. (as in Section 19.3 (Figure 19.7)) is shown in Figure 19.13.

For the negative sequence, the slip changes from $S^+ = f/(f\text{-}v)$ to $S^- = f/(f + v)$.

Also, as the slip frequency is different for the two sequences, $f_s^+ = S^+ \cdot f \cdot f_{1b}$; $f_s^- = S^- \cdot f \cdot f_{1b}$, the skin effect in the rotor will change R_2 in the negative sequence circuit accordingly. To a first approximation, this aspect may be neglected. The superposition of effects in the (+,–) method implies that X_m is the same also, though still influenced by magnetic saturation.

Core loss is neglected for simplicity.

19.8.1 The delta-connected SEIG

Let us consider a delta connected IG tied to a delta connected admittance network Y_{ab}, Y_{ac}, Y_{bc} (Figure 19.14).

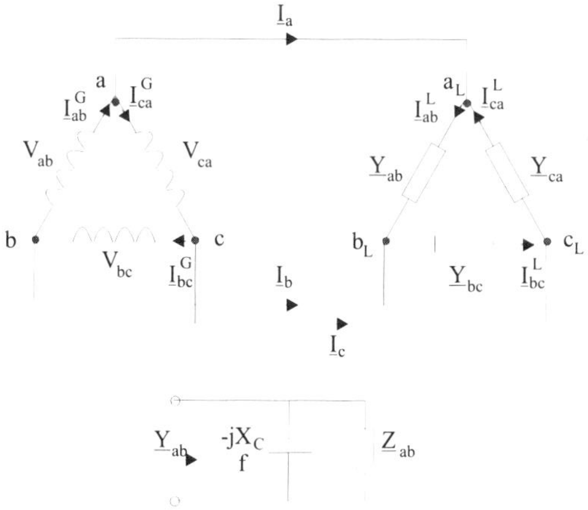

Figure 19.14 SEIG with delta connected unbalanced load and excitation capacitors

The expression of load and excitation capacitive admittance is

$$Y_{ab} = \frac{1}{\underline{Z}_{ab}} + j\frac{f}{X_C} \tag{19.29}$$

The total load-capacitor supposedly unbalanced admittances are first decomposed into the sequence components as

$$\begin{vmatrix} \underline{Y}^0 \\ \underline{Y}^+ \\ \underline{Y}^- \end{vmatrix} = \frac{1}{3}\begin{vmatrix} 1 & 1 & 1 \\ 1 & a & a^2 \\ 1 & a^2 & a \end{vmatrix} \cdot \begin{vmatrix} \underline{Y}_{ab} \\ \underline{Y}_{bc} \\ \underline{Y}_{ca} \end{vmatrix} \tag{19.30}$$

The total load-capacitor phase currents $\underline{I}_{ab}{}^L$, $\underline{I}_{bc}{}^L$, $\underline{I}_{ca}{}^L$, can also be transformed into their sequence components as

$$\begin{vmatrix} \underline{I}_{ab}{}^{L0} \\ \underline{I}_{ab}{}^{L+} \\ \underline{I}_{ab}{}^{L-} \end{vmatrix} = \begin{vmatrix} \underline{Y}^0 & \underline{Y}^- & \underline{Y}^+ \\ \underline{Y}^+ & \underline{Y}^0 & \underline{Y}^- \\ \underline{Y}^- & \underline{Y}^+ & \underline{Y}^0 \end{vmatrix} \cdot \begin{vmatrix} \underline{V}_{ab}{}^0 \\ \underline{V}_{ab}{}^+ \\ \underline{V}_{ab}{}^- \end{vmatrix} \tag{19.31}$$

In the delta connection, there is no zero sequence voltage $\underline{V}_{ab}{}^0 = 0$. Consequently, the line currents $\underline{I}_a$ components $\underline{I}_a{}^+$ and $\underline{I}_a{}^-$ are

$$\begin{aligned} \underline{I}_a{}^+ &= \underline{I}_{ab}{}^{L+} - \underline{I}_{ca}{}^{L+} = (1-a)\underline{I}_{ab}{}^{L+} = (1-a)\left(\underline{Y}^0\underline{V}_{ab}{}^+ + \underline{Y}^-\underline{V}_{ab}{}^-\right) \\ \underline{I}_a{}^- &= \underline{I}_{ab}{}^{L-} - \underline{I}_{ca}{}^{L-} = \left(1-a^2\right)\underline{I}_{ab}{}^{L-} = \left(1-a^2\right)\left(\underline{Y}^+\underline{V}_{ab}{}^+ + \underline{Y}^0\underline{V}_{ab}{}^-\right) \end{aligned} \tag{19.32}$$

From Figure 19.13, the generator phase current components $I_1{}^+$ and $I_1{}^-$ are

$$\underline{I}_1{}^+ = \underline{Y}_G{}^+\underline{V}_1{}^+; \quad \underline{I}_1{}^- = \underline{Y}_G{}^-\underline{V}_1{}^- \tag{19.33}$$

Similar to (19.32), we may write the line current $\underline{I}_a$ components based on generator phase current components as

$$\begin{aligned} \underline{I}_a{}^+ &= (1-a)\underline{I}_{ab}{}^{G+} = -(1-a)V_{ab}{}^+Y_G{}^+ \\ \underline{I}_a{}^- &= (1-a)\underline{I}_{ab}{}^{G-} = -\left(1-a^2\right)V_{ab}{}^-Y_G{}^- \end{aligned} \tag{19.34}$$

Note that the generator phases are balanced. Equating (19.32) to (19.34) and then dividing them leads to the elimination of $V_{ab}{}^+$ and $V_{ab}{}^-$ to yield

$$\left(\underline{Y}_G{}^+ + \underline{Y}_0\right)\left(\underline{Y}_G{}^- + \underline{Y}_0\right) - \underline{Y}^+\underline{Y}^- = 0 \tag{19.35}$$

Equation (19.35) represents the self-excitation condition and leads to two real coefficient nonlinear equations. In general, the solution to (19.35) for a given capacitance, frequency (or speed) and load is similar to the solution for balanced condition, though a bit more involved. For balanced operation $\underline{Y}^+ = \underline{Y}^-$ $= \underline{Y}$ and $\underline{Y}^0 = 0$. So (19.35) degenerates to $Y_G{}^+ - \underline{Y} = 0$, as expected.

If the load is Y connected, it should be first transformed into an equivalent delta connection and then apply (19.35). Notice that in general the capacitors are delta connected to exploit the higher voltage available to them.

19.8.2 Star-connected SEIG

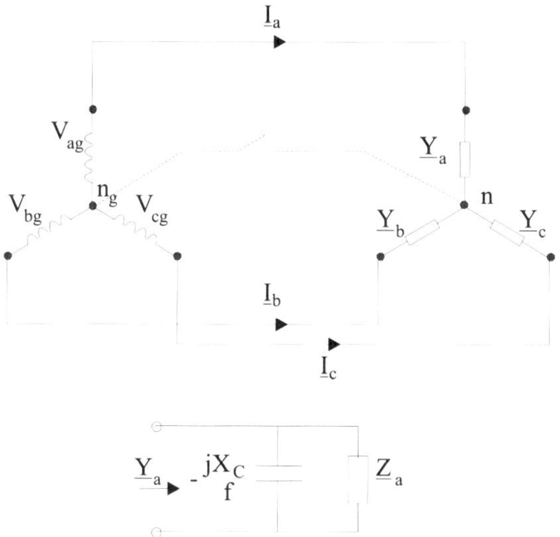

Figure 19.15 Star connected SEIG

This time the capacitors are star connected to the load. If they are delta connected, X_C for delta connection is replaced by $3X_C$ for star connection ($X_{CY} = 3X_{C\Delta}$).

In this case, $\underline{Y}^0$, $\underline{Y}^+$ and $\underline{Y}^-$, the total capacitive load admittance components, refer to $\underline{Y}_a$, $\underline{Y}_b$, $\underline{Y}_c$.

The symmetrical components of the line current $\underline{I}_a$ calculated from the load are

$$\begin{vmatrix} \underline{I}_a^0 \\ \underline{I}_a^+ \\ \underline{I}_a^- \end{vmatrix} = \begin{vmatrix} \underline{Y}^0 & \underline{Y}^- & \underline{Y}^+ \\ \underline{Y}^+ & \underline{Y}^0 & \underline{Y}^- \\ \underline{Y}^- & \underline{Y}^+ & \underline{Y}^0 \end{vmatrix} \cdot \begin{vmatrix} \underline{V}_a^0 \\ \underline{V}_a^+ \\ \underline{V}_a^- \end{vmatrix} \tag{19.36}$$

with $\underline{I}_a^0 = 0$ (no neutral connection)

$$\underline{V}_a^0 = -\frac{\left(\underline{Y}^-\underline{V}_a^- + \underline{Y}^+\underline{V}_a^+\right)}{\underline{Y}^0} \tag{19.37}$$

Making use of $\underline{V}_a^0$ in the star two equations of (19.36), we obtain

$$\underline{I}_a^{\,+} = \left(\underline{Y}^0 - \frac{\underline{Y}^+\underline{Y}^-}{\underline{Y}^0}\right)\underline{V}_a^{\,+} + \left(\underline{Y}^- - \frac{\underline{Y}^+\underline{Y}^+}{\underline{Y}^0}\right)\underline{V}_a^{\,-}$$
$$\underline{I}_a^{\,-} = \left(\underline{Y}^+ - \frac{\underline{Y}^-\underline{Y}^-}{\underline{Y}^0}\right)\underline{V}_a^{\,+} + \left(\underline{Y}^0 - \frac{\underline{Y}^+\underline{Y}^-}{\underline{Y}^0}\right)\underline{V}_a^{\,-} \qquad (19.38)$$

The same line current components, calculated from the generator side, are simply

$$\underline{I}_a^{\,+} = -\underline{Y}_G^{\,+}\underline{V}_{ag}^{\,+};\ \underline{I}_a^{\,-} = -\underline{Y}_G^{\,-}\underline{V}_{ag}^{\,-} \qquad (19.39)$$

As the generator and the load line voltages have their sum equal to zero, even with unbalanced load, it may be shown that

$$\underline{V}_{ab}^{\,+} = \left(1 - a^2\right)\underline{V}_a^{\,+} = \left(1 - a^2\right)V_{ag}^{\,+}$$
$$\underline{V}_{ab}^{\,-} = \left(1 - a\right)\underline{V}_a^{\,-} = \left(1 - a\right)V_{ag}^{\,-} \qquad (19.40)$$

Consequently, $$V_a^{\,+} = V_{ag}^{\,+} \text{ and } V_a^{\,-} = V_{ag}^{\,-} \qquad (19.41)$$

From (19.38)-(19.39), with (19.41), we obtain the self-excitation equation

$$\left(\underline{Y}_G^{\,+} + \underline{Y}^0 - \frac{\underline{Y}^+\underline{Y}^-}{\underline{Y}^0}\right)\left(\underline{Y}_G^{\,-} + \underline{Y}^0 - \frac{\underline{Y}^+\underline{Y}^-}{\underline{Y}^0}\right) - \left(\underline{Y}^+ - \frac{\underline{Y}^-\underline{Y}^-}{\underline{Y}^0}\right)\left(\underline{Y}^- - \frac{\underline{Y}^+\underline{Y}^+}{\underline{Y}^0}\right) = 0$$

(19.42)

Note that if the SEIG is star-connected and the capacitor load combination is delta connected with original +-0 sequence admittances as $\underline{Y}^0$, $\underline{Y}^+$, $\underline{Y}^-$, then a self-excitation equation similar to (19.35) is obtained

$$\left(\frac{1}{3}\underline{Y}_G^{\,+} + \underline{Y}^0\right)\left(\frac{1}{3}\underline{Y}_G^{\,-} + \underline{Y}^0\right) - \underline{Y}^+\underline{Y}^- = 0 \qquad (19.43)$$

Also when n_g and n are connected, the zero sequence current $I_a^{\,0}$ is not zero, generally. However there is no zero sequence generator voltage $\underline{V}_a^{\,0} = \underline{V}_{ag}^{\,0} = 0$.

Consequently, the situation is in this case similar to that of delta/delta connection, but the total load admittances $\underline{Y}^0$, $\underline{Y}^+$, $\underline{Y}^-$correspond to the star connected load admittances $\underline{Y}_a$, $\underline{Y}_b$, $\underline{Y}_c$.

Whereas the above treatment is fairly standard and general, a few examples are in order.

19.8.3. Two phase open

Let us consider the case of Δ/Δ connection (Figure 19.16) with a single capacitor between a_L and b_L and a single load

$$\underline{Y}_{ab} = \underline{Y} = \frac{1}{\underline{Z}}; \quad \underline{Y}_{bc} = \underline{Y}_{ca} = 0 \tag{19.44}$$

For this case

$$\underline{Y}^0 = \underline{Y}^+ = \underline{Y}^- = \frac{\underline{Y}}{3} \tag{19.45}$$

With (19.45) the selfexcitation equation (19.35) becomes

$$3\underline{Y}_G^+ \underline{Y}_G^- + \left(\underline{Y}_G^+ + \underline{Y}_G^-\right)\underline{Y} = 0 \tag{19.46}$$

or

$$\frac{3}{\underline{Y}} + \frac{1}{\underline{Y}_G^+} + \frac{1}{\underline{Y}_G^-} = 0 \tag{19.47}$$

Equation (19.47) illustrates the series connection of + and-generator sequence circuits to $\frac{3}{\underline{Y}} = 3\underline{Z}$ (Figure 19.16).

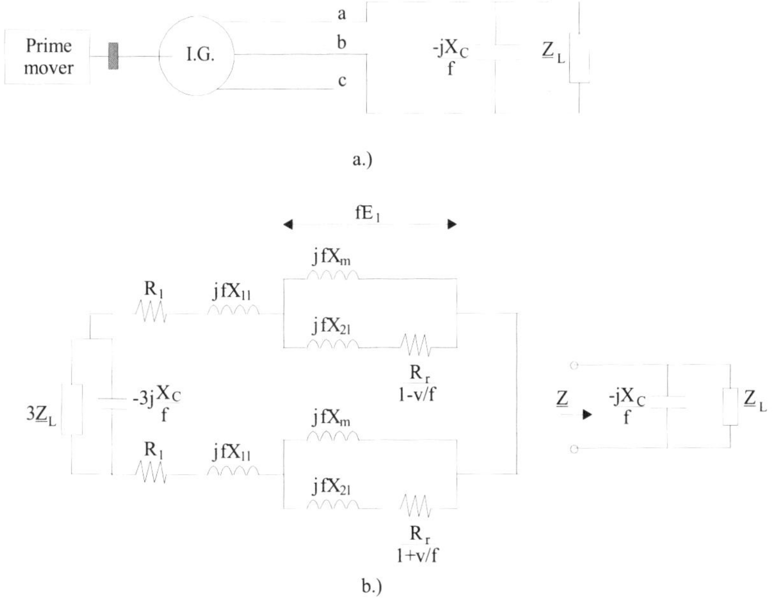

Figure 19.16 Unbalanced SEIG with Δ/Δ connection single capacitor and single load

Other unbalanced conditions may occurs. For example, if phase C opens for some reason, while the load $\underline{Z}$ is balanced, the SEIG performs as with a single capacitor and load impedance but with new load admittance $Y_{ab} = 1.5Y$ and $Y_{bc} = Y_{ca} = 0$ [15].

For a short-circuit across phase a (Δ/Δ connection) $\underline{Y}_{ab} = \infty$ and thus $\underline{Y}^0 = \underline{Y}^+ = \underline{Y}^- = \infty$. Consequently (19.35) becomes (Figure 19.16a).

$$\underline{Y}_G^{\ +} + \underline{Y}_G^{\ -} = 0 \tag{19.48}$$

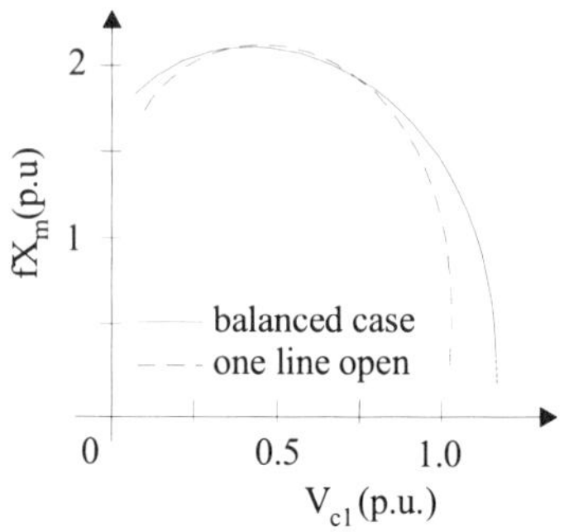

Figure 19.17 Positive sequence magnetizing characteristics

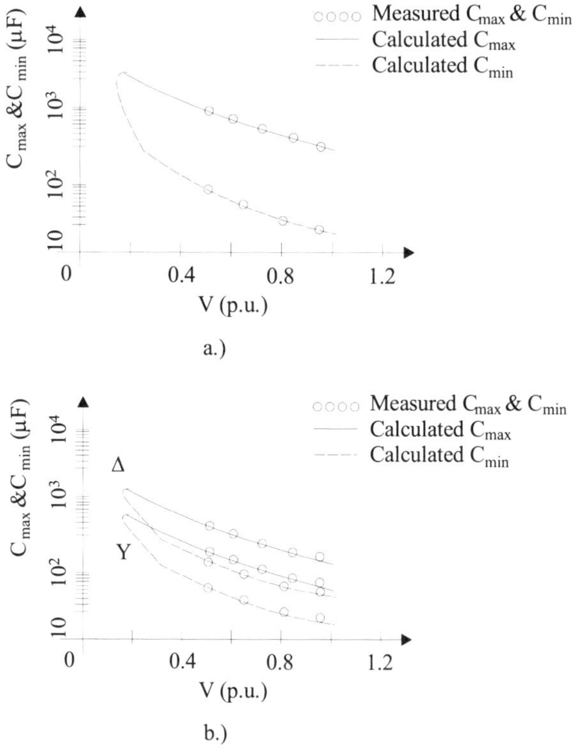

Figure 19.18 Min-max capacitors on no load
a.) balanced capacitor; b.) one capacitor only

As the capacitor is short-circuited there will be no selfexcitation. The same is true for a three phase short-circuit.

In essence, the computation of steady state performance under unbalanced conditions includes solving first the self-excitation nonlinear equations (19.35) or (19.42) or (19.43), (19.47) for given speed (frequency), load and capacitance, to find the frequency (speed) and the magnetization reactance X_m. Then, as for the balanced conditions, equivalent circuits are solved, provided the magnetization curve $E_1(X_m)$ is known. It was proved in Reference [15] that the positive sequence magnetization curves obtained for balanced no load and one-line-open load differ from each other in the saturation zone (Figure 19.17).

It turns out that, whenever a zero line current occurs, the one-line-open characteristic in Figure 19.17 should be applied. [15]

Sometimes the range of self-excitation capacitance is needed. Specifying the maximum magnetization reactance X_{max}, the speed (frequency), machine parameters and load, the solution of self-excitation equations developed in this paragraph produces the required frequency (speed) and capacitances.

There are two solutions for the capacitances, C_{max} and C_{min}, and they depend strongly on load and speed. For maximum load, the two values become equal to each other. Beyond that point the voltage collapses.

Sample results obtained for C_{max} and C_{min} on no load under balanced and unbalanced (one capacitor only) conditions are shown in Figure 19.18 [15] for $r_1 = 0.09175$, $r_2 = 0.06354$, $p_1 = 2$, $f_{1b} = 60Hz$, $x_{1l} = x_{2l} = 0.2112$, $x_{max} = 2.0$.

Meticulous investigation of various load and capacitor unbalances on load has proved [15] that the theory of symmetrical components works well.

The range of C_{min} to C_{max} for delta connection is 3 times smaller for the star connection, In terms of parameter-sensitivity analysis, the trends for unbalanced load are similar to those for balanced load. However, the efficiency is notably reduced and stays low with load variation.

19.9 TRANSIENT OPERATION OF SEIG

SEIGs work alone or in parallel on variable loads. Load and speed perturbations, voltage build-up on no load, excitation capacitance changes during load operation are all causes for transients. To investigate the transients the d-q model is generally preferred. Also stator coordinates are used to deal easily with eventual parallel operation or active loads (a.c. motors).

The d-q model (in stator coordinates) for the SEIG alone, including cross-coupling saturation [16, 17], is given below

$$[V_G] = -[R_G][i_G] - [L_G]p[i_G] - \omega_{rG}[G_G][i_G] \tag{19.49}$$

with

$$[V_G] = [V_d, V_q, 0, 0] \tag{19.50}$$

$$[i_G] = [i_d, i_q, i_{dr}, i_{qr}] \tag{19.51}$$

The torque

$$T_{eG} = \frac{3}{2} p_{1G} L_m \left(i_q i_{dr} - i_d i_{qr} \right) \tag{19.52}$$

$$\frac{J}{p_{1G}} \frac{d\omega_{rG}}{dt} = T_{PM} - T_{eG} \tag{19.53}$$

$$pV_d = \frac{1}{C} i_{dc}; \; pV_q = \frac{1}{C} i_{qc} \tag{19.54}$$

$$i_{dc} = i_d + i_{dL}; \quad i_{qc} = i_q + i_{qL} \tag{19.55}$$

i_{dc}, i_{qc}, i_{dL}, i_{qL} are d, q components of capacitor and load currents, respectively.

The association of signs in (19.49) corresponds to generator mode.

Now the parameter matrices R_G, L_G, G_G are

$$R_G = \text{Diag}\left[R_1, R_1, R_2, R_2 \right]$$

$$L_G = \begin{bmatrix} L_{1d} & L_{dq} & L_{md} & L_{dq} \\ L_{dq} & L_{1q} & L_{dq} & L_{mq} \\ L_{md} & L_{dq} & L_{2d} & L_{dq} \\ L_{dq} & L_{mq} & L_{dq} & L_{2q} \end{bmatrix} \tag{19.56}$$

$$G_G = \begin{bmatrix} 0 & 0 & 0 & 0 \\ 0 & 0 & 0 & 0 \\ 0 & L_m & 0 & L_2 \\ -L_m & 0 & -L_2 & 0 \end{bmatrix} \tag{19.57}$$

L_m is the magnetisation inductance and i_m the magnetisation current

$$i_{md} = i_d + i_{dr}; \qquad i_{mq} = i_q + i_{qr}; \; i_m = \sqrt{i_{md}^2 + i_{mq}^2} \tag{19.58}$$

The crosscoupling transient inductance L_{dq} [16,17] is

$$L_{dq} = \frac{i_{md} i_{mq}}{i_m} \frac{dL_m}{di_m} \tag{19.59}$$

Also the transient magnetisation inductances along the two axes L_{md} and L_{mq} [16, 17] are

$$L_{md} = L_m + \frac{i_{md}^2}{i_m} \frac{dL_m}{di_m} \tag{19.60}$$

$$L_{mq} = L_m + \frac{i_{mq}^{\;2}}{i_m}\frac{dL_m}{di_m} \tag{19.61}$$

Also

$$\begin{aligned} L_{1d} &= L_{11} + L_{md}; \quad L_{1q} = L_{11} + L_{mq} \\ L_{2d} &= L_{21} + L_{md}; \quad L_{2q} = L_{21} + L_{mq} \end{aligned} \tag{19.62}$$

The load equations are to be added. The load may be passive or active. A series $R_L L_L C_L$ load would have the equations

$$\begin{aligned} [V_G] &= [R_L][i_{sL}] + [L_L]\frac{d[i_{sL}]}{dt} + [V_{cL}] \\ \frac{d[V_{cL}]}{dt} &= \frac{1}{C_L}[i_{sL}]; \quad [V_{cL}] = [V_{cd} \;\; V_{cq}]^T \\ [i_{sL}] &= [i_{dL} \;\; i_{qL}]^T \end{aligned} \tag{19.63}$$

In contrast, an induction motor load (in stator coordinates) is characterized by a set of equations such as

$$[V_G] = [R_M]\,[i_M] + [L_M]\frac{d[i_M]}{dt} + \omega_{rM}[G_M]\,[i_M]$$

$$\frac{J_M}{P_{1M}}\frac{d\omega_{rM}}{dt} = T_{eM} - T_{Load}$$

$$\begin{aligned} T_{eM} &= \frac{3}{2}P_{1M}L_{mM}\left(i_{qM}i_{drM} - i_{dM}i_{qrM}\right) \\ [i_M] &= [i_{dM} \;\; i_{qM} \;\; i_{drM} \;\; i_{qrM}]^T \end{aligned} \tag{19.64}$$

Motor equations are similar to generator equations, as are the parameter definitions in (19.56)-(19.62).

We also should note that $i_{dM} = i_{dL}$, $i_{qM} = i_{qL}$. When multiple passive and active loads are connected to the generator, the generator current $[i_L] = [i_{dL}, i_{qL}]$.

Solving for the transients in the general case is done through numerical methods. The unknowns are the various currents and the capacitor voltages. The capacitor currents are dummy variables.

19.10 SEIG TRANSIENTS WITH INDUCTION MOTOR LOAD

There are practical situations, in remote areas, when the SEIG is supplying an induction motor which in turns drives a pump or similar load.

A pump load is characterized by increasing torque with speed such that

$$T_{load} \approx T_{L0} + K_L\omega_{rM}^{\;2} \tag{19.65}$$

The most severe design problem of such a system is the starting (transient) process. The rather low impedance of IM at low speeds translates into high transient currents from the SEIG. On the other hand, a large excitation capacitor is needed for rated voltage with the motor on load. When the motor is turned off, this large capacitor produces large voltage transients. Part of it must be disconnected or an automatic capacitance reduction to control the voltage is required. An obvious way out of this difficulty would be to have a smaller (basic) capacitance (C_G) connected to the generator and one (C_M) to the motor (Figure 19.19).

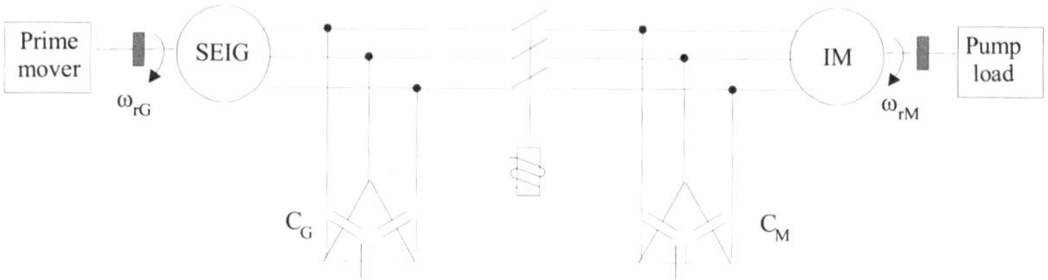

Figure 19.19 SEIG with IM load and splitted excitation capacitor

First, the SEIG is accelerated by the prime mover until it establishes a certain speed with a certain output voltage. Then the IM is connected. Under rather notable voltage and current transients, the motor accelerates and settles at a speed corresponding to the case when the motor torque equals the pump torque.

To provide safe starting, the generator rated power should be notably in excess of the motor rated power. In general, even an 1/0.6 ratio marginally suffices.

Typical such transients for a 3.7kW SEIG and a 2.2kW motor with C_G = 18μF, C_M = 16μF, are shown in Figure 19.20 [18].

Voltage and current transients are evident, but the motor accelerates and settles on no load smoothly. Starting on pump load is similar but slightly slower.

On the other hand, the voltage build-up in the SEIG takes about 1-2 seconds. Sudden load changes are handled safely up to a certain level which depends on the ratio of power rating of the motor of the SEIG.

For the present case, a zero to 40% motor step load is handled safely but an additional 60% step load is not sustainable. [18] Very low overloading is acceptable unless the SEIG to motor rating is larger than 3 ÷ 4.

In case of overloading, an over current relay set at 1.2-1.3p.u. for 300-400ms can be used to trip the circuit.

During a short-circuit at generator terminals, surges of 2 p.u. in voltage and 5 p.u. in current occur for a short time (up to 20 ms) before the voltage collapses.

Though the split capacitor method (Figure 19.19) is very practical, for motors rated less than 20% of SEIG rating, a single capacitor will do.

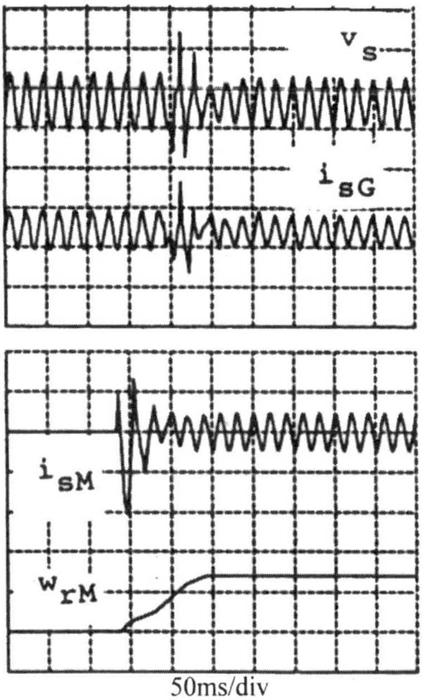

Figure 19.20 Motor load starting transients for a SEIG, $C_G = 18\mu F$, $C_m = 16\mu F$

19.11 PARALLEL OPERATION OF SEIGS

An isolated power system made of a few SEIGs is a practical solution for upgrading the generating power rating upon demand. Such a stand alone power system may have a single excitation capacitor but most probably, it has to be variable to match the load variations which inevitably occur. Capacitor change may be done in steps through a few power switches or continuously through power electronics control of a paralleled inductor current (Figure 19.21).

The operation of such a power grid on steady state is to be approached as done for the single SEIG. However, the magnetization curves and parameters of all SEIGs are required. Finally, the self-excitation condition will be under the form of an admittance summation equal to zero. Neglecting the "de-excitation" inductance L_{ex}, the terminal voltage equation is

$$\underline{V}_s\left(\frac{1}{R_L} - \frac{j}{fX_L} + \frac{jf}{X_C}\right) = \sum_{i=1}^{N} \underline{I}_{si} \tag{19.66}$$

The SEIG's equations (section 19.3 and Figure 19.7) may be expressed as

$$\underline{I}_{mi} + \underline{I}_{si} = \underline{I}_{2i} = \frac{-f\underline{E}_{1i}}{\dfrac{R_{2i}}{1-\dfrac{v_i}{f}} + jfX_{2li}} \tag{19.67}$$

v_i is the p.u. speed of the ith generator. $X_{mi}(E_{1i})$ represents the magnetization curve of the ith machine. All these curves have to be known. Again for a given excitation capacitance C, motor parameters and speeds of various SEIGs and load, the voltage, frequency, SEIG currents, capacitor currents and load currents and load powers (active and reactive) could be iteratively determined from (19.66)-(19.67). A starting value of f is in general adopted to initiate the iterative computation. For sample results, see [19].

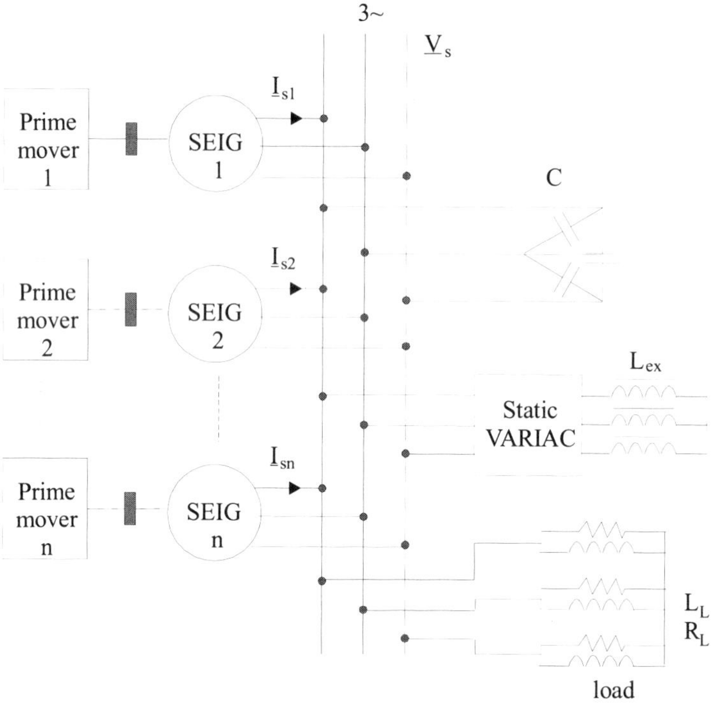

Figure 19.21 Isolated power grid with paralleled SEIGs

The SEIGs speeds in p.u. have to be close to each other to avoid that some of them switch to motoring mode. Transient operation may be approached by using the d-q models of all SEIGs with the summation of the stator phase currents equal to the capacitor plus load current. So the solution of transients is similar to that for a single SEIG, as presented earlier in this chapter, but the number of equations is larger. Only numerical methods such as Runge-Kutta, Gear, etc. may succeed in solving such stiff nonlinear systems.

19.12 THE DOUBLY-FED IG CONNECTED TO THE GRID

As shown in Figure 19.1, a wound-rotor induction generator (dubbed as doubly - fed) may deliver power not only through the stator but also through the rotor; provided a bi-directional power flow converter is used in the rotor circuit. The main merit of the doubly-fed IG is its capability to deliver constant voltage and frequency output for ±(20-40)% speed variation around conventional synchronous speed. If the speed variation range is (20-40)% so is the rating of static power converter and transformer in the rotor circuit (Figure 19.1). For bi-directional power flow in the rotor circuit, single stage converters (cycloconverters or matrix converters) or dual-stage back-to-back voltage source PWM inverters have to be used.

Independent reactive and active power flow control-with harmonics elimination-may be achieved this way for both super and subsynchronous speed operation. Less costly solutions use a unidirectional power flow converter in the rotor when only super or subsynchronous operation as generator is possible.

Here we are interested mainly in the IG behaviour when doubly fed. So we will consider sinusoidal variables with rotor voltage amplitude, frequency and phase open to free change through an ideal bi-directional power electronics converter.

19.12.1. Basic equations

To investigate the doubly-fed IG (DFIG) the d-q (space - phasor) model as presented in Chapter 13 on transients is used.

In synchronous coordinates

$$\begin{aligned}\overline{V}_1 &= \left(R_1 + (p + j\omega_1)L_1\right)\bar{i}_1 + (p + j\omega_1)L_m\bar{i}_2 \\ \overline{V}_2 &= (p + jS\omega_1)L_m\bar{i}_1 + \left(R_2 + (p + jS\omega_1)L_2\right)\bar{i}_2\end{aligned} \tag{19.68}$$

with

$$\overline{V}_1 = V_d + jV_q;\ \ \overline{V}_2 = V_{dr} + jV_{qr}$$

$$\bar{i}_1 = i_d + ji_q;\ \ \bar{i}_2 = i_{dr} + ji_{qr}$$

$$L_1 = L_{1l} + L_m;\ \ L_2 = L_{2l} + L_m$$

$$T_e = \frac{3}{2}p_1 L_m \,\mathrm{Imag}\left[\bar{i}_1\bar{i}_2^{\,*}\right];\ \ \frac{Jp\omega_r}{p_1} = T_{PM} - T_e$$

We should add the Park transformation

$$\begin{vmatrix} V_d \\ V_q \end{vmatrix} = \frac{2}{3}\begin{bmatrix} \cos\theta_{1,2} & \cos\left(-\theta_{1,2} + \frac{2\pi}{3}\right) & \cos\left(-\theta_{1,2} - \frac{2\pi}{3}\right) \\ \sin\left(-\theta_{1,2}\right) & \sin\left(-\theta_{1,2} + \frac{2\pi}{3}\right) & \sin\left(-\theta_{1,2} - \frac{2\pi}{3}\right) \end{bmatrix}\begin{vmatrix} V_a \\ V_b \\ V_c \end{vmatrix} \tag{19.69}$$

The transformation is valid for stator and rotor

$$\theta_1 = \int \omega_1 t = \omega_1 t \quad \text{- for constant frequency stator;}$$

$$\theta_2 = \int (\omega_1 - \omega_r) t = \theta_1 - \int \omega_r dt \quad \text{- for the rotor with variable speed;}$$

So,

$$\frac{d\theta_2}{dt} = \omega_1 - \omega_r = S\omega_1 \tag{19.70}$$

For constant speed (ω_r = constant)

$$\theta_2 = S\omega_1 t + \delta \tag{19.71}$$

Also, for an infinite power bus

$$\begin{aligned} V_{1abc} &= V_1\sqrt{2}\cos\left(\omega_1 t - (i-1)\frac{2\pi}{3}\right) \\ V_{2abc} &= V_2\sqrt{2}\cos\left(\omega_1 t - \delta - (i-1)\frac{2\pi}{3}\right) \end{aligned} \tag{19.72}$$

In (19.72) the rotor phase voltages V_{2abc} are expressed in stator coordinates. This explains the same frequency in V_{1abc} and V_{2abc} ($\theta_1 = 0$, $\theta_2 = -\omega_r t$). Using (19.72) and (19.71) in (19.69) we obtain

$$V_{1d} = \sqrt{2}V_1; \quad V_{1q} = 0 \tag{19.73}$$

For constant speed

$$V_{2d} = \sqrt{2}V_2\cos\delta; \quad V_{2q} = -\sqrt{2}V_2\sin\delta$$

As seen from (19.73), in synchronous coordinates, the DFIG voltages in the stator are dc quantities. Under steady state-constant δ-the rotor voltages are also d.c. quantities, as expected. The angle δ may be considered as the power angle and thus DFIG operates as a synchronous machine. The rotor voltage $\sqrt{2}V_2$ and its phase δ with respect to stator voltage in same coordinates (synchronous in our case) and slip S are thus the key factors which determine the machine operation mode (motor or generator) and performance.

The proper relationship between V_2, δ for various speeds (slips) represents the fundamental question for DFIG operation.

The various phase displacements in (19.71)-(19.72) are shown in Figure 19.22.

The d-q model equations serve to solve most transients in the time domain by numerical methods with $\overline{V}_1$, $\overline{V}_2$ and prime mover torque T_{PM} as inputs and the various currents and speed as variables.

However to get an insight into the DFIG behaviour we here investigate steady state operation and its stability limits.

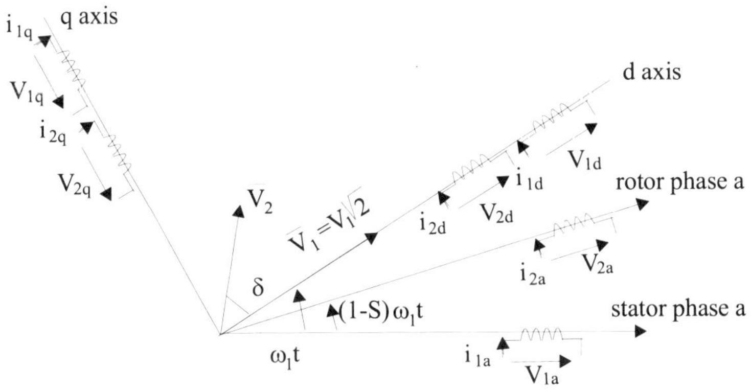

Figure 19.22 The d-q axis model and the voltage vectors $\overline{V}_1$, $\overline{V}_2$ in synchronous coordinates

19.12.2 Steady state operation

The steady state performance may be calculated with $S = S_0$ and $\delta = \delta_0$ and $\overline{V}_1$, $\overline{V}_2$ from (19.73)

$$
\begin{aligned}
\overline{V}_1\sqrt{2} &= R_1 I_{d0} - \omega_1 \psi_{q0}; \quad \psi_{q0} = L_m I_{qr0} - L_1 I_{q0} \\
0 &= R_1 I_{q0} + \omega_1 \psi_{d0}; \quad \psi_{d0} = L_m I_{dr0} + L_1 I_{d0} \\
V_2\sqrt{2}\cos\delta &= R_2 I_{dr0} - S_0\omega_1 \psi_{qr0}; \quad \psi_{qr0} = L_2 I_{qr0} + L_m I_{q0} \\
-V_2\sqrt{2}\sin\delta &= R_2 I_{qr0} + S_0\omega_1 \psi_{dr0}; \quad \psi_{dr0} = L_2 I_{dr0} + L_m I_{d0}
\end{aligned}
\tag{19.74}
$$

with the torque T_{e0} given by

$$T_{e0} = \frac{3}{2} p_1 L_m \left(I_{q0} I_{dr0} - I_{d0} I_{qr0}\right) \tag{19.75}$$

The main characteristic is $T_{e0}(\delta)$ for the rotor voltage amplitude V_2 as a parameter. The frequency of rotor voltage system is in the real machine $S\omega_1$, while its phase shift with respect to the stator voltages at time zero is δ.

Though an analytical solution at $T_{e0}(\delta)$ is obtainable from (19.74)-(19.75) as they are, for $R_1 = R_2 = 0$ a much simpler solution is obtained. Such a solution is acceptable for large power and slip values

$$\left(T_{e0}\right)_{\substack{R_1=R_2=0 \\ S_0\neq 0}} = 3P_1\left(\frac{V_1}{\omega_1}\right)\left(\frac{V_2}{S\omega_1}\right)\frac{\sin\delta}{\omega_1 L_{sc}}\left(\frac{L_m}{L_1}\right); \quad L_{sc} = \left(L_1 L_2 - L_m^{\ 2}\right) L_1 \tag{19.76}$$

Positive torque means motoring. Positive δ means V_2 lagging $\overline{V}_1$. It is now evident that the behavior of DFIG resembles that of a synchronous machine. As long as the ratio $V_2/S_0\omega_1$ is constant, the maximum torque value remains the same. But this is to say that when the rotor flux is constant, the maximum torque is the same and thus the stability boundaries remain large.

As $R_1 = R_2 = 0$ assumption does not work at zero slip ($S_0 = 0$), we may derive this case separately from (19.74) with $R_1 = 0$, $R_2 \neq 0$, but with $S_0 = 0$. Hence,

$$\left(T_{e0}\right)_{\substack{R_1=0\\S_0=0}} = 3\left(\frac{p_1}{\omega_1}\right)\left(\frac{V_1V_2}{R_2}\right)\left(\frac{L_m}{L_1}\right)\cos\delta \tag{19.77}$$

The two approximations of the torque, one valid at rather large slips and the other valid at zero slip (conventional synchronism) are shown in Figure 19.23.

The statically stable operation zones occur between points A_M and A_G.

For $S_0 \neq 0$ motor and generator operation is in principle feasible both for super and subsynchronous speeds ($S_0 < 0$ and $S_0 > 0$). However, the complete expression of torque for $S_0 \neq 0$, which includes the resistances, will alter the simple expression of torque shown in (19.76). A better approximation would be obtained for $R_2 \neq 0$ with $R_1 = 0$. For low values of slip S_0 the rotor resistance is important even for large power induction machines.

After some manipulations (19.74) and (19.75) yield

$$\left(T_{e0}\right)_{R_1=0} = 3\left(\frac{p_1V_1V_2}{\omega_1}\right)\left(\frac{L_m}{L_1}\right)\frac{R_2\cos\delta + S_0\omega_1L_{sc}\sin\delta - \frac{V_1}{V_2}R_2S_0\frac{L_m}{L_1}}{\left(R_2^{\,2} + \left(S_0\omega_1L_{sc}\right)^2\right)} \tag{19.78}$$

As expected, when $R_2 = 0$ expression (19.76) is reobtained and when $S_0 = 0$, expression (19.77) follows from (19.78).

Expression (19.78) does not collapse either for $V_2 = 0$ or for $S_0 = 0$. It only neglects the stator resistances, for simplicity (Figure 19.24).

For $\delta_0 = 0$ the torque is

$$\left(T_{e0}\right)_{\substack{\delta_0=0\\R_1=0}} = \frac{3p_1V_1V_2}{\omega_1}\frac{\left(\frac{L_m}{L_1} - S_0\frac{V_1}{V_2}\right)}{R_2^{\,2} + \left(S_0\omega_1L_{sc}\right)^2} \tag{19.79}$$

So $\left(T_{e0}\right)_{\delta_0=0}$ may be either positive or negative

$$\frac{V_2}{V_1} <> \frac{L_1S_0}{L_m};\ \left(T_{e0}\right)_{\delta_0=0} <> 0 \tag{19.80}$$

For S_0 negative the situation in (19.80) is reversed. As seen from Figure 19.24 the stability zones of motoring and generating modes are no longer equally large due to rotor resistance R_2 and also due to the ratio V_2/V_1 and the sign of slip $S_0 <> 0$.

The extension of the stability zone A_G-A_M in terms of δ_{KG} to δ_{0KM} may be simply found by solving the equation

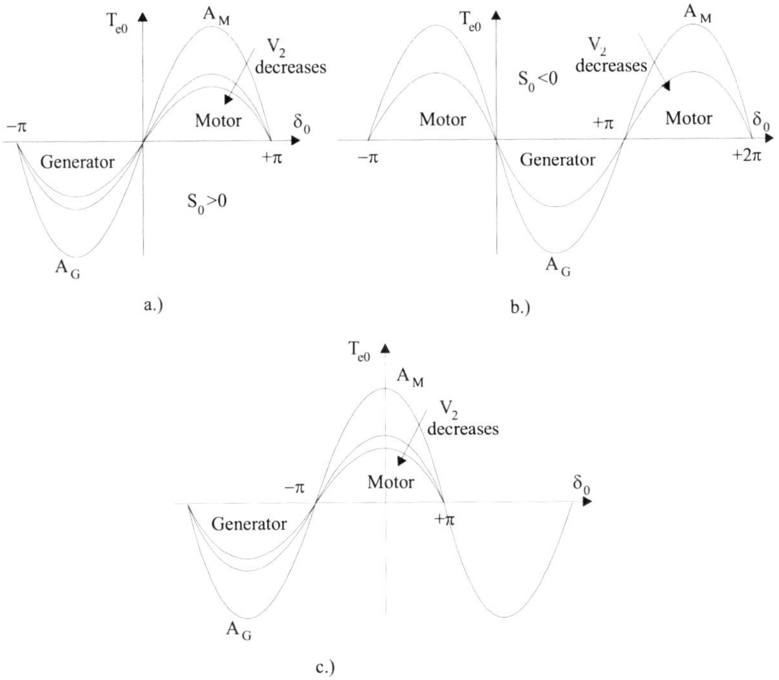

Figure 19.23 Ideal torque-angle curves
a.) & b.) at non zero (rather large) slips $|S| > 0.1$-$R_1 = R_2 = 0$ and c.) at zero slip ($R_1 = 0$, $R_2 \neq 0$)

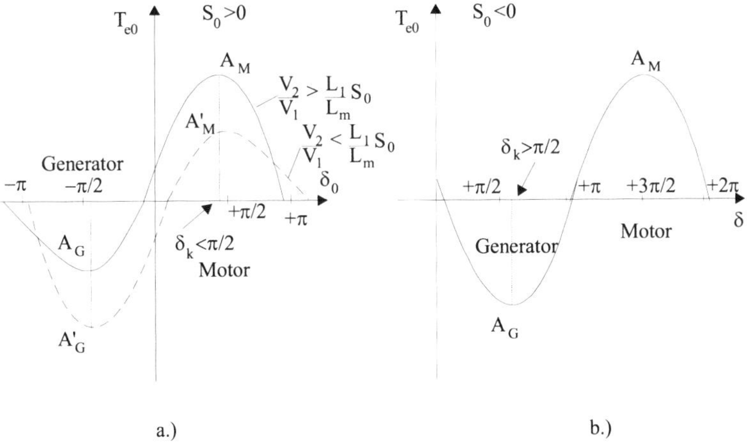

Figure 19.24 Realistic torque-angle curves ($R_2 \neq 0$, $R_1 = 0$)
a.) $S_0 > 0$; b.) $S_0 < 0$

$$\frac{\partial T_{e0}}{\partial \delta_0} = 0; \quad -R_2 \sin\delta_K + S_0\omega_1 L_{sc}\cos\delta_K = 0 \tag{19.81}$$

$$\tan\delta_{K_{G,M}} = \frac{S_0\omega_1 L_{sc}}{R_2} \tag{19.82}$$

The lower the value of the rotor resistance, the larger the value of $\delta_{K_{G,M}}$ and thus the larger the total stability angle zone travel (A_G - A_M).

Slightly larger stability zones for generating occur for negative slips ($S_0 < 0$). As the slip decreases the band of the stability zones decreases also (see 19.82).

The static stability problem may be approached through the eigenvalues method [20] of the linearized system only to obtain similar results even with nonzero stator resistance.

The complete performance of DFIG may be calculated by directly solving the system (19.74) for currents. Saturation may be accounted for by noting that $L_1 = L_{1l} + L_m$, $L_2 = L_{2l} + L_m$ and $L_m(i_m)$-magnetization curve-is given, with

$$i_m = \sqrt{i_{dm}^2 + i_{qm}^2}; \; i_{dm} = i_d + i_{dr}; \; i_{qm} = i_q + i_{qr}$$

The core losses may be added also, both in the stator and rotor as slip frequency is not necessarily small.

The stator and rotor power equations are

$$\underline{S}_1 = \frac{3}{2}\left(V_d i_{d0} + V_q i_{q0}\right) + j\frac{3}{2}\left(V_d i_{q0} - V_q i_{d0}\right) \tag{19.83}$$

$$\underline{S}_2 = \frac{3}{2}\left(V_{dr} i_{dr0} + V_{qr} i_{qr0}\right) + j\frac{3}{2}\left(V_{dr} i_{qr0} - V_{qr} i_{dr0}\right) \tag{19.84}$$

In principle, the active and reactive power in the stator or rotor may be positive or negative depending on the capabilities with respect of the power electronics converter connected in the rotor circuit.

Advanced vector control methods with decoupled active and reactive power control of DFIG has been recently proposed. [21, Chapter 14]

Slip recovery systems, which use unidirectional power flow power electronics converters in the rotor, lead to limited performance in terms of power factor, harmonics and flexibility. Also, they tend to have equivalent characteristics different from those discussed here. They are well documented in the literature. [22] Special doubly-fed IMs-with dual stator windings (p_1 and p_2 pole pairs) and nested cage rotor with $p_1 + p_2$ poles-have been recently proposed [23]. Also dual stator windings with p_1 and p_2 pole pairs and a regular cage-rotor may be practical. The lower power winding is inverter fed. With the wind energy systems on the rise and low power hydropower plants gaining popularity, the IG systems with variable speed are expected to have a dynamic future.

19.13 SUMMARY

- Three phase induction machines are used as generators both in wound rotor (doubly - fed) or the cage-rotor configurations.
- IGs may work in isolation or grid-connected.
- IGs may be used in constant or variable speed, constant or variable frequency, and constant or variable output voltage systems.
- As a rule less frequency sensitive isolated systems use cage rotor IGs with self-excitation (SEIG).
- Grid connected systems use both cage-rotor IGs (CRIGs) or doubly fed (wound rotor) IGs (DFIGs).
- Selfexcited IGs (SEIGs) use parallel capacitors at the terminals. For a given speed, IG, and a given no load voltage, there is a specific value of the capacitor that can produce it. Magnetic saturation is a must in SEIGs.
- When SEIGs with constant speed prime mover are loaded the frequency and voltage decrease with load. The slip (negative) increases with load.
- The basic problem for steady state operation starts with a given frequency, load, IG parameters and selfexcitation capacitance to find the speed (slip), voltage, currents, power, efficiency, etc.
- Impedance and admittance methods are used to study steady state. While the impedance methods lead to $4(5)^{th}$ order equations in S (or X_m), an admittance method that leads to a second order equation in slip is presented. This greatly simplifies the computational process as X_m (the magnetisation reactance) is obtained from a first order equation.
- There is a minimum load resistance which still allows SEIG selfexcitation which is inversely proportional to the selfexcitation capacitance C (19.22).
- The ideal maximum power slip is $S_{max} = -R_2/(fX_{2l})$, with R_2, X_{2l} as rotor resistance and leakage reactance at rated frequency and f, stator frequency in p.u. This expression is typical for constant airgap flux conditions in vector controlled IM drives.
- The SEIG steady state curves are voltage/current, voltage/power, frequency/ power for given speed and selfexcitation capacitance.
- SEIG parameter sensitivity studies show that voltage and maximum output power increases with capacitance.
- For no load the minimum capacitor that can provide selfexcitation is inversely proportional to speed squared.
- When feeding an IM from a SEIG, in general the rating of SEIG should be 2-3 times the motor rating though for pump-load starting even 1.6 times overrating will do. To avoid high overvoltages when the IM load is disconnected the excitation capacitor may be split between the IM and the SEIG.

- Pole changing stator SEIGs are used for wide speed range harnessing of SEIG's prime mover energy more completely (wind turbines for example).
- Pole changing winding design has to maintain large enough airgap flux densities, that saturate the cores to secure selfexcitation over the entire design speed range.
- Unbalanced operation of SEIGs may be analysed through the method of symmetrical components and is basically similar to that of balanced load but efficiency is notably lower and the phase voltages may differ from each other notably. When one line current is zero the single capacitor no load curve should be considered.
- Transients of SEIG occur for voltage build up or load perturbation. They may be investigated by the d-q model. At shortcircuit, for example, with IM-load, the SEIG experiences for a few periods 2 p.u. voltage and up to 5 p.u.current transients before the voltage collapses.
- Parallel operation of SEIGs in isolated systems makes use of a single but variable capacitor to control the voltage for variable load.
- The slip of all SEIGs in parallel has to be negative to avoid motoring. So the speeds (in p.u.) have to be very close to each other.
- Full usage of doubly fed induction generators (DFIG) occurs when it allows bidirectional power flow in the rotor circuit.
  Generating is thus feasible both sub and over conventional synchronous speed. Separate active and reactive power control may be commanded.
- The study of DFIG requires the d-q model which allows for an elegant treatment of steady state with rotor voltage V_2 and power angle δ as variables and slip S_0 as parameter. The DFIG works as a synchronous machine with constant stator voltage V_1 and frequency. The rotor voltage a.c. system has the slip frequency. Thus variable speed is handled implicitly.
- The stability angle δ_K extension increases with slip and leakage reactance and decreases with rotor resistance. Also V_2/V_1 (voltage ratio) plays a crucial role in stability. Constant $(V_2/S\omega_1)$ ratio-constant rotor flux-tends to hold the peak torque less variable with slip and thus enhances stability margins.
- Eigen values of the linearized d-q model may serve well to investigate the DFIG stability boundaries. Vector control with separate active and reactive rotor power control change the behaviour of the DFIG for the better in terms of both steady state and transients [21, Chapter 14].
- New topologies such as dual pole stator winding (p_1, p_2 pole pairs) with $p_1 + p_2$ rotor cage nests [23] and, respectively, conventional cage rotors [24] have been recently proposed. The IG subject seems far from exhausted.

19.14 REFERENCES

1. P. K. S. Khan, J. K. Chatterjee, Three-Phase Induction Generators, A Discussion on Performance, EMPS, vol. 27, no. 8, 1999, pp. 813-832.
2. E. D. Basset, F. M. Potter, Capacitive Excitation of Induction Generators, Electrical Engineering, vol. 54, 1935, pp. 540-545.
3. S. S. Murthy, O. P. Malik, A. K. Tandon, Analysis of Self-excited Induction Generators, Proc IEE, vol. 129C, no. 6, 1982, pp. 260-265.
4. L. Shridhar, B. Singh, C. S. Jha, A Step Toward Improvements in the Characteristics of Self-excited Induction Generators, IEEE Trans. vol. EC-vol. 8, 1993, pp. 1, pp. 40-46.
5. S. P. Singh, B. Singh, B. P. Jain, Steady State Analysis of Self-excited Pole-Changing Induction Generator, The Inst. of Engineering (India), Journal-EL, vol. 73, 1992, pp. 137-144.
6. S. P. Singh, B. Singh, M. P. Jain, A New Technique for the Analysis of Self-excited Induction Generator, EMPS, vol. 23, no. 6, 1995, pp. 647-656.
7. L. Quazene, G. McPherson Jr, Analysis of Isolated Induction Generators, IEEE Trans. vol. PAS-102, no. 8, 1983, pp. 2793-2798.
8. N. Ammasaigounden, M. Subbiah, M. R. Krishnamurthy, Wind Driven Self-Excited Pole-Changing Induction Generators, Proc. IEE, vol. 133B, no. 5., 1986, pp. 315-321.
9. K. S. Sandhu, S. K. Jain, Operational Aspects of Self-excited Induction Generator Using a New Model, EMPS vol. 27, no. 2, 1999, pp. 169-180.
10. I.Boldea, S. A. Nasar, Vector control of a.c. drives, book, CRC Press, 1992, Boca Raton, Florida.
11. N. N. Malik, A. A. Mazi, Capacitance Requirements for Isolated Self-Excited Induction Generators, IEEE Trans, vol. EC-2, no. 1, 1987, pp. 62-68.
12. A. Kh.Al-Jabri, A. I. Alolah, Limits On the Performance of the Three-Phase Self-Excited Induction Generators, IEEE Trans, vol. EC-5, no. 2, 1990, pp. 350-356.
13. E. Bim, J. Szajner, Y. Burian, Voltage Compensation of an Induction Generator with Long Shunt Connection, IBID vol. 4, no.3, 1989, pp. 506-513.
14. S. M. Alghumainen, "Steady state analysis of an induction generator self-excited by a capacitor in parallel with a saturable reactor", EMPS vol.26, no.6, 1998, pp.617-625.

15. A. H. Al-Bahrani, Analysis of Self-Excited Induction Generators Under Unbalanced Conditions, EMPS, vol. 24, no. 2, 1996, pp.117-129.

16. I. Boldea, S. A. Nasar, Unified Treatment of Core Losses and Saturation in The Orthogonal-Axis Model of Electric Machines, Proc. IEE-vol. 134B, no. 6, 1987, pp. 355-363.

17. K. E. Hallenius, P. Vas, J. E. Brown, The Analysis of Saturated Self-excited Asynchronous Generator, IEEE Trans., vol. EC-6, no. 2, 1991, pp. 336-345.

18. B. Singh, L. Shridhar, C. B. Iha, Transient Analysis of Self-excited Induction Generator Supplying Dynamic Load, EMPS vol. 27, no. 9, 1999, pp. 941-954.

19. Ch. Chakraborty, S. N. Bhadra, A. K. Chattopadhyay, Analysis of Parallel-Operated Self-Excited Induction Generators, IEEE Trans. vol EC-14, no. 2, 1999, pp. 209-216.

20. A. Masmoudi, A. Tuomi, M. B. A. Kamoun, M. Poloujadoff, Power Flow Analysis and Efficiency Optimisation of a Double Fed Synchronous Machine, EMPS vol.21, no. 4, 1993, pp. 473-491.

21. I. Boldea, S. A. Nasar, Electric Drives, book, CRC Press, Boca Raton, Florida, 1998.

22. M. M. Eskander, T. El-Hagri, Optimal Performance of Double Output Induction Generators WECS, EMPS, vol. 25, no. 10, 1997, pp.1035-1046.

23. A. Wallace, P. Rochelle, R.Spée, Rotor Modelling Development for Brushless Double-Fed Machines, EMPS vol. 23, no. 6, 1995, pp. 703-715.

24. O. Ojo, I. E. Davidson, PWM-VSI Inverter-Assisted Stand-Alone Dual Stator Winding Induction Generator, IEEE Trans. vol. IA-36, no. 6, 2000, pp. 1604-1610.

Chapter 20

LINEAR INDUCTION MOTORS

20.1 INTRODUCTION

For virtually every rotary electric machine, there is a linear motion counterpart. So is the case with induction machines. They are called linear induction machines (LIMs).

LIMs directly develop an electromagnetic force, called *thrust,* along the direction of the travelling field motion in the airgap.

The imaginary process of "cutting" and "unrolling" rotary counterpart is illustrated in Figure 20.1.

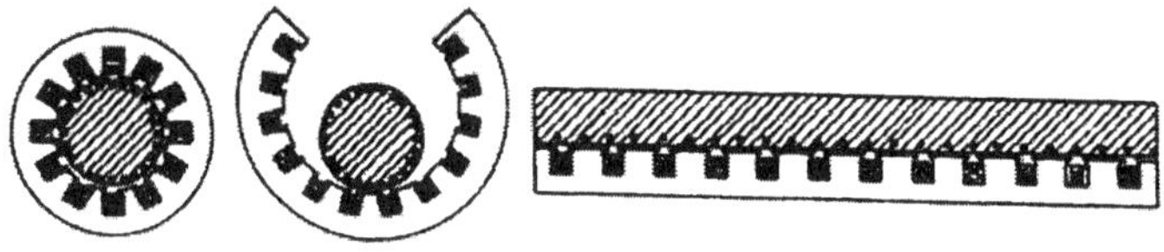

Figure 20.1 Imaginary process of obtaining a LIM from its rotary counterpart

The *primary* usually contains a three phase winding in the uniform slots of the laminated core.

The *secondary* is either made of a laminated core with a ladder cage in the slots or of an aluminum (copper) sheet with (or without) a solid iron back core.

Apparently the LIM operates as its rotary counterpart does, with *thrust* instead of torque and linear speed instead of angular speed, based on the principle of travelling field in the airgap.

In reality there are quite a few differences between linear and rotary IMs such as [1 - 8]

- The magnetic circuit is open at the two longitudinal ends (along the travelling field direction). As the flux law has to be observed, the airgap field will contain additional waves whose negative influence on performance is called *dynamic longitudinal* end effect (Figure 20.2a).
- In short primaries (with 2, 4 poles), there are current asymmetries between phases due to the fact that one phase has a position to the core longitudinal ends which is different from those of the other two. This is called static longitudinal effect (Figure 20.2b).
- Due to same limited primary core length, the back iron flux density tends to include an additional nontravelling (ac) component which should be considered when sizing the back iron of LIMs (Figure 20.2c).
- In the LIM on Figure 20.1 (called single sided, as there is only one primary along one side of secondary), there is a normal force (of attraction or

repulsion type) between the primary and secondary. This normal force may be put to use to compensate for part of the weight of the moving primary and thus reduce the wheel wearing and noise level (Figure 20.2d).

- For secondaries with aluminum (copper) sheet with (without) solid back iron, the induced currents (in general at slip frequency Sf_1) have part of their closed paths contained in the active (primary core) zone (Figure 20.2c). They have additional-longitudinal (along OX axis)-components which produce additional losses in the secondary and a distortion in the airgap flux density along the transverse direction (OY). This is called the transverse edge effect.

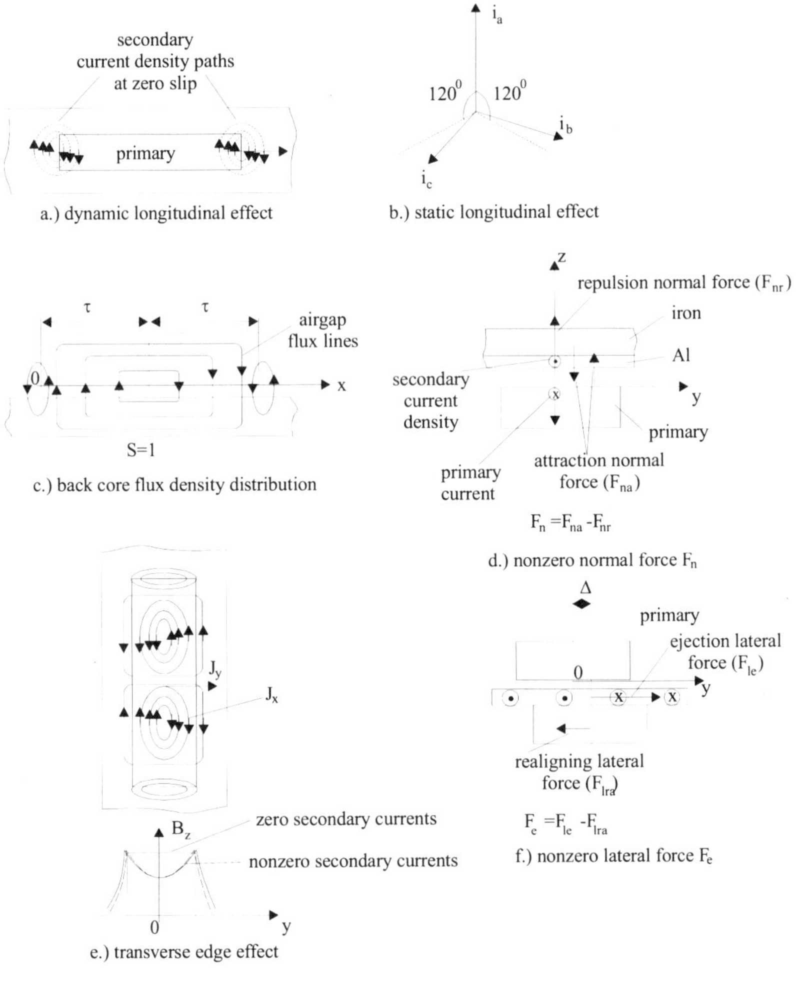

Figure 20.2 Panoramic view of main differences between LIMs and rotary IMs

- When the primary is placed off center along OY, the longitudinal components of the current density in the active zone produce an ejection type lateral force. At the same time, the secondary back core tends to realign the primary along OY. So the resultant lateral force may be either decentralizing or centralizing in character (Figure 20.2.f).

All these differences between linear and rotary IMs warrant a specialized investigation of field distribution and performance in order to limit the adverse effects (longitudinal end effects and back iron flux distortion, etc.) and exploit the desirable ones (normal and lateral forces, or transverse edge effects).

The same differences suggest the main merits and demerits of LIMs.

Merits

- Direct electromagnetic thrust propulsion (no mechanical transmission or wheel adhesion limitation for propulsion)
- Ruggedness; very low maintenance costs
- Easy topological adaptation to direct linear motion applications
- Precision linear positioning (no play (backlash) as with any mechanical transmission)
- Separate cooling of primary and secondary
- All advanced drive technologies for rotary IMs may be applied without notable changes to LIMs

Demerits

- Due to large airgap to pole pitch (g/τ) ratios–$g/\tau > 1/250$–the power factor and efficiency tend to be lower than with rotary IMs. However, the efficiency is to be compared with the combined efficiency of rotary motor + mechanical transmission counterpart. Larger mechanical clearance is required for medium and high speeds above 3m/s. The aluminum sheet (if any) in the secondary contributes an additional (magnetic) airgap.
- Efficiency and power factor are further reduced by longitudinal end effects. Fortunately these effects are notable only in high speed low pole count LIMs and they may be somewhat limited by pertinent design measures.
- Additional noise and vibration due to uncompensated normal force, unless the latter is put to use to suspend the mover (partially or totally) by adequate close loop control.

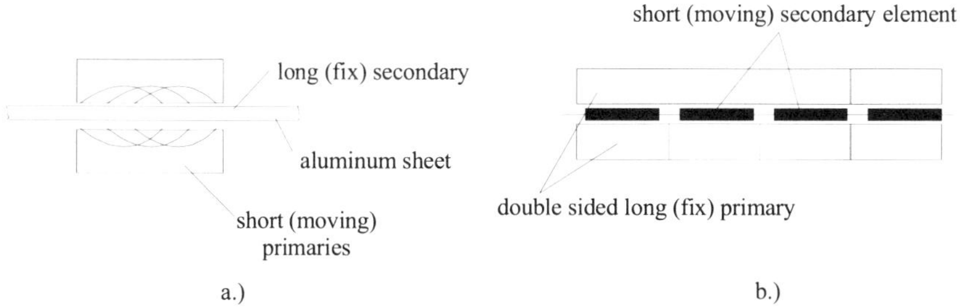

Figure 20.3 Double sided flat LIMs
a.) double sided short (moving) primary LIM; b.) double sided short (moving) secondary LIM for conveyors

As sample LIM applications have been presented in Chapter 1, we may now proceed with the investigation of LIMs, starting with classification and practical construction aspects.

20.2 CLASSIFICATIONS AND BASIC TOPOLOGIES

LIMs may be built single sided (Figure 20.1) or double sided (Figure 20.3a), with moving (short) primary (Figure 20.1) or moving (short) secondary (Figure 20.3b).

As single sided LIMs are more rugged, they have found more applications. However (Figure 20.3b) shows a double sided practical short-moving secondary LIM for low speed short travel applications.

LIMs on Figure 20.1-20.3 are flat. For flat single sided LIMs the secondaries may be made of aluminum (copper) sheets on back solid iron (for low costs), ladder conductor in slots of laminated core (for better performance), and a pure conducting layer in electromagnetic metal stirrers (Figure 20.4a, b, c).

In double sided LIMs, the secondary is made of an aluminum sheet (or structure) or from a liquid metal (sodium) as in flat LIM pumps.

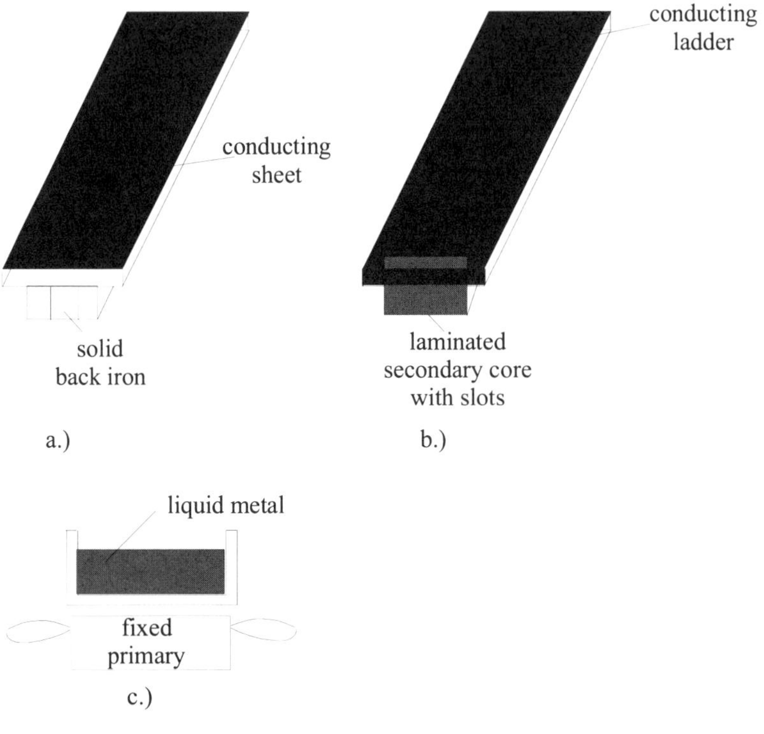

Figure 20.4 Flat single-sided secondaries
a.) sheet on iron ; b.) ladder conductor in slots; c.) liquid metal

Besides flat LIMs, tubular configurations may be obtained by rerolling the flat structures along the transverse (OY) direction (Figure 20.5). Tubular LIMs or tubular LIM pumps are in general single sided and have short fix primaries and moving limited length secondaries (except for liquid metal pumps).

The primary core may be made of a few straight stacks (Figure 20.5a) with laminations machined to circular stator bore shape. The secondary is typical aluminum (copper) sheet on iron. The stator coils have a ring shape.

While transverse edge effect is absent and coils appear to lack end connections, building a well centered primary is not easy.

An easier solution to build is obtained with only two-size disk shape laminations both on primary and secondary (Figure 20.5b). The secondary ring shape conductors are also placed in slots.

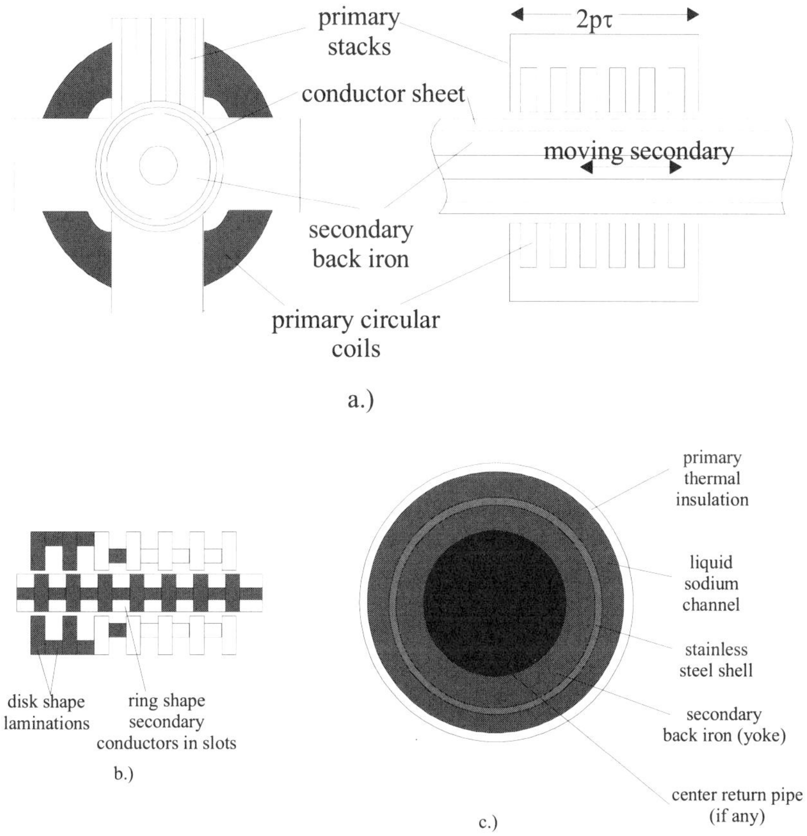

Figure 20.5 Tubular LIMs
a.) with longitudinal primary lamination stacks; b.) with disk-shape laminations
c.) liquid metal tubular LIM pump the secondary

Better performance is expected by the fact that, in the back cores, the magnetic field goes perpendicular to laminations and tends to produce additional core losses. The interlamination insulation leads to an increased magnetization m.m.f. and thus takes back part of this notable improvement.

The same rationale is valid for tubular LIM liquid metal pumps (Figure 20.5c) in terms of primary manufacturing process. Pumps allow for notably higher speeds (u = 15m/s or more) when the fixed primaries may be longer and have more poles ($2p_1$ = 8 or more). The liquid metal (sodium) low electrical conductivity leads to a smaller dynamic longitudinal effect which at least has to be checked to see if it is negligible.

20.3 PRIMARY WINDINGS

In general, three phase windings are used as three phase PWM converters and are widely available for rotary induction motor drives. Special applications which require only $2p_1$ = 2 pole might benefit from two phase windings as they are both placed in the same position with respect to the magnetic core ends. Consequently, the phase currents are fully symmetric at very low speeds.

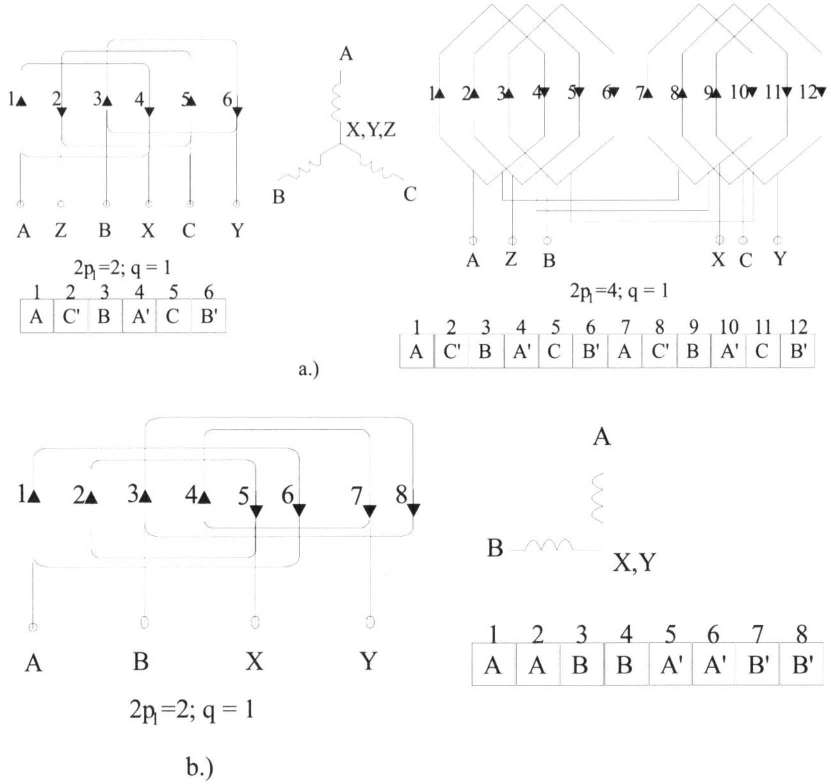

Figure 20.6 Single layer windings
a.) three phase; b.) two phase

LIM windings are similar to those used for rotary IMs and ideally they produce a pure travelling m.m.f.. However, as the magnetic circuit is open along the direction of motion, there are some particular aspects of LIM windings.

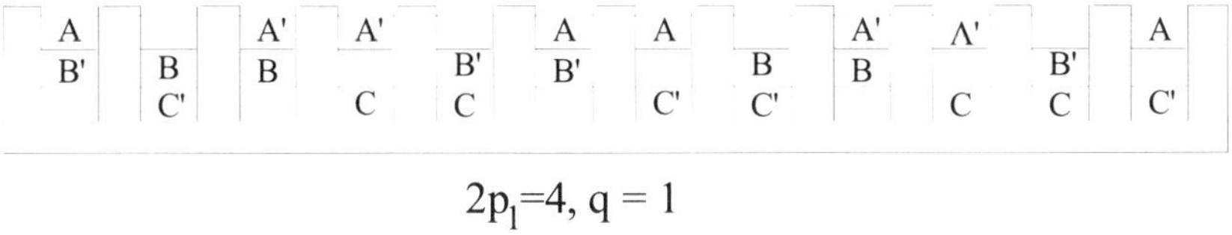

Figure 20.7 Triple layer winding with chorded coils ($y/\tau = 2/3$) (very short end connections)

A C B
X Z Y
1 2 3 4 5 6 7 8 9 10 11 12 13 14
A C' B A' C B' A C' B A' C B'
A' C B' A C' B A' C B' A C' B
2p1+1 = 5; q = 1, y/τ = 2/3
1 2 3 4 5 6 7 8 9 10 11 12 13 14 15 16 17 18 19 20 21 22 23 24 25 26 27 28 29 30 31 32 33 34 35 36 37 38 39 40
A A C' C' B B A' A' C C B' B' A A C' C' B B A' A' C C B' B' A A C' C' B B A' A' C C B' B'
A' A' C C B' B' A A C' C' B B A' A' C C B' B' A A C' C' B B A' A' C C B' B' A A C' C' B B
2p1+1 = 7; q = 2, y/τ = 5/6

Figure 20.8 Double layer chorded coil windings with $2p_1 + 1$ poles

Among the possible winding configurations we illustrate a few

- Single layer full pitch ($y = \tau$) windings with an even number of poles $2p_1$, Figure 20.6 three phase and two phase.
- Triple layer chorded coil ($y/\tau = 2/3$) winding with an even number of poles $2p_1$ (Figure 20.7).
- Double layer chorded coil ($2/3 < y/\tau < 1$) coil winding with an odd number of poles $2p_1 + 1$ (the two end poles have half-filled slots)-Figure 20.8.
- Fractionary winding for miniature LIMs (Figure 20.9).

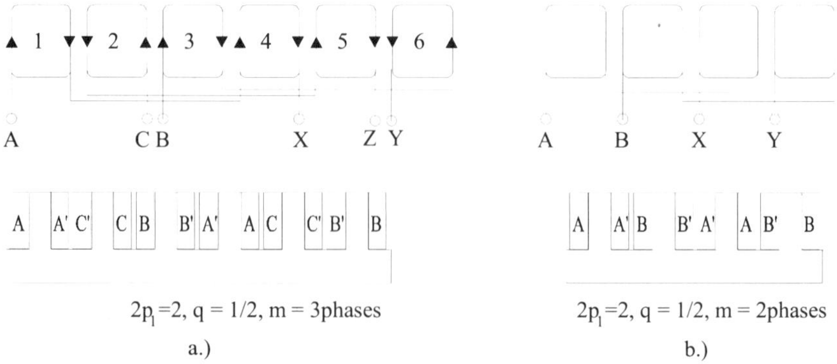

Figure 20.9 Fractionary single layer winding (low low end connections)
a.) three phase; b.) two phase

A few remarks are in order

- The single layer winding with an even number of poles makes better usage of primary magnetic core but it shows rather large coil end connections. It is recommended for $2p_1 = 2, 4$.
- The triple layer chorded coil windings is easy to manufacture automatically; it has low end connections but it also has a rather low winding (chording) factor $K_y = 0.867$.
- The double layer chorded coil winding with an odd total number of poles (Figure 20.8) has shorter end connections and is easier to build but it makes a poorer use of primary magnetic core as the two end poles are halfwound. As the number of poles increases above 7, 9 the end poles influence becomes small. It is recommended for large LIMs ($2p_1 + 1>5$).
- The fractionary winding (Figure 20.9)-with q = ½ in our case-is characterised by very short end connections but the winding factor is low. It is recommended only in miniature LIMs where volume is crucial.
- When the number of poles is small, $2p_1 = 2$ especially, and phase current symmetry is crucial (low vibration and noise) the two phase LIM may prove the adequate solution.
- Tubular LIMs are particularly suitable for single-layer even-number of poles windings as the end connections are nonexistent with ring-shape coils.

In the introduction we mentioned the transverse edge and longitudinal end effects as typical to LIMs. Let us now proceed with a separate analysis of transverse edge effect in double sided and in single sided LIMs with sheet-secondary.

20.4 TRANSVERSE EDGE EFFECT IN DOUBLE-SIDED LIM

A simplified single dimensional theory of transverse edge effect is presented here.

The main assumptions are

- The stator slotting is considered only through Carter coefficient K_c.

$$K_c = \frac{1}{1-\gamma g/\tau_s}; \quad \gamma = \frac{\left(\frac{g}{b_{os}}\right)^2}{5+\gamma\frac{g}{b_{os}}} \tag{20.1}$$

- The primary winding in slots is replaced by an infinitely this current sheet traveling wave J_1 (x, t)

$$J_1(x,t) = J_m e^{j\left(S\omega_1 t - \frac{\pi}{\tau}x\right)}; \quad J_m = \frac{3\sqrt{2}W_1K_{w1}I_1}{p_1\tau} \tag{20.2}$$

$2p_1$-pole number, W_1-turns/phase, K_{w1}-winding factor; τ-pole pitch, I_1-phase current (RMS). Coordinates are attached to secondary.

- The skin effect is neglected or considered through the standard correction coefficient.

$$K_{skin} \approx \frac{d}{2d_s}\left[\frac{\sinh\left(\frac{d}{d_s}\right)+\sin\left(\frac{d}{d_s}\right)}{\cosh\left(\frac{d}{d_s}\right)-\cos\left(\frac{d}{d_s}\right)}\right] \tag{20.3}$$

$$\frac{1}{d_s} \approx \sqrt{\frac{\mu_0\omega_1 S\sigma_{Al}}{2}} \tag{20.4}$$

For single sided LIMs, d/d_s will replace $2d/d_s$ in (20.2); d_s-skin depth in the aluminum (copper) sheet layer. Consequently, the aluminum conductivity is corrected by $1/K_{skin}$,

$$\sigma_{Als} = \frac{\sigma_{Al}}{K_{skin}} \tag{20.5}$$

- For a large airgap between the two primaries, there is a kind of flux leakage which makes the airgap look larger g_l [1].

$$g_l = gK_cK_{leakage}$$

$$K_{leakage} = \frac{\sinh\left(\frac{g}{2\tau}\right)}{\frac{g}{2\tau}} > 1 \tag{20.6}$$

- Only for large g/τ ratio $K_{leakage}$ is notably different from unity. The airgap flux density distribution in the absence of secondary shows the transverse fringing effect (Figure 20.10).

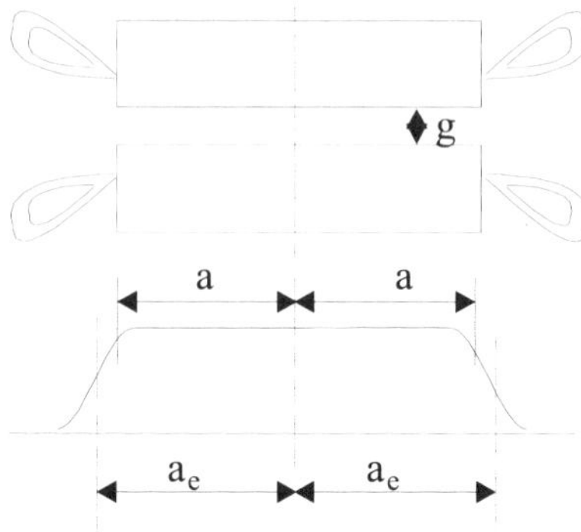

Figure 20.10 Fringing and end connection flux considerations

The transverse fringing effect may be accounted for by introducing a larger (equivalent) stack width $2a_e$ instead of 2a

$$2a_e = 2a + (1.2 \div 2.0)\, g \tag{20.7}$$

For large airgap in low thrust LIMs this effect is notable.

As expected, the above approximations may have been eliminated provided a 3D FEM model was used. The amount of computation effort for a 3D FEM model is so large that it is feasible mainly for special cases rather than for preliminary or optimisation design.

- Finally, the longitudinal effect is neglected for the time being and space variations along thrust direction and time variations are assumed to be sinusoidal.

In the active region ($|z| \le a_e$) Ampere's law along contour 1 (Figure 20.11) yields

$$g_e \frac{\partial \underline{H}_y}{\partial z} = \underline{J}_{2x} d;\ \underline{J}_1 = J_m e^{j\left(\omega_1 t - \frac{\pi}{\tau} x\right)} \tag{20.8}$$

Figure 20.11 shows the active and overhang regions with current density along motion direction.

The same law applied along contour 2, Figure 20.11.b, in the longitudinal plane gives

$$-g_e \frac{\partial}{\partial x}\left(\underline{H}_x + \underline{H}_0\right) = J_m + \underline{J}_{2z} d \tag{20.9}$$

Faraday's law also yields

$$\frac{\partial \underline{J}_{2z}}{\partial x} - \frac{\partial \underline{J}_{2x}}{\partial z} = -j\mu_0 S\omega_1\sigma_{Als}\left(\underline{H}_y + \underline{H}_0\right) \tag{20.10}$$

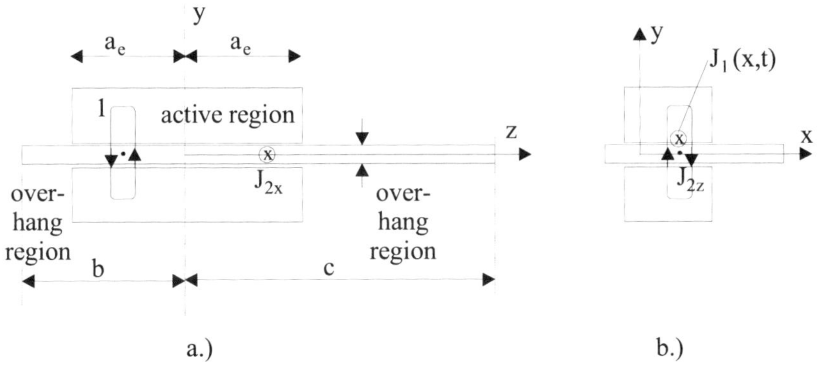

Figure 20.11 Transverse cross-section of a double sided LIM a.), and longitudinal view b.)

Equations (20.8-20.10) are all in complex terms as sinusoidal time variation was assumed

$\underline{H}_0$ is the airgap field in absence of secondary

$$\underline{H}_0 = \frac{jJ_m}{\frac{\pi}{\tau} g_e} \tag{20.11}$$

The three equations above combine to yield

$$\frac{\partial^2 \underline{H}_y}{\partial x^2} + \frac{\partial^2 \underline{H}_y}{\partial z^2} - -j\mu_0 S\omega_1\sigma_{Als}\frac{d}{g_e}\underline{H}_y = j\mu_0 S\omega_1\sigma_{Als}\frac{d}{g_e}\underline{H}_0 \tag{20.12}$$

The solution of (20.12) becomes

$$\underline{H}_y = -\frac{j\underline{H}_0 SG_e}{1 + jSG_e} + A\cosh \underline{\alpha} z + B\sinh \underline{\alpha} z \tag{20.13}$$

with

$$G_i = \frac{\tau^2\omega_1\mu_0\sigma_{Als}}{\pi^2}\frac{d}{g_e} \tag{20.14}$$

$$\underline{\alpha}^2 = \left(\frac{\pi}{\tau}\right)^2\left(1 + jsG_e\right) \tag{20.15}$$

G_i is called the rather ideal goodness factor of LIM, a performance index we will return to in what follows frequently.

In the overhang region ($|z| > a_e$) we assume that the total field is zero, that is J_{2z}, J_{2x} satisfy Laplace's equation

$$\frac{\partial \underline{J}_{2z}}{\partial x} - \frac{\partial \underline{J}_{2x}}{\partial z} = 0 \tag{20.16}$$

Consequently

$$\begin{aligned} \underline{J}_{2zr} &= \underline{C}\sinh\frac{\pi}{\tau}(c-z) \qquad a_e < z \le c \\ \underline{J}_{2xr} &= j\underline{C}\cosh\frac{\pi}{\tau}(c-z) \end{aligned} \tag{20.17}$$

$$\begin{aligned} \underline{J}_{2zl} &= \underline{D}\sinh\frac{\pi}{\tau}(b+z) \qquad -b < z < -a_e \\ \underline{J}_{2xl} &= j\underline{D}\cosh\frac{\pi}{\tau}(b+z) \end{aligned} \tag{20.18}$$

where r and l refer to right and left, respectively.

From the continuity boundary conditions at $z = \pm a_e$, we find

$$\underline{A} = \frac{\underline{H}_0 jSG_i}{1 + jSG_i}\left[\frac{(\underline{C}_1 + \underline{C}_2)\cosh(\underline{\alpha} a_e) + 2\sinh(\underline{\alpha} a_e)}{(1 + \underline{C}_1\underline{C}_2)\sinh(2\underline{\alpha} a_e) + (\underline{C}_1 + \underline{C}_2)\cosh(2\underline{\alpha} a_e)}\right] \tag{20.19}$$

$$\underline{B} = \frac{\underline{A}(\underline{C}_2 - \underline{C}_1)\sinh(\underline{\alpha} a_e)}{(\underline{C}_1 + \underline{C}_2)\cosh(\underline{\alpha} a_e) + 2\sinh(\underline{\alpha} a_e)}$$

$$\underline{C}_1 = \frac{\alpha\tau}{\pi}\tanh\frac{\pi}{\tau}(c - a_e); \quad \underline{C}_2 = \frac{\alpha\tau}{\pi}\tanh\frac{\pi}{\tau}(b - a_e) \tag{20.20}$$

$$\underline{C} = \frac{-jg_e\underline{\alpha}}{d\cosh\frac{\pi}{\tau}(c - a_e)}\left[\underline{A}\sinh(\underline{\alpha} a_e) + \underline{B}\cosh(\underline{\alpha} a_e)\right] \tag{20.21}$$

$$\underline{D} = \frac{-jg_e\underline{\alpha}}{d\cosh\frac{\pi}{\tau}(b - a_e)}\left[-\underline{A}\sinh(\underline{\alpha} a_e) + \underline{B}\cosh(\underline{\alpha} a_e)\right] \tag{20.22}$$

Sample computation results of flux and current densities distributions for a rather high speed LIM with the data of $\tau = 0.35$m, $f_1 = 173.3$Hz, $S = 0.08$, $d = 6.25$mm, $g = 37.5$mm, $J_m = 2.25\cdot10^5$A/m are shown in Figure 20.12.

The transverse edge effect produces a "deep" in the airgap flux density transverse (along OZ) distribution. Also if the secondary is placed off center, along OZ, the distribution of both secondary current density is nonsymmetric along OZ.

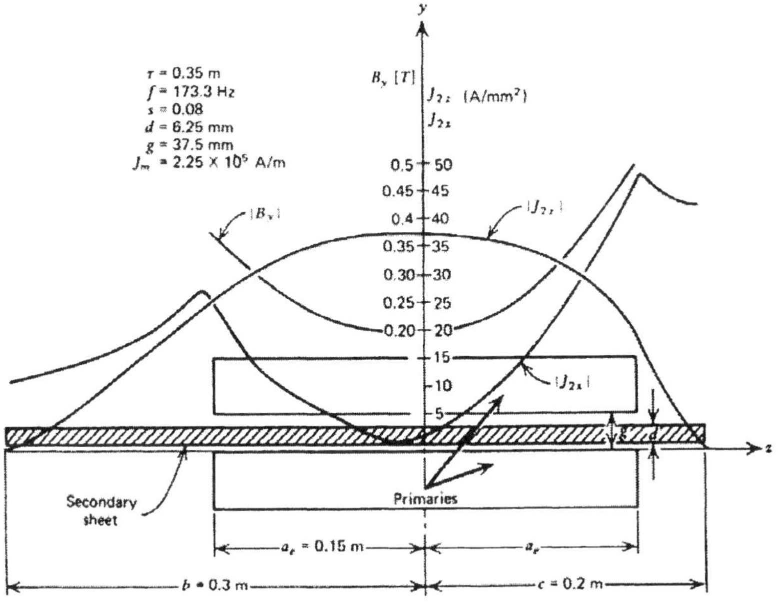

Figure 20.12 Transverse distribution of flux and secondary current densities

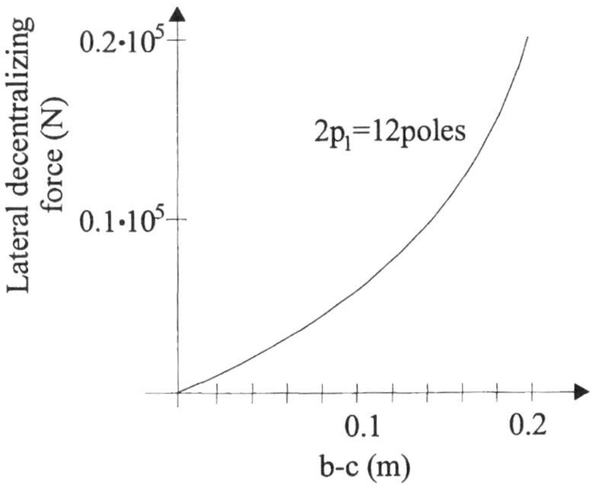

Figure 20.13 Decentralizing lateral force F_z

The main consequences of transverse edge effect are an apparent increase in the secondary equivalent resistance R_2' and a decrease in the magnetisation inductance (reactance X_m). Besides, when the secondary is off center in the transverse (lateral) direction, a lateral decentralising force F_{za} is produced

$$F_{za} = \mu_0 d\tau p_1 \,\mathrm{Re}\int_{-a_e}^{a_e}\left(\underline{J}_{2x}\underline{H}_y^*\right)dz \tag{20.23}$$

Again sample results for the same LIM as above are given in Figure 20.13.

The lateral force F_z decreases as a_e/τ decreases or the overhangs c-a_e, b-a_e > τ/π. In fact it is of no use to extend the overhangs of secondary beyond τ/π as there are few currents for $|z| > |\tau/\pi + a_e|$.

The transverse edge effect correction coefficients

In the absence of transverse edge effect, the magnetization reactance X_m has the conventional expression (for rotary induction machines)

$$X_m = \frac{6\mu_0\omega_1}{\pi^2}\frac{(W_1K_{w1})^2\tau(2a_e)}{p_1 g_1} \tag{20.24}$$

The secondary resistance reduced to the primary R'_2 is

$$R_2' = \frac{X_m}{G_i} = \frac{12(W_1K_{w1})^2 a_e}{\sigma_{Als} p_1 \tau d} \tag{20.25}$$

Because of the transverse edge effect, the secondary resistance is increased by $K_t > 1$ times and the magnetization inductance (reactance) is decreased by K_m < 1 times.

$$K_t = \frac{K_X^2}{K_R}\left[\frac{1+S^2G_i^2K_R^2/K_X^2}{1+S^2G_i^2}\right] \geq 1 \tag{20.26}$$

$$K_m = \frac{K_R K_t}{K_X} \leq 1 \tag{20.27}$$

For b = c [9],

$$K_R = 1 - \mathrm{Re}\left[(1 - SG_i)\frac{\underline{\lambda}}{\underline{\alpha} a_e}\tanh(\underline{\alpha} a_e)\right] \tag{20.28}$$

$$K_X = 1 + \mathrm{Re}\left[(j + SG_i)\frac{SG_i\underline{\lambda}}{\underline{\alpha} a_e}\tanh(\underline{\alpha} a_e)\right] \tag{20.29}$$

$$\underline{\lambda} = \frac{1}{1 + (1 + jSG_i)^{1/2}\tanh(\underline{\alpha} a_e)\cdot\tanh\frac{\pi}{\tau}(c - a_e)} \tag{20.30}$$

For LIMs with narrow primaries ($2a_e/\tau < 0.3$) and at low slips, $K_m \approx 1$ and

$$K_t \approx \frac{1}{1 - \frac{\tanh\frac{\pi}{\tau}a_e}{\frac{\pi}{\tau}a_e}\left[1 + \tanh\left(\frac{\pi}{\tau}a_e\right)\cdot\tanh\frac{\pi}{\tau}(c - a_e)\right]^{-1}} \quad (20.31)$$

Transverse edge effect correction coefficients depend on the goodness factor G_i, slip, S, and the geometrical type factors a_e/τ, $(c - a_e)/\tau$.

For $b \neq c$ the correction coefficients have slightly different expressions but they may be eventually developed based on the flux and secondary current densities transverse distribution.

The transverse edge effect may be exploited for developing large thrusts with lower currents or may be reduced by large overhangs (up to τ/π) and an optimum a_e/τ ratio.

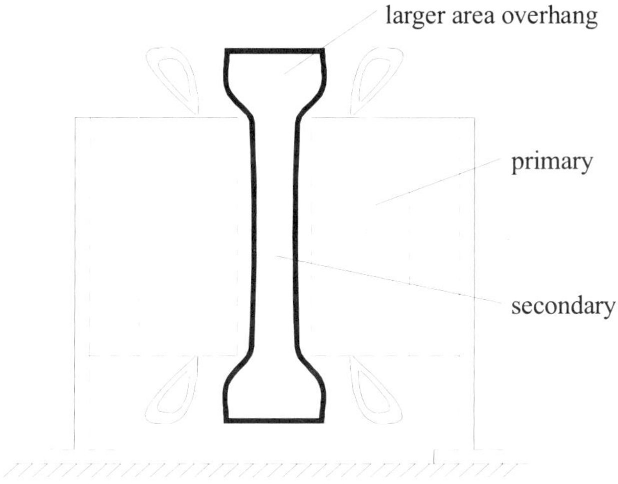

Figure 20.14 Reduced transverse edge effect secondary

On the other hand, the transverse edge effect may be reduced, when needed, by making the overhangs of a larger cross-section or of copper (Figure 20.14).

In general, the larger the value of SG_i, the larger the transverse edge effect for given a_e/τ and c/τ. In low thrust (speed) LIMs as the pole pitch τ is small, so is the synchronous speed u_s

$$u_s = 2\tau f_1 = \tau\frac{\omega_1}{\pi} \quad (20.32)$$

Consequently, the goodness factor G_i is rather small below or slightly over unity.

Now, we may define an equivalent-realistic-goodness factor G_e as

$$G_e = G_i \frac{K_m}{K_t} < G_i \tag{20.33}$$

where the transverse edge effect is also considered.

Alternatively, the combined airgap leakage (K_{leak}), skin effect (K_{skin}) and transverse edge effect (K_m, K_t) may be considered as correction coefficients for an equivalent airgap g_e and secondary conductivity σ_e,

$$g_e = g \frac{K_c K_{leak}}{K_m} > g \tag{20.34}$$

$$\sigma_e = \frac{\sigma_{Al}}{K_{skin} K_t} < \sigma_{Al} \tag{20.35}$$

We notice that all these effects contribute to a reduction in the realistic goodness factor G_i.

20.5 TRANSVERSE EDGE EFFECT IN SINGLE-SIDED LIM

For the single sided LIM with conductor sheet on iron secondary (Figure 20.15), a similar simplified theory has been developed where both the aluminum and saturated back iron contributions are considered [5,7].

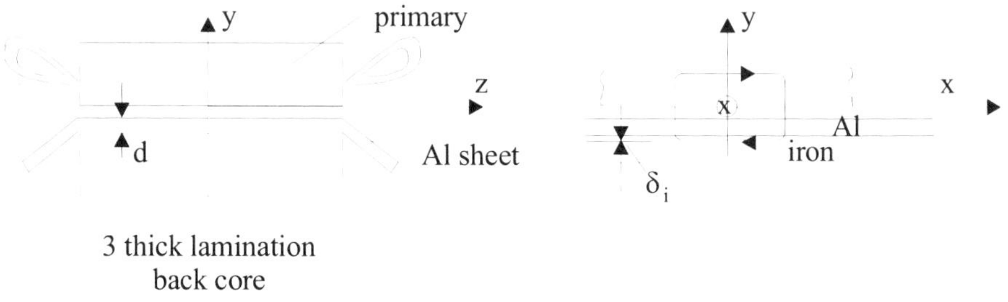

Figure 20.15 Single side LIM with 3 thick lamination secondary back core

The division of solid secondary back iron into three (i = 3) pieces along transverse direction leads to a reduction of eddy currents. This is due to an increase in the "transverse edge effect" in the back iron.

As there are no overhangs for secondary back iron ($c - a_e = 0$) the transverse edge effect coefficient K_t of (20.31) becomes

$$K_{ti} = \frac{1}{1 - \frac{\tan\frac{\pi}{\tau}\frac{i}{a_e}}{\frac{\pi}{\tau}\frac{i}{a_e}}} \tag{20.36}$$

Now an equivalent iron conductivity σ_{ti} may be defined as

$$\sigma_{ti} = \frac{\sigma_{iron}}{K_{ti}} \tag{20.37}$$

The depth of field penetration in the secondary back iron δ_i is thus

$$\delta_i^{-1} = \mathrm{Real}\left[\sqrt{\left(\frac{\pi}{\tau}\right)^2 + jS\omega_1\mu_{iron}\frac{\sigma_{iron}}{K_{ti}}}\right] \tag{20.38}$$

The iron permeability μ_{iron} depends mainly on the tangential (along OX) flux density B_{xi}, in fact on its average over the penetration depth δ_i

$$B_{xi} = \frac{\tau}{\pi}\frac{B_g}{\delta_i}K_{pf}; \quad 1 \le K_{pf} < 2 \tag{20.39}$$

where B_g is the given value of airgap flux density and K_{pf} accounts for the increase in back core maximum flux density in LIMs due to its open magnetic circuit along axis x [2, 7]; $K_{pf} = 1$ for rotary IMs.

Now we may define an equivalent conductivity σ_e of the aluminum to account for the secondary back iron contribution

$$\sigma_e = \frac{\sigma_{Al}}{K_{skinal}}\left[\frac{1}{K_{ta}} + \frac{\sigma_{iron}\delta_i K_{skinal}}{\sigma_{Al}K_{ti}d}\right] \tag{20.40}$$

K_{skinal}-is the skin effect coefficient for aluminum (20.3). K_{ta}-the transverse edge effect coefficient for aluminum.

In a similar way an equivalent airgap may be defined which accounts for the magnetic path in the secondary path by a coefficient K_p

$$g_e = (1 + K_p)g\frac{K_c K_{leak}}{K_{ma}} \tag{20.41}$$

$$K_p = \frac{\tau^2}{\pi^2}\frac{\mu_0}{2gK_c\delta_i\mu_{iron}} \tag{20.42}$$

K_p may not be neglected even if the secondary back iron is not heavily saturated.

Now the equivalent goodness factor G_e may be written as

$$G_e = \frac{\mu_0\omega_1\tau^2\sigma_e d}{\pi^2 g_e} \tag{20.43}$$

The problem is that G_e depends on ω_1, S, and μ_{iron}, for given machine geometry.

An iterative procedure is required to account for magnetic saturation in the back iron of secondary (μ_{iron}). The value of the resultant airgap flux density $B_g = \mu_0(H + H_0)$ may be obtained from (20.13) and (20.11) by neglecting the Z dependent terms

$$B_g \approx \frac{\mu_0 \tau}{\pi g_e}\frac{J_m}{(1+jSG_e)}; \quad J_m = \frac{3\sqrt{2}W_1 K_{w1} I_1}{p_1 \tau} \tag{20.44}$$

For given value of B_g, S, W_1, K_t, K_{ma}, K_{ti} are directly calculated. Then from (20.38)-(20.39) and the iron magnetization curve μ_{iron} (and B_{xi}, and δ_i) are iteratively computed.

Then σ_e and g_e are calculated from (20.40)-(20.42). Finally G_e, is determined. The primary phase current I_1 is computed from (20.44).

All above data serve to calculate the LIM thrust and other performance indices to be dealt with in the next paragraph in a technical longitudinal effect theory of LIMs.

20.6 A TECHNICAL THEORY OF LIM LONGITUDINAL END EFFECTS

Though we will consider the double sided LIM, the results to be obtained here are also valid for single-sided LIMs with same equivalent airgap g_e and secondary conductivity σ_e.

The technical theory as introduced here relies on a quasi-one dimensional model attached to the short (moving) primary. Also, for simplicity, the primary core is considered infinitely long but the primary winding is of finite length (Figure 20.16). All effects treated in previous paragraph enter the values of σ_e, g_e and G_e and the primary m.m.f. is replaced by the travelling current sheet $\underline{J}_1$ (20.2). Complex variables are used as sinusoidal time variations are considered.

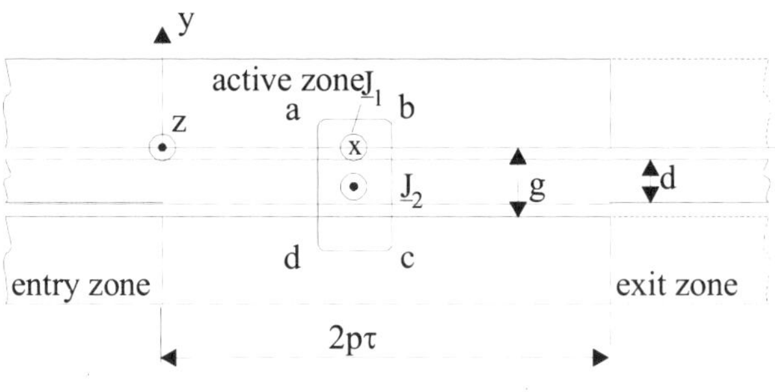

Figure 20.16 Double-sided LIM with infinitely long primary core

In the active zone ($0 \le x \le 2p_1\tau$) Ampere's law along abcd (Figure 20.16) yields

$$g_e \frac{\partial H_t}{\partial x} = J_1 e^{j\left(\omega_1 t - \frac{\pi}{\tau}x\right)} + J_2 d \tag{20.45}$$

H_t is the resultant magnetic field in airgap. It varies only along OY axis. Also the secondary current density J_2 has only one component (along OZ).

Faraday's law applied to moving bodies

$$\operatorname{curl} E = -\frac{\partial B}{\partial x} + \overline{u} \times \overline{B}; \quad J = \sigma_e E \tag{20.46}$$

yields

$$\frac{\partial J_2}{\partial x} = j\omega_1 \mu_0 \sigma_e H_t + \mu_0 u \sigma_e \frac{\partial H_t}{\partial x} \tag{20.47}$$

where u is the relative speed between primary and secondary.

Equations (20.45) and (20.47) may be combined into

$$\frac{\partial^2 H_t}{\partial x^2} - \mu_0 u \sigma_e' \frac{\partial H_t}{\partial x} - j\omega_1 \mu_0 \sigma_e' H_t = -j\frac{\pi}{\tau g_e} J_m e^{-j\frac{\pi}{\tau}x} \tag{20.48}$$

$$\sigma_e' = \sigma_e \frac{d}{g_e} \tag{20.49}$$

The characteristic equation of (20.48) becomes

$$\gamma^2 - \mu_0 \sigma_e' u \gamma - j\omega_1 \mu_0 \sigma_e' = 0 \tag{20.50}$$

Its roots are

$$\gamma_{1,2} = \pm\frac{a_1}{2}\left[\sqrt{\frac{b_1+1}{2}} \pm 1 + j\sqrt{\frac{b_1-1}{2}}\right] = \gamma_{1,2r} \pm j\gamma_i \tag{20.51}$$

with

$$a_1 = \mu_0 \sigma_e' u = \frac{\pi}{\tau} G_e (1 - S) \tag{20.52}$$

$$b_1 = \sqrt{1 + \left(\frac{4}{G_e (1-S)^2}\right)^2}$$

The complete solution of H_t within active zone is

$$H_{ta}(x) = \underline{A}_t e^{\gamma_1 x} + \underline{B}_t e^{\gamma_2 x} + \underline{B}_n e^{-j\frac{\pi}{\tau}x} \tag{20.54}$$

with

$$\underline{B}_n = \frac{j\tau J_m}{\pi g_e (1 + SG_e)}; \quad S = \frac{u_s - u}{u_s} \tag{20.55}$$

The coefficient $e^{j\omega_1 t}$ is understood.

In the entry ($x < 0$) and exit ($x > 2p_1\tau$) zones there are no primary currents. Consequently

$$H_{entry} - \underline{C}_e e^{\gamma_1 x}; \quad x \leq 0 \tag{20.56}$$

$$H_{exit} = \underline{D}_e e^{\gamma_2 x}; \quad x > 2p_1\tau \tag{20.57}$$

At $x = 0$ and $x = 2p_1\tau$, the magnetic fields and the current densities are continuous

$$\left(H_{entry}\right)_{x=0} = \left(H_{ta}\right)_{x=0} \tag{20.58}$$

$$\left(H_{exit}\right)_{x=2p_1\tau} = \left(H_{ta}\right)_{x=2p_1\tau} \tag{20.59}$$

$$\left(\frac{\partial H_{entry}}{\partial x}\right)_{x=0} = \left(J_2\right)_{x=0}; \quad \left(\frac{\partial H_{exit}}{\partial x}\right)_{x=2p_1\tau} = \left(J_2\right)_{x=2p_1\tau} \tag{20.60}$$

These conditions lead to

$$\underline{A}_t = -j\frac{J_m}{\underline{D}}\left(\gamma_2\frac{\tau}{\pi} + SG_e\right)e^{-2p_1\tau\gamma_1} \tag{20.61}$$

$$\underline{B}_t = j\frac{J_m}{\underline{D}}\left(\gamma_1\frac{\tau}{\pi} + SG_e\right) \tag{20.62}$$

$$\underline{D} = g_e\left(\gamma_2 - \gamma_1\right)\left(1 + jSG_e\right) \tag{20.63}$$

$$G_e = \frac{\tau^2}{\pi^2}\omega_1\mu_0\sigma_e\frac{d}{g_e} \tag{20.64}$$

20.7 LONGITUDINAL END-EFFECT WAVES AND CONSEQUENCES

The above field analysis enables us to investigate the dynamic longitudinal effects.

Equation (20.54) reveals the fact that the airgap field H_t and its flux density B_t has, besides the conventional unattenuated wave, two more components a forward and a backward travelling wave, because of longitudinal end effects. They are called end effect waves

$$\underline{B}_{backward} = -j\mu_0\frac{J_m}{\underline{D}}\left(\gamma_2\frac{\tau}{\pi} + SG_e\right)e^{(\gamma_{1r} + j\gamma_i)(x - 2p_1\tau)} \tag{20.65}$$

$$\underline{B}_{forward} = j\mu_0 \frac{J_m}{\underline{D}}\left(\gamma_1 \frac{\tau}{\pi} + SG_e\right)e^{(\gamma_{2r} - j\gamma_i)x} \tag{20.66}$$

They may be called the exit and entry end-effect waves respectively. The real parts of $\gamma_{1,2}$ (γ_{1r}, γ_{2r}) determine the attenuation of end-effect waves along the direction of motion while the imaginary part $j\gamma_i$ determines the synchronous speed (u_{se}) of end effect waves

$$u_{se} = \frac{\omega_1}{\gamma_i}; \quad \tau_e = \frac{\pi}{\gamma_i} \tag{20.67}$$

The values of $1/\gamma_{1r}$ and $1/\gamma_{2r}$ may be called the depths of the end effect waves penetration in the (along) the active zone.

Apparently from (20.51)

$$\frac{1}{\gamma_{1r}} << \frac{1}{\gamma_{2r}}; \quad \frac{1}{\gamma_{1r}} < (8 \div 10)\cdot 10^{-3}\text{m} \tag{20.68}$$

Consequently, the effect of backward (exit) end effect wave is negligible. Not so with the forward (entry) end effect wave which attenuates slowly in the airgap along the direction of motion.

The higher the value of goodness factor G_e and the lower the slip S, the more important the end effect waves are. High G_e means implicitly high synchronous speeds.

The pole pitch ratio of end-effect waves (τ_e/τ) is

$$\frac{\tau_e}{\tau} = \frac{2\sqrt{2}}{G_e(1-S)\left(\sqrt{1+\left(\frac{4}{G_e(1-S)^2}\right)^2} - 1\right)} \geq 1 \tag{20.69}$$

It may be shown that $\tau_e/\tau \geq 1$ and is approaching unity (at S = 0) for large goodness factor values G_e.

The conventional thrust F_{xc} is

$$F_{xc} = \mu_0 a_e J_m \,\text{Re}\left(\int_0^{2p_1\tau} B_a^* dx\right) \tag{20.70}$$

The end-effect force F_{xe} has a similar expression

$$F_{xe} = \mu_0 a_e J_m \,\text{Re}\left(\int_0^{2p_1\tau} B_t^* e^{\gamma_2^* x} e^{-j\frac{\pi}{\tau}x} dx\right) \tag{20.71}$$

The ratio f_e of these forces is a measure of end effect influence.

$$f_e = \frac{F_{xe}}{F_{xc}} = \frac{\mathrm{Re}\left[B_t^* \frac{-1+\exp 2p_1\tau\left(\gamma_2^* - j\frac{\pi}{\tau}\right)}{\left(\gamma_2^* - j\frac{\pi}{\tau}\right)}\right]}{\mathrm{Re}\left[B_n^* 2p_1\tau\right]} \tag{20.72}$$

Or, finally,

$$f_e = \frac{1+S^2G_e^{\ 2}}{SG_e}\mathrm{Re}\left\{\frac{j\left(\frac{\gamma_1\tau}{\pi}+SG_e\right)\left(\exp\left[2p_1\tau\left(\gamma_2^*\frac{\tau}{\pi}-j\right)\right]-1\right)}{\frac{\tau}{\pi}\left(\gamma_2^*-\gamma_1^*\right)\left(1-jSG_e\right)2p_1\pi\left(\gamma_2^*\frac{\tau}{\pi}-j\right)}\right\} \tag{20.73}$$

As seen from (20.73), f_e depends only on slip S, realistic goodness factor G_e, and on the number of poles $2p_1$.

Quite general p.u. values $(F_{xe})_{p.u.}$ of F_{xe} may be expressed as

$$\left(F_{xe}\right)_{p.u.} = F_{xe}\frac{\pi^2 g_e}{a_e\mu_0 J_m^{\ 2}} \tag{20.74}$$

$(F_{xe})_{p.u.}$ depends only on $2p_1$, S and G_e and is depicted in Figure 20.17a, b, c. The quite general results on Figure 20.17 suggest that

- The end effect force at zero slip may be either propulsive-positive-(for low G_e values or (and) large number of poles) or it may be of braking character (negative)-for high G_e or (and) smaller number of poles.
- For a given number of poles and zero slip, there is a certain value of the realistic goodness factor G_{eo}, for which the end effect force is zero. This value of G_e is called the optimum goodness factor.
- For large values of G_e, the end effect force changes sign more than once as the slip varies from 1 to zero.
- The existence of the end effect force at zero slip is a distinct manifestation of longitudinal end effect.

Further on the airgap flux density $B_g = \mu_o H_{ta}$ (see (20.54)) has a nonuniform distribution along OX that accentuates if S is low, goodness factor G_e is high and the number of poles is low.

Typical qualitative distributions are shown in Figure 20.18.

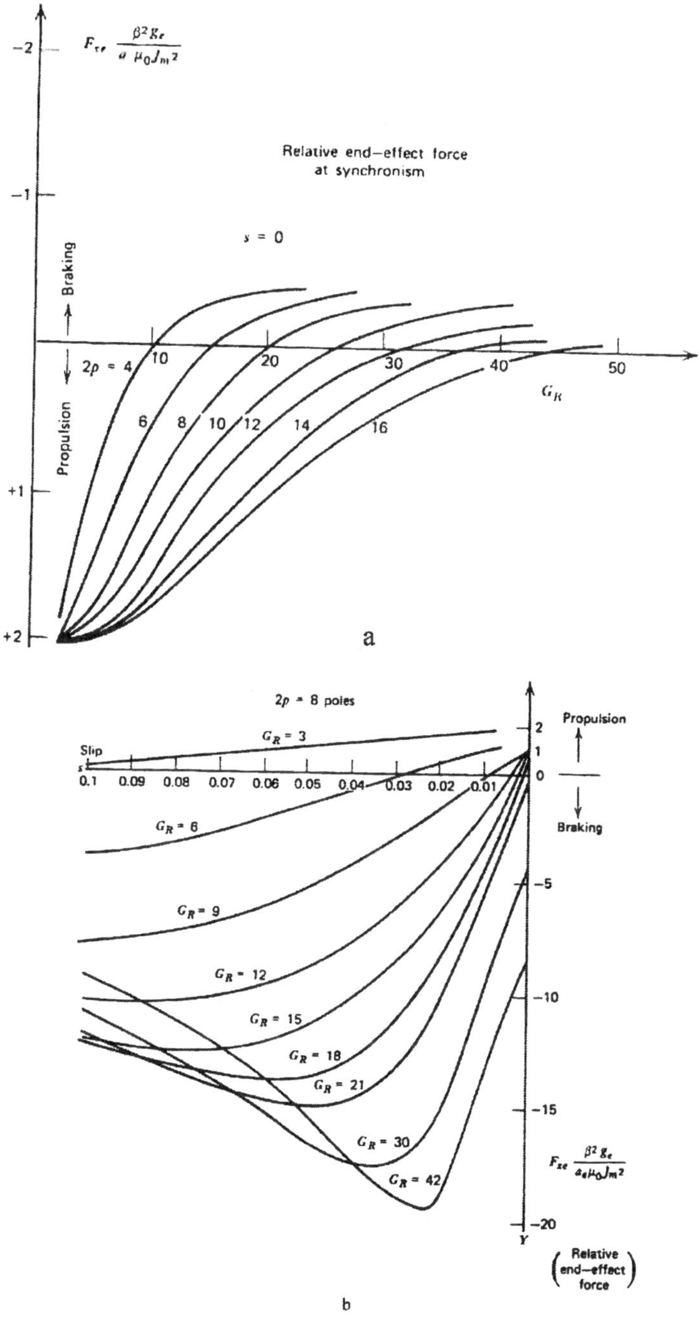

Figure 20.17 End effect force F_{xe} in p.u.
a.) at zero slip; b.) at small slips; c.) versus normalized speed (continued)

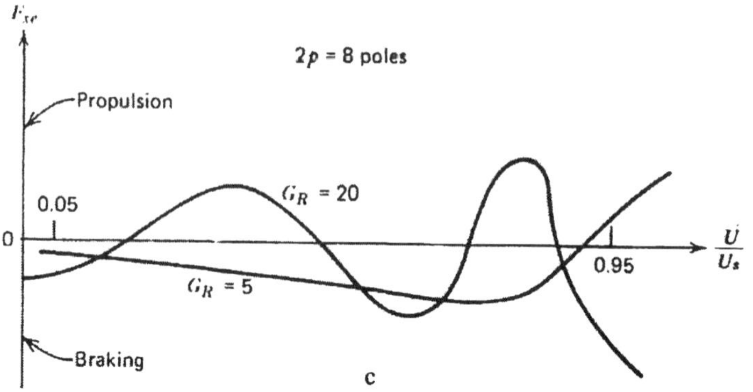

Figure 20.17 (continued)

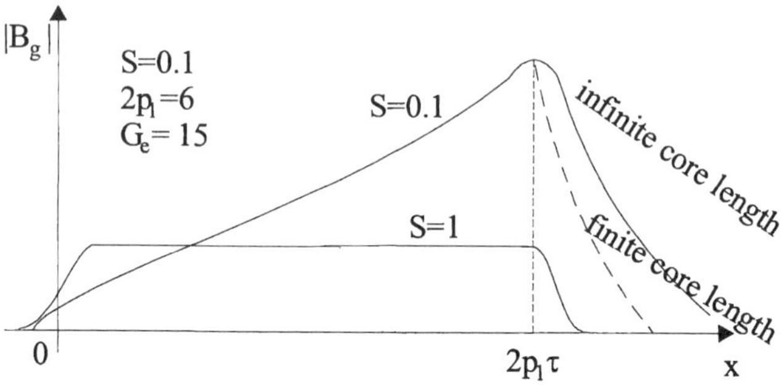

Figure 20.18 Airgap flux density distribution along OX

The problem is similar in single sided LIMs but there the saturation of secondary solid iron requires iterative computation procedures.

A nonuniform distribution shows also the secondary current density J_2

$$\underline{J}_2 = \frac{g_e}{d}\left[\underline{\gamma}_1\underline{A}_t e^{\gamma_1 x} + \underline{\gamma}_2\underline{B}_t e^{\gamma_2 x} - \frac{jSG_e J_m e^{-j\frac{\pi}{\tau}x}}{g_e(1 + jSG_e)}\right] \tag{20.76}$$

Higher current density values are expected at the entry end (x = 0) and/or at the exit end at low values of slip for high goodness factor G_e and low number of poles $2p_1$ [5, pp. 271].

Consequently, the secondary plate losses are distributed nonuniformly along the direction of motion in the airgap [5, pp. 231].

Also, the propulsion force is not distributed uniformly along the core length. In presence of large longitudinal end effects, the thrust at entry end goes down to zero, even to negative values [5, pp.232].

Similar aspects occur in relation to normal forces in double-sided and single-sided LIMs [5, pp. 232].

20.8 SECONDARY POWER FACTOR AND EFFICIENCY

Based on the secondary current density $\underline{J}_2$ distribution (20.76), the power losses in the secondary P_2 are

$$P_2 = \frac{a_e d^2}{\sigma_e g_e} \int_{-\infty}^{\infty} \left(\underline{J}_2 \underline{J}_2^*\right) dx \tag{20.77}$$

Similarly, the reactive power Q_2 in the airgap is

$$Q_2 = a_e \omega_1 \mu_0 g_e \int_{-\infty}^{\infty} \left(\underline{H}_{ta} \underline{H}_{ta}^*\right) dx \tag{20.78}$$

The secondary efficiency η_2 is

$$\eta_2 = \frac{uF_x}{uF_x + P_2}; \quad F_x = F_{xc} + F_{xe} \tag{20.79}$$

The secondary power factor $\cos\varphi_2$ is

$$\cos\varphi_2 = \frac{\left(uF_x + P_2\right)}{\sqrt{\left(uF_x + P_2\right)^2 + Q_2^2}} \tag{20.80}$$

Longitudinal end-effects deteriorate both the secondary efficiency and power factor. Typical numerical results for a super-high speed LIM are shown in Figure 20.19.

So far, we considered the primary core as infinitely long. In reality, this is not the case. Consequently, the field in the exit zone decreases more rapidly (Figure 20.18) and thus the total secondary power losses are in fact, smaller than calculated above.

However, due to the same reason, at exit end there will be an additional reluctance small force [3, pp.74 - 79].

Numerical methods such as FEM would be suitable for a precise estimation of field distribution in the active, entry, and exit zones. However to account for transverse edge effect also, 3D FEM is mandatory.

Alternatively, 3D multilayer analytical methods have been applied successfully to single-sided LIMs with solid saturated and conducting secondary back iron [10 - 12] for reasonable computation time.

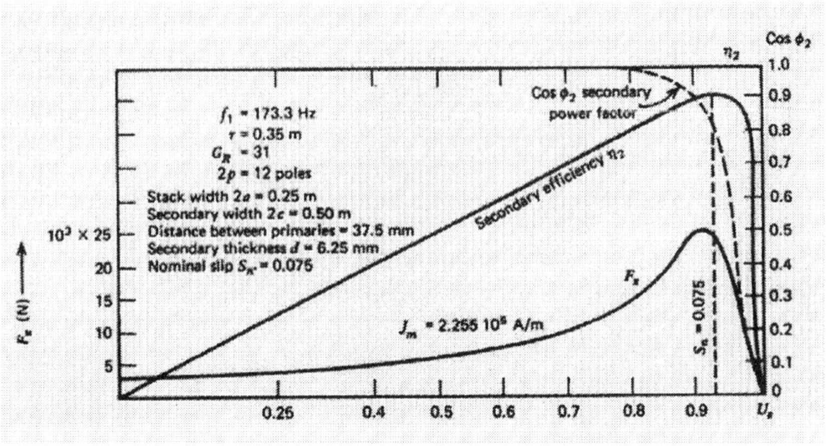

Figure 20.19 LIM secondary efficiency η_2 and power factor $\cos\varphi_2$

20.9 THE OPTIMUM GOODNESS FACTOR

As we already noticed, the forward end effect wave has a longitudinal penetration depth of $\delta_{end} = 1/\gamma_{2r}$. We may assume that if

$$\frac{(\delta_{end})_{S=0}}{2p_1\tau} \leq 0.1 \tag{20.81}$$

the longitudinal end-effect consequences are negligible for $2p_1 \geq 4$. Condition (20.81) involves only the realistic goodness factor, G_o, and the number of poles.

Indirectly condition (20.81) is related to frequency, pole pitch (synchronous speed), secondary sheet thickness, conductivity, total airgap etc.

Consequently two LIMs of quite different speeds and powers may have the same longitudinal effect relative consequences if G_e and $2p_1$ and slip S are the same. [13]

End effect compensation schemes have been introduced early [2,3] but they did not prove to produce overall (global) advantages in comparison with well designed LIMs.

By well designed LIMs for high speed, we mean those designed for zero longitudinal end-effect force at zero slip. That is, designs at optimum goodness factor G_{eo} [5, pp. 238] which is solely dependent on the number of poles $2p_1$ (Figure 20.20).

G_{eo} is a rather intuitive compromise as higher G_e leads to both conventional performance enhancement and increase in the longitudinal effect adverse influence on performance.

LIMs where the dynamic longitudinal end effect may be neglected are called low speed LIMs or linear induction actuators while the rest of them are called high speed LIMs.

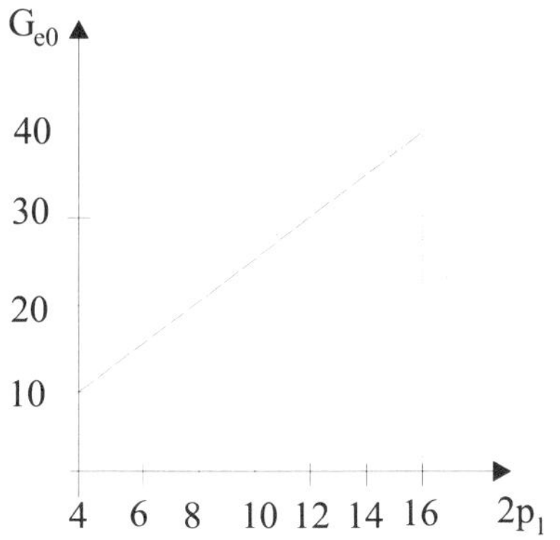

Figure 20.20 The optimum goodness factor

High speed LIMs are used for transportation-urban and inter-urban. In urban (suburban) transportation the speed seldom goes above 20(30) m/s but this is enough to make the longitudinal end effects worth considering, at least by global thrust correction coefficients. [14]

20.10 LINEAR FLAT INDUCTION ACTUATORS

Again, we mean by linear induction actuators (LIAs) low speed short travel, linear induction motor drives for which the dynamic longitudinal end effect may be neglected (20.81).

Most LIAs are single-sided (flat and tubular) with short primary and long conductor-sheet-iron or ladder secondary in a laminated slotted core. For double sided LIA, the long primary and short (moving) secondary configuration is of practical interest.

a. The equivalent circuit

All specific effects-airgap leakage, aluminum plate skin effect and transverse edge effects–have been considered and their effects lumped into equivalent airgap g_e (2.41) and aluminum sheet conductivity σ_e (2.40).

These expressions also account for the solid secondary back iron contribution in eddy currents and magnetic saturation. The secondary resistance R_2' reduced to the primary and the magnetizing reactance X_m (the secondary leakage reactance is neglected) can be adapted from (20.24)-(20.25) as

$$X_m = \frac{\mu_0\omega_1(W_1K_{w1})^2\tau(2a_e)}{\pi^2 p_1 g_e(S\omega_1, I_1)} = K_m W_1^2 \tag{20.82}$$

$$R_2' = \frac{X_m}{G_e} = \frac{12(W_1K_{w1})^2 a_e}{p_1\tau d\sigma_e(S\omega_1, I_1)} = K_{R2} W_1^2 \tag{20.83}$$

So in fact both X_m and R_2' vary with slip frequency $S\omega_1$ and the stator current due to skin effect and transverse edge effect in both the conducting sheet and the back iron (if it is made of solid iron).

The equivalent circuit of rotary IM is now valid for LIAs, but with the variable parameters X_m and R_2' (Figure 20.21).

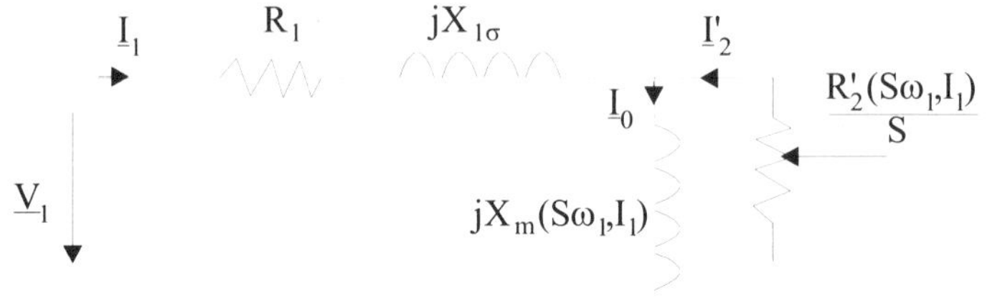

Figure 20.21 Linear induction actuator equivalent circuit

To fully exploit the equivalent circuit, we still need the expressions of primary resistance and leakage reactance R_1 and $X_{1\sigma}$.

$$R_1 = \frac{1}{\sigma_{Co}}\frac{(4a+2l_{ec})}{W_1 I_1}W_1^2 j_{con} = K_{R1}W_1^2 \tag{20.84}$$

where l_{ec} is the coil end connection length, j_{con}-rated current density, σ_{Co}-copper electrical conductivity.

$$X_{1\sigma} = \frac{2\mu_0\omega_1}{p_1 q}\left[(\lambda_s+\lambda_d)2a+\lambda_f l_{ec}\right]W_1^2 = K_{1\sigma}W_1^2 \tag{20.85}$$

$$\lambda_s \approx \frac{1}{12}\frac{h_s}{b_s}(1+3\beta');\ \ \beta' = \frac{y}{\tau}\ \text{(coil span)} \tag{20.86}$$

$$\lambda_d = \frac{5\frac{g}{b_s}}{5+4\frac{g}{b_s}};\ \ b_s, h_s - \text{open slot width and height} \tag{20.87}$$

$$\lambda_f \approx 0.3q(3\beta-1);\ l_{ec} \approx K_f\tau;\ \ K_f = 1.3 \div 1.6 \tag{20.88}$$

λ_s, λ_d, λ_f-slot, differential, end connection specific geometrical permeances.

The primary phase m.m.f. W_1I_1 (RMS) is

$$W_1 I_1 = j_{con}\left(\frac{\tau}{3q}\right)^2 K_d^{\,2}\frac{h_s}{b_s}K_{fill} \tag{20.89}$$

where $K_d = \frac{3qb_s}{\tau}$, the open slot width/slot pitch, is 0.5 to 0.7 for LIAs; hs/bs is the slot aspect ratio and varies between 3 and 6(7) for LIAs; K_{fill} is the slot fill factor and varies in the interval K_{fill} = 0.35-0.45 for round wire random wound coils and is 0.6-0.7 for preformed rectangular wire coils.

b. Performance computation

Making use of the equivalent circuit (Figure 20.21), with all parameters already calculated, we may simply determine the thrust F_x,

$$F_x = \frac{3I_2'^2 R_2'}{S \cdot 2\tau f_1} = \frac{3I_1^2 R_2'}{S \cdot 2\tau f_1\left[\left(\frac{1}{SG_e}\right)^2 + 1\right]} \tag{20.90}$$

The efficiency η_1 and power factor $\cos\varphi_1$ are

$$\eta_1 = \frac{2\tau f_1(1-S)F_x}{2\tau f_1 F_x + 3I_1^2 R_1} \tag{20.91}$$

$$\cos\varphi_1 = \frac{2\tau f_1 F_x + 3I_1^2 R_1}{3V_{1f}I_1} \tag{20.92}$$

As expected, the realistic goodness factor G_e plays an important role in thrust production (for given stator current I_1) and in performance (Figure 20.22).

The peak thrust for given stator current follows from (20.90) with

$$\frac{\partial F_x}{\partial S} = 0 \rightarrow S_k G_e = \pm 1;\; F_{xk} = \frac{\pm 3I_1^2 R_2' G_e}{2\tau f_1 \cdot 2} \tag{20.93}$$

Evidently (20.93) is valid only if R_2', X_m, G_e are constant.

In many low speed applications, the realistic goodness is around unity. If peak thrust/ current is needed at standstill, then S_k = 1 and thus the design goodness factor at start is

$$G_e = \omega_1 \sigma_e \mu_0 \frac{d}{g_e}\frac{\tau^2}{\pi^2} = 1 \;\text{ for }\; S_k = 1 \tag{20.94}$$

Here $\omega_1 = 2\pi f_1$ is the primary frequency at start. If a variable frequency converter is used, the starting frequency is reduced.

The thrust/secondary losses is, from 20.90,

$$\frac{F_x}{P_2} = \frac{F_x}{3I_2'^2 R_2'} = \frac{1}{2\tau f_1 \cdot S} \tag{20.95}$$

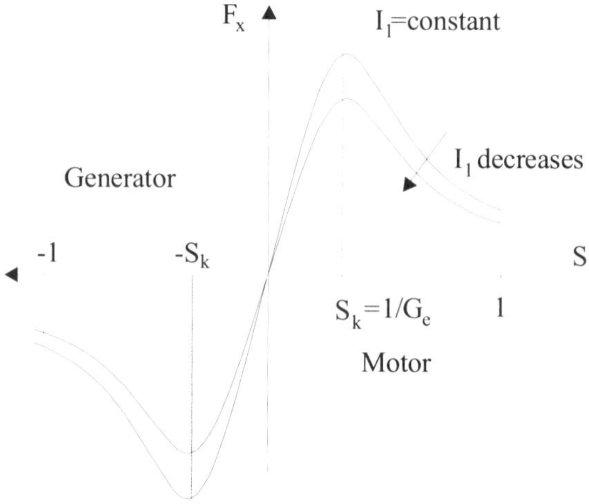

Figure 20.22 Thrust versus slip

For S = 1 the smaller the frequency f_1 at start, for given τ, the larger the thrust/ secondary conductor losses.

However as the value of τ decreases, to reduce the back core height, for given airgap flux density, the starting frequency does not fall below (5-6)Hz for most practical cases.

c. Normal force in single-sided configurations

The normal force between single-sided LIAs primary and secondary has two components

- An attraction force F_{na} between the primary core and the secondary back iron produced by the normal airgap flux density component.
- A repulsion force F_{nr} between the primary and secondary current m.m.f.s.

As the longitudinal end effect is absent, F_{na} is

$$F_{na} = \frac{2a_e\left|\underline{B}_g\right|^2 2p_1\tau}{2\mu_0} \tag{20.96}$$

Making use of (20.44) for $\underline{B}_g$, (20.96) becomes

$$F_{na} \approx \frac{a_e\mu_0 J_{1m}^{2}\tau^2 2p_1\tau}{\pi^2 g_e^{2}\left(1+S^2G_e^{2}\right)} \tag{20.97}$$

To calculate the repulsive force F_{nr}, the tangential component (along OX) of airgap magnetic field H_x is needed. The value of H_x along the primary surface is

$$H_x = J_m; \quad B_x = \mu_0 J_m \tag{20.98}$$

$$F_{nr} = 4\frac{a_e}{2}\tau p_1 d\mu_0 \, \mathrm{Re}\left[\underline{J}_2^* J_m\right] \tag{20.99}$$

But

$$\underline{J}_2 = S\omega_1\sigma_e \frac{\pi}{\tau}\underline{B}_g \tag{20.100}$$

Finally,

$$F_{nr} \approx -2a_e\tau S\omega_1 a_e d p_1 \frac{\tau^2}{\pi^2}\frac{(\mu_0 J_m)^2 S G_e}{g_e\left(1+S^2G_e^{\,2}\right)} \tag{20.101}$$

The net normal force F_n is

$$F_n = F_{na} + F_{nr} = 2a_e p_1\tau \frac{\mu_0 J_m^{\,2}\tau^2}{\pi^2 g_e^{\,2}\left(1+S^2G_e^{\,2}\right)}\left[1-\left(\frac{g_e\pi}{\tau}\right)^2 S^2G_e^{\,2}\right] \tag{20.102}$$

From (20.102), the net normal force becomes repulsive if

$$SG_e > \frac{\tau}{g_e\pi} \tag{20.103}$$

which is not the case in most low speed LIMs (LIAs) even at zero speed (S = 1).

Core losses in the primary core have been neglected so far but they may be added as in rotary IMs. Anyway, they tend to be relatively smaller as the airgap flux density is only B_{gn} = (0.2-0.45)T, because of the rather large magnetic airgap (air plus conductor sheet thickness).

d. A numerical example

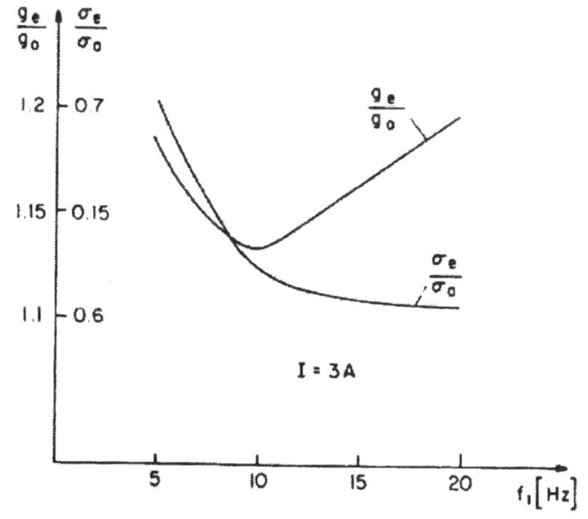

Figure 20.23 Equivalent airgap (g_e/g) and secondary conductivity (σ_e/σ_{Al}) thrust F_x and goodness factor G_e versus frequency at standstill (continued)

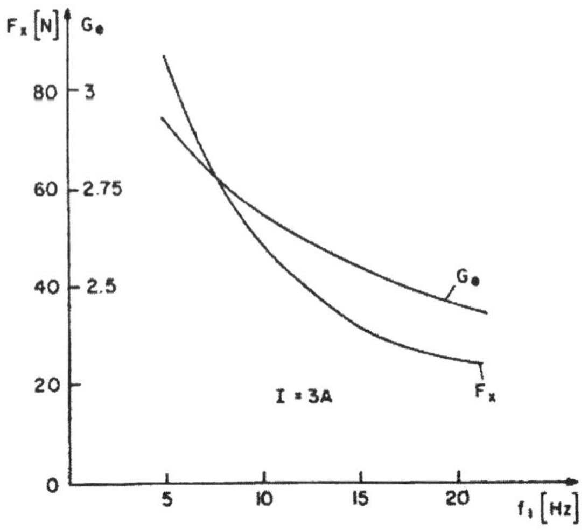

Figure 20.23 (continued)

Let us consider a sheet on iron secondary (single sided) LIA. The initial data are pole pitch τ = 0.084 m, $2a_e$ = 0.08 m, the number of secondary back iron "solid" laminations is i = 3, number of poles $2p_1$ = 6, q = 2, W_1 = 480 turns/phase, g = 0.012 m (out of which d_{Al} = 0.006 m), aluminum plate width 2c = 0.12 m, $\sigma_{Al} = 3.5 \cdot 10^7 (\Omega m)^{-1}$, $\sigma_{iron} = 3.55 \cdot 10^6 (\Omega m)^{-1}$, slot depth h_s = 0.045 m, slot width b_s = 0.009 m, $\beta' = y/\tau = 5/6$ (coil span ratio).

The equivalent conductivity and airgap ratios versus frequency at start (S = 1) and the thrust and goodness factor for given (constant) stator current I_1 = 3A (RMS) are shown in Figure 20.23.

Figure 20.23 summarizes the influence of skin effect, saturation, and eddy currents in the secondary back iron.

As the frequency increases G_e slowly deriorates (decreases) because the skin depth δ_i in iron decreases and so the back iron contribution to the equivalent airgap increases.

The airgap ratio shows a minimum. It is the σ_e/g_e ratio that counts in the goodness factor. For our case, approaching f_1 = 50Hz at start seems good choice.

e. Design methodology by example

In a single sided LIA, the back iron in the secondary is necessary to provide a low magnetic reluctance path for the magnetic flux which crosses the airgap. Usage of a solid back iron is imposed by economic reasons. If a laminated secondary core is to be used to step-up performance, then a ladder secondary in slots seems the adequate choice. This case will be investigated in a separate paragraph.

For the time being, we consider the case of secondary solid back iron. Due to larger equivalent airgap per pole pitch ratio (g_e/τ) both, the efficiency and power factor are not particularly high, they do not seem the best design criteria to start with.

We also need some data from past experience mechanical gap g-d = 1 to 4 mm, airgap flux density B_g = 0.15 to 0.35T, aluminum sheet thickness d = 2 to 4 mm, slot depth/slot width $K_{slot} = h_s/b_s$ = 3 to 6.

The number of poles should be $2p_1$ = 4, 6, 8. The case of $2p_1$ = 2 should be used only in very small thrust applications. The secondary overhangs (c-a) should not surpass τ/π. Let us consider c-a = τ/π.

The basic design criterion is chosen as maximum thrust/given stator current (20.93) which yields

$$S_k G_e = 1 \tag{20.104}$$

This case corresponds approximately to maximum thrust per conductor losses.

Design specifications are considered here to be

- Peak thrust up to rated speed F_{xk} = 1500N
- Rated speed u_n = 6m/s
- Rated frequency f_{1n} = 50Hz
- Rated phase voltage V_1 = 220V (RMS) star connection
- Variable frequency converter supply
- Excursion length L_1 = 50m

The design algorithm unfolds as follows.

The peak thrust F_{xk} (from 20.90 with R'_2 from (20.83)) is

$$F_{xn} = \left(F_{xk}\right)_{S_k G_e = 1} = \frac{3I_1^2 \left(W_1 K_{w1}\right)^2 12 a_e}{d_{Al} \tau p_1 \sigma_e \left(2\tau f_1 - u_n\right)} \frac{1}{2} = 1500\text{N} \tag{20.105}$$

Also from (20.104) with G_e from (20.43)

$$S_k G_e = 1 = 2\mu_0 \frac{\tau^2 f_1}{\pi g_e} S_k d_a \sigma_e \tag{20.106}$$

The rated speed

$$u_n = 2 f_1 \tau \left(1 - S_k\right) \tag{20.107}$$

Also from (20.43) the airgap flux density for $S_k G_e = 1$ is

$$B_{gn} = \left(B_g\right)_{S_k G_e = 1} = \frac{3\sqrt{2} W_1 K_{w1} \mu_0 I_1}{g_e p_1 \pi \sqrt{2}}; \ B_{gn} = 0.3\text{T} \tag{20.108}$$

The secondary back iron tangential flux density B_{xin} should also be limited to a given value, say B_{xin} = 1.7T (even higher values may be accepted). From (20.39) the secondary iron penetration depth required δ_{in} is

$$\delta_{in} = \frac{\tau B_{gn} K_{pf}}{\pi B_{xin}}; \; K_{pf} = 1.6 \qquad (20.109)$$

We may also choose the rated current density j_{con} = 4A/mm^2 (for no cooling), the number of slots per pole and phase q = 2, slot aspect ratio K_{slot} = (3-6) and slot/slot pitch ratio $K_d = 3qb_s/\tau$ = 0.5-0.7, slot fill factor K_{fill} = 0.35-0.45. With these data the phase m.m.f. $W_1 I_{1n}$ is

$$W_1 I_{1n} = C_s p_1 \tau^2; \; C_s = \frac{1}{9q} K_d{}^2 K_{slot} K_{fill} j_{con} \qquad (20.110)$$

$$\text{with } K_d = \frac{3b_s q}{\tau} = 1 - \frac{B_{gn}}{B_{tn}}; \; B_{tn} - \text{primary tooth design flux density} \qquad (20.111)$$

The unknowns of the problem as defined so far are τ, a_e/ τ, p_1, $S_k\omega_1$ and d_{Al} and only an iterative procedure can lead to solutions.

However, by choosing a_e/τ as a parameter the problem simplifies considerably.

From (20.109) and (20.38), we get

$$\tau^2 S_k \omega_1 = \frac{K_{ti}\pi^2}{\mu_0 \sigma_i}\sqrt{\left(\frac{B_{xin}}{B_{gn}}\right)^4 - 4} = c_\tau \qquad (20.112)$$

The thrust F_{xk}, from (20.105) with (20.108) and (20.106), is

$$F_{xn} = 6C_s B_{gn} K_{w1} p_1 \tau^3 \left(\frac{a_e}{\tau}\right) \qquad (20.113)$$

As the primary frequency f_1 is given we may choose also the number of poles as a parameter $2p_1$ = 4, 6 .... In this case from (20.113), the pole pitch τ may be calculated as

$$\tau = \sqrt[3]{\frac{F_{xn}}{6C_s B_{gn} K_{w1} p_1 \left(\frac{a_e}{\tau}\right)}} \qquad (20.114)$$

Now $S_k\omega_1$ is

$$S_k \omega_1 = 2\pi f_1 - u_n \frac{\pi}{\tau} \qquad (20.115)$$

Alternatively, we may have used (20.112) to find $S_k\omega_1$ in which case even f_1 from (20.115) could have been calculated. This would be the case with f_{1n} not specified. Anyway (20.112) is to be checked.

Now from (20.105-20.108)

$$\sigma_e d_{Al} = \frac{3C_s \tau \pi}{S_k \omega_1 B_{gn}} \tag{20.116}$$

From (20.40) the value $\sigma_{Alskin} d_a$ is

$$\sigma_{Alskin} d_{Al} = \left(\frac{3C_s \tau \pi}{S_k \omega_1 B_g} - \frac{\sigma_i \delta_i}{K_{ti}} \right) K_{ta} \tag{20.117}$$

As K_{ta} comes from Equations (20.14), (20.26-20.30), where G_e (which contains $\sigma_{Alskin} d_a$) is included, the computation of $\sigma_{Alskin} d_{Al}$ may be done only iteratively from these equations.

With $\sigma_{Alskin} d_{Al}$ known, again iteratively, d_{Al} may be determined (the skin effect depends on d_{Al}). A valid solution is found if $d_{Al} < g_e$.

With the above procedure the equivalent parameters entering the equivalent circuit, X_m, R_2', R_1, $X_{1\sigma}$ from (20.92-20.88), for rated speed u_n and thrust F_{xn}, may be calculated. The only unknown is now the number of turns per phase W_1

$$W_1 = \frac{V_1}{W_1 I_{1n} \left[\left(K_{R1} + \frac{K_{R2}'}{S_k \left(1 + (1/S_k G_e)^2\right)} \right)^2 + \left(K_{x1\sigma} + \frac{K_m}{\left(1 + (S_k G_e)^2\right)} \right)^2 \right]} \tag{20.118}$$

with $W_1 I_{1n}$ from (20.110), (20.118) provides a unique value for W_1.

To choose the adequate a_e/τ ratio, which will lead to the best design, the cost C_t of both primary (C_p) and secondary (C_{sec}) are evaluated

$$C_t = C_p + C_{sec} \tag{20.119}$$

$$C_p = \frac{6W_1 I_{1n}}{J_{con}} (2a + K_1 \tau) \gamma_{Co} C_{Co} + \left[2p_1 \tau (1 - K_d) \cdot 2aK_{slot} \frac{K_d \tau}{3q} + 3\delta_i 2p_1 \tau \cdot 2a \right] \gamma_i C_{ii} \tag{20.120}$$

where γ_{Co}, γ_i, C_{Co}, C_{ii} are the specific weights and specific prices of copper and core respectively

$$C_{sec} = \left[\left(2a + \frac{2\tau}{\pi} \right) d_{Al} \gamma_{Al} C_{Al} + 3\delta_i 2a \gamma_i C_i \right] L_l \tag{20.121}$$

γ_{Al}, C_{Al} are aluminum specific weight and specific cost, respectively, and L_l is the total LIA secondary length (excursion length + primary length).

When $L_l >> 2p_1\tau$, the secondary cost becomes an important cost factor. Thus a high a_e/τ ratio leads to a smaller length primary but also to a wider (large cost) secondary. The final choice is left to the designer. The apparent power S_1 may be used as an indicator of converter costs.

For our numerical case, the results in Figure 20.24 are obtained. The total cost/iron specific cost (C_t/C_{ii}) has a minimum for $a_e/\tau = 0.7$ but this corresponds to a rather high S_1 (KVA). A high S_1 means a higher cost frequency converter. A good compromise would be $a_e/\tau = 0.9$, $S_k\omega_1 = 93$rad.sec, $\tau = 0.085$m, $d_{Al} = 1.8 \cdot 10^{-3}$m, $W_1 = 384$turns/phase, $I_{1n} = 32.757$A, $S_1 = 21.92$KVA.

Now that the LIAs dimensions are all know, the performance may be calculated iteratively for every (V_1, f_1) pair and speed U.

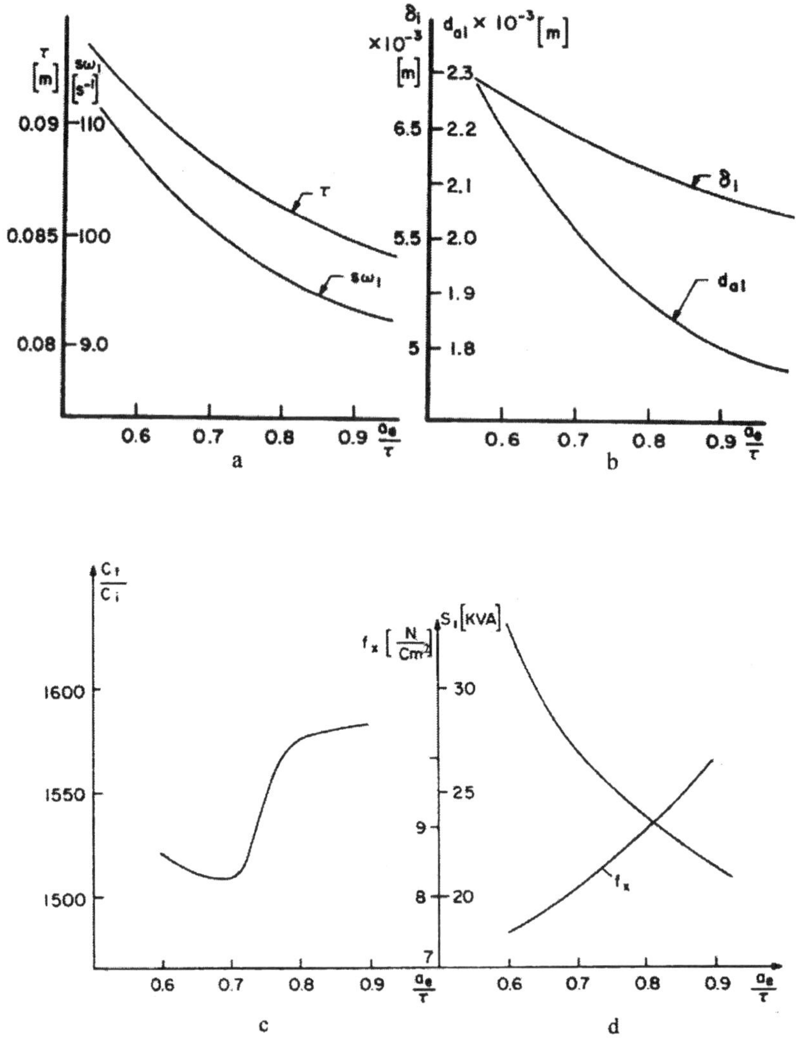

Figure 20.24 Design numerical results
a.) $S_k\omega_1$, b.) iron penetration depth δ_i aluminum thickness d_{Al},
c.) cost versus a_e/τ d.) S_1 (KVA), specific thrust f_x(N/cm^2)

For a double-sided LIA with aluminum sheet secondary the same design procedure may be used, after its drastic simplification due to the absence of solid iron in the secondary. Also the two-sided windings may be lumped into an equivalent one.

f. The ladder secondary

Lowering the airgap to g = 1mm or even less in special applications with LIAs may be feasible. In such cases the primary slots should be semiclosed to reduce Carter's coefficient value and additional losses. The secondary may then be built from a laminated core with, again, semiclosed slots, for same reasons as above (Figure 20.25).

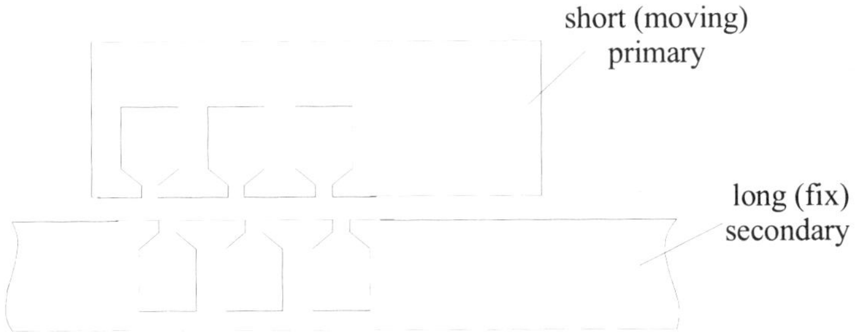

Figure 20.25 LIA with ladder secondary

To calculate the performance, the conventional equivalent circuit is used. The secondary parameter expressions are identical to those derived for the rotary IM. The adequate number of combinations of primary/secondary slots (per primary length) are the same as for rotary IM. In fact, the whole design process is the same as for the rotary IM.

The only difference is that the electromagnetic power is

$$P_{elm} = F_x 2\tau f_1 = \frac{3R_2' I_2'}{S} \quad (20.122)$$

instead of

$$P_{elm} = T_e \frac{\omega_1}{p_1} = \frac{3R_2' I_2'^2}{S} \quad (20.123)$$

for rotary IMs.

For short excursion (less than a few meters) and g ≈ 1mm quite good performance may be obtained with ladder secondary

20.11 TUBULAR LIAs

There are two main tubular LIA configurations one with longitudinal primary laminations stack and the other with ring shape laminations and secondary ring-shape conductors in slots (Figure 20.26).

Due to manufacturing advantages we consider here-except for liquid pump applications-disk shape laminations (Figure 20.26).

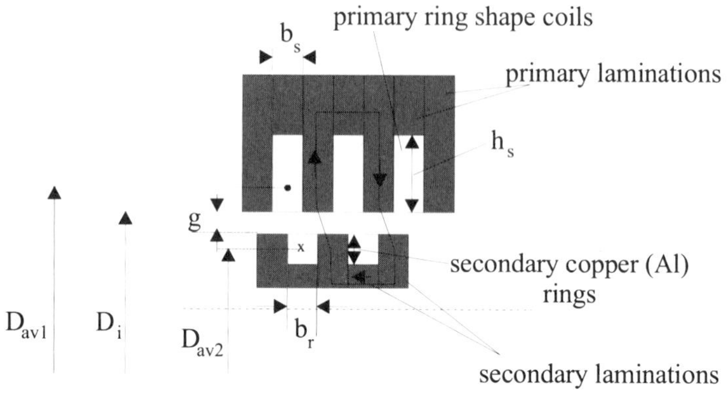

Figure 20.26 Tubular LIA with secondary conductor rings

The main problem of this configuration is that it has open slots on both sides of the airgap.

To limit the airgap flux harmonics (with all their consequences in "parasitic" losses and forces), for an airgap g (g ≈ 1mm) the primary slot width b_s should not be larger than (4-6)g. Fortunately, this suffices for many practical cases.

The secondary ring width b_r should not be larger than (3-4)g. Another secondary effect originates in the fact that the back core flux path goes perpendicular to laminations.

Consequently, the total airgap is increased at least by the insulation total thickness along say ≈ 1/3τ. Fortunately, this increase by, g_{ad}, is not very large

$$g_{ad} = 0.03\frac{1}{3}\tau \approx 0.01\tau \tag{20.124}$$

For a pole pitch τ = 30mm g_{ad} = 0.3-0.4mm, which is acceptable. A very good stacking factor of 0.97 has been assumed in (20.124). Also eddy current core losses will be increased notably in the back iron due to the perpendicular field (Figure 20.27).

Performing slits in the back-core disk shape laminations during stamping leads to a notable reduction of eddy current losses. Finally, the secondary is to be coated with mechanically resilient, nonconducting, nonmagnetic material for allowing the use of linear bearings.

Once these peculiarities are taken care of, the tubular LIM performance computation runs smoothly. The absence of end connections, both in the primary and secondary coils, is a definite advantage of tubular configurations.

So no transverse edge effect occurs in the secondary. The primary back iron is not likely to saturate but the secondary back iron may do so as its area for the half-pole flux is much smaller than in the primary.

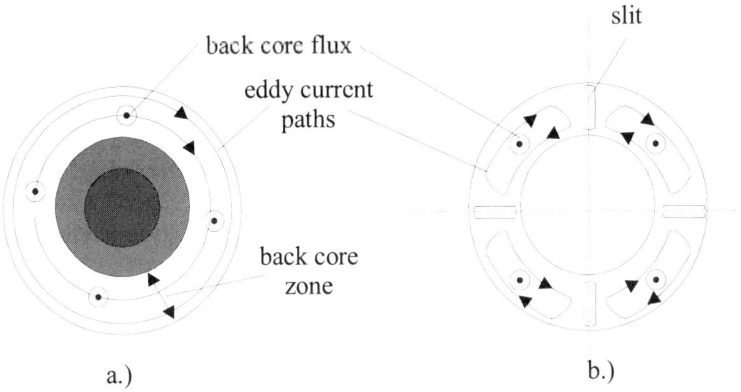

Figure 20.27 Tubular LIA back core disk-shape lamination a.) and its slitting b.) to reduce eddy current losses

The equivalent circuit parameters are

$$R_{1t} = \frac{1}{\sigma_{Co}} \frac{\pi D_{av1} J_{con}}{W_1 I_{1n}} W_1^2 \tag{20.125}$$

$$X_{1t} = 2\mu_0 \omega_1 \frac{\pi D_{av1}}{p_1 q} (\lambda_{s1} + \lambda_{d1}) W_1^2 \tag{20.126}$$

We have just "marked" in (20.125-20.126) that the primary coil average length is πD_{av1}. For the secondary with copper rings

$$X_{2r}' = 2\mu_0 \omega_1 \pi D_{av2} \cdot 12 \frac{(W_1 K_{w1})^2}{N_{s2}} (\lambda_{s2} + \lambda_{d2}) \tag{20.127}$$

$$R_{2r}' = \frac{1}{\sigma_{Co}} \frac{\pi D_{av2}}{A_{ring}} \cdot 12 \frac{(W_1 K_{w1})^2}{N_{s2}} \tag{20.128}$$

λ_s and λ_d are defined in (20.86)-(20.87), N_{s2} is the number of secondary slots along primary length, A_{ring}-copper ring cross-section.

The magnetisation reactance X_m is

$$X_m = \frac{6\mu_0 \omega_1}{\pi^2} (W_1 K_{w1})^2 \frac{\pi D_i \tau}{p_1 g K_c \left(1 + \frac{g_{ad}}{g} + K_{sat}\right)} \tag{20.129}$$

K_{sat}-iron saturation coefficient should be kept below 0.4-0.5.

For an aluminum sheet-secondary on laminated core

$$R_2' = \frac{6\pi(D_i - 2g - d_{Al})(W_1K_{w1})^2}{p_1\tau d_{Al}\sigma_{Al}}; \quad X_2' = 0 \tag{20.130}$$

The equivalent circuit (Figure 20.28) now contains also the secondary leakage reactance which is however small as the slots are open and not very deep.

Based on the equivalent circuit, all performance may be obtained at ease.

The tubular LIA is used for short stroke applications which implicitly leads to maximum speeds below 2-3m/s. Consequently, the design of tubular LIAs, in terms of energy conversion, should aim to produce the maximum thrust at standstill per unit of apparent power if possible.

The primary frequency may be varied (and controlled) through a static power converter, and chosen below the industrial one.

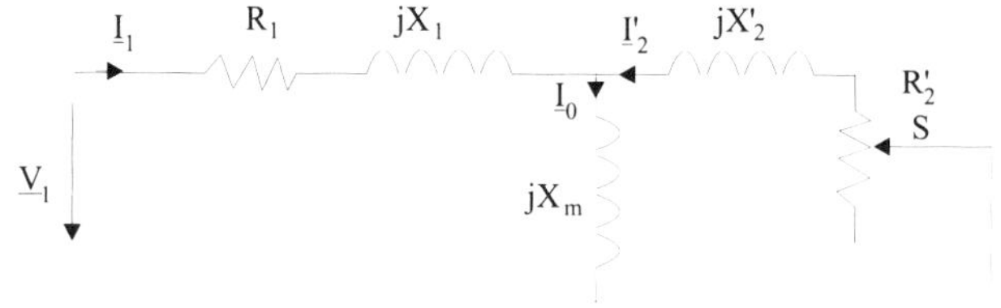

Figure 20.28 Equivalent circuit of tubular LIA with secondary copper rings in slots

In terms of costs both the LIA and power converter costs are to be considered. Hereby, however, we illustrate the criterion of maximum thrust at standstill.

The electromagnetic power P_{elm} is

$$P_{elm} = F_x 2\tau f_1 = 3R_e I_1^2 \tag{20.131}$$

$$R_e = \frac{\frac{R_2'}{S}X_m^2}{\left(\frac{R_2'}{S}\right)^2 + (X_m + X_2')^2} \tag{20.132}$$

or with $G_e = \dfrac{X_m}{R_2'}$

$$R_e = \frac{R_2'}{S\left[\left(\frac{1}{SG_e}\right)^2 + \left(1 + \frac{X_2'}{X_m}\right)^2\right]} \tag{20.133}$$

Now the maximum thrust is obtained for

$$(S_k G_e)_{opt} = \frac{1}{1 + \frac{X_2'}{X_m}} \tag{20.134}$$

For standstill, $S_k = 1$. When initiating the design process, the ratio X_2'/X_m is not known and can only be assigned a value to be adjusted iteratively later on. A good start would be $X_2'/X_m = 0.2$.

From now on, the design methodology developed for the flat LIA may be used here.

a. A numerical example

Let us consider a tubular LIA with copper ring secondary and the data primary bore diameter $D_i = 0.15$ m, primary external diameter $D_e = 0.27$ m, pole pitch $\tau = 0.06$ m, q = 1 slots/pole/phase, primary slot depth $h_s = 0.04$ m, primary slot width $b_s = 0.0125$ m, number of poles $2p_1 = 8$, the secondary slot pitch $\tau_{s2} = 10^{-2}$ m, secondary slot width $b_{s2} = 6\cdot10^{-3}$ m, secondary slot depth $h_{s2} = 4\cdot10^{-3}$ m, airgap $g = 2\cdot10^{-3}$ m, current density $j_{con} = 3\cdot10^6$ A/m^2, slot fill factor $K_{fill} = 0.6$.

Let us determine the frequency f_2 for maximum thrust at standstill and the corresponding thrust, apparent power and conductor losses.

First, with the available data, we may calculate from (20.127) the secondary leakage reactance as a function of $f_1 W_1 K_{w1}$

$$X_2' = 4\mu_0\pi D_{av2}\cdot\frac{12}{N_{s2}}(\lambda_{s2}+\lambda_{d2})f_1 W_1^2 K_{w1}^2 = 1.97\cdot10^{-6} f_1 W_1^2 K_{w1}^2 \quad (20.135)$$

with

$$\lambda_{s2} = \frac{h_{s2}}{3b_{s2}} = \frac{4}{3\cdot6} = 0.33;\ \ \lambda_{d2} = \frac{\frac{5g}{b_{s2}}}{5+\frac{4g}{b_{s2}}} = \frac{\frac{5\cdot3}{6}}{5+\frac{4\cdot3}{6}} = 0.57 \quad (20.136)$$

The secondary resistance (20.128) is

$$R_2' = \frac{1}{\sigma_{Co}}\frac{\pi D_{av2}}{h_{s2}b_{s2}}\cdot12(W_1K_{w1})^2 = 0.876\cdot10^{-4}(W_1K_{w1})^2 \quad (20.137)$$

The magnetic reactance X_m (20.129) becomes

$$X_m = \frac{6\mu_0\omega_1}{\pi^2}(W_1K_{w1})^2\frac{\pi D_i\tau}{p_1 g\left(1+\frac{g_{ad}}{g}+K_{sat}\right)} = 10^{-5} f_1(W_1K_{w1})^2 \quad (20.138)$$

Let assign $\frac{g_{ad}}{g}+K_{sat} = 0.5$.

Now the optimum goodness factor is (from (20.134))

$$G_0 = \frac{X_m}{R_2'} = 0.114 f_{1s} = \frac{1}{1+\frac{X_2'}{R_2'}} = \frac{1}{1+2.24\cdot10^{-2} f_{1s}} \quad (20.139)$$

From (20.139), the primary frequency at standstill f_{1s} is $f_{1s} \approx 7.5Hz$. The primary phase m.m.f. W_1I_{1n} is

$$W_1I_{1n} = pqn_sI_{1n} = p_1q\left(h_sb_sK_{fill}J_{con}\right) = 4\cdot 4\cdot 10^{-2}\cdot 1.2\cdot 10^{-2}\cdot 0.6\cdot 3\cdot 10^{-6} = 3456At \tag{20.140}$$

Hence, the primary phase resistance and leakage reactance R_{1t}, X_{1t} are (20.125-20.126)

$$R_{1t} = \frac{1}{\sigma_{Co}}\frac{\pi D_{av1}J_{con}}{W_1I_{1n}}W_1^2 = 1.035\cdot 10^{-5}W_1^2 \tag{20.141}$$

with $D_{av1} = D_i + h_s = 0.15 + 0.04 = 0.19$ m.

$$X_{1t} = 2\mu_0\omega_1\frac{\pi D_{av1}}{p_1q}\left(\lambda_{s1} + \lambda_{d1}\right)W_1^2 = 2.483\cdot 10^{-5}W_1^2 \tag{20.142}$$

The number of turns W_1 may be determined once the rated current I_{1n} is assigned a value. For $I_{1n} = 20A$, from (20.140), $W_1 = 3456/20 \approx 172$turns/phase.

The apparent power S_1 does not depend on voltage (or on W_1) and is

$$\left(S_1\right)_{S=1} = 3I_1^2\left|R_{1t} + jX_{1t} + \frac{jX_m\left(R_2' + jX_2'\right)}{j\left(X_m + X_2'\right) + R_2'}\right| \tag{20.143}$$

with W_1I_{1n} and R_{1t}, X_{1t}, X_m, R_2', X_2' proportional to W_1^2 and with $K_{w1} = 1.00$ (q = 1) we obtain

$$S_1 \approx 3000VA;\quad \cos\varphi_1 \approx 0.52 \tag{20.144}$$

The corresponding phase voltage at start is

$$V_1 = \frac{S_1}{3I_{1n}} = \frac{3000}{3\cdot 20} = 50.00V(RMS) \tag{20.145}$$

The rated input power $P_1 = S_1\cos\varphi_1 = 3000\cdot 0.52 = 1560W$.

The primary conductor losses p_{cos} are

$$p_{cos} = 3R_1I_{1n}^2 = 3\cdot 0.135\cdot 10^{-4}\left(3456\right)^2 = 370.86W \tag{20.146}$$

The electromagnetic power P_{elm} writes

$$P_{elm} = P_1 - p_{cos} = 1560 - 370.86 = 1189W \tag{20.147}$$

The thrust at standstill F_{xk} is

$$F_{xk} = \frac{P_{elm}}{2\tau f_1} = \frac{1189}{2\cdot 0.06\cdot 7.5} = 1321.26N \tag{20.147'}$$

Thus the peak thrust at standstill per watt is

$$f_{xW} = \frac{F_{xk}}{P_1} = \frac{1321.26}{1560} = 0.847 N/W \quad (20.148)$$

Also

$$f_{xVA} = \frac{F_{xk}}{S_1} = \frac{1321.26}{3000} = 0.44 N/VA \quad (20.149)$$

Such specific thrust values are typical for tubular LIAs.

20.12 SHORT-SECONDARY DOUBLE-SIDED LIAs

In some industrial applications such as special conveyors, the vehicle is provided with an aluminum (light weight) sheet which travels guided between two layer primaries (Figure 20.29).

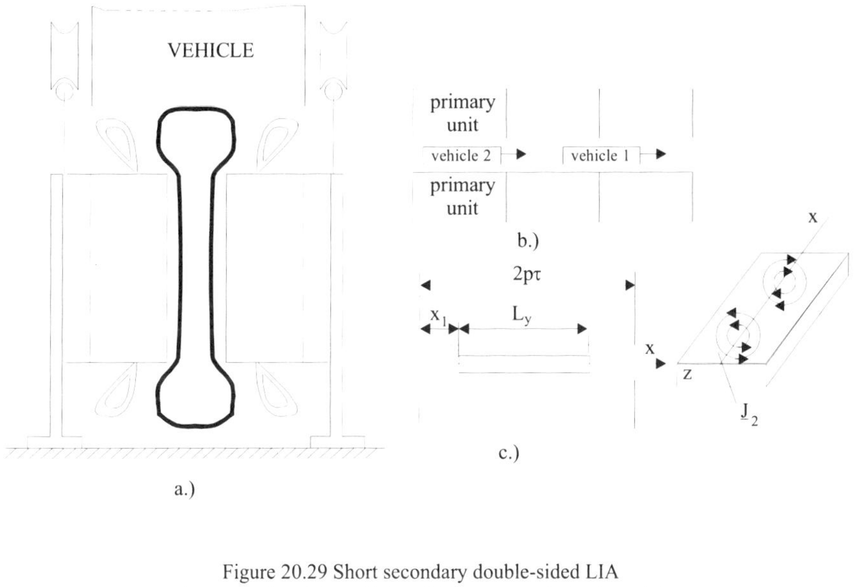

Figure 20.29 Short secondary double-sided LIA
a.) cross section b.) multiple vehicle system c.) single vehicle

The primary units may be switched on and off as required by the presence of a vehicle in the nearby unit.

The fact that we deal with a secondary which is shorter than the primary leads to a redistribution of secondary current density along the direction of motion. Let us consider the simplified case of a single short vehicle inside a single longer primary.

The second current density paths (Figure 20.29c) reveal the transverse edge effect which may be accounted for as in previous paragraphs. Also the skin effect may be considered as done before in the chapter.

The dynamic longitudinal end effect is considered negligible. So the current density in the secondary is simply having the conventional travelling component and a constant which provides for the condition

$$\int_0^{L_v}\left(\underline{J}_{2z}dx\right)=0 \tag{20.150}$$

$\underline{J}_{2z}$ (from (20.76) with $A_t = B_t = 0$) is

$$\underline{J}_{2z}=\frac{-jSG_eJ_me^{-j\frac{\pi}{\tau}x}}{d_{Al}(1+jSG_e)}+A \tag{20.151}$$

From (20.150)-(20.151) $\underline{A}$ is

$$\underline{A}=\frac{-jSG_eJ_m}{d_{Al}(1+jSG_e)}\frac{\tau}{j\pi}\frac{\left(1-e^{j\frac{\pi}{\tau}L_v}\right)}{L_v} \tag{20.152}$$

In fact $\underline{A}$ represents an alternative current density which produces additional losses and a pulsation in thrust. Its frequency is the slip frequency Sf_1, the of entire $\underline{J}_{2z}$.

As expected when $L_v = 2K\tau$, $\underline{A} = 0$.

Care must be exercised when defining the equivalent parameters of the equivalent circuit as part of the primary "works" on empty airgap while only the part with the vehicle is active. 3D FEM approaches seem most adequate to solve this case.

Linear induction pumps are also complex systems characterised by various phenomena. For details see Reference 5, Chapter 5.

20.13 LINEAR INDUCTION MOTORS FOR URBAN TRANSPORTATION

LIMs when longitudinal effects have to be considered are typically applied to people movers on airports (Chicago, Dallas-Fortworth), in metropolitan areas (Toronto, Vancouver, Detroit) or even for interurban transportation. Only single sided configurations have been applied so far.

As in previous paragraphs, we did analyse all special effects in LIMs including the optimal goodness factor definition corresponding to zero longitudinal end effect force at zero slip (Figure 20.20). We will restrict ourselves here to discuss only some design aspects through a rather practical numerical example. More elaborated optimization design methods including some based on FEM are reported in [15-17]. However, they seem to ignore the transverse edge effect and saturation and eddy currents in the secondary back iron.

Specifications

- Peak thrust at standstill $F_{xk} = 12.0kN$

- Rated speed; $U_n = 34$m/s
- Rated thrust, at U_n $F_{xn} = 3.0$kN
- Pole pitch $\tau = 0.25$m (from 0.2 to 0.3m)
- Mechanical gap, $g_m = 10^{-2}$m
- Starting primary current $I_K = 400\text{-}500$A

<u>Data from past experience</u>

- The average airgap flux density $B_{gn} = 0.25\text{-}0.40$T for conductor sheet on solid iron secondary and $B_{gn} = 0.35\text{-}0.45$T for ladder type secondary.
- The primary current sheet fundamental $J_{mk} = 1.5\text{-}2.5 \cdot 10^5$A/m.
- The effective thickness of aluminum (copper) sheet of the secondary $d_{Al} = (4\text{-}6) \cdot 10^{-3}$m.
- The pole pitch $\tau = 0.2\text{-}0.3$m to limit the secondary back iron depth and reduce end connections of coils in the primary winding.
- The frequency at standstill and peak thrust should be larger than $f_{1sc} = 4\text{-}5$Hz to avoid large vibration and noise during starting.
- The primary stack width/pole pitch $2a/\tau = 0.75\text{-}1$. The lower limit is required to reduce the too large influence of end connection losses, and the upper limit to obtain reasonable secondary costs.

<u>Objective functions</u>

Typical objective functions to minimise are

- Inverse efficiency $F_1 = 1/\eta$
- Secondary costs $F_2 = C_{sec}$
- Minimum primary weight $F_3 = G_1$
- Primary KVA $F_4 = S_{1k}$
- Capitalised cost of losses $F_c = (P_{loss})_{av}$

A combination of these objectives may be used to obtain a reasonable compromise between energy conversion performance and investment and loss costs. [15]

<u>Typical constraints</u>

- Primary temperature $T_1 < 120^0$ (with forced cooling)
- Secondary temperature in stations during peak traffic hours T_2
- Core flux densities < (1.7-1.9)T
- Mechanical gap $g_m \geq 10^{-2}$m

Some of the unused objective functions may be taken as constraints.

<u>Typical variables</u>

Integer variables

- Number of poles $2p_1$
- Slots per pole per phase q
- Number of turns per coils n_c ($W_1 = 2p_1qn_c$)

Real variables

- Pole pitch τ
- Airgap g_m
- Aluminum thickness d_{Al}
- Stack width 2a

- Primary slot height h_s
- Primary teeth flux density B_{t1}
- Back iron tangential flux density B_{xin}
- Rated frequency f_1

The analysis model

The analysis model has to make the connection between the variables, the objective functions and the constraints. So far in this chapter we have developed a rather complete analytical model which accounts for all specific phenomena in LIMs such as transverse edge effect, saturation and eddy currents in the secondary back iron, airgap leakage, skin effect in the aluminum sheet of secondary and longitudinal end effect.

So, from this point of view, it would seem reasonable to start with a feasible variable vector and then calculate the objective function(s) and constraints and use a direct search method to change the variables until sufficient convergence of the objective function is obtained (see Hooks–Jeeves method presented in Chapter 18).

To avoid the risk of being trapped in a local minimum, a few starting variable vectors should be tried. Only if the final variable vector and the objective function minimum are the same in all trials, we may say that a global optimum has been obtained. Finding a good starting variable vector is a key factor in such approaches.

On the other hand, Genetic Algorithms are known for being able to reach global optima as they start with a population of variable vectors (chromosomes)– see Chapter 18. Neural network-FEM approaches have been presented in [16].

To find a good initial design, the methodologies developed for rotary IM still hold but the thrust expression includes an additional term to account for longitudinal effect.

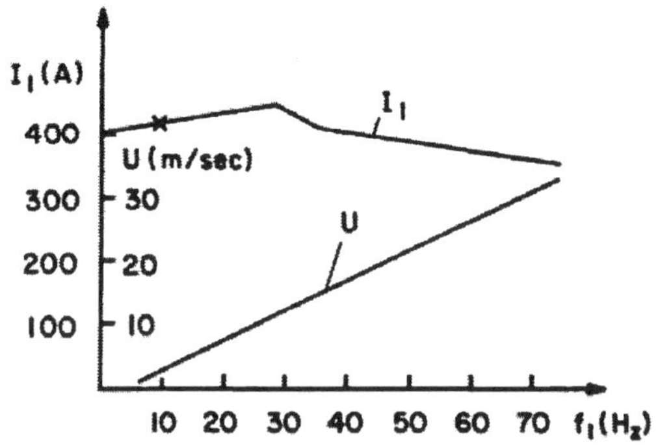

Figure 20.30 Imposed current and speed dependence of frequency

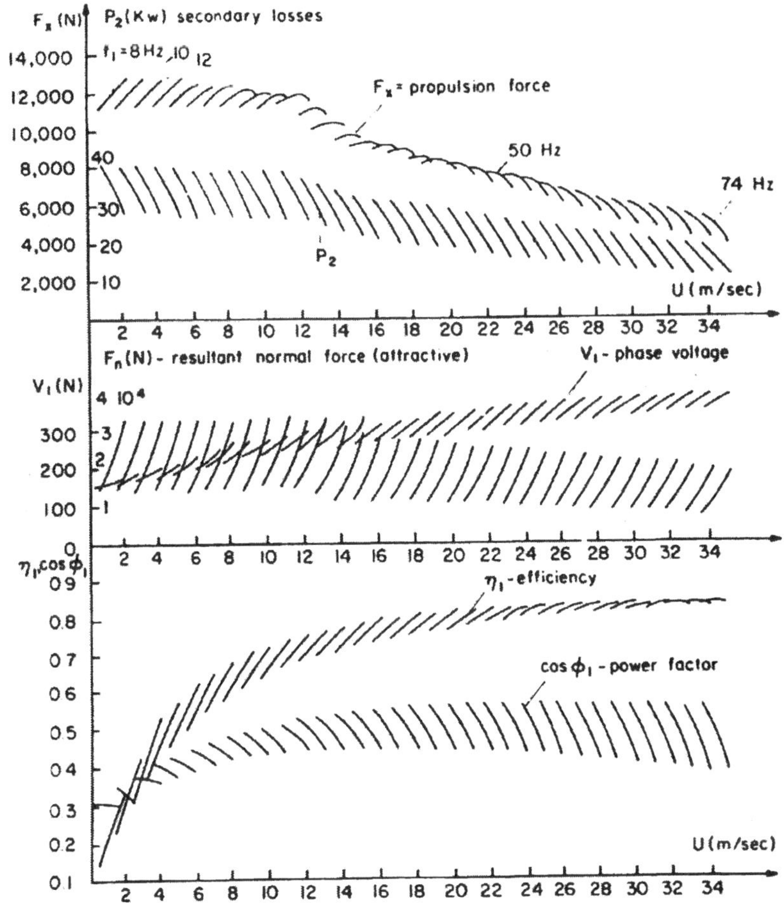

Figure 20.31 Motoring performance

Following the analysis model developed in this chapter, the following results have been obtained for the above initial data $\tau = 0.25$ m, $2p_1 = 10$, $W_1 = 90$, $I_{1sc} = 420$ A (for peak thrust), $f_{1sc} = 6$ Hz, $2a = 0.27$ m, $V_{1sc} = 220$ V (at start), phase voltage at 20 m/s and rated thrust is $V_{1n} = 200$ V, efficiency at rated speed and thrust $\eta_1 = 0.8$, rated power factor $\cos\varphi_1 = 0.55$. Performance characteristics are shown in Figures 20.30-20.32.

<u>Discussion of numerical results</u>

- Above 80% of rated speed (34 m/s) the efficiency is quite high 0.8.
- The power factor remains around 0.5 above 50% of rated speed.
- The thrust versus speed curves satisfy typical urban traction requirements.
- Regenerative braking, provided the energy retrieval is accepted through the static converter, is quite good down to 25% of maximum speed. More

regenerative power can be produced at high speeds by adequate control (full flux in the machine) but at higher voltages in the d.c. link.

- Though not shown, a substantial net attraction force is developed (2-3 times or more the thrust) which may produce noise and vibration if not kept under control.
- These results may be treated as an educated starting point in the optimisation design. Minimum cost of secondary objective function would lead to a not so wide stack (2a < 0.27m) with a lower efficiency in the motor. The motor tends to be larger. The pole pitch may be slightly smaller. [15]

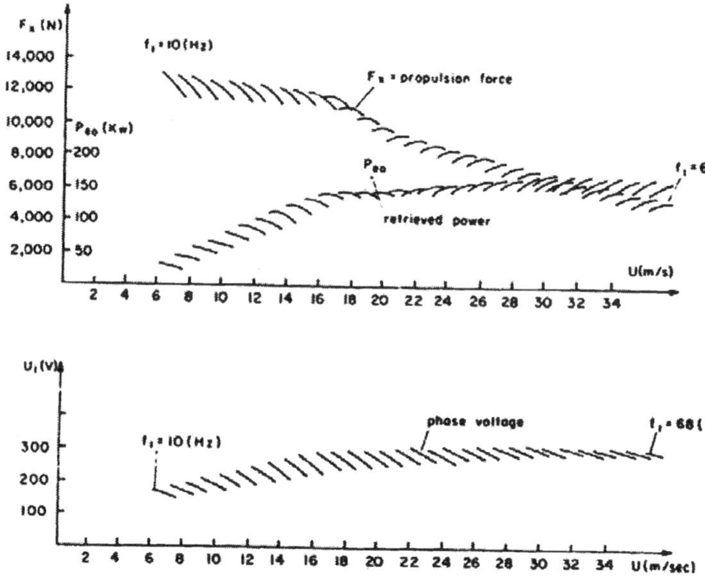

Figure 20.32 Generator braking performance

- The thermal design of secondary, especially around the stop stations, during heavy traffic, is a particular constraint which needs a special treatment. [5, pp. 248 - 250]

20.14 TRANSIENTS AND CONTROL OF LIMs

In the absence of longitudinal end effects, the theory of transients as developed for rotary induction motors may be used. The influence of transverse edge effect and secondary eddy currents may be modelled by an equivalent double cage (Chapter 13). The ladder secondary with broken bars, may need a bar to bar simulation as done for rotary IMs (Chapter 13).

A fictitious ladder [18] (or dq pole by pole [19]) model of transients may be used to account for longitudinal end effects.

Also FEM-circuit coupled models have been introduced recently for the scope. [17] The control of LIMs by static power converters is very similar to that presented in Chapter 8 for rotary IMs. Early control systems in use today apply slip frequency control in PWM-voltage source inverters such as in the UTDC2 system. [20]

Flux control may be used to tame the normal force while still providing for adequate flux and thrust combinations for good efficiency over the entire speed range.

Advanced vector control techniques-such as direct torque and flux control (DTFC) have been proposed for combined levitation-propulsion of a small vehicle for indoor transport in a clean room. [21]

20.15 ELECTROMAGNETIC INDUCTION LAUNCHERS

The principle of electromagnetic induction may be used to launch by repulsion an electrically conducting armature at a high speed by quickly injecting current in a primary coil placed in its vicinity (Figure 20.33).

A multicoil arrangement is also feasible for launching a large weight vehicle.

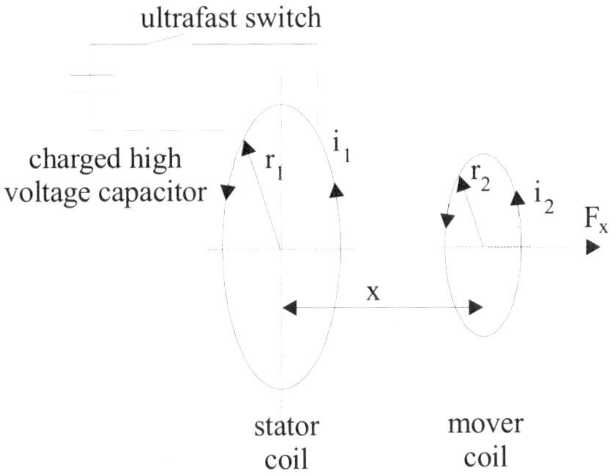

Figure 20.33 Coaxial filamentary coil structure

The principle is rather simple. The "air core" stator coil gets a fast injection of current i_1 from a high voltage capacitor through a very fast high voltage switch. The time varying field created produces an e.m.f. and thus a current i_2 in the secondary (mover) coil(s). The current in the mover coil i_2 decreases slowly enough in time to produce a high repulsion force on the mover coil. That is to launch it.

A complete study of this problem requests a 3D FEM eddy current approach. As this is time consuming, some analytical methods have been introduced for preliminary design purposes.

The main assumption is that the two currents (m.m.f.s) are constant in time during the launching process and equal to each other $W_1I_1 = - W_2 I_2$, as in an ideal short-circuited transformer.

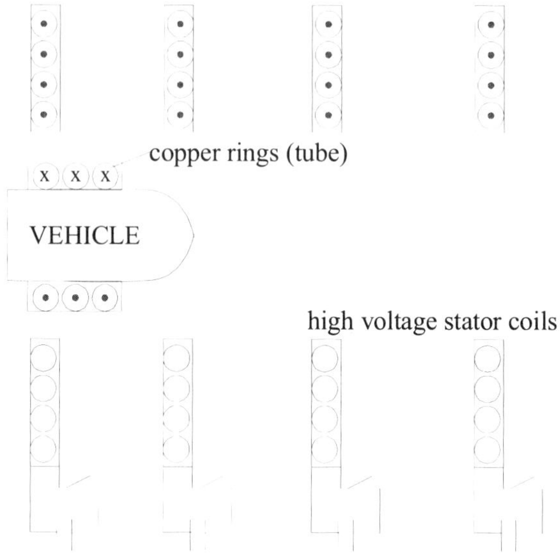

Figure 20.34 Multiple coil structure

In this case, the force F_x in the mover may be calculated from the known formula.

$$F_x = i_1 i_2 \frac{\partial M(x)}{\partial x} \tag{20.153}$$

Now the problem retorts to the computation of the mutual inductance M(x) between stator and mover coil versus position.

For concentric filamentary coils, analytical expressions for M(x) making use of elliptic functions exist. [22] For more involved configurations (Figure 20.34) again FEM may be used to calculate the mutual inductance M and, for given constant currents, the force F_x, even the force along radial direction which is related to motion stability. With known (given) currents the FEM computation effort is reasonably low.

The design of such a system faces two limitations"

- the numerical maximum stress on the stator coils, in the radial direction mainly, σ_s
- the mover coil maximum admitted temperature T_m.

For given stator coil base diameter D_i, mover coil weight m_α (total mover weight is $m(1 + \alpha)$) and barrel length L, the maximum speed of the mover is pretty much determined for given σ_s and T_m constraints. [22]

There is rich literature on this subject (see IEEE Trans. vol. MAG, no.1, 1999, 1997, 1995–the symposium on electromagnetic launch technology (EML)).

The radial force in concentric coils may be used to produce "magnetic" compression of magnetic powders, for example, with higher permeability, to be used in permanent magnetic electric motor with complex geometry.

20.16 SUMMARY

- Linear induction motors (LIMs) develop directly an electromagnetic force along the travelling field motion in the airgap.
- Cutting and unrolling a rotary IM leads to a single sided LIM with a ladder type secondary.
- A three phase winding produces basically a travelling field in the airgap at the speed Us = $2\tau f_1$ (τ-pole pitch, f_1-primary frequency).
- Due to the open character of the magnetic circuit along the travelling field direction, additional currents are induced at entry and exit ends.
- They die out along the active part of LIM producing additional secondary losses, and thrust, power factor and efficiency deterioration. All these are known as longitudinal end effects.
- The longitudinal end effects may be neglected in the so-called low speed LIMs called linear induction actuators.
- When the secondary ladder is replaced by a conducting sheet on solid iron, a lower cost secondary is obtained. This time the secondary current density has longitudinal components under the stack zone. This is called transverse edge effect which leads to an equivalent reduction of conductivity and an apparent increase in the airgap. Also the transverse airgap flux density is nonuniform, with a minimum around the middle position.
- The ratio between the magnetization reactance X_m and secondary equivalent resistance R_2' is called the goodness factor $G_e = X_m/R_2$'. Airgap leakage, skin effect, and transverse edge effects are accounted for in G_e.
- The longitudinal end effect is proven to depend only on G_e, number of poles $2p_1$ and the value of slip S.
- The longitudinal effect introduces a backward and a forward travelling field attenuated wave in the airgap. Only the forward wave dies out slowly and, when this happens along a distance shorter than 10% of primary length, the longitudinal end effect may be neglected.
- Compensation windings to destroy longitudinal effect have not yet proved practical. Instead, designing the LIM with an optimum goodness factor G_0 produces good results. G_0 has been defined such as the longitudinal end effect force at zero slip be zero. G_0 increases rather linearly with the number of poles.
- In designing LIAs for speed short travel applications a good design criterion is to produce maximum thrust per conductor losses (N/W) at zero

speed. This condition leads to $(G_e)_{S=1}$ and it may be met with a rather low primary frequency f_{1sc} = 6-15 Hz for most applications.

- In designing LIA systems the costs of primary + secondary + power converter tend to be a more pressing criterion than energy conversion ratings which are, rather low generally for very low speeds ($\eta_1\cos\varphi_1 < 0.45$).
- The ladder secondary leads to magnetic airgaps of 1 mm or so, for short travels (in the meter range) and thus better energy conversion performance is to be expected ($\eta_1\cos\varphi_1$ up to 0.55).
- Tubular LIMs with disk shape laminations and secondary copper rings in slots are easy to build and produce satisfactory energy conversion ($\eta_1\cos\varphi_1$ up to 0.5) for short travels (below 1m long) at rather good thrust densities (up to 1 N/W and 1.5 N/cm^2).
- LIMs for urban transportation have been in operation for more than a decade. Their design specifications are similar to a constant torque + constant power torque speed envelope rotary IM drive. The peak thrust at standstill is obtained at $G_e = 1$ for a frequency in general larger than (5–6)Hz to avoid large vibrational noise during starting.
  For base speed (continuous thrust at full voltage) and up to a maximum speed (at constant power) energy conversion (η, KVA) or minimum costs of primary and secondary or primary weight objective functions are combined for optimization design. All deterministic and stochastic optimisation methods presented in Chapter 18 for rotary IMs are also suitable for LIMs.
- Electromagnetic induction launchers based on the principle of repulsion force between opposite sign currents in special single and multistator ring-shape stator coils fed from precharged high voltage capacitors and concentric copper rings, may have numerous practical applications. They may be considered as peculiar configurations of air core linear induction actuators. [22, 23] Also they may be used to produce huge compression stresses for various applications.

20.17 SELECTED REFERENCES

1. E. R.Laithwaite, Induction Machines for Special Purposes, Chemical Publish. Comp. New York, 1966.

2. S. Yamamura, Theory of Linear Induction Motors, Wiley Interscience, 1972.

3. S. A.Nasar and I. Boldea, Linear Motion Electric Machines, Wiley Interscience, 1976.

4. M. Poloujadoff, Theory of Linear Induction Machines, Oxford University Press, 1980.

5. I. Boldea, and S. A. Nasar, Linear Motion Electromagnetic Systems, Wiley, 1985.

6. S.A.Nasar and I.Boldea, Linear Electric Motors, Prentice Hall, 1987.

7. I. Boldea and S. A. Nasar, Linear Electric Actuators and Generators, Cambridge University Press, 1997.

8. I.Boldea and S.A.Nasar, Linear Motion Electromagnetic Devices, Gordon & Breach, 2001.

9. H. Bolton, Transverse Edge Effect in Sheet Rotor Induction Motors, Proc. IEE, vol. 116, 1969, pp. 725-739.

10. K. Oberretl, Three Dimensional Analysis of the Linear Motor Taking Into Account Edge Effects and the Distribution of the Windings, Ach. für Electrotechnik, vol.55, 1973, pp. 181-190.

11. I. Boldea and M. Babescu, Multilayer Approach to the Analysis of Single-Sided Linear Induction Motors, Proc. IEE, vol. 125, no. 4, 1978, pp. 283-287.

12. I. Boldea and M. Babescu, Multilayer Theory of DC Linear Brakes with Solid Iron Secondary, IBID, vol. 123, no. 3, 1976, pp. 220-222.

13. I. Boldea, and S. A. Nasar, Simulation of High Speed Linear Induction Motor end Effects in Low Speed Tests", Proc.IEE, vol. 121, no. 9, 1974, pp. 961-964.

14. J. F. Gieras, Linear Induction Drives, O.U.P. Clarendon Press, 1994.

15. T. Higuchi, K. Himeno, S. Nonaka, Multiobjective Optimisation of Single-Sided Linear Induction Motor for Urban Transit, Record of LDIA-95, Nagasaki, pp.45-48.

16. D. Ho Im, S-Ch. Park, Il. Ho Lee, Inverse Design of Linear Induction Motor for Subway Using NN and FEM, IBID, pp. 61-64.

17. D. Ho Im, S-Ch Park, Ki-Bo Jang, Dynamic Characteristic Prediction of Linear Motor Car by NN and FEM, IBID, pp. 65-68.

18. K. Oberretl, Single-Sided Linear Motor With Ladder Secondary, (in German), Arch. für Electrotechnik, vol. 56, 1976, pp. 305-319.

19. T. A. Lipo and T. A. Nondahl, Pole by Pole d-q Model of a LIM", IEEE-PES Winter Meeting, New York, 1978.

20. A. K. Wallace, J. M. Parker, G. E. Dawson, Slip Control for LIM Propelled Transit Vehicle, IEEE Trans. vol. MAG-16, no. 5, 1980, pp. 710-712.

21. I. Takahashi and Y. Ide, Decoupling Control of Thrust and Attractive Force of a LIA-Using Space Vector Controlled Inverter, Record of IEE-IAS, 1990, Annual Meeting.

22. S. Williamson and A.Smith, Pulsed Coilgun Limits, IEEE Trans. vol. MAG-33-vol. 1, 1997, pp. 201-207.

23. A. Musolino, M. Raugi, B. Tellini, 3D Field Analysis in Tubular Induction Launchers With Armature Transverse Motion, IBID vol. 35, no. 1, 1999, pp. 154-159.

24. K. Oberretl, General Harmonic Field Theory for a Three Phase LIM with Ladder Secondary Accounting for Multiple Secondary Reaction and Slot Openings" (in German), Arch. für Electrotechnik, vol. 76, 1993, part I-II, pp. 111-120 and 201-212.

Chapter 21

SUPER-HIGH FREQUENCY MODELS AND BEHAVIOUR OF IMs

21.1 INTRODUCTION

Voltage strikes and restrikes produced during switching operations of induction motors fed from standard power grid may cause severe dielectric stresses on the stator IM windings, leading, eventually, to failure.

In industrial installations high dielectric stresses may occur during second and third pole circuit breaker closure. The second and the third pole closure in electromagnetic power circuit breakers has been shown to occur within 0 to 700 μs [1].

For such situations, electrical machine modelling in the frequency range of a few KHz is required. Steep fronted waves with magnitudes up to 5 p.u. may occur at the machine terminals under certain circuit breaker operating conditions.

On the other hand, PWM voltage source inverters produce steep voltage pulses which are applied repeatedly to induction motor terminals in modern electric drives.

In IGBT inverters, the voltage switching rise times of 0.05–2 μs, in presence of long cables, have been shown to produce strong winding insulation stresses and premature motor bearing failures.

With short rise time IGBTs and power cables longer than a critical l_c, repetitive voltage pulse reflection may occur at motor terminals.

The reflection process depends on the parameters of the feeding cable between motor and inverter, the IGBTs voltage pulse time t_r, and the motor parameters.

The peak line to line terminal overvoltage (V_{pK}) at the receiving end of an initially uncharged transmission line (power cable) subjected to a single PWM pulse with rise time t_r [2] is

$$\begin{aligned}\left(V_{pK}\right)_{l\geq l_c} &= \left(1+\Gamma_m\right)V_{dc}\\ \Gamma_m &= \frac{Z_m - Z_0}{Z_m + Z_0}\end{aligned} \tag{21.1}$$

where the critical cable length l_c corresponds to the situation when the reflected wave is fully developed; V_{dc} is the dc link voltage in the voltage source inverter and Γ_m is the reflection coefficient ($0 < \Gamma_m < 1$).

Z_0 is the power cable and Z_m – the induction motor surge impedance. The distributed nature of a long cable L–C parameters favor voltage pulse reflection, besides inverter short rising time. Full reflection occurs along the power cable if the voltage pulses take longer than one-third the rising time to travel from converter to motor at speed u* ≈ 150–200 m/μs.

The voltage is then doubled and critical length is reached. [2] (Figure 21.1)

The receiving (motor) end may experience $3V_{dc}$ for cable lengths greater than l_c when the transmission line (power cable) has initial trapped charges due to multiple PWM voltage pulses.

Inverter rise times of 0.1–0.2 μs lead to equivalent frequency in the MHz range.

Consequently, investigating super–high frequency modelling of IMs means frequency from kHz to 10MHz or so.

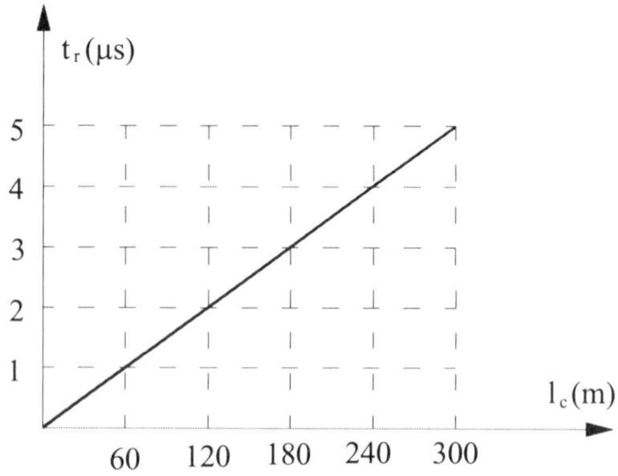

Figure 21.1 Critical power cable length l_c

The effects of such fast voltage pulses on the machine windings include:

- Non-uniform voltage distribution along the windings, with the first 1–2 coils (connected to the terminals) experiencing most of the voltage drop. Especially with random coils, the inter-turn voltage between the first and, say, the last turn, which may be located nearby, may become high enough to produce premature insulation aging.
- The common mode voltage PWM inverter pulses, on the other hand, produce parasitic capacitive currents between stator windings and motor frame and, in parallel, through airgap parasitic capacitance and through bearings, to motor frame. The common mode circuit is closed through the cable capacitances to ground.
- Common mode current may unwarrantly trip the null protection of the motor and damage the bearing by lubricant electrostatic breakdown.

The super-high frequency or surge impedance of the IM may be approached either globally when it is to be identified through direct tests or as a complex distributed parameter (capacitor, inductance, resistance) system, when identification from tests at the motor terminals is in fact not possible.

In such a case, either special tests are performed on a motor with added measurement points inside its electric (magnetic) circuit, or analytical or FEM methods are used to calculate the distributed parameters of the IM.

IM modelling for surge voltages may then be used to conceive methods to attenuate reflected waves and change their distribution within the motor so as to reduce insulation stress and bearing failures.

We will start with global (lumped) equivalent circuits and their estimation and continue with distributed parameter equivalent circuits.

21.2 THREE HIGH FREQUENCY OPERATION IMPEDANCES

When PWM inverter-fed, the IM terminals experience three pulse voltage components.

- Line to line voltages (example: phase A in series with phase B and C in parallel): V_{ab}, V_{bc}, V_{ca}
- Line (phase) to neutral voltages: V_{an}, V_{bn}, V_{cn}
- Common mode voltage V_{oin} (Figure 21.2)

$$V_{oin} = \frac{V_a + V_b + V_c}{3} \tag{21.2}$$

Assume that the zero sequence impedance of IM is Z_0.

The zero sequence voltage and current are then

$$V_0 = \frac{(V_a - V_n) + (V_b - Vn) + (V_c - V_n)}{3} = \frac{V_a + V_b + V_c}{3} - V_n \tag{21.3}$$

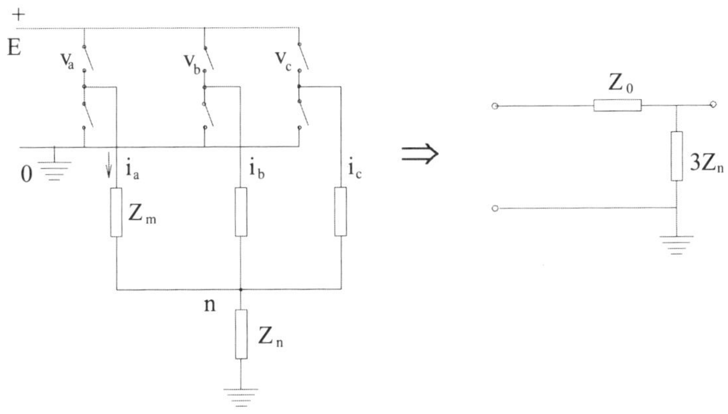

Figure 21.2 PWM inverter with zero sequence impedance of the load (insulated neutral point)

Also, by definition,

$$
\begin{aligned}
V_n &= Z_n I_n \\
V_0 &= Z_0 I_0 \\
I_0 &= \frac{I_a + I_b + I_c}{3} = \frac{I_n}{3}
\end{aligned}
\tag{21.4}
$$

Eliminating V_n and V_0 from (21.3) with (21.4) yields

$$
I_n = \frac{3}{Z_0 + 3Z_n} \cdot \frac{(V_a + V_b + V_c)}{3}
\tag{21.5}
$$

and again from (21.3),

$$
\begin{aligned}
V_{oin} &= \frac{V_a + V_b + V_c}{3} = Z_0 I_0 + V_n \\
V_n &= 3Z_n I_0 = Z_n I_n
\end{aligned}
\tag{21.6}
$$

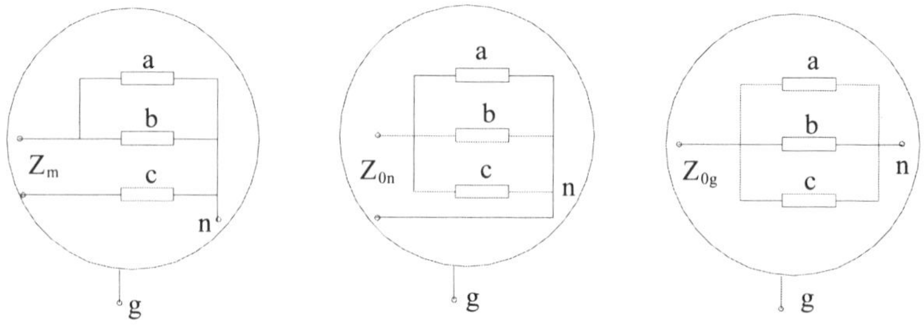

Figure 21.3 Line (Z_m) neutral (Z_{on}) and ground (common mode Z_{og})

This is how the equivalent lumped impedance (Figure 21.2) for the common voltage evolved.

It should be noted that for the differential voltage mode, the currents flow between phases and thus no interference between the differential and common modes occurs.

To measure the lumped IM parameters for the three modes, terminal connections as in Figure 21.3 are made.

Impedances in Figure 21.3, dubbed here as differential (Z_m), zero sequence – neutral – (Z_{on}), and common mode (Z_{og}), may be measured directly by applying single phase ac voltage of various frequencies. Both the amplitude and the phase angle are of interest.

21.3 THE DIFFERENTIAL IMPEDANCE

Typical frequency responses for the differential impedance: Z_m are shown in Figure 21.4. [3]

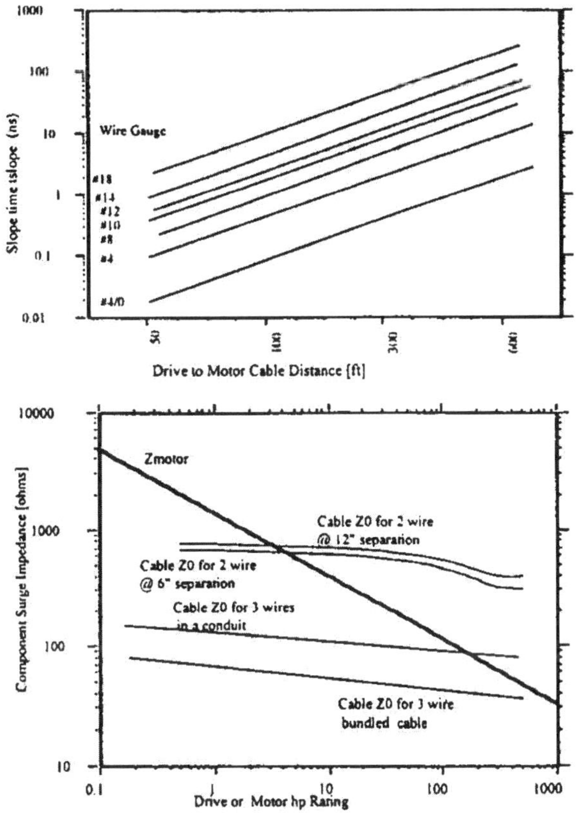

Figure 21.4 Differential mode impedance (Z_m)

Z_m has been determined by measuring the line voltage V_{ac} during the PWM sequence with phase a,b together and c switched from the + d.c. bus to the – d.c. bus. The value Δe is the transient peak voltage V_{ac} above d.c. bus magnitude during t_{rise} and ΔI is phase *c* peak transient phase current.

$$Z_m = \frac{\Delta e}{\Delta I} \tag{21.7}$$

As seen in Figure 21.4, Z_m decreases with IM power. Also it has been found out that Z_m varies from manufacturer to manufacturer, for given power, as much as 5/1 times. When using an RLC analyser (with phases a and b in parallel, connected in series with phase c) the same impedance has been measured by a frequency response test (Figure 21.5). [3]

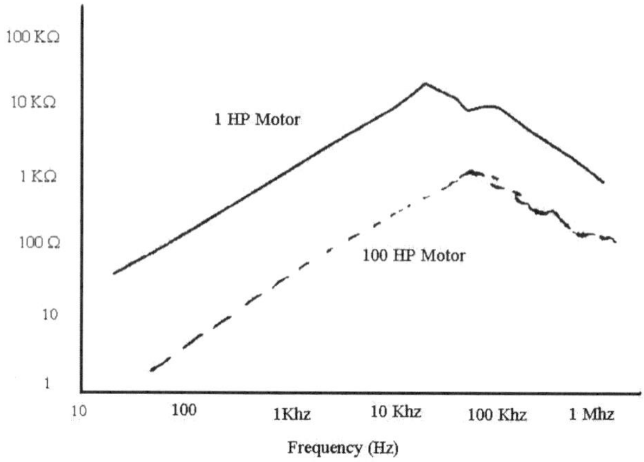

a.)

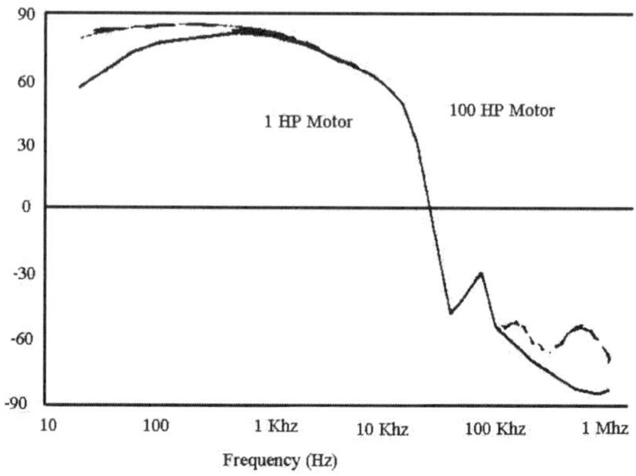

b.)

Figure 21.5 Differential impedance Z_m versus frequency
a.) amplitude b.) phase angle

The phase angle of the differential impedance Z_m approaches a positive maximum between 2 kHz and 3 kHz (Figure 21.4 b). This is due to lamination skin effect which reduces the iron core ac inductance.

At critical core frequency f_{iron}, the field penetration depth equals the lamination thickness d_{lam}.

$$d_{lam} = \sqrt{\frac{1}{\pi f_{iron} \sigma_{iron} \cdot \mu_{rel} \cdot \mu_0}} \tag{21.8}$$

The relative iron permeability μ_{rel} is essentially determined by the fundamental magnetization current in the IM. Above f_{iron} eddy current shielding becomes important and the iron core inductance starts to decrease until it approaches the wire self-inductance and stator aircore leakage inductance at the resonance frequency f_r = 25 kHz (for the 1 HP motor) and 55 kHz for the 100 HP motor.

Beyond f_r, turn to turn and turn to ground capacitances of wire perimeter as well as coil to coil and phase to phase capacitances prevail such that the phase angle approaches now – 90^0 degrees (pure capacitance) around 1 MHz.

So, as expected, with increasing frequency the motor differential impedance switches character from inductive to capacitive. Z_m is important in the computation of reflected wave voltage at motor terminals with the motor fed from a PWM inverter through a power (feeding) cable. Many simplified lumped equivalent circuits have been tried to model the experimentally obtained wide-band frequency response [4,5].

As very high frequency phenomena are confined to the stator slots, due to the screening effect of rotor currents – and to stator end connection conductors, a rather simple line to line circuit motor model may be adopted to predict the line motor voltage surge currents.

The model in [3] is basically a resonant circuit to handle the wide range of frequencies involved (Figure 21.6)

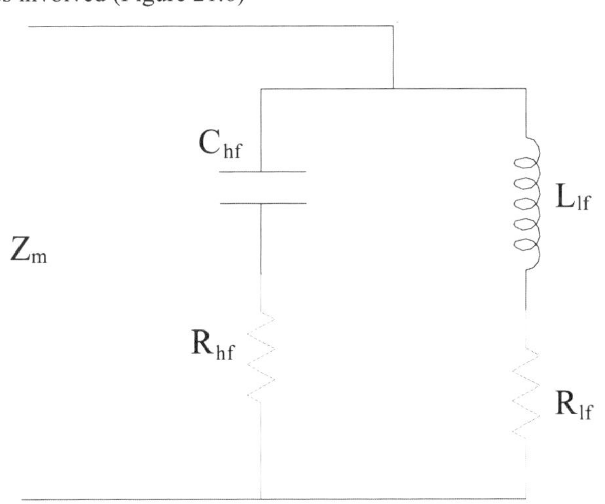

Figure 21.6 Differential mode IM equivalent circuit

C_{hf} and R_{hf} determine the model at high frequencies (above 10kHz in general) while L_{lf}, R_{lf} are responsible for lower frequency modelling. The identification of the model in Figure 21.6 from frequency response may be done through optimization (regression) methods.

Typical values of C_{hf}, R_{hf}, L_{lf}, R_{lf} are given in Table 21.1 after [3].

Table 21.1 Differential mode IM model parameters

	1 kW	10 kW	100 kW
C_{hf}	250pF	800pF	8.5nF
R_{hf}	18Ω	1.3Ω	0.13Ω
R_{lf}	150Ω	300Ω	75Ω
L_{lf}	190mH	80mH	3.15μH

Table 21.1 suggests a rather linear increase of C_{hf} with power and a rather linear decrease of R_{hf}, R_{lf} and L_{lf} with power.

The high frequency resistance R_{hf} is much smaller than the low frequency resistance R_{lf}.

21.4 NEUTRAL AND COMMON MODE IMPEDANCE MODELS

Again we start our inquiries from some frequency response tests for the connections in Figure 21.3 b (for Z_{on}) and in Figure 21.3 c (for Z_{og}) [6]. Sample results are shown in Figure 21.7 a,b.

Many lumped parameter circuits to fit results such as those in Figure 21.7 may be tried.

Such a simplified phase circuit is shown in Figure 21.8 [6].

The impedances Z_{on} and Z_{og} with the three phases in parallel from Figure 21.8 are

$$Z_{on} = \frac{pL_d}{3\left[1 + p\frac{L_d}{R_e} + p^2 L_d \frac{C_g}{2}\right]} \tag{21.9}$$

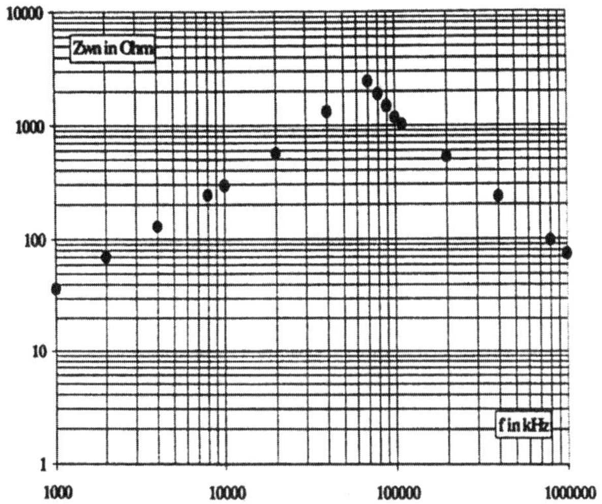

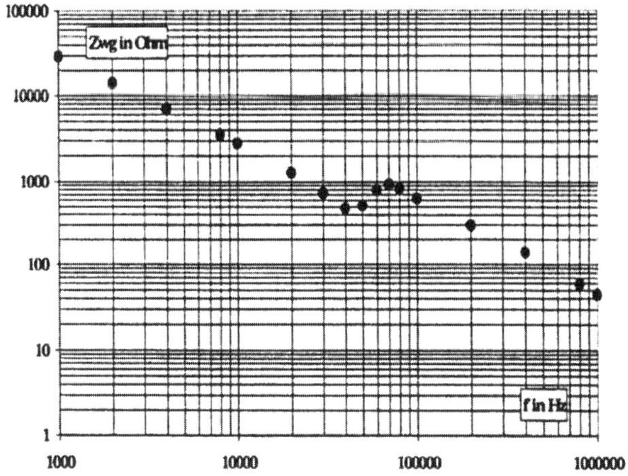

Figure 21.7. Frequency dependence of Z_{on} and Z_{og}

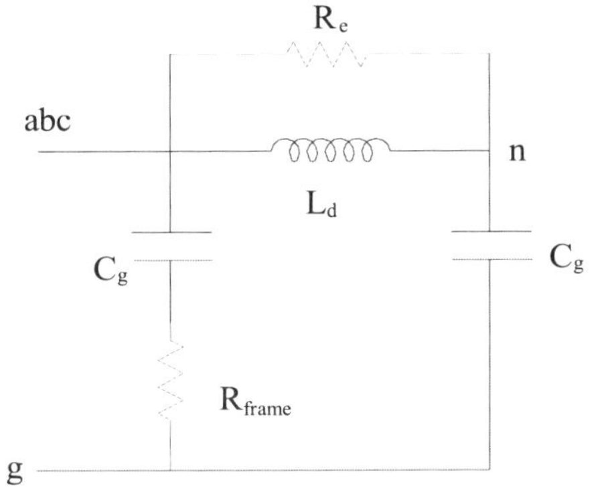

Figure 21.8 Simplified high frequency phase circuit

$$Z_{og} = \frac{1 + p\frac{L_d}{R_e} + p^2 L_d C_g}{6pC_g\left[1 + p\frac{L_d}{R_e} + p^2 L_d \frac{C_g}{2}\right]} \tag{21.10}$$

The poles and zeros are

$$f_p\left(Z_{on}, Z_{og}\right) = \frac{1}{2\pi}\sqrt{\frac{2}{L_d C_g}} \tag{21.11}$$

$$f_z\left(Z_{og}\right) = f_p\left(Z_{on}, Z_{og}\right)/\sqrt{2} \tag{21.12}$$

At low frequency (1 kHz – 10 kHz), Z_{og} is almost purely capacitive.

$$C_g = \frac{1}{6 \times 2\pi f \times (Z_{og})_f} \tag{21.13}$$

With f = 1 kHz, and Z_{og} (1 kHz) from Figure 21.7 b, the capacitance C_g may be determined.

With the pole frequency f_p (Z_{on}, Z_{og}), corresponding to peak Z_{on} value, the resistance R_e is

$$R_e = 3(Z_{on})_{f_p} \tag{21.14}$$

Finally, L_d is obtained from (21.11) with f_p known from Figure 21.7 and C_g calculated from (21.13).

This is a high frequency inductance.

It is thus expected that at low frequency (1 – 10 kHz) fitting of Z_{on} and Z_{og} with the model will not be so good.

Typical values for a 1.1 kW 4 pole, 50 Hz, 220/380 V motor are C_g = 0.25 nF, $L_d(h_f)$ = 28 mH, R_e = 17.5 kΩ.

For 55 kW C_g = 2.17 nF, $L_d(h_f)$ = 0.186 mH, R_e = 0.295 kΩ.

On a logarithmic scale, the variation of C_g and $L_d(h_f)$ with motor power in kW may be approximated [6] to

$$C_g = 0.009 + 0.53\ln(P_n(kW)), \ [nF] \tag{21.15}$$

$$\ln(L_d(h_f)) = 2.36 - 0.1P_n(kW), \ [mH] \tag{21.16}$$

for 220/380 V, 50 Hz, 1 – 55 kW IMs.

To improve the low frequency between 1 kHz and 10 kHz fitting between the equivalent circuit and the measured frequency response, an additional low frequency R_e, L_e branch may be added in parallel to L_d in Figure 21.8.

Alternatively, the dq model may be placed in parallel and thus obtain a general equivalent circuit acceptable for digital simulations at any frequency.

The addition of an eddy current resistance representing the motor frame R_{frame} (Figure 21.8) may also improve the equivalent circuit precision.

Test results with triangular voltage pulses and with a PWM converter and short cable have proven that the rather simple high frequency phase equivalent circuit in Figure 21.8 is reliable.

More elaborated equivalent circuits for both differential and common voltage modes may be adopted for better precision. [5] For their identification however, regression methods have to be used.

The lumped equivalent circuits for high frequency presented here are to be identified from frequency response tests. Their configuration retains a large degree of approximation. They serve only to assess the impact of differential and common mode voltage surges at the induction motor terminals.

The voltage surge paths and their distribution within the IM will be addressed below.

21.5 THE SUPER-HIGH FREQUENCY DISTRIBUTED EQUIVALENT CIRCUIT

When the power grid connected, IMs undergo surge-connection or atmospherical surge voltage pulses. When PWM voltage surge-fed, IMs experience trains of steep front voltage pulses. Their distribution, in the first few microseconds, along the winding coils, is not uniform.

Higher voltage stresses in the first 1–2 terminal coils, and their first turns occur.

This non uniform initial voltage distribution is due to the presence of stray capacitors between turns (coils) and the stator frame.

The complete distributed circuit parameters should contain individual turn to turn and turn to ground capacitances, self – turn, turn-to-turn and coil-to-coil inductances and self-turn, eddy current resistances.

Some of these parameters may be measured through frequency response tests only if the machine is tapped adequately for the purpose. This operation may be practical for a special prototype to check design computed values of such distributed parameters.

The computation process is extremely complex even impossible in terms of turn-to-turn parameters in random wound coil windings.

Even via 3D–FEM, the complete set of distributed circuit parameters valid from 1 kHz to 1 MHz, is not yet feasible.

However, at high frequencies (in the MHz range), corresponding to switching times in the order of tens of a microsecond, the magnetic core acts as a flux screen and thus most of the magnetic flux will be contained in air as leakage flux.

The high frequency eddy currents induced in the rotor core will confine the flux within the stator. In fact the magnetic flux will be non-zero in the stator slot and in the stator coil end connections zone.

Let us suppose that the end connections resistances, inductances, and capacitances between various turns and to the frame can be calculated separately. The skin effect is less important in this zone and the capacitances between the end turns and the frame may be neglected, as their distance to the frame is notably larger in comparison with conductors in slots which are much closer to the slot walls.

So, in fact, the FEM may be applied within a single stator slot, conductor by conductor (Figure 21.9).

The magnetic and electrostatic field is zero outside the stator slot perimeter Γ and non zero inside it.

The turn in the middle of the slot has a lower turn-to-ground capacitance than turns situated closer to the slot wall.

The conductors around a conductor in the middle of the slot act as an eddy current screen between turn and ground (slot wall).

In a random wound coil, positions of the first and of the last turns (n_c) are not known and thus may differ in different slots.

Computation of the inductance and resistance slot matrixes is performed separately within an eddy current FEM package. [7]

The first turn current is set to 1A at a given high frequency while the current is zero in all other conductors. The mutual inductances between the active turn and the others is also calculated.

The computation process is repeated for each of the slot conductors as active.

With I = 1A (peak value), the conductor self inductance L and resistance R are

$$L = \frac{2W_{mag}}{\left(\frac{I_{peak}}{\sqrt{2}}\right)^2}; R = \frac{P}{\left(\frac{I_{peak}}{\sqrt{2}}\right)^2} \tag{21.17}$$

The output of the eddy current FE analysis per slot is an $n_c \times n_c$ impedance matrix.

The magnetic flux lines for $n_c = 55$ turns/coil (slot), $f \approx \frac{1}{t_{rise}}$, from 1 to 10 MHz, with one central active conductor ($I_{peak} = 1A$) (Figure 21.9, [7]) shows that the flux lines are indeed contained within the slot volume.

The eddy current field solver calculates A and Φ in the field equation

$$\nabla x \frac{1}{\mu_r}(\nabla \times A) = (\sigma + j\omega\varepsilon_r)(-j\omega A - \Delta\Phi) \tag{21.18}$$

with:

A(x,y) – the magnetic potential (Wb/m)

Φ(x,y) – the electric scalar potential (V)

μ_r – the magnetic permeability

ω – the angular frequency

σ – the electric conductivity

ε_r – the dielectric permitivity

The capacitance matrix may be calculated by an electrostatic FEM package.

This time each conductor is defined as a 1V (dc) while the others are set to zero volt. The slot walls are defined as a zero potential boundary. The electrostatic field simulator now solves for the electric potential Φ(x,y).

$$\nabla(\varepsilon_r\varepsilon_0\nabla\Phi(x,y)) = -\rho(x,y) \tag{21.19}$$

ρ(x,y) is the electric charge density.

The result is an $n_c \times n_c$ capacitance matrix per slot which contains the turn-to-turn and turn-to-slot wall capacitance of all conductors in slot. The computation process is done n_c times with always a different single conductor as the 1 V(dc) source.

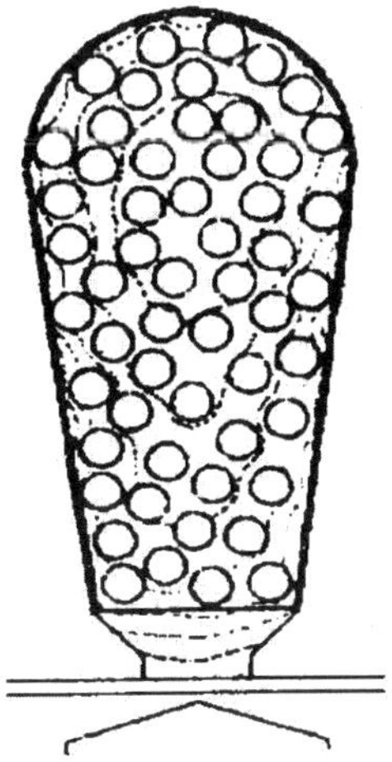

Figure 21.9 Flux distribution in a slot for eddy current FE analysis

The matrix terms C_{ij} are

$$C_{ij} = 2W_{elij} \tag{21.20}$$

W_{elij} is the electric field energy associated to the electric flux lines that connect charges on conductor "i" (active) and j (passive).

As electrostatic analysis is performed the dependence of capacitance on frequency is neglected.

A complete analysis of a motor with 24, 36, 48,...stator slots with each slot represented by $n_c \times n_c$ (for example, 55×55) matrixes would hardly be practical.

A typical line-end coil simulation circuit, with the first 5 individual turns visualised, is shown in Figure 21.10 [7]

To consider extreme possibilities, the line end coil and the last five turns of the first coil are also simulated turn-by-turn. The rest of the turns are simulated by lumped parameters. All the other coils per phase are simulated by lumped parameters. Only the diagonal terms in the impedance and capacitance matrixes are non zero. Saber-simulated voltage drops across the line-end coil and in its first three turns are shown in Figure 21.11 a and b for a 750 V voltage pulse with $t_{rise} = 1$ μs. [7]

The voltage drop along the line-end coil was 280 V for $t_{rise} = 0.2$ μs and only 80 V for $t_{rise} = 1$ μs. Notice that there are 6 coils per phase. Feeder cable tends to lead to 1.2–1.6kV voltage amplitude by wave reflection and thus dangerously high electric stress may occur within the line-end coil of each phase, especially for IGBTs with $t_{rise} < 0.5$ μs, and random wound machines.

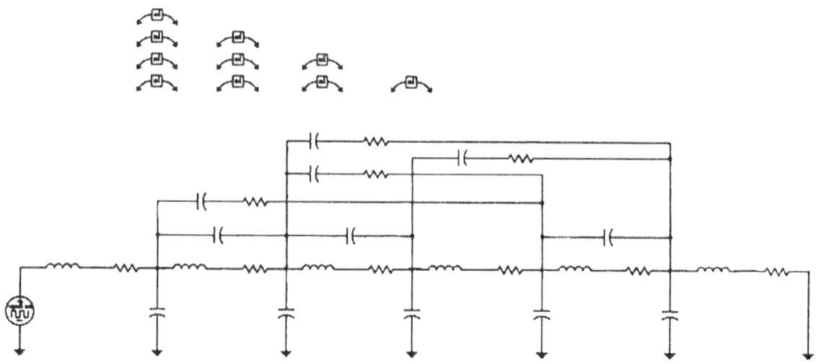

Figure 21.10 Equivalent circuit of the line – end coil

Thorough frequency response measurements with a tapped winding IM have been performed to measure turn–turn and turn to ground distributed parameters. [8]

Further on, the response of windings to PWM input voltages (rise time: 0.24μs) with short and long cables, have been obtained directly and calculated through the distributed electric circuit with measured parameters. Rather satisfactory but not very good agreement has been obtained. [8]

It was found that the line-end coil (or coil 01) takes up 52% and the next one (coil 02) also takes a good part of input peak voltage (42%). [8] This is in contrast to FE analysis results which tend to predict a lower stress on the second coil [7].

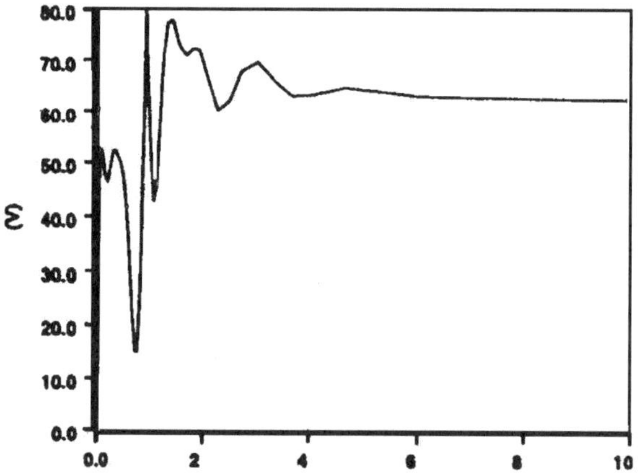

a.)

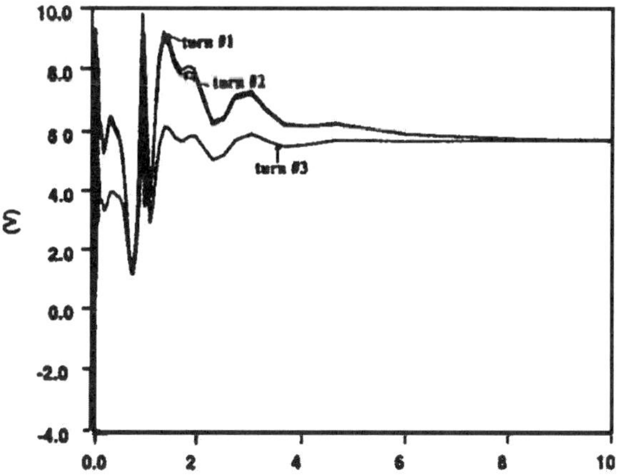

b.)

Figure 21.11 Voltage drop versus time for f_{rise} = 1μs
a) line-end coil; b)Turns 1,2,3 of line- end coil

As expected, long power cables tend to produce higher voltage surges at motor terminals. Consequently, the voltage drop peaks along the line-end coil and its first turn are up to (2–3) times higher. The voltage distribution of PWM voltage surges along the winding coils, especially along its line-end first 2 coils and the line-end first 3 turns thus obtained, is useful to winding insulation design.

Also preformed coils seem more adequate than random coils.

It is recommended that in power-grid fed IMs, the timing between the first, second, and third pole power switch closure be from 0–700μs. Consequently, even if the commutation voltage surge reaches 5 p.u. [9], in contrast to 2–3 pu for PWM inverters, the voltage drops along the first two coils and their first turns is not necessarily higher because the t_{rise} time of commutation voltage surges is much larger than 0.2–1μs. As we already mentioned, the second main effect of voltage surges on the IM is the bearing early failures with PWM inverters. Explaining this phenomenon however requires special lumped parameter circuits.

21.6 BEARING CURRENTS CAUSED BY PWM INVERTERS

Rotor eccentricity, homopolar flux effects or electrostatic discharge are known causes of bearing (shaft) ac currents in power grid fed IMs. [10,11] The high frequency common mode large voltage pulse at IM terminals, when fed from PWM inverters, has been suspected to further increase bearing failure. Examination of bearing failures in PWM inverter-fed IM drives indicates fluting, induced by electrical discharged machining (EDM). Fluting is characterized by pits or transverse grooves in the bearing race which lead to premature bearing wear.

When riding the rotor, the lubricant in the bearing behaves as a capacitance. The common mode voltage may charge the shaft to a voltage that exceeds the lubricant's dielectric field rigidity believed to be around $15V_{peak}/\mu m$. With an average oil film thickness of 0.2 to 2 μm, a threshold shaft voltage of 3 to 30 V_{peak} is sufficient to trigger electrical discharge machining (EDM).

As already shown in Section 21.1, a PWM inverter produces zero sequence besides positive and negative sequence voltages.

These voltages reach the motor terminals through power cables, online reactors, or common mode chokes. [2] These impedances include common mode components as well.

The behavior of the PWM inverter IM system in the common voltage mode is suggested by the three phase schemata in Figure 21.12. [12]

The common mode voltage, originating from the zero sequence PWM inverter source, is distributed between stator and rotor neutral and ground (frame):

C_{sf} – is the stator winding-frame stray capacitor,
C_{sr} – stator – rotor winding stray capacitor (through airgap mainly)
C_{rf} – rotor winding to motor frame stray capacitor
R_b – bearing resistance
C_b – bearing capacitance
Z_l – nonlinear lubricant impedance which produces intermittent shorting of capacitor C_b through bearing film breakdown or contact point.

With the feeding cable represented by a series/parallel impedance Z_s, Z_p, the common mode voltage equivalent circuit may be extracted from 21.12, as shown in Figure 21.13. R_0, L_0 are the zero sequence impedances of IM to the inverter voltages. Calculating C_{sf}, C_{sr}, C_{rf}, R_b, C_b, Z_l is still a formidable task.

Consequently, experimental investigation has been performed to somehow segregate the various couplings performed by C_{sf}, C_{sr}, C_{rf}.

The physical construction to the scope implies adding an insulated bearing support sleeve to the stator for both bearings. Also brushes are mounted on the shaft to measure V_{rg}.

Grounding straps are required to short outer bearing races to the frame to simulate normal (uninsulated) bearing operation. (Figure 21.14) [12]

In region A, the shaft voltage V_{rg} charges to about $20V_{pk}$. At the end of region A, V_{sng} jumps to a higher level causing a pulse in V_{rg}. In that moment, the oil film breaks down at 35 V_{pk} and a 3 A_{pk} bearing current pulse is produced. At high temperatures, when oil film thickness is further reduced, the breakdown voltage (V_{rg}) pulse may be as low as 6–10 volts.

Region B is without bearing current. Here, the bearing is charged and discharged without current.

Region C shows the rotor and bearing (V_{rg} and V_{sng}) charging to a lower voltage level. No EDM occurs this time. $V_{rg} = 0$ with V_{sng} high means that contact asperities are shorting C_b.

The shaft voltage V_{rg}, measured between the rotor brush and the ground, is a strong indicator of EDM potentiality. Test results on Figure 21.15 [2] show V_{rg}, the bearing current I_b and the stator neutral to ground voltage V_{sng}.

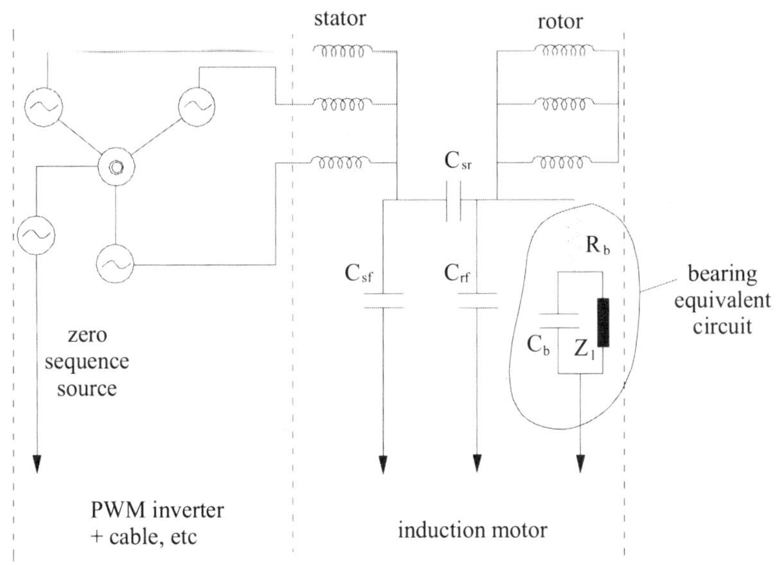

Figure 21.12 Three phase PWM inverter plus IM model for bearing currents

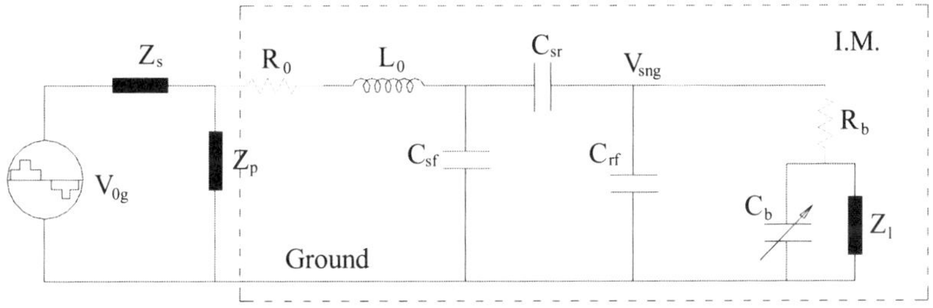

Figure 21.13 Common mode lumped equivalent circuit

An indicator of shaft voltage is the bearing voltage ratio (BVR):

$$\mathrm{BVR} = \frac{V_{rg}}{V_{sng}} = \frac{C_{sr}}{C_{sr} + C_b + C_{rf}} \tag{21.21}$$

With insulated bearings, neglecting the bearing current (if the rotor brush circuits are open and the ground of the motor is connected to the inverter frame), the ground current I_G refers to stator winding to stator frame capacitance C_{sf} (Figure 21.16 a). With an insulated bearing, but with both rotor brushes connected to the inverter frame, the measured current I_{AB} is related to stator winding to rotor coupling (C_{sr}). (Figure 21.16 b)

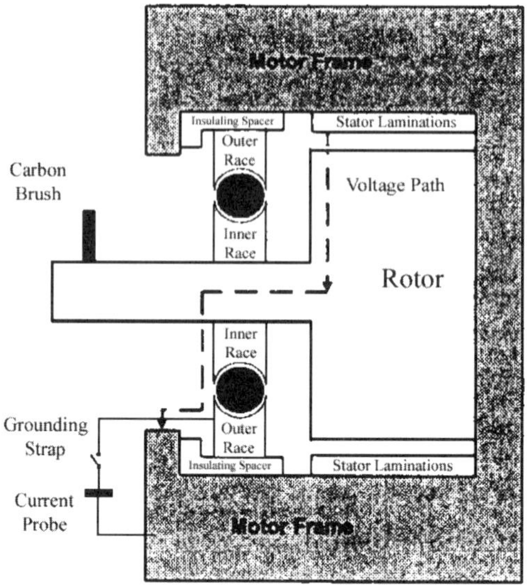

Figure 21.14 The test motor

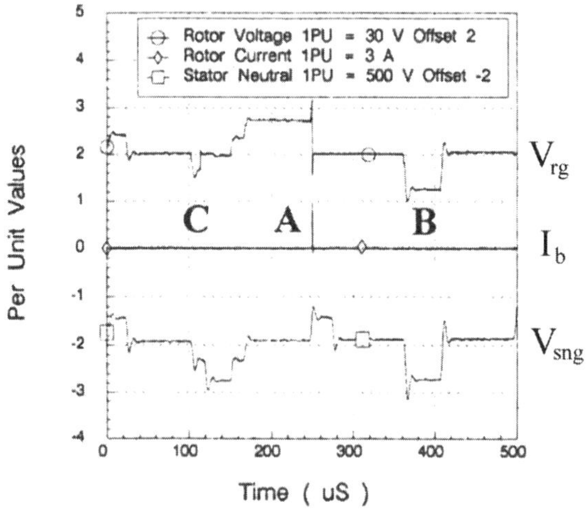

Figure 21.15 Bearing breakdown parameters

In contrast, shortcircuiting the bearing insulation sleeve allows the measurement of initial (uninsulated) bearing current I_b (Figure 21.16 c).

Experiments as those suggested in Figure 21.16 [13] may eventually lead to C_{sr}, C_{sf}, C_{rf}, R_b, C_b identification.

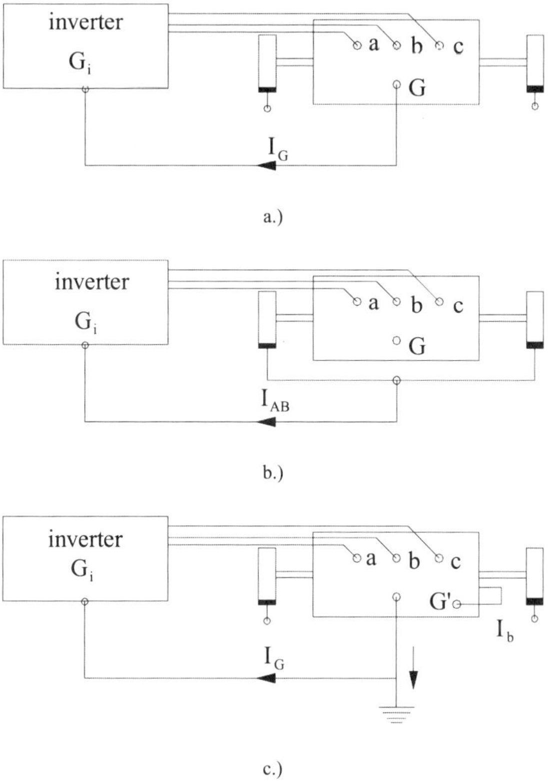

Figure 21.16 Capacitive coupling modes
a.) stator winding to stator
b.) stator winding to rotor
c.) uninsulated bearing current I_b

21.7 WAYS TO REDUCE PWM INVERTER BEARING CURRENTS

Reducing the shaft voltage V_{rg} to less than 1–1.5 V_{pk} is apparently enough to avoid EDM and thus eliminate bearing premature failure.

To do so, bearing currents should be reduced or their path be bypassed by larger capacitance path; that is, increasing C_{sf} or decreasing C_{sr}.

Three main practical procedures have evolved [12] so far

- Properly insulated bearings (C_b decrease)
- Conducting tape on the stator in the airgap (to reduce C_{sr})
- Copper slot stick covers (or paint) and end windings shielded with nomex rings and covered with copper tape and all connected to ground (to reduce C_{sr}) – Figure 21.17 b

Various degrees of shaft voltage attenuation rates (from 50% to 100%) have been achieved with such methods depending on the relative area of the shields. Shaft voltages close to NEMA specifications have been obtained.

The conductive shields do not notably affect the machine temperatures.

Besides the EDM discharge bearing current, a kind of circulating bearing current that flows axially through the stator frame has been identified. [14] Essentially a net high frequency axial flux is produced by the difference in stator coil end currents due to capacitance current leaks along the stack length between conductors in slots and the magnetic core. However, the relative value of this circulating current component proves to be small in comparison with EDM discharge current.

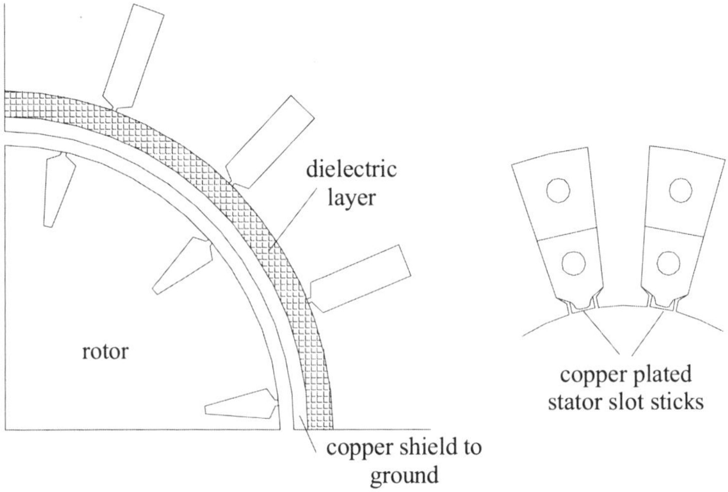

Figure 21.17 Conductive shields to bypass bearing currents (C_{sr} is reduced)

21.8 SUMMARY

- Super-high frequency models for IMs are required to assess the voltage surge effects due to switching operations or PWM inverter-fed operation modes.
- The PWM inverter produces 0.2–2μs rise time voltage pulses in both differential of common modes. These two modes seem independent of each other and, due to multiple reflection with long feeder cables, may reach up to 3 p.u. dc voltage levels.
- The distribution of voltage surges within the IM-repetitive in case of PWM inverters – along the stator windings is not uniform. Most of the voltage drops along the first two line-end coils and particularly along their first 1-3 turns.

- The common mode voltage pulses seem to produce also premature bearing failure through electric discharge machining (EDM). They also produce most of the electromagnetic interference effects.
- To describe the global response of IM to voltage surges, the differential impedance Z_m, the neutral impedance Z_{ng}, and the common voltage (or ground) impedance Z_{og} are defined through pertinent stator winding connections (Figure 21.3) in order to be easily estimated through direct measurements.
- The frequency response tests performed with RLC analyzers suggest simplified lumped parameter equivalent circuits.
- For the differential impedance Z_m, a resonant parallel circuit with high frequency capacitance C_{hf} and eddy current resistance R_{hf} in series is produced. The second R_{lf}, L_{lf} branch refers to lower frequency modelling (1 kHz to 50 kHz or so). Such parameters are shown to vary almost linearly with motor power (Table 21.1). Care must be exercised, however, as there may be notable differences between $Z_m(p)$ from different manufacturers, for given power.
- Th neutral (Z_{ng}) and ground (common voltage) Z_{og} impedances stem from the phase lumped equivalent circuit for high frequencies (Figure 21.8). Its components are derived from frequency responses through some approximations.
- The phase circuit boils down to two ground capacitors (C_g) with a high frequency inductance in between, L_d. C_d and L_d are shown to vary simply with motor power ((21.15)-(21.16))
- To model the voltage surge distribution along the stator windings, a distributed high frequency equivalent circuit is necessary. Such a circuit should visualize turn-to-turn and turn-to-ground capacitances, turn-to-turn inductances and eddy current resistances. The computation of such distributed parameters has been attempted by FEM, for super-high frequencies (1 MHz or so). The electromagnetic and electrostatic field is non zero only in the stator slots and in and around stator coil end connections.
- FEM-derived distributed equivalent circuits have been used to predict voltage surge distribution for voltage surges within 0-2 μs rising time. The first coil takes most of the voltage surge. Its first 1-3 turns particularly so. The percentage of voltage drop on the line-end coil decreases from 70% for 0.1μs rising time voltage pulses to 30% for 1μs rising time. [15]
- Detailed tests with thoroughly tapped windings have shown that as the rising time t_{rise} increases not only the first but also the second coil takes a sizeable portion of voltage surge. [8] Improved discharge resistance (by adding oxides) insulation of magnetic wires [1], together with voltage surge reduction are the main avenues to long insulation life in IMs.

- Bearing currents, due to common mode voltage surges, are deemed responsible for occasional bearing failures in PWM inverter-fed IMs, especially when fed through long power cables.
- The so-called shaft voltage V_{rg} is a good indicator of bearing-failure propensity. In general, V_{rg} values of 15 V_{pk} at room temperature and 6–10 V_{pk} in hot motors are enough to trigger the electrostatic discharge machining (EDM) which produces bearing fluting. In essence, the high voltage pulses reduce the lubricant non linear impedance and thus a large bearing current pulse occurs which further advances the fluting process towards bearing failure.
- Insulated bearings or Faraday shields (in the airgap or on slot taps, covering also the end connections) are practical solutions to avoid large bearing EDM currents. They increase the bearings life.
- Given the complexity and practical importance of super-high frequency behavior of IMs, much progress is expected in the near future, including pertinent international test standards.

21.9 REFERENCES

1. K. J. Cormick and T. R. Thompson, Steep Fronted Switching Voltage Transients and Their Distribution in Motor Windings, Part 1, Proc IEEE, vol. 129, March 1982, pp. 45-55.
2. I. Boldea and S. A. Nasar, Electric Drives, book, CRC Press, Florida, 1998 Chapter 13, pp. 359-370.
3. G. Skibinski, R. Kerman, D. Leggate, J. Pankan, D. Schleger, Reflected Wave Modelling Techniques for PWM AC Motor Drives, Record of IEEE IAS-1998 Annual Meeting, vol. 2, pp 1021-1029.
4. G. Grandi, D. Casadei, A. Massarini, High Frequency Lumped Parameter Model for AC Motor Windings, EPE ’97, pp. 2578-2583.
5. I. Dolezel, J. Skramlik, V.Valough, Parasitic Currents in PWM Voltage Inverter-Fed Asynchronous Motor Drives, EPE ’99, Lausanne, p1-p10.
6. A. Boglietti and E. Carpaneto, An Accurate Induction Motor High Frequency Model for Electromagnetic Compatibility Analysis, EMPS 2001.
7. G. Suresh, H. A. Toliyat, D. A. Rendussara, P. N. Enjeti, Predicting the Transient Effects of PWM Voltage Waveform on the Stator Windings of Random Wound Induction Motors, Record of IEEE-IAS-Annual Meeting, 1997, vol. 1, pp. 135-141.
8. F. H. Al-Ghubari, Annete von Jouanne, A. K. Wallace, The Effects of PWM Inverters on the Winding Voltage Distribution in Induction Motors, EMPS vol. 29, 2001.

9. J. Guardado and K. J. Cornick, Calculation of Machine Winding Parameters at High Frequencies for Switching Transients Study, IEEE Trans, vol. EC-11, no. 1, 1996, pp. 33-40.

10. F. Punga and W. Hess, Bearing currents, Electrotechnick und Maschinenbau, vol. 25, August 1907, pp. 615-618 (in German).

11. M. J. Costello, Shaft Voltage and Rotating Machinery, IEEE Trans, vol. IA-29, no. 2, 1993, pp. 419-426.

12. D. Busse, J. Erdman, R. J. Kerkman, D. Schlegel, G. Skibinski, An Evaluation of the Electrostatic Shielded Induction Motor: A Solution to Rotor Shaft Voltage Buildup and Bearing Current, Record of IEEE-IAS-1996, Annual Meeting, vol. 1, pp. 610-617.

13. S. Chen, T. A. Lipo, D. Fitzgerald, Source of Induction Motors Bearing Caused by PWM Inverters, IEEE Trans. vol. EC-11, no. 1, 1996, pp. 25-32.

14. S. Chen, T. A. Lipo, D. W. Novotny, Circulating Type Motor Bearing Current in Inverter Drives, Record of IEEE-IAS-1996, Annual Meeting, vol. 1, pp. 162-167

15. G. Stone, S. Campbell, S. Tetreault, Inverter-fed drives: which motor stators are at risk ?, IEEE-IA magazine, vol. 6, no. 5, 2000, pp. 17-22

Chapter 22

TESTING OF THREE-PHASE IMs

Experimental investigation or testing of induction machines at the manufacturer's and user's site may be considered an engineering art in itself.

It is also an indispensable tool in research and development of new induction machines in terms of new materials, sizing, topologies or power supply and application requirements.

There are national and international standards on the testing of IMs of low and large power with cage or wound rotor, fed from sinusoidal or PWM converter, and working in various environments.

We mention here the International Electrotechnical Committee (IEC) and the National Electrical Manufacturers Association (NEMA) with their standards on induction machines (IEC–34 series and NEMA MG1–1993 for large IMs).

Temperature, losses and efficiency, starting, unbalanced operation, overload, dielectric, cooling, noise, surge capabilities, and electromagnetic compatibility tests are all standardized.

A description of the standard tests is not considered here as the reader may study the standards for himself; the space required would be too large and the diversity of different standards prescriptions is so pronounced that it may create confusion for a newcomer in the field.

Instead, we decided to present the most widely accepted tests and a few non standardized ones which have recently been promoted with strong international vigor.

They refer to

- Loss segregation/power and temperature based methods
- Load testing/direct and indirect approaches
- Machine parameters estimation methods
- Noise testing methodologies

22.1 LOSS SEGREGATION TESTS

Let us first recall here the loss breakdown (Figure 22.1) in the induction machine as presented in detail in Chapter 11.

For sinusoidal (power grid) supply, the time harmonics are neglected. Their additional losses in the stator and rotor windings and cores are considered zero.

However besides the fundamental, stator core and stator and rotor fundamental winding losses, additional losses occur. There are additional core losses (rotor and stator surface and tooth pulsation losses) due to slotting, slot-openings for different stator and rotor number of slots. Saturation adds new losses. Also, there are space harmonics produced time harmonic rotor-current losses which tend to be smaller in skewed rotors where surface iron additional losses are larger. The lack of insulation between the rotor cage-bars and the rotor core allow for inter-bar currents and additional losses which are not negligible for high frequency rotor current harmonics.

All these nonfundamental losses are called either additional or stray load losses.

In fact, the correct term would be "additional" or "nonfundamental" as they exist, to some extent, even under no mechanical load. They accentuate with load and are, in general, considered proportional with current (or torque) squared.

On top of that, time-harmonics additional losses (Figure 22.1), both in copper and iron, occur with nonsinusoidal voltage (current) power supplies such as PWM power electronics converters.

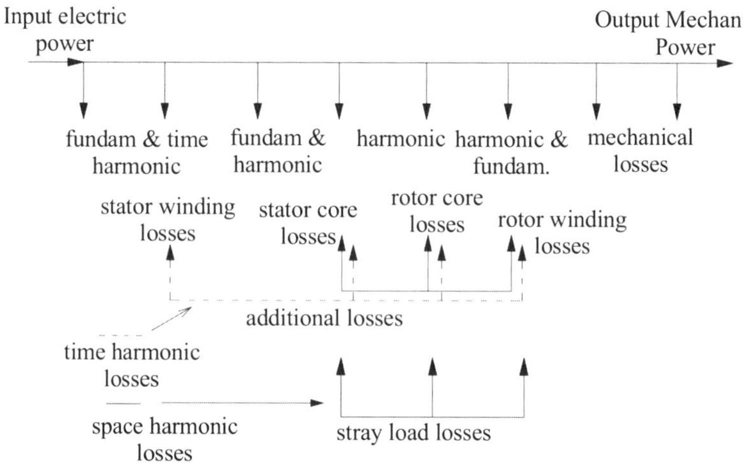

Figure 22.1 Loss breakdown in IMs

As a rather detailed analysis of such a complex loss composition has been done in Chapter 11, here we present only one sequence of testing for loss segregation believed to be coherent and practical. A few additional methods are merely suggested. This line of testing contains only the standard no load test at variable voltage, but extended well above rated voltage, and the stall rotor test.

The no load and shortcircuit (stall rotor) variable-voltage tests are known to allow for the segregation of mechanical losses, p_{mec}, the no load core losses and, respectively, the fundamental copper losses.

The extension of no load test well above rated voltage (as suggested in [1]) is used as a basis to derive an expression for the stray load losses considered proportional to current squared. The same tests are recommended for the PWM power electronics converter IM drives when the inverter is used in all tests.

22.1.1 The no-load test

A variable voltage transformer, with symmetric phase voltages, supplies an induction motor whose rotor is free at shaft. A data acquisition system acquires three currents and three voltages and if available, a power analyzer, the power per each phase.

The method of two wattmeters leads to larger errors as the power factor on no load is low and the total power is obtained by subtracting two large numbers.

It is generally accepted that at no load, up to rated voltage, the loss composition is approximately

$$P_0 = 3R_1 I_{10}^2 + p_{iron} + p_{mec} \tag{22.1}$$

If hysteresis losses are neglected or measured as the jump in input power when the motor on no load is driven through the synchronous speed, the iron losses may be assimilated with eddy current losses which are known to be proportional to flux and frequency squared. This is to say that p_{iron} is proportional to voltage squared:

$$p_{iron} = K_{iron} V_1^2 \tag{22.2}$$

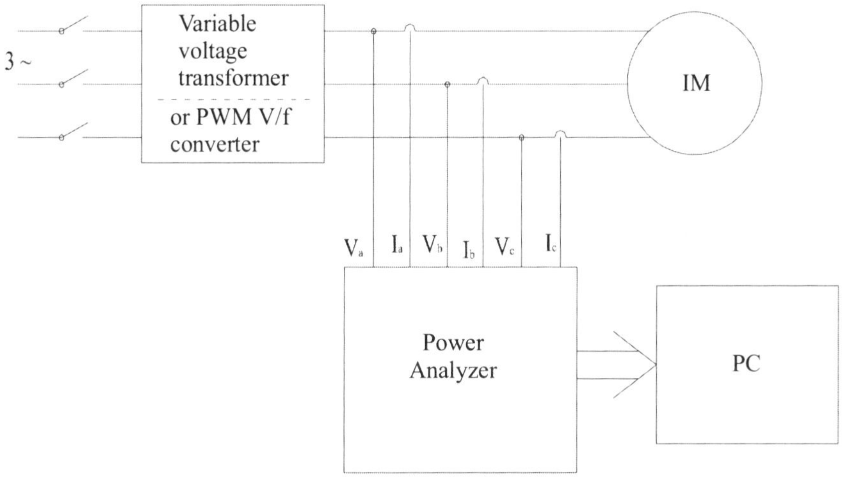

Figure 22.2 No-load test arrangement

When reducing the voltage to 25-30% of rated value, the speed decreases very little so the mechanical losses are independent of voltage V_1. The voltage reduction is stopped when the stator current starts rising.

Consequently,

$$P_0 - 3R_1 I_{10}^2 = K_{iron} V_1^2 + p_{mec} \tag{22.3}$$

The stator resistance may be measured through a d.c. voltage test with two phases in series.

Alternatively, when all six terminals are available, the a.c. test with all phases in series is preferable as the airgap field is very low, so the core loss is negligible (Figure 22.3).

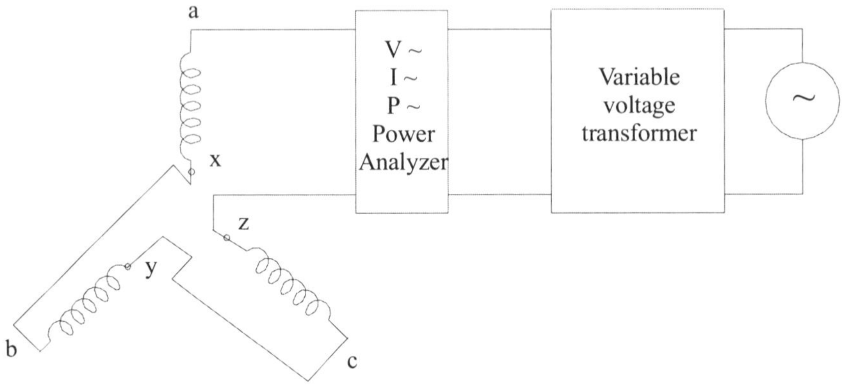

Figure 22.3 AC resistance measurement when six terminals are available

With voltage, current and power measured, the stator resistance R_s and homopolar reactance $X_0^{'}$ (lower or equal to stator leakage reactance) are:

$$R_{1\sim} = \frac{V_\sim}{3I_\sim} \tag{22.4}$$

$$X_0 \le X_{l1} = \sqrt{\left(\frac{V_\sim}{3I_\sim}\right)^2 - R_1^2} \tag{22.5}$$

With full pitch coil windings, $X_0 = X_{l1}$. However, $X_0 < X_{l1}$ for chorded windings. Low voltage is required to avoid over-currents in this test.

For large machines with skin effect in the stator, even at fundamental frequency, $R_{1\sim}$ is required.

The same is valid with IMs fed from PWM power converters. In the latter case, the Variac is replaced by the PWM converter, triggered for two power switches only, with a low modulation index.

The graphical representation of (22.3) is shown on Figure 22.4.

The intercept of the graph on the vertical axis is the mechanical losses.

The fundamental iron losses p^1_{iron} result from the graph in Figure 22.4 for various values of voltage.

We may infer that in most IMs the stator impedance voltage drop is small so the fundamental core loss determined at no load is valid as well for on load conditions.

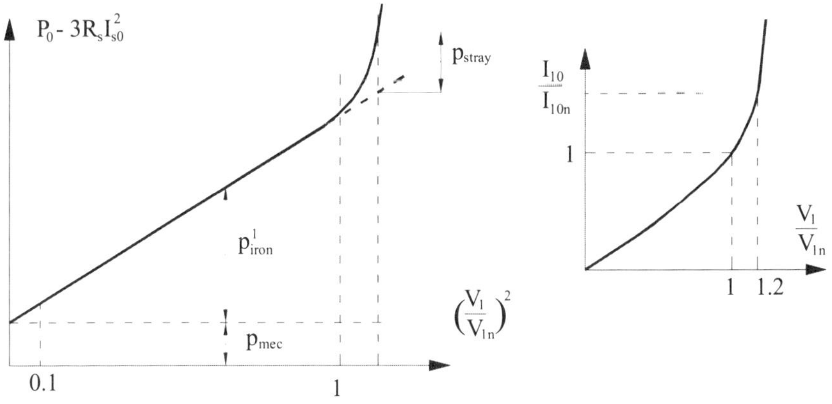

Figure 22.4 No load input (less stator copper loss) versus voltage squared

In reality under load conditions, the value of $p^{1}{}_{iron}$ slightly decreases as the e.m.f. does the same.

22.1.2 Stray losses from no-load overvoltage test

When increasing the voltage over its rated (design) value, Equation (22.3) depart from a straight line (Figure 22.4), at least for small power induction machines. [1]

Apparently in this case, the iron saturates so the third flux harmonic due to saturation causes more losses. This is not so in most IMs as explained in paragraph 5.4.4. However, the stator current increases notably above rated no load current. Space harmonics induced currents in the rotor will increase also. Not so for the fundamental rotor current. So the stator core surface and tooth pulsation losses are not notable. Fortunately, they are not large even under load conditions.

All in all, it is very tempting to use this extention of no load test to find the stray loss coefficient as

$$K_{stray} = \frac{P_{stray}}{3I_{10}^2} \tag{22.6}$$

With the voltage values less than 115-120%, the difference in power on Figure 22.4 is calculated, together with the respective current I_{10} measured. A few readings are done and an average value for the stray load coefficient K_{stray} is obtained.

Test results on three low power IMs (250 W, 550 W, 4.2 kW) have resulted in stray losses at rated current of 2%, 0.9%, and 2.5%. [1] It is recognized that stray load losses, generalizations have to be made with extreme caution.

On load tests are required to verify the practicality of this rather simple method for wide power ranges. It is today recognized that stray load losses are much larger than 0.5% or 1% as stipulated in some national and in IEC standards. [3] Their computation from on load tests, as in IEEE 112B standard, seems a more realistic approach. As the on load testing is rather costly, other simpler methods are considered.

Historically, the reverse rotation test at low voltage and rated current has been considered an acceptable way of segregating stray load losses. [2]

22.1.3 Stray load losses from the reverse rotation test

Basically, the IM is rotated in the opposite direction of its stator travelling field (Figure 22.5). The stator is fed at low voltage through a Variac.

With the speed $n = -f_1/p$, the value of slip $S = 1 - np_1/f_1 = 2$, and thus the frequency of rotor currents is $2f_1$.

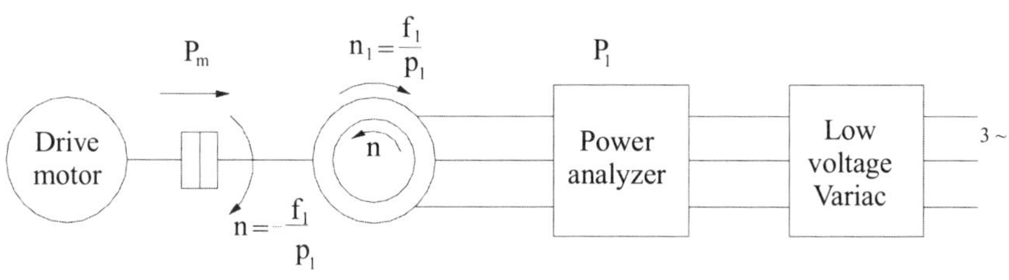

Figure 22.5 Reverse rotation test

The mechanical input will cover the mechanical losses p_{mec}, the rotor stray losses plus the term due to $3I_2^2R_2(1 - S)/S$. For S = 2 this term is $-3I_2^2R_2/2$. Consequently, the difference between the stator electric input P_1 and the mechanical input P_m is

$$P_1 - P_m = 3R_1I_1^2 + P_{iron} + 3\frac{I_2^2R_2}{2} - P_{stray} - \frac{3I_2^2R_2}{2} - P_{mec} \tag{22.7}$$

The rotor winding loss terms cancel in (22.7).

The fundamental iron losses in P_{iron} are different from those for rated power motoring as they tend to be proportional to voltage squared. Also P_{iron} contains the stator flux pulsation losses, which again may be considered to depend on voltage squared.

So,

$$p_{iron} \approx P(_{iron})_{rated} \cdot \left(\frac{V_1}{V_{1n}}\right)^2 \tag{22.8}$$

The method has additional precision problems as detailed in [4] besides the need for a drive with measurable shaft torque (power).

By comparison, the extension of no load test above rated voltage is much more practical. But is it a satisfactory method? Only time will tell as many other attempts to segregate the stray load losses have not yet gotten universal acceptance.

22.1.4 The stall rotor test

Traditionally, the stall rotor (shortcircuit) test is done with a three phase supply and mechanical blockage to stall the rotor. With single phase supply, however, the torque is zero and thus the rotor remains at standstill by itself (Figure 22.6).

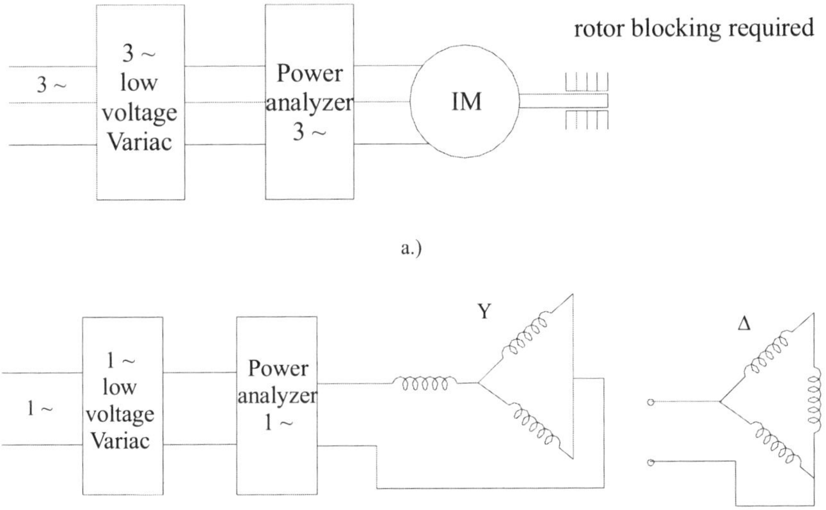

a.)

b.)

Figure 22.6 Stall rotor tests
a.) three phase supply
b.) single phase supply

The tests are done for low voltages until the current reaches its rated value. If the test is done at rated frequency f_{1n}, the skin effect in the rotor is pronounced and thus the rotor resistance is notably larger than in load operation when the slip frequency $Sf_{1n} << f_{1n}$. Only for low power IMs (in the kW range), of general design (moderate starting torque), the assumption of low rotor skin effect is true.

It is true also for all wound rotor IMs.

With voltage, current, and power measured, the stall rotor (shortcircuit) resistance R_{sc} and reactance X_{sc} are found.

We use the superscript s to emphasize that these values have been obtained at stall. The values of R^s_{sc} and X^s_{sc} are less practical than they seem for load conditions. Skin effect and leakage saturation, at high values of stall currents, for rated voltage, lead to different values of R^s_{sc}, X^s_{sc} at stall.

$$R_1 + R_2^s = R_{sc}^s = \frac{P_{sc3}}{3I_{sc3}^2};$$

$$X_{sc}^s = \sqrt{\left(\frac{V_{sc3}}{I_{sc3}}\right)^2 - R_{sc}^2} = X_{l1}^s + X_{l2}^s$$

$$R_1^s + R_2^s = R_{sc}^s = 2\frac{P_{sc1}}{3I_{sc1}^2};$$

$$X_{sc}^s = \sqrt{\frac{2}{3}\left(\frac{V_{sc1}}{I_{sc1}}\right)^2 - R_{sc}^2} = X_{l1}^s + X_{l2}^s \tag{22.9}$$

The shortcircuit test is also supposed to provide for fundamental winding losses computation at rated current,

$$P_{Co} = P_{co1n} + P_{co2n} = 3R_{sc}^s \cdot I_{1n}^2 \tag{22.10}$$

That is, to segregate the winding losses for rated currents. It manages to do so correctly only for IMs with no skin effect at rated frequency in the rotor ($Sf_1 = f_1$) and in wound rotor IMs.

However, if a low frequency low voltage power source is available (even 5% of rated frequency will do), the test will provide correct data of rated copper losses (and R_1+R_2, X_{l1}+X_{l2}) to be used in fundamental winding losses for efficiency calculation attempts.

22.1.5 No-load and stall rotor tests with PWM converter supply

The availability of PWM converters for IM drives raises the question if their use in no – load and stall tests as variable voltage and frequency power sources is not the way of the future.

It is evident that for IMs destined for variable speed applications, with PWM converter supplies, the no load and stall tests are to be performed with PWM converter rather than Variac (transformer) supplies.

As the voltage supplied to the motor has time harmonics, the stator and rotor currents have time harmonics. So, in the first place, even if $S \approx 0$, the time harmonics produced rotor currents are non zero as their slip is $S_\upsilon \approx 1$. The space harmonics produced rotor harmonics currents, existing also with sinusoidal voltage supply, are augmented by the presence of stator voltage time harmonics. They are, however, load independent ($S_\upsilon \approx 1$).

So the loss breakdown with PWM converter supply at no load is

$$P_0' = 3R_1' I_{10}'^2 + 3R_2' I_{20}'^2 + p_{iron}' + p_{mec} \tag{22.11}$$

It has been shown that, despite the fact that the stator and rotor no load currents show time harmonics due to PWM converters, the airgap e.m.f. is quasi-sinusoidal. [5] And so is the magnetization current I_m.

So, performing the no load test for the same fundamental voltage and frequency, once with sinusoidal power source and once with PWM converter, the current relationships are

$$I_{10}^2 = I_m^2$$
$$I_{10}^{'2} \approx I_{20}^{'2} + I_m^2 \tag{22.12}$$

Thus the rotor current under no load I'_{20} for PWM converter supply is

$$I_{20}^{'} = \sqrt{I_{10}^{'2} - I_{10}^2} \tag{22.12'}$$

The stator and rotor resistances R'_1 and R'_2 valid for the given current harmonics spectrum are not known.

However, with the rotor absent, a slightly overestimated value of R'_1 may be obtained. From a stall rotor test at low 5% fundamental frequency, $R'_1 + R'_2$ may be found.

From Equation (22.11), we may represent graphically $p'_{iron}+p'_{mec}$ as a function of fundamental voltage squared for various fundamental frequencies. Finally, we obtain $P_{iron}(f_1,(V_1/V_{1n})^2)$ and $p_{mec}(f_1)$.

In general, for PWM converter supply, both the no load losses and the stall rotor losses are larger than for sinusoidal supply with same fundamental voltage and frequency.

With high switching frequency PWM converters, the time harmonics content of voltage and current is less important and thus smaller no load and stall rotor additional losses occur [5] (Figure 22.7-22.8). For the cage rotor IM stall tests, the difference in losses is negligible, though the current (RMS) is larger (Figure 22.8)

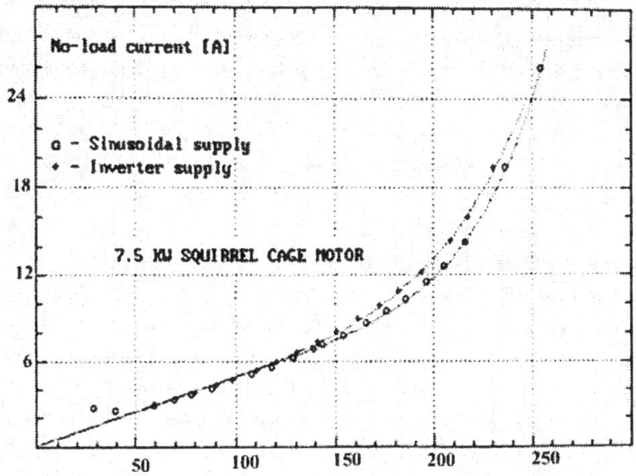

a.)

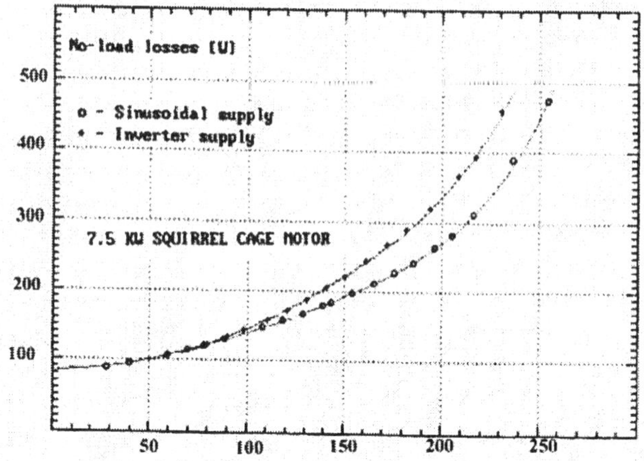

b.)

Figure 22.7 No load testing
a.) current; b.) losses

It may be argued that for given a.c. power grid parameters, in general, PWM converters cannot produce rated voltage fundamental to the motor.

A 3-5% voltage fundamental reduction due to converter limits is accepted. So either correction of all losses, proportional to voltage ratio squared, is applied to PWM converter tests or the power grid is set to provide 5% more than rated voltage fundamental.

As for the extension of the voltage beyond rated voltage by 10-15% for the no load test, to calculate the stray load loss coefficient mentioned in a previous paragraph, a voltage up/down transformer is required. Alternatively, the ratio V_{1n}/f_1 may be increased by decreasing the frequency.

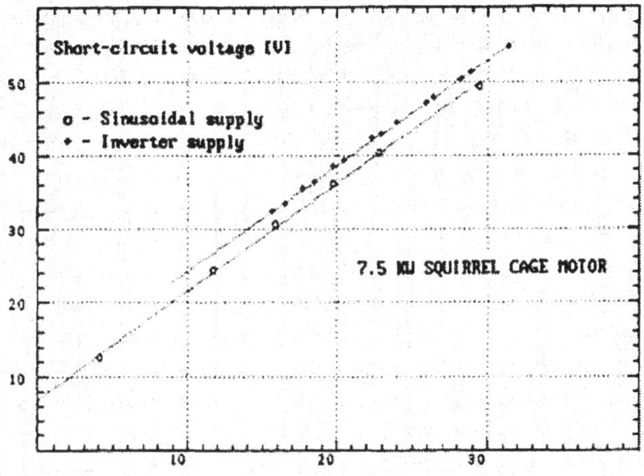

a.)

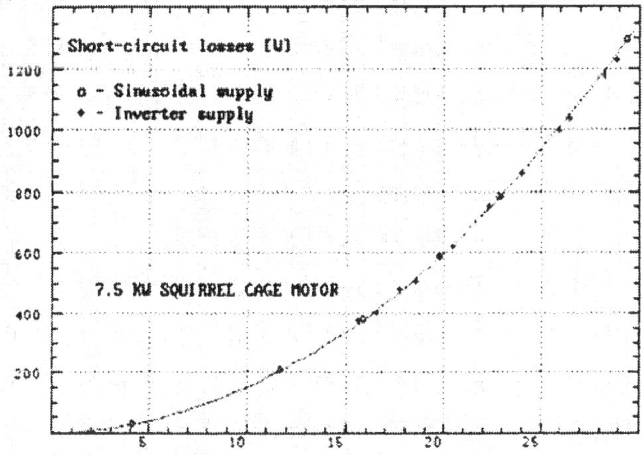

b.)

Figure 22.8 Stall rotor test
a.) current b.) losses

All this being said, it appears that the no load and stall rotor tests for loss segregation may be extended to PWM converter supplied induction motors.

Pertinent data acquisition and processing systems are required with non-sinusoidal voltages and currents to calculate fundamentals and losses.

In parallel with the electrical methods to segregate the losses in IM, presented so far, temperature/time and calorimetric methods have also been developed to determine the various losses in the IM.

Temperature-time methods require numerous temperature sensors to be planted in key points within the IM. If the loss distribution is known and IM is disconnected from the source and kept at constant speed, the temperature/time derivative, at the time of disappearance of the loss source, is

$$\frac{dT}{dt} = -\frac{1}{C} \cdot Q \tag{22.13}$$

Where C is the thermal capacity of the body volume considered and Q – the respective whole losses in that region.

The tests may be done both on load and no load. [6,7] The intrusive character of the method and the requirement of knowing the thermal capacity of various parts of IM body seem to limit the use of this method to prototyping.

22.1.6 Loss measurement by calorimetric methods

The calorimetric method [8] is based on the principle that the temperature rise in a wall insulated chamber is proportional to the losses dissipated inside.

For steady state (thermal equilibrium), the rate of heat transfer by the air coolant, Q, which in fact represents the IM losses, is

$$Q = MC_p\Delta T \tag{22.14}$$

Where M(kg/s) is the mass flow rate of coolant, C_p(J/KgK) – the specific heat of the air; ΔT temperature rise (K) of coolant at exit with respect to entrance. For better precision, the dual chamber calorimetric approach has been introduced. [9], Figure 22.9

In the first chamber, the IM is placed. In the second chamber, a known power heater is located. This way the motor losses Σp_{IM} are

$$\sum p_{IM} = P_{heater} \times \frac{\Delta T_1}{\Delta T_2} \tag{22.15}$$

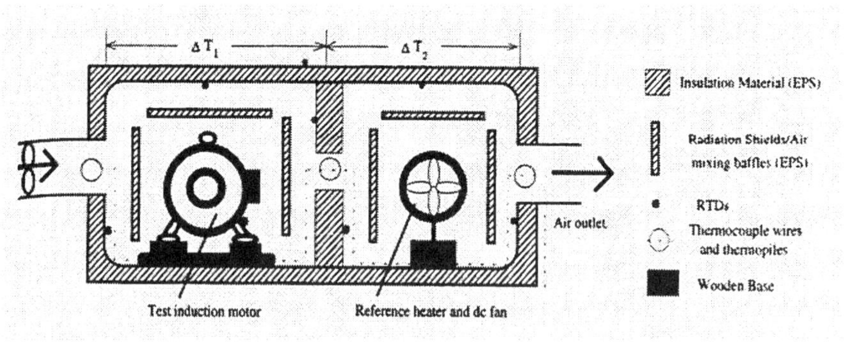

Figure 22.9 The dual chamber calorimeter method

The air properties, not easy to find with variable temperature, are not required. Also the errors of temperature sensors tend to cancel each other, as we need only $\Delta T_1/\ \Delta T_2$.

There is some heat leakage from the calorimeter. This loss may be determined with a precision of ±0.5W. [9] Also, as an order of magnitude, 60-100 litres/sec of air is required for loss measurements of 0.2-1 kW. A proper temperature rise per chamber should be around 10C.

The overall uncertainty of the method is about ±15 W. The friction losses due to stuffing the motor shaft through the chamber walls have to be considered.

For IMs with losses above 150 W, an error in the loss measurements less than 10% is expected. The method works for any load levels and with any kind of IM supply. However, it takes time – until thermal steady state is reached and implies notable costs. For prototyping, however, it seems appropriate.

22.2 EFFICIENCY MEASUREMENTS

Due to its impact on energy costs, efficiency is the single most important parameter index in electric machines.

As induction motors are fabricated and used worldwide, national and international standards for efficiency measurements have been introduced. We mention here some of them which are deemed to be highly representative

- IEEE Standard 112–1996 [10]
- IEC–34–2 and IEC–34–2A [11]
- JEC 37

NEMA MG1–1993 and the Canadian standard C390 correspond to IEEE–112 while most European countries abide by IEC standard. JEC holds in Japan, mainly.

The definition of efficiency η is basically unique

$$\eta = \frac{output\_power}{input\_power} = 1 - \frac{overall\_losses}{input\_power} \tag{22.16}$$

However, the main difference between the above standards involves how the stray load losses are defined and treated as part of overall losses.

Direct measurement of stray losses is obtained from

$$P_{stray} = P_{input} - \left(P_{output} + p_{iron} + 3R_1 I_1^2 + 3R_2 I_2^2 + p_{mec}\right) \tag{22.17}$$

$$\text{with: } I_2 \approx \sqrt{I_1^2 - I_m^2}$$

where: I_m, the magnetization current, is equal to the no load current at the respective voltage.

The various losses in (22.17) are obtained through the loss segregation methods. The input and output powers are measured directly.

Let us now summarize how the efficiency and stray load losses are handled in the IEEE–112 and IEC–34–2 standards.

22.2.1 IEEE Standard 112–1996

This rather complete standard consists of five methods to determine efficiency. They are called methods A, B, C, E and F.

Method A is, in fact, a direct method where the input and output powers are measured directly.

Method B directly measures the stray load losses.

Linear regression is used to reduce measurement errors. The stray load losses are considered proportional to torque squared.

$$\left(P_{stray}\right)_{corrected} = AT_{shaft}^2 \tag{22.18}$$

The correlation coefficient of linear regression has to be larger than 0.9 to secure good measurements.

Method C introduces a back to back (motor/generator) test. The stray load losses are obtained by loss segregation (22.16) but the measured power is, in fact, the difference between the input and output of the two identical machines. So the total losses in the two machines are measured. Consequently, better precision is expected. The total stray load losses thus segregated are divided between the motor and generator considering that they are proportional to current squared.

Method E and E_1 are indirect methods. So the output power is not measured directly.

Table 22.1 Assumed stray load losses/rated output/IEEE–112 method E1

Rated power (kW)	Stray load losses
0.75 – 90	1.8%
91 – 375	1.5%
376 – 1800	1.2%
≥1800	0.9%

In fact, in method E, the stray load losses are measured via the reverse rotation test (see paragraph 22.1.3). In method E_1, the stray load losses are simply assumed at a certain value (Table 22.1).

Methods F and F_1 make use of the equivalent circuit with stray losses directly measured (F) or assigned a certain value (F_1).

22.2.2 IEC standard 34–2

The efficiency is estimated by determining all loss components of the IM.

The losses are to be determined by the loss segregation methods or from the measurement of overall losses

- Direct load test with torque measurement
- Calibrated torque load machine
- Mechanical back to back test as for IEEE–112 C
- Electrical back to back test

The preferred method in the standard is however the segregation of losses with stray load losses having a fix value of 0.5% of rated power. The Japanese standard JEC neglects the stray load losses altogether.

Stray load losses in Table 22.1 (IEEE–112E1) are notably higher than the 0.5% in IEC–34–2 standard and the zero value in JEC standard. Consequently, the same induction motor would be labelled with highest efficiency in the JEC standard, then in IEC–34–2 and finally the lowest, and most realistic in the IEEE–112E1 standard.

22.2.3 Efficiency test comparisons

Typical tests according to the three standards run on the same induction motor of 75 kW are shown in Table 22.2. [12]

A few remarks are in order

- The first test, the direct method, involves direct measurement of input electrical power, shaft torque and speed with a total possible error in efficiency of maximum 1%.
- IEEE–172B tests imply that linear regression is used on the measured stray losses, considered proportional to torque squared.
- For IEC–34–2 the stray load losses are considered 0.5% of rated output and proportional to torque squared.

Table 22.2 Efficiency testing of a 75 kW standard IM according to direct measurements: IEEE-112B, IEC-34-2 and JEC

Load	25%	50%	75%	100%
$U_{average}$ [V]	407.8	404.7	400.2	391.3
$I_{average}$ [A]	71.4	91.2	118.1	149.6
P_{el} [W]	22270	41820	61850	81360
Power Factor [-]	0.44	0.65	0.76	0.80
T [Nm]	182.2	364.0	548.3	723.3
n [rpm]	997	995	994	991
P_{shaft} [W]	19021	37924	57078	75064
slip [%]	0.3	0.5	0.6	0.9
P_{loss} measured directly [W]	3249	3896	4772	6296
$P_{RI^2,1}$ [W]	354	578	968	1553
P_{core} [W]	1774	1728	1660	1527
$P_{RI^2,2}$ [W]	60.4	198	355	705
$P_{w,fr}$ [W]	910	910	910	910
P_{stray} [W]	151	482	879	1601
P_{shaft}/P_{rated} [%]	25.4	50.6	76.1	100.1
Efficiency Direct [%]	85.4	90.7	92.3	92.3
Efficiency using IEEE-112 Linear regression coefficient. Slope: 2.802 10^{-3} Correlation: 0.99				
$P_{stray\ corr.}$ [W]	93	371	843	1466
$P_{loss\ core}$ [W]	3191	3785	4736	6161
$P_{shaft\ core}$ [W]	19079	38035	57114	75199
P_{shaft}/P_{rated} [%]	25.4	50.7	76.2	100.3
Efficiency IEEE – 112 [%]	85.7	90.9	92.3	92.4
Efficiency using IEC 34-2				
P_{stray} [W]	93	151	253	407
P_{loss} [W]	31.91	3565	4146	5102
P_{shaft} [W]	19079	38255	57704	76258
P_{shaft}/P_{rated} [%]	25.4	51.0	76.9	101.7
Efficiency IEC 34 – 2 [%]	85.7	91.5	93.3	93.7
Efficiency using JEC				
P_{stray} [W]	0	0	0	0
P_{loss} [W]	3098	3414	3893	4695
P_{shaft} [W]	19172	38406	57957	76665
P_{shaft}/P_{rated} [%]	25.6	51.2	77.3	102.2
Efficiency JEC [%]	86.1	91.8	93.7	94.2

The main conclusion is that differences in efficiency, with respect to direct measurements of more than 2% may be encountered with the IEC–34–2 and JEC, while IEEE–112B is much closer to reality.

The evident suggestion then is to use the IEEE–112B method as often as possible whenever the direct method is feasible.The problem is that the direct method requires almost one day testing time per motor, notable man power and energy

Providing and mounting a brushless torque-meter with integrated speed sensor both of high precision 1% and, respectively, 1 rpm speed error, is not an easy task as torque (power) increases. Also the power grid rating increases with the power of the tested IM.

22.2.4 The motor/generator slip efficiency method

Difficulties related to the torque-meter acquisition mounting and frequent calibration and excessive energy consumption may be tamed by eliminating the torque-meter and loading the IM with another IM fed from the now available bi-directional PWM voltage-source converter supply used for high performance variable speed drives (Figure 22.10). This method may be seen as an extension of IEEE–112C back to back method. A very precise speed sensor or a slip frequency measuring device is still required.

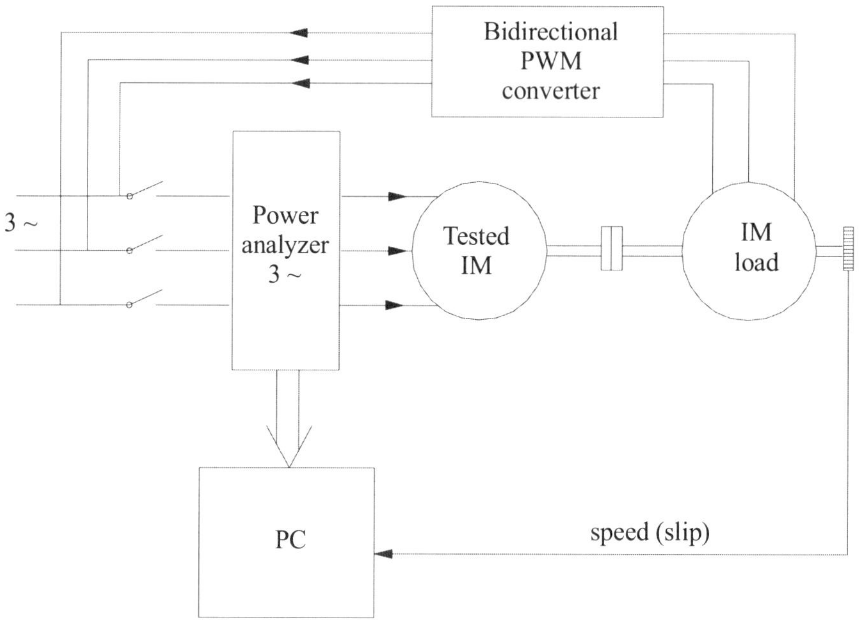

Figure 22.10 Motor/generator slip efficiency method [13]

In principle, the IM is tested for ±0.25, ±0.5, ±0.75, ±1.0, ±1.25, ±1.5 times rated slip speeds as a motor and as a generator, respectively, with the voltages, currents, and input power measured via a power analyzer.

It is admitted that the power level is proportional to slip so the 25%, 50%, 75%, 100%, 125%, 150% loads are approximately obtained. Alternatively, the slip is adjusted to the required power. The power balance equations for the motor and generator modes and equal (or known) slip values are

$$P_{in/m} = 3R_1 I_{1m}^2 + p_{iron/m} + K_{stray} \cdot I_{1m}^2 + 3R_2 \frac{I_{2m}^2}{S_m} \tag{22.19}$$

$$P_{out/g} = 3\frac{R_2 I_{2g}^2}{|S_g|} - K_{stray} I_{1g}^2 - p_{iron/g} - 3R_1 I_{1g}^2 \tag{22.20}$$

For the 25%-150% load range the rotor circuit is highly resistive and thus,

$$I_{2m}^2 = I_{1m}^2 - I_m^2 \tag{22.21}$$

$$I_{2g}^2 = I_{1g}^2 - I_m^2 \tag{22.22}$$

With I_m equal to the no load current at rated voltage, the rotor currents for motor and generator modes, I_{2m} and I_{2g} are obtained.

The core losses are considered with their value at no load and rated voltage, the same for motor and generator,

$$p_{iron/m} = p_{iron/g} \tag{22.23}$$

Now in Equations (22.19)–(22.20) we have only two unknowns: R_2–rotor resistance and K_{stray} (stray load coefficient). As the equations are linear, the solution is straightforward.

It may be argued that while R_2 enters large power components, K_{stray} enters loss(small) power components. While this is true, the direct method again subtracts output from input to compute efficiency. Good instrumentation should deal with this properly.

A few remarks on this method now follow:

- No torque-meter is required
- Motor/generator slip loading is performed
- The motor/generator slip levels may be rather equal but accuracy is not a must
- The core losses and the no load current at rated voltage is required
- The mechanical losses p_{mec}, as determined from variable voltage no load test, are required to calculate the efficiency

$$\eta = 1 - \left(3R_1 I_{1m} + p_{iron/m} + 3K_{stray} I_{1m}^2 + 3R_2 I_{2m}^2 + p_{mec}\right) / P_{in/m} \tag{22.24}$$

Efficiency could also be calculated from generator mode test results with

$$p_{mecg} = p_{mecm} \cdot \left(1 + |S|\right)^2 \tag{22.25}$$

The rotor resistance R_2 at low (slip) frequency may be determined eventually only at 25% rated slip and considered constant over the loading tests at 50%, 75%, 100%, 125%, 150% as the temperature also has to stabilize for each test. This way for all loads except for 25% only the difference between Equations (22.19)–(22.20) will be used. This should enhance precision notably. This estimation of R_2, independent of skin effect, should be preferred to that from stall rotor tests at rated frequency. To cut the testing time, we may acquire the temperature of the stator frame at the time of every load level testing and correct the stator and rotor resistances accordingly. The stator resistance is to be found from a d.c. test at a known temperature.

If time permits, the machine may be stopped quickly, by fast braking it with the load machine which is supplied by a bi-directional power flow converter, for the stator resistance to be d.c. (or miliohmeter) measured.

Test results in [13] proved satisfactory when compared with IEEE–112B tests.

Still the coupling of a loading machine at the IM shaft is required. For vertical shaft IMs as well as for large power IMs, this may be incurring too high costs.

Recently, the mixed frequency method, stemming from the classical two-frequency method [14], has been revived by using a static power converter to supply the IM. [15,16] The key problem in this artificial loading test is the equivalence of losses with direct load testing.

22.2.5 The PWM mixed frequency temperature rise and efficiency tests

The two-frequency test relies on the principle of supplying the IM with voltages V_1, V_2 of two different frequencies f_1 and f_2

$$V(t) = V_1 \cos\omega_1 t + V_2 \cos\omega_2 t \tag{22.26}$$

The rated frequency f_1 differs from f_2 such that the IM speed oscillates very little. Consequently, the IM works as a motor for f_1 and as a generator for f_2. The RMS value of stator current depends on the amplitude and frequency of second voltage: V_2, f_2.

Traditionally, the test uses a synchronous generator of frequency $f_2 \approx (0.8–0.85)f_1$ and $V_2 < V_1$ connected in series with the power grid voltage $V_1(f_1)$ to supply the IM. V_2, f_2 are modified until the RMS value of stator current has the desired value (around or equal to rated value in general).

The test has been traditionally used to determine the temperature rise transients and steady state value, to replicate direct load loss conditions. The exceptional advantage is that no loading machine is attached to the shaft. Consequently, the testing costs and time are drastically reduced, especially for vertical shaft IMs.

However, due to the lack of correct loss equivalence such tests were reported to yield, in general 6–10^{0}C higher temperature than direct load tests for same RMS current.

This is one reason why the mixed frequency traditional tests did not get enough acceptance.

The availability of PWM converters should change the picture to the point that such tests could be applied, at least for IMs destined for variable speed drives, to determine the efficiency. There are quite a few ways to control the PWM converter to obtain artificial loading.

Among them we mention here,

- The accelerating–decelerating method [15]
- Fast primary frequency oscillation method [15]
- PWM dual frequency method [16]

• The accelerating–decelerating method

The IM is supplied from an off the shelf PWM converter – even one with unidirectional power flow will the do. The reference speed is ramped up and down linearly or sinusoidally around the rated frequency synchronous speed so that the IM with free shaft can follow the path.

The speed variation range $n_{min} - n_{max}$ and its time variation:

$$\frac{dn}{dt} \approx \frac{n_{max} - n_{min}}{T} \tag{22.27}$$

are used as parameters to try to provide loss equivalence with the direct load test.

Common sense indicates that for equivalent stator copper loss, the same RMS current as in direct load testing is required.

As the speed has to go above rated (base) speed, the inverter voltage ceiling is causing constant voltage to be applied above rated speed with only frequency as a variable.

As $(n_{max} - n_{min})/n_{rated} \approx 0.2 - 0.4$, the mechanical losses vary. As they are, in general, proportional to speed squared, their equivalencing should not be very difficult.

$$\left(P_{mec}\right)_{rated} = K_m n_n^2 = \frac{1}{\left(n_{max} - n_{min}\right)} \int_{n_{min}}^{n_{max}} K_m n^2 dn = \frac{n_{max}^3 - n_{min}^3}{3\left(n_{max} - n_{min}\right)} \cdot K_m \tag{22.28}$$

Consequently,

$$n_{max}^2 + n_{max} \cdot n_{min} + n_{min}^2 = 3n_n^2$$

As the frequency varies, all core losses are considered proportional to voltage squared, and the torque is also proportional with voltage squared (constant slip during the tests), the average core loss is slightly smaller than for direct load tests. In general, n_{max} and n_{min} are rather symmetric with respect to rated speed.

Consequently, n_{max} and n_{min} are found easily.

Now the pace of acceleration is left to be adjusted so as to produce the equivalent RMS stator current over the speed oscillation cycle.

An RMS current estimator based on current acquisition over a few periods can be used as current feedback signal. The reference RMS current will be input to a slow current close loop that outputs the reference speed oscillation frequency between n_{min} and n_{max}.

The voltages V_a, V_b and the currents I_a and I_b are acquired and transformed to synchronous coordinates

$$\begin{vmatrix} V_d(I_d) \\ V_q(I_q) \end{vmatrix} = \frac{2}{3}\begin{vmatrix} \cos(-\theta_{es}) & \cos\left(-\theta_{es} + \frac{2\pi}{3}\right) & \cos\left(-\theta_{es} - \frac{2\pi}{3}\right) \\ \sin(-\theta_{es}) & \left(-\theta_{es} + \frac{2\pi}{3}\right) & \sin\left(-\theta_{es} - \frac{2\pi}{3}\right) \end{vmatrix} \cdot \begin{vmatrix} V_a(I_a) \\ V_b(I_b) \\ V_c(I_c) \end{vmatrix}; \tag{22.29}$$

$$V_c = -V_a - V_b;$$
$$I_c = -I_a - I_b;$$
$$\theta_{es} = \int \omega_1 dt;$$

The frequency f_1 varies with time as the machine accelerates and decelerates.

The input power

$$P_1 = \frac{3}{2}\left(V_d I_d + V_q I_q\right) \tag{22.30}$$

is in fact the instantaneous active power.

If the reference frame axis d falls along phase a axis at time zero, then $V_q = 0$ and thus only one term appears in (22.30).

The reactive input power Q_1 is

$$Q_1 = \frac{3}{2}\left(V_d I_q - V_q I_d\right) \tag{22.30'}$$

and retains also one term only. Notice that the current has both components.

The main advantage of this variable transformation – to be done off line through a PC – is that it provides "instantaneous" power values without requiring averaging over a few integral electrical periods. [15]

Consequently, while the machine travels from n_{min} to n_{max} and back, the active and (reactive) instantaneous power travel a circle when shown as a function of speed. (Figure 22.11) [15] Basically, the machine works as a motor during acceleration from n_{min} to n_{max} and as a generator from n_{max} to n_{min} and thus the input instantaneous active power is at times positive or negative.

The average power per cycle represents the average total losses in the IM per cycle P_{loss}. By the same token, the average of positive values yields the average input power P_{in1m}. So the efficiency η is

$$\eta = 1 - P_{loss} / P_{in1m} \tag{22.31}$$

Ref. [15] reports less than 1% difference in efficiency between this method and the direct loading method.

The main demerit of the method is the speed continuous dynamics which gives rise to additional noise and vibration losses in the IM.

If the frequency is changed up and down quickly, the IM speed cannot follow it and thus speed dynamics is avoided. Care must be exercised in this case not to go beyond the d.c. line capacitor threshold voltage during inevitable fast switching from motor to generator mode.

• The PWM dual frequency test [16]

An inverter with an open control architecture is prepared to produce, through PWM, the V(t) of (22.26) which contains two distinct frequencies.

This may be chosen around the rated synchronous speed of the IM:

$$f_1 = f_{1n} + \Delta f$$
$$f_2 = f_{1n} - \Delta f \tag{22.32}$$

Now Δf is the output of a slow RMS current close loop controller, which will result in the required current load.

Again, Δf is so large that the IM speed oscillates very little, somewhere below the average of the two frequencies. The switching from motoring to generating is now done, by the electromagnetic field, with a frequency equal to the difference between f_1 and f_2.

To provide for the same core losses, they are considered proportional to voltage squared and superposition is applied.

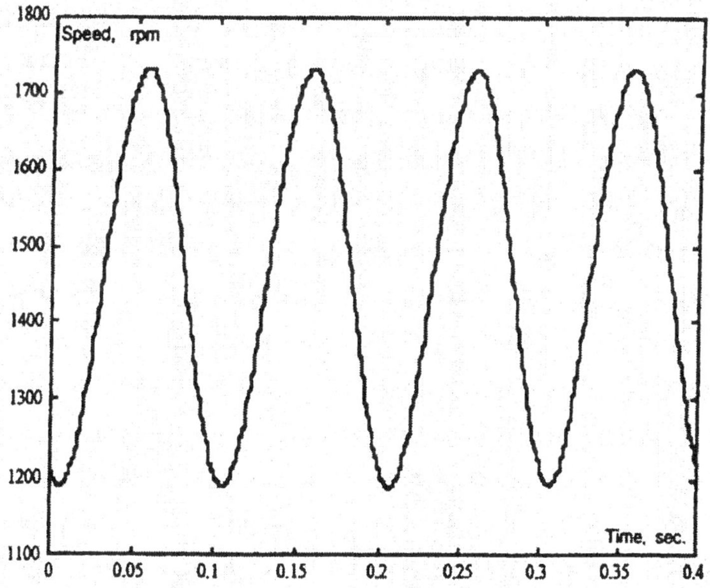

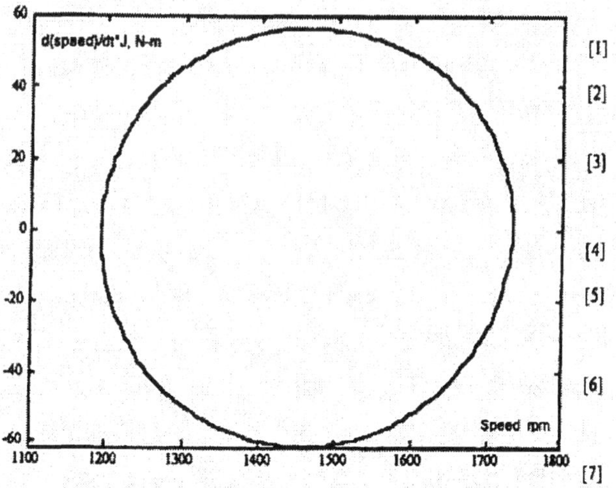

Figure 22.11 Speed oscillation and instantaneous active power variation ($P_n = 7.5$ kW)

Consequently,

$$V_{1n}^2 = V_1^2 + V_2^2 \tag{22.33}$$

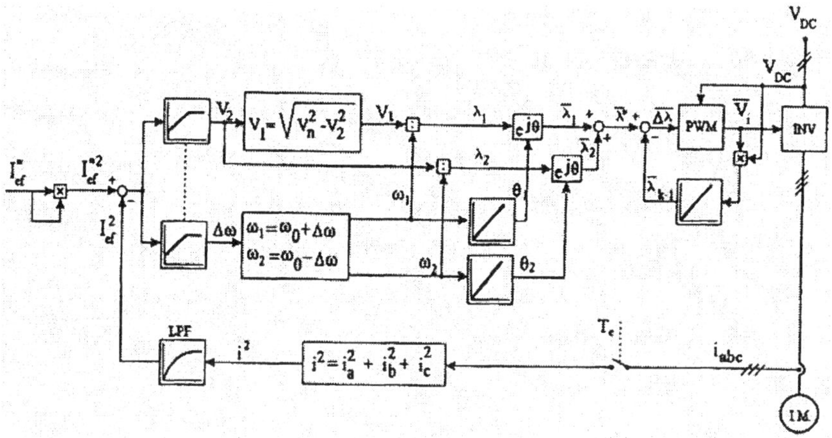

Figure 22.12 PWM dual frequency test system: the block diagram

Still, to determine V_1 and V_2 from (22.33), one of them should be adopted. Alternatively, the amplitude of the magnetic flux may be conserved to provide about the same saturation level

$$\left(\frac{V_{1n}}{f_{1n}}\right)^2 = \left(\frac{V_1}{f_1}\right)^2 + \left(\frac{V_2}{f_2}\right)^2 \tag{22.34}$$

with $\Delta f = 5\times10^{-2}f_{1n}$ from (22.33) and (22.34) $V_2 = 0.5383V_{1n}$ and $V_1 = 0.8427V_{1n}$. A rather complete block diagram of the entire system is shown in Figure 22.12. As evident in Figure 22.12, the whole testing process is "mechanized." This includes starting the IM with a single frequency voltage, which is ramped slowly until the rated frequency is reached. Then gradually, the dual frequency voltage PWM enters into action and produces the recalculated voltage amplitudes and frequencies. It should be noted that as the load is varied, through the RMS current, not only Δf varies but, also the voltage amplitudes V_1 and V_2 vary according to (22.33)–(22.34). Typical test results are shown in Figure 22.13. [16]

As expected the voltage, flux, and current are modulated. Also the capacitor voltage in the d.c. link varies. When it increases, generating mode is encountered.

The instantaneous active power may be calculated as in the previous method, in synchronous coordinates. This time a few electrical cycles may be averaged but in essence the average of instantaneous active power represents the total losses of the machine. The average of the positive values yields the average input power P_{in1m}. So the efficiency may be calculated again as in (22.32) for various reference stator RMS currents. Ultimately, the efficiency may be plotted versus output power, which is the difference between input power and losses.

Also temperature measurements may be added either to check the validity of the loss equivalence or, for a given temperature, to calibrate the reference RMS current value.

Temperature measurements [16] seem to confirm the method though verifications on many motors are required for full validation.

It should be emphasized that the IM current has been raised up to 150% and thus the complete efficiency test up to 150% load may be replicated by the PWM dual frequency test.

Notice that the slip frequency is around the rated value and thus the skin effect is not different from the case of direct load testing.

The mixed frequency methods may be used also for the temperature rise tests.

One more indirect method for temperature rise testing, used extensively by a leading manufacturer for the last 40 years for powers up to 21MW, is presented in what follows. [17]

22.3 THE TEMPERATURE-RISE TEST VIA FORWARD SHORTCIRCUIT (FSC) METHOD

For the large power IMs, direct loading to evaluate efficiency and temperature rise under load involves high costs. To avoid direct (shaft) loading, the superposition and equivalent (artificial) loading methods have been used for large powers, non-standard frequency or vertical shaft IMs.

The superposition test relies on the fact that temperature rise due to various losses adds up and thus a few special tests could be used to "simulate" the actual temperature rise in the IM. Scaling of results and the rather impossibility to consider correctly the stray load losses renders such methods as less practical for IMs.

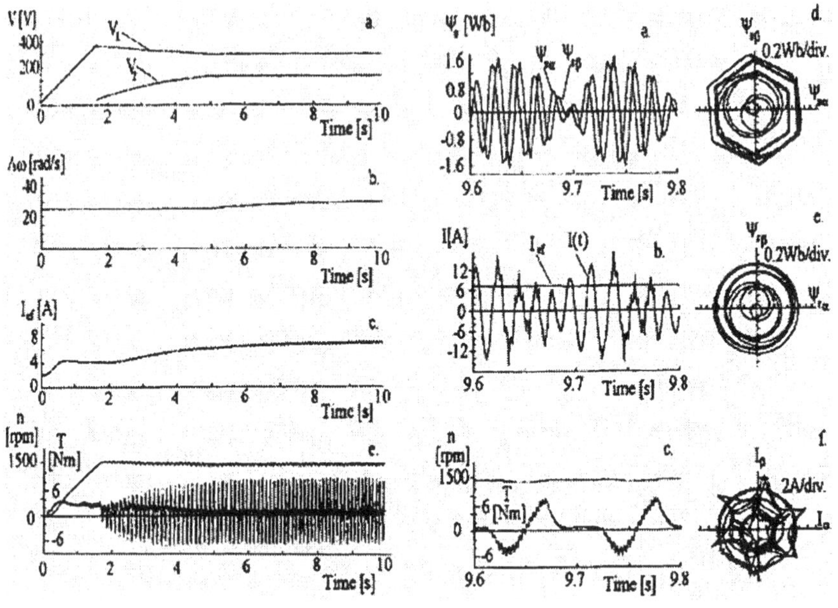

Figure 22.13. Sample test results with the PWM dual frequency method

The forward shortcircuit test (FSC) [17] performs a replica of load testing from the point of view of losses. It is not standardized yet but it uses standard hardware. In essence the tested IM is driven to rated speed slowly by drive motor 1 rated at 10% the rated power of the tested IM (Figure 22.14).

Then a synchronous generator is rotated by drive motor 2, rated at 10% of generator rated power, and then excited to produce low voltages of frequency f'_1.

$$f_1^{'} \approx (0.8 - 0.85)f_{1n} \tag{22.35}$$

As the generator excitation current increases, so do the generator voltages supplying the tested motor. The tested motor starts operating as a generator feeding active power to the generator, which becomes now a motor (albeit at low power) and the drive motor 2 works as a generator. The frequency of the currents in the rotor of the tested motor Sf'_1 is

$$Sf_1^{'} = f_1^{'} - n_n p_1 = -(0.15 - 0.2)f_{1n} \tag{22.36}$$

To secure low active power delivery from the tested motor, the slip frequency f_2 is such that the generator operates beyond peak torque slip frequency.

$$\left|Sf_1^{'}\right| > f_{2k} \approx \frac{R_2}{2\pi L_{sc}} \tag{22.37}$$

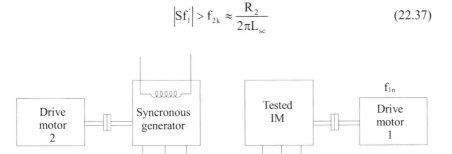

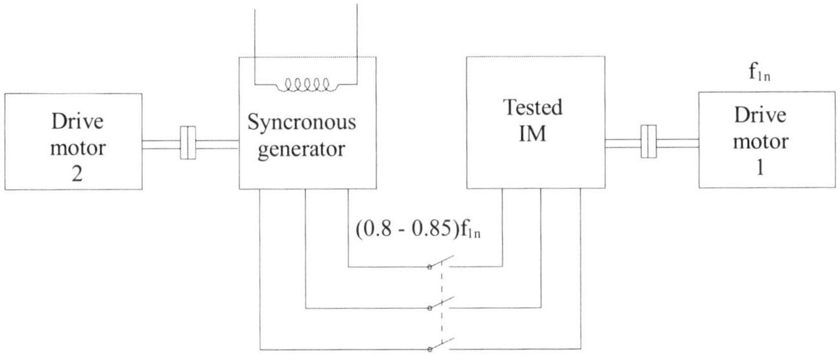

Figure 22.14 Test arrangement for forward shortcircuit test

Consequently, the equivalent impedance of the tested IM is small (large slip values), and, to limit the currents around rated value, the voltage at its terminals has to be low. The synchronous generator excitation current controls the voltage at the tested motor terminals. So low active power is delivered by the IM, albeit at rated current. However, operating beyond the peak torque point may result in instabilities unless the excitation current of the generator is not regulated to "freeze" the operation into a point on the descending torque/speed curve.

As low active powers are handled, the drive motors rating may be as low as 10% of tested motor power. So, in fact, the tested motor experiences rated current but low power (torque) in the generator regime at rather large slip values (0.15–0.25) and low voltage.

The fundamental problem in using this method intended to replicate temperature rise for rated loading is, again, equivalence of losses.

But before that, choosing the generator frequency f'_1 is crucial.

When the test f'_1 frequency decreases the copper losses in the tested motor tend to increase (slip increases) and the required rating of drive motor 1 decreases.

In general, as the slip is rather large the rotor current, for a given stator current, is higher than for rated load conditions (low slip). The difference is generated by the no load (magnetization) current, as the rotor current tends to get closer to stator current with increasing slip.

For low no load current IMs (low number of poles and/or high power), the increase in rotor current in the test with respect to rated power conditions is rather small. Not so for large number of poles and (or) low power IMs.

For $f'_1 = 40$ Hz and $f_{1n} = 50$ Hz the slip frequency $(Sf_1) = 10$ Hz and thus, at least for medium and large power cage rotor IMs, the skin effect is notable in contrast to rated power conditions.

This is an inherent limitation of the method. Let us now discuss the various loss components status in the FST:

- The stator copper losses

As the current is kept at rated value only, the lower skin effect, due to lower frequency f'_1 ($f'_1 < f_{1n}$), will lead to slightly lower copper losses than in rated load conditions in large power IMs. If the design data are known, with f'_1 given, the skin effect resistance coefficient may be calculated for f_{1n} and f'_1 (Chapter 9). Corrections then may be added.

- The fundamental iron losses are definitely lower than for rated conditions as f'_1 is

lower than for rated conditions and so is the voltage, even the flux.
To a first approximation, the fundamental core losses are proportional to voltage squared (irrespective of frequency). Also, there are some fundamental core losses in the rotor as Sf'_1 is (0.15–0.25) $f'1$.
The no load iron losses at rated voltage and frequency conditions may be compared with the low iron losses estimated for the FSC with the difference added by increasing stator current to compensate for the difference in the loss balance.

- Rotor cage fundamental losses

As indicated earlier, the rotor cage losses are larger and dependent on the generator frequency f'_1, but, to calculate their difference, would require the magnetization reactance and the rotor resistance R_2 and leakage reactance X_{l2} at the rather large frequency Sf'_1 (with skin effect considered).

- Stray load losses

A good part of stray load losses occur in the rotor bars. The frequency $S_v f'_1$ of the rotor bar harmonics currents due to stator m.m.f. space harmonics is

$$S_v f_1' = [1 - \upsilon(1 - S)]f_1' \tag{22.38}$$

For phase belt harmonics,

$$\nu = \pm 6K + 1 \tag{22.39}$$

with ν= –5, +7, –11, 13… (22.40)

If for FST f'_1= 40 Hz and S = -0.2375, for rated load conditions f_{1n} = 50 Hz and S_n = 0.01. The rotor current harmonics frequency $S_v f'_1$ are very close in the two cases (287.5/297.5, 306.5/296.5, 584.5/594.5, 603.5/593.5).

So, from this point of view, the two cases are rather equivalent. That is, the FST method is consistent with the direct load method. However, the rotor current harmonics are also influenced by slot permeance harmonics, which are voltage dependent. So in general, the stator m.m.f. caused rotor bar stray losses are smaller for the FST than for rated conditions.

On the other hand, tooth flux pulsation core losses have to be considered. As the saturation is not present (low voltage; low main flux level), the flux pulsation in the tooth, for given currents, tends to be larger in the FST, than for rated load conditions.

The rotor slot harmonic pole pairs p_μ is

$$p_\mu = KN_r \pm p_1 \tag{22.41}$$

N_r – number of rotor slots, p_1 – IM pole pairs.

The frequency f_μ of the eddy currents induced in the stator tooth is

$$f_\mu' = [p_\mu(1-S)+S]f_1' \tag{22.42}$$

It suffices to consider K = 1,2.

For the same example as above (f'_1 = 40 Hz, S = -0.2375 and f_1 = 50 Hz, S_n = 0.01) and K = 1,2, again frequencies f_μ very close to each other are obtained for the two situations.

So the stator harmonics core losses are equivalent.

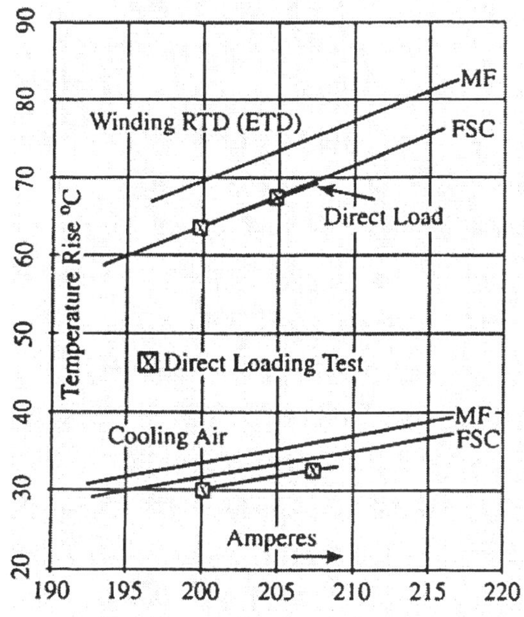

Figure 22.15 Steady state temperature rises with three tests

Still we have to remember that the rotor cage fundamental losses tend to be somewhat larger (because of much larger slip and skin effect) while the fundamental core losses are much lower.

The degree to which these two effects neutralize each other is an issue to be tackled for every machine separately.

Final results related to temperature rise for three tests – FSC, IEEE-112F and the two frequency methods (MF) – are shown on Figure 22.15 for a 1960 kW 4 pole, 50 Hz, 6600 V IM. [17] The FSC and direct loading tests produce very close results while the mixed frequency (MF) method overestimates the temperature by 5–7^{0}C.

The standard mixed–frequency method overestimation of temperature rise (Figure 22.15) is due to lower speed, speed oscillations, and due to higher core losses as the supply frequency voltage V_1 comes in with 100% with V_2 added to it. It appears that the PWM dual frequency test is free from such limitations and would produce reliable results in temperature rise as well as in efficiency tests. Returning to FSC, we notice that if the core losses are neglected (the voltage is reduced), the difference between the generated power P_G and the shaft input power P_{shaft} for the tested motor is

$$P_{shaft} - P_G = 3I_1^2R_1 + 3I_2^2R_2 + p_{stray} + p_{mec} + p_{iron}\left(\frac{V}{V_{1n}}\right)^2 \tag{22.43}$$

The mechanical losses p_{mec} are considered to be known while, P_{shaft}, P_g, I_1 have to be measured. When a d.c. machine is used as drive motor 1, its calibration is easy and thus P_{shaft} may be estimated without a torque-meter. The iron losses at rated voltage are p_{iron}.

Now if we consider the stray losses located in the stator, then the electromagnetic power concept yields

$$\frac{3I_2^2R_2}{|S|} = P_G + 3R_1I_1^2 + p_{stray} + p_{iron}\left(\frac{V}{V_{1n}}\right)^2 \tag{22.44}$$

Eliminating $I^2{}_2R_2$ from (22.43)–(22.44) yields

$$p_{stray} = \frac{(P_{shaft} - p_{mec})}{1+|S|} - P_G - 3R_1I_1^2 - p_{iron}\left(\frac{V}{V_{1n}}\right)^2 \tag{22.45}$$

As all powers involved are rather small, the error of calculating p_{stray} is not expected to be large. With p_{stray} calculated, from (22.44) the rotor fundamental cage losses $3R_2I^2{}_2$ are determined. To a first approximation, with $I_2 \approx \sqrt{I_1^2 - I_m^2}$ (I_m – the no load current at low voltage levels that is under unsaturated conditions), the rotor resistance R_2 at Sf'_1 frequency may be determined as a bonus. Comparing results on parameter calculation from FSC and IEEE–112F shows acceptable correlation for the 1960 kW IM investigated in Reference 17

The level of voltage at IM terminals during FSC is around 20% rated voltage. A few remarks on FSC are in order

- The loss distribution is changed with more losses in the rotor and less in the stator.
- The skin effect is rather notable in the rotor.
- It is possible to replace the drive motor 2 and the a.c. generator by a bi-directional power PWM converter sized around 20% the rated power of the tested IM (Figure 22.16).

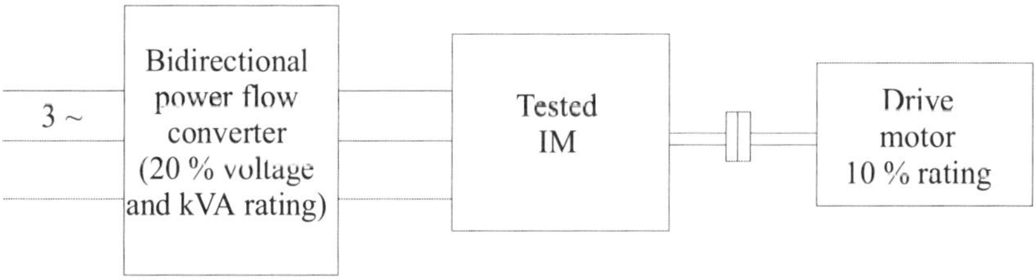

Figure 22.16 The forward shortcircuit (FSC) test with PWM converter

The control options, for adequate loss equivalence and stable operation, are improved by the presence and the capabilities of the PWM converter.

22.4 PARAMETER ESTIMATION TESTS

By induction machine parameters, we mean the resistances and inductances in the adapted circuit model and the inertia.

There are many ways we may develop a circuit model, which duplicates better or worse the performance of IM in different operation modes. Magnetic saturation, skin effect and the space and time m.m.f. harmonics make the problem of parameter estimation very difficult. Traditionally, the no load and stall rotor tests at rated frequency are used to estimate the parameters of the single cage circuit model, with skin effect neglected.

On the other hand for IM with deep rotor bars, or double cage rotors, shortcircuit tests at rated frequency produce too a high value for the rotor resistance for the IM running at load and rated frequency.

Standstill frequency response tests are recommended for such cases.

Still the large values of currents during IM direct starting at the power grid produce saturation in the leakage path of both stator and rotor magnetic field. Consequently, the leakage inductances are notably reduced. This effect adds to the rotor leakage inductance reduction due to skin effect.

The closed rotor cage slot leakage field, on the other hand, saturates the upper iron bridge for rotor currents above 10% of rated current.

This means that for such a case the rotor leakage inductance does not vary essentially at currents above 10% rated current, that is during starting or on load conditions.

For variable speed IM operation, the level of the airgap flux varies notably. Consequently, the magnetization inductance varies also. The apparent way out of these difficulties may seem to use the traditional single-rotor circuit model with all parameters as variables (Figure 22.17) with primary (f_1), rotor slip frequency Sf_1, and stator, rotor and magnetization currents (I_1, I_2, I_m). The trouble is that most parameters depend on more than one variable and thus such an equivalent circuit is hardly practical (Figure 22.17). Also, if space and time harmonics are to be considered, the general equivalent circuit becomes all but manageable.

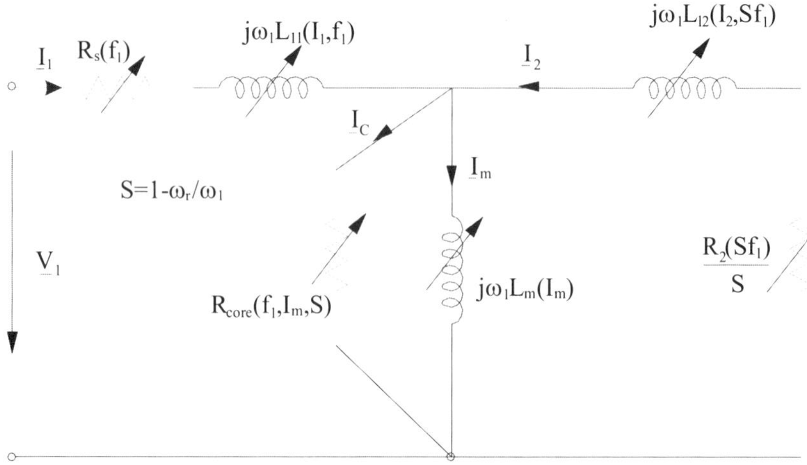

Figure 22.17 Fundamental single cage rotor equivalent circuit with variable parameters

For the space harmonics, the slip S changes to S_v while for time harmonics both f_K and S_K vary with the harmonics order K (see chapter 10 on harmonics). We should remember that, while spinning the rotor, slip frequency Sf_1 is different from stator frequency f_1 and thus is full reality only tests done in such conditions match the performance. Also, FEM is now approaching enough maturity to be useful in the computation of all parameters in any conditions of operation.

Finally, for IM control for drives a rather simplified equivalent circuit (model) with some parameter variation is practical. This is how simplified solutions for IM parameter estimation off line and on line evolved.

Among them, in fact only the no load and shortcircuit single frequency test derived parameter estimation method is standardized worldwide. The two frequency and the frequency response standstill tests have gained some momentum lately. Modified regression methods to estimate double cage model parameters based on acquiring data over a direct slow starting period or directly from transients during a fast start, have been developed recently. Finally for variable speed drives commissioning and their adaptive control, off and on line simplified methods to determine some parameters detuning are currently used. In what follows, a synthesis of these trends is presented.

22.4.1 Parameter calculation from no-load and standstill tests

The no-load motor test – paragraph 22.1.1. – is hereby used to calculate the electric parameters appearing in the equivalent circuit for zero rotor currents (Figure 22.18).

With voltage V_1, current I_{10}, and power P_0 measured, we may calculate two parameters:

$$R_1 + R_{cs} = \frac{(P_0 - p_{mec})}{3I_{10}^2}$$
$$L_{l1} + L_m = \frac{1}{\omega_1}\sqrt{\left(\frac{V_1}{I_{10}}\right)^2 - (R_1 + R_{cs})^2} \tag{22.46}$$

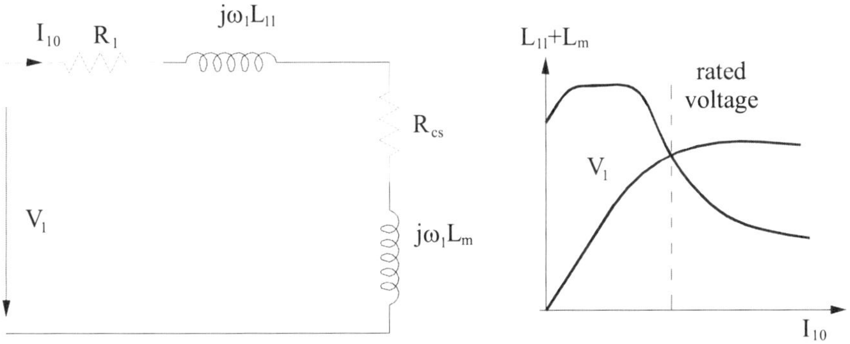

Figure 22.18 Fundamental no load equivalent circuit

The stator phase resistances may be d.c. measured or, if possible, in a.c. without the rotor in place. In this case, if a small three phase voltage is applied, the machine will show a very small R_{cs} (core losses are very small). The calculated phase inductance slightly over estimates the stator leakage inductance.

The overestimation is related to the magnetic energy in the air left by the missing rotor.

The equivalent inductance of the "air" is, in fact, corresponding to the magnetization induction of an equivalent airgap of τ/π (τ the pole pitch).

$$L_g = (L_m)_{unsat} \cdot \frac{g\pi}{\tau} \tag{22.47}$$

Suppose we do these measurements (without the rotor) and obtain from power, voltage, and current measurements,

$$R_{1\sim} = \frac{P_{3\sim}}{3I_{10\sim}^2}$$
$$L_{l1} + L_g = \frac{1}{\omega_1}\sqrt{\left(\frac{V_{1\sim}}{I_{10\sim}}\right)^2 - R_{1\sim}^2} \tag{22.48}$$

From the no-load test at low voltage, the unsaturated value of L_m is obtained. Then from, (22.47) L_g is easily calculated, provided the airgap is known. Finally from (22.48) L_{l1} is found with L_g already calculated.

In general, however, the stator leakage inductance is not considered separable from the no load and shortcircuit tests.

The shortcircuit (standstill or stall) tests, on the other hand, are performed at low voltage with three phase or single-phase supply (paragraph 22.1.4), and R_{sc} and $X_{sc} = \omega_1 L_{sc}$ are calculated from (22.9)–(22.10) and averaged over a few current values up to about rated current. When voltage versus current is plotted, the curve tends to be linear but, for the closed-slot rotor, it does not converge into the origin (Figure 22.19).

The residual voltage E_{bridge}, corresponding to the rapidly saturating iron bridge above the rotor closed slot, may be added to the equivalent circuit in the rotor part. All the other components of the rotor leakage inductance occur in L_{l2}. Such an approximation appears to be practical for small induction machines.

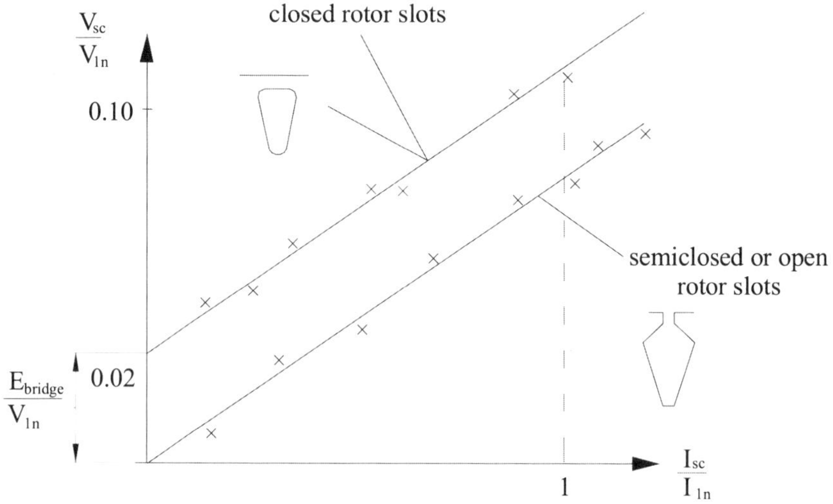

Figure 22.19 Typical voltage/current curves at standstill

Now, with R_l known, R_{2sc} may be calculated from (22.9)–(22.10). That is, the rotor resistance for the rated frequency, with the skin effect accounted for.

The stator and rotor leakage inductances at shortcircuit come together in $X_{sc} = (L_{l1} + L_{l2sc})\omega_1$. Unless L_{l1} is measured as indicated above or its design value is known, L_{l1} is taken as $L_{l1} \approx L_{sc}/2$.

The no load and shortcircuit tests for the wound rotor IM are straightforward as both the stator and rotor currents may be measured directly. Also the tests may be run with stator or rotor energized. Even both stator and rotor circuits may be energized. Changing the amplitude and phase of rotor voltage until the airgap field is zero, leads to a complete separation of the two circuits. The same tests have been proved acceptable to derive torque/current load characteristics for superhigh speed PWM converter fed IMs. [18] The current is limited in this case by the converter rating. Neglecting the magnetization inductance in the equivalent circuit at standstill (S = 1) – Figure 22.20 – is valid as long as $\omega_1 L_m >> R_{2sc}$.

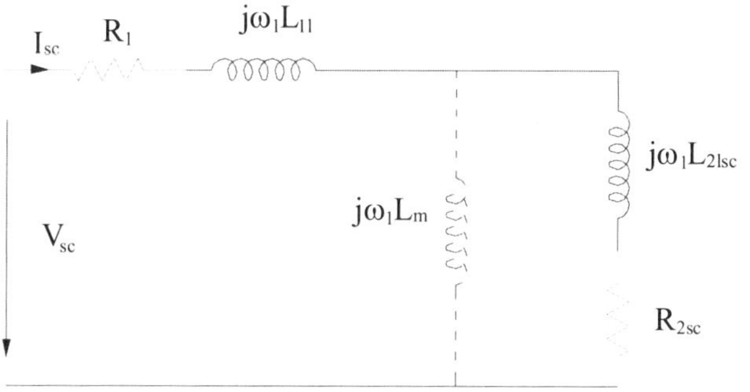

Figure 22.20 Equivalent circuit at standstill

It is known that at full voltage, the airgap flux is only about (50–60%) of its no load value due to the large voltage drop along the stator resistance R_1 and leakage reactance $\omega_1 L_{l1}$. So, in any case, the main magnetic circuit is nonsaturated. That is to say that the nonsaturated value of L_m – $(L_m)_{unsat}$ – has to be introduced in the equivalent circuit at low frequencies when $\omega_1 L_m$ is not much larger than R'_{2sc}. If the L_m branch circuit is left out in low frequency standstill tests, the errors become unacceptable.

The presence of skin effect suggests that the single rotor circuit model is not sufficient to accommodate a wide spectrum of operation modes from standstill to no load.

The first easy step is to perform one more standstill test, but at low frequency: less than 5% of rated frequency. This is how the so-called two-frequency standstill test was born.

22.4.2 The two frequency standstill test

The second shortcircuit test, at low frequency, requires a low frequency power source. The PWM converter makes a good (easily available) low frequency power supply.

This time the equivalent circuit contains basically two loops (Figure 22.21)

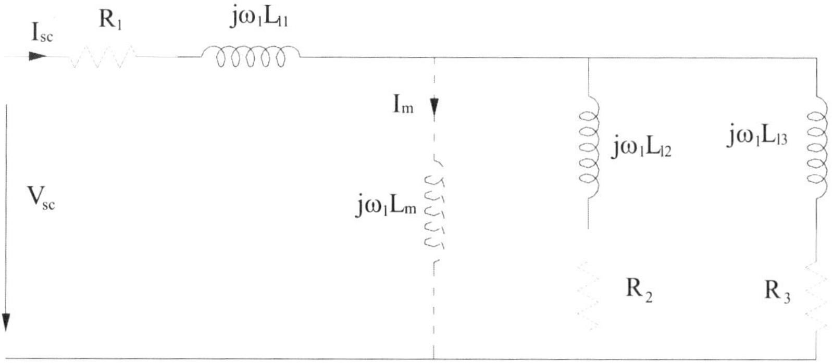

Figure 22.21 Dual rotor – loop equivalent circuit at standstill

To calculate the four rotor parameters R_2, L_{l2}, R_3, L_{l3}, it is assumed that at rated frequency $f_{1n}(\omega_{1n})$ only the second (starting) rotor loop exists. That is, only L_{l3} and R_3 remain as unknowns and the L_m branch may be eliminated. At the second frequency $f'_1 \leq 0.05f_{1n}$, only the first (working) cage acts. Thus, only R_2, L_{l2} are to be calculated with the L_m branch accounted for. L_m is taken as the nonsaturated value of L_m from the no load test. An iterative procedure may be used with $L_m(I_m)$ from the no load test as the main circuit may get saturated unless $V_{sc}/V_{1n}<f'_1/f_{1n}$.

The rotor parameters of rotor loops are constant. One dominates the performance at large slip frequencies Sf_1 (R_3, L_{l3}) and the other one prevails at low slip frequency.

The double-cage rotor IM may be assimilated with this case. However, the two cages measured parameters may not overlap with the actual double-cage parameters.

As of now, with the above parameter estimation methods, skin effect has been accounted for approximately. Also by the $L_m(I_m)$ function, the magnetic saturation of the main circuit is accounted for.

Still, the leakage saturation at high currents has not been present in tests. Consequently, calculating the torque, current, power factor from S = 1 to S_n (S_n – rated slip) with the above parameters is bound to produce large errors. And so it does!

The dissatisfaction of IM users with this situation led to quite a few simplified attempts to calculate the parameters, basically R_2, R_{2sc}, L_{l2}, L_{l2sc}, based on catalogue data to fit both the starting current, and torque and the rated current and power factor for given efficiency, rated speed and torque. These approaches may be dubbed as catalogue based methods.

22.4.3 Parameters from catalogue data

Let us consider that the starting current and torque I_s, T_{es}, rated speed, frequency, current, power factor, efficiency and rated power are all catalogue data for the IM. It is well understood that the logic solution to parameter estimation for two extreme situations – start and full load – is to use the equivalent circuit. First from the no-load current I_0,

$$I_0 \approx \frac{V_{1n}}{\omega_1 L_m \left(1 + L_{l1} / L_m\right)} \tag{22.49}$$

We obtain an approximate value of L_m. For start $L_{l1}/L_m = 0.02–0.05$. At the end of the estimation process, L_{l1} is obtained.

Then we may return with a corrected value of L_{l1}/L_m to (22.49) and perform the whole process again until sufficient convergence is reached.

This saturated value is valid for rated load calculations.

From the starting torque,

$$T_{es} = \frac{3R_{2sc} \cdot I_{2s}^2 \cdot p_1}{\omega_1}$$
$$I_{2s} \approx \sqrt{I_s^2 - I_0^2} \tag{22.50}$$

From (22.50), R_{2sc} may be determined. The equivalent circuit at start yields

$$\left(L_{l1sc} + L_{l2sc}\right) = \frac{1}{\omega_1}\sqrt{\left(\frac{V_{1n}}{I_s}\right)^2 - \left(R_1 + R_{2sc}\right)^2} \tag{22.51}$$

The stator resistance R_1 may be d.c. measured and thus (L_{l1sc}+L_{l2sc}) as influenced by skin effect and leakage saturation, is determined.

On full load, the mechanical losses are assigned an initial value $p_{mec} = 0.01P_n$.

Thus the electromagnetic torque is equal to the shaft torque plus the mechanical loss torque.

$$T_{en} = p_1\left(1 + \frac{p_{mec}}{P_n}\right)\frac{P_n}{2\pi f_1(1 - S_n)} = \frac{3R_2 I_{2n}^2}{S_n\omega_1}p_1$$
$$I_{2n}^2 \approx \sqrt{I_{1n}^2 - I_0^2} \tag{22.52}$$

With R_2 found easily from (22.52), we may write the balance of powers at rated slip,

$$Q_1 = 3V_{1n}I_{1n}\sin\varphi_{1n} = 3\omega_1 L_{l1}I_{1n}^2 + 3\omega_1 L_{l2}I_{2n}^2 + 3L_m\omega_1 \cdot I_0^2 \tag{22.53}$$

$$P_1 = 3V_{1n}I_{1n}\cos\varphi_{1n} = 3R_1 I_{1n}^2 + p_{iron} + p_{stray} + 3\frac{R_2 I_{2n}^2}{S_n} \tag{22.54}$$

$$P_n\left(1 - \frac{1}{\eta}\right) = 3R_1 I_1^2 + p_{iron} + p_{stray} + 3R_2 I_{2n}^2 + p_{mec} \tag{22.55}$$

There are 5 unknowns – L_{l1}, L_{l2}, p_{iron}, p_{stray}, p_{mec} – in Equations (21.53-21.55).

However from (21.54), p_{iron} + p_{stray} can be calculated directly. They may be introduced in (21.55) to calculate mechanical loss p_{mec}. If they differ from the initial value of 1% of rated power P_n, their value is corrected in (21.52) and p_{core}+p_{stray} and p_{mec} are recalculated again until sufficient convergence in p_{mec} is obtained.

We thus remain with Equation (21.53) where in fact we do have two unknowns: the stator and rotor leakage inductances. They may not be separated and thus we may consider them equal to each other: $L_{l1} = L_{l2}$.

This way

$$L_{l1} = L_{l2} = \frac{\left(3V_{1n}I_m \sin\varphi_{1n} - 3\omega_1 L_m I_0^2\right)}{3\omega_1\left(2I_{1n}^2 - I_0^2\right)} \tag{22.56}$$

Now going back to (22.51), if the IM has open stator slots, there will be no leakage flux path saturation in the stator and thus $L_{l1sc} = L_{l1} = L_{l2}$. Consequently, from (22.51), L_{l2sc} is found. With semiclosed stator slots, both stator and rotor leakage flux paths may saturate. So, we might as well consider $L_{l1sc} = L_{l2sc}$. Again $L_{l2sc} = L_{l1sc}$ is found from (22.51).

Attention has to be paid to (21.56) as computation precision is important because most of the reactive power is "spent" in the airgap, in L_m. In general, $2L_{l1} > L_{l1sc}+L_{l2sc}$. If $2L_{l1} \le L_{l1sc}+L_{l2sc}$ is obtained the skin and saturation effects are mild and the shortcircuit values hold as good for all slip values.

An approximate dependence of rotor resistance and leakage inductances with slip, between S = 1 and S = S_n may be admitted

$$R_2(S) = R_2 + \left(R_{2sc} - R_2\right)\sqrt{\frac{\left(S - S_n\right)}{1 - S_n}}; S \ge S_n \tag{22.57}$$

$$L_{l2}(S) = L_{l2} - \left(L_{l2} - L_{l2sc}\right)\frac{S - S_n}{1 - S}; S \ge S_n \tag{22.58}$$

$$L_{l1}(S) = L_{l1} - \left(L_{l1} - L_{l1sc}\right)\frac{S - S_n}{1 - S_n}; S \ge S_n \tag{22.59}$$

This way the parameters dependence on slip is known, so torque, current, power factor versus slip may be calculated. Alternatively, we may consider that the rotor parameters at standstill R_{2sc}, L_{l2sc} represent the starting cage and R_{2l}, L_{l2} the working cage. Such an attitude may prove practical for the investigation of IM transients and control.

It should be emphasized that catalogue data methods provide for complete agreement between the circuit model and the actual machine performance only at standstill and at full load. There is no guarantee that the rotor parameters vary linearly or as in (21.57) with slip for all rotor (stator) slot geometries.

In other words, the agreement of the torque and current versus slip curves for all slips is not guaranteed to any definable error.

From the need to secure good agreement between the circuit model and the real machine, for ever wider operation modes, more sophisticated methods have been introduced.

We introduce here two of the most representative ones: the standstill frequency response method (SSFR) and the step-wise regression general method.

22.4.4 Standstill frequency response method

Standstill tests, especially when a single phase ac supply is used (zero torque), require less manpower, equipment and electrical energy to perform than running tests. To replicate the machine performance with different rotor slip frequency Sf_1, tests at standstill with a variable voltage and frequency supply have been proposed [19-20]. Dubbed as SSFR, such tests have been first used for synchronous for generator parameter estimation.

In essence, the IM is fed with variable voltage and frequency from a PWM converter supply.

Single phase supply is preferred as the torque is zero and thus no mechanical fixture to stall the rotor is required. However, as three phase PWM converters are available up to high powers and the tests are performed at low currents (less than 10% of rated current), three phase testing seems more practical (Figure 22.22)

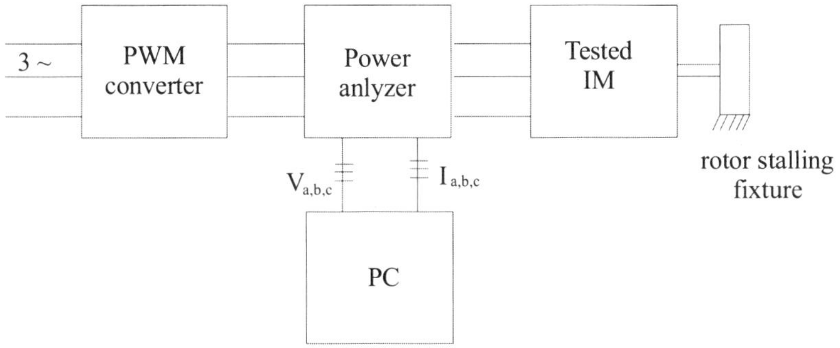

Figure 22.22 Standstill frequency test arrangement

The test provides for the consideration of skin effect, which is frequency dependent. However, as the current is low, to avoid IM overheating, the SSFR test methods should not be expected to produce good agreement with the real machine in the torque/slip or current/slip curves at rated voltage. This is so because the leakage saturation effect of reducing the leakage inductances is not considered. A way out of this difficulty is to adopt a certain functional for stator leakage inductance variation with stator current.

Also at low frequencies, 0.01 Hz, 2–3 periods acquisition time means 200–300 seconds. So, in general, acquiring data for 50–100 frequencies is time prohibitive. The voltages and currents are acquired and their fundamental wave amplitudes and ratio of amplitudes and phase lag are calculated.

A two (even three) rotor loop circuit model may be adopted (Figure 22.23).

The operational inductance L(p) for a double cage model (see Chapter 13 on transients) in rotor coordinates, is:

$$L(p) = (L_{l1} + L_m)\frac{(1 + pT')(1 + pT'')}{(1 + pT_0')(1 + pT_0'')} \tag{22.60}$$

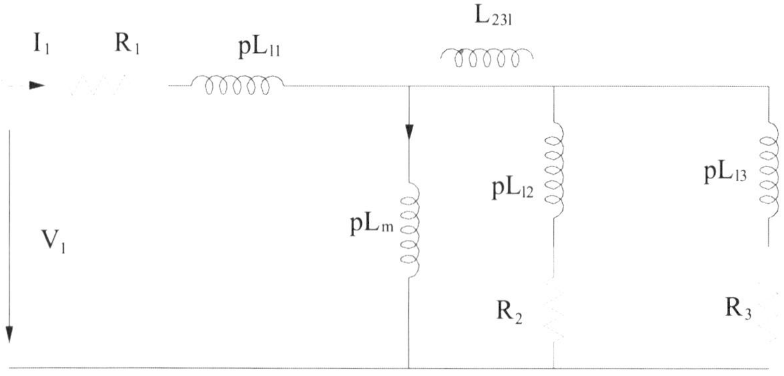

Figure 22.23 Double cage operational equivalent circuit at standstill

The equivalent circuit does not show the mutual leakage inductance between the two virtual cages (L_{l23}) because it can not be measured directly.

The time constants in (22.60) are related to the equivalent circuit parameters as follows

$$T' = \frac{1}{R_3}\left[L_{3l} + \frac{L_m \cdot L_{l1}}{L_m + L_{l1}}\right] \quad (22.61)$$

$$T'' = \frac{1}{R_2}\left[L_{l2} + \frac{L_m L_{l1} L_{l3}}{L_{l3}L_m + L_{l3}L_{l1} + L_m L_{l1}}\right] \quad (22.62)$$

$$T_0' = \frac{1}{R_3}\left(L_m + L_{l1} + L_{l3}\right) \quad (22.63)$$

$$T_0'' = \frac{1}{R_2}\left[L_{l2} + \frac{L_m L_{l3} \cdot}{L_m + L_{l3}}\right] \quad (22.64)$$

The unsaturated value of the no load inductance $L_s = L_{l1} + L_m$ has to be determined separately from the no load test at lower than rated voltage.

Also the stator leakage inductance value has to be known from design or is being assigned a "reasonable" value of 2 – 5% of no load inductance L_s.

The measurements will result in impedance estimation for every frequency test performed.

$$\underline{Z}_{sc}(\omega) = \frac{\underline{V}_{sc}}{\underline{I}_{sc}} = R_{sc}(\omega) + jX_{sc}(\omega) \quad (22.65)$$

When the IM is single-phase supply, fed between 2 phases, a factor of 2 occurs in the denominator of (22.65).

We may now calculate L(jω) from (22.65).

$$L(j\omega)=\frac{(Z_{sc}-R_1)}{j\omega_1}=\mathrm{Re}(\omega)+j\,\mathrm{Im}(\omega) \tag{22.66}$$

Equating the real and imaginary parts in (22.60) and (22.66), we obtain

$$\mathrm{Re}(\omega)=\omega L_s+\omega\,\mathrm{Im}(\omega)\,(T_0'+T_0'')-\omega^3 L_s T' T''+\omega^2\,\mathrm{Re}(\omega)\cdot T_0'\cdot T_0'' \tag{22.67}$$

$$\mathrm{Im}(\omega)=\omega^2 L_s (T'+T'')-\omega\,\mathrm{Re}(\omega)\,(T_0'+T_0'')-\omega^2\,\mathrm{Im}(\omega) T_0'\cdot T_0'' \tag{22.68}$$

A new set of unknowns, instead of T', T'', T_0', T_0'', is now defined. [21]

$$\begin{aligned} X_1 &= L_s \\ X_2 &= T_0'+T_0'' \\ X_3 &= L_s T' T'' \\ X_4 &= T_0' T_0'' \\ X_5 &= L_s (T'+T'') \end{aligned} \tag{22.69}$$

L_s may be discarded by introducing the known value of the non-saturated no load inductance.

A linear system is obtained.

In matrix form,

$$|Z_i|=|H_i||X|;i=1,2,....,m \tag{22.70}$$

with

$$|Z_i|=|\mathrm{Re}(\omega),\mathrm{Im}(\omega)\,|_I^T \tag{22.71}$$

$$|X|=|X_1,X_2,X_3,X_4,X_5,| \tag{22.72}$$

$$|H_i|=\begin{vmatrix}\omega & \omega\,\mathrm{Im}(\omega) & -\omega^3 & \omega^2\,\mathrm{Re}(\omega) & 0\\ 0 & -\omega\,\mathrm{Re}(\omega) & 0 & \omega^2\,\mathrm{Im}(\omega) & \omega^2\end{vmatrix} \tag{22.73}$$

The measurements are taken for $m \geq 5$ distinct frequencies.

As the number of equations is larger than the number of unknowns, various approximation methods may be used to solve the problem.

The solution of (22.70) in the least squared error method sense is

$$|X|=\left[[H]^T\cdot[H]\right]^{-1}\cdot[H]^T\cdot[Z] \tag{22.74}$$

Once the variable vector $|X|$ is known, the actual unknowns L_s, T', T'', T_0', T_0'' are

$$L_s = X_1$$

$$T', T'' = \frac{1}{2X_1}\left(X_5 \pm \sqrt{X_5^2 - 4X_3X_1}\right) \tag{22.75}$$

$$T_0', T_0'' = \frac{1}{2}\left(X_2 \pm \sqrt{X_2^2 - 4X_4}\right) \tag{22.76}$$

The least squared error method has been successfully applied to an 1288 kW IM in Ref. [21].

Typical direct impedance results are shown on Figure 22.24. [21]

Comparative results [20,21] of parameter values estimated with Bode diagram and, respectively, by the least squared method, for the same machine, are given in Table 22.3. [21]

Good agreement between the two methods of solution is evident.

However, as Ref. [20] shows, when applying these parameters to the torque/slip or current/slip curves at rated load, the discrepancies in torque and current at high values of slip are inadmissible. This may be explained by the neglect of leakage saturation, which in the real machine with higher starting torque and current is notable.

Table 22.3 Comparative parameter estimates for an 1288 kW IM

Parameter	Bode diagram [10]	Least squared method [21]
l_s p.u	5.3	5.299
T' sec.	88×10^{-3}	87.99×10^{-3}
T'' sec.	3.6×10^{-3}	3.6×10^{-3}
T_0' sec.	2.0	2.00
T_0'' sec.	4.55×10^{-3}	4.55

As always in such cases, an empirical expression of the stator leakage inductance, which decreases with stator current, may be introduced to improve the model performance at high currents.

Still, there seem to be some notable discrepancies in the breaking torque region, which hold in there even after correcting the stator leakage inductance for leakage saturation. A three-circuit cage seems necessary in this particular case.

This raises the question of when the SSFR parameter estimation may be used with guaranteed success. It appears that only for small periodic load perturbations around rated speed or for estimating the machine behaviour with respect to high frequency time harmonics so typical in PWM converter-fed IMs.

Step response methods to different levels of voltage and current peak values may be the response to the leakage saturation at high currents inclusion into the IM models.

Short voltage pulses may be injected in the IM at standstill between two phases until the current peaks to a threshold, predetermined value. A pair of positive and negative voltage pulses may be applied to eliminate the influence of hysteresis.

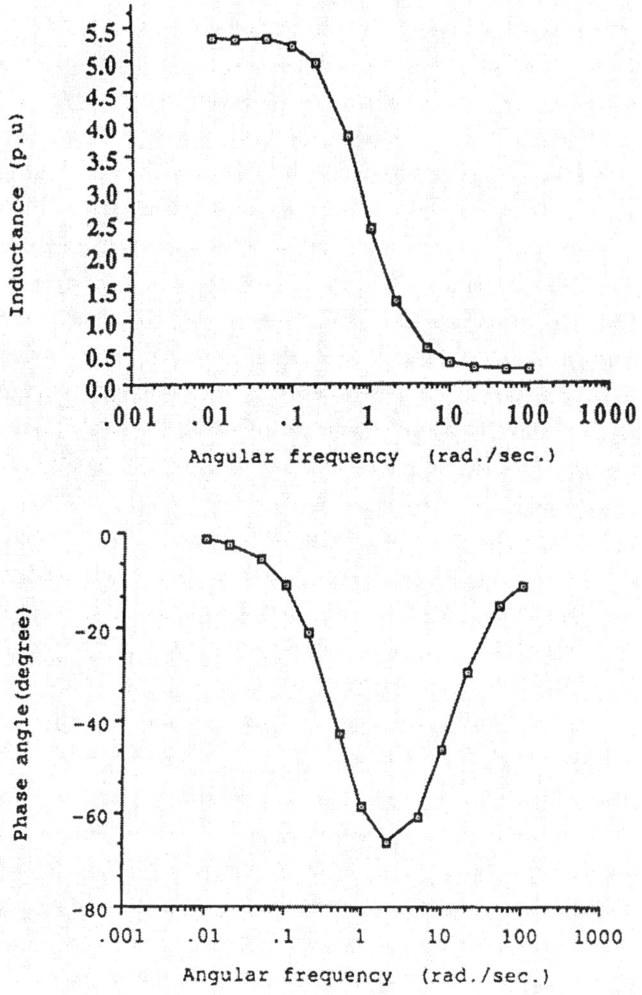

Figure 22.24 Standstill frequency response of an 1288 kW IM

The current and voltage are acquired and processed using the same operational inductance (22.60).

$$v(p) = 2 \cdot i(p)[R_1 + pL(p)] \tag{22.77}$$

The two cage model may be used, with its parameters determined to fit the current/time transient experimental curve via an approximation method (least squared method, etc.).

The voltage signals will stop, via a fast static power switch, at values of current up to the peak starting current to explore fully the leakage saturation phenomenon. The pulses are so short that the motor does not get warm after 10–15 threshold current level tests. Also the testing will last only 1–2 minutes in all and requires simplified power electronics equipment. It may be argued that the step voltage is too a fast signal and thus the rotor leakage inductances and resistances are too much influenced by the skin effect. Only extensive tests hold the answer to this serious question.

The method to estimate the parameters from this test is somehow included in the general regression method that follows.

22.4.5 The general regression method for parameters estimation

As variable parameters, with current and (or frequency) or rotor slip frequency, are difficult to handle, constant parameters are preferred. By using a two rotor cage circuit model, the variable parameters of a single cage with slip frequency are accounted for. Not so for magnetic saturation effects.

The deterministic way of defining variable parameters for the complete spectrum of frequency and current has to be apparently abandoned for constant parameter models which, for a given spectrum of operation modes, give general satisfaction according to an optimization criterion, such as the least squared error over the investigated operation mode spectrum.

For transient modes, the transient model has to be used while for steady state modes the steady state equations are applied.

The d.q. model of double cage IM in synchronous coordinates in space phasor quantities, is (Chapter 13 on transients)

$$V_1 = R_1\overline{I_1} + \frac{d\overline{\Psi_1}}{dt} + j\omega_1\overline{\Psi_1}; \overline{\Psi_1} = \overline{\Psi_m} + L_{l1}\bar{I}_1 \quad (22.78)$$

$$0 = R_2\overline{I_2} + \frac{d\overline{\Psi}_2}{dt} + jS\omega_1\overline{\Psi_2}; \overline{\Psi_2} = \overline{\Psi_m} + L_{l2}\overline{I_2} \quad (22.79)$$

$$0 = R_3\overline{I_3} + \frac{d\overline{\Psi_3}}{dt} + jS\omega_1\overline{\Psi_3}; \overline{\Psi_3} = \overline{\Psi_m} + L_{l3}\overline{I_3} \quad (22.80)$$

$$\overline{\Psi}_m = L_m\left(\bar{I}_1 + \bar{I}_2 + \bar{I}_3\right); \bar{I}_2 = I_{d2} + jI_{q2}; \bar{I}_3 = I_{d3} + jI_{q3} \quad (22.81)$$

$$\frac{J}{p_1}\frac{d\omega_r}{dt} = T_e - T_{load}; T_e = \frac{3}{2}p_1L_m \cdot \mathrm{Imag}\left[\bar{I}_1 \cdot \left(\bar{I}_2^* + \bar{I}_3^*\right)\right] \quad (22.82)$$

For sinusoidal voltages

$$\overline{V}_1 = V\sqrt{2}\cos\gamma - jV\sqrt{2}\sin\gamma; \bar{I}_1 = I_{d1} + jI_{q1} \quad (22.83)$$

Equation (22.78)–(22.82) represent a seven order system with 7 variables (I_{d1}, I_{q1}, I_{d2}, I_{q2}, I_{d3}, I_{q3},ω_r).

For steady state d/dt = 0 in Equations (22.78)–(22.80) – synchronous coordinates.

When the IM is fed from a non-sinusoidal voltage source, $\overline{V}_s$ has to be calculated according to Park transformation.

$$\overline{V_s} = \frac{2}{3}\left[V_a(t) + V_b(t)e^{j\frac{2\pi}{3}} + V_c(t)e^{-j\frac{2\pi}{3}}\right]e^{-j\theta_1} \quad (22.84)$$

$$\frac{d\theta_1}{dt} = \omega_1 \quad (22.85)$$

Equation (22.85) represents the case when the fundamental frequency varies during the parameter estimation testing process.

For steady state (22.79)–(22.81) may be solved for torque T_e, stator current, I_1, and input power P_1. [22]

$$T_e(S) = 3V^2 \cdot \frac{p_1}{\omega_1} \cdot \frac{R_r}{S(A_2 + B_2)} \quad (22.86)$$

$$I_1(S) = V\sqrt{\frac{C^2 + D^2}{A^2 + B^2}} \quad (22.87)$$

$$P(S) = 3V^2 \frac{AC - BD}{A^2 + B^2} \quad (22.88)$$

$$R_r = \frac{R_2R_3(R_2 + R_3) + (R_2X_{13}^2 + R_3X_{12}^2)S^2}{(R_2 + R_3)^2 + (X_{12} + X_{13})^2 S^2} \quad (22.89)$$

$$X_{rl} = \frac{(X_{12}X_{13}(X_{12} + X_{13})S^2 + R_2^2X_{13} + R_3^2X_{12})}{(R_2 + R_3)^2 + (X_{12} + X_{13})^2} \quad (22.90)$$

$$C = 1 + X_{rl}/X_m; D = R_r/(SX_m); E = 1 + X_{l1}/X_m \quad (22.91)$$

$$A = R_1C + \frac{R_r}{S}E; B = X_{l1} + X_{rl} \cdot E - R_1D \quad (22.92)$$

$$X_i = \omega_1 L_i \quad (22.93)$$

The steady state, and transient models have as inputs the phase voltage RMS value V, its phase angle γ, p_1 pole pairs, and frequency ω_1.

For steady state, the slip value is also a given. The stator resistance R_l, stator leakage and magnetization inductances, L_{l1} and L_m, are to be known apriori.

The stator current or torque or power under steady state, are measured.

Using torque, or current or input power measured functions of slip (from S_l to S_2), the rotor double cage parameters that match the measured curve have to be found. This is only an example for steady state. In a similar way, the observation vector may be changed for transients into stator currents d–q components versus time: $I_{d1}(t)$, $I_{q1}(t)$. For a number of time steps $t_1,\dots.t_n$, the two currents are calculated from the measured stator current after transformation through Park transformation (22.87).

To summarize, the parameters vector $|X|$,

$$|X| = |R_2, R_3, L_{l2}, L_{l3}|^T \tag{22.94}$$

may be estimated from a row $|Y|$ of test data at steady state or transients,

$$|Y| = [Y(S_1), Y(S_2), \dots\dots, Y(S_n)] \tag{22.95}$$

$$Y(S_K) = T_e(S_K) \text{ or } I_1(S_K) \text{ or } P_1(S_K) \tag{22.96}$$

where $S_1,\dots.,S_n$ are the n values of slip considered for steady state. For transients, the test data $I_{d1}(t)$, $I_{q1}(t)$ are acquired at n time instants $t_1,\dots.t_n$. The stator current d–q components in synchronous coordinates now constitute the observation vector:

$$[Y] = [Y(t_1), Y(t_2), \dots\dots, Y(t_n)] \tag{22.97}$$

Table 22.4 Estimated parameters for various observation vectors [22]

	Estimated parameters [Ω]				
	R_2	R_3	X_{l2}	X_{l3}	X_{l1}
Values used to calculate the input (theoretical) characteristics and runs	3.1144	0.2717	3.0187	4.6098	3.1316
Values obtained after estimation from $T_e = f(S)$	3.1225	0.2717	3.032	4.6145	3.1261
Values obtained after estimation from $I_s = f(S)$	3.0221	0.2722	2.868	4.5552	3.1946
Values obtained after estimation from $i_a = f(S)$	3.0842	0.2718	2.9706	4.5931	3.1513
Values obtained after estimation from $\|i_s\| = f(t)$	3.0525	0.2721	2.9099	4.5671	3.1762
Values obtained after estimation from $T_e = f(t)$	3.1114	0.2664	3.0693	4.5831	3.093

$$Y(t_K) = I_{d1}(t_K) \text{or} I_{q1}(t_K) \tag{22.98}$$

In general, the problem to solve may be written as

$$Y = f(\zeta_K, X) + \xi_K = 1,...,n \tag{22.99}$$

with $\zeta = S_K$ for steady-state estimation and $\zeta = t_K$ for transients estimation. The parameters ξ_K is the error for the Kth value of S_K (or t_K).

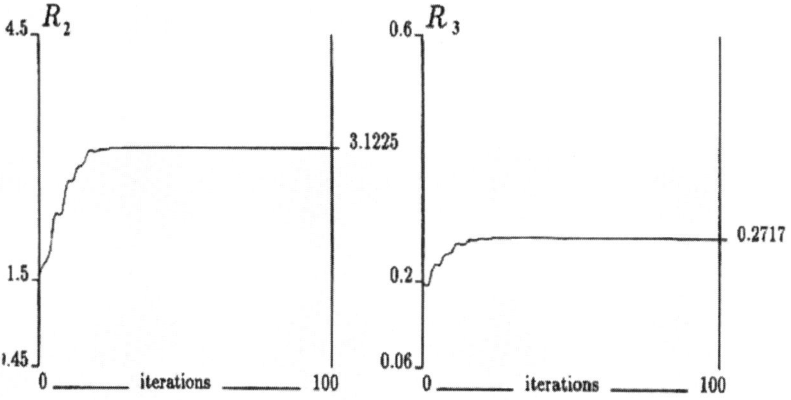

Figure 22.25 Convergence progress in rotor resistance estimation from $T_e(S)$ [22]

There are many ways to solve such a problem iteratively. Gauss–Newton method, when used after linearization, tends to run into almost singularity matrix situations. On the contrary, the step-wise regression method, based on Gauss – Jordan pivot concept [21], seems to produce safely good results across the board. For details on the computation algorithm, see Reference [22]. The method's convergence should not depend on the width of change limits of parameters or on their assigned start values.

Typical results obtained in [22] are given in Table 22.4 for steady state and, respectively, for dynamic (direct starting) operation modes.

Sample results on convergence progress for $T_e(S)$ observation vector show fast convergence rates in Figure 22.25. [22]

The corresponding torque/slip curve fitting is, as expected, very good (Figure 22.26).

Similarly good agreement for the torque–time curve during starting transients is shown in Figure 22.27 when the $T_e(t)$ function is observed.

A few remarks are due.

- The curve fitting of the observed vectors over the entire explored zone is very good and the rate of convergence is safe and rather independent of the extension of parameter existence range.
- The estimated parameter vectors (Table 22.4) have only up to 3% variations depending on the observation vector; torque or current. This

is a strong indication that if the torque/slip is observed, the current/slip curve will also fit rather well, etc.

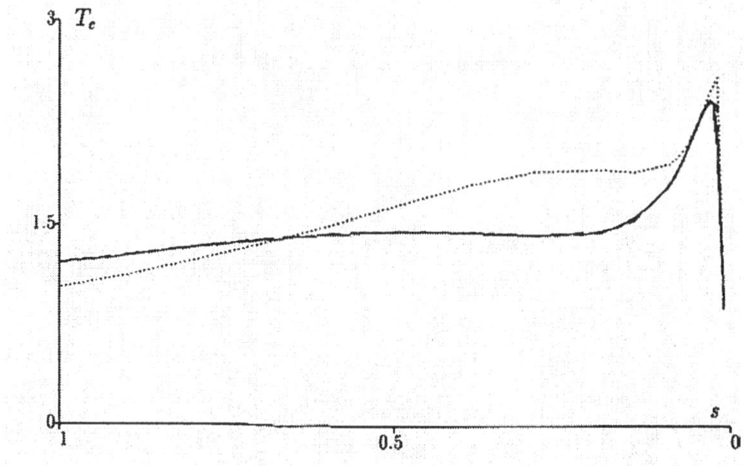

Figure 22.26 The static characteristic of electromagnetic torque $T_e = f(S)$: [22]
____ – the theoretical characteristic of $T_e = f(S)$
...... – the characteristic calculated for starting values of parameters
------ – the characteristic calculated for final values of estimated parameters

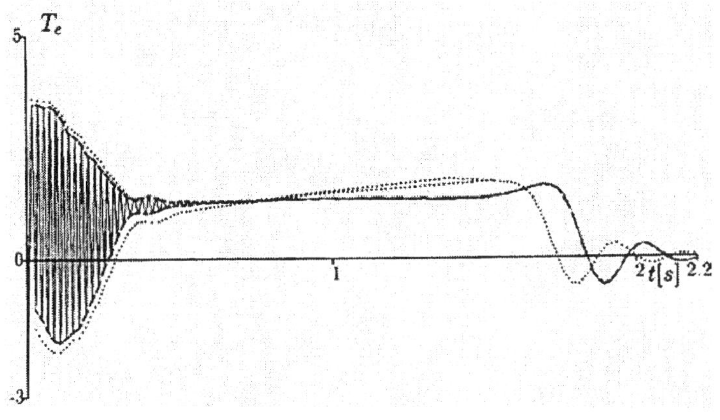

Figure 22.27 Runs of the electromagnetic torque T_e during the start – up: [22]
____ – the theoretical run of $T_e = f(t)$
...... – the run calculated for starting values of parameters
------ – the run calculated for final values of estimated parameters

- The phase current measurement is straightforward but the torque measurement is not so. We may however calculate the torque off line from measured voltages and currents V_a, V_b, I_a, I_b,

$$Te = \frac{3}{2} p_1 \left(\Psi_\alpha I_\beta - \Psi_\beta I_\alpha \right) \tag{22.100}$$

$$\begin{aligned} V_\alpha &= V_a \\ V_\beta &= \frac{1}{\sqrt{3}} \left(V_a + 2V_b \right) \\ I_\alpha &= I_a \\ I_\beta &= \frac{1}{\sqrt{3}} \left(I_a + 2I_b \right) \end{aligned} \tag{22.101}$$

$$\begin{aligned} V_\alpha &= \int \left(V_\alpha - R_1 I_\alpha \right) dt \\ V_\beta &= \int \left(V_\beta - R_1 I_\beta \right) dt \end{aligned}$$

The precision of torque estimates by (22.100) is good at rated frequency but smaller at low frequencies f_1. The integrator should be implemented as a filter to cancel the offset. In (22.100), the reduction of the electromagnetic torque due to core loss, which is smaller than 1% of rated torque in general, is neglected. If core loss p_{core} is known from the no load test, the torque of (22.100) may be reduced by ΔT_{ecore}.

$$\Delta T_{ecore} = p_{core} \times \frac{p_1}{\omega_1} \tag{22.102}$$

- The direct start is easy to perform, as only two voltages and two currents are to be acquired and the data is processed off line. It follows that this test may be the best way to estimate the IM parameters for a large spectrum of operation modes. Such a way is especially practical for medium–large power IMs.
- Very large power machines direct starting is rather slow as the relative inertia is large and the power grid voltage drops notably during the process. In such a case, the method of parameter estimation data may be simplified as shown in the following paragraph where also the inertia is estimated.

22.4.6 Large IM inertia and parameters from direct starting acceleration and deceleration data

Testing large power IMs takes time and notable human effort. Any attempt to reduce it while preserving the quality of the results should be considered.

Direct connection to the power grid, even if through a transformer at lower than rated voltage in a limited power manufacturer's laboratory, seems a practical way to do it. The essence of such a test consists of the fact that the acceleration would be slow enough to validate steady state operation mode approximation all throughout the process. In large power IMs, reducing the supply voltage to 60% would do it in general. We assume that the core loss is proportional to voltage squared. By performing no load tests at decreasing voltages, we can separate first the mechanical from iron losses by the standard loss segregation method. The mechanical losses vary with speed. To begin, we may suppose a linear dependence

$$p_{mec} \approx p_{mec0} \cdot \frac{n}{f_1/p}$$
$$p_{iron} = p_{iron0} \cdot \left(\frac{V}{V_0}\right)^2 \tag{22.103}$$

During the no load acceleration test, the voltage and current of two phases are acquired together with the speed measurement. Based on this, the average power is calculated. To simplify the process of active power calculation, synchronous coordinates may be used.

$$\begin{vmatrix} V_d \\ V_q \end{vmatrix} = \frac{2}{3} \begin{vmatrix} \cos(-\omega_1 t) & \cos\left(-\omega_1 t + \frac{2\pi}{3}\right) & \cos\left(-\omega_1 t - \frac{2\pi}{3}\right) \\ \sin(-\omega_1 t) & \sin\left(-\omega_1 t + \frac{2\pi}{3}\right) & \sin\left(-\omega_1 t + \frac{2\pi}{3}\right) \end{vmatrix} \cdot \begin{vmatrix} V_a \\ V_b \\ -(V_a + V_b) \end{vmatrix} \tag{22.104}$$

The same formula is valid for the currents I_d and I_q. Finally, the instantaneous active power P_1 is

$$P_1 = \frac{3}{2}\left(V_d I_d + V_q I_q\right) \tag{22.105}$$

As V_d, V_q, I_d, I_q, for constant frequency ω_1 are essentially d.c. variables the power signal P_1 is rather clean and no averaging is required.

We might define now the kinetic useful power P_K as

$$P_K(n) = P_{elm}(1 - S) + p_{mec}$$
$$P_{elm} = P_1 - 3R_s I_1^2 - p_{iron} \tag{22.106}$$

As all terms in (22.106) are known for each value of speed, the kinetic useful power $P_K(n)$, dependent on speed, may be found. The time integral of $P_K(n)$ gives the kinetic energy of the rotor E_c.

$$E_c = \int_0^t P_K(n) dt = \frac{J}{2} 4\pi^2 n_f^2 \tag{22.107}$$

where n_f is the final speed considered.

We may terminate the integration at any time (or speed) during acceleration. Alternatively,

$$P_K = J 4\pi^2 n \frac{dn}{dt} \tag{22.108}$$

At every moment during the acceleration process, we may calculate P_K, measure speed n and build a filter to determine dn/dt.

Consequently, the inertia J may be determined either from (22.107) or from (22.108) for many values of speed (time) and take an average.

However, (22.107) avoids the derivative of n and performs an integral and thus the results are much smoother.

Sample measurements results on a 7500 kW, 6000 V, 1490 rpm, 50 Hz, IM with $R_1 = 0.0173\Omega$, $P_{mec0} = 44.643$ kW and $P_{iron} = 78.628$ kW are shown in Figure 22.28. [23]

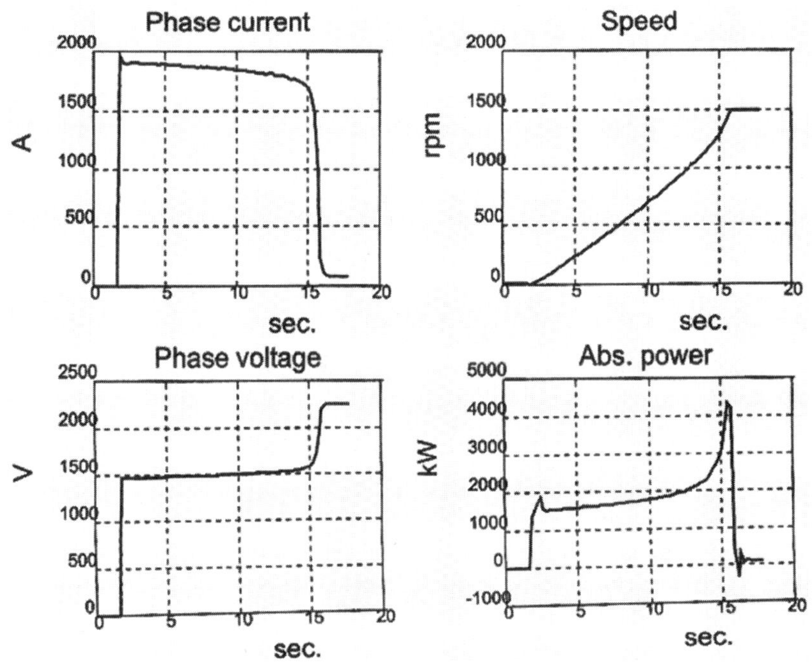

Figure 22.28 Current, phase, voltage, power and speed during direct starting of a 7500kW IM

The moment of inertia, calculated during the acceleration process, by (22.108), is shown in Figure 22.29.

Even with dn/dt calculated, J is rather smooth until we get close to the settling speed zone, which has to be eliminated when the average J is calculated.

The torque may also be calculated from two different expressions (Figure 22.30).

$$T = 2\pi J \frac{dn}{dt} \tag{22.109}$$

$$T_e = \frac{P_K}{2\pi n} \tag{22.110}$$

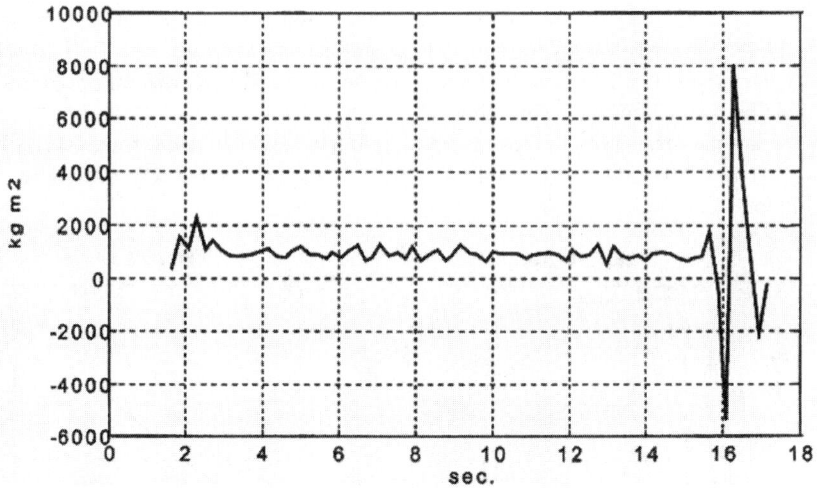

Figure 22.29 Inertia calculated during the acceleration process [23] using the speed derivative (22.108)

Again, Equation (22.110), which avoids dn/dt, produces a smoother torque/time curve.

The torque may be represented as a function of speed also, as speed has been acquired.

The torque in Figure 22.30 is scaled up to 6kV line voltage (star connection), assuming that the torque is proportional to voltage squared.

The testing may add a free deceleration test from which, with inertia known from the above procedure, the more exact mechanical loss dependence on speed $p_{mec}(n)$ may be determined.

$$P_{mec}(n) = -J2\pi\frac{dn}{dt} \tag{22.111}$$

Then $P_{mec}(n)$ from (22.111) may be used in the acceleration tests processing to get a better inertia value. The process proved to converge quickly. Finally, an exponential regression of mechanical losses has been adopted.

$$P_{mec} = \alpha n^{2.4}; \alpha = 0.001533 \tag{22.112}$$

The stator voltage attenuation during free deceleration has also been acquired.

Sample deceleration test results are shown in Figure 22.31 a,b,c.

The mechanical losses, determined from the deceleration tests, have been consistently smaller (that is for quite a few machines tested: 800 kW, 2200 kW, 2800 kW, 7500 kW) than those obtained from loss segregation method.

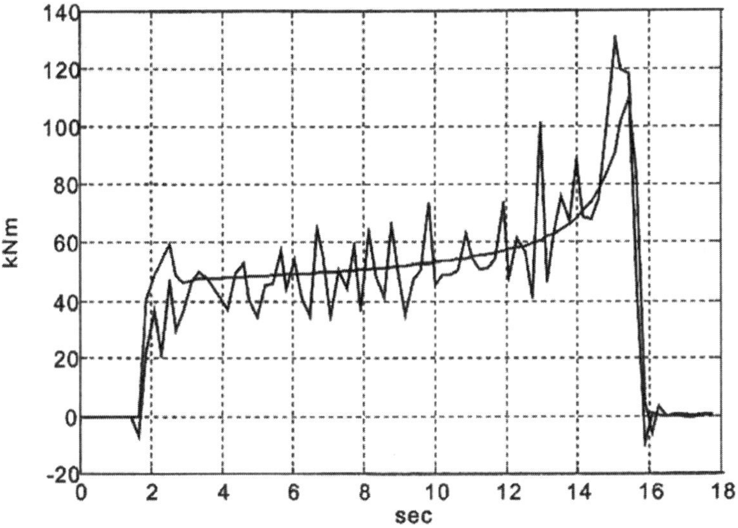

Figure 22.30 Torque versus time during acceleration [23] from (22.109) and from (22.110)

In our case at no load speed P_{mec} = 64.631kW from the free deceleration tests. That is, 14 kW less than from the loss segregation method.

This difference may be attributed to additional (stray losses).

To calculate the stray load losses, we may adopt the formula

$$P_{stray}(I) = \left[(P_{mec})_{l.s.m.} - (P_{mec})_{f.d.m.}\right] \cdot \left(\frac{I}{I_0}\right)^2 \tag{22.113}$$

l.s.m. – loss segregation method

f.d.m. – free deceleration method

where I is the stator current and I_0 the no load current.

To complete loss segregation and secure performance assessment, all IM parameters have to be determined. We may accomplish this goal avoiding shortcircuit test due to both time (work) needed and also due to the fact that large IMs tend to have strong rotor skin effect at standstill and thus a double cage rotor circuit has to be adopted.

The IM parameters may be identified from the acceleration and free deceleration test data alone through a regression method. [23]

22.5. NOISE AND VIBRATION MEASUREMENTS: FROM NO-LOAD TO LOAD

Noise in induction machine has electromagnetic and mechanical origins. The noise level accepted depends on the environment in which the IM works. At the present time, standards (ISO1680) prescribe noise limits for IMs, based on sound power levels measured at no load conditions, if the noise level does not vary with load. It is not as clear how to set the sufficient conditions for no load noise tests.

The load machine produces additional noise and thus, it is simpler to retort to IM no load noise tests only.

If the electromagnetic noise is relatively large, it is very likely that under load it will increase further. [22] Also, when PWM converter fed, the IM noise will change from no-load to load conditions. Skipping the basics of sound theory [26] in electrical machines, we will synthesize here ways to measure the noise under load in IMs.

22.5.1 When on-load noise tests are necessary?

Noise tests with the machine on no-load (free at shaft) are much easier to perform, eventually in an acoustic chamber (anechoic room for free field or reverberant room for diffuse field). So it is natural to take the pains and perform noise measurements on load only when the difference in noise level between the two situations is likely to be notable.

An answer to this question may be that the on load noise measurements are required when the electromagnetic noise at no-load is relatively important. Under load, space harmonics fields accentuate and are likely to produce increased noise, aside from the noise related to torque pulsations (parasitic torques) and uncompensated radial forces.

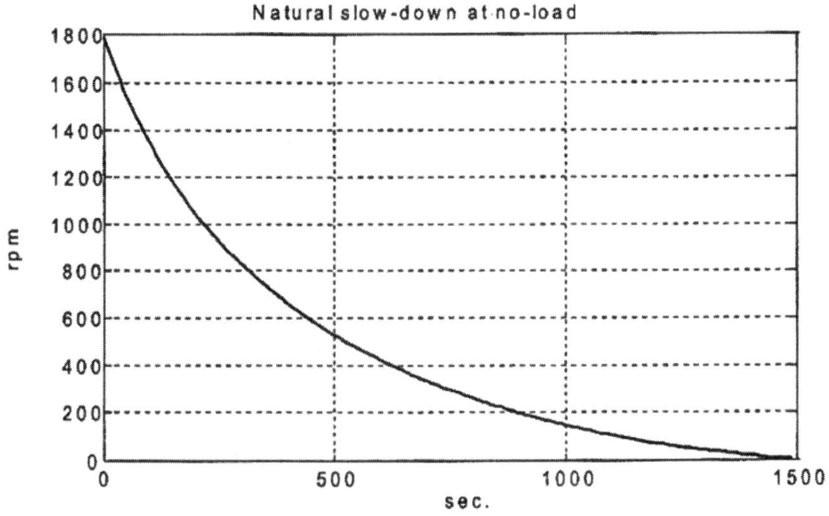

a.)

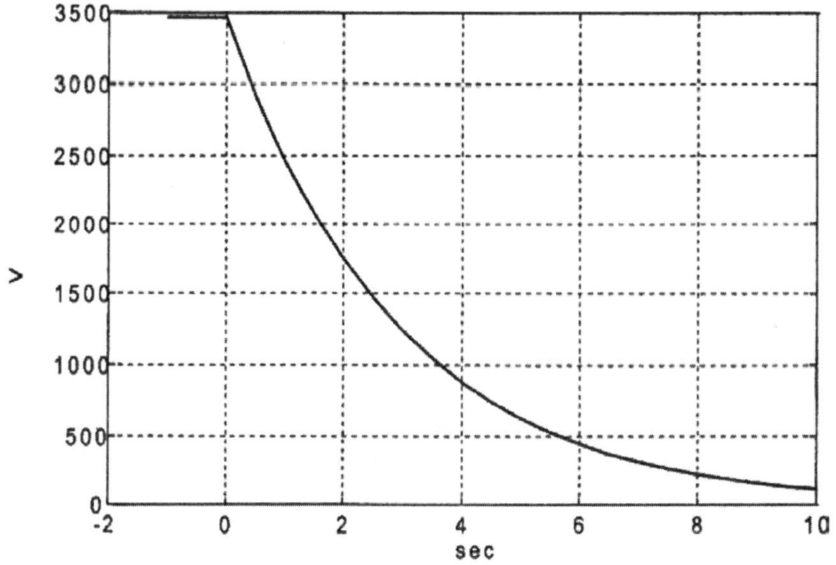

b.)

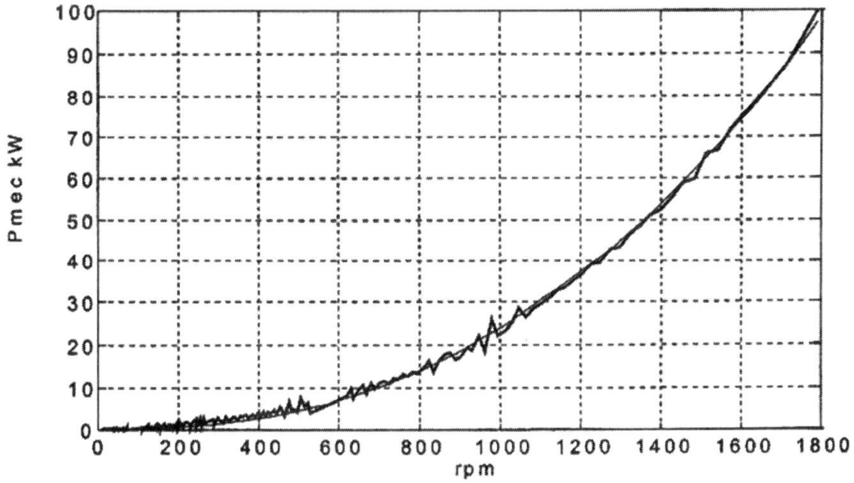

c.)

Figure 22.31 Free deceleration tests [23]
a.) speed versus time; b.) stator voltage amplitude versus time
c.) mechanical losses versus speed

To separate the no-load electromagnetic noise sound power level $L_{W,em}$, three tests on no load are required [27,29]:

a) The no-load sound power L_{wo} is measured according to standards.

b) After taking off the ventilator to sound power level L_{Wf} is measured. Subtracting L_{Wf} from L_{Wo}, we determine the ventilator system sound power level $L_{W,vent}$.

$$L_{W,vent} = 10\log\left(10^{0.1L_{Wo}} - 10^{0.1L_{Wf}}\right) \tag{22.114}$$

c) Without the ventilator, the no-load noise measurements are carried out at ever smaller voltage level before the speed decreases. By extrapolating for zero voltage the mechanical sound power level $L_{W,mec}$ is obtained. Consequently, the electromagnetic sound level at no load $L_{W,em}$ is

$$L_{W,em} = 10\log\left(10^{0.1L_{Wo}} - 10^{0.1L_{Wmec}}\right) \tag{22.115}$$

Now if the electromagnetic sound level $L_{W,em}$ is more than 8dB smaller than the total mechanical sound power, the noise measurements on load are not necessary.

$$L_{W,em} < 8\text{dB} + 10\log\left(10^{0.1L_{W,mec}} + 10^{0.1L_f}\right) \tag{22.116}$$

22.5.2 How to measure the noise on-load

There are a few conditions to meet for noise tests

- The sound pressure has to be measured by microphones placed in the acoustic far field (about 1m from the motor)
- The background sound pressure level radiated by any sound source has to be 10 dB lower than the sound pressure generated by the motor itself
- Every structure-based vibration has to be eliminated

Acoustical chambers fulfil these conditions but large machine shops, especially in the after hours, are also suitable for noise measurements.

Under on-load conditions, the loading machine and the mechanical transmission (clutch) produce additional noise. As the latter is located close to the IM, its contribution to the resultant noise power level is not easy to segregate.

Separation of the loading machine may be performed through putting it outside the acoustic chamber which contains the tested motor.

Also, an acoustic capsule may be built around the coupling plus the load machine with the tested motor placed in the machine shop.

Both solutions are costly and hardly practical unless a special “silent” motor is to be tested thoroughly.

Thus, in situ noise measurements, in large machine shops, after hours, appears as the practical solution for noise measurements on load.

The sound pressure or power is measured in the far field 1 m from the motor in general. However, to reduce the influence of the loading machine noise, the measurements are taken at near field 0.2 m from the motor.

However, in this case, near field errors occur mainly due to motor vibration modes.

To eliminate the near field errors, comparative measurements in the same points are made.

For example, in a near field point (0.2m from the motor), the sound pressure is measured on no-load and with the load coupled to obtain $L_{W,on}$ and $L_{W,n}$. The difference between the two represents the load contribution to noise and is not affected by the near field error.

$$\Delta L_n = L_{W,n} - L_{W,on} \tag{22.117}$$

This method of near field measurements is most adequate for tests at the working place of the IMs.

Noise, in the far or near field, may be measured directly by sound pressure microphones (Figure 22.32) or calculated based on vibration speed measurements.

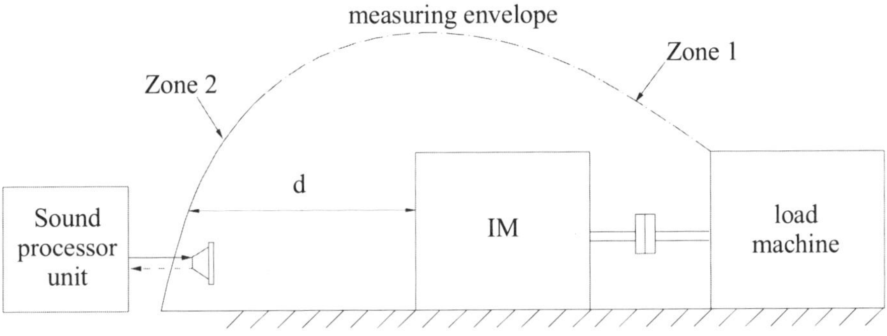

Figure 22.32 Sound intensity measurement arrangements

The new microphone method measures sound pressure and produces very good results if the microphone placement avoids the aggregation of load machine noise contributions. Their placements in zone 2 (Figure 22.32) at d = 1 m from motor, though the best available solution may still lead to about 3 dB increase in noise as loading machine contribution. This may be considered a systematic error at worst.

The artificial loading by PWM dual frequency methods avoids the presence of the load machine and it may constitute a viable method to verify IM noise measurements under load.

The noise level may also be calculated from vibration mode speeds measurement. Measurements of IM frame vibration mode speeds are made on its frame. This is the traditional method and is described in detail in the literature. [24–28]

Experimental results reported in Ref [27] lead to conclusions such as:

- The sound power level may be determined directly by sound intensity (pressure) measurement in the far field (1 m away from IM) through microphones (Figure 22.32)
- The sound power level of the IM under load may be determined indirectly via measurements by microphones, on no-load and on-load, made in the near field (0.2 m) of the IM (Equations (22.125)–(22.126))
- Finally, the sound power may be calculated from noise or vibration measurements depending on which the mechanical or electromagnetic noise, dominates
- All these procedures seem worthy of further consideration, as their results are close. In [27], the far field microphone method departured from the other two by 0.80 dB (from 78.32 dB to 79.12 dB on-load)
- A ΔL_n of 6.84dB increase due to load was measured for a 5kW IM in Ref [27]

For details on noise and vibration essentials, see [28].

In the power electronics era, IM noise theory testing and reduction are expected to prompt new R & D efforts.

22.6 SUMMARY

- IM testing refers essentially to loss segregation, direct and indirect load testing, parameter estimation and noise and vibration measurements.
- Loss segregation is performed to verify design calculations and avoid full load testing for efficiency calculation.
- The no-load and standstill tests are standard in segregating the mechanical losses, fundamental core losses, and winding losses.
- The key point in loss segregation is however the so-called stray load losses.
- Stray load losses may be defined by identifying their origins: rotor and stator surface and teeth flux pulsation space harmonics core losses, the rotor cage non-fundamental losses. The latter are caused by stator m.m.f. space harmonics, augmented by airgap permeance harmonics, and/or inter-bar rotor core losses. Separating these components in measurements is not practical.
- In face of such a complex problem, the stray load losses have been defined as the difference between the total losses calculated from load tests and the ones obtained from no-load and standstill (loss segregation) tests.
- Though many tests to directly measure the stray load losses have been proposed, so far only two seem to hold.
- In reverse motion test, the IM fed at low voltage is driven at rated speed but in reverse direction to the stator voltage system. The slip $S \approx 2$. Though the loss equivalence with the rated load operation is far from complete, the elimination of rotor cage fundamental loss from calculations "saves" the method for the time being.

- In the no-load test extension over rated voltage – so far performed only on small IMs – the power input minus stator winding losses depart from its proportionality to voltage squared. This difference is attributed to additional "stray load" losses considered then proportional to current squared.
- The stator resistance is traditionally d.c. measured. However, for large IMs, a.c. measurements would be useful. The a.c. test at low voltage with the rotor outside the machine is considered to be pertinent for determining both the stator resistance and leakage inductance.
- The no-load and standstill tests may be performed by using PWM converter supplies, especially for IM destined for variable speed drives. Additional losses, in comparison with sinusoidal voltage supply, occur at no-load, not so at standstill test as the switching frequency in the converter is a few kHz.
- If the distribution of the losses is known, in the moment their source is eliminated by turning off the stator winding, the temperature gradient is proportional to those losses. Consequently, measuring temperature versus time leads to loss value. Unfortunately, loss distribution is not easy to find unless FEM is used and the temperature sensors, placed at sensitive points, represent an intrusion which may be accepted only for prototyping.
- On the contrary, the calorimetric method – in the dual chamber version – leads to the direct determination of overall losses, with an error of less than 5% generally.
- Load tests are performed to determine efficiency and temperature rise for various loads.
- Efficiency may be calculated based on loss segregation methods or by direct (or indirect: artificial) loading tests. In the latter case, both the input and output powers are measured.
- Standard methods to determine efficiency are amply presented in the international standards IEEE–112 (USA), IEC–34–2 (Europe) and JEC (Japan). The essential difference between these three standards lies in the treatment of non-fundamental (or stray load) losses. JEC apparently considers the stray load losses as zero, IE–34–2 takes them as a constant: 0.5% of rated power. IEEE-112, through its many alternative methods, determines them as the difference between overall losses in direct load tests and no-load losses at rated voltage plus standstill losses at rated current. Alternatively, the latter sets constant values for stray load losses, all higher than 0.5% - from 1.8% below 90 kW to 0.9% above 1800 kW. So, in fact, the same IM tested according to these representative standards would have different labelled efficiency. Or, conversely, an IM with the same label efficiency and power will produce most power if the efficiency was measured according to IEEE–112 and least power for JEC efficiency definition. The differences may run as high as 2%.

- IEEE–112B by its provisions seems much closer to reality.
- The direct torque measurement through a brushless torque-meter in direct load tests may be avoided either by a calibrated d.c. generator loading machine or by estimating the torque of an inverter–fed IM machine load. Alternatively, the tested IM may be run both as a motor and generator at equal slip values. Rather, simple calculations will allow stray load loss and efficiency computation based on input power, current, voltage, and slip precise measurements.
- Avoiding altogether the burden of coupling a second machine at the tested IM shaft led to artificial (indirect) loading. The two frequency method introduced in 1921 has been used only for temperature rise calculations as loss equivalence with direct load tests was not done equitably.
- The use of PWM converter to mix frequency and provide indirect (artificial) loading apparently changes the picture. Equivalence of losses may be now pursued elegantly and the whole process may be mechanized through PC control. Good agreement in efficiency with the direct methods has been reported recently. Is this the way of the future? More experience with this method will only tell.
- A rather conventional indirect method, applied by a leading manufacturer for 40 years is the forward shortcircuit approach. In essence, the unsupplied IM is first driven at rated speed by a 10% rating IM of the same number of poles. Then another 10% rating motor drives a synchronous generator to produce about 80–85% rated frequency of the tested IM. Finally, the a.c. generator is excited and its terminals are connected to the tested motor. The voltage is raised to reach the desired current. The tested IM works at high slip as a generator and at rated speed and thus the active power output is low: 10% of rated power. Despite the fact that the distribution of losses differs from that in direct load tests, with more losses on rotor and less on stator and skin rotor effect presence, good stator temperature rise agreement with direct load tests has been reported for large IMs. The introduction of bi-directional power electronics to replace the second drive motor plus a.c. generator might prove a valuable way to improve on this genuine and rather practical method for large power IMs.
- Parameter estimation is essential for IM precise modelling over a wide spectrum of operation modes.
- Single and multiple rotor loop circuit models are used for modelling.
- Basically, only no-load and shortcircuit tests are standardized for parameter estimation.
- No-load tests provide reliable results on the core loss resistance and magnetization inductance dependance on magnetization current.
- Shortcircuit tests performed at standstill and at rated frequency contain too much skin effect influence on rotor resistance and leakage

inductance, which precludes their usage for calculating on-load IM performance.

- Even for starting torque and current estimation at rated voltage, they are not proper as they are performed up to rated current and thus the leakage flux path saturation is not considered. Large errors at full voltage – up to 60–70% – may be expected in torque in extreme cases.
- Two frequency shortcircuit tests at rated current – one at rated frequency and the other at 5% rated frequency – could lead to a dual cage rotor model good both for large and low slip values, provided the current is not larger than 150–200% of rated current; most PWM converter–fed general variable speed IM drives qualify for these conditions.
- The standstill frequency response (SSFR) method is an improvement on the two-frequency method but, being performed at low current and many frequencies, excludes leakage saturation effect as well and time prohibitive. Its validity is restricted to small deviation transients and stability analysis in large IMs.
- Even catalogue – data – based methods have been proposed to match (by a single model with dual virtual cage rotor) both the starting and rated torque and currents. However, in between, around breakdown torque, large errors persist.
- The general regression method, based on steady state or transients measurements, for wide speed range, from standstill to no-load speed, has been successfully used to estimate the parameters of an equivalent dual cage rotor model according to the criterion of least squared error. Thus, a wide spectrum of operation modes can be handled by a single but adequate model. A slow acceleration start and a direct (fast) one are required to cover large ranges of current and rotor slip frequency. The slow start may be obtained by mounting an inertia disk on the shaft. However direct fast start transients tend to produce, by the general regression method, about the same parameters as the steady state test and may be preferred.
- A simplified version of the naturally slow acceleration (quasi-steady state electromagnetic-wise) for very large IMs has been successfully used to separate the losses, calculate the inertia and dual cage rotor model parameters that fit the whole slip range.
- Noise and vibration tests are environmental constraints. Only no-load noise tests are standardized but sometimes on-load noise is larger by 6-8 dB or more
- When and how on-load in situ noise tests are required (done) is discussed in detail. For PWM converter fed IMs, variable speed noise tests seem necessary. [29]
- We did not exhaust, rather prioritize, the subject of IM testing. Many other methods have been proposed and will continue to be introduced

as the IM experimental investigations may see a new boost by making full use of PWM converter variable voltage and frequency supplies.

- An interesting example is the revival of the method of driving the IM on no-load through the synchronous speed and detect the power input jump at that speed. This power jump corresponds basically to hysteresis losses. The test is made easier to do with a PWM converter supply with a precise speed control loop.

22.7 REFERENCES

1. P. Bourne, No-Load Method for Estimating Stray Load Loss in Small Cage Induction Motors, Proc. IEE, vol. 136–B, no. 2, 1989, pp.92–95.

2. A. A. Jimoh, S. R. D.Findlay, M. Poloujadoff , Stray Losses in Induction Machines, Parts 1 and 2, IEEE Trans. Vol. PAS–104, no. 6, 1985, pp.1500–1512.

3. C. H. Glen, Stray Load Losses in Induction Motors, – A Challenge to Academia, Record of EMD Conf. 1997, IEEE Publ. no. 444, pp.180–184.

4. B. J. Chalmers and A. C. Williamson, Stray Losses in Squirrel-Cage Motors, Proc. IEE, vol.110, Oct.1963.

5. A. Boglietti, P. Ferraris, M. Lazari, F. Profumo, Loss Items Evaluation in Induction Motors Fed by Six-step VSI, EMPS–vol. 19, no. 4, 1991, pp. 513–526.

6. N. Benamrouche et.al, Determination of Iron Loss Distribution in Inverter-Fed Induction Motors, IBID, vol. 25, no. 6, 1997 pp. 649–660.

7. N. Benamrouche et.al., Determination of Iron and Stray Load Losses in IMs Using Thermometric Method, IBID, vol. 26, no. 1, 1998, pp.3–12.

8. IEC Publication 34–2A, Methods for Determining Losses and Efficiency of Rotating Electric Machinery from Tests, Measurement of Losses by the Calorimetric Method, 1974 First supplement to Publication 34–2 (1972), Rotating Electric Machines, Part 2.

9. A. Jalilian, V. J. Gosbell, B. S. P. Perera, Double Chamber Calorimeter (DCC): A New Approach to Measure Induction Motor Harmonic Losses, IEEE Trans. vol. EC–14, no. 3, 1999, pp. 680–685.

10. IEEE Standard Procedure for Polyphase Induction Motors and Generators, IEEE Std. 112–1996, IEEE–PES, New York, NY.

11. Rotating Electrical Machines Methods for Determining Losses and Efficiency of Rotating Electrical Machines from Tests, IEC Std. 34–2:1972.

12. R. Reiner, K. Hameyer, R. Belmans, Comparison of Standards for Determining Efficiency of Three Phase Induction Motors, IEEE Trans. vol. EC–14, no. 3, 1999, pp. 512–517.

13. L. Kis, I. Boldea, A. K. Wallace, Load Testing of Induction Machines Without Torque Measurements, Record of IEEE–IEMDC–1999, Seattle, U.S.A.

14. A. Ytterberg, Ny Method for Fullbelasting av Electriska Maskiner Utan Drivmotor Eller Avlastningsmaskin, S. K. Shakprov Technisk Lidskrift, no. 79, 1921, pp. 42–46.

15. C. Grantham, Full Load Testing of Three Phase IMs Without the Use of a Dynamometer, ICEMA 14–16, Sept. 13, 1994, Adelaide–Australia pp. 147–152.

16. L. Tutelea, I. Boldea, E. Ritchie, P. Sandholdt, F. Blaabjerg, Thermal Testing for Inverter–fed IMs Using Mixed Frequency Method, Record of ICEM–1998, vol. 1, pp. 248–253.

17. D. H. Plevin, C. H. Glew, J. H. Dymond, Equivalent Load Testing for Induction Machines–The Forward Shortcircuit Test, IEEE Trans vol. EC–14, no. 3, 1999, pp. 419–425.

18. A. Boglietti, P. Ferraris, M. Lazzari, F. Profumo, Test Procedure for Very High Speed Spindle Motors, EMPS vol. 23, no. 4, 1995, pp. 443–458.

19. J. R. Willis, G. J. Brock, J. S. Edmonds, Derivation of Induction Motor Models From Standstill Frequency Response Tests, IEEE Trans, vol. EC–4, no. 4, 1989, pp. 608–615.

20. S. A. Soliman, G. S. Christensen, Modelling of Induction Motors From Standstill Frequency Response Tests and a Parameter Estimation Algorithm, EMPS vol. 20, no. 2, 1992, pp. 123–136.

21. R. I. Jennrich and P. F. Sampson, Application of Stepwise Regression to Non–Linear Estimation, Technometrics, 10(1), 1968.

22. K. Macek–Kaminska, Estimation of the Induction Machine Parameters, EMPS vol. 23, no. 3, 1995, pp. 329–344.

23. R.Babau, Inertia, Parameter, and Loss Identification of Large Induction Motors Through Free Acceleration and Deceleration Tests Only, Ph.D Thesis, 2001, University "Politehnica" Timisoara (Prof. I.Boldea).

24. R.Brozek, No-Load to Full Load Airborne Noise Level Change in High Speed Polyphase Induction Motors, IEEE Trans. Vol. IA–9, 1973, pp. 1973, pp. 180-184.

25. P. L. Timar, Noise Measurement of Electrical Machines Under Load, Record of ICEM–90, Boston, Part 2, pp. 911–916.

26. P. L. Timar, Noise and Vibration of Electrical Machines, Elsevier, 1989, New York.

27. P. L. Timar, Noise Test in Rotating Electrical Motors Under Load, EMPS vol. 20, no. 4, 1992, pp. 339–353.

28. S. A. Nasar (editor), Handbook of Electric Machines, Mc.Graw Hill, 1987, Chapter 14 by S. J Jang and A. J. Ellison.

29. R. P. Lisner, P. L. Timar, A New Approach to Electric Motor Acoustic Noise Standards and Test Procedures, IEEE Trans. vol. EC–14, no. 3, 1998, pp. 692–697.

Chapter 23

SINGLE-PHASE INDUCTION MACHINES: THE BASICS

23.1 INTRODUCTION

Most small power (generally below 2 kW) induction machines have to operate with single-phase a.c. power supplies that are readily available in homes, and remote rural areas. When power electronics converters are used three phase a.c. output is produced and thus three phase induction motors may still be used.

However, for constant speed applications (the most frequent situation), the induction motors are fed directly from the available single-phase a.c. power grids. In this sense, we call them single phase induction motors.

To be self-starting, the induction machine needs a travelling field at zero speed. This in turn implies the presence of **two windings** in the stator, while the rotor has a standard squirrel cage. The first winding is called the **main winding** while the second winding (for start, especially) is called **auxiliary winding**.

Single phase IMs may run only on the main winding once they started on two windings. A typical case of single phase single-winding IM occurs when a three IM ends up with an open phase (Chapter 7). The power factor and efficiency degrade while the peak torque also decreases significantly.

Thus, except for low powers (less than ¼ kW in general), the **auxiliary winding** is active also during running conditions to improve performance.

Three types of single-phase induction motors are in use today:

- Split-phase induction motors
- Capacitor induction motors
- Shaded-pole induction motors

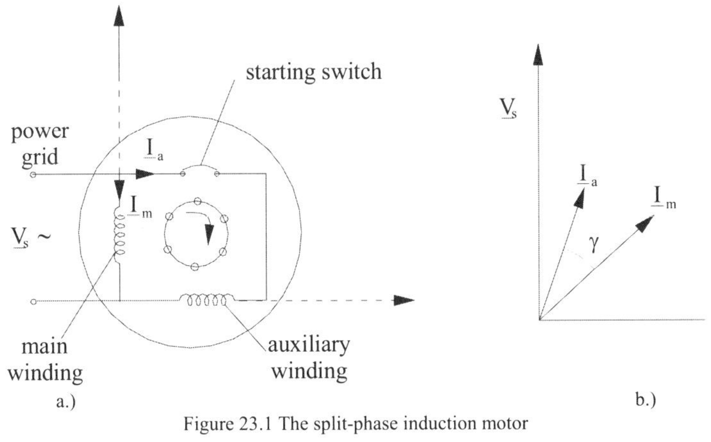

Figure 23.1 The split-phase induction motor

23.2 SPLIT-PHASE INDUCTION MOTORS

The split phase induction motor has a main and an auxiliary stator winding displaced by 90^0 or up to $110\text{-}120^0$ degrees (Figure 23.1a).

The auxiliary winding has a higher ratio between resistance and reactance, by designing it at a higher current density, to shift the auxiliary winding current $\underline{I}_a$ ahead of main winding current $\underline{I}_m$ (Figure 23.1b).

The two windings-with a 90^0 space displacement and a $\gamma \approx 20\text{-}30^0$ current time phase shift-produce in the airgap a magnetic field with a definite forward travelling component (from m to a). This travelling field induces voltages in the rotor cage whose currents produce a starting torque which rotates the rotor from m to a (clockwise on Figure 23.1).

Once the rotor catches speed, the starting switch is opened to disconnect the auxiliary winding, which is designed for short duty. The starting switch may be centrifugal, magnetic, or static type. The starting torque may be up to 150% rated torque, at moderate starting current, for frequent starts long-running time applications. For infrequent starts and short running time, low efficiency is allowed in exchange for higher starting current with higher rotor resistance.

During running conditions, the split-phase induction motor operates on one winding only and thus it has a rather poor power factor. It is used below 1/3 kW, generally, where the motor costs are of primary concern.

23.3 CAPACITOR INDUCTION MOTORS

Connecting a capacitor in series with the auxiliary winding causes the current in that winding $\underline{I}_a$ to lead the current in the main winding $\underline{I}_m$ by up to 90^0.

Complete symmetrization of the two windings m.m.f. for given slip may be performed this way.

That is a pure travelling airgap field may be produced either at start ($S = 1$) or at rated load ($S = S_n$) or somewhere in between.

An improvement in starting and running torque density, efficiency and especially in power factor is brought by the capacitor presence. Capacitor motors are of quite a few basic types:

- Capacitor-start induction motors
- Two-value capacitor induction motors
- Permanent-split capacitor induction motors
- Tapped-winding capacitor induction motors
- Split-phase capacitor induction motors
- Capacitor three phase induction motors (for single-phase supply)

23.3.1 Capacitor-start induction motors

The capacitor-start IM (Figure 23.2a) has a capacitor and a start switch in series with the auxiliary winding.

The starting capacitor produces an almost 90^0 phase advance of its current $\underline{I}_a$ with respect to $\underline{I}_m$ in the main winding (Figure 23.2b).

This way a large travelling field is produced at the start. Consequently, the starting torque is large. After the motor starts, the auxiliary winding circuit is opened by the starting switch, leaving on only the main winding. The large value of the starting capacitor is not adequate for running conditions. During running conditions, the power factor and efficiency are rather low.

Capacitor start motors are built for single or dual voltage (115 V and 230 V). For dual voltage, the main winding is built in two sections connected in series for 230 V and in parallel for 115 V.

23.3.2 The two-value capacitor induction motor

The two-value capacitor induction motor makes use of two capacitors in parallel-one for starting, C_s, and one for running C_n ($C_s >> C_n$).

The starting capacitor is turned off by a starting switch while the running capacitor remains in series with the auxiliary winding (Figure 23.3).

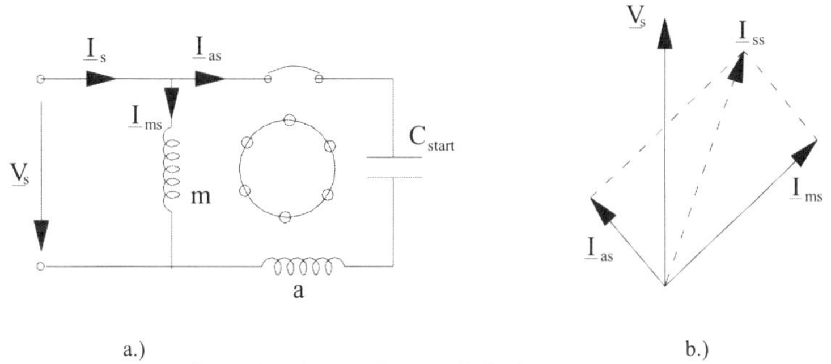

Figure 23.2 The capacitor-start induction motor
a.) schematics b.) phasor diagram at zero speed

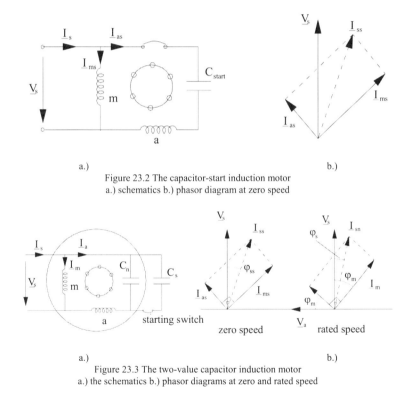

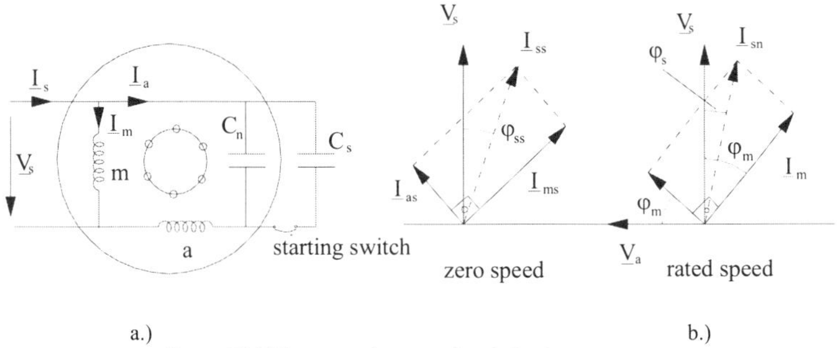

Figure 23.3 The two-value capacitor induction motor
a.) the schematics b.) phasor diagrams at zero and rated speed

The two capacitors are sized to symmetrize the two windings at standstill (C_s) and at rated speed (C_n).

High starting torque is associated with good running performance around rated speed (torque). Both power factor and efficiency are high.

The running capacitor is known to produce effects such as [1]

- 5 to 30^0 increase of breakdown torque
- 5-10% improvement in efficiency
- Power factors values above 90%
- Noise reduction at load
- 5-20% increase in the locked-rotor torque

Sometimes, the auxiliary winding is made of two sections (a_1 and a_2). One section works at start with the starting capacitor while both sections work for running conditions to allow a higher voltage (smaller) running capacitor (Figure 23.4).

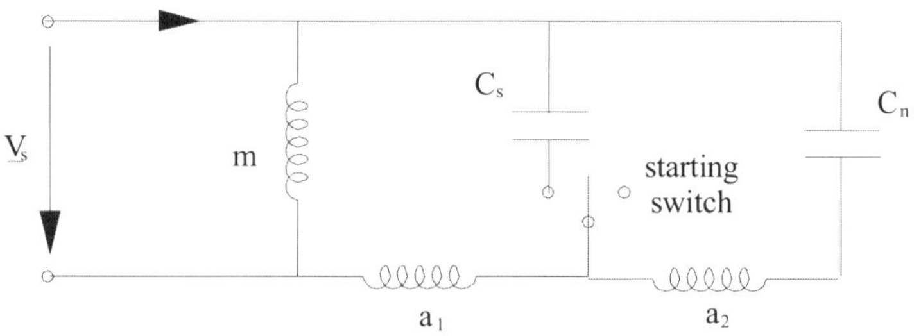

Figure 23.4 Two-value capacitor induction motor with two-section auxiliary winding

23.3.3 Permanent-split capacitor induction motors

The permanent-split capacitor IMs have only one capacitor in the auxiliary winding which remains in action all the time.

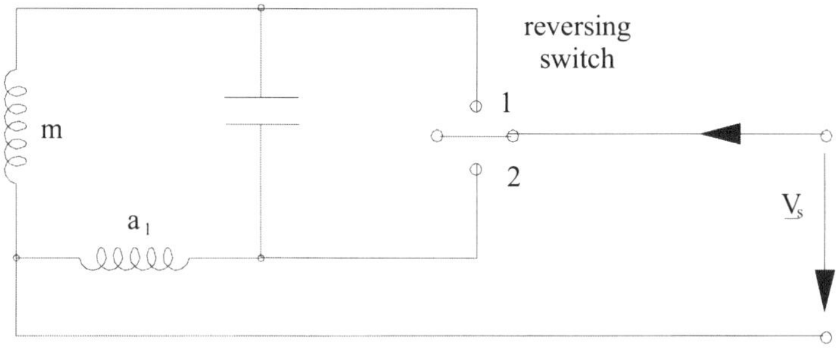

Figure 23.5 The reversing permanent-split capacitor motor

As a compromise between starting and running performance has to be performed, this motor has a rather low starting torque but good power factor and efficiency under load. This motor is not to be used with belt-transmissions due to its low starting torque. It is however suitable for reversing, intermittent-duty service, and powers from 1 W to 200 W. For speed reversal the two windings are identical and the capacitor is "moved" from the auxiliary to the main winding circuit (Figure 23.5)

23.3.4 Tapped-winding capacitor induction motors

Tapped-winding capacitor induction motors are used when two or more speeds are required.

For two speeds the T and L connections are highly representative (Figure 23.6 a,b). The main winding contains two sections m_1 and m_2.

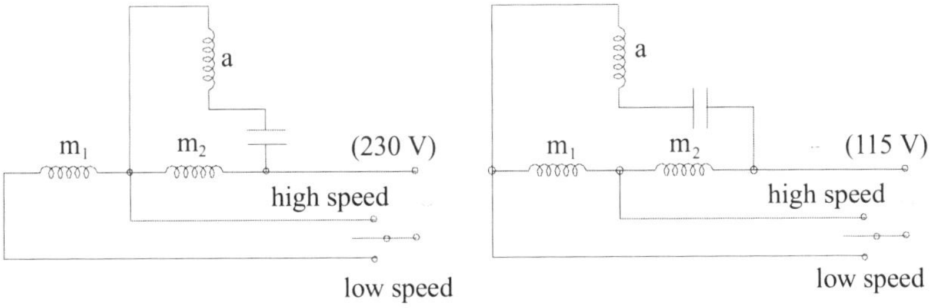

Figure 23.6 T(a) and L(b) connections of tapped winding capacitor induction motors

The T connection is more suitable for 230 V while the L connection is more adequate for 115V power grids since this way the capacitor voltage is higher and thus its cost is lower.

The difference between high speed and low speed is not very important unless the power of the motor is very small, because only the voltage is reduced. The locked rotor torque is necessarily low (less than the load torque at the low speed). Consequently, they are not to be used with belted drives. Unstable low speed operation may occur as the breakdown torque decreases with voltage squared.

A rather general tapped-winding capacitor induction motor is shown in Figure 23.7 [2]

In this configuration, each coil of the single layer auxiliary winding has a few taps corresponding to the number of speeds required.

23.3.5 Split-phase capacitor induction motors

Split-phase capacitor induction motors start as split-phase motors and then commute to permanent-capacitor motors.

This way both higher starting torque and good running performance is obtained. The auxiliary winding may also contain two sections to provide higher capacitance voltage. (Figure 23.8)

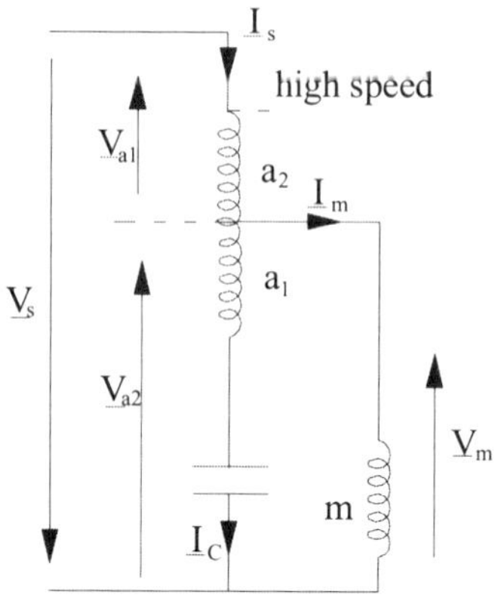

Figure 23.7 General tapped-winding . Capacitor induction motor

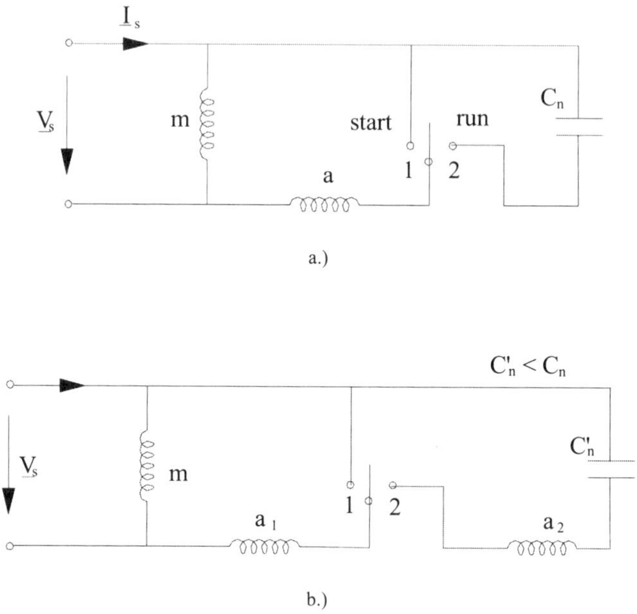

Figure 23.8 The split-phase capacitor induction motor
a.) with single section auxiliary winding; b.) with two section auxiliary winding

23.3.6 Capacitor three-phase induction motors

The three phase winding induction motors may be connected and provided with a capacitor to comply with single-phase power supplies. The typical connections also called Steinmetz star and delta connections-are shown in Figure 23.9. As seen in Figure 23.9 c), with an adequate capacitance C_n,the voltage of the three phases A,B,C may be made symmetric for a certain value of slip ($S = 1$ or $S' = S_n$ or a value in between).

The star connection prevents the occurrence of third m.m.f. space harmonic and thus its torque-speed curve does not show the large dip at 33% synchronous speed, accentuated by saturation and typical for low power capacitor single phase induction motors with two windings. Not so for delta connection (Figure 23.9 b) which allows for the third m.m.f. space harmonic. However, the delta connection provides $\sqrt{3}$ times larger voltage per phase and thus, basically, 3 times more torque in comparison with the star connection, for same single phase supply voltage.

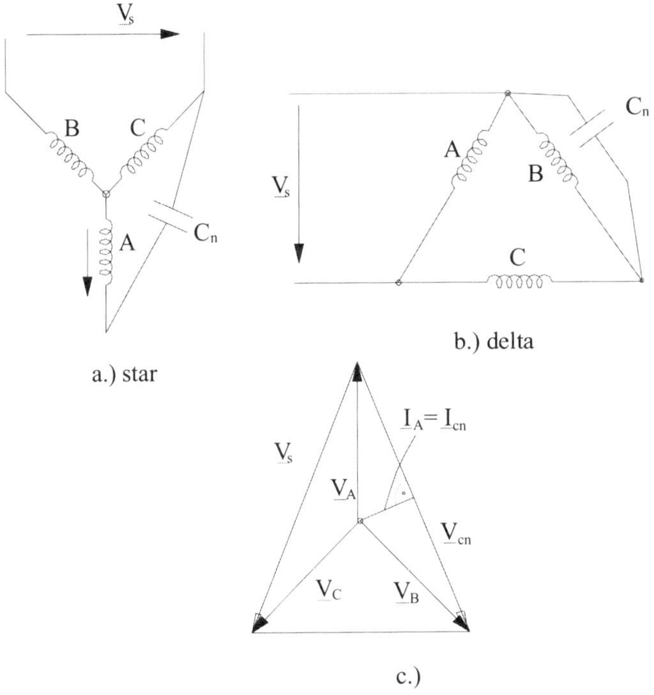

Figure 23.9 Three phase winding IM with capacitors for single phase supply a.) star; b.) delta; c.) phasor diagram (star)

The three phase windings also allow for pole changing to reduce the speed in a 2:1 or other ratios (see Chapter 4)-Figure 23.10.

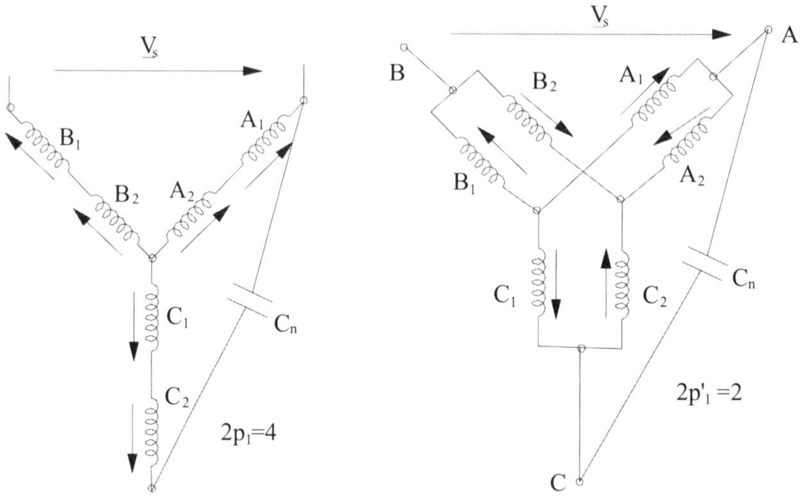

a.) b.)
Figure 23.10 Pole-changing single phase induction motor with three phase stator winding a.) $2p_1 = 4$ poles; b.) $2p_1' = 2$ poles

23.3.7 Shaded-pole induction motors

The shaded-pole induction motor has a single, concentrated coil, stator winding, with $2p_1 = 2, 4, 6$ poles, connected to a single phase a.c. power grid. To provide for selfstarting a displaced shortcircuited winding is located on a part of main winding poles (Figure 23.11).

The main winding current flux induces a voltage in the shaded-pole coil which in turn produces a current which affects the total flux in the shaded area. So the flux in the un-shaded area $\underline{\Phi}_{mm}$ is both space-wise and time-wise shifted ahead with respect to the flux in the shaded area ($\underline{\Phi}_a$).

These two fluxes will produce a travelling field component that will move the rotor from the unshaded to the shaded area. Some further improvement may be obtained by increasing the airgap at the entry end of stator poles (Figure 23.11). This beneficial effect is due to the third space harmonic field reduction.

As the space distribution of airgap field is far from a sinusoid, space-harmonic parasitic asynchronous torques occur. The third harmonic is in general the largest and produces a notable dip in the torque-speed curve around 33% of synchronous speed $n_1 = f_1/p_1$.

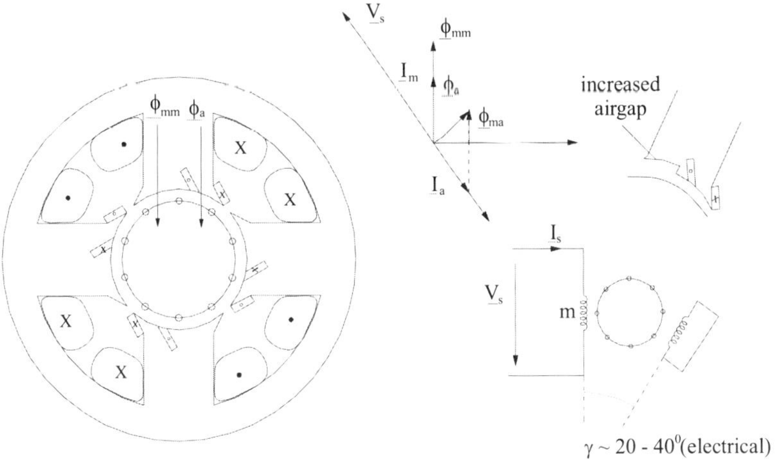

Figure 23.11 The shaded-pole induction motor

The shaded-pole induction motor is low cost but also has low starting torque, torque density, low efficiency (less than 30%), and power factor. Due to these demerits is used only at low powers between 1/25 to 1/5 kW in round frames (Figure 23.11), and less and less.

For even lower power levels (tens of mW), an unsymmetrical (C-shape) stator configuration is used (Figure 23.12).

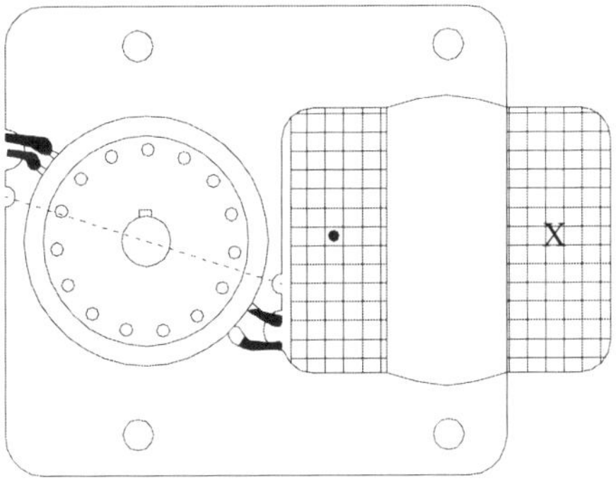

Figure 23.12 Two-pole C-shape shaded-pole induction motor

23.4 THE NATURE OF STATOR-PRODUCED AIRGAP FIELD

The airgap field distribution in single phase IMs, for zero rotor currents, depends on the distribution of the two stator windings in slots and the amplitude ratio I_m/I_a and phase shift γ of phase currents.

When the current waveforms are not sinusoidal in time, due to magnetic saturation or due to power grid voltage time harmonics, the airgap field distribution gets even more complicated. For the time being, let us neglect the stator current time harmonics and consider zero rotor current.

Chapter 4 presents in some detail typical distributed windings for capacitor single phase induction motors.

Let us consider here the first and the third space harmonic of stator winding m.m.fs.

$$F_m(\theta_{es},t) = F_{1m}\cos\omega_1 t\left[\cos\theta_{es} + \frac{F_{3m}}{F_{1m}}\cos(3\theta_{es})\right] \tag{23.1}$$

$$F_a(\theta_{es},t) = -F_{1a}\cos(\omega_1 t + \gamma_i)\left[\sin\theta_{es} + \frac{F_{3a}}{F_{1a}}\sin(3\theta_{es})\right] \tag{23.2}$$

θ_{es}-stator position electrical angle.

For windings with dedicated slots, and coils with same number of turns, expressions of F_{1m}, F_{3m}, F_{1a}, F_{3a} (see Chapter 4) are

$$F_{1m\nu} = \frac{2N_m K_{Wm\nu} I_m\sqrt{2}}{\pi\nu p_1};\nu = 1,3 \tag{23.3}$$

$$F_{1a\nu} = \frac{2N_a K_{Wa\nu} I_a\sqrt{2}}{\pi\nu p_1};\nu = 1,3$$

The winding factors $K_{Wm\upsilon}$, $K_{Wa\upsilon}$ depend on the relative number of slots for the main and auxiliary windings (S_m/S_A) and on the total number of stator slots S_1 (see Chapter 4).

For example in a 4 pole motor with $S_1 = 24$ stator slots, 16 slots for the main winding and 8 slots for the auxiliary winding, both of single layer type, the coils are full pitch. Also $q_m = 4$ slots/pole/phase for the main winding and $q_a = 2$ slots/pole/phase for the auxiliary winding. Consequently,

$$K_{Wm\nu} = \frac{\sin q_m \frac{\alpha_{c\nu}}{2}}{q_m \sin\frac{\alpha_{c\nu}}{2}} = \frac{\sin 4\cdot\frac{\pi}{24}\nu}{4\sin\frac{\pi}{24}\nu} \tag{23.4}$$

$$K_{Wa\nu} = \frac{\sin q_a \frac{\alpha_{c\nu}}{2}}{q_a \sin\frac{\alpha_{c\nu}}{2}} = \frac{\sin 2\cdot\frac{\pi}{24}\nu}{2\sin\frac{\pi}{24}\nu} \tag{23.5}$$

$$F_{3m}/F_{1m} = 0.226;\ F_{3a}/F_{1a} = 0.3108$$

Let us first neglect the influence of slot openings while the magnetic saturation is considered to produce only an "increased apparent airgap."

Consequently, the airgap magnetic field $B_{go}(\theta_{er}, t)$ is:

$$B_{go}(\theta_{es}, t) = \frac{\mu_0}{gK_c(1+K_s)}\left[F_m(\theta_{es}, t) + F_a(\theta_{es}, t)\right] \tag{23.6}$$

K_s-saturation coefficients, K_c-Carter coefficient.

So, even in the absence of slot-opening and saturation influence on airgap magnetic permeance, the airgap field contains a third space harmonic. The presence of slot openings produces harmonics in the airgap permeance (see Chapter 10, paragraphs 10.3). As in single phase IM both stator and rotor slots are semiclosed, these harmonics may be neglected in a first order analysis.

On the other hand, magnetic saturation of slot neck leakage path produces a second order harmonic in the airgap permeance (Chapter 10; paragraph 10.4). This, in turn, is small unless the currents are large (starting conditions).

The main flux path magnetic saturation tends to create a second order airgap magnetic conductance harmonic (Chapter 10, paragraph 10.5),

$$\lambda_{gs}(\theta_{es}) = \frac{1}{gK_C}\left[\frac{1}{1+K_{s1}} + \frac{1}{1+K_{s2}}\sin\left(2\frac{\theta_{es}}{p_1} - \omega_{1t} - \gamma_s\right)\right] \tag{23.7}$$

K_{s1}, K_{s2} are saturation coefficients that result from the decomposition of main flux distribution in the airgap.

We should mention here that rotor eccentricity also "induces" airgap magnetic conductance harmonics (Chapter 10, paragraph 10.7).

For example, the static eccentricity $\varepsilon = e/g$ yields

$$\lambda_{ge}(\theta_{es}) \approx \frac{1}{g}\left(C_0 + C_1\cos\frac{\theta_{es}}{p_1}\right)$$
$$C_0 = \frac{1}{\sqrt{1-\xi^2}}; C_1 = 2(C_0 - 1)/\xi \tag{23.8}$$

Now the airgap flux density $B_{g0}(\theta_{es},t)$, for zero rotor currents, becomes

$$B_{g0}(\theta_{es}, t) = \mu_0\lambda_{gs}(\theta_{es})\left[F_m(\theta_{es}, t) + F_a(\theta_{es}, t)\right] \tag{23.9}$$

As seen from (23.9) and (23.1)-(23.2) the main flux path saturation introduces an additional third space harmonic airgap field, besides that produced by the m.m.f. third space harmonic. The two third harmonic field components are phase shifted with an angle γ_3 dependent on the relative ratio of the main and auxiliary windings m.m.f. amplitudes and on their time phase lag.

The third space harmonic of airgap field is expected to produce a third order notable asynchronous parasitic torque whose synchronism occurs around 1/3 of ideal no load speed $n_1 = f_1/p_1$. A rather complete solution of the airgap field distribution, for zero rotor currents (the rotor does not have the cage in slots) may be obtained by FEM.

In face to such complex field distributions in time and space, simplified approaches have traditionally become widely accepted.

Let us suppose that magnetic saturation and m.m.f. produced third harmonics are neglected. That is, we return to (23.6).

23.5 THE FUNDAMENTAL M.M.F. AND ITS ELLIPTIC WAVE

Neglecting the third harmonic in (23.1)-(23.2), the main and auxiliary winding m.m.fs are:

$$F_m(\theta_{es}, t) = F_{1m} \cos \omega_{1t} \cdot \cos \theta_{es} \tag{23.10}$$

$$F_a(\theta_{es}, t) = -F_{1a} \cos(\omega_{1t} + \gamma_i) \sin \theta_{es} \tag{23.11}$$

F_{1a}/F_{1m} is, from (23.3)

$$F_{1a} / F_{1m} = \frac{N_m K_{wm1}}{N_a K_{wa1}} \cdot \frac{I_m}{I_a} = \frac{1}{a} \frac{I_m}{I_a} \tag{23.12}$$

Given the sinusoidal spatial distribution of the two m.m.fs, their resultant space vector F(t) may be written as

$$F(t) = F_{1m} \sin \omega_{1t} + jF_{1a} \sin(\omega_{1t} + \gamma_i) \tag{23.13}$$

So the amplitude F(t) and its angle γ(t) are

$$\begin{aligned} F(t) &= \sqrt{F_{1m}^2 \cos^2 \omega_{1t} + F_{1a}^2 \sin^2(\omega_{1t} + \gamma_i)} \\ \gamma(t) &= \tan^{-1}\left[-a \frac{I_a}{I_m} \frac{\cos(\omega_{1t} + \gamma_i)}{\cos \omega_{1t}} \right] \end{aligned} \tag{23.14}$$

A graphical representation of (23.13) in a plane with the real axis along the main winding axis and the imaginary one along the auxiliary winding axis, is shown in Figure 23.13. The elliptic hodograph is evident.

Both the amplitude F(t) and time derivative (dγ/dt) (speed) vary in time from a maximum to a minimum value.

Only for symmetrical condition, when the two m.m.f.s have equal amplitudes and are time-shifted by $\gamma_i = 90^0$, the hodograph of the resultant m.m.f. is a circle. That is, a pure **travelling** wave ($\gamma = \omega_1 t$) is obtained. It may be demonstrated that the m.m.f. wave speed dγ/dt is positive for $\gamma_i > 0$ and negative for $\gamma_i < 0$.

In other words, the motor speed may be reversed by changing the sign of the time phase shift angle between the currents in the two windings. One way to do it is to switch the capacitor from auxiliary to main winding (Figure 23.5).

The airgap field is proportional to the resultant m.m.f. (23.6), so the elliptic hodograph, of m.m.f. stands valid for this case also.

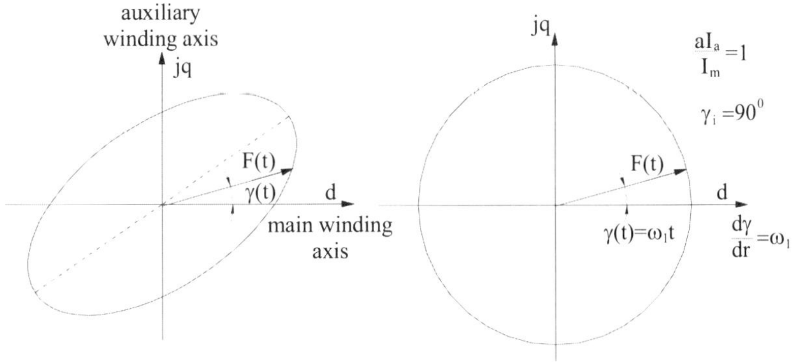

Figure 23.13 The hodograph of fundamental m.m.f. in single phase IMs
a.) in general b.) for symmetrical conditions

When heavy magnetic saturation occurs for sinusoidal voltage supply, zero rotor currents, the stator currents become nonsinusoidal, while the e.m.f. phase flux linkages, also depart from sinusoids.

The presence of rotor currents further complicates the picture. It is thus fair to say that the computation of field distribution, currents, power, torque for the single phase IMs is more complicated than for three-phase IMs. A saturation model will be introduced in Chapter 24.

23.6 FORWARD-BACKWARD M.M.F. WAVES

The resultant stator m.m.f.. $F(\theta_{es},t)$-(23.10-23.11)-may be decomposed in two waves

$$F(\theta_{es},t) = F_{1m}\cos\omega_1 t\cos\theta_{es} + F_{1a}\cos(\omega_1 t+\gamma_i)\sin\theta_{es} = \\ = \frac{1}{2}F_m C_f \sin(\theta_{es}-\omega_1 t-\beta_f) + \frac{1}{2}F_m C_b\sin(\theta_{es}+\omega_1 t-\beta_b) \tag{23.15}$$

$$C_f = \sqrt{\left(1+\frac{F_{1a}}{F_{1m}}\sin\gamma_i\right)^2 + \left(\frac{F_{1a}}{F_{1m}}\right)^2 \cos\gamma_1^2} \\ \sin\beta_f = \frac{(1+F_{1a}\sin\gamma_i/F_{1m})}{C_f} \tag{23.16}$$

$$C_b = \sqrt{\left(1-\frac{F_{1a}}{F_{1m}}\sin\gamma_i\right)^2 + \left(\frac{F_{1a}}{F_{1m}}\cos\gamma_i\right)^2}$$

$$\sin\beta_b = \frac{(1-F_{1a}\sin\gamma_i/F_{1m})}{C_b} \tag{23.17}$$

Again, for $F_{1a}/F_{1m} = 1$ and $\gamma_i = 90^0$, $C_b = 0$ and thus the backward (b) travelling wave becomes zero (the case of circular hodograph in Figure 23.13 b).

Also, the forward (f) travelling amplitude, is in this case, equal to $F_{1m} = F_{1a}$ (it is 3/2 F_{1m} for a three phase symmetrical winding).

For $F_{1a} = 0$, the f.b. waves have the same magnitude: ½ F_{1m}.

The m.m.f.. decomposition in forward and backward waves may be done even with saturation while the superposition of the forward and backward field waves is correct only in the absence of magnetic saturation.

The fb decomposition is very practical as it leads to simple equivalent circuits with the slip $S_f = S$ and, respectively, $S_b = 2\text{-}S$. This is how the travelling (fb)-revolving field theory of single phase IMs evolved.

$$S_f = \frac{\omega_1 - \omega_r}{\omega_r} = S; S_b = \frac{(-\omega_1) - \omega_r}{(-\omega_1)} = 2 - S \qquad (23.18)$$

For steady state, the fb model is similar to the symmetrical components model. On the other hand, for the general case of transients or nonsinusoidal voltage supply, the dq cross-field model has become standard.

23.7 THE SYMMETRICAL COMPONENTS GENERAL MODEL

For steady state and sinusoidal currents, time phasors may be used. The unsymmetrical m.m.f., currents and voltages corresponding to the two windings of single phase IMs may be decomposed in two symmetrical systems (Figure 23.14) which are in fact the forward and backward components introduced in the previous paragraph.

It goes without saying that the two windings are 90^0 electrical degrees phase shifted spatially.

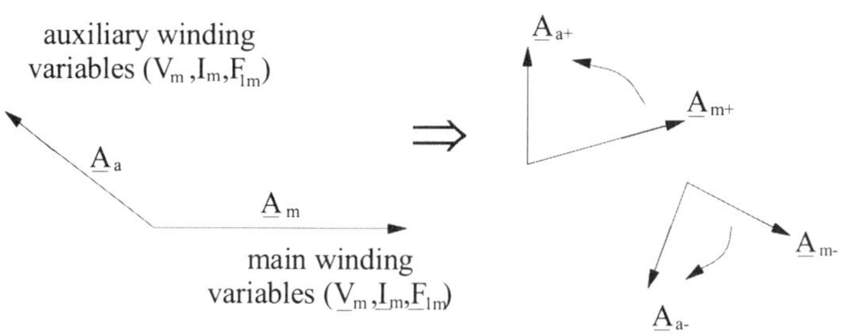

Figure 23.14 Symmetrical (fb or + -) components of a two phase winding motor

From Figure 23.14

$$\underline{A}_{a+} = j\underline{A}_{m+}; \underline{A}_{a-} = -j\underline{A}_{m-} \qquad (23.19)$$

The superposition principle yields

$$\begin{aligned}\underline{A}_m &= \underline{A}_{m+} + \underline{A}_{m-}\\ \underline{A}_f &= \underline{A}_{a+} + \underline{A}_{a-}\end{aligned} \tag{23.20}$$

From (23.19)-(23.20),

$$A_{m+} = \frac{1}{2}(\underline{A}_m - j\underline{A}_a) \tag{23.21}$$

$$A_{m-} = \frac{1}{2}\left(A_m + j\underline{A}_a\right) = \underline{A}_{m+}^{*} \tag{23.22}$$

Denoting by $\underline{V}_{m+}$, $\underline{V}_{m-}$, $\underline{I}_{m+}$, $\underline{I}_{m-}$ and respectively $\underline{V}_{a+}$, $\underline{V}_{a-}$, $\underline{I}_{a+}$, $\underline{I}_{a-}$, - the voltage and current components of the two winding circuits, we may write the equations of the two fictitious 2-phase symmetric machines (whose travelling fields are opposite (f,b)) as

$$\begin{aligned}\underline{V}_{m+} &= \underline{Z}_{m+}\underline{I}_{m+};V_{m-} = \underline{Z}_{m-}\underline{I}_{m-}\\ \underline{V}_{a+} &= \underline{Z}_{a+}\underline{I}_{a+};V_{a-} = \underline{Z}_{a-}\underline{I}_{a-}\\ \underline{V}_m &= \underline{V}_m + \underline{V}_{m-};\underline{V}_a = \underline{V}_{a+} + \underline{V}_{a-}\end{aligned} \tag{23.23}$$

The $\underline{Z}_{m+}$, $\underline{Z}_{m-}$ and respectively $\underline{Z}_{a+}$, $\underline{Z}_{a-}$ represent the resultant forward and backward standard impedances of the fictitious induction machine, per phase, with the rotor cage circuit reduced to the m, and respectively, a stator winding (Figure 23.15).

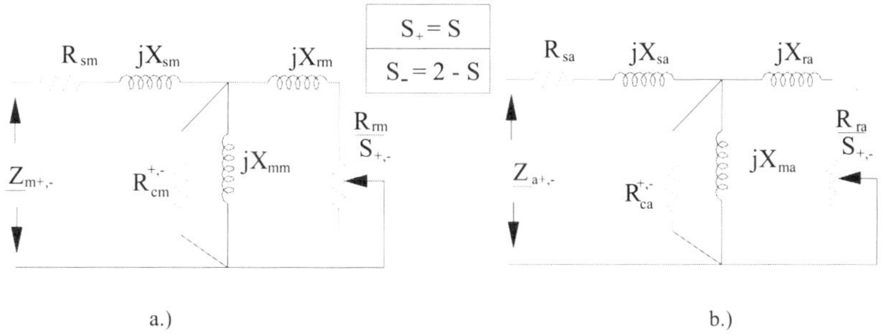

Figure 23.15. Equivalent symmetrical (+,-) impedances

a.) reduced to the main stator winding m; b.) reduced to the auxiliary stator winding a

We still need to solve Equations (23.33), with given slip S value and all parameters in $\underline{Z}_{m+,-}$ and $\underline{Z}_{a+,-}$. A relationship between the $\underline{V}_m$, $\underline{V}_a$ and the source voltage $\underline{V}_s$ is needed.

When an impedance ($\underline{Z}_a$) is connected in series with the auxiliary winding and then both windings are connected to the power grid given voltage $\underline{V}_s$ we obtain a simple relationship:

$$\underline{V}_m = \underline{V}_s;\underline{V}_a + \left(I_{a+} + I_{a-}\right)\underline{Z}_a = \underline{V}_s \tag{23.24}$$

It is generally more comfortable to reduce all windings on rotor and on stator to the main winding.

The reduction ratio of auxiliary to main winding is the coefficient a in Equation (23.12)

$$a = \frac{N_a}{N_m}\frac{K_{w1a}}{K_{W1m}} \tag{23.25}$$

The rotor circuit is symmetric when $a^2X_{rm} = X_{ra}, a^2R_{rm} = R_{ra}$.

Consequently, when

$$\begin{aligned} \Delta X_{sa} &= X_{sa} - a^2X_{sm} = 0 \\ \Delta R_{sa} &= R_{sa} - a^2R_{sm} = 0 \end{aligned} \tag{23.26}$$

the two equivalent circuits on Figure 23.15, both reduced to the main winding, become identical.

This situation occurs in general when both windings occupy the same number of uniform slots and the design current density for both windings is the same. When the auxiliary winding occupies 1/3 of stator periphery, its design current density may be higher to fulfill (23.26).

The voltage and current reduction formulas from the auxiliary winding to main winding are

$$\underline{V}'_a = \underline{V}_a / a; \underline{I}'_a = a\underline{I}_a \tag{23.27}$$

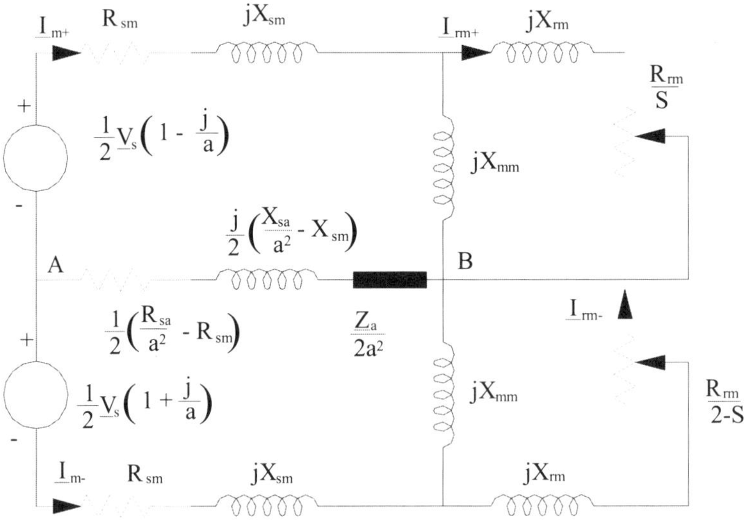

Figure 23.16 General equivalent circuit of split-phase and capacitor induction motor

In such conditions, when $\Delta X_{sa} \neq 0$ and $\Delta R_{sa} \neq 0$, their values are introduced in series with the capacitor impedance $\underline{Z}_a$ after reduction to the main winding and division by 2. The division by 2 is there to conserve powers (Figure 23.16).

Note again that $\underline{Z}_a$ contains the capacitor. In split-phase IMs, $\underline{Z}_a = 0$. When the auxiliary winding is open: $\underline{Z}_a = \infty$. Also note that the + - voltages are now defined at the grid terminals ($V_a = V_s/a$) rather than at the winding terminals.

This is how the A to B section impedance on Figure 23.16 takes its place inside the equivalent circuit. The equivalent circuit in Figure 23.16 is fairly general. It has only two unknowns I_{m+} and I_{m-} and may be solved rather easily for given slip, machine parameters and supply voltage $\underline{V}_s$.

Tapped winding (multiple-section winding or three phase winding) capacitor IMs may be represented directly by the equivalent circuit in Figure 23.16. Core loss resistances may be added in Figure 23.16 in parallel with the magnetization reactance X_{mm} or with the leakage reactances (for additional losses) but the problem is how to calculate (measure) them.

Main flux path saturation may be included by making the magnetization reactance X_{mm} a variable. The first temptation is to consider that only the forward (direct component) X_{mm} (the upper part of Figure 23.16) saturates, but, as shown earlier, the elliptic m.m.f. makes the matters far more involved.

This is why the general equivalent circuit in Figure 23.16 is suitable when the saturation level is constant or it is neglected.

To complete the picture we add here the average torque expression

$$T_e = T_{e+} + T_{e-} = \frac{2p_1}{\omega_1}\left[I_{rm+}^2 \cdot \frac{R_{rm}}{S} - \frac{I_{rm-}^2 \cdot R_{rm}}{2-S}\right] \tag{23.28}$$

For the case when the auxiliary winding is open ($Z_a = \infty$), $I_{rm+} = I_{rm-}$. In such a case for S = 1 (start), the total torque T_e is zero, as expected.

It may be argued that as the rotor currents experience two frequencies: $f_{2+} = Sf_1$ and $f_2=(2 - S)f_1 > f_1$, the rotor cage parameters differ for the two components due to skin effects.

To account for skin effect, two fictitious cages in parallel may be considered in Figure 23.18. Their parameters are chosen to fit the rotor impedance variation from $f_2 = 0$ to $f_2 = 2f$ according to a certain optimization criterion.

23.8 THE d-q MODEL

As the stator is provided with two orthogonal windings and the rotor is fully symmetric, the single phase IM is suitable for the direct application of dq model in stator coordinates (Figure 23.17).

The rotor will be reduced here to the main winding, while the auxiliary winding is not reduced to the main winding.

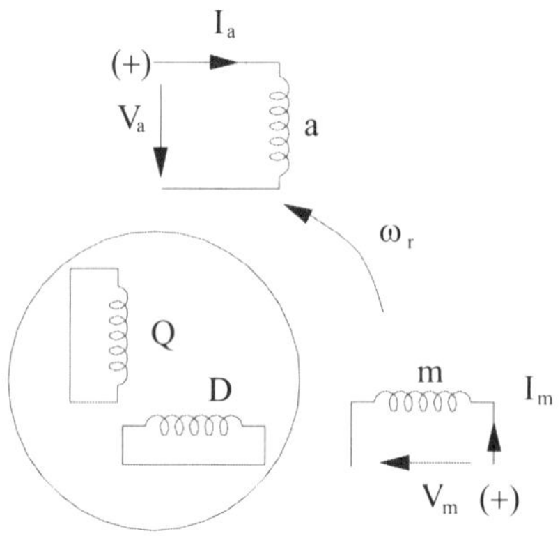

Figure 23.17 The d-q model of single phase IM

The dq model equations in stator coordinates are straightforward

$$
\begin{aligned}
I_m R_{sm} - V_m &= -\frac{d\psi_m}{dt} \\
I_a R_{sa} - V_a &= -\frac{d\psi_a}{dt} \\
i_D R_r &= -\frac{d\psi_D}{dt} - \omega_r \psi_a \\
I_Q R_r &= -\frac{d\psi_a}{dt} + \omega_r \psi_D
\end{aligned}
\tag{23.29}
$$

The flux current relationships are

$$
\begin{aligned}
\psi_D &= L_{rm} I_D + L_{mm}(I_m + I_D) \\
\psi_Q &= L_{rm} I_Q + L_{am}(aI_q + I_Q)/a^2 \\
\psi_m &= L_{sm} I_m + L_{mm}(I_m + I_D) \\
\psi_a &= L_{sa} I_a + L_{am}\left(I_a + \frac{I_Q}{a}\right)
\end{aligned}
\tag{23.30}
$$

For nonlinear saturation conditions, it is better to define a the reduction ratio as

$$
a = \sqrt{\frac{L_{am}}{L_{mm}}}
\tag{23.31}
$$

This aspect will be treated in a separate chapter on testing (Chapter 28). The magnetization inductances L_{mm} and L_{am} along the two axes (windings) are dependent on the resultant magnetization current I_μ

$$I_\mu(t)=\sqrt{(I_m+I_D)^2+(aI_a+I_Q)^2} \tag{23.32}$$

The torque and motion equations are straightforward.

$$T_e=-p_1(\psi_D I_Q-\psi_Q I_D)=p_1 L_{mm}\left(\sqrt{\frac{L_{am}}{L_{mm}}}i_a I_D-I_m I_Q\right) \tag{23.33}$$

$$\frac{J}{p_1}\frac{d\omega_r}{dt}=T_e-T_{load}(\omega_r,\theta_{er},t) \tag{23.34}$$

Still missing are the relationships between the stator voltages V_m, V_a and the source voltage V_s. In a capacitor IM

$$\begin{aligned} V_m(t)&=V_s(t)\\ V_a(t)&=V_s(t)-V_c(t) \end{aligned} \tag{23.35}$$

$$C\frac{dV_C(t)}{dt}=I_a(t) \tag{23.36}$$

V_C-is the capacitor voltage.

A six order nonlinear model has been obtained. The variables are: I_m, I_a, I_D, I_Q, ω_r, V_C, while the source voltage $V_s(t)$ and the load torque T_{load} (ω_r , θ_{er} , t) constitute the inputs.

With known $L_{am}(I_\mu)$ and $L_{mm}(I_\mu)$ functions even the saturation presence may be handled elegantly, both during transients and steady state.

23.9 THE d-q MODEL OF STAR STEINMETZ CONNECTION

The three phase winding IM, in the capacitor motor connection for single phase supply, may be reduced simply to the d-q (m,a) model as

$$V_m+jV_a=\sqrt{\frac{2}{3}}\left(Va(t)+V_b(t)e^{j\frac{2\pi}{3}}+V_c(t)e^{-j\frac{2\pi}{3}}\right)e^{-j\theta_0} \tag{23.37}$$

The same transformation is valid for currents while the phase resistances and leakage inductances stand unchanged.

The voltage relationships depend on the connection of phases. For the Steinmetz star connection (Figure 23.18),

$$\begin{aligned} V_B(t) - V_C(t) &= V_S(t) \\ V_a(t) + V_b(t) + V_c(t) &= 0 \\ V_B(t) - V_A(t) &= V_{cap}(t) \\ I_A(t) &= I_{cap}(t) \\ C\frac{dV_{cap}}{dt} &= I_{cap}(t) \end{aligned} \tag{23.38}$$

With $\theta_0 = -\pi/2$, d axis falls along the axis of the main winding

$$\begin{aligned} V_m(t) &= \frac{1}{\sqrt{2}}(V_B - V_C) = \frac{V_S(t)}{\sqrt{2}} \\ V_a(t) &= +\sqrt{\frac{3}{2}}V_A(t) \end{aligned} \tag{23.39}$$

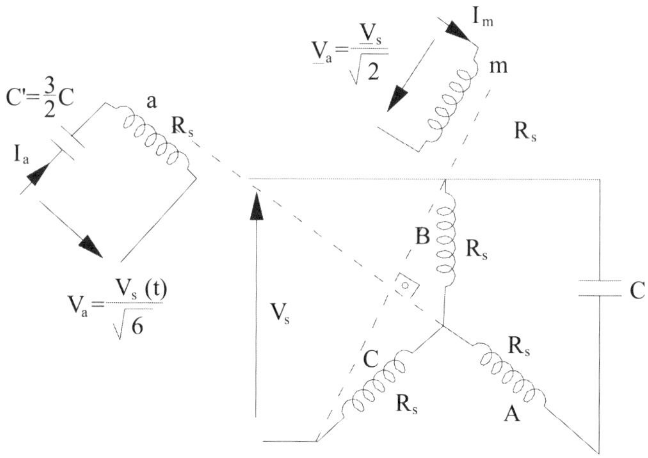

Figure 23.18 Three (two) phase equivalence

From (23.38) with (23.39)

$$\sqrt{\frac{2}{3}}V_{cap} + V_a = \frac{V_S(t)}{\sqrt{6}} \tag{23.40}$$

Notice that

$$I_a = \sqrt{\frac{3}{2}}I_{cap} \tag{23.41}$$

For the d-q model equation (23.40) becomes

$$V'_{cap} + V_a = \frac{V_S(t)}{\sqrt{6}}$$
$$\frac{dV'_{cap}}{dt} = \frac{1}{C} I_a \qquad (23.42)$$

Also,

$$V'_{cap} = \frac{1}{C} \cdot \int I_a dt = \sqrt{\frac{2}{3}} \frac{1}{C} \int I_A dt = \frac{2}{3C} \int I_a dt = \frac{1}{\left(\frac{3}{2} C\right)} \int I_a dt \qquad (23.43)$$

So,

$$C' = \frac{3}{2} C \qquad (23.44)$$

The equivalent capacitance circuit C' in the d-q (m.a)-2 phase –model is 1.5 times higher than the actual capacitance C. The source voltage to the main winding m(d) is $V_s / \sqrt{2}$ instead of V_s as it is for the usual model. For the auxiliary winding the "virtual source" voltage is $V_s / \sqrt{6}$ (Equation 23.40). Finally the resistance, leakage inductance and magnetization inductance are those of the three phase machine per phase. Consequently, for the m&a model stator windings are now identical.

The steady state performance may be approached by running transients from start (or given initial conditions) until conditions stabilize. As the state-space system of equations is solvable only via numerical methods, its solution takes time. With zero initial values for currents, capacitor voltage and given (final) slip (speed), this time is somewhat reduced.

Alternatively, it is possible to use $j\omega_1$ instead of d/dt in Equations (23.29) and to extract the real part of torque expression, after solving the algebraic system of equations for the currents and the capacitor voltage.

This approach may be more suitable than the travelling (revolving) field model when the magnetic saturation is to be accounted for, for the resultant (elliptical) magnetic field in the machine may be accounted for directly. However, it does not exhibit the intuitive attributes of the revolving field theory. In presence of strong computation facilities, this steady state approach to the dq model may revive for usage when magnetic saturation is heavy.

Finally, the reduction of the three-phase Steinmetz star connection to the m,a model performed above may also be used in the symmetrical components model (Figure 23.16).

23.10 SUMMARY

- Single phase supply induction motors have a cage rotor and two phase stator windings: the main winding and the auxiliary winding.
- The two windings are phaseshifted spatially by 90^0 electrical degrees.

- The auxiliary winding may have a larger resistance/reactance ratio or a series capacitance to push ahead its current phase angle with respect to that of the main winding.
- Alternatively, the auxiliary winding may be shortcircuited, but in this case its coils should occupy only a part of the main winding poles. This is the shaded-pole induction motor. It is low cost but also low performance. It is fabricated for powers below (1/5) kW and down to 10-15W.
- The capacitor induction motor may be built in capacitor-start,dual-capacitor permanent-capacitor, tapped winding capacitor, split-phase capacitor configurations. Each of these types is suitable for certain applications. Except for the capacitor-start IMs, all the other capacitor IMs have good efficiency and over 90% power factor but only capacitor-start, dual-capacitor and split-phase configurations exhibit large starting torque (up to 250% rated torque)
- Speed reversal is obtained by "moving" the capacitor from the auxiliary to the main winding circuit. In essence, this move reverses the predominant travelling field speed in the airgap.
- The three phase stator winding motor may be single-phase supply fed, adding a capacitor mainly in the so-called Steinmetz star connection. Notable derating occurs unless the three phase motor was not designed for delta connection.
- The nature of the airgap field in capacitor IMs may be easily investigated for zero rotor currents (open rotor cage). The two stator winding m.m.f.s and their summation contains a notable third space harmonic. So does the airgap flux. This effect is enhanced by the main flux path magnetic saturation which creates a kind of second space harmonic in the airgap magnetic conductance. Consequently, a notable third harmonic asynchronous parasitic torque occurs. A dip in the torque/speed curve around 33% of ideal no load speed shows up, especially for low powers motors (below 1/5 kW)
- Even in absence of magnetic saturation and slot opening effects, the stator m.m.f. fundamental produces an elliptic hodograph. This hodograph degenerates into a circle when the two phase m.m.f.s are perfectly symmetric (90^0 (electrical) winding angle shift, 90^0 current phase shift angle).
- The main (m) and auxiliary (a) m.m.f. fundamentals produce together elliptic wave which may be decomposed into a forward (+) and a backward (-) travelling wave.
- The backward wave is zero for perfect symmetrization of phases.
- For zero auxiliary m.m.f. (current), the forward and backward m.m.f. waves have equal magnitudes.
- The (+) (-) m.m.f. waves produce, by superposition, corresponding airgap field waves (saturation neglected). The slip is S for the forward wave and 2-S for the backward wave. This is a phenomenological basis for applying

symmetrical components to assess capacitor IM performance under steady state.

- The symmetrical components method leads to a fairly general equivalent circuit (Figure 23.16) which shows only two unknowns-the + and - components of main winding current, I_{m+} and I_{m-}, for given slip, supply voltage and machine parameters. The average torque expression is straightforward (23.18).
- When the auxiliary winding parameters R_{sa}, X_{sa}, reduced to the main winding R_{sa}/a^2, X_{sa}/a^2, differ from the main winding parameters R_{sm}, X_{sm}, the equivalent circuit (Figure 23.16) contains, in series with the capacitance impedance $\underline{Z}_a$ (reduced to the main winding $\underline{Z}_a/a^2$) after division by 2, ΔR_{sa} and ΔX_{sa}. This fact brings a notable generality to the equivalent circuit in Figure 23.16.
- To treat the case of open auxiliary winding ($I_a = 0$), it is sufficient to put Za $= \infty$ in the equivalent circuit in Figure 23.16. In this case, the forward and backward main winding components are equal to each other and thus at zero speed (S = 1), the total torque (23.18) is zero, as expected.
- The dq model seems ideal for treating the transients of single phase induction motor. Cross-coupling saturation may be elegantly accounted for in the dq model. The steady state is either obtained after solving the steady state 6 order system via a numerical method at given speed for zero initial variable values. Alternatively, the solution may be found in complex numbers by putting $d/dt = j\omega_1$. The latter procedure though handy on the computer, in presence of saturation, does not show the intuitive attributes (circuits) of the travelling (revolving) field theory of symmetrical components.
- The star Steinmetz connection may be replaced by a 2 winding capacitor motor provided the capacitance is increased by 50%, the source voltage is reduced $\sqrt{2}$ times for the main winding and is $V_s/\sqrt{6}$ for the auxiliary one. The two winding parameters are equal to phase parameters of the three phase original machine. The star Steinmetz connection is lacking the third m.m.f. space harmonic. However, the third harmonic saturation produced asynchronous torque still holds.
- Next chapter deals in detail with steady state performance in detail.

23.11 REFERENCES

1. S. A. Nasar, editor, Handbook of Electric Machines, (book) McGraw-Hill, 1987, Chapter 6 by C. G. Veinott.
2. T. J. E. Miller, J. H. Gliemann, C. B. Rasmussen, D. M. Ionel, Analysis of a Tapped-Winding Capacitor Motor, Record of ICEM-1998, Istambul, Turkey, vol. 1, pp. 581-585.
3. W. J. Morrill, Revolving Field Theory of the Capacitor Motor, Trans. AIEEE, vol. 48, 1929, pp. 614-632.

Chapter 24

SINGLE-PHASE INDUCTION MOTORS: STEADY STATE

24.1 INTRODUCTION

Steady state performance report in general on the no-load and on-load torque, efficiency and power factor, breakdown torque, locked-rotor torque, and torque and current versus speed.

In Chapter 23, we already introduced a quite general equivalent circuit for steady state (Figure 23.28) with basically only two unknowns-the forward and backward current components in the main windings with space and time harmonics neglected.

This circuit portrays steady state performance rather well for a basic assessment. On the other hand, the d-q model/cross field model (see Chapter 23) may alternatively be used for the scope with $d/dt = j\omega_1$.

The cross field model [1, 2] has gained some popularity especially for T-L tapped and generally tapped winding multi-speed capacitor motors. [3,4] We prefer here the symmetrical component model (Figure23.18) for the split phase and capacitor IMs as it seems more intuitive.

To start with, the auxiliary phase is open and a genuine single phase winding IM steady state performance is investigated.

Then, we move on to the capacitor motor to investigate both starting and running steady state performance.

Further on, the same investigation is performed on the split phase IM (which has an additional resistance in the auxiliary winding).

Then the general tapped winding capacitor motor symmetrical component model and performance are treated in some detail.

Steady state modelling for space and time harmonics is given special attention.

Some numerical examples are meant to give a consolidated feeling of magnitudes.

24.2. STEADY STATE PERFORMANCE WITH OPEN AUXILIARY WINDING

The auxiliary winding is open, in capacitor start or in split-phase IMs, after the starting process is over (Figure 24.1).

The auxiliary winding may be kept open intentionally for testing. So this particular connection is of practical interest.

According to (23.21)-(23.22) the direct and inverse main winding current components $\underline{I}_{m+}$ and $\underline{I}_{m-}$ are

$$\underline{I}_{m+,-} = \frac{1}{2}(\underline{I}_m \mp j\underline{I}_a) = \frac{1}{2}\underline{I}_m \tag{24.1}$$

The equivalent circuit in Figure 23.18 gets simplified as $\underline{Z}_a = \infty$ (Figure 24.2)

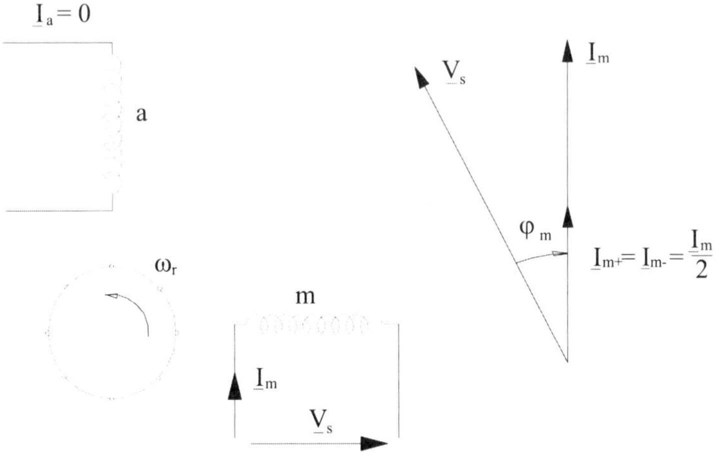

Figure 24.1 The case of open auxiliary winding: a.) schematics, b.) phasor diagram

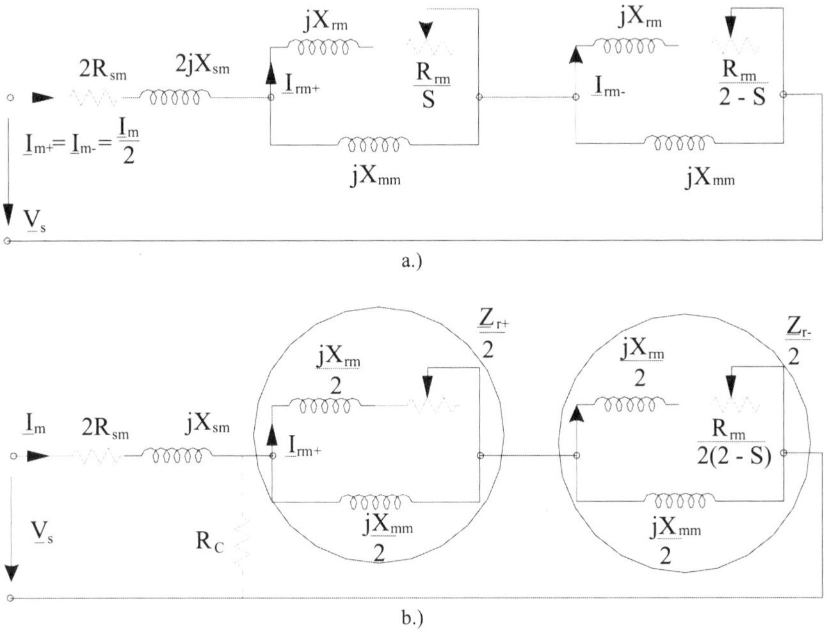

Figure 24.2 Equivalent circuit with open auxiliary phase: a.) with forward current $\underline{I}_{m+} = \underline{I}_{m-} = \underline{I}_m/2$; b.) with total current $\underline{I}_m$

As the two current components I_{m+} and I_{m-} are equal to each other, the torque expression T_e (23.18) becomes

$$T_e = 2\frac{R_r \cdot p_1}{\omega_1}\left(\frac{I_{rm+}^2}{S} - \frac{I_{rm-}^2}{2-S}\right) \tag{24.2}$$

with

$$\underline{I}_m = \frac{\underline{V}_s}{R_{sm} + jX_m + \left(\frac{\underline{Z}_{r+}}{2}\right) + \left(\frac{\underline{Z}_{r-}}{2}\right)} \tag{24.3}$$

$$\frac{\underline{Z}_{r+}}{2} = \frac{j\frac{X_{mm}}{2}\left(\frac{R_{rm}}{2S} + j\frac{X_{rm}}{2}\right)}{\frac{R_{rm}}{2S} + j\frac{(X_{mm} + X_{rm})}{2}}$$

$$\frac{Z_{r-}}{2} = \frac{j\frac{X_{mm}}{2}\left(\frac{R_{rm}}{2(2-S)} + j\frac{X_{rm}}{2}\right)}{\frac{R_{rm}}{2(2-S)} + j\frac{(X_{mm} + X_{rm})}{2}} \tag{24.4}$$

It is evident that for S = 1 (standstill), though the current $\underline{I}_m$ is maximum, I_{ms}

$$I_{ms} \approx \frac{V_s}{\sqrt{(R_{sm} + R_{rm})^2 + (X_{sm} + X_{rm})^2}} \tag{24.5}$$

The total torque (24.2) is zero as S = 2-S = 1.

The torque versus slip curve is shown on Figure 24.3.

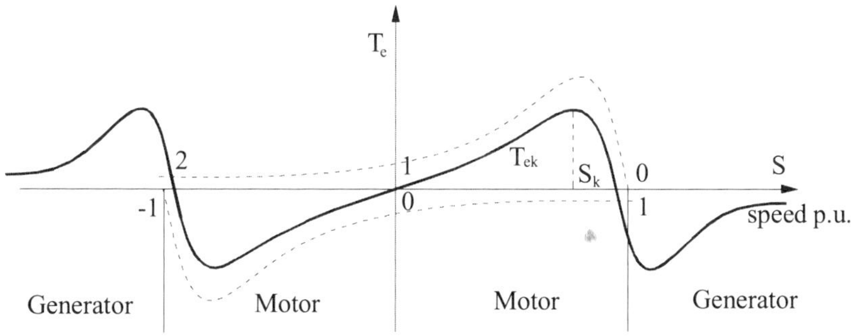

Figure24.3 Torque /speed curve for open auxiliary winding (single phase winding operation)

At zero slip (S = 0) the torque is already negative as the direct torque is zero and the inverse torque is negative. The machine exhibits only motoring and generating modes.

The peak torque value T_{ek} and its corresponding slip (speed) S_K are mainly dependent on stator and rotor resistances R_{sm}, R_{rm} and leakage reactances X_{sm}, X_{rm}.

As in reality, this configuration works for split-phase and capacitor-start IMs, the steady state performance of interest relate to slip values $S \le S_K$.

The loss breakdown in the machine is shown on Figure 24.4.

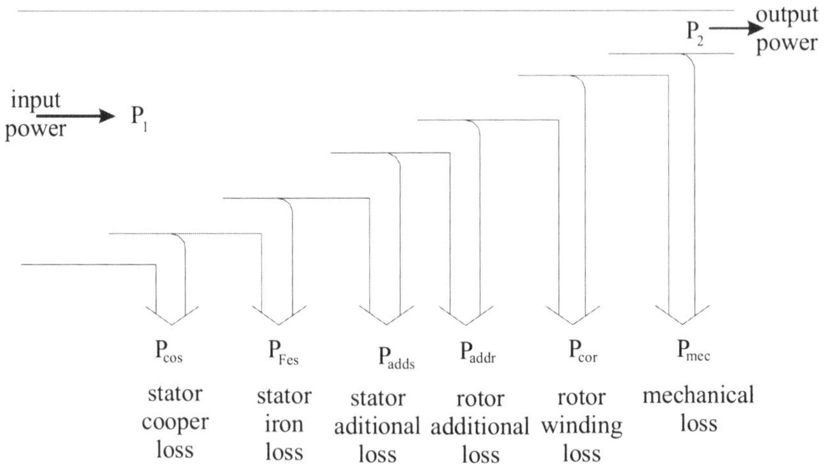

Figure 24.4 Loss breakdown in the single phase IM

Winding losses p_{cos}, p_{cor} occur both in the stator and in the rotor. Core losses occur mainly in the stator, p_{Fes}.

Additional losses occur in iron and windings and both in the stator and rotor p_{adds}, p_{addr}.

The additional losses are similar to those occurring in the three phase IM. They occur on the rotor and stator surface and as tooth-flux pulsation in iron, in the rotor cage and in the rotor laminations, due to transverse interbar currents between the rotor bars.

The efficiency η is, thus,

$$\eta = \frac{P_2}{P_1} = \frac{P_1 - \sum p}{P_1} \tag{24.6}$$

$$\sum p = p_{cos} + p_{Fes} + p_{cor} + p_{adds} + p_{addr} + p_{mec} \tag{24.7}$$

A precise assessment by computation of efficiency (of losses, in fact) constitutes a formidable task due to complexity of the problem caused by magnetic saturation, slot openings, skin effect in the rotor cage, transverse (interbar) rotor currents, etc.

In a first approximation, only the additional losses $p_{add} = p_{adds} + p_{addr}$ and the mechanical losses p_{mec} may be assigned a percentage value of rated output power. Also, the rotor skin effect may be neglected in a first approximation and the interbar resistance may be lumped into the rotor resistance.

Moreover, the fundamental core loss may be calculated by simply "planting" a core resistance R_c in parallel, as shown in Figure24.2b.

A numerical example follows.

Example 24.1.

Let us consider a permanent capacitor (C = 8 μF) 6 pole IM fed at 230 V, 50 Hz, whose rated speed n_n = 940 rpm. The main and auxiliary winding and rotor parameters are: R_{sm} = 34 Ω, R_{sa} = 150 Ω, a = 1.73, X_{sm} = 35.9 Ω, $X_{sa} = a^2X_{sm}$, X_{rm} = 29.32 Ω, R_{rm} = 23.25 Ω, X_{mm} = 249 Ω.

The core loss is p_{Fes} = 20 W and the mechanical loss p_{mec} = 20 W; the additional losses p_{add} = 5 W.

With the auxiliary phase open, calculate

a) The main winding current and power factor at n = 940 rpm
b) The electromagnetic torque at n = 940 rpm
c) The mechanical power, input power, total losses and efficiency at 940 rpm
d) The current and torque at zero slip (S = 0)

Solution

a) The no load ideal speed n_1 is

$$n_1 = \frac{f_1}{p_1} = \frac{50}{3} = 16.66\text{rps} = 1000\text{rpm}$$

The value of rated slip S_n is

$$S_n = \frac{n_1 - n_n}{n_1} = \frac{1000 - 940}{1000} = 0.06$$

$$2 - S_n = 2 - 0.06 = 1.94!$$

The direct and inverse component rotor impedances $Z_{r+}/2$, $Z_{r-}/2$ (24.3)-(24.4) are

$$\frac{\underline{Z}_{r+}}{2} = \frac{1}{2}\frac{j24.9(23.25/0.06 + j29.32)}{23.25/0.06 + j(29.32 + 249)} \approx 52.7 + j87$$

$$\frac{\underline{Z}_{r-}}{2} = \frac{1}{2}\frac{j249(23.25/1.94 + j29.32)}{23.25/1.94 + j(29.32 + 249)} \approx 4.8 + j13.5$$

The total impedance at S = 0.06 in the circuit of Figure 24.2 b) $\underline{Z}_{11}$ is

$$\underline{Z}_{11} = R_{sm} + jX_{sm} + \frac{\underline{Z}_{r+}}{2} + \frac{\underline{Z}_{r-}}{2} = 92.8 + j137$$

Now the RMS stator current $\underline{I}_m$, (24.3), is

$$\left|I_m\right| = \left|\frac{V_s}{\underline{Z}_{11}}\right| = \frac{230}{165.19} \approx 1.392\text{A}$$

The power factor $\cos\varphi_{11}$ is

$$\cos\varphi_{11} = \frac{\text{Re al}(\underline{Z}_{11})}{\left|\underline{Z}_{11}\right|} = \frac{92.8}{165.19} \approx 0.5618 \text{ !}$$

b) The torque T_e is

$$T_e = \frac{p_1}{\omega_1} I_m^2 [\text{Re}(\underline{Z}_{r+}) - \text{Re}(\underline{Z}_{r-})] = \frac{3}{314} \cdot 1.392^2 [52.7 - 4.8] = 0.8867\text{Nm} \quad (24.8)$$

Notice that the inverse component torque is less than 10% of direct torque component.

c) The iron and additional losses have been neglected so far. They should have occurred in the equivalent circuit but they did not.

So,

$$p_{cos} + p_{cor} = P_1 - T_e \cdot 2\pi n = I_m^2 \text{Re}[\underline{Z}_{11}] - T_e \cdot 2\pi n_1 =$$
$$= 1.392^2 \cdot 92.8 - 0.8867 \cdot 2\pi \cdot 16.6 = 87.1\text{W}$$

The mechanical power P_{mec} is

$$P_{mec} = T_e \cdot 2\pi n - p_{mec} = 0.8867 \cdot 2\pi \cdot 16.666(1 - 0.06) - 20 = 85.20\text{W}$$

If we now add the core and additional losses to the input power, the efficiency η is

$$\eta = \frac{85.2}{87.1 + 85.2 + 2 + 20 + 5} = 0.4275$$

The input power P_1 is

$$P_1 = \frac{P_{mec}}{\eta} = \frac{85.2}{0.4975} = 199.30\text{W}$$

The power factor, including core and mechanical losses, is

$$\cos\varphi = P_1/(V_s I_m) = 199.30/(230 \cdot 1.392) = 0.6225 \text{ !}$$

d) The current at zero slip is approximately

$$(I_m)_{S=0} \approx \left|\frac{\underline{V}_s}{R_{sm} + j\left(X_{sm} + \frac{X_{mm}}{2}\right) + \frac{(\underline{Z}_{r-})_{S=0}}{2}}\right| \approx \left|\frac{230}{34 + j(35.9 + 249) + 4.8 + j13.5}\right|$$
$$= 0.764\text{A}$$

Only the inverse torque is not zero

$$\left(T_e\right)_{S=0} = -(I_m)^2_{s=0} \operatorname{Re}\frac{\left(\underline{Z}_{r-}\right)_{S=0}}{2}\cdot\frac{P_1}{\omega_1} \approx -0.764^2\cdot 4.8\cdot\frac{3}{314} = -0.0268\text{Nm}$$

Negative torque means that the machine has to be shaft-driven to maintain the zero slip conditions.

In fact, the negative mechanical power equals half the rotor winding losses as $S_b = 2$ for $S = 0$.

So the rotor winding loss $(p_{cor})_{s=0}$ is

$$\left(p_{cor}\right)_{S=0} = 2\times\left(T_e\right)_{S=0}\cdot 2\pi n_1 = 2\times 0.0268\times 2\pi\cdot 16.6 = 5.604\text{W}$$

The total input power $(P_1)_{S=0}$ is

$$P_{10} = \left(R_{sm} + \operatorname{Re}(\underline{Z}_b)_{s=0}\right)\left(I_m\right)^2_{s=0} \approx \left(34 + 4.8\right)\cdot 0.764^2 = 22.647\text{W}$$

The stator winding losses $p_{cos} = 34\times 0.764^2 = 19.845\text{W}$

Consequently half of rotor winding losses at $S = 0$ (2.8W) are supplied from the power source while the other half is supplied from the shaft power.

The frequency of rotor currents at $S = 0$ is $f_{2b} = S_b f_1 = 2f_1$.

The steady state performance of IM at $S = 0$ with an open auxiliary phase is to be used for loss segregation and some parameters estimation.

24.3 THE SPLIT PHASE AND THE CAPACITOR IM: CURRENTS AND TORQUE

For the capacitor IM the equivalent circuit of Figure 23.18 remains as it is with

$$\underline{Z}_a = -j/\omega_1 C \tag{24.9}$$

as shown in Figure 24.5.

For the split phase IM, the same circuit is valid but with $C = \infty$ and $R_{sa}/a^2 - R_{sm} > 0$.

Also for a capacitor motor with a permanent-capacitor in general: $R_{sa}a^2 - R_{sm} = 0$; $X_{sa}a^2 - X_{sm} = 0$; that is, it uses about the same copper weight in both, main and auxiliary windings.

Notice that all variables are reduced to the main winding (see the two voltage components). This means that the main and auxiliary actual currents $\underline{I}_m$ and $\underline{I}_a$ are now

$$\underline{I}_m = \underline{I}_{m+} + \underline{I}_{m-};\ \underline{I}_a = j\left[\underline{I}_{m+} - \underline{I}_{m-}\right]/a \tag{24.10}$$

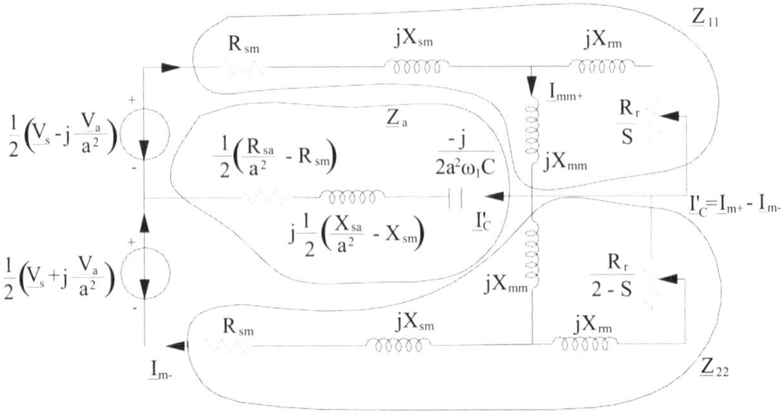

Figure 24.5 Equivalent circuit for the capacitor IM (V_a=V_s) in the motor mode

Let us denote

$$\underline{Z}_+ = R_{sm} + jX_{sm} + \frac{JX_{mm}(jX_{rm} + R_{rm}/S)}{R_{rm}/S + j(X_{mm} + X_{rm})} \tag{24.11}$$

$$\underline{Z}_a^m = \Delta R_{sm} + j\Delta X_{sm} - \frac{j}{2a^2}X_c; X_c = \frac{1}{\omega_1 C} \tag{24.12}$$

$$\Delta R_{sm} = \frac{1}{2}(R_{sa}/a^2 - R_{sm}); \Delta X_{sm} = \frac{1}{2}(X_{sa}/a^2 - X_{sm}) \tag{24.13}$$

$$\underline{Z}_- = R_{sm} + jX_{sm} + \frac{jX_{mm}(jX_{rm} + R_{rm}/(2-S))}{R_{rm}/(2-S) + j(X_{rm} + X_{mm})} \tag{24.14}$$

The solution of the equivalent circuit in Figure 23.19, with (24.11)-(24.14), leads to matrix equations

$$\begin{vmatrix} \dfrac{\underline{V}_s(1 - j/a)}{2} \\ \dfrac{\underline{V}_s(1 + j/a)}{2} \end{vmatrix} = \begin{vmatrix} \underline{Z}_+ + \underline{Z}_a^m & -\underline{Z}_a^m \\ -\underline{Z}_a^m & \underline{Z}_- + \underline{Z}_a^m \end{vmatrix} \cdot \begin{vmatrix} \underline{I}_{m+} \\ \underline{I}_{m-} \end{vmatrix} \tag{24.15}$$

with the straightforward solution

$$\underline{I}_{m+} = \frac{\underline{V}_s}{2}\frac{(1 - j/a)(\underline{Z}_- + 2\underline{Z}_a^m)}{\underline{Z}_+ \cdot \underline{Z}_- + \underline{Z}_a^m(\underline{Z}_+ + \underline{Z}_-)} \tag{24.16}$$

$$\underline{I}_{m-} = \frac{\underline{V}_s}{2}\frac{(1 + j/a)(\underline{Z}_+ + 2\underline{Z}_a^m)}{\underline{Z}_+ \cdot \underline{Z}_- + \underline{Z}_a^m(\underline{Z}_+ + \underline{Z}_-)} \tag{24.17}$$

The electromagnetic torque components T_{e+} and T_{e-} retain the expressions (24.8) under the form

$$T_{e+} = \frac{2p_1}{\omega_1} I_{m+}^2 \left[Re(\underline{Z}_+) - R_{sm} \right] \tag{24.18}$$

$$T_{e-} = \frac{2p_1}{\omega_1} I_{m-}^2 \left[Re(\underline{Z}_-) - R_{sm} \right] \tag{24.19}$$

The factor 2 in (24.18)-(24.19) is due to the fact that the equivalent circuit on Figure 23.19 is per phase and it refers to a two phase winding model.

Apparently, the steady state performance problem is solved once we have the rather handy Equations. (24.10)-(24.18) at our disposal.

We now take a numerical example to get a feeling of magnitudes.

Example 24.2. Capacitor motor case

Let us consider again the capacitor motor from example 24.1.; this time with the auxiliary winding on and R_{sa} changed to $R_{sa} = a^2 R_{sm}$, $C' = 4\mu F_0$ and calculate

a) The starting current and torque with $C = 8\ \mu F$
b) For rated slip $S_n = 0.06$: supply current and torque
c) The main phase I_m and auxiliary phase current I_a and the power factor at rated speed ($S_n = 0.06$)
d) The phasor diagram

Solution

a) For S = 1, from (24.10)-(24.13),

$$\begin{aligned}(\underline{Z}_+)_{S=1} = (\underline{Z}_-)_{S=1} = \underline{Z}_{sc} &\approx R_{sm} + R_{rm} + j(X_{sm} + X_{rm}) \\ &= (34 + 23.25) + j(35.9 + 29.32) \\ &= 57.5 + j65.22\end{aligned}$$

Note that $\Delta R_{sm} = \frac{R_{sa}}{a^2} - R_{sm} = 0$ and $\Delta X_{sm} = \frac{X_{sa}}{a^2} - X_{sm} = 0$ and thus

$$\underline{Z}_a^m = \frac{-jX_c}{2a^2} = \frac{-j}{2\cdot 1.73^2 \cdot 2\pi \cdot 50 \times 8 \cdot 10^{-6}} \approx -j66.34$$

From (24.16)-(24.17)

$$\begin{aligned}(\underline{I}_{m+})_{S=1} &= \frac{230}{2} \frac{\left[(1 - j/1.73)(57.5 + j65.22) - 2j66.7\right]}{(57.5 + j65.22)^2 - j66.7 \cdot (57.5 + j65.22) \times 2} \\ &= 1.444 - j1.47\end{aligned}$$

$$(\underline{I}_{m-})_{S=1} = \frac{230}{2}\frac{[(1+j/1.73)(57.5+j65.22)-2j\cdot 66.7]}{(57.5+j65.22)^2 - j66.7\cdot(57.5+j65.22)\times 2}$$
$$= 0.3 - j0.5$$

Consequently, the direct and inverse torques are ((24.17)-24.18)

$$(T_{e+})_{S=1} = \frac{2\times 3}{314}\cdot 2.06^2(57.5-34) = 1.9066\text{Nm}$$

$$(T_{e-})_{S=1} = \frac{-2\times 3}{314}\cdot 0.583^2(57.5-34) = -0.1526\text{Nm}$$

The inverse torque T_{e-} is much smaller than the direct torque T_{e+}. This indicates that we are close to symmetrization through adequate a and C.

The main and auxiliary winding currents $\underline{I}_m$ and $\underline{I}_a$ are (24.10)

$$\underline{I}_m = \underline{I}_{m+} + \underline{I}_{m-} = 1.444 - j1.47 + 0.30 - j0.50 \approx 1.744 - j1.97$$

$$\underline{I}_a = \frac{j(\underline{I}_{m+} - \underline{I}_{m-})}{a} = \frac{j(1.44 - j1.47 - 0.3 + j0.5)}{1.73} = j0.6613 + 0.56$$

$$\underline{I}_s = \underline{I}_m + \underline{I}_a = 1.744 - j1.97 + j0.6693 + 0.56 = 2.304 - j1.308$$

At the start, the source current is lagging the source voltage (which is a real variable here) despite the influence of the capacitance.

The capacitance voltage,

$$(V_c)_{s=1} = \frac{(I_a)_{s=1}}{\omega C} = \frac{0.866}{314\cdot 8\cdot 10^{-8}} = 344.75\text{V}$$

b) For rated speed (slip)–S_n = 0.06, the same computation routine is followed. As the capacitance is reduced twice

$$\underline{Z}_a^m = -j133.4$$

$$(\underline{Z}_+)_{S=0.06} = R_{sm} + jX_{sm} + \frac{jX_{mm}\left(\frac{R_{rm}}{S} + jX_{rm}\right)}{\frac{R_{rm}}{S} + j(X_{mm} + X_{rm})} =$$

$$= 34 + j23.25 + \frac{j249\left(\frac{23.25}{0.06} + j29.32\right)}{\frac{23.25}{0.06} + j(249 + 29.32)} = 139.4 + j197.75$$

$$(\underline{Z}_-)_{S=0.06} = 340 + j23.25 + \frac{j249\left(\frac{23.25}{1.94} + j29.32\right)}{\frac{23.25}{1.94} + j(249 + 29.32)} = 46.6 + j50.25\Omega$$

$$(\underline{I}_{m+})_{S_n=0.06} = \frac{230}{2}\frac{[(1 - j/1.73)(46.6 + j50.2) - 2j133.4]}{(139.4 + j197)(46.6 + j50.25) - j133.4 \cdot (139.4 + 46.66 + j(197 + 50.25))}$$
$$= 0.525 - j0.794$$

$$(\underline{I}_{m-})_{S_n=0.06} = \frac{230}{2}\frac{[(1 + j/1.73)(139 + j197.5) - 2j133.4]}{(139.4 + j197)(46.6 + j50.25) - j133.4 \cdot (139.4 + 46.66 + j(197 + 50.25))}$$
$$= 0.1016 + j0.0134$$

$$(T_{e+})_{S=1} = \frac{2 \times 3}{314} \cdot (0.9518)^2 (139.4 - 34) = 1.8245\text{Nm}$$

$$(T_{e-})_{S=1} = -\frac{2 \times 3}{314} \cdot (0.1025)^2 (46.6 - 34) = -2.53 \times 10^{-3}\text{Nm}$$

Both the inverse current and torque components are very small. This is a good sign that the machine is almost symmetric. Perfect symmetry is obtained for $\underline{I}_{m-} = 0$.

The main and auxiliary winding currents are

$$I_m = I_{m+} + I_{m-} = 0.525 - j0.794 + 0.1016 + j0.0134 = 0.6266 - j0.7806$$
$$I_a = j\frac{(I_{m+} - I_{m-})}{a} = \frac{j(0.525 - j0.794 - 0.1016 - j0.0134)}{1.73} = j0.274 + 0.466$$

The phase shift between the two currents is almost 90^0.

Now the source current $(\underline{I}_s)_{s=0.06}$ is

$$(I_s)_{S=0.06} = 0.6266 - j0.7806 + j0.274 + 0.466$$
$$= 1.0926 - j0.5066$$

The power factor of the motor is

$$\cos\varphi_s = \frac{\text{Re}(\underline{I}_s)}{|\underline{I}_s|} = \frac{1.0926}{1.2043} = 0.9072$$

The power factor is good. The 4 μF capacitance has brought the motor close to symmetry and providing for good power factor.

The speed torque curve may be calculated, for steady state, point by point as illustrated above. (Figure 24.6)

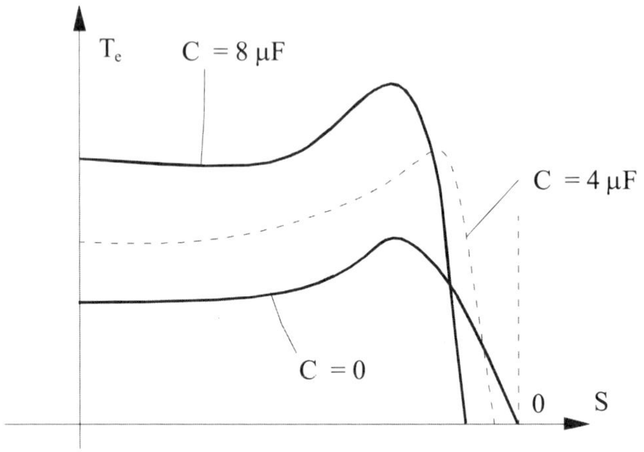

Figure 24.6 Torque speed curve of capacitor motors from example 24.2 (qualitative)

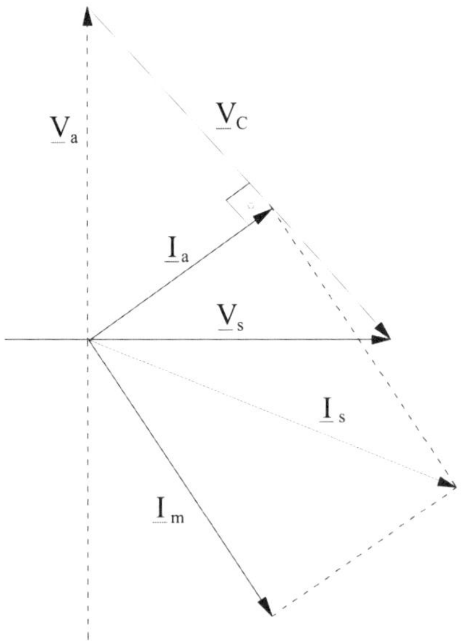

Figure 24.7 Phasor diagram for S = 0.06, C = 4 μF and the data of example 24.2

A large capacitor at start leads to higher starting torque due to symmetrization tendency at s = 1. At low slip values, however, there is too much inverse torque (Figure 24.6) as the motor departures from symmetrization conditions.

A proper capacitor at rated load would lead again the symmetrization conditions (zero inverse current) and thus good performance is obtained.

This is how the dual-capacitor IM evolved. The ratio of the two capacitance is only 2/1 in our case but as power goes up in the KW range, this ratio may go up to 6/1.

The phasor diagram at S = 0.06 is shown on Figure 24.7.

As the equivalent circuit (Figure 24.5) and the current and torque components expressions (24.16)-(24.19) remain unchanged, the case of split-phase IM (without a capacitor, but with an additional resistance: $\Delta R_{sm} \neq 0$ by design or for real) may be treated the same way as the capacitor IM, except for the content of $\underline{Z}_a^m$.

24.4 SYMMETRIZATION CONDITIONS

By symmetrisation we mean zero inverse current component I_{m-} for a given slip. This requires by necessity a capacitance C dependent on slip. From (14.17) with $\underline{I}_{m-} = 0$, we obtain

$$(1 + j/a)\underline{Z}_+ + 2\underline{Z}_a^m = 0; \underline{Z}_+ = R_+ + jX_+ \tag{24.20}$$

with

$$\underline{Z}_a^m = -j\frac{1}{2\omega Ca^2} + \frac{1}{2}\left(\frac{R_{sa}}{a^2} - R_{sm}\right) + \frac{1}{2}j\left(X_{sa}/a^2 - X_{sm}\right) = R_a^m + jX_a^m \tag{24.21}$$

The direct total impedance $\underline{Z}_+$ may be written as

$$\underline{Z}_+ = Z_+ e^{j\varphi+}; \varphi_+ = \tan^{-1}(X_+/R_+) \tag{24.22}$$

With (24.21-24.22), Equation (24.20) yields two symmetrization conditions,

$$R_+ - \frac{X_+}{a} = -2R_a^m = -\left(R_{sa}/a^2 - R_{sm}\right) \tag{24.23}$$

$$\frac{R_+}{a} + X_+ = +\frac{1}{\omega Ca^2} - \left(X_{sa}/a^2 - X_{sm}\right) \tag{24.24}$$

The two unknowns of Equations (24.23)-(24.24) are

- C and *a* for the capacitor IM
- R_{sa} and *a* for the split-phase IM

The eventual additional resistance in the auxiliary winding of the split-phase IM is lumped into R_{sa}.

For the permanent capacitor motor, in general, $\Delta R_{sm} = R_{sa}/a^2 - R_{sm} = 0$ and $\Delta X_{sm} = X_{sa}/a^2 - X_{sm} = 0$. So we obtain:

$$a = \frac{X_+}{R_+} = \tan\varphi_+ \tag{24.25}$$

$$X_c = \frac{1}{\omega C} = \frac{X_+}{\cos^2\varphi_+} = Z_+ \cdot a\sqrt{a^2+1} \tag{24.26}$$

φ_+ is the direct component power factor angle.

A few remarks are in order

- The turn ratio *a* for symmetrization is equal to the tangent of the direct component of power factor angle. This angle varies with slip for a given motor.
- So, in dual capacitor-motor, only, it may be possible to match a and C both for start and running conditions when and only when the value of turn ratio *a* changes also from low to high speed running when the capacitor value is changed. This change of *a* means a dual section main or auxiliary winding.

For the split phase motor $C = \infty$ and $\Delta R_{sm} \neq 0$ and $\Delta X_{sm} \neq 0$ and thus from (24.23)-(24.24), even if

$$X_{sa}/a^2 < X_{sm} \tag{24.27}$$

we obtain

$$a = \frac{-R_+ \pm \sqrt{R_+^2 - 4(X_+ - X_{sm})X_{sa}}}{X_+ - X_{sm}} \tag{24.28}$$

It is now clear that as $X_+ > X_{sm}$, there is no *a* to symmetrize the split/phase IM, as expected.

Choosing a and R_{sa} should thus follow other optimisation criterion.

For the dual-capacitor motor IM in example 24.2, the value of a and C for symmetrization, according to (24.25)-(24.26) are

$$(a)_{S=1} = \frac{(X_+)_{S=1}}{(R_+)_{S=1}} = \frac{65.22}{57.5} = 1.134$$

$$(C)_{S=1} = \frac{1}{314\cdot\left(\sqrt{57.5^2+65.22^2}\right)\times 1.134\sqrt{1+1.134^2}} = 21.36\times10^{-6}\,F$$

$$(a)_{S=0.06} = \frac{(X_+)_{S=0.06}}{(R_+)_{S=0.06}} = \frac{197.25}{139.4} = 1.415$$

$$(C)_{S=0.06} = \frac{1}{314\left(\sqrt{139.25^2+197.25^2}\right)\cdot 1.415\sqrt{1+1.415^2}} = 5.378\cdot10^{-6}\,F$$

The actual values of the capacitor were 8 μF for starting (S = 1) and 4 μF for rated slip (S = 0.06). Also the ratio of equivalent turns ratio was a = 1.73.

It is now evident that with this value of a (a = 1.73) no capacitance C can symmetrize perfectly the machine for S = 1 or S = 0.06. Once the machine is made, finding a capacitor to symmetrize the machine is difficult. In practice, in general, coming close to symmetrization conditions for rated load is recommendable, while other criteria for a good design have to be considered..

24.5 STARTING TORQUE AND CURRENT INQUIRIES

For starting conditions (S = 1) $\underline{Z}_+ = \underline{Z}_- = \underline{Z}_{sc}$, so (24.16) and (24.17) become

$$\underline{I}_{m+} = \frac{\underline{V}_s}{2\underline{Z}_{sc}}\left[1 - \frac{j}{a}\frac{\underline{Z}_{sc}}{Z_{sc} + 2\underline{Z}_a^m}\right] \tag{24.29}$$

$$\underline{I}_{m+} = \frac{\underline{V}_s}{2\underline{Z}_{sc}}\left[1 + \frac{j}{a}\frac{\underline{Z}_{sc}}{Z_{sc} + 2\underline{Z}_a^m}\right] \tag{24.30}$$

The first components of + - currents in (24.29)-(24.30) correspond to the case when the auxiliary winding is open $\left(\left|\underline{Z}_a^m\right| = \infty\right)$

We may now write

$$Z_{sc} = Z_{sc}e^{j\varphi_{sc}}; \underline{Z}_a^m = Z_a^m e^{j\varphi_a}; K_a = \frac{2\left|\underline{Z}_a^m\right|}{\left|\underline{Z}_{sc}\right|} \tag{24.31}$$

This way (24.29) may be written as

$$\underline{I}_{m+} = \underline{I}_{sc}(1 + \alpha + j\beta)/2 \tag{24.32}$$

$$\underline{I}_{m-} = \underline{I}_{sc}(1 - \alpha - j\beta)/2 \tag{24.33}$$

with

$$\alpha + j\beta = \frac{-j}{a(1 + K_a \exp(j\gamma))} \tag{24.34}$$

with

$$\gamma = \varphi_{sc} - \varphi_a$$

Let us denote by T_{esc} the starting torque of two phase symmetrical IM.

$$T_{esc} = 2\frac{p_1}{\omega_1} \cdot I_{sc}^2 \operatorname{Re}(Z_{r+})_{S=1} \tag{24.35}$$

The p.u. torque at start t_{es} is

$$t_{es} = \frac{(T_e)_{S=1}}{T_{esc}} = \frac{|\underline{I}_{m+}|^2 - |\underline{I}_{m-}|^2}{|\underline{I}_{sc}|^2} \quad (24.36)$$

Making use of (24.32)-(24.33) t_{es} becomes

$$t_{es} = \alpha = \frac{K_a \sin\gamma}{a\left(1 + K_a^2 + 2K_a \cos\gamma\right)} \quad (24.37)$$

$$\beta = -\left(1 + K_a \cos\gamma\right)/\left(1 + K_a^2 + 2K_a \cos\gamma\right)/a \quad (24.38)$$

A few comments are in order

- The relative starting torque depends essentially on the ratio K_a and on the angle γ.
- We should note that for the capacitor motor with windings using the same quantity of copper ($\Delta R_{sa} = 0$ $\Delta X_{sa} = 0$),

$$(K_a)_C = \frac{1}{\omega C a^2 |\underline{Z}_{sc}|} \quad (24.39)$$

Also for this case $\varphi_a = -\pi/2$ and thus

$$(\gamma)_c = \varphi_{sc} + \pi/2 \quad (24.40)$$

- On the other hand, for the split phase capacitor motor with, say, $\Delta X_{sa} = 0$,

$C = \infty$.

$$(K_a)_{\Delta R_{sa}} = \frac{\Delta R_{sa}}{|\underline{Z}_{sc}|} = \frac{\left(R_{sa} - a^2 R_{sm}\right)}{a^2 |\underline{Z}_{sc}|} \quad (24.41)$$

Also $\varphi_a = 0$ and $\gamma = \varphi_{sc}$.

For the capacitor motor, with (24.39)-(24.40), the expression (24.37) becomes

$$t_{es} = \frac{\cos\varphi_{sc}}{a\left[a^2 Z_{sc} \cdot \omega_1 C + \frac{1}{\omega_1 C Z_{sc} a^2} - 2\sin\varphi_{sc}\right]} \quad (24.42)$$

The maximum torque versus the turns ratio *a* is obtained for

$$a_K = \sqrt{\frac{\sin\varphi_{sc}}{Z_{sc}\omega C}}$$

$$(t_{es})_{max} = \frac{\sqrt{Z_{sc}\omega_1 C \sin\varphi_{sc}}}{\cos\varphi_{sc}} \quad (24.43)$$

For the split phase motor (with ΔR_{sa} replacing $1/\omega C$) and $\gamma = \varphi_{sc}$, no such maximum starting torque occurs.

The maximum starting torque conditions are not enough for an optimum starting.

At least the starting current has to be also considered.

From (24.32)-(24.33)

$$\underline{I}_m = \left(\underline{I}_{m+} + \underline{I}_{m-}\right) = \underline{I}_{sc} \tag{24.44}$$

$$\underline{I}_a = j\frac{\left(\underline{I}_{m+} - \underline{I}_{m-}\right)}{a} = j\frac{(\alpha + j\beta)}{a}\underline{I}_{sc} \tag{24.45}$$

The source current $\underline{I}_s$ is

$$\begin{aligned} \underline{I}_s &= \left[1 + \frac{(-\beta + j\alpha)}{a}\right]\underline{I}_{sc} \\ \frac{I_s}{I_{sc}} &= \sqrt{\left(1 - \frac{\beta}{a}\right)^2 + \frac{\alpha^2}{a^2}} \end{aligned} \tag{24.46}$$

Finally for the capacitor motor

$$\frac{I_s}{I_{sc}} = \sqrt{\left[1 + \frac{t_{es}}{a}\frac{\left(\omega C Z_{sc} a^2 - \sin\varphi_{sc}\right)}{\cos\varphi_{sc}}\right]^2 + \frac{t_{es}^2}{a^2}} \tag{24.47}$$

The p.u. starting torque and current expressions reflect the strong influence of: capacitance C, the shortcircuit impedance Z_{sc}, its phase angle, and the turns ratio a. As expected, a higher rotor resistance would lead to higher starting torque.

In general $1.5 < a < 2.0$ seems to correspond to most optimization criteria. From (24.42), it follows that in this case $Z_{sc}\omega C < 1$.

For the split phase induction motor with $\gamma = \varphi_{sc}$ from (24.38)

$$\beta = \frac{-\left(1 + K_a \cos\varphi_{sc}\right)}{a\left(1 + K_a^2 + 2K_a \cos\varphi_{sc}\right)} \tag{24.48}$$

with α (t_{es}) from (24.37)

$$\beta = \frac{-\left(1 + K_a \cos\varphi_{sc}\right)}{K_a \sin\varphi_{sc}} \cdot t_{es} \tag{24.49}$$

With K_a from (24.41) the starting current (24.46) becomes

$$\frac{I_s}{I_{sc}} = \sqrt{\left[1 + \frac{t_{es}}{a}\frac{\left(\frac{|\underline{Z}_{sc}|a^2}{(R_{sa} - a^2 R_{sm})} + \cos\varphi_{sc}\right)}{\sin\varphi_{sc}}\right]^2 + \left(\frac{t_{es}}{a}\right)^2} \quad (24.50)$$

Expressions (24.47)-(24.50) are quite similar except for the role of φ_{sc}, the shortcircuit impedance phase angle. They allow to investigate the existence of a minimum I_s/I_{sc} for given starting torque for a certain turns ratio a_{ki}.

Especially for the capacitor start and split-phase IM the losses in the auxiliary winding are quite important as, in general, the design current density in this winding (which is to be turned off after start) is to be large to save costs.

Let us notice that for the capacitor motor, the maximum starting torque conditions (24.43) lead to a simplified expression of I_s/I_{sc} (24.47).

$$\left(\frac{I_s}{I_{sc}}\right)_{t_{es}\max} = \sqrt{1 + \left(\frac{(t_{es})_{max}}{a_K}\right)^2} = \sqrt{1 + \frac{(Z_{sc}\omega C)^2}{\cos^2\varphi_{sc}}} \quad (24.51)$$

For $Z_{sc}\omega C < 1$, required to secure $a_K > 1$, both the maximum starting torque (24.43) and the starting current are reduced.

For an existing motor, only the capacitance C may be changed. Then directly (24.43) and (24.47) may be put to work to investigate the capacitor influence on starting torque and on starting losses (squared starting current).

As expected there is a capacitor value that causes minimum starting losses (minimum of $(I_s/I_{sc})^2$) but that value does not also lead to maximum starting torque.

The above inquires around starting torque and current show clearly that this issue is rather complex and application-dependent design optimization is required to reach a good practical solution.

24.6 TYPICAL MOTOR CHARACTERISTIC

To evaluate capacitor or split-phase motor steady-state performance, the general practice rests on a few widely recognized characteristics.

- Torque (T_e) versus speed (slip)
- Efficiency (η) versus speed
- Power factor ($\cos\varphi_s$) versus speed
- Source current (I_s) versus speed
- Main winding current (I_m) versus speed
- Auxiliary winding current (I_a) versus speed
- Capacitor voltage (V_c) versus speed
- Auxiliary winding voltage (V_a) versus speed

All these characteristics may be calculated via the circuit model in Figure 24.2 and Equations (24.16-24.19) with (24.11)-(24.14), (24.44)-(24.45), for given motor parameters, capacitance C, and speed (slip).

Typical such characteristics for a 300 W, 2 pole, 50 Hz permanent capacitor motor are shown in Figure 24.8.

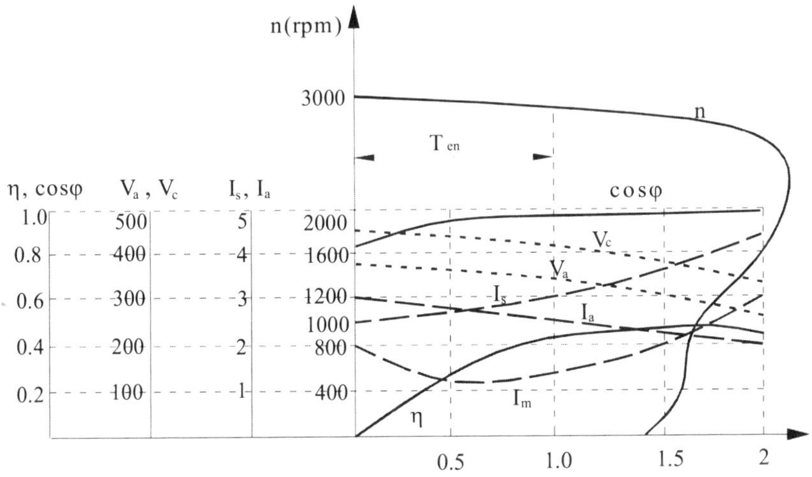

Figure 24.8 Typical steady-state characteristics of a 300W, 2 pole, 50Hz

The characteristics on Figure 24.8 prompt remarks such as

- The torque is zero for a speed smaller than $n_1 = f_1/p_1 = 50$ rps (3000 rpm).
- The efficiency is only moderate as a rather large rotor resistance has been considered to secure good starting torque for moderate starting current.
- The power factor is high all over the torque (speed) range.
- The capacitor and auxiliary winding voltages (V_c, V_a) decrease with torque increase (above breakdown torque speed).
- Notice that data on Figure 24.8 correspond to possible measurements above breakdown torque speed and not from the start. Only the torque-speed curve covers the entire speed range (from the start).
- From zero torque to rated torque (about 1 Nm), the source current increases only a little while I_a decreases and I_m has a minimum below rated torque.
- The characteristics on Figure 24.8 are typical for a permanent capacitor IM where the designer has to trade efficiency for starting torque.

24.7 NON-ORTHOGONAL STATOR WINDINGS

It was shown that for the split-phase motor, with a more resistive auxiliary winding, there is no way to even come close to symmetry conditions with orthogonal windings.

This is where the non-orthogonal windings may come into play.

As the time phase angle shift $(\varphi_a-\varphi_m) < 90^0$, placing spatially the windings at an angle $\xi > 90^0$ intuitively leads to the idea that more forward travelling field might be produced.

Let us write the expressions of two such m.m.f.s

$$F_m(\theta_{es},t) = F_{m1}\cos\theta_{es}\cdot\sin\omega_1 t \tag{24.52}$$

$$F_a(\theta_{es},t) = F_{a1}\cos(\theta_{es}+\xi)\cdot\sin(\omega_1 t+(\varphi_a-\varphi_m)) \tag{24.53}$$

To the simple mathematics let us suppose that $F_{m1} = F_{a1}$ and that $(\xi > 90^0)$,

$$\xi + (\varphi_a - \varphi_m) = 180^0 \tag{24.54}$$

$$\begin{aligned} F(\theta_{es},t) &= F_m(\theta_{es},t) + F_a(\theta_{es},t) = \\ &= \frac{F_{m1}}{2}\left[\sin(\omega_1 t - \theta_{es}) + \sin(\omega_1 t - \theta_{es} + 180 - 2\xi)\right] \end{aligned} \tag{24.55}$$

For $\xi = 90^0$, the symmetry condition of capacitor motor is obtained. This corresponds to the optimum case.

For $\xi > 90^0$, still, a good direct travelling wave can be obtained. Its amplitude is affected by $\xi \neq 90^0$.

It is well understood that even with a capacitor motor, such an idea may be beneficial in the sense of saving some capacitance. But it has to be remembered that placing the two stator windings nonorthogonally means the using in common of all slots by the two phases.

For a split phase IM condition, (24.54) may not be easily met in practice as $\varphi_a-\varphi_m = 30^0$-$35^0$ and $\xi = 145^0$-150^0 is a rather large value.

To investigate quantitatively the potential of nonorthogonal windings, a transformation of coordinates (variables) is required (Figure 24.9)

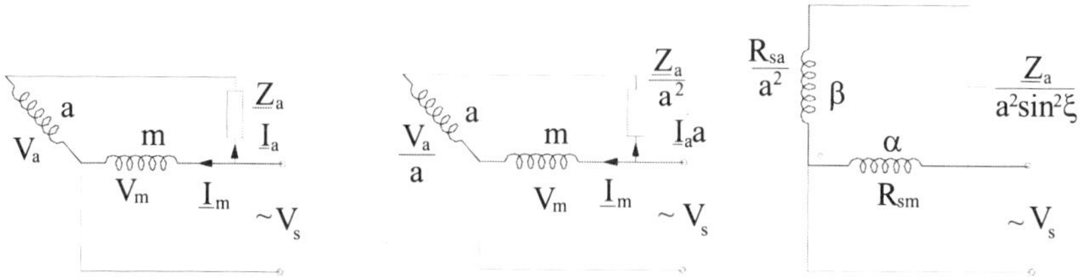

Figure 24.9 Evolutive equivalent schematics of nonorthogonal winding IM

The m & a nonorthogonal (real) windings are gradually transformed into orthogonal α-β windings. The conservation of m.m.f.s and apparent power leads to :

$$\begin{aligned}\underline{I}_\alpha &= \underline{I}_m + a\underline{I}_a\cos\xi\\ \underline{I}_\beta &= a\underline{I}_a\sin\xi\\ \underline{V}_\alpha &= \underline{V}_m\\ \underline{V}_\beta &= \frac{V_a/a - V_m\cos\xi}{\sin\xi}\end{aligned} \tag{24.56}$$

The total stator copper losses for the m & a and α & β model are

$$P_{m\&a} = R_{sm}I_m^2 + R_aI_a^2 \tag{24.57}$$

$$\begin{aligned}P_{\alpha\&\beta} &= R_{sm}I_\alpha^2 + \frac{R_a}{a^2}I_\beta^2 =\\ &= R_{sm}I_a^2 + R_aI_a^2 + R_{sm}\cdot 2I_a\cdot a\cdot I_m\cos\xi\cos(\varphi_m - \varphi_a)\end{aligned} \tag{24.58}$$

φ_m and φ_a are the phase angles of $\underline{I}_m$ and $\underline{I}_a$. It is now evident that only with φ_m-φ_a = 90^0 (the capacitor motor) the a and m and α and β are fully equivalent in terms of losses.

As long as ξ is close to 90^0 (up to 100-110^0) and φ_m-φ_a = 70^0-80^0, the two models are practically equivalent.

The additional impedance in the β circuit is obtained from

$$\underline{Z}_aI_a^2 = Z_{a\beta}I_\beta^2 \tag{24.59}$$

With (24.56)-(24.59) becomes

$$\underline{Z}_{a\beta} = \frac{\underline{Z}_a}{a^2\sin^2\xi} \tag{24.60}$$

as shown on Figure 24.9 c.

The relationship between the main and auxiliary terminal voltages is

$$\begin{aligned}\underline{V}_s &= \underline{V}_a + \underline{Z}_a\underline{I}_a = \underline{V}_{a0} + (R_{sa} + jX_{sa})\underline{I}_a + \underline{Z}_a\underline{I}_a\\ \underline{V}_s &= \underline{V}_m = \underline{V}_{m0} + \underline{I}_m(R_{sm} + jX_{sm})\end{aligned} \tag{24.61}$$

To consider the general case of stator windings when $R_{sa} / a^2 \neq R_{sm}$ and $X_{sa} / a^2 \neq X_{sm}$, we have introduced the total e.m.f.s $\underline{V}_{a0}$ and $\underline{V}_{m0}$ in (24.61).

Also, we introduce a coefficient m which is unity if $R_{sa} / a^2 \neq R_{sm}$ and $X_{sa} / a^2 \neq X_{sm}$ and zero, otherwise,

$$\begin{aligned}\underline{V}_s &= \underline{V}_{a0} + [\underline{Z}_a + (R_{sa} + jX_{sa})(1-m)]\underline{I}_a\\ \underline{V}_s &= \underline{V}_m = \underline{V}_{m0} + (R_{sm} + jX_{sm})(1-m)\underline{I}_m\end{aligned} \tag{24.62}$$

We now apply the symmetrical components method to V_α, V_β

$$\begin{vmatrix}\underline{V}_+ \\ \underline{V}_-\end{vmatrix} = \frac{1}{2}\begin{vmatrix}1-j \\ 1+j\end{vmatrix}\begin{vmatrix}\underline{V}_\alpha \\ \underline{V}_\beta\end{vmatrix}$$

$$\underline{V}_\alpha = \underline{V}_+ + \underline{V}_- \tag{24.63}$$

$$\underline{V}_\beta = j(\underline{V}_+ - \underline{V}_-)$$

With (24.56), (24.60), (24.62), Equation (24.63) yields only two equations,

$$\underline{V}_s = a\left[j(\underline{V}_+ - \underline{V}_-)\sin\xi + (\underline{V}_+ + \underline{V}_-)\cos\xi\right] - \frac{\left[\underline{Z}_a + (R_{sa} + jX_{sa})(1-m)\right]}{ja\sin\xi}(\underline{I}_+ - \underline{I}_-)$$

$$\underline{V}_s = \underline{V}_+ + \underline{V}_- + (R_{sm} + jX_{sm})\left[\underline{I}_+ + \underline{I}_- - \frac{\cos\xi}{\sin\xi}j(\underline{I}_+ - \underline{I}_-)\right] \tag{24.64}$$

The + - components equations are

$$\underline{V}_1 = \underline{Z}_+\underline{I}_+ ; \underline{V}_2 = \underline{Z}_-\underline{I}_- \tag{24.65}$$

with

$$\underline{Z}_+ = (R_{sm} + jX_{sm})m + \frac{jX_{mm}\left(\frac{R_{rm}}{S} + jX_{rm}\right)}{\frac{R_{rm}}{S} + j(X_{mm} + X_{rm})} \tag{24.66}$$

$$\underline{Z}_- = (R_{sm} + jX_{sm})m + \frac{jX_{mm}\left(\frac{R_{rm}}{2-S} + jX_{rm}\right)}{\frac{R_{rm}}{2-S} + j(X_{mm} + X_{rm})} \tag{24.67}$$

Notice again the presence of factor m (with values of m = 1 and respectively 0 as explained above).

The electromagnetic torque components are

$$T_{e+} = 2\left[\mathrm{Re}\left(\underline{V}_+\underline{I}_+^*\right) - R_{sm}(\underline{I}_+)^2 m\right]\frac{p_1}{\omega_1}$$

$$T_{e+} = 2\left[\mathrm{Re}\left(\underline{V}_-\underline{I}_-^*\right) - R_{sm}(\underline{I}_-)^2 m\right]\frac{p_1}{\omega_1} \tag{24.68}$$

$$T_e = T_{e+} + T_{e-}$$

Making use of (24.65), $\underline{I}_+$ and $\underline{I}_-$ are eliminated from (24.64) to yield

$$\underline{V}_+ = \frac{\underline{a}_0}{\underline{a}_1}\underline{V}_s \tag{24.69}$$

$$\underline{V}_- = -\frac{V_s - \underline{V}_+\left[1 - (R_{sm} + jX_{sm})(1-m)\left(\frac{1 - j\frac{\cos\xi}{\sin\xi}}{\underline{Z}_+}\right)\right]}{1 + \frac{(R_{sm} + jX_{sm})}{\underline{Z}_-}\left(1 + j\frac{\cos\xi}{\sin\xi}\right)} \tag{24.70}$$

$$\underline{C}_2 = -aj\sin\xi + a\cos\xi + \frac{[\underline{Z}_a + (R_{sa} + jX_{sa})(1-m)]}{ja\underline{Z}_- \cdot \sin\xi} \tag{24.71}$$

$$\underline{a}_0 = 1 - \frac{\underline{C}_2}{1 + \frac{(R_{sm} + jX_{sm})}{\underline{Z}_-}(1-m)\left(1 + j\frac{\cos\xi}{\sin\xi}\right)} \tag{24.72}$$

$$\begin{aligned} a_1 &= aj\sin\xi + a\cos\xi - \frac{\underline{Z}_a + (R_{sa} + jX_{sa})(1-m)}{jaZ_+ \sin\xi} - \\ &- (1 - \underline{a}_0)\left[+1 + \frac{(R_{sm} + jX_{sm})}{\underline{Z}_+}(1-m)\left(1 - j\frac{\cos\xi}{\sin\xi}\right)\right] \end{aligned} \tag{24.73}$$

24.8 SYMMETRIZATION CONDITIONS FOR NON-ORTHOGONAL WINDINGS

Let us consider the case of m = 1 ($R_{sa} = a^2R_{sm}$, $X_{sa} = a^2X_{sm}$), which is most likely to be used for symmetrical conditions.

In essence

$$\underline{V}_- = 0 \tag{24.74}$$

$$\underline{V}_+ = \underline{V}_\alpha = \underline{V}_m = \underline{V}_s \tag{24.75}$$

Further on, from (24.69)

$$\underline{a}_0 = \underline{a}_1 \tag{24.76}$$

With m = 1, from (24.71)-(24.73),

$$\underline{C}_2 = -aj\sin\xi + a\cos\xi + \frac{\underline{Z}_a}{ja\underline{Z}_- \cdot \sin\xi} \tag{24.77}$$

$$\underline{a}_0 = 1 - \underline{C}_2 \tag{24.78}$$

$$\underline{a}_1 = aj\sin\xi + a\cos\xi - \frac{\underline{Z}_a}{ja\underline{Z}_+ \cdot \sin\xi} + \underline{a}_0 - 1 \tag{24.79}$$

Finally, from (24.79) with (24.76)

$$\underline{Z}_a = aZ_+ \sin\xi[j(a\cos\xi - 1) - a\sin\xi] \tag{24.80}$$

For the capacitor motor, with orthogonal windings ($\xi = \pi/2$)

$$\underline{Z}_a = -a\underline{Z}_+[a + j] = \frac{-j}{\omega C} \tag{24.81}$$

Condition (24.81) is identical to (24.26) derived for orthogonal windings.

For a capacitor motor, separating the real and imaginary parts of (24.80) yields:

$$a = \frac{X_+}{R_+ \sin\xi + X_+ \cos\xi} = \frac{\sin\varphi_+}{\sin(\varphi_+ + \xi)} \tag{24.82}$$

$$\frac{1}{\omega C} = a^2 X_+ \sin\xi + aR_+ \sin\xi(1 - a\cos\xi) \tag{24.83}$$

with

$$\tan\varphi_+ = X_+ / R_+ \tag{24.84}$$

Table 24.1. Symmetrization conditions for $\varphi_+ = 45^0$

ξ	60^0	70^0	80^0	90^0	100^0	110^0	120^0
a	0.732	0.78	0.863	1.00	1.232	1.232	2.73
$\frac{1}{\omega C X_+}$	0.732	1.109	1.456	2.00	2.967	4.944	12.046!

The results on Table 24.1 are fairly general.

When the windings displacement angle ξ increases from $60^0$to 120^0 (for the direct component power factor angle $\varphi_+ = 45^0$), both the turns ratio a and the capacitance reactance steadily increase.

That is to say, a smaller capacitance is required for say $\xi = 110^0$ than for $\xi = 90^0$ but, as expected, the capacitance voltage V_c and the auxiliary winding voltage are higher.

Example 24.3 Performance: orthogonal versus non-orthogonal windings

Let us consider a permanent capacitor IM with the parameters $R_{sm} = 32\ \Omega$ $R_{sa} = 32\ \Omega$, $R_{rm} = 35\ \Omega$, $L_{rm} = 0.2$ H, $L_{sm} = L_{sa} = 0.1$ H, $L_{mm} = 2$ H, $p_1 = 1$, a = 1, $V_s = 220$ V, $\omega_1 = 314$ rad/sec, $\xi = 90^0$ (or 110^0); C = 5 μF, R_{core} (parallel resistance) = 10000 Ω.

Let us find the steady-state performance versus slip for orthogonal ($\xi = 90^0$) and non-orthogonal ($\xi = 110^0$) placement of windings.

A C++ code was written to solve this problem and the results are shown on Figures. 24.10-24.11.

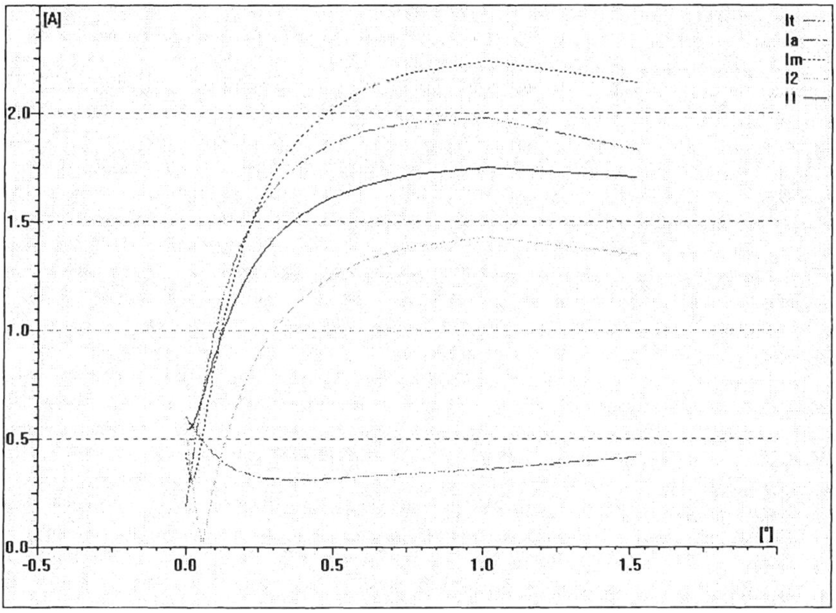

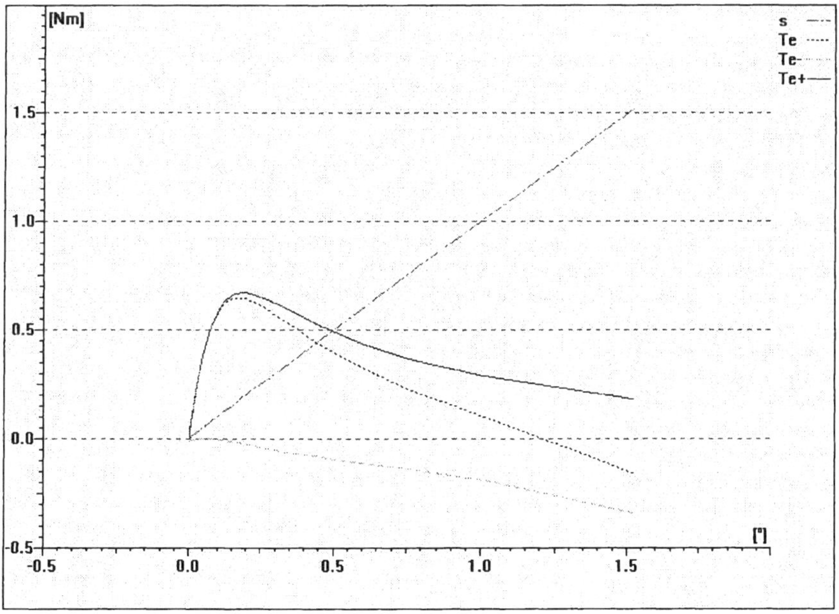

Figure 24.10 Performance with orthogonal windings ($\xi = 90^0$) (continued)

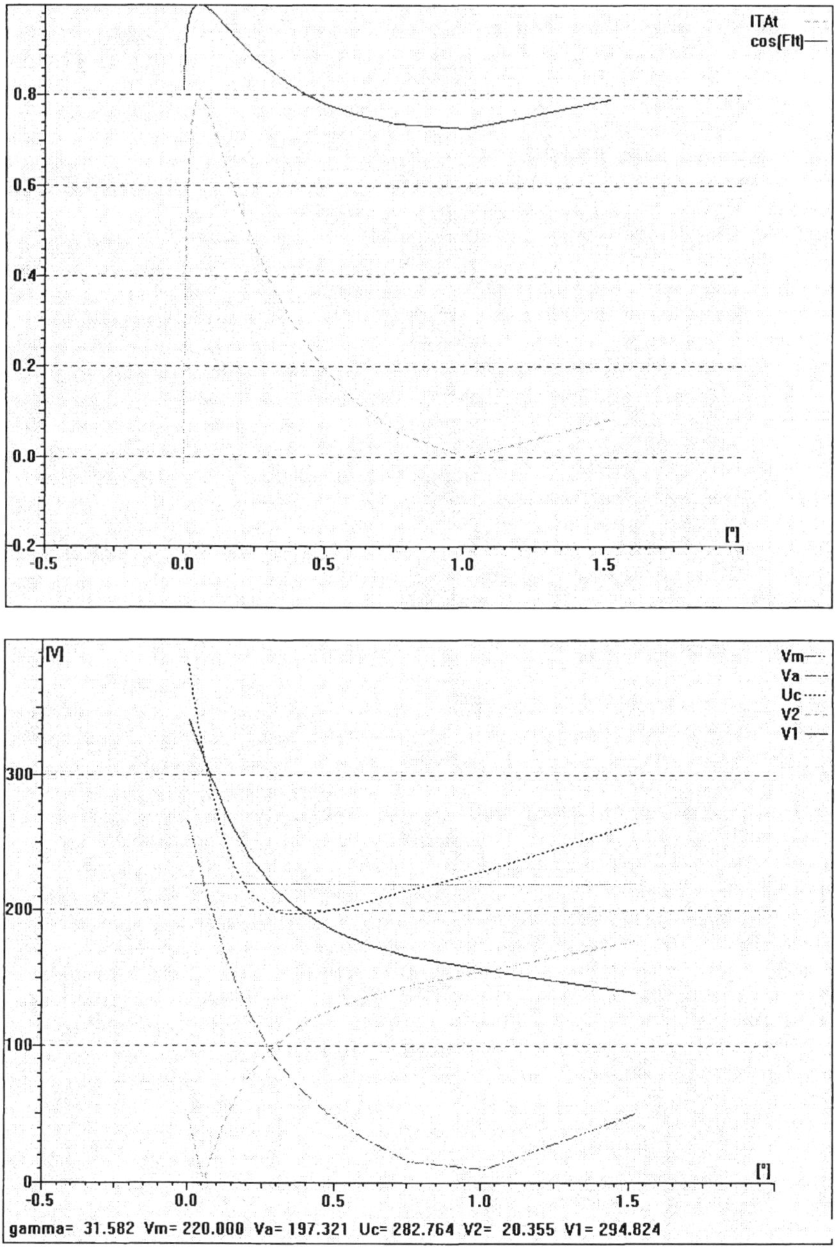

Figure 24.10 (continued)

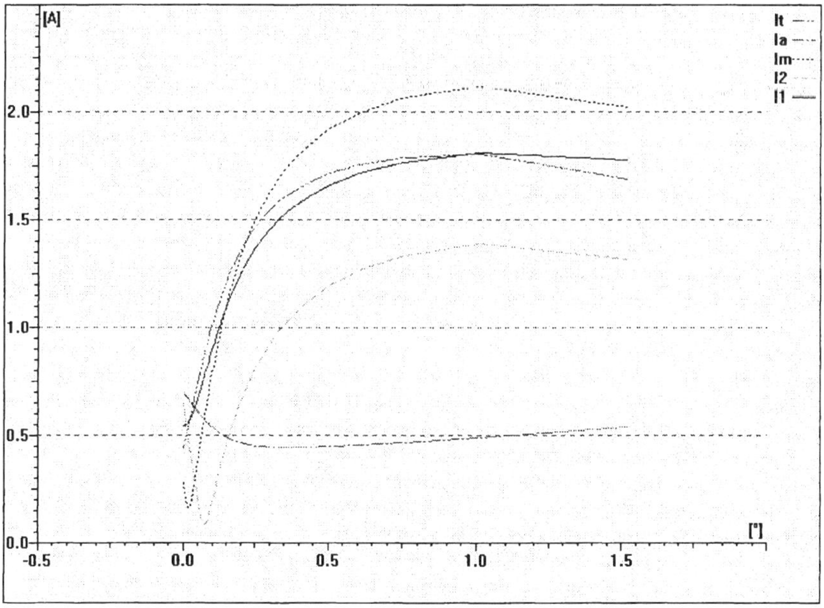

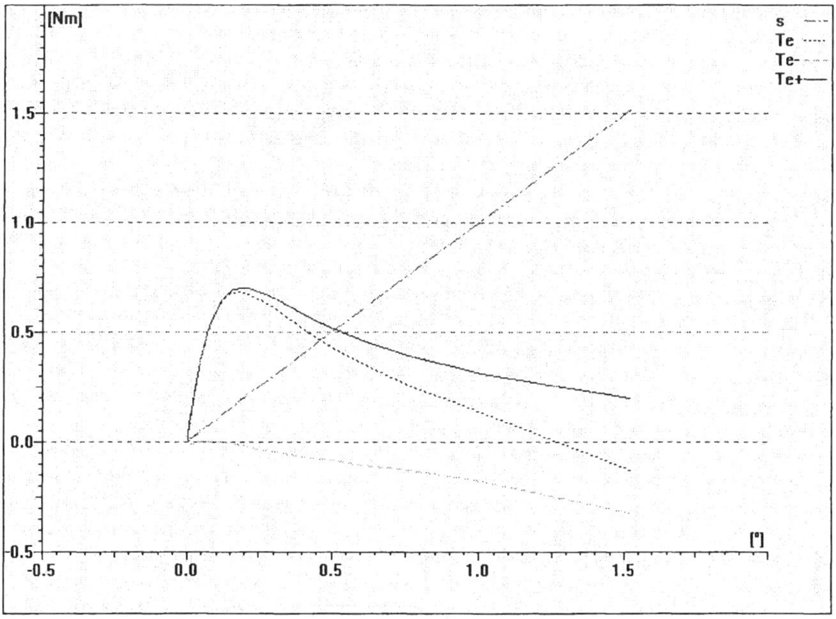

Figure 24.11 Performance with non-orthogonal windings ($\xi = 110^0$) (continued)

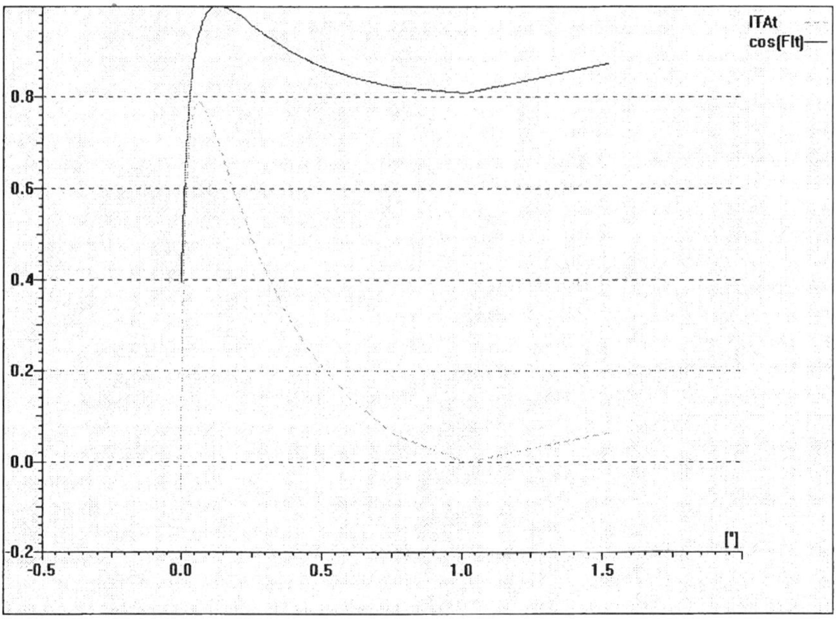

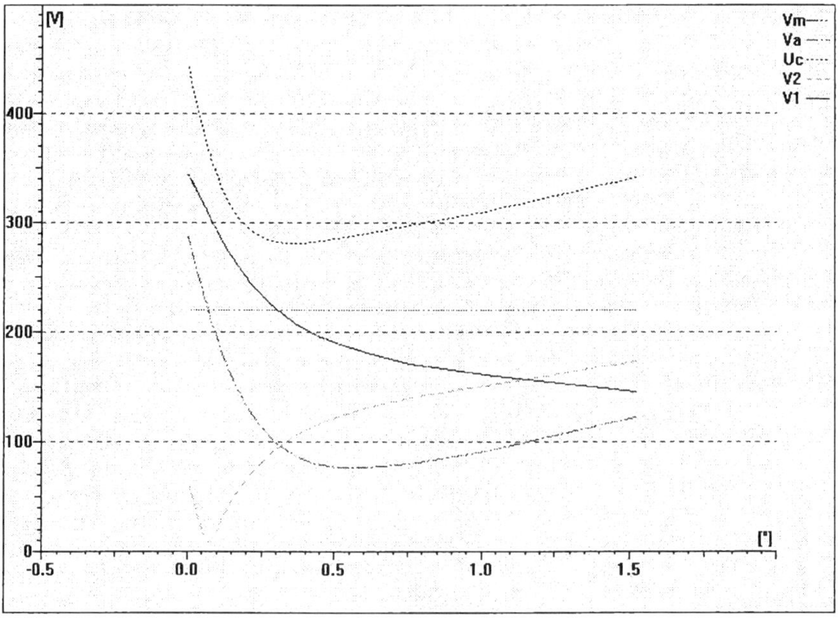

Figure 24.11 (continued)

The performance illustrated on Figures 24.10-24.11 leads to remarks such as:

- The torque/speed curves are close to each other but more starting torque is obtained for the non-orthogonal windings motor.

- The symmetrization (zero I_- and V_-) slip is larger for the non-orthogonal winding motor.
- The power factor of the non-orthogonal winding motor is larger at 20% slip but smaller at slip values below 10%.
- The efficiency curves are very close.
- The capacitor voltage $\underline{V}_c$ is larger for the non-orthogonal winding motor.

It is in general believed that non-orthogonal windings make a more notable difference in split/phase rather than in capacitor IMs.

The mechanical characteristics ($T_e(S)$) for two different capacitors ($C = 5\times10^{-6}$F and $C_a = 20\times10^{-6}$F) are shown on Figure 24.12 for orthogonal windings.

As expected, the larger capacitor produces larger torque at start but it is not adequate at low slip values because it produces a too large inverse torque component.

The same theory can be used to investigate various designs in the pursuit of chosen optimization criterion such as minimum auxiliary winding loss during starting, etc.

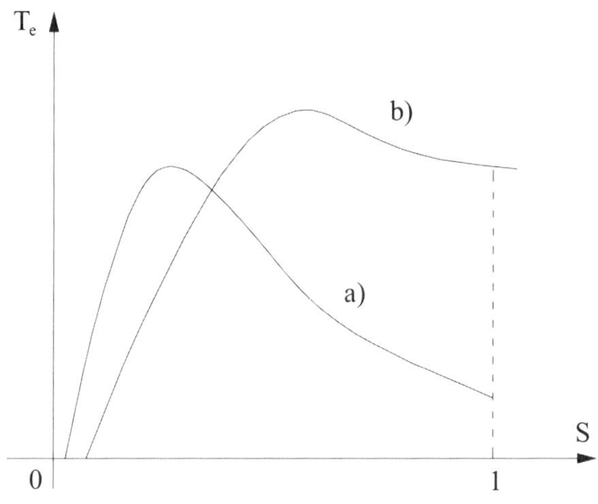

Figure 24.12 Torque/slip curves
a.) $C = 5\times10^{-6}$ F; b.) $C_a = 20\times10^{-6}$ F

The split phase motor may be investigated as a particular case of capacitor motor with either $\underline{Z}_a = 0$ and increased R_{sa} value or with the original (real) R_{sa} and $\underline{Z}_a = R_a$. Above a certain slip, the auxiliary phase is turned off which means $|\underline{Z}_a| \approx \infty$.

A general circuit theory to handle 2 or 3 stator windings located at random angles has been developed in [5,6].

24.9 M.M.F. SPACE HARMONIC PARASITIC TORQUES

So far only the fundamental of m.m.f. space distribution along rotor periphery has been considered for the steady-state performance assessment.

In reality, the placement of stator conductors in slots leads to space m.m.f. harmonics. The main space harmonic is of the third order. Notice that the current is still considered sinusoidal in time.

From the theory of three phase IMs, we remember that the slip S_υ for the υth space harmonics is

$$S_\nu = 1 \mp \nu(1 - s) \qquad (24.85)$$

For a single phase IM (the auxiliary phase is open) the space harmonic m.m.f.s are stationary, so they decompose into direct and inverse travelling components for each harmonic order.

Consequently, the equivalent circuit in Figure 24.2 may be adjusted by adding rotor equivalent circuits for each harmonic (both direct and inverse waves)-Figure 24.13.

Once the space harmonic parameters $X_{rm\upsilon}$, $X_{mn\upsilon}$ are known, the equivalent circuit on Figure 24.13 allows for the computation of the additional (parasitic) asynchronous torques.

In the case when the reaction of rotor current may be neglected (say for the third harmonic), their equivalent circuits on Figure 24.14 are reduced to the presence of their magnetization reactances $X_{mm\upsilon}$. These reactances are not difficult to calculate as they correspond to the differential leakage reactances already defined in detail in Chapter 6, paragraph 6.1.

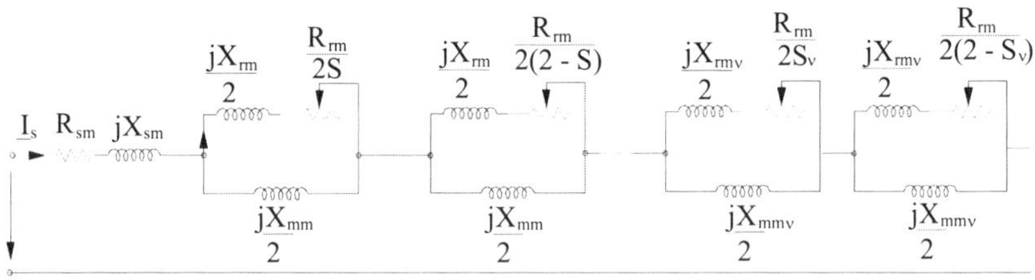

Figure 24.13 Equivalent circuit of single phase IM with m.m.f. space harmonics included

The presence of slot openings and of magnetic saturation enhances the influence of m.m.f. space harmonics on additional losses in the rotor bars and on parasitic torques.

Too large a capacitor would trigger deep magnetic saturation which produces a large 3rd order space harmonic. Consequently, a deep saddle in the torque/slip curve, around 33% of ideal no-load speed, occurs (Figure 24.14).

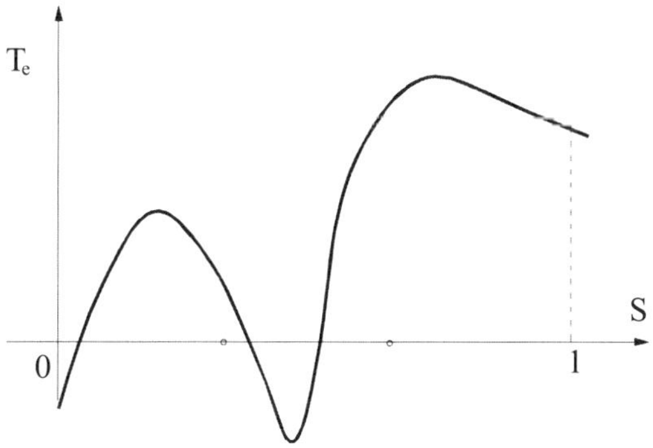

Figure 24.14 Grossly oversized capacitor IM torque-speed curve

Besides asynchronous parasitic torques, synchronous parasitic torques may occur as they do for three phase IMs. The most important speeds at which such torques may occur are zero and $2f_1/N_r$ (N_r-rotor slots).

When observing the conditions

$$N_s \neq N_r \tag{24.86}$$

$$\left|N_r - N_s\right| \neq 2p_1 \tag{24.87}$$

the synchronous torques are notably reduced (Figure 24.15).

The reduction of parasitic synchronous torques, especially at zero speed, is required to secure a safe start and reduce vibration, noise, and additional losses.

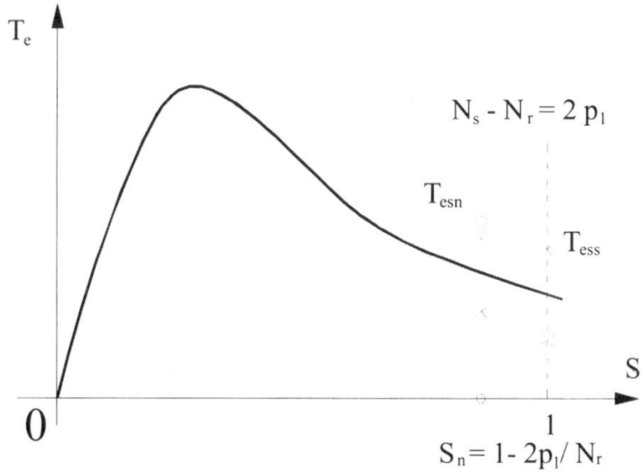

Figure 24.15 Synchronous parasitic torques

24.10 TORQUE PULSATIONS

The speed of the elliptical travelling field, present in the airgap of single phase or capacitor IMs, varies in time in contrast to the symmetrical IM where the speed of the circular field is constant.

As expected, torque pulsations occur.

The equivalent circuits for steady state are helpful to calculate only the time average torque.

Besides the fundamental m.m.f. produced torque pulsations, the space harmonics cause similar torque pulsations.

The d-q models for transients, covered in the next chapter, will deal with the computation of torque pulsations during steady state also.

Notable radial uncompensated forces occur, as in three phase IMs, unless the condition (24.88) is met

$$N_s \pm p_1 \neq N_r \tag{24.88}$$

24.11 INTER-BAR ROTOR CURRENTS

In most single phase IMs, the rotor cage bars are not insulated and thus rotor currents occur between bars through the rotor iron core. Their influence is more important in two pole IMs where they span over half the periphery for half a period (Figure 24.16).

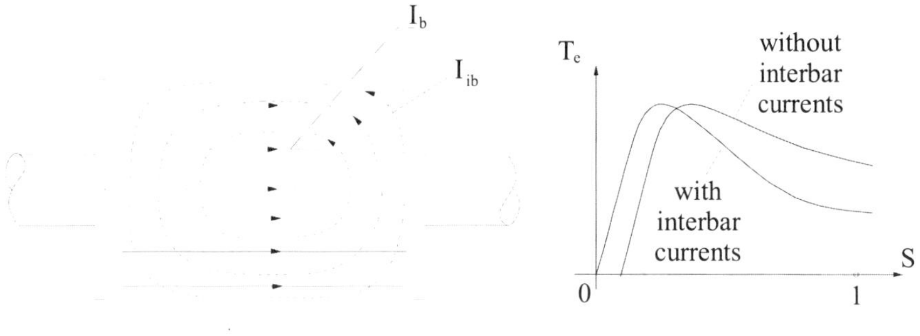

Figure 24.16 Interbar rotor currents and their torque ($2p_1 = 2$ poles)

The interbar current ($\underline{I}_{ib}$) m.m.f. axis is displaced by 90^0(electrical) with respect to the rotor bar current ($\underline{I}_b$) m.m.f.

This is why the interbar currents are called transverse currents.

The interbar currents depend on the rotor skewing, besides the contact resistance between rotor bars and rotor iron and on rotor slip frequency.

The torque/speed curve may be influenced notably (as in Figure 24.16) by 2 pole IMs.

The slot harmonics may further increase the interbar currents as the frequency of their produced rotor currents is much higher than for the fundamental ($f_2 = Sf_1$).

Special thermal treatments of the rotor may increase the bar-core rotor resistance to reduce the interbar currents and thus increase the torque for large slip values.

The influence of rotor skewing and rotor stack length/pole ratio on interbar current losses are as for the three phase IMs (see Chapter 11, paragraph 11.12). Larger stacks tend to accentuate the interbar current effects. Also a large number of rotor slots ($N_r >> N_s$) tend to lead to large interbar currents.

24.12 VOLTAGE HARMONICS EFFECTS

Voltage time harmonics occur in a single phase IM either when the motor is fed through a static power converter or when the power source is polluted with voltage (and current) time harmonics from other static power converters acting alone.

Both odd and even time harmonics may occur if there are imbalances within the static power converters.

High efficiency low rotor resistance single phase capacitor IMs are more sensitive to voltage time harmonics than low efficiency single phase induction motors with open auxiliary phase in running conditions.

For some voltage time harmonics, frequency resonance conditions may be met in the capacitor IM and thus high time harmonics currents occur. They in turn lead to marked efficiency reduction.

As the frequency of time voltage harmonics increases, so does the skin effect. Consequently, the leakage inductances decrease and the rotor resistance increases markedly. A kind of leakage flux "saturation" occurs. Moreover, main flux path saturation also occurs.

So the investigation of voltage time harmonics influence presupposes known, variable, parameters in the single phase capacitor IM.

Such parameters may be calculated or measured. Correctly measured parameters are the safest way, given the nonlinearities introduced by magnetic saturation and skin effect.

Once these parameters are known, for each voltage time harmonic, equivalent forward and backward circuits for the main and auxiliary winding may be defined.

For a capacitor IM, the equivalent circuit of Figure 24.5 may be simply generalized for the time harmonic of order h (Figure 24.16).
The method of solution is similar to that described in paragraph 24.3.

The main path flux saturation, may be considered to be related only to the forward (+) component of the fundamental. X_{mm} in the upper part of Figure 24.5 depends on the magnetization current I_{mm+}. Results obtained in a similar way (concerning harmonic losses), for distinct harmonic orders, are shown in Figure 24.18, for a 1.5 kW, 12.5 μF capacitor single phase induction motor. [9]

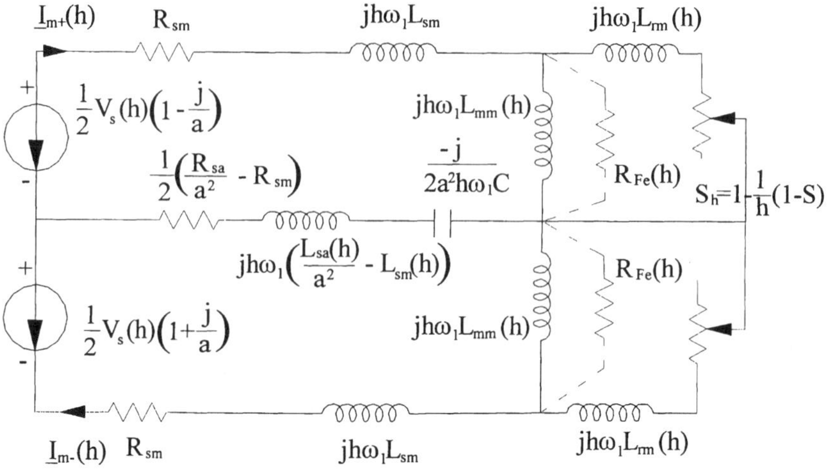

Figure 24.17 Time harmonics voltage equivalent circuit of capacitor IMs

Resonance conditions are met somewhere between the 11th and 13th voltage time harmonics. An efficiency reduction of 5% has been measured for this situation [9].

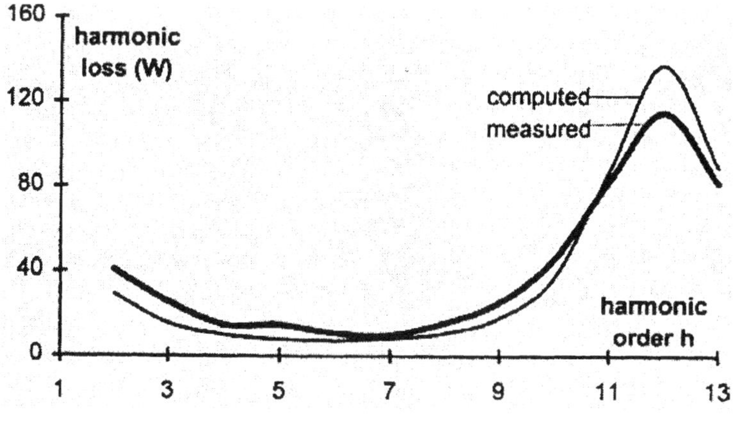

Figure 24.18 Harmonic losses at rated load and 10% time voltage harmonics

It is generally argued that the computation of total losses-including stray load losses, etc., is very difficult to perform as visible in Figure 24.18. Experimental investigation is imperative, especially when the power decreases and the number of poles increases and severe interbar rotor currents may occur.

The neglect of space harmonics in the equivalent circuit on Figure 24.5 and 24.17 makes the latter particularly adequate only from zero to breakdown torque slip.

24.13 THE DOUBLY TAPPED WINDING CAPACITOR IM

As already alluded to in Chapter 23 tapped windings are used with capacitor multispeed IMs or with dual-capacitor IMs (Figure 23.4) or with split-phase capacitor motors (Figure 23.8).

In the L or T connections (Figure 23.6), the tapping occurs in the main winding. For dual-capacitor or split phase capacitor IMs (Figures 23.4 and 23.8), the tapping is placed in the auxiliary winding. Here we are presenting a fairly general case (Figure 24.19).

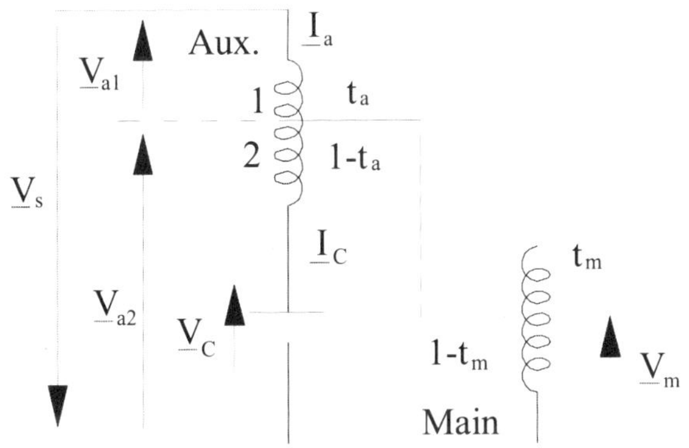

Figure 24.19 Doubly tapped winding capacitor IM

The two taps are t_a, for the auxiliary winding, and t_m for the main winding. In practice, only one tap is used at any time.

When only the tapping on the main winding is used $t_a = 0$ and $0 < t_m < 0.5$ in general. On the contrary, when the tapping is done in the auxiliary winding $0 < t_a < 0.5$ with $t_m = 0$. [8]

As it is evident from Figure 24.19, the main winding is present with a single circuit even for $t_m \neq 0$ (main winding tapping). Meanwhile, the tapped auxiliary winding is present with both sections aux 1 and aux 2.

Consequently, between the two sections there will be a mutual inductance.

Denoting by X the total main self inductance of the auxiliary winding, $X_1 = t_a^2 X$ represents aux 1 and $X_2 = (1 - t_a^2)X$ refers to auxiliary 2. The mutual inductance X_{12} is evidently $X_{12} = Xt_a(1 - t_a)$.The coupling between the two auxiliary winding sections may be represented as in Figure 24.20.

Besides the main flux inductances, the leakage inductance components and the resistances are added. What is missing in Figure 24.20 is rotor and main windings e.m.f.s in the auxiliary winding sections and in the main winding.

As part of main winding is open (Figure 24.19) when tapping is applied, the equivalent circuit is much simpler (no coupling occurs).

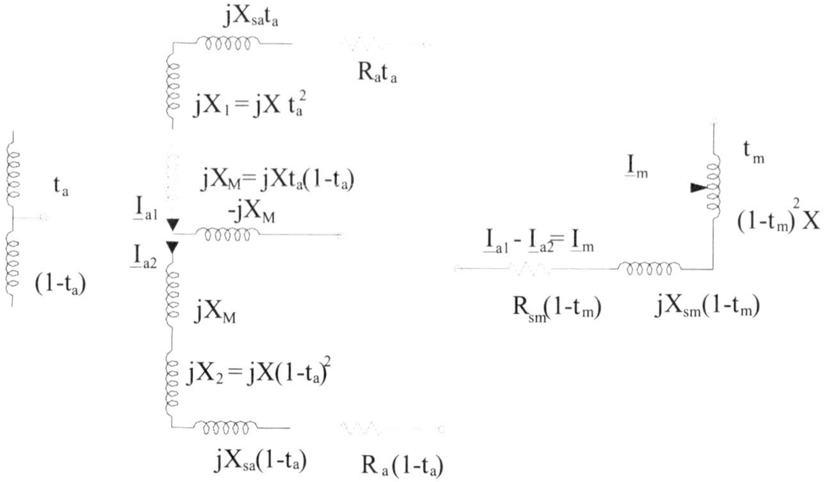

Figure 24.20 The tapped auxiliary a) and main b) winding equivalent (noncoupled) circuit

Let us denote as forward and backward $\underline{Z}_f$, $\underline{Z}_b$, the rotor equivalent circuits

$$\underline{Z}_f = \frac{1}{2}\frac{jX_{mm}(R_{rm}/S + jX_{rm})}{R_{rm}/S + j(X_{mm} + X_{rm})} \tag{24.89}$$

$$\underline{Z}_b = \frac{1}{2}\frac{jX_{mm}(R_{rm}/(2-S) + jX_{rm})}{R_{rm}/(2-S) + j(X_{mm} + X_{rm})} \tag{24.90}$$

We may now apply the revolving field theory (Morrill's theory) to this general case, observing that each of the two sections of the auxiliary winding and the main winding are producing self induced forward and backward e.m.f.s $\underline{E}_{fa1}$, $\underline{E}_{ba1}$, $\underline{E}_{fa2}$, $\underline{E}_{ba2}$, $\underline{E}_{fm}$, $\underline{E}_{bm}$. The main winding tapping changes the actual turn ratio to $a/(1-t_m)$; also $\underline{Z}_f$, and $\underline{Z}_b$ to $\underline{Z}_f(1-t_m)^2$ and $\underline{Z}_b(1-t_m)^2$.

$$\begin{aligned}
\underline{E}_{fa1} &= t_a^2 a^2 \cdot \underline{Z}_f \underline{I}_{a1}\\
\underline{E}_{ba1} &= t_a^2 a^2 \cdot \underline{Z}_b \underline{I}_{a1}\\
\underline{E}_{fa2} &= (1-t_a)^2 a^2 \cdot \underline{Z}_f \underline{I}_{a2}\\
\underline{E}_{ba2} &= (1-t_a)^2 a^2 \cdot \underline{Z}_b \underline{I}_{a2}\\
\underline{E}_{fm} &= \underline{Z}_f \cdot (1-t_m)^2 \underline{I}_m\\
\underline{E}_{bm} &= \underline{Z}_b \cdot (1-t_m)^2 \underline{I}_m
\end{aligned} \tag{24.91}$$

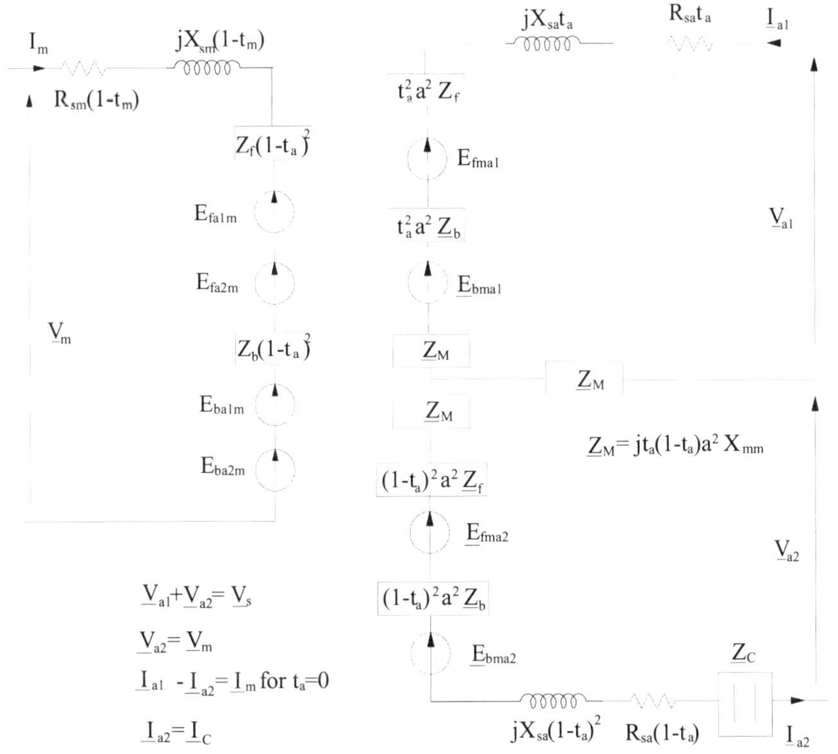

Figure 24.21 The equivalent circuit of dualy tapped winding capacitor induction motor

Besides these self-induced e.m.f.s, there are motion induced e.m.f.s from one axis (m axis) to the other (orthogonal) axis (a axis), produced both by the forward and backward components; in all, there are four such such e.m.f.s in the main winding and two for each of the two auxiliary windings: $\underline{E}_{fa1m}$, $\underline{E}_{fa2m}$, $\underline{E}_{ba1m}$, $\underline{E}_{ba2m}$, $\underline{E}_{fma1}$, $\underline{E}_{bma1}$, $\underline{E}_{fma2}$, $\underline{E}_{bma2}$.

$$\begin{aligned} \underline{E}_{fa1m} &= -jat_a \underline{Z}_f \underline{I}_{a1}(1-t_m) \\ \underline{E}_{ba1m} &= jat_a \underline{Z}_b \underline{I}_{a1}(1-t_m) \\ \underline{E}_{fa2m} &= -ja(1-t_a)\underline{Z}_f \underline{I}_{a2}(1-t_m) \\ E_{ba2m} &= -ja(1-t_a)\underline{Z}_b \underline{I}_{a2}(1-t_m) \end{aligned} \tag{24.92}$$

Now we may easily construct the equivalent circuit model of the entire machine (Figure 24.21).

It is now only a matter of algebraic manipulations to solve the equivalent circuit on Figure 24.21. After eliminating $\underline{V}_m$, $\underline{V}_{a1}$, $\underline{V}_{a2}$, and $\underline{I}_m$ two equations are obtained. They have $\underline{I}_{a1}$ and $\underline{I}_{a2}$ as variables

$$\begin{aligned} \underline{Z}_{11}\underline{I}_{a1} + \underline{Z}_{12}\underline{I}_{a2} &= \underline{V}_s \\ \underline{Z}_{21}\underline{I}_{a1} + \underline{Z}_{22}\underline{I}_{a2} &= 0 \end{aligned} \tag{24.93}$$

with:

$$\underline{Z}_{11} = (1 - t_m)(R_{sm} + jX_{sm}) + (\underline{Z}_f + \underline{Z}_b)(1 - t_m)^2 + t_a(R_{sa} + jX_{sa}) + \\ + t_a^2 a^2 (\underline{Z}_f + \underline{Z}_b)$$

$$\underline{Z}_{22} = -(1 - t_m)(R_{sm} + jX_{sm}) - (\underline{Z}_f + \underline{Z}_b)(1 - t_m)^2 - \\ - (1 - t_a)(R_{sa} + jX_{sa}) - (1 - t_a)^2 a^2 (\underline{Z}_f + \underline{Z}_b) - \underline{Z}_c$$

$$\underline{Z}_{12} = -ja(1 - t_m)(\underline{Z}_f - \underline{Z}_b) - (1 - t_m)(R_{sm} + jX_{sm}) - (\underline{Z}_f + \underline{Z}_b)(1 - t_m)^2 + \\ + jt_a(1 - t_a)a^2 X_{mm}$$

$$\underline{Z}_{21} = -ja(1 - t_m)(\underline{Z}_f - \underline{Z}_b) + (1 - t_m)(R_{sm} + jX_{sm}) + (\underline{Z}_f + \underline{Z}_b)(1 - t_m)^2 - \\ - jt_a(1 - t_a)a^2 X_{mm}$$

For no tapping in the main winding ($t_m = 0$), the results in Ref. 8 are obtained. Also with no tapping at all ($t_m = t_a = 0$), (24.91-24.94) degenerate into the results obtained for the capacitor motor ((24.16)-(24.17) with (24.10); $I_{a2} = I_a$, $V_m = V_{a2} = V_s$).

The equivalent circuit on Figure 24.20, for no tapping ($t_m = t_a = 0$) degenerates into the standard revolving theory circuit of the capacitor IM (Figure 24.22).

The torque may be calculated from the power balance

$$T_e = [\mathrm{Re}(\underline{V}_m \underline{I}_m^*) - R_{sm}(1 - t_m)I_m^2 + \mathrm{Re}(\underline{V}_{a1}\underline{I}_{a1}^*) + \\ \mathrm{Re}(V_{a2}I_{a2}^*) - t_a R_{sa} I_{a1}^2 - (1 - t_a)R_{sa} I_{a2}^2] \cdot p_1/\omega_1 \tag{24.94}$$

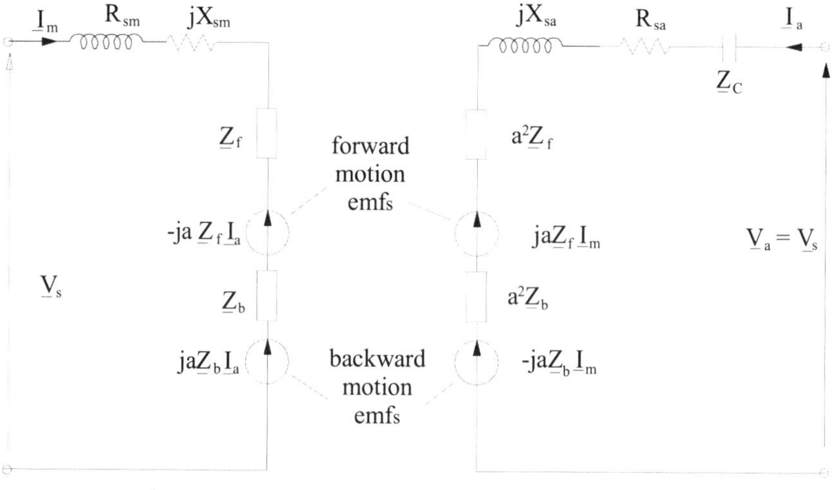

Figure 24.22 The revolving theory equivalent circuit of the capacitor motor

Symmetrization conditions

The general symmetrization may be determined by zeroing the backward (inverse) current components. However, based on geometrical properties, it is easy to see that, for orthogonal windings, orthogonal voltages in the main and auxiliary winding sections, with same power factor angle in both windings, correspond to symmetry conditions (Figure 24.23).

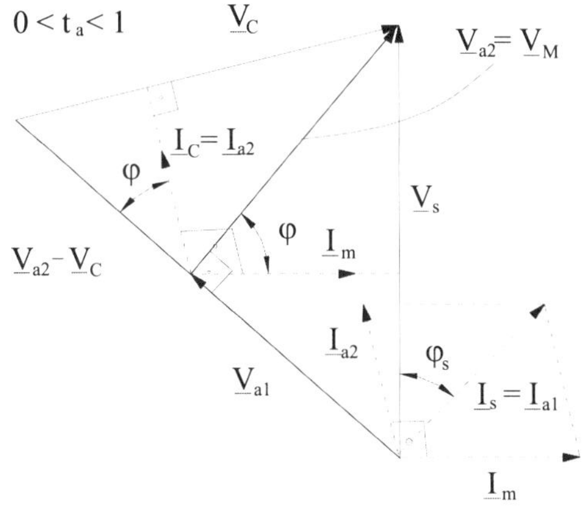

Figure 24.23 The phasor diagram for symmetrization conditions of the generally tapped winding capacitor IM

The two currents I_{a2}, I_m are 90^0 phase shifted as the voltages are. On the other hand the current I_{a1} in the first section of the auxiliary winding a_1 is highly reactive and it serves basically for voltage reduction. This way some speed reduction is produced.

When no tapping is performed ta = 0, Va1 = 0, and thus $V_{a2} = V_m = V_s$ and simpler symmetrization conditions are met (Figure 24.24).

This case corresponds to highest $\underline{V}_{a2}$ - $\underline{V}_c$ voltage; that is, for highest auxiliary active winding voltage. Consequently, this is the highest speed situation expected.

Note on magnetic saturation and steady-state losses

As mentioned previously (in Chapter 23), magnetic saturation produces a flattening of the airgap flux density distribution. That is, a third harmonic e.m.f. and thus a third harmonic current may occur. The elliptic character of the magnetic field in the airgap makes the accounting of magnetic saturation even more complicated. Analytical solutions to this problem tend to use an equivalent sinusoidal current to handle magnetic saturation. It seems however that only FEM could offer a rather complete solution to magnetic saturation.

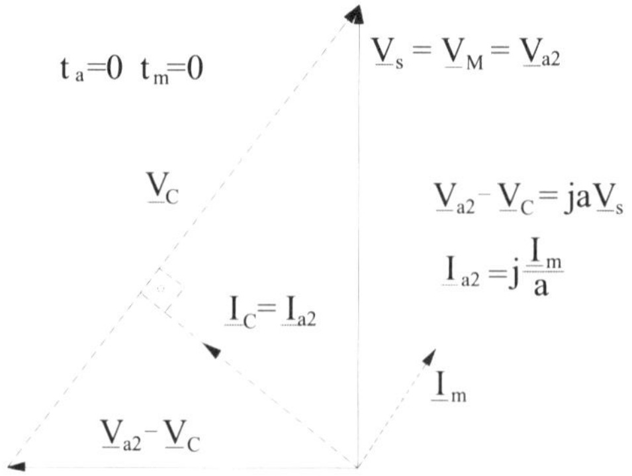

Figure 24.24 The phasor diagram for symmetrization conditions with no tapping

This appears even more so if we need to calculate the iron losses. On the other hand, experiments could be useful to asses magnetic saturation and fundamental and additional losses both in the core and in the rotor cage or losses caused by the interbar rotor iron currents. More on these complicated issues in the next two chapters.

24.14 SUMMARY

- The steady state performance of single phase IM is traditionally approached by the revolving field theory or the + – symmetrical components model. In essence, the voltages and currents in the two phases are decomposed into + and – components for which the slip is S and, respectively 2-S. Superposition is applied. Consequently, magnetic saturation cannot be accounted for unless the backward component is considered not to contribute to the level of magnetic saturation in the machine.
- On the other hand, the d-q (or crossfield) model (in stator coordinates), with $d/dt = j\omega_1$ may be used for steady state. In this case, keeping the current sinusoidal, magnetic saturation may be iteratively considered. This model has gain some popularity with T-L or generally tapped winding multispeed capacitor IMs.
- The revolving theory (or + – or fb) model leads, for the single phase IM with open auxiliary winding, to a very intuitive equivalent circuit composed of + and – equivalent circuits in series. The torque is made of + and – components and is zero at zero speed (S = 1) as T_+ and T_- are equal to each other. At zero slip (S = 0) $T_+ = 0$ but $T_- \neq 0$ (and negative). Also at S = 0, with the auxiliary phase open, half of the rotor cage losses are covered electrically from stator and half mechanically from the driving motor.
- The split-phase and capacitor IM travelling field (+ –) model yields a special equivalent circuit with two unknowns I_{m+}, I_{m-} (the main winding current components). The symmetrization conditions are met

when $I_{m-} = 0$. As the speed increases above breakdown torque speed so do the auxiliary winding and capacitor voltages for the permanent capacitor IM. Capacitor IMs have a good power factor and a moderate efficiency in general.

- Besides the configuration with orthogonal windings, it is feasible to use non-orthogonal windings especially with split-phase IMs in order to increase the output. It has be shown that symmetrization conditions correspond to

$$\xi + (\varphi_a - \varphi_m) = 180^0$$

where ξ is the electrical spatial angle of the two windings and $\varphi_a - \varphi_m$ is the time phase angle shift between $\underline{I}_a$ and $\underline{I}_m$.
The above conditions cannot be met in split-phase IMs but they may be reached in capacitor motors. Apparently less capacitance is needed for $\xi > 90^0$ and slightly more starting torque may be produced this way. However, building non-orthogonal windings means implicitly to use most or all slots, for both main and auxiliary windings.

- The split phase IM may be considered as a particular case of the capacitor motor with $Z_a = R_{aux}$. Above a certain speed the auxiliary phase is open ($R_{aux} = \infty$).
- Space harmonics occur due to the winding placement in slots. The third harmonic and the first slot harmonic are especially important. The equivalent circuit may be augmented with the space harmonics contribution (Figure 24.13). Asynchronous, parasitic, forward and backward (+, –) torques occur.
- For the permanent capacitor IM, the symmetrization conditions ($I_{m-} = 0$) yield rather simple expressions for the turns ratio a between auxiliary and main winding and the capacitor reactance X_c

$$a = \tan\varphi_+; X_c = \frac{X_+}{\cos^2\varphi_+}$$

where φ_+ is the power factor angle for the + component and X_+ is the + component equivalent reactance.

- Starting torque is a major performance index. For a capacitor motor, it is obtained for a turn ratio a_K

$$a_K = \sqrt{\frac{\sin\varphi_{sc}}{Z_{sc}\omega C}}$$

Unfortunately, the maximum starting torque condition does not provide, in general, for overall optimum performance.
As for an existing motor only the capacitance C may be changed, care must be exercised when the turns ratio *a* is chosen.

- Starting current is heavily dependent on the turns ratio *a* and the capacitor C. A large capacitor is needed to produce sufficient starting torque but starting current limitation must also be observed.
- In general, for good starting performance of a given motor, a much larger capacitor C_s is needed in comparison with the condition of good running performance C ($C_s >> C$).
- Torque, main and auxiliary currents, I_m, I_a, auxiliary winding voltage V_a, capacitor voltage V_c, efficiency and power factor versus speed, constitute standard steady state performance characteristics.
- The third harmonics may produce, with a small number of slots/pole, a large deep in the torque/speed curve around 33% of ideal no load speed ($n_1 = f_1/p_1$). Magnetic saturation tends to accentuate this phenomenon.
- When $N_s = N_r$ synchronous parasitic torques occur at zero speed. For N_s-$N_r = 2p_1$, such torques occur at $S = 1\text{-}2p_1/N_r$. So, in general

$$N_s \neq N_r; N_s - N_r \neq 2p_1$$

  to eliminate some very important synchronous parasitic torques.
- To avoid notable uncompensated radial forces $N_s \pm p_1 \neq N_r$, as for three phase IMs.
- As the rotor bars are not insulated from rotor core, interbar currents occur. In small ($2p_1 = 2$) motors, they notably influence the torque/speed curves acting as an increased rotor resistance. That is, the torque is reduced at low slips and increased at high slip values when the bar-core contact resistance is increased.
- Voltage time harmonics originating from static power converters may induce notable additional losses in high efficiency capacitor IMs, especially around electrical resonance conditions, when the efficiency may be reduced to 3-5%. Such resonance voltage time harmonics have to be filtered out.
- Winding tapping is traditionally used to reduce speed for various applications. A fairly general revolving field model for dual winding tapping was developed. Symmetrisation conditions are illustrated (Figures 24.23, 24.24). They show the orthogonality of main and auxiliary windings voltages and currents for orthogonally placed stator windings.
- Magnetic saturation is approachable through the + (forward) magnetization inductance functional $L_{mm}(I_{mm+})$, but it seems to lead to large errors, especially when the backward current component is notable. [10] A more complete treatment of saturation is performed in Chapter 25 on transients.

24.15 REFERENCES

1. P. H. Trickey, Capacitor Motor Performance Calculations by the Cross Field Theory" Trans. AIEE, Feb. 1957, pp. 1547-1552.
2. C. G. Veinott, Theory and Design of Small Induction Motors, Mc.Graw-Hill, New York, 1959.
3. B. S. Guru, Performance Equations and Calculations on T and L Connected Tapped-Winding Capacitor Motors by Cross Field Theory, Electric Machines & Electromechanics, vol. 1, no. 4, 1977, pp. 315-336.
4. B. S. Guru, Performance Equations and Calculations on L - and T - Connected Multispeed Capacitor Motors by Cross Field Theory, Electric Machines & Electromechanics, IBID, vol. 2, 1977, pp. 37-48.
5. J. Stepina, The Single Phase Induction Motors, paragraph 3.7 (book), Springer Verlag, 1982 (in German).
6. S. Williamson and A. C. Smith, A Unified Approach to the Analysis of Single Phase Induction Motors, IEEE-Trans. Vol. IA-35, no. 4, 1999, pp. 837-843.
7. I. Boldea, T. Dumitrescu, S. A. Nasar, Steady State Unified Treatment of Capacitor A.C. Motors, IEEE Trans. vol. EC-14, no. 3, 1999, pp. 557-582.
8. T. J. E. Miller, J. H. Gliemann, C. B. Rasmussen, D.M. Ionel, Analysis of Tapped Winding Capacitor Motor, Record of ICEM-1998, Istambul, Turkey, vol. 2, pp. 581-585.
9. D. Lin, T. Batan, E. F. Fuchs, W. M. Grady, Harmonic Losses of Single Phase Induction Motors Under Nonsinusoidal Voltages, IEEE Trans. vol. EC-11, no. 2, 1996, pp. 273-282.
10. C. B. Rasmussen, T. J. E. Miller, Revolving-field Polygon Technique for Performance Prediction of Single Phase Induction Motors, Record of IEEE-IAS-2000, Annual meeting.

Chapter 25

SINGLE-PHASE IM TRANSIENTS

25.1 INTRODUCTION

Single-phase induction motors undergo transients during starting, load perturbation or voltage sags etc. When inverter fed, in variable speed drives, transients occur even for mechanical steady state during commutation mode.

To investigate the transients, for orthogonal stator windings, the cross field (or d-q) model in stator coordinates is traditionally used. [1]

In the absence of magnetic saturation, the motor parameters are constant. Skin effect may be considered through a fictitious double cage on the rotor.

The presence of magnetic saturation may be included in the d-q model through saturation curves and flux linkages as variables. Even for sinusoidal input voltage, the currents may not be sinusoidal. The d-q model is capable of handling it. The magnetisation curves may be obtained either through special flux decay standstill tests in the d-q (m.a) axes (one at a time) or from FEM-in d.c. with zero rotor currents. The same d-q model can handle nonsinusoidal input voltages such as those produced by a static power converter or by power grid polluted with harmonics by other loads nearby.

To deal with nonorthogonal windings on stator, a simplified equivalence with a d-q (orthogonal) winding system is worked out. Alternatively a multiple reference system + - model is used [3]

While the d - q model uses stator coordinates, which means a.c. during steady state, the multiple-reference model uses + - synchronous reference systems which imply d.c. steady state quantities. Consequently, for the investigation of stability, the frequency approach is typical to the d-q model while small deviation linearization approach may be applied with the multiple-reference + - model.

Finally, to consider the number of stator and rotor slots-that is space flux harmonics-the winding function approach is preferred. [4] This way the torque/speed deep around 33% of no load ideal speed, the effect of the relative numbers of stator and rotor slots, broken bars, rotor skewing may be considered. Still saturation remains a problem as superposition is used.

A complete theory of single phase IM, valid both for steady-state and transients, may be approached only by a coupled FEM-circuit model, yet to be developed in an elegant computation time competitive software. In what follows, we will illustrate the above methods in some detail.

25.2 THE d-q MODEL PERFORMANCE IN STATOR COORDINATES

As in general the two windings are orthogonal but not identical, only stator coordinates may be used directly in the d-q (cross field) model of the single phase IMs.

True, when one phase is open any coordinates will do it but this is only the case of the split-phase IM after starting. [5]

First, all the d-q model variables are reduced to the main winding

$$\begin{aligned} V_{ds} &= V_s(t) \\ V_{qs} &= \left(V_s(t) - V_c(t)\right)/a \\ I_{ds} &= I_m \\ I_{qs} &= I_a \cdot a \end{aligned} \tag{25.1}$$

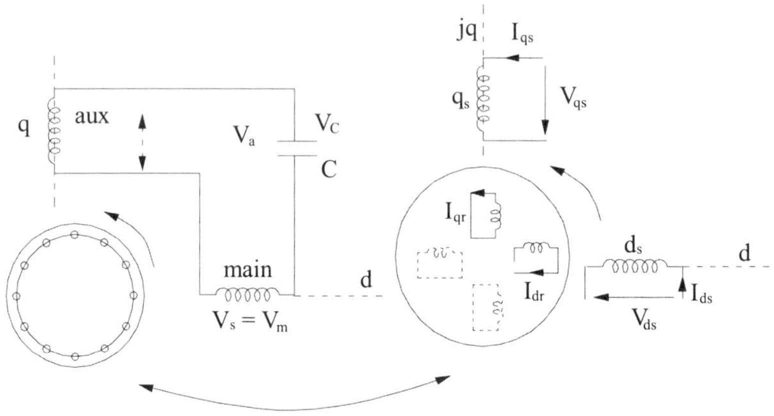

Figure 25.1 The d-q model

Where a is the equivalent turns ratio of the auxiliary to main winding.
Also

$$\frac{dV_c}{dt} = \frac{1}{C}\frac{I_{qs}}{a} \tag{25.2}$$

The d-q model equations, in stator coordinates, are now straightforward

$$
\begin{aligned}
\frac{d\Psi_{ds}}{dt} &= V_s(t) - R_{sm} I_{ds} \\
\frac{d\Psi_{qs}}{dt} &= \left(V_s(t) - V_c(t)\right)/a - \frac{R_{sa}}{a^2} I_{qs} \\
\frac{d\Psi_{dr}}{dt} &= I_{dr} R_{rm} - \omega_r \Psi_{qr} \\
\frac{d\Psi_{qr}}{dt} &= I_{qr} R_{rm} + \omega_r \Psi_{dr}
\end{aligned}
\tag{25.3}
$$

The flux/current relationships are

$$
\begin{aligned}
\Psi_{ds} &= L_{sm} I_{ds} + \Psi_{dm}; \Psi_{dm} = L_m\left(I_{ds} + I_{dr}\right) = L_m I_{dm} \\
\Psi_{dr} &= L_{rm} I_{dr} + \Psi_{dm} \\
\Psi_{qs} &= \frac{L_{sa}}{a^2} I_{qs} + \Psi_{qm}; \Psi_{qm} = L_m\left(I_{qs} + I_{qr}\right) = L_m I_{qm} \\
\Psi_{qr} &= L_{rm} I_{qr} + \Psi_{qm}
\end{aligned}
\tag{25.4}
$$

The magnetization inductance L_m depends both on I_{dm} and I_{qm} when saturation is accounted for because of cross-coupling saturation effects. As the magnetic circuit looks nonisotropic even with slotting neglected, the total flux vector Ψ_m is a function of total magnetization current I_m:

$$
\begin{aligned}
\Psi_m &= L_m(I_m) \cdot I_m \\
I_m &= \sqrt{I_{dm}^2 + I_{qm}^2}; \Psi_m = \sqrt{\Psi_{dm}^2 + \Psi_{qm}^2}
\end{aligned}
\tag{25.5}
$$

Once $L_m(I_m)$ is known, it may be used in the computation process with its previous computation step value, provided the sampling time is small enough.

We have to add the motion equation

$$
\begin{aligned}
\frac{J}{p_1}\frac{d\omega_r}{dt} &= T_e - T_{load}\left(\theta_r, \omega_r\right) \\
\frac{d\theta_r}{dt} &= \omega_r / p_1 \\
T_e &= \left(\Psi_{ds} I_{qs} - \Psi_{qs} I_{ds}\right) p_1
\end{aligned}
\tag{25.6}
$$

Equations (25.2)-(25.6) constitute a nonlinear 7th order system with V_c, I_{ds}, I_{qs}, I_{dr}, I_{qr} and ω_r, θ_r as variables. The inputs are $V_s(t)$-the source voltage and T_{load}-the load torque.

The load torque may vary with speed (for pumps, fans) or with rotor position (for compressors).

In most applications, θ_r does not intervene as the load torque depends on speed, so the order of the system becomes six.

When the two stator windings do not occupy the same number of slots, there may be slight differences between the magnetization curves along the two axes, but they are in many cases small enough to be neglected in the treatment of transients.

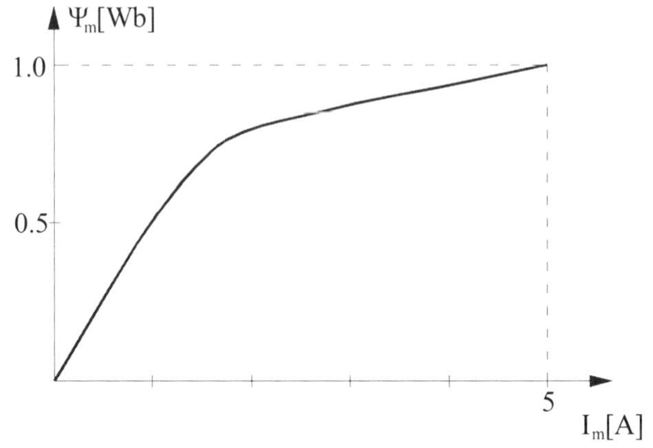

Figure 25.2 The magnetization curve

A good approximation of the magnetization curve may be obtained through FEM, at zero speed, with d.c. excitation along one axis and contribution from the second axis. Dc current decay tests at standstill in one axis, with dc current injection in the second axis should provide similar results. After many "points" in Figure 25.2 are calculated, curve fitting may be used to yield a univoque correspondance between Ψ_m and I_m.

As expected, if enough digital simulation time for transients is allowed, steady state behaviour may be reached.

In the absence of saturation, for sinusoidal power source voltage, the line current is sinusoidal for no-load and under load.

In contrast, when magnetic saturation is considered, the simulation results show nonsinusoidal line current under no-load and load conditions (Figure 25.3). [2]

The influence of saturation is visible. It is even more visible in the main winding current RMS computation versus experiments. (Figure 25.4) [2]

The differences are smaller in the auxiliary winding for the case in point: P_n = 1.1 kW, U_n = 220 V, I_n = 6.8 A, C = 25 μF, f_n = 50 Hz, a = 1.52, Rsm = 2.6 Ω, R_{sa} = 6.4 Ω, R_{rm} = 3.11 Ω, L_{sm} = L_{rm} = 7.5 mH, J = 1.1×10^{-3} Kgm^2, L_{sa} = 17.3 mH. The magnetization curve is as in Figure 25.2. The motor power factor, calculated for the fundamental, at low loads, is much less than predicted by the nonsaturated d-q model. On the other hand it looks like the saturated d-q model produces results which are very close to the experimental ones, for a wide range of loads.

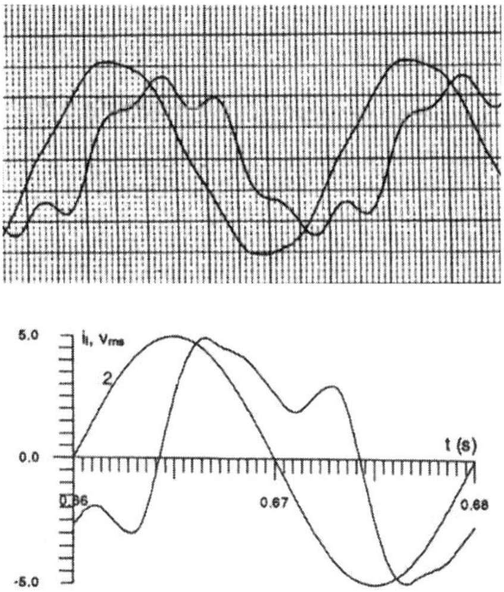

a.) b.)

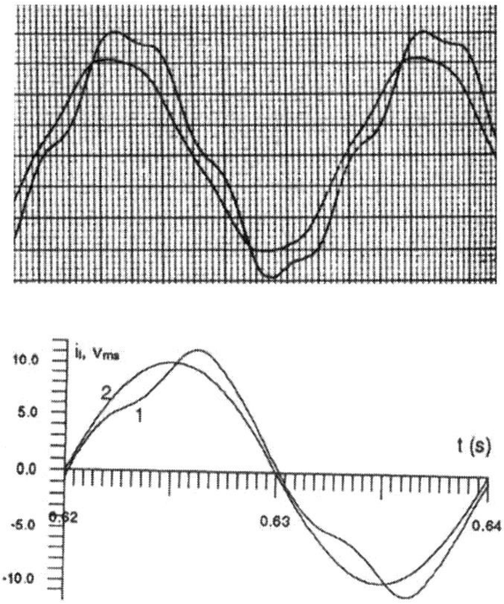

c.) d.)

Figure 25.3 Steady state by the d-q model with saturation included
a.) and b.)-no load
c.) and d.)-on load
a.) and c.)-test results
b.) and d.)-digital simulation

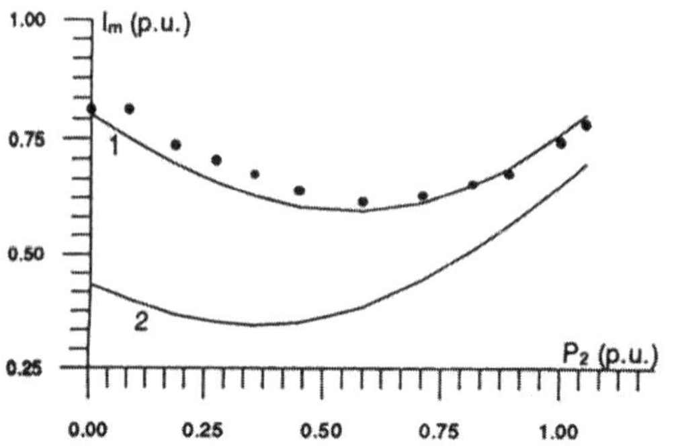

Figure 25.4. RMS of main winding current versus output power
1-saturated model
2-unsaturated model
•-test results

This may be due to the fact that the saturation curve has been obtained through tests and thus the saturation level is tracked for each instantaneous value of magnetization current (flux).

The RMS current correct prediction by the saturated d-q model for steady state represents notable progress in assessing more correctly the losses in the machine.

Still, the core losses-fundamental and additional-and additional losses in the rotor cage (including the interbar currents) are not yet included in the model. Space harmonics though apparently somehow included in the $\Psi_m(I_m)$ curve are not thoroughly treated in the saturated d-q model.

25.3 STARTING TRANSIENTS

It is a known fact that during severe starting transients, the main flux path saturation does not play a crucial role. However it is there embedded in the model and may be used if so desired.

For a single phase IM with the data: Vs = 220 V, f_{1n} = 50 Hz, $R_{sm} = R_{sa}$ = 1 Ω, a = 1, L_m = 1.9 H, $L_{sm} = L_{sa}$ = 0.2 H, R_{rm} = 35 Ω, L_{rm} = 0.1 H, J = 10^{-3} Kgm^2, C_a=5 μF, $\zeta = 90^0$, the starting transients are presented on Figure 25.5. The load torque is zero from start to t = 0.4 s, when a 0.4 Nm load torque is suddenly applied.

The average torque looks negative because of the choice of signs.

The torque pulsations are large and so evident on Figure 25.5 b), while the average torque is 0.4 Nm, equal to the load torque. The large torque pulsations reflect the departure from symmetry conditions. The capacitor voltage goes up to a peak value of 600 V, another sign that a larger capacitor might be needed.

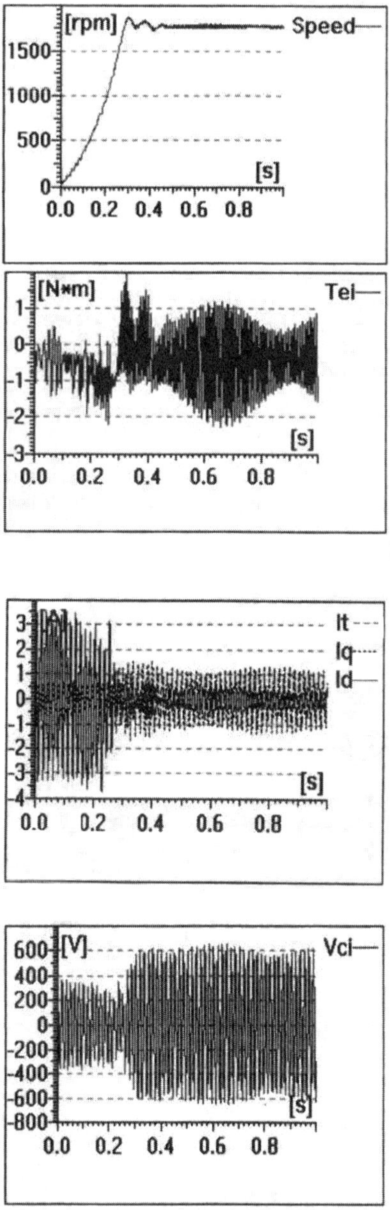

a.) b.)

c.) d.)

Figure 25.5 Starting transients
a.) speed versus time, b.) torque versus time
c.) input current versus time, d.) capacitor voltage versus time

The hodograph of stator current vector

$$I_s = I_d + jI_q \tag{25.7}$$

is shown on Figure 25.6.

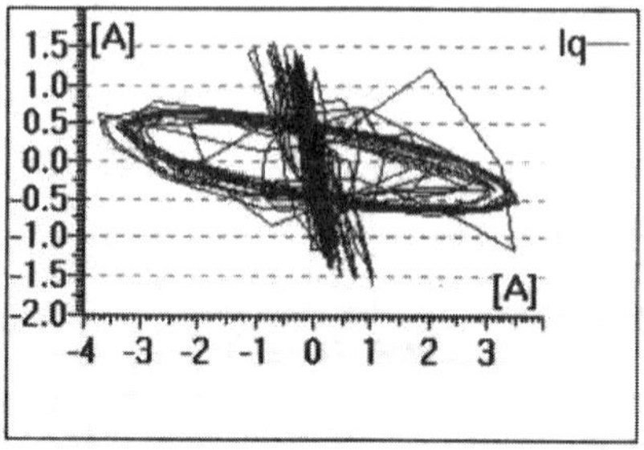

Figure 25.6 Current hodograph during starting

For symmetry, the current hodograph should be a circle. Not so in Figure 25.6.

To investigate the influence of capacitance on the starting process, the starting capacitor is Cs = 30μF, and then at t = 0.4s the capacitance is reduced suddenly C_a = 5μF (Figure 25.7)

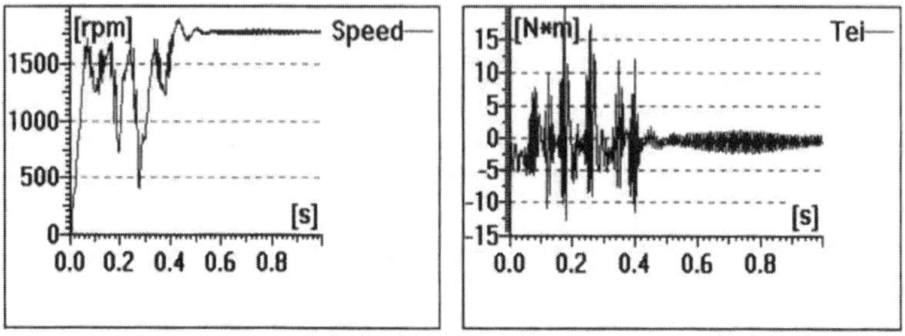

a.) b.)

Figure 25.7 Starting transients for C_s = 30 μF at start and Ca = 5 μF from t = 0.4 s on a.) speed; b.) torque

The torque and current transients are too large with C_S = 30 μF, a sign that the capacitor is now too large.

25.4 THE MULTIPLE REFERENCE MODEL FOR TRANSIENTS

The d-q model has to use stator coordinates as long as the two windings are not identical. However if the d-q model is applied separately for the instantaneous + and-(f, b) components then for synchronous coordinates: $+\omega_1$ (for the +(f) component) and $-\omega_1$ (for the - (b) component), the steady state means d.c. variables. [3]

Consequently, the dynamic (stability) analysis may be performed via the system linearization (small deviation theory).

Stability to sinusoidal torque pulsation such as in compressor loads, can thus be treated. The superposition principle precludes the inclusion of magnetic saturation in the model. [3]

25.5 INCLUDING THE SPACE HARMONICS

The existence of space harmonics causes torque pulsations, cogging, and crawling.

The space harmonics have three origins, as mentioned before in Chapter 23,

- Stator m.m.f. space harmonics
- Slot opening permeance pulsations
- Main flux path saturation

A general method to deal with space harmonics of all these three origins is presented in Reference 6 based on the multiple magnetic circuit approach. The rotor is considered bar by bar.

Reference 4 treats space harmonics produced only by the stator m.m.f. but deals also with the effect of skewing. Based on the winding function approach, the latter method also considers m windings in the stator and N_r bars in the rotor. Magnetic saturation is neglected.

The deep(s) in the torque/speed curve may be predicted by this procedure. Also the influence of the ratio of slot numbers N_s/N_r of stator and rotor and of cogging (parasitic, asynchronous) torques is clearly evidentiated. As an example, Figure 25.8 illustrates [4] the starting transients of a single-phase capacitor motor with the same number of slots per stator and rotor, without skewing. As expected, the motor is not able to start due to the strong parasitic synchronous torque at stall (Figure 25.8 a)

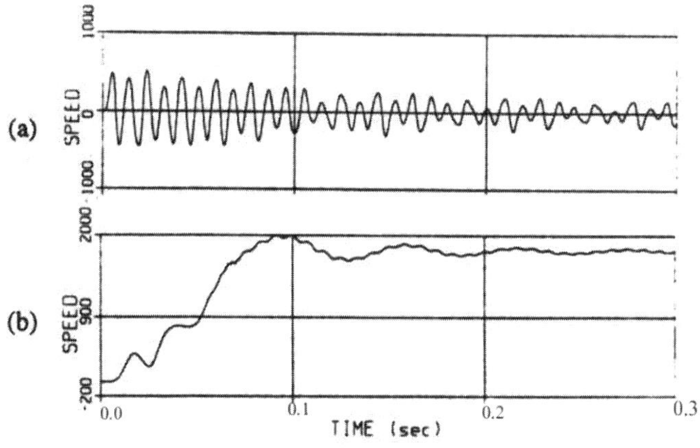

Figure 25.8 Starting speed transients for $N_s = N_r$
a) without skewing b) with skewing

With skewing of one slot pitch, even with $N_s = N_r$ slots, the motor can start (Figure 25.8 b).

25.6 SUMMARY

- Transients occur during starting or load torque perturbation operation modes.
- Also power source voltage (and frequency) variation cause transients in the single phase IM.
- When power converter (variable voltage, variable frequency)-fed, the single phase IM experiences time harmonics also.
- The standard way to handle the single phase IM is by the cross-field (d-q) model when magnetic saturation, time harmonics and skin effect may be considered simultaneously.
- Due to magnetic saturation the main winding and total stator currents depart from sinusoidal waveforms, both at no-load and under load conditions. The magnetic saturation level depends on load, machine parameters and on the capacitance value. It seems that neglecting saturation leads to notable discrepancy between the main and total currents measured and calculated RMS values. This is one of the reasons why losses are not predicted correctly, especially for low powers.
- Switching the starting capacitor off produces important speed, current, and torque transients.
- The d-q model has a straightforward numerical solution for transients but, due to the stator winding asymmetry, it has to make use of stator coordinates to yield rotor position independent inductances. Steady state means a.c. variables at power source frequency. Consequently, small deviation theory, to study stability, may not be used, except for the case when the auxiliary winding is open.
- The multiple reference system theory, making use of + - (f,b) models in their synchronous coordinates ($+\omega_1$ and $-\omega_1$) may be used to investigate stability after linearization.
- Finally, besides FEM-circuit coupled model, the winding function approach or the multiple equivalent magnetic circuit method simulate each rotor bar and thus asynchronous and synchronous parasitic torques may be detected.
- A FEM-circuit couple model (software), easy to handle, and computation time competitive, for single-phase IM, is apparently not available as of this writing.

25.7 REFERENCES

1. I. Boldea and S. A. Nasar, Electrical Machine Dynamics, Macmillan Publishing Company, New York 1986.
2. K. Arfa, S. Meziani, S. Hadji, B. Medjahed, Modelization of Single Phase Capacitor Run Motor Accounting for Saturation, Record of ICEM-1998, Vol.1, pp.113-118.
3. T. A. Walls and S. D. Sudhoff , Analysis of a Single-Phase Induction Machine With a Shifted Auxiliary Winding, IEEE Trans., vol. EC-11, no. 4, 1996, pp. 681-686.
4. H. A. Toliyat and N. Sargolzaei, Comprehensive Method for Transient Modelling of Single Phase Induction Motors Including the Space Harmonics, EMPS Journal vol. 26, no. 3, 1998, pp. 221-234.
5. S. S. Shokralla, N. Yasin, A. M. Kinawy, Perturbation Analysis of a Split Phase Induction Motor in Time and Frequency Domains, EMPS Journal vol. 25, no. 2, 1997, pp. 107-120.
6. V. Ostovic, Dynamics of Saturated Electric Machines (book), Springer Verlag, New York, 1985.

Chapter 26

SINGLE-PHASE INDUCTION GENERATORS

26.1 INTRODUCTION

Small portable single-phase generators are built for up to 10-20 kW.

Traditionally they use a synchronous single-phase generator with rotating diodes.

Self excited, self-regulated single-phase induction generators (IGs) provide, in principle, good voltage regulation, more power output/weight and a more sinusoidal output voltage.

In some applications, where tight voltage control is required, power electronics may be introduced to vary the capacitors "seen" by the IM. Among the many possible configurations [1,2] we investigate here only one, which holds a high degree of generality in its analysis and seems very practical in the same time. (Figure 26.1)

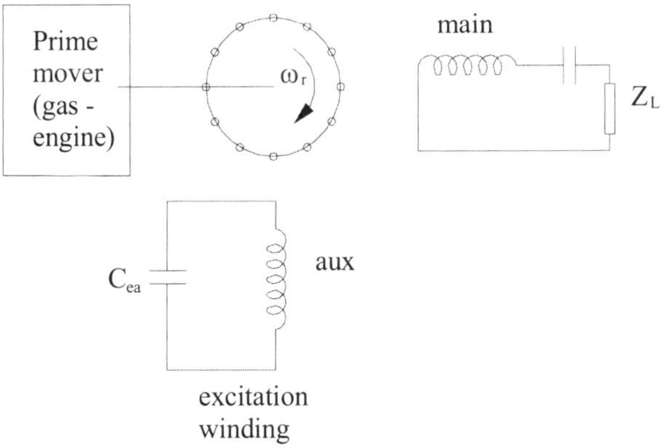

Figure 26.1 Self-excited self-regulated single-phase induction generator

The auxiliary winding is connected over a self-excitation capacitor C_{ea} and constitutes the excitation winding.

The main winding has a series connected capacitor C_{sm} for voltage self-regulation and delivers output power to a given load.

With the power (main) winding open the IG is rotated to the desired speed. Through self-excitation (in presence of magnetic saturation) it produces a certain no load voltage. To adjust the no load voltage the self-excitation capacitor may be changed accordingly, for a given IG.

After that, the load is connected and main winding delivers power to the load.

The load voltage / current curve depends on the load impedance and its power factor, speed, IG parameters and the two capacitors C_{ea} and C_{sm}. Varying C_{sm} the voltage regulation may be reduced to desired values.

In general increasing C_{sm} tends to increase the voltage at rated load, with a maximum voltage in between. This peak voltage for intermediate load may be limited by a parallel saturable reactor.

To investigate the steady state performance of single-phase IGs the revolving theory seems to be appropriated. Saturation has to be considered as no self-excitation occurs without it. On the other hand, to study the transients, the d-q model, with saturation included, as shown in Chapter 25, may be used.

Let us deal with steady state performance first.

26.2 STEADY STATE MODEL AND PERFORMANCE

Examining carefully the configuration on Figure 25.1 we notice that:

- The self-excitation capacitor may be lumped in series with the auxiliary winding whose voltage is then $V_a = 0$.
- The series (regulation) capacitor C_{sm} may be lumped into the load

$$\underline{Z}_L^{'} = \underline{Z}_L - \frac{jX_C}{F} \tag{26.1}$$

F is the P.U. frequency with respect to rated frequency. In general

$$\underline{Z}_L = R_L + jX_L \cdot F \tag{26.2}$$

Now with $V_a = 0$, the forward and backward voltage components, reduced to the main winding, are ($V_a = 0$, $V_m = V_s$)

$$V_{m+} = V_{m-} \tag{26.3}$$

$$V_{m+} = V_s/2 = -Z_L(I_{m+} + I_{m-})/2 \tag{26.4}$$

Equation (26.4) may be written as

$$\begin{aligned}\underline{V}_{AB} = V_{m+} &= -\frac{\underline{Z}_L^{'}}{2}(\underline{I}_{m+} + \underline{I}_{m-}) = -\underline{Z}_L^{'}\underline{I}_{m+} + \frac{\underline{Z}_L^{'}}{2}(\underline{I}_{m+} - \underline{I}_{m-}) \\ \underline{V}_{AB} = V_{m-} &= -\frac{\underline{Z}_L^{'}}{2}(\underline{I}_{m+} + \underline{I}_{m-}) = -\underline{Z}_L^{'}\underline{I}_{m-} - \frac{\underline{Z}_L^{'}}{2}(\underline{I}_{m+} + \underline{I}_{m-})\end{aligned} \tag{26.5}$$

Consequently, it is possible to use the equivalent circuit in Figure 24.5 with $\underline{Z}'_L$ in place of both $\underline{V}_{m+}$ (V_{AB}) and $\underline{V}_{m-}$(V_{BC}) as shown in Figure 26.2. Notice that $-Z'_L/2$ also enters the picture, flowed by (I_{m+}-I_{m-}), as suggested by Equations (26.5).

All parameters in Figure 26.2 have been divided by the P.U. frequency F. Denoting

$$\underline{Z}_{1mL} = \frac{R_{sm}}{F} + jX_{sm} + \frac{R_L}{F} + jX_L - j\frac{X_{csm}}{F^2}$$

$$\underline{Z}_{aL}^{'} = \frac{R_{sa}}{2Fa^2} + j\frac{X_{sa}}{2a^2} - j\frac{X_{cea}}{2F^2a^2} - \frac{\underline{Z}_{1mL}}{2}$$

$$\underline{Z}_{+} = \frac{jX_{mm}\left(\frac{R_{rm}}{F-U} + jX_{rm}\right)}{\frac{R_{rm}}{F-U} + j(X_{mm} + X_{rm})} \qquad (26.6)$$

$$\underline{Z}_{-} = \frac{jX_{mm}\left(\frac{R_{rm}}{F+U} + jX_{rm}\right)}{\frac{R_{rm}}{F+U} + j(X_{mm} + X_{rm})}$$

the equivalent circuit of Figure 26.2 may be simplified as in Figure 26.3.

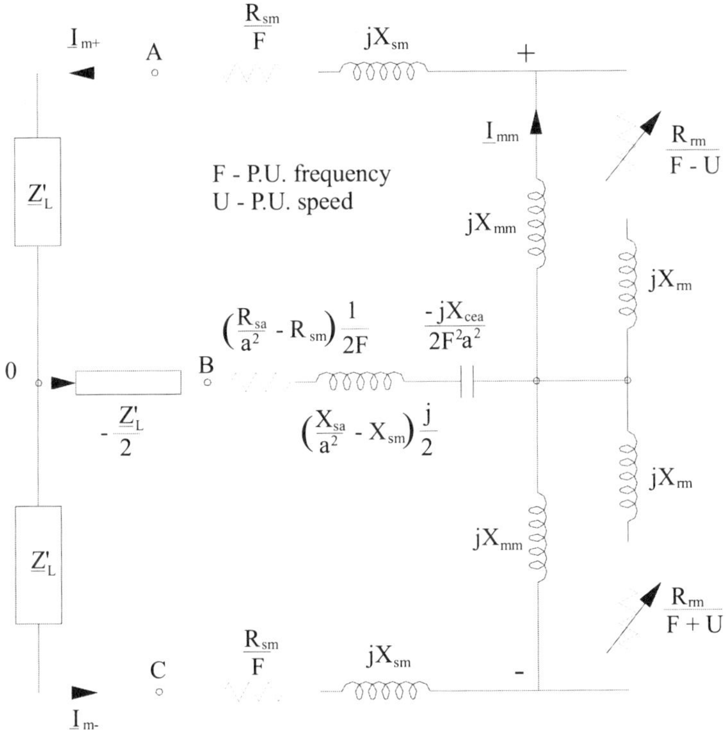

Figure 26.2 Self-excited self-regulated single-phase IG (Figure 26.1): equivalent circuit for steady state

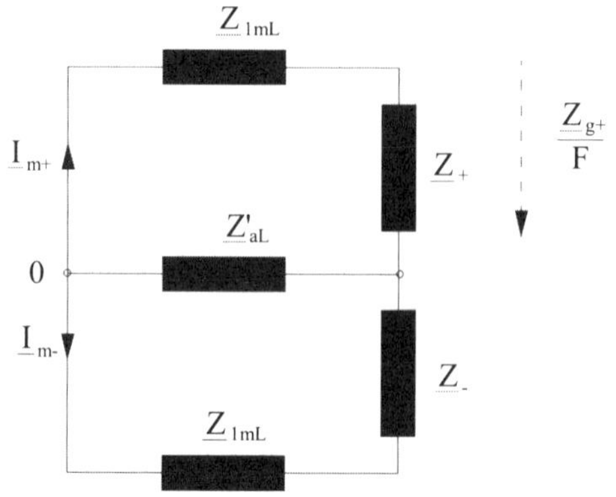

Figure 26.3 Simplified equivalent circuit of single-phase generator

The self-excitation condition implies that the sum of the currents in node 0 is zero:

$$\frac{1}{\underline{Z}_{1mL}+\underline{Z}_+}=\frac{1}{\underline{Z}_{1mL}+\underline{Z}_-}+\frac{1}{\underline{Z}'_{aL}} \tag{26.7}$$

Two conditions are provided by (26.7) to solve for two unknowns. We may choose F and X_{mm}, provided the magnetisation curve: $V_{g+}(X_{mm})$ is known from the measurements or from FEM calculations.

In reality, X_{mm} is a known function of the magnetisation current: I_{mm}: $X_{mm}(I_{mm})$ (Figure 26.4).

To simplify the computation process we may consider that $\underline{Z}_-$ is

$$\underline{Z}_- \approx \frac{R_{rm}}{F+U}+jX_{rm} \tag{26.8}$$

Except for X_{mm}, as all other parameters are considered constant, we may express $\underline{Z}_+$ from (26.7) as:

$$\underline{Z}_+ = \frac{(\underline{Z}_{1m}+\underline{Z}_-)\cdot\underline{Z}'_{aL}}{\underline{Z}_{1mL}+\underline{Z}_-+\underline{Z}'_{aL}}-\underline{Z}_{1mL} \tag{26.9}$$

All impedances on the right side of Equations (26.9) are solely dependent on frequency F, if all motor parameters, speed n and C_{ae}, C_{sm}, X_L are given, for an adopted rated frequency f_{1n}.

For a row of values for F we may simply calculate from (26.9) $\underline{Z}_+ = f(F)$ for given speed n, capacitors, load

$$\underline{Z}_+(F, X_{mm}) = \frac{jX_{mm}\left(\frac{R_{rm}}{F-U}+jX_{rm}\right)}{j(X_{mm}+X_{rm})+\frac{R_{rm}}{F-U}} \tag{26.10}$$

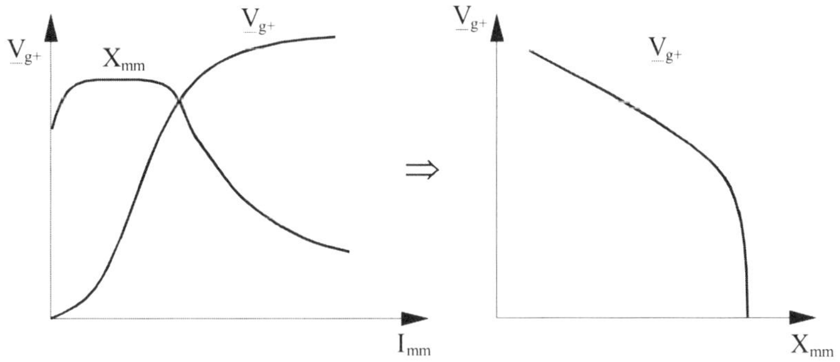

Figure 26.4 The magnetization curves of the main winding

We may use only the imaginary part of (26.10) and determine rather simply the $X_{mm}(F)$ function.

Now, from Figure (26.4), we may determine for each F, that is for every X_{mm} value, the airgap voltage value $\underline{V}_{g+}$ and thus the magnetization current

$$I_{mm} = \frac{\underline{V}_{g+}}{X_{mm}} \tag{26.11}$$

As we now know F, X_{mm} and $\underline{V}_{g+}$, the equivalent circuit of Figure 26.3 may be solved rather simply to determine the two currents I_{m+} and I_{m-}.

From now on, all steady state characteristics may be easily calculated.

$$\underline{I}_{m+}^{(F)} = \frac{\underline{V}_{g+}}{F} \cdot \frac{1}{\left(\underline{Z}_{1mL} + \frac{\underline{Z}'_{aL} \cdot \underline{Z}_{1mL}}{\underline{Z}'_{aL} + \underline{Z}_{1mL}}\right)} \tag{26.12}$$

$$\underline{I}_{m-}^{(F)} = \underline{I}_{m+}^{(F)} \cdot \frac{\underline{Z}'_{aL}}{\underline{Z}_{-} + \underline{Z}_{1mL}} \tag{26.13}$$

The load current $\underline{I}_m$ is

$$\underline{I}_m = \underline{I}_{m+} + \underline{I}_{m-} \tag{26.14}$$

The auxiliary winding current writes

$$\underline{I}_a = j(\underline{I}_{m+} - \underline{I}_{m-}) \tag{26.15}$$

The output active power P_{out} is

$$P_{out} = I_m^2 \cdot R_L \tag{26.16}$$

The rotor + current component $\underline{I}_{r+}$ becomes

$$\underline{I}_{r+} = -\underline{I}_{m+} \cdot \frac{jX_{mm}}{\frac{R_{rm}}{F-U} + j(X_{rm} + X_{mm})} \tag{26.17}$$

The total input active power from the shaft P_{input} is

$$P_{input} \approx 2I_{r+}^2 \cdot \frac{R_{rm} \cdot U}{F-U} - 2I_{m-}^2 \cdot \frac{R_{rm}U}{F+U} \tag{26.18}$$

For a realistic efficiency formula, the core additional and mechanical losses p_{iron}+p_{stray}+p_{mec} have to be added to the ideal input of (26.18).

$$\eta = \frac{P_{out}}{P_{input} + p_{iron} + p_{stray} + p_{mec}} \tag{26.19}$$

As the speed is given, varying F we change the slip. We might change the load resistance with frequency (slip) to yield realistic results from the beginning.

As $X_{mm}(V_{g+})$ may be given as a table, the values of $X_{mm}(F)$ function may be looked up simply into another table. If no X_{mm} is found from the given data it means that either the load impedance or the capacitors, for that particular frequency and speed, are not within the existence domain.

So either the load is modified or the capacitor is changed to reenter the existence domain.

The above algorithm may be synthesized as in Figure 26.5.

The IG data obtained through tests are: P_n = 700 W, n_n = 3000 rpm, V_{Ln} = 230 V, f_{1n} = 50 Hz, R_{sm} = 3.94 Ω, R_{sa} = 4.39 Ω, R_{rm} = 3.36 Ω, X_{rm} = X_{sm} = 5.48 Ω, X_{sa} = 7.5 Ω, unsaturated X_{mm} = 70 Ω, C_{ea} = 40 μF, C_{sm} = 100 μF [1].

The magnetization curves $V_{g+}(I_m)$ has been obtained experimentally, in the synchronous bare rotor test. That is, before the rotor cage was located in the rotor slots, the IG was driven at synchronism, n = 3000 rpm (f = 50 Hz), and was a.c.-fed from a Variac in the main winding only. Alternatively it may be calculated at standstill with d.c. excitation via FEM. In both cases the auxiliary winding is kept open. More on testing of single-phase IMs in Chapter 28. The experimental results in Figure 26.6 warrant a few remarks

- The larger the speed, the larger the load voltage
- The lower the speed, the larger the current for given load
- Voltage regulation is very satisfactory: from 245 V at no load to 230 V at full load
- The no load voltage increases with C_{ea} (the capacitance) in the auxiliary winding
- The higher the series capacitor (above C_{sm} = 40 μF) the larger the load voltage
- It was also shown that the voltage waveform is rather sinusoidal up to rated load
- The fundamental frequency at full load and 3,000 rpm is f_{1n} = 48.4 Hz, an indication of small slip

A real gas engine (without speed regulation) would lose some speed when the generator is loaded. Still the speed (and additional frequency) reduction from no load to full load is small. So aggregated voltage regulation is, in these conditions, at full load, slightly larger but still below 8% with a speed drop from 3000rpm to 2920rpm [1].

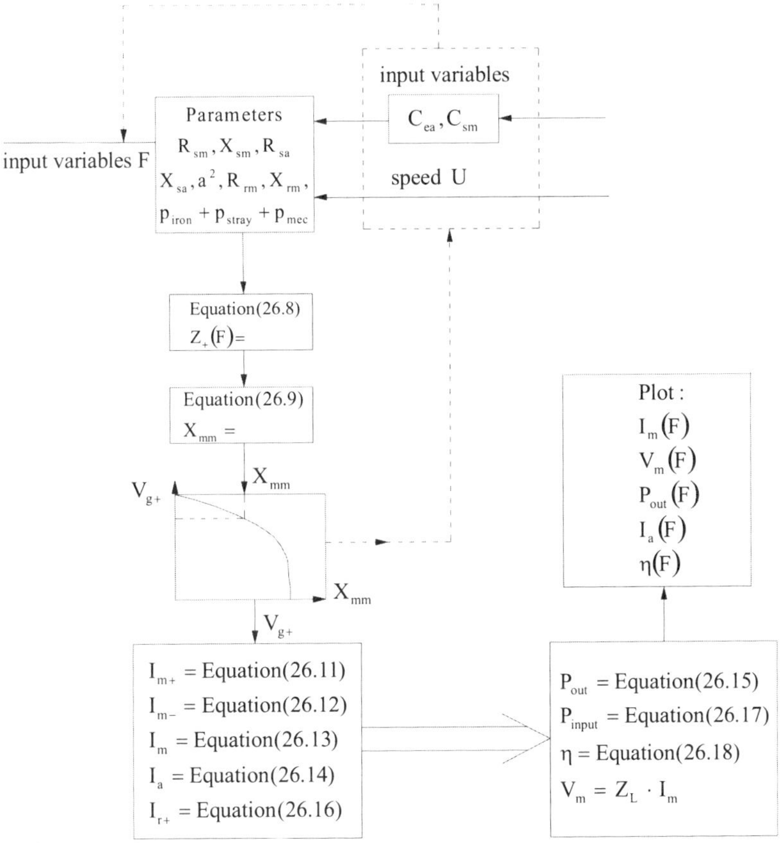

Figure 26.5 Performance computation algorithm

Typical steady state performance obtained for such a self-regulated single-phase IG are shown in Figure 26.6. [1]

26.3 THE d-q MODEL FOR TRANSIENTS

The transients may be treated directly via d-q model in stator coordinates with saturation included (as done for motoring).

$$V_{ds} = -V_{csm} - R_L I_{ds} - L_L \frac{dI_{ds}}{dt} \tag{26.20}$$

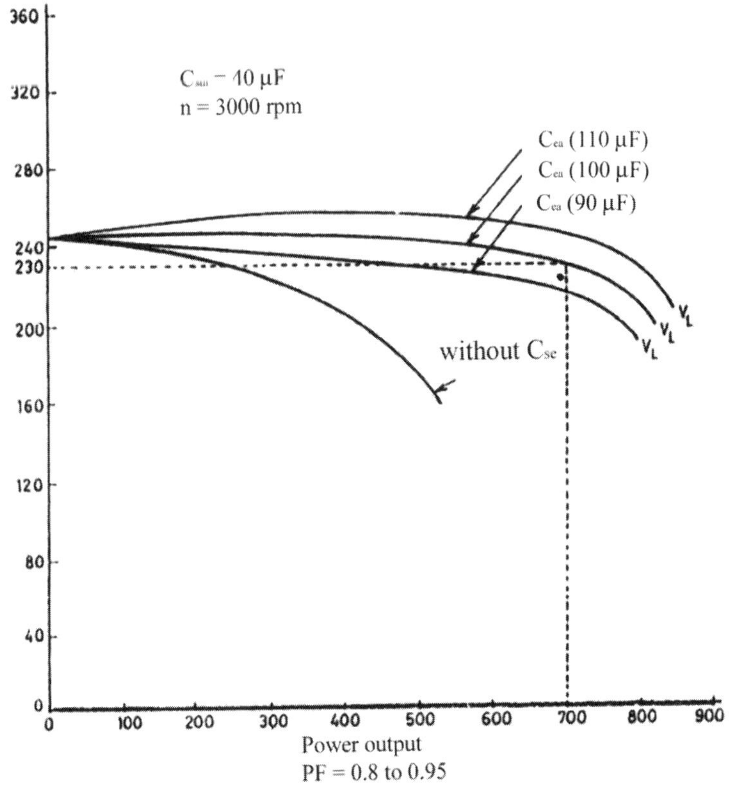

Figure 26.6 Steady state performance of a self-excited self-regulated single-phase IG

$$\frac{dV_{csm}}{dt} = \frac{1}{C_{sm}} I_{ds}; I_{ds} = I_m \tag{26.21}$$

$$V_q = -V_{cea} \tag{26.22}$$

$$\frac{dV_{cea}}{dt} = \frac{1}{a^2 C_{ea}} I_{qs}; I_{qs} = I_a \cdot a \tag{26.23}$$

The d-q model in paragraph (25.2) is:

$$\begin{aligned}
\frac{d\Psi_{ds}}{dt} &= V_{ds}(t) - R_{sm}I_{ds}\\
\frac{d\Psi_{qs}}{dt} &= -V_{cea} - \frac{R_{sa}}{a^2}I_{qs}\\
\frac{d\Psi_{dr}}{dt} &= I_{dr}R_{rm} - \omega_r\Psi_{qr}\\
\frac{d\Psi_{qr}}{dt} &= I_{qr}R_{rm} - \omega_r\Psi_{dr}
\end{aligned} \tag{26.24}$$

$$\begin{aligned}
\psi_{ds} &= L_{sm}I_{ds} + \Psi_{dm}; \Psi_{dm} = L_{mm}I_{dm}\\
\psi_{dr} &= L_{rm}I_{dr} + \Psi_{qm}; \Psi_{qm} = L_{mm}I_{qm}\\
\psi_{qs} &= \frac{L_{sa}}{a^2}I_{qs} + \Psi_{qm}; I_{dm} = I_{ds} + I_{dr}\\
\psi_{qr} &= L_{rm}I_{qr} + \Psi_{qm}; I_{qm} = I_{qs} + I_{qr}\\
\text{and}: \psi_m &= L_{mm}(I_m)\cdot I_m; I_m = \sqrt{I_{dm}^2 + I_{qm}^2}
\end{aligned} \tag{26.25}$$

To complete the model the motion equation is added

$$\begin{aligned}
\frac{J}{p_1}\frac{d\omega_r}{dt} &= T_{pmover} + T_e\\
T_e &= p_1\left(\psi_{ds}I_{qs} - \Psi_{qs}I_{ds}\right) < 0
\end{aligned} \tag{26.26}$$

The prime mover torque may be dependent on speed or on the rotor position also. The prime mover speed governor (if any) equations may be added.

Equation (26.20) shows that when the load contains an inductance L_L (for example a single-phase IM), I_{ds} has to be a variable and thus the whole d-q model (Equation 26.24) has to be rearranged to accommodate this situation in presence of magnetic saturation.

However, with resistive load (R_L)-$L_L = 0$-the solution is straightforward with: V_{csm}, V_{cea}, Ψ_{ds}, Ψ_{qs}, V_{qs}, Ψ_{dr}, Ψ_{qr} and ω_r as variables.

If the speed ω_r is a given function of time the motion equation (26.26) is simply ignored. The self-excitation under no load, during prime mover start-up, load sudden variations, load dumping, or sudden shortcircuit are typical transients to be handled via the d-q model.

26.4 EXPANDING THE OPERATION RANGE WITH POWER ELECTRONICS

Power electronics can provide more freedom to the operation of single-phase IMs in terms of load voltage and frequency control. [3] An example is shown on Figure 26.7 [4].

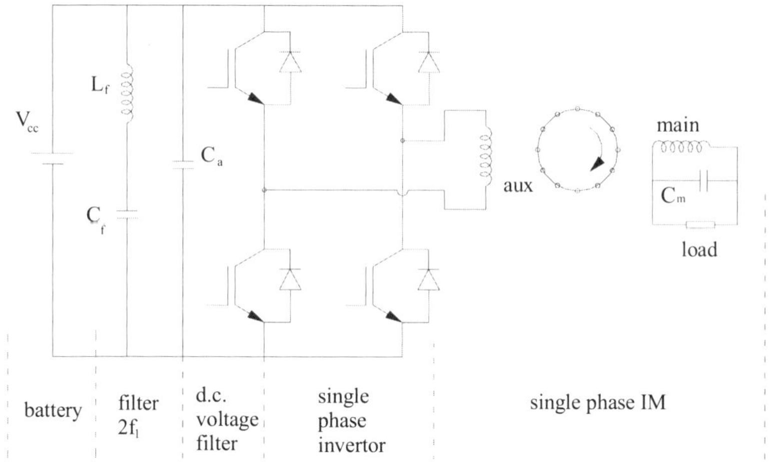

Figure 26.7 Single-phase IG with battery-inverter-fed auxiliary winding

The auxiliary winding is now a.c. fed at the load frequency f_1, through a single-phase inverter, from a battery.

To filter out the double frequency current produced by the converter, the L_FC_F filter is used. C_a filters the d.c. voltage of the battery.

The main winding reactive power requirement may be reduced by the parallel capacitor C_m with or without a short or a long shunt series capacitor.

By adequate control in the inverter, it is possible to regulate the load frequency and voltage when the prime mover speed varies.

The inverter may provide more or less reactive power.

It is also possible that, when the load is large, the active power is contributed by the battery. On the other hand, when the load is low, the auxiliary winding can pump back active power to recharge the battery.

This potential infusion of active power from the battery to load may lead to the idea that, in principle, it is possible to operate as a generator even if the speed of the rotor ω_r is not greater than $\omega_1 = 2\pi f_1$. However, as expected, more efficient operation occurs when $\omega_r > \omega_1$.

The auxiliary winding is 90^0 (electrical) ahead of the main winding and thus no pulsation type interaction with the main winding exists. The interaction through the motion e.m.f.s is severely filtered for harmonics by the rotor currents. So the load voltage is practically sinusoidal. The investigation of this system may be performed through the d-q model, as presented in the previous paragraph.

For details, see Reference 4.

26.5 SUMMARY

- The two winding induction machine may be used for low power autonomous single-phase generators.

- Amongst many possible connections it seems that one of the best connects the auxiliary winding upon an excitation capacitors C_{ea}, while the main winding (provided with a self-regulation series capacitors C_{sm}) supplies the load.
- The steady state modelling may be done with the revolving theory (+,-or f, b) model.
- The saturation plays a key role in this self–excited self–regulated configuration
- The magnetic saturation is related, in the model, to the direct (+)-forward-component.
- A rather simple computer program can provide the steady state characteristics: output voltage, current, frequency versus output power for given speed, machine parameters and magnetization curve.
- Good voltage regulation (less than 8%) has been reported.
- The sudden shortcircuit apparently does not threaten the IG integrity.
- The transients may be handled through the d-q model in stator coordinated via some additional terminal voltage relationships.
- More freedom in the operation of single-phase IG is brought by the use of a fractional rating battery-fed inverter to supply the auxiliary winding. Voltage and frequency control may be provided this way. Also, bi-direction power flow between inverter and battery can be performed. So the battery may be recharged when the IG load is low.
- In the power range (10-20kW) the single-phase IG represents a strong potential competitor to existing gensets using synchronous generators.

26.6 REFERENCES

1. S. S. Murthy, A Novel Self-Excited Self-Regulated Single Phase Induction Generator, Part I+II, IEEE Trans, vol. EC-8, no. 3, 1993, pp. 377-388.

2. O. Ojo, Performance of Self-Excited Single-Phase Induction Generators with Short Shunt and Long Shunt Connections, IEEE Trans, vol. EC-11, no. 3, 1996, pp. 477-482.

3. D. W. Novotny, D. J. Gritter, G. H. Studmann, Self-Excitation in Inverter Driven Induction Machines, IEEE Trans, vol. PAS-96, no. 4, 1977, pp. 1117-1125.

4. O. Ojo, O. Omozusi, A. A. Jimoh, Expanding the Operating Range of a Single-Phase Induction Generator with a PWM Inverter, Record of IEEE-IAS-1998, vol. 1, pp. 205-212.

Chapter 27

SINGLE-PHASE IM DESIGN

27.1 INTRODUCTION

By design we mean "dimensioning." That is, the finding of a suitable geometry and manufacturing data and performance indexes for given specifications. Design, then, means first dimensioning (or synthesis), sizing, then performance assessment (analysis). Finally if the specifications are not met the process is repeated according to an adopted strategy until satisfactory performance is obtained.

On top of this, optimization is performed according to one or more objectives functions, as detailed in Chapter 18 in relation to three phase induction machines. Typical specifications (with a case study) are

- Rated power P_n = 186.5 W (1/4 HP)
- Rated voltage V_{sn} = 115 V
- Rated frequency f_{1n} = 60 Hz
- Rated power factor $\cos\varphi_n$ = 0.98 lagging service continuous or shortduty
- Breakdown p. u. torque 1.3-2.5
- Starting p. u. torque 0.5-3.5
- Starting p. u. current 5-6.5
- Capacitor p. u. maximum voltage 0.6-1.6

The breakdown torque p.u. may go as high as 4.0 for the dual capacitor configuration and special-service motors.

Also starting torques above 1.5 p. u. are obtained with a starting capacitor.

The split-phase IM is also capable of high starting and breakdown torques in p. u. as during starting both windings are active, at the expense of rather high resistance, both in the rotor and in the auxiliary stator winding.

For two (three) speed operation the 2 (3) speed levels in % of ideal synchronous speed have to be specified. They are to be obtained with tapped windings. In such a case it has to be verified that for each speed there is some torque reserve up to the breaking torque of that tapping.

Also for multispeed motors the locked rotor torque on low speed has to be less than the load torque at the desired low speed. For a fan load, at 50% as the low speed, the torque is 25 % and thus the locked rotor torque has to be less than 25%.

We start with the sizing of the magnetic circuit, move on to the selection of stator windings, continue with rotor slotting and cage sizing. The starting and (or) permanent capacitors are defined. Further on the parameter expressions are given and steady state performance is calculated. When optimization design is performed, objective (penalty) functions are calculated and constraints are verified. If their demands are not met the whole process is repeated, according to a deterministic or stochastic optimization mathematical method, until sufficient convergence is reached.

27.2 SIZING THE STATOR MAGNETIC CIRCUIT

As already discussed in Chapter 14, when dealing with design principles of three phase induction machines, there are basically two design initiation constants, based on past experience

- The machine utilisation factor C_u in W/m^3 D_0^2L
- The rotor tangential stress f_t in N/cm^2 or N/m^2

D_0 is the outer stator diameter and L-the stator stack length.

As design optimization methods advance and better materials are produced, C_u and f_t tend to improve slowly. Also, low service duty allows for improved in C_u and f_t.

However, in general, better efficiency requires larger C_u and lower f_t.

Figure 27.1 [1] presents standard data on C_u

$$C_u = D_0^2L \tag{27.1}$$

for the three phase small power IMs.

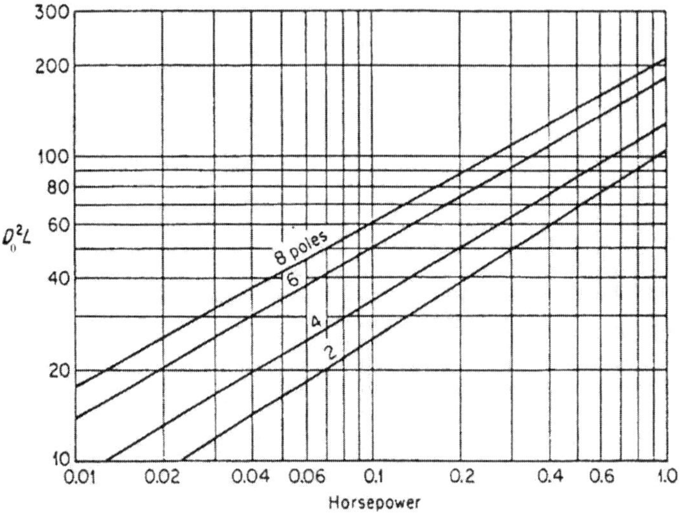

Figure 27.1 Machine utilization factor C_u=D_0^2L (cubic inches) for fractional/horsepower three-phase IMs.

Concerning the rated tangential stress there is not yet a history of its use but it is known that it increases with the stator interior (bare) diameter D_i, with values of around f_t = 0.20 N/cm^2 for D_i = 30 mm to f_t = 1 N/cm^2 for D_i = 70 mm or so , and more.

In general, any company could calculate f_t (D_i) for the 2, 4, 6, 8 single phase IMs fabricated so far and then produce its own f_t database.

The rather small range of f_t variation may be exploited best in our era of computers for optimization design.

The ratio between stator interior (bore) diameter D_i and the external diameter D_o depends on the number of poles, on D_o and on the magnetic (flux densities) and electric (current density) loadings.

Figure 27.2. presents standard data [1] from three sources T. C. Lloyd, P. M. Trickey and Reference 2. In Reference 2, the ratio D_i/D_o is obtained for maximum airgap flux density in the airgap per given stator magnetisation m.m.f. in three phase IMs ($D_i/D_o = 0.58$ for $2p_1 = 2$, 0.65 for $2p_1 = 4$, 0.69 for $2p_1 = 6$, 0.72 for $2p_1 = 8$).

The D_i/D_o values of Reference 2 are slightly larger than those of P. M. Trickey, as they are obtained from a contemporary optimization design method for 3 phase IMs.

In our case study, from Figure 27.1, for 186.5 W (1/4 HP), $2p_1 = 4$ poles, we choose

$$C_u = D_o^2 L = 3.5615 \cdot 10^{-3} m^3$$

with $L/D_o = 0.380$, $D_o = 0.137$ m, $L = 0.053$ m.

The outer stator punching diameter D_o might not be free to choose, as the frames for single phase IMs come into standardized sizes [3].

For 4 poles we choose from fig. 27.2 a kind of average value of the three sets of data $D_i/D_o = 0.60$. Consequently the stator bore diameter $D_i = 0.6 \times 0.137 = 82.8 \times 10^{-3}$ m.

The airgap $g = 0.3$ mm and thus the rotor external diameter $D_{or} = D_i - 2g = (82.8 - 2\times0.3)\times10^{-3} = 82.2 \times 10^{-3}$ m.

The number of slots of stator N_s is chosen as for three phase IMs (the rules for most adequate combinations N_s and N_r established for three phase IMs hold in general also for single phase IMs see chapter 10). Let us consider $N_s = 36$ and $N_r = 30$.

The theoretical peak airgap flux density $B_g = 0.6$-0.75 T.Let us consider $B_g = 0.705$ T. Due to saturation it will be somewhat lower (flattened). Consequently, the flux per pole in the main winding Φ_m is

$$\Phi_m \approx \frac{2}{\pi} \cdot K_{dis} \cdot B_g \cdot \tau \cdot L \tag{27.3}$$

The pole pitch,

$$\tau = \pi \cdot \frac{D_i}{2p_1} = \frac{\pi}{4} \cdot 82.2 \times 10^{-3} \approx 64.57 \times 10^{-3} m \tag{27.4}$$

With $K_{dis} = 0.9$

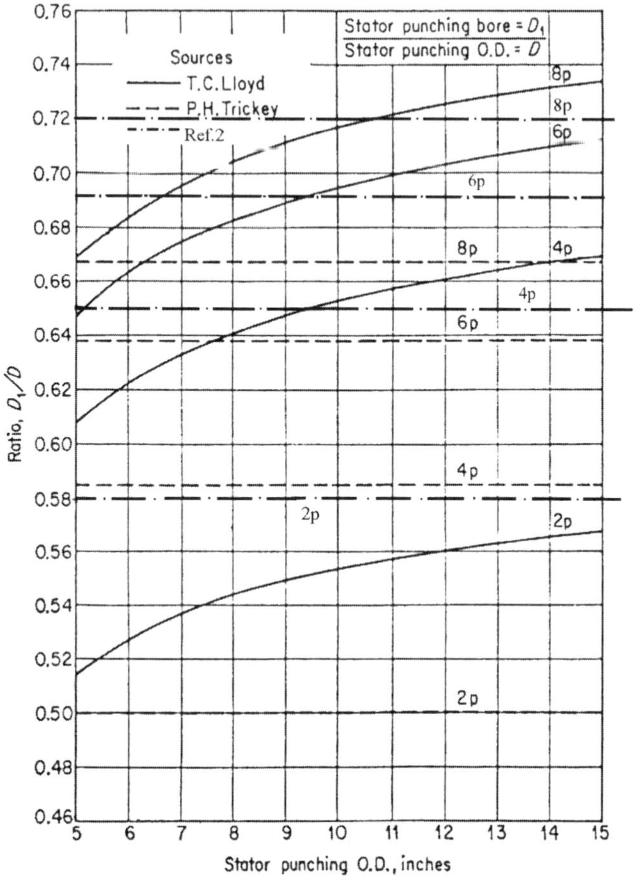

Figure 27.2 Interior/outer diameter ratio D_i/D_o versus D_o.

$$\Phi_m = \frac{2}{\pi} 0.9 \cdot 0.705 \cdot 64.57 \times 10^{-3} \cdot 0.053 \approx 1.382 \times 10^{-3}\,\text{Wb}$$

Once we choose the design stator back iron flux density B_{cs} = 1.3-1.7 T, the back iron height h_{cs} may be computed from

$$h_{cs} = \frac{\Phi_m}{2 \cdot B_{cs} \cdot L} = \frac{1.382 \times 10^{-3}}{2 \cdot 1.5 \times 0.053} = 8.69 \times 10^{-3} \approx 9 \times 10^{-3}\,\text{m} \qquad (27.4)$$

The stator slot geometry is shows on Figure 27.3.

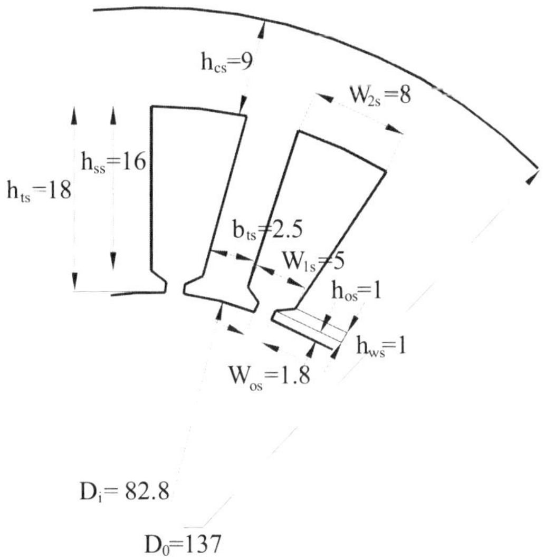

Figure 27.3 Stator slot geometry in mm

The number of slots per pole is $N_s/2p_1 = 36/(2\times2) = 9$. So the tooth width b_{ts} is

$$b_{ts} = \frac{\Phi_m \cdot 2p_1}{N_s \cdot B_{ts} \cdot L} \tag{27.5}$$

With the tooth flux density $B_{ts} \cong (0.8\text{-}1.0)\ B_{cs}$

$$b_{ts} = \frac{1.382\times10^{-3}\times4}{36\cdot1.3\cdot0.053} \approx 2.55\times10^{-3}\,\mathrm{m} \tag{27.6}$$

This value is close to the lowest limit in terms of punching capabilities. Let us consider $h_{os} = 1\times10^{-3}$ m, $w_{os} = 6g = 6\times0.3\times10^{-3} = 1.8\times10^{-3}$ m. Now the lower and upper slot width w_{1s} and w_{2s} are

$$W_{1s} = \frac{\pi(D_i + 2(h_{os} + h_{ws}))}{N_s} - b_{ts} = \left(\frac{\pi(82.8 + 2(1+1))}{36} - 2.55\right)\times10^{-3} \approx 5.00\times10^{-3}\,\mathrm{m} \tag{27.7}$$

$$W_{2s} = \frac{\pi(D_o - 2h_{cs})}{N_s} - b_{ts} = \left(\frac{\pi(137 - 2\times9)}{36} - 2.5\right)\times10^{-3} \approx 8.00\times10^{-3}\,\mathrm{m} \tag{27.8}$$

The useful slot height is

$$h_{ts} = \frac{D_o}{2} - h_{cs} - h_{ws} - h_{os} - \frac{D_i}{2} = \left(\frac{137}{2} - 9 - 2 - \frac{82.8}{2}\right) \times 10^{-3} \quad (27.9)$$

$$\approx 16.0 \times 10^{-3} m$$

So the "active" stator slot area A_s is

$$A_0 = \frac{(W_{1s} + W_{2s})}{2} h_{ss} - \frac{(5.00 + 8.00)}{2} \times 16 \times 10^{-6} = 104 \times 10^{-6} m^2 \quad (27.9)$$

For slots which host both windings or in split phase IMs some slots may be larger than others.

27.3 SIZING THE ROTOR MAGNETIC CIRCUIT

The rotor slots for single phase IMs are either round or trapezoidal or in between (Figure 27.4).

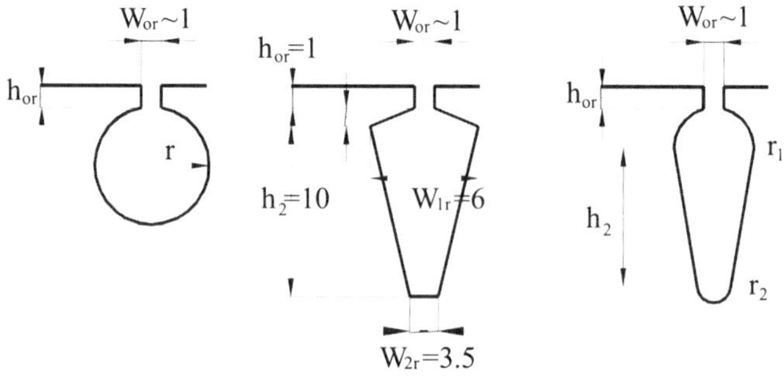

Figure 27.4 Typical rotor slot geometries

With 30 rotor slots, the rotor slot pitch τ_{sr} is

$$\tau_{sr} = \frac{\pi \cdot D_r}{N_r} = \frac{\pi \cdot 82.2 \times 10^{-3} \cdot 4}{30} = 8.6 \times 10^{-3} \quad (27.10)$$

With a rotor tooth width $b_{tr} = 2.6 \times 10^{-3}$ m the tooth flux density B_{tr} is

$$B_{tr} = \frac{\Phi \cdot 2p_1}{N_r \cdot b_{tr} \cdot L} = \frac{1.388 \times 10^{-3} \cdot 4}{30 \cdot 2.6 \cdot 10^{-3} \cdot 0.053} = 1.343T \quad (27.11)$$

The rotor slots useful area is in many cases (35–60) % of that of the stator slot

$$A_r = 0.38 \frac{A_s \cdot N_s}{N_r} = 0.38 \times 104 \times 10^{-6} \times \frac{36}{30} = 47.42 \times 10^{-6} m^2 \quad (27.12)$$

The trapezoidal rotor slot (Figure 27.4), with

$$h_{or} = W_{or} = 1\times 10^{-3}\,m; h_{orl} = 1\times 10^{-3}\,m$$
$$\text{has } W_{1r} \approx \tau_{sr} - b_{tr} = (8.6 - 2.6)\times 10^{-3} = 6\times 10^{-3}\,m \quad (27.13)$$

Adopting a slot height $h_2 = 10 \times 10^{-3}$ m we may calculate the rotor slot bottom width W_{2r}

$$W_{2r} = \frac{2A_r}{h_2} - W_{1s} = \frac{2\cdot 47.40\times 10^{-3}}{10\times 10^{-3}} - 6\times 10^{-3} = 3.48\times 10^{-3}\,m \quad (27.14)$$

A "geometrical" verification is now required.

$$W_{2r} = \frac{\pi[D_r - (h_{or} + h_{orl} + h_2)\cdot 2]}{N_r} - b_{tr} =$$
$$= \frac{\pi[82.2 - 2(1+1+10)]\times 10^{-3}}{30} - 2.6\times 10^{-3} = 3.49\times 10^{-3}\,m \quad (27.15)$$

As (27.14) and (27.15) produce the same value of slot bottom width, the slot height h_2 has been chosen correctly. Otherwise h_2 should have been changed until the two values of W_{2r} converged.

To avoid such an iterative computation (27.14)–(27.15) could be combined into a second order equation with h_2 as the unknown after the elimination of W_{2r}.

A round slot with a diameter $d_r = W_{2r} = 6 \times 10^{-3}$ m would have produced an area $A_r = \frac{\pi}{4}(6\times 10^{-3})^2 = 28.26\times 10^{-6}\,m^2$.

This would have been too small a value unless copper bars are used instead of aluminum bars.

The end ring area A_{ring} is

$$A_{ring} \approx A_r \times \frac{1}{2\sin\frac{\pi\cdot p_1}{N_r}} = 47.42\times 10^{-6}\frac{1}{2\sin\frac{\pi\cdot 2}{30}} = 114\times 10^{-6}\,m^2 \quad (27.16)$$

With a radial height $b_r = 15 \times 10^{-3}$ m, the ring axial length $a_r = \frac{A_{ring}}{b_r} = 7.6\times 10^{-3}\,m$.

27.4 SIZING THE STATOR WINDINGS

By now the number of slots in the stator is known $N_s = 36$. As we do have a permanent capacitor motor the auxiliary winding is always operational. It thus seems natural to allocate both windings about same number of slots in fact 20/16. In general single layer windings with concentrated coils are used. (Figure 27.5)

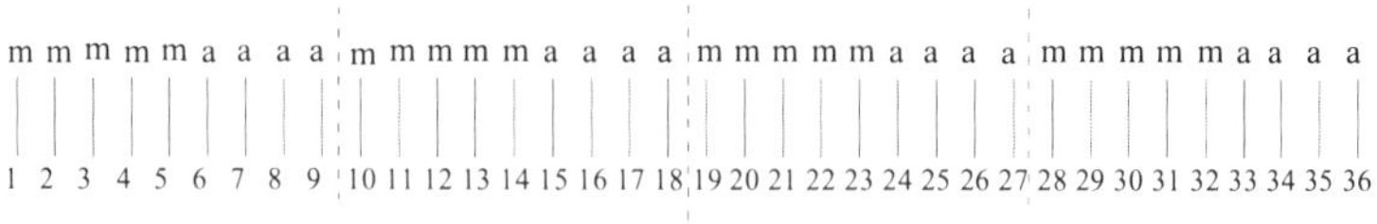

Figure 27.5 Stator winding m-main phase; a-auxiliary phase.

With identical coils per slot and phase; the winding factors for the two phases (Chapter 4) are

$$K_{wm} = \frac{\sin\left(q_m \frac{\alpha_s}{2}\right)}{q_m \sin\frac{\alpha_s}{2}} = \frac{\sin\left(5\times\frac{\pi}{36}\right)}{5\sin\frac{\pi}{36}} = 0.9698$$

$$K_{wa} = \frac{\sin\left(q_a \frac{\alpha_s}{2}\right)}{q_a \sin\frac{\alpha_s}{2}} = \frac{\sin\left(4\times\frac{\pi}{36}\right)}{4\sin\frac{\pi}{36}} = 0.9810 \quad (27.17)$$

There are 5 ($q_m = 5$) slots per pole per phase for the main winding and 4 ($q_a = 4$) for the auxiliary one.

In case the number of turns/phase in various slots is not the same, the winding factor can be calculated as shown below.

As detailed in chapter 4, two types of sinusoidal windings may be built (Figure 27.6)

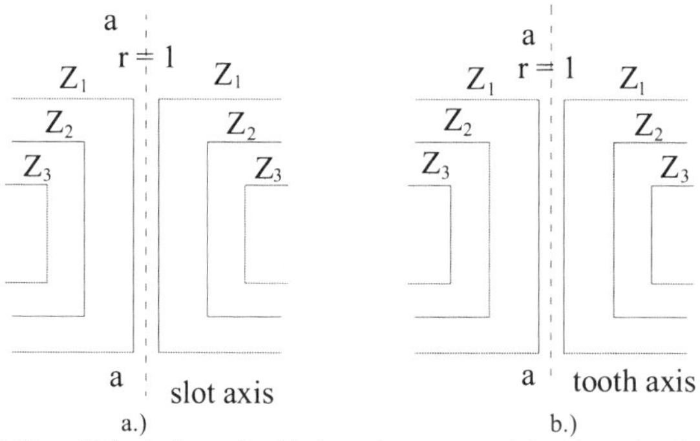

Figure 27.6 Sinusoidal windings a.) with slot axis symmetry; b.) with tooth axis symmetry.

When the total number of slots per pole per phase is an odd number, concentrated coils with slot axis symmetry seem adequate. In contrast, for even number of slots/pole/phase, concentrated coils with tooth axis symmetry are recommended.

Should we have used such windings for our case with $q_m = 5$ and $q_a = 4$, slot axis symmetry would have applied to the main winding and tooth axis symmetry to the auxiliary winding. A typical coil group is shown on Figure 27.7.

From Figure 27.7 the angle between the axis a-a (Figure 27.6-27.7) and the kth slot, $\beta_{k\nu}$ (for the νth harmonic), is

$$\beta_{k\nu} = \gamma \frac{\pi}{N_s} \nu + (k-1)\frac{2\pi\nu}{N_s} \tag{27.18}$$

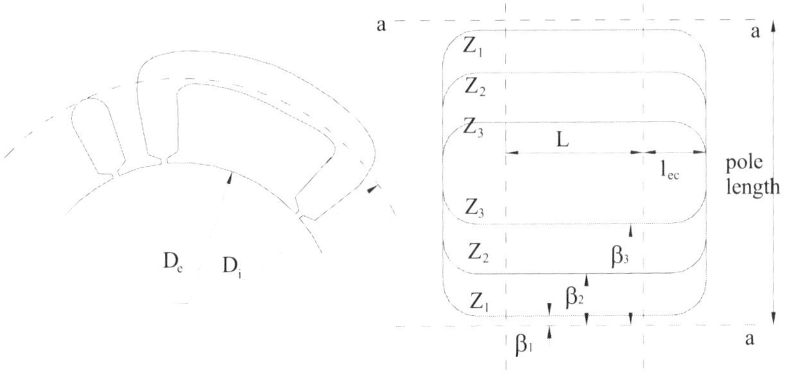

Figure 27.8 Typical "overlapping" winding

The effective number of conductors per half a pole Z_ν' is [4]

$$Z_\nu' = \sum_{k=1}^{n} Z_k \cos\beta_{k\nu}$$
$$N_\nu = 4p_1 \cdot Z_\nu' \tag{27.19}$$

N_ν is the effective number of conductors per phase.
The winding factor $K_{w\nu}$ is simply

$$K_{w\nu} = \frac{Z_\nu'}{\sum_{k=1}^{n} Z_k} = \frac{\sum_{k=1}^{n} Z_k \cos\beta_{k\nu}}{\sum_{k=1}^{n} Z_k} \tag{27.20}$$

The length of the conductors per half a pole group of coil, l_{cn}, is [3]

$$l_{cn} \approx \sum_{k=1}^{n} Z_k \left[L + (\pi - 2)L_{ec} + D_e\left(\frac{\pi}{2p_1} - \beta_{k1}\right)\right] \tag{27.21}$$

So the resistance per phase R_{phase} is

$$R_{phase} = \rho_{Co} \frac{4p_1 l_{cn}}{A_{con}} \tag{27.22}$$

A_{con} is the conductor (magnet wire) cross-section area. A_{con} depends on the total number of conductors per phase, slot area and design current density.

The problem is that, though the supply current at rated load may be calculated as

$$I_s = \frac{P_n}{\eta_n V_s \cos\varphi} = \frac{186.5}{0.7 \times 115 \times 0.98} = 2.36A\ , \qquad (27.23)$$

with an assigned value for the rated efficiency, η_n, still the rated current for the main and auxiliary currents $\underline{I}_m$, $\underline{I}_a$ are not known at this stage.

$\underline{I}_m$ and $\underline{I}_a$ should be almost 90^0 phase shifted for rated load and

$$\underline{I}_s = \underline{I}_m + \underline{I}_a \qquad (27.24)$$

Now as the ratio $a = \frac{I_m}{I_a} = \frac{N_a}{N_m}$, for symmetry, we can assume that

$$I_m \approx \frac{I_s}{\sqrt{1+\left(\frac{1}{a}\right)^2}} \qquad (27.25)$$

With a the turns ratio in the interval a = 1.0-2, in general $I_m = I_s \cdot (0.700 - 0.90)$.

The number of turns of the main winding N_m is to be determined by observing that the e.m.f. in the main winding $E_m = (0.96 - 0.98) \cdot V_s$ and thus with (27.3)

$$E_m = \pi\sqrt{2}\Phi_m N_m k_{wm} f_{1n} \qquad (27.26)$$

Finally

$$N_m = \frac{0.97 \times 115}{\pi\sqrt{2}1.382 \times 10^{-3} \times 0.9698 \times 60} = 313\text{turns}$$

In our case, the main winding has 10 ($p_1 q_m = 2 \times 5 = 10$) identical coils. So the number of turns per slot n_{sm} is

$$n_{sm} = \frac{N_m}{q_m p_1} = \frac{313}{5 \times 2} = 31\text{turns / coil} \qquad (27.27)$$

Assuming the turn ratio is a = 1.5, the number of turns per coil in the auxiliary winding is

$$n_{sa} = \frac{N_m \times a \times k_{wm}}{q_a \times p_1 \times k_{wa}} = \frac{310 \times 1.5 \times 0.9698}{4 \times 2 \times 0.9810} \approx 57\text{turns / coil} \qquad (27.28)$$

As the slots are identical and their useful area is (from 27.9) $A_s = 104 \times 10^{-6}$ m^2, the diameters of the magnetic wire used in two windings are

$$d_m = \sqrt{\frac{4}{\pi} \cdot \frac{A_s K_{fill}}{n_{sm}}} = 10^{-3}\sqrt{\frac{4}{3.14} \cdot \frac{104 \times 0.4}{31}} \approx 1.3 \times 10^{-3} m$$
$$d_a = \sqrt{\frac{4}{\pi} \cdot \frac{A_s K_{fill}}{n_{sa}}} = 10^{-3}\sqrt{\frac{4}{3.14} \cdot \frac{104 \times 0.4}{57}} \approx 1.0 \times 10^{-3} m \quad (27.29)$$

The filling factor was considered rather large $K_{fill} = 0.4$.
The predicted current density in the two windings would be

$$j_{Com} = \frac{I_m}{\frac{\pi d_m^2}{4}} = \frac{2.36}{\left(\sqrt{1+\left(\frac{1}{1.5}\right)^2}\right) \cdot \pi \cdot \frac{1.3^2}{4} \cdot 10^{-6}} = 1.4847 \times 10^6 \, A/m^2$$
$$j_{Coa} = \frac{I_m}{\frac{a \cdot \pi \cdot d_a^2}{4}} = \frac{1.9696}{\frac{1.5 \times \pi \cdot 1^2 \times 10^{-6}}{4}} = 1.6726 \times 10^6 \, A/m^2 \quad (27.30)$$

The forecasted current densities, for rated power, are rather small so good efficiency (for the low power considered here) is expected.

An initial value for the permanent capacitor has to be chosen.

Let us assume complete symmetry for rated load, with $I_{an} = \frac{I_{mn}}{a} = \frac{1.96}{1.5} = 1.3A$.

Further on

$$V_{an} = V_{mn} \cdot a = 115 \times 1.5 = 172V \quad (27.31)$$

The capacitor voltage V_c is

$$V_c = \sqrt{V_s^2 + V_{an}^2} = 115\sqrt{1 + 1.5^2} = 207.3V \quad (27.32)$$

Finally the permanent capacitor C is

$$C = \frac{I_a}{\omega_1 V_C} = \frac{1.3}{2 \cdot \pi \cdot 60 \cdot 207.3} \approx 16\mu F \quad (27.33)$$

We may choose a higher value of C than 16 μF (say 25 μF) to increase somewhat the breakdown and the starting torque.

27.5 RESISTANCES AND LEAKAGE REACTANCES

The main and auxiliary winding resistances R_{sm} and R_{sa} (27.22) are

$$R_{sm} = \rho_{Co} \cdot 4 \cdot \pi \frac{l_{cnm}}{\frac{\pi d_m^2}{4}}$$
$$R_{sa} = \rho_{Co} \cdot 4 \cdot \pi \frac{l_{cna}}{\frac{\pi d_a^2}{4}} \quad (27.34)$$

The length of the coils/half a pole for main and auxiliary windings l_{cnm} (27.21) are

$$l_{cnm} \approx n_{cm}\left[\left(2+\frac{1}{2}\right)L + 3(\pi-2)L_{ec} + D_e\left(\frac{3\pi}{2p_1} - \frac{2\pi}{36} - \frac{4\pi}{36}\right)\right] =$$
$$= 31\times\left[\left(2+\frac{1}{2}\right)\times 0.053 + 3(\pi-2)\times 0.025 + 0.14\left(\frac{3\pi}{4} - \frac{2\pi}{36} - \frac{4\pi}{36}\right)\right] = 9.805\text{m} \quad (27.35)$$

$$l_{cna} \approx n_{ca}\left[2L + 3(\pi-2)L_{ec} + D_e\left(\frac{2\pi}{2p_1} - \frac{2\pi}{36} - \frac{4\pi}{36}\right)\right] =$$
$$= 57\times\left[2\times 0.053 + 3(\pi-2)\times 0.025 + 0.14\left(\frac{2\pi}{4} - \frac{\pi}{18} - \frac{\pi}{9}\right)\right] = 11.38\text{m} \quad (27.36)$$

Finally

$$R_{sm} = \frac{2.1\times 10^{-8}\times 4\times 2\times 9.805}{\frac{\pi}{4}\times 1.3^2\times 10^{-6}} = 1.2417\Omega$$
$$R_{sa} = \frac{2.1\times 10^{-8}\times 4\times 2\times 11.38}{\frac{\pi}{4}\times 1^2\times 10^{-6}} = 2.435\Omega$$

The main winding leakage reactance X_{sm}

$$X_{sm} = X_{ss} + X_{se} + X_{sd} + X_{skew} \quad (27.37)$$

where X_{ss}-the stator slot leakage reactance
X_{se}-the stator end connection leakage reactance
X_{sd}-the stator differential leakage
X_{skew}-skewing reactance

$$X_{se} + X_{ss} = 2\mu_0\omega_1 \frac{W_1^2 \cdot L}{p \cdot q}(\lambda_{ss} + \lambda_{se}) \quad (27.38)$$

with (chapter 6 and Figure 27.3)

$$\lambda_{ss} = \frac{2h_{ss}}{3(W_{1s} + W_{2s})} + \frac{2h_{ws}}{W_{1s} + W_{2s}} + \frac{h_{os}}{W_{os}} = \frac{2\times16}{3(5+8)} + \frac{2\times1}{5+8} + \frac{1}{1.8} = 1.53 \quad (27.39)$$

Also (chapter 6, equation (6.28))

$$\lambda_{se} = \frac{0.67\cdot\frac{q_m}{2}}{L}(l_{ec} - 0.64\cdot\tau) \quad (27.40)$$

It is $q_m/2$ as the q coil/pole are divided in two sections whose end connections go in opposite directions (Figure 27.5).

The end connection average length l_{ec} is

$$l_{ec}^m = \frac{l_{cn}\times 2}{q_m n_{cm}} - L \quad (27.41)$$

With l_{cn} from (27.21)

$$l_{ec}^m = \frac{9.805\times2}{5\times31} - 0.053 = 0.0735\text{m}$$

$$\lambda_{se} = \frac{0.67\times\frac{5}{2}}{0.053}(0.0735 - 0.64\times0.065) = 1.008$$

The differential leakage reactance X_{sd} (chapter 6, Equation (6.2) and (6.9) and (Figure 6.2-Figure 6.5)) is

$$\frac{X_{sd}}{X_{mm}} = \sigma_{ds}\cdot\Delta_d = 2.6\times10^{-2}\times0.92 = 2.392\times10^{-2} \quad (27.42)$$

for $q_m = 5$, $N_s = 36$, $N_r = 30$, one slot pitch skewing of rotor.

X_{mm} is the nonsaturated magnetization reactance of the main winding (Chapter 5, Equation (5.115))

$$X_{mm} = \omega_1\frac{4\mu_0}{\pi^2}(W_m\cdot K_{wm1})^2\frac{L\cdot\tau}{p_1\cdot g\cdot K_C} =$$

$$= 2\pi\cdot60\frac{4\cdot4\pi\times10^{-7}}{\pi^2}(10\times31\times0.9698)^2\frac{0.053\cdot0.065}{2\cdot0.3\times10^{-3}\cdot1.3} = 72.085\Omega$$

K_C-Carter coefficient; g-the airgap.

$$\frac{X_{skew}}{X_{mm}} = (1 - K_{skew}^2) = 1 - \frac{\sin^2\frac{\alpha_{skew}}{2}}{(\alpha_{skew})^2} = 1 - \frac{\sin^2\frac{\pi}{18}}{\left(\frac{\pi}{18}\right)^2} = 9.108\times10^{-3} \quad (27.43)$$

Finally the main winding leakage reactance X_{sm} is

$$X_{sm} = 2\times 4\pi\times 10^{-7}\frac{2\pi\cdot 60\times 310^2\cdot 0.053\cdot(1.53+1.008)}{2\times 5}+$$
$$+72.085\cdot(23.92+9.108)\cdot 10^{-3} = 4.253\Omega$$

The auxiliary winding is placed in identical slots and thus the computation process is similar. A simplified expression for X_{sa} in this case is

$$X_{sa} \approx X_{sm}\left(\frac{W_a}{W_m}\right) = 4.253\times\left(\frac{57\times 8}{31\times 10}\right)^2 = 9.204\Omega \quad (27.44)$$

The rotor leakage reactance, after its reduction to the main winding (chapter 6, equations (6.86)-(6.87)) is

$$X_{rm} = X_{rmd} + X'_{bem} = \omega_1\left(L_{rmd} + L'_{bem}\right) \quad (27.45)$$

The rotor-skewing component has been "attached" to the stator and the zig-zag component is lumped into differential one X_{rmd}.

From Equation (6.16) (Chapter 6)

$$X_{rmd} = \sigma_{dr0}X_{mm} = 2.8\times 10^{-2}X_{mm} \quad (27.46)$$

With $N_r/p_1 = 30/2 = 15$ and one slot pitch skewing, from Figure 6.4

$$\sigma_{dr0} = 2.8\times 10^{-2}$$

The equivalent bar-end ring leakage inductance L'_{ben} is (Equation 6.86)

$$L'_{bem} = L_{bem}\frac{12\cdot K_{wm}^2 W_m^2}{N_r} \quad (27.47)$$

From (6.91)-(6.92)
$$L_{ben} = \mu_0 l_b\lambda_b + 2\mu_0\lambda_{ei}\cdot 2\pi\cdot\frac{D_{ir}}{N_r} \quad (27.48)$$

With the rotor bar (slot) and end ring permeance coefficients, λ_b and λ_{ei} (from (6.18) and (6.46) and Figure 27.5)

$$\lambda_b = \frac{h_{or}}{W_{or}} + \frac{2h_{or1}}{(W_{o1}+W_{o2})} + \frac{2h_2}{3(W_{1r}+W_{2r})} = \frac{1}{1} + \frac{1\times 2}{1+6} + \frac{2\times 10}{3(6+3.5)} = 1.987 \quad (27.49)$$

$$\lambda_{ei} = \frac{2.3\times D_{ir}}{4\cdot N_r\cdot L\cdot\sin^2\left(\frac{\pi p_1}{N_r}\right)}\log\frac{4.7\cdot D_{ir}}{a_r + 2\cdot b_r} \quad (27.50)$$

With $D_{ir} \approx D_r - b = (82.8-15)\times 10^{-3} = 67.8\times 10^{-3}\text{m}$, $a_r = 7.6\times 10^{-3}\text{m}$ and $b_r = 15\times 10^{-3}\text{m}$ from (27.16)

$$\lambda_{ei} = \frac{2.3 \times 67.8 \times 10^{-3}}{4 \cdot 30 \cdot 0.053 \cdot \sin^2\left(\frac{\pi}{15}\right)} \log\left(\frac{4.7 \times 67.8}{2 \times 15 + 7.6}\right) = 0.579$$

so from (27.48)

$$L_{ben} = 1.256 \times 10^{-6}\left(0.053 \times 1.987 + 0.579\frac{2\pi \times 67.8 \times 10^{-3}}{30}\right) = 0.1425 \times 10^{-6}\,H$$

Now from (27.45) and (24.76)

$$X_{rm} = 2.8 \times 10^{-2} X_{mm} + \omega_1 L_{ben} \frac{12K_{wm}^2 W_m^2}{N_r} =$$

$$= 2.8 \times 10^{-2} \times 72.085 + 2\pi \times 60 \times 0.1425 \times 10^{-6} \frac{12 \times 0.9698^2 \times 310^2}{30} = \tag{27.51}$$

$$= 3.961\Omega$$

The rotor cage resistance

$$R_{rm} = R_{be} \frac{12K_{wm}^2 W_m^2}{N_r} \tag{27.52}$$

With R_{be} (equation (6.63))

$$R_{be} = R_b + \frac{R_{ring}}{2 \cdot \sin^2\left(\frac{\pi \cdot p_1}{N_r}\right)} \tag{27.53}$$

$$\begin{aligned} R_b &= \rho_{Co} \frac{l_b}{A_r} = 3 \cdot 10^{-8} \frac{0.065}{47.42 \times 10^{-6}} = 4.112 \times 10^{-5}\Omega \\ R_{ring} &= \rho_b \frac{l_{ring}}{A_{ring}} = 3 \times 10^{-8} \frac{2\pi \times 67.8 \times 10^{-3}}{1.14 \times 10^{-4} \times 30} = 3.735 \times 10^{-6}\Omega \end{aligned} \tag{27.54}$$

Finally

$$R_{rm} = \left(4.112 \times 10^{-5} + \frac{3.735 \times 10^{-6}}{2 \cdot \sin^2 \frac{\pi \cdot 2}{30}}\right) \cdot 12 \times \frac{0.9698^2 \cdot 310^2}{30} = 3.048\Omega$$

Note The rotor resistance and leakage reactance calculated above are not affected by the skin effects.

27.6 THE MAGNETIZATION REACTANCE X_{mm}

The magnetisation reactance X_{mm} is affected by magnetic saturation which is dependent on the resultant magnetisation current I_m.

Due to the symmetry of the magnetisation circuit it is sufficient to calculate the functional X_m (I_m) for the case with current in the main winding (I_a = 0, bare rotor).

The computation of $X_m(I_m)$ may be performed analytically (see Chapter 6 or Ref. 4) or by F.E.M. For the present case, to avoid lengthy calculations let us consider X_{mm} constant for a moderate saturation level $(1+K_s) = 1.5$

$$X_{mm} = \left(X_{mm}\right)_{unsat}\frac{1}{1+K_s} = \frac{72.085}{1.5} = 48.065\Omega \tag{27.55}$$

27.7 THE STARTING TORQUE AND CURRENT

The theory behind the computation of starting torque and current is presented in chapter 24, paragraph 24.5

$$\left(I_s\right)_{S=1} = I_{sc}\sqrt{\left[1+\frac{t_{es}}{a}\frac{\left(\omega C\left|\underline{Z}_{sc}\right|\cdot a^2 - \sin\varphi_{sc}\right)}{\cos\varphi_{sc}}\right]^2 + \frac{t_{es}^2}{a^2}} \tag{27.56}$$

$$t_{es} = \frac{T_{es}}{\left(\frac{2p_1}{\omega_1}I_{sc}^2\left(R_{r+}\right)_{S=1}\right)} = \frac{K_a\cdot\sin\varphi_{sc}}{a\left(1+K_a^2+2K_a\cos\varphi_{sc}\right)}$$

$$K_a = \frac{1}{\omega\cdot C\cdot a^2\cdot\left|\underline{Z}_{sc}\right|};\ I_{sc} = \frac{V_s}{2\cdot\left|\underline{Z}_{sc}\right|}$$

$$Z_{sc} = \sqrt{\left(R_{sm}+R_{rm}\right)^2+\left(X_{sm}+X_{rm}\right)^2} = \\ = \sqrt{\left(1.2417+3.048\right)^2+\left(4.235+3.96\right)^2} = 9.26\Omega \tag{27.57}$$

$$\cos\varphi_{sc} = 0.463$$
$$\sin\varphi_{sc} = 0.887$$

$$K_a = \frac{1}{2\pi\cdot 60\times 25\times 10^{-6}\times 1.5^2\times 9.26} = 5.095$$

$$I_{sc} = \frac{115}{2\times 9.26} = 6.2095\text{A};\ \left(R_{r+}\right)_{S=1} = R_{rm} = 3.048\Omega$$

$$T_{sc} = \frac{2\times 2}{2\pi\cdot 60}\times 6.2095^2 \times 3.048 = 2.495\text{Nm}$$

The relative value of starting torque t_{es} is

$$t_{es} = \frac{5.095\times 0.887}{1.5\left(1+5.095^2+2\times 5.095\times 0.463\right)} = 0.095$$

$$T_{es} = t_{es}T_{sc} = 0.095\times 2.495 = 0.237\text{Nm}$$

The rated torque T_{en}, for an alleged rated slip $S_n = 0.06$, would be

$$T_{en} \approx \frac{P_n p_1}{\omega_1\cdot(1-S_n)} = \frac{186.5\times 2}{2\pi\cdot 60\cdot(1-0.06)} = 1.053\text{Nm} \tag{27.58}$$

So the starting torque is rather small as the permanent capacitance $C_a = 25 \times 10^{-6}$ F is too small.

The starting current (27.56) is

$$(I_s)_{S=1} = 6.2095\sqrt{\left[1+\frac{0.095}{1.5}\left(\frac{\frac{1}{5.095}-0.887}{0.463}\right)\right]^2+\left(\frac{0.095}{1.5}\right)^2} \approx 5.636\text{A}$$

The starting current is not large (the presumed rated source current I_{sn} = 2.36 A,(Equation 27.23)) but the starting torque is small.

The result is typical for the permanent (single) capacitor IM.

27.8 STEADY STATE PERFORMANCE AROUND RATED POWER

Though the core losses and the stray load losses have not been calculated, the computation of torque and stator currents for various slips for S = 0.04-0.20 may be performed as developed in chapter 24, paragraph 24.3.

To shorten the presentation we will illustrate this point by calculating the currents and torque for S = 0.06.

First the impedances $\underline{Z}_+^m, \underline{Z}_-^m$ and $\underline{Z}_a^m$ ((24.11)-(24.14)) are calculated

$$\underline{Z}_+ = R_{sm} + j\cdot X_{sm} + j\frac{X_{mm}\left(jX_{rm}+\frac{R_{rm}}{S}\right)}{\frac{R_{rm}}{S}+j(X_{mm}+X_{rm})} =$$

$$= 1.2417 + j\cdot 4.253 + \frac{j\cdot 48.056\left(j\cdot 3.96+\frac{3.048}{0.06}\right)}{\frac{3.048}{0.06}+j\cdot(48.056+3.96)} = 23.43 + j\cdot 29.583 \tag{27.59}$$

$$\underline{Z}_{-} = R_{sm} + j\cdot X_{sm} + j\cdot\frac{X_{mm}\left(j\cdot X_{rm} + \frac{R_{rm}}{(2-S)}\right)}{\frac{R_{rm}}{(2-S)} + j\cdot(X_{mm} + X_{rm})} \approx$$

$$\approx 1.2417 + j\cdot 4.253 + j\cdot\frac{48.056\cdot\left(j\cdot 3.96 + \frac{3.048}{(2-0.06)}\right)}{\frac{3.048}{(2-0.06)} + j\cdot(48.056 + 3.96)} \approx 2.7 + j\cdot 7.912 \quad (27.60)$$

$$\underline{Z}_{a}^{m} = \frac{1}{2}\left(\frac{R_{sa}}{a^2} - R_{sm}\right) + j\cdot\frac{1}{2}\left(\frac{X_{sa}}{a^2} - X_{sm}\right) - j\cdot\frac{1}{2\cdot a^2\cdot\omega\cdot C} =$$

$$= \frac{1}{2}\left(\frac{2.435}{1.5^2} - 1.2417\right) + 0 - \frac{j}{2\times 1.5^2\times 2\pi\times 60\times 25\times 10^{-6}} =$$

$$= -0.08 - j\cdot 23.6 \approx -j\cdot 23.6$$

The current components $\underline{I}_{m+}$ and $\underline{I}_{m-}$ ((24.16)-(24.17)) are

$$\underline{I}_{m+} = \frac{V_s}{2}\cdot\frac{\left[\left(1-\frac{j}{a}\right)\underline{Z}_{-} + 2\underline{Z}_{a}^{m}\right]}{\underline{Z}_{+}\cdot\underline{Z}_{-} + \underline{Z}_{a}^{m}(\underline{Z}_{+} + \underline{Z}_{-})} = 1.808 - j\cdot 2.414 \quad (27.62)$$

$$\underline{I}_{m-} = \frac{V_s}{2}\cdot\frac{\left[\left(1+\frac{j}{a}\right)\underline{Z}_{+} + 2\underline{Z}_{a}^{m}\right]}{\underline{Z}_{+}\cdot\underline{Z}_{-} + \underline{Z}_{a}^{m}(\underline{Z}_{+} + \underline{Z}_{-})} = 0.304 - j\cdot 8.197\times 10^{-3} \quad (27.63)$$

Now the torque components T_+ and T_- are computed from (24.18)-(24.19)

$$T_{e+} = \frac{2p_1}{\omega_1} I_{m+}^2\left[R_e(\underline{Z}_{+}) - R_{sm}\right] \quad (27.64)$$

$$T_{e-} = -\frac{2p_1}{\omega_1} I_{m-}^2\left[R_e(\underline{Z}_{-}) - R_{sm}\right] \quad (27.65)$$

$$T_e = T_{e+} + T_{e-}$$

The source current I_s is

$$\underline{I}_{s} = \underline{I}_{m} + \underline{I}_{a} = \underline{I}_{m+} + \underline{I}_{m-} + j\cdot\frac{(\underline{I}_{m+} - \underline{I}_{m-})}{a} = 3.721 - j\cdot 1.422 \quad (27.66)$$

The source power factor becomes

$$\cos\varphi_1 = \frac{R_e(I_s)}{I_s} = 0.934 \tag{27.67}$$

$$T_{e+} = \frac{2\times 2}{2\pi\times 60}\times 9.096\cdot[23.43-1.2417] = 2.1425\text{Nm}$$

$$T_{e-} = -\frac{2\times 2}{2\pi\times 60}\times 0.09248\cdot[2.7-1.2417] = -0.967\times 10^{-3}\text{Nm}$$

Note The current is about 60 % higher than the presumed rated current (2.36 A) while the torque is twice the rated torque $T_{en} \approx 1\text{Nm}$.

It seems that when the slip is reduced gradually, perhaps around 4 % (S = 0.04) the current goes down and so does the torque, coming close to the rated value, which corresponds, to the rated power P_n (27.58).

On the other hand, if the slip is gradually increased the breakdown torque region is reached.

To complete the design the computation of core, stray load and mechanical losses is required. Though the computation of losses is traditionally performed as for the three phase IMs, the elliptic travelling field of single phase IM leads to larger losses [5]. We will not follow this aspect here in further detail.

We can now consider the preliminary electromagnetic design finished. Thermal model may than be developed as for the three-phase IM (Chapter 12).

Design trials may now start to meet all design specifications. The complexity of the nonlinear model of the single phase IM makes the task of finding easy ways to meet, say, the starting torque and current, breakdown torque and providing for good efficiency, rather difficult.

This is where the design optimization techniques come into play.

However to cut short the computation time of optimization design, a good preliminary design is useful and so are a few design guidelines based on experience (see [1]).

27.9 GUIDELINES FOR A GOOD DESIGN

- In general a good value for the turn ratio a, lies in the interval between 1.5 to 2.0 except for reversible motion when a = 1 (identical stator windings).
- The starting and breakdown torques may be considered proportional to the number of turns of main winding squared.
- The maximum starting torque increases with the turn ratio a.
- The flux densities in various parts of the magnetic circuit are inversely proportional to the number of turns in the main winding, for given source voltage.
- The breakdown torque is almost inversely proportional to the sum $R_{sm} + R_{rm} + X_{sm} + X_{rm}$.
- When changing gradually the number of turns in the main winding, the rated slip varies with W^2_m.
- The starting torque may be increased, up to a point, in proportion to rotor resistance R_{rm} increase.

- For a given motor, there is a large capacitor C_{ST} which could provide maximum starting torque and another one C_{SA} to provide, again at start, maximum torque/current. A value between C_{ST} and C_{SA} is recommended for best starting performance. For running conditions a smaller capacitor C_a is needed. In permanent capacitor IMs a value C'_a closer to C_a ($C'_a > C_a$) is generally used.
- The torque varies with the square root of stack length. If stack length variation ratio K is accompanied by the number of turns variation by $1/\sqrt{K}$, the torque remains almost unchanged.

27.10 OPTIMIZATION DESIGN ISSUES

Optimization design implies

- A machine model for analysis (as the one described in previous paragraphs);
- Single or multiple objective functions and constraints;
- A vector of initial independent variables (from a preliminary design) are required in nonevolutionary optimization methods.
- A method of optimization (search of new variable vectors until the best objective function value is obtained).

Typical single objective functions F are

F1-maximum efficiency without excessive material cost

F2-minimum material (iron, copper, aluminum, capacitor) cost for an efficiency above a threshold value.

F3-maximum starting torque

F4-minimum global costs (materials plus loss capitalized costs for given duty cycle over the entire life of the motor).

A combination of the above objectives functions could also be used in the optimization process.

A typical variable vector X might contain

1. Outer rotor diameter D_r
2. Stator slot depth h_{st};
3. Stator yoke height h_{cs};
4. Stator tooth width b_{ts};
5. Stack length L;
6. Airgap length g;
7. Airgap flux density B_g;
8. Rotor slot depth h_{rt};
9. Rotor tooth width b_{tr};
10. Main winding wire size d_m;
11. Auxiliary winding wire size d_a;
12. Capacitance C_a;
13. Effective turns ratio a.

Variables 10-12 vary in steps while the others are continuous.

Typical constraints are

- Starting torque T_{es}

- Starting current $(I_s)_{S=1}$
- Breakdown torque T_{eb}
- Rated power factor $\cos\varphi_n$
- Stator winding temperature rise ΔT_m
- Rated slip S_n
- Slot fullness k_{fill}
- Capacitor voltage V_C
- Rotor cage maximum temperature T_{rmax}

Chapter 18 of this book presented in brief quite a few optimization methods. For more information, see Reference 6.

Table 27.1 Design constraints

Constraint	Limit	Standard	F_1	F_2	F_3
Power factor	> 0.92	0.96	0.96	0.95	0.94
Main winding temperature [^{0}C]	< 90	86.2	82.2	86.0 (+)	84.8
Max. torque [Nm]	> 1.27	1.29	1.27 (+)	1.28 (+)	1.49
Start torque [Nm]	0.84	0.878	0.876	0.84 (+)	1.06
Start current [A]	< 11.5	10.5	10.9	11.5 (+)	11.5 (+)
Slot fullness $k_{fill} \times \frac{4}{\pi}(\text{to } d^2)$	< 0.8	0.78	0.8 (+)	0.66	0.8 (+)
Capacitor voltage [V]	< 280	263	264	246	261
Efficiency	> 0.826	0.826	0.859	0.828 (+)	0.836

According to most of them the objective function F (x) is augmented with the SUMT, [7]

$$P(X_k, r_k) = F(X_k) + r_k \sum_{j=1}^{m} \frac{1}{G_j(X_k)} \tag{27.68}$$

where the penalty factor r_k is gradually decreased as the optimization search counter k increases.

There are many search engines which can change the initial variable vector towards a global optimum for P (X_k, r_k) (Chapter 18).

Hooke-Jeeves modified method is a good success example for rather moderate computation time efforts.

Reference 8 presents such an optimization design attempt for a 2 pole, 150 W, 220 V, 50 Hz motor with constraints as shown in Table 27.1.

Efficiency (F_1) and material cost (F_2) evolution during the optimization process is shown in Figure 27.8 a). [8] Stack length evolution during F_1 and F_2 optimization process in Figure 27.8 b) shows an increase before decreasing towards the optimum value.

Also, it is interesting to note that criteria F_1 (max. efficiency) and F_2 (minimum material costs) lead to quite different wire sizes both in the main and auxiliary windings (Figure 27.9 a, b).

A 3.4% efficiency increase has been obtained in this particular case. This means about 3.4% less energy input for the same mechanical work.

It has to be noted that reducing the core losses by using better core material and thermal treatments (and various methods of stray load loss reduction) could lead to further increases in efficiency.

Though the power/unit of single phase IMs is not large, their number is.

Consequently increases in efficiency of 3% or more have a major impact both on world's energy consumption and on the environment (lower temperature motors, less power from the power plants and thus less pollution).

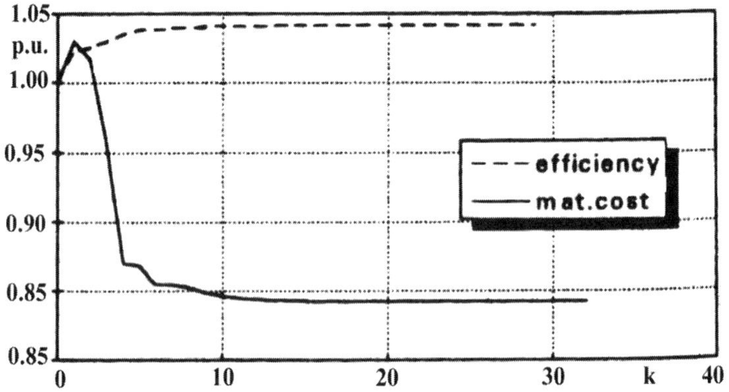

a.)

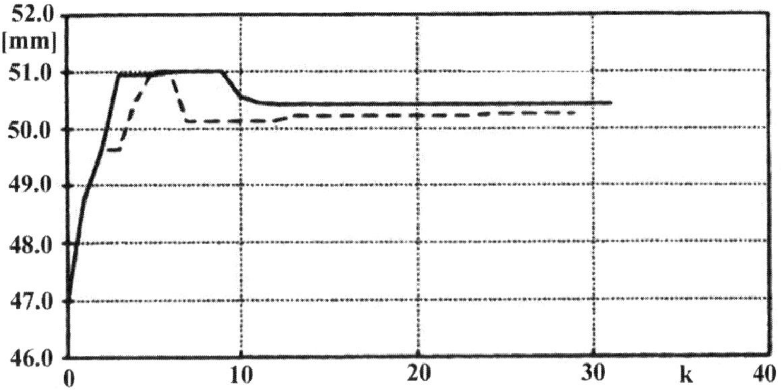

b.)

Figure 27.8 Evolution of the optimization process a.) Efficiency and material costs F_1 and F_2. b.) Stack length for F_1 and F_2.

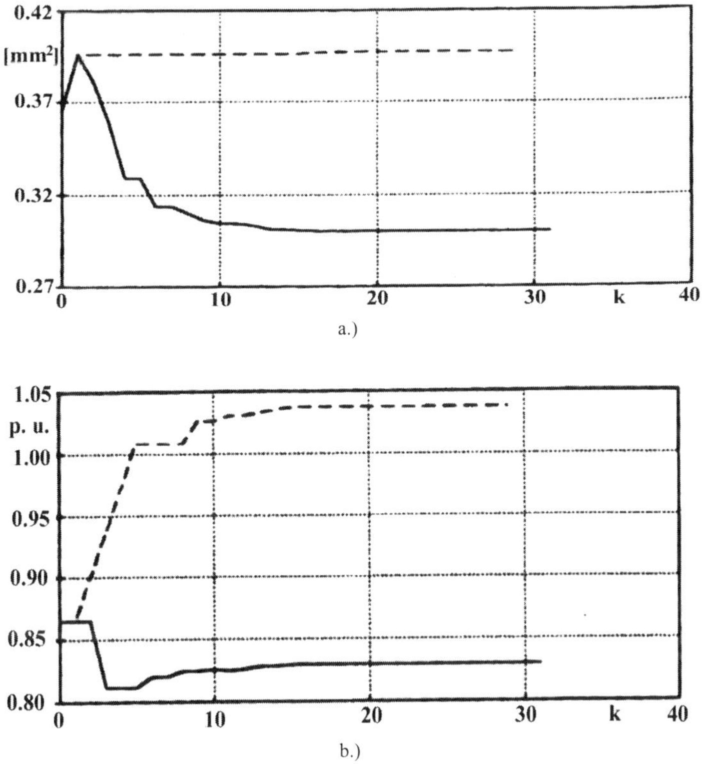

Figure 27.9 Main (a) and auxiliary (b) winding wire sizes evolution during the optimization process

27.11 SUMMARY

- The design of single phase IM tends to be more involved as the magnet field is rather elliptical, in contrast to being circular for three phase IMs.
- The machine utilization factor C_u and the tangential force density (stress) f_t tend to be higher and, respectively, smaller than for three phase IMs. They also depend on the type of single phase IM split phase, dual-capacitor, permanent capacitor, or split-phase capacitor.
- For new specifications, the tangential force density f_t (0.1-1.5)N/cm^2 is recommended to initiate the design process. For designs in given standardized frames the machine utilisation factor C_u seems more practical as an initiation constant.
- The number of stator N_s and rotor slots N_r selection goes as for three phase IMs in terms of parasitic torque reduction. The larger the number of slots, the better in terms of performance.

- The stator bore to stator outer diameter ratio D_i/D_o may be chosen as variable in small intervals, increasing with the number of poles above 0.58 for $2p_1 = 2$, 0.65 for $2p_1 = 4$, 0.69 for $2p_1 = 6$ and 0.72 for $2p_1 = 8$. These values correspond to laminations which provide maximum no-load airgap flux density for given stator m.m.f. in three phase IMs [2].
- By choosing the airgap, stator, rotor tooth and core flux densities, the sizing of stator and rotor slotting becomes straightforward.
- The main winding effective number of turns is then calculated by assuming the e. m. f. $E_m/V_s \approx 0.95\text{-}0.97$.
- The main and auxiliary windings can be made with identical coils or with graded turns coils (sinusoidal windings). The effective number of turns (or the winding factor) for sinusoidal windings is computed by a special formula. In such a case the geometry of various slots may differ. More so for the split-phase IMs where the auxiliary winding occupies only 33 % of stator periphery and is active only during starting.
- The rotor cage cross section total area may be chosen for start as 35% to 60% of the area of stator slots.
- The turns ratio between auxiliary and main winding a = 1.5-2.0, in general. It is equal to unity (a = 1) for reversible motors which have identical stator windings.
- The capacitance initial value is chosen for symmetry conditions at an assigned value of rated slip and efficiency (for rated power).
- Once the preliminary sizing is done the resistance and leakage reactance may be calculated. Then, either by refined analytical methods or by FEM (with bare rotor and zero auxiliary winding current), the magnetisation curve $\Psi_m (I_{mm})$ or reactance $X_{mm}(I_{mm})$, is obtained.
- To estimate the starting and steady state running performance (constraints) such as starting torque and current, rated slip, current, efficiency and power factor and breakdown torque, revolving field or cross field models are used.
- The above completes a preliminary electromagnetic design. A thermal model is then used to estimate stator and rotor temperatures.
- Based on such an analysis model an optimization design process may be started. A good initial (preliminary) design is useful in most nonevolutionary optimization methods [5].
- The optimization design (chapter 18) is a constrained nonlinear programming problem. The constraints can be lumped into an augmented objective function by procedures such as SUMT [7]. Better procedures for safe global optimization are currently proposed [9].
- Increases in efficiencies of a few percent can be obtained by optimization design. Given the immense number of single phase IMs, despite the small power/unit, their total power is significant. Consequently, any improvement of efficiency of more than 1-2% is relevant both in energy costs and environmental effects.

27.12 REFERENCES

1. C. G. Veinott, Small electric motors, Chapter 6, Handbook of Electric Machines, Editor S. A. Nasar, McGraw-Hill Company, 1987.
2. G. Madescu, I. Boldea, T. J. E. Miller, The Optimal Lamination Approach (OLA) to Induction Machine Design Global Optimization, Record of IEEE-IAS-1996, volume 2, pp. 574-580.
3. E.S. Hamdi, Design of Small Electrical Machines, Wiley & Sons, 1994, pp. 140.
4. J. Stepina, Single Phase Induction Motors, Chapter 5, Springer Verlag, New York, 1981 (in German).
5. C. B. Rasmoussen and T. J. E. Miller, Revolving Field Polygon Technique for Performance Prediction of Single Phase Induction Motors, Record of ICEM-2000, Helsinki.
6. H. Huang, E. F. Fuchs, Z. Zak, Optimization of Single Phase Induction Motor Design, Part I + II, IEEE Trans. Vol. EC-3, 1988, pp. 349-366.
7. G. I. Hang, S. S. Shapiro, Statistical Models in Engineering, John Wiley & Sons, 1967.
8. F. Parasiliti, M. Villani, Design Procedure for Single Phase Capacitor Motors Performance Improvement, Record of ICEM-1994, Paris, France, pp. 193-197.
9. X. Liu and W. Xu, Global Optimization of Electrical Machines with the Filled Function Methods, EMPS, vol. 29, 2001.

Chapter 28

SINGLE-PHASE IM TESTING

28.1 INTRODUCTION

The elliptic magnetic field in the airgap of single phase IMs in presence of space m.m.f. harmonics, magnetic saturation, rotor skin effect, and interbar rotor currents makes a complete theoretical modelling a formidable task.

In previous chapters, we did touch all these subjects through basically refined analytical approaches. Ideally, a 3D-FEM, with eddy currents computation and circuit model coupling should be used to tackle simultaneously all the above phenomena.

However, such a task still requires a prohibitive amount of programming and computation effort.

As engineering implies intelligent compromises between results and costs, especially for single-phase IMs, characterized by low powers, experimental investigation is highly recommended. But, again, it is our tendency to make tests under particular operation modes such as locked rotor (shortcircuit) and no-load tests to segregate different kinds of losses and then use them to calculate on-load performance.

Finally, on-load tests are used to check the loss segregation approach.

For three phase IMs, losses from segregation methods and direct on-load tests are averaged to produce safe practical values of stray load losses and efficiency (IEEE Standard 112B).

The presence of rotor currents even at zero slip (S = 0)-due to the backward field component-in single phase IM makes the segregation of losses and equivalent circuit parameter computation rather difficult. Among many potential tests to determine single phase IM parameters and loss segregation two of them have gained rather large acceptance.

One is based on single phase supplying of either main or auxiliary winding of the single phase IM at zero speed (S = 1) and on no-load. The motor may be started as capacitor or split-phase motor and then the auxiliary phase is turned off with the motor free at shaft. [1]

The second method is based on the principle of supplying the single phase IM from a symmetrical voltage supply. The auxiliary winding voltage V_a is 90^0 ahead of the main winding voltage and $V_a = V_m a$, such that the current I_a is $Ia = I_m/a$; a is the ratio between main and auxiliary winding effective turns. [2]

This means in fact that pure forward travelling field conditions are provided. The 90^0 shifted voltage source is obtained with two transformers with modified Scott connection and a Variac.

Again shortcircuit (zero speed)-at low voltage-and no load testing is excersized. Moreover, ideal no-load operation (S = 0) is performed by using a drive at synchronous speed $n_1 = f_1/p_1$

Then both segregation methods are compared with full (direct) load testing. [2] For the case studies considered both methods claim superior results. [1,2]

The two methods have a few common attributes

- They ignore the stray losses (or take them as additional core losses already present under no load tests).
- They consider the magnetization inductance as constant from shortcircuit (S = 1) to no-load and load conditions.
- They ignore the space m.m.f. harmonics, and, in general, consider the current in the machine as sinusoidal in time.
- They neglect the skin effect in the rotor cage. Though the value of the rotor slot depth is not likely to go over 15×10^{-3}m (the power per unit is limited to a few kW), the backward field produces a rotor current component whose frequency $f_{2b} = f_1(2-S)$ varies from f_1 to $2f_1$ when the motor accelerates from zero to rated speed. The penetration depth of electromagnetic field in aluminum is about 12×10^{-3}m at 50Hz and $\left(12/\sqrt{2}\right)\times10^{-3}$ m at 100Hz. The symmetrical voltage method [2] does not have to deal with the backward field and thus the rotor current has a single frequency ($f_{2f} = Sf_1$). Consequently the skin effect may be neglected. The trouble is that during variable load operation, the backward field exists and thus the rotor skin effect is present. In the single phase voltage method [1] the backward field is present under no load and thus it may be claimed that somehow the skin effect is accounted for. It is true that, as for both methods the shortcircuit tests (S = 1) are made at rated frequency, the rotor resistance thus determined already contains a substantial skin effect. This may explain why both methods give results which are not far away from full load tests.
- Avoiding full load tests, both methods measure under no-load tests, smaller interbar currents losses in the rotor than under load. Though there are methods to reduce the interbar currents, there are cases when they are reported to be important, especially with skewed rotors. However, in this case, the rotor surface core additional losses are reduced and thus some compensation of errors may occur to yield good overall loss values.
- Due to the applied simplifications, neither of the methods is to be used to calculate the torque/speed curve of the single-phase IM beyond the rated slip ($S>S_n$). Especially for tapped winding or split-phase IMs which tend to have a marked third order m.m.f. space harmonic that causes a visible deep in the torque/speed curve around 33% of ideal no load speed (S = 0).

As the symmetrical voltage method of loss segregation and parameter computation is quite similar to that used for three phase IM testing (Chapter 22), we will concentrate here on Veinott's method [1] as it sheds more light on single phase IM peculiarities, given by the presence of backward travelling field. The presentation here will try to retain the essentials of Veinott's method while making it show a simple form, for the potential user.

28.2 LOSS SEGREGATION THE SPLIT PHASE AND CAPACITOR START IMs

The split phase and capacitor start IMs start with the auxiliary winding on but end up operating only with the main stator winding connected to the power grid.

It seems practical to start with this case by exploring the shortcircuit (zero speed) and no load operation modes with the main winding only on, for parameter computation and loss segregation.

We will make use of the cross field model (see Chapter 24, Figure 24.21), though the travelling field (+, -) model would give similar results as constant motor parameters are considered.

For the zero speed test (S = 1) $Z_f = Z_b$ and, thus,

$$R_{sm} + R_{rm} \approx \frac{P_{sc}}{I_{msc}^2} \tag{28.1}$$

$$\left|\underline{Z}_{sc}\right| = \frac{V_s}{I_{msc}} \tag{28.2}$$

$$X_{sm} + X_{rm} \approx \sqrt{\left(\frac{V_s}{I_{msc}}\right)^2 - \left(R_{sm} + R_{rm}\right)^2} \tag{28.3}$$

With R_{sm} d.c. measured and temperature-corrected, and $X_{sm} \approx X_{rm}$ for first iteration the values of R_{sm}, $X_{sm} = X_{rm}$ and R_{rm} are determined. For the no-load test (still $I_a = 0$) we may measure the slip value S_0 or we may not. If we do, we make use of it. If not, $S_0 \approx 0$.

Making use of the equivalent circuit of Figure 28.1, for $S = S_0 = 0$ and with $I_{m0}(A)$, $V_s(V)$, $P_m(W)$ and E_a measured, we have the following mathematical relations

$$Z_f \approx \frac{1}{2} jX_{mm} \tag{28.4}$$

$$Z_b \approx \frac{1}{2}\left[\frac{R_{rm}}{2} + jX_{rm}\right] \tag{28.5}$$

$$Z_{f,b} = \frac{1}{2} \frac{j\,X_{mm}\left(\dfrac{R_{rm}}{S_{f,b}}+jX_{rm}\right)}{\dfrac{R_{rm}}{S_{f,b}}+j(X_{mm}+X_{rm})} \qquad \begin{aligned} S_f &= S \\ S_b &= 2-S \end{aligned}$$

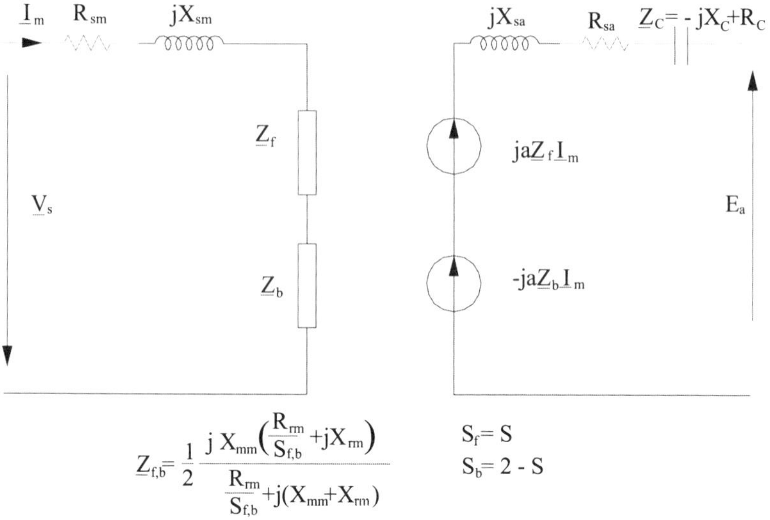

Figure 28.1 The cross field single phase IM with auxiliary winding open ($I_a = 0$)

From
$$\frac{E_a}{a} \approx \left|\left(\underline{Z}_f - \underline{Z}_b\right)\right|I_{m0} \tag{28.6}$$

with R_{rm} and X_{sm} already determined from the shortcircuit test and I_{m0} and E_a measured, we need the value of a in Equation 28.6 to determine the magnetization reactance X_{mm}

$$X_{mm} = 2\sqrt{\frac{E_a^2}{a^2 I_{m0}^2} - \frac{R_{rm}^2}{16}} - X_{rm} \tag{28.7}$$

A rather good value of a may be determined by running on no load the machine additionally, with the main winding open and the auxiliary winding fed from the voltage $V_{a0} \approx 1.2E_a$. With E_m measured, [1]

$$a = \sqrt{\frac{E_a V_{a0}}{V_s E_m}} \tag{28.8}$$

Once X_{mm} is known, from (28.7) with (28.8), we may make use of the measured P_{m0}, I_{m0}, V_s (see Figure 28.1) to determine the sum of iron and mechanical losses

$$p_{mec} + p_{iron} = P_{m0} - \left(R_{sm} + \frac{R_{rm}}{4}\right)I_{m0}^2 \tag{28.9}$$

The no load test may be performed at different values of V_s, below rated value, until the current I_{m0} starts increasing; a sign that the slip is likely to increase too much.

As for the three phase IM, the separation of mechanical and core losses may be done by taking the ordonate at zero speed of the rather straight line dependence of ($p_{mec}+p_{iron}$) of V_s^2 (Figure 28.2). Alternatively, we may use only the results for two voltages to segregate p_{mec} from p_{iron}.

A standard straight-line curve fitting method may be used for better precision.

The core loss is, in fact, dependent on the e.m.f. E_m (not on input voltage V_m) and on the magnetic field ellipticity.

The field ellipticity decreases with load and this is why, in general, the core losses are attributed to the forward component.

Consequently, the core loss resistance R_{miron} may be placed in series with the magnetization reactance X_{mm} and thus (Figure 28.3)

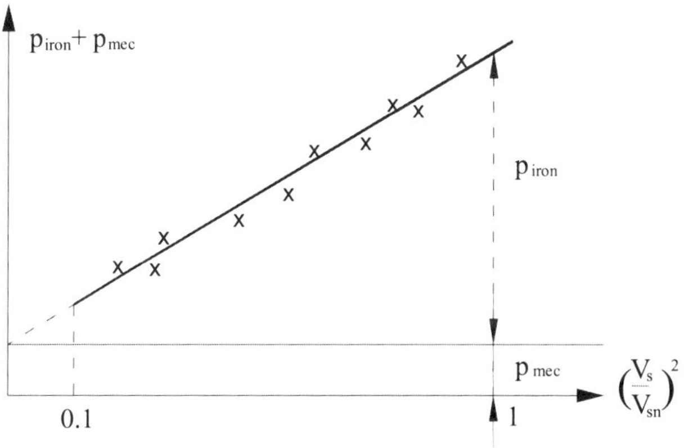

Figure 28.2 Mechanical plus core losses at no load (open auxiliary winding)

$$R_{miron} \approx \frac{p_{iron}}{2I_{m0}^2} \tag{28.10}$$

The impedance at no load $\underline{Z}_{om}$ is

$$\left|\underline{Z}_{om}\right| = \frac{V_s}{I_{m0}} = \sqrt{\left(R_{sm} + R_{miron} + \frac{R_{rm}}{4}\right)^2 + \left(\frac{X_{mm} + X_{rm}}{2} + X_{sm}\right)^2} \tag{28.11}$$

Equation (28.11) allows us to calculate again X_{mm}. An average of the value obtained from (28.7) and (28.11) may be used for more confidence.

Now that all parameters are known, it is possible to refine the results by introducing the magnetization reactance X_{mm} in the shortcircuit impedance, while still $X_{sm} = X_{rm}$, to improve the values of R_{rm} and X_{sm} until sufficient convergence is obtained.

Today numerical methods available through many software programs on PCs allow for such iterative procedures to be applied rather easily.

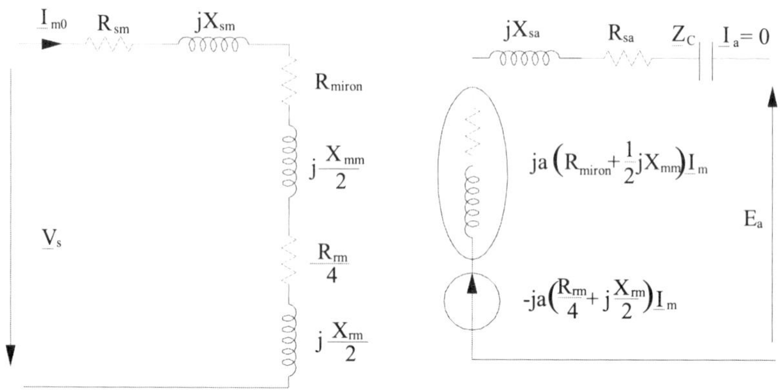

Figure 28.3 Simplified no load (S = 0) equivalent circuit with series core loss resistance R_{miron} in both circuits (the auxiliary winding is open)

Example 28.1

A 123 W (1/6 hp), 6 poles, 60 Hz, 110 V, split phase motor was tested as follows [1]

$R_{sm} = 2.54\ \Omega$ after locked rotor reading, $R_{sm} = 2.65\ \Omega$ after no-load single phase running.

Locked rotor watts at 110 V is $P_{sc} = 851$ W.

No-load current I_{m0} at $V_{s0} = 105$ V is $I_{m0} = 2.68$ A. Locked rotor current is $I_{sc} = 11.65$A.

$$V_{s0} = V_{sn} - I_{mn}R_{sm}^{h}\cos\varphi_n = 110 - 3.17\times 2.85\times 0.588 \approx 105\text{V} \qquad (28.12)$$

where $I_{mn} = 3.17$ A, $R_{sm}^{h} = 2.85\Omega$, $\cos\varphi_n = 0.588$, full load slip $S_n = 0.033$. Full load input $P_{1n} = 205$ W and rated efficiency $\eta_n = 60.5$ % have been obtained from a direct load (brake) test.

The no-load power versus applied volts V_s, produces $(p_{iron})_{Vs0} = 24.7$ W and mechanical losses $p_{mec} = 1.5$ W.

The no-load auxiliary winding voltage $E_a = 140.0$ V.

The value of turns ratio a is found from an additional no-load test with the auxiliary winding fed at 1.2 E_a (V). With the measured no-load main winding voltage $E_{m0} = 105$ V, a is (28.8)

$$a = \sqrt{\frac{140\times 140\times 1.2}{105\times 105}} = 1.46$$

Let us now find the motor parameters and directly check the efficiency measured by the loss segregation method.

The rotor resistance (referred to the main winding) R_{rm} is (28.1)

$$R_{rm} = \frac{851}{11.65^2} - 2.54 = 3.73\Omega$$

The stator and rotor leakage reactances $X_{rm} = X_{sm}$ from (28.3) are

$$X_{sm} = X_{rm} = \frac{1}{2}\sqrt{\left(\frac{105}{11.65}\right)^2 - (3.73 + 2.54)^2} = 3.237\Omega$$

R_{rm} is getting larger during the no load test, due to heating, to the same extent that R_{sm} does

$$R_{rm}^{o} = R_{rm} \times \frac{R_{sm}^{o}}{R_{sm}} = 3.73 \times \frac{2.65}{2.54} = 3.89\Omega$$

Now the magnetisation reactance X_{mm}, from (28.7), is

$$X_{mm} = 2\sqrt{\left(\frac{140.0}{1.46 \times 2.58}\right)^2 - \left(\frac{3.89}{4}\right)^2} - 3.237 = 77.57\Omega$$

with the core resistance R_{miron} (28.10)

$$R_{miron} = \frac{p_{iron}}{2I_{m0}^2} = \frac{24.7}{2 \cdot 2.58^2} = 1.87\Omega$$

Now from (28.11) we may recalculate X_{mm}

$$X_{mm} = 2\sqrt{\left(\frac{105}{2.58}\right)^2 - \left(2.65 + 1.87 + \frac{3.89}{4}\right)^2} - 3.237 = 74.71\Omega$$

An average of the two X_{mm} values would be 76.1235 Ω.

By now the parameter problem has been solved. A few iterations may be used with the complete circuit at S = 1 to get better values for R_{rm} and $X_{sm} = X_{rm}$. However,we should note that $X_{mm} / X_{rm} \approx 20$, and thus not much is to be gained from these refinements.

For efficiency checking we need to calculate the winding losses at $S_n = 0.033$ with the following parameters

$$R_{sm}^{h} = 2.85\Omega, R_{rm}^{h} = R_{rm}^{o} \times \frac{R_{sm}^{h}}{R_{sm}^{o}} = 3.89 \times \frac{2.85}{2.65} = 4.1835\Omega$$

$$X_{sm} = X_{rm} = 3.237\Omega, X_{mm} = 76.12\Omega$$

The core resistance R_{miron} may be neglected when the rotor currents are calculated

$$I_{rmf} = I_m \left| \frac{j \cdot X_{mm}}{\frac{R^h_{rm}}{S_n} + j \cdot (X_{mm} + X_{rm})} \right| = 3.117 \cdot \left| \frac{j \cdot 76.12}{\frac{4.1835}{0.033} + j \cdot (76.12 + 3.237)} \right| = 1.586A$$

$$I_{rmb} \approx I_m = 3.117A$$

So the total rotor winding losses $p_{Corotor}$ are

$$p_{Corotor} = I^2_{rmf} \frac{R^h_{rm}}{2} + I^2_{rmb} \frac{R^h_{rm}}{2} = \left(3.117^2 + 1.586^2\right)\frac{4.1835}{2} = 25.58W$$

The stator copper losses p_{cos} are

$$p_{cos} = I^2_{m0} R^h_{sm} = 3.117^2 \times 2.85 = 27.69W$$

The total load losses Σp from loss segregation, are

$$\sum p = p_{cos} + p_{Corotor} + p_{iron} + p_{mec} = 27.69 + 25.58 + 24.7 + 1.5 = 79.47W$$

The losses calculated from the direct load test are

$$\left(\sum p\right)_{load} = P_{in} - P_{out} = 205 - 123 = 82W$$

There is a small difference of 2 W between the two tests, which tends to validate the methods of loss segregation. However, it is not sure that this is the case for most designs.

It is recommended to back up loss segregation by direct shaft loading tests whenever possible.

It is possible to define the stray load losses as proportional to stator current squared and then to use this expression to determine total losses at various load levels

$$\text{stray load losses} = (\Sigma p)_{load} - \Sigma p_{segregation} \approx R_{stray}\left[I^2_m + (I_a a)^2\right]$$

R_{stray} may then be lumped into the stator resistances R_{sm} and R_{sa} / a^2.

28.3 THE CASE OF CLOSED ROTOR SLOTS

In some single phase IMs (as well as three phase IMs) closed rotor slots are used to reduce noise.

In this case the rotor slot leakage inductance varies with rotor current due to the magnetic saturation of the iron bridges above the rotor slots.

The shortcircuit test has to be done now for quite a few values of voltage (Figure 28.4)

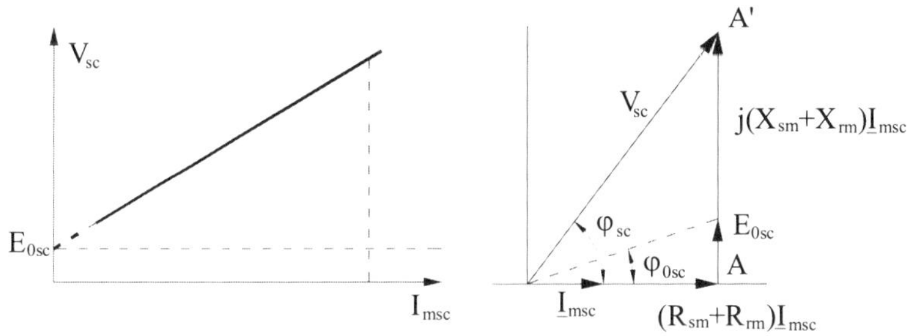

Figure 28.4 Shortcircuit characteristics

In this case, the equivalent circuit should additionally contain a constant e.m.f. E_{0sc}, corresponding to the saturated closed rotor upper iron bridge (Figure 28.5).

Only the segment AA′-E_{0sc} (on Figure 28.5) represents the voltage drop on the rotor and stator constant leakage reactances.

$$X_{sm} + X_{rm} = \frac{V_{sc} \sin \varphi_{sc} - E_{0sc}}{I_{msc}} \tag{28.13}$$

28.4 LOSS SEGREGATION THE PERMANENT CAPACITOR IM

The loss segregation for the permanent capacitor IM may be performed as for the single phase IM, with only the main winding ($I_a = 0$) activated.

The auxiliary winding resistance and leakage reactance R_{sa} and X_{sa} may be measured, in the end, by a shortcircuit (zero speed) test performed on the auxiliary winding

$$P_{sca} \approx \left(R_{sa} + R_{ra}\right) I_{sca}^2 \tag{28.14}$$

$$X_{sa} + X_{ra} = \sqrt{\left(\frac{V_{sca}}{I_{sca}}\right)^2 - \left(R_{sa} + R_{ra}\right)^2} \tag{28.15}$$

When $X_{ra} = a^2 X_{rm}$ (with a and X_{rm} known, from (28.13)) it is possible to calculate X_{sa} with measured input power and current. With R_{sa} d.c. measured, R_{ra} may be calculated from 28.14.

The capacitor losses may be considered through a series resistance R_C (see Figure 28.1). The value of R_C may be measured by separately supplying the capacitor from an a.c. source.

The active power in the capacitor P_C and the current through the capacitor I_C are directly measured

$$R_C = \frac{P_C}{I_C^2} \tag{28.16}$$

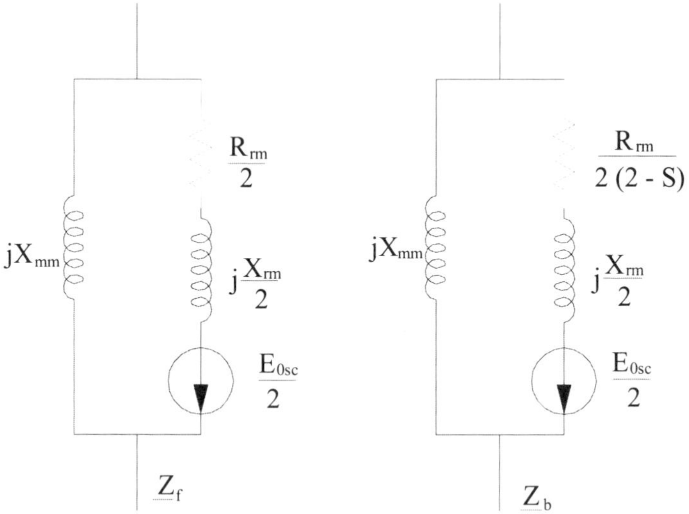

Figure 28.5 The forward (f, b) impedances for the closed slot rotor

Once all parameters are known, the equivalent circuit in Figure 28.1 allows for the computation of both stator currents, I_m and I_a, under load with both phases on, and given slip value, S. Consequently, the stator and rotor winding losses may then be determined.

From this point on, we repeat the procedure in the previous paragraph to calculate the total losses by segregation method and from direct input and output measurements, with the machine shaft loaded.

28.5 SPEED (SLIP) MEASUREMENTS

One problem encountered in the load tests is the slip (speed) measurement.

Unless a precision optical speedometer (with an error in the range of 2 rpm or less) is available, it is more convenient to measure directly the slip frequency Sf_1.

The "old" method of using a large diameter circular shortcircuited coil with a large number of turns-to track the axial rotor leakage flux by a current Hall probe (or shunt), may be used for the scope.

The coil is placed outside the motor frame at the motor end which does not hold the cooling fan. Coil axis is concentric with the shaft.

The acquired current signal contains two frequencies Sf_1 and $(2\text{-}S)f_1$. An off-line digital software (or a low pass hardware) filter may be applied to the current signal to extract the Sf_1 frequency component.

The slip computation error is expected to be equivalent to that of a 2 rpm precision speedometer or better.

28.6 LOAD TESTING

There are two main operation modes to test the single phase IM on load the motor mode and the generator mode.

Under the motor mode the electric input power, P_{1e}, and the output mechanical power, P_{2m}, are measured. P_{2m} is in fact calculated indirectly from the measured torque T_{shaft} and speed n

$$P_{2m} = T_{shaft} \cdot 2\pi n \tag{28.16}$$

The torque is measured by a torquemeter (Figure 28.6). Alternatively the load machine may have the losses previously segregated such that at any load level the single-phase IM mechanical power P_{2m} may be calculated

$$P_{2m} = P_{3e} - \sum p_{loadmachine} \tag{28.17}$$

The load machine may be a PM d.c. or a.c. generator with resistive load, a hysteresis or an eddy current d.c. brake.

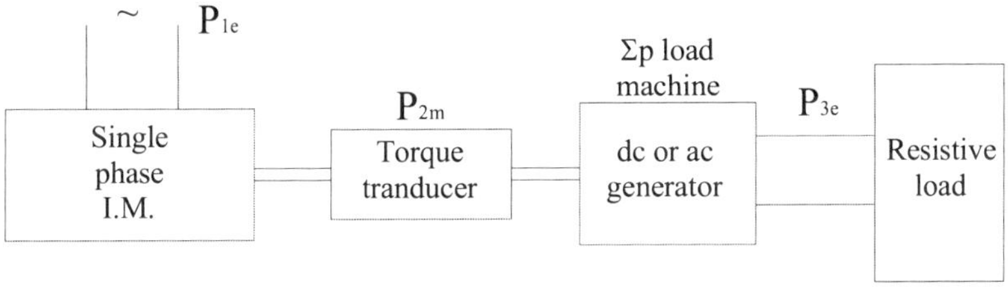

Figure 28.6 Load tests with a torquemeter

28.7 COMPLETE TORQUE-SPEED CURVE MEASUREMENTS

Due to magnetic saturation m.m.f. space harmonics, slotting, and skin effect influences on harmonic rotor currents, it seems that direct torque measurements at various speeds (below rated speed) is required.

Such a measurement may be performed by using a d.c. or an a.c. generator with power-converter energy retrieval to the power grid (Figure 28.7).

Alternatively a slow acceleration test on no load may be used to calculate the torque speed curve. For slow acceleration, the supply voltage may be lowered from V_{sn} to V_s.

The core losses are considered proportional to voltage squared. They are measured by the loss segregation method.

$$p_{iron} = \left(p_{iron}\right)_{V_{sn}} \left(\frac{V_s}{V_{sn}}\right)^2 \quad (28.18)$$

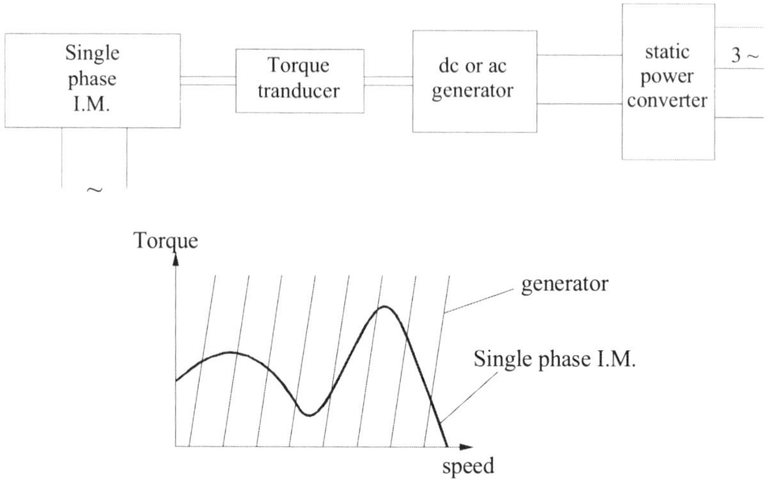

Figure 28.7 Torque/speed measurements with a torquemeter

The mechanical losses proportional to speed squared

$$p_{mec} = \left(p_{mec}\right)_{n_0} \left(\frac{n}{n_0}\right)^2 \quad (28.19)$$

With R_{sm}, R_{rm}, R_{sa}, X_{sm}, X_{ra}, X_{rm}, X_{mm} known and with I_m and I_a and input power measuredm P_1, torque may be calculated as for steady state around rated speed.

$$T_e = \frac{p_{mec}}{2\pi n} + T_{shaft} = \frac{\left[P_1 - p_{iron} - R_{sm} I_m^2 - R_{sa} I_a^2 - R_{rm}\left(I_{rmf}^2 + I_{rmb}^2\right) - R_C I_a^2\right]}{2\pi n} \quad (28.20)$$

The computation of T_{shaft} is to be done offline. An optical speedometer could be used to measure the speed during the slow acceleration test and during a free deceleration after turn-off.

With $p_{mec}(n)$ known, the free deceleration test yields

$$J = -\frac{p_{mec}(n)}{2\pi \frac{dn}{dt}} \quad (28.21)$$

The moment of inertia J is thus obtained.

With J known and speed n acquired during the no-load slow acceleration test, the torque is

$$T_e(n) = J2\pi\frac{dn}{dt} + \frac{p_{mec}}{2\pi n} \tag{28.22}$$

The torque for rated voltage is considered to be

$$(T_e(n))_{V_{sn}} \approx (T_e(n))_{V_s}\left(\frac{V_{sn}}{V_s}\right)^2 \tag{28.23}$$

The speed derivative in (28.22) may be obtained offline, with an appropriate software filter, from the measured speed signal.

The two values of torque, (28.20) and (28.22), are then compared to calculate a measure of stray losses

$$p_{stray} \approx \left[(T_e)_{Equation(28.22)} - (T_e(n))_{Equation(28.20)}\right]\cdot\left(\frac{V_{sn}}{V_s}\right)^2 2\pi n \tag{28.24}$$

Temperature measurements methods are very similar to those used for three phase IMs. Standstill d.c. current decay tests may also be used to determine the magnetization curve $\Psi_{mm}(I_{mm})$ and even the resistances and leakage reactances, as done for three phase IMs.

28.8 SUMMARY

- The single phase IM testing aims to determine equivalent circuit parameters to segregate losses and to measure the performance on load; even to investigate transients.
- Due to the backward field (current) component, even at zero slip, the rotor current (loss) is not zero. This situation complicates the loss segregation in no load tests.
- The no load test may be done with the auxiliary phase open $I_a' = 0$, after the motor starts.
- The shortcircuit (zero speed) test is to be performed, separately for the main and auxiliary phases, to determine the resistances and leakage reactances.
- Based on these results the no load test (with $I_a = 0$) furnishes data for loss segregation and magnetization curve ($X_{mm}(I_{mm})$), provided it is performed for quite a few voltage levels below rated voltage.
- The calculation of rotor resistance R_{rm} from the zero speed test (at 50 (60) Hz) produces a value that is acceptable for computing on load performance, because it is measured at an average of forward and backward rotor current frequencies $\frac{Sf_1 + (2-S)f_1}{2} = f_1$.
- The difference between total losses by segregation method and by direct input / output measurements under load is a good measure of stray load losses. The stray load losses tend to be smaller than in three phase motors because the ratio between full load and no load current is smaller.

- Instead of single (main) phase no load and shortcircuit tests, symmetrical two voltage supplying for same tests has also been proved to produce good results.
- The complete torque-speed curve is of interest also. Below the breakdown torque speed value there may be a deep in the torque speed curve around 33 % of no load ideal speed (f_1 / p_1) due to the third space m.m.f. harmonic. A direct load method may be used to obtain the entire torque versus speed curve. Care must be exercised that the load machine had a rigid torque / speed characteristic to handle the statically unstable part of the single-phase IM torque-speed curve down to standstill.
- To eliminate direct torque measurements and the load machine system, a slow free acceleration at reduced voltage and a deceleration test may be performed. With the input power, speed and stator currents and voltage measured and parameters already known (from the shortcircuit and single phase no load tests), the torque may be computed after loss subtraction from input at every speed (slip). The torque is also calculated from the motion equation with inertia J determined from free deceleration test. The mechanical power difference in the two measurements should be a good measure of the stray load losses caused by space harmonics. The torque in the torque / speed curve thus obtained is multiplied by the rated to applied voltage ratio squared to obtain the full voltage torque-speed. It is recognized that this approximation underestimates the influence of magnetic saturation on torque at various speeds. This is to say that full voltage load tests from S = 0 to S = 1 are required for magnetic saturation complete consideration.
- DC flux (current) decay tests at standstill, in the main and auxiliary winding axes, may also be used to determine resistances, leakage inductances and the magnetization curve. Frequency response standstill tests may be applied to single phase IM in a manner very similar to one applied to three phase IMs (Chapter 22).
- Temperature measurement tests are performed as for three phase IMs, in general (Chapter 22).

28.9 REFERENCES

1. C. G. Veinott, Segregation of Losses in Single Phase Induction Motors, Trans AIEE, Volume 54, December 1935, pp. 1302-1306.
2. C. Van der Merwe and F. S. Van der Merwe, Study of Methods to Measure the Parameters of Single Phase Induction Motors, IEEE Trans. vol. EC-10, no. 2, 1995, pp. 248-253.

INDEX